Biometrical Genetics: The Study of Continuous Variation.

K. Mather and J. L. Jinks

1st Edn 1949

2nd Edn 1971

Reprinted 1974

Chapman + Hall Ltd.
11, New Fetter Lane
London EC4

Classic on analysis of quantitative characters

An Introduction to Genetic Analysis

An Introduction to Genetic Analysis

SECOND EDITION

David T. Suzuki
The University of British Columbia

Anthony J. F. Griffiths
The University of British Columbia

Richard C. Lewontin
Harvard University

W. H. Freeman and Company
San Francisco

Developmental Editor: Nancy Flight
Copy Editor: Lawrence W. McCombs
Designer: Gary A. Head
Project Editor: Larry Olsen
Production Coordinator: Linda Jupiter
Illustration Coordinator: Cheryl Nufer
Artists: John and Judy Waller, Tim Keenan, Donna Salmon, Kelly Solis-Navarro, John and Jean Foster, Georg Klatt
Photo Researcher: Kay James
Compositor: Typothetae
Printer and Binder: Kingsport Press

The cover illustration is a representation of the garden pea, *Pisum sativum,* based on a lino cut print by the renowned botanical illustrator Henry Evans. Mendel's experiments using this species were the beginning of modern genetics. Cover design by Gary A. Head.

Library of Congress Cataloging in Publication Data

Suzuki, David T 1936–
An introduction to genetic analysis.

Bibliography: p.
Includes index.
1. Genetics. I. Griffiths, Anthony J. F., joint author. II. Lewontin, Richard C., 1929– joint author. III. Title.
[DNLM: 1. Genetics. QH 430 S968i]
QH430.S94 1981 575.1 80-24522
ISBN 0-7167-1263-6

Printed in the United States of America
9 8 7 6 5 4 3 2

Contents

Preface

This book attempts to teach the reader how to do genetic analysis. It both describes the findings of genetics and provides a set of methods and exercises to help the student understand the way genetic inference is made. We believe this is the best way to develop an appreciation of the discipline of genetics, which plays such a central role in modern biology. The book is designed for an introductory course in college genetics. It could serve equally well as a one-semester or two-semester course, with additions or deletions at the discretion of the instructor.

Rather than beginning by describing our current knowledge of molecular genetics, we present genetic concepts and discoveries for the most part in historical sequence. We chose this approach because it seems to us that a student begins much as biologists did at the turn of the century, asking general questions about the laws governing the inheritance of traits. Only after one understands these laws can one logically proceed to analyses at the cellular and molecular levels of hereditary organization. In the text we have used a question and answer format to simulate this step-by-step evolution of genetic understanding.

We emphasize the quantitative aspects of genetics throughout the text because progress in genetics, in particular the formulation of its abstract ideas, has been based largely on numerical experimental data. Various kinds of quantitative analyses are introduced in the historical discussion. The problem sets at the ends of chapters provide the student with the opportunity to apply these analytical methods to experimental situations. The act of working a problem effectively simulates the modus operandi of genetic analysis. The problems are arranged in the order of the discussion of topics in each chapter, and in order of increasing difficulty. These problem sets constitute an essential aspect of the teaching of the quantitative nature of genetic analysis. Despite this emphasis on quantitative analysis, the only mathematics required for understanding the text and working the problems is arithmetic and basic algebra.

Today geneticists can be divided into two broad groups, those concerned with the mechanisms of heredity and those using the techniques of genetics to probe other fundamental biological processes or phenomena. Both of these approaches are emphasized and recur as themes throughout the text.

This Second Edition contains several special features to aid the student. Throughout the text, key concepts are summarized in special Message sections. These sections provide convenient stopping points from which the reader may orient himself within each chapter, and they may be used as a convenient way of reviewing the material. Summaries at the end of each chapter also provide a means for reviewing the important points of discussion. Key terms are set in boldface type, and most of these are defined in the Glossary at the end of the book. Further Readings and Answers to Selected Problems are also provided for each chapter in a section at the end of the book.

In the Second Edition, three chapters (Chapter 13, Manipulation of DNA, Chapter 14, The Structure and Function of Chromosomes, and Chapter 15, Organelle Genes) and the last section of Chapter 16, "Transposable Genetic Elements," are devoted almost entirely to new material. Since the preparation of the First Edition, geneticists have acquired an arsenal of revolutionary new analytical techniques, largely molecular in their focus, which are now used on a routine basis in genetic research. These new techniques have proved to be satisfyingly complementary to the traditional techniques of genetic analysis. We have tried to convey in the new chapters a sense of this immensely powerful combined approach to the study of life and its processes.

In addition, some material has been completely rewritten by our new coauthor, Richard C. Lewontin, resulting in three substantially new chapters (Chapter 1, Gene and Organism, Chapter 18, Quantitative Genetics, and Chapter 19, Population Genetics). The order of chapters has been changed somewhat in relation to the First Edition—Chapter 7, Gene Mutation, now precedes Chapter 8, Chromosome Mutation, and Chapter 9, Recombination in Bacteria and Their Viruses, now follows these two chapters. We have extensively modified and expanded the remaining chapters, often in response to the specific suggestions of users. Many examples have been added from medicine and agriculture. Many new figures have been added, especially photographs, and the figure legends have been greatly expanded. The number of problems has been doubled to a total of over four hundred. An Instructor's Guide is available for this edition, containing chapter by chapter commentaries and sets of suggested exam questions. A separate Solutions Manual is also available, containing answers to all problems in the text. With the instructor's approval, the Solutions Manual will be available to students. Both of these manuals were prepared by Robert J. Robbins (Michigan State University).

Thanks are due to the following people: To Cedric I. Davern (University of Utah), Robert G. Fowler (San Jose State University), Edward W. Hanley (University of Utah), Ross J. MacIntyre (Cornell University), and Robert J. Robbins for invaluable reviews and helpful modifications of the manuscript. Of these, special credit is due to Bob Robbins, whose suggested modifications have been adopted without change in several places, notably at the beginning of Chapter 9, and whose careful attention to detail and pedagogy has contributed in a major way throughout the text. To Jack von Borstel (University of Alberta), David P. Campbell (California State Polytechnic University), Thomas W. Cline (Princeton University), Leonard E. Kelly (University of California, Berkeley), Mary R. Murnick (Western Illinois University), Allen F. Sherald (George Mason University), Wendell E. Wall (University of Alabama, Birmingham), and Jerry Woolpy (Earlham College) for constructive reviews of the proposed outline for the Second Edition. To Lawrence McCombs, who edited the final manuscript. To Clayton Person and Tom Kaufman for many helpful discussions and suggestions. To Joan Griffiths, Glynis Kirby, Chris Poirier-Skelton, Shirley Macaulay, and Valerie Kohn for help with the typing. To Tara Cullis and Dorienne Jaffe for help with correspondence. To the long list of colleagues who generously contributed material for the figures. To Kay James for photograph research. And finally to our friends at W. H. Freeman and Company, who have contributed so much through their professional competence and enthusiasm. The present structure of the book is attributable in large part to the sound ideas and tenacity of the developmental editor, Nancy Flight. The production team of Gary Head, Linda Jupiter, Cheryl Nufer, and Larry Olsen showed exceptional dedication and creativity.

We hope this book will stimulate the reader to do some first-hand experimental genetics, whether as professional scientist, student, or amateur gardener or animal fancier. Failing this, we hope some lasting impression will be formed of the precision, elegance, and power of genetic analysis.

November 1980

David T. Suzuki
Anthony J. F. Griffiths
Richard C. Lewontin

An Introduction to Genetic Analysis

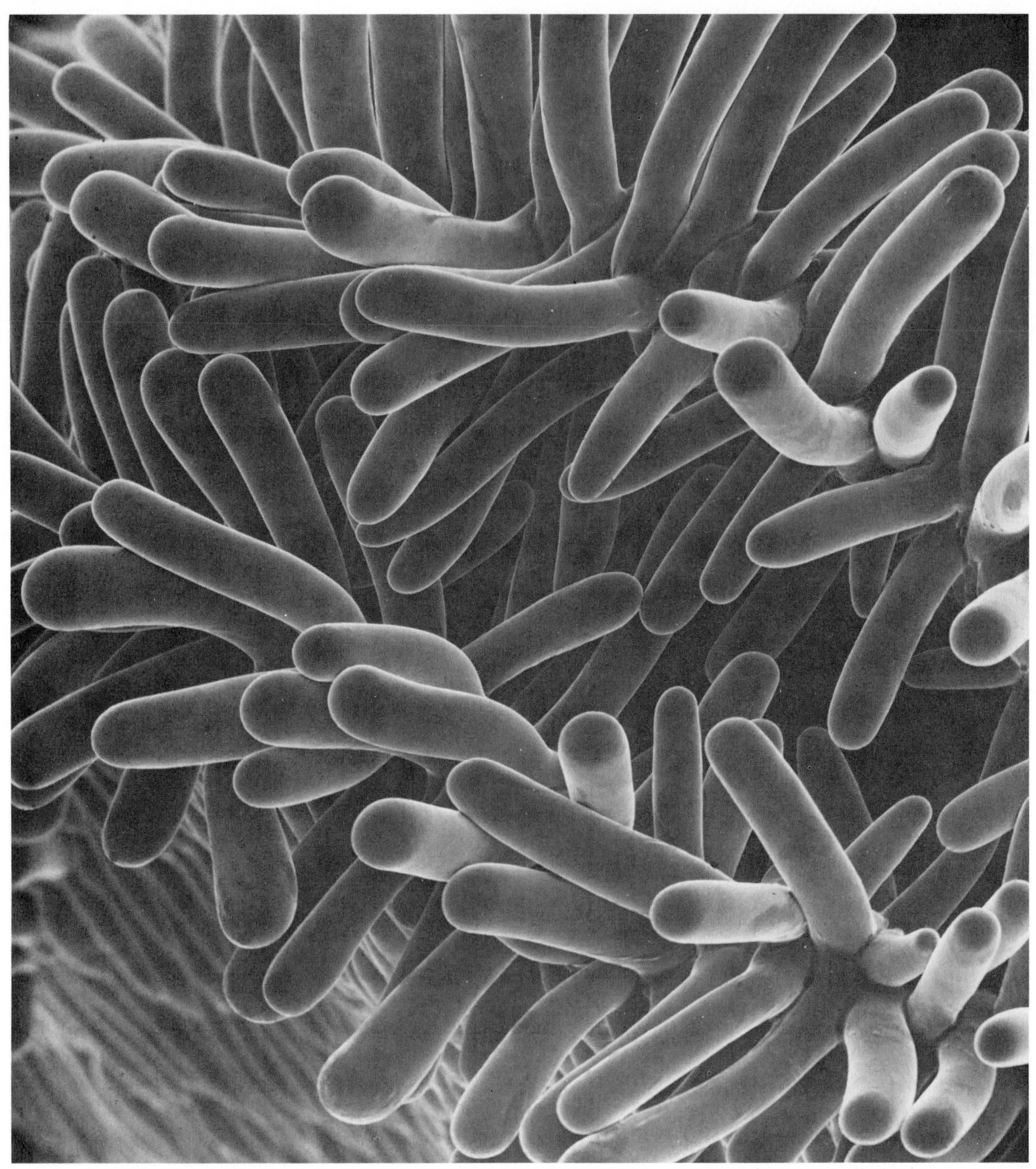

Order and symmetry in life: Gladiola *stigmas (×375). (Copyright © David Scharf/Peter Arnold, Inc.)*

Introduction

Genetics in Biology

The universe naturally tends toward disorder or disarray. This river of chaos does contain a few isolated eddies of order, and one of the most interesting is called **life.** Living systems are highly ordered. In fact, life can be viewed as a secret or trick for taking disorderly components and assembling them into highly symmetrical structures—structures that are extremely unlikely as far as the rest of the universe is concerned. The key to this unlikely phenomenon of life is found in the chemical behavior of biological structures called **genes,** the fundamental units that control heredity. The branch of science called **genetics,** as the name implies, is devoted to the study of the genes. Because the genes play such a central role in the processes of life, genetics has become a keystone that supports and ties together many areas of biology.

Under the appropriate conditions, genes are able to replicate—to generate faithful copies of themselves. Replication gives life its continuity from one generation to the next. Each living system possesses instructions, handed down from its remote ancestors, for the battle against disorder. This continuity through time, in its turn, permits evolution—the diversification and increasing complexity of life forms. Replication also is the basis for the development, from a single egg cell, of a relatively massive and complex multicellular organism, such as a person or a flowering plant.

Two major findings of genetics have been monumentally significant in the unification of biology into a coherent science. First, consider the staggering variety of life forms on earth—a spectrum including some 286,000 species of flowering plants, 500,000 species of fungi, 750,000 species of insects, and so forth. Genetics has shown that all these life forms probably share a common ancestor, and genetics has revealed precise molecular mechanisms that provide a model to explain this evolutionary process. Second, genetics has shown that all life forms—from bacteria to humans—are based on a common system for storing, duplicating, copying, and translating information. Once again, the details of this mechanism have been revealed down to the level of the molecules involved.

Finally, genetics has provided a major tool for the study of biological phenomena. This approach, which we call **genetic dissection** of biological systems, is a central theme of this book. Clues provided by genetic dissection have triggered much of the progress in biology in recent decades.

Genetics and Human Affairs

Knowledge about hereditary phenomena has been important to humans for a very long time. Civilization itself became possible when nomadic tribes learned to domesticate plants and animals. Long before biology existed as a scientific discipline, people selected grains with higher yields and greater vigor and animals with better fur or meat. They also puzzled about the inheritance of desirable and undesirable traits in the human population. Despite this long-standing concern with heredity and the practice of selective breeding, it was not until the last century that we were able to elucidate the actual basis for inheritance.

As in many areas of science, this new knowledge has produced new challenges as well as solutions to some human problems. For example, early in this century a new wheat strain called Marquis was developed in Canada. This high-quality strain is resistant to disease; furthermore, it matures two weeks earlier than other commercially used strains—a very important factor where the growing season is short. At the time of its introduction, the use of Marquis wheat opened up millions of square miles of fertile soil to cultivation in such northern countries as Canada, Sweden, and the USSR. Table I-1 shows how geneticists have bred a wide range of desirable characteristics into another commercial crop, rice. In addition to improving crop varieties, geneticists have learned to alter the genetic systems of insects to reduce their fertility. This technique is providing an important new weapon in the age-old struggle to keep insects out of human crops and habitations.

Such successes in recent years led to the concept of the "Green Revolution" as the scientific answer to the problem of human hunger. Using sophisticated breeding techniques based on new knowledge about genes, geneticists created high-yield varieties of dwarf wheat (Figure I-1) and rice (Figure I-2). Extensive planting of these crops around the world did provide new food supplies, but new problems quickly became apparent. These specialized crops require extensive cultivation and costly fertilizers, and their use produced a wide range of

Table I-1. Development of pest-resistant strains of rice

Strain	Year developed	Diseases					Insects		
		Blast fungus	Bacterial blight	Leaf-streak virus	Grassy stunt virus	Tungro virus	Green leaf-hopper	Brown hopper	Stem borer
IR 8	1966	MR	S	S	S	S	R	S	MS
IR 5	1967	S	S	MS	S	S	R	S	S
IR 20	1969	MR	R	MR	S	R	R	S	MS
IR 22	1969	S	R	MS	S	S	S	S	S
IR 24	1971	S	S	MR	S	MR	R	S	S
IR 26	1973	MR	R	MR	MR	R	R	R	MR

NOTE: The table entries describe each strain's susceptibility or resistance to each pest as follows: S = susceptible; MS = moderately susceptible; MR = moderately resistant; R = resistant. (From *Research Highlights*, I.R.R.I, 1973, p. 11.)

Figure I-1. A specially bred strain of dwarf wheat (right) *resists crop damage from "lodging" far better than the normal strain* (left). *(Courtesy of The Rockefeller Foundation.)*

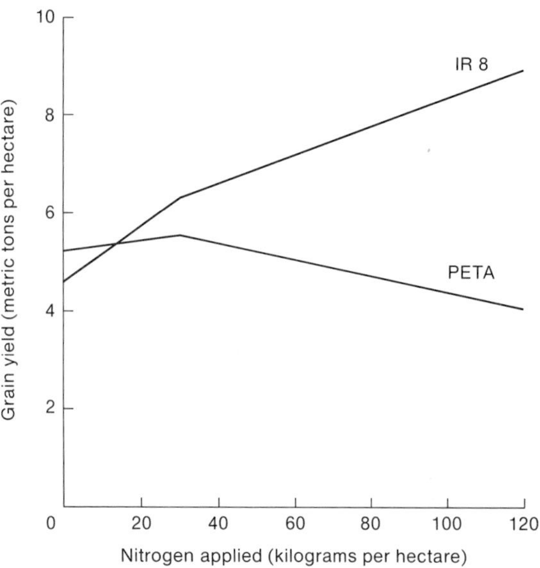

Figure I-2. The specially bred strain of dwarf rice called IR 8 owes part of its success to its remarkable response to the application of fertilizer. The strain PETA is an older nondwarf strain showing a more typical response. (From Peter R. Jennings, "The Amplification of Agricultural Production." Copyright © 1976 by Scientific American, Inc. All rights reserved.)

social and economic problems in the impoverished countries where they were most needed. Furthermore, the spread of monoculture (the extensive reliance on a single plant variety) left vast areas at the mercy of some newly introduced or newly evolved form of pathogen—say, a plant disease or an insect pest. With the huge population of humans on earth, our dependence on high-yield varieties of crop plants and domestic animals is becoming increasingly obvious. In a very real sense, the stability of human society depends on the ability of geneticists to juggle the inherited traits that shape life forms, keeping the crops a jump ahead of the destructive parasites and predators (Figure I-3).

Other genetic advances in recent years include the isolation of special strains of fungi and bacteria, which have greatly increased yields of antibiotics and other drugs derived from these organisms. The use of specially bred microorganisms offers even wider possibilities for the future. They may be used to clean up pollutants, to form a significant source of food, to transfer the ability to fix atmospheric nitrogen into important crop plants (thereby reducing the need for fertilizers), to provide functions that are missing in people suffering from disease—and on through a stunning list of ideas that are no longer simply the wild dreams of science fiction writers.

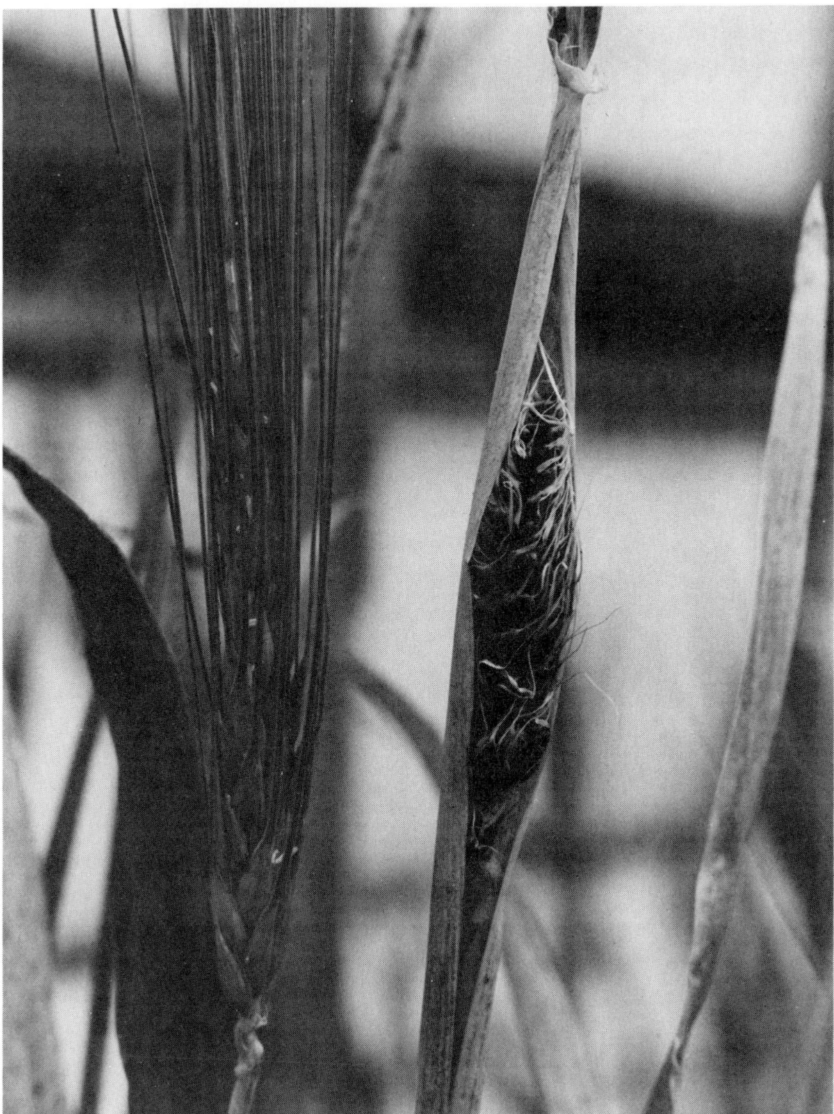

Figure I-3. Seed heads of barley. The plant at left has genetic resistance to the barley smut fungus Ustilago hordei. *The plant at right has no such resistance and has become infected with fungus, seen as a mass of black-colored spores.*

But the most exciting and frightening application of genetic knowledge is to the human species itself. Genetic discoveries have had major effects on medicine. We can now diagnose hereditary diseases before or soon after birth, and in some cases we can provide secondary cures. Using family pedigrees, a genetic counselor can give prospective parents the information they need to make intelligent decisions about the risks of genetic disease in their offspring. Such refined techniques as amniocentesis and fetoscopy provide information about possible genetic disease at early stages of pregnancy. A battery of postnatal chemical tests can detect problems in the newborn infant, so that some corrective techniques can be applied immediately to alleviate the effects of many genetic diseases.

Our new ability to recognize genetic disease poses an important moral dilemma. An estimated five percent of our population survives with severe physical or mental genetic defects. This percentage probably will increase with extended exposure to various environmental factors—and, paradoxically, with improved medical technology. As geneticist Theodosius Dobzhansky has remarked,

> If we enable the weak and the deformed to live and propagate their kind, we face the prospect of a genetic twilight. But if we let them die or suffer when we can save or help them, we face the certainty of a moral twilight.

Thirty percent of the patients admitted to pediatric hospitals in North America are estimated to have diseases that can be traced to genetic causes. The financial burden to society is already significant. Are we prepared to shoulder this genetic burden? How much money are we willing to spend to keep the genetically handicapped alive and to enable them to lead as normal a life as possible?

Many other significant issues are raised by the potential applications of genetic knowledge to human beings. Obviously, the human brain is subject to the same rules of genetic determination as the rest of the body. Does this mean that our thoughts and behavior are extensively determined by inherited predispositions? Or can we view the mind as a clean slate at birth, written upon only by individual experience? The nature of inborn constraints on thought and personality—the implications for present sociological problems—have fascinated many geneticists and other scientists. Such books as *African Genesis, The Territorial Imperative, On Aggression,* and other popular titles on sociobiology have stimulated widespread public interest. A bitter debate has raged about the differences in intelligence among various racial and social groups. Of course, this topic is not new. It was debated by Lycurgus in Sparta against Plato in Athens, and dreams of producing a pure race of superior humans have motivated many important figures in history. As the problem of human overpopulation becomes obvious to almost everyone, there is increased talk of legislated sterilization and the planned selection of human offspring. There is very serious talk, even among geneticists, about the ability of the human race to take control of its own evolution. Others are frightened by the possibilities for disastrous error or unpleasant sociological consequences.

The sophisticated technology of molecular genetics has given us a wide range of new techniques for shaping our genetic makeup. Even more bizarre procedures loom in the near future. We have moved beyond conventional breeding techniques to the ability to make chemical and molecular modifications in the genetic apparatus. While some emphasize the promised benefits of such research, others raise disturbing questions about possible dangers. Could there be an

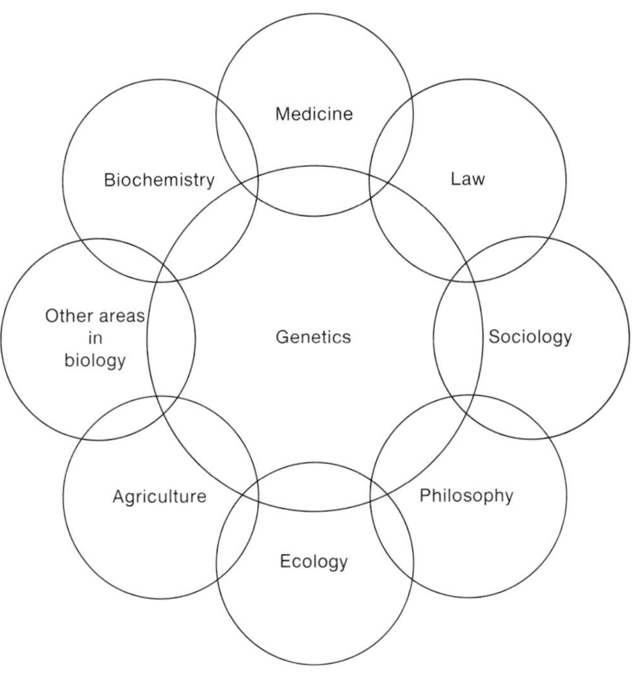

Figure I-4. The findings of genetics have powerful impacts on many interacting areas of human endeavor.

accidental release from some laboratory of an artificial pathogen that has never existed on this planet before? Such worries have led to calls for a complete moratorium on such research—or at least for legislation that requires ruthlessly efficient containment facilities. A number of popular books warn us of a *Genetic Fix,* a *Biological Time Bomb,* the *Genetic Revolution,* a *Fabricated Man,* and the *Biocrats.* Knowledge of genetic mechanisms has made us aware of other new dangers as well. Some geneticists fear that increased exposure to chemical food additives and to the vast array of chemicals in other commercial products may be changing our genetic makeup in a very undesirable and haphazard way. This type of random genetic change can also be caused by such environmental agents as fallout from nuclear weapons, radioactive contamination from nuclear reactors, and radiation from X-ray machines. These agents may be contributing to inherited disease, but they almost certainly are contributing to the incidence of cancer, a genetic disease of the somatic cells.

The study of genetics is relevant not only to the biologist but to any thinking member of today's complex technological society. A working knowledge of the principles of genetics is essential for making informed decisions on many scientific, political, and personal levels (Figure I-4). Such a working knowledge can come only through an understanding of the way that genetic inferences are made—that is, from an understanding of genetic analysis, the subject of this book.

Identical twins: identical genotypes but different people. (Copyright © Charles Harbutt/Magnum Photos, Inc.)

1

Gene and Organism

Genetics is the study of two contradictory aspects of nature: offspring resemble their parents, yet they are not identical to their parents. The offspring of lions are lions and never lambs, yet no two lions are identical, even if they come from the same litter. We have no trouble recognizing the differences between sisters—even "identical" twins are recognized as distinctive individuals by their parents and close friends. But we also see subtle similarities between parents and children.

As we shall see, **heredity** (the similarity of offspring to parents) and **variation** (the difference between parents and offspring, and between the offspring themselves) turn out to be two aspects of the same fundamental mechanism (Figure 1-1). This book explains the underlying mechanism and the methods of analysis that have revealed the mechanism. This chapter introduces some important concepts and discusses the scope of genetic analysis.

When humans began domesticating plants and animals (around 10,000 years ago), they necessarily became involved with both heredity and variation, because they had to choose among the organisms at their disposal those with advantageous characters and then seek to propagate those advantageous characters in future generations. References to sound breeding practices in Egyptian tomb inscriptions and in the Bible convince us that conscious concern with genetic phenomena is at least as old as civilization. The farmers and shepherds involved with such concerns could quite deservedly have claimed to be called geneticists. But what we now call genetics—a coherent and unified theory of heredity and variation—is little more than a century old.

Figure 1-1. *Generations. (Photo by Bruce Davidson, © Magnum Photos, Inc.)*

Modern genetics as a set of principles and analytic rules began with the work of an Augustinian monk, Gregor Mendel, who worked in a monastery in the middle of the nineteenth century in what is now Brno in Czechoslovakia. Mendel was taken into the monastery by its director, Abbot Knapp, with the expressed purpose of trying to discover a firm mathematical and physical foundation underlying the practice of plant breeding. Knapp and others in Brno were interested in fruit breeding. They believed that recent advances in mathematics in the physical sciences could be a model for building a science of variation. Mendel was recommended to Knapp as a good scholar of mathematics and physics, although a rather mediocre student of biology!

Mendel's methods, which he developed in the monastery garden, are the ones still used today (in an extended form), and they form an integral part of genetic analysis. (Mendel's work is considered in detail in Chapter 2.) Mendel realized that both the similarities and the differences among parents and their offspring can be explained by a mechanical passage of discrete hereditary units, which we now call genes, from parent to offspring. We now know that in all organisms —whether bacteria, fungi, animals, or plants—there is a regular passage of hereditary information from parent to offspring by means of the genes. The regularities we observe in heredity and variation are a consequence of the regularities of the mechanical lanes of passage and activity of these genes. Genes, however, are only part of the story, as we see in the following sections.

Gene and Environment

It is a characteristic of living organisms that they mobilize the components of the world around themselves and convert these components into their own living material, or into artifacts that are extensions of themselves. An acorn becomes an oak tree, using in the process only water, oxygen, carbon dioxide, some inorganic materials from the soil, and light energy.

The seed of an oak tree develops into an oak, while the spore of a moss develops into a moss, although both are growing side by side in the same forest (Figure 1-2). The two plants that result from these developmental processes

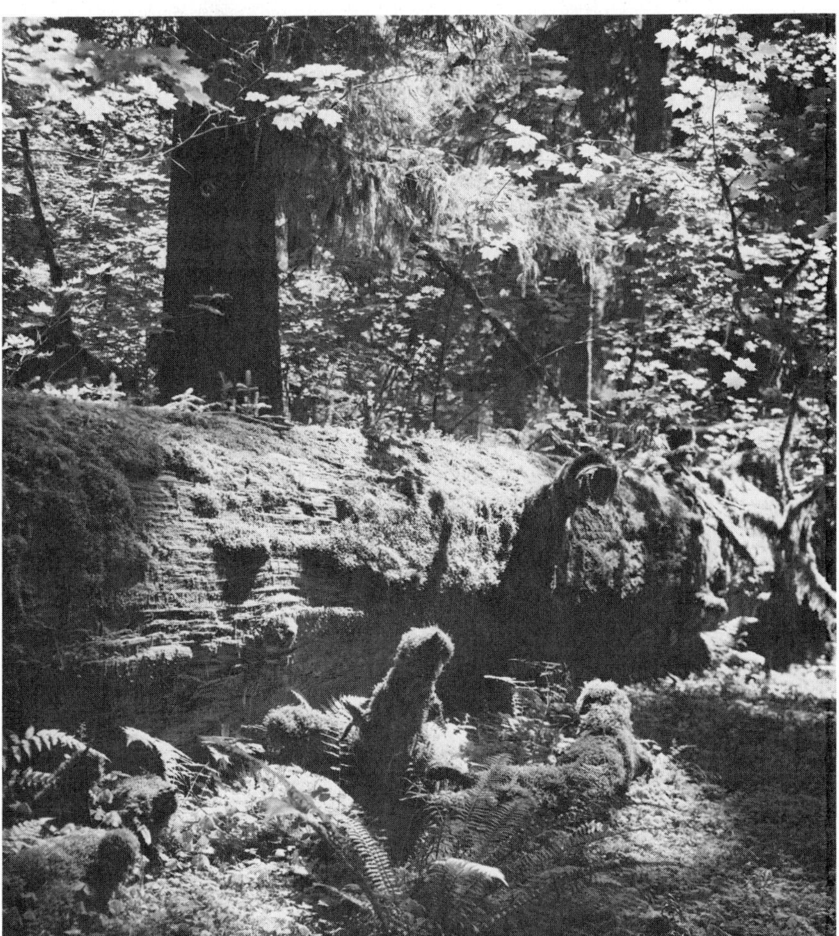

Figure 1-2. The genes of a moss direct environmental components to be shaped into a moss, whereas the genes of a tree cause a tree to be constructed from the same components. (From Grant Heilman.)

resemble their parents and differ from each other, even though they have access to the same narrow range of inorganic materials from the environment. The specifications for building living protoplasm from the environmental materials are passed in the form of genes from parent to offspring through the physical materials of the fertilized egg. As a consequence of the information in the genes, the seed of the oak develops into an oak, and the moss spore becomes a moss.

What is true for the oak and moss is also true within species. Consider plants of the species *Plectritis congesta,* the sea blush. Two forms of this species are found wherever the plants grow in nature; one form has winged fruits, and the other has wingless fruits (Figure 1-3). These plants will self-pollinate, and we can observe the offspring that result from such "selfs" when these are grown in a greenhouse under uniform conditions. It is commonly observed that the progeny of a winged-fruited plant are all winged-fruited, and the progeny from a wingless-fruited plant all have wingless fruits. Since all the progeny were grown in an identical environment, we can safely conclude that the difference between the original plants must result from the different genes they carry.

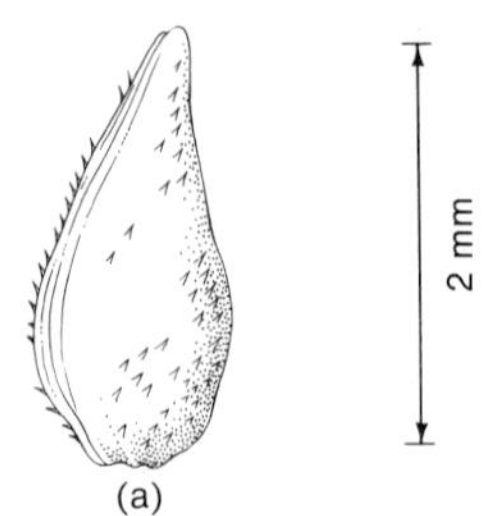

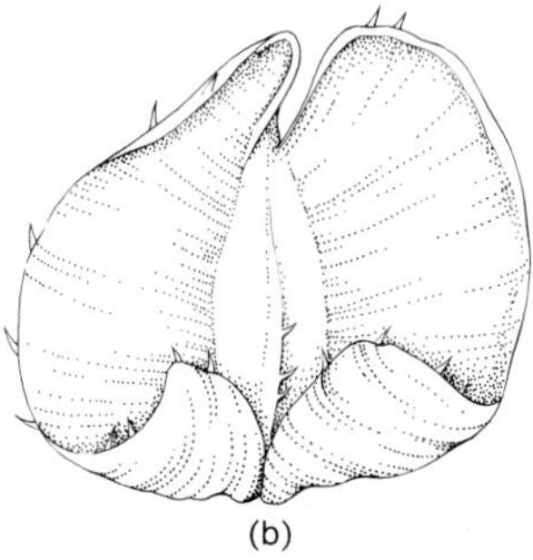

Figure 1-3. (a) Wingless and (b) winged fruits of Plectritis congesta, *the sea blush. Any one plant has either all wingless or all winged fruits, and only the fruits are different. The striking difference in appearance is determined by a simple genetic difference.*

Observations like these lead to a model of the interaction of genes and environment like that shown in Figure 1-4. In this view, the genes act as a set of instructions for turning more or less undifferentiated environmental materials into a specific organism, much as blueprints specify what form of house is to be built from basic materials. The same bricks, mortar, wood, and nails can be made into an A-frame or a flat-roofed house, according to different plans. Such a model implies that the genes are really the dominant elements in the determination of organisms; the environment simply supplies the undifferentiated raw materials.

But now consider two monozygotic ("identical") twins, the product of a single fertilized egg that divided and produced two complete individuals with identical genes. Suppose that the twins are born in England but separated at birth and taken to different countries. If one is raised in China by Chinese-speaking foster parents, she will speak perfect Chinese, while her sister raised in Budapest will speak fluent Hungarian. Each will absorb the cultural values and customs of her environment. Although the twins begin life with identical genetic properties, the different cultural environments in which they live pro-

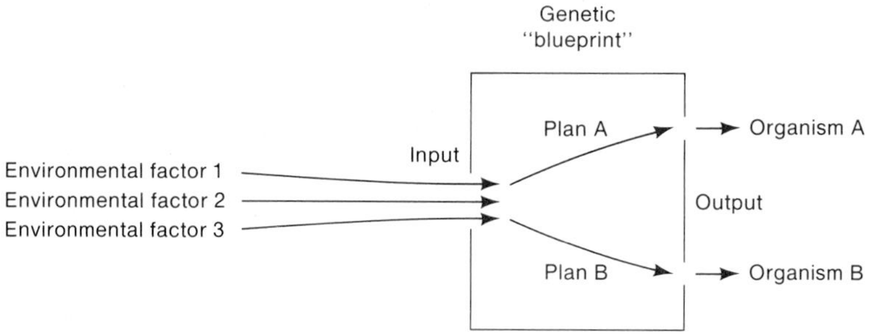

Figure 1-4. A model of determination that emphasizes the role of genes.

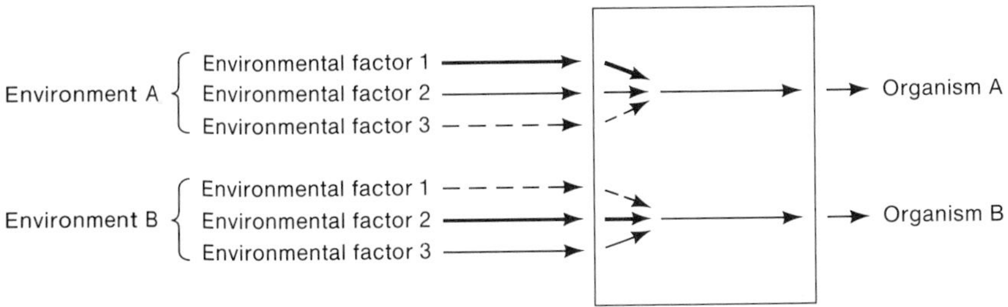

Figure 1-5. A model of determination that emphasizes the role of the environment.

duce differences between the sisters (and differences from their parents). Obviously, the difference in this case is due to the environment, and genetic effects are of little importance.

This example suggests the model of Figure 1-5, which is the opposite of that shown in Figure 1-4. Here the genes impinge on the system, giving certain general signals for development, but the environment determines the actual course of change. Imagine a set of specifications for a house that simply calls for "a floor that will support 30 pounds per square foot" or "walls with an insulation factor of 15"; the actual appearance and nature of the structure would be determined by the available building materials.

Our two different sets of examples lead to two very different models. Given a pair of seeds and a uniform growth environment, we would be unable to predict future growth patterns solely from a knowledge of the environment. In any environment we can imagine, if growth occurs at all, then the acorn will become an oak and the spore will become a moss. On the other hand, considering the twins, no information about the set of genes they inherit could possibly enable us to predict their ultimate languages and cultures. Two individuals that are *genetically different* may develop differently in the *same environment,* but two *genetically identical* individuals may develop differently in *different environments.*

In general, of course, we deal with organisms that differ in both genes and environment. If we wish to understand and predict the outcome of the development of a living organism, we must first know the genetic constitution that it inherits from its parents. Then we must know the *historical sequence* of environments to which the developing organism is exposed. We emphasize the historical sequence of environments rather than simply the general environment. Every organism has a developmental history from birth to death. What an organism will become in the next moment depends critically both on the environment it encounters during that moment and on its present state. It makes a difference to an organism not only what environments it encounters but in what sequence it encounters them. A fruitfly (*Drosophila*) develops normally at 20°C. If the

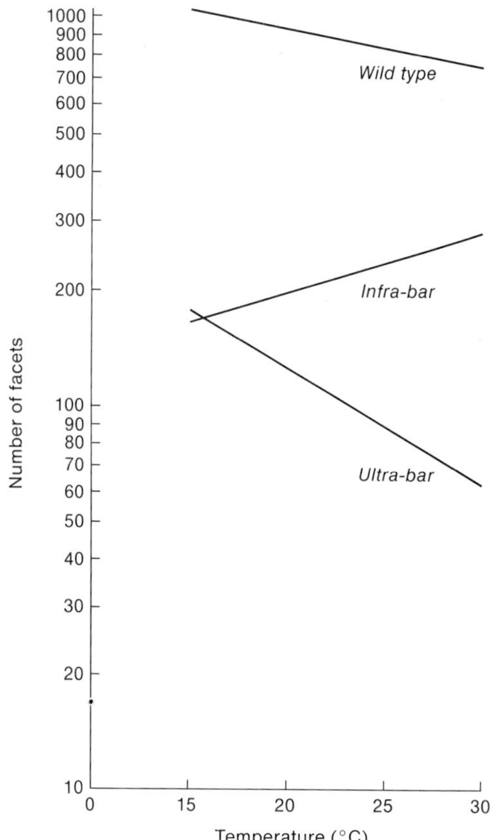

Figure 1-7. Norms of reaction to temperature for three different eye-size genotypes of Drosophila melanogaster: wild type, infra-bar, *and* ultra-bar. *Eye size is measured by the number of facets in the eye.*

of more extensive tabulated data. The size of the fly eye is measured by counting its individual facets, or cells. The vertical axis of the graph shows the number of facets (on a logarithmic scale); the horizontal axis shows the constant temperature at which the flies develop.

Three norms of reaction are shown on the graph. Flies of the *wild-type* genotype (characteristic of flies in natural populations) show somewhat smaller eyes at higher temperatures of development. Flies of the abnormal *ultra-bar* genotype have smaller eyes than *wild-type* flies at any particular temperature of development. Higher temperatures also have a stronger effect upon eye size in the *ultra-bar* flies (the *ultra-bar* line slopes more steeply). Any fly of the other abnormal genotype, *infra-bar,* also has smaller eyes than any *wild-type* fly, but higher temperatures have the opposite effect on flies of this genotype. *Infra-bar* flies raised at a higher temperature tend to have larger eyes than those raised at lower temperatures. These norms of reaction indicate that the relationship between genotype and phenotype is complex rather than simple.

Message
A single genotype can produce many different phenotypes, depending on the environment. A single phenotype may be produced by various different genotypes, depending on the environment.

If we know that a fruitfly has the *wild-type* genotype, this information alone does not tell us whether its eye has 800 or 1000 facets. On the other hand, the knowledge that a fruitfly's eye has 170 facets does not tell us whether its genotype is *ultra-bar* or *infra-bar.* We cannot even make a general statement about the effect of temperature on eye size in *Drosophila,* because the effect is opposite in two different genotypes. We see from Figure 1-7 that some genotypes do differ unambiguously in phenotype, no matter what the environment: any *wild-type* fly has larger eyes than any *ultra-bar* or *infra-bar* fly. But other genotypes overlap in phenotypic expression: the eyes of *ultra-bar* flies may be larger or smaller than those of *infra-bar* flies, depending on the temperatures at which the individuals developed.

To obtain a norm of reaction like those in Figure 1-7, we must allow different individuals of identical genotype to develop in many different environments. To carry out such an experiment, we must be able to obtain or produce many zygotes with identical genotypes. For example, to test a human genotype in ten environments, we would have to obtain identical decuplets and raise each individual in a different milieu. Obviously, that is possible neither biologically nor socially. At the present time, we do not know the norm of reaction of any human genotype for any character in any set of environments. Nor is it clear how we can ever acquire such information without unacceptable manipulation of human individuals.

For a few experimental organisms, special genetic methods make it possible to replicate genotypes and thus determine norms of reaction. Such studies are particularly easy in plants that can be propagated vegetatively—that is, by cuttings. The pieces cut from a single plant all have the same genotype, so all offspring produced in this way have identical genotypes. Figure 1-8 shows the results of a study using such vegetative offspring of the plant *Achillea.* Many plants were collected, and three cuttings were taken from each plant. One cutting was planted at low elevation (30 meters above sea level), one at medium elevation (1400 meters), and one at high elevation (3050 meters). Figure 1-8 shows the mature individuals that developed from cuttings of seven collected (parental) plants; the three plants of identical genotype are aligned vertically in the figure for comparison.

First, we note an average effect of environment: in general, the plants grew poorly at the medium elevation. This is not true for every genotype, however; the cutting of plant 24 grew best at the medium elevation. Second, we note that no genotype is unconditionally superior in growth to all others. Plant 4 showed the best growth at low and high elevations but showed the poorest

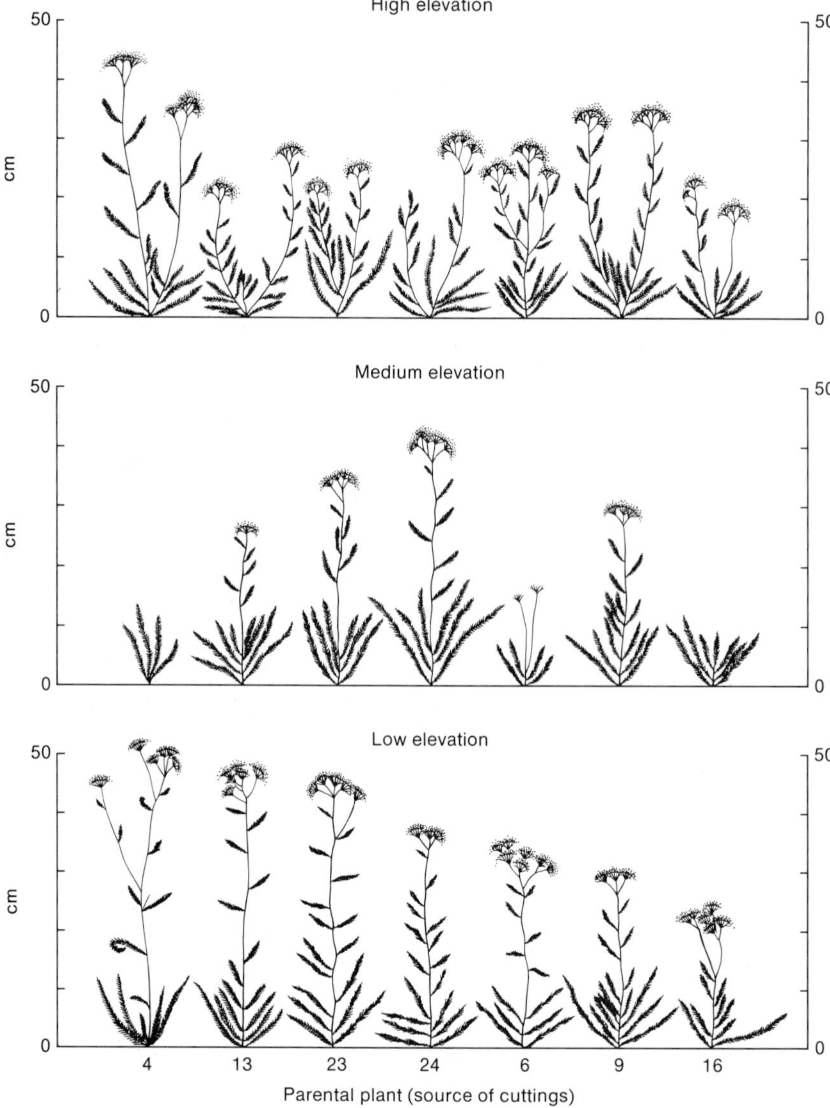

Figure 1-8. Norms of reaction to elevation for seven different Achillea *plants (seven different genotypes). A cutting from each plant was grown at low, medium, and high elevations. (Carnegie Institution of Washington.)*

growth at the medium elevation. Plant 9 showed the second-worst growth at low elevation and the second-best at high elevation. Once again we see the complex relationship between genotype and phenotype. Figure 1-9 graphs the norms of reaction derived from the results shown in Figure 1-8. Each genotype has a different norm of reaction, and the norms cross one another so that we

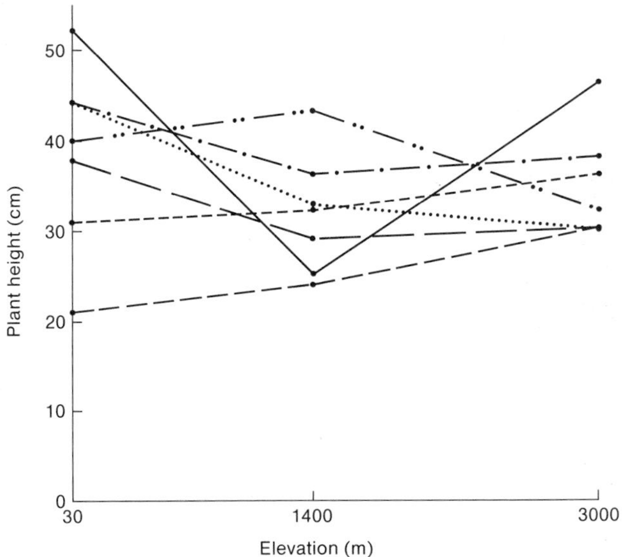

Figure 1-9. Graphic representation of the results in Figure 1-8. Each line represents the norm of reaction of a separate plant. Notice that the norms of reaction cross one another, so that no sharp distinction is apparent.

cannot identify either a "best" genotype or a "best" environment for *Achillea* growth.

We have seen two different patterns of reaction norms. The difference between *wild type* and the abnormal eye-size genotypes in *Drosophila* is such that the corresponding phenotypes show a consistent difference, regardless of the environment. Any fruitfly of *wild-type* genotype has larger eyes than any fruitfly of the abnormal genotypes, so we could (imprecisely) speak of "large-eye" and "small-eye" genotypes. In this case, the differences in phenotype between genotypes is much greater than the variation within a genotype for different environments. In the case of *Achillea,* however, the variation for a single genotype in different environments is so great that the norms of reaction cross one another and form no consistent pattern. In this case, it makes no sense to identify a genotype with a particular phenotype except in terms of response to particular environments.

Developmental Noise

Thus far, we have assumed that the phenotype is uniquely determined by the interaction of a specific genotype and a specific environment. But a closer look shows some further unexplained variation. According to Figure 1-7, a *Dro-*

sophila of *wild-type* genotype raised at 16°C has 1000 facets in each eye. In fact, this is only an average value; individual flies studied under these conditions commonly have 980 or 1020 facets. Perhaps these variations are due to slight fluctuations in the local environment or slight differences in genotypes? However, a typical count may show that a fly has, say, 1017 facets in the left eye and 982 in the right eye. In another fly, the left eye has slightly fewer facets than the right eye. Yet the left and right eyes of the same fly are genetically identical. Furthermore, the fly has developed as a larva a few millimeters long burrowing in homogeneous artificial food in a laboratory bottle, and then completed its development as a pupa (also a few millimeters long) glued vertically to the inside of the glass high up off the food surface. Surely, the environment has not differed significantly from one side of the fly to the other! But if the two eyes have experienced the same sequence of environments and are identical genetically, then why is there any phenotypic difference between left and right eyes?

Differences in shape and size are partly dependent on the process of cell division that turns the zygote into the multicellular organism. Cell division, in turn, is sensitive to molecular events within the cell, and these may have a relatively large random component. For example, the vitamin biotin is essential for growth, but its *average* concentration is only one molecule per cell! Obviously, any process that depends on the presence of this molecule will necessarily be subject to fluctuations in rate because of random variations in concentration. But if a cell division is to produce a differentiated eye cell, it must occur within a relatively short developmental period during which the eye is being formed. Thus we would expect random variation in such phenotypic characters as the number of eye cells, the number of hairs, the exact shape of small features, and the variation of neurons in a very complex central nervous system—even when the genotype and the environment are precisely fixed. Even in such structures as the very simple nervous system of nematodes, these random variations do occur.

Message

Random events in development lead to an uncontrollable variation in phenotype; this variation is called ***developmental noise.*** *In some characters, such as eye cells or bristles in* Drosophila, *developmental noise is a major source of the observed variations in phenotype.*

Like noise in a verbal communication, developmental noise adds small random variations to the predictable process of development governed by norms of reaction. Adding developmental noise to our model of phenotype development, we obtain something like Figure 1-10. With a given genotype and environment, there is a range of possible outcomes of each developmental

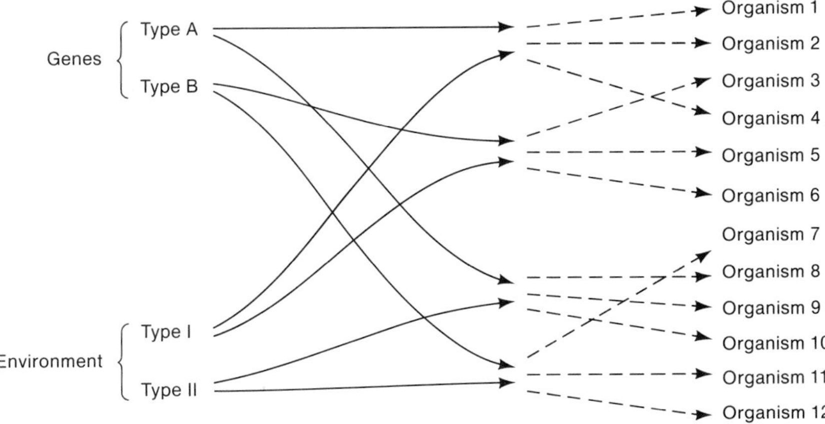

Figure 1-10. A model of phenotypic determination that shows how genes, environment, and developmental noise interact to produce any given phenotype.

step. The development process does contain feedback systems that tend to hold the deviations within certain bounds, so that the range of deviation does not increase indefinitely through the many steps of development. However, this feedback is not perfect. For any given genotype, developing in any given sequence of environments, there remains some uncertainty in the exact phenotype that will result.

Variation Is the Raw Material of Genetic Analysis

We have seen that heredity and variation are results of the same basic genetic process. Simple observations of biological and environmental variation have led us to some important general statements about the similarities and differences among organisms. Genetic analysis goes further. In genetic analysis, biological variants are experimentally manipulated in a variety of standard ways. From the results of these manipulations, the geneticist makes specific inferences about the nature of the genetic material, the mechanism of its passage between generations, and its role in influencing the phenotype.

Recall that a set of variants of any given biological character can fall into either of two broad categories: (1) those who underlying genotypes show non-overlapping norms of reaction, and (2) those whose underlying genotypes show overlapping norms of reaction. The techniques of genetic analysis are different in each case.

Analysis of Genotypes with Nonoverlapping Norms of Reaction

The analysis of genotypes with nonoverlapping norms of reaction has produced most of the phenomenal success of genetics in elucidating the cellular and molecular processes of organisms. The appeal of such systems from the experimenter's viewpoint is that the environment can virtually be ignored, because each genotype produces a discrete, identifiable phenotype. In fact, geneticists have deliberately sought and selected just such variants. Experiments become much easier—in fact, they only become possible—when the phenotype observed is an unambiguous indication of a particular genotype. After all, one cannot observe genotypes, only phenotypes.

The availability of such variants made Mendel's experiments possible. He used genotypes that were established horticultural varieties. For example, in one set of variants, one genotype always produces tall pea plants, and another always produces dwarf plants. Had Mendel chosen variants whose genotypes have overlapping norms of reaction, he would have obtained complex and patternless results like those in the *Achillae* experiments. He could never have identified the simple relationships that became the foundation of genetic understanding. Much of modern molecular genetics is based on research with variants of bacteria that similarly show distinctive characters whose genotypes have nonoverlapping norms of reaction.

In this category of nonoverlapping norms of reaction, the variants tend to be rather drastic. Many of these variants could never survive in nature, simply because they are so developmentally extreme. In other cases, such as the winged and wingless fruits of *Plectritis,* the strikingly different forms do appear regularly in natural populations.

Reflecting the importance of discrete variants in developing our present understanding of the genetic basis of life, the next 16 chapters of this book are devoted to analysis of this kind of variation. The story begins with Mendel's research and theories, proceeds through classical genetics, and ends with the startling discoveries of molecular biology. We must emphasize that the entire development of this knowledge critically depended on the availability of phenotypes having simple relationships to genotypes.

Analysis of Genotypes with Overlapping Norms of Reaction

We have seen that simple one-to-one relationships between genotype and phenotype dominate the world of experimental genetics. In the natural world, such relationships are quite rare. When norms of reaction for size, shape, color, metabolic rate, reproductive rate, and behavior have been obtained for organisms taken from natural populations, they almost always turn out to be like those of *Achillea.* The relationships of genotype to phenotype in nature are almost always one-to-many rather than one-to-one. This is the underlying cause of the rarity of discrete phenotypic classes in natural populations.

Obviously, the analysis of these one-to-many relationships is far more complex. The researcher is confronted with a bewildering range of phenotypes. Special statistical techniques must be used in an attempt to disentangle the genetic, environmental, and noise components. Chapters 18 and 19 deal with the special techniques needed for such analyses.

There is another problem that is distinct from the purely analytical difficulties. The major discoveries of genetics were founded on work with discrete norms of reaction. Although these discoveries have revolutionized pure and applied biology, great care must be taken in extrapolating such ideas to systems that are based on complex interaction of gene and environment, especially those found in nature. The difference between yellow-bodied and gray-bodied *Drosophila* in laboratory stocks can be shown to be a simple genetic difference, but it does not follow that skin color in humans (for example) obeys similar genetic laws. In fact, even in *Drosophila,* the various intensities of black pigment in flies from natural populations turn out to be a consequence of the interaction of temperature with a complex genetic system.

Message

In general, the relationship between genotype and phenotype cannot be extrapolated from one species to another or even between phenotypic traits that seem superficially similar within a species. A genetic analysis must be carried out for each particular case.

Our discussion of genetics thus far has been based on the wisdom of hindsight. With the wealth of genetic knowledge we now share, we can talk of genes, phenotypes, and genotypes as though these concepts are self-evident. Obviously, this was not always the case. Mendel almost certainly was completely in the dark at the beginning of his experiments. How did he infer the existence of genes? How was an entirely new scientific discipline born? That is the subject of the next chapter.

Summary

Genetics has two fundamental aspects: heredity, or the similarity of offspring to parents, and variation, or the differences among parents and their offspring. Both the similarity and the differences are based on the transmission of discrete hereditary units called genes.

Genes do not act in a vacuum, however. It is the interaction of genes and environment that determines how an organism develops. Sometimes people confuse an individual's genes with the individual's developmental outcome. The distinction between the two is expressed in the terms *genotype,* which refers to gene makeup, and *phenotype,* which refers to manifest characters.

The correspondence between environment and phenotype may be tabulated as the norm of reaction of a genotype. There are two basic categories of reaction norms. Genotypes with nonoverlapping norms of reaction have a simple, one-to-one relation to phenotype. Genotypes with overlapping norms of reaction have a one-to-many relation to phenotype. Thus, a particular genotype may have many different phenotypes, depending on environment; conversely, a particular phenotype may have several different genotypes underlying it. Phenotype, however, is not solely determined by the interaction of genotype and environment. Unexplained variation may occur as a result of developmental noise.

Variation is essential to genetic analysis. By manipulating biological variants in a variety of ways, geneticists can make specific inferences about the nature of the genetic material, the mechanics of its passage between generations, and its role in influencing the phenotype. Analysis of genotypes with nonoverlapping norms of reaction is relatively easy because of the simple relation between genotype and phenotype. Analysis of genotypes with overlapping norms of reaction is more complex because of the one-to-many relation.

The garden pea, Pisum sativum, *Mendel's experimental organism. (Copyright © John Colwell/Grant Heilman.)*

2

Mendelism

The gene is the focal point of the discipline of modern genetics. In all lines of genetic research, it is the gene that provides the common unifying thread to a great diversity of experimentation. Geneticists are concerned with the transmission of genes from generation to generation, with the nature of genes, with the variation in genes, and with the ways that genes function to dictate the features that constitute any given species.

In this chapter we trace the birth of the gene as a concept. We shall see that genetics is, in one sense, an abstract science: most of its entities began as hypothetical constructs in the minds of geneticists and were later identified in physical form if the reasoning was sound.

The concept of the gene (but not the word) was first set forth in 1865 by Gregor Mendel. Until then, little progress had been made in understanding heredity. The prevailing notion was that the spermatozoon and egg contain a sampling of essences from the various parts of the parental body. At conception, these essences are somehow blended to form the pattern for the new individual. This idea of **blending inheritance** accounts for the fact that offspring typically show some characteristics similar to those of both parents. However, attempts to expand and improve this theory never led to much progress.

As a result of his research with pea plants, Mendel proposed instead a theory of **particulate inheritance.** A genetic determinant of a specific character is passed on from one generation to the next as a unit, without any blending of the units. This model explained many observations that could not be explained by blending inheritance. It also proved a very fruitful framework for further progress in understanding the mechanism of heredity.

For many reasons, the importance of Mendel's ideas was not recognized until about 1900 (after his death). His report was then rediscovered by three scientists after each independently had reached the same conclusions. Historically, then, Mendel's work was irrelevant to the development of ideas about heredity. However, his achievement is important, and his analysis provides a good example of basic genetic reasoning, so we now examine Mendel's work.

Mendel's Experiments

Mendel's studies provide an outstanding example of good scientific technique. He chose research material well suited to study of the problem at hand, designed his experiments carefully, collected large amounts of data, and used mathematical analysis to show that the results were consistent with his explanatory hypothesis. The predictions of the hypothesis were then tested in a new round of experimentation. (Some historians of science believe that Mendel may have "fudged" his data to fit his hypothesis, but we shall accept his reports at face value in our discussion here.)

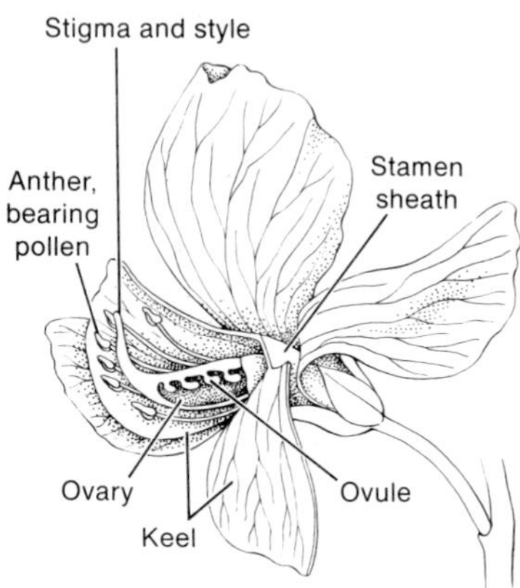

Figure 2-1. Cutaway view of reproductive parts of a pea flower. (After J. B. Hill, H. W. Popp, and A. R. Grove, Jr., Botany, *McGraw-Hill, 1967.)*

Mendel studied the garden pea (*Pisum sativum*) for two main reasons. First, peas were available through a seed merchant in a wide array of distinct shapes and colors that are very easily identified and analyzed. Second, peas left to themselves will self-pollinate (or **self**) because the male parts (anthers) and female parts (ovaries) of the flower—which produce the pollen and the eggs, respectively—are enclosed in a petal box, or keel (Figure 2-1). The gardener or experimenter can cross-pollinate (or **cross**) any two plants at will. The anthers from one plant are clipped off to prevent selfing; pollen from the other plant is then transferred to the receptive area with a paintbrush or on transported anthers (Figure 2-2). Thus the experimenter can readily choose to self or to cross the pea plants.

Plants Differing in One Character

First Mendel chose several characters to study, and he established pure lines of plants for these characters. This was a clever beginning because it amounts to a control experiment; he had to be sure of the constant behavior of his research material. A **pure line** is a population that breeds true for the particular character being studied; that is, all offspring produced by selfing or crossing within this population show the same form for this character.

For example, Mendel obtained two lines, each of which bred true for the character of flower color. One line bred true for purple flowers, and the other

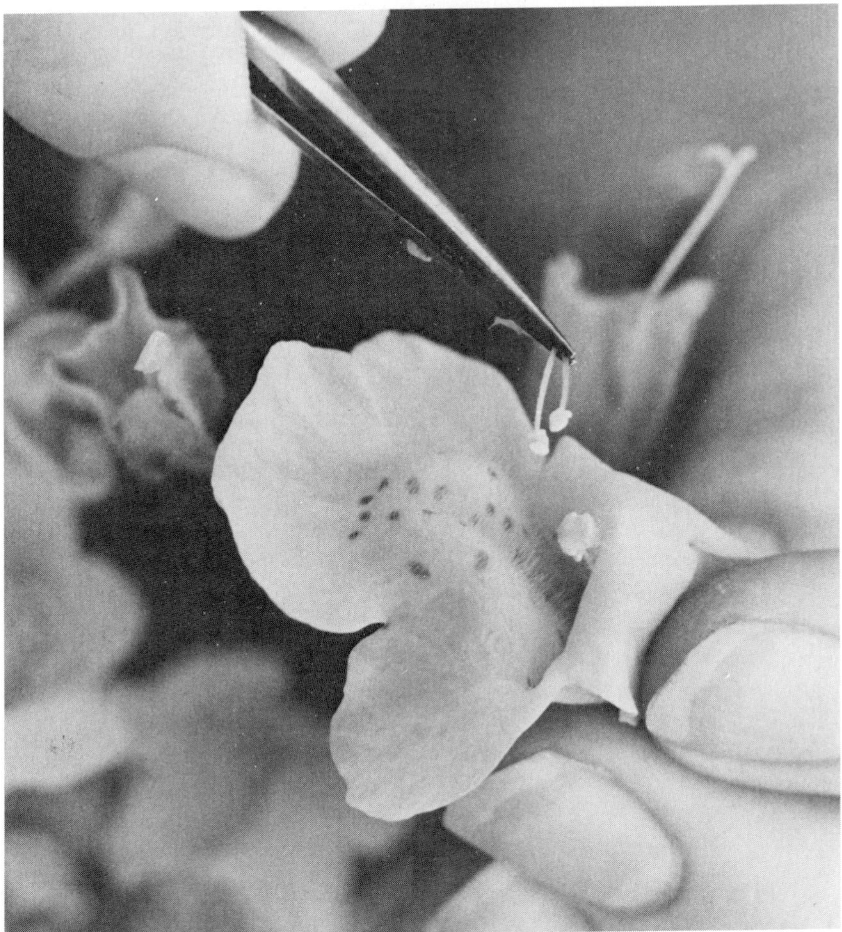

Figure 2-2. One technique of artificial cross-pollination, demonstrated with Mimulus guttatus, *the yellow monkey flower. Anthers from the male parent are applied to the stigma of an emasculated flower, which acts as the female parent.*

for white flowers. Any plant in the purple-flowered line—when selfed, or when crossed with others from the same line—produced seeds that all grew into plants with purple flowers. When these plants in turn were selfed or crossed within the line, their progeny also had purple flowers, and so on. The white-flowered line similarly produced only white flowers through all generations. In all, Mendel obtained seven pairs of pure lines for seven characters, with each pair differing in respect to only one character (Figure 2-3).

It is important to clarify the meaning of the term **character.** A character is a specific property of an organism; we use it as a synonym for the terms *characteristic* or *trait*. Each pair of Mendel's plant lines can be said to show a **character difference,** a contrasting difference between two lines of organisms

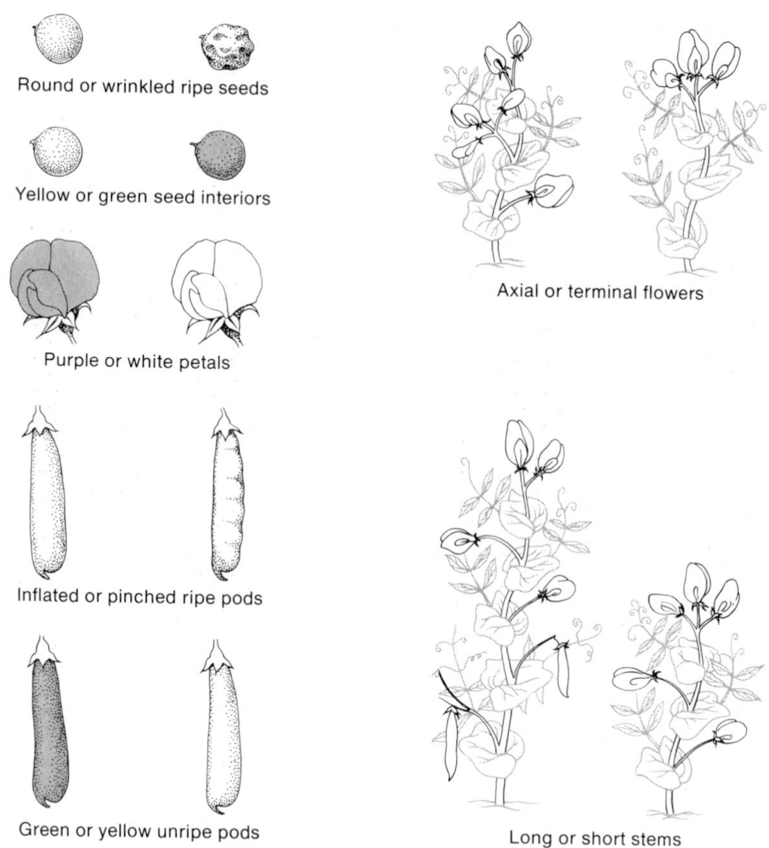

Figure 2-3. The seven character differences studied by Mendel. (After S. Singer and H. Hilgard, The Biology of People. *Copyright © 1978, W. H. Freeman and Company.)*

(or between two organisms) with respect to one particular character. The differing lines (or individuals) represent different forms that the character may take: they can be called character forms, character variants, or phenotypes. The useful term **phenotype** (derived from Greek) literally means "the form that is shown." The term is extensively used in genetics, and we use it in this discussion, even though such words as *gene* and *phenotype* were not coined or used by Mendel. We shall describe Mendel's results and hypotheses in terms of more modern genetic language.

Figure 2-3 shows seven characters, each represented by two contrasting phenotypes. Contrasting phenotypes for a particular character are the starting point for any genetic analysis, by Mendel or by the modern geneticist. Of course, the definition of characters is somewhat arbitrary; there are many different ways to "split up" an organism into characters. For example, consider the following different ways of stating the same character and character difference.

Character	Phenotypes
Flower color	Red versus white
Flower redness	Presence versus absence
Flower whiteness	Absence versus presence

In many cases, the description chosen is a matter of convenience (or chance). Fortunately, the choice of definition does not alter the final conclusions of the analysis, except in the names used.

We turn now to some of Mendel's specific experimental results. In our discussion, we shall follow his analysis of the lines breeding true for flower color.

In one of his early experiments, Mendel used pollen from a white-flowered plant to pollinate a purple-flowered plant. These plants from the pure lines are called the **parental generation** (P). All the plants resulting from this cross had purple flowers (Figure 2-4). This progeny generation is called the **first filial generation** (F_1). (The subsequent generations in such an experiment are called F_2, F_3, and so on.)

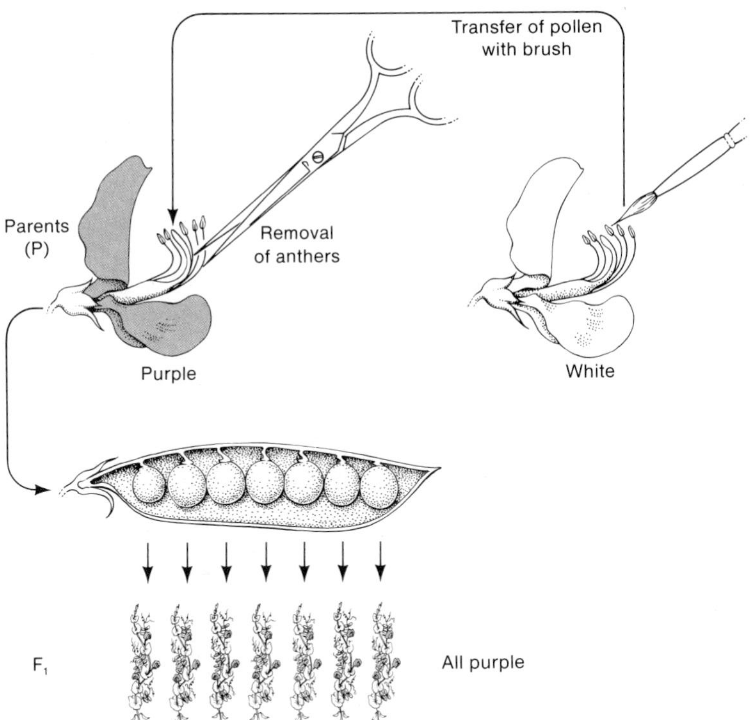

Figure 2-4. Mendel's cross of purple-flowered ♀ × white-flowered ♂.

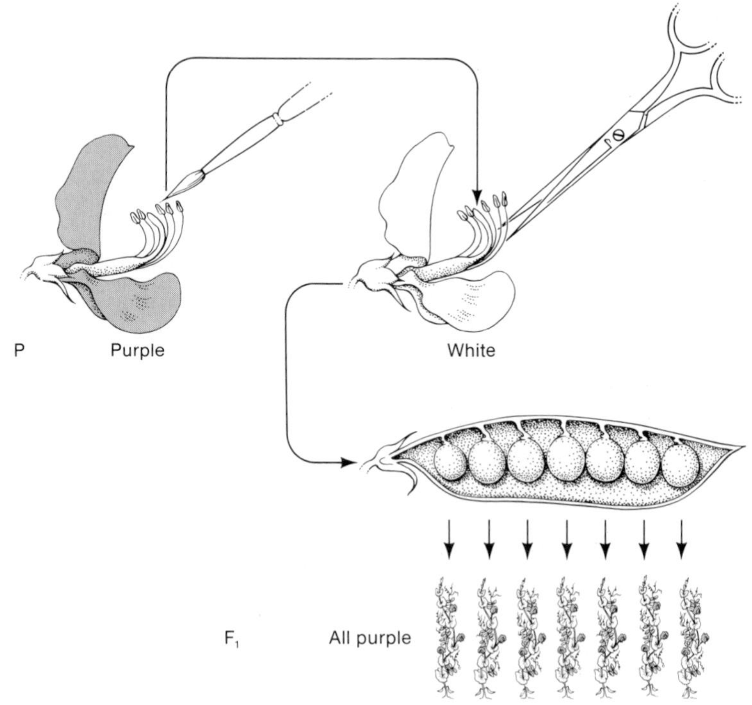

Figure 2-5. *Mendel's cross of white-flowered ♀ × purple-flowered ♂.*

The **reciprocal cross**—that is, a white flower pollinated by a purple-flowered plant—produced the same result in the F_1 (Figure 2-5). Mendel concluded that it makes no difference which way the cross is made. If one parent is purple-flowered and the other white-flowered, all plants in the F_1 are purple-flowered. The purple flower color in the F_1 generation is identical to that in the purple-flowered parental plants. In this case, the inheritance obviously is not a simple blending of purple and white colors to produce some intermediate color. To maintain a theory of blending inheritance, we would have to assume that the purple color is somehow "stronger" than the white color, completely overwhelming any trace of the white phenotype in the blend.

Next, Mendel selfed the F_1 plants, allowing the pollen of each flower to fall on the stigma within its petal box. He obtained 929 pea seeds from this selfing (the F_2 individuals) and planted them. Amazingly, some of the resulting plants were white-flowered; the white phenotype had reappeared! Mendel then did something that, more than anything else, marks the birth of modern genetics: he *counted* the numbers of plants with each phenotype. This procedure seems obvious to us after another century of quantitative scientific research, but it had seldom if ever been used in genetic studies before Mendel's work. There

Table 2-1. Results of all Mendel's crosses in which parents differed for one character

Parent phenotypes	F_1	F_2	F_2 ratio
1. Round × wrinkled seeds	All round	5474 round; 1850 wrinkled	2.96:1
2. Yellow × green seeds	All yellow	6022 yellow; 2001 green	3.01:1
3. Purple × white petals	All purple	705 purple; 224 white	3.15:1
4. Inflated × pinched pods	All inflated	882 inflated; 299 pinched	2.95:1
5. Green × yellow pods	All green	428 green; 152 yellow	2.82:1
6. Axial × terminal flowers	All axial	651 axial; 207 terminal	3.14:1
7. Long × short stems	All long	787 long; 277 short	2.84:1

SOURCE: From *The Biology of People* by S. Singer and H. H. Hilgard. Copyright © 1978 by W. H. Freeman and Company.

were 705 purple-flowered plants and 224 white-flowered plants. Mendel observed that the ratio of 705:224 is almost a 3:1 ratio (in fact, it is 3.1:1).

Mendel repeated this breeding procedure for six other pairs of pea character differences. He found the same 3:1 ratio in the F_2 generation for each pair (Table 2-1). By this time, he was undoubtedly beginning to believe in the significance of the 3:1 ratio and seeking an explanation. The white phenotype is completely absent in the F_1 generation, but it reappears (in its full original form) in one-fourth of the F_2 plants. It is very difficult to devise an explanation of this result in terms of blending inheritance. Even though the F_1 flowers are purple, the plants still must carry the *potential* to produce progeny with white flowers.

Mendel inferred that the F_1 plants receive from their parents the ability to produce both the purple phenotype and the white phenotype, and that these abilities are retained and passed on to future generations rather than blended. Why is the white phenotype not expressed in the F_1 plants? Mendel invented the terms **dominant** and **recessive** to describe this phenomenon without explaining the mechanism. In modern terms, the purple phenotype is dominant to the white phenotype, and the white phenotype is recessive to the purple. Thus, the phenotype of the F_1 provides the operational definition of dominance.

Mendel made another important observation when he individually selfed the F_2 plants. In this case, he was working with the character of seed color. Because this character can be observed without growing plants from the peas, much larger numbers of individuals can be counted. (In this species, the color of the seed is characteristic of the offspring—the seed itself—not of the parent plant.) Mendel used two pure lines of plants with yellow and green seeds, respectively. In a cross between one plant from each line, he observed that all of the F_1 peas were yellow. Symbolically,

$$
\begin{array}{ll}
\text{P} & \text{Yellow} \times \text{Green} \\
 & \quad\quad\downarrow \\
\text{F}_1 & \quad\text{All yellow}
\end{array}
$$

Therefore, in this character, yellow is dominant and green is recessive.

Mendel grew F_1 plants from these yellow F_1 peas and selfed the plants. Of the resulting F_2 peas, 3/4 were yellow and 1/4 were green—the 3:1 ratio again (see the second line of Table 2-1). He then grew plants from 519 of the yellow F_2 peas and selfed each of these F_2 plants. When the peas appeared (the F_3 generation), he found that 166 of the plants had only yellow peas. The remaining 353 plants bore both yellow and green peas on the same plant. Counting all the peas from these plants, he again obtained a 3:1 ratio of yellow to green peas. The green F_2 peas all proved to be pure-breeding green peas; that is, selfing produced only green peas in the F_3 generation.

In summary, all of the F_2 green peas were pure-breeding greens like one of the parents. Of the F_2 yellows, about 2/3 were like the F_1 yellows (producing yellow and green seeds in a 3:1 ratio when selfed), and the remaining 1/3 were like the pure-breeding yellow parent. Thus the study of the F_3 generation revealed that the apparent 3:1 ratio in the F_2 generation could be more accurately described as a 1:2:1 ratio.

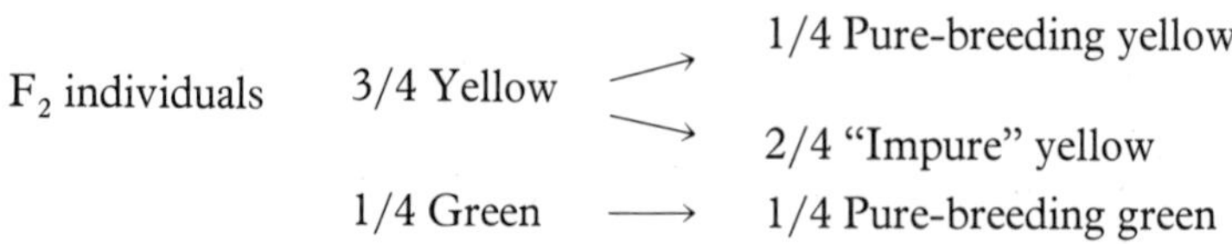

Further studies showed that these 1:2:1 ratios existed in all of the apparent 3:1 ratios that Mendel had observed. Thus the problem really was to explain the 1:2:1 ratio.

Mendel's explanation was a classical example of a model or hypothesis derived from observation, a model that was well-suited for testing by further experimentation. Mendel deduced the following explanation.

1. There are hereditary determinants of a particulate nature. (He saw no blending of phenotypes, so he was forced to this particulate notion.) We now call these determinants genes.

2. Each adult pea plant has two genes (a **gene pair**) in each cell for each character studied. The reasoning here was obvious: the F_1 plants, for example, must have had one gene responsible for the dominant phenotype (called the **dominant gene**) and one gene for the recessive phenotype, which only showed up in later generations (called the **recessive gene**).

3. During sex-cell formation, the members of each gene pair segregate (separate) equally into the sex cells.

4. Consequently, each sex cell (pollen or egg cell) carries only one gene for each character.

5. The union of sex cells, to form the first cell (or zygote) of a new progeny individual, is random and occurs irrespective of which member of a gene pair is carried.

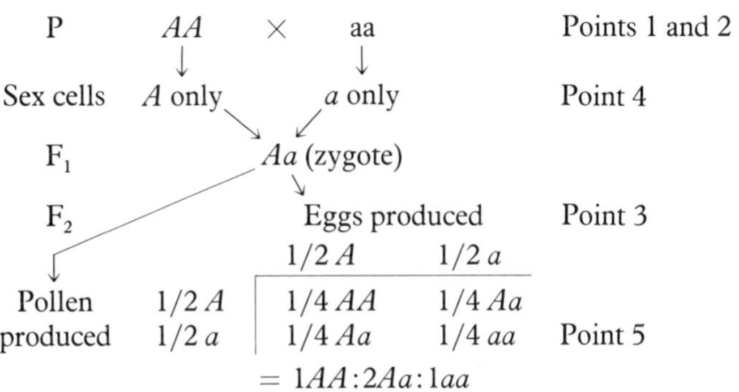

Figure 2-6. Symbolic representation of the P, F_1, and F_2 generations in Mendel's system involving a character difference determined by one gene difference. The five points are those listed in the text.

These points can be illustrated diagrammatically for a general case, using *A* to represent a dominant gene, and *a* the recessive gene (as Mendel did), much as a mathematician uses symbols to represent abstract entities of various kinds. This is shown in Figure 2-6, which illustrates how these five points explain the 1:2:1 ratio.

The whole model made beautiful sense of the data. However, many beautiful models have been knocked down under test: Mendel's next job was to test it. He did this by taking (for example) an F_1 yellow and crossing it with a green. A 1:1 ratio of yellow to green seeds could be predicted in the next generation. If we use *Y* to stand for the dominant gene causing yellow seeds and *y* to stand for the recessive gene causing green seeds, we can diagram Mendel's predictions as in Figure 2-7. In this experiment, he obtained 58 yellow and 52 green seeds, a very close approximation to the predicted 1:1 ratio, and confirming

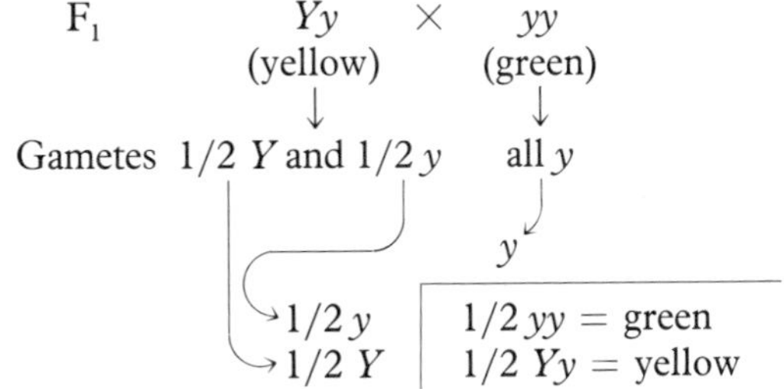

Figure 2-7. Predicted consequences of crossing an F_1 yellow with any green.

the equal segregation of Y and y in the F_1 individual. The third point in our version of Mendel's explanation has been given formal recognition as Mendel's first law.

Mendel's First Law
The two members of a gene pair segregate (separate) from each other during sex-cell formation, so that one-half of the sex cells carry one member of the pair and the other one-half of the sex cells carry the other member of the gene pair.

Now we need to introduce some more terms. The individuals represented by *Aa* are called **heterozygotes** (sometimes, **hybrids**), whereas those in pure lines are called **homozygotes**. Thus an *AA* plant is said to be homozygous for the dominant gene (sometimes called **homozygous dominant**), and an *aa* plant is homozygous for the recessive gene (**homozygous recessive**). Sex cells in the organisms most familiar to us are the **gametes**. The designated genetic constitution with respect to the character or characters under study is called the **genotype.** Thus, *YY* and *Yy*, for example, are different genotypes even though the seeds of both types are of the same phenotype (that is, yellow). You can see that the phenotype can be thought of simply as the outward manifestation of the underlying genotype.

Note that the expressions *dominant* and *recessive* have been used in conjunction with both the phenotype *and* the gene. This is accepted usage. The dominant phenotype is established in analysis by the appearance of the F_1. Obviously, however, a phenotype (which is merely a description), cannot really exert dominance. Mendel showed that the dominance of one phenotype over another is in fact due to the dominance of one member of a gene pair over another.

Let's pause to let the significance of this work sink in. What Mendel had done was to develop an analytical scheme for the identification of major genes regulating any biological character or function. Starting with two different phenotypes (purple and white) of one character (petal color), he was able to show that the difference was caused by differences centering on one gene pair. Let's take the two forms of the petal-color character as an example. Modern geneticists would say that Mendel's analysis had identified a major gene for petal color. What does this mean? It means that in these organisms there is a *kind* of gene that has a profound effect on the color of the petals. This gene can exist in different forms: the dominant form of the gene (represented by *C*) causes purple petals, and the recessive form of the gene (represented by *c*) causes white petals. The forms *C* and *c* are called **alleles** (or alternative forms) of that gene for petal color. They are given the same letter symbol to show that they are forms of the same kind of gene. One could express this another way by saying that there is a kind of gene, called phonetically a "see" gene, with alleles *C* and *c*. Any individual pea plant will always have two "see" genes,

Table 2-2. Summary of the modus operandi for establishing simple Mendelian inheritance

Experimental procedure:	1. Choose pure lines showing a character difference (purple versus white flowers). 2. Intercross the lines. 3. Self the F_1 individuals.
Results:	F_1 is all purple; F_2 is 3/4 purple and 1/4 white.
Inferences:	1. The character difference is controlled by a major gene for flower color. 2. The dominant allele of this gene causes purple petals; the recessive allele causes white petals.

Symbolic interpretation:

Character	Phenotypes	Genotypes	Alleles	Gene
Flower color	Purple (dominant)	*CC* (homozygous dominant)	*C* (dominant)	Flower-color gene
		Cc (heterozygous)	*c* (recessive)	
	White (recessive)	*cc* (homozygous) recessive)		

forming a gene pair, and the actual members of the gene pair can be either *CC*, *Cc*, or *cc*. Notice that although the members of a gene pair can produce different effects, they obviously both affect the same character, and presumably in nature they both do exactly the same job, in a double dose.

Students often find the term *allele* confusing, and the reason is probably that the words *allele* and *gene* are used interchangeably in some situations. For example, *dominant gene* and *dominant allele* both refer to the same thing in an interchangeable way. This stems from the fact that the forms (alleles) of any type of gene are of course genes themselves.

The basic route of Mendelian analysis for a single character is summarized in Table 2-2.

Message
Genes originally were inferred (and still are today) by observing precise mathematical ratios in the filial generations issuing from two parental individuals.

Plants Differing in Two Characters

The experiments described thus far dealt with a single gene pair (a **monohybrid** system); we considered the alleles of only one gene affecting one character. The next obvious question is what happens when a **dihybrid** cross is made, involving two pairs of genes affecting two different characters. We

Figure 2-8. Round (R) *and wrinkled* (r) *garden peas. (Grant Heilman.)*

can use the same symbolism that Mendel used to indicate the genotype of seed color (*Y* and *y*) and seed shape (*R* and *r*).

A pure-breeding line of *RR yy* plants, on selfing, produces seeds that are round and green. Another pure-breeding line is *rr YY*; on selfing, this line produces wrinkled yellow seeds (*r* is a recessive allele of the seed-shape gene and produces a wrinkled seed; see Figure 2-8). When Mendel crossed plants from these two lines, he obtained round yellow F_1 seeds, as expected. The results in the F_2 are complex (Figure 2-9). Mendel performed similar experiments using other pairs of characters in many other dihybrid crosses; in each case, he obtained 9:3:3:1 ratios. So, he had another phenomenon to explain, some more numbers to turn into an idea.

He first checked to see whether the ratio for each gene pair in the dihybrid cross is the same as that for a monohybrid cross. If you look at only the round and wrinkled phenotypes and add up all the seeds falling into these two classes, the totals are $315 + 108 = 423$ round, and $101 + 32 = 133$ wrinkled. Hence the monohybrid 3:1 ratio still prevails. Similarly, the ratio for yellow to green is $(315 + 101):(108 + 32) = 416:140 \cong 3:1$. From this clue, Mendel concluded that the two systems of heredity are independent. He was mathematically astute enough to realize that the 9:3:3:1 ratio is nothing more than a random combination of two independent 3:1 ratios.

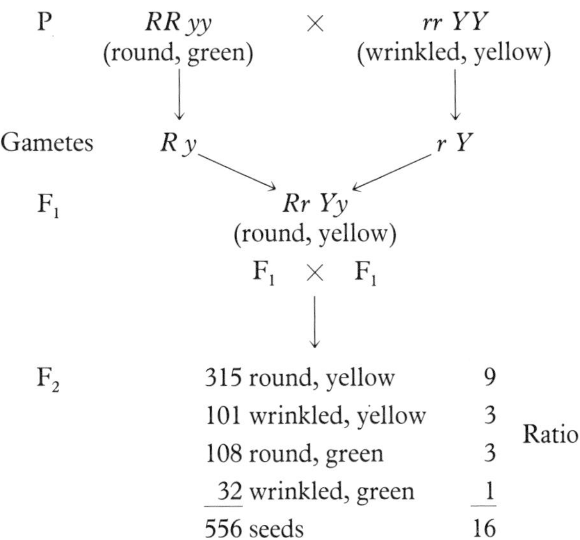

Figure 2-9. The F_2 generation resulting from parents differing in two characters.

This is a convenient point to introduce some elementary rules of probability that we will often use throughout this book.

The Rules of Probability

1. *Definition of probability.*

$$\textbf{Probability} = \frac{\text{The number of times an event happens}}{\text{The number of opportunities for it to happen (or the number of trials)}}$$

For example, the probability of rolling a "four" on a die in a single trial is written p(of one four) $= 1/6$ because the die has six sides. If each side is equally likely to turn up, then the average result should be one "four" for each six trials.

2. *The product rule.* The probability of two independent events occurring simultaneously is the product of each of their probabilities. For example, with two dice we have independent objects, and

$$p(\text{of two fours}) = 1/6 \times 1/6 = 1/36$$

3. *The sum rule.* The probability of *either one* of two mutually exclusive events occurring is the sum of their individual probabilities. For example, with two dice,

$$p(\text{of two fours } or \text{ two fives}) = 1/36 + 1/36 = 1/18$$

In the pea example, the F_2 of a dihybrid cross can be predicted if the mechanism for putting R or r into a gamete is *independent* of the mechanism for putting Y or y into a gamete. The frequency of gamete types can be calculated by determining their probabilities (see box on page 39.) That is, if you pick a gamete at random, the *probability* of picking a certain type of gamete is the same as the frequency of that type in the population.

We know from Mendel's first law that

$$Y \text{ gametes} = y \text{ gametes} = 1/2$$
$$R \text{ gametes} = r \text{ gametes} = 1/2$$

Therefore, in a $Rr\ Yy$ plant,

$$p(\text{of gamete being } R \text{ and } Y) = p(R\ Y) = 1/2 \times 1/2 = 1/4 \text{ (by product rule)}$$
$$p(\text{of gamete being } R \text{ and } y) = p(R\ y) = 1/2 \times 1/2 = 1/4 \text{ (by product rule)}$$
$$p(\text{of gamete being } r \text{ and } y) = p(r\ y) = 1/2 \times 1/2 = 1/4 \text{ (by product rule)}$$
$$p(\text{of gamete being } r \text{ and } Y) = p(r\ Y) = 1/2 \times 1/2 = 1/4 \text{ (by product rule)}$$

Thus we can represent the F_2 generation by a giant grid named (after its inventor) a Punnett square (Figure 2-10).

♀ \ ♂	R Y 1/4	R y 1/4	r y 1/4	r Y 1/4
R Y 1/4	RR YY 1/16	RR Yy 1/16	Rr Yy 1/16	Rr YY 1/16
R y 1/4	RR Yy 1/16	RR yy 1/16	Rr yy 1/16	Rr Yy 1/16
r y 1/4	Rr Yy 1/16	Rr yy 1/16	rr yy 1/16	rr Yy 1/16
r Y 1/4	Rr YY 1/16	Rr Yy 1/16	rr Yy 1/16	rr YY 1/16

Key:

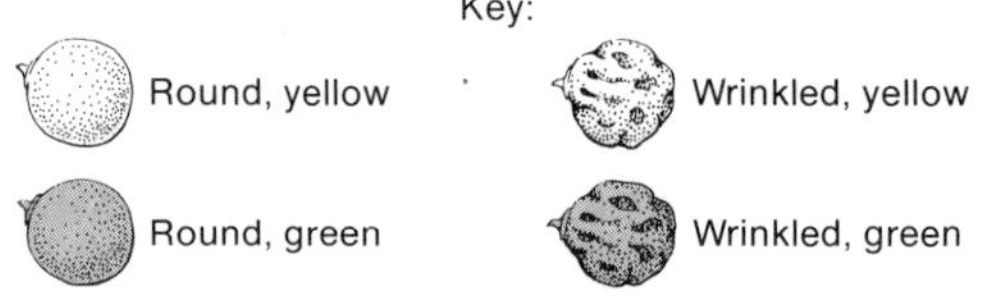

Figure 2-10. Symbolic representation of the genetic and phenotypic constitution of the F_2 generation resulting from parents differing in two characters. (Figure 2-9 shows the P and F_1 generations.)

The probability of 1/16 shown for each box in the square is derived by further application of the product rule. The assortment of alleles into the male gametes and that into the female gametes are independent events. Thus, for example, the probability (or frequency) of *RR YY* (combining an *R Y* male gamete with an *R Y* female gamete) will be $1/4 \times 1/4 = 1/16$. Grouping all the types that will look the same from Figure 2-9, we find the 9:3:3:1 ratio (now not so mysterious) in all its beauty.

Round, yellow	9/16 or 9
Round, green	3/16 or 3
Wrinkled, yellow	3/16 or 3
Wrinkled, green	1/16 or 1

The concept of independence of the two systems (round/wrinkled and yellow/green) is important. Mendel's insight has been generalized to give the statement now known as Mendel's second law.

Mendel's Second Law

Different segregating gene pairs behave independently. (This statement sometimes is called the law of independent assortment.)

Note how Mendel's counting led to the discovery of such unexpected regularities as the 9:3:3:1 ratio, and how a few simple assumptions (such as equal segregation and independent assortment) can explain this ratio that initially seems so baffling. Although it was unappreciated at the time, Mendel's approach proved to provide the key to an understanding of genetic mechanisms.

Of course, Mendel went on to test this second law. For example, he crossed an F_1 dihybrid *Rr Yy* with a double-homozygous recessive strain *rr yy*. He predicted that the dihybrid *Rr Yy* should produce the gametic types *R Y*, *r Y*, *R y*, and *r y* in equal frequency—that is, as shown along one edge of the Punnett square, in the frequencies 1/4, 1/4, 1/4, and 1/4. On the other hand, because it is homozygous, the *rr yy* plant should produce only one gamete type (*r y*), regardless of equal segregation or independent assortment. Thus the progeny phenotypes should be a direct reflection of the gametic types from the *Rr Yy* parent (because the *r y* contribution from the *rr yy* parent does not alter the phenotype indicated by the other gamete). Mendel predicted a 1:1:1:1 ratio of *Rr Yy*, *rr Yy*, *Rr yy*, and *rr yy* progeny from this cross, and his prediction was confirmed. (A cross with a homozygous recessive is now known as a **testcross**. We shall see this kind of cross many times.) Mendel tested the concept of independent assortment on a large number of gene pairs and found that it applied to every combination in his published work.

Of course, the deduction of equal segregation and independent assortment as abstract concepts that explain the observed facts leads immediately to the question of what structures or forces are responsible for generating them. The

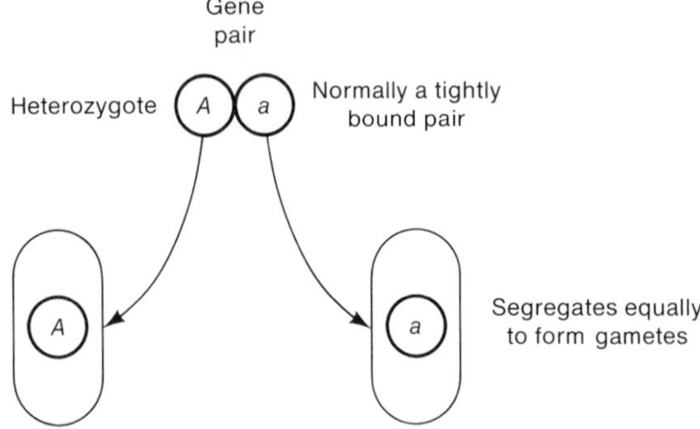

Figure 2-11. Diagrammatic visualization of the equal segregation of one gene pair into gametes.

idea of equal segregation seems to indicate that both alleles of a pair actually exist in a paired configuration, from which they can separate cleanly during gamete formation (Figure 2-11). If any other gene pair behaves independently in the same way, then we have independent assortment (Figure 2-12). But this is all speculation at this stage of our discussion, as it was after the rediscovery of Mendel's work. The actual mechanisms are now known, and we shall discuss them later. (We shall see that it is the chromosomal location of genes that is responsible for their equal segregation and independent assortment.)

The key point to appreciate at this stage is that the ground rules for genetic analysis were established by Mendel. His work made it possible to infer the existence and nature of hereditary particles and mechanisms without ever seeing them. All such theories were based on the analysis of phenotype frequencies in controlled crosses; this is the experimental approach still used in modern genetics.

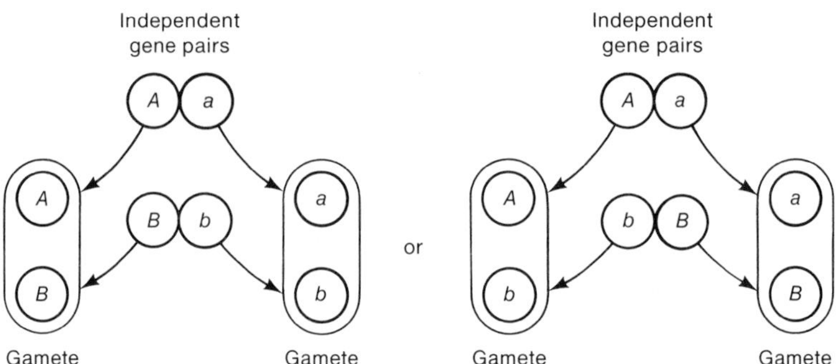

Figure 2-12. Diagrammatic visualization of the segregation of two independent gene pairs into gametes.

Methods for Working Problems

We pause here for a few words on the working of problems. The Punnett square is sure and graphic, but it is unwieldy; it is suited only for illustration, not for efficient calculation.

A branch diagram is useful for solving some problems. For example, the 9:3:3:1 ratio can be derived by drawing a branch diagram and applying the product rule to determine the frequencies. (Note the use of the convention that *R*– represents both *RR* and *Rr*. That is, either allele can occupy the space indicated by the dash.)

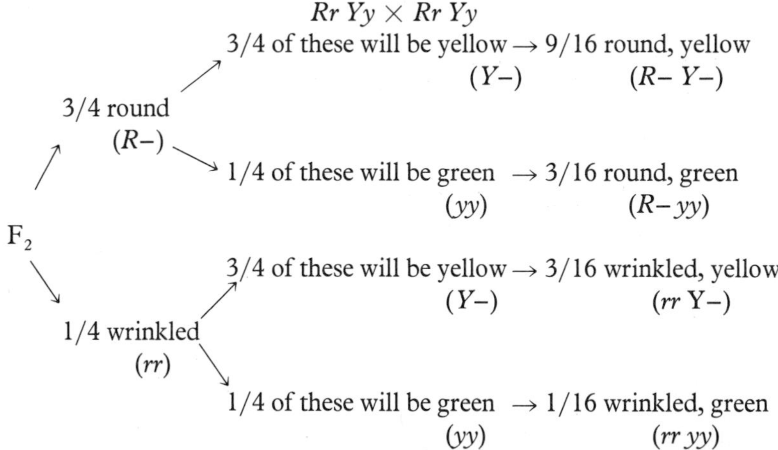

The diagram can be extended to a trihybrid ratio (such as *Aa Bb Cc* × *Aa Bb Cc*) by drawing another set of branches on the end. However, as the number of gene pairs increases, the number of identifiable phenotypes rises startlingly, and the number of genotypes climbs even more steeply (Table 2-3). With such large class numbers, even the branch method becomes unwieldy.

In such cases, one must resort to devices based directly on the product and sum rules. For example, what proportion of progeny from the cross *Aa Bb Cc Dd Ee Ff* × *Aa Bb Cc Dd Ee Ff* will be *AA bb Cc DD ee Ff*? The answer is easily obtained if the gene pairs all assort independently, thereby allowing use of the product rule. That is, 1/4 of the progeny will be *AA*, 1/4 will be *bb*, 1/2 will be *Cc*, 1/4 will be *DD*, 1/4 will be *ee*, and 1/2 will be *Ff*, and so we obtain the answer by multiplying these frequencies:

$$p(AA\ bb\ Cc\ DD\ ee\ Ff) = 1/4 \times 1/4 \times 1/2 \times 1/4 \times 1/4 \times 1/2 = 1/1024$$

Table 2-3. Rise in number of genotypic classes as the power of the number of segregating gene pairs

Number of segregating gene pairs	Number of phenotypic classes	Number of genotypic classes
1	2	3
2	4	9
3	8	27
4	16	81
⋮	⋮	⋮
n	2^n	3^n

When Mendel's results were rediscovered in 1900, his principles were tested in a wide spectrum of higher organisms. The results of these tests showed that Mendelian genetics (sometimes with extensions that we shall discuss in the next three chapters) is universally applicable. Mendelian ratios (such as 3:1, 1:1, 9:3:3:1 and 1:1:1:1) were extensively reported (Figure 2-13), suggesting that equal segregation and independent assortment are fundamental heredi-

Figure 2-13. A 9:3:3:1 ratio in the phenotype of kernels of corn. Each kernel represents a progeny individual. The progeny result from a self individual of genotypes Aa Bb *where* A = *dark,* a = *light,* B = *smooth, and* b = *wrinkled. Two representative ears of corn are shown.*

tary processes found throughout nature. The experimental approach used by Mendel can be extensively applied in plants. However, in some plants, and in animals, the technique of selfing is impossible. This problem can be circumvented by intercrossing identical genotypes. For example, an F_1 animal resulting from the mating of parents from differing pure lines can be mated to its F_1 siblings (brothers or sisters) and an F_2 produced. The F_1 individuals are genetically identical, so the F_1 cross amounts to a selfing.

Simple Mendelian Genetics in Humans

Other systems present some special problems in the application of Mendelian methodology. One of the most difficult, yet most interesting, is the human species. Obviously, controlled crosses cannot be made, so human geneticists must resort to a scrutiny of established matings in the hope that informative matings have been made by chance. The scrutiny of established matings is called **pedigree analysis.** A member of a family who first comes to the attention of a geneticist is called the **propositus.** Usually the phenotype of the propositus is exceptional in some way—for example, a dwarf. The investigator then traces the history of the character shown to be interesting in the propositus back through the history of the family, and a family tree or pedigree is drawn up using certain standard symbols (Figure 2-14). (The terms *autosomal* and *sex-linked* will be explained later; they are included here to make the table complete.)

Many human diseases and other exceptional conditions are determined by simple Mendelian recessive alleles. There are certain clues in the pedigree that must be sought. Characteristically the condition appears in progeny of unaffected parents; on average, if all such progenies are totaled, about one-fourth of the siblings are affected. Furthermore, two affected individuals cannot have an unaffected child. Quite often such recessive alleles are revealed by consanguinous matings—for example, cousin marriages. This is particularly true of rare conditions where chance matings of heterozygotes are expected to be extremely rare. It has been estimated, for example, that first-cousin marriages account for about 18% to 24% of albino children and 27% to 53% of children with Tay–Sachs disease; both are rare recessive conditions. Some other examples of disease-causing recessive alleles in humans are those for cystic fibrosis and phenylketonuria (PKU). A typical pedigree for a rare recessive condition is shown in Figure 2-15. Of course, variants that are not regarded as diseases also may be caused by recessive alleles. These may be rare as in albinism (Figure 2-16) or common as in light eye color (blue or green) in North American populations.

There are also examples of exceptional conditions caused by dominant alleles. (This is in contrast to the situation for such conditions as PKU, where the normal condition is attributable to the dominant allele, and the recessive allele causes PKU.) Once again, there are some simple rules to follow to discern from pedigrees a condition caused by a dominant allele: the condition

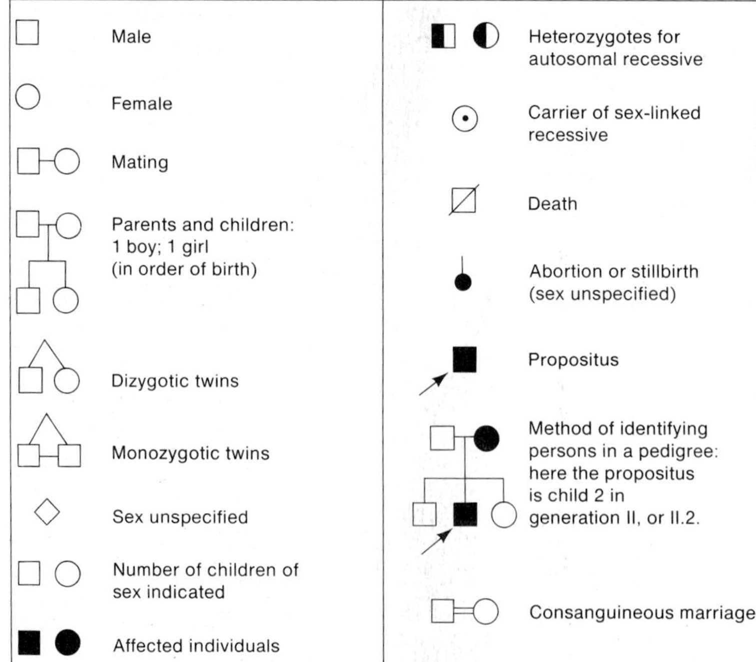

Figure 2-14. Symbols used in human pedigree analysis. (After W. F. Bodmer and L. L. Cavalli-Sforza, Genetics, Evolution, and Man. *Copyright © 1976, W. H. Freeman and Company.)*

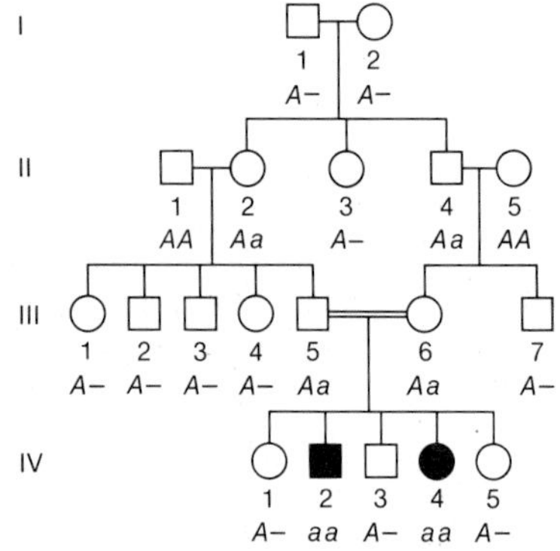

Figure 2-15. Illustrative pedigree, involving an exceptional recessive phenotype determined by the recessive Mendelian allele a. *Gene symbols normally are not included in pedigree charts, but genotypes are inserted here for reference. Note that individuals II.1 and II.5 marry into the family; they are assumed to be normal because the heritable condition under scrutiny in a pedigree typically is rare. Note also that it is not possible to be certain of the genotype in some individuals with normal phenotype; such individuals are indicated by* A–.

typically occurs in every generation; unaffected individuals never transmit the condition to their offspring; two affected parents may have unaffected children; and the condition is passed, on average, to one-half of the children of an affected individual. As with recessive alleles acting in a Mendelian manner, both sexes may be equally affected. Achondroplasia (a kind of dwarfism), white forelock, and brachydactyly (very short fingers) are examples of exceptional conditions in man caused by dominant alleles. A typical pedigree of a rare dominant condition is shown in Figure 2-17.

Notice that, in this kind of Mendelian analysis, there is again the notion of identification of genes affecting major biological function, this time in humans. Pedigrees for PKU, for example, demonstrate that there is a kind of gene controlling the character we might call "normal PKU function." The two alleles are presence and absence. This identification is an important step toward discovery of the precise way in which the abnormal allele is failing and its possible correction. The medical applications of such genetic analysis obviously are far reaching. In fact, medical genetics is today a key part of medical training. Simple pedigree analyses have extensive use, not only in such medical research but also in the day-to-day counseling of prospective parents who fear genetic disease in their children.

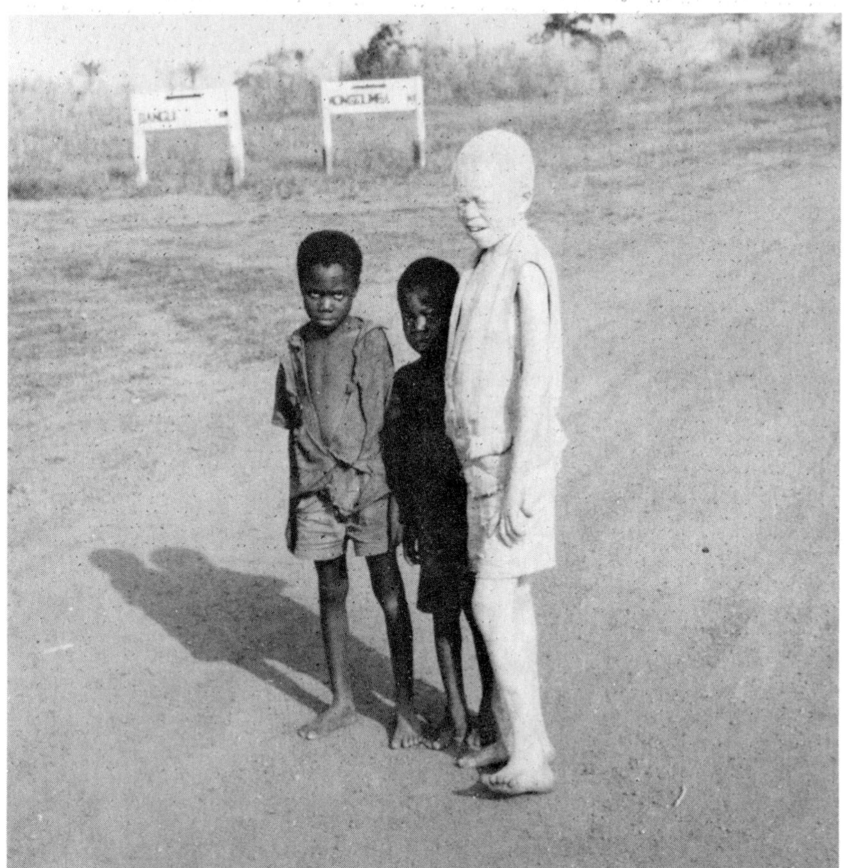

Figure 2-16. Albinism in an African child. This phenotype is caused by homozygosity for a recessive Mendelian gene. (Courtesy of W. F. Bodmer and L. L. Cavalli-Sforza, Genetics, Evolution, and Man. *Copyright © 1976, W. H. Freeman and Company.)*

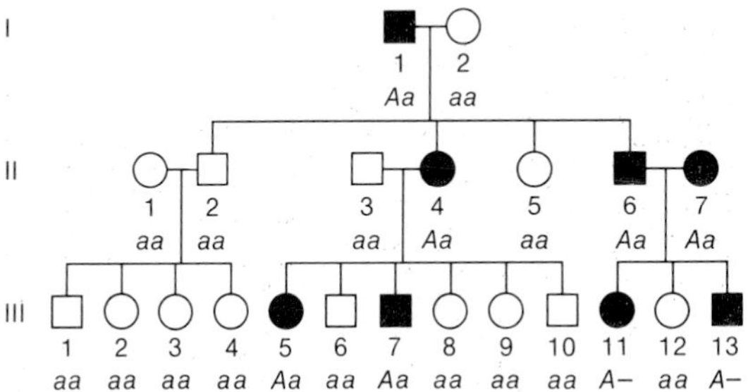

Figure 2-17. Illustrative pedigree involving an exceptional dominant phenotype determined by the dominant Mendelian allele A. *In this pedigree, most of the genotypes can be deduced.*

Simple Mendelian Genetics in Agriculture

As mentioned in the Introduction, there has been an interest in plant breeding since prehistoric times. The methods used by Neolithic farmers were probably the same as those used until the discovery of Mendelian genetics. Basically the approach was to select superior phenotypes from seeds or plants derived from natural populations. Particularly desirable were pure lines of favorable phenotype, because these lines produced constant results over generations of planting. Without the knowledge of Mendelian genetics, how is it possible to develop pure lines? It so happens that self-pollinating plants, such as many crop plants, naturally tend to be homozygous, and any heterozygous gene pairs become homozygous over generations of selfing (Figure 2-18). Thus pure lines have developed automatically over the years.

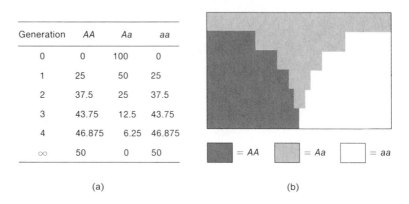

Generation	*AA*	*Aa*	*aa*
0	0	100	0
1	25	50	25
2	37.5	25	37.5
3	43.75	12.5	43.75
4	46.875	6.25	46.875
∞	50	0	50

(a)

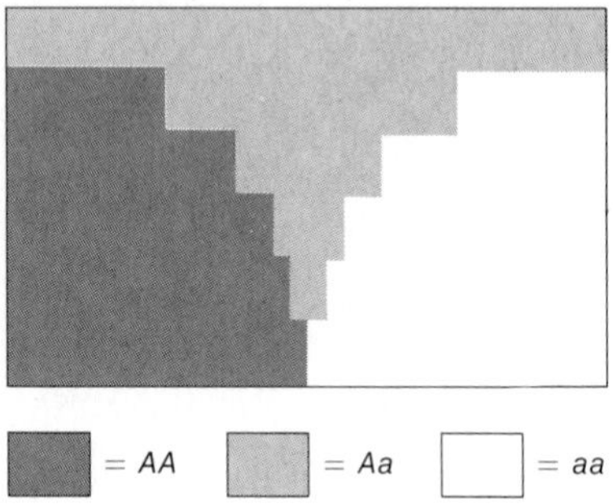

(b)

Figure 2-18. Selfing (and inbreeding in general) produces an increasing proportion of homozygous individuals with time. (a) The relative proportions of the three genotypes are shown through several generations, assuming that all individuals in generation 0 are Aa. *At each generation,* AA *and* aa *breed true, but* Aa *individuals produce* Aa, AA, *and* aa *progeny in a 2:1:1 ratio. (b) A graphical depiction of the process.*

However, these less sophisticated breeding practices suffered from a major problem: the breeder was forced to rely on favorable combinations of genes that occurred in nature. With the advent of Mendelian genetics, it became evident that favorable qualities in different lines could be combined through hybridization and subsequent gene reassortment. This procedure forms the basis of modern plant breeding.

For naturally self-pollinating plants, such as rice or wheat, two pure lines (each of different favorable genotype) are hybridized by manual cross-pollina-

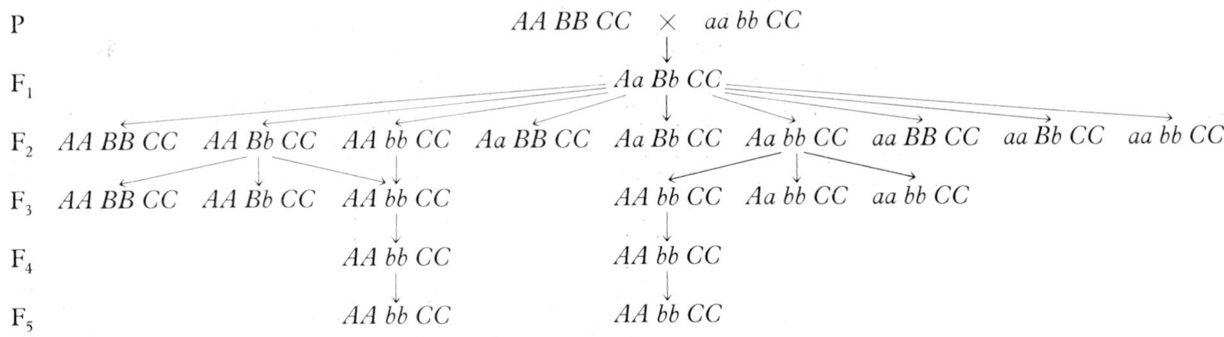

Figure 2-19. *The basic technique of plant breeding. A hybrid is generated, and breeding stock is selected in subsequent generations to obtain pure-breeding improved types. For simplicity, we assume here that the parental lines differ in only two genes* (A *and* B). *In this example, the* AA bb CC *genotype is the one desired—that is, tests on the F_2 generation show this to be the most desirable phenotype. A pure-breeding line of this phenotype may be obtained in the F_2. It can also be obtained in the F_3 either from a nonpure parent of similar phenotype* (right) *or from a nonpure parent of different phenotype* (left). *In typical cases of plant breeding, far larger numbers of genes, individuals, and generations are involved. Furthermore, the breeder usually knows little or nothing about the precise genotypes involved.*

tion, and hence an F_1 is developed. The F_1 is then allowed to self, and its heterozygous gene pairs assort to produce many different genotypes, some of which represent desirable new combinations of the parental genes. A small proportion of these new genotypes will be pure breeding already, but if not, several generations of selfing will produce homozygosity of the relevant genes. Figure 2-19 summarizes this method; Figure 2-20 shows an example of the complex series of hybridizations that have taken place in rice.

An example of genetic improvement in a species more familiar to most is the tomato. Anyone who has read a recent seed catalog will be familiar with the abbreviations V, F, or N next to a tomato variety. These represent resistance to the pathogens *Verticillium, Fusarium,* and nematodes—resistance that has been crossed into the tomatoes, typically from wild forms of tomato. Another familiar phenomenon in tomatoes is determinate as opposed to indeterminate growth pattern. Determinate plants are bushier and more compact, and they do not need as much staking (Figure 2-21). Determinate growth is caused by a recessive allele *sp* (self-pruning), which has been crossed into modern varieties. Another useful allele is *u* (uniform ripening); this allele eliminates the green patch or shoulder around the stem on the ripe fruit.

Such examples could be listed for many pages. The point is that simple Mendelian genetics (as in this chapter) has provided agricultural plant breeding with its rationale and its modern methods. Entire complex genotypes may be constructed from an array of ancestral lines, each showing some desirable

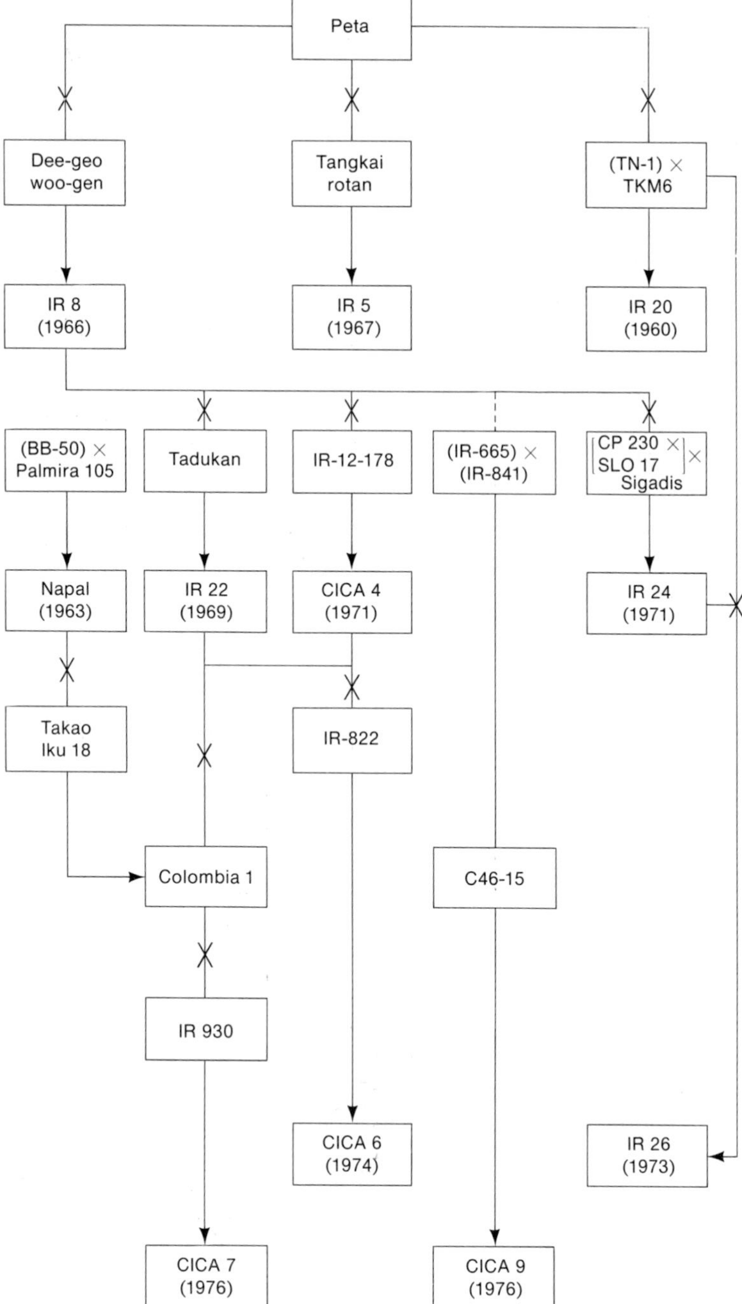

Figure 2-20. The complex pedigree of modern rice varieties. The progenitor of most modern dwarf types was IR 8, selected from a cross between the vigorous Peta (Indonesian) and the dwarf Dee-geo woo-gen (from Taiwan). Most of the other crosses represent progressive improvement of IR 8. The diagram illustrates the extensive plant breeding that produced modern crop varieties. (From Peter R. Jennings, "The Amplification of Agricultural Production." Copyright © 1976 by Scientific American, Inc. All rights reserved.)

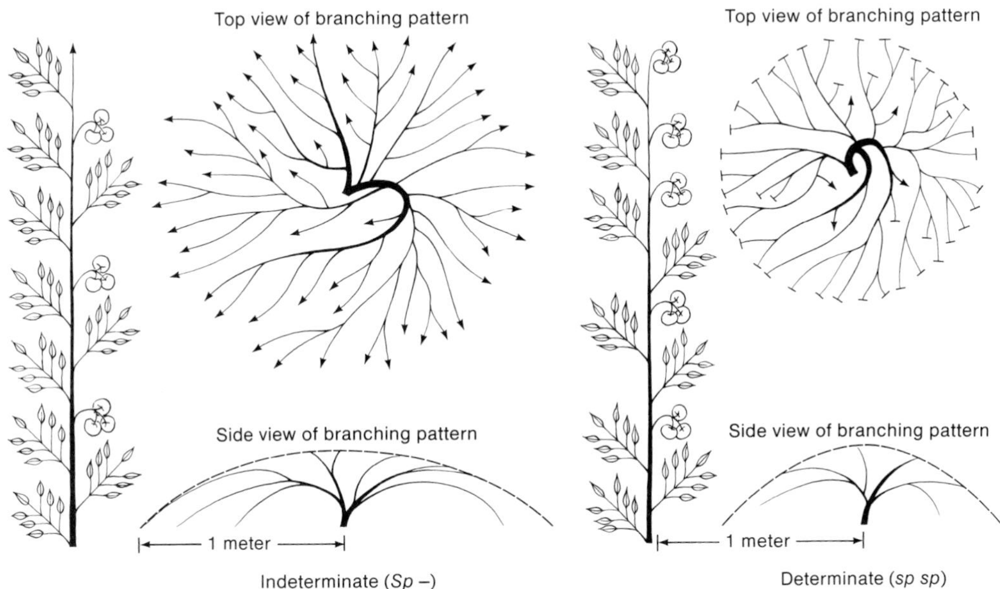

Figure 2-21. *Growth characteristics of indeterminate* (Sp–) *and determinate* (sp sp) *tomatoes. Determinate has progressively fewer leaves between inflorescences, and the determinate stems end with a size-arresting inflorescence. Individual stem branches are shown, plus typical branching patterns for mature plants in top and side views. Note the size differences. (Note: the symbol* sp *represents a single gene; many gene symbols involve more than one letter.) (From Charles M. Rick, "The Tomato." Copyright © 1978 by Scientific American, Inc. All rights reserved.)*

feature. When we think of genetic engineering, we think of the genetic techniques of the 1970s, but genetic engineering for plant improvement began long ago. Similar statements could be made for animal breeding.

The Origin of Variants

Genetic analysis, as we have seen, must start with parental differences. Without variants, no genetic analysis is possible. Where do these variants, this raw material for genetic analysis, come from? This is a question that can be answered in full only in later chapters. Briefly, most of the variants used by Mendel (and by ancestral and modern breeders of plants and animals) arise spontaneously, without the deliberate action of geneticists, in nature or in the breeders' populations. Undoubtedly, some of these variants would be "weeded out" by natural selection, but they may be kept alive through their deliberate nurture by geneticists, either for some specific breeding program or for the elucidation of the basic biological mechanisms of which the different forms are variants.

Genetic Dissection

We have seen that the genetic analysis of variants can identify a gene involved in an important biological process. This is a central aspect of modern genetics; throughout the book we shall refer to this process as genetic dissection of biological systems. Mendel was the first genetic surgeon. Using genetic analysis, he was able to identify and distinguish among the several components of the hereditary process in a way as convincing as if he had microdissected those components. The fact that the genes he was using were for pea shape, pea color, and so on was largely irrelevant. Those genes were being used simply as **genetic markers,** which enabled Mendel to trace the hereditary processes of segregation and assortment. It is almost as though he were able to "paint" two alleles different colors and send them through a cross to see how they behaved! Genetic markers are now routinely used in genetics and in all of biology to study all sorts of processes that the marker genes themselves do not directly affect.

Probably without realizing it, Mendel had also invented another aspect of genetic dissection in which the precise genes used *were* important. In this type of analysis, genetic variants are used in such a way that, by studying the variant gene function, we can make inferences about the "normal" operation of the gene and the process it controls. Let's consider the petal-color gene in peas again. This is a major gene affecting an important biological function, which is the color of the petals. Undoubtedly the success or failure of a plant in nature would largely hinge on its having the right kind of petal color, presumably to attract the appropriate pollinator. So Mendel had pinpointed a gene for a process of central importance to the biology of peas, thus opening the way to extensive further study. Genetic dissection is a versatile tool of modern biological research.

Once a gene has been identified, affecting (say) petal color in peas, we have called it a major gene, but what does this really mean? It is major in that it is obviously having a profound effect on the color of the petals. But can we conclude that it is *the* single most important step in the determination of petal color? The answer is no, and the reason may be seen in an analogy. If we were trying to discover how a car engine works, we might investigate this by pulling out various parts and observing the effect on the running of the engine. If a battery cable were disconnected, the engine would stop; one might erroneously conclude that this cable is *the* most important part of the running of the engine. Obviously, other parts are equally necessary, and their removal could also stop or seriously cripple the engine. In a similar way it can be shown (see Chapter 4) that several genes can be identified, all of which have a major and similar effect on petal coloration.

Mendel's work has withstood the test of time and has provided us with the basic groundwork for all modern genetic study. Yet his work went unrecognized and neglected for 35 years following its publication. Why? There are many possible reasons, of which we will mention just one. Perhaps it was because biological science at that time could not provide evidence for any real physical units within cells that might correspond to Mendel's genetic particles.

Chromosomes had certainly not yet been studied, meiosis not yet described, and even the full details of plant life cycles had not been worked out. Without this basic knowledge, it may have seemed that Mendel's ideas were mere numerology.

Message
Mendel's work is significant for the following reasons.

1. *It showed how it is possible to study some central processes of heredity and biology in general through the use of genetic markers.*
2. *It showed how the biological functions of genes themselves can be elucidated from the study of variant alleles.*
3. *It had far-reaching ramifications in agriculture and medicine.*

In the next chapter, we focus on the physical locations of genes in cells and on the consequences of these locations.

Summary

Modern genetics is based on the concept of the gene, the fundamental unit of heredity. In his experiments with the garden pea, Mendel was the first to recognize the existence of genes. For example, by crossing a pure line of purple-flowered pea plants with a pure line of white-flowered pea plants and then selfing the F_1 generation, which was entirely purple, Mendel produced an F_2 generation of purple plants and white plants in a 3:1 ratio. In crosses such as those of pea plants bearing yellow seeds and pea plants bearing green seeds, he discovered that a 1:2:1 ratio underlies all 3:1 ratios. From these precise mathematical ratios Mendel concluded that there are hereditary determinants of a particulate nature (now known as genes). In higher plant and animal cells, genes exist in pairs. The forms of a gene are called alleles. Individual alleles can be either dominant or recessive.

In a cross of heterozygous yellow plants with homozygous green plants, a 1:1 ratio of yellow to green plants was produced. From this ratio Mendel confirmed his so-called first law, which states that two members of a gene pair segregate from each other during sex cell formation into equal numbers of sex cells. Thus, each sex cell carries only one gene for each gene pair. The union of sex cells to form a zygote is random and occurs irrespective of which member of a gene pair is carried.

The foregoing conclusions came from Mendel's work with monohybrid crosses. In dihybrid crosses Mendel found 9:3:3:1 ratios in the F_2, which are really two 3:1 ratios combined at random. From these ratios Mendel inferred that the two gene pairs studied in a dihybrid cross behave independently. This concept has been stated as Mendel's second law.

Although controlled crosses cannot be made in human beings, Mendelian genetics has great significance for humans. Many diseases and other exceptional conditions in humans are determined by simple Mendelian recessive alleles; other exceptional conditions are caused by dominant alleles. In addition, Mendelian genetics is widely used in modern agriculture. By combining favorable qualities from different lines through hybridization and subsequent gene reassortment, plant geneticists are able to produce new lines of superior phenotype.

Finally, Mendel was responsible for the basic techniques of genetic dissection still in use today. One such technique is the use of genes as genetic markers to trace the hereditary processes of segregation and assortment. The other is the study of genetic variants to discover how genes operate normally.

Problems

1. Holstein cattle normally are black and white. A superb black and white bull, Charlie, was purchased by a farmer for $100,000. The progeny sired by Charlie were all normal in appearance. However, certain pairs of his progeny, when interbred, produced red and white progeny at a frequency of about 25%. Charlie was soon removed from the stud lists of the Holstein breeders. Explain precisely why, using symbols.

2. Maple-syrup-urine disease is a rare inborn error of metabolism. It derives its name from the odor of the urine of affected individuals. If untreated, affected children die soon after birth. The disease tends to recur in the same family, but the parents of the affected individuals are always normal. What does this information suggest about the transmission of the disease: is it dominant or recessive?

3. On a hike into the mountains, you notice a beautiful harebell plant that has white flowers instead of the usual blue. Assuming this effect to be caused by a single gene, outline *precisely* what you would do to find out whether the allele causing white flowers is dominant or recessive to that causing blue.

4. Mother and father both find the taste of a chemical called phenylthiourea very bitter. However, two of their four children find the chemical tasteless. Assuming the inability to taste this chemical to be a monogenic trait, is it dominant or recessive?

5. In humans, the disease galactosemia is inherited as a monogenic recessive trait in a simple Mendelian manner. A woman whose father had galactosemia intends to marry a man whose grandfather was galactosemic. They are worried about having a galactosemic child. What is the probability of this outcome?

6. Huntington's chorea is a rare, fatal disease that usually develops in middle age. It is caused by a dominant allele. A phenotypically normal man in his early twenties learns that his father has developed Huntington's chorea.

   a. What is the probability that he will himself develop the symptoms later on?

   b. What is the probability that his son will develop the symptoms in later life?

7. Achondroplasia is a form of dwarfism inherited as a simple monogenic trait. Two achondroplastic dwarfs working in a circus marry and have a dwarf child; later they have a second child who is normal.
   a. Is achondroplasia produced by a recessive or a dominant allele?
   b. What are the genotypes of the two parents in this mating?
   c. What is the probability that their next child will be normal? a dwarf?

8. Suppose that a husband and wife are both heterozygous for a recessive gene for albinism. If they have dizygotic (two-egg) twins, what is the probability that both of the twins will have the same phenotype with respect to pigmentation?

9. The plant blue-eyed Mary grows on Vancouver Island and on the lower mainland of British Columbia. Near Nanaimo, one plant was observed in nature that had blotched leaves. This plant, which had not yet flowered, was dug up and taken to a laboratory, where it was allowed to self. Seeds were collected and grown into progeny. Figure 2-22 shows one randomly selected (but typical) leaf from each of the progeny.

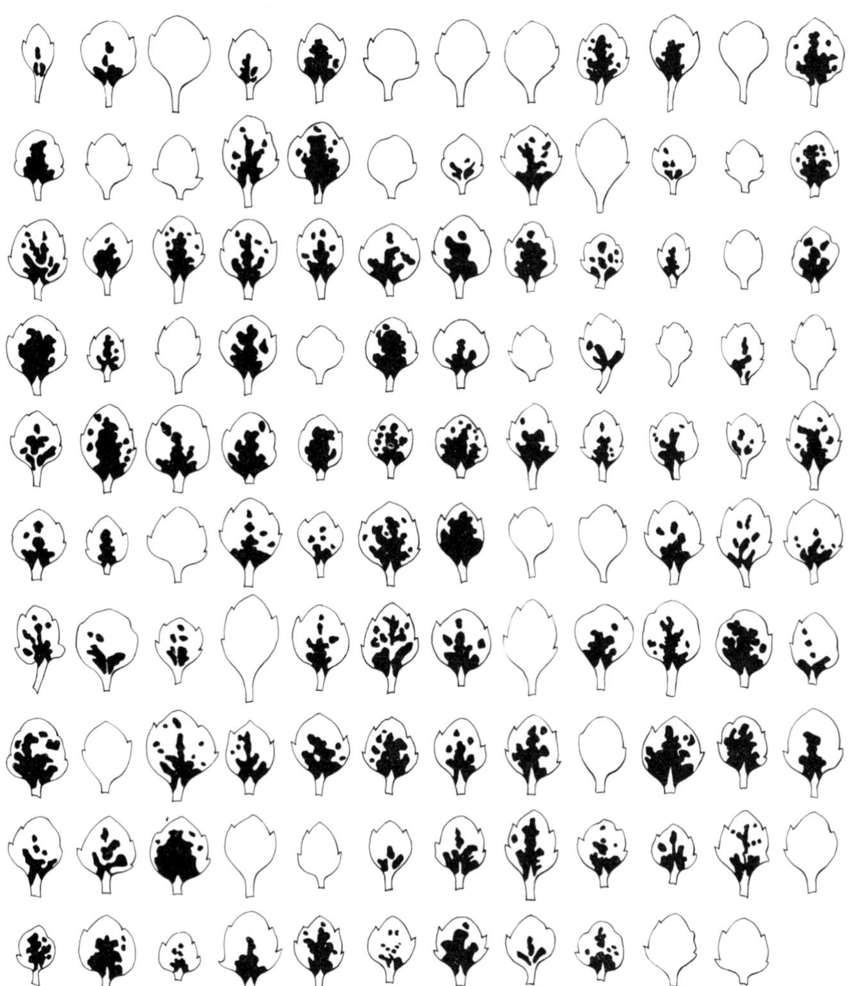

Figure 2-22.

a. Formulate a concise genetic hypothesis to explain these results. Explain all symbols and show all genotypic classes (and the genotype of the original plant).

b. How would you test your hypothesis? Be specific.

10. Some species of plants (called heterostylous) produce two forms that differ in their flowers (Figure 2-23). In the Lompoc fiddleneck (*Amsinckia spectabilis*), a heterostylous species, the following four pollinations are made:

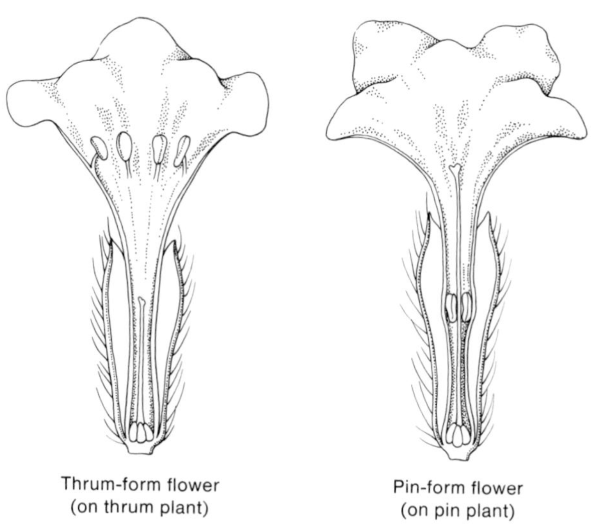

Figure 2-23.

Parents	Progeny
Pin plant #1 × Pin plant #2	→ 37 pin plants
Thrum plant #3 × Thrum plant #3	→ 28 thrum plants
Thrum plant #3 × Pin plant #1	→ 29 thrum plants
Thrum plant #4 × Pin plant #2	→ 19 pins, 16 thrums

Represent the dominant allele of the heterostyle gene by *H* and the recessive allele by *h*.

a. What are the genotypes of (1) Pin plant #1, (2) Pin plant #2, (3) Thrum plant #3, and (4) Thrum plant #4?

b. What proportions of pins and thrums would be expected in the following crosses? (1) Thrum plant #3 × Thrum plant #4. (2) Thrum plant #3 × Pin Plant #2. (3) Thrum plant #4 × Thrum plant #4.

(Problem 10 courtesy of F. R. Ganders.)

11. Can it ever be proved that an animal is *not* a carrier of a recessive allele (that is, not a heterozygote for a given gene)? Explain.

12. A green seed is planted in a greenhouse and grows into a plant. When the seeds on this plant are examined, they are all yellow! What must have happened?

13. In nature, individual plants of *Plectritis congesta* bear either winged or wingless fruits (Figure 2-24). Plants were collected from nature before flowering and were crossed or selfed with the results shown in Table 2-4. Interpret these results, and derive the mode of inheritance of these fruit-shape phenotypes. Use symbols. (NOTE: the progeny marked by asterisks probably have a nongenetic explanation. What do you think it is?)

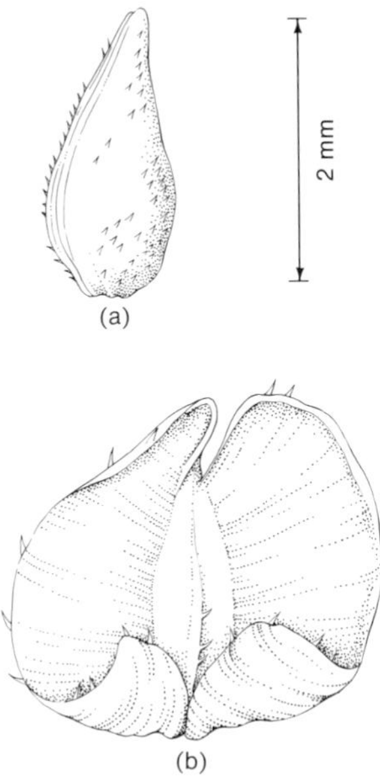

Figure 2-24. (a) Wingless. (b) Winged.

Table 2-4.

Pollination	Number of progeny plants	
	Winged	Wingless
Winged (selfed)	91	1*
Winged (selfed)	90	30
Wingless (selfed)	4*	80
Winged × wingless	161	0
Winged × wingless	29	31
Winged × wingless	46	0
Winged × winged	44	0
Winged × winged	24	0

14. Figure 2-25 shows four human pedigrees. The black symbols represent an abnormal condition (phenotype) inherited in a simple Mendelian manner.

1

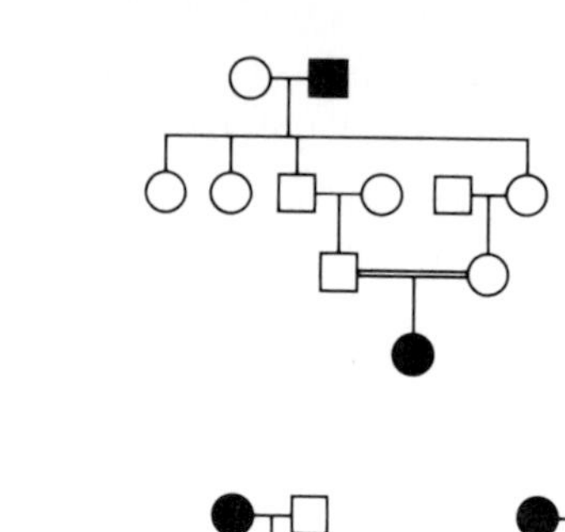

2

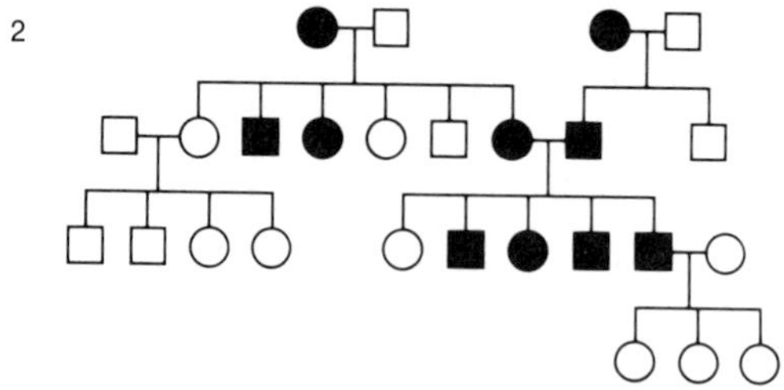

3

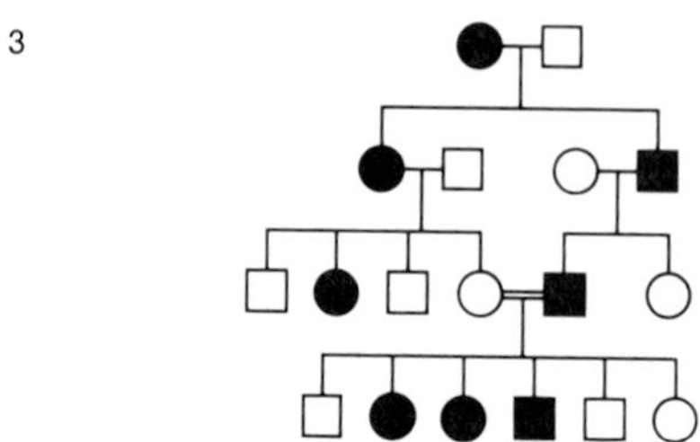

4

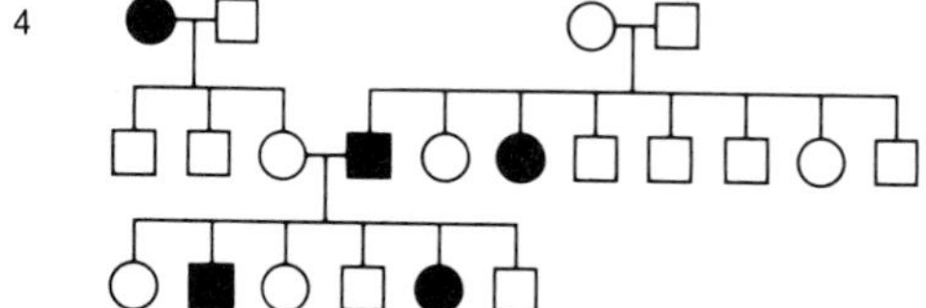

Figure 2-25.

a. For each pedigree, state whether the abnormal condition is dominant or recessive. Try to state the logic behind your answer.

b. In each pedigree, describe the genotypes of as many individuals as possible.

15. When a pea plant of genotype *Aa Bb* produces gametes, what proportion will be *A b*? (Assume that the two genes are independent.) Choose the correct answer from the following possible answers: 3/4, 1/2, 9/16, None, or 1/4.

16. When a fruitfly of genotype *Mm Nn Oo* is mated to another fly of identical genotypes, what proportion of the progeny flies will be *MM nn Oo*? (Assume that the three genes are independent.) Choose the correct answer from the following possible answers: 1/2, 1/8, 3/8, 1/32, or 1/64.

17. Suppose that you have two strains of plants—one is *AA BB*, the other *aa bb*. You cross the two and self the F_1 plants. With respect to these two genes, what

is the probability that an F_2 plant will obtain half its alleles from one grandparent and half from the other? What is the probability that it will obtain all its alleles from a single grandparent?

18. In dogs, dark coat color is dominant over albino, and short hair is dominant over long hair. If these effects are caused by two independently assorting genes, write the genotypes of the parents in each of the crosses shown in Table 2-5. Use the symbols *C* and *c* for the dark and albino coat-color alleles, and *S* and *s* for the short-hair and long-hair alleles, respectively. Assume homozygosity unless there is evidence otherwise.

(Problem 18 reprinted with the permission of Macmillan Publishing Co., Inc., from *Genetics* by M. Strickberger. Copyright © Monroe W. Strickberger, 1968.)

Table 2-5.

	Number of progeny			
Parental phenotypes	Dark, short	Dark, long	Albino, short	Albino, long
a. Dark, short × dark, short	89	31	29	11
b. Dark, short × dark, long	18	19	0	0
c. Dark, short × albino, short	20	0	21	0
d. Albino, short × albino, short	0	0	28	9
e. Dark, long × dark, long	0	32	0	10
f. Dark, short × dark, short	46	16	0	0
g. Dark, short × dark, long	30	31	9	11

SOURCE: Reprinted by permission of Macmillan Publishing Co., Inc., from *Genetics* by M. Strickberger. Copyright © Monroe W. Strickberger, 1968.

19. In tomatoes, two alleles of one gene determine the character difference of purple versus green stems, and two alleles of a separate independent gene determine the character difference of "cut" versus "potato" leaves. Table 2-6 gives the results for five separate matings of tomato plant phenotypes.

Table 2-6.

		Number of progeny			
Mating	Parental phenotypes	Purple, cut	Purple, potato	Green, cut	Green, potato
1	Purple, cut × green, cut	321	101	310	107
2	Purple, cut × purple, potato	219	207	64	71
3	Purple, cut × green, cut	722	231	0	0
4	Purple, cut × green, potato	404	0	387	0
5	Purple, potato × green, cut	70	91	86	77

SOURCE: A. M. Srb, R. D. Owen, and R. S. Edgar, *General Genetics*, 2nd ed. San Francisco: W. H. Freeman and Company. Copyright © 1965.

a. Determine which alleles are dominant.

b. What are the most probable genotypes for the parents in each cross?

(Problem 19 from A. M. Srb, R. D. Owen, and R. S. Edgar. *General Genetics*, 2nd ed. Copyright © 1965 by W. H. Freeman and Company.)

20. In the mountains of British Columbia, a small group of Sasquatches was discovered. A study of four matings that occurred in the group in the course of several years produced the results shown in Table 2-7.

Table 2-7.

Mating	Parent #1	Parent #2	Progeny
1	Bowlegs, hairy knees	Bowlegs, hairy knees	3/4 bowlegs, hairy knees 1/4 knock-knees, hairy knees
2	Bowlegs, smooth legs	Knock-knees, smooth legs	1/2 bowlegs, smooth legs 1/2 knock-knees, smooth legs
3	Bowlegs, hairy knees	Knock-knees, smooth legs	1/4 bowlegs, smooth legs 1/4 bowlegs, hairy knees 1/4 knock-knees, hairy knees 1/4 knock-knees, smooth legs
4	Bowlegs, hairy knees	Bowlegs, hairy knees	3/4 bowlegs, hairy knees 1/4 bowlegs, smooth legs

a. How many genes are involved in these phenotypes?

b. Which character differences are controlled by which alleles of these genes?

c. Which alleles are dominant or recessive?

d. There were in fact only five parents participating in these matings. Give the genotypes of these five individuals.

21. Peach trees have fuzzy fruits. Nectarine trees have smooth fruits (no fuzz). Most commercial varieties of peaches and nectarines have yellow-fleshed fruits, but some have white-fleshed fruits. Suppose that you have some peach and some nectarine trees and plan to live a long time, so you decide to do some genetic experiments with them. Table 2-8 shows the crosses you perform and the progenies you obtain. Represent the allele for fuzzy fruits by f^+, the allele for smooth fruits by f^0, the allele for yellow flesh by y^+, and the allele for white flesh by y^0.

Table 2-8.

Cross	Parents	Progeny
1	white peach #1 × yellow nectarine #1	12 yellow peach trees and 10 white peach trees
2	yellow nectarine #1 × yellow nectarine #2	15 yellow nectarine trees
3	white peach #1 × yellow nectarine #2	14 yellow peach trees

a. What are the genotypes of white peach #1? yellow nectarine #1? yellow nectarine #2?

b. What proportions of the four possible phenotypes would you expect in the progeny if you selfed yellow nectarine #1?

c. What proportions of phenotypes would you expect in the progeny if you selfed one of the yellow peaches in the progeny from cross 3?

d. What can you conclude from these experiments about the genetic difference between peaches and nectarines?

(Problem 21 courtesy of F. R. Ganders.)

22. We have dealt mainly with only two pairs of genes, but the same principles hold for more than two at a time. In the cross *AA Bb Cc Dd Ee* × *aa Bb CC Dd ee*, what proportion of progeny will phenotypically resemble

   a. the first parent?

   b. the second parent?

   c. either parent?

   d. neither parent?

   What proportion of progeny will *genotypically* resemble

   e. the first parent?

   f. the second parent?

   g. either parent?

   h. neither parent?

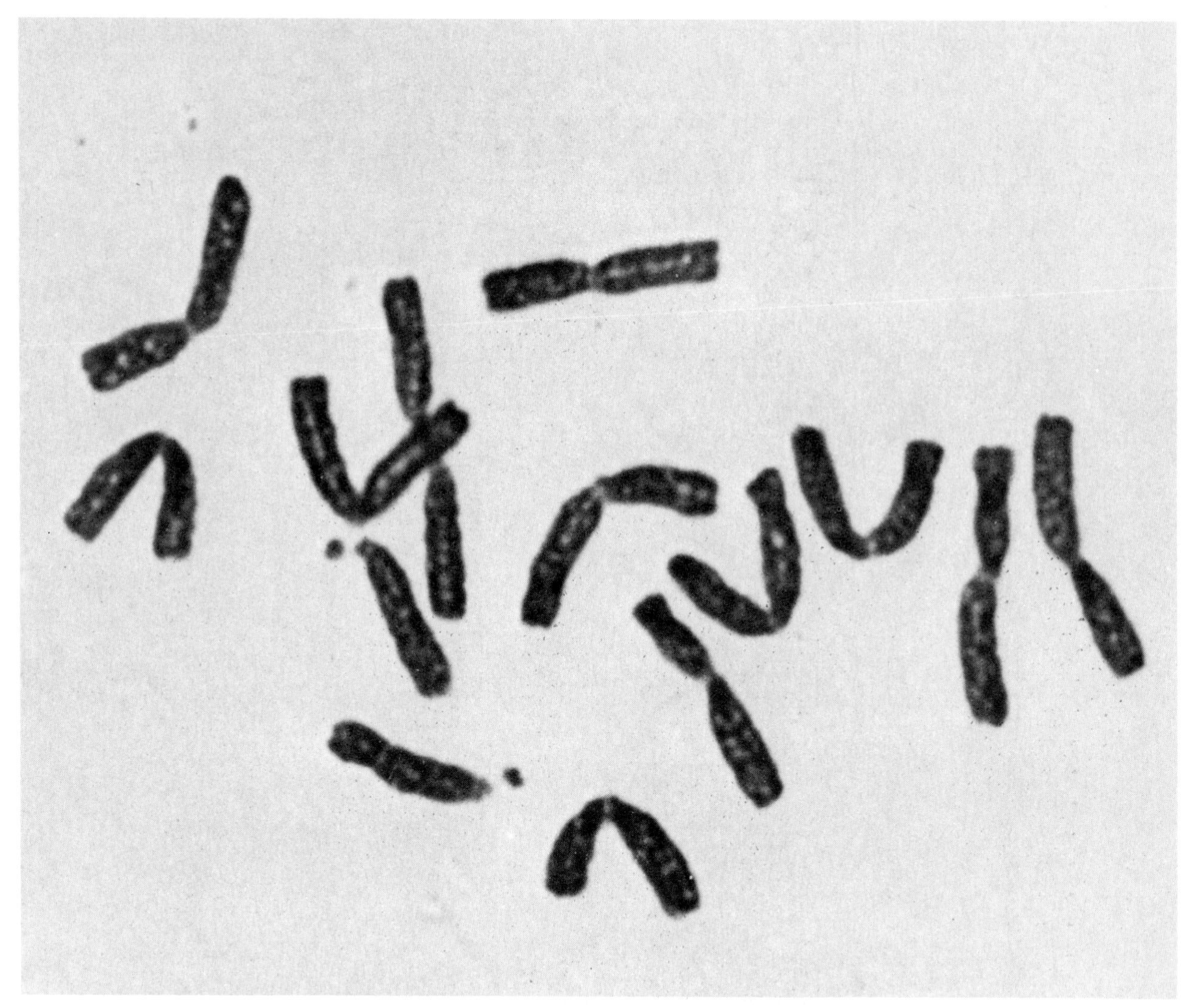

The 14 chromosomes of Hepatica acutiloba, *a higher plant. (C. J. Marchant and A. M. Adamovich.)*

3

Chromosome Theory of Inheritance

The beauty of Mendel's analysis is that data derived from genetic crosses can be interpreted by means of the laws of segregation and independent assortment. Furthermore, it then becomes possible to make predictions about the outcome of later crosses. All this is possible simply by representing abstract hypothetical factors of inheritance (or genes) by symbols—without any concern about their physical nature or their location in a cell. Nevertheless, although the validity of Mendelian principles is verified in many different organisms and genotypes, it is obvious that the next question is: What structure (or structures) within cells correspond to these hypothetical genes?

Mitosis and Meiosis

Little was known about the process of cell division when Mendel's work was published. Before its rediscovery in 1900, however, cytologists had carefully recorded the sequence of events visible under the microscope as cells divide. They noted the apparent distinction within the cell between the nucleus and the cytoplasm, and they observed the appearance of thread-like structures called **chromosomes** within the nucleus at the time of division.

Division in somatic (body) cells proceeds by a process called **mitosis.** A somewhat different division process called **meiosis** occurs in reproductive cells. A single fertilized human egg will ultimately produce, by mitosis, an adult consisting of 60 trillion (plus or minus a few billion) cells! Within the gonads of a postpubertal male, millions of sperm are produced daily as the result of meiosis. Figures 3-1 and 3-2 show these two processes of cell division. You have undoubtedly studied these processes in other courses; at this point, the fine details of the changes within the cell are not important. Just recognize that the processes are dynamic and that it is only for convenience of description that arbitrary stages are selected and defined. For our discussion here, we need only consider the following features of the processes.

Before mitosis, each chromosome duplicates along its entire length into **sister chromatids.** Initially, the chromatids are attached to one another at a point called the **centromere,** or **kinetochore.** Fibers attached to each centromere extend away to anchor points at opposite "poles" of the cell. Because these fibers form a shape like a spindle, they are called **spindle fibers.** The fibers exert force on the centromeres of the sister chromatids, pulling one chromatid of each pair to each pole, with the chromatid arms dragging behind them. The result is the formation of two nuclei—one at each pole—each consisting of the same number and type of chromosomes as the other (and as the original cell before division).

In meiosis, on the other hand, the same material (the chromosomes) is processed in a different way. Before meiosis, each chromosome duplicates along its entire length, except for the centromere, to form a joined pair of sister chromatids called a **dyad.** At some stage, each dyad unites (**synapses**) with another dyad like itself, forming bundles of four similar chromatids called **tetrads.** This synapsis becomes apparent when the chromosomes become visible, which is during the first meiotic division. The alikeness of chromosomes or chromatids is termed **homology** in genetics, and the alike units are said to be **homologs** of each other. This homology is manifested as similarity of size and shape. One can infer (and this is supported by cytological examination) that a cell destined to go through meiosis has many pairs of homologous chromosomes, initially unsynapsed, and that duplication and synapsis occur (probably in that order) to produce the tetrad structures.

Meiosis consists of two cell divisions, usually following in rapid sequence. In the first division, spindle fibers from opposite poles attach to the two centromeres of each tetrad. In contrast to mitosis, the centromeres do not split, so it is the dyads that are pulled to opposite poles and end up in two separate daughter cells. This first division is often called **reduction division** because the number of centromeres per cell is halved in each daughter cell.

In the second division of meiosis, centromeres divide, spindle fibers from each pole attach to these divided centromeres, and the sister chromatids are pulled apart in a manner reminiscent of mitosis. The second division is often called the **equational division,** because the number of centromeres per cell remains constant. Because there are two cell divisions, there are nearly always $2 \times 2 = 4$

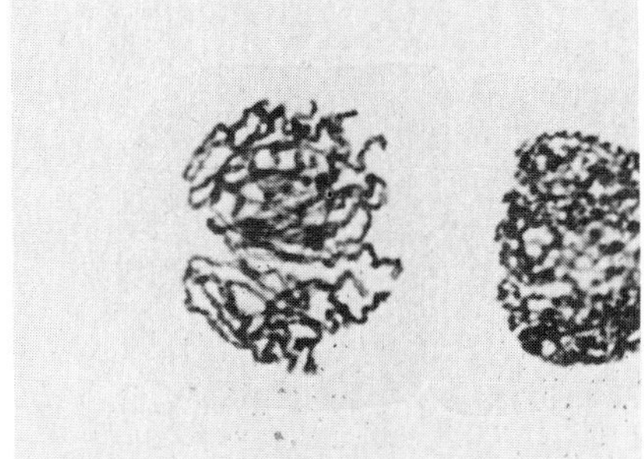

(a) *Prophase.* Chromosomes are contracting and becoming visible. Each chromosome has already duplicated into two sister chromatids, but these will not be clearly visible until metaphase.

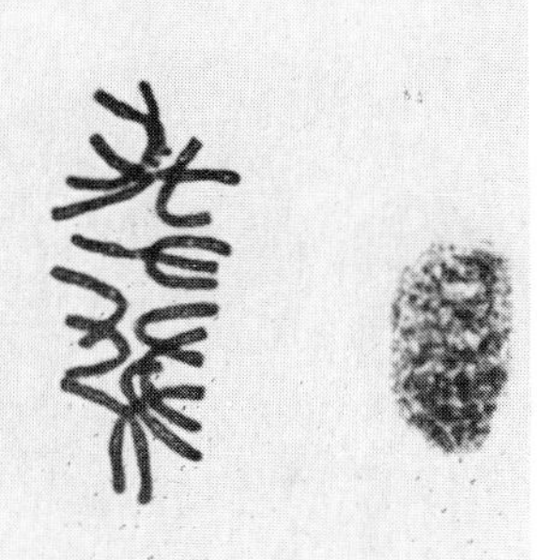

(b) *Metaphase.* All chromosomes assemble on the equatorial plane of the cell. Chromatids are now visible, and centromeres have also duplicated. Note that the chromosomes do *not* pair during mitosis.

(c) *Anaphase.* Centromeres segregate (or disjoin), pulling one sister chromatid to each cell pole.

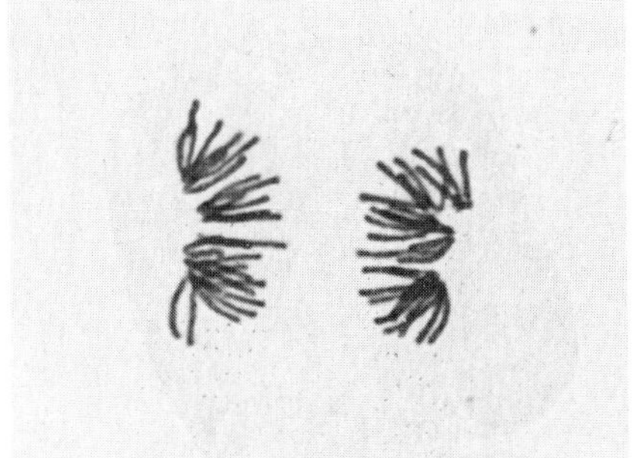

(d) *Anaphase.* Disjunction is complete. Compare the number of chromosomes in each cell with that in the original cell.

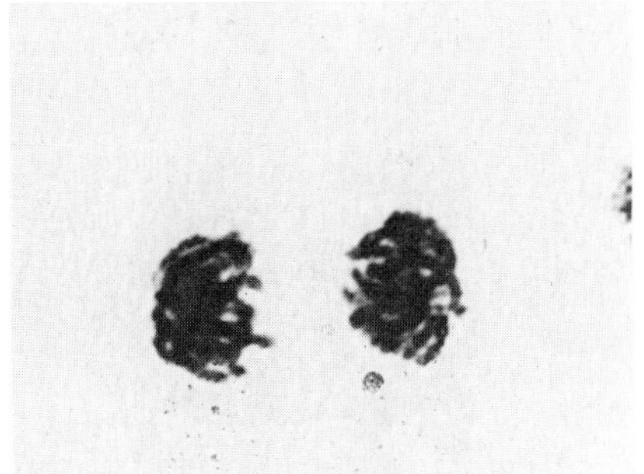

(e) *Telophase.* The nucleus reorganizes and cell division is almost complete.

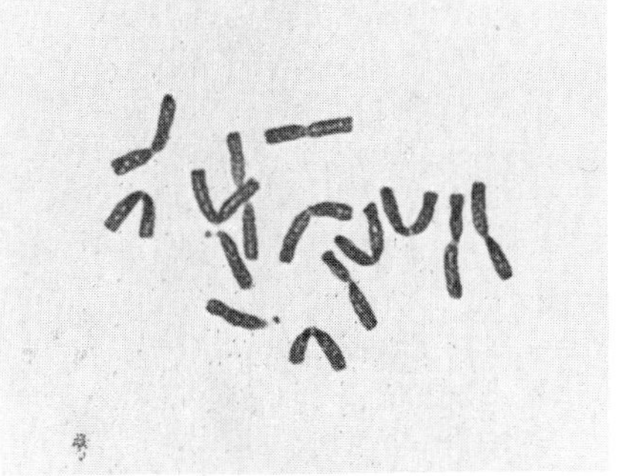

(f) Metaphase that has been disrupted by chemical treatment to show the individual chromosomes more clearly. Can you see any homologous ones?

Figure 3-1. Mitosis in Hepatica acutiloba *(2*n *= 14). (Photographs by C. J. Marchant and A. M. Adamovich.)*

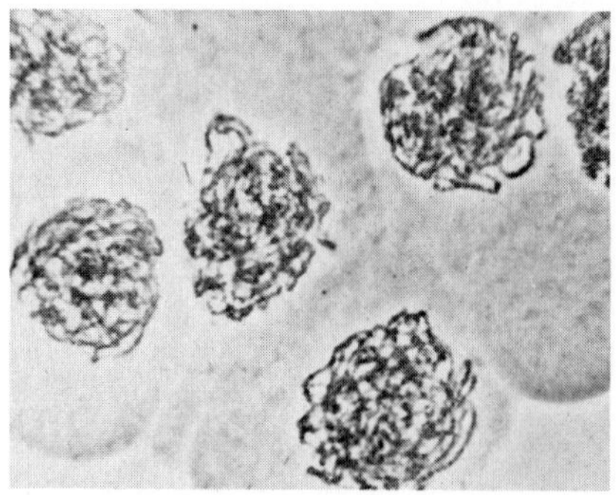

(a) *Early prophase I.* The chromosomes in several cells are becoming visible as long threads. Each chromosome is already duplicated into two daughter chromatids, but in this set of photographs the chromatids are not visible until anaphase I.

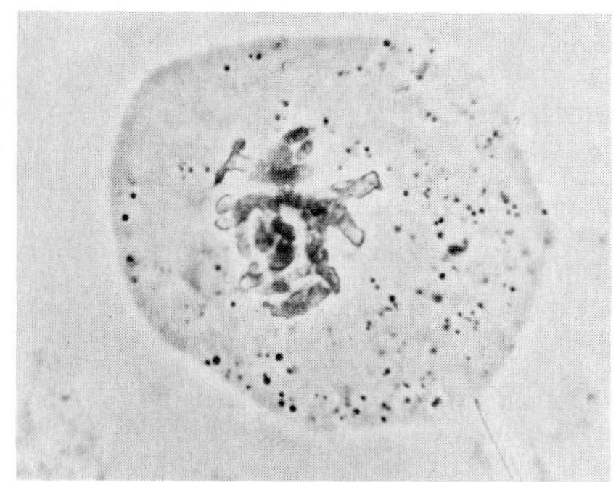

(b) *Prophase I.* The chromosomes have shortened (in this species the chromosomes become very squat), and the pairing of homologs can be discerned.

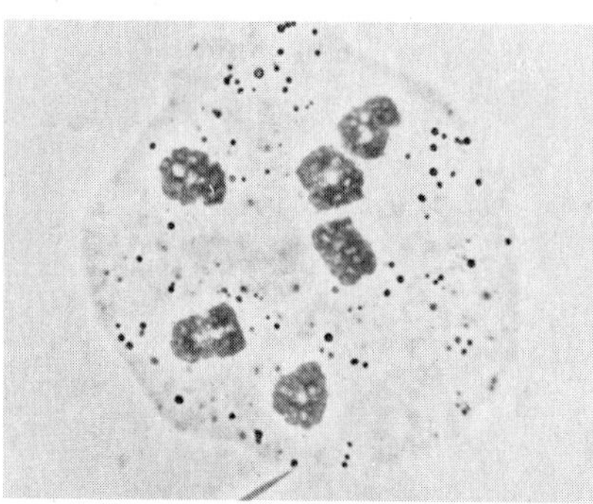

(c) *Late prophase I.* The pairing of homologous units is now clearly visible. Note that each of the six groups consists of four homologous chromatids and two undivided centromeres.

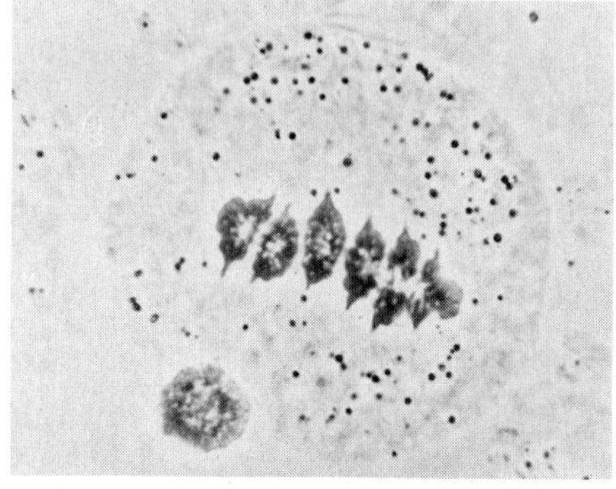

(d) *Metaphase I.* The groups of homologs are lined up on the equatorial plane of the cell, and segregation (or disjunction) of the centromeres is beginning.

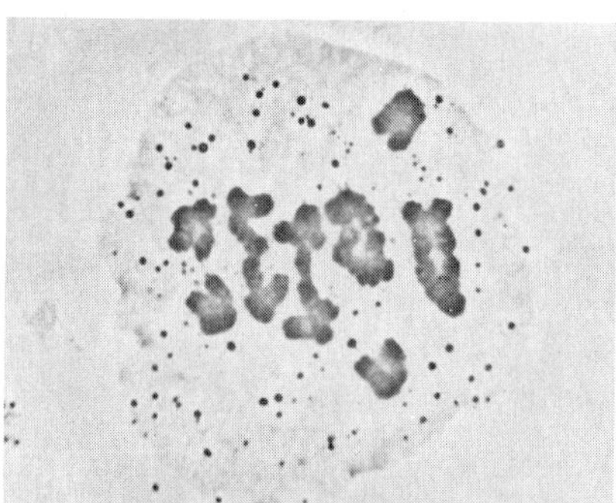

(e) *Anaphase I.* Different stages of disjunction are visible in different groups. Note that sister chromatids are now visible, still attached at their centromere.

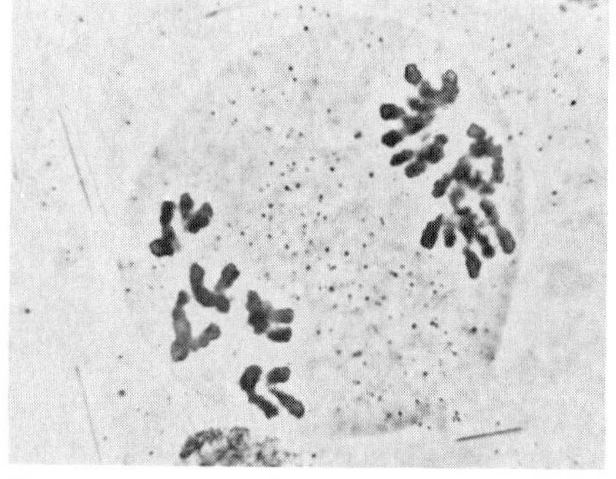

(f) *Late anaphase I.* The division of the nuclear contents is now almost complete. How many centromeres does each new nucleus have?

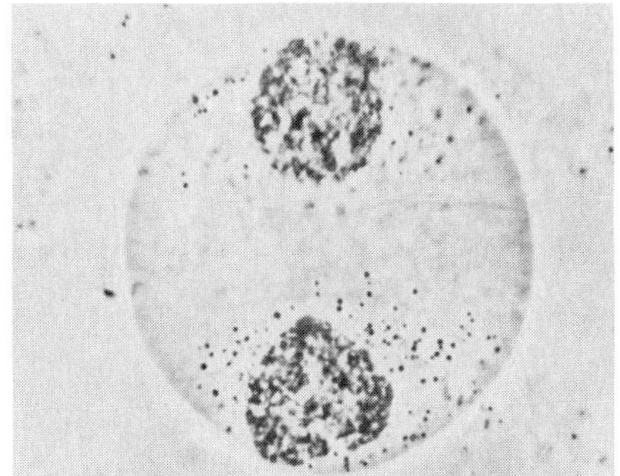

(g) *Telophase I.* Reorganization of the nuclei has occurred, and cell division is almost complete. The chromosomes elongate again.

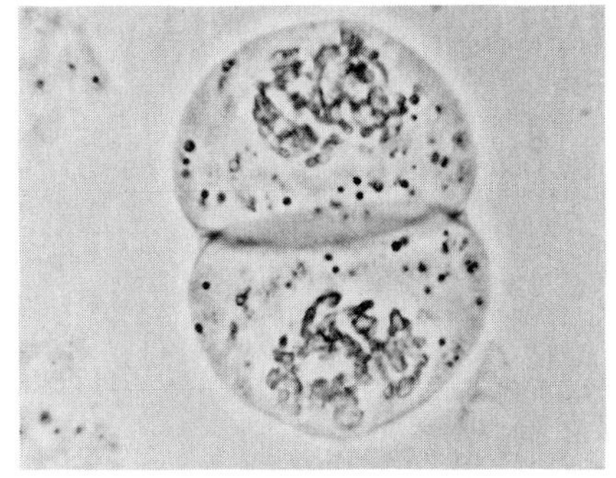

(h) *Prophase II.* Both cells begin the second meiotic division. The chromosomes contract.

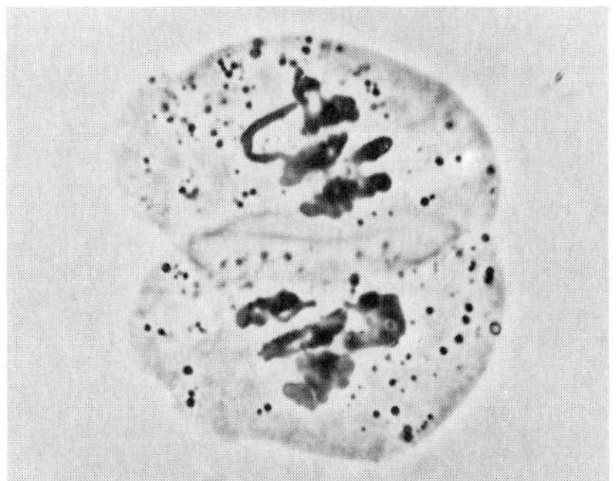

(i) *Metaphase II.* The centromeres have divided and the sister chromatids are now disjoining.

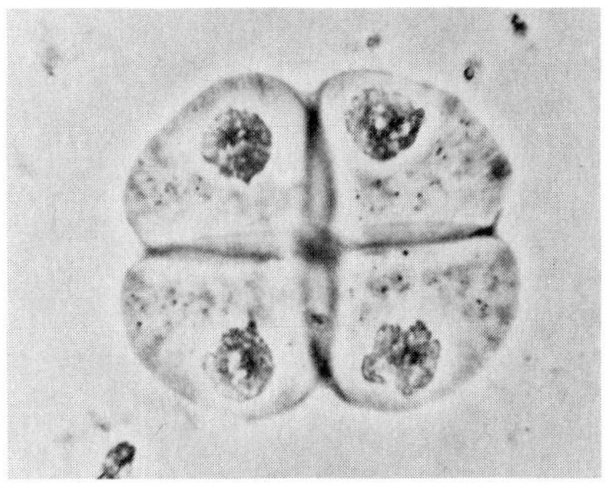

(j) *The four products of meiosis.* Each nucleus has one chromosome set and has six chromatids, which are now the new chromosomes. Each cell will develop into a pollen grain, or male gamete.

Figure 3-2. Meiosis in the anthers of Tradescantia paludosa *(2*n *= 12). (Photographs by C. J. Marchant and A. M. Adamovich.)*

cells representing the products of one meiosis. The important end result is that each of the four meiotic product cells has one-half the chromosome number of the initial premeiotic cell. (You can see that each cell ends up with one of the bundle of four chromatids we called a tetrad, and this is true for each tetrad present.) Meiotic products are, in many organisms, the gametes, and it is the fertilization of one gamete by another that restores the original chromosome number present in premeiotic cells.

In summary, then, mitosis produces two cells, each with the same number of chromosomes as the parental cell. Meiosis produces four cells, each with one-half the number of chromosomes of the parental cell. Chromosome pairing, which occurs only at meiosis, is the main cause of this difference in outcome.

The Chromosome Theory of Heredity

The regularity with which chromosomes duplicate, separate, and are restored in number after fertilization suggested to cytologists (before 1900) that chromosomes must be biologically important structures. Indeed, if we look at large numbers of individuals *within* a given species, the chromosome number is constant from individual to individual (with very rare exceptions). (In some species, the number varies between sexes, but in a very regular way.) On the other hand, the chromosome number varies remarkably *between* different species (Table 3-1). The entire chromosome complement of a single somatic cell is called a **genome.**

Soon after the rediscovery of Mendel's laws in 1900, Walter Sutton (an American graduate student) and Theodor Boveri (a great German biologist) recognized that the behavior of chromosomes during meiosis parallels the behavior of Mendel's hypothetical units of heredity, which we now call genes. Thus Sutton and Boveri in 1902 postulated that Mendel's hereditary particles "are on chromosomes" (Figure 3-3).

Message
The parallel behavior of Mendelian genes and chromosomes with respect to segregation and independent assortment led to the suggestion that genes are on chromosomes.

Table 3-1. Chromosome numbers of various species

Organism	Total chromosome number
Human	46
Dog	78
Horse	64
Cat	38
Mallard	80
Chicken	78
Alligator	32
Cobra	38
Bullfrog	26
Goldfish	94
Starfish	36
Fruitfly	8
Housefly	12
Neurospora	7
Sphagnum moss	23
Field horsetail	216
Giant sequoia	22
Tobacco	48

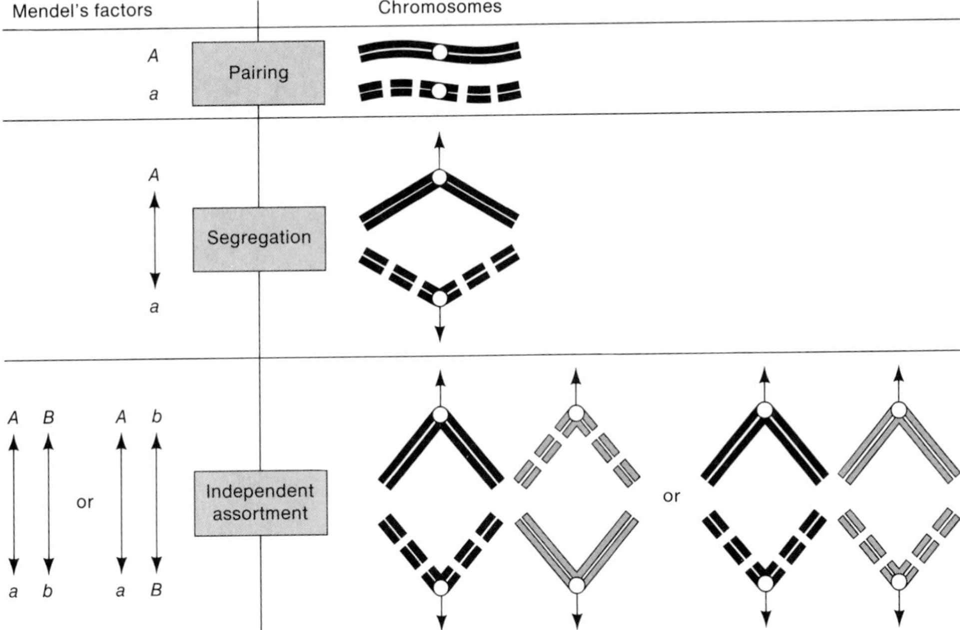

Figure 3-3. Parallels in the behavior of Mendel's hypothetical particles (genes) and chromosomes at meiosis. Black is used to represent one homologous pair and gray to represent another. Solid and dashed lines are used to distinguish one homolog from another.

To modern biology students, this suggestion may not seem very earthshaking. However, early in the twentieth century, this hypothesis (which potentially united cytology and the infant field of genetics) was a bombshell. Of course, the first response to such an important hypothesis is to try to pick holes in it. For years after, there was a raging controversy over the validity of what became known as the Sutton–Boveri chromosome theory of heredity.

It is worth considering some of the objections raised to the Sutton–Boveri theory. For example, at the time, chromosomes could not be detected during interphase (between cell divisions). Boveri had to make some very diligent studies of chromosome position before and after interphase before he could argue persuasively that chromosomes retain their physical integrity through interphase, even though they are cytologically invisible during interphase. It was also pointed out that all chromosomes look quite a lot alike in many organisms, so that they might be pairing randomly, whereas Mendel's laws absolutely require segregation of alleles. However, in species where chromosomes do differ in size and shape, it was verified that similar chromosomes do occur in pairs and that the homologs pair and segregate during meiosis.

Others argued that all chromosomes appear as stringy structures, so that one cannot detect any qualitative differences between them. Perhaps all chromosomes are just more or less of the same stuff. Alfred Blakeslee countered this

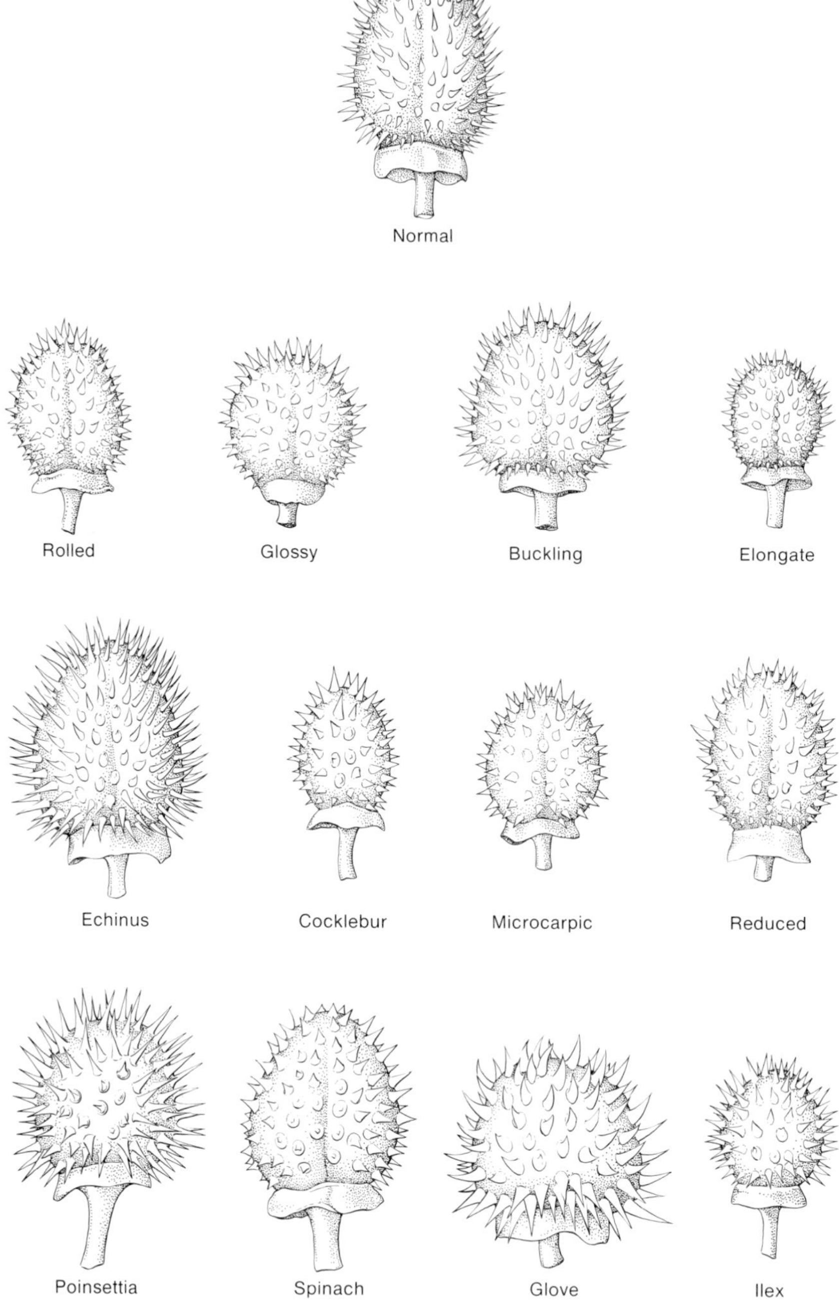

Figure 3-4. Fruits from Datura *plants, each having one different extra chromosome. Their characteristic appearances show that each chromosome produces a unique effect. (From E. W. Sinnott, L. C. Dunn, and T. Dobzhansky,* Principles of Genetics, *5th ed., McGraw-Hill Book Co.)*

objection through a study of the jimsonweed (*Datura*), which has 12 pairs of chromosomes. He obtained 12 different strains, each of which had the normal 12 chromosome pairs plus an extra representative of one pair. Blakeslee showed that each strain was phenotypically distinct from the others (Figure 3-4). This result would not be expected if the nonhomologous chromosomes were all alike.

Finally, Elinor Carothers found an unusual chromosomal situation in a certain species of grasshopper, a situation that permitted a direct test of whether different chromosome pairs do indeed segregate independently. Studying grasshopper testes, she found one chomosome pair (that regularly synapses) whose members are nonidentical; this is called a **heteromorphic pair,** and the chromosomes presumably show only partial homology. Futhermore, another chromosome, apparently unrelated to the heteromorphic pair, has no pairing partner at all. Carothers was able to use these unusual chromosomes as visible cytological markers of the behavior of chromosomes during assortment. By looking at anaphase nuclei, she could count the number of times that each dissimilar chromosome of the heteromorphic pair migrated to the same pole as the chromosome with no pairing partner (Figure 3-5). She found that the two patterns of chromosome behavior occur with equal frequency. Although these unusual chromosomes obviously are not typical, the results do suggest that nonhomologous chromosomes assort independently.

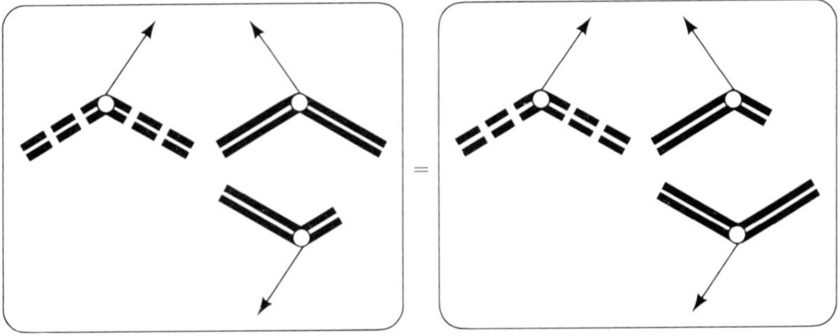

Figure 3-5. The two segregation patterns involving a heteromorphic pair (solid lines) *and an unpaired chromosome* (dashed lines).

All these results indicate that the behavior of chromosomes closely parallels that of genes. This of course makes the Sutton-Boveri theory attractive, but we do not yet have any *proof* that genes are on chromosomes. Further observations, however, did provide such proof.

The Discovery of Sex Linkage

In the crosses discussed thus far, it does not matter which sex of parent is selected from which strain under study. That is, reciprocal crosses (such as strain-A female × strain-B male, and strain-A male × strain-B female) yield similar progeny. The first exception to this pattern was discovered in 1906 by L. Doncaster and G. H. Raynor. They were studying wing color in the magpie moth (*Abraxas*), using two different lines—one with light wings, the other with dark. If light-winged females are crossed with dark-winged males, all the progeny have dark wings, thus showing that the allele for light wings is recessive. However, in the reciprocal cross (dark female × light male), all the female progeny have light wings, and all the male progeny have dark wings. Thus, this pair of reciprocal crosses does not give similar results, and in the second cross the wing phenotypes are associated with the sex of the moths. Note that the female progeny of this second cross are phenotypically similar to their fathers, and the males to their mothers. This is sometimes called **criss-cross inheritance.** How can we explain these results? Before attempting an explanation, let's consider another example.

William Bateson had been studying the inheritance of feather pattern in chickens. One line had feathers with alternating stripes of dark and light coloring, a phenotype called barred. Another line, nonbarred, had feathers of uniform coloring. In the cross of barred male × nonbarred female, all the progeny were barred, thus showing that nonbarred is recessive. However, the reciprocal cross (barred female × nonbarred male) gave barred males and nonbarred females. Again, the result is criss-cross inheritance. Can we find an explanation for these similar results with moths and with chickens?

An explanation came from the laboratory of Thomas Hunt Morgan, who in 1909 began studying inheritance in a fruitfly (*Drosophila melanogaster*). Because this organism has played a key role in the study of inheritance, a brief digression about the creature is worthwhile. The life cycle of *Drosophila* is typical of the life cycles of many insects (Figure 3-6).

The flies grow vigorously in the laboratory. In the egg, the early embryonic events lead to the production of a larval stage called the first "instar." Growing rapidly, the larva molts twice, and the third instar larva then pupates. In the pupa, the larval carcass is replaced by adult structures, and an "imago" (or adult) emerges from the pupal case, ready to mate within 12 to 14 hours. The adult fly is about 2 mm in length, so it takes up very little space. The life cycle is very short (12 days at room temperature) in comparison with that of a human, a mouse, or a corn plant; thus, many generations can be reared in a year. Moreover, the flies are extremely prolific—a single female is capable of laying several hundred eggs. Perhaps the beauty of the animal when observed through a microscope added to its early allure. In any case, as we shall see, the choice of *Drosophila* was a very fortunate one for geneticists—and especially for Morgan, whose work earned him a Nobel Prize in 1934.

The normal eye color of *Drosophila* is bright red. Early in his studies, Morgan discovered a male with completely white eyes. When he crossed this male with red-eyed females, all the F_1 progeny had red eyes, showing that the allele for white is recessive. Crossing the red-eyed F_1 males and females, Morgan obtained a 3:1 ratio of red-eyed to white-eyed flies, but all the white-eyed flies were males. Among the red-eyed flies, the ratio of females to males was 2:1. What is going on?

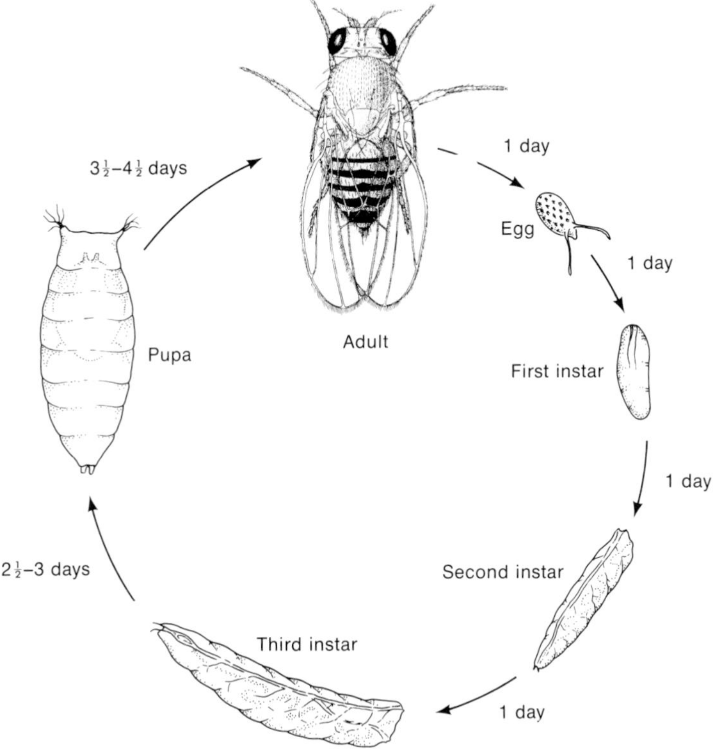

Figure 3-6. Life cycle of Drosophila melanogaster, *the common fruit fly.*

Morgan gathered further data. When he crossed white-eyed males with red-eyed female progeny of the cross of white males × red females, he obtained red males, red females, white males, and white females in equal numbers. Finally, in a cross of white females and red males (which is the reciprocal of the cross of the original white male with a normal female), all the females were red and all the males were white. This is criss-cross inheritance again. However, note that criss-cross inheritance was observed in the experiments on chickens and moths when the parental males carried the recessive genes; in the *Drosophila* cross, it is seen when the female parent carries the recessive genes.

Before turning to Morgan's explanation of the *Drosophila* results, we should look at some of the cytological information he was able to use in his interpretations. In 1891, working with males of a species of Hemiptera (the true bugs), H. Henking observed that meiotic nuclei contained 11 pairs of chromosomes and an unpaired element that moved to one of the poles during the first meiotic division. Henking called this unpaired element an "X body"; he interpreted it as a nucleolus, but later studies showed it to be a chromosome. Similar unpaired elements were later found in other species. In 1905, Edmond Wilson noted that females of *Protenor* (another Hemipteran bug) have seven pairs of chromosomes, whereas males have six pairs and an unpaired chromosome, which Wilson called (by analogy) the X chromosome. The females, in fact, have a pair of X chromosomes.

Also in 1905, Nettie Stevens found that males and females of the beetle *Tenebrio* have the same number of chromosomes, but one of the pairs in males is heteromorphic (of different shapes, as in the grasshoppers studied by Carothers). One member of the heteromorphic pair appears identical to the members of a pair in the female; she called this the X chromosome. The other member of the heteromorphic pair is never found in females; she called this the Y chromosome (Figure 3-7). Stevens found a similar situation in *Drosophila melanogaster,* which has four pairs of chromosomes, with one of the pairs being heteromorphic in males. Figure 3-8 summarizes the two basic situations we have described. (You may be wondering about the male grasshoppers studied by Carothers, which had both a heteromorphic pair and an unpaired chromosome. This situation is very unusual, and we needn't worry about it at this point of our discussion.)

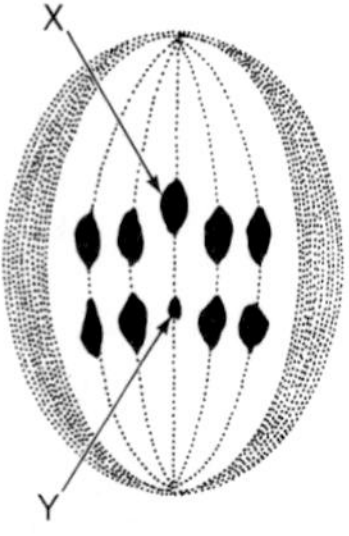

Figure 3-7. Segregation of the heteromorphic pair (X and Y) during meiosis in a Tenebrio *male. The X and Y chromosomes are being pulled to opposite poles during anaphase I. (From A. M. Srb, R. D. Owen, and R. S. Edgar,* General Genetics, *2nd ed. Copyright © 1965, W. H. Freeman and Company.)*

With this background information, Morgan constructed an interpretation of his genetic data. First, it appears that the X and Y chromosomes determine the sex of the fly. *Drosophila* females have four chromosome pairs, whereas males have three normal pairs plus a heteromorphic pair. Thus, meiosis in the female produces eggs that each bear one X chromosome. Although the X and Y chromosomes in males are heteromorphic, they seem to synapse and segregate like homologs (Figure 3-9). Thus, meiosis in the male produces two types of sperm, one type bearing an X chromosome and the other bearing a Y chromosome. According to this explanation, union of an egg with an X-bearing sperm produces an XX (female) zygote, and union with a Y-bearing sperm produces an XY (male) zygote. Furthermore, approximately equal numbers of males and females are expected because of the equal segregation of X and Y.

Morgan next turned to the problem of eye color. Assume that the alleles for red or white eye color are present on the X chromosome, with no counterpart

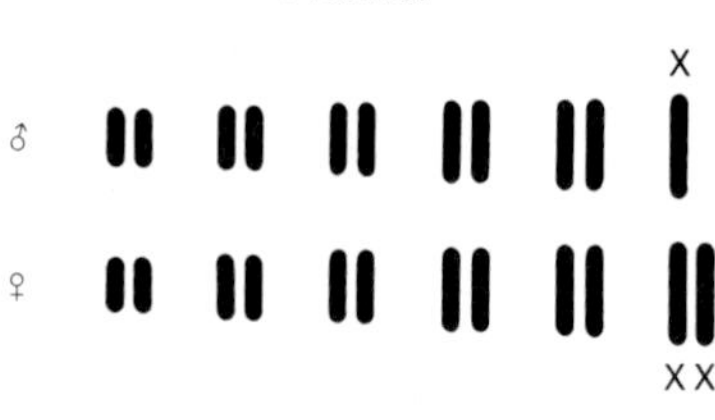

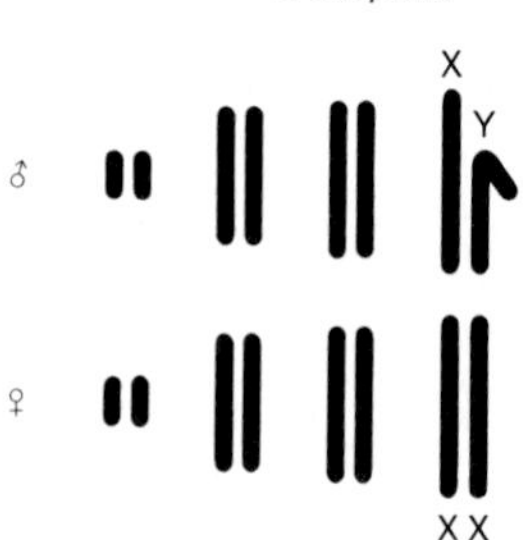

Figure 3-8. Diagrammatic representation of the chromosomal constitutions of males and females in two different insect species.

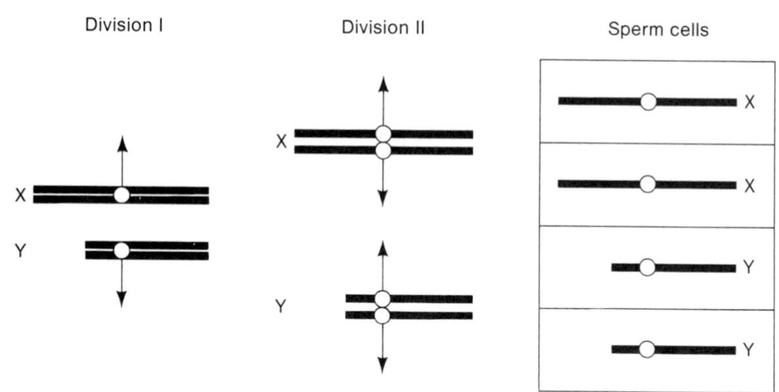

Figure 3-9. Meiotic pairing and the segregation of the X and Y chromosomes into equal numbers of sperms.

on the Y. Thus, females would have two copies of this gene, whereas males would have only one. This highly unexpected situation proves to fit the data. In the original cross of the white-eyed male with red-eyed females, all F_1 progeny had red eyes, showing that the gene for red eyes is dominant. Therefore, we can represent the two alleles as W (red) and w (white). If we designate the X chromosomes as X^W and X^w to indicate the alleles supposedly carried by them, we can diagram the two reciprocal crosses as shown in Figure 3-10.

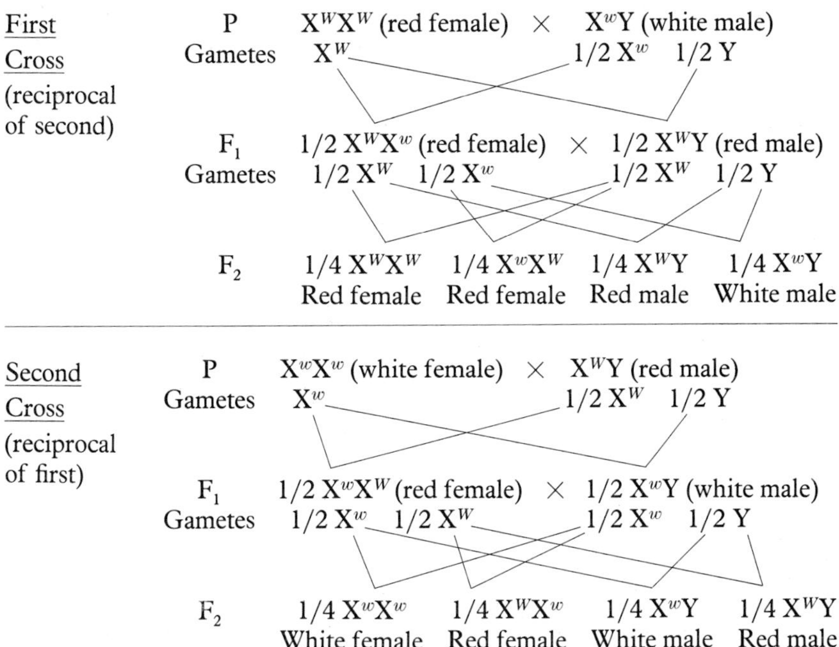

Figure 3-10. Explanation of the different results obtained from reciprocal crosses between red-eyed and white-eyed Drosophila.

As you can see from the figure, the genetic results of the two reciprocal crosses are completely consistent with the known meiotic behavior of the X and Y chromosomes. This experiment strongly supports the notion of chromosomal location of genes; however, it is only a correlation, and it does not constitute a definitive proof of the Sutton–Boveri theory.

Can the same XX and XY chromosome theory be applied to the results of the earlier crosses made with chickens and moths? You will find that it cannot. However, Richard Goldschmidt recognized immediately that these results can be explained with a similar hypothesis, simply assuming that the *males* have pairs of identical chromosomes, whereas the *females* have a heteromorphic pair. To distinguish this situation from the X–Y situation in *Drosophila*, Morgan suggested that the heteromorphic chromosomes in chickens and moths be called W–Z, with males being ZZ and females WZ. Thus, if the genes in the chicken and moth crosses are on the Z chromosome, the crosses can be diagrammed as shown in Figure 3-11.

CHICKENS				
First cross (reciprocal of second)	P	Z^BZ^B barred males	×	Z^bW nonbarred females
	F_1	Z^BZ^b barred males		Z^BW barred females
Second cross (reciprocal of first)	P	Z^bZ^b nonbarred males	×	Z^BW barred females
	F_1	Z^BZ^b barred males		Z^bW nonbarred females

MOTHS				
First cross (reciprocal of second)	P	Z^LZ^L dark males	×	Z^lW light females
	F_1	Z^LZ^l dark males		Z^LW dark females
Second cross (reciprocal of first)	P	Z^lZ^l light males	×	Z^LW dark females
	F_1	Z^lZ^L dark males		Z^lW light females

Figure 3-11. Inheritance pattern of genes on the sex chromosomes of two species having the ZW mechanism of sex determination.

Again, the interpretation is consistent with the genetic data. In this case, cytological data provided a confirmation of the genetic hypothesis. In 1914, J. Seiler verified that both chromosomes are identical in all pairs in male moths, whereas females have one heteromorphic pair.

Message
The special inheritance pattern of some genes makes it extremely likely that they are borne on the chromosomes associated with sex, which show a parallel pattern of inheritance.

An Aside on Genetic Symbols

These correlations were consistent with the supposition that genes do indeed reside on chromosomes, but they did not settle the matter. The heated debate continued for many years after these experiments. The critical proof of the Sutton–Boveri theory came from experiments performed by one of Morgan's most brilliant students, Calvin Bridges. Before we look at his work, however, we should introduce some symbols.

In *Drosophila*, a special symbolism for allele designation was introduced defining the alleles in relation to a "normal" allele. This system is now used by many geneticists, and is especially useful in genetic dissection. For a given *Drosophila* character, the allele that is found most frequently in natural populations (or, alternatively, that found in standard laboratory stocks) is designated as the standard, or **wild type.** All other alleles are then nonwild type. Sometimes the establishment of what is the wild-type allele can present a problem, but geneticists find this a useful approach in most situations. The symbolic designation of a gene is provided by the first nonwild allele found. In the *Drosophila* situation elucidated by Morgan, this was white eyes, so the nonwild allele is symbolized by w. The wild-type counterpart allele is conventionally represented by a + superscript, so the normal red-eye allele is written w^+

The wild-type allele is not always dominant over a nonwild-type allele. For the two alleles w and w^+, the use of the lower-case letter indicates that the wild type is dominant over white (that is, w is recessive to w^+). In another case, the wild-type condition of a fly's wing is straight and flat. A nonwild-type allele causes the wing to be curled. Because this allele is dominant over the wild-type allele, this gene is called Curly, and the nonwild-type allele is written *Cy*, while the wild-type allele is written Cy^+. Here, note that the capital letter indicates that *Cy* is dominant over Cy^+. (Also note that this two-letter symbol represents a single gene, not two genes.)

This kind of symbolism is useful because it helps geneticists to focus on the procedure of genetic dissection. The wild-type allele is defined as the normal functioning situation, and nonwild-type alleles (whether recessive or dominant) can be regarded as abnormal. The abnormal alleles then become probes to investigate how the normal situations work, though an examination of the ways in which the normal mechanism can go wrong. Most geneticists interested in using genetics to explore biological processes use this kind of symbolism. You will note that Mendel's convention (*A* and *a*, or *B* and *b*) does not define or emphasize normality. In a pea flower, for example, is red or white normal? However, the Mendelian symbolism is useful for some purposes; it is used extensively in plant and animal breeding. Table 3-2 summarizes the two systems of symbolism.

Table 3-2. Summary of two systems for assigning symbols to genes

	Recessive variant allele, *a*		Dominant variant allele, *A*	
Symbolic system	Symbol for wild-type allele	Symbol for variant allele	Symbol for wild-type allele	Symbol for variant allele
Normal/abnormal	a^+ (or $+^a$ or $+$)	a (or a^-)	A^+ (or $+^A$ or $+$)	A (or A^-)
Mendelian	A	a	a	A

Proof of the Chromosome Theory

We return now to Bridges' work. Consider a fruitfly cross we have discussed before (now represented in our new symbolism), X^wX^w (white ♀) + $X^{w^+}Y$ (red ♂). We know that the progeny are $X^{w^+}X^w$ (red ♀♀) and X^wY (white ♂♂). However, Bridges discovered that rare exceptions occur when the cross is made on a large scale. About 1 of every 2000 F_1 progeny is a white-eyed female or a red-eyed male. Because these exceptional progeny resemble their parents of the same sex, the phenotype of females is said to be **matroclinous,** and that of the males is called **patroclinous;** collectively, they are called **primary exceptional progeny.** All the patroclinous males proved to be sterile. However, when Bridges crossed the primary exceptional white-eyed females with normal red-eyed males, 4% of the progeny were matroclinous white-eyed females and patroclinous red-eyed males that were fertile. Thus, exceptional offspring were again recovered, but at a higher frequency, and the males were fertile. These exceptional progeny of primary exceptional mothers are called **secondary exceptional offspring** (Figure 3-12). How do we explain the exceptional progeny?

It is obvious that the matroclinous females—which, like all females, have two X chromosomes—must get both of them from their mothers because they are homozygous for *w*. Similarly, patroclinous males must derive their X chromosomes from their fathers because they carry w^+. Bridges hypothesized that rare mishaps occur during meiosis in the female, whereby the paired X chromosomes fail to separate during either the first or second division. This would result in meiotic nuclei containing either two X chromosomes or no X at all. Such a failure to separate is called nondisjunction; it produces an XX nucleus and a nullo-X nucleus (containing no X). Fertilization of these two types of nuclei will produce four zygotic classes (Figure 3-13).

It is important to note that, in these diagrams, as in many other meiotic diagrams, the lines representing the chromosomes are in fact not single chro-

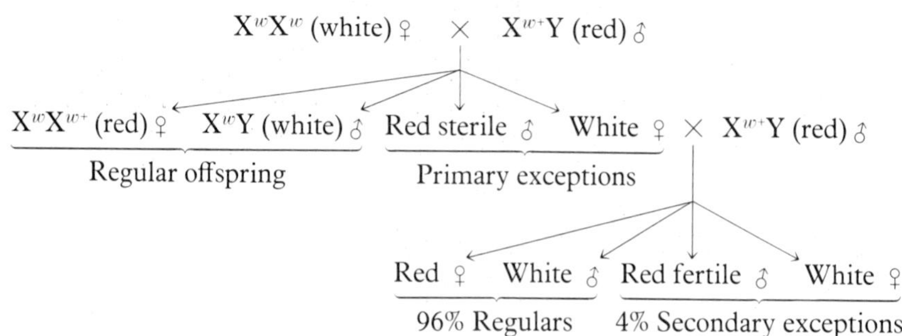

Figure 3-12. Drosophila *crosses from which primary and secondary exceptional progeny were originally obtained.*

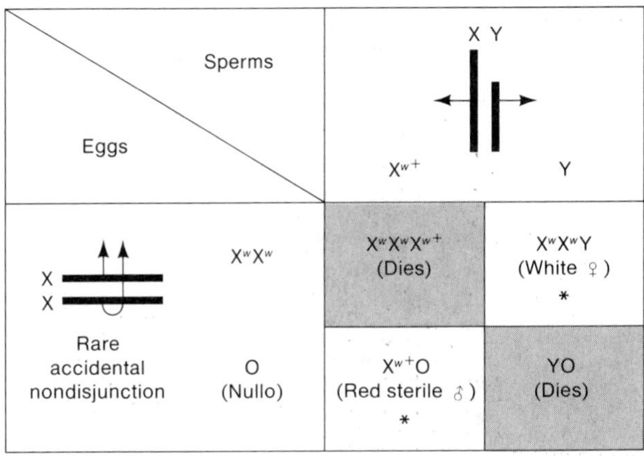

Figure 3-13. Proposed explanation of primary exceptional progeny through nondisjunction of the X chromosomes in the maternal parent.

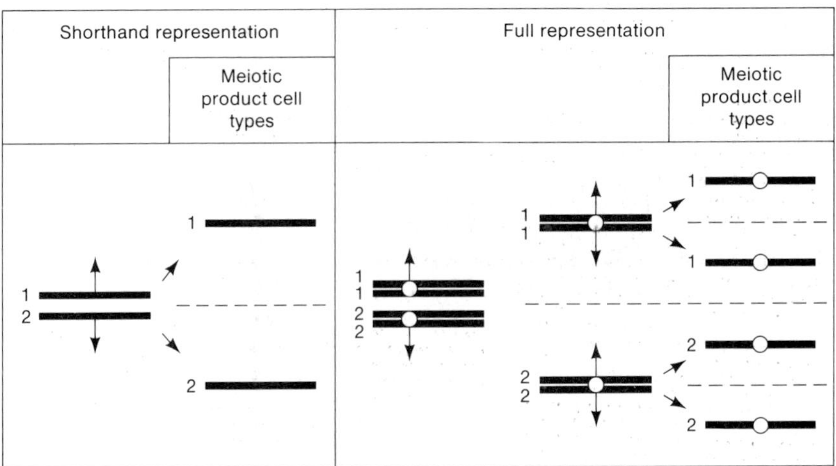

Figure 3-14. Two different representations of chromosomes used in genetics.

mosomes but dyads. This makes the diagrams simpler, because the second meiotic division need not be drawn. For many cases, this is an adequate shorthand technique, but sometimes there will be a need to draw both sister chromatids of the dyads. Figure 3-14 summarizes the shorthand convention.

If we assume that the XXX and YO classes die, then the two types of exceptional progeny can be expected to be X^w X^w Y (♀) and X^{w+} 0 (♂) if their

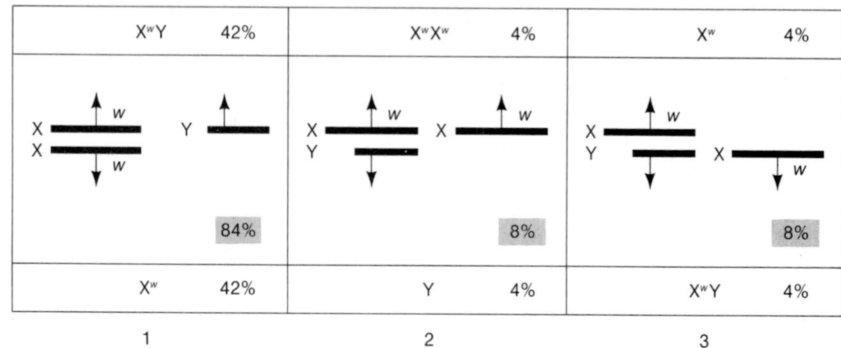

Figure 3-15. Three different segregation patterns in an XXY female fruitfly.

chromosomes are examined. Notice that it is implicit in Bridges' model that the sex of *Drosophila* is determined not by the presence or absence of the Y but by the number of X chromosomes. Two X chromosomes produce a female, and one X produces a male, in most situations.

What about the sterility of the primary exceptional males? This is explicable if we assume that the Y chromosome must be present in order to have male fertility.

			Sperms: X^{w+} (50%)	Sperms: Y (50%)	
Eggs	X–Y pairing (16%)	X^wX^w (4%)	$X^wX^wX^{w+}$ (dies) (2%)	X^wX^wY (white ♀) (2%)	Secondary exceptional progeny phenotype (4%) (4% die)
		Y (4%)	$X^{w+}Y$ (red fertile ♂) (2%)	YY (dies) (2%)	
		X^w (4%)	$X^{w+}X^w$ (red ♀) (2%)	X^wY (white ♂) (2%)	"Regular" (expected) progeny phenotypes (92%)
		X^wY (4%)	$X^wX^{w+}Y$ (red ♀) (2%)	X^wYY (white ♂) (2%)	
	X–X pairing (84%)	X^wY (42%)	$X^wX^{w+}Y$ (red ♀) (21%)	X^wYY (white ♂) (21%)	
		X^w (42%)	X^wX^{w+} (red ♀) (21%)	X^wY (white ♂) (21%)	

Figure 3-16. The proposed origin of secondary exceptional progeny as a result of specific gamete types produced by the XXY parent.

How can we explain the secondary exceptional offspring? During meiosis in the XXY females, if the two X chromosomes pair and disjoin most of the time, leaving the Y chromosome unpaired, then we should expect equal numbers of X-bearing and XY-bearing eggs. However, we know that the X and Y chromosomes can pair and segregate, because normal males produce equal numbers of X-bearing and Y-bearing sperms. To explain the observed results, we must assume that the Y successfully pairs with an X^w in approximately 16% of the pairings in X^wX^wY females, leaving the other X^w free to separate to either pole. One-half (8%) of these pairings will result in X^w and X^wY eggs, and the other one-half (8%) will result in X^wX^w and Y eggs (Figure 3-15).

We can now look at the results of fertilization by equal numbers of X^{w+} and Y sperms (Figure 3-16). We find that one-half of the fertilized X^wX^w and Y eggs will produce $X^wX^wX^{w+}$ and YY zygotes, which we presume die. The other one-half of these fertilized eggs produce the secondary exceptions, X^wX^wY and $X^{w+}Y$. Now we see why the secondary exceptional males are fertile: each of them receives a Y chromosome from the XXY mother.

So far, this is all a model—an intellectual edifice. We have made assumptions about the chromosome location of w and w^+, and we have hypothesized nondisjunction to explain the exceptional progeny. However, if this model is correct, we can now make testable predictions.

1. Cytological study of the primary exceptional progeny (which we have identified through the genetic study) should show that the females are XXY and the males are XO. Bridges confirmed this prediction.

2. Cytological study of the secondary exceptional progeny (which we have identified genetically) should show that the females are XXY and the males are XY. Bridges confirmed this prediction.

3. One-half of the red-eyed daughters of exceptional white-eyed females should be XXY and one-half should be XX. Bridges confirmed this prediction.

4. One-half of the white-eyed sons of exceptional white-eyed females should themselves give exceptional progeny, and all of those that do should be XYY. Bridges confirmed this prediction.

Thus Bridges verified all the testable predictions arising from the assumptions that w and w^+ are indeed on the X chromosome and that an unexplained process of nondisjunction occurs in infrequent cases of meiosis. These confirmations provide unequivocal evidence that genes are associated with chromosomes.

Message

When Bridges used the chromosome theory to predict successfully the outcome of certain genetic analyses, the chromosome location of genes was established beyond reasonable doubt.

Sex Chromosomes in Other Species

Humans and all mammals also show an X–Y sex-determining mechanism, with males XY and females XX. Unlike *Drosophila*, however, it is the presence of the Y that determines maleness in humans. This difference is demonstrated by the sexual phenotypes of the abnormal chromosome types XXY and XO (Table 3-3). However, we postpone a full discussion until a later chapter.

Table 3-3. Chromosomal determination of sex in *Drosophila* and humans

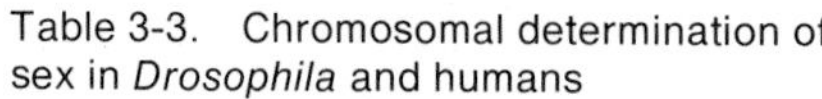

Species	Sex chromosomes			
	XX	XY	XXY	XO
Drosophila	♀	♂	♀	♂
Humans	♀	♂	♂	♀

Higher plants show a variety of sexual situations. Some species have both male and female sex organs on the same plant, often combined into the same flower (called **hermaphroditic;** the rose is an example), or in separate flowers on the same plant (called **monoecious;** corn is an example). These would be the equivalent of hermaphroditic animals. **Dioecious** species, however, show plants of separate sexes, with male plants bearing flowers containing only anthers, and female plants bearing flowers containing only ovaries (Figure 3-17). Some, but not all, dioecious plants have a heteromorphic pair of chromosomes associated with (and almost certainly determining) the sex of the plant. Of the species with heteromorphic sex chromosomes, a large proportion have an X–Y system. Critical experiments in a few species suggest a *Drosophila*-like system. Table 3-4 lists some examples of dioecious plants with heteromorphic sex-determining chromosomes. Some dioecious plants have no visibly heteromorphic pair of chromosomes; they might still have sex chromosomes, but not visibly distinguishable types. Other dioecious plants have heteromorphic

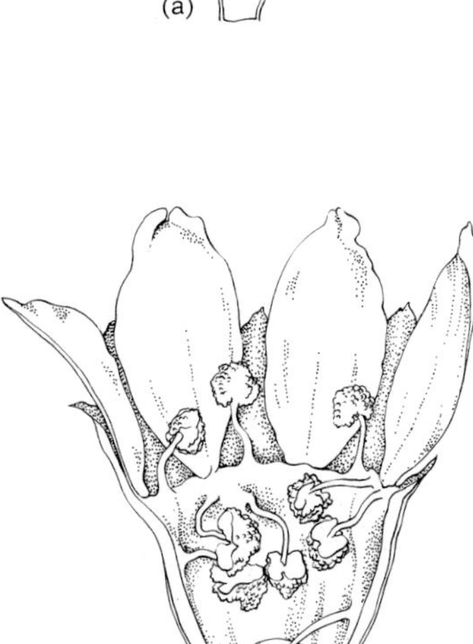

Figure 3-17. Female (a) and male (b) forms of flowers in a dioecious species (Osmaronia).

Table 3-4. Sex-chromosome situations in some dioecious plant species

Species	Chromosome number	Sex-chromosome constitution	
		Female	Male
Cannabis sativa (hemp)	20	XX	XY
Humulus lupulus (hop)	20	XX	XY
Rumex angiocarpus (dock)	14	XX	XY
Melandrium album (campion)	22	XX	XY

chromosomes that determine sex, but the constitution of the different sexes is much more complex than the simply X–Y system of the *Drosophila* type. We shall not consider these systems.

The sex chromosomes can be tentatively divided into pairing and differential regions (Figure 3-18). These regions are based on studies of meiosis in males. The pairing regions of the X and Y chromosomes are thought to be homologous. The differential region of each chromosome appears to hold genes that have no counterparts on the other kind of sex chromosome. These genes, whether dominant or recessive, show their effects in the male phenotype. Genes in the differential regions are called **hemizygous** ("half-zygous"). Hemizygous genes in the differential region of the X show an inheritance pattern called **X linkage;** those in the differential region of the Y show **Y linkage.** Genes in the pairing region show what might be called **X-and-Y linkage.** In general, genes on the sex chromosomes show **sex linkage.**

We can introduce some other common technology here. The sex showing only one kind of sex chromosome (XX ♀ or ZZ ♂) is called the **homogametic** sex; the other (XY ♂ or ZW ♀) is called **heterogametic** sex. Thus, human and *Drosophila* females (and male birds and moths) are homogametic; that is, they produce only one type of gamete with respect to sex-chromosome constitution. The nonsex chromosomes (what we might call the "regular" chromosomes) are called **autosomes.** Humans have 46 chromosomes per cell: 44 autosomes plus two sex chromosomes. The plant *Melandrium album* has 22 chromosomes per cell: 20 autosomes plus two sex chromosomes.

Genes on the autosomes show the typical kind of inheritance pattern—the type discovered and studied by Mendel. The genes on the different regions of the sex chromosomes show their own typical patterns of inheritance, as follows.

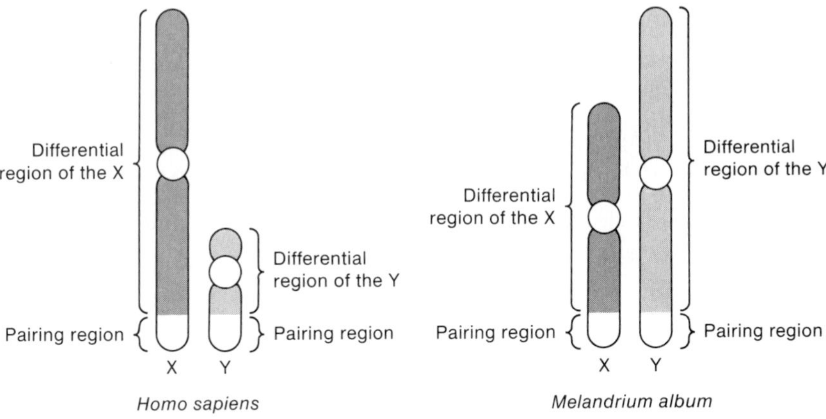

Figure 3-18. Differential and pairing regions of sex chromosomes of humans and of the plant Melandrium album. *The pairing regions have been located chiefly through cytological examination.*

X-Linked Inheritance

We have already seen an example of X-linked inheritance in *Drosophila:* the inheritance pattern of white eye and its wild-type allele. Of course, eye color is not concerned with sex determination, so we see that not all genes on the sex chromosomes are involved with sexual function. The same is true in humans, where many X-linked genes have been discovered through pedigree analysis, of which hardly any could be construed to have any connection to sexual function. Just as we earlier listed rules for detecting autosomal inheritance patterns in humans, we can list clues for detecting X-linked genes in human pedigrees. (We shall soon see that Y-linked genes are very rare in humans, so we can usually ignore them.)

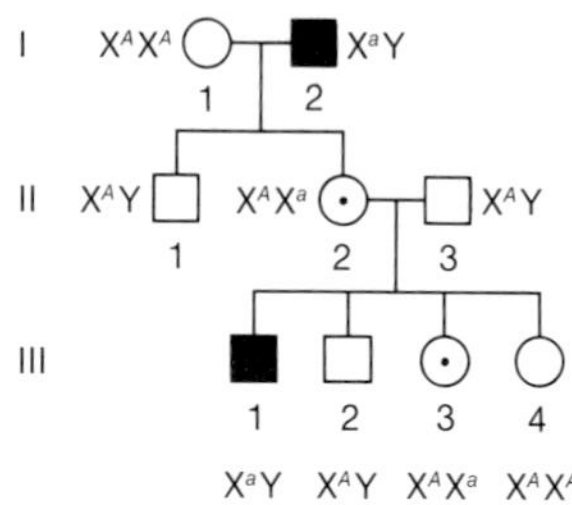

Figure 3-19. Illustrative pedigree showing how X-linked recessives are expressed in males, then carried unexpressed by females in the next generation, to be expressed in their sons. (Note that III-3 and III-4 cannot be distinguished phenotypically.)

Recessive genes showing X-linked inheritance can be detected in human pedigrees through the following clues.

1. Typically, many more males than females show the recessive phenotype. This is because an affected female can result only when both mother and father bear the gene (for example, $X^AX^a \times X^aY$), whereas an affected male can result when only the mother carries the gene. If the recessive gene is very rare, almost all observed cases will occur in males.

2. Usually none of the offspring of an affected male will be affected, but all his daughters will carry the gene in masked heterozygous condition, so one-half of their sons will be affected (Figure 3-19).

3. None of the sons of an affected male will inherit the gene, so not only will they be free of the phenotype, but they will not pass the gene along to their offspring.

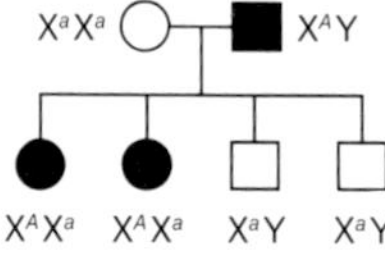

Figure 3-20. Illustrative pedigree showing how X-linked dominants are expressed in all the daughters of affected males.

Some good examples of X-linked recessive genes in humans are hemophilia, red–green color blindness, and Duchenne's muscular dystrophy.

Dominant genes showing X-linked inheritance can be detected in human pedigrees through the following clues.

1. The most important clue here is that affected males pass the condition on to all of their daughters but to none of their sons (Figure 3-20).

2. Females, on the other hand, usually pass the condition on to one-half of their sons and daughters (Figure 3-21).

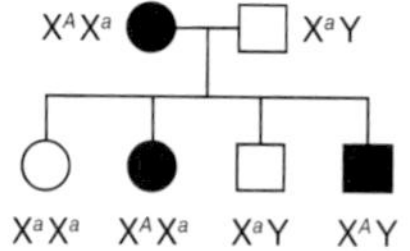

Figure 3-21. Illustrative pedigree showing how females affected by an X-linked dominant condition usually are heterozygous and pass the condition to one-half of their progeny.

Y-Linked Inheritance

In humans, genes on the differential region of the Y would be inherited only by males, with transmission from father to son. No examples of such inheritance have been confirmed, although hairy ear rims has been suggested as a likely possibility (Figure 3-22). However, in humans (and in other species with similar sex-determination systems), the presence of the Y and its unique regions

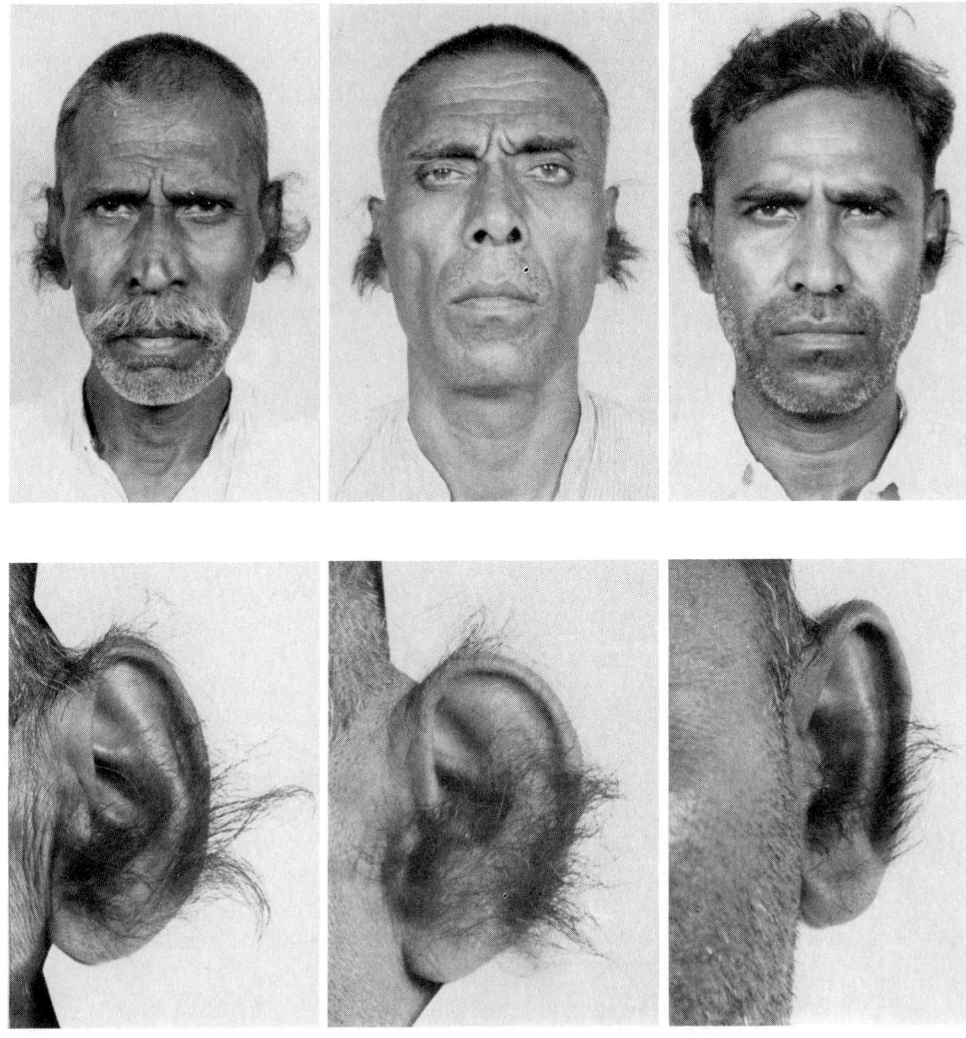

Figure 3-22. Hairy ear rims are thought to be due to a Y-linked gene. (From C. Stern, W. R. Centerwall, and S. S. Sarkar, The American Journal of Human Genetics *16 (1964):467. By permission of Grune & Stratton, Inc.)*

determines maleness. Therefore, it seems safe to speculate that "maleness" genes of some kind must exist on the differential region of the Y.

In the fish *Lebistes*, the Y chromosome carries a gene called maculatus that determines a pigmented spot at the base of the dorsal fin. This phenotype is passed only from father to son, and females never carry or express the gene, so it seems to be a clear example of Y linkage.

X-and-Y-Linked Inheritance

Are there ways of identifying genes on homologous regions of the sex chromosomes? The existence of homologous regions is itself somewhat in doubt, being in part inferred from pairing association. However, there is a gene pair in *Drosophila* that is inherited in such a way that an X-and-Y location is likely. Curt Stern found that a certain nonwild recessive allele in *Drosophila* causes a phenotype of shorter and more slender bristles; he called it bobbed (*b*). If a bobbed X^bX^b female is crossed with a wild-type X^+Y^+ male, all the F_1 progeny are wild type: X^+X^b ♀♀ and X^bY^+ ♂♂. The same result, of course, would be expected from an autosomal gene pair. However, the X-and-Y linkage is revealed in the F_2:

Males:	All wild-type phenotype	→ Inferred X^+Y^+ and X^bY^+ genotypes
Females:	1/2 bobbed phenotype	→ Inferred X^bX^b genotype
	1/2 wild-type phenotype	→ Inferred X^+X^b genotype

The Chromosome Theory in Review

The patterns of inheritance centering on the normal and nondisjunctional behavior of the sex chromosomes provide satisfying confirmation of the chromosome theory of inheritance, which was originally suggested by the parallel behavior of Mendelian genes and autosomal chromosomes. Now we should pause and state clearly the situation for the regular (autosomal) genes, because these are the genes most commonly encountered. Such a summary is best achieved in a diagram; Figure 3-23 illustrates the passage of a hypothetical cell through meiosis. Two gene pairs are shown on two chromosome pairs. The hypothetical cell type has four chromosomes: a pair of homologous long chromosomes and a pair of homologous short ones. (Such size differences between pairs are common.) The genotype of the cell is assumed to be *Aa Bb*.

As the diagram shows, two equally frequent kinds of spindle attachment (4a and 4b) result in two basic kinds of segregation patterns of gene pairs. Meiosis then produces four cells of the genotypes shown from each of these segregation patterns. Because the segregation patterns 4a and 4b are equally common, the meiotic product cells of genotypes *Ab*, *ab*, *Ab*, and *aB* are produced in equal frequencies. In other words, each of the four genotypes occurs with frequency 1/4. This, of course, is the distribution postulated in Mendel's model and is the one we have written along one edge of the standard Punnett square (Figure 2-9). We can now understand the exact chromosomal mechanism that produces the Mendelian ratios.

Notice that Mendel's first law (equal segregation) is a direct result of the separation of a pair of homologs (bearing the gene pair under study) into

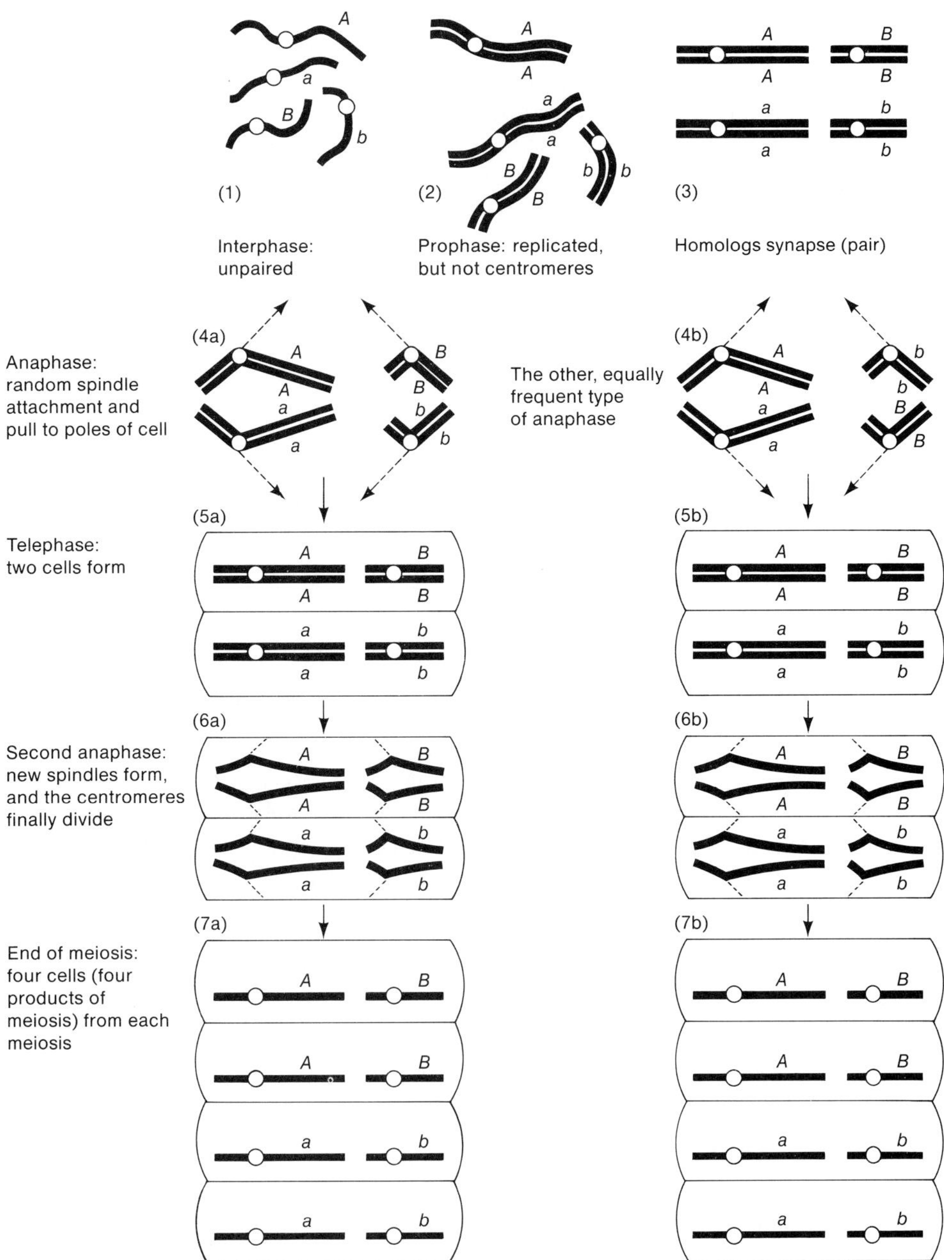

Figure 3-23. The route of two heterozygous gene pairs (on separate chromosome pairs) through meiosis. Steps 1 through 7 represent the various events of meiosis. The a and b series result from the two equally frequent spindle-attachment arrangements in the first meiotic division. In stage 7, note that AB, Ab, aB, *and* ab *occur with equal frequency.*

opposite cells at the first division. Notice also that Mendel's second law (independent assortment) results from the independent behavior of separate pairs of homologous chromosomes.

The chromosome theory was important in many ways. At this point in our discussion, we can stress the importance of the theory in centering attention on the role of the cell genotype in determining the organism phenotype. The phenotype of an organism is determined by the phenotypes of all its individual cells. The cell phenotype, in turn, is determined by the genes (and alleles of those genes) present on the chromosomes of the cell. When we say that an organism is (for example) *AA*, we really mean that each cell of the organism is *AA*. This cell genotype determines how the cell functions, thus controlling the phenotype of the cell and hence the phenotype of the organism.

Message
An organism's phenotype is determined by the kinds of genes it has in its cells and by the forms (alleles) of those genes that are present.

Mendelian Genetics in Less Familiar Sexual Cycles

Thus far we have been discussing diploid organisms—organisms with two identical chromosome sets in each cell. The diploid is designated $2n$, where n stands for one chromosome set (for example, the pea cell contains two sets of seven chromosomes, so $2n = 14$). Most of the organisms we encounter in our daily existence are diploid; these are the so-called higher plants (including flowering plants) and higher animals (including humans). In fact, evolution seems to have produced a trend toward diploidy (perhaps you can speculate on reasons for this). Nevertheless, a vast proportion (probably the major proportion) of the biomass on earth is composed of organisms that spend most of their life cycles in a haploid condition, in which each cell has only one set of chromosomes. Haploids are designated n. Important here are the fungi and algae, most of which are predominantly haploid. Also important are organisms that spend part of their life cycles as haploid and another part as diploid. Such organisms are said to show alternation of generations (a poor term—alternation of $2n$ and n would be more descriptive); familiar examples are mosses and ferns. Bacteria could be considered haploid, but they form a special case because their cells contain no nuclei (they are called prokaryotic cells), whereas most other life-forms (eukaryotic cells) do have nuclei. Bacteria are discussed in Chapter 9.

Do these unconventional life cycles show Mendelian genetics? Or is Mendelian inheritance observed only in the "higher" organisms? The answer is that Mendelian inheritance patterns appear in any organism with meiosis as part of its cycle, because Mendelian laws are based on the process of meiosis. All the groups of organisms mentioned, except bacteria, utilize meiosis as part of their cycles. We next consider the inheritance patterns shown by these less familiar forms, and we compare them to more familiar cycles. This is important because it shows how universal Mendelian genetics is and because some of these "exotic" organisms have been the subject of extensive genetic research. We describe here three major types of cycle, beginning with the more familiar diploid types.

Diploids

Figure 3-24 summarizes the diploid cycle in skeletal form. This is the cycle shown, with minor variations, by animals (including humans) and by higher plants (including flowers and trees). Meiosis occurs in specialized diploid cells,

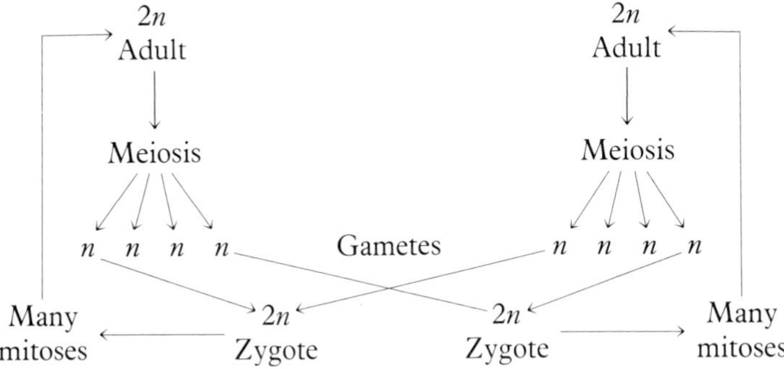

Figure 3-24. The diploid life cycle.

which are set aside for the purpose but which are part of the diploid adult organism. In most cases, the cells that are the products of meiosis can be called gametes (eggs or sperm). Fusion of haploid gametes forms a diploid zygote, which (through mitosis) produces a multicellular organism. Mitosis in a diploid proceeds in the fashion outlined in Figure 3-25.

Haploids

Figure 3-26 shows the basic haploid cycle. Here the "adult" (either multicellular or unicellular) is haploid. How can meiosis possibly occur in a haploid? After all, meiosis involves pairing of two homologous chromosome sets! The answer is that all haploid organisms that undergo meiosis employ a *transient* diploid stage. In some cases, unicellular haploid adult individuals fuse to form a diploid cell, which then undergoes meiosis. In other cases, specialized (or sometimes representative) haploid cells from different parents fuse to form diploid cells so that meiosis can occur. (These fusing cells are properly called the gametes,

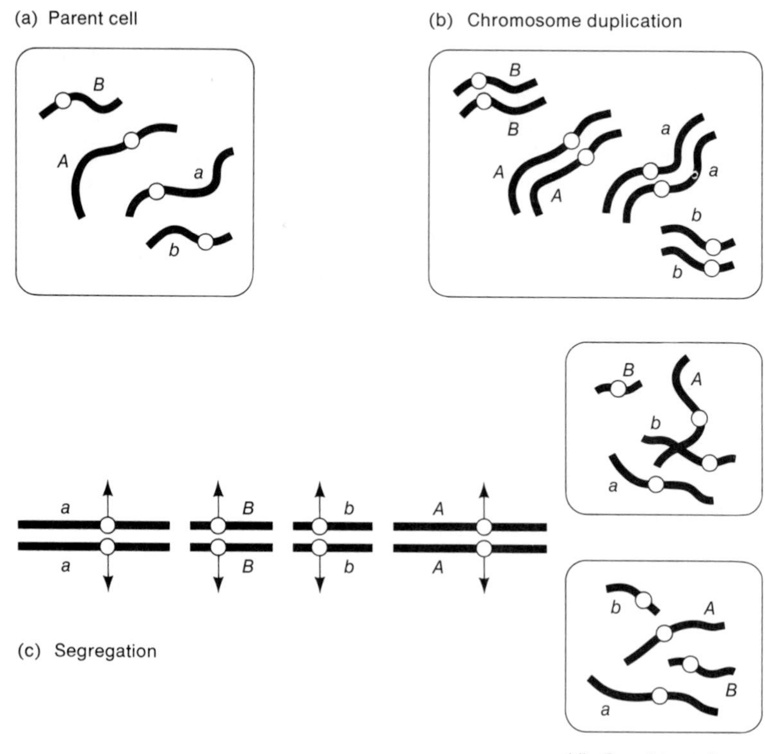

Figure 3-25. Mitosis in a diploid cell of genotype Aa Bb.

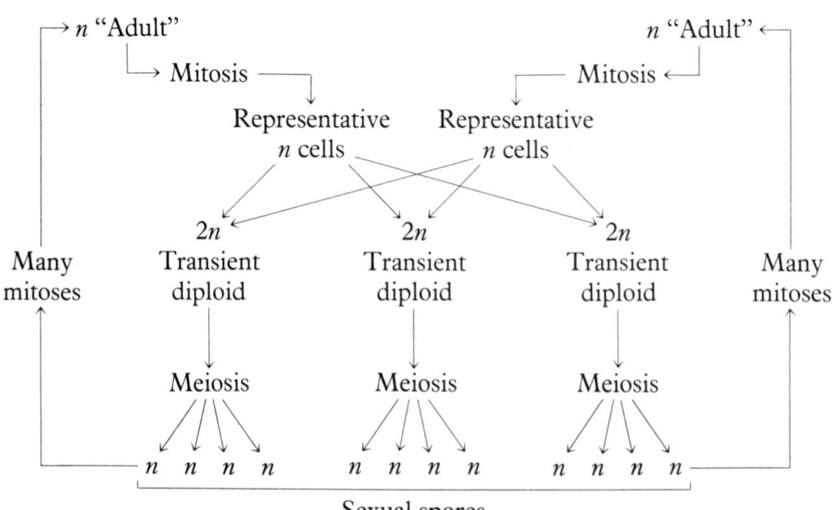

Figure 3-26. The haploid life cycle.

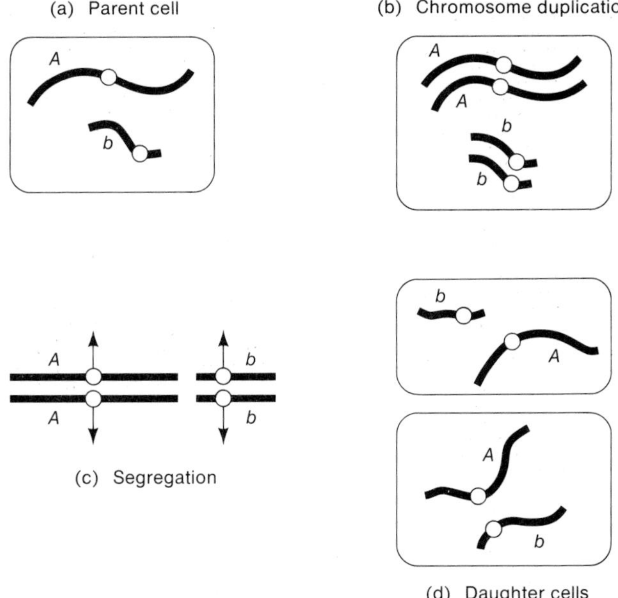

Figure 3-27. Mitosis in a haploid cell of genotype A b.

so you see that in these cases gametes arise from mitosis.) Meiosis produces haploid products of meiosis, which are called **sexual spores.** The sexual spores can become new unicellular adults; in other species, they develop through mitosis into a multicellular haploid individual. In haploids, mitosis proceeds as shown in Figure 3-27, which happens to illustrate a cell of genotype *A b*. Notice that a cross between two adult haploid organisms involves only one meiosis, whereas a cross between two diploid organisms involves a meiosis in each organism. As we shall see, this simplification makes haploids very useful for genetic analysis.

Let's consider a cross in a haploid. A convenient organism is the pink bread mold **Neurospora.** This fungus is a multicellular haploid in which the cells are joined end to end to form **hyphae,** or threads of cells. The individual cells can become detached, in which case they are known as **asexual spores.** Asexual spores can disperse to form new colonies; alternatively, they can act as male gametes (Figure 3-28). Another specialized cell (which develops inside a knot of hyphae) can be regarded as the female gamete.

What characters can be studied in such an organism? One is the color of the cells. Variants of the normal pink color can be found—for example, an albino. Figure 3-29 shows a normal and an albino culture. Another character is the morphological nature of the culture—perhaps fluffy (normal) versus colonial. We can make a cross by allowing the asexual spores to act as male gametes.

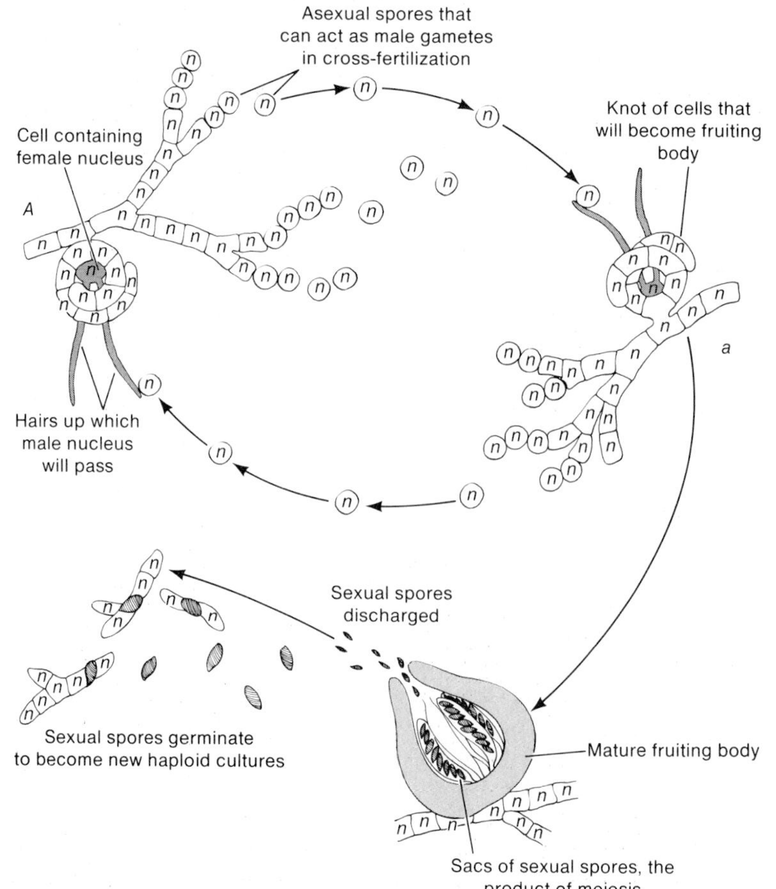

Figure 3-28. Simplified representation of the life cycle of Neurospora crassa, *the pink bread mold. Self-fertilization is not possible in this species: there are two mating types, determined by the alleles* A *and* a *of one gene. A cross will succeed only if it is* A × a. *Male and female nuclei divide mitotically; then fusion occurs to form many diploid nuclei, each of which undergoes meiosis. The sexual spores are the meiotic products.*

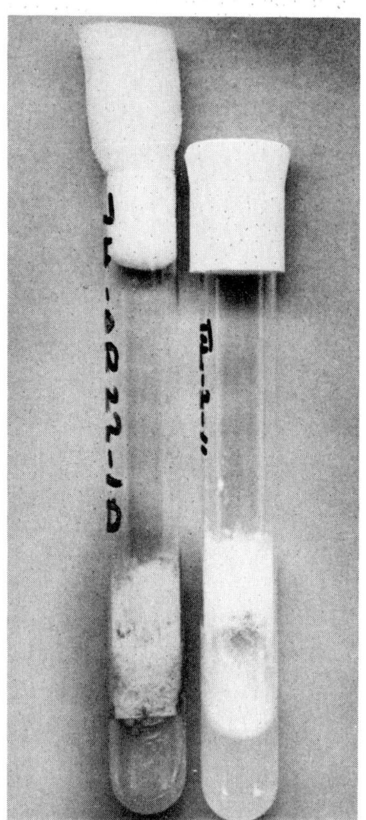

Figure 3-29. A pink wild-type Neurospora *culture* (left) *and an albino mutant culture* (right) *lacking the reddish carotenid pigment (genotypically the cultures are* al^+ *and* al^-, *respectively).*

A culture cannot self in *Neurospora,* so crosses are carried out by adding asexual spores of one culture to another, or vice versa. The asexual spores will fuse with the female gametes; then meiosis occurs, and sexual spores (called **ascospores**) are formed. These ascospores are black and football-shaped; they are shot out of the knot of hyphae, which is now known as a **fruiting body.** The ascospores can be isolated, each into a culture tube, where each ascospore will grow into a new culture by mitosis (Figure 3-30).

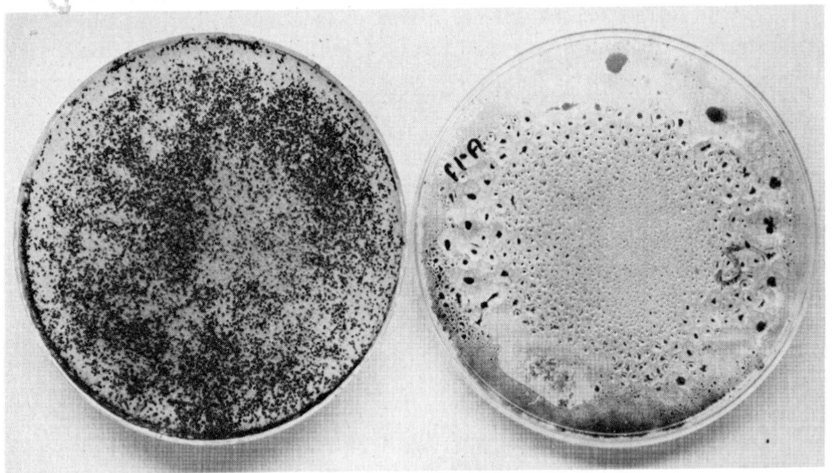

(a)

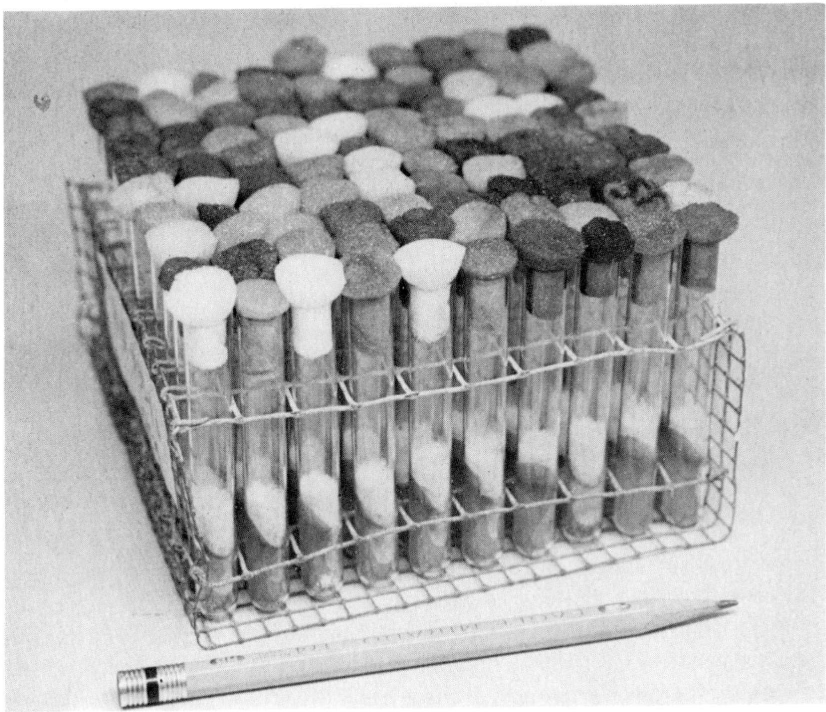

(b)

Figure 3-30. (a) A Neurospora *cross made in a Petri plate. The many small black spheres are fruiting bodies in which meiosis has occurred, shooting the ascospores (sexual spores) as a fine dust into the condensed moisture on the lid (which has been removed and is sitting on the right side of the plate). (b) A rack of progeny cultures, each resulting from one isolated ascospore.*

If we cross a fluffy pink culture with a colonial albino culture and isolate and culture 100 ascospores, the resulting cultures (on average) would be

25 fluffy pink cultures
25 colonial albino cultures
25 fluffy albino cultures
25 colonial pink cultures

In total, one-half of the "progeny" are fluffy and one-half are colonial. Thus these phenotypes are determined by one gene pair that has segregated equally at meiosis. The same is true of the other character: one-half are pink and one-half are albino, so color is determined by a different gene pair. We could represent the four culture types as

$col^+ \; al^+$ (fluffy pink)
$col \; al$ (colonial albino)
$col^+ \; al$ (fluffy albino)
$col \; al^+$ (colonial pink)

The 25%:25%:25%:25% ratio is a result of independent assortment, as illustrated in the following branch diagram:

$$1/2\ col^+ \begin{cases} \nearrow 1/2\ al^+ \rightarrow col^+\ al^+ \\ \searrow 1/2\ al \rightarrow col^+\ al \end{cases}$$

$$1/2\ col \begin{cases} \nearrow 1/2\ al \rightarrow col\ al \\ \searrow 1/2\ al^+ \rightarrow col\ al^+ \end{cases}$$

So we see that, even in such a lowly organism, Mendel's laws are still in operation.

Alternating Haploid/Diploid

In the organism with alternating generations, there are two multicellular stages that could be called adult stages; in most cases, one form is more visible. For example, what we call a fern plant is a diploid stage (the **sporophyte**), but the organism does have a small underground haploid stage (the **gametophyte**). In mosses, the green plant actually is a haploid gametophyte, and the brown stalk that grows up out of it is a diploid sporophyte. (The term *gametophyte* means "gamete-producing plant," and *sporophyte* means "sexual spore-producing plant.")

The cycle of these organisms is outlined in Figure 3-31. The haploid gametophyte produces haploid gametes by mitosis. Gametes (often from different gametophytes) fuse to form a diploid cell, which grows into the diploid sporophyte by mitosis. Specialized cells in the sporophyte undergo meiosis,

13. Duchenne's muscular dystrophy is sex-linked and usually affects only boys. Victims of the disease become progressively weaker, starting early in life.
    a. What is the probability that a woman whose brother has Duchenne's disease will have an affected child?
    b. If your mother's brother (your uncle) had Duchenne's disease, what is the probability that you will receive the gene?
    c. If your father's brother had the disease, what is the probability that you will receive the gene?
14. The pedigree shown in Figure 3-33 is concerned with an inherited dental abnormality, amelogenesis imperfecta.

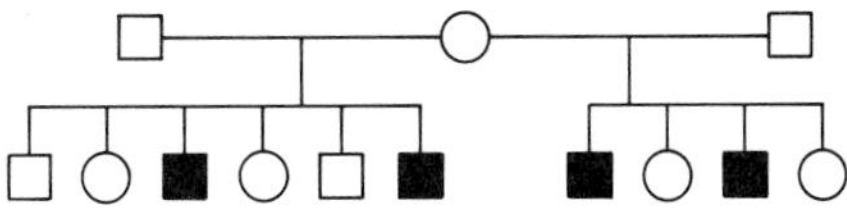

Figure 3-33.

    a. What mode of inheritance *best* accounts for the transmission of this trait?
    b. Write the genotypes of the individual members according to your hypothesis.
15. A sex-linked recessive gene *c* produces a red-green color blindness in humans. A normal woman whose father was color blind marries a color-blind man.
    a. What genotypes are possible for the mother of the color-blind man?
    b. What are the chances that the first child from this marriage will be a color-blind boy?
    c. Of the girls produced by these parents, what percentage is expected to be color blind?
    d. Of all the children (sex unspecified) of these parents, what proportion can be expected to have normal color vision?
16. Male house cats are either black or yellow; females are black, tortoise-shell pattern, or yellow.
    a. If these colors are governed by a sex-linked gene, how can these observations be explained?
    b. Using appropriate symbols, determine the phenotypes expected in the progeny of a cross between a yellow female and a black male.
    c. Repeat part b for the reciprocal of the cross described there.
    d. One-half of the females produced by a certain kind of mating are tortoise-shell and one-half are black; one-half of the males are yellow and one-half are black. What colors are the parental males and females in this kind of mating?
    e. Another kind of mating produces progeny in the following proportions: 1/4 yellow males, 1/4 yellow females, 1/4 black males, and 1/4 tortoise-shell females. What colors are the parental males and females in this kind of mating?

17. Mr. Brown is heterozygous for one autosomal gene pair *Bb*, and he carries a recessive X-linked gene *d*. What proportion of his sperms will be *bd*? (a) 0; (b) 1/2; (c) 1/8; (d) 1/16; (e) 1/4.

18. A woman has the rare (hypothetical) disease called quackerlips. She marries a normal man, and all of their sons and none of their daughters are quackerlipped. What is the mode of inheritance of quackerlips? (a) Autosomal recessive; (b) Autosomal dominant; (c) X-linked recessive; (d) X-linked dominant.

19. What do you think are the advantages and disadvantages for the organism of having genes organized into chromosomes? Why don't genes float free in the nucleus or the cell? (Try to remember to reconsider this question after finishing the book, and see if your answers have changed.)

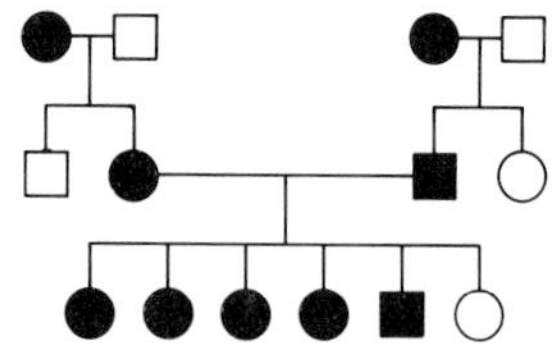

Figure 3-34.

20. Assume the pedigree in Figure 3-34 to be straightforward, with no complications such as illegitimacy. Trait W, found in individuals represented by the shaded symbols, is rare in the general population. Which of the following patterns of transmission for W are consistent with this pedigree, and which are excluded? (a) Autosomal recessive; (b) Autosomal dominant; (c) X-linked recessive; (d) X-linked dominant; (e) Y-linked.

(Problem 20 from A. M. Srb, R. D. Owen, and R. S. Edgar, *General Genetics*, 2nd ed. Copyright © 1965 by W. H. Freeman and Company.)

21. In mice, there is a mutant allele that causes a bent tail. From the cross results given in Table 3-5, deduce the mode of inheritance of this trait.

    a. Is it recessive or dominant?

    b. Is it autosomal or sex-linked?

    c. What are the genotypes of parents and progeny in all crosses shown in the table?

Table 3-5.

	Parents		Progeny	
Cross	Female	Male	Female	Male
1	Normal	Bent	All bent	All normal
2	Bent	Normal	1/2 bent, 1/2 normal	1/2 bent, 1/2 normal
3	Bent	Normal	All bent	All bent
4	Normal	Normal	All normal	All normal
5	Bent	Bent	All bent	All bent
6	Bent	Bent	All bent	1/2 bent, 1/2 normal

22. In the pedigree of Figure 3-35, a dot represents the occurrence of an extra finger, and a shaded symbol represents the occurrence of an eye disease.

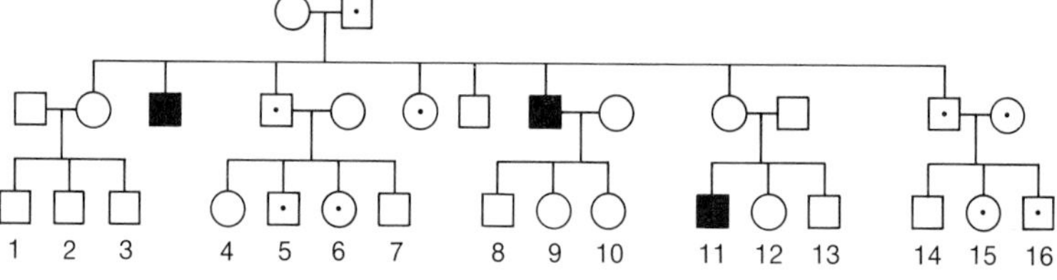

Figure 3-35.

a. What can you tell about the inheritance of an extra finger?

b. What can you tell about the inheritance of the eye disease?

c. What were the genotypes of the original parents?

d. What is the probability that a child of individual 4 will have an extra finger? will have the eye disease?

e. What is the probability that a child of individual 12 will have an extra finger? will have the eye disease?

23. In the ovaries of female mammals (including humans), the first meiotic division produces two cells (as normal), but one cell becomes a small nonfunctional cell called a polar body. Consequently, a second meiotic division occurs in only one of the first-division products. Furthermore, one of the products of the second meiotic division also becomes a polar body. Thus, the net result of one meiosis is one ovum (egg) plus two polar bodies.

a. Diagram this system at the cell level and at the chromosome level.

b. In this system, Mendelian genetics obviously does not operate: the Mendelian ratios observed in mammals are due solely to segregation in the male. Is this statement true or false? Explain your answer.

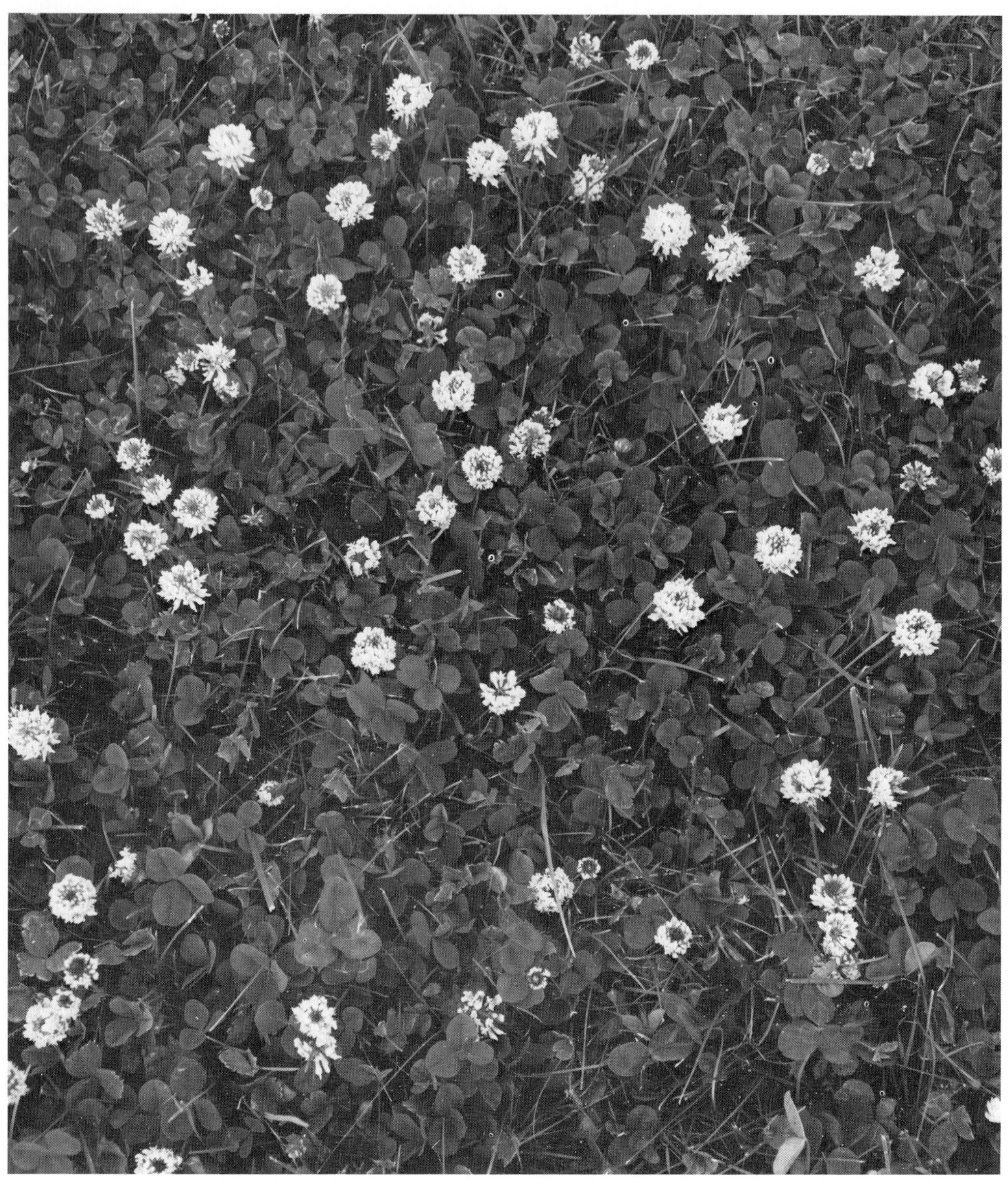

*White clover (*Trifolium repens*) showing leaf chevron markings in a natural population. (Grant Heilman.)*

4

Extensions to Mendelian Analysis

We have seen that Mendel's laws seem to hold across the entire spectrum of eukaryotic organisms—that is, we can identify analogous phenomena that reveal segregation and independent assortment. These laws form a base for predicting the outcome of simple crosses. However, it is only a base; the real world of genes and chromosomes is more complex than is revealed by Mendel's laws, and exceptions and extensions abound. These situations do not invalidate Mendel's laws. Rather, they show that there are more situations than can be explained by segregation and independent assortment of gene pairs and that these situations must be accommodated into the fabric of genetic analysis. This is the challenge we now must meet. Of course, one extension has already been accommodated—sex linkage. This chapter presents a grab bag of other extensions, and the two following chapters discuss two major extensions. We shall see that, rather than creating a hopeless and bewildering situation, these complexities combine to form a precise and unifying set of principles for the genetic analyst. These principles interlock and support each other in a highly satisfying way that has provided great insight into the mechanics of inheritance.

Variations on Dominance Relations

Dominance is a good place to start. Mendel observed (or at least reported) full dominance (and recessiveness) for all the seven gene pairs he studied. He may have been selective in his choice of pea characters to study, because variations on the basic theme crop up quite often in analysis. The problems center on the phenotype of the heterozygote. Some examples will illustrate this.

In four-o'clock plants, when a pure line with red petals is crossed to a pure line with white petals, the F_1 have *not* red petals but pink! If an F_2 is produced, the result is

1/4 red petals	1*RR*
1/2 pink petals	2*Rr*
1/4 white petals	1*rr*

The explanation is that the red phenotype (and its determining allele *R*) is **incompletely dominant** over the white phenotype (and its allele *r*). Incomplete dominance is quite common. The operational key to its identification is a heterozygote whose phenotype is intermediate between those of the homozygotes.

The behavior of the heterozygote is also the key in the phenomenon of **codominance,** in which the heterozygote shows the phenotypes of *both* the homozygotes. In a sense, then, codominance is no dominance at all! A good example is found in the M–N blood-group gene pair in humans. Three blood groups are possible—M, N, and MN—and these are determined by the genotypes $L^M L^M$, $L^N L^N$, and $L^M L^N$, respectively.

Blood groups actually represent the presence of an immunological antigen on the surface of red blood cells. People of genotype $L^M L^N$ have both antigens, and their blood transfusions must be arranged accordingly. This example shows how the heterozygote can show both phenotypes.

Another good example of codominance is found in the disease sickle-cell anemia. The gene pair concerned effects the oxygen-transport molecule hemoglobin, the major constituent of red blood cells. The three genotypes have different phenotypes, as follows:

$Hb^A\ Hb^A$:	normal
$Hb^S\ Hb^S$:	sickle-cell anemia (a severe, often fatal anemia)
$Hb^A\ Hb^S$:	sickle-cell trait (a mild anemia)

One might be tempted to call this incomplete dominance, at first sight, but further study reveals that it also shows the characteristics of codominance. Luckily, the two types of hemoglobin concerned, dictated by the two alleles, have different chemical charges, and this property can be used in a technique called **electrophoresis** to make a more refined distinction.

In electrophoresis, mixtures of proteins can be separated on the basis of their charges. A small sample of protein (here hemoglobin from red blood cells) is placed in a small well, cut in a slab of gel. A powerful electric field is applied across this gel, and the hemoglobin moves according to the degree of electrostatic charge. Figure 4-1 is a sketch of a typical electrophoresis apparatus. The gel is actually a supporting web containing electrolyte solution in its spaces; hence the hemoglobin can move easily through the field. After an appropriate time, the gel can be stained for protein, with the results shown in Figure 4-2. The blotches are the stained hemoglobin. We see that homozygous normal people have one type of hemoglobin (A), and anemics have another slower-moving type (S). The heterozygotes have both types, A and S.

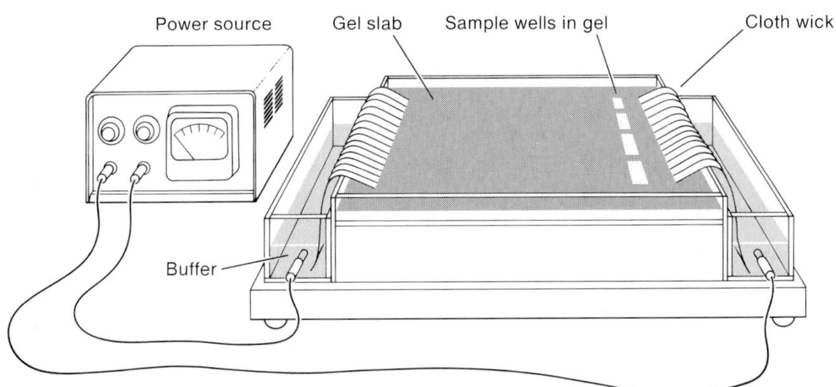

Figure 4-1. Equipment for electrophoresis. Each sample mixture is placed in a well in the gel. The components of the mixture migrate different distances on the gel because of their different electric charges. Several sample mixtures can be tested at the same time (one in each well). The positions of the separated components are later revealed by staining.

Phenotype	*Genotype*	*Hemoglobin electrophoretic pattern* Origin ↓	+4 ↓	+2 ↓	0 ↓	*Hemoglobin types present*
Sickle-cell trait	$Hb^S Hb^A$					S and A
Sickle-cell anemia	$Hb^S Hb^S$					S
Normal	$Hb^A Hb^A$					A

Figure 4-2. Electrophoresis of hemoglobin from a normal individual, an individual with sickle-cell anemia, and an individual with sickle-cell trait. The smudges show the positions to which the hemoglobins migrate on the starch gel. (After D. L. Rucknagel and R. K. Laros, Jr., "Hemoglobinopathics: Genetics and Implications for Studies of Human Reproduction." Clinical Obstetrics and Gynecology *12, 1969, 49-75.)*

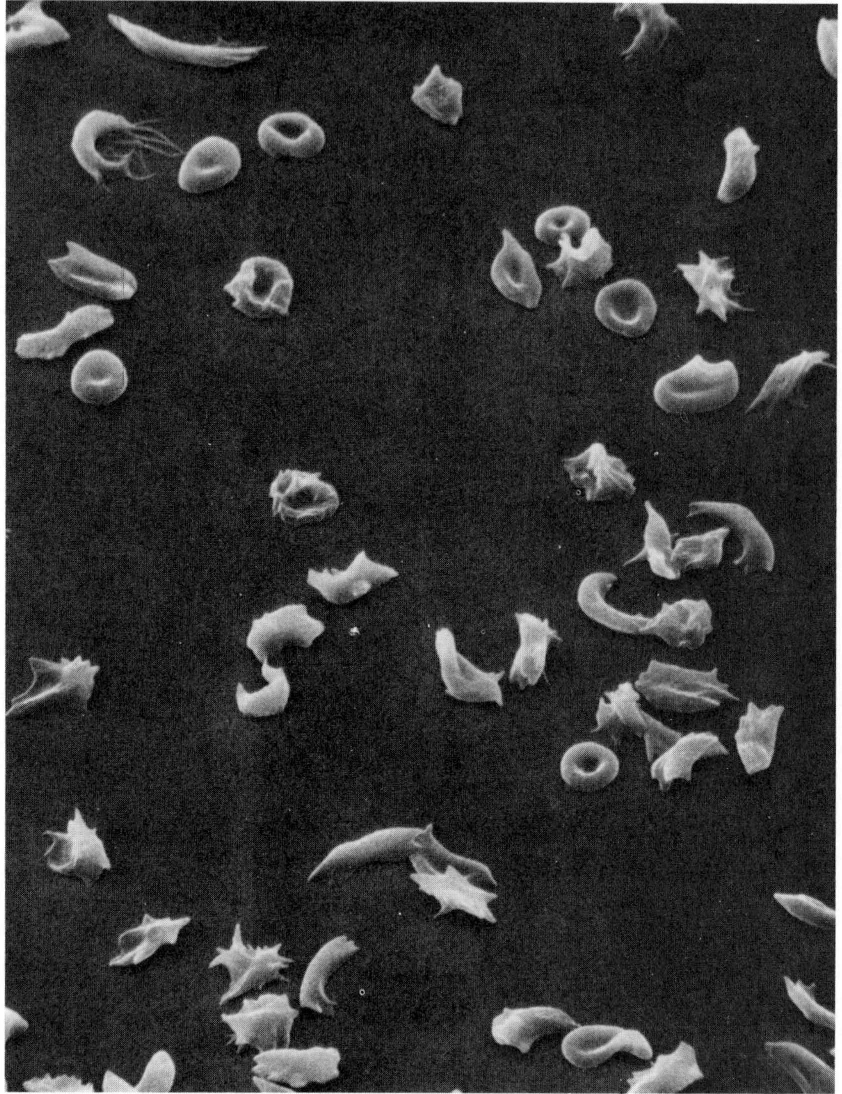

Figure 4-3. Electron micrograph of red blood cells from an individual with sickle-cell anemia. A few rounded cells appear almost normal. (Courtesy of Patricia N. Farnsworth.)

You may be wondering why the term *sickle-cell* is used in connection with the anemia. In $Hb^S Hb^S$ homozygotes, the altered hemoglobin causes a characteristic distortion of their transporting cells, the red blood cells, so that a variety of sickle-like shapes are seen (Figure 4-3).

Sickle-cell anemia illustrates the somewhat arbitrary nature of the terms *incomplete dominance* and *codominance*. Also, in the four o'clock, isn't pinkness a joint manifestation of redness and whiteness? The real distinction is probably best made in terms of gene function. Briefly (for now), codominance is observed if both alleles are active in some way, and incomplete dominance is observed in cases where one allele is completely inactive (a **null allele**) and cannot be compensated by the one active allele. (With full dominance, such compensation is made.) We shall later find chemical explanations of this distinction.

Message
Complete dominance and recessiveness are not essential aspects of Mendel's laws; those laws deal more with the inheritance patterns of genes than with their nature or function.

Multiple Alleles

Early in the history of genetics, it became clear that it is possible to have more than two forms of one kind of gene. Although only two actual alleles of a gene can exist in a diploid cell (and only one in a haploid cell), the total number of possible different allelic forms that might exist in a population of individuals is often quite large. This situation is called **multiple allelism,** and the set of alleles itself is called an **allelic series.** The concept of allelism is a crucial one in genetics, so we consider several examples. The examples themselves serve also to introduce important areas of genetic investigation.

ABO Blood Group in Humans

The human ABO blood-group alleles afford a modest example of multiple allelism. There are four blood types (or phenotypes) in the ABO system (Table 4-1). The allelic series includes three major alleles, which can be present in any pairwise combination in one individual; thus only two of the three alleles can be present in any one individual. Note that the series include cases of both complete dominance and codominance. In this allelic series, the alleles I^A and I^B each determine a unique form of one type of antigen; the allele i determines a failure to produce either form of that type of antigen.

Table 4-1. ABO Blood Group in Humans

Blood phenotype	Genotype
O	ii
A	I^AI^A or I^Ai
B	I^BI^B or I^Bi
AB	I^AI^B

C Gene in Rabbits

Another example of multiple allelism involves a larger allelic series determining coat color in rabbits. The alleles in this series are C (full color), c^{ch} (chinchilla, a light greyish color), c^h (Himalayan, albino with black extremities), and c (albino). You will have noted the use of a superscript to indicate an allele of a type of gene; this symbolism often is necessary because more than the two symbols C and c are needed to describe multiple alleles. In this series, dominance is in the order of the alleles listed; verify this by careful study of Table 4-2.

Table 4-2. C Gene in Rabbits

Coat-color phenotype	Genotype
Full color	CC or Cc^{ch} or Cc^h or Cc
Chinchilla	$c^{ch}c^{ch}$ or $c^{ch}c^h$ or $c^{ch}c$
Himalayan	c^hc^h or c^hc
Albino	cc

Operational Test of Allelism

Now that we have seen two examples of allelic series, it is a good time to pause and consider a question. How do we know that a set of phenotypes is determined by alleles of the same type of gene? In other words, what is the operational test for allelism? For now, the answer is simply the observation of Mendelian single-gene-pair ratios in all combinations crossed. For example, consider three pure-line phenotypes in a hypothetical plant species. Line 1 has round spots on the petals; line 2 has oval spots on the petals; and line 3 has no spots on the petals. Suppose that crosses yield the following results:

line 1 × line 2	F_1 all round-spotted	F_2 3/4 round, 1/4 oval
line 1 × line 3	F_1 all round-spotted	F_2 3/4 round, 1/4 unspotted
line 2 × line 3	F_1 all oval-spotted	F_2 3/4 oval, 1/4 unspotted

These results prove that we are dealing with three alleles of a single gene that affects petal spotting. We can choose any symbols we wish. Because we don't know which phenotype is the wild type, we could follow the rabbit system and use S for the round-spotted allele, s^o for the oval-spotted allele, and s for the unspotted allele. Alternatively, we could use S^r for round, S^o for oval, and s for unspotted. The convention provides no firm rules about whether to use capital or small letters, particularly for the alleles in the middle of the series that are dominant to some of their alleles but recessive to others.

Clover Chevrons

Clover is the common name for plants of the genus *Trifolium.* There are many species—some native to North America and some that grow here as introduced weeds. The red-flowered and white-flowered clovers are familiar to most people. These are *different* species, but each shows considerable variation among individuals in the curious V or "chevron" pattern on the leaves. Much genetic research has been done with white clover; Figure 4-4 shows that the different chevron forms (and the absence of chevrons) constitute an allelic series in this species. Furthermore, several types of dominance relations may be seen in the allele combinations shown in the figure.

Incompatability Alleles in Plants

It has been known for millennia that some plants just will not self-pollinate. A single plant may produce both male and female gametes, but no seeds will ever be produced. The same plants, however, will cross with certain other plants, so obviously they are not sterile. This phenomenon is called **self-incompatibility.** Of course, the pea plants used by Mendel were not self-

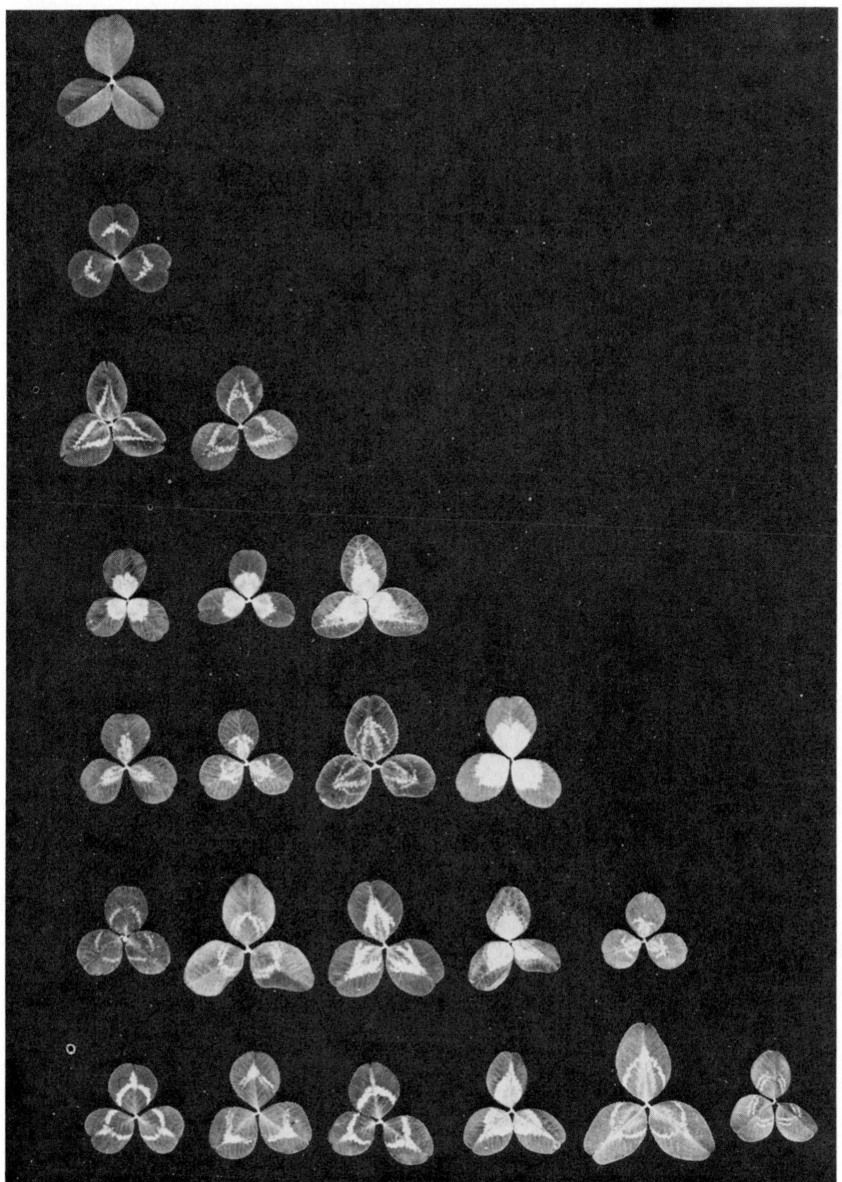

Figure 4-4. Multiple alleles are involved in determining the chevron pattern on the leaves of white clover. The genotype of each plant is shown below. Inspect these data for dominance and codominance. (Courtesy of W. Ellis Davies.)

vv

$V^{l}V^{l}$

$V^{h}V^{h}$ $V^{l}V^{h}$

$V^{f}V^{f}$ $V^{l}V^{f}$ $V^{h}V^{f}$

$V^{ba}V^{ba}$ $V^{l}V^{ba}$ $V^{h}V^{ba}$ $V^{f}V^{ba}$

$V^{b}V^{b}$ $V^{l}V^{b}$ $V^{h}V^{b}$ $V^{f}V^{b}$ $V^{ba}V^{b}$

$V^{by}V^{by}$ $V^{l}V^{by}$ $V^{h}V^{by}$ $V^{f}V^{by}$ $V^{ba}V^{by}$ $V^{b}V^{by}$

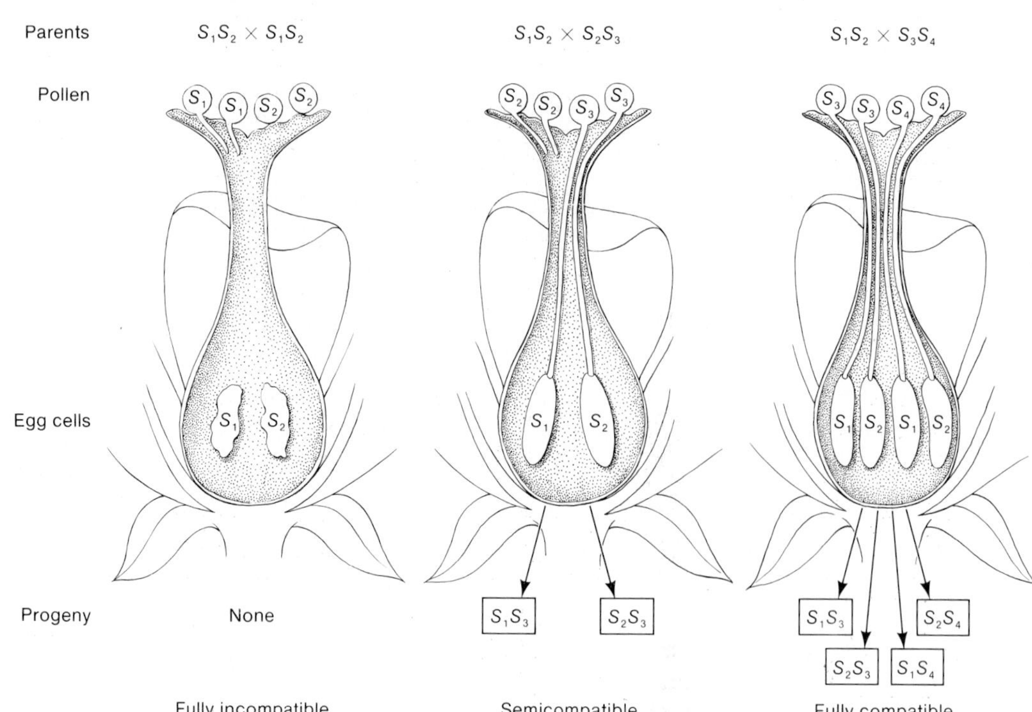

Figure 4-5. Diagram showing how multiple alleles control incompatibility in certain plants. A pollen tube will not grow if the S *allele that it contains is present in the female parent. This diagram shows only four multiple alleles, but many plant incompatibility systems use far larger numbers of alleles. Such systems act to promote exchange of genes between plants by making selfing impossible and crosses between near relations very unlikely. (From A. M. Srb, R. D. Owen, and R. S. Edgar,* General Genetics, *2nd ed. Copyright © 1965, W. H. Freeman and Company.)*

incompatible, for he was able to self his plants with ease. We now know that incompatibility has a genetic basis and that there are several different genetic systems acting in different self-incompatible species. These systems form very nice examples of multiple allelism.

One of the most common systems is found in many dicot and monocot plants, including sweet cherries, tobacco, petunias, and evening primroses. In each of these species, one gene determines compatibility/incompatibility relations, with many different allelic forms of the gene possible in different plants of any one species. Figure 4-5 shows an example. The figure shows a fully incompatible reaction, a semicompatible reaction, and a fully compatible reaction.

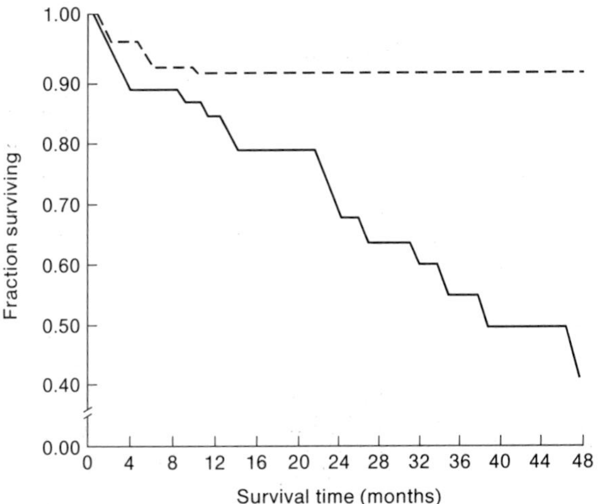

Figure 4-6. The success of kidney transplants in HLA-matched (dashed line) *and HLA-mismatched* (solid line) *sibling pairs. HLA refers to genes related to tissue compatibility. (From D. P. Singal et al.,* Transplantation *7:246, copyright © 1965, The Williams & Wilkins Co., Baltimore.)*

If a pollen grain bears an *S* allele that is also present in the maternal parent, then it will not grow; however, if that allele is not in the maternal tissue, then the pollen grain produces a pollen tube containing the male nucleus, and this tube effects fertilization. The number of *S* alleles in a series in one species can be very large—over 50 in the evening primrose and clover, and cases of more than 100 alleles have been reported in some species.

Tissue Incompatibility in Humans

When an organ or tissue graft is medically necessary in a human, the success or failure of the graft depends on the genotypes of the host and the donor. If the two are mismatched, rejection of the graft eventually occurs, often accompanied by death of the host (Figure 4-6).

The rejection system is based on two important genes called *HLA-A* and *HLA-B*, which determine the immunological acceptability of the introduced tissue. Each gene has a series of allelic forms. There are eight alleles for *HLA-A*, designated *A1*, *A2*, *A3*, *A9*, *A10*, *A11*, *A28*, and *A29*. By coincidence, there are also eight alleles for *HLA-B*, designated *B5*, *B7*, *B8*, *B12*, *B13*, *B14*, *B18*, and *B27*. (The two genes do not show independent assortment, but this need not concern us now. For the moment we shall simply use these allelic series as excellent examples of multiple allelism in humans.)

The HLA type of an individual is determined by testing his or her lymphocytes (white blood cells) with a battery of standard antibodies from donors

Table 4-3. Success or failure of transplants determined by HLA types

Transplant	Recipient genotype	Donor genotype	Result
1	*A1 A2, B5 B5*	*A1 A1, B5 B7*	Rejected (because of *B7*)
2	*A2 A3, B7 B12*	*A1 A2, B7 B7*	Rejected (because of *A1*)
3	*A1 A2, B7 B5*	*A1 A2, B7 B7*	Accepted
4	*A2 A3, B7 B5*	*A3 A3, B5 B5*	Accepted

of known HLA type. In a transplant, the recipient's immune system will recognize the graft as "foreign" and reject it if an allele is present in the donor tissue that is not present in the recipient. In the typing test, this rejection reaction is observed as lymphocyte death. The immune system will not reject tissue that lacks some alleles present in the recipient—it only rejects "strange" alleles. Table 4-3 shows examples.

You might wonder what function these HLA genes normally serve—obviously, tissue transplants have not been a normal aspect of human evolution! One possible answer is that some HLA alleles seem to confer resistance or susceptibility to certain specific diseases (Figure 4-7). Thus these alleles may have played some role in our evolutionary history, in response to some environmental selection pressure. This supposition is supported by the finding that certain ethnic groups and races show a preponderance of specific alleles in their populations. The study of such alleles is important both in deciphering the history of human evolution and in developing modern medical techniques—not only for transplants, but also for the general study of resistance to disease.

Another very interesting possibility emerging from current research is that the HLA genes may be intimately involved in cancer. Presumably, one of the normal functions of the HLA genes is to recognize cancer cells as a kind of "foreign" agent within the body. The development of cancer cells may be a fairly common event in healthy individuals, with these cells being recognized and destroyed before they cause noticeable damage. A cancerous tumor may be the result of a failure of this detection-and-destruction system at some stage.

Message

A gene can exist in several different states or forms—a situation called multiple allelism. The alleles are said to constitute an allelic series, and the members of a series can show any type of dominance relationship with one another.

As a postscript to this section, it is worth noting that several different "kinds" or "species" of cell-surface antigens have been discovered. Each of these kinds is, of course, specified by one kind of gene. The different forms one kind of antigen can take are determined by the multiple alleles of the gene concerned.

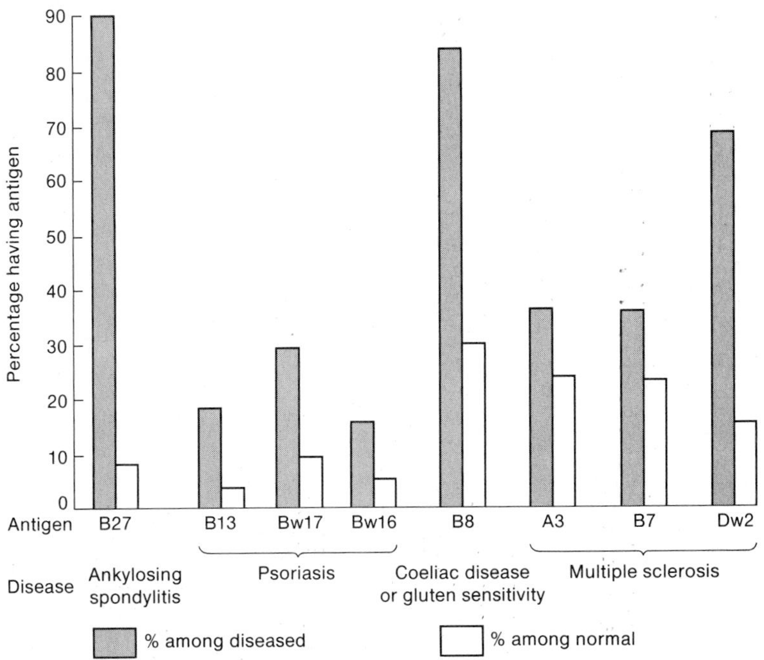

Figure 4-7. Histograms showing effects of specific HLA alleles on predisposition to certain diseases in humans. For example, the allele B27 *is found in 90% of individuals with ankylosing spondylitis (a kind of severe rheumatism), but in only 9% of normal individuals. Most of the diseases that show such HLA correlations are known to involve the immune system. (From W. F. Bodmer and L. L. Cavalli-Sforza,* Genetics, Evolution, and Man. *Copyright © 1976, W. H. Freeman and Company. Based on data from H. O. McDevitt and W. F. Bodmer,* Lancet *1:1267, 1974.)*

Lethal Genes

Normal wild-type mice have coats with a rather dark overall pigmentation. In 1904, Lucien Cuenot studied mice having a lighter coat color called yellow. Mating a yellow mouse to a normal mouse from a pure line, Cuenot observed a 1:1 ratio of yellow to normal mice in the progeny. This observation suggests that a single gene determines this aspect of coat color and that an allele for yellow is dominant to an allele for normal color. However, the situation became more confusing when he crossed yellow mice with one another. The result was always the same, no matter which yellow mice were used:

$$\text{yellow} \times \text{yellow} \begin{array}{l} \nearrow 2/3 \text{ yellow} \\ \searrow 1/3 \text{ normal color} \end{array}$$

Two things are of interest in these results. First, the 2:1 ratio is a departure from Mendelian expectations. Second, Cuenot failed to detect (in any generation of breeding) a homozygous yellow mouse.

Cuenot suggested the following explanation for these results. A cross between two heterozygotes would be expected to yield a Mendelian genotype ratio of 1:2:1. If one of the homozygous classes died before birth, the live births would then show a 2:1 ratio of heterozygotes to the surviving homozygotes. In other words, the allele A^Y for yellow is dominant to the normal allele A with respect to its effect on color, and it also acts as a **recessive lethal** allele with respect to a character we could call "survival." Thus a mouse with the homozygous genotype A^YA^Y dies before birth and is not observed among the progeny. All surviving yellow mice must be heterozygous A^YA, so a cross between yellow mice will always yield the following results:

$$A^YA \times A^YA \begin{array}{l} \nearrow 1/4\,AA \quad \text{(normal)} \\ \rightarrow 2/4\,A^YA \quad \text{(yellow)} \\ \searrow 1/4\,A^YA^Y \quad \text{(die before birth)} \end{array}$$

The expected Mendelian ratio of 1:2:1 would be observed among the zygotes, but it is altered to a 2:1 ratio of viable progeny because of the lethality of the A^YA^Y genotype. This hypothesis was confirmed by the removal of uteri from pregnant females of the yellow × yellow cross; one-fourth of the embryos were found to be dead. Figure 4-8 shows a typical litter from a cross between yellow mice.

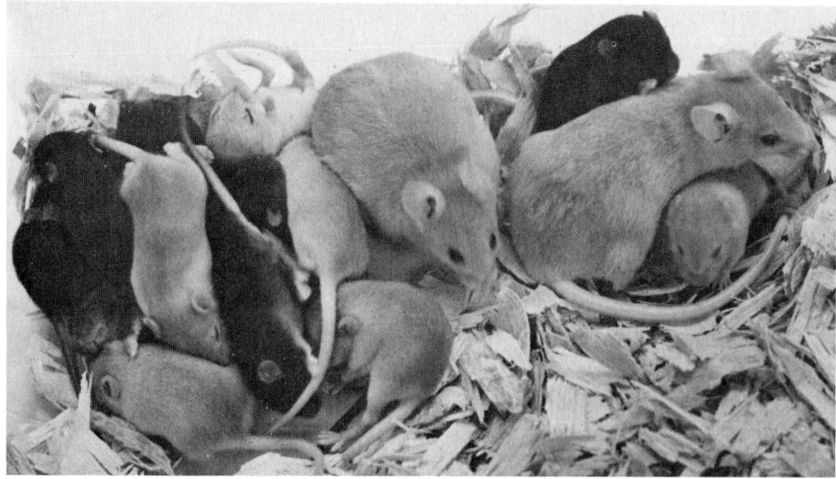

Figure 4-8. A mouse litter from two parents heterozygous for the yellow coat-color allele, which is lethal in a double dose. The larger mice are the parents. Not all progeny are visible.

The A^Y allele has effects on two characters: coat color and survival. (Genes that have more than one distinct phenotypic effect are called **pleiotropic** genes.) It is entirely possible that both effects are the result of the same basic cause, which promotes yellowness of coat in a single dose and death in a double dose.

Lethal genes are quite common, even in humans. Some lethal effects occur in utero, others much later—in infancy, in childhood, or even in adulthood. Only rarely is a lethal gene associated with a distinguishable heterozygous phenotype, as in yellow mice. Typically, the only detectable effect of a lethal gene is on survival of the individual. In some cases, a specific abnormality can be pinpointed as the cause of death. For example, a recessive lethal gene in rats may have its pleiotropic effects traced to a basic problem in the nature of the rat's cartilage (Figure 4-9).

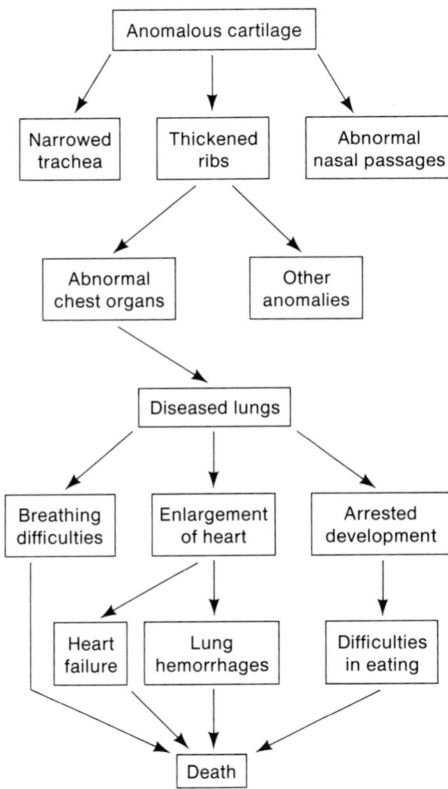

Figure 4-9. Diagram showing how one specific lethal gene causes death in rats. (From I. M. Lerner and W. J. Libby, Heredity, Evolution, and Society, *2nd ed. Copyright © 1976, W. H. Freeman and Company; after H. Grüneberg.)*

The lethality of an allele of a gene is often dependent on the environment in which the organism develops. Whereas certain alleles would be lethal in virtually any environment, others are viable in one environment but lethal in another. For example, remember that many of the phenotypes favored and selected by agricultural breeders would almost certainly be wiped out in nature in competition with the normal members of their population. Petal-color and petal-form variants are good examples; it is only the careful nurturing of the gardener that has maintained such phenotypes for our study.

In practical genetics, we very commonly encounter situations where perfect expected Mendelian ratios are consistently skewed in one direction by one allele. For example, in the cross $Aa \times aa$, we predict a progeny phenotypic ratio of 50% $A-$ and 50% aa, but we might consistently observe a ratio such as 55%:45% or 60%:40%. In such a case, the a allele is said to be **subvital**—a kind of partial lethality, expressed in only some individuals. The partial lethality may range from 0% to 100%, depending on the rest of the genome and on the environment. We shall return to this topic.

Several Genes Affecting the Same Character

We pointed out earlier that, in a genetic dissection, the identification of a major gene affecting a character does not mean that this is the *only* gene affecting that character. An organism is a highly complex machine in which all functions interact to a greater or lesser degree. At the level of genetic determination, the genes likewise can be regarded as interacting. A gene does not act in isolation; its effects depend not only on its own functions but also on the functions of other genes (and on the environment). In many cases, the complex interactions of major genes are detectable through genetic analysis, and we now look at some examples. Typically, the mode of interaction is revealed by **modified Mendelian ratios.**

Coat Color in Mammals

A character that has been extensively studied at the genetic level is coat color in mammals. These studies have revealed a beautiful set of examples of the interplay between different genes in the determination of one character. The best-studied mammal in this regard is the mouse, because of its small size and its short reproductive cycle. However, the genetic determination of coat color in mice has direct parallels in other mammals, and we shall look at some of these as our discussion proceeds. There are at least five interacting major genes involved in determining the coat color of mice: A, B, C, D, and S.

The A Gene. The wild-type allele *A* produces a phenotype called agouti. Agouti is an overall grayish color with a "brindled" or "mousy" appearance. It is common in many mammals in nature. The effect is produced by a band of yellow on the hair shaft. In the nonagouti phenotype (determined by the allele *a*), the yellow band is absent, so the coat color appears solid (Figure 4-10).

The lethal yellow A^Y allele is another form of this gene. Another form is a^t, which gives a "black-and-tan" effect involving a cream-colored belly with dark pigmentation elsewhere. We shall not include these two alleles in the following discussion.

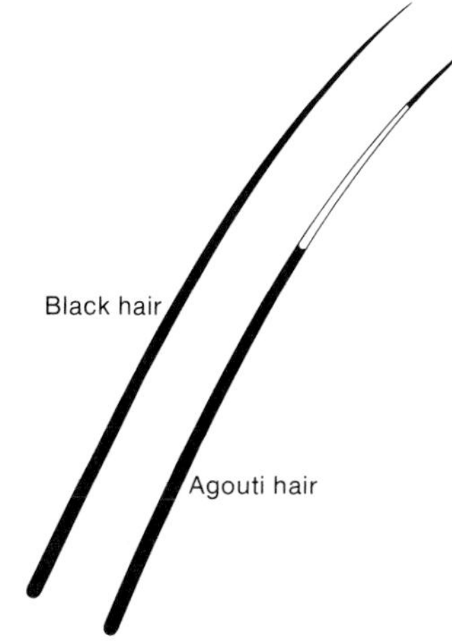

Figure 4-10. Individual hairs from an agouti (wild-type) *and a black mouse. The yellow band on each hair gives the agouti pattern its "mousy" appearance.*

The B Gene. There are two major alleles of the *B* gene. The allele *B* gives the normal agouti color in combination with *A*, but with *aa* it gives solid black. The genotype *A– bb* gives a color called cinnamon ("mousy" brown), and *aa bb* gives solid brown.

The following sample cross illustrates the inheritance pattern of the *A* and *B* genes.

AA bb (cinnamon) × *aa BB* (black)
or *AA BB* (agouti) × *aa bb* (brown)
↓
F_1 all *Aa Bb* (agouti)
↓
F_2 9 *A– B–* (agouti)
3 *A– bb* (cinnamon)
3 *aa B–* (black)
1 *aa bb* (brown)

In horses, no *A* agouti gene seems to have survived the generations of breeding, although such a gene does exist in certain wild relatives of the horse. The color we have called brown in mice is called chestnut in horses, and this phenotype also is recessive to black.

The C Gene. The wild-type allele *C* permits color expression, and the allele *c* prevents color expression. The *cc* constitution is said to be **epistatic** to the other color genes. The word *epistatic* literally means "standing upon"; the *c* allele in homozygous condition "stands on" (blots out) the expression of other genes concerned with coat color. The *cc* animals, lacking coat color, are called albino. Albinos are common in many mammalian species; in fact, albinos have been occasionally reported among birds, snakes, and fish. Epistatic genes produce interesting modified ratios, as seen in the following sample cross (where we assume that both parents are *aa*):

$$\begin{array}{ll} BB\,cc \text{ (albino)} & \times\ bb\,CC \text{ (brown)} \\ \text{or } BB\,CC \text{ (black)} & \times\ bb\,cc \text{ (albino)} \end{array}$$

$$\downarrow$$

F_1 all *Bb Cc* (black)

$$\downarrow$$

$$F_2 \quad \begin{array}{ll} 9\ B{-}\,C{-} \text{ (black)} & 9 \\ 3\ bb\,C{-} \text{ (brown)} & 3 \\ \left.\begin{array}{l} 3\ B{-}\,cc \text{ (albino)} \\ 1\ bb\,cc \text{ (albino)} \end{array}\right\} & 4 \end{array}$$

A phenotypic ratio of 9:3:4 is observed. This ratio is the signal for inferring gene interaction of the type called recessive epistasis. In some other organisms, dominant epistasis is observed. (What ratio is produced in dominant epistasis?)

We have already encountered the c^h (Himalayan) allele in rabbits. It exists also in other mammals, including mice (also called Himalayan) and cats (called Siamese).

The D Gene. The *D* gene controls the intensity of pigment specified by the other coat-color genes. The genotypes *DD* and *Dd* permit full expression of color in mice, but *dd* "dilutes" the pigment to a milky appearance. Dilute agouti, dilute cinnamon, dilute brown, and dilute black coats all are possible. A gene of this nature is called a **modifier gene**. In the following sample cross, we assume that both parents are *aa CC*.

$$\begin{array}{ll} BB\,dd \text{ (dilute black)} & \times\ bb\,DD \text{ (brown)} \\ \text{or}\quad BB\,DD \text{ (black)} & \times\ bb\,dd \text{ (dilute brown)} \end{array}$$

$$\downarrow$$

F_1 all *Bb Dd* (black)

$$\downarrow$$

$$F_2 \quad \begin{array}{l} 9\ B{-}\,D{-} \text{ (black)} \\ 3\ B{-}\,dd \text{ (dilute black)} \\ 3\ bb\,D{-} \text{ (brown)} \\ 1\ bb\,dd \text{ (dilute brown)} \end{array}$$

In horses, the *D* allele shows incomplete dominance. Figure 4-11 shows how dilution affects the appearance of chestnut and bay horses.

The S Gene. The *S* gene controls the presence or absence of spots. The genotype *S–* results in no spots, and *ss* produces a spotting pattern called

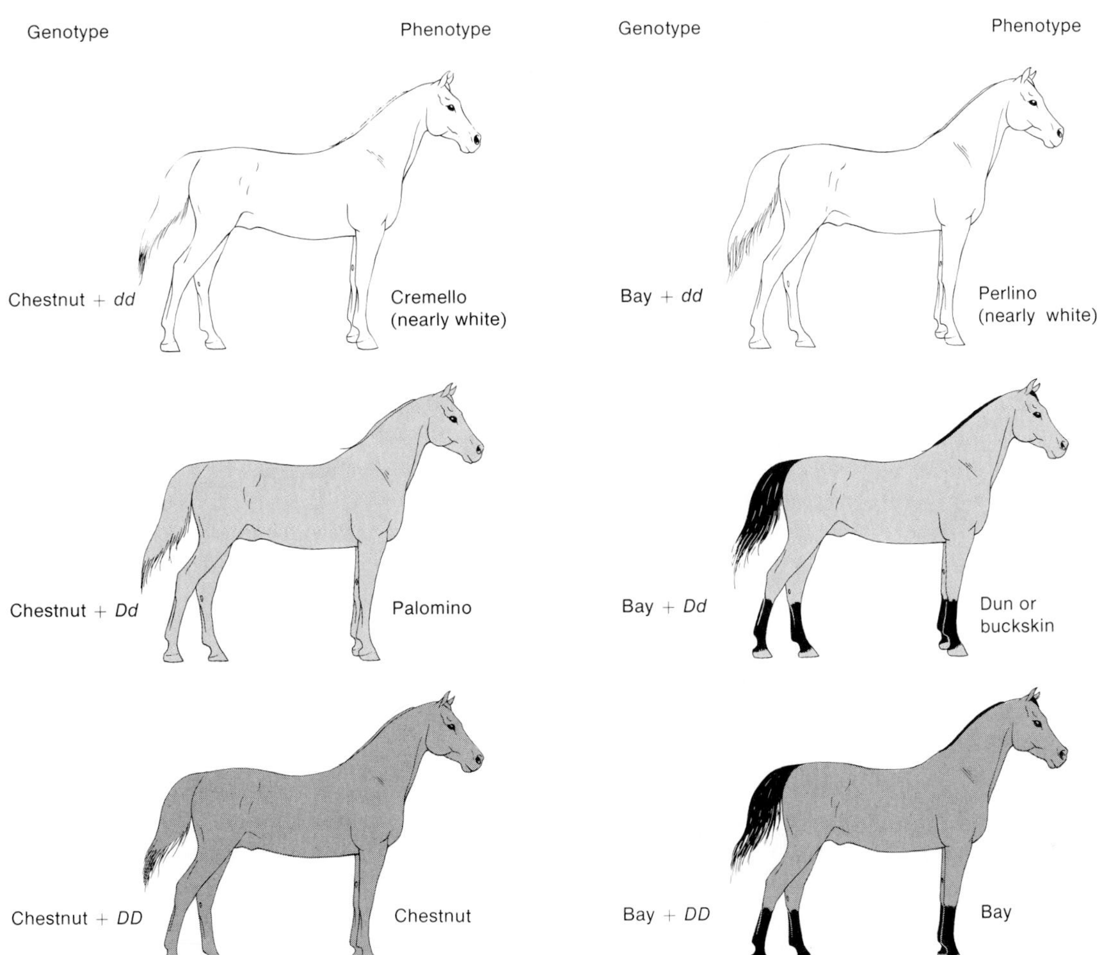

Figure 4-11. The modifying effect of dilution genes on basic chestnut and bay genotypes in horses. Note the incomplete dominance shown by D. *(From J. W. Evans et al.,* The Horse. *Copyright © 1977, W. H. Freeman and Company.)*

piebald in both mice and horses. This pattern can be superimposed on any of the coat colors discussed earlier—with the exception of albino, of course.

By this time, the point of the discussion should be obvious. Normal coat appearance in wild mice is produced by a complex set of interacting genes determining pigment type, pigment distribution in the individual hairs, pigment distribution on the animal's body, and the presence or absence of pigment. Similar situations exist for any character in any organism.

Figure 4-12 illustrates some pigment patterns in mice.

Figure 4-12. *Some coat phenotypes in mice.*

Examples of Gene Interaction in Other Organisms

Some other kinds of gene interaction are best illustrated in other organisms. Peas provide a good example of one important situation. Two different, independently obtained pure lines of pea plants are both white-petaled. When these lines are crossed, all of the F_1 have purple flowers. The F_2 shows both purple and white plants in a ratio of 9:7. How can this result be explained? By now, you should immediately suspect that the 9:7 ratio is a modification of the Mendelian 9:3:3:1 ratio.

Two different genes in the pea have similar effects on petal color. Let us represent the alleles of these genes by A, a, B, and b.

white strain 1 × white strain 2

$AA\,bb \times aa\,BB$

↓

F_1 all $Aa\,Bb$ (purple)

↓

F_2 9 $A\!-\!B\!-$ (purple) 9

3 $A\!-\!bb$ (white)
3 $aa\,B\!-$ (white) } 7
1 $aa\,bb$ (white)

Both gene pairs affect petal color. Whiteness can be produced by a recessive allele of either gene pair, but purpleness is a phenotype produced by a combination of the dominant alleles of *both* gene pairs. This phenomenon is called **complementary gene action,** a term that satisfactorily describes how the two dominant alleles are uniting to produce a specific phenotype—in this example, purple pigment. (Note that we might have mistakenly inferred purpleness to be a specific phenotype of one gene pair if we had only one white strain available for our crosses.)

Another important kind of interaction is **suppression** of one gene by another. This interaction is illustrated by the inheritance of the production of a chemical called malvidin in the plant genus *Primula*. Malvidin production is determined by a single dominant gene *K*. However, the action of this dominant gene may be suppressed by a nonallelic dominant suppressor *D*. The following pedigree is informative:

$KK\,dd$ (malvidin) × $kk\,DD$ (no malvidin)

↓

F_1 all $Kk\,Dd$ (no malvidin)

↓

F_2 9 $K\!-\!D\!-$ (no malvidin)
3 $kk\,D\!-$ (no malvidin) } 13
1 $kk\,dd$ (no malvidin)
3 $K\!-\!dd$ (malvidin) 3

Recessive suppression also is known of both dominant and recessive genes. The suppressor gene may have its own associated phenotype or (as in the malvidin example) have no known phenotypic effect other than the suppression.

Our final example of gene interaction introduces a concept that we shall develop in a later chapter. It concerns the genes that control fruit shape in the plant called shepherd's purse. Two different lines have fruits of different shapes: one is "round," the other "narrow." Are these two phenotypes determined by two alleles of a single gene? A cross between the two lines reveals an F_1 with round fruit; this result is consistent with the hypothesis of a single gene pair. However, the F_2 shows a 15:1 ratio of round to narrow. Again, this ratio immediately suggests a modification of the 9:3:3:1 Mendelian ratio, and it can be explained in terms of two **duplicate genes** (Figure 4-13). Apparently, round fruits are produced as a result of the presence of at least one dominant allele of either gene. The two genes appear to be identical in function. (Contrast this 15:1 ratio with the 9:7 ratio obtained from complementary genes, where *both* dominant genes are necessary to produce a specific phenotype.)

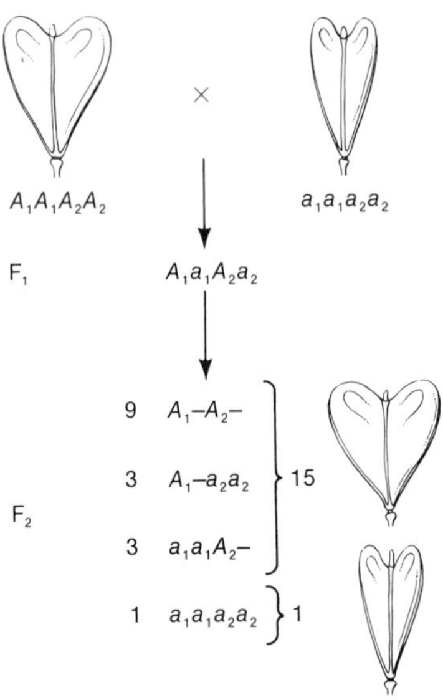

Figure 4-13. Inheritance pattern of duplicate genes controlling fruit shape in shepherd's purse. Either A_1 *or* A_2 *can cause a round fruit.*

Gene interactions also can be detected in haploid organisms. We have already discussed a gene in the fungus *Neurospora* where one allele causes albino asexual spores, as opposed to the normal pinkish-orange color produced by the wild-type allele of the same gene. A cross *al* × al^+ gives 1/2 *al* and 1/2 al^+ progeny. Another interesting gene, *ylo*, gives yellow asexual spores, and the cross *ylo* × ylo^+ gives 1/2 *ylo* and 1/2 ylo^+ progeny. When an *al* culture is crossed with a *ylo* culture, the resulting progeny are 1/4 yellow, 1/4 wild type, and 1/2 albino. How can we explain this result? The answer is a kind of epistasis: *al* and *ylo* are forms of separate genes, each of which can affect the normal production of pink pigment. The genotypes in the cross are the following:

haploid parental cultures	*al* ylo^+ (albino) × al^+ *ylo* (yellow)	
	↓	
transient diploid	$al^+/al, ylo^+/ylo$	
	↓ meiosis	
progeny (cultures from sexual spores)	1/4 *al ylo* (albino)	1/2
	1/4 *al* ylo^+ (albino)	
	1/4 al^+ *ylo* (yellow)	1/4
	1/4 al^+ ylo^+ (normal)	1/4

Message
Modified Mendelian ratios reveal that a character is determined by the complex interaction of different genes.

Penetrance and Expressivity

We have made the point that genes do not act in isolation. A gene does *not* determine a phenotype by acting alone; it does so only in conjunction with other genes and with the environment. Although geneticists do routinely ascribe a particular phenotype to an allele of a gene they have identified, you should remember that this is merely a convenient kind of jargon designed to facilitate genetic analysis. This jargon arises from the ability of geneticists to isolate individual components of a biological process and to study them as part of genetic dissection. Although this logical isolation is an essential aspect of genetics, the message of this chapter is that a gene cannot act by itself.

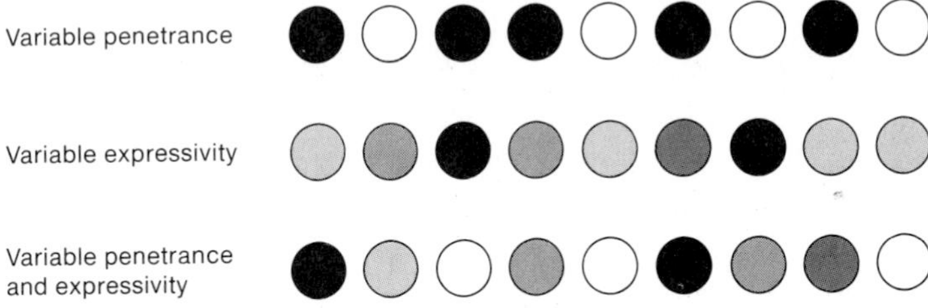

Figure 4-14. Diagram representing effects of penetrance and expressivity through a hypothetical character "pigment intensity." In each row, all individuals have the same genotype (say, PP*), giving them the same "potential to produce pigment." However, the effects of the rest of the genome and of the environment may suppress or modify pigment production in each individual. (Be sure to recall this kind of variation when we discuss polygenes in Chapter 18.)*

In the preceding examples, the genetic basis of the dependence of one gene on another has been worked out. In other situations, where the phenotype ascribed to a gene is known to be dependent on other factors but the precise nature of those factors has not been established, the terms *penetrance* or *expressivity* may be very useful in describing the situation.

We have already encountered penetrance in our discussion of lethal alleles. **Penetrance** is defined as the percentage of individuals with a given genotype who exhibit the phenotype associated with that genotype. For example, an organism may be of genotype *aa* or *A*– but may not express the phenotype normally associated with its genotype—because of the presence of modifiers, epistatic genes, or suppressors in the rest of the genome or because of a modifying effect of the environment. Penetrance can be used to describe such an effect when the exact cause is not known.

Expressivity, on the other hand, describes the degree or extent to which a given genotype is expressed phenotypically in an individual. Again, the lack of full expression may be due to the rest of the genome or to environmental factors. Figure 4-14 diagrams the distinction between penetrance and expressivity.

Human pedigree analysis and predictions in genetic counseling can often be thwarted by the phenomena of penetrance and expressivity. For example, if a disease-causing allele is not fully penetrant (as usually is the case), it is difficult to give a clean genetic bill of health to any individual involved as part of a disease pedigree—for example, individual R in Figure 4-15. On the other hand, pedigree analysis can sometimes identify individuals who do not express but almost certainly have a disease genotype—for example, individual Q in Figure 4-15.

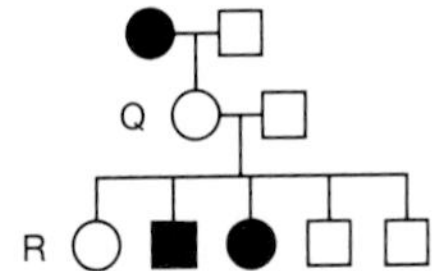

Figure 4-15. Lack of penetrance illustrated by a pedigree for a dominant gene. Individual Q must have the gene (because it was passed on to the progeny), but it was not expressed in the phenotype. An individual such as R cannot be sure that his or her genotype lacks the gene.

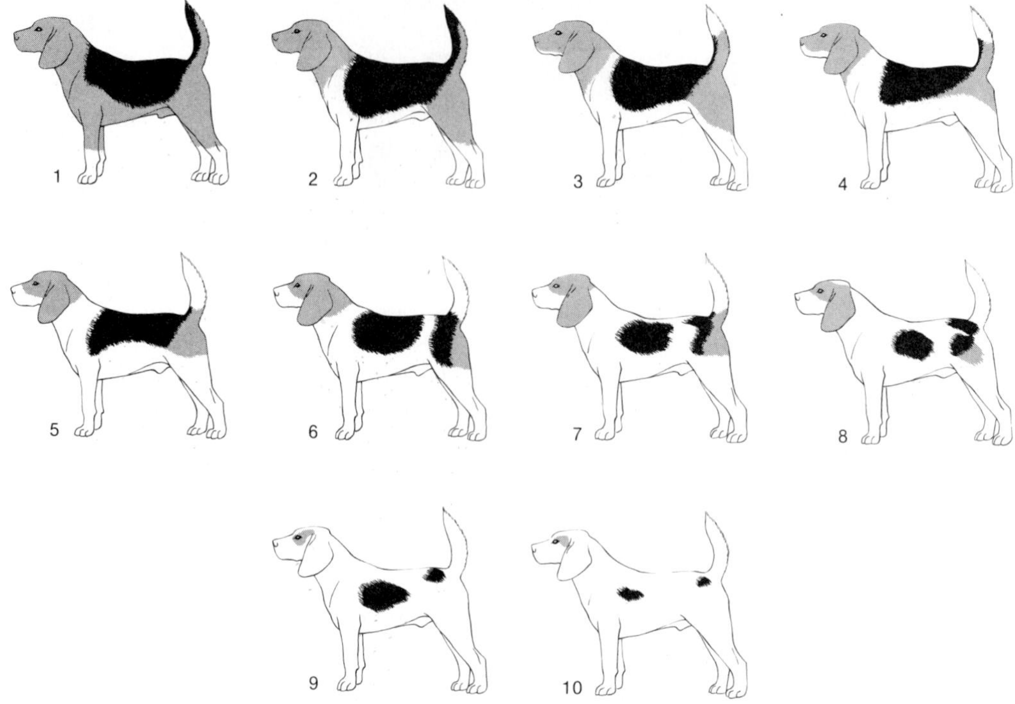

Figure 4-16. Ten grades of piebald spotting in beagles. Each of these dogs has the causative gene S^P*; the variation in expressivity is attributed to modifying genes. (Adapted from Clarence C. Little:* The Inheritance of Coat Color in Dogs. *Copyright © 1957 by Cornell University. Used by permission of the publisher, Cornell University Press.)*

Further examples of variable expressivity are found in Figure 4-16 and in Problem 6 of Chapter 2.

Message

The impact of a gene at the phenotype level depends not only on its dominance relations but also on the conditions of the rest of the genome and the condition of the environment.

The exceptions to Mendelian analysis covered in this chapter and following chapters are not the only ones encountered in routine genetic analysis; they are simply among the most common. Each exception—whether, for example, multiple allelism, incomplete dominance, epistasis, or variable expressivity—

has its key recognition features at the experimental level. The analyst must be constantly on the lookout for these and any other results that might indicate the uniqueness of any given situation. Such signals often lead to the discovery of new phenomena and to the opening up of new research areas.

Summary

Although it has been shown that Mendel's laws apply to all eukaryotic organisms, these laws are only a base for understanding heredity. The real world of genes and chromosomes is much more complex.

In addition to the full dominance that Mendel observed in his experiments, incomplete dominance and codominance may exist. In incomplete dominance, the phenotype of a heterozygote is intermediate between those of the homozygotes. In codominance, the heterozygote shows the phenotypes of both homozygotes.

In his experiments, Mendel reported genes with two forms. It was later discovered that in fact a gene may have more than two forms. This situation is known as multiple allelism. The members of an allelic series may exhibit any type of dominance relationship with the other members. The genes controlling the rejection of incompatible tissues in humans are an example of multiple allelism.

One gene may affect more than one character. Such genes are known as pleiotropic genes. An example is the A^Y gene in mice, which affects both coat color and survival.

The identification of a major gene affecting a character does not mean that it is the only gene affecting that character; several genes may be interacting. A good example of gene interaction is the coat color of mice, which is produced by a complex set of interacting genes that determine pigment type, pigment distribution in the hair, pigment distribution on the animal, and the presence or absence of pigment.

Gene interaction often produces modified Mendelian ratios in the F_2. Some kinds of interaction have specific names, such as complementary gene action, epistasis, suppression, and duplicate gene action. Gene interaction occurs in both diploid and haploid organisms.

Two other important extensions to Mendelian analysis are the concepts of penetrance and expressivity. Penetrance is the percentage of individuals of a specific genotype who express the phenotype associated with that genotype. Expressivity refers to the degree of expression, or severity, of a particular genotype at the phenotypic level.

Problems

1. In *Clarkia elegans* plants, the allele for white flowers is recessive to the allele for pink flowers. Pollen from a heterozygous pink flower is placed on the pistil of a white flower. What is the expected ratio of phenotypes in the progeny? (a) 1 pink:1 white; (b) 1 red:2 pink:1 white; (c) all pink; (d) 3 pink:1 white; (e) all white.

   (Problem 1 courtesy of F. R. Ganders.)

2. "Erminette" fowls have mostly light-colored feathers with an occasional black one, giving a "flecked" appearance. A cross of two erminettes produced a total of 46 progeny consisting of 22 erminettes, 14 blacks, and 12 pure whites. What genetic basis of the erminette pattern is suggested? How would you test your hypothesis?

3. In a moss, a certain protein is studied through electrophoresis. We consider two electrophoretic protein forms (P1 and P2), determined by two alleles, *P1* and *P2*. In this moss, the protein is found in both sporophyte and gametophyte. A gametophyte showing P1 is crossed with a gametophyte showing P2. What protein forms will be detected upon electrophoresis of the sporophyte? When this sporophyte produces spores, what gametophytes will result in what proportions?

4. In the plant incompatibility system described in Figure 4-5, progeny from the cross *S1S3* × *S2S4* are intercrossed in all combinations as males and as females. What proportion of the crosses will be (a) fully fertile; (b) fully sterile; (c) partially fertile?

5. In some compatibility systems in plants, pollen compatibility type is determined by the diploid genotype of the male parent. Often in these systems there is dominance of *S* alleles, not only in the male but also in the female. Assume the dominance series $S1 > S2 > S3 > S4$. Thus, *S1S2* is type 1, *S2S3* is type 2, and so on. Assume that compatibility is possible only between different types. From the cross *S1S2* × *S3S4*, the progeny are intercrossed in all combinations. What will be the proportion of compatible matings? What progeny will be produced from the compatible matings?

6. In the multiple allelic series $C^+ > C^{ch} > C^h$, dominance is from left to right as shown. In a cross of $C^+ C^{ch} \times C^{ch} C^h$, what proportion of the progeny will be Himalayan? (a) 100%; (b) 3/4; (c) 1/2; (d) 1/4; (e) 0%.

7. a. The extremities of mammals show lower temperature than the rest of the body. What do you suppose this fact might have to do with the action of Himalayan alleles in mice and other animals?

   b. If a small patch of hair is shaved from the back of a Himalayan mouse and the area is kept covered with an ice pack, what color would you expect the hair to be when it grows back in this area? (1) white; (2) agouti; (3) black; (4) cinnamon; (5) yellow.

8. In a maternity ward, the babies became accidentally mixed up. The ABO types of the four babies are known to be O, A, B, and AB. The ABO types of the four sets of parents are determined. Indicate which baby belongs to each set of parents: (a) AB × O; (b) A × O; (c) A × AB; (d) O × O.

9. The M, N, and MN blood groups are determined by two alleles, L^M and L^N. The Rh^+ (rhesus positive) blood group is caused by a dominant allele R of a different gene. In a court case concerning a paternity dispute, each of two men claimed three children to be his own. The blood groups of the men, the children, and their mother were as follows:

Husband	O	M	Rh^+
The wife's lover	AB	MN	Rh^-
Wife	A	N	Rh^+
Child 1	O	MN	Rh^+
Child 2	A	N	Rh^+
Child 3	A	MN	Rh^-

From this evidence, can the paternity of the children be established?

10. On a fox ranch in Wisconsin, a mutation arose that gave a "platinum" coat color. The platinum color proved very popular with buyers of fox coats, but the breeders could not develop a pure-breeding platinum strain. Every time two platinums were crossed, some normal foxes appeared in the progeny. For example, in repeated matings of the same pair of platinums, a total of 82 platinums and 38 normals was produced. This result was typical of all other such matings. State a *concise* genetic hypothesis to account for these results.

11. Over a period of several years, Hans Nachtsheim investigated an inherited anomaly of the white blood cells of rabbits. This Pelger anomaly, in its usual condition, involves an arrest of the typical segmentation of the nuclei of certain of the white cells. This anomaly does not appear to seriously inconvenience the rabbits.
    a. When rabbits showing the typical Pelger anomaly were mated with rabbits from a true-breeding normal stock, Nachtsheim counted 217 offspring showing the Pelger anomaly and 237 normal progeny. What appears to be the genetic basis of the Pelger anomaly?
    b. When rabbits with the Pelger anomaly were mated to each other, Nachtsheim found 223 normal progeny, 439 showing the Pelger anomaly, and 39 extremely abnormal progeny. These very abnormal progeny not only had defective white blood cells but also showed severe deformities of the skeletal system; almost all of them died soon after birth. In genetic terms, what do you suppose these extremely defective rabbits represented? Why do you suppose that there were only 39 of them?
    c. What additional experimental evidence might you collect to support or disprove your answers to part b?
    d. About one human in a thousand (in Berlin) shows a Pelger anomaly of white blood cells very similar to that described in rabbits. It is inherited as a simple dominant, but the homozygous type has not been observed in humans. Can you suggest why, if you are permitted an analogy with the condition in rabbits?
    e. Again by analogy with rabbits, what genetic situations might be expected among the children of a man and woman each showing the Pelger anomaly?

(Problem 11 is from A. M. Srb, R. D. Owen, and R. S. Edgar, *General Genetics*, 2nd ed., W. H. Freeman and Company, 1965.)

12. You have been given a single virgin *Drosophila* female. You notice that the bristles on her thorax are much shorter than normal. You mate her with a normal male (with long bristles) and obtain the following F_1 progeny: 1/3 short-bristle females, 1/3 long-bristle females, and 1/3 long-bristle males. A cross of the F_1 long-bristle females with their brothers gives only long-bristle F_2. A cross of short-bristle females with their brothers gives 1/3 short-bristle females, 1/3 long-bristle females, and 1/3 long-bristle males. Explain these data.

13. In *Drosophila*, a dominant allele *H* reduced the number of body bristles, giving rise to a "hairless" condition. In the homozygous condition *H* is lethal. An independently assorting dominant allele *S* has no effect on bristle number except in the presence of *H*, in which case a single dose of *S* suppresses the hairless phenotype, thus restoring the hairy condition. However, *S* also is lethal in the homozygous (*SS*) condition.

    a. What ratio of hairy to hairless individuals would you find in the live progeny of a cross between two hairy flies both carrying *H* in the suppressed condition?

    b. If the hairless progeny are backcrossed with a parental fly, what phenotypic ratio is expected in the live progeny?

14. A panther fancier has developed two pure-breeding lines of panther: one is pink and spotted (spotted pink), the other is blue and without spots (solid blue). Reciprocal crosses between these lines always give the same result: 1/2 solid pink and 1/2 solid blue. The solid pink panthers from this progeny then are intercrossed, and a large number of progeny are reared. They are of four kinds: 1/2 solid pink, 1/6 spotted pink, 1/4 solid blue, and 1/12 spotted blue. Explain this peculiar phenotypic ratio and assign genotypes, using symbols of your own choosing.

    (Problem 14 courtesy of Clayton Person.)

15. In squash plants, white and yellow are fruit colors, and disk and sphere are fruit shapes. The cross of white, disk × yellow, sphere produces an F_1 of all white, disk. Selfing the F_1 produces an F_2 of 9/16 white, disk; 3/16 white, sphere; 3/16 yellow, disk; and 1/16 yellow, sphere. Interpret, using symbols of your own choosing.

16. Two albinos marry and have four normal children. How is this possible?

17. In corn, pure lines are obtained that have either sun-red, pink, scarlet, or orange kernels when exposed to sunlight (normal kernels remain yellow in sunlight). Table 4-4 gives the results of some crosses between these lines. Analyze the results of each cross, and provide a unifying hypothesis to account for *all* the results. (Explain all symbols you use.)

Table 4-4.

Cross	Parental phenotypes	F_1 phenotypes	F_2 phenotypes
1	Sun-red × pink	All sun-red	66 sun-red, 20 pink
2	Orange × sun-red	All sun-red	998 sun-red, 314 orange
3	Orange × pink	All orange	1300 orange, 429 pink
4	Orange × scarlet	All yellow	182 yellow, 80 orange, 58 scarlet

18. Many kinds of wild animals have the agouti coloring pattern, in which each hair has a yellow band around it (Figure 4-10).

    a. In black mice and other black animals, the yellow band is not present, and the hair is all black. This absence of wild agouti pattern is called nonagouti. When mice of a true-breeding agouti line are crossed with nonagoutis, the F_1 is all agouti, and the F_2 has a 3:1 ratio of agoutis to nonagoutis. Diagram this cross, letting A = agouti and a = nonagouti. Show the phenotypes and genotypes of the parents, their gametes, the F_1, their gametes, and the F_2.

    b. Another inherited color deviation in mice substitutes brown for the black color in the wild-type hair. Such brown-agouti mice are called cinnamons. When wild-type mice are crossed with cinnamons, the F_1 is all wild type, and the F_2 has a 3:1 ratio of wild type to cinnamon. Diagram this cross as in part a, letting B = wild-type black and b = the brown of the cinnamon.

    c. When mice of a true-breeding cinnamon line are crossed with mice of a true-breeding nonagouti (black) line, the F_1 is all wild type. Use a genetic diagram to explain this result.

    d. In the F_2 of the cross in part c, a fourth color called chocolate appears in addition to the parental cinnamon and nonagouti and the wild type of the F_1. Chocolate mice have a solid, rich-brown color. What is the genetic constitution of the chocolates?

    e. Assuming that the A–a and B–b allelic pairs assort independently of each other, what would you expect to be the relative frequencies of the four color types in the F_2 described in part d? Diagram the cross of parts c and d, showing phenotypes and genotypes (including gametes).

    f. What phenotypes (and in what proportions) would be observed in the progeny of a backcross of F_1 mice from part c to the cinnamon parent stock? to the nonagouti (black) parent stock? Diagram these backcrosses.

    g. Diagram a testcross for the F_1 of part c. What colors would result, and in what proportions?

    h. Albino (pink-eyed white) mice are homozygous for the recessive member of an allelic pair C–c, which assorts independently of the A–a and B–b pairs. Suppose that you have four different highly inbred (and therefore presumably homozygous) albino lines. You cross each of these lines with a true-breeding wild-type line, and you raise a large F_2 progeny from each cross. What genotypes for the albino lines would you deduce from the F_2 phenotypes shown in Table 4-5?

Table 4-5.

	Numbers of F_2 progeny				
F_2 of line	Wild type	Nonagouti (black)	Cinnamon	Chocolate	Albino
1	87	0	32	0	39
2	62	0	0	0	18
3	96	30	0	0	41
4	287	86	92	29	164

(Problem 18 is adapted from A. M. Srb, R. D. Owen, and R. S. Edgar, *General Genetics*, 2nd ed., W. H. Freeman and Company, 1965.)

19. In rats, yellow coat color is determined by an allele A that is not lethal when homozygous. At a separate gene that assorts independently, the allele R produces a black coat. Together, A and R produce a grayish coat, whereas $aa\ rr$ is white. A gray male is crossed with a yellow female, and the F_1 is 3/8 yellow, 3/8 gray, 1/8 black, and 1/8 white. Deduce the genotypes of the parents.

20. Normal *Drosophila melanogaster* have deep red eyes. You are able to establish two homozygous lines, one having bright scarlet eyes and the other having dark brown eyes. When you cross scarlet-eyed flies with brown-eyed flies, all of the F_1 progeny have deep red eyes. An $F_1 \times F_1$ cross produces the following progeny: 432 red eyes, 158 scarlet eyes, 139 brown eyes, and 52 white eyes. Explain these results, using symbols of your own devising.

21. In the fowl, the genotype $rr\,pp$ gives single comb, $R\!-\ P\!-$ gives walnut comb, $rr\ P\!-$ gives pea comb, and $R\!-pp$ gives rose comb.
    a. What comb types will appear (and in what proportions) in F_1 and in F_2 if single-combed birds are crossed with birds of a true-breeding walnut-combed strain?
    b. What are the genotypes of the parents in a walnut-combed × rose-combed mating from which the progeny are 3/8 rose-combed, 3/8 walnut-combed, 1/8 pea-combed, and 1/8 single-combed?
    c. What are the genotypes of the parents in a walnut-combed × rose-combed mating from which all the progeny are walnut-combed?
    d. How many genotypes will produce a walnut phenotype? Write them out.

22. F_2 phenotypic ratios of 27:37 have been observed. What is their genetic basis? (HINT: You may find it useful to consider trihybrids.)

    (Problems 21 and 22 are adapted from A. M. Srb, R. D. Owen, and R. S. Edgar, *General Genetics,* 2nd ed., W. H. Freeman and Company, 1965.)

23. *Answer the parts of this problem in sequence.* You have two homozygous lines of *Drosophila,* one found in Vancouver (line A) and the other found in Los Angeles (line B). Both lines have bright scarlet eyes, a phenotype quite distinct from the deep red eyes of the wild type.
    a. A cross of line-A males with line-B females produces an F_1 of 200 wild-type males and 198 wild-type females. From this result, what can you say about the inheritance of eye color in the two lines?
    b. A cross of line-B males with line-A females produces an F_1 of 197 scarlet-eyed males and 201 wild-type females. What does this result indicate about the inheritance of eye color?
    c. When you intercross the F_1 progeny from part a, you obtain the following F_2 progeny: 151 wild-type females, 49 scarlet-eyed females, 126 scarlet-eyed males, and 74 wild-type males. Diagram the genotypes of the parents and of the F_1 offspring. Indicate the expected ratios of F_2 genotypes and phenotypes.

24. *Answer the parts of this problem in sequence.* In *Drosophila melanogaster,* wild-type eyes are deep red in color. You have obtained two lines of *Drosophila,* line A and line B, in which the eyes are white.

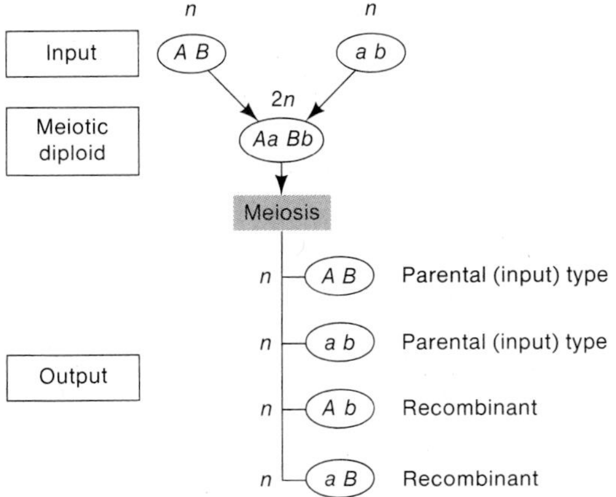

Figure 5-4. *Recombination. This diagram shows a meiosis, comparing its input haploid genotypes to its output haploid genotypes.*

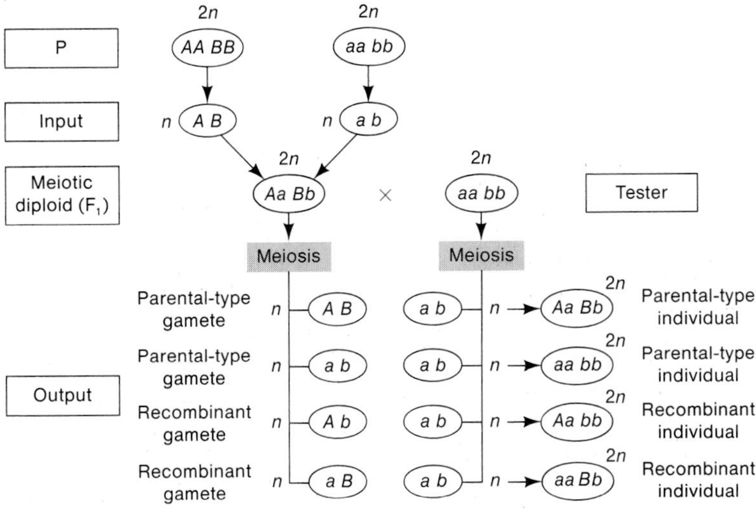

Figure 5-5. *An expansion of Figure 5-4 to clarify the process of recombination in diploid organisms. Note that Figure 5-4 is actually a part of this figure. In a diploid meiosis, the recombinant meiotic products are most readily detected by a cross to a recessive tester.*

product of meiosis, it too is called a recombinant. Notice again that the testcross allows us to concentrate on *one* meiotic system, avoiding ambiguity. In a self of the F_1 in Figure 5-5, for example, an *AA Bb* offspring (which would demonstrate recombination in F_1) cannot be distinguished from *AA BB* without further extensive crosses.

There are two kinds of recombination, and they produce recombinants by completely different methods. But a recombinant is a recombinant, so how can we decide which type of recombination has occurred in any particular progeny? The answer lies in the *frequency* of recombinants, as we shall soon see. First, though, let us consider the two types of recombination.

Interchromosomal Recombination

Interchromosomal recombination is achieved by Mendelian independent assortment. The two recombinant classes always make up 50% of the progeny; that is, there are 25% of each recombinant type among the progeny (Figure 5-6). If we observe this frequency, we can infer that the gene pairs under study are assorting independently.

The simplest interpretation of these frequencies is that the gene pairs are on

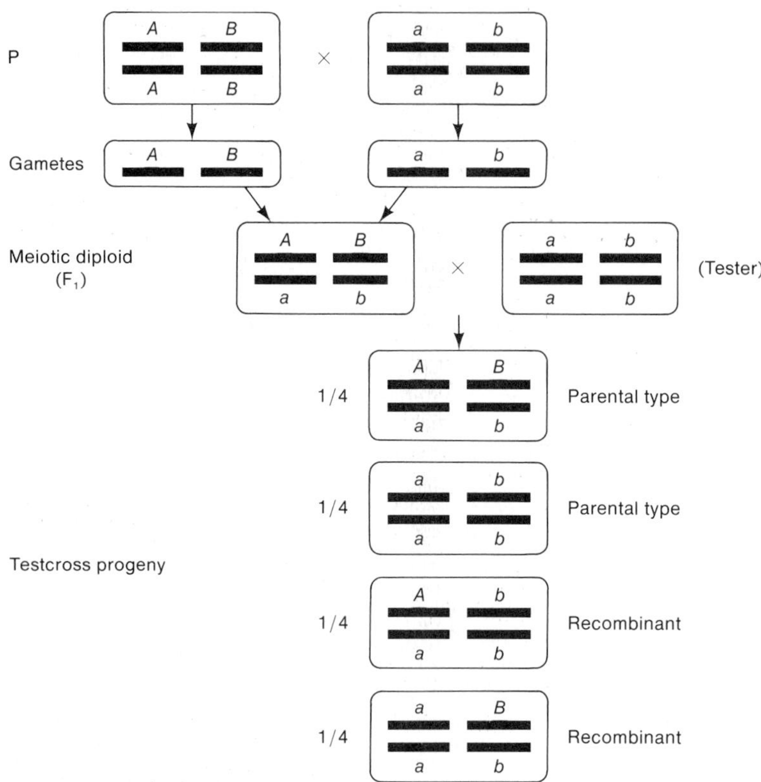

Figure 5-6. Interchromosomal recombination, which always produces a recombinant frequency of 50%. This diagram shows a diploid organism, but we can see the haploid situation by removing the part marked P and the testcross.

separate chromosome pairs. However, as we have already hinted, gene pairs that are far apart on the *same* chromosome pair can act virtually independently, producing the same result. We shall see in the next chapter how a large amount of crossing over effectively unlinks the distantly linked gene pairs.

Intrachromosomal Recombination

The sign of intrachromosomal recombination is a recombinant frequency of less than 50%. The physical linkage of parental gene combinations prevents the free assortment of gene pairs (Figure 5-7). We saw an example of this situation in Morgan's data, where the recombinant frequency was (151 + 154) ÷ 2839 = 10.7%. This is obviously much less than the 50% we would expect with independent assortment. What is the significance of recombinant frequencies greater than 50%? Such frequencies are *never* observed, as we shall prove in Chapter 6.

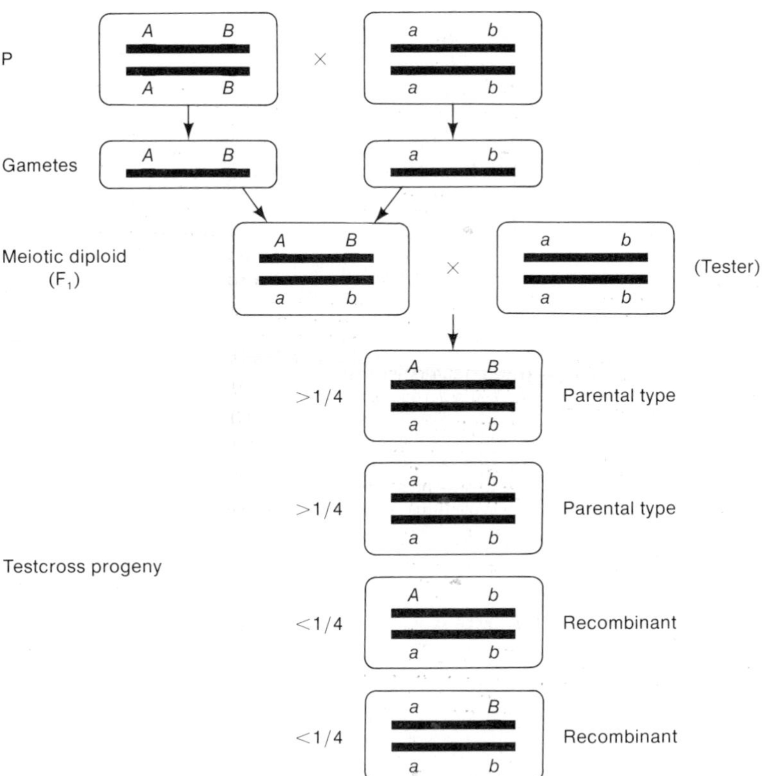

Figure 5-7. Intrachromosomal recombination, which always produces a recombinant frequency equal to or (usually) less than 50%. Again, a diploid organism is shown.

Linkage Symbolism

Our symbolism for describing crosses becomes cumbersome with the introduction of linkage. We can show the genetic constitution of each chromosome in the *Drosophila* cross as

$$\frac{pr \qquad vg}{\overline{pr^+ \qquad vg^+}}$$

where each line represents a chromosome, the genes above being on one chromosome and those below on the other. A crossover is represented by × between the two chromosomes so that

$$\begin{array}{c} pr \qquad vg \\ \hline \times \\ \hline pr^+ \qquad vg^+ \end{array}$$

is the same as

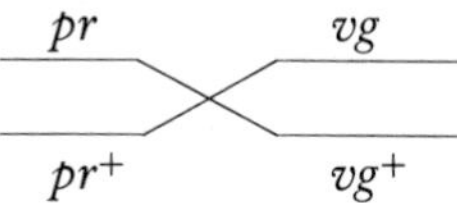

We can simplify the genotypic designation of linked genes by drawing a single line, with the genes on each side being on the same chromosome; now our symbol is

$$\frac{pr \qquad vg}{pr^+ \qquad vg^+}$$

But this gives us problems in typing and writing, so let's tip the line to give us *pr vg*/*pr*$^+$ *vg*$^+$, still keeping the genes on one chromosome on one side of the line, and those of its homolog on the other. We always designate linked genes on each side in the same order, so it is always *a b*/*a b*, never *a b*/*b a*. That being the case, we can indicate the wild-type allele with a plus sign (+), and *pr vg*/*pr*$^+$ *vg*$^+$ becomes *pr vg*/+ +. From now on, we'll use this kind of designation unless its use creates an ambiguity.

Linkage of Genes on the X Chromosome

If we now reconsider the results obtained by Bateson and Punnett, we can easily explain the coupling phenomenon by means of the concept of linkage. Their results are complex because they did not do a testcross. See if you can derive estimated numbers for recombinant and parental types in the gametes.

Until now, we have been considering recombination in autosomal genes. If, however, we were to look at sex-linked genes, a testcross would not be necessary. (Don't confuse linkage with sex linkage. Linkage describes the association of genes on the same chromosome, whereas sex linkage describes the association of genes with an X chromosome or other sex-determining chromosome.) Consider what happens when recombination involves the X chromosome. A female, heterozygous for two different sex-linked genes, will produce male progeny hemizygous for those genes, so that the mother's gametic genotype will be the son's phenotype. Let's take an example, using the genes *y* (yellow body) and y^+ (brown body), and *w* (white eye) and w^+ (red eye). In the following cross, Y represents the Y chromosome; do not confuse it with the y/y^+ gene.

P $\quad y+/y+ ♀ \times +w/\text{Y} ♂$ $\qquad$ (that is, $y\,w^+/y\,w^+ ♀ \times y^+\,w/\text{Y}♂$)

$F_1 \quad y+/+w♀ \times y+/\text{Y}♂$ $\qquad$ (that is, $y\,w^+/y^+w♀ \times y\,w^+/\text{Y}♂$)

The number of F_2 males in each phenotypic class is

y w	43
+ *w*	2146
y +	2302
+ +	22
	4513

These classes reflect the products of meiosis in the F_1 female. The system is in effect a testcross, because the F_2 males obtain only a Y chromosome from the F_1 $y+/\text{Y}$ males, and thus they have only one allele (obtained from the mother) for each of the genes we are considering. The recombinant frequency in this case is $(43 + 22) \div 4513 = 1.4\%$.

Linkage Maps

The frequency of recombinants for the *Drosophila* autosomal genes we studied (*pr* and *vg*) was 10.7% of the progeny, much greater than that for the linked genes on the X chromosome. Apparently, the amount of crossing over between linked gene pairs is not constant. Indeed, there is no reason to expect that crossing over between different linked gene pairs would occur with the same frequency. As Morgan studied more linked genes, he saw that the proportion of recombinant progeny varied considerably, depending on which linked gene pairs were being studied, and he thought that these variations in crossover frequency might somehow reflect the actual distances separating genes on the chromosomes. Morgan assigned the study of this problem to a student, Alfred Sturtevant, who (like Bridges) became one of the great geneticists. Morgan asked Sturtevent, still an undergraduate at the time, to make some sense of the data on crossing over between different linked genes. In one night, Sturtevant developed a method for describing relationships between genes that is still used today.

For example, consider a testcross from which we obtain the following results:

pr vg/pr vg	165	Parental (noncrossover) types
+ +/pr vg	191	
pr +/pr vg	23	Recombinant (crossover) types
+ vg/pr vg	21	
	400	

The progeny in this example represent 400 female gametes, of which 44 (or 11%) are recombinant. Sturtevant suggested that we can use this percentage of recombinants as a quantitative index of the distance between two gene pairs on a **genetic map.**

The basic idea here is quite simple. Imagine two specific gene pairs a certain fixed distance apart (Figure 5-8). Now imagine crossovers occurring randomly along the paired homologs. In some meiotic divisions, a crossover occurs by chance in the chromosomal region between these gene pairs; from these meioses, the recombinants are produced. In other meiotic divisions, no crossover occurs in the intervening region; no recombinants result from these

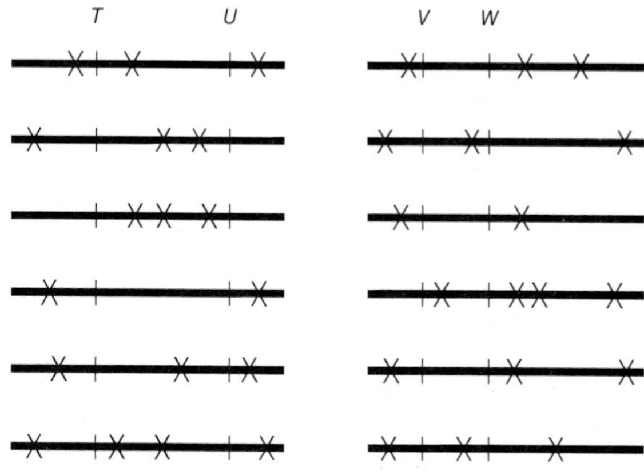

Figure 5-8. Proportionality between chromosomal distance and recombinant frequency. Each line represents a chromosome pair (four chromatids) during meiosis, and the crosses represent the positions of the crossovers that are taking place. On the left, two distant *gene pairs* (T *and* U) *are shown; on the right (a different chromosome) are two* close *gene pairs* (V *and* W). *Crossovers occur at random along the chromosomes, and there are more of them between* T *and* U *than between* V *and* W, *so the recombinant frequency for* T *and* U *will be higher than that for* V *and* W.

meioses. Sturtevant postulated a rough proportionality: the greater the distance between the linked genes, the greater the chance of a crossover occurring in the intervening region and hence the greater the proportion of meioses in which a crossover occurs here. Thus, by measuring the frequency of recombinants, one obtains a measure of the map distance between the gene pairs.

Message
One ***genetic map unit (m.u.)*** *is the distance between gene pairs for which one product of meiosis out of one hundred is recombinant. Put another way, a* ***recombinant frequency (RF)*** *of 0.01 (or 1%) is defined as 1 m.u.*

The place on the map that represents a gene pair is called the **gene locus** (plural, **loci**). The locus of the eye-color gene pair and the locus of the wing-shape gene pair, for example, are 11 map units apart. They are usually diagrammed this way:

pr 11.0 *vg*

although they could be diagrammed equally well like this:

pr^+ 11.0 vg^+

or like this:

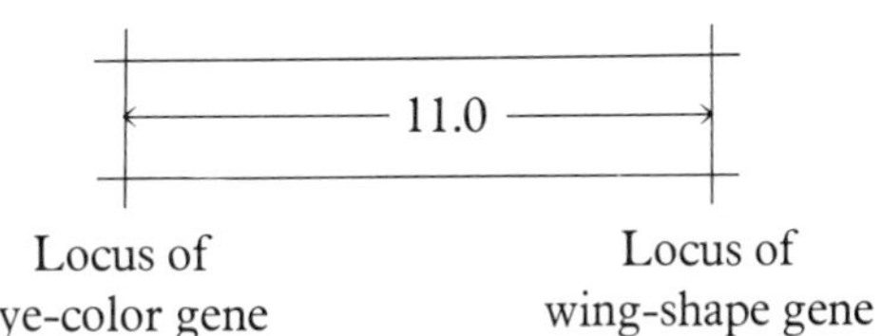

Usually the locus of this eye-color gene is referred to in shorthand as the "*pr* locus," after the most famous nonwild allele, but it means the place on the chromosome where any allele of this gene will be found.

There is a strong implication that the "distance" on a genetic map is a physical distance along a chromosome, and Morgan and Sturtevant certainly intended to imply just that. But we should realize that the genetic map is another example of an entity constructed from a purely genetic analysis. The genetic map (or **linkage map,** as it is sometimes called) could have been derived without even knowing that chromosomes existed. Furthermore, at this point in our discussion, we cannot say whether the "genetic distances" calculated by means of recombinant frequencies in any way reflect actual physical distances on chromosomes. However, clever cytogenetic analysis of *Drosophila* has shown that genetic distances are, in fact, largely proportional to chromosome distances.

Given a genetic distance in map units, one can predict frequencies of progeny in different classes. For example, of the progeny from a testcross of a *pr vg*/+ + heterozygote, 5½% will be *pr* +/*pr vg* and 5½% will be + *vg*/*pr vg*; of the progeny from a testcross of a *pr* +/+ *vg* heterozygote, 5½% will be *pr vg*/*pr vg* and 5½% will be + +/*pr vg*.

Let's take a few other examples showing different map distances. In corn, there are two linked loci controlling seed color (*A*, colored, and *a*, colorless) and shape (*Sh*, plump, and *sh*, shrunken). In the cross *A Sh*/*a sh* × *a sh*/*a sh*, we get

A Sh/*a sh*	5020
a sh/*a sh*	4960
A sh/*a sh*	12
a Sh/*a sh*	8
	10000

In this example, the number of recombinants is very small; consequently the genetic distance is short—0.2 map unit, or [(12 + 8)/10,000] × 100. On the other hand, in chickens, two genes affecting feather color (*I*, white, and *i*, colored) and texture (*F*, frizzle, and *f*, normal) are less tightly linked. In the cross *i F*/*I f* × *i f*/*i f*, we obtain

i F/*i f*	63
I f/*i f*	63
I F/*i f*	18
i f/*i f*	13
	157

In this example, the distance between the gene pairs is 19 map units, or [(18 + 13)/157] × 100.

Message
The genetic map is another example of a hypothetical entity based on genetic analysis. Its construction does not depend on the knowledge of any cytological structure.

Three-Point Testcross for Linkage Detection

How do we analyze linkage in crosses involving more than two heterozygous gene pairs? When organisms are extensively investigated, it becomes possible to study such multiple heterozygotes and to derive a great deal of interlocking linkage information from a single testcross. We confine our attention to triple heterozygotes in this section. When these are crossed to testers, the procedure is called a **three-point testcross.**

Let's follow some examples from *Drosophila.* In our first example, the non-wild-type alleles are *sc* (scute, or loss of certain thoracic bristles), *ec* (echinus, or roughened eye surface), and *vg* (vestigial wing). If we take *sc/sc, ec/ec, vg/vg* triple-recessive flies and cross these with wild types, a triple heterozygote *sc*/+, *ec*/+, *vg*/+ is established in the F_1. We can take F_1 females of this genotype and testcross them with *sc/sc, ec/ec, vg/vg*. (By drawing the genotypes this way, we don't commit ourselves to any particular linkage arrangement at this stage of the analysis.) The progeny, listed as the gametic types from the triple heterozygote, will be as follows. The numbers are invented for the purpose of illustration, but they are based on real linkage values for these particular genes.

sc	*ec*	*vg*	235
+	+	+	241
sc	*ec*	+	243
+	+	*vg*	233
sc	+	*vg*	12
+	*ec*	+	14
sc	+	+	14
+	*ec*	*vg*	16
			1008

There is considerable deviation from the 1:1:1:1:1:1:1:1 ratio we expect for a trihybrid cross of independently assorting genes, so linkage must be involved. The four classes with the largest numbers must consist of those individuals that are not the result of a crossover between linked loci. Which loci are linked? Taking only the four largest classes, we can determine whether *sc* and *ec* are linked. (Remember that by *sc* and *ec* we really mean the gene sites, or loci, represented by *sc*/+ and *ec*/+.) Ignoring *vg* then, we get:

sc	*ec*	235
+	+	241
sc	*ec*	243
+	+	233

There are only two classes, *sc ec* and + +; so we know they are linked and that these are the two parental classes. What about *vg*? Let's ask whether *ec* and *vg* are linked and ignore *sc* (because we know it is linked to *ec*). The data then are

ec	*vg*	235
+	+	241
ec	+	243
+	*vg*	233

which is a nearly perfect 1:1:1:1 ratio, showing that *vg* is assorting independently of *ec* (and *sc*). So we know *sc* and *ec* are on one chromosome, and *vg* assorts independently. We can calculate the genetic distance between *sc* and *ec* by looking at the smaller classes, which (if we ignore *vg* because it is assorting independently) are obviously derived from crossovers.

sc	+	12
+	*ec*	14
sc	+	14
+	*ec*	16

The distance between *sc* and *ec* is $[(12 + 14 + 14 + 16)/1008] \times 100 = 5.5$ map units, and we can diagram the genetic maps as

sc —— 5.5 —— *ec*   —*vg*—

Now let's take another cross, in which the genes are *sc*, *ec*, and another nonwild type, *cv* (crossveinless, or absence of a crossvein on the wing). We cross flies homozygous for *sc*, *ec*, and *cv* with homozygous wild type to get triple-heterozygote females, which we then testcross to obtain (again, invented numbers)

sc	*ec*	*cv*	417
+	+	+	430
sc	+	+	25
+	*ec*	*cv*	29
sc	*ec*	+	44
+	+	*cv*	37
			982

First, do these genes assort independently? We see that there are two classes, *sc ec cv* and + + +, that occur much more often than the others, thereby showing that we are not getting a 1:1:1:1 ratio of independent assortment of any two pairs. Moreover, they are reciprocal classes (that is, the alleles in one are all different from those in the other); so *sc*, *ec*, and *cv* must have been on one of the homologous chromosomes and +, +, and + on the other. We now know the three genes are linked; so we can determine how far apart they are. Looking at one pair of loci at a time, let's take *sc* and *ec* first (we can cover up the *cv* and consider only the *sc* and *ec* pair of loci).

sc	*ec*	~~*cv*~~	417
+	+	~~+~~	430
sc	+	~~+~~	25
+	*ec*	~~*cv*~~	29
sc	*ec*	~~+~~	44
+	+	~~*cv*~~	37
			982

There are $417 + 430 + 44 + 37 = 928$ parentals and $25 + 29 = 54$ recombinants with respect to these two gene pairs. Therefore, the genetic distance between *sc* and *ec* is 5.5 map units [(54/982) × 100], which is the same distance found earlier for another cross involving these loci. The map can be diagrammed as

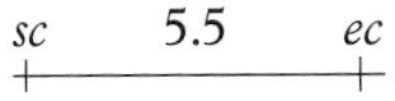

sc —— 5.5 —— *ec*

Now let's take *ec* and *cv*; the data are

~~*sc*~~	*ec*	*cv*	417
~~+~~	+	+	430
~~*sc*~~	+	+	25
~~+~~	*ec*	*cv*	29
~~*sc*~~	*ec*	+	44
~~+~~	+	*cv*	37
			982

There are 417 + 430 + 25 + 29 = 901 parentals and 44 + 37 = 81 recombinants. The distance between *ec* and *cv*, then, is (81/982) × 100 = 8.2, and the map is

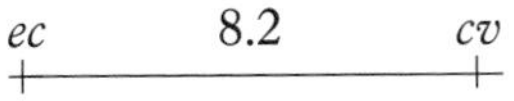

ec —— 8.2 —— *cv*

The distance between *sc* and *ec* being 5.5 and that between *ec* and *cv* being 8.2, we can draw a map as either

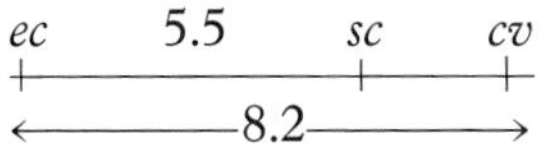

ec —— 5.5 —— *sc* —— *cv*
←—— 8.2 ——→

or

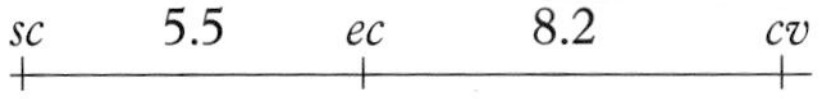

sc —— 5.5 —— *ec* —— 8.2 —— *cv*

Now let's see how far *sc* and *cv* are apart, forgetting *ec*:

sc	~~*ec*~~	*cv*	417
+	~~+~~	+	430
sc	~~+~~	+	25
+	~~*ec*~~	*cv*	29
sc	~~*ec*~~	+	44
+	~~+~~	*cv*	37
			982

There are 417 + 430 = 847 parentals and 25 + 29 + 44 + 37 = 135 recombinants for a distance of (135/982) × 100 = 13.7 map units. We know then that the map should be drawn as

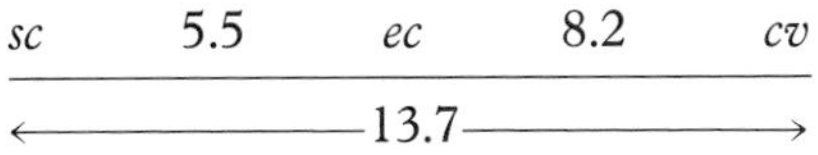

sc —— 5.5 —— *ec* —— 8.2 —— *cv*
←—— 13.7 ——→

Double Crossovers

Let's turn to another cross in which the genes are *cv* and the two more non-wild types, *ct* (cut, or snipped wing edges) and *v* (vermilion, or bright scarlet eye color). A homozygous *cv ct v* fly is crossed with a homozygous wild-type fly to yield a triply heterozygous female, which is then testcrossed to give

cv	*ct*	*v*	580
+	+	+	592
cv	+	+	45
+	*ct*	*v*	40
cv	*ct*	+	89
+	+	*v*	94
cv	+	*v*	3
+	*ct*	+	5
			1448

Again we recognize a pretty obvious linkage of all three gene pairs by the absence of Mendelian ratios, the *cv ct v* and + + + classes being the parental chromosomes. Unlike the previous cross, this cross yields six classes of recombinant chromosomes. We can calculate map distances between two gene pairs at a time, ignoring the third again. Let's ignore *v* and look at *cv* and *ct*, for which the recombinant classes are *cv* + and + *ct*; the genetic distance is 6.4 map units, or [(45 + 40 + 3 + 5)/1448] × 100. Doing the same for *ct* and *v* (and ignoring *cv*), we get a distance of 13.2 map units, or [(89 + 94 + 3 + 5)/1448] × 100. Now if we ignore *ct* and look at *cv* + and + *v* recombinants, we find the distance to be 18.5 map units, or [(45 + 40 + 89 + 94)/1448] × 100. This shows that *cv* and *v* are the farthest apart and that *ct* must be between them. However, if we add the distances between *cv* and *ct* (6.4) and between *ct* and *v* (13.2), we get a value of 19.6, which is larger than the computed value of 18.5. Why? We see that eight of the flies (*cv* + *v* and + *ct* +) were included in calculations for both the *cv*-to-*ct* and the *ct*-to-*v* intervals, but *not* for the calculations for the *cv*-to-*v* interval, even though *two crossovers* had taken place between *cv* and *v* in each of these chromosomes. The reason becomes obvious if we look at a representation of this **double-crossover** event (Figure 5-9).

Although two crossovers have taken place between *cv* and *v*, the genetic result is that *cv* and *v* remain linked in the parental combination. Only when *ct* and its wild-type allele are present can we recognize the double-crossover

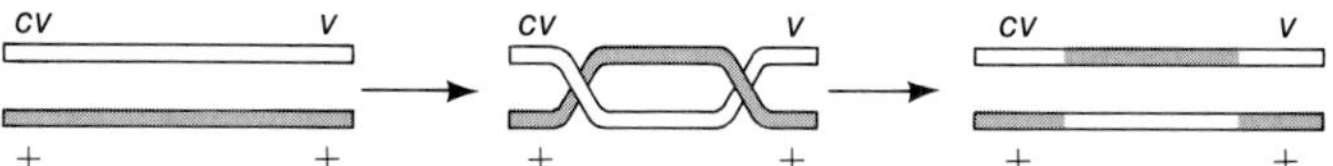

Figure 5-9. Double crossover chromatids involving two linked gene pairs are normally not detected genetically because recombinant genotypes are not produced.

chromosome. So, in a calculation of the *cv*-to-*v* distance, we should add in *twice* the number of double-recombinant types ($8 \times 2 = 16$). When this is done, the *cv*-to-*v* distance becomes $[(45 + 40 + 89 + 94 + 16)/1448] \times 100 = 19.6$ m.u., which is the same as the combined values.

In almost any crossover study of three linked genes, the double-recombinant classes will be the most rare. Knowing that, we could have determined the gene order (which gene is in the middle) without calculating map distances. We know from the nonrecombinant classes that the parental arrangement of genes was *cv ct v*/+ + +. If we do not know the proper gene order, then it could be any of *cv v ct*, *v cv ct*, and *cv ct v* (as shown in Figure 5-10). Of these squences, only the one shown in part c would generate the genotypes of the double-crossover classes *cv* + *v* and + *ct* +.

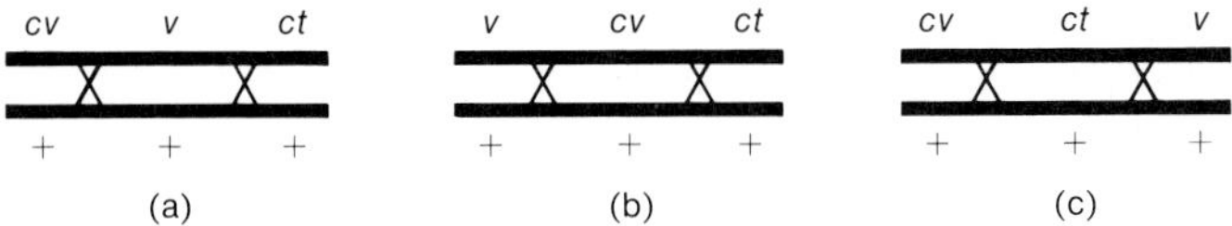

Figure 5-10. Three different gene-pair orders are possible. Each case leads to a different double-recombinant genotype.

Interference

The detection of the double-exchange class allows us to ask another question. Are exchanges occurring in each of two different regions of the same chromosome independently of each other? We know that we get recombinants between *cv* and *ct* 6.4% of the time and between *ct* and *v* 13.2% of the time. If a crossover in one of these regions in no way affects the probability of a crossover occurring in the other region (that is, if they are independent), then the probability that an individual will be produced that is recombinant in both regions at once is the product of the individual recombinant frequencies. If there is independence, then the expected frequency of double-recombinant types is $0.064 \times 0.132 = 0.0084$ (0.84%). In a sample of 1448, we would expect $1448 \times 0.0084 = 12$ double crossovers, whereas we actually observed 8. Had we obtained a million progeny, the observed number of doubles would still have been lower than the expected. What does that mean? It means that crossing over in one region is not independent of crossing over in the other; in fact, when a crossover does occur in one region, it decreases the likelihood that another will occur in the other region. The effect of a crossover in one region on the likelihood of a crossover in another region is called **interference.** We may quantify this effect by the **coefficient coincidence (c.c.),** which is simply the number of double recombinants observed divided by the number expected.

An **interference value (I)** is calculated by subtracting the c.c. from 1:

$$I = 1 - \text{c.c.} = 1 - \frac{\text{observed doubles}}{\text{expected doubles}}$$

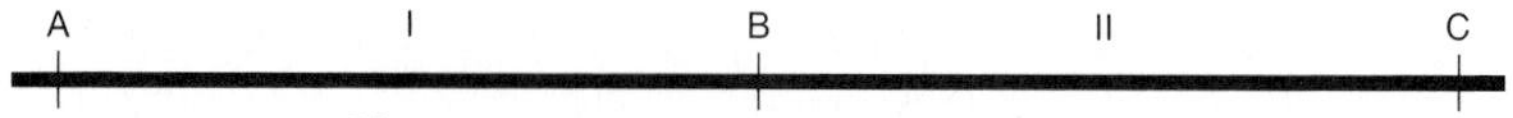

A — I — B — II — C

20 m.u. (A–B) | 20 m.u. (B–C)

p(of recombinant individual) = 0.2 | p(of recombinant individual) = 0.2

	No interference	With 10% interference
p(that recombinant for I but not II)	0.2 × 0.8 = 0.16	0.164 (up 0.004)
p(that recombinant for II but not I)	0.2 × 0.8 = 0.16	0.164 (up 0.004)
p(that recombinant for I and II)	0.2 × 0.2 = 0.04	0.036 (down 0.004)
p(that recombinant for neither I nor II)	0.8 × 0.8 = 0.64	0.636 (down 0.004)
	1.00	1.000

Figure 5-11. How to calculate the expected frequencies of nonrecombinant, single-recombinant, and double-recombinant types, with and without interference. The gene pairs A *and* B *are separated by distance I (20 m.u.); gene pairs* B *and* C *are separated by distance II (20 m.u.).*

When there are no double crossovers, the c.c. is zero, and so I is one; interference is complete. (Interference is complete in the *sc–ec–cv* interval in the example on page 150.) When there are fewer doubles than expected, the c.c. is less than one, making I a positive number; there is interference. When the c.c. is one, $I = 0$; there is no interference. Finally, when there are more doubles than expected, the c.c. is greater than one, making I a negative number; interference is negative. Figure 5-11 further explores the effects of interference by showing the redistribution of progeny types due to interference. (Note that the map distances I and II remain the same.)

Table 5-2 shows an explanatory analogy in which 1000 people are given at random 200 buns and 200 cookies. A socialistic impulse leads some of those people with both to give up one or the other to someone who has neither. The four people who give up either cookie or bun generate four additional one-bun people and four additional one-cookie people, but also generate four less bun-and-cookie people (their original selves) and four less people having neither (who have now received one item). A measure of the extent of the socialistic impulse (in this case, 10%) would correspond to the interference value.

You may have noticed that we always used heterozygous *females* for testcrosses in our crossover studies in *Drosophila.* When *pr vg*/+ + males are crossed with *pr vg*/*pr vg* females, only *pr vg*/+ + and *pr vg*/*pr vg* progeny are recovered. This result shows that crossing over does not occur in *Drosophila* males. However, this absence of crossing over in one sex is limited to certain

Table 5-2. A bun-and-cookie analogy to interference

Individuals with	After random distribution	After redistribution
One bun only	160	164 (up 4)
One cookie only	160	164 (up 4)
Bun and cookie	40	36 (down 4)
Neither	640	636 (down 4)
	1000	1000

species; it is not the case for males of all species (or for the heterogametic sex). In other organisms, crossing over can occur in XY males or in WZ females. The reason for this sex difference is completely unknown; you will just have to remember that male *Drosophila* have this special characteristic.

Linkage maps are an essential aspect of the experimental genetic study of any organism. They are the prelude to any serious piece of genetic manipulation. Many organisms have had their genes intensively mapped in this way. The resultant maps represent a tremendous amount of basic applied genetic analysis. Figures 5-12 and 5-13 show two examples of linkage maps, from the *Drosophila* and from the tomato.

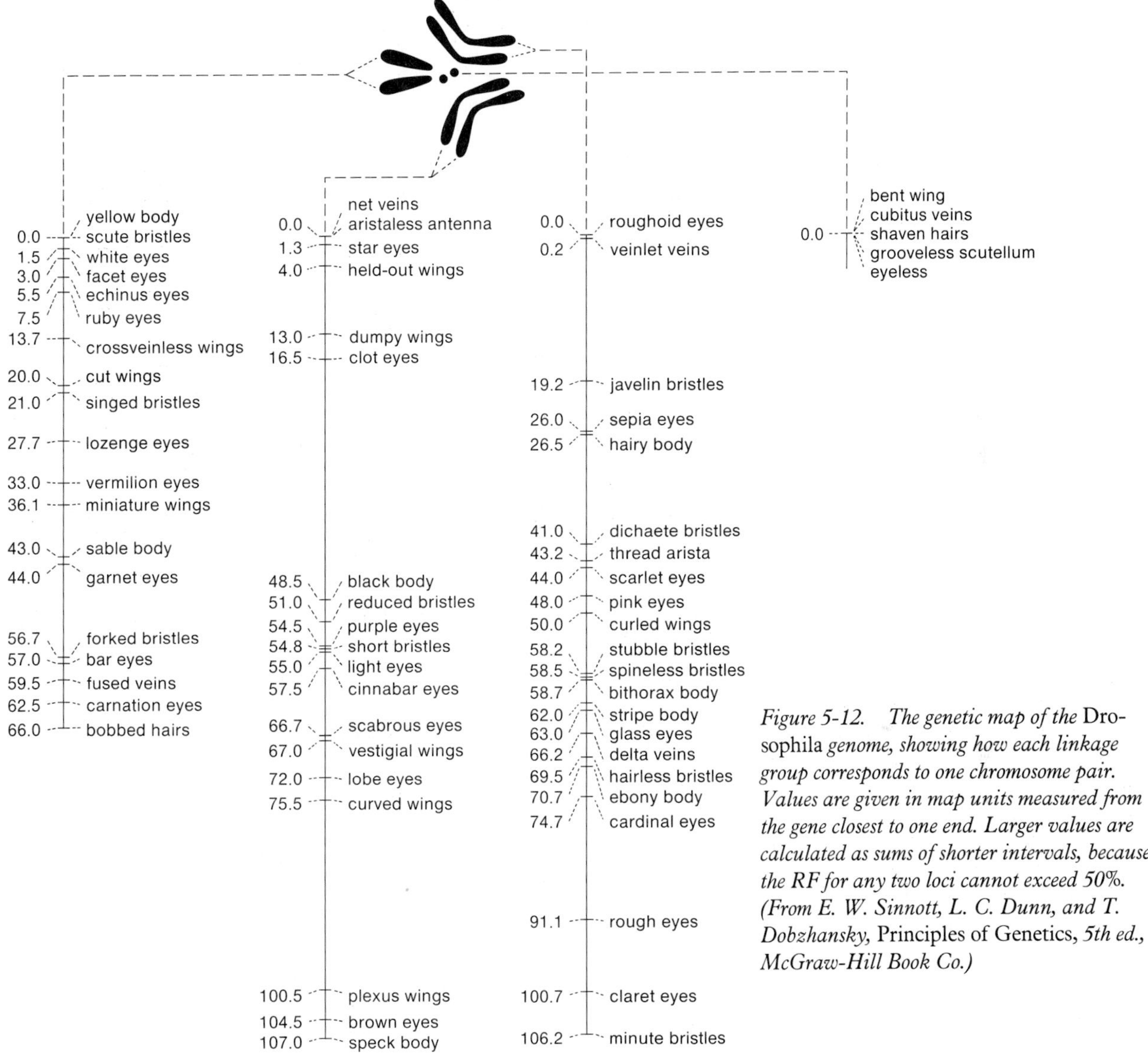

Figure 5-12. The genetic map of the Drosophila *genome, showing how each linkage group corresponds to one chromosome pair. Values are given in map units measured from the gene closest to one end. Larger values are calculated as sums of shorter intervals, because the RF for any two loci cannot exceed 50%. (From E. W. Sinnott, L. C. Dunn, and T. Dobzhansky,* Principles of Genetics, *5th ed., McGraw-Hill Book Co.)*

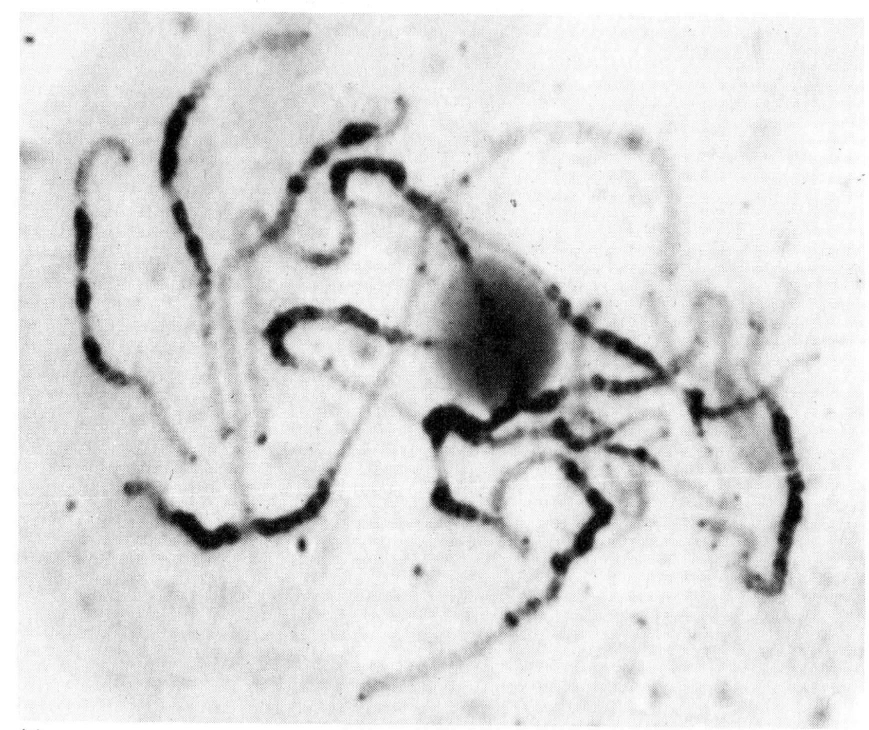
(a)

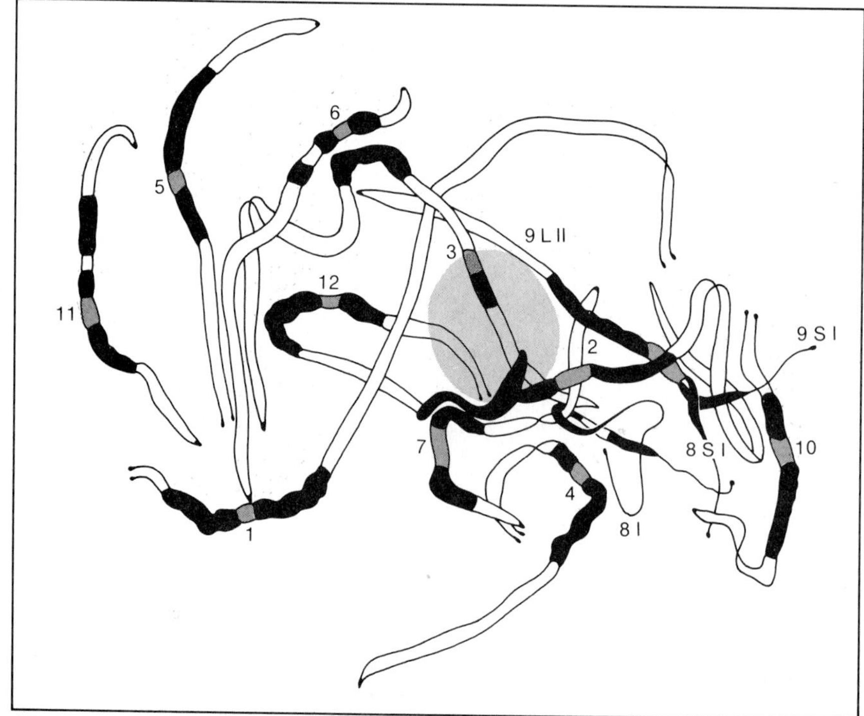

(b)

Figure 5-13. *The tomato genome. (a) Photomicrograph of a meiotic pachytene (prophase I) from anthers. (b) Diagram corresponding to part a, identifying the twelve chromosome pairs. Grayish centromeres are flanked by densely staining chromosome regions called heterochromatin. Heterochromatin is thought to be genetically inert. (c) Genetic map of the tomato genome. The twelve linkage groups correspond to the twelve chromosome pairs of part b. Centromeres and heterochromatin regions are indicated. (Part a courtesy of Charles M. Rick. Parts b, c from Charles M. Rick, "The Tomato." Copyright © 1978 by Scientific American, Inc. All rights reserved.)*

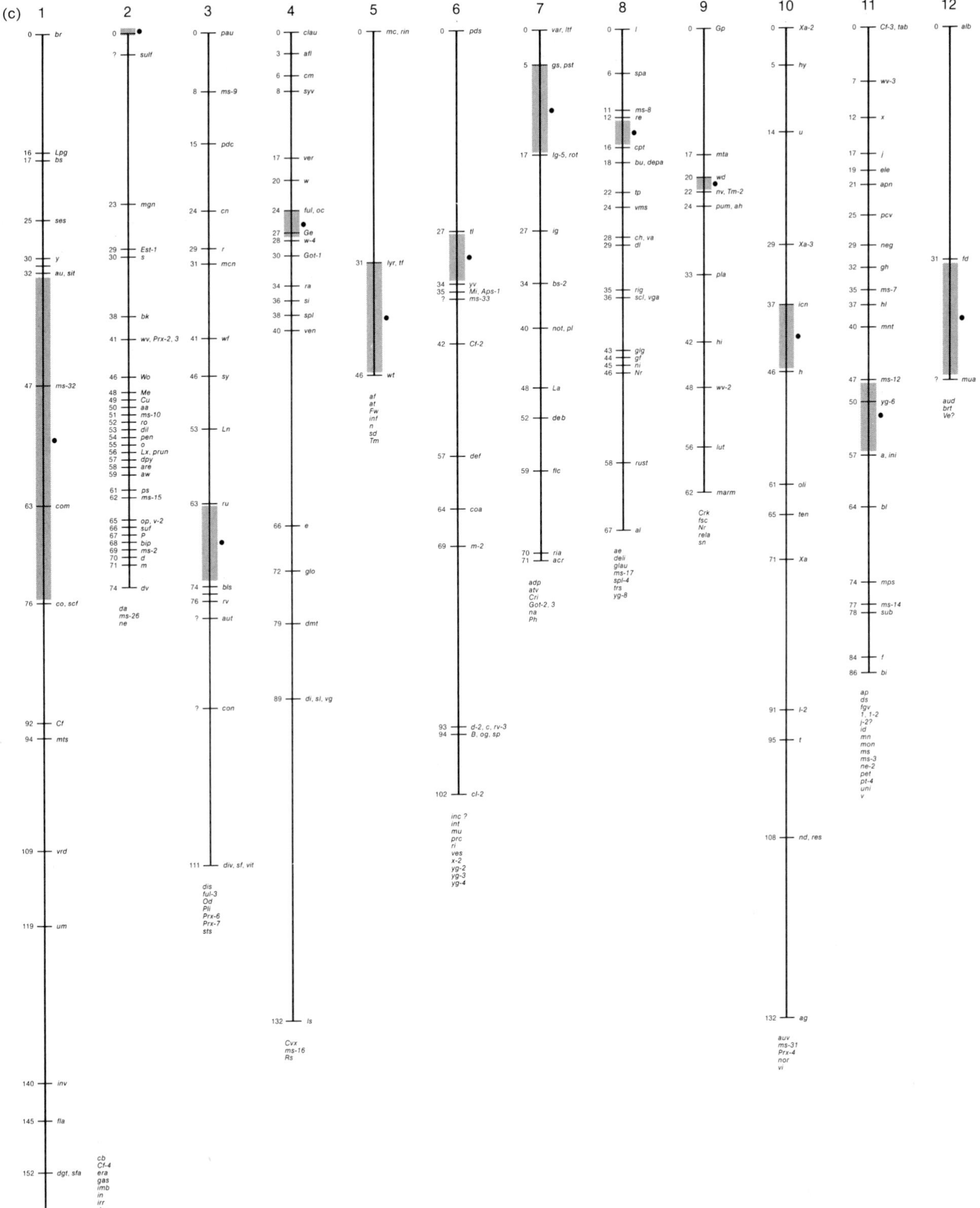
(c)
1 br Lpg bs ses y au, sit ms-32 com co, scf Cf mts vrd um inv fla dgt, sfa dt
cb Cf-4 era gas imb in irr Jau lu ms-6 per pr
2 sulf mgn Est-1 s bk wv, Prx-2, 3 Wo Me Cu aa ms-10 ro dil pen o Lx, prun dpy are aw ps ms-15 op, v-2 suf P bip ms-2 d m dv
da ms-26 ne
3 pau ms-9 pdc cn r mcn wf sy Ln ru bls rv aut con div, sf, vit
dis ful-3 Od Pli Prx-6 Prx-7 sts
4 clau afl cm syv ver w ful, oc Ge w-4 Got-1 ra si spl ven e glo dmt di, sl, vg ls
Cvx ms-16 Rs
5 mc, rin lyr, tf wt
af at Fw inf n sd Tm
6 pds tl yv Mi, Aps-1 ms-33 Cf-2 def coa m-2 d-2, c, rv-3 B, og, sp cl-2
inc ? int mu prc ri ves x-2 yg-2 yg-3 yg-4
7 var, ltf gs, pst lg-5, rot ig bs-2 not, pl La deb flc ria acr
adp atv Cri Got-2, 3 na Ph
8 l spa ms-8 re cpt bu, depa tp vms ch, va dl rig scl, vga glg gf ni Nr rust al
ae deli glau ms-17 spl-4 trs yg-8
9 Gp mta wd nv, Tm-2 pum, ah pla hi wv-2 lut marm
Crk fsc Nr rela sn
10 Xa-2 hy u Xa-3 icn h oli ten Xa l-2 t nd, res ag
auv ms-31 Prx-4 nor vi
11 Cf-3, tab wv-3 x j ele apn pcv neg gh ms-7 hl mnt ms-12 yg-6 a, ini bl mps ms-14 sub f bi
ap ds fgv I, I-2 j-2? id mn mon ms ms-3 ne-2 pet pt-4 uni v
12 alb fd mua
aud brt Ve?

The χ^2 Test

We come now to a subject that emerges naturally at this point in the discussion, a subject that is particularly relevant to the detection of linkage. It has been stated that the functional test for the presence or absence of linkage is based on the relative frequencies of the meiotic product types. If there is no linkage, the four product-cell types *A B*, *a b*, *A b*, and *a B* are produced in a 1:1:1:1 ratio, and this fixes the recombinant frequency at 50%. If there is linkage, there is deviation from the 1:1:1:1 ratio, and two types (the recombinants) are present in a minority ($<$50%). In practical terms, the following question must be answered in any particular case: "Is this a 1:1:1:1 ratio?" An answer of "yes" indicates absence of linkage, and "no" indicates linkage. "Obvious" departures from the 1:1:1:1 ratio present no decision problems, but smaller departures are trickier to handle, and they require further analytical approaches.

What is the precise problem here? An example will be helpful. A double heterozygote in coupling conformation produces 500 meiotic products distributed as follows:

A B	140
a b	135
A b	110
a B	115

By the application of the recombinant-frequency test, we find 225 recombinants, or 45%. This is admittedly less than 50%, but not convincingly so. The skeptic would say, "This is merely a chance deviation from a 1:1:1:1 ratio! If you repeatedly grabbed samples of 500 marbles from a dark sack containing equal numbers of red, blue, yellow, and green marbles, you would fairly often get a variation this great from the 1:1:1:1 ratio." What should we decide? The χ^2 test can help in this kind of predicament.

The χ^2 test in general tells how often deviations from expectations will occur purely on the basis of chance. The procedure is as follows.

1. *State a simple hypothesis that gives a precise expectation.* Obviously, in our example, the best hypothesis is "lack of linkage," which gives an expected 1:1:1:1 ratio. This **null hypothesis** is obviously better than a hypothesis of linkage, which is not precise because (as we have seen) the recombinant frequency can be large or small.
2. *Calculate* χ^2. This value *always* is calculated from actual numbers—never from percentages, fractions, or decimal fractions. In fact, part of the usefulness of the χ^2 test is that it takes sample size into consideration. The sample is composed of several operational classes, with O the observed number in any class and E the expected number based on the null hypothesis. The formula for calculating χ^2 is as follows:

$$\chi^2 = \text{total of } \frac{(O - E)^2}{E} \text{ over all classes.}$$

In our example, we would set up the calculation as shown in Table 5-3.

Table 5-3. Calculation of χ^2

Class	O	E	$(O - E)^2$	$\frac{(O - E)^2}{E}$
AB	140	125	225	1.8
ab	135	125	100	0.8
Ab	110	125	225	1.8
aB	115	125	100	0.8
	500	500		$\chi^2 = 5.2$

3. Using χ^2, calculate the probability p of obtaining the present results if the null hypothesis is correct. Before this can be done, another item must be computed: the number of degrees of freedom (d.f.). In the present context, the number of degrees of freedom can be simply defined as

$$\text{d.f.} = (\text{number of classes} - 1)$$

In our example,

$$\text{d.f.} = (4 - 1) = 3$$

We now turn to the table of χ^2 values (Table 5-4), which will give us our p values if we plug in our computed χ^2 and d.f. values. Looking along the d.f. = 3 line, we find our χ^2 value of 5.2 lies between the p values of 0.5 (50%) and 0.1 (10%). Hence we can conclude that our p value is just greater than 10% (0.1). What this means is that, if our null hypothesis is correct, a deviation *at least* as great as that observed is expected a little more than 10% of the time.

4. *Reject or accept the null hypothesis.* How do we know which p value is too low to be acceptable? Scientists in general arbitrarily use the 5% level. Any p value less than 5% results in rejection of the null hypothesis, and any value greater than 5% results in acceptance of the null hypothesis. In the case of acceptance, of course, the null hypothesis is not proved, merely possible. Now that we have accepted our null hypothesis of absence of linkage, we must live with it and acknowledge that the skeptic was right!

In this way, the χ^2 test helps decide between linkage and absence of linkage. In fact, probably the major use of the χ^2 test in genetics is in the determination

Table 5-4. Critical values of the χ^2 distribution

d.f. \ p	0.995	0.975	0.9	0.5	0.1	0.05	0.025	0.01	0.005	d.f.
1	.000	.000	0.016	0.455	2.706	3.841	5.024	6.635	7.879	1
2	0.010	0.051	0.211	1.386	4.605	5.991	7.378	9.210	10.597	2
3	0.072	0.216	0.584	2.366	6.251	7.815	9.348	11.345	12.838	3
4	0.207	0.484	1.064	3.357	7.779	9.488	11.143	13.277	14.860	4
5	0.412	0.831	1.610	4.351	9.236	11.070	12.832	15.086	16.750	5
6	0.676	1.237	2.204	5.348	10.645	12.592	14.449	16.812	18.548	6
7	0.989	1.690	2.833	6.346	12.017	14.067	16.013	18.475	20.278	7
8	1.344	2.180	3.490	7.344	13.362	15.507	17.535	20.090	21.955	8
9	1.735	2.700	4.168	8.343	14.684	16.919	19.023	21.666	23.589	9
10	2.156	3.247	4.865	9.342	15.987	18.307	20.483	23.209	25.188	10
11	2.603	3.816	5.578	10.341	17.275	19.675	21.920	24.725	26.757	11
12	3.074	4.404	6.304	11.340	18.549	21.026	23.337	26.217	28.300	12
13	3.565	5.009	7.042	12.340	19.812	22.362	24.736	27.688	29.819	13
14	4.075	5.629	7.790	13.339	21.064	23.685	26.119	29.141	31.319	14
15	4.601	6.262	8.547	14.339	22.307	24.996	27.488	30.578	32.801	15
16	5.142	6.908	9.312	15.338	23.542	26.296	28.845	32.000	34.267	16
17	5.697	7.564	10.085	16.338	24.769	27.587	30.191	33.409	35.718	17
18	6.265	8.231	10.865	17.338	25.989	28.869	31.526	34.805	37.156	18
19	6.844	8.907	11.651	18.338	27.204	30.144	32.852	36.191	38.582	19
20	7.434	9.591	12.443	19.337	28.412	31.410	34.170	37.566	39.997	20
21	8.034	10.283	13.240	20.337	29.615	32.670	35.479	38.932	41.401	21
22	8.643	10.982	14.042	21.337	30.813	33.924	36.781	40.289	42.796	22
23	9.260	11.688	14.848	22.337	32.007	35.172	38.076	41.638	44.181	23
24	9.886	12.401	15.659	23.337	33.196	36.415	39.364	42.980	45.558	24
25	10.520	13.120	16.473	24.337	34.382	37.652	40.646	44.314	46.928	25
26	11.160	13.844	17.292	25.336	35.563	38.885	41.923	45.642	48.290	26
27	11.808	14.573	18.114	26.336	36.741	40.113	43.194	46.963	49.645	27
28	12.461	15.308	18.939	27.336	37.916	41.337	44.461	48.278	50.993	28
29	13.121	16.047	19.768	28.336	39.088	42.557	45.722	49.588	52.336	29
30	13.787	16.791	20.599	29.336	40.256	43.773	46.979	50.892	53.672	30
31	14.458	17.539	21.434	30.336	41.422	44.985	48.232	52.192	55.003	31
32	15.135	18.291	22.271	31.336	42.585	46.194	49.481	53.486	56.329	32
33	15.816	19.047	23.110	32.336	43.745	47.400	50.725	54.776	57.649	33
34	16.502	19.806	23.952	33.336	44.903	48.602	51.966	56.061	58.964	34
35	17.192	20.570	24.797	34.336	46.059	49.802	53.203	57.342	60.275	35
36	17.887	21.336	25.643	35.336	47.212	50.998	54.437	58.619	61.582	36
37	18.586	22.106	26.492	36.335	48.363	52.192	55.668	59.893	62.884	37
38	19.289	22.879	27.343	37.335	49.513	53.384	56.896	61.162	64.182	38
39	19.996	23.654	28.196	38.335	50.660	54.572	58.120	62.428	65.476	39
40	20.707	24.433	29.051	39.335	51.805	55.758	59.342	63.691	66.766	40
41	21.421	25.215	29.907	40.335	52.949	56.942	60.561	64.950	68.053	41
42	22.139	25.999	30.765	41.335	54.090	58.124	61.777	66.206	69.336	42
43	22.860	26.786	31.625	42.335	55.230	59.304	62.990	67.460	70.616	43
44	23.584	27.575	32.487	43.335	56.369	60.481	64.202	68.710	71.893	44
45	24.311	28.366	33.350	44.335	57.505	61.656	65.410	69.957	73.166	45
46	25.042	29.160	34.215	45.335	58.641	62.830	66.617	71.202	74.437	46
47	25.775	29.956	35.081	46.335	59.774	64.001	67.821	72.443	75.704	47
48	26.511	30.755	35.949	47.335	60.907	65.171	69.023	73.683	76.969	48
49	27.250	31.555	36.818	48.335	62.038	66.339	70.222	74.920	78.231	49
50	27.991	32.357	37.689	49.335	63.167	67.505	71.420	76.154	79.490	50

SOURCE: Values from 1 to 30 degrees of freedom from C. M. Thompson, *Biometrika* 32(1941):188–189, with permission of the publisher. Values from 31 to 50 degrees of freedom from *Statistical Tables*, 2nd ed., by F. James Rohlf and Robert R. Sokal, W. H. Freeman and Company.

of linkage. But there are several other situations in which the χ^2 test is useful, and these all involve testing actual results against simple expectations of a hypothesis.

For example, the test is ideal (as we have seen) for testing deviations from any genetic ratio: 3:1, 9:3:3:1, 9:7, 1:1, and so on. But there are pitfalls here. Suppose, for example, you believe you have identified a major gene affecting a specific biological function. You testcross a presumed heterozygote *A a* to a homozygous recessive individual *a a*, expecting a 1:1 phenotypic ratio in the progeny. Out of 200 progeny, 116 are *Aa* and 84 are *aa*. Here $\chi^2 = (16^2 \div 100) + (16^2 \div 100) = 0.512$ with 1 d.f. The p value is 2.5%, so you reject the null hypothesis that there is a single gene pair segregating. Actually, however, you must recall that the 1:1 ratio is expected if there is a single pair of alleles *of equal viability* segregating. With the rejection of the null hypothesis, you must now reject *either* the notion of a single allele pair *or* that of their equal viability (or both). The χ^2 test cannot tell you which portion of a compound null hypothesis to reject, so care must be taken to determine all the hidden assumptions in a null hypothesis if error is to be avoided in the application of statistical testing.

Message
The χ^2 test is used to test experimental results against the expectations derived from a null hypothesis. The test generates a p *value that is the probability of obtaining a specific deviation at least as great as that observed, assuming that the null hypothesis is correct*

Early Thoughts on the Nature of Crossing Over

The production of recombinant chromosomes is an interesting phenomenon that cries out for explanation. How are they generated? At least three different hypotheses were initially put forward (around 1930) to account for this phenomenon.

1. Morgan's explanation was that crossing over involves a **physical exchange** of homologous chromosome parts to produce the crossover chromosome, through a mechanism on which he did not speculate.

2. Goldschmidt suggested that genes during interphase (when chromosomes disappear) become more or less loose elements like unstrung beads. He suggested that the beads are restrung in the same sequence during meiotic prophase but that the members of each pair of alleles can exchange places during this restringing.

3. H. Winkler suggested that some interaction between different alleles of the same gene in a heterozygous cell may (at a low frequency)

convert one allele into the form of the other. Thus in $a/+$, sometimes a would be converted into $+$, and other times $+$ would be converted into a. In a heterozygote for linked genes, $a\ b/+\ +$, this **gene conversion** (say, of b to $+$) would give a chromosome ($a\ +$) that would be recorded genetically as a recombinant.

Figure 5-14 diagrams these three hypotheses, using beads to represent genes, with each pair of alleles consisting of a black bead and a white bead.

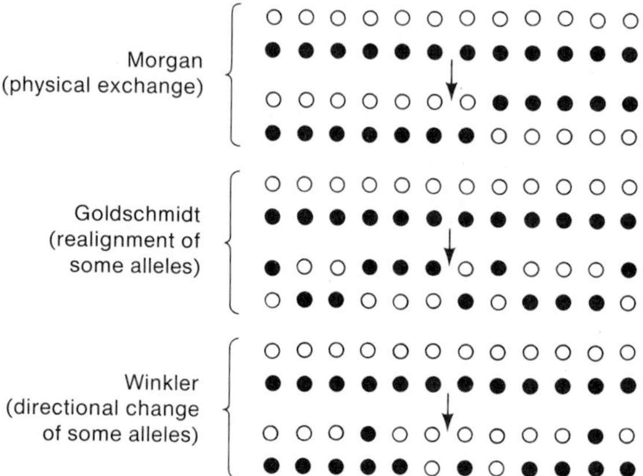

Figure 5-14. Diagrammatic representation of three different hypotheses about the general mechanism of intrachromosomal recombination.

Is there any way to test these hypotheses? In 1931, Stern performed an elegant experiment on *Drosophila*, and Harriet Creighton and Barbara McClintock did the same on corn, to determine the method of crossover production. Stern studied two X chromosomes that differed cytologically in appearance, one having an apparent discontinuity in it

and the other carrying an extra long piece on one side of the centromere

He knew (we'll see how later in the book) that two genes, *car* (carnation, or light eye color) and *B* (bar, or smaller eye), are linked near the centromere and have this appearance:

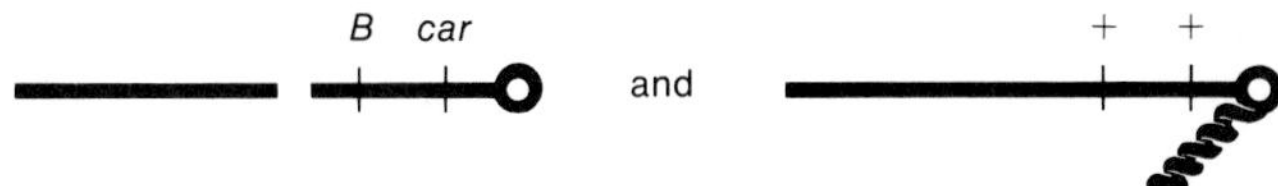

A female heterozygous for the two chromosome types and *B* and *car* would be

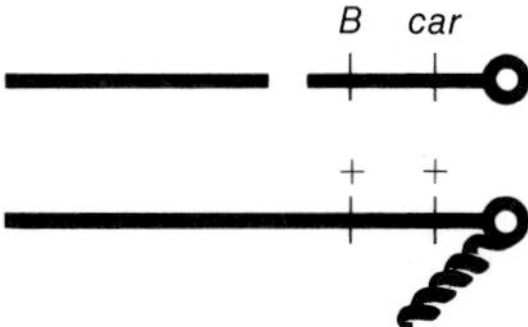

and, on testcrossing, Stern identified both parental and recombinant types with respect to *B* and *car*. He found that the parental types determined genetically (as *B car*, and + +) always had parental chromosome configurations. That is, *B car* flies had

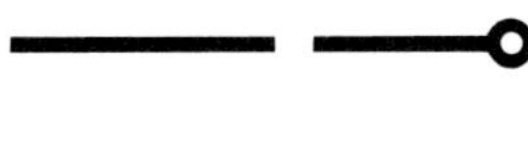

and + + flies carried

When he looked at the chromosomes of the recombinant progeny, he found new chromosome arrangements: *B* + flies had

and + *car* flies had

Thus, he showed a direct relationship between crossovers detected genetically and a physical rearrangement of the chromosome parts. This rules out both Goldschmidt's and Winkler's models, although gene conversion can rarely be detected using specialized systems.

It is worth repeating these observations by describing the work done by Creighton and McClintock. In order to correlate crossing over with chromosome exchange, it is important to have chromosomes that are cytologically different on both sides of the region in which crossing over occurs. Creighton and McClintock were studying chromosome 9 of corn, which has two loci: one affects seed color (*C*, colored, and *c*, colorless) and the other, plant composition (*Wx*, waxy, and *wx*, starchy). Furthermore, the chromosome carrying *C* and *Wx* was distinguished from its homolog by the presence of a large, densely staining element (called a knob) on the *C* end and by a longer piece of chromosome on the *Wx* end; thus, a heterozygote is

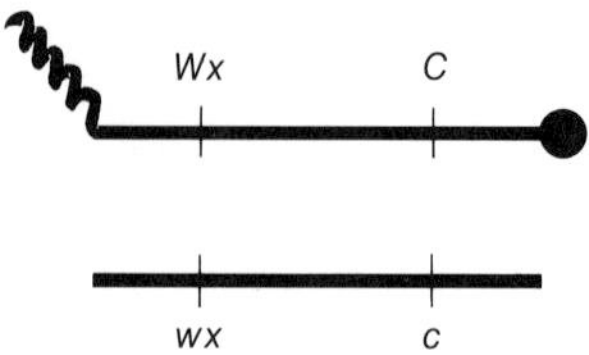

When they separated recombinants from nonrecombinants, they showed that the nonrecombinants retained the parental chromosome arrangements, whereas the recombinants were

and

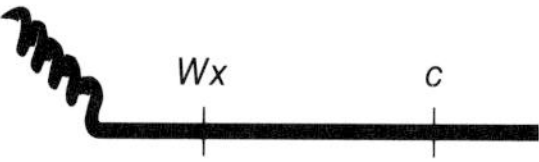

again confirming Morgan's suggestion that crossovers are a result of physical exchange consisting of an apparent breakage and a reunion of chromosome parts.

Message
Crossing over seems to be the result of a physical breakage and a reunion of chromosome parts.

How might the breakage and reunion be produced? Is it a "cut-and-paste" process like splicing sound tapes? Or is it a process that only *appears* to consist of a breakage and a reunion? An idea that favors the latter was proposed in 1928 by John Belling, who studied meiosis in plant chromosomes and observed bumps along the chromosome called **chromomeres,** which he thought may correspond to genes (Figure 5-15). Belling visualized the genes as beads strung

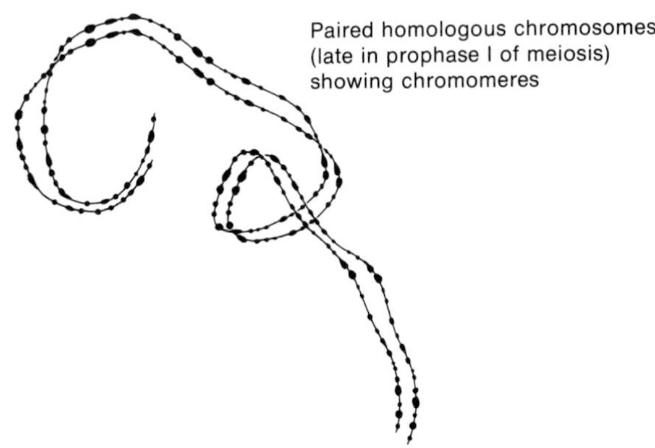

Figure 5-15. Diagrammatic representation of the chromomeres that are important in Belling's copy-choice model.

together with some nongenic linking substance. He reported that, during prophase of meiosis, chromomeres duplicate so that newly made chromomeres are stuck to the originals (Figure 5-16). After duplication, the newly formed chromomeres are fastened together, but, because all the chromomeres are

Figure 5-16. Division of the chromomeres according to Belling's model.

tightly juxtaposed, the linking elements could switch from a newly made chromomere on one homologous chromosome to an adjacent one on the other homolog. This model became known as the **copy-choice model** (or switch model) of crossing over. You can see how it can generate a crossover chromatid that would seem to have arisen from a physical breakage and the reunion of chromosomes (Figure 5-17). You can also see that it suggests that the event producing recombinant chromosomes can take place only between newly made chromomeres (and hence newly made chromatids), so that any multiple crossover could involve only two chromatids.

Figure 5-17. The hooking together of the newly synthesized chromomeres according to Belling's model. Joining usually forms parental combinations, but sometimes a switch can occur.

Thus, the copy-choice model predicts that, in every meiosis in which multiple crossovers occur (for example, double crossovers and even higher multiples), only two chromatids out of the four would ever be involved. The prediction could be tested if only we had some way of recovering all four products from individual meiotic divisions. Each group of four could then be examined for the presence of multiple recombinants. In any group of four that contains at least one multiple recombinant, there must be two accompanying parental types for the copy-choice model to be true. Luckily, several haploid organisms are suited to the recovery of all four products of a single meiosis. These organisms are some fungi and some unicellular algae. We discuss the full analysis of these groups of four (called *tetrad analysis*) at length in Chapter 6. But we need them right now to answer one specific question:

What companion genotypes are found in tetrads containing multiple recombinants? Several types of such tetrads are possible, but one interesting situation (often encountered in tetrad analysis) is illustrated in the following diagram, in which the genes are linked in the order shown:

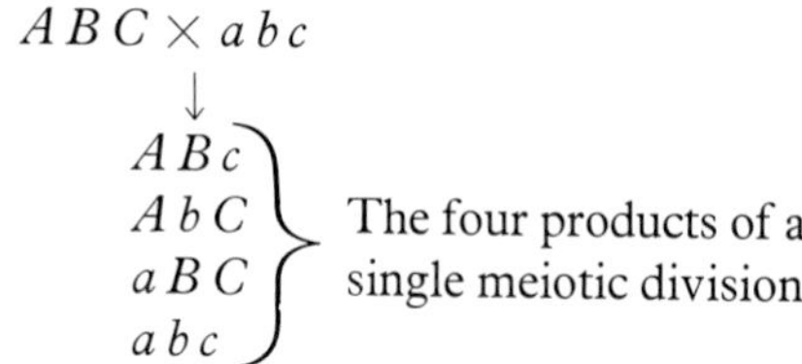

Note that a recombination event has occurred between the first two loci (*A*/*a* and *B*/*b*) *and* between the second two loci (*B*/*b* and *C*/*c*) in the same meiosis; a double-exchange event has occurred (and one of the two products, *A b C*, is a double recombinant). *But* more than two chromatids must have been involved; in fact, three *must* have been involved in this case (Figure 5-18). Thus, Belling's suggestion (which would predict that only two chromatids could ever be involved) is most unlikely. (We shall see some *positive* evidence for the model of breakage and reunion in Chapter 16.)

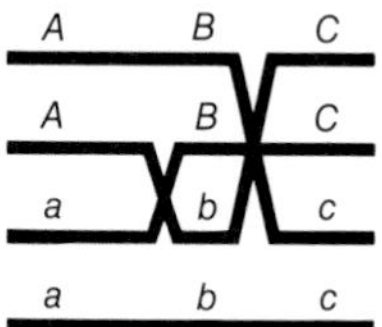

Figure 5-18. One of the several possible types of double-crossover tetrads that are regularly observed, showing how more than two chromatids must be involved. This evidence makes Belling's copy-choice (switch) model very unlikely.

Message
The "cut-and-paste" method of chromosome crossing over wins by default.

The ability to isolate the four products of meiosis in fungi and algae also clears up another mystery, about whether crossing over occurs at the two-strand (two-chromosome) stage or at the four-strand (four-chromatid) stage. If it occurs at the two-strand stage, there can never be more than two different products of a given meiosis. If it occurs at the four-strand stage, up to four different products of meiosis are possible (Figure 5-19). In fact, four different products of a single meiosis are regularly observed, showing that crossing over occurs at the four-strand stage of meiosis.

Crossing over is a remarkably precise process. The synapsis and exchange of chromosomes is such that no segments are lost or gained, and four complete chromosomes emerge in a tetrad. At present, however, the actual mechanisms of crossing over and interference are still an enigma, although several attractive models do exist, and a lot is known about the sorts of chemical reactions that might ensure precision at the molecular level. We shall return to these points in Chapter 16.

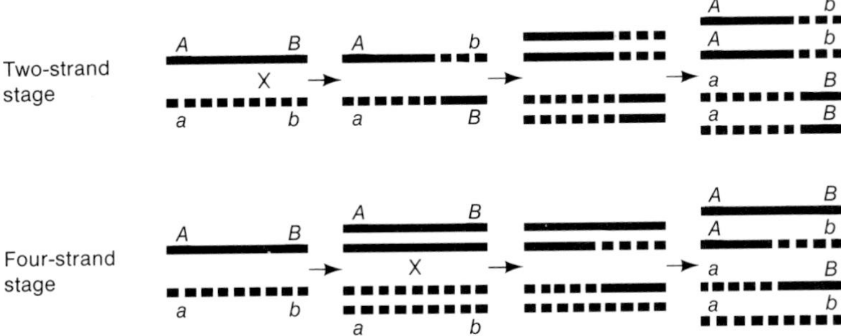

Figure 5-19. Tetrad analysis provides evidence to decide whether crossing over occurs at the two-strand (two-chromosome) or at the four-strand (four-chromatid) stage of meiosis. Because more than two different products of a single meiosis can be seen in some tetrads, crossing over cannot occur at the two-strand stage.

Linkage Mapping by Recombination in Humans

Humans, partly because of their relatively large chromosome number and partly because of the lack of suitable pedigrees containing what amount to testcrosses, have revealed very few examples of autosomal linkage through recombination analysis. The X chromosome, however, has been far more amenable to analysis because of hemizygosity in males. Consider the following situation involving the rare recessive genes for defective sugar processing (*g*) and for color blindness (*c*). A doubly affected male, $c\,g/\mathrm{Y}$, marries a normal woman (who is almost certainly $C\,G/C\,G$). The daughters of this mating are coupling-conformation heterozygotes. When they marry, they will almost certainly marry normal men $C\,G/\mathrm{Y}$, and their male children provide an opportunity for the study of the frequency of recombinants (Figure 5-20). Using such pedigrees, the total frequency of recombinants may be estimated. A map based on such techniques is shown in Figure 5-21. Further data on mapping in humans have been provided by less conventional techniques, as we shall see in Chapter 6.

Linkage studies occupy a large proportion of the routine day-to-day activities of geneticists. When a new variant gene is discovered, one of the first questions to be asked concerning it is "Where does it map"? Not only is this knowledge a necessary component in the engineering and maintenance of genetic stocks for research use, but it is also of fundamental importance in the piecing together of an overall view of the architecture of the chromosome—and in fact of the entire genome. The operational key for the detection of linkage is the same in studies from viruses to humans, and that key is the nonindependence of genes in their transmission from generation to generation. Look for this key in discussions of linkage in following chapters.

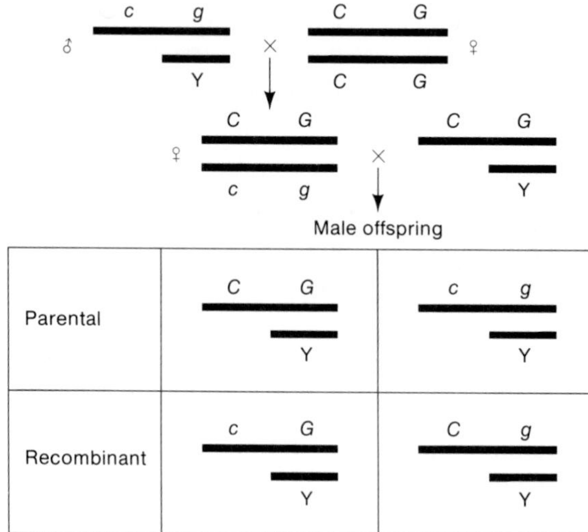

Figure 5-20. The male children of women heterozygous for two X-linked genes can be used to calculate recombinant frequency. Thus, mapping of the X chromosome is possible by studying certain selected pedigrees.

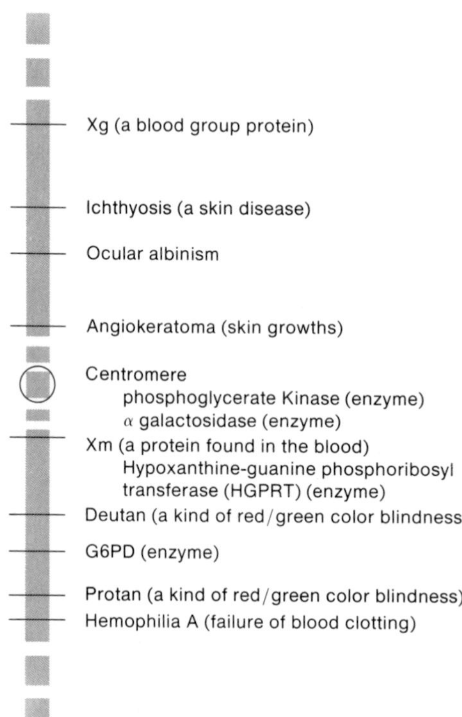

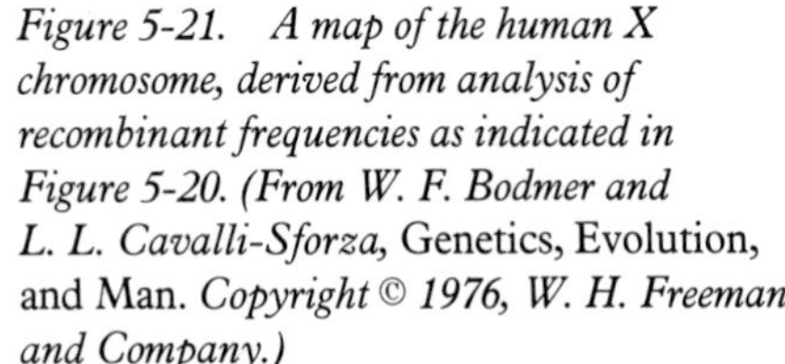

Figure 5-21. A map of the human X chromosome, derived from analysis of recombinant frequencies as indicated in Figure 5-20. (From W. F. Bodmer and L. L. Cavalli-Sforza, Genetics, Evolution, and Man. *Copyright © 1976, W. H. Freeman and Company.)*

Summary

In dihybrid crosses of sweet pea plants, Bateson and Punnett discovered deviations from the 9:3:3:1 ratio of phenotypes expected in the F_2 generation. The parental gametic types occurred with much greater frequency than the other two classes. Later, in his studies of two different autosomal gene pairs in *Drosophila,* Morgan found a similar deviation from Mendel's law of independent assortment. He postulated that the two gene pairs were located on the same pair of homologous chromosomes. This phenomenon is called linkage.

Linkage explains why the parental gene combinations stay together but not how the nonparental combinations arise. Morgan postulated that during meiosis there may be a physical exchange of chromosome parts by a process called crossing over, or meiotic recombination. Thus there are two types of meiotic recombination. Interchromosomal recombination is achieved by Mendelian independent assortment and results in a recombination frequency of 50%. Intrachromosomal recombination occurs when the physical linkage of the parental gene combinations prevents full assortment of the gene pairs, and the recombinant frequency is less than 50%.

As Morgan studied more linked genes, he discovered considerable variation in crossover frequency and wondered if this variation reflected the actual distance between genes. Sturtevant, a student of Morgan's, developed a method of determining the distance between gene pairs on a genetic map based on the percentage of recombinants. A genetic map is another example of a hypothetical entity based on genetic analysis.

Although the basic test for linkage is deviation from the 1:1:1:1 ratio of progeny types in a testcross, deviation may not be all that obvious. The χ^2 test, which tells how often deviations from expectations will occur purely by chance, can help determine whether linkage exists or not. The χ^2 test has other applications in genetics, in the testing of observed against expected events.

Several theories about how recombinant chromosomes are generated have been set forth. We now know that crossing over is the result of physical breakage and a reunion of chromosomal parts and that it occurs at the four-strand stage of meiosos. The actual mechanism of crossing over, however, remains largely a mystery.

Problems

1. A geneticist studying bloops (an exotic organism found only in textbook problems) has been unable to find any examples of linkage. Suggest some explanations for this situation.

2. A strain of *Neurospora* with the genotype *H I* is crossed with a strain with the genotype *h i*. One-half of the progeny are *H I*, and one-half are *h i*. Explain how this is possible.

3. The cross *RR SS* × *rr ss* is made, and the F_1 is backcrossed to the double-recessive parent. If the gene pairs involved are closely linked, which two genotypes will be in the minority in the progeny? (a) *Rr Ss* and *rr ss*; (b) *Rr Ss* and *Rr ss*; (c) *Rr ss* and *rr Ss*; (d) *rr Ss* and *rr ss*; (e) *rr Ss* and *Rr Ss*.

4. The genes for yellow (*Y*) and green (*y*) peas are linked to the genes for red (*R*) and white (*r*) pea flowers. A pure-breeding yellow-pea, red-flower plant is crossed to a green-pea, white-flower plant, and the F_1 is backcrossed to the double-recessive parent. In the progeny of the backcross, which genotype will be equal in frequency to the *Yy Rr* genotype? (a) *Yy rr*; (b) *yy rr*; (c) *yy RR*; (d) *YY RR*; (e) *yy Rr*.

5. A female animal with genotype *Aa Bb* is crossed with a double-recessive male (*aa bb*). Their progeny includes 442 *Aa Bb*, 458 *aa bb*, 46 *Aa bb*, and 54 *aa Bb*. Explain these results.

6. In a haploid organism, the loci *C/c* and *D/d* are linked 8 m.u. apart. In a cross *C d* × *c D*, what proportion of the progeny will be (a) *C D*? (b) *c d*? (c) *C d*? (d) recombinants?

7. A fruitfly of genotype *B R/b r* is testcrossed to *b r/b r*. In 84% of the meioses, no chiasmata occur between the linked gene pairs; in 16% of the meioses, one chiasma occurs between them. What proportion of the progeny will be *Bb rr*? (a) 50%; (b) 4%; (c) 84%; (d) 25%; (e) 16%.

8. There is an autosomal gene *N* in humans that causes abnormalities in nails and patellae (kneecaps), called the nail–patella syndrome. In marriages of people with the phenotypes

   nail–patella syndrome, blood type A × normal nail–patella, blood type O

   some children are born with the nail–patella syndrome and blood type A. When marriages between such children (unrelated, of course) take place, their children are of the following types:

   66% nail–patella syndrome, blood type A
   16% normal nail–patella, blood type O
   9% normal nail–patella, blood type A
   9% nail–patella syndrome, blood type O

   Fully analyze these data.

9. You obtain two lines of *Drosophila*, one having light yellow eyes and the other having bright scarlet eyes. (Remember that wild-type *Drosophila* have deep red eyes.) When you cross a yellow female with a scarlet male, you obtain 251 wild-type females and 248 yellow males in the F_1 generation. When you cross F_1 males and females, you obtain the following F_2 phenotypes; 260 wild-type females, 253 yellow females, 77 wild-type males, 179 yellow males, 183 scarlet males, and 80 brown-eyed males (brown is a new phenotype). Explain these results, using diagrams where possible.

10. In the ovaries of higher plants, a haploid nucleus resulting from meiosis undergoes several mitotic divisions without cell division, forming a multinucleate cell called the female gametophyte. (The male nucleus fuses with just one of these nuclei, which acts as the egg nucleus.) In some conifers, the female gametophyte can be quite large—in fact, large enough to be removed and analyzed electrophoretically. One tree of lodgepole pine (*Pinus contorta*) is heterozygous for fast and slow "electrophoretic alleles" of the enzymes alcohol dehydrogenase (ADH^F/ADH^S) and phosphoglucomutase (PGM^F/PGM^S). A sample of 237 female gametophytes is studied, and the following phenotypes are found: 21 $ADH^F\ PGM^F$, 19 $ADH^S\ PGM^S$, 95 $ADH^F\ PGM^S$, and 102 $ADH^S\ PGM^F$.

    a. Explain these results.

    b. Can you think of any other uses in genetics for this female-gametophyte system?

11. Using the data obtained by Bateson and Punnett (Table 5-1), calculate the map distance (in m.u.) separating the color and shape genes. (This will require some trial and error.)

12. You have a homozygous *Drosophila* line carrying the autosomal recessive genes *a*, *b*, and *c* linked in that order. You cross females of this line with males of a homozygous wild-type line. You then cross the F_1 heterozygous males with their heterozygous sisters, and you obtain the following F_2 phenotypes: 1364 + + +, 365 *a b c*, 87 *a b* +, 84 + + *c*, 47 *a* + +, 44 + *b c*, 5 *a* + *c*, and 4 + *b* +.

    a. What is the recombinant frequency between *a* and *b*? between *b* and *c*?

    b. What is the coefficient of coincidence?

13. R. A. Emerson crossed two different pure-breeding parental lines of corn and obtained an F_1 that was heterozygous for three recessive genes: *an* (anther), *br* (brachytic), and *f* (fine). He testcrossed the F_1 to a completely homozygous recessive tester and obtained the following progeny phenotypes: 355 anther, 339 brachytic and fine, 88 completely wild type, 55 anther and brachytic and fine, 21 fine, 17 anther and brachytic, 2 brachytic, and 2 anther and fine.

    a. What were the genotypes of the parental lines?

    b. Draw a linkage map to illustrate the linkage arrangement of the three genes (include map distances).

    c. Calculate the interference value.

14. In corn, the following allelic pairs have been identified in chromosome 3: +/*b* (plant-color booster versus nonbooster), +/*lg* (liguled versus liguleless), and +/*v* (green plant versus virescent). A testcross involving triple recessives and F_1 plants heterozygous for the three gene pairs yields the following progeny phenotypes: 305 + *v lg*, 275 *b* + +, 128 *b* + *lg*, 112 + *v* +, 74 + + *lg*, 66 *b v* +, 22 + + +, and 18 *b v lg*. Give the gene sequence on the chromosome, the map distances between genes, and the coefficient of coincidence.

15. In *Drosophila*, Dichaete is a third-chromosome allele with a dominant effect on wing shape. Pink and ebony are third-chromosome recessive mutations that affect eye color and body color, respectively. Dichaete male flies are crossed to pink, ebony flies. The F_1 females with the Dichaete phenotype are then crossed to pink, ebony homozygotes. The progeny of this cross show the following phenotypes:

800 pink, ebony; 761 Dichaete; 184 pink; 175 Dichaete, ebony; 31 wild type; 28 Dichaete, ebony, pink; 11 Dichaete, pink; and 10 ebony.

a. What is the map distance between each pair of these genes?

b. Draw a map of the third chromosome showing these three loci with map distances.

c. Is there any interference? If so, how much?

d. If the coefficient of coincidence is 0.66, how many ebony flies would you expect among 10,000 F_2 progeny from the cross described?

16. Groodies are useful (but fictional) haploid organisms that are pure genetic tools. A wild-type groody has a fat body, a long tail, and flagella. Nonwild-type lines are known that have thin bodies, or are tailless, or do not have flagella. Groodies can mate with each other (although they are so shy that we do not know how) and produce recombinants. A wild-type groody is crossed with a thin-bodied groody lacking both tail and flagella. The 1000 baby groodies resulting are classified as shown in Figure 5-22. Assign genotypes, and map the three genes.

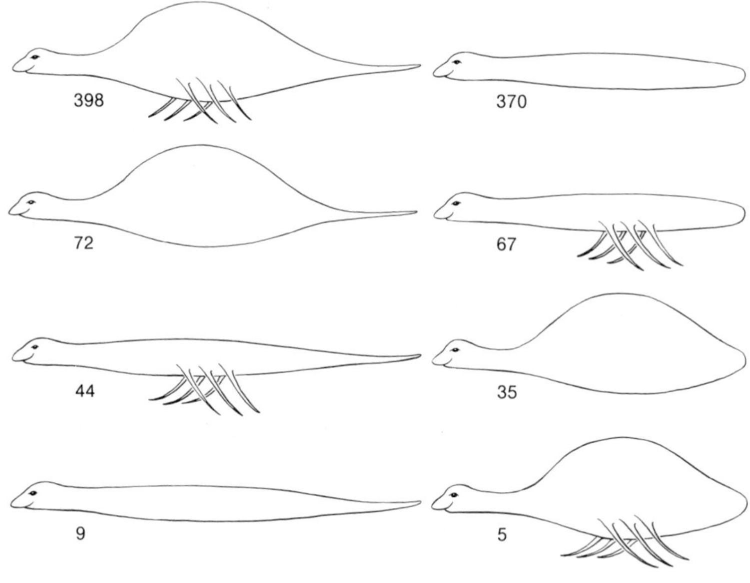

Figure 5-22.

(Problem 16 courtesy of Burton S. Guttman.)

17. Assume that three pairs of alleles are found in *Drosophila:* $+/x$, $+/y$, and $+/z$. As shown by the symbols, each nonwild allele is recessive to its wild-type allele. A cross between females heterozygous at these three loci and wild-type males yields the following progeny phenotypes: 1010 $+ + +$ females; 430 $x + z$ males; 441 $+ y +$ males; 39 $x y z$ males; 32 $+ + z$ males; 30 $+ + +$ males; 27 $x y +$ males; 1 $+ y z$ male; and 0 $x + +$ males.

a. How were members of the allelic pairs distributed in the members of the appropriate chromosome pair of the heterozygous female parents?

b. What is the sequence of these linked genes in their chromosome?

c. Calculate the map distances between the genes, and the coefficient of coincidence.

d. In what chromosome of *Drosophila* are these genes carried?

18. You have two homozygous lines of *Drosophila.* Line 1 has bright scarlet eyes and a wild-type thorax. Line 2 has dark brown eyes and a humpy thorax. When you cross virgin females of line 2 with males of line 1, you obtain 232 wild-type males and 225 wild-type females in the F_1 generation. You then cross F_1 males with virgin F_1 females and obtain the following F_2 phenotypes: 283 completely wild-type females; 145 completely wild-type males; 139 wild-thorax, scarlet-eyed males; 78 humpy-thorax, brown-eyed females; 40 humpy-thorax, white-eyed males; 39 humpy-thorax, brown-eyed males; 20 wild-thorax, brown-eyed females; 19 humpy-thorax, wild-eyed females; 11 humpy-thorax, scarlet-eyed males; 10 wild-thorax, white-eyed males; 9 wild-thorax, brown-eyed males; and 8 humpy-thorax, wild-eyed males. Explain these results as fully as possible (using symbols wherever possible).

19. The mother of a family with ten children has blood type Rh^+ and has a very rare condition (elliptocytosis, E) with no adverse clinical effect, in which the red cells are oval rather than round. The father is Rh^- (lacks the Rh^+ antigen) and has normal red cells (symbol: e). The children are 1 Rh^+ e, 4 Rh^+ E, and 5 Rh^- e. Information is available on the mother's parents, who are Rh^+ E and Rh^- e. One of the ten children (who is Rh^+ E) marries someone who is Rh^+ e, and they have an Rh^+ E child.

a. Draw the pedigree of this whole family.

b. Is the pedigree in agreement with the hypothesis that Rh^+ is dominant and Rh^- is recessive?

c. What is the mechanism of transmission of elliptocytosis?

d. Could the genes for E and Rh be on the same chromosome? If so, estimate the map distance between them, and comment on your result.

20. The father of Mr. Spock, second officer of the Starship Enterprise, came from the planet Vulcan; his mother came from Earth. A Vulcanian has pointed ears (P), adrenals absent (A), and a right-sided heart (R). All of these alleles are dominant over normal Earth alleles. These genes are autosomal, and they are linked as shown in this linkage map:

P/p ——+—— A/a ——+—— R/r

If Mr. Spock marries an Earth woman and there is no (genetic) interference, what porportion of their children

a. will show Vulcanian appearance for all three characters?

b. will show Earth appearance for all three characters?

c. will have Vulcanian ears and heart but Earth adrenals?

d. will have Vulcanian ears but Earth heart and adrenals?

(Problem 20 is from D. Harrison, *Problems in Genetics,* Addison-Wesley, 1970.)

21. In a certain diploid plant, the three loci A/a, B/b, and C/c are linked as follows:

A/a —— 20 m.u. —— B/b —— 30 m.u. —— C/c

One plant is available to you (call it the parental plant). It has the constitution $A\,b\,c/a\,B\,C$.

a. Assuming no interference, if the plant is selfed, what proportion of the progeny will be of the genotype $a\,b\,c/a\,b\,c$?

b. Again assuming no interference, if the parental plant is crossed with the $a\,b\,c/a\,b\,c$ plant, what genotypic classes will be found in the progeny? What will be their frequencies if there are 1000 progeny?

c. Repeat part b, this time assuming 20% interference between the regions.

22. From several crosses of the general type *AA BB* × *aa bb*, the F_1 individuals of type *Aa Bb* were testcrossed to *aa bb*. The results are shown in Table 5-5. In each case, use the χ^2 test to decide if there is evidence of linkage.

Table 5-5.

Testcross of F_1 from cross	Number of individuals of genotype			
	Aa Bb	*aa bb*	*Aa bb*	*aa Bb*
1	310	315	287	288
2	36	38	23	23
3	360	380	230	230
4	74	72	50	44

23. Certain varieties of flax show different resistances to specific races of the fungus called flax rust. For example, the flax variety 77OB is resistant to rust race 24 but susceptible to rust race 22, whereas flax variety Bombay is resistant to rust race 22 and susceptible to rust race 24. When 77OB and Bombay were crossed, the F_1 hybrid was resistant to both rust races. When selfed, it produced an F_2 containing the phenotypic proportions shown in Table 5-6.

Table 5-6.

		Rust race 22	
		Resistant	Susceptible
Rust race 24	Resistant	184	63
	Susceptible	58	15

a. Propose a hypothesis to account for the genetic basis of resistance in flax for these particular rust races. Make a concise statement of the hypothesis, and define any gene symbols you use. Show your proposed genotypes of the 77OB, Bombay, F_1, and F_2 flax plants.

b. Test your hypothesis, using the χ^2 test. Give the expected values, the value of χ^2 (to two decimal places), and the appropriate probability value. Explain exactly what this value is the probability of. Do you accept or reject your hypothesis on the basis of the χ^2 test?

(Problem 23 is adapted from M. Strickberger, *Genetics,* Macmillan, 1968.)

Asci of Neurospora. *Top: normally maturing asci. Bottom: asci from a cross of wild type by a mutant affecting ascospore color (note different segregation patterns). (Namboori B. Raju,* European Journal of Cell Biology, *1980.)*

6

Advanced Transmission Genetics

The analytical structure we have developed in preceding chapters is called **transmission genetics.** Transmission genetics is the set of rules that governs the transmission, or handing on, of the genetic material from one generation to the next. (Transmission genetics is normally not concerned with the chemical nature of the genetic material, a broad area called **molecular genetics** that is discussed later in the book.)

The present chapter, with its rather awesome title, introduces a number of topics concerned with transmission genetics that are more sophisticated treatments of (or in some cases more recent additions to) the basic theory already developed. These subjects are mapping functions (an accurate treatment of mapping by RF), tetrad analysis, and mitotic genetics.

Mapping Functions

In Chapter 5, we defined a map unit (m.u.) as a recombinant frequency (RF) of 1%. We have seen that this definition leads to "reasonable" estimates of map distances. However, when *larger* locus-to-locus intervals are being examined, the estimate becomes very imprecise, as in the following example:

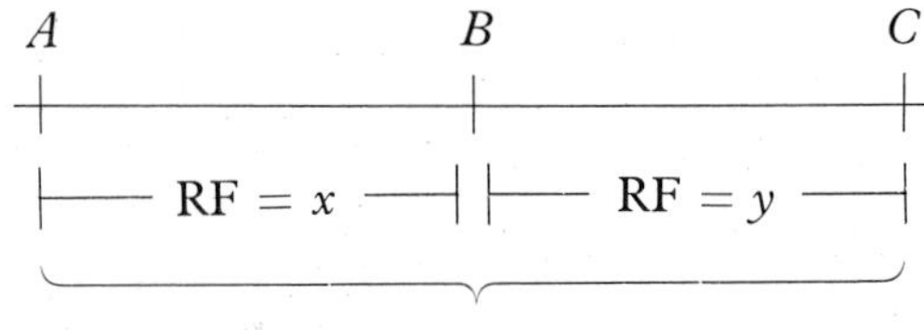

We have seen that we can improve the situation by including the double-recombinant types (twice). However, the point is that map distance (or the "true" distance) between loci is not linearly related to recombinant frequency throughout the range of values possible in mapping experiments. We have the clue provided by double crossovers, but now we must come to grips with the notion that double crossovers are part of a much larger problem of multiple crossovers and their effect on recombinant frequency. It might be pertinent to ask, for example, whether the RF value of x for the A-to-B interval in the example is an accurate reflection of the true distance between the A and B loci. Perhaps some double crossovers occur in this interval alone—we cannot detect them because we have no markers between A and B, but we should account for them if RF is to reflect physical distance at all well. This section describes a mathematical treatment of the problem that does give a way of correcting for multiple exchanges without actually detecting them at the phenotypic level.

Any relationship between one variable entity and another is called a function: the relationship between real map distance and recombination frequency is called the **mapping function.**

Message
The relationship between real map distance and RF is not linear. The accurate relationship is called the mapping function.

To calculate the mapping function, we need a mathematical tool that is widely used in genetic analysis because it describes many genetic phenomena well. It is called the **Poisson distribution.** A distribution is merely a device that describes the frequencies of various classes of samples. The Poisson distribution describes the frequency of classes of samples containing 0, 1, 2, 3, 4, . . . , n events when the average frequency (occurrence) of the event is small in relation to the total number of times that the event could occur. For example, the number of tadpoles one *could* get in a single dip of a net in a pond is quite large, but most dips yield only one or two or none. The number of dead birds on the side of the highway is potentially very large, but in a sample kilometer the number is usually small. Such samplings are described well by the Poisson distribution.

Let's consider a numerical example. We'll randomly distribute 100 one-dollar bills to 100 students in a lecture room, perhaps by scattering them over the class from some vantage point near the ceiling. The average (or mean) number of bills per student is 1.0, but common sense tells us that it is very unlikely that each of the 100 students will capture one bill. We expect a few lucky students to grab three or four bills each, and quite a few should come up with two bills each. However, we would expect most students to get either one bill or none. The Poisson distribution provides a quantitative prediction of the results.

In this example, the event being considered is capture of one bill by a student. We want to divide the students into classes according to the number of events occurring (number of bills captured per student) and find the frequency of each class. Let m represent the mean number of events (here $m = 1.0$ bill per student). Let i represent the number of events for a particular class (say, $i = 3$ for those students who get three bills each). Let $f(i)$ represent the frequency of the class with i events—that is, the proportion of the 100 students who each capture i bills. The general expression for the Poisson distribution states that

$$f(i) = \frac{e^{-m}m^i}{i!}$$

where e is the base of natural logarithms ($e \cong 2.7$), and ! is the factorial symbol (for example, $3! = 3 \times 2 \times 1 = 6$, and $4! = 4 \times 3 \times 2 \times 1 = 24$; by definition, $0! = 1$). When computing $f(0)$, recall that any number raised to the power of zero is defined as one. Table 6-1 gives values of e^{-m} for m values from 0.000 to 1.000.

In our example, $m = 1.0$. Using Table 6-1, let's compute the frequencies of the classes of students who each capture 0, 1, 2, 3, and 4 bills.

$$f(0) = \frac{e^{-1}\,1^0}{0!} = \frac{e^{-1}}{1} = 0.368$$

$$f(1) = \frac{e^{-1}\,1^1}{1!} = \frac{e^{-1}}{1} = 0.368$$

$$f(2) = \frac{e^{-1}\,1^2}{2!} = \frac{e^{-1}}{2 \times 1} = \frac{e^{-1}}{2} = 0.184$$

$$f(3) = \frac{e^{-1}\,1^3}{3!} = \frac{e^{-1}}{3 \times 2 \times 1} = \frac{e^{-1}}{6} = 0.061$$

$$f(4) = \frac{e^{-1}\,1^4}{4!} = \frac{e^{-1}}{4 \times 3 \times 2 \times 1} = \frac{e^{-1}}{24} = 0.015$$

Figure 6-1 shows a histogram of this distribution. We predict that about 37 students will capture no bills, about 37 will capture one bill, about 18 will capture two bills, about 6 will capture three bills, and about 2 will capture four bills. This accounts for all 100 of the students; in fact, you can verify that the Poisson distribution yields $f(5) = 0.003$, indicating the probability that no student in this sample will capture five bills.

Table 6-1. Values of e^{-m} for m values of 0 to 1

m	e^{-m}	m	e^{-m}	m	e^{-m}	m	e^{-m}
0.000	1.00000	0.250	0.77880	0.500	0.60653	0.750	0.47237
0.010	0.99005	0.260	0.77105	0.510	0.60050	0.760	0.46767
0.020	0.98020	0.270	0.76338	0.520	0.59452	0.770	0.46301
0.030	0.97045	0.280	0.75578	0.530	0.58860	0.780	0.45841
0.040	0.96079	0.290	0.74826	0.540	0.58275	0.790	0.45384
0.050	0.95123	0.300	0.74082	0.550	0.57695	0.800	0.44933
0.060	0.94176	0.310	0.73345	0.560	0.57121	0.810	0.44486
0.070	0.93239	0.320	0.72615	0.570	0.56553	0.820	0.44043
0.080	0.92312	0.330	0.71892	0.580	0.55990	0.830	0.43605
0.090	0.91393	0.340	0.71177	0.590	0.55433	0.840	0.43171
0.100	0.90484	0.350	0.70469	0.600	0.54881	0.850	0.42741
0.110	0.89583	0.360	0.69768	0.610	0.54335	0.860	0.42316
0.120	0.88692	0.370	0.69073	0.620	0.53794	0.870	0.41895
0.130	0.87810	0.380	0.68386	0.630	0.53259	0.880	0.41478
0.140	0.86936	0.390	0.67706	0.640	0.52729	0.890	0.41066
0.150	0.86071	0.400	0.67032	0.650	0.52205	0.900	0.40657
0.160	0.85214	0.410	0.66365	0.660	0.51685	0.910	0.40252
0.170	0.84366	0.420	0.65705	0.670	0.51171	0.920	0.39852
0.180	0.83527	0.430	0.65051	0.680	0.50662	0.930	0.39455
0.190	0.82696	0.440	0.64404	0.690	0.50158	0.940	0.39063
0.200	0.81873	0.450	0.63763	0.700	0.49659	0.950	0.38674
0.210	0.81058	0.460	0.63128	0.710	0.49164	0.960	0.38289
0.220	0.80252	0.470	0.62500	0.720	0.48675	0.970	0.37908
0.230	0.79453	0.480	0.61878	0.730	0.48191	0.980	0.37531
0.240	0.78663	0.490	0.61263	0.740	0.47711	0.990	0.37158
						1.000	0.36788

SOURCE: F. James Rohlf and Robert R. Sokal, *Statistical Tables*, 2nd ed. (San Francisco: W. H. Freeman and Company). Copyright © 1981 by W. H. Freeman and Company.

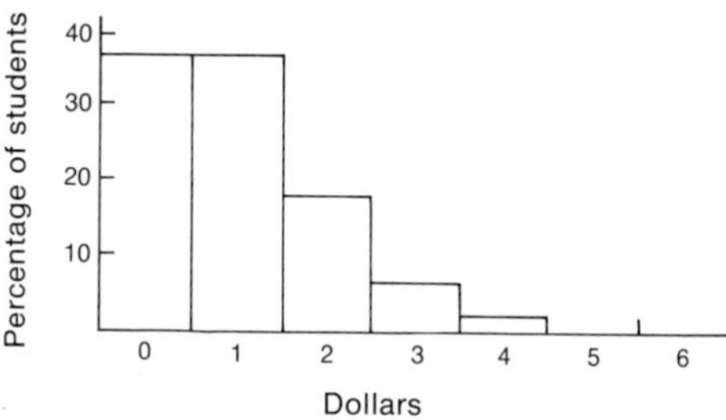

Figure 6-1. Poisson distribution for a mean of 1.0, illustrated in terms of a random distribution of dollar bills to students.

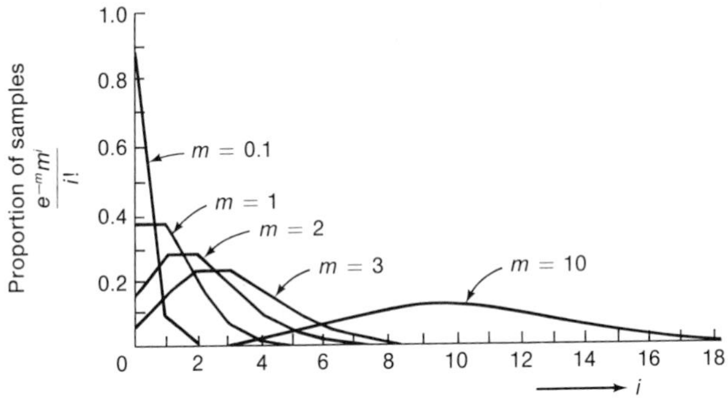

Figure 6-2. Poisson distributions for five different mean values: m *is the mean number of events per sample, and* i *is the actual number of events per sample. (From R. R. Sokal and F. J. Rohlf,* Introduction to Biostatistics. *Copyright © 1973, W. H. Freeman and Company.)*

Similar distributions may be developed for other m values. Some are shown in Figure 6-2, which uses curves instead of bar histograms.

The occurrence of crossovers along a chromosome during meiosis also can be described by the Poisson distribution. In any given genetic region, the actual number of crossovers occurring is probably small in relation to the total number of opportunities for such a crossover in that stretch. If we knew the *mean* number of crossovers in the region per meiosis, we could calculate the distribution of meioses with zero, one, two, three, four, and more multiple crossovers. This is unnecessary in the present context because, as we shall see, the only class we are really interested in is the zero class. The reason for this is that we want to correlate real distances with observable RF values, and it turns out that meioses in which there are one, two, three, four, or *any* finite number of crossovers per meiosis all behave similarly in that they produce an RF of 50% *among the products of those meioses,* whereas the meiosis with no crossovers produce an RF of 0%. Consequently, the determining force in actual RF values is the ratio of class zero to the rest!

The truth of these statements can be illustrated by considering meioses in which zero, one, and two crossovers occur between nonsister chromatids. (Try the three-crossover class yourself.) Figure 6-3 tells us that the only way we can get a recombinant product of meiosis is from a meiosis with at least one crossover in the marked region, and then *always* precisely one-half of the products of such meioses will be recombinant. (Note that in Figure 6-3 we consider only crossovers that occur between nonsister chromatids. **Sister-chromatid exchange** is thought to be very rare at meiosis. If it does occur, it can be shown to have no net effect in most meiotic analyses; see Problem 22 for this chapter.)

At last we can derive the map function. Recombinants will make up one-half of the products of those meioses in which at least one crossover occurs in the

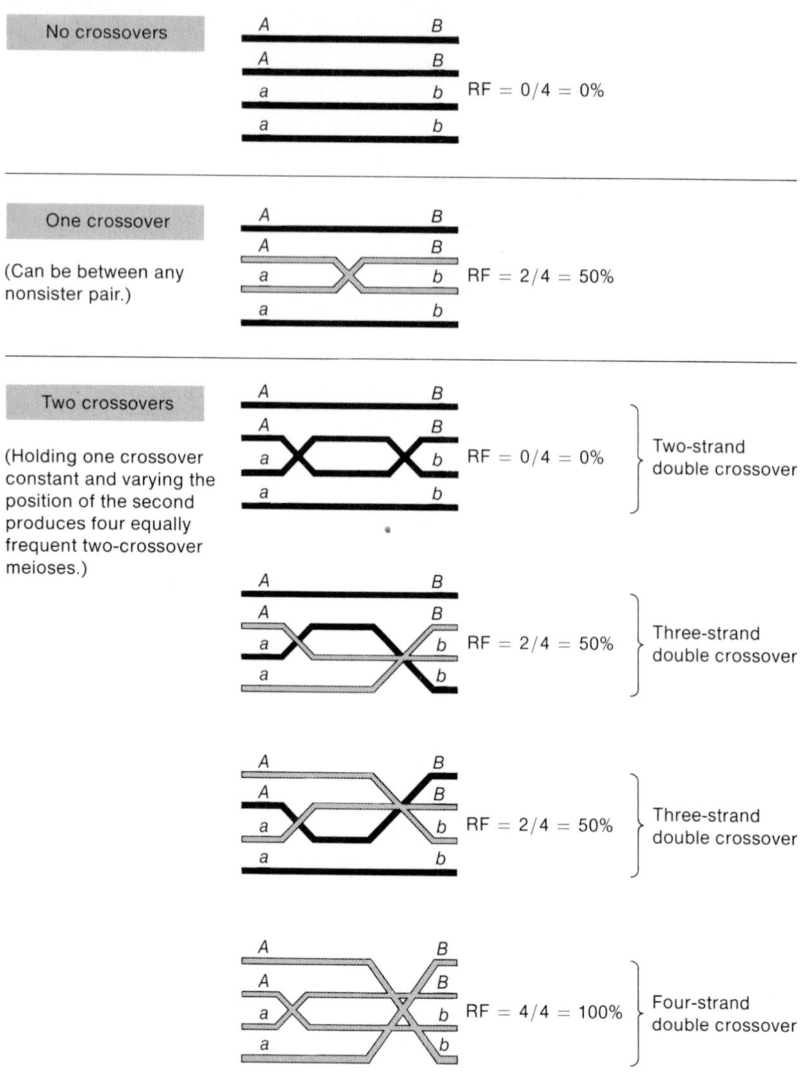

Figure 6-3. Demonstration that the overall RF is 50% for meioses in which any nonzero number of crossovers occurs. The lighter shade represents a recombinant chromatid.

region of interest. The proportion of meioses with at least one crossover is one minus the fraction with zero crossovers. Hence RF $= \frac{1}{2}(1 - e^{-m})$, because the zero-class frequency is $e^{-m}m^0/0! = e^{-m}$. When you think about it, m is the best measure we can have of *true* genetic distance—the *actual* average number of crossovers in a region.

If we know an RF value, we can calculate m by solving the equation. When we plot the function as a graph (Figure 6-4), several interesting points emerge.

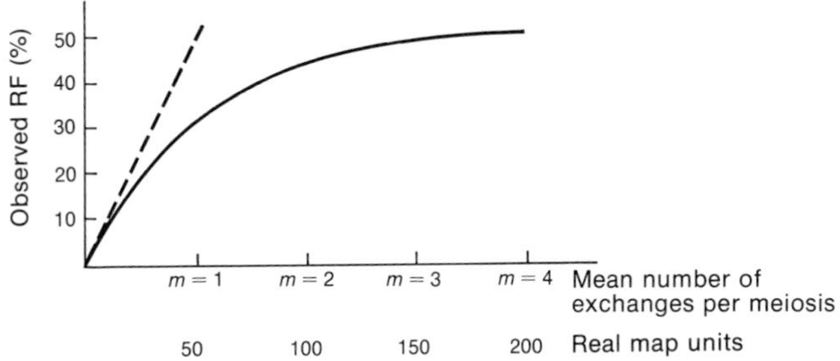

Figure 6-4. The mapping function in graphic form (solid line). *The dashed line represents the linear relationship found for small values of* m.

1. No matter how far apart two loci are on a chromosome, we never observe an RF value of greater than 50%. Consequently, an RF value of 50% would leave us in doubt about whether two loci are linked or are on separate chromosomes. Put another way, as m gets larger, e^{-m} gets smaller, and RF approaches $\frac{1}{2}(1 - 0) = \frac{1}{2} \times 1 = 50\%$. This may be surprising to you: RF values of 100% can never be observed, no matter how far apart the loci are!
2. The function is linear for a certain range corresponding to very small m values (genetic distances). Therefore RF is a good measure of distance if the distance is small and no multiple exchanges are likely. In this region, the map unit defined as 1% RF has real meaning. Therefore, let's use this region of the curve to define real map units. For small m, such as $m = 0.05$, $e^{-m} = 0.95$, and RF $= \frac{1}{2}(1 - 0.95) = \frac{1}{2}(0.05) = \frac{1}{2} \times m$; for $m = 0.10$, $e^{-m} = 0.90$, and RF $= \frac{1}{2}(1 - 0.90) = \frac{1}{2}(0.10) = \frac{1}{2} \times m$. We see that RF $= m/2$, and this relation defines the dotted line on the graph in Figure 6-4. It allows us to translate m values into real (low-end-of-the-curve-style) map units.

So an m value of 1 is the equivalent of 50 real map units, and we can express the horizontal axis in map units, as indicated on the lower scale. Here we see that two loci separated by 150 real map units show only 50% RF. For regions of the graph in which the line is not horizontal, we can use the function to convert the RF into map distance simply by drawing a horizontal line to the curve and dropping a perpendicular to the map-unit axis (a process equivalent to using the formula).

Example of Calculation of m

Suppose we get an RF of 27.5%. How many real map units does this represent?

$$0.275 = \tfrac{1}{2}(1 - e^{-m})$$
$$0.55 = 1 - e^{-m}$$

therefore,

$$e^{-m} = 1 - 0.55 = 0.45$$

From e^{-m} tables (or by solving the hard way using logarithms), we find that $m = 0.8$, which is 40 real map units. If we had been happy to accept 27.5% RF as meaning 27.5 map units, we would have been considerably underestimating the true distance between the loci.

A note to calculator owners: the mapping function may be rearranged for easy solution by calculator as follows:

$$e^{-m} = 1 - 2\,\text{RF}$$
$$-m = \ln(1 - 2\,\text{RF})$$
$$m = -\ln(1 - 2\,\text{RF})$$

Message
To get good estimates of map distance, put RF values through the map function, or stick to small regions in which the graph is linear.

Tetrad Analysis

We have already hinted (in Chapter 5) at the existence of marvelous organisms in which the four products of a single meiosis are recoverable and testable. The group of four is called a **tetrad.** Tetrad analysis has been possible only in those fungi and single-celled algae in which the products of each meiosis are held together in a kind of bag.

These organisms are all haploid. There are many advantages to using haploids for genetic analysis; a few are listed here.

1. Because the organisms are haploid, there is no complication of dominance. The nuclear genotype is expressed directly in the phenotype.
2. There is only one meiosis to analyze at a time (refer to the discussion of life cycles in Chapter 3), whereas, in diploids, gametes from two different meiotic events fuse to form the zygote. In diploids, the testcross is an attempt to achieve the same end, but the procedure is technically more laborious and sometimes not possible.

Random-Meiotic-Product-Linkage Analysis in Haploids

Before embarking on a study of tetrad analysis, it is worthwhile pointing out that a conventional analysis of the random products of meiosis is possible in haploids and is much easier than in diploids. For example, the products of the cross $+ + \times a\ b$ might be

$+ +$	45%
$a\ b$	45%
$a\ +$	5%
$+\ b$	5%

This result permits a direct calculation of RF as 10%, so 10 map units separate the *a* and the *b* loci. Note how easy it is to compare product-of-meiosis genotypes with the haploid genotypes that constituted the diploid cells in which meiosis occurred.

3. Because the organisms are small, fast growing, and inexpensive to culture, it is possible to produce very large numbers of progeny from a cross. Thus, good statistical accuracy is possible; also, very rare events at frequencies as low as 10^{-8} are detectable.
4. In several well-studied species (such as *Neurospora*), the structure and behavior of the chromosomes is very similar to that found in higher organisms. Thus, these simpler forms represent very useful, easily analyzed **eukaryotic models.**
5. Last, but not least, there is the possibility of tetrad analysis.

Tetrad analysis itself has proved useful for several reasons.

1. It provides an opportunity to test *directly* some of the assumptions of the chromosome theory of heredity. Our analyses so far have been essentially **random-meiotic-product** analyses. In these studies, individuals are examined and inferences are made about the meioses that produced them. This was the basic approach used by Mendel and by most eukaryotic geneticists since. The following example illustrates this kind of inference. A testcross of *Aa* to *aa* produces a 1:1 ratio, from which the equal segregation of *A* and *a* in a single meiosis is inferred. The use of tetrads provides a far more direct test of this notion because the direct products of a single meiosis are examined. Furthermore, each meiotic product in the tetrad reinforces inferences about the other three products; the data are interlocking, and they increase the confidence with which we can make judgments on meiotic behavior. In research, this is an extremely useful facility.
2. It makes possible the mapping of centromeres as genetic loci.

3. It permits examination of the distribution of crossovers between the four chromatids, and hence investigation of the possibility of **chromatid interference.**

4. It permits several approaches to studying the mechanism of chromosome exchange (crossing over). We have already used it to deduce the stage at which crossing over occurs (four strands) and to rule out Belling's copy-choice hypothesis. But perhaps the most significant use has been in the analysis of gene conversion (see Chapter 16).

5. It provides a unique approach to the study of abnormal chromosome sets (see Chapter 8).

Centromere Mapping in Linear Tetrads

Tetrads can take different forms in different organisms. In some species, there are four sexual spores containing the four nuclei, which are the products of meiosis. In other species, there are eight sexual spores: in these cases each of the four meiotic product nuclei undergoes a further mitotic division, producing eight nuclei that are then enclosed in the eight spores. These groups of eight may be called **octads,** but most geneticists also call them tetrads because they represent simply doubled tetrads.

The sexual spores, whether four or eight in number, may be enclosed in two different ways. In some species, the spores are always found in a jumbled arrangement called a **nonlinear tetrad** (Figure 6-5a). In other species, the

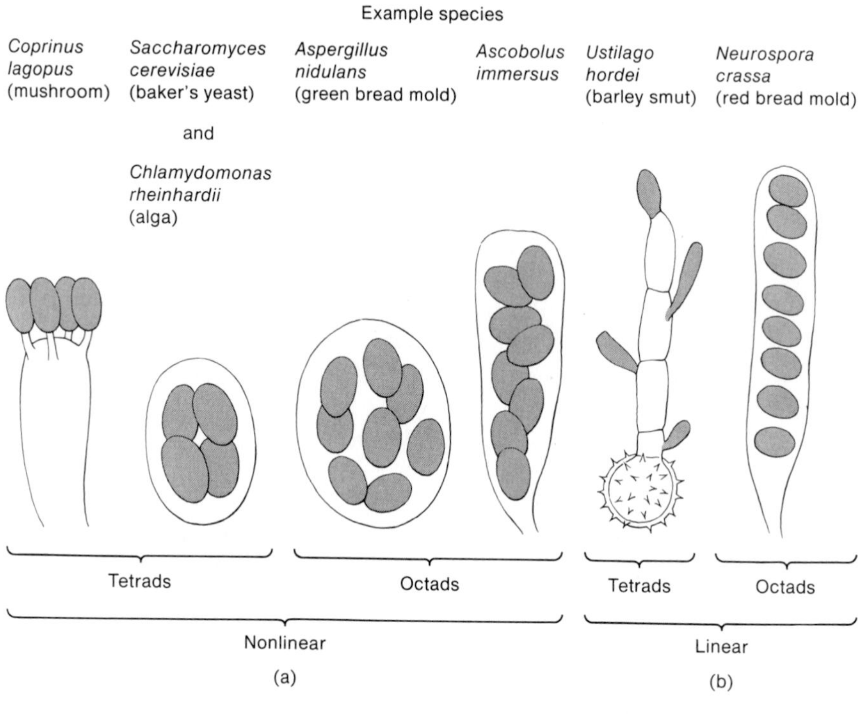

Example species

Coprinus lagopus (mushroom)

Saccharomyces cerevisiae (baker's yeast) and *Chlamydomonas rheinhardii* (alga)

Aspergillus nidulans (green bread mold)

Ascobolus immersus

Ustilago hordei (barley smut)

Neurospora crassa (red bread mold)

Tetrads | Octads | Tetrads | Octads

Nonlinear (a) | Linear (b)

Figure 6-5. Various forms of tetrads and octads found in different organisms. (a) Nonlinear. (b) Linear.

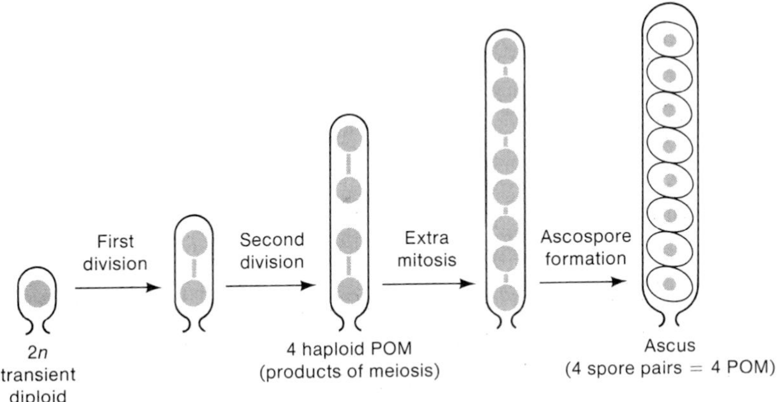

Figure 6-6. A linear meiosis and subsequent mitosis. There is no nuclear passing (change in sequence of nuclei) because there is no spindle overlap. The vertical bars represent spindles.

spores are arranged in a line, an arrangement called a **linear tetrad** (Figure 6-5b). Species with linear tetrads are particularly interesting because in these species it is possible to map the centromeres and include them on a linkage map. A knowledge of the position of the centromere can be extremely useful in a variety of genetic analyses. This ability is particularly useful in study of the lower eukaryotes, whose small chromosomes make cytological identification of centromeres difficult.

How are linear tetrads produced? The linear array of sexual spores is a result of the lack of spindle overlap in the meiotic divisions or in any subsequent mitotic divisions. Because no spindle overlap occurs, the nuclei do not pass each other in the long sac, and a linear array of nuclei (and hence of spores) is produced (Figure 6-6). This particular quirk allows us to map centromeres.

One Locus. Let's look at a specific cross to see how this works. The species we use is the fungus *Neurospora crassa,* which has a linear octad. In the group of fungi to which *Neurospora* belongs, the spores and the bag surrounding them are together called an **ascus.** The locus under study in this example is the one that controls crossing ability, the so-called **mating-type locus.** There are two alleles, *A* and *a,* each determining a mating type. For a cross to be fertile, one parent must be *A* and the other *a.* But here we are using these genes merely as genetic markers.

We can cross *A* and *a* cultures by mixing them on a medium that promotes the sexual cycle. We can then isolate the octads. From each octad, we can then carefully pick out the eight spores one at a time, in the order found. We can test the culture arising from each spore to see if it is *A* or *a.* C. Lindegren performed this experiment in the 1930s, obtaining the following results. The bottom of the ascus as written represents the "stalk" where the ascus sac is joined to the rest of the fungal tissue.

Cross $A \times a$

Asci						
	A	a	A	a	A	a
	A	a	A	a	A	a
	A	a	a	A	a	A
	A	a	a	A	a	A
	a	A	A	a	a	A
	a	A	A	a	a	A
	a	A	a	A	A	a
	a	A	a	A	A	a
Number	126	132	9	11	10	12	Total = 300

What do these patterns and their relative proportions mean? The first two types on the left are called **first-division segregation patterns,** or M_I patterns. These asci result from meioses in which there has been no detectable crossover between the mating-type locus and the corresponding centromere locus of that chromosome. Because of this lack of crossover, the two alleles

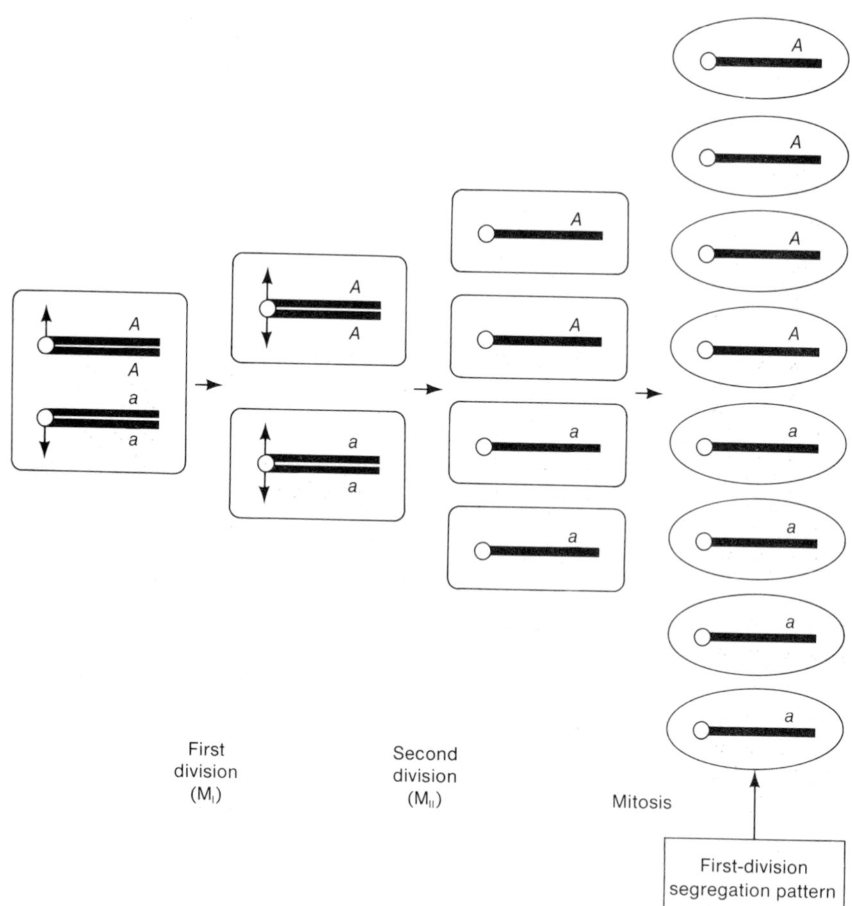

Figure 6-7. Segregation of A *and* a *into separate nuclei at the first meiotic division. The resultant allele pattern in the octad is called a first-division segregation pattern.*

A and a segregate into separate nuclei at the first meiotic division (M_I) (Figure 6-7). The approximate equality of the first two types ($126 \simeq 132$) reflects random spindle attachment to the centromeres at the first meiotic division. One pattern is simply an inverted copy of the other.

The four ascus types on the right show **second-division segregation patterns,** or M_{II} patterns. These asci result from meioses in which there *has* been a crossover between the mating-type locus and the centromere locus. As a result of the crossover, A and a alleles appear in the same nucleus at the end of the first meiotic division, and they do not segregate until the second division (Figure 6-8). The M_{II} pattern in Figure 6-8 is only one of the four observed, so how are the others explained? We must conclude that the attachment of the spindle fibers to the centromere at the second meiotic division must be random. This random attachment produces four patterns, only one of which is shown in Figure 6-8, but all of which must be interpreted as manifestations of the same event, a crossover. The rough equivalence of these classes ($9 \simeq 11 \simeq 10 \simeq 12$) is a satisfying confirmation of the random spindle-attachment idea. Random spindle attachment is illustrated in Figure 6-9.

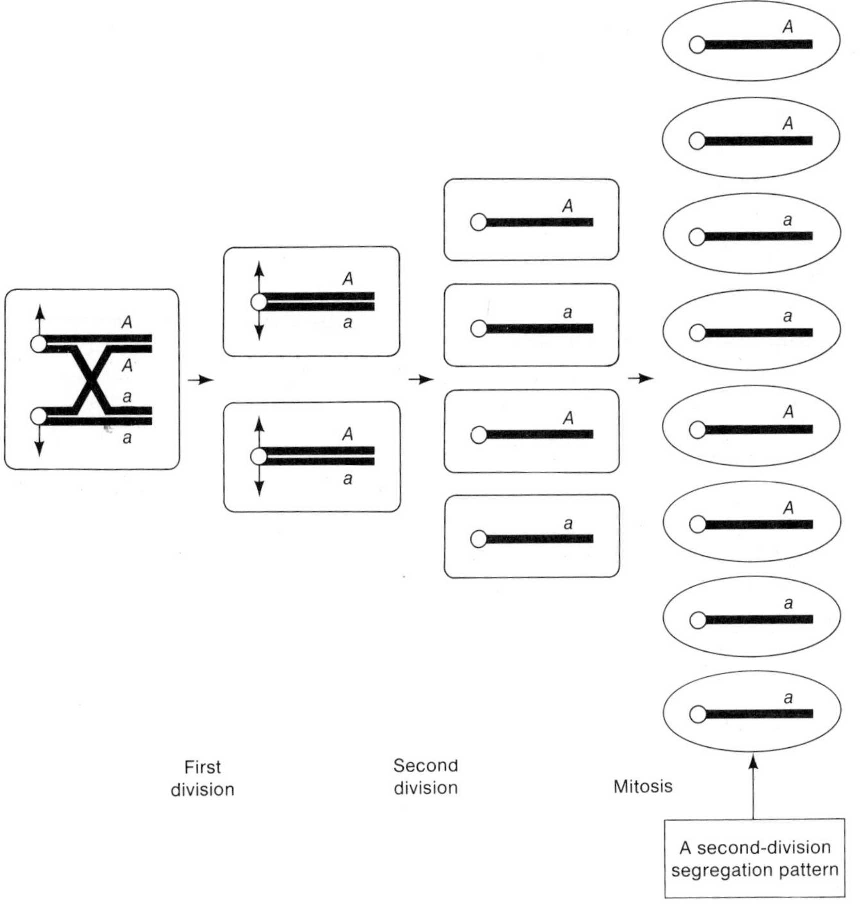

Figure 6-8. Segregation of A *and* a *into separate nuclei at the second meiotic division because of a crossover. The allele pattern in the octad is called a second-division segregation pattern. (Other second-division patterns are possible.)*

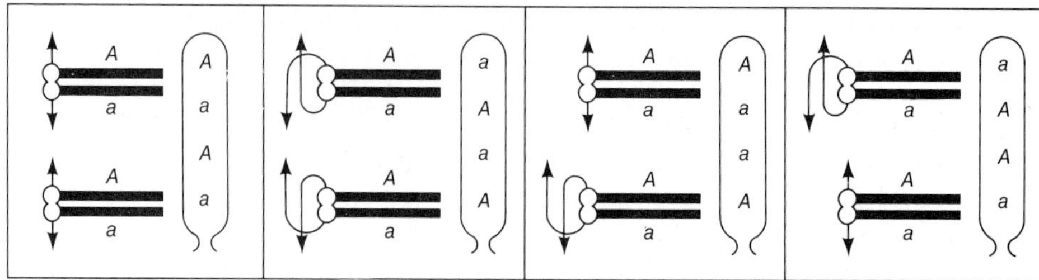

Figure 6-9. Four second-division segregation ascus patterns are equally frequent in linear asci. These are produced by random spindle-to-centromere attachments at the second meiotic division.

Notice in these data that the pairing of spores in the octad reflects the lack of nuclei passing each other in the mitotic division following meiosis. It is obvious that the octad in this species is nothing more than a doubled tetrad, so the asci of *Neurospora* often are informally called tetrads.

We instinctively feel that the relative abundance of M_I versus M_{II} types must reflect the map distance between the centromere locus and the mating-type locus. The total M_{II} pattern frequency is $42/300 = 14\%$; can we state this simply as a distance of 14 map units (14 m.u.)? The answer is a resounding NO! The reason is an important one: map units are defined as the percentage of *recombinant products* of meiosis. The 14% obtained here is the percentage of *meioses* in which a crossover has occurred. One crossover between two loci in a meiosis—in the present case, between a gene locus and a centromere locus—produces only 50% recombinant meiotic products in that meiosis (Figure 6-10). Because the recombinant chromatids represent recombinant meiotic products, and because only two of the four products in this kind of meiosis are recombinant, the M_{II} frequency obviously must be halved to obtain a good estimate of the number of map units between the loci. The mating-type locus therefore is 7 m.u. from its centromere.

Message

To calculate the distance of a locus from its centromere, simply multiply the frequency of M_{II} asci by $\frac{1}{2}$.

Figure 6-10. Only one-half of the chromatids from a meiosis with a single crossover are recombinant.

In this way, we can locate the centromere on a genetic map with units consistent with other mapping analyses.

Two Loci. In a cross involving linked marker loci, (say, *a*/+ and *b*/+), either or both of the gene pairs may show an M_I (or an M_{II}) pattern in a tetrad analysis. The combined pattern obtained will depend of course on the exact location of the crossover. If the locus order is centromere–*a*/+–*b*/+, then a crossover between *a*/+ and *b*/+ will produce an M_{II} pattern for *b*/+ and an M_I pattern for *a*/+ (Figure 6-11).

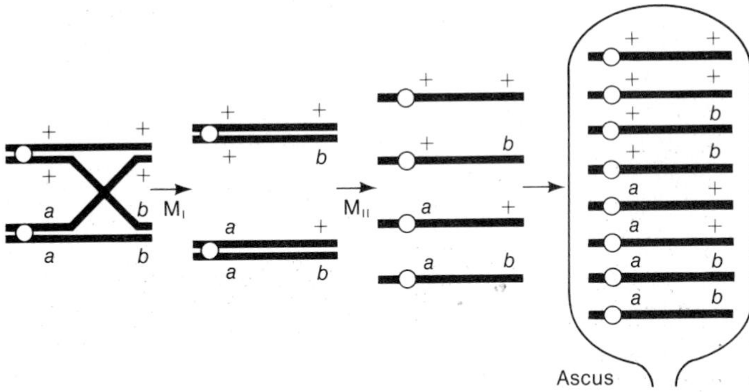

Figure 6-11. A first-division segregation pattern for a/+ *and a second-division segregation pattern for* b/+ *will result from a single crossover with the linkage arrangment shown.*

If the relation of several gene loci with respect to the centromere locus (or loci if the genes are unlinked) is unknown, then a linear tetrad analysis can be used to provide this information. We look next at an example of such an analysis.

The cross is *nic* + × + *ad*, in which *nic* is an allele that causes the fungus to require nicotinic acid in order to grow, and *ad* is an allele of another gene and confers a requirement for adenine. Normal strains have no medium-supplementation requirements. (Don't worry about these phenotypes for now; they are being used only as genetic markers.) In a cross involving two marker loci, only seven basic linear octad classes are possible (Figure 6-12). The octads are written as tetrads for simplicity, because members of each spore pair are identical (except in very rare cases, discussed in Chapter 16). Note also that in arriving at these seven classes, the order of genotypes within the half-ascus has been ignored because it simply reflects random spindle attachment; for example, class 5 also includes

nic +	+ *ad*	*nic* +
+ *ad*	*nic* +	+ *ad*
+ *ad*	*nic* +	*nic* +
nic +	+ *ad*	+ *ad*

1	2	3	4	5	6	7
+ *ad*	+ +	+ +	+ *ad*	+ *ad*	+ +	+ +
+ *ad*	+ +	+ *ad*	*nic ad*	*nic* +	*nic ad*	*nic ad*
nic +	*nic ad*	*nic* +	+ +	+ *ad*	+ +	+ *ad*
nic +	*nic ad*	*nic ad*	*nic* +	*nic* +	*nic ad*	*nic* +
M_I M_I	M_I M_I	M_I M_{II}	M_{II} M_I	M_{II} M_{II}	M_{II} M_{II}	M_{II} M_{II}
(PD)	(NPD)	(T)	(T)	(PD)	(NPD)	(T)
808	1	90	5	90	1	5

Figure 6-12. All the seven possible tetrad types for a cross in which two loci are heterozygous. Some numbers are given as an example, and these are interpreted in the text. The results shown are for a total of 1000 acsi. PD = parental ditype (the asci contain only two genotypes, both parental); NPD = nonparental genotype (there are only two genotypes and both are nonparental, or recombinant); and T = tetratype (there are four genotypes, two parental and two nonparental).

The asci in Figure 6-12 are labeled to indicate which segregation pattern is shown by each locus. The asci also are labeled according to another classification:

1. **parental ditypes** (PD), in which there are only two genotypes (hence *di*type) with respect to the marker loci, and both are parental (classes 1 and 5);

2. **nonparental ditypes** (NPD), in which there are only two genotypes, and both are nonparental or recombinant (classes 2 and 6);

3. **tetratype** (T) in which there are four genotypes—two parental and two nonparental (classes 3, 4, and 7).

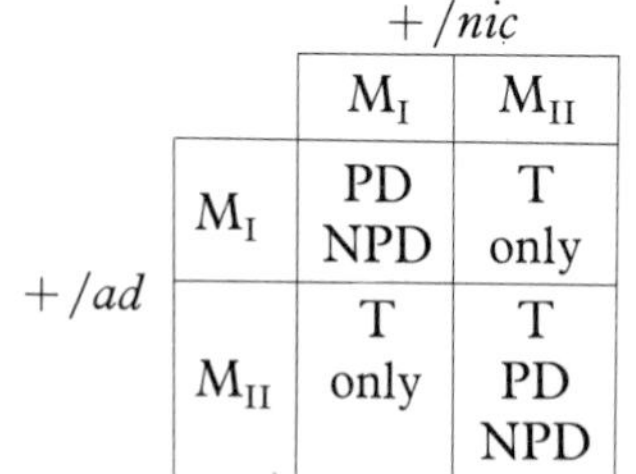

Figure 6-13. Tetratypes can be produced only if there is a second-division segregation for at least one gene.

These classifications are the result of the combinations of M_I and M_{II} segregations (Figure 6-13). Note that a tetratype can occur only when there is a crossover between at least one locus and its centromere. Now for the calculations.

First, we calculate the distance of each locus from its centromere. For +/*nic*, it is

$$\frac{5 + 90 + 1 + 5}{1000} = \frac{101}{1000} = 10.1\%\ M_{II} = 5.05 \text{ m.u.}$$

For +/*ad*, it is

$$\frac{90 + 90 + 1 + 5}{1000} = \frac{186}{1000} = 18.6\%\ M_{II} = 9.30 \text{ m.u.}$$

However, we still have three linkage possibilities (Figure 6-14). Most of the asci are M_IM_I parental ditypes (808/1000). Therefore most genomes are parental, and independent assortment cannot have occurred; so possibility 1 can be ruled out.

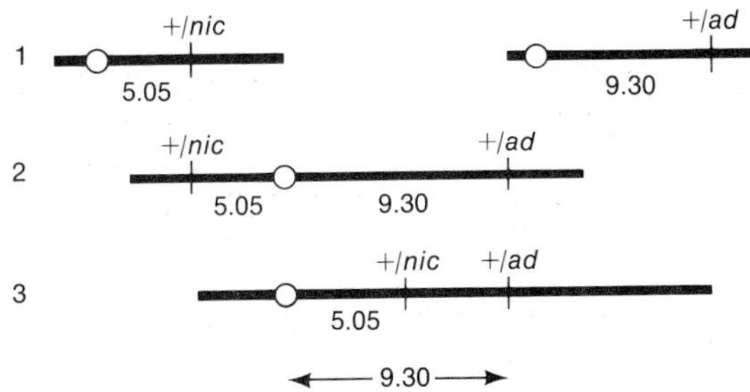

Figure 6-14. The three linkage possibilities for + /ad *and* + /nic *in the example discussed.*

If we arrange the data in a different way, then possibility 2 also is ruled out as follows:

+ / *nic*	+ / *ad*	
M_I	M_I	809
M_I	M_{II}	90
M_{II}	M_I	5
M_{II}	M_{II}	96
		1000

Now we can clearly see that a crossover in the centromere-to-*nic* region is almost always (96/101 times) accompanied by a crossover in the centromere-to-*ad* region. We conclude that the *same crossover* simultaneously generates an M_{II} pattern for + / *nic* and for + / *ad*. This is very powerful evidence in favor of possibility 3, with the following crossover producing the observed products:

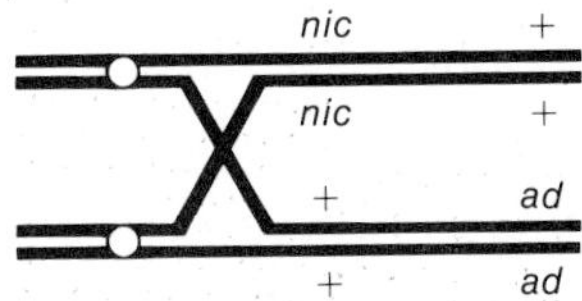

Having ascertained that possibility 3 is correct, we can calculate the recombinant frequency between + / *nic* and + / *ad*. The simple subtraction 9.30 − 5.05 does not give an accurate value for the following reason. We arrived at the figure of 9.30 m.u. by calculating the M_{II} frequency and dividing by 2. In adding up the total M_{II} frequency, we now know that we missed quite a few asci containing crossovers between the centromere and the *ad* locus. For example, class 4 was scored as M_I for + / *ad*, but we know now that *two* crossovers occurred in these asci:

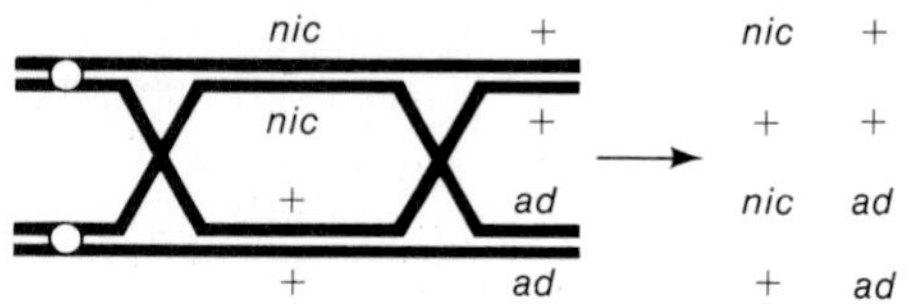

We can better calculate the distance between $+/nic$ and $+/ad$ by using the formula

$$RF = \frac{NPD + \frac{1}{2}T}{\text{total asci}} \times 100$$

because we know that the NPD asci are full of recombinant genotypes, and the T asci are "half-full" of recombinant genotypes. Therefore,

$$RF = \frac{2 + \frac{1}{2}(100)}{1000} \times 100 = 5.2 \text{ m.u.}$$

Now we can redraw the best map obtainable from these data:

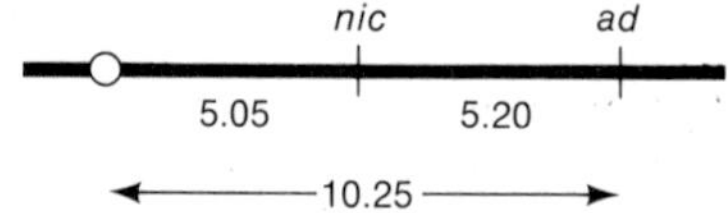

(NOTE: Class 6 is a very complex ascus. Draw the crossovers needed to produce it, right now!)

Message
Linear tetrad analysis can be used to map loci in relation to their centromere(s) and to each other.

Analysis of Nonlinear Tetrads

Because isolation of linear tetrads is a laborious process, nonlinear tetrads are sometimes used as a less informative but simpler way to study meiotic genetics. The groups of four or eight are recovered clustered together, but not in linear sequence. In *Neurospora,* the fungus shoots the spores out of the ascus sac, and they may be caught as a somewhat scattered group of eight on a collection surface, such as a slide. In many other species, of course, the tetrad is always nonlinear. Nonlinear tetrads can be scored only as PD, NPD, or T. Although centromere mapping is not normally possible, useful linkage information is still available. Furthermore, an analysis of unordered tetrads introduces some concepts that are central to a full appreciation of meiotic genetics. Although these specific analyses can be performed only in fungi or algae, the mechanisms revealed (and the analytical thinking involved) are highly relevant and are of direct application to any eukaryotic meiotic system.

Remember that PD asci contain no recombinants, NPD asci are full of

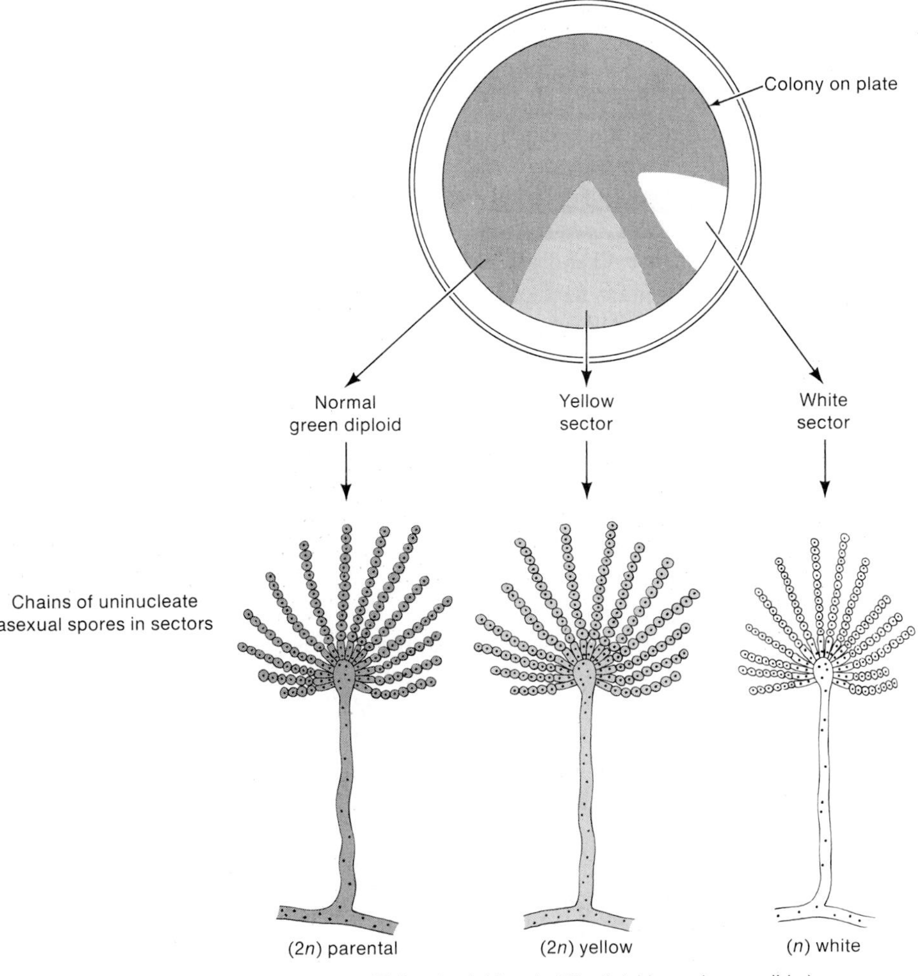

Figure 6-24. Some sectors showing segregation in an Aspergillus *diploid of genotype* +/w +/y, *where* w *is white and* y *is yellow asexual spores. Haploids are identified by smaller cell size.*

show the genotype *w ad* + *paba y* +. Thus, the original diploid nucleus has somehow become haploid (a process known as **haploidization**), presumably by a process involving the progressive loss of one member of each chromosome pair. By looking at only white haploid spores, we automatically selected for the *w*-bearing chromosome. In half of the *w* sectors, the + *pro* + + *bi* chromosome is retained; in the other half, the *ad* + *paba y* + chromosome is retained. In general, the recessive spore-color genes can be used to derive linkage information because haploidization is similar to independent assortment. You can see

There is a virus called Sendai that has a useful property. Normally, a virus has a specific point for attachment to and penetration of a host cell. Each Sendai virus has several points of attachment, so that it can simultaneously attach to two different cells if they happen to be close together. A virus, though, is very small in comparison with a cell (similar to the comparison between the planet Earth and the sun), so that the two cells to which it is attached are held very close together indeed. In fact, in many cases, the membranes of the two cells fuse together, and the two cells become one, a binucleate heterokaryon.

If suspensions of human and mouse cells are mixed together in the presence of Sendai virus (which has been inactivated by ultraviolet light), the virus can mediate fusion of the two kinds of cells (Figure 6-27). Once the cells fuse, the nuclei can fuse to form a uninucleate cell line. Because the mouse and human chromosomes are very different in number and shape, the hybrid cells can be readily recognized. However, for some unknown reason, the human chromosomes are gradually eliminated from the hybrid as the cells divide (perhaps

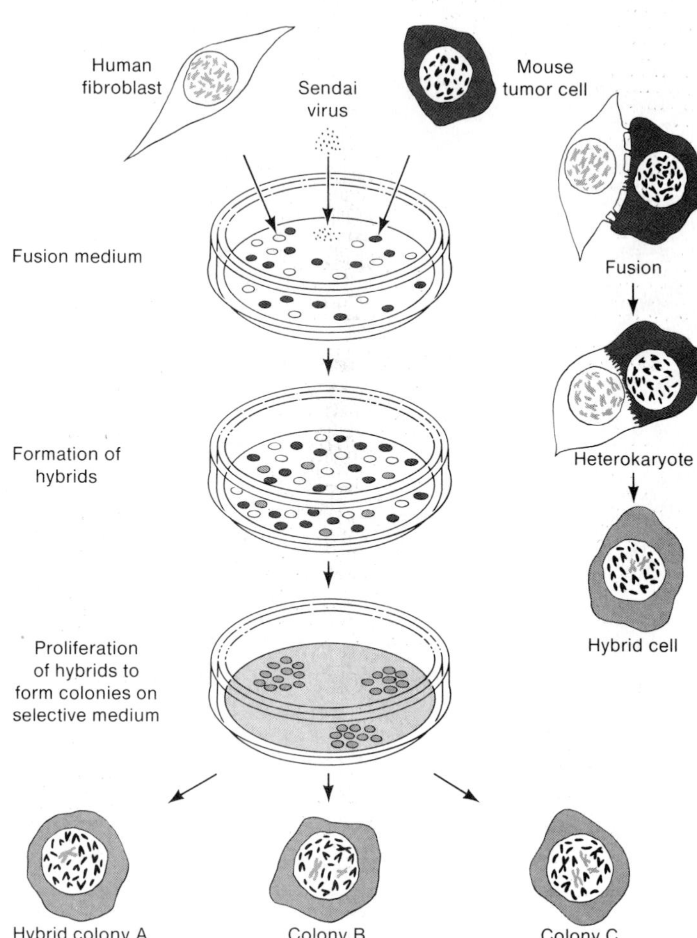

Figure 6-27. Cell-fusion techniques applied to human and mouse cells, producing colonies each containing a full mouse genome plus a few human chromosomes (lighter shade). *A fibroblast is a cell from fibrous connective tissue. (From F. H. Ruddle and R. S. Kucherlapati, "Hybrid Cells and Human Genes." Copyright © 1974 by Scientific American, Inc. All rights reserved.)*

this is analogous to haploidization in *Aspergillus*). This process can be arrested to encourage the formation of a stable partial hybrid in the following way. The mouse cells that are used can be made genetically deficient in some function (usually a nutritional one) so that, for growth of cells to occur, the function must be supplied by the human genome. This selective technique usually results in the maintenance of hybrid cells that have a complete set of mouse chromosomes and a small number of human chromosomes, which vary in number and type from hybrid to hybrid but always include the nutritionally sufficient human chromosome.

Luckily, this process can be followed under the microscope because mouse chromosomes can easily be distinguished from human chromosomes. Recently, this procedure has been made a lot easier by the development of fluorescent stains (such as quinacrine and giemsa) that reveal a pattern of "banding" within the chromosomes. The size and the position of these bands vary from chromosome to chromosome, but the banding patterns are highly specific and constant for each chromosome. Thus, for any hybrid, it is relatively easy to identify the human chromosomes that are present (Figure 6-28).

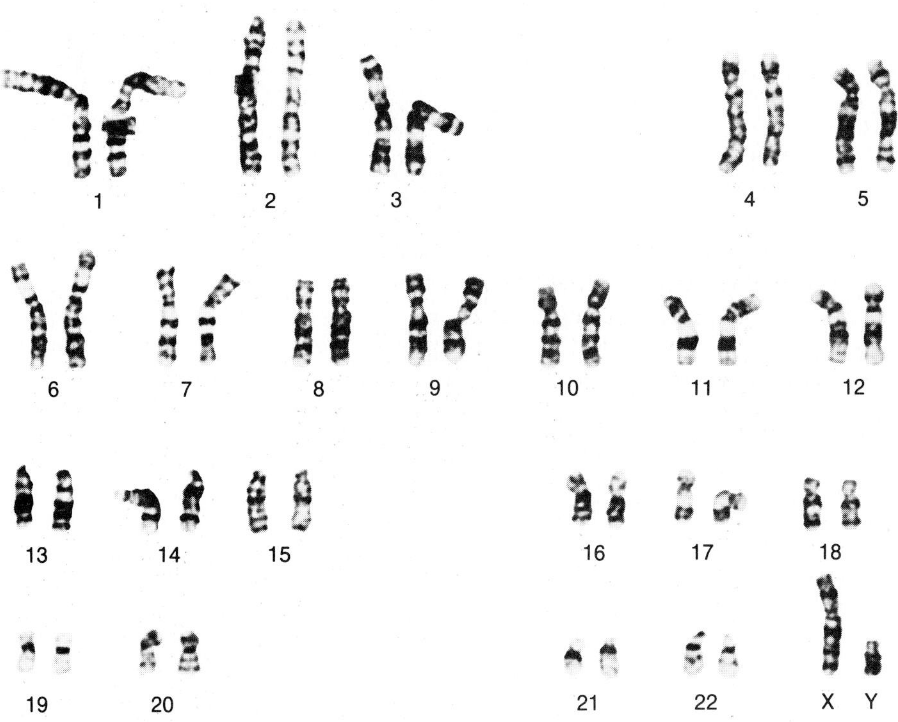

Figure 6-28. Karyotype of a human male, showing how size, centromere position, and banding pattern (produced by trypsin–giemsa treatment) can be used to recognize specific human chromosomes. (Photograph by Fred Dill.)

The mapping technique works as follows. If the human chromosome set contains a genetic marker (such as a gene that controls a specific cell-surface antigen, drug resistance, a nutritional requirement, or a protein variant), then the presence or absence of the genetic marker in each line of hybrid cells can be correlated with the presence or absence of certain human chromosomes in each line (Table 6-3). We can see that, in the different hybrid cell lines, genes 1 and 3 are always present or absent together. We conclude then that they are linked. Furthermore, the presence or absence of genes 1 and 3 is directly correlated with the presence or absence of chromosome 2, so we assume these genes are located on chromosome 2. By the same reasoning, gene 2 must be on chromosome 1, but the location of gene 4 cannot be assigned.

Table 6-3. Comparison of five hybrid lines

		Hybrid cell lines				
		A	B	C	D	E
Human genes	1	+	−	−	+	−
	2	−	+	−	+	−
	3	+	−	−	+	−
	4	+	+	+	−	−
Human chromosomes	1	−	+	−	+	−
	2	+	−	−	+	−
	3	−	−	−	+	+

Large numbers of human genes have now been localized to specific chromosomes in this way, but of course we cannot derive a linkage map showing the order and distances between genes. Other tricks are needed—for example, the loss or gain of variously sized bits of a specific chromosome might be correlated with the presence or absence of genetic markers. A problem at the end of Chapter 8 encourages you to think through the kind of logic involved. However, the results of this kind of intrachromosomal mapping are so far not nearly as extensive as those on simple chromosome location (Figure 6-29).

Message
Mitotic as well as meiotic phenomena can give information on gene location in fruit flies, fungi, and humans.

In this chapter, we have discussed more advanced treatments of transmission genetics, including mapping functions, tetrad analysis, and mitotic genetics. These genetic tools have enabled geneticists to test some of the assumptions of the chromosome theory of heredity and to learn more about how genetic material is passed from one generation to the next.

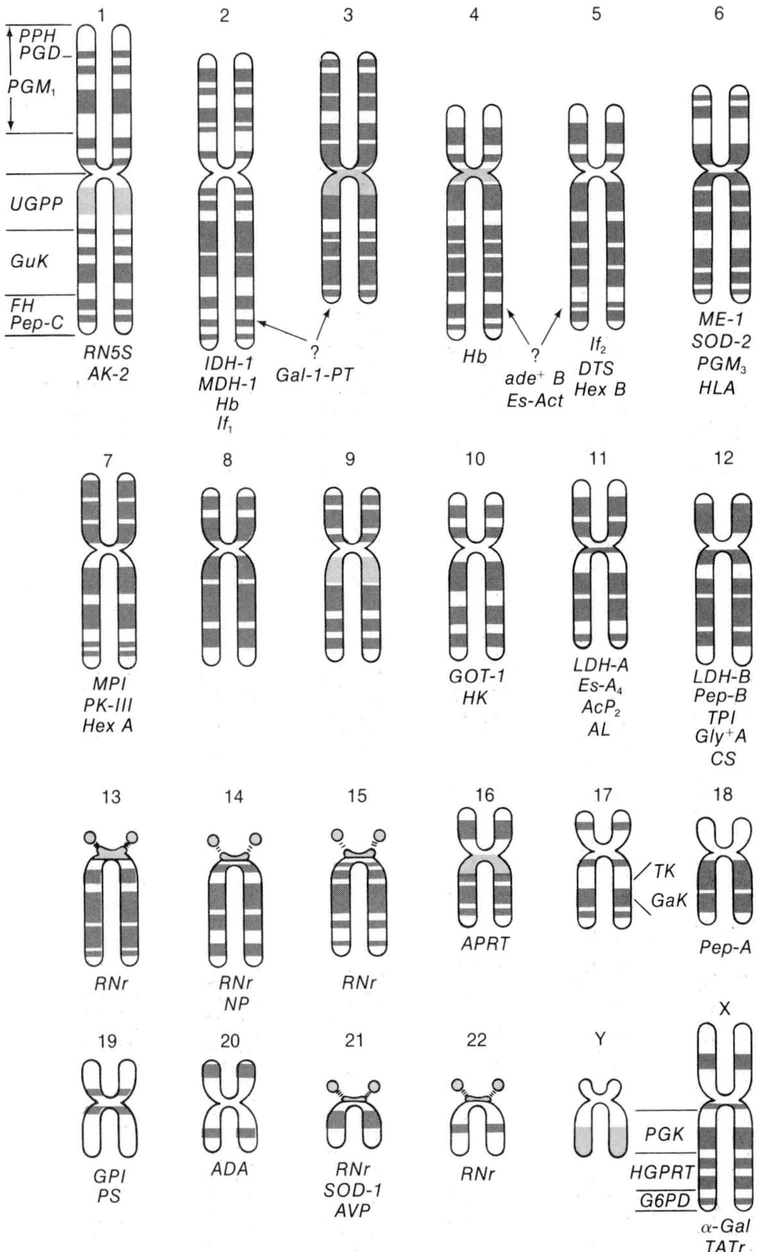

Figure 6-29. Genes that have been assigned to specific human chromosomes by the cell-hybridization technique. Characteristic quinacrine banding patterns of each chromosome are shown. The shorthand gene designations relate to their phenotypes. Note that only a few approximate loci have been assigned. (From F. H. Ruddle and R. S. Kucherlapati, "Hybrid Cells and Human Genes." Copyright © 1974 by Scientific American, Inc. All rights reserved.)

Summary

Because multiple crossovers are likely to occur when great distances separate loci, map distance between such loci is not linearly related to recombinant frequency. The true relationship between map distance and recombinant frequency is called the mapping function, which can be calculated by using the Poisson distribution.

Another useful genetic tool is tetrad analysis, which analyzes the four products of meiosis in those fungi and single-celled algae in which the four products of each meiosis are held together in a kind of bag. Tetrad analysis provides the opportunity to test directly some of the assumptions of the chromosome theory of heredity, to map centromeres as genetic loci, to investigate the possibility of chromatid interference, to examine the mechanisms of chromosome exchange, and to study abnormal chromosome sets.

Tetrads may be linear or nonlinear. Analysis of linear tetrads is particularly useful because it is possible to map loci in relation to their centromeres and to each other. In crosses involving two linked loci, the asci in a linear or nonlinear tetrad may be classified as parental ditype (PD), nonparental ditype (NPD), and tetratype (T). The number of PD asci in relation to the number of NPD asci provides a critical test for linkage; if the number of PD asci is greater than the number of NPD asci, the recombinant frequency must be less than 50%, and there must be linkage between the genes. If PD equals NPD, the proportion of T asci can be used to distinguish independent assortment from loose linkage.

Although segregation and recombination are normally thought of as meiotic phenomena, segregation and recombination do occasionally occur during mitosis. Mitotic segregation was first identified in the 1930s, when Bridges observed patches of M^+ bristles on the body of a female *Drosophila* of predominantly M phenotype. He concluded that the patches were the result of abnormal chromosome segregation at the mitotic level. Around the same time, Stern observed twin spots in *Drosophila* and assumed they must be the reciprocal products of mitotic crossing over. Fungi are also extensively used to study mitotic segregation and recombination.

Humans are generally unsuitable subjects for traditional genetic analysis. However, by using the Sendai virus to fuse human and mice cells, geneticists have been able to locate human genes on specific chromosomes.

Problems

1. In haploid yeast, a cross between $arg^- \; ad^- \; nic^+ \; leu^+$ and $arg^+ \; ad^+ \; nic^- \; leu^-$ produces haploid sexual spores, and 20 of these are isolated at random. The resulting cultures are tested on various media as shown in Table 6-4 (where + means growth and − means no growth).

Table 6-4.

Culture	Minimal medium plus: Arginine, adenine, nicotinamide	Arginine, adenine, leucine	Arginine, nicotinamide, leucine	Adenine, nicotinamide, leucine
1	+	+	−	−
2	−	−	+	+
3	−	+	−	+
4	+	−	+	−
5	−	−	+	+
6	+	+	−	−
7	+	+	−	−
8	−	−	+	+
9	+	−	+	−
10	−	+	−	+
11	−	+	−	+
12	+	−	+	−
13	+	+	−	−
14	+	−	+	−
15	−	+	−	+
16	+	−	−	−
17	+	+	−	−
18	−	−	+	+
19	+	+	−	−
20	−	+	−	+

a. What can you say about the linkage arrangement of these genes?

b. What is the origin of culture 16?

2. Every Friday night, genetics student Jean Allele, exhausted by her studies, goes to the student's union bowling lane to relax. But even there she is haunted by her genetics studies. The rather modest bowling lane has only four bowling balls: two red and two blue. These are bowled at the pins, then collected and returned down the chute in random order, coming to rest at the end stop. Over the evening, Jean notices familiar patterns of the four balls as they come to rest at the stop. Compulsively, she counts the different patterns that occur. What patterns did she see, what were their frequencies, and what is the relevance of this matter to genetics? (This is not a trivial question.)

3. a. By using the map function, calculate how many real map units are indicated by a recombinant frequency of 20%. Remember that a mean of 1 equals 50 real map units.

   b. If you obtain an RF value of 45% in one experiment, what can you say about linkage? (The actual figures are 58, 52, 47, and 43 out of 200 progeny.)

4. Complete Table 6-5 for a situation in *Neurospora* in which the *mean* number of crossovers between the *a* locus ($a/+$) and its centromere is equal to one per meiosis.

Table 6-5.

	Number of exchanges				
	0	1	2	3	4
Probability of this kind of meiosis?					
What proportion of each of these kinds of meiosis will result in an M_{II} pattern for $a/+$?					
What proportion of all asci from this cross will show an M_{II} pattern as a result of each of these kinds of meiosis?					

a. What is the total M_{II} frequency if the mean = 1?

b. Complete similar tables for means of 0.5, 2.0, and 4.0, and hence draw a mapping function for M_{II} frequency. (That is, plot the total M_{II} frequency against mean crossover frequency per meiosis.)

c. Why does the curve bend downward? How would you correct this?

5. In the fungus *Aspergillus nidulans,* there are no mating types. A cross is attempted by mixing two lines, whose genotypes are $y^- paba^+$ and $y^+ paba^-$. Asci are produced, and they are of three types:

(a) $y^- paba^+ / y^- paba^+ / y^+ paba^- / y^+ paba^-$

(b) $y^- paba^+ / y^- paba^+ / y^- paba^+ / y^- paba^+$

(c) $y^+ paba^- / y^+ paba^- / y^+ paba^- / y^+ paba^-$

Explain these results.

6. In *Neurospora,* the cross $+ + + + \times a\,b\,c\,d$ is made (*a, b, c,* and *d* are linked in the order written). Draw crossover diagrams to illustrate how the following unordered (nonlinear) ascus patterns could arise:

+ + *c* +	+ *b* *c* *d*	+ + *c* +	+ + *c* +	+ *b* + *d*
a *b* *c* +	+ + + *d*	+ *b* + *d*	+ *b* + *d*	+ *b* + *d*
+ *b* + *d*	*a* *b* *c* +	*a* + *c* +	*a* + *c* +	*a* + *c* +
a + + *d*	*a* + + +	*a* *b* + *d*	*a* *b* + *d*	*a* + *c* +

+ *b* *c* *d*	+ *b* + *d*	+ *b* + +	+ *b* + *d*
a *b* *c* +	+ + + *d*	+ *b* + +	*a* + *c* +
+ + + *d*	*a* *b* *c* +	*a* + *c* *d*	+ *b* + *d*
a + + +	*a* + *c* +	*a* + *c* *d*	*a* + *c* *d*

7. In *Neurospora,* crosses $a\,b \times + +$ (in which *a* and *b* represent different loci in each cross) are made. From each cross, 100 linear asci are analyzed, and the results shown in Table 6-6 are obtained. For each cross, map the genes in relation to each other and to their respective centromere(s).

7

Gene Mutation

Genetic analysis would not be possible without **variants**—organisms that differ in a particular character. We have considered many examples where analyses have been performed in organisms that have different phenotypes connected with a particular character. Now we consider the origin of the variants. How in fact do genetic variants arise?

The simple answer to this question is that organisms have an inherent tendency to promote change from one hereditary state to another. This process is called **mutation.** We can recognize two basic levels of mutation.

1. **Gene mutation.** A gene can mutate from one allelic form to another. These changes occur at or within a single gene, so they sometimes are called **point mutations.**

2. **Chromosomal mutation.** Segments of chromosomes, whole chromosomes, or even entire sets of chromosomes may be involved in genetic change, and this process is collectively called chromosomal mutation. Gene mutation is not necessarily involved in such a process; the effects of chromosomal mutation are due more to the new arrangements of chromosomes and of the genes they contain.

In this chapter we explore gene mutation; in Chapter 8 we consider chromosomal mutation.

In any consideration of the subject of change, a fixed reference point, or standard, is necessary. In genetics, that point is provided by the so-called wild type. Remember that the wild-type gene may be a form actually isolated from nature or a form commonly used as a standard laboratory stock. Any change away from the standard form is called **forward mutation;** any change toward the standard form is called **reverse mutation, reversion,** or **back mutation.** For example,

$$\left.\begin{array}{l} a^+ \rightarrow a \\ D^+ \rightarrow D \end{array}\right\} \text{forward mutation}$$

$$\left.\begin{array}{l} a \rightarrow a^+ \\ D \rightarrow D^+ \end{array}\right\} \text{reverse mutation}$$

The nonwild-type form of a gene usually is called a mutation. (To use the same word for the process and the product may sound wanton to you, but in practice little confusion arises!) Thus one can speak of a dominant mutation such as D above) or a recessive mutation (such as a). Bear in mind how arbitrary these gene states are: the wild type of today may have been a mutation in the evolutionary past, and vice versa.

Another useful term is **mutant.** This is, strictly speaking, an adjective and should properly precede a noun. A mutant individual or cell is one whose changed phenotype is attributable to the possession of a mutation. Sometimes the noun is left unstated; in this case, a mutant always means an individual or cell whose phenotype shows that it bears a mutation.

One final useful term is **mutation event,** the actual occurrence of a mutation.

Somatic Versus Germinal Mutation

Mutation can occur in either somatic or germinal tissue; these are called somatic and germinal mutations, respectively. The two types are diagrammed in Figure 7-1.

Somatic Mutation

Somatic mutation may lead to a sector, or **clone,** of identical mutant cells that can be recognized in the background of normal cells. The genetic system for the detection of somatic mutation must be such as to rule out the possibility that the sector might be due to mitotic segregation or recombination. If the individual was a homozygous diploid, such sectoring is almost certainly due to mutation. The timing of a somatic mutation during development will determine the proportion of cells affected (Figure 7-2).

What about sexual transmission to progeny? In the organism in which the original mutation arises, there is little chance of transmission unless germinal tissue is involved. But in a cutting-propagated plant, the gamete will be either normal or mutant, depending on what cell line forms the germinal tissue.

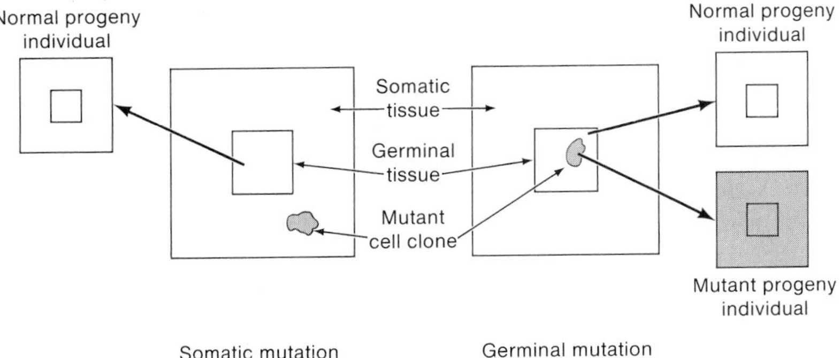

Figure 7-1. Diagrammatic representation of the differing consequences of somatic and germinal mutation.

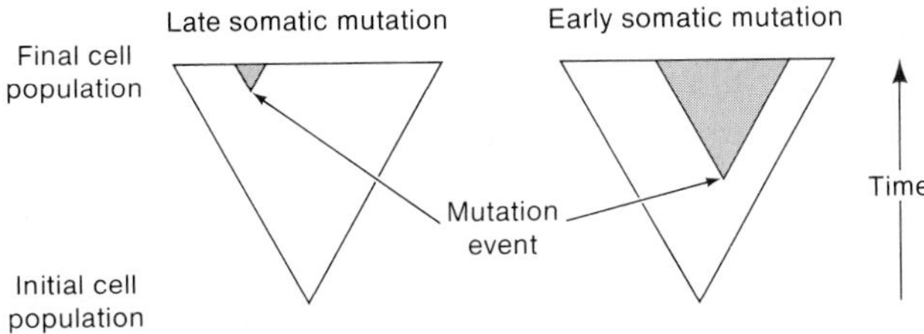

Figure 7-2. Early mutation produces a larger proportion of mutant cells in the growing population than does later mutation.

Germinal Mutation

Germinal mutation occurs in tissue that ultimately will form sex cells. Then, if these mutant sex cells act in fertilization, the mutation will be passed on to the next generation. Of course, an individual of perfectly normal phenotype and of normal ancestry can harbor undetected mutant sex cells. These mutations can be detected only if they turn up in the zygote. Such a situation is thought to have occurred in Queen Victoria, who was almost certainly the origin of a germinal mutation to the X-linked recessive mutant allele for hemophilia (failure of the blood to clot). This mutation showed up in some of her male descendants.

At the operational level, the detection of germinal mutation depends on the ability to rule out meiotic segregation and recombination as possible causes.

Message
At the operational level, before any new variant hereditary state can be attributed to mutation, both segregation and *recombination must be ruled out. This is true for somatic and for germinal mutation.*

Types of Mutations

What kinds of mutants are there? The phenotypic consequences of mutation may be so subtle as to require refined biochemical techniques to detect a difference from wild type or so severe as to produce gross morphological defects or death. A rough classification follows, based only on the ways in which the mutations are recognized. This is by no means an attempt at a complete classification.

Morphological Mutations. *Morph* means "form." In this class are the mutations that affect the visible properties of an organism, usually the outward properties, such as shape, color or size. Albino ascospores in *Neurospora,* curly wings in *Drosophila,* and dwarf peas are all considered morphological mutations. Some examples of morphological mutants are shown in Figures 7-3 and 7-4.

Lethal Mutations. Here the new allele is recognized through its lethal effects on the organism. Sometimes a primary cause of death is easy to identify—for example, in certain blood abnormalities. But often the gene is recognizable *only* by its effects on mortality.

Figure 7-3. A rare morphological mutant in the thimbleberry, Rubus parviflorus, *which arose spontaneously in nature. Left, three normal wild-type flowers; right, a flower from the mutant. The mutation causes a spectacular increase in the number of petals per flower. Such mutants have been used extensively in horticulture to increase the showiness of flowers.*

(a)

(b)

Figure 7-4. *(a) Normal and (b) a rare plumage mutant which arose spontaneously in a laboratory population of Japanese quail. The mutation is an autosomal recessive. It interferes with the normal development of the feathers. (Courtesy of Janet Fulton.)*

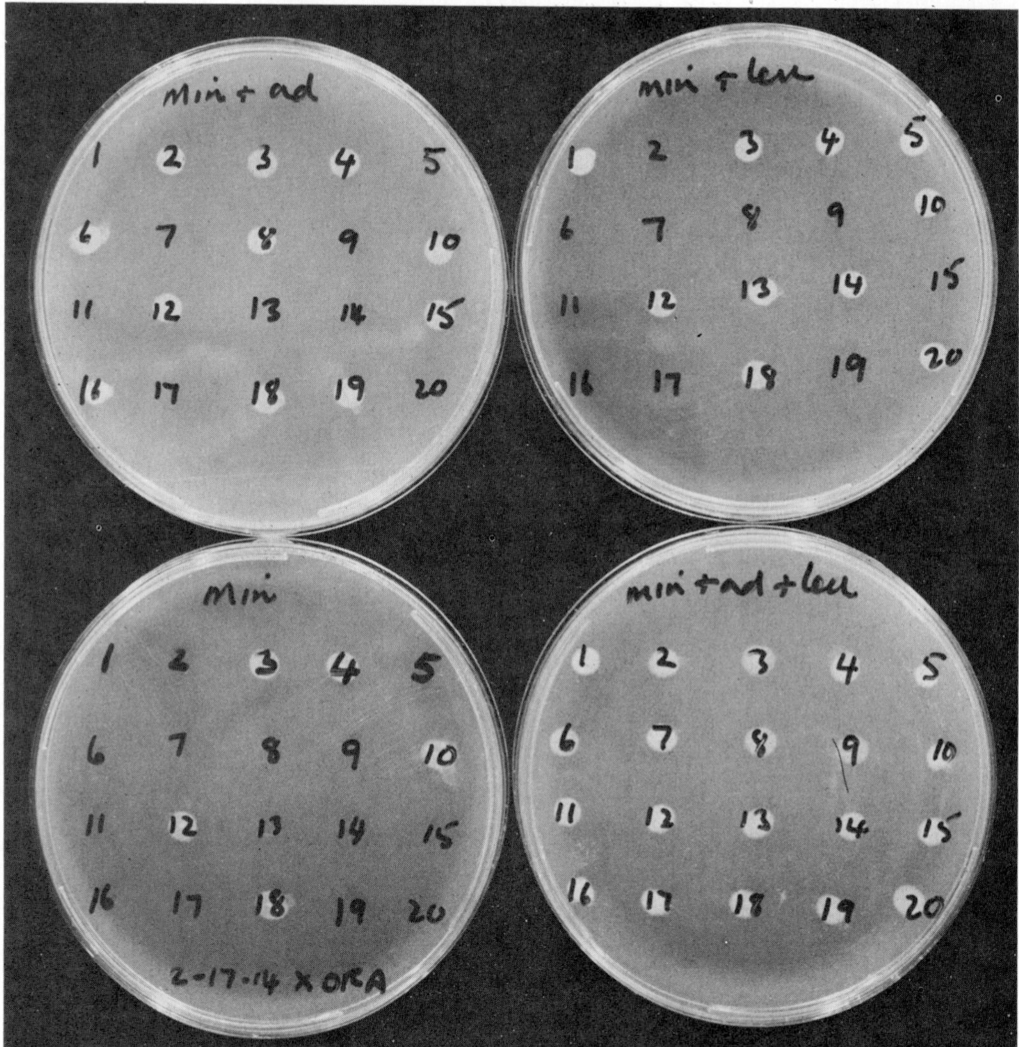

Figure 7-5. Testing for auxotrophy and prototrophy in the fungus Neurospora crassa. *The process is illustrated here by testing the genotypes of twenty haploid cultures derived from a cross between an adenine-requiring auxotrophic mutant and a leucine-requiring auxotrophic mutant (thus the cross was* $ad^- \ leu^+ \times ad^+ \ leu^-$*). The cultures are tested by dipping a sterile needle into a culture tube (say, tube 1) and then touching the needle to the surface of each of four different media, where a few cells will adhere. If the cells proliferate, a colony is formed, and these appear as white circles on the photograph. The four plates contain different combinations of adenine and leucine, where different genotypes produce different results. For example, culture 1 obviously requires leucine, but not adenine, so its genotype must be* $ad^+ \ leu^-$*. Four different genotypes are present in the twenty cultures. Min means the basic "minimal" medium of wild-type* Neurospora, *consisting of simple inorganic salts, an energy (carbon) source, and a nonnutritive gel called agar.*

Conditional Mutations. In this class, a mutant allele expresses the mutant phenotype under a certain condition (called the **restrictive** condition) but expresses a normal phenotype under another condition (called the **permissive** condition). Temperature-conditional mutants have been the most frequently studied. For example, a certain class of mutations in *Drosophila* is known as "dominant heat-sensitive lethal." Heterozygotes in this class (say, H^+/H) are normal at 20° C (the permissive condition) but die if the temperature is raised to 30° C (the restrictive condition).

Many mutant organisms are less vigorous than normal forms. For this reason, conditional mutants are handy in that many of them can be grown in permissive conditions and then shifted to restrictive conditions for study. There are other advantages to conditional mutations, as we shall see in following chapters.

Biochemical Mutations. This class is identified by the loss of, or a change in, some biochemical function of the cell. This change typically results in an inability to grow and proliferate. In many cases, however, growth of a mutant can be restored by supplementing the growth medium with a specific nutrient. Biochemical mutants have been extensively analyzed in microorganisms. Microorganisms, by and large, are **prototrophic**—that is, they are nutritionally self-sufficient and can exist on a substrate of simple inorganic salts and a soluble carbon source (called a **minimal medium**). Biochemical mutants, however, often are **auxotrophic**—that is, they require supplementation with complex nutrients in order to grow. For example, in fungi, a certain class of biochemical mutants is recognized by the fact that they will not grow unless specifically supplemented with the important cellular chemical adenine. These auxotrophic mutants are called *ad* or "adenine-requiring." The method of auxotroph testing is shown in Figure 7-5.

Resistant Mutations. Here, the mutant cell or organism acquires the ability to grow in the presence of some specific inhibitor, such as cycloheximide or a pathogen, to which wild types are susceptible. Such mutants have been extensively used because they are relatively easy to select for, as we shall see.

Table 7-1 illustrates detection of three of these mutation types. Clearly, the five classes are not mutually exclusive, nor do they cover all mutation types. Nevertheless, they are useful components in the vocabulary of mutation.

Table 7-1. Operational basis for detection of mutant phenotype in three different types of mutations

	Conditional mutation (temperature-sensitive)		Auxotrophic mutation		Resistant mutation	
Genotype	Low temperature	High temperature	Without supplement	With supplement	Without agent	With agent
Wild type	Normal	Normal	Growth	Growth	Growth	**No growth**
Mutant	Normal	**Mutant**	**No growth**	Growth	Growth	Growth

The Usefulness of Mutations

Mutation, as a biological process that has been occurring as long as there has been life on this planet, is certainly fascinating and worthy of study. Mutant alleles such as those mentioned in the previous section obviously are invaluable in the study of the process of mutation itself. In this connection, they are used as **genetic markers**—they are used as representative genes, and their precise function is not particularly important except as a way to detect them.

In modern genetics, however, mutant genes have another important role, in which their precise function *is* important. We have already referred (in Chapter 2) to genetic dissection as an established approach to biological analysis. Mutant genes are like probes, which can be used to disassemble the constituent parts of a biological function and to examine their workings and interrelationships. Thus, it is of considerable interest to a biologist studying function A to have as many mutant forms affecting that function as possible. This has led to "mutant hunts" as an important prelude to any genetic dissection in biology. To identify a genetic variant is to identify a component of the process.

Message
Mutations can be used for two purposes: (1) to study the process of mutation itself and (2) to permit genetic dissection of biological function. The two main tools of genetic dissection are recombination dissection and mutational dissection.

Mutation Detection Systems

The tremendous stability and constancy of form of species from generation to generation suggest that mutation must be a rare process. This supposition has been confirmed, creating a problem for the geneticist trying to demonstrate mutation.

The prime need is for a **detection system,** designed in such a way that a mutant allele will make its presence known at the phenotypic level. Such a system ensures that any of the rare mutations that might occur will not be missed.

One of the main considerations here is that of dominance. The system must be set up so that recessive mutations will not be masked by a paired dominant normal allele. (Dominant mutations are less of a problem.) As an example, we can use one of the first detection systems ever set up—that used by Lewis Stadler in the 1920s to study mutation in corn from C (colored kernel) to c (white kernel). He crossed $CC \times cc$ and simply examined thousands of individual kernels on the corn ears that resulted from this cross. Each kernel represents a progeny individual. In the absence of mutation, every kernel would be Cc and show the colored phenotype. Therefore, the presence of a white kernel indicates mutation from C to c in the CC parent. Although laborious, this is a very straightforward and reliable approach (Figure 7-6).

This basic system can be extended to as many loci as can be conveniently made heterozygous in the same cross. For example,

$$\begin{array}{ccc} ++\ ++\ ++\ ++ & \times & aa\ bb\ cc\ dd \\ & \swarrow \quad \downarrow \quad \searrow & \\ \text{Normal} & \text{Mutant 1} & \text{Mutant 2} \\ +a\ +b\ +c\ +d & +a\ bb\ +c\ +d & +a\ +b\ +c\ dd \\ & \uparrow & \uparrow \end{array}$$

This type of test is called a **specific locus test** for detecting mutation. It has been used extensively in corn and in mammalian genetics.

In *Tradescantia* plants of the vegetatively propagated strain called 02, there is a simple detection system for somatic mutations. These plants are heterozygous for dominant blue and recessive pink pigmentation alleles of one gene. The pigmentation is expressed in the flower parts, petals, and stamens. In this plant, the stamens have hairs that are chains of single cells. Millions of single cells can be screened for pink cells representing somatic mutation of the blue to the pink allele (Figure 7-7).

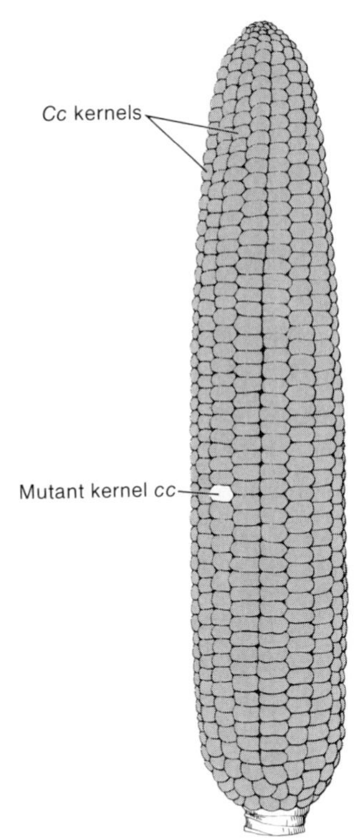

Figure 7-6. Detecting mutants by screening large numbers of progeny (kernels) in a corn cross of CC *(mutagenized)* × cc.

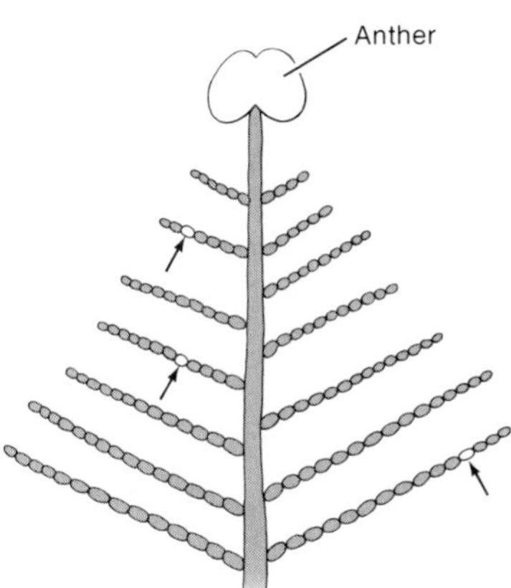

Figure 7-7. A Tradescantia *stamen. In the chains that constitute the lateral hairs, some cells (marked by arrows) are mutant. Shading represents a blue color; the lack of shading represents a pink color.*

In humans, mutation detection is theoretically simple but difficult in practice. For example, dominant mutations arising anew in a pedigree should be clearly recognizable. However, there are complications resulting from the possibility that a parent may have carried the gene but not expressed it. Thus the most convincing cases of dominant mutations are for conditions known to have high penetrance and expressivity (see Chapter 4 for definitions of these terms). Autosomal recessive mutations, on the other hand, may never be manifested in a phenotype.

Haploids have a great advantage over diploids in mutation studies. Here the detection system is quite straightforward; any newly arising allele will announce its presence unhampered by any dominant partner allele. In fact, the question of dominance or recessiveness need never arise: there is what amounts to a built-in detection facility. Let's look at some examples. In some cases, a direct identification of mutants is possible. In *Neurospora,* for example, auxotrophic adenine-requiring mutants have been found to map at several loci. One of these gene loci (*ad-3*) is unique, in that auxotrophic mutants accumulate a purple pigment in their cells when grown on a low concentration of adenine. Thus auxotrophic mutants of this gene may be detected simply by allowing single asexual spores to grow into colonies on medium with limited adenine. The purple colonies may be identified easily among the normal white colonies.

What about other auxotrophs? Usually there are no visual pleiotrophic effects such as that with *ad-3.* The most commonly used detection technique is called replica plating, but we postpone discussion of this technique until later in this chapter.

How Common Are Mutations?

If a detection system is available, one can set out to find mutations. One thing will become apparent: mutations are in general very rare. This is shown in some data collected by Stadler working with several corn loci (Table 7-2). Mutation studies of this sort are a lot of work! Counting a million of *anything* is no small task. Another feature shown by these data is that different genes seem to generate different frequencies of mutations; a 500-fold range is seen in the corn results. Obviously, one of the prime requisites of mutation analysis is to be able to measure the tendency of different genes to mutate. Two terms are commonly used to quantify mutation.

1. **Mutation rate.** This is a number that represents an attempt to measure the probability of a specific kind of mutation *event* over a specific unit of time. This is obviously getting close to the intrinsic mutation

Table 7-2. Forward mutation frequencies at some specific corn loci

Gene	Number of gametes tested	Number of mutations	Average per million gametes
$R \rightarrow r$	554,786	273	492.0
$I \rightarrow i$	265,391	28	106.0
$Pr \rightarrow pr$	647,102	7	11.0
$Su \rightarrow su$	1,678,736	4	2.4
$Y \rightarrow y$	1,745,280	4	2.2
$Sh \rightarrow sh$	2,469,285	3	1.2
$Wx \rightarrow wx$	1,503,744	0	0.0

tendency of a gene. The unit of time can be one of several. Instead of an actual time unit, such as hours or days, a unit such as an organismal generation, cell generation, or cell division is normally used. Consider the lineage of cells in Figure 7-8.

Obviously, only one mutation event (M) has occurred, so the numerator of a mutation rate is established. But what can be used as the denominator? The total opportunity for mutation, the "time" element, may be represented either by the total number of the lines in the diagram (14 total generations) or alternatively by the total number of actual cell divisions (7). Either is acceptable if stated clearly—for example, 1/7 cell divisions. (Note that we are dealing with generations, not generation cycles, of which there are three.)

In practice, mutation rates are not easy to obtain in most organisms. The reason is that the precise number of proliferative steps, which generated a population of cells, must be very carefully monitored in some way. We shall see a specific example later in the chapter.

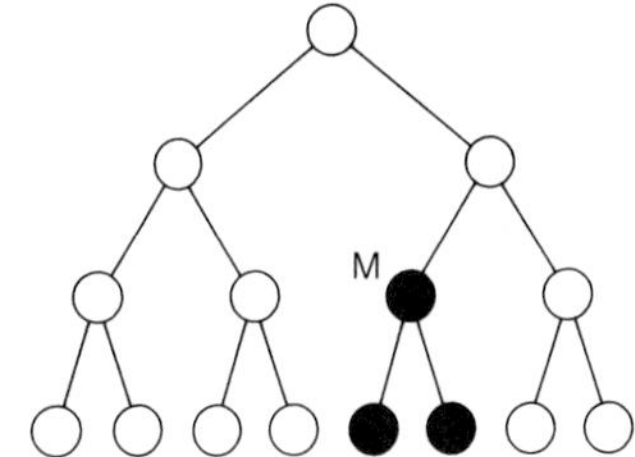

Figure 7-8. A simple cell pedigree showing a mutation at M.

2. **Mutation frequency.** This is the frequency at which a specific kind of mutation (or mutant) is found in a population of cells or individuals. The cell population can be of gametes (in higher organisms), or of asexual spores, or of almost any other cell type. This variable is much easier to measure than mutation rate. In our example in Figure 7-7, the mutation frequency in the final population of 8 cells would be $2/8 = 0.25$.

Some mutation rates and frequencies are shown in Table 7-3.

Selective Systems

The rarity of mutations is a problem, if you are trying to amass a collection of a specific type for genetic study. Geneticists respond to this problem in two ways. One approach is to use **selective systems.** These are techniques specially designed to facilitate the picking out of the desired mutant types from among the rest of the individuals. The other approach is to try to increase the mutation rate using **mutagens,** agents that have the biological effect of inducing mutations above the background (or spontaneous) rate.

Message
Obtaining rare mutations is facilitated by the use of selective recovery systems and/or mutagens.

Selective systems are many and varied. Their scope is immense and is limited only by the ingenuity of the experimenter. But they all have one thing in common, and that is elevated **resolving power.** That is, a selective system can

Table 7-3. Mutation rates or frequencies in various organisms

Organism	Mutation	Value	Units
Bacteriophage T2 (bacterial virus)	Lysis inhibition $r \to r^+$	1×10^{-8}	*Rate*: mutant genes per gene replication
	Host range $h^+ \to h$	3×10^{-9}	
Escherichia coli (bacterium)	Lactose fermentation $lac \to lac^+$	2×10^{-7}	*Rate*: mutant cells per cell division
	Histidine requirement $his^- \to his^+$	4×10^{-8}	
	$his^+ \to his^-$	2×10^{-6}	
Chlamydomonas reinhardi (alga)	Streptomycin sensitivity *str-s* → *str-r*	1×10^{-6}	
Neurospora crassa (fungus)	Inositol requirement $inos^- \to inos^+$	8×10^{-8}	*Frequency* per asexual spore
	adenine requirement $ad^- \to ad^+$	4×10^{-8}	
Corn	See Table 7-2		
Drosophila melanogaster (fruitfly)	Eye color $W \to w$	4×10^{-5}	*Frequency* per gamete
Mouse	Dilution $D \to d$	3×10^{-5}	
Humans: *to autosomal dominants*			
	Huntington's chorea	0.1×10^{-5}	
	Nail–patella syndrome	0.2×10^{-5}	
	Epiloia (predisposition to type of brain tumor)	$0.4\text{–}0.8 \times 10^{-5}$	
	Multiple polyposis of large intestine	$1\text{–}3 \times 10^{-5}$	
	Achondroplasia (dwarfism)	$4\text{–}12 \times 10^{-5}$	
	Neurofibromatosis (predisposition to tumors of nervous system)	$3\text{–}25 \times 10^{-5}$	
to X-linked recessives			
	Hemophilia A	$2\text{–}4 \times 10^{-5}$	
	Duchenne's muscular distrophy	$4\text{–}10 \times 10^{-5}$	
bone-marrow tissue-culture cells	Normal → azaguanine resistance	7×10^{-4}	*Rate*: mutant cells per cell division

SOURCE: R. Sager and F. J. Ryan, *Heredity,* John Wiley and Sons, 1961.

automatically distinguish between (or resolve) two alternative states—in this case, mutant and nonmutant. In other words, the selective system lets the material, instead of the experimenter, do the resolving work. We shall see that resolving power is an important aspect of many areas of genetic analysis. It is especially important in allowing the experimenter to select rare events of any kind. (You will remember a previous message that rare exceptions are often the key to understanding the normal situation.) Most examples we give are from microorganisms. This doesn't mean that selective systems are impossible in higher organisms, but merely that selection can be used to much better advantage in microbes. A million spores or bacterial cells are easy to produce—but a million mice, or even a million fruitflies involves a large-scale commitment of money, time, and laboratory space. Microbes appear frequently in the discussion that follows, so a few words on culturing and routine microbial manipulation are appropriate here.

Microbial Techniques

Microbes that we consider in this book are bacteria, fungi, and unicellular algae. All these can be regarded as haploid, but whereas fungi and algae are **eukaryotic** (having their chromosomes in a nucleus surrounded by a nuclear membrane), bacteria are **prokaryotic** (which means that their chromosomes are not enclosed in a separate compartment of any kind).

In liquid culture, these organisms proliferate as suspensions of individual cells. Each starting cell goes through repeated cell divisions so that, from each, a series of 2, 4, 8, 16, . . . descendant cells is produced. This exponential growth is limited by the availability of nutrients in the medium, but a dense suspension of millions of cells is soon produced. Fungi such as yeasts follow this pattern exactly, but a slightly different situation occurs in the mycelial fungi. Here the descendant cells remain attached as long chains called hyphae; thus a liquid culture started from asexual spores tends to look like tapioca pudding, with small fuzzy balls of hyphae in suspension, each originating from one spore.

In solid culture, usually on an agar-gel surface, descendant cells tend again to stay together, so that colonies are produced—one colony from each original cell in the suspension that is spread on the surface of the culture medium.

It is usually necessary to know how many cells you have in a culture. A suspension of cells may be counted by several methods.

1. **Microscope counts.** The suspension is placed in a chamber of known depth that is marked off in a grid of known dimensions. This device is known as a haemocytometer. The haemocytometer is placed under the microscope, and cells are counted directly.

2. **Turbidity.** Bacteria are too small to be counted conveniently under a light microscope, so they often are counted using turbidity measurements. The amount of light passing through suspensions of known density is measured, and a turbidity calibration curve is drawn up. Suspensions of unknown density can then be checked off on this calibration curve.

3. **Colony-forming units.** Suspensions of known volume and appropriate dilution are **plated** (spread) on the surface of nutrient medium solidified with agar (Figure 7-9). Each cell will form a colony, and the colonies can be counted with the unaided eye.

4. **Electronic cell counters.**

Figure 7-10 summarizes these methods.

What kinds of phenotypes may be examined in microorganisms? Morphological mutations affecting color, shape, and size of colony are useful but of limited occurrence. Other characters have been far more useful—for example, auxotrophic mutations (can these cells grow without this specific supplement?), resistance mutations (can these cells grow on this growth inhibitor?), and

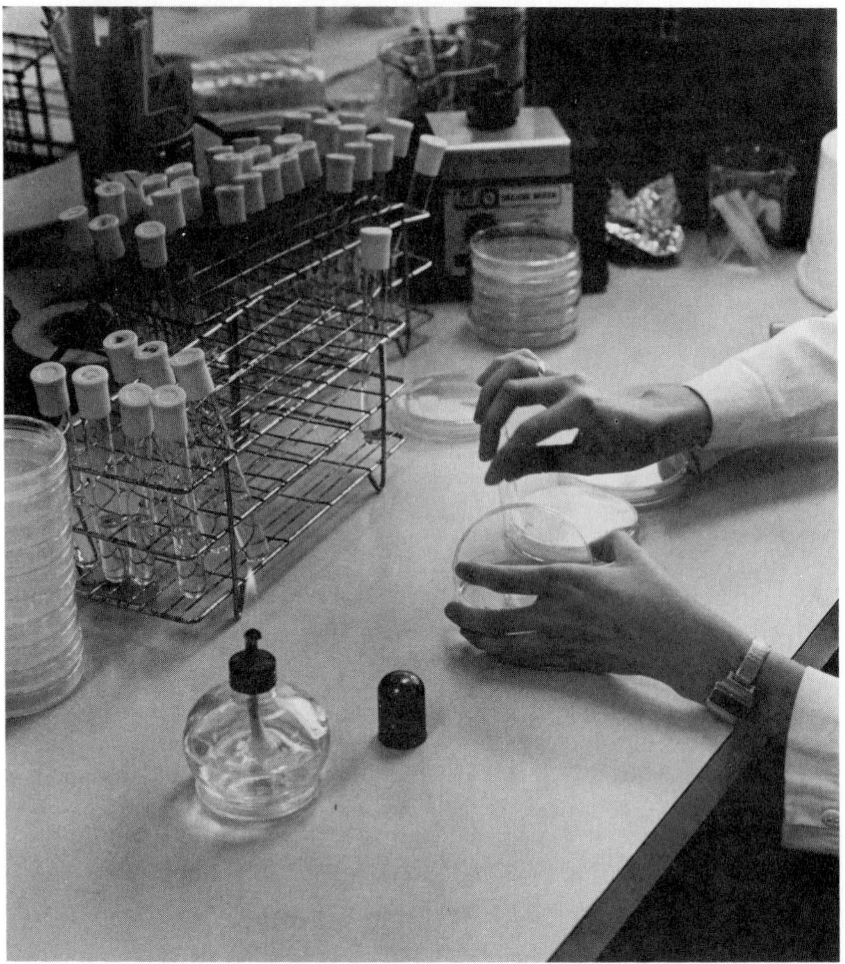

Figure 7-9. Plating microbial cells. A cell suspension of appropriate density is being poured over the surface of a plate of medium. Each cell will ultimately produce a visible colony. This is one of the routine techniques used by microbial geneticists, referred to several times in later sections of the book.

substrate-utilization mutants (can these cells utilize this sugar as an energy source, as wild types can?)

We now return to selective systems, and some examples thereof.

Reversion of Auxotrophs

For the detection of reversion of auxotrophy to prototrophy, there is a direct selection system. Take an adenine auxotroph, for example. A culture of the auxotrophic mutant is grown on adenine-containing medium. The cells are then

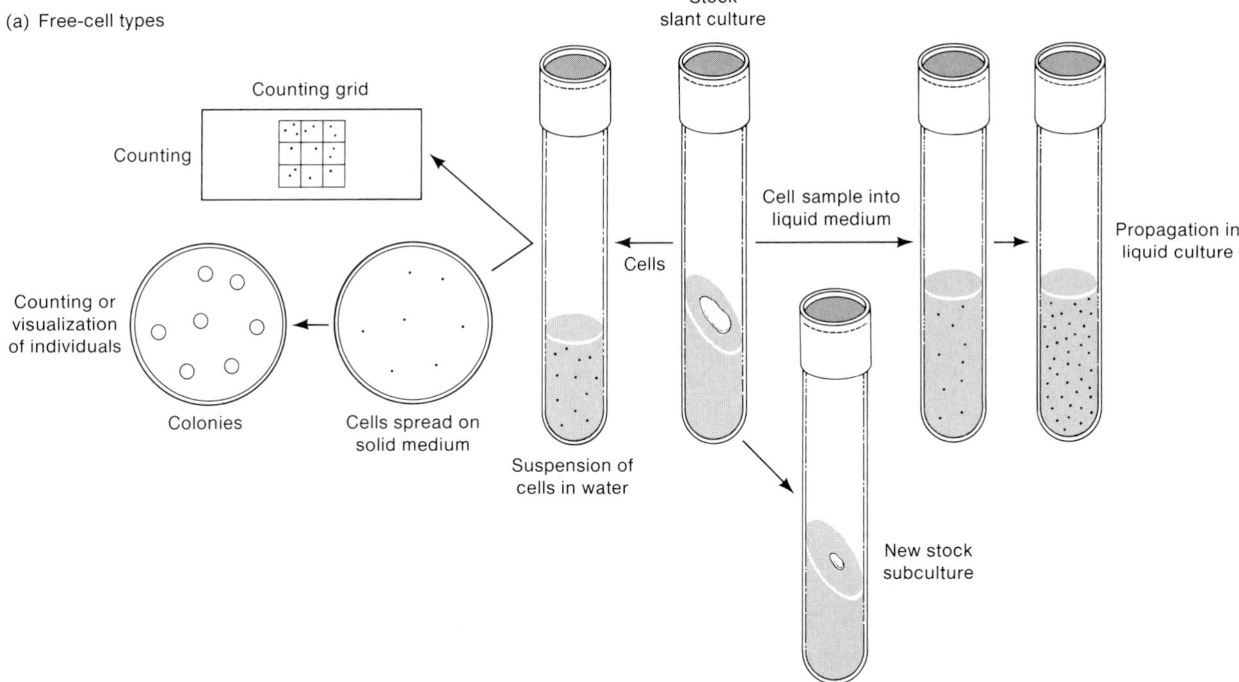

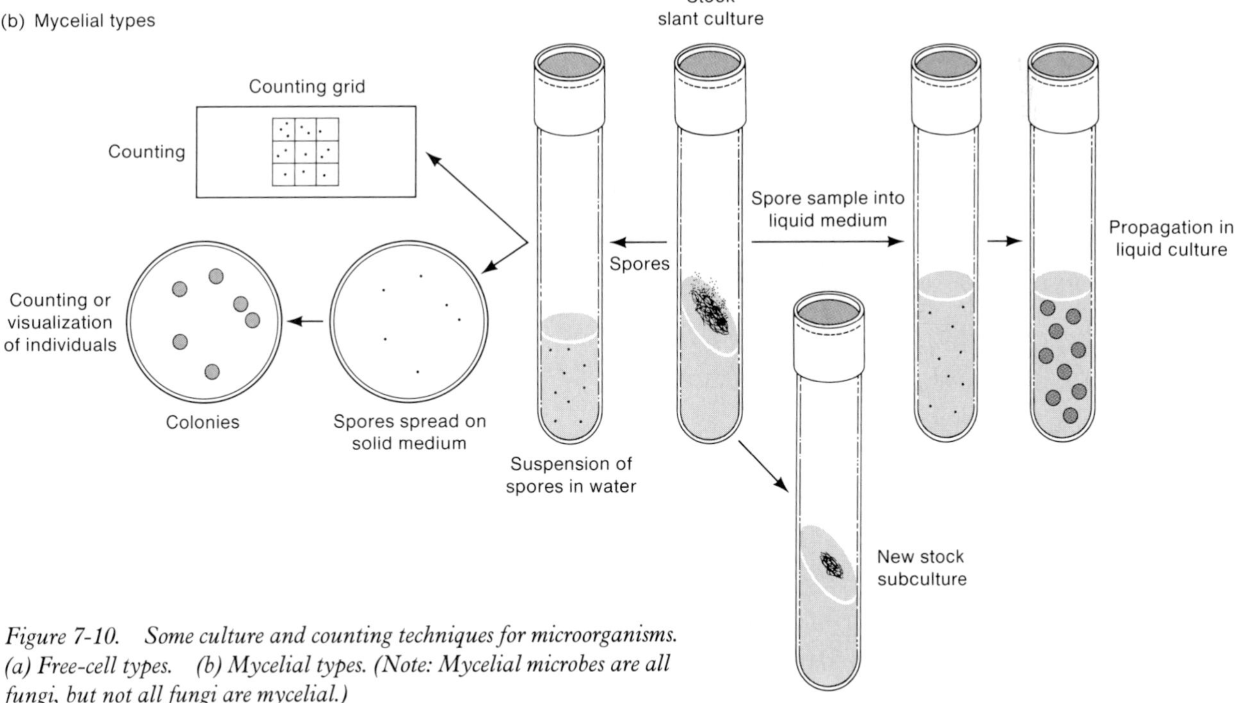

Figure 7-10. Some culture and counting techniques for microorganisms. (a) Free-cell types. (b) Mycelial types. (Note: Mycelial microbes are all fungi, but not all fungi are mycelial.)

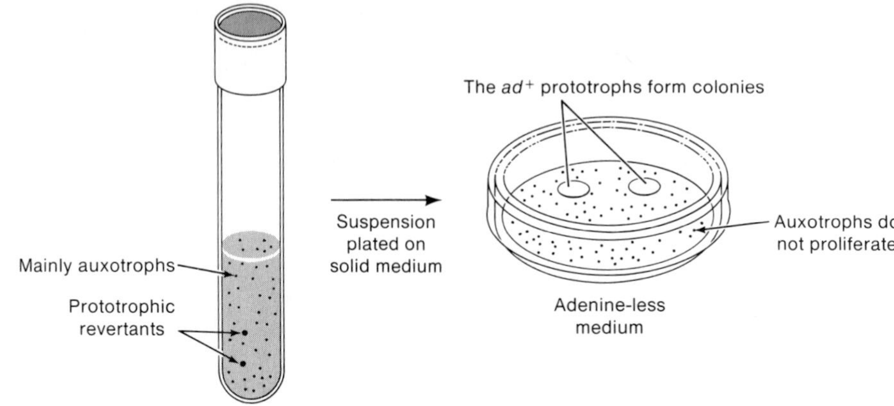

Figure 7-11. Selection for prototrophic revertants of auxotrophic mutants.

plated on solid medium containing no adenine. The only cells that can proliferate (grow and divide) on this medium are adenine prototrophs, which must have arisen by back mutation in the original culture (Figure 7-11). For most genes (not only those concerned with nutrition), the rate of reversion is generally much lower that the rate of forward mutation. (We shall explore the reason for this later.)

Filter Enrichment

Filter enrichment is used in mycelial fungi to select specific auxotrophic mutants. A suspension of (predominantly) prototrophic spores is grown in liquid culture with no growth supplements. (We have already mentioned this minimal medium, consisting of inorganic salts and an energy source like sugar.) Any auxotrophic mutants that had arisen in the original culture will not grow in this medium, but prototrophs of course will. The prototrophic colonies may be filtered off using a glass-fiber filter, allowing the auxotrophic cells to pass through. Of course, these auxotrophs will be a mixed bag of types—some requiring A, some B, and so on. If we are specifically interested in adenine auxotrophs, then we plate the heterogeneous suspension that comes through the filter on medium supplemented with adenine. Only adenine auxotrophs will respond to this medium and form colonies (Figure 7-12).

Penicillin Enrichment

An analogous technique is available for auxotroph selection in bacteria. Bacteria are in the main highly sensitive to the antibiotic penicillin, but only the proliferating cells are sensitive. If penicillin is added to a suspension of cells in

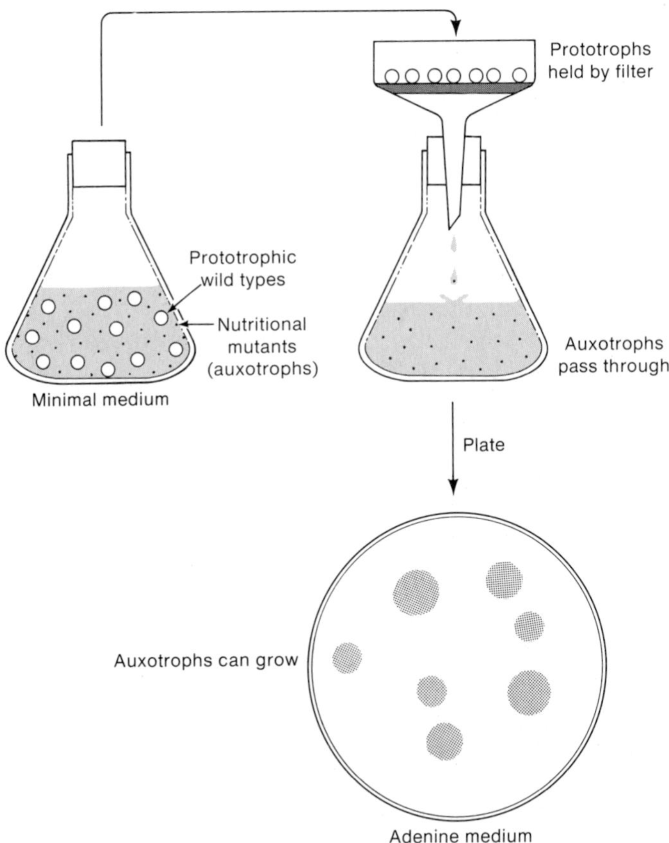

Figure 7-12 The filter-enrichment method for selecting forward mutations to auxotrophy in filamentous organisms. (This example involves mutation to adenine requirement.)

liquid culture, all the prototrophs are killed because they proliferate, but the auxotrophs survive! The penicillin can be removed by washing the cells on a filter. Then plating on medium supplemented with a specific chemical will reveal colonies of the auxotrophs specifically requiring that compound.

Resistance

Resistance to specific environmental agents not normally tolerated by wild types is easily demonstrated in microorganisms. We use an example that will serve to introduce the next subject.

Viruses, like most parasites, are highly specific with regard to the hosts they will parasitize. Some viruses are specific to bacteria—these are called either bacterial viruses, or **bacteriophages,** or **phages.** Phages have had a major role in the elucidation of our present level of understanding of molecular genetics. They appear frequently throughout the rest of the book. Although they are introduced properly in the chapter on bacterial genetics, we need them at this point to demonstrate resistance.

The gut bacterium *Escherichia coli* has many specific phages. One of these, called T1, was used early in bacterial mutation studies. T1 will attack and kill most *E. coli* cells, and a host of fresh viruses are liberated from the dead cell. T1 is a subcellular lollipop-shaped particle that can be seen only under an electron microscope, but the progress of phage infection can be followed on a plate by its effects. If a plate is spread with large numbers of bacteria (around 10^9) and phages, most of the bacteria will be killed. However, T1 phage-resistant bacterial cells survive and produce colonies that can be isolated. These individuals are called T-one resistant (*Ton*^{r}).

During these early studies on the selection of variants, the origin of these *Ton*^{r} bacterial mutants was questioned by Salvadore Luria and Max Delbrück (in 1943) in a classical experiment that was highly relevant to the study of mutation in general and, more specifically, to all selection experiments to follow. Although *Ton*^{r} individuals were obviously genuine mutants—after all, they represent a stable inherited phenotype—there was doubt about how they originated. Were the *Ton*^{r} colonies derived from cells that were genetically *Ton*^{r} *before* the exposure to T1 phage? (The *Ton*^{r} cells would have originated through random genetic change.) Or did the *Ton*^{r} genotype arise in response to the T1 exposure in a kind of physiological adaptation?

How is it possible to distinguish between these two alternatives? Let's assume that the resistant cells result from a random genetic change that can occur at any time during growth. If we initiate a bacterial culture with a small number of cells, and then the population increases, random genetic change to resistance to T1 phage could occur in any cell at any time. If we take a large number of small populations of cells and let each one expand into a large population, which is then exposed to T1 phage, the number of mutants in each population will vary considerably, depending on *when* the change occurred. On the other hand, if each cell has the same probability of becoming resistant by physiological adaptation, then in each population of cells there should be the same general frequency of survivors and little variation of that value (Figure 7-13).

Luria and Delbrück carried out such a test. Into each of twenty cultures containing 0.2 ml of medium and into one containing 10 ml, they introduced 10^3 *E. coli* cells per milliliter and incubated them until they obtained about 10^8 cells per milliliter. Each of the twenty 0.2 ml cultures were spread on plates that had a dense layer of T1 phages. From the 10-ml "bulk" culture, ten 0.2 ml volumes were withdrawn and plated. Many colonies were T1-resistant, as shown in Table 7-4.

A tremendous amount of variation from plate to plate was seen in the individual 0.2 ml cultures, but not in the samples from the bulk culture (which represented a kind of control). This situation cannot be explained by physiological adaptation, because all the samples spread had the same approximate number of

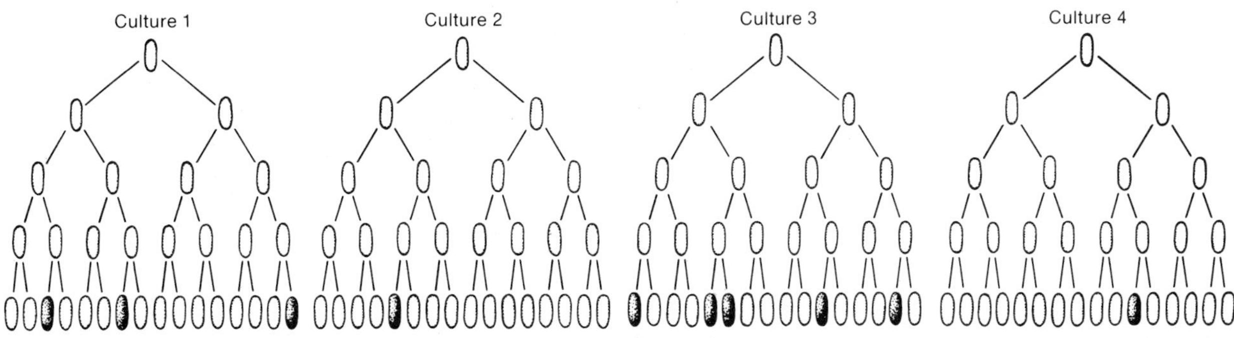

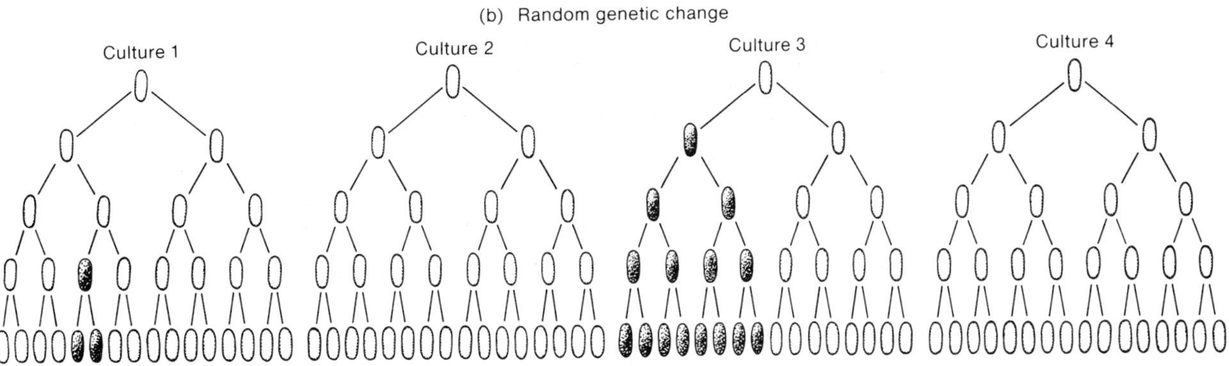

Figure 7-13. Cell pedigrees illustrating the expectations from two contrasting theories about the origin of resistant cells. (a) Physiological adaptation. (b) Random genetic change. (From G. S. Stent, and R. Calendar, Molecular Genetics, *2nd ed. Copyright © 1978, W. H. Freeman and Company.)*

cells. The simplest explanation is random genetic change, occurring either early (large number of resistant cells), or late (few resistant cells), or not at all (no resistant cells) in the 0.2 ml cultures.

This elegant analysis suggests that the resistant cells are *selected* by the environmental agent (here, phages) rather than produced by it. Can the existence of mutants in a population *before* selection be directly demonstrated? This was done by means of a **replica-plating** technique developed by Joshua Lederberg and Esther Lederberg in 1952. A sterile piece of velvet placed lightly on the surface of the petri plate will pick up cells wherever there is a colony (Figure 7-14). (Under a microscope, velvet looks like a series of needles, which explains why lint adheres to it so well and why it picks up colonies of cells.) On touching the velvet to another sterile plate, some of the cells clinging to the velvet will be in-

Table 7-4. Results of Luria and Delbrück's test

Individual cultures		Bulk culture	
Number of culture	T1-resistant	Number of culture	T1-resistant
1	1	1	14
2	0		
3	3	2	15
4	0		
5	0	3	13
6	5		
7	0	4	21
8	5		
9	0	5	15
10	6		
11	107	6	14
12	0		
13	0	7	26
14	0		
15	1	8	16
16	0		
17	0	9	20
18	64		
19	0	10	13
20	35		
Mean	11.3	Mean	16.7

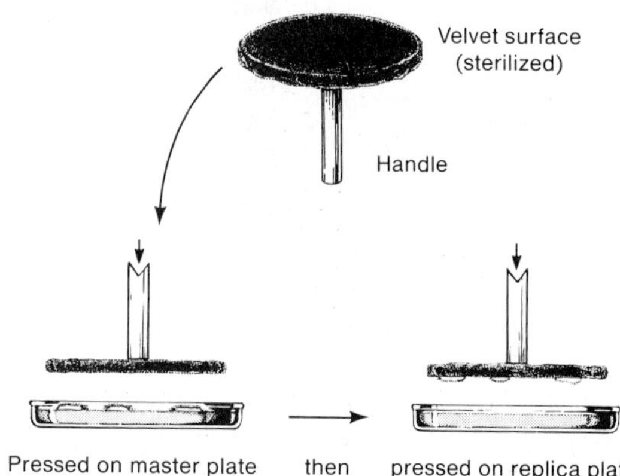

Figure 7-14. Replica-plating methodology. Replica plating is used to identify mutant colonies on a master plate, through their behavior on selective replica plates. (From G. S. Stent and R. Calendar, Molecular Genetics, *2nd ed. Copyright © 1978, W. H. Freeman and Company.)*

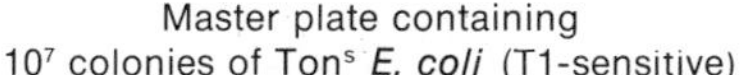

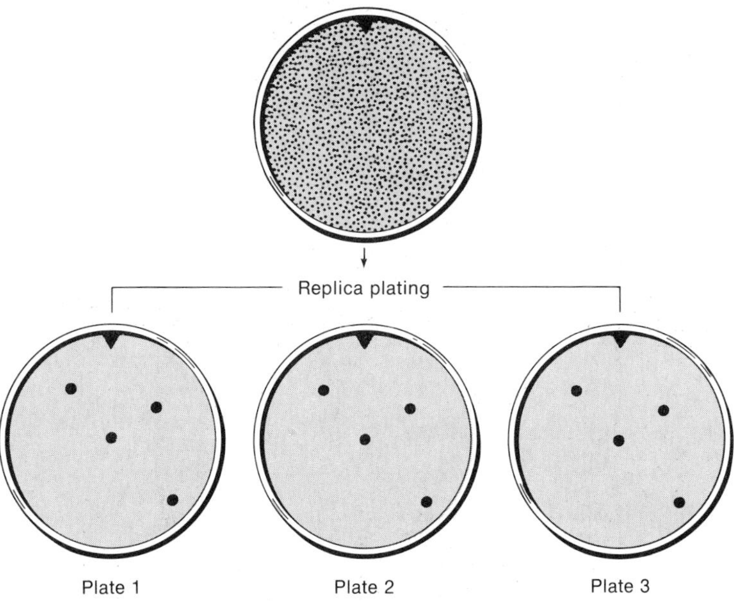

Figure 7-15. Replica plating to demonstrate the presence of mutants before selection. The identical patterns on the replicas show that the resistant colonies are from the master. (From G. S. Stent and R. Calendar, Molecular Genetics, *2nd ed. Copyright © 1978, W. H. Freeman and Company.)*

oculated onto the plate in the same relative positions as the colonies on the original "master" plate. This simple technique allows us to grow cells on a nonselective medium (either complete nutrients or free of antibiotics or phages) and then to transfer a copy of the colonies to a selective medium (either minimal or plus antibiotics or phages), as shown in Figure 7-15. The cells from some of the colonies on the master plate form colonies when transferred to the selective plate, in which case cells from those colonies on the master plate can be retested. If they are found to be mutant (in this example, resistant), we have proof that the mutation occurred *before* any selection was applied for the mutant. (It should be noted that there are some situations in which physiological adaptation can occur. The point for now is that mutations do not arise by physiological adaptation.)

Message
Mutation is a random process that can occur in any cell at any time.

A Way of Calculating Mutation Rate

In passing, we would like to show how Luria and Delbrück's test (called a **fluctuation test**) provides a way of calculating mutation rate, according to its strict definition. Consider the 20 cultures that were tested for T1 resistance. Here we have a situation that is well described by the Poisson distribution—a great opportunity for mutations to be found, but a large class having no mutations at all in the culture (11/20). Look back at Figure 7-8, and convince yourself that the number of cell divisions necessary to produce n cells is n minus the original number in the culture (it would be $8 - 1 = 7$ in the example in Figure 7-8). If n is very large, which it is in the fluctuation-test cultures, and the original number of cells was relatively very small, then an accurate estimate of the number of cell divisions is given by n itself. If the mutation rate per cell division is μ, each culture tube may be expected to have entertained an averge of μn mutation events. The Poisson distribution then tells us that the zero class (no mutants detectable) will equal $e^{-\mu n}$. Because we know that the zero class $= 11/20 = 0.55$ and that $n = 0.2 \times 10^8$, we can solve the following equation for μ:

$$0.55 = e^{-\mu(0.2 \times 10^8)}$$

We find that $\mu = 3 \times 10^{-8}$ mutation events per cell division. [Note that in the same data, the mutation frequency is $(1 + 3 + 5 + 5 + 6 + 107 + 1 + 64 + 35)/(20 \times 0.2 \times 10^8)$, which equals $227/(4 \times 10^8)$, or 5.7×10^{-7}.]

Microbial-Like Selection Techniques in Mammalian Cell Culture

There are techniques for growing cells of higher organisms (animals and plants) in culture. A few cells taken from a specific tissue, including cancerous tissue, will proliferate much like microbial cells in culture. Furthermore, many of the techniques used for mutant induction and selection in microbes can also be used on these higher cells, permitting the flourishing of **somatic cell genetics** of higher organisms as an active area of research. The combination of these mutation–selection techniques with sophisticated techniques of cell fusion and hybridization now makes possible a wide variety of *in vitro* manipulation on higher cells, including human cells.

One extensively worked cell-culture system uses "CHO" cells, derived from the hamster. In this system (and in others), many mutant cell types have been isolated. Let's have a look at some examples. In the first place, dominant mutations are common. Dominance is defined in terms of the phenotype observed when the mutant cell line is fused with normal cells. Resistance to such drugs as ouabain, α-amanitin, methotrexate, and colchicine are all examples of dominant mutations. Dominant mutations are not surprising: one might expect to find *only* dominant mutations because the cells are diploid.

Nevertheless, recessive mutants are common, too! Resistance to lectins, 8-azaguanine, or methotrexate, auxotrophic requirement for glycine, proline, or lysine, and various temperature-sensitive lethals are all examples of recessive mutants. (Note that certain kinds of mutant phenotypes can be either dominant or recessive in different mutants.)

How is it possible that recessive mutants show up in diploid cells? No one has the complete answer to this question. However, there is some evidence that, although the cells are predominantly diploid, they are hemizygous for very small sections of the genome. Recessive mutations would manifest themselves if they occurred in such regions. But whatever the mechanism of such mutations, their existence provides an invaluable tool for the higher-cell somatic geneticist.

Selection Systems in *Drosophila:* The ClB test

Compared with finding specific gene changes in microorganisms, the task of finding such changes in multicellular organisms is tremendously complex. In 1928 Hermann J. Müller devised a method for searching for any lethal mutation on the X chromosome in *Drosophila*. He was able to do this by constructing a chromosome called ClB, which carries a chromosomal inversion (labeled *C* for crossover suppressor; see Chapter 8), a lethal (*l*), and the dominant eye marker *Bar*. *Clb*/Y males die because of the hemizygosity for the lethal gene, but the chromosome can be maintained in heterozygous *ClB*/ + females. By mating wild-type males with *ClB*/ + females, one can test for lethal mutations anywhere on the X chromosomes in samples of the male gametes by mating single *ClB*/ + F_1 females with other wild-type males (Figure 7-16). You can see that,

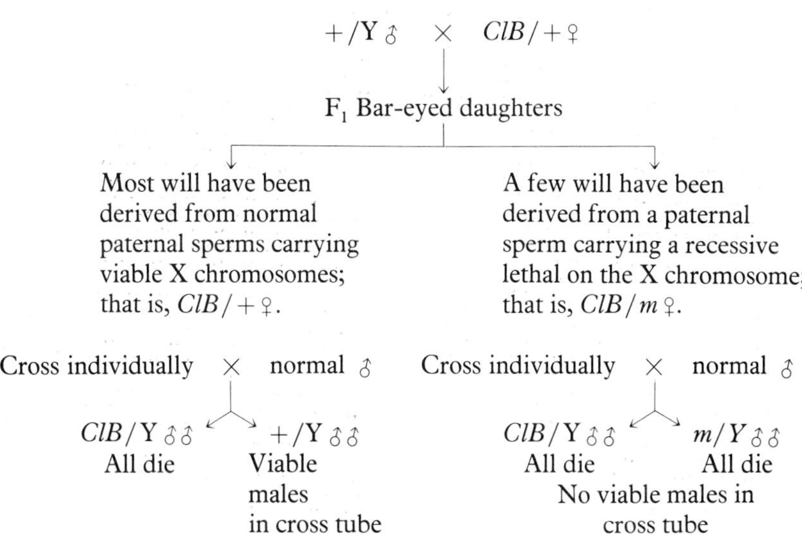

Figure 7-16. The ClB test for mutant detection in Drosophila. *The symbol m represents a recessive lethal mutation anywhere on the X chromosome.*

if a mutation has occurred on an X chromosome in one of the original male gametes sampled, then the F_1 female carrying that chromosome will not produce any viable male progeny. Note that only rarely will the new lethal mutation on the X chromosome be an allele of the lethal gene on the ClB chromosome (of course, when it is, the *ClB/l* female will die). The absence of males in a vial is very easily seen under low-power magnification and readily allows one to score for the presence of lethal mutations on the X chromosome. Müller found their frequency to be about 1.5×10^{-3}, still a relatively low value for an entire chromosome with its many different genes.

Mutation Induction by Mutagens

Müller's experiments showed that mutation is a rare event that occurs spontaneously at most loci. He then asked whether there were any agents that would increase the rate of mutation. Using the ClB test, he scored for sex-linked lethal frequencies after irradiating males with X rays and discovered a striking increase. This was the first indication of a **mutagen** (an agent that induces mutations). Mutagens have been invaluable tools not only for studying the process of mutation itself but for increasing the yield of mutants for other genetic studies. It is now known that many kinds of radiation will increase mutations (Table 7-5).

The harnessing of nuclear energy for weapons and fuel has become a social issue because of the mutagenic effect of radiation. We leave it to you to consider the pros and cons of the use of nuclear energy, but let's address one point that is often poorly understood. For any organism, the vast majority of newly formed mutations are deleterious. But you may ask, if organisms evolved through an advantage conferred by a mutant condition, why aren't many mutations an improvement? To answer this question, let's use an analogy often cited in this connection, in which a cell is compared to a highly complicated watch that has

Table 7-5. Relative efficiencies of various types of radiation in producing mutations in *Drosophila*

Type of radiation	Sex-linked recessive lethals per 1,000 roentgens*	Percentage of irradiated male X chromosomes
Visible light (spontaneous)	0.0015	0.15
X rays (25 Mev)	0.0170	1.70
β rays, γ rays, hard X rays	0.0290	2.90
Soft X rays	0.0250	2.50
Neutrons	0.0190	1.90
α rays	0.0084	0.84

*The roentgen (r) is a unit of radiation energy.

many delicate parts. This watch evolved through generations of minor changes in the design of the machine. If we expose the cogs and wheels by removing the back casing, close our eyes, and plunge a thick needle (analogous to a mutagen) into the workings, there is a remote possibility that this random hit will improve the efficiency of the watch, but the chances are overwhelmingly high that any change inflicted in this way will damage it. So it is with a cell or an organism. (We consider good molecular reasons for the detrimental nature of mutations in Chapter 16.)

Message
Most mutations are deleterious. Evolution is possible because of those very few that are not.

Many people have studied and are still studying the genetic effects of radiation. It is clear that, within a certain range of radiation dosage, point-mutation induction is linear—that is, if we double or halve the radiation level, the number of mutants produced will vary accordingly. From the kind of graph shown in Figure 7-17, it is possible to extrapolate to very low radiation levels and to infer very low frequencies of mutation induction. Because exposure to radiation from X-ray machines, to radioactive fallout from bomb testing, and to contamination from nuclear plants is very low, it is possible to conclude that the effects are negligible. Yet every year 200 million new gametes form 100 million new babies in the world. In this very large annual "mutation experiment," even low mutation frequencies are potentially translatable into large numbers of mutant babies.

Radiation doses are cumulative. If a population of organisms is repeatedly exposed to radiation, the frequency of mutations induced will be in direct proportion to the *total amount* of radiation absorbed. However, there are exceptions to both additivity and cumulativeness. For example, if mice given x rads (a biologi-

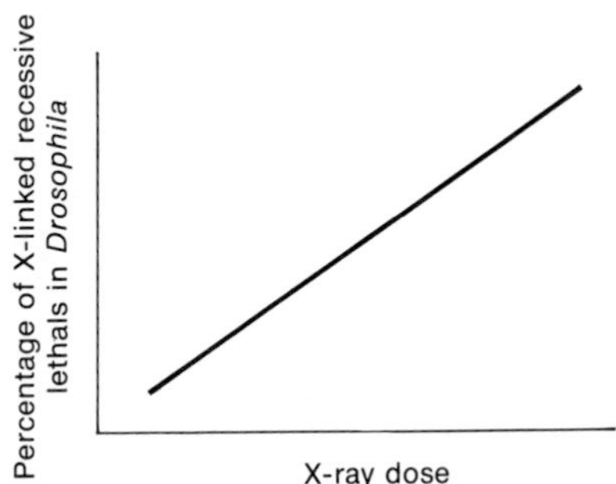

Figure 7-17. Linear relationship between X-ray dose and mutation.

cal measure of a dose of radiation) in one short burst (called an "acute" dose) are compared with those given the same dose over a protracted period of weeks or months (a "chronic" dose), significantly fewer mutations will be found in the chronically exposed group. This has been interpreted to mean that there is some form of repair of radiation-induced genetic damage with time.

We have not considered *how* ionizing radiation (the type discussed so far) causes mutation. Perhaps the simplest way to explain it is to compare a particle of radiation to a bullet; when the particle enters living cells, it strikes and breaks the genetic material directly. Or, it breaks molecules (such as water) into entities that themselves become highly reactive chemically and interact secondarily with the genetic material. We consider ultraviolet radiation, a nonionizing type, in detail in Chapter 16.

Of even greater importance for geneticists was the discovery that certain chemicals also may be mutagenic. The first demonstration of chemical mutagenesis was made by Charlotte Auerbach and J. M. Robson, who conducted experiments on mustard gases that were used in gas warfare. (You might be interested to know that their discovery was kept from publication by the British military for several years.) This initial study opened a floodgate to research into the mutagenic effects of a wide variety of chemicals.

The mutagenicity of chemicals is an important phenomenon because, in many cases, the chemical reactions responsible for the mutagenic action of a compound can be determined, thereby providing a clue to the molecular basis of the mutation. Furthermore, many chemicals are much less toxic to an organism that radiation is and yet give much higher frequencies of mutation. So, as a tool, chemical mutagens have been used to induce much of the wide array of mutants now available for genetic studies. Finally, great controversy now exists over the potential mutagenic effects of a host of molecules in the human environment, ranging from caffeine to pollutants, pesticides, and LSD. The molecular basis of mutation is considered in Chapter 16.

In recent years, mutation studies have taken on a new dimension. Whereas geneticists previously worried about environmental mutagens inducing germinal mutations that would increase our society's load of inherited diseases, it is now feared that somatic mutation may present equally fearsome consequences. It has been known for some time that there are agents like X rays, ultraviolet radiation, and various chemicals that are **carcinogenic.** In other words, these agents cause cancer if applied to the tissue of humans or experimental mammals. It has recently been shown that a very large proportion of carcinogens are also mutagens, and this has led to the speculation that these agents are causing cancer by causing somatic mutation. However, the situation is more complex than this. There are other obvious determining factors in some cancers, notably viruses and hereditary predispositions. Furthermore, the correlation doesn't work the other way: many potent mutagens are not carcinogenic. This is an extremely active area of mutation research, and we return to this topic in Chapter 16. Mutation has been implicated in other unexpected areas of medicine. For example, atherosclerosis, one of the major causes of death in western society, has been attributed to mutation in a cell of the arterial wall, which proliferates to form a fatty plaque. Mutation has also been implicated in the generation of antibody diversity as part of the body's immune system.

The Mutant Hunt

Mutations are very useful. In the same way that we can learn how the engine of a car works by tinkering with its parts one at a time to see what effect each has, we can see how a cell works by altering its parts one at a time by means of induced mutations. The **mutation analysis** is a further aspect of the process we have called genetic dissection of living systems.

The first stage of any mutational dissection is the **mutant hunt.** Before embarking on a mutant hunt, one usually has some specific biological process in mind. This process must be associated with some recognizable character of an organism. For example, you might be interested in phototrophic responses in algae—their tendency to swim toward the light. It would be comparatively easy to establish a selective system to recover mutants defective in this particular response. These mutants would then be tested for their single-gene inheritance, the number of different loci involved in the mutant set, and finally for specific cellular effects. In this way, the components of the response could be determined unambiguously.

So, armed with a biological phenomenon to study, a selection system, and a mutagen, you embark on a mutant hunt. How do you proceed? If you are going to induce mutations, the first thing needed is a "survival curve" in order to judge which dose to use for the best yield. Some standard survival curves after irradiation are shown in Figure 7-18. As you can see, some curves have a shoulder. The meaning of this shoulder has been the topic of much debate, but a formal ex-

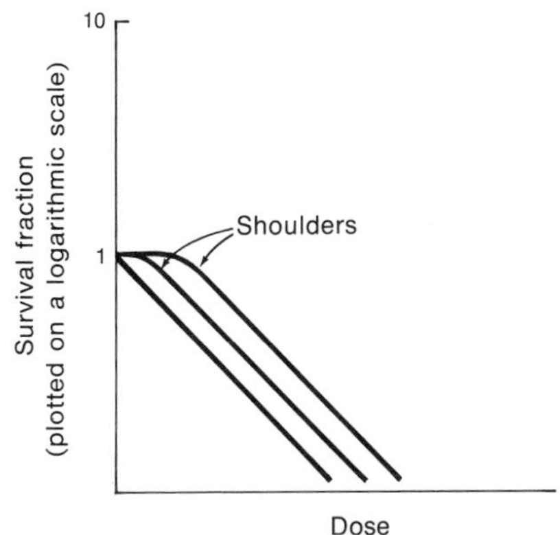

Figure 7-18. Representative simple survival curves after mutagen treatment.

planation is found in the target theory, in which the mutagen is compared to a bullet fired from a gun at targets. The "hits" are mutations that inactivate the targets and cause a decrease in survival percentage. What are the targets? Are they genes? chromosomes? nuclei? cells? This also has been the subject of great debate. The fact is that it is difficult to state precisely what the targets are, but let's develop the theory anyway and worry about the targets later.

We consider a situation in which each cell has two targets. We define dose (d) very simply as the average number of hits per target. The Poisson distribution will describe the actual distribution of hits, and the fraction of targets with *no* hits will be e^{-d}. Say that we have two targets, each of which must be hit at least once to cause death. We can calculate the probability of death as follows:

$$p(\text{that first target is hit}) = 1 - e^{-d}$$
$$p(\text{that second target is hit}) = 1 - e^{-d}$$
$$p(\text{that both targets are hit}) = (1 - e^{-d})^2$$

Let N_0 be the original population, and let N be the population that survives. The survival fraction, then, is N/N_0, and

$$\begin{aligned} N/N_0 &= 1 - (1 - e^{-d})^2 \\ &= 1 - [1 - 2e^{-d} + (e^{-d})^2] \end{aligned}$$

For high doses, the final term is negligibly small, so we have $N/N_0 = 2e^{-d}$.

Taking logarithms of both sides, we have $\ln (N/N_0) = -d + \ln 2$. Applying this to the general graph equation $y = mx + c$, we see that the slope of a plot of $\ln (N/N_0)$ against dose will be -1 (not of interest), but the intercept on the $\ln (N/N_0)$ axis will be the target number. This intercept is obtained by extrapolating the straight-line part of the curve backwards. Therefore, it is called the extrapolation number, and it is equal to the number of targets (in our example, it is 2; see Figure 7-19).

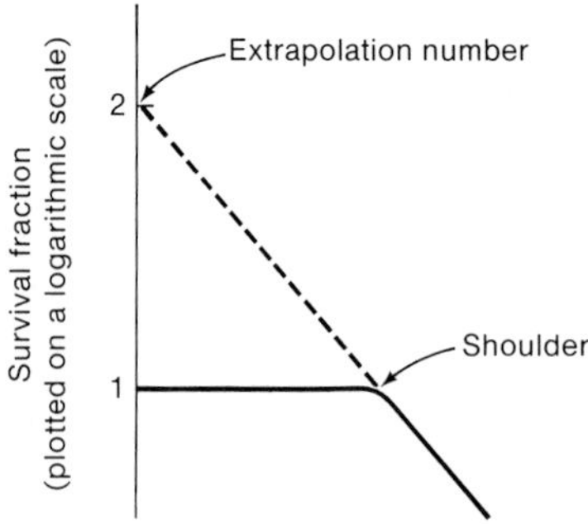

Figure 7-19. Using the survival curve to determine extrapolation number. By extrapolating the curve back to the vertical axis, the intercept point provides target number.

Thus we see that the shoulder represents a situation in which survival does not drop off immediately because there is more than one target. The presence of the shoulder gives us an extrapolation number that tells us the number of targets.

Well, what *are* targets in a genetic system? We have reached a level of understanding of the genome that allows us to think about this. In a haploid cell, one can see that a mutation in virtually any gene will inactivate (kill) the cell, and so it seems that the genome is the target. In a diploid, however, the situation is more complex because an inactivated gene on one chromosome will nearly always be complemented by an untouched gene on its homolog. Only simultaneous hits in the *same locus* of both homologs or a single hit that produces a dominant effect will cause inactivation of diploids. So, is the *gene* the target in a diploid cell? This is obviously closer to the truth, but it is still not fair to state that genes are the targets because *any* two hits in homologous loci will cause inactivation. It is like rolling two dice and saying *any pair* of numbers will cause inactivation. Thus, in diploids, the concept of the target is difficult to translate into real entities. Suffice it to say that in haploids or prokaryotes an extrapolation number of 1 is common, whereas in diploids a shoulder is nearly always seen, often with an extrapolation number of 2. The situation is further complicated by repair systems that can fix up the targets after a hit before death occurs. Also, the type of mutagen and the genetic ancestry of the strains often greatly affect the shape of the survival curve. So, target theory may be difficult to translate literally, but it does provide one of the few interesting and useful models for explaining survival curves.

Once a survival curve is obtained, a dose is chosen that will not give too low a survival. The best dose varies with mutagenic agent and organism. At low survival levels there is always the possibility of picking up multiple mutations, and these are to be avoided. A *double* mutant is not always immediately recognized, and if the stock is treated as though it carried a single mutation, all sorts of strange results can be obtained. (For example, what would happen if you tried to map a sex-linked lethal in *Drosophila* that really is two lethals?) Using the optimal dose, a battery of mutants may be amassed for study.

Mutation Breeding

Mutation has uses other than in mutational dissection of biological systems. We saw in Chapter 2 that one way of breeding a better crop plant is to make a hybrid and then select desired recombinants from the progeny generations. That approach makes use of the variation naturally found between available stocks or isolates from nature. Another way to generate variability for selection is to treat with a mutagen. In this way, the variability is produced through human intervention.

A variety of procedures may be used. Pollen may be mutagenized, then used in pollination. Dominant mutations will appear in the next generation, and further generations of selfing will reveal recessives. Alternatively, seeds may be mutagenized. A cell in a seed's enclosed embryo may become mutant, and then it may become part of germinal tissue or somatic tissue. If the mutation is in somatic tissue, any dominant mutations will show up in the plant derived from that seed (Figure 7-20), but this will be the end of the road for such mutations.

(a)

(b) (c)

Figure 7-20. Irradiating seeds of Collinsia grandiflora *with gamma rays to produce mutations in the embryo. (a) The effects of increasing dose* (left to right) *on survival and vigor. (b) A somatic mutation in a survivor, causing leaf-shape abnormalities. (c) Another somatic mutation causing a mutant sector. Germinal mutations also were recovered in the next generation.*

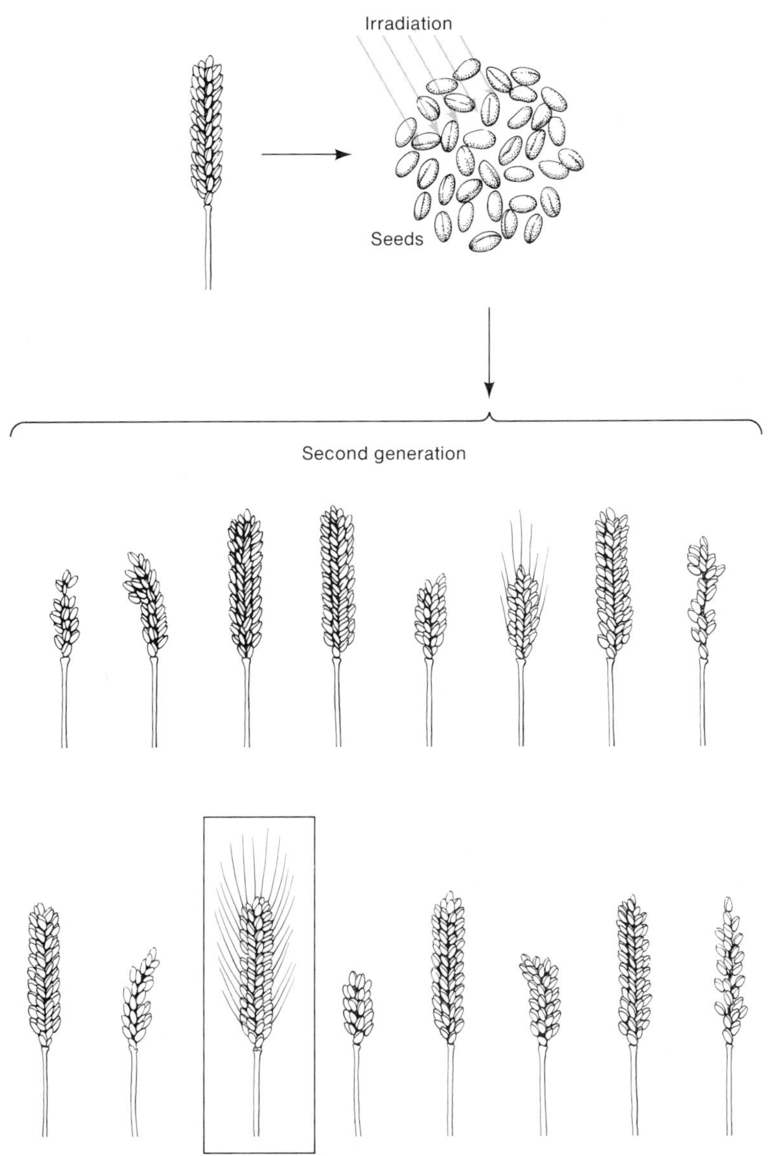

Figure 7-21. Mutation breeding in crops. (Courtesy of Björn Sigurbjörnsson, "Induced Mutations in Plants." Copyright © 1970 by Scientific American, Inc. All rights reserved.)

Germinal mutations will show up in later generations where they can be selected as appropriate. Figure 7-21 summarizes mutation breeding.

In this chapter, we have seen how gene mutation is not only of biological interest in itself but can also be put to a variety of experimental uses. In the next chapter, we will see that much the same kind of statement can be made about chromosome mutation.

Summary

Gene mutation may occur in somatic cells or germinal cells. Within these two categories there are several kinds of mutations, including morphological, lethal, conditional, biochemical, and resistant mutations.

Mutations can be used to study the process of mutation itself or to permit genetic dissection of biological functions. In order to carry out such studies, however, it is necessary to have a system for detecting mutant alleles at the phenotypic level. In diploids, dominant mutations should be easily detected; recessive mutations, on the other hand, may never be manifested in the phenotype. For this reason detection systems are much more straightforward in haploids, where the question of dominance does not arise.

Because mutations are rare, geneticists use selective systems or mutagens (or both) to obtain mutations. Selective systems automatically distinguish between mutant and nonmutant states and have been used mainly in microbes. Mutagens are valuable not only in studying the mechanisms of mutation but also for inducing mutations to be used in other genetic studies. In addition, mutagens are frequently used in crop breeding.

Problems

1. More than 10,000 new molecules are synthesized every year. Many of them could be mutagenic. How would you readily screen large numbers of compounds for their mutagenicity?

2. How would you use the replica-plating technique to select arginine-requiring mutants of haploid yeast?

3. Using the filtration-enrichment technique, you do all your filtering with minimal medium and do your final plating on complete medium that contains every known nutritional compound. How would you find out what *specific* nutrient is required? After replica plating onto every kind of medium supplement known to science, you still can't identify the nutritional requirement of your new yeast mutant. What could be the reason(s)?

4. *Meiotic* (*mei*) mutations are known in several organisms. The observable effect of this mutation on the phenotype is to cause a drastic disruption of the meiotic process, often producing only aborted meiotic products. In *Neurospora,* a cross is made between two strains called 1 (mating type A) and 2 (mating type a). The cross is perfectly fertile, with no spore abortion. Seven randomly selected progeny labeled 3 through 9 are intercrossed and crossed to the parental strains. Table 7-6 shows the fertility of the resulting progeny. In the table, F means fully fertile, and O means

almost sterile (with many aborted white ascospores, and with the few normal spores showing reduced recombination throughout the genome). Explain these results in terms of a recessive *mei* mutation in one of the parental strains. Suggest a possible mode of action for the *mei* mutation at the cellular level.

Table 7-6.

Strains of mating type *a*	Strains of mating type *A*				
	1	3	4	5	6
2	F	F	O	F	O
7	F	F	O	F	O
8	F	F	O	F	O
9	F	F	F	F	F

5. An experiment is initiated to measure the reversion rate of an *ade-3* mutant allele in haploid yeast cells. One hundred tubes of liquid adenine-containing medium are each inoculated with a very small number of mutant cells and incubated until there are 10^6 cells per tube. Then the contents of each tube are spread over a separate plate of solid medium containing no adenine. The plates are observed after one day, and colonies are seen on 63 plates. Calculate the reversion rate of this allele per cell division.

6. Suppose that you want to determine whether caffeine induces mutations in higher organisms. Describe how you might do this (include control tests).

7. Some mealybugs called coccids have a diploid number of 10. In cells of males, 5 of the chromosomes are always seen to be heterochromatic, and 5 are euchromatic. In female cells, all 10 are always euchromatic. Spencer Brown and Walter Nelson-Rees gave large doses of X radiation to males and females and obtained the following results:

   Female (X-rayed) × Male (non-X-rayed)
   ↓
   No surviving progeny

   Female (non-X-rayed) × Male (X-rayed)
   ↓
   Lots of male progeny
   but no female progeny

   Interpret these results. (NOTE: *Heterochromatic* means densely staining, and *euchromatic* means not densely staining.)

8. Assume that, in a leaf, an albino mutation is present in the inner meristem layers but not in the epidermal layer. During development, a few cells of the epidermis migrate into the photosynthetic layers of the leaf. What do you predict will be the appear of the leaf? (NOTE: The epidermis normally has no chlorophyll.)

9. In corn, a single gene determines presence (*Wx*) or absence (*wx*) of amylose in the cell's starch. Cells that have *Wx* stain blue with iodine, and those that have only

wx stain red. Design a system for studying frequency of rare mutations from $WX \rightarrow wx$ without using acres of plants. (HINT: You might start by thinking of an easily studied cell type.)

10. A new high-yielding strain of wheat suffers from the disadvantages that it tends to "lodge" (fall over) during storms. You have an X-ray source (and a couple of years) to correct this defect. How would you proceed? State which part of the plant (or which stage of its life cycle) you would treat, what results you would look for, which generation you might expect your results in, and so forth.

11. Suppose that you cross a single male mouse from a homozygous wild-type stock with several homozygous wild-type, virgin females. The F_1 consists of 38 wild-type females and males, and 5 black males and females. How can you explain this result?

12. Joe Smith accidentally receives a heavy dose of radiation in the gonadal region. Nine months later, his girlfriend has a daughter, Mary. Mary appears perfectly normal, and she eventually marries a homozygous normal man. Figure 7-22 shows their pedigree.

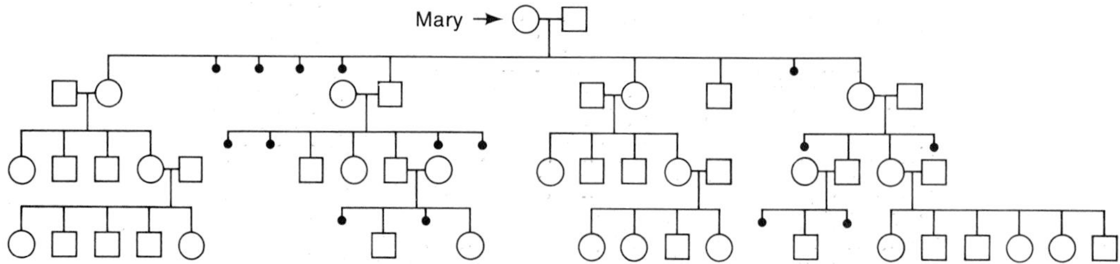

Figure 7-22.

a. What possible genetic mechanisms could explain these results?

b. How would you prove which explanation is correct?

13. A man employed for several years in a nuclear power plant becomes the father of a hemophiliac boy, the first in the extensive family pedigrees of both his own and his wife's ancestry. Another man, also employed for several years in the same plant, has an achondroplastic dwarf child, the first occurrence in his ancestry or in that of his wife. Both men sue their employer for damages. As a geneticist, you are asked to testify in court; what do you say about each situation? (NOTE: Hemophilia is an X-linked recessive, and achondroplasia is an autosomal dominant.)

14. In a large maternity hospital in Copenhagen, there were 94,075 births. Ten of the infants were achondroplastic dwarfs. (Achondroplasia is an autosomal dominant showing virtually full penetrance.) Only two of the dwarfs had a dwarf parent. What is the mutation frequency to achondroplasia in gametes? Do you have to worry about reversion rates in this queston? Explain.

15. One of the jobs of the Hiroshima/Nagasaki Atomic Bomb Casualty Commission was to assess the genetic consequences of the blast. One of the first things they studied was the sex ratio in the offspring of the survivors. Why do you suppose they did this?

16. The government wants to build a nuclear reactor near the town of Poadnuck. The townspeople are very upset about the possibility of an accidental release of radioactivity and are putting up a stiff fight to keep it out. They call on you, a geneticist, to inform them about the biological hazards of an accident. What would you say in your speech?

17. The nuclear plant has been built and you are now a professor at Poadnuck State College. A radical group manages to infiltrate the plant and blow it up, releasing a large amount of radioactivity in the area. Fortunately, a thunderstorm washes most of the radioisotopes to the ground, preventing widespread contamination. You set out immediately to determine the genetic effects of the radiation. How do you go about it? (Remember, you must have controls.)

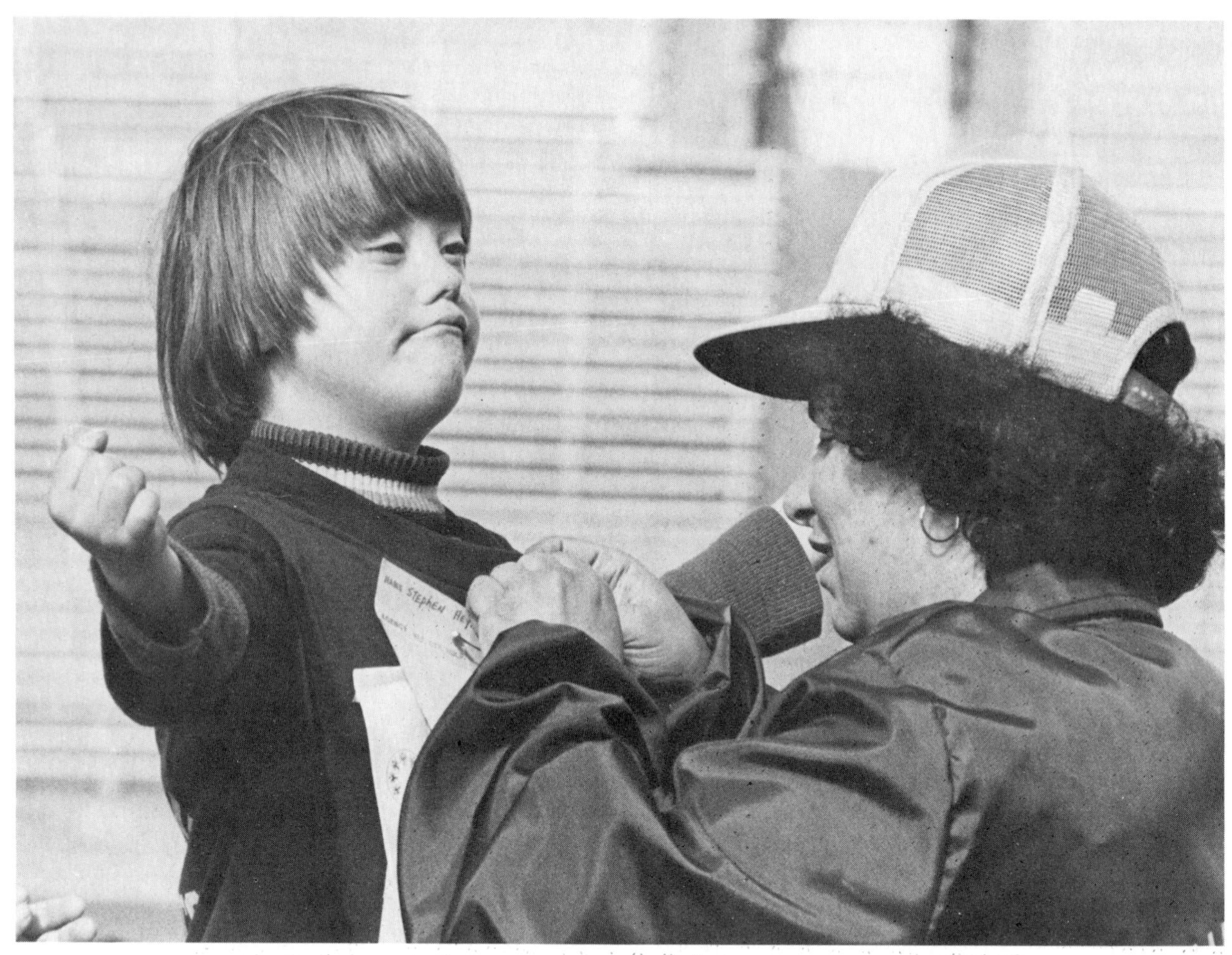

Young athlete with Down's syndrome (trisomy-21). (Copyright © Cheryl A. Traendly/Icon.)

8

Chromosome Mutation

The gene mutations discussed in Chapter 7 are not detectable by examination of chromosomes. A chromosome bearing a gene mutation looks the same as one bearing the normal allele. However, visible changes in the genome do occur regularly, and these changes are called **chromosome mutations** or **chromosome aberrations**. We see in this chapter that such changes may be detected not only with a microscope but also by standard genetic analysis.

Chromosome mutations are changes in the genome involving chromosome parts, whole chromosomes, or whole chromosome sets. Thus far we have treated the eukaryotic chromosome rather naively as merely a string of genes. A great deal is now known about the genetic organization of chromosomes (see Chapter 14), but there still are major gaps in our understanding at the molecular level. Luckily, the simplistic approach is not a major impediment in our understanding of chromosomes at the behavior level. A large amount of information has been amassed about normal chromosome behavior and about how this behavior can go awry in the creation of chromosome mutations. This information has come from cytological and genetic studies, whose union constitutes the discipline of **cytogenetics.**

Genetics, perhaps more than any other field of biology, makes extensive use of deviations from the norm, and the study of chromosomes is no exception.

Although this means of investigation may seem very esoteric to some, the findings have proven to be very important in applied biology—especially in agriculture, animal husbandry, and medicine. Many genetic tricks and devices that geneticists routinely use to build certain genotypes are discussed in this chapter. Finally, although we shall not give this point the attention it deserves, many of the aberrations we discuss here have been important in building the theories of evolution and speciation.

Cytological Properties of Chromosomes

First, let's consider some purely descriptive aspects of chromosomes. From pictures you have seen, chromosomes may appear to be skinny, wormlike squiggles with few distinctive landmarks for identification other than size. However, closer inspection can and often does show considerable variation. Chromosomes may vary immensely in size from species to species, from less than one micron (1 μ, or 10^{-6} m) to several hundred microns in length. Even within a single organism, chromosomes may vary in size from tissue to tissue. Amphibian oocytes, for example, contain giant so-called lampbrush chromosomes, which may be as much as 800 μ in length, whereas their somatic cell chromosomes are only a few microns long. Techniques of electron microscopy and chemistry have revealed the chromosome as a highly condensed complex of nucleic acids and proteins.

At any given moment in cell division, segments of chromosomes may exhibit staining differences. Some regions (generally found near the centromere and at the tips, or **telomeres**) are darkly stained and have been designated **heterochromatin,** whereas other areas are less densely stained and are termed **euchromatin.** The distribution of heterochromatin and euchromatin within chromosomes is usually constant from cell to cell and is thus, in a sense, hereditary. Heterochromatin found in the same area of both homologs in a cell is called **constitutive heterochromatin.** There are other cases, such as the X chromosomes of mammalian females and the paternally derived chromosomes in some insects, in which one chromosome may be heterochromatic whereas its homolog is not. This is called **facultative heterochromatin.** Although the genetic difference between heterochromatin and euchromatin is not yet clear, there is good evidence that heterochromatic regions are genetically inactive in the cells in which they are seen.

Within a chromosome, there may be small areas of intense staining connected by lightly stained areas. These correspond to the chromomeres on which Belling built his crossover model (Chapter 5). Recently, staining procedures have been developed that result in patterns of lightly and darkly stained regions or bands. The banding patterns are highly specific to each chromosome pair and have allowed unequivocal identification of all 23 chromosome pairs in humans. They are equally useful in identifying specific chromosomes in a wide range of other organisms. We have already encountered this banding in an earlier chapter (see Figures 6-27 and 6-28).

There are other landmarks for identification of specific chromosomes. The position of the centromere (alias spindle-fiber attachment, or kinetochore) within a chromosome determines the length of the **arms** on each side and the chromosome's shape during anaphase. Thus, the terms **telocentric, acrocentric,** and **metacentric** describe chromosomes in which the centromere is located anywhere from either end to the middle, producing differences in shape at anaphase that range from a simple rod through a J to a V (Figure 8-1).

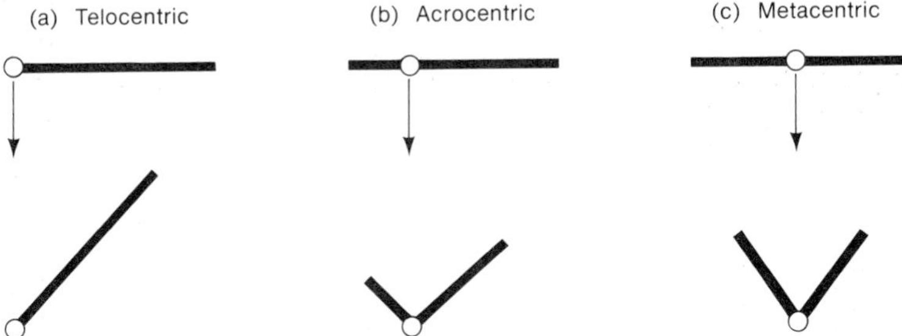

Figure 8-1. The classification of chromosomes by the position of the centromere. A telocentric chromosome has its centromere at one end. When the chromosome moves toward one pole of the cell during anaphase of cellular division, it appears as a simple rod. An acrocentric chromosome has its centromere somewhere between the end and the middle of the chromosome. During anaphase movement the chromosome will appear as a J. A metacentric chromosome has its centromere in the middle and will appear as a V during anaphase.

The region of the centromere may appear pinched and is called the **primary constriction.** Other constrictions, the **secondary constrictions,** indicate the position of the nucleolus-organizer region, the region on a specific chromosome seen to be physically associated with the nucleolus. In some organisms, such as Lepidoptera, centromeres are "diffuse," so that spindle fibers attach all along the chromosome, in which case the chromosome's migration to one of the poles is parallel to the metaphase plate. When such a chromosome is broken, both parts can still migrate to the pole because they are both attached by spindle fibers. In contrast, a break in a chromosome that has a single centromere results in the inability of the **acentric** (without a centromere) fragment to move owing to the loss of the point of spindle-fiber attachment (Figure 8-2).

An important tool for the cytological study of chromosomes was supplied by the rediscovery of so-called giant chromosomes in certan organs of the Diptera. In 1881, E. G. Balbiani had recorded peculiar structures in the nuclei of certain secretory cells of two-winged flies. These structures were long and sausage-shaped and marked by swellings and cross striations. Unfortunately, he did not recognize them as chromosomes, and the report remained buried in the literature. It was not until 1933 that Theophilus Painter, Ernst Heitz, and H. Bauer rediscovered them and realized they are chromosomes.

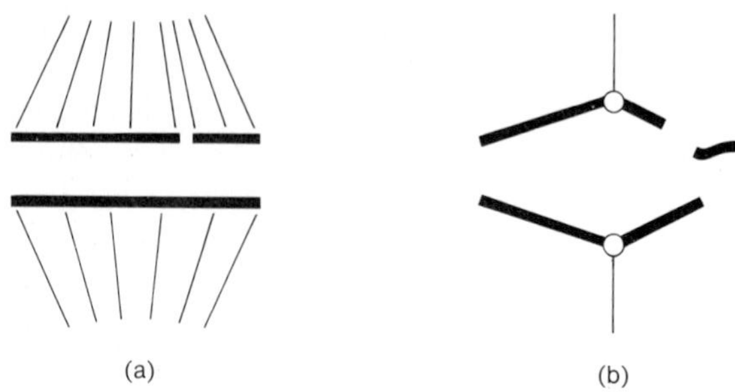

Figure 8-2. The spindle fiber attachment to chromosomes. On the left is an example of a relatively rare, diffuse centromere, with many spindle fibers attaching along the length of a single chromosome. A break in the chromosome will not result in the loss of any chromosomal material. The diagram on the right shows spindle-fiber attachment to a single centromere; this is the more usual form. In this case, a break in the chromosome will result in an acentric fragment, which will be lost during cellular division.

In secretory tissues—such as Malpighian tubules, rectum, gut, footpads, and salivary glands of Diptera—the chromosomes apparently replicate their genetic material many times without actual separation into distinct chromatids. Thus, as the chromosome increases in replicas, it elongates and thickens. *Drosophila* has an n number of 4, and only 4 chromosomes are seen in the cells of such tissues because, for some reason, homologs are tightly paired. Furthermore, all four are joined at the **chromocenter,** which represents a coalescence of the heterochromatic areas around the centromeres of all four chromosome pairs. Salivary gland chromosomes are shown in Figure 8-3, in which L and R stand for arbitrarily assigned left and right arms.

The formation of such giant chromosomes is a process called **endomitosis.** The bundles of multiple replicas are called **polytene chromosomes.** The most commonly examined polytene chromosomes are in the salivary gland nuclei. These nuclei never divide. Along the chromosome length, characteristic stripes called **bands,** which vary in width and morphology, can be observed and identified. In addition, there are regions that may at times appear swollen (**puffs**) or greatly distended (**Balbiani rings**); these are presumed to correspond to regions of genetic activity. Recently, it has been found in *Drosophila* that, in general, each band contains the genetic material of a single gene. (However, the significance of the bands in human and other chromosomes is not known.) The polytene chromosomes corresponding to linkage groups have been identified through the use of chromosome aberrations. Such aberrations, as we shall see, have also been useful in specific localization of genes along the chromosomes.

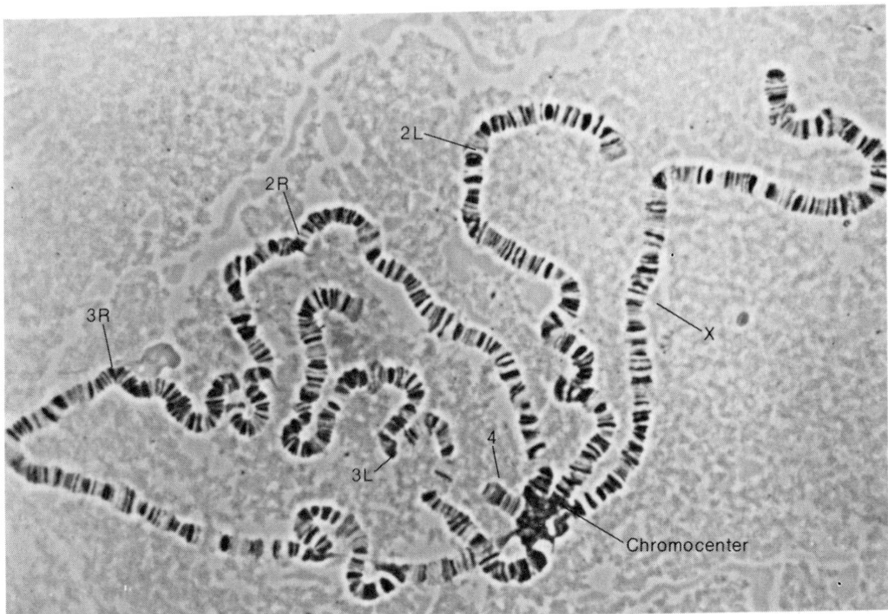

Figure 8-3. Drosophila *larval salivary chromosomes. Note that the two homologs of each chromosome pair have fused to form just one banded unit. The centromeres of all chromosomes are united at the common chromocenter. Thus, the telocentric X chromosome appears as a single unit, while the metacentric second and third chromosomes are seen as left (2L or 3L) and right (2R and 3R) portions. (Photograph by Tom Kaufman.)*

So much for some cytological properties of normal chromosomes. The rest of this chapter deals with what has been termed *chromosome mechanics.*

First, let's classify chromosome aberrations.

1. Aberrations concerning **chromosome structure:**
   a. deletions
   b. duplications
   c. inversions
   d. translocations
2. Aberrations concerning **chromosome number:**
   a. Euploids: variation in the number of sets of chromosomes. The abnormal situation can be one set (monoploid, or sometimes haploid), two sets (diploid), three sets (triploid), four sets (tetraploid), and so on. Sets of three or more are called **polyploid.** In the case of some polyploids, a different set nomenclature is used, in which the letter x stands for the number of chromosomes in the basic set. Wheat, for example, is hexaploid, or $6x$, where $x = 7$. Because the

plant follows basically a diploid type of life cycle, its 42 chromosomes are still known as the diploid (or $2n$) number. Hence, for wheat, $x = 7$ and $n = 21$.

b. Aneuploids: addition or subtraction of chromosomes from entire sets. For example, in a diploid organism, $2n - 1$ is monosomic, $2n + 1$ is trisomic, $2n + 1 + 1$ is double trisomic, and $2n - 2$ is nullosomic (here the two chromosomes lost are homologs). In a haploid organism, $n + 1$ would be called disomic.

We must emphasize again that meiotic pairing is incredibly precise. There is good evidence that homologous regions generally pair gene for gene, and chromosomes will go through remarkable contortions to fulfill the demands of homology. However, in the presence of more than two homologous regions, only two of them can pair at any one time or position. Bear this in mind throughout the following discussion.

Changes in Chromosome Structure

It is possible that a segment of a chromosome might be lost.

This type of change is called a **deletion,** or a **deficiency.** The reciprocal of such a change would be its **duplication.**

a b b c d e f

We can also conceive of a segment of a chromosome that has rotated 180° and rejoined the chromosome as an **inversion.**

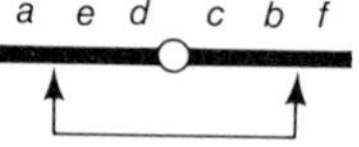

Finally, parts of two nonhomologous chromosomes might be exchanged to produce a **translocation.**

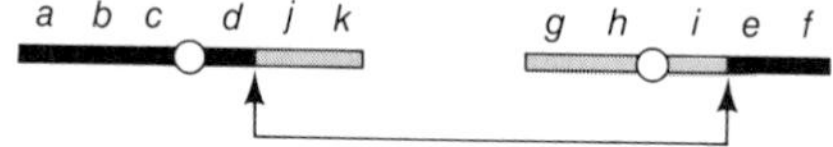

In fact, all of these types of chromosomal aberrations do occur, and so we can examine their genetic and cytological properties. Collectively, this class of aberrations is known as **chromosome rearrangements.**

Deletions

Obviously, the occurrence of chromosomal rearrangements results from a break or disruption in the linear continuity of a chromosome. Deletions and duplications can be produced by the same event if breaks occur simultaneously at different points in two homologs, which can be visualized as occurring when the homologs overlap (Figure 8-4). This does not mean, however, that duplications and deletions are *always* reciprocal products of the same event.

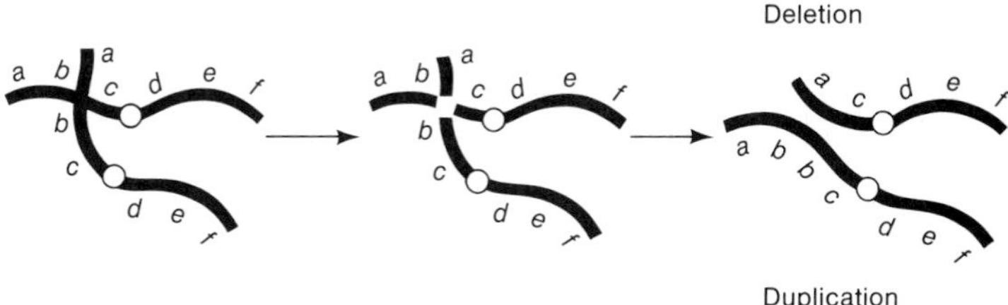

Figure 8-4. One possible way of producing deletions and duplications, by the reunion of broken homologs.

In general, if a deletion is made homozygous, it is lethal. This suggests that most regions of the chromosomes are essential for normal viability and that complete elimination of any segment from the genome is deleterious. Even individuals heterozygous for deletions usually do not survive because the genome has been finely "tuned" during evolution to require a specific balance or ratio of most genes; the presence of the deletion upsets this balance. Nevertheless, in some cases, individuals with relatively large deficiencies can survive if those deletions are heterozygous with normal chromosomes. If meiotic chromosomes in such heterozygotes can be examined, the region of the deletion can be detected by the failure of the corresponding segment on the normal chromosome to pair properly. You can see that deletions can be located on the chromosome by this technique (Figure 8-5).

A good mutagen for inducing chromosome rearrangements of all kinds is ionizing radiation. This kind of radiation, of which X rays and γ (gamma) rays are examples, is highly energetic and causes chromosome breaks. The way in which the breaks rejoin will determine the kind of rearrangement produced. In the case of deletion, a single break can cause a **terminal deletion,** and two breaks can produce an **interstitial deletion** (Figure 8-6).

But what of the genetic properties of deletions? Cytological detection of a "deletion loop" (for example, in salivary gland chromosomes of *Drosophila*)

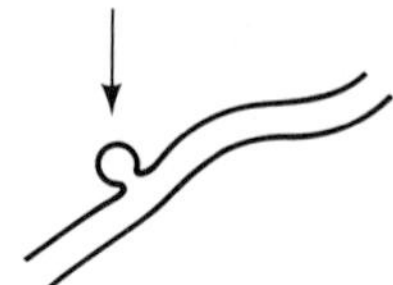

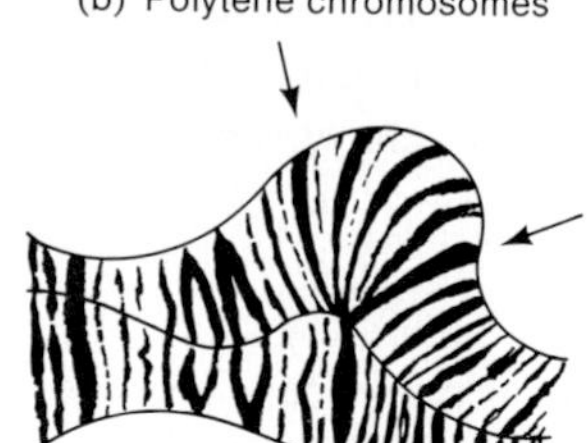

Figure 8-5. Cytogenetic configurations in a deletion heterozygote in Drosophila. *The "looped-out" portion of the meiotic chromosome is the normal form. The genes in this deletion loop have no alleles with which to pair during synapsis. Since polytene chromosomes in* Drosophila *have specific banding patterns, the missing bands of the deleted chromosome can be observed in the deletion loop of the normal chromosome.*

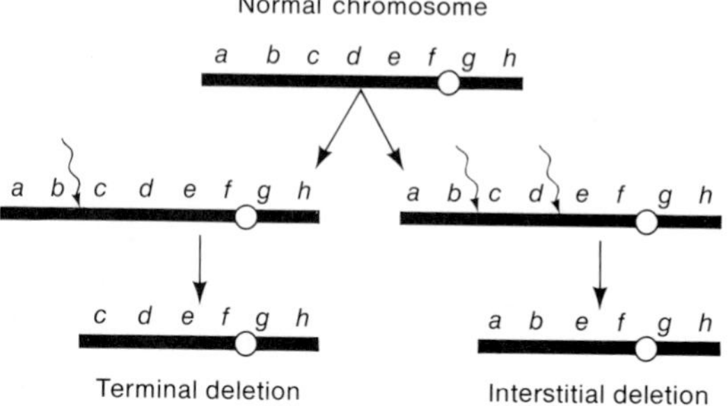

Figure 8-6. Production of terminal and interstitial deletions. Chromosomes can be broken when struck by ionizing radiation (wavy arrows). *A terminal deletion is formed when the end piece of a chromosome is lost. An interstitial deletion is formed when two breaks are induced. The acentric fragment* (cd) *is lost, while the terminal portion* (ab) *rejoins the main body of the chromosome.*

confirms its presence, but there are genetic criteria for inferring the presence of a deletion. One is the failure of the chromosome to survive as a homozygote, but that, of course, could also be produced by any lethal gene. Another is the suppression of crossing over in the region spanning the deficiency, but again this could occur with other aberrations, and small deficiencies may have only minor effects on crossing over. The best criterion is that chromosomes with deletions can never revert to a normal condition. Another reliable criterion is the phenotypic expression of a recessive gene on a normal chromosome when the region in which it is located has been deleted from the homolog.

a b c d e f
\+ + + + +
Phenotype is + b + + + +

Such **pseudodominance** (the expression of a recessive gene when present in a single dose) also allows the cytological location of that gene when coupled with the chromosomal positioning of the deficiency loop. This technique permits a correlation between the genetic map (based on linkage analysis) and the cytological map (devised by marking the position of deficiency loops in specific cases of pseudodominance). By and large, the maps correspond well—a satisfying cytological endorsement of a purely genetic creation.

Message
Deletion analysis proves that linkage maps are in fact reflections of chromosome maps.

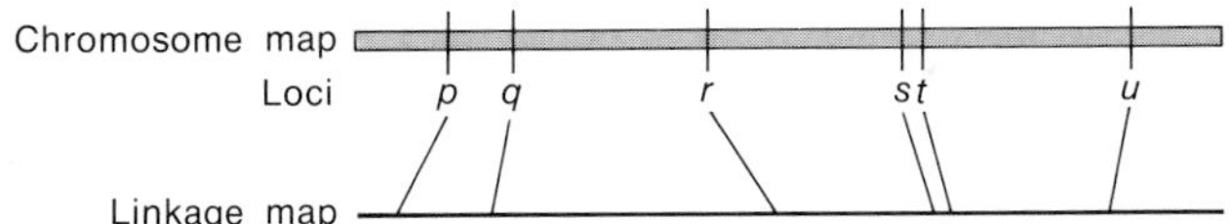

Sometimes it is possible to map deletions *genetically* using a combination of linkage and pseudodominance analysis. As an example of how this works, suppose that you have a strain of flies having an X chromosome (which we'll designate as X?) that, in the cross X?/X♀ × X/Y♂, gives progeny in a ratio of 2 female:1 male. This suggests the presence of a sex-linked lethal factor that causes the death of X?/Y males. We can first ask: Where is the lethal factor genetically? We can find this out by crossing X?/X females with males carrying X chromosomes that have conveniently placed markers from one tip to the centromere (let's label them *a*–*b*–*c*–*d*). The cross is then X?/X♀ × *a b c d*/Y♂, which gives us X?/*a b c d* ♀♀, which can be used to map the position of the lethal factor when crossed with wild males. Because we have a factor that is lethal in males, any parental or cross-over chromosome that carries the lethal factor will die in the male progeny of these females. Suppose that the defect is at the tip of the X chromosome between *a* and *b*. We will be able to map it immediately by the genotypes of the surviving males (Figure 8-7). You can see that one of each pair of reciprocal crossover products will always die in males. Nevertheless, map

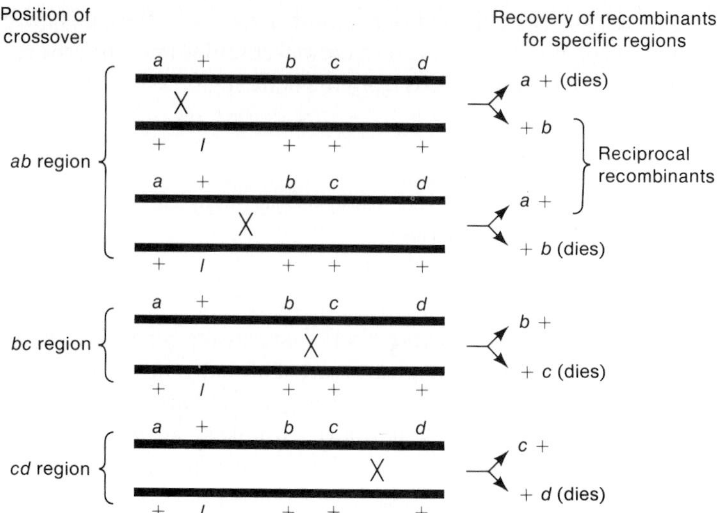

Figure 8-7. Genetic location of a recessive lethal on the X chromosome. A female is made heterozygous for an X chromosome with a lethal mutation (l) *and an X chromosome containing a conveniently spaced array of markers* (a, b, c, d). *She is crossed to a normal male, and the male offspring are scored. If the recessive lethal is between* a *and* b, *one-half of the male progeny with crossovers between* a *and* b *will be* a + + +, *and one-half will be* + b c d. *However, for crossovers occurring between* b *and* c, *only* a b + + *offspring will live. All* + + c d *progeny will die. The relative proportions of reciprocal recombinants will thus locate the position of* l *in the* ab *region.*

distances can be determined because all classes, even parentals, are halved by the lethal factor. If the lethal factor does not lie within a genetic region in which a crossover occurs, then only one of the two reciprocal crossovers in that region will live. On the other hand, if the lethal factor does lie within a crossover region, *both* reciprocal recombinants survive, but in fact they come from crossovers on either side of the lethal. Therefore, the percentage of these two classes will give us the genetic position of the lethal factor within the region; that is, the percentages of + *b* and *a* + male recombinants are the genetic distances from *a* to lethal and from lethal to *b*, respectively.

If the total *a-to-b* recombinant frequency is much lower than the control or normal value, then we can suspect that the lethal factor is a deletion, because the total physically paired region would be shorter. If our X? chromosome carries a lethal factor between *a* and *b* that reduces the genetic interval in this way, we can then cross X?-bearing females with males carrying different recessive genes known to reside in that interval. A map of genes in the tip region is

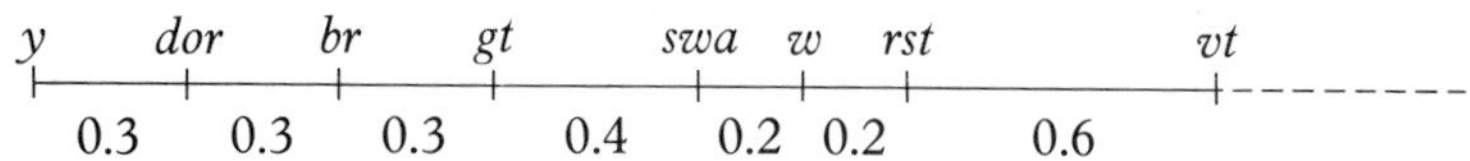

Suppose we get all wild-type flies in crosses between X?/X females and males carrying *y*, *dor*, *br*, *gt*, *rst*, or *vt* but get pseudodominance of *swa* and *w* with X? (that is, X?/*swa* is swa and X?/*w* is w). Then we have good genetic evidence for a deletion of the chromosome including at least the *swa* and *w* loci but not *gt* or *rst*.

Message

Deletions are recognized genetically by (1) lack of revertability, (2) pseudodominance, and (3) recessive lethality, and cytologically by (4) deletion loops.

An interesting difference between animals and plants is revealed by deletions. In animals, a male that is heterozygous for a deletion chromosome and a normal one will produce functional sperm carrying each of the two chromosomes in approximately equal numbers. In other words, sperm seem to function to some extent regardless of their genetic content. In diploid plants, on the other hand, the pollen produced by a deletion heterozygote is of two types: functional pollen carrying the normal chromosome and nonfunctional (or aborted) pollen carrying the deficient homolog. Thus, pollen cells seem to be sensitive to changes in *amount* of chromosome material, and this might act to weed out any large changes in the genome. The situation is somewhat different for polyploid plants, which are far more tolerant of pollen deletions. This is because there are several chromosome sets even in the pollen, and the loss of a segment in one of these sets is less crucial than it would be in a haploid pollen cell. Ovules, in either diploid or polyploid plants, also are quite tolerant of deletions. Presumably this is because of the nurturing effect of the surrounding maternal tissues.

Duplications

Duplications are very important chromosomal changes from the standpoint of evolution because they supply additional genetic material potentially capable of assuming new functions. Adjacent duplicated segments may occur in **tandem sequence** with respect to each other (*a bc bc d*) or in **reverse order** (*a bc cb d*). The pairing patterns obtained in these two sequences are different and illustrate the high affinity of homologous regions for pairing. Thus, chromosomes in meiotic nuclei containing a normal chromosome and a homolog with a duplication are seen to pair in the configurations shown in Figure 8-8. Alternatively, duplicated segments may be nonadjacent, either in the same chromosome or in separate chromosomes.

An organism that has evolved with a fixed number of genetic units would presumably be able to "spare" duplicated copies when they arise. Thus, loss or alteration of gene function due to changes in one of the duplicated genes would be covered by the duplicate copies. This provides an opportunity for divergence in function of genes, which could be potentially advantageous. Indeed, in situations in which different gene products with related functions can be compared, such as the globins (which are discussed in a later chapter), there is good evidence that they indeed arose as duplicates of each other. Interestingly, once an

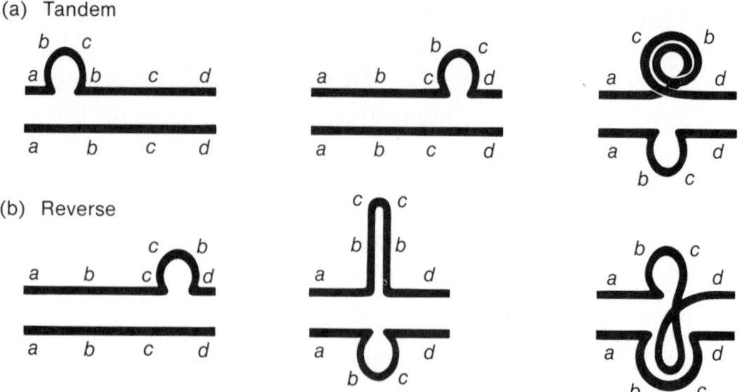

Figure 8-8. Possible pairing configurations in duplication heterozygotes. Duplicated segments may occur in tandem (a bc bc d) or in reverse order (a bc cb d). There are several ways in which the homologs of duplicated heterozygotes may pair, illustrating the high affinity of homologous regions for pairing.

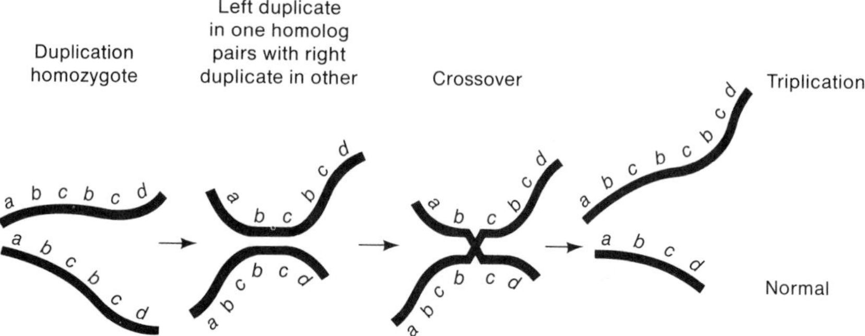

Figure 8-9. Generation of higher orders of duplications by asymmetric pairing and recombination in a duplication homozygote.

adjacent duplication arises in a population, homozygosity for such a duplication can result in higher orders of duplication by crossing over when the chromosomes are **asymmetrically paired** (Figure 8-9).

Message
Duplications are important chromosomal alterations that supply additional genetic material capable of evolving new functions.

Duplications of certain genetic regions may produce specific phenotypes and act like a gene mutation. For example, the dominant mutation Bar in *Drosophila*

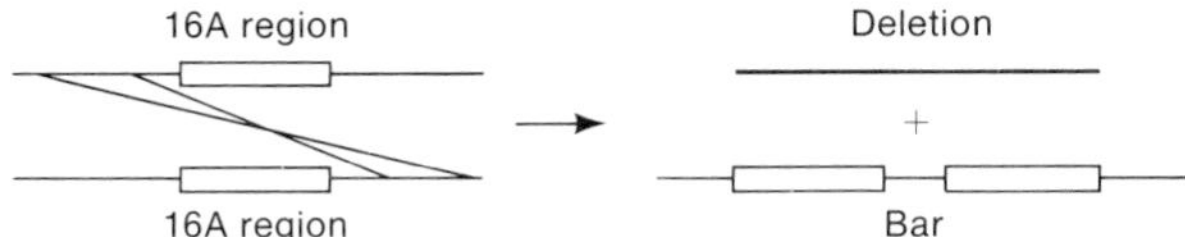

Figure 8-10. Production of bar-eye duplication by nonreciprocal crossing over. Since this event will occur during meiosis, gametes containing the deletion *chromosome will presumably die or produce an inviable zygote. The gamete containing the* Bar *duplication, however, will produce a male offspring with severely reduced eye size* (Bar) *or a female offspring with a slightly reduced eye size* (Bar *heterozygote*).

produces a slitlike eye instead of the normal oval one (see Figure 10-1). Cytologically, in the polytene chromosomes Bar was found to be in fact a tandem duplication that probably resulted from an **unequal crossover** (Figure 8-10). Evidence for the asymmetrical pairing and crossing over in *Drosophila* comes from studying homozygous Bar females. Occasionally, such females produce offspring with extremely small eyes called "double bar." They are found to carry three doses of the Bar region in tandem (Figure 8-11).

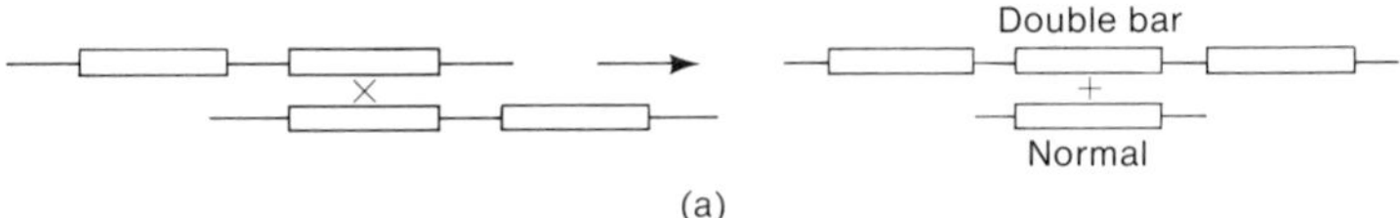

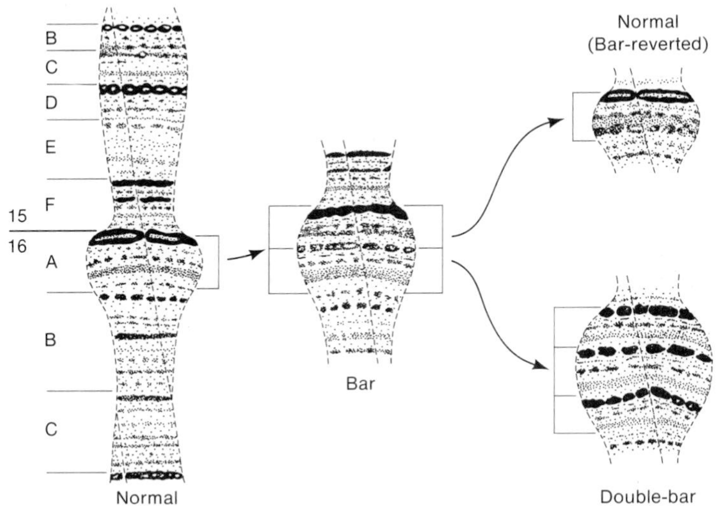

Figure 8-11. Production of double-bar (triplication) and bar-revertant (normal) chromosomes by asymmetric pairing and recombination in a duplication homozygote: (a) diagrammatic and (b) cytological representation. (Part b is from Bridges, Science *83:210, 1936.)*

In general, duplications are hard to detect and are rare. However, they are useful tools and can be generated from other aberrations by tricks that we shall learn later.

Duplications and deficiencies are detected in human chromosomes. As a matter of fact, some of the best evidence for the unequal crossover origin of tandem duplications (and their reciprocal deletions) comes from studies of the genes that determine the structure of the oxygen-transport molecule, hemoglobin, in humans. The hemoglobin molecule is always composed of two different kinds of subunits or components. Furthermore, there are always two of each kind of subunit, so that there is a total of four subunits. The kinds of subunits that constitute hemoglobin are different at different stages of development. For example, the fetus has two alpha subunits and two gamma subunits, ($\alpha_2\gamma_2$), whereas the adult has $\alpha_2\beta_2$. The structures of these subunits are determined by different genes, some of which are linked and some not. The situation is summarized in Figure 8-12. It is the linked γ–δ–β group that provides the data we need on unequal crossover.

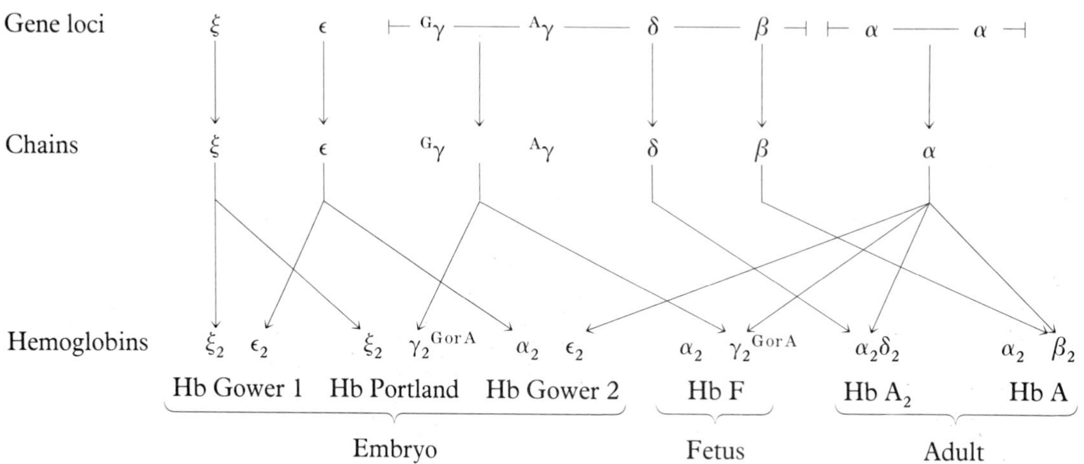

Figure 8-12. Hemoglobin genes in humans. Each Greek letter represents a different gene locus and its type of hemoglobin subunit chain. These combine in different ways at different stages in development, as indicated, to yield functioning hemoglobin molecules. (From D. J. Weatherall and J. B. Clegg, "Recent Developments in Molecular Genetics of Human Hemoglobin," Cell *16, 1979. Copyright © 1979, M.I.T. Press.)*

Some thalassemias (a kind of inherited blood disease) proved to involve hemoglobin subunits that are part δ and part β (Lepore hemoglobin) or part γ and part β (Kenya hemoglobin). The origin of these rare forms can be explained by the unequal crossover models shown in Figure 8-13. Also represented in the diagram are the reciprocal crossover products called anti-Lepore and anti-Kenya. It can be seen that the deletion types led to the anemic blood-disease symptoms.

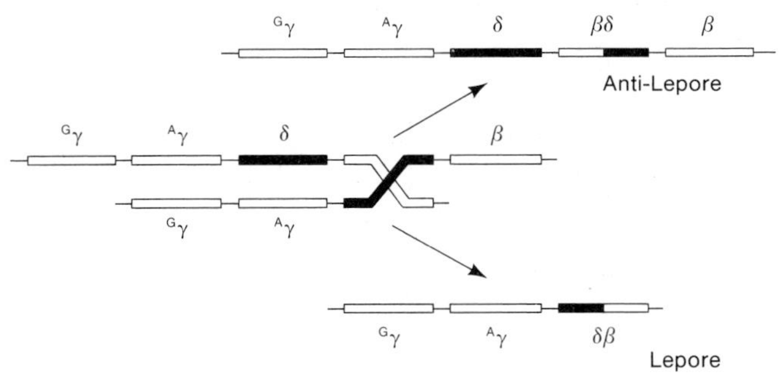

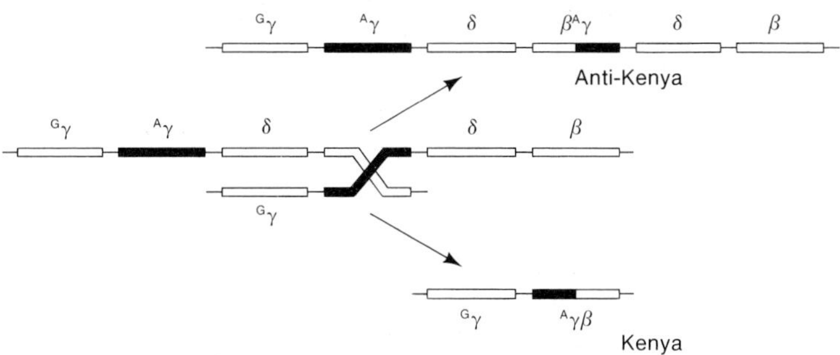

Figure 8-13. Proposed generation of the variant human hemoglobin chains "Lepore" and "Kenya" by unequal crossing over in the γ-δ-β genetic region. The deletion types, and not the "Anti-Lepore" and "Anti-Kenya," result in the anemic blood diseases, a type of thalassemia. (From D. J. Weatherall and J. B. Clegg, "Recent Developments in Molecular Genetics of Human Hemoglobin." Cell *16, 1979. Copyright © 1979, M.I.T. Press.)*

Visible deletions of chromosome segments in humans are always associated with major incapacities. For example, **cri du chat** syndrome is caused by a deletion of one-half of the long arm of chromosome 5 and involves severe mental retardation. (The syndrome was named after the cat-like mewing cry of infants suffering from the condition.) Figure 8-14 shows a pedigree involving cri du chat syndrome; examine it now, but don't worry about the details of the production of the deletion until after we have dealt with translocations. Deletions for certain other chromosomes also are found in humans, always associated with severe incapacity or with intrauterine lethality.

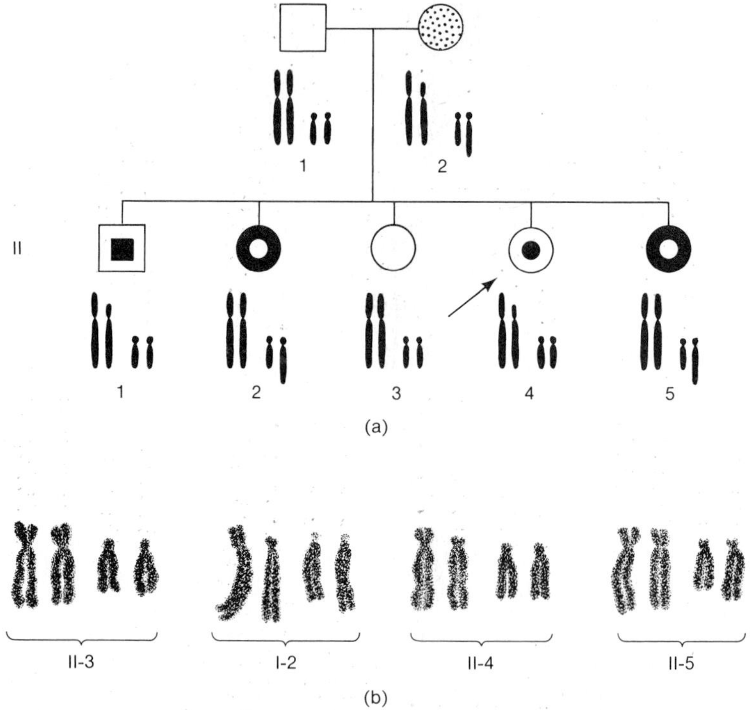

Figure 8-14. *(a) Pedigree with cases of* cri du chat *syndrome:*
I. Both parents are phenotypically normal.
II. Of the offspring, II-1 and II-4 have cri du chat *syndrome, II-2 and II-5 have the reciprocal chromosome abnormality, and II-3 is normal. Symbols represent different karyotypes, and the arrow points to the* proband *or affected individual with whom the study began.*
(b) Chromosomes 5 and 13 of four persons in the pedigree. (From J. Lejeune, J. Lafourcade, H. Berger, and R. Turpin, Comptes Rendius. L'Academie des Sciences, Paris *258, 1964.)*

Inversions

Homozygosity for a chromosome carrying an inverted gene sequence will result in a linkage map with a different gene order. In a heterozygote having a chromosome that contains an inversion and one that is normal, there are important genetic and cytological effects. Because there is no net loss or gain of material, heterozygotes usually are perfectly viable. The location of the inverted segment can be recognized cytologically in the meiotic nuclei of such heterozygotes by the presence of an "inversion loop" in the paired homologs (Figure 8-15).

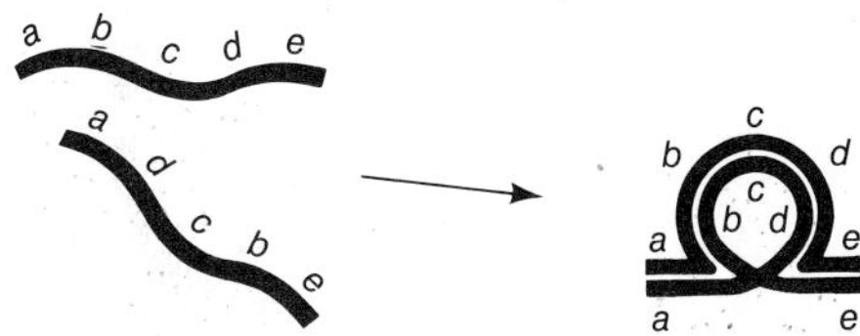

Figure 8-15. Inversion loop in paired homologs of an inversion heterozygote. Tight meiotic pairing produces this cytological configuration.

The location of the centromere relative to the inverted segment determines the genetic behavior of the chromosomes. If the centromere is not included in the inversion, the inversion is **paracentric,** whereas inversions spanning the centromere are **pericentric** (Figure 8-16).

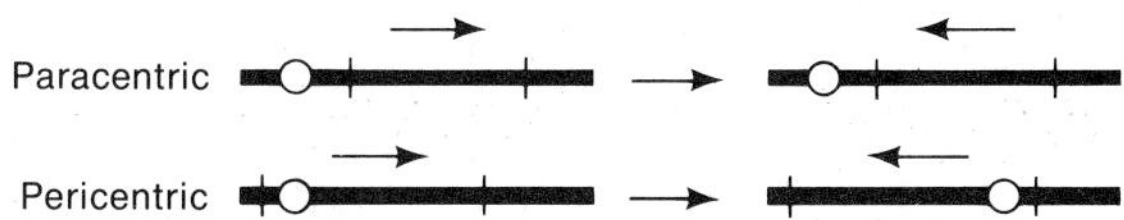

Figure 8-16. The location of the centromere relative to the inverted segment. If the centromere is not included in the inversion, the inversion is paracentric. Inversions including the centromere are pericentric.

What do inversions do genetically? In a heterozygote for a paracentric inversion, crossing over within the inversion loop has the effect of connecting homologous centromeres in a **dicentric bridge,** as well as producing an acentric piece of chromosome—that is, one without a centromere (Figure 8-17).

Thus, as the chromosomes separate during anaphase I, the disjoining centromeres will remain linked by means of the bridge. This orients the centromeres so that the noncrossover chromatids lie farthest apart. The acentric fragment cannot align itself or move, and consequently it will be lost. Remarkably, in *Drosophila* and in plant eggs, the dicentric bridge may remain intact long after anaphase I and, as the second meiotic division begins, the noncrossover chromatids are directed to the outermost nuclei (Figure 8-18). Thus, the two inner nuclei either will be linked by the dicentric bridge or will contain fragments of the bridge if it breaks, whereas the outer nuclei contain the noncrossover chromatids.

Fertilization of a nucleus carrying the broken bridge should produce defective zygotes that die because they have an unbalanced set of genes. Consequently, in a testcross the recombinant chromosomes would end up in dead zygotes, and

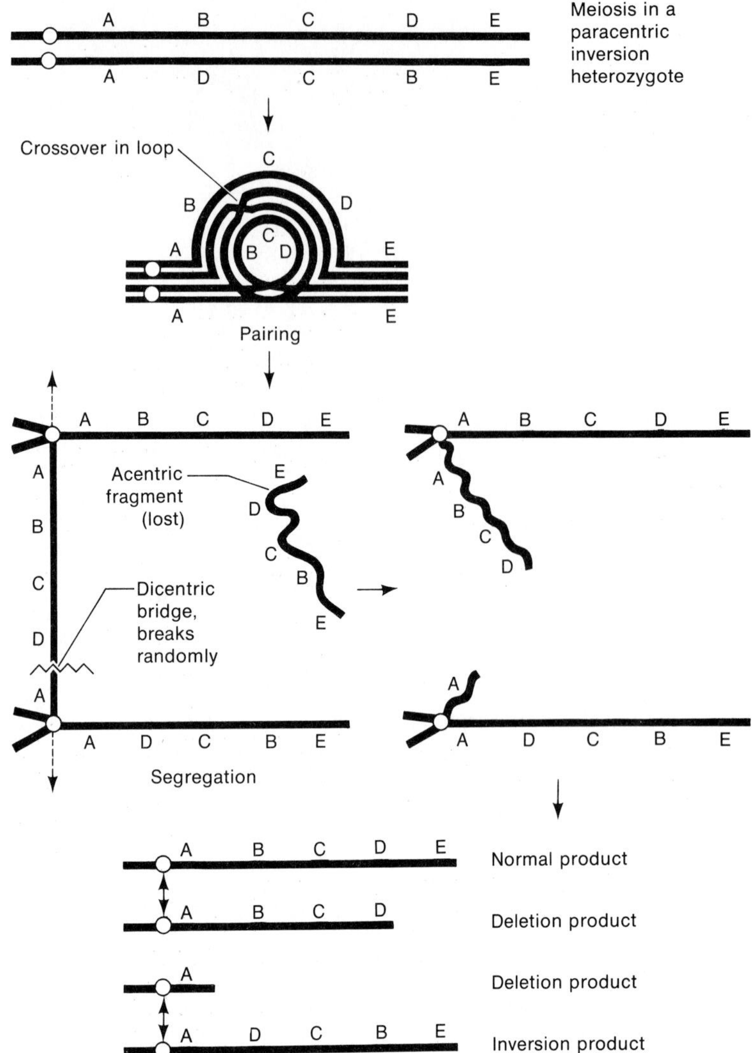

Figure 8-17. Meiotic products resulting from a single crossover within a heterozygous paracentric inversion loop. Crossing over occurs in the four-strand stage (two identical chromatids connected to each centromere).

recombinant frequency would be lowered. However, in *Drosophila,* the presence of large inversions does not result in a large increase in zygotic mortality. Consequently, it has been inferred that the inner nuclei never participate in fertilization and that only one of the outer nuclei can be the egg nucleus. Thus, you can see that the chromatids participating in a crossover event will be selectively

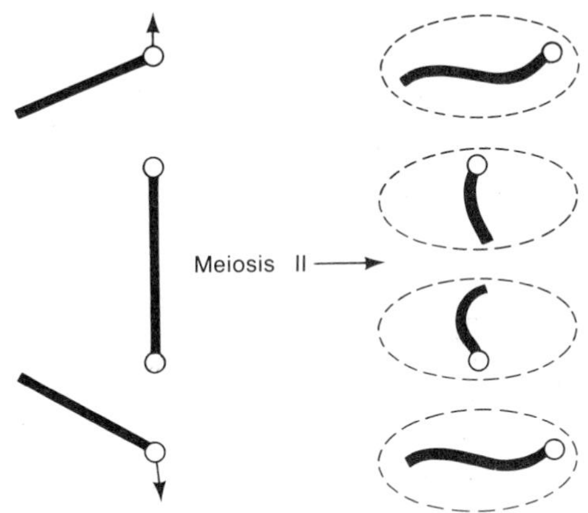

Figure 8-18. In some organisms, such as Drosophila *and some plants, the dicentric bridge resulting from a single crossover within a heterozygous paracentric inversion loop will not break during anaphase I. As the second meiotic division begins, the noncrossover chromatids are directed to the outermost nuclei. Thus, the two inner nuclei either are linked by the dicentric bridge or contain fragments of the bridge if it breaks.*

retained in the central nuclei, thereby allowing recovery of the noncrossover chromatids in the egg nuclei. This remarkable suggestion based on genetic results was later confirmed cytologically in *Drosophila*. It has also been shown in plants. (How would you test this suggestion in *Neurospora* using tetrad analysis?) Nevertheless, the genetic consequence of inversion heterozygosity is the same—that is, the selective recovery of noncrossover chromatids from exchange tetrads. In addition, inversion heterozygotes often have mechanical pairing problems in the area of the inversion and this also reduces crossing over and recombinant frequency in the vicinity.

Message

Although inversion heterozygosity does reduce the number of recombinants recovered, it in fact does so by two mechanisms: one by inhibiting the process of chromosome pairing in the vicinity of the inversion, and the other by selectively eliminating the products of crossovers in the inversion loop.

It is worth noting that paracentric inversions have been useful in studying crossing over in organisms for which no genetic markers are known but which do have chromosomes that can be studied. Dicentric bridges are the consequence of crossovers within the inversion loop, and so their frequency is related to the amount of crossing over.

The net genetic effect of a pericentric inversion is the same as that of a paracentric one—that is, crossover products are not recovered—but for different reasons. In a pericentric inversion, because the centromeres are contained within the inverted region, disjunction of crossover chromosomes is normal. However, a crossover within the inversion produces chromatids that contain a duplication and a deficiency for different parts of the chromosome (Figure 8-19). In this case, fertilization of a nucleus carrying a crossover chromosome generally

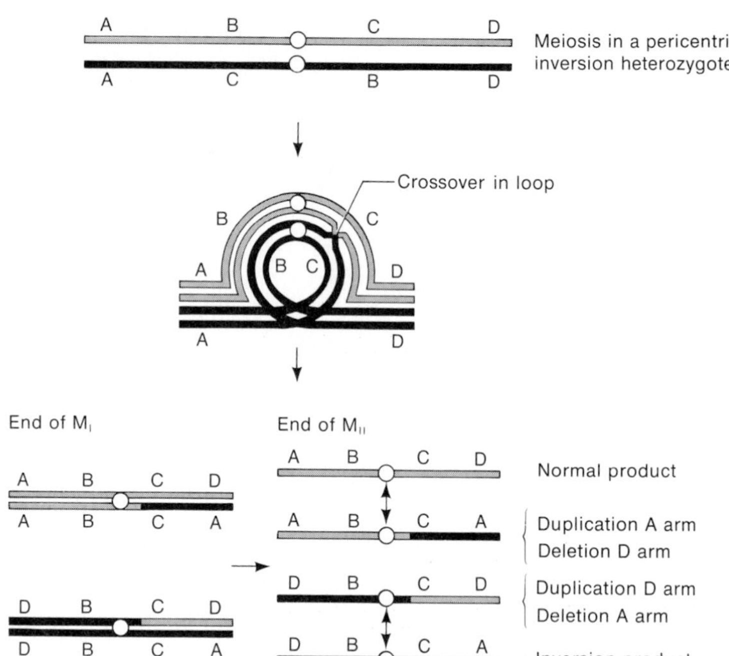

Figure 8-19. Meiotic products resulting from a meiosis with a single crossover within a heterozygous pericentric inversion loop.

results in its elimination through zygotic mortality caused by an imbalance of genes. Again, the result is the selective recovery of noncrossover chromosomes as viable progeny.

To generate a duplication "on purpose," it is possible to use a pericentric inversion having one breakpoint at the tip of the chromosome (Figure 8-20). A crossover in the loop produces a chromatid type in which the entire left arm is duplicated, and, if the tip is nonessential, a duplication stock is generated for investigation. Another way to make a duplication (and a deficiency) uses two paracentric inversions whose breakpoints overlap (Figure 8-21). (These tricks are possible only in genetically well-marked organisms.)

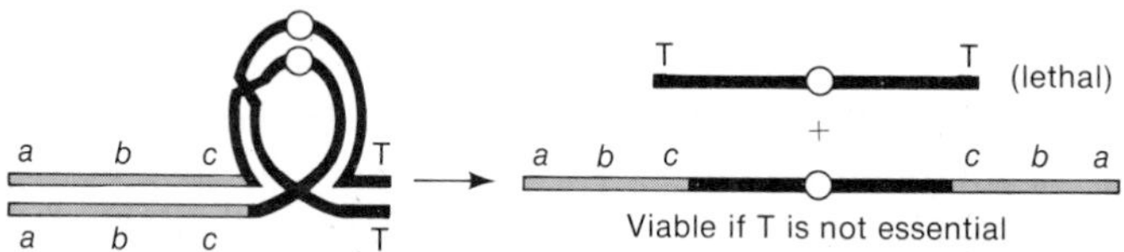

Figure 8-20. Generation of a viable nontandem duplication from a pericentric inversion close to a dispensible chromosome tip.

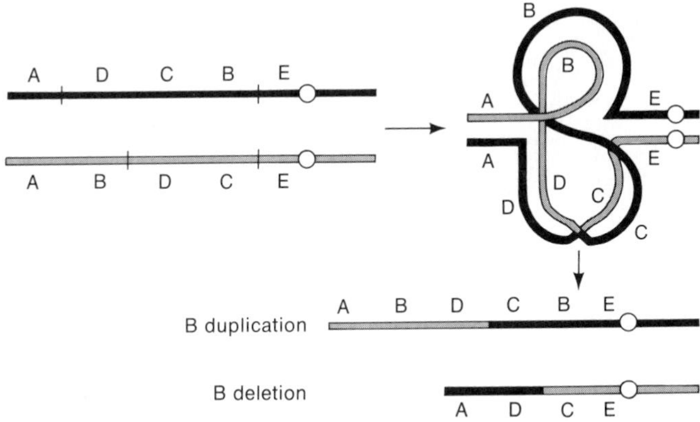

Figure 8-21. Generation of a nontandem duplication from two overlapping inversions.

So far, we have considered only single crossovers. You may ask: What happens in double-exchange tetrads? The consequences are predictable from a tetrad analysis of normal chromosomes (discussed in Chapter 6). We would expect two-, three-, and four-strand double exchanges within the inversion loop in a 1:2:1 ratio. Let's consider such tetrads in a paracentric inversion (Figure 8-22). In this case, viable crossover chromatids can be recovered, but only if they are the products of both exchanges. Thus, comparison of the frequency of double crossovers within inversion loops of heterozygotes and in the same region of structurally normal chromosomes provides a measure of how much effect the inversion has on the occurrence of crossing over.

The reduction in recovery of recombinants caused by the presence of inversions is a useful property in laboratory experiments. Thus, if a particularly useful or interesting combination of genes has been linked by crossing over but cannot be made homozygous (because some of the genes cause sterility, reduced viability, or lethality), an inversion of its homolog can be introduced to prevent disruption of that sequence by crossing over. The use of inversions in this way may, in fact, be an imitation of nature. In natural populations of organisms, inversions are commonly found. Thus, it has been inferred that inversions may be of value in reducing the recovery of crossovers in chromosomes that carry particularly advantageous combinations of genes that had arisen previously.

We have seen that genetic analysis and meiotic chromosome cytology are both good ways of detecting inversions. As with most rearrangements, there is also the possibility of detection through mitotic chromosome analysis. One of the key operational features is to look for new arm ratios (Figure 8-23). Note that the ratio of the long to the short arm has been changed from about 4 to about 1 by the pericentric inversion. Paracentric inversions are more difficult to detect, but they may be detected if banding or other chromosomal landmarks are available.

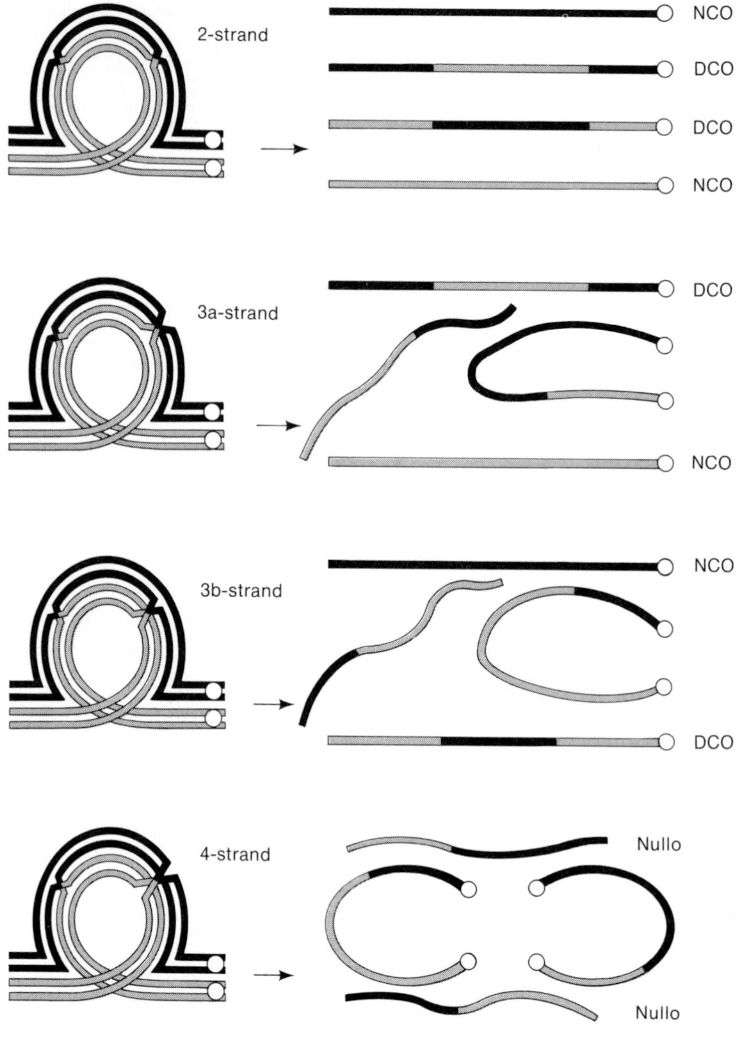

Figure 8-22. *The consequences of a double crossover within a heterozygous paracentric inversion loop. Since the crossover occurs in the four-strand stage, two, three, or all four strands may be involved in a double crossover. N = nonrecombinant; D = double recombinant. Unlabeled chromosomes are not viable.*

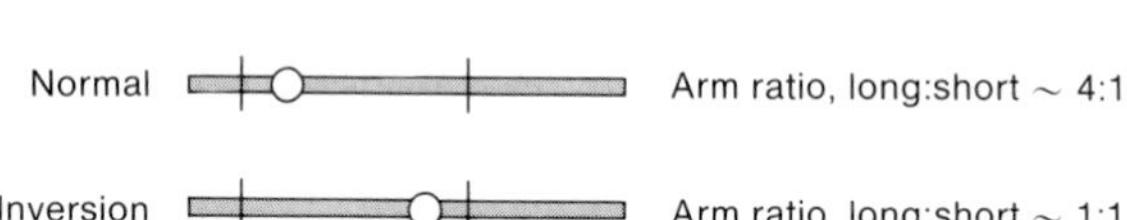

Figure 8-23. *The ratio of the lengths of the left and right arms of a chromosome can be changed by a pericentric inversion. Thus, changes in arm ratios can be used to detect pericentric inversions.*

Translocations

Here we discuss **reciprocal translocations,** the most common kind of translocation. A segment from one chromosome is exchanged with a segment from another nonhomologous one, so that in reality two translocations are simultaneously achieved.

The exchange of chromosome parts between nonhomologs establishes new linkage relationships if the translocated chromosomes are homozygous. Furthermore, translocations may drastically alter the size of a chromosome as well as the position of its centromere. For example,

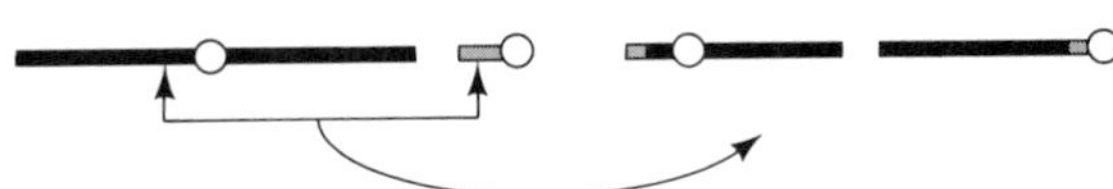

Here a large metacentric chromosome is shortened by one-half its length to an acrocentric one, whereas the small chromosome becomes a large one. Examples in natural populations are known in which chromosome numbers have actually been changed by translocation between acrocentric chromosomes and the subsequent loss of the resulting small chromosome elements (Figure 8-24).

In heterozygotes having translocated and normal chromosomes, the genetic and cytological effects are important. Again, the pairing affinities of homologous regions dictate a characteristic configuration when all chromosomes are synapsed in meiosis. In Figure 8-25, which illustrates meiosis in a reciprocally translocated heterozygote, the configuration is that of a cross.

Remember, this configuration lies on the metaphase plate with the spindle fibers perpendicular to the page. Thus the centromeres would actually migrate

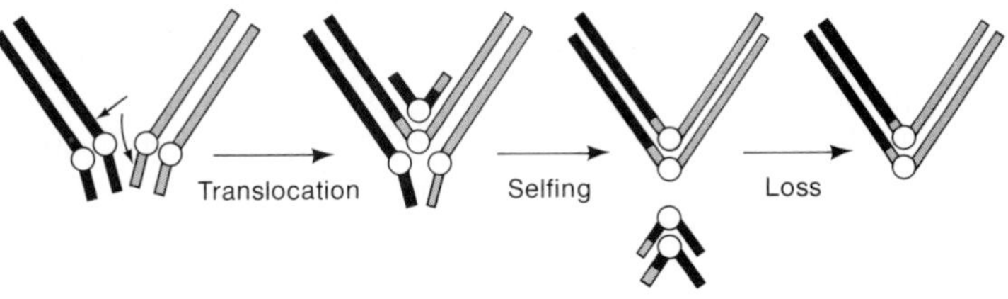

Figure 8-24. Genome restructuring by translocations. Small arrows indicate break points in one homolog of each of two pairs of acrocentric chromosomes. The resulting fusion of the breaks yields one short metacentric and one long metacentric. If, as in plants, self-fertilization takes place (selfing), an offspring could be formed with only one pair of long metacentrics and one pair of short metacentrics. Under appropriate conditions, the short metacentric may be lost. Thus, we see a conversion from two acrocentric pairs of chromosomes to one pair of metacentrics.

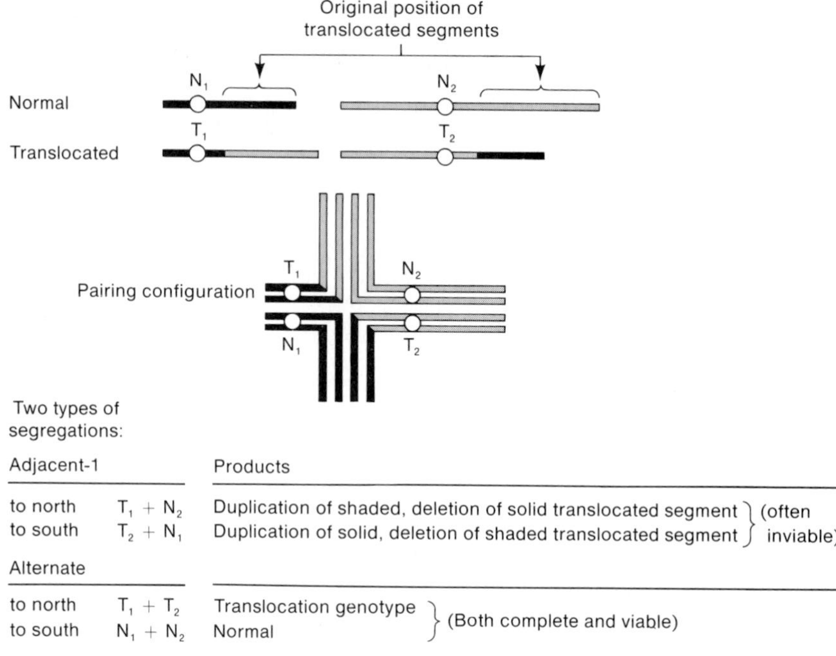

Adjacent-1		Products	
to north	$T_1 + N_2$	Duplication of shaded, deletion of solid translocated segment	(often inviable)
to south	$T_2 + N_1$	Duplication of solid, deletion of shaded translocated segment	
Alternate			
to north	$T_1 + T_2$	Translocation genotype	(Both complete and viable)
to south	$N_1 + N_2$	Normal	

Figure 8-25. The meiotic products resulting from the two most commonly encountered chromosome segregation patterns in a reciprocal translocation heterozygote.

up out of the page or down into it. Homologous paired centromeres disjoin, translocation or not. Because Mendel's second law still applies to *different paired centromeres,* there are two common patterns of disjunction. The segregation of each of the structurally normal chromosomes with one of the translocated ones (T_1 with N_2 and T_2 with N_1) is called **adjacent-1 segregation.** Both meiotic products are duplicated and deficient for different regions. On the other hand, the two normal chromosomes may segregate together, as do the reciprocal parts of the translocated ones to produce $T_1 + T_2$ and $N_1 + N_2$ products. This is called **alternate segregation.** There is another event called adjacent-2 segregation in which homologous centromeres migrate to the same pole, but in general this is a rare occurrence.

Semisterility in Plants. Once again, in the study of rearrangements, a careful distinction must be made between animals and plants. In animals, the unbalanced products of adjacent-1 segregation in reciprocal translocation heterozygotes usually produce viable gametes. Thus, in *Drosophila,* for example, the gamete population is composed of approximately equal numbers of alternate and adjacent segregation meioses. However, in diploid plants, the T_1N_2 and T_2N_1 gametes normally abort, giving a situation known as semisterility. (You will remember that semisterility in plants also can be produced in gametes containing deletions and in the products of crossovers in inversion loops.) Semisterility is often directly identifiable through the observation of a mixture of shriveled, abnormal pollen grains with normal pollen grains (Figure 8-26).

Even in *Drosophila* and other tolerant animals, the unbalanced gametes (when fertilized by normal gametes) give rise to unbalanced zygotes that tend not to survive unless the imbalance is for very small translocated regions.

Let's consider a cross between two reciprocal translocation heterozygotes in *Drosophila* and examine the viable progeny types. We assume equal numbers of alternate and adjacent segregations for simplicity (Figure 8-27). You can see that a complete diploid set of the chromosomes is obtained only in certain gametic combinations. In this example, 6/16 of the zygotes carry a complete diploid chromosome set. Of these six, one will be homozygous for the normal chromosomes, and one will be homozygous for the translocation.

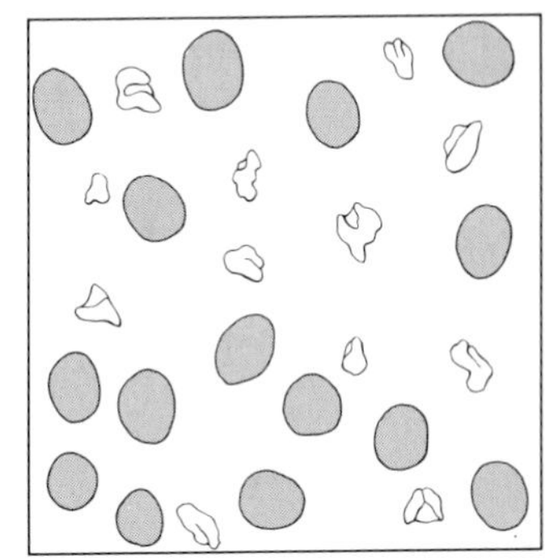

Figure 8-26. Sketch of normal and aborted pollen in a semisterile corn plant. The small, shriveled pollen grains contain aneuploid meiotic products of a reciprocal translocation heterozygote. The normal-shaped pollen grains, containing either the complete translocation genotype or normal chromosomes, will be functional in fertilization and development.

		Male			
		T_1N_2	T_2N_1	T_1T_2	N_1N_2
Female	T_1N_2	−	+	−	−
	T_2N_1	+	−	−	−
	T_1T_2	−	−	+	+
	N_1N_2	−	−	+	+

Figure 8-27. Possible gametic combinations resulting from a cross between two reciprocal translocation heterozygotes in Drosophila *(assuming all gametes are viable). T_1 and T_2 represent the translocated chromosomes, while N_1 and N_2 represent normal chromosomes. Pluses indicate zygotes that have whole genomes and that should survive to adulthood. Minuses indicate zygotes with incomplete genomes; these usually die as zygotes or during early embryogenesis.*

In a cross between translocation heterozygotes and structurally normal homozygotes, inviable and viable zygotes are produced in equal frequency:

	T_1N_2	T_2N_1	T_1T_2	N_1N_2
N_1N_2	−	−	+	+

Of the viable progeny, one-half will be homozygous for structurally normal chromosomes, and the other one-half will remain heterozygous.

Message
Translocations, inversions, and deletions produce semisterility by generating unbalanced meiotic products that may themselves be lethal or that may result in lethal zygotes.

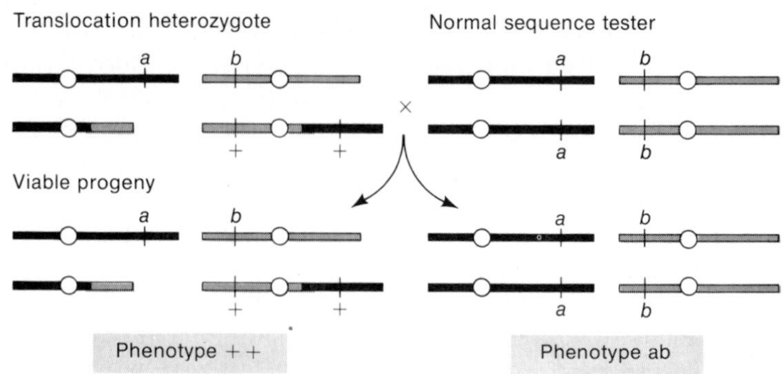

Figure 8-28. Inviability in some translocation progeny produces apparent genetic linkage because only parental types are recovered in the progeny.

Genetically, markers on nonhomologous chromosomes will appear to be linked if these chromosomes are involved in a translocation. Figure 8-28 shows a situation where a translocation heterozygote has been established by crossing an *aa bb* individual with a translocation bearing the wild-type genes. Upon test-crossing the heterozygote, the only viable progeny are those bearing the parental genotypes, so linkage is seen between loci on different chromosomes. In fact, if all four arms of the meiotic pairing structure are genetically marked, a cross-shaped linkage map will result. Such interlinkage group linkage is often a genetic giveaway for the presence of a translocation.

The Importance of Translocations for Humans. Translocations are economically important as well. In agriculture, the occurrence of translocations in certain crop strains can reduce yields considerably because of the number of unbalanced zygotes formed. On the other hand, translocations are potentially useful: it has been proposed that the high incidence of inviable zygotes could be used to control insect pests by the introduction of translocations into the wild. Thus, 50% of the offspring of crosses between insects carrying the translocation and wild types would die, and 10/16 of the progeny of crosses between translocation-bearing insects would die.

Translocations occur in humans, usually in association with a normal chromosome set in a translocation heterozygote. Down's syndrome (previously referred to as Mongolism) can arise in the progeny of an individual heterozygous for a translocation involving chromosome 21. The heterozygous person is phenotypically normal (and is called a carrier), but during meiosis an adjacent-1 segregation will produce gametes carrying duplicated parts of chromosome 21, and possibly a deficiency for some part of the other chromosome involved in the translocation. For unknown reasons, the extra chunk of chromosome 21 is the cause of Down's syndrome. Of the normal children of a carrier, one-half will themselves be carriers (Figure 8-29).

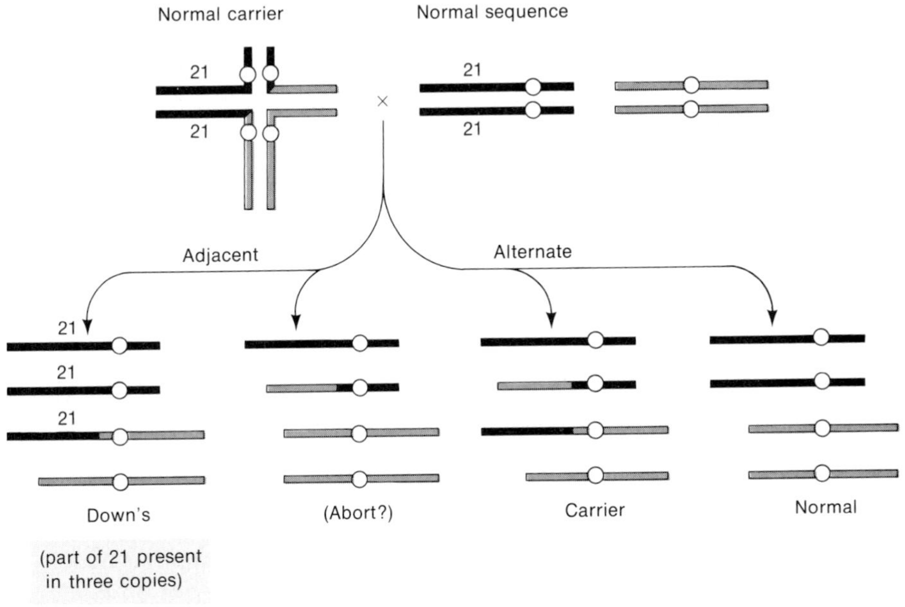

Normal carrier
21
21
×
Normal sequence
21
21
Adjacent
Alternate
21
21
21
Down's
(part of 21 present in three copies)
(Abort?)
Carrier
Normal

Figure 8-29. Production of the translocation form of Down's syndrome. This form is much rarer than that discussed on p. 300.

Later we shall discuss another way in which Down's syndrome is generated. Bear in mind for now that under the present method there should be a high recurrence rate in the family pedigree involved: a translocation heterozygote or carrier can repeatedly produce children with Down's syndrome, and of course the carriers can do the same in subsequent generations. The other method does *not* show this recurrence within a family. The factors responsible for numerous other hereditary disorders have been traced to translocation heterozygosity in the parents—as, for example, in the cri du chat pedigree in Figure 8-14.

The Use of Translocations in Producing Duplications and Deficiencies. Another method of producing duplications and deficiencies for specific chromosome regions makes use of translocations. Let's take *Drosophila* as an example. For reasons that are yet unclear, heterochromatin near the centromere, although very extensive physically, contains very few genes. In fact, for a long time heterochromatin was considered useless and inert material. In any case, for our purposes, *Drosophila* can tolerate a loss or an excess of heterochromatin with little effect on viability or fertility.

Let's now select two different reciprocal translocations involving the same two chromosomes. Each of them has a breakpoint somewhere in heterochromatin, and each has a euchromatic break on one side or the other of the region we want to be duplicated and deleted (Figure 8-30). You can see that, if we have a

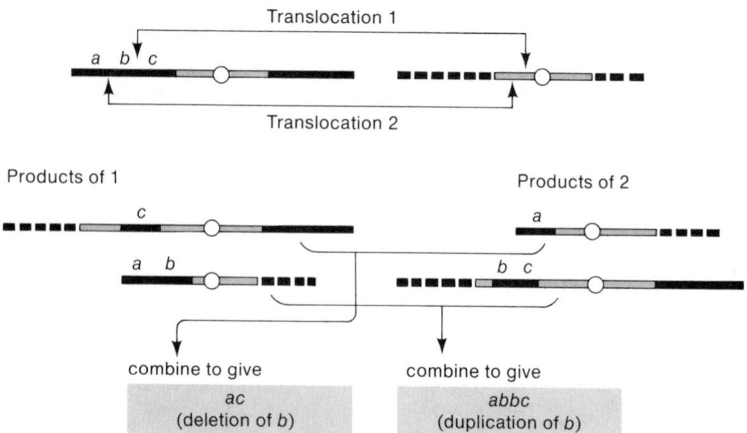

Figure 8-30. Using translocations with one breakpoint in heterochromatin to produce a duplication and a deletion. (Heterochromatin is lightly shaded.) If the upper product of translocation 1 is combined with the upper product of translocation 2 by means of an appropriate mating, a deletion of b *results. If the lower products of the two translocations are combined, the genotype* a b b c *is produced.*

large collection of translocations having one heterochromatic break and euchromatic breaks at many different sites, then duplications and deletions for many parts of the genome can be produced at will.

Position-Effect Variegation. In previous chapters, we have discussed several ways of generating variegation in the somatic cells of a multicellular organism. You will recall somatic segregation, somatic crossover, and somatic mutation as causes—and mosaics, chimeras, or sectors as descriptive terms. To clarify those terms, **variegation** is the mere existence of different-looking sectors of somatic tissue; a **mosaic** is a mixture of genetically different tissue types; a **chimera** is probably best thought of as a self-perpetuating kind of mosaic. For example, if a plant meristem is a mosaic, then the mosaic may be perpetuated vegetatively, resulting in a chimera.

A further cause of variegation is associated with translocations and is called **position-effect variegation.** The following example is from *Drosophila.*

The locus for white eye color in *Drosophila* is near the tip of the X chromosome. If the tip of a chromosome carrying w^+ is translocated to a heterochromatic region—say, of chromosome 4—then in a heterozygote for the translocation and a normal chromosome 4, and with the normal X chromosome carrying w, the eye color is a mosaic of white and red cells. Because the translocation carries the dominant w^+ allele, we would expect the eye to be completely red. How have the white areas come about? We could suppose that, when the translocation was formed, the w^+ gene itself was somehow changed to a state that made it more mutable in somatic cells; so the white eye tissue reflects cells in which w^+ has mutated to w.

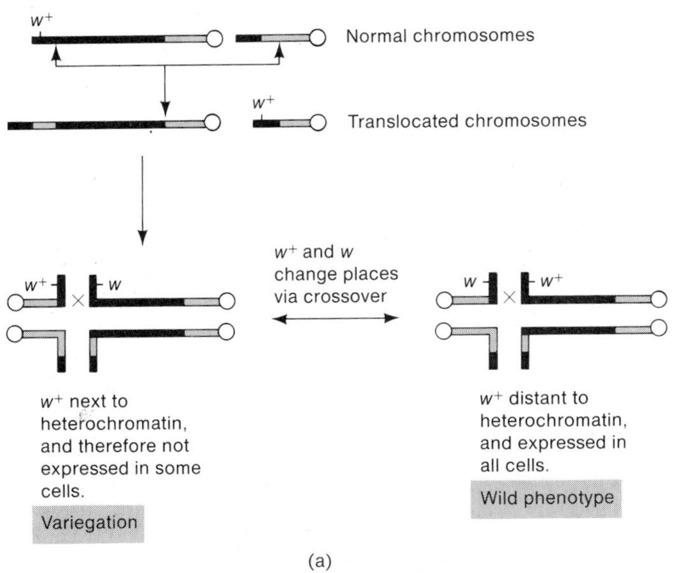

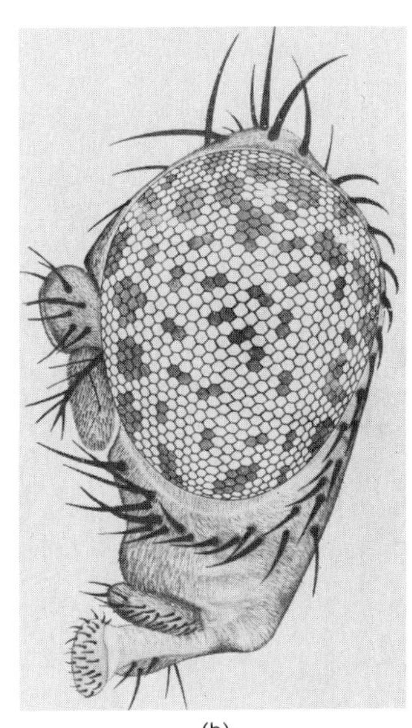

Figure 8-31. (a) Translocation of w$^+$ *to a position next to heterochromatin* (shaded lightly) *causes* w$^+$ *function to fail in some cells, producing position-effect variegation. (b) Appearance of a* Drosophila *eye showing position-effect variegation. Dark = red, unshaded = white. (Part b courtesy of Randy Mottus.)*

Burke Judd tested this hypothesis by breeding the w^+ allele out of the translocation and onto a normal X chromosome and by breeding a w gene from the normal X onto the translocation (Figure 8-31). He found that, when the w^+ gene on the translocation was crossed onto a normal X chromosome and w was inserted into the translocation, the eye color was red; so obviously the w^+ was not defective. A w^+ gene that is crossed back onto the translocation again variegates. We can conclude, then, that, for some reason, the w^+ gene in the translocation is not expressed in some cells, thereby allowing expression of w. This kind of variegation was called position-effect variegation because the unstable expression of a gene is a reflection of its position in a rearrangement.

Message
The expression of a gene can be affected by its position in the genome.

In fungi, a tetrad analysis can be very useful in detecting chromosome aberrations in general. In *Neurospora,* for example, any genome containing a deletion will produce an ascospore that does not ripen to the normal black color, and parents of crosses with high proportions of such "aborted" white ascospores usually contain rearrangements (duplications are generally recovered as black *viable* ascospores). Specific spore-abortion patterns sometimes identify specific rearrangements. The pattern resulting from a translocation is an equal number

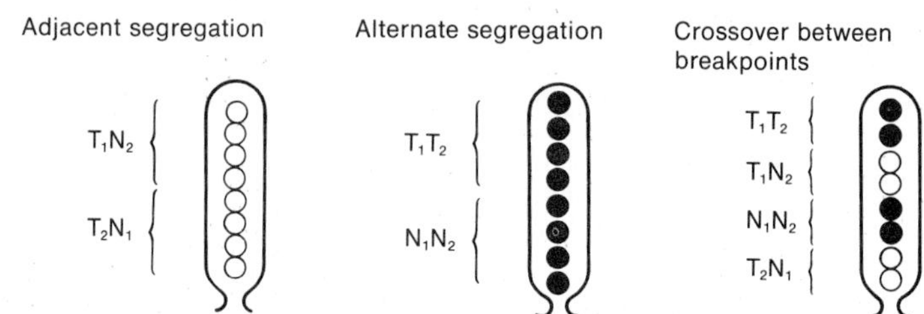

Figure 8-32. Consequences of various meioses in a cross of Neurospora *heterozygous for a reciprocal translocation. White sexual spores abort; black are viable.* T_1 *and* T_2 *represent the respective translocated chromosomes;* N_1 *and* N_2 *represent the normal chromosomes. The 4 black–4 white spores are produced by crossing over between either centromere and the translocation breakpoint.*

of asci having 8 black and 0 white spores (alternate-segregation meioses) and 0 black and 8 white spores (adjacent-segregation meioses) and some asci having 4 black and 4 white. The 4:4 asci are produced by crossing over between either centromere and the translocation breakpoint (Figure 8-32).

Changes in Chromosome Number

Euploidy

Monoploids. A **monoploid** contains one chromosome set in a kind of organism that is usually associated with a diploid type of life cycle, thus being distinguished from true haploids. Monoploids can arise spontaneously in natural populations as rare aberrations, but in several forms (such as bees, wasps, and ants) the males are normally monoploid, having been derived from unfertilized eggs.

In the germ cells of a monoploid, meiosis cannot occur normally because the chromosomes have no pairing partners. Thus monoploids are characteristically sterile. (However, meiosis can be bypassed in some monoploid animals, such as male honeybees, which produce gametes essentially by mitotic division.) If meiosis occurs and the single chromosomes segregate randomly, then the probability of them all going to one pole is $(1/2)^{x-1}$, where x is the number of chromosomes. This will determine the frequency of viable (whole-set) gametes, obviously a vanishingly small number if x is large.

Monoploids have a major role in modern approaches to plant breeding. Diploidy is an inherent nuisance in the induction and selection of new favorable plant mutations and of new combinations of genes already present. Monoploids

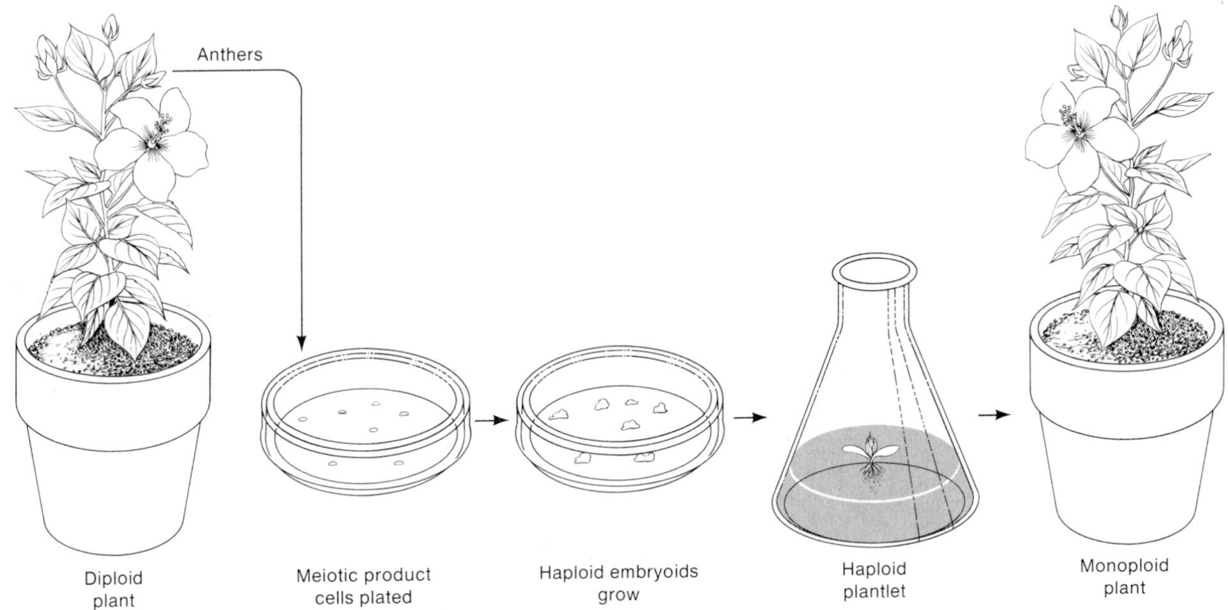

Figure 8-33. Generating a monoploid plant by tissue culture. Appropriately treated pollen grains (haploid) can be plated on agar containing certain plant hormones. Under these conditions, haploid embryoids will grow into monoploid plantlets. With another change in plant hormones, these plantlets will grow into mature monoploid plants with roots, stems, leaves, and flowers.

provide a way around some of these problems. Monoploids may be artificially derived from the products of meiosis in the plant's anthers. A cell destined to become a pollen grain may be induced by cold treatment to grow instead into an **embryoid,** a small dividing mass of cells. The embryoid may be grown on agar to form a monoploid plantlet, which can then be potted in soil to mature (Figure 8-33).

Monoploids may be exploited in several ways. First, they may be examined for favorable traits or gene combinations. These may arise from heterozygosity already present in the parent or induced in the parent by mutagens. The monoploid can then be subjected to chromosome doubling to achieve a completely homozygous diploid with a normal meiosis, capable of providing seed. How is this achieved? Quite simply, by the application of a compound called **colchicine** to somatic tissue, perhaps a meristem. Colchicine inhibits the mitotic spindle, so that cells with two chromosome sets are produced (Figure 8-34). These may proliferate to form a sector of diploid tissue that can be identified cytologically.

Another way the monoploid may be used is to treat its cells, basically like a population of haploid organisms, in a mutagenesis-and-selection procedure. The cells are isolated, their walls are removed by enzyme treatment, and they

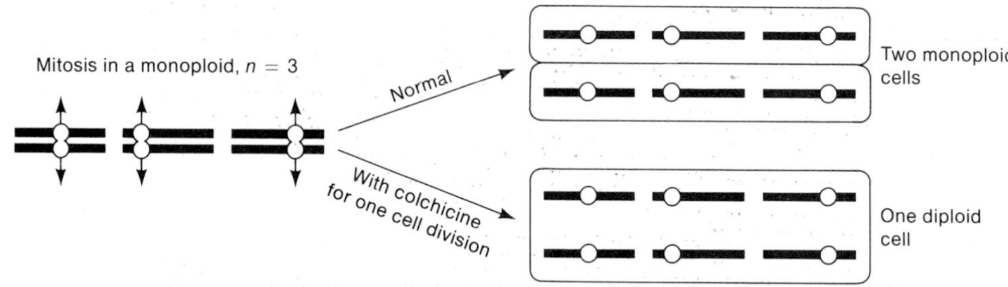

Figure 8-34. Using colchicine to generate a diploid from a monoploid. Colchicine added to mitotic cells during metaphase and anaphase disrupts spindle-fiber formation, thus preventing separation of chromatids after the centromere is split. A single cell is created containing pairs of identical chromosomes homozygous at all loci.

are treated with mutagen. They are then plated on selective medium—perhaps a toxic compound normally produced by one of the plant's parasites, or an insecticide—to select resistant cells. Resistant plantlets grow eventually into haploid plants, which can then be doubled (using colchicine) into a pure-breeding resistant type (Figure 8-35). These are potentially powerful techniques that can circumvent the normally slow process of what is basically meiotic plant breeding. The techniques have been successfully applied in several important crop plants, such as soybeans and tobacco. This is of course another aspect of somatic-cell genetics in higher organisms.

Triploids. Once into the realm of polyploids (having more than two chromosome sets), we must distinguish between autopolyploids and allopolyploids. **Autopolyploids** are formed from sets within one single species, whereas **allopolyploids** are formed from sets from different species.

Triploids are usually autopolyploids. They are constructed from the cross of a $4x$ (tetraploid) with a $2x$ (diploid). The $2x$ and the x gametes unite to form a $3x$ triploid.

Triploids also are characteristically sterile. The problem again is pairing at meiosis. Pairing can occur in several ways, but always between only two chromosomes at a time (Figure 8-36). The net result is always the same, an unbalanced segregation of one of the following types:

$$\frac{1+2}{3} \text{ or } \frac{1+3}{2} \text{ or } \frac{2+3}{1}$$

This happens for every chromosome threesome, and the probability of obtaining either a $2x$ or an x gamete is $(\frac{1}{2})^{x-1}$, where x is the number of chromosomes in a set. The others will be unbalanced gametes, having two of one chromosome type, one of another, two of another, and so on, and most will be nonfunctional. Even if the gametes are functional, the resulting zygotes will be unbalanced. A practical application of the sterility associated with triploidy lies in the production of seedless varieties of watermelons and bananas.

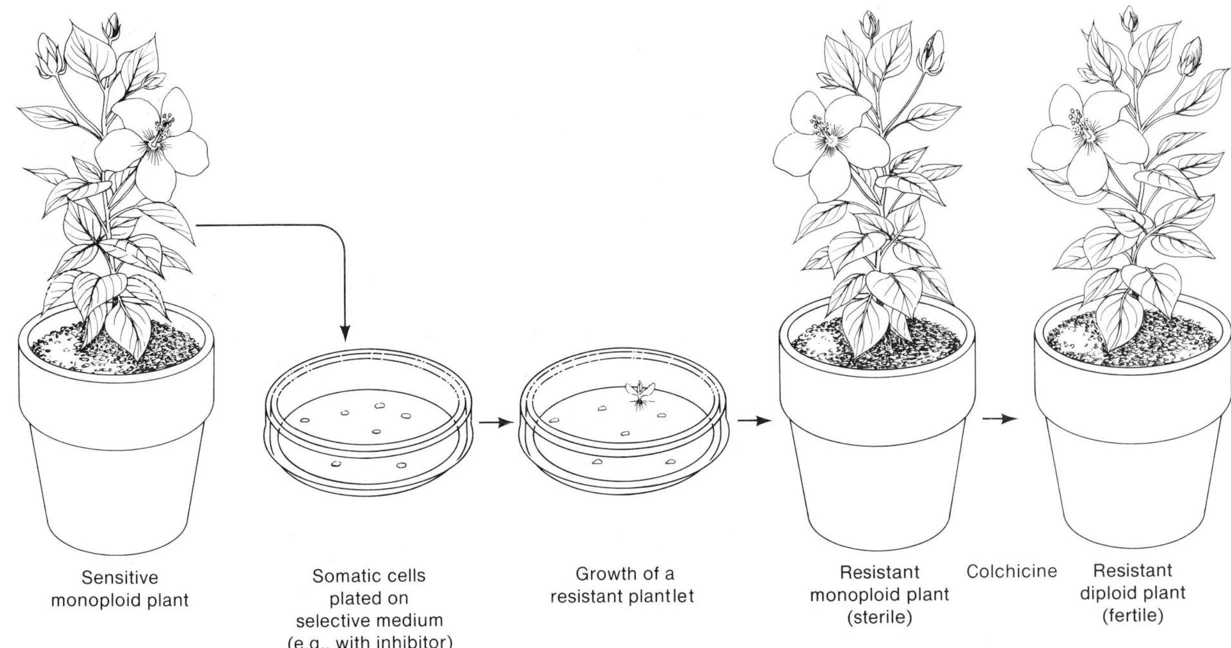

Figure 8-35. *Using microbial techniques in plant engineering. Haploid cells have their cell walls removed enzymatically. The cells are then exposed to a mutagen and plated on an agar medium containing a selective agent, such as a toxic compound produced by a plant parasite. Only those cells containing a resistance mutation that allows the cells to live within the presence of this toxin will grow. After treatment with the appropriate plant hormones, they will grow into mature monoploid plants and, with proper colchicine treatment, can be converted into homozygous diploid plants.*

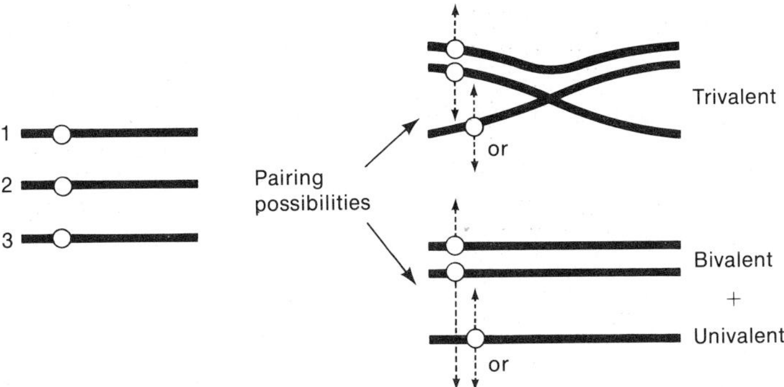

Figure 8-36. *Meiotic pairing possibilities in a triploid. (Each chromosome is really two chromatids.) Pairing in meiosis with the resultant segregation always occurs between only two of the three homologs. The possibility that one gamete will receive the univalent of all chromosome sets is very low. Consequently, gametes of unbalanced chromosome number are produced, and sterility results.*

Autotetraploids. Autotetraploids occur either naturally, by the spontaneous accidental doubling of a $2x$ genome to $4x$, or artificially, through the use of colchicine. Autotetraploids are evident in many commercially important crop plants because, as with other polyploids, the larger number of chromosome sets is often associated with increased size of the plant. This is manifested in increased cell size, fruit size, stomata size, and so on (Figure 8-37).

Because four is an even number, autotetraploids can have a regular meiosis, although this is by no means always the case. The crucial factor is how the four chromosomes of one type pair and segregate. There are several possibilities (Figure 8-38). The two bivalent and the quadrivalent pairing modes tend to be most regular in segregation, but even here there is no guarantee of a 2↔2 segregation. If a regular 2↔2 segregation is achieved at each chromosome type, as is the case in some species, then a formal genetic analysis can be developed for autotetraploids.

Let's use colchicine to double the chromosomes of an *Aa* plant into an *AAaa* autotetraploid, which we will assume shows 2↔2 segregation. We now have a further worry because autotetraploids give different genetic results in their progeny, depending on whether or not the locus concerned is tightly linked to the centromere. First we consider a centromeric gene. Three pairing and segregation patterns are possible (Figure 8-39); these occur by chance and with equal frequency. As the figure shows, the $2x$ gametes will be *Aa*, *AA*, or *aa*, and these will be produced in a ratio of 8:2:2, or 4:1:1. If such a plant is selfed, the probability of an *aaaa* phenotype appearing in the offspring is obviously $1/6 \times 1/6 = 1/36$. In other words, a 35:1 phenotypic ratio will be observed.

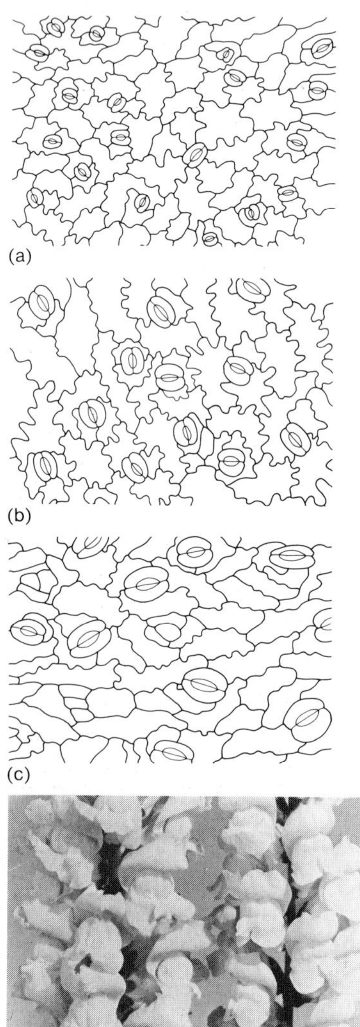

Figure 8-37. Epidermal leaf cells of tobacco plants showing increase in cell size, particularly evident in stomata size, with increase in autopolyploidy: (a) diploid, (b) tetraploid, (c) octoploid. (From W. Williams, Genetic Principles and Plant Breeding, *Blackwell Scientific Publications, Ltd.) (d) Diploid* (right) *and tetraploid* (left) *snapdragons. (W. Atlee Burpee Company.)*

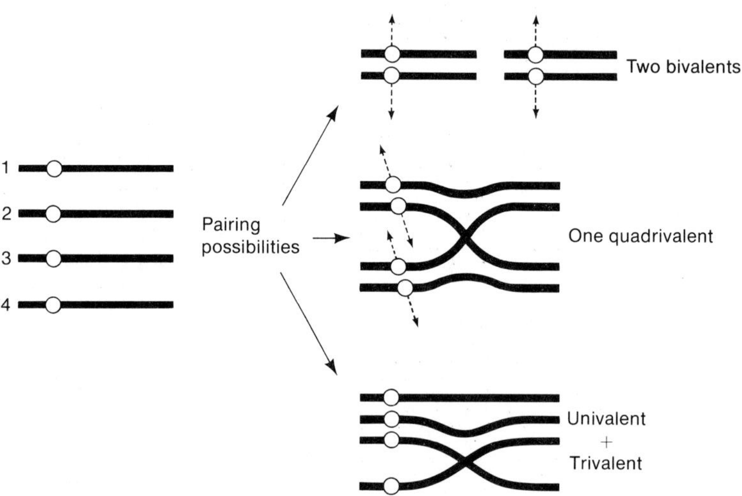

Figure 8-38. Meiotic pairing possibilities in tetraploids. The four chromosomes of one type may pair as two bivalents or as a quadrivalent. Both of these possibilities can yield functional gametes. However, the four chromosomes may also pair in a univalent:trivalent combination, yielding nonfunctional gametes. Some tetraploids have adopted the first or second method as a routine in meiotic segregation and thus can function normally.

1 A
2 A
3 a
4 a

Pairing	1, 2 (A, A)	3, 4 (a, a)	1, 3 (A, a)	2, 4 (A, a)	1, 4 (A, a)	2, 3 (A, a)
Gametes produced by random spindle attachment	$\frac{1+3}{2+4}$	Aa Aa	$\frac{1+2}{3+4}$	AA aa	$\frac{1+2}{3+4}$	AA aa
	$\frac{1+4}{2+3}$	Aa Aa	$\frac{1+4}{2+3}$	Aa Aa	$\frac{1+3}{2+4}$	Aa Aa

Figure 8-39. Genetic consequences in a tetraploid showing orderly pairing by bivalents. The locus is assumed to be close to the centromere. Self-fertilization could yield a variety of genotypes, including aaaa.

If, in the same kind of plant, a genetic locus B/b is very far removed from the centromere, crossing over must be considered. This forces us to think in terms of chromatids instead of chromosomes, and we have 4 B chromatids and 4 b chromatids (Figure 8-40). Because the number of crossovers in such a long region will be large, the genes will become effectively unlinked from their original centromeres, and the packaging of genes two at a time into gametes is very much like grabbing two balls at random from a bag of eight balls, four of one kind and four of another. The probability of picking two b genes then is

$$4/8 \text{ (the first one)} \times 3/7 \text{ (the second one)} = 12/56 = 3/14$$

Then, in a selfing, the probability of a *bbbb* phenotype will equal $3/14 \times 3/14 = 9/196 \cong 1/22$. Hence there will be a 21:1 phenotype ratio of *B*---: *bbbb*. For genetic loci of intermediate position, intermediate ratios will, of course, result.

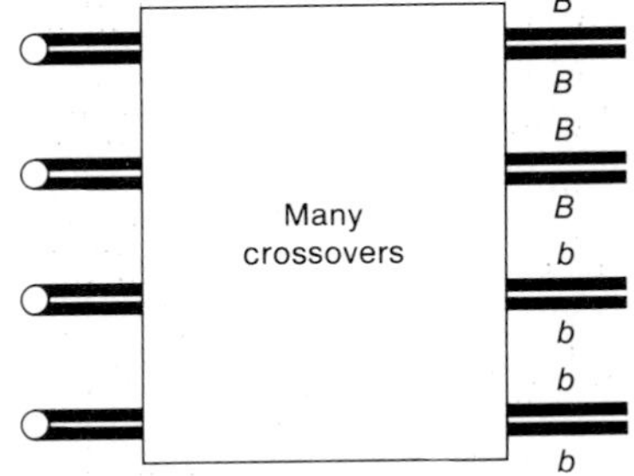

Figure 8-40. Highly diagrammatic representation of a tetraploid meiosis involving a heterozygous locus distant from the centromere. Although tetravalents may not form, the net effect of multiple crossovers in such a long region will be large, and the genes become effectively unhooked from their original centromeres. Genes are packaged two at a time into gametes, much as two balls may be grabbed at random from a bag containing eight balls, four of one kind and four of another.

Allopolyploids. The "classical" allopolyploid was synthesized by G. Karpechenko in 1928. He wanted to make a fertile hybrid between the cabbage (*Brassica*) and the radish (*Raphanus*) that would have the leaves of the former and roots of the latter. Each of these species has 18 chromosomes. They are closely enough related that they could be intercrossed, and a viable hybrid progeny individual was produced from seed. However, this hybrid was sterile because the nine chromosomes from the cabbage parent were different enough from the radish chromosomes that homology was insufficient for normal synapsis and disjunction.

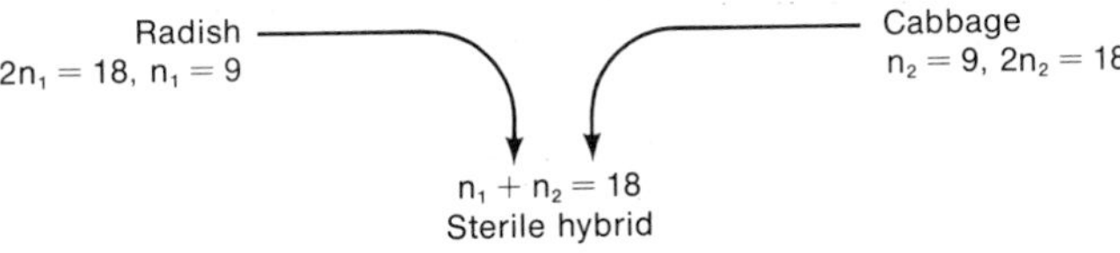

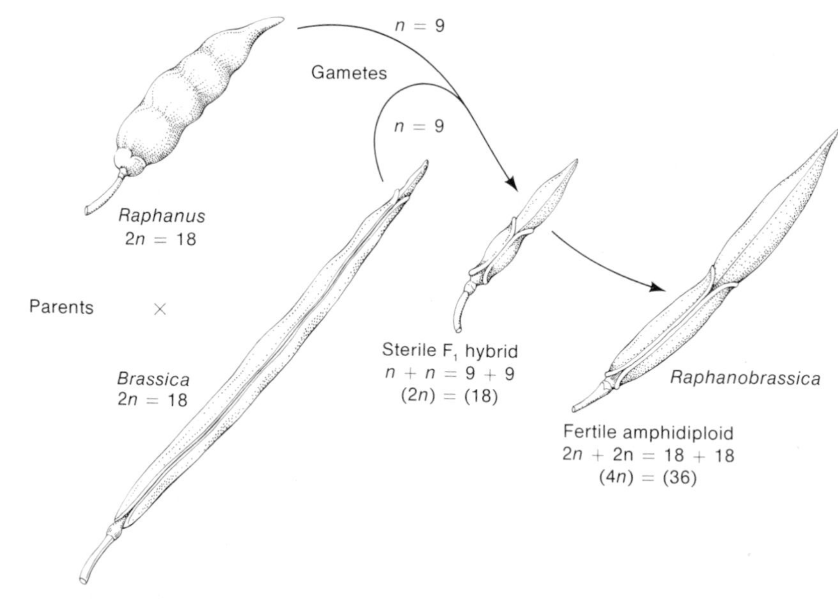

Figure 8-41. The origin of the amphidiploid (Raphanobrassica) *formed from cabbage* (Brassica) *and radish* (Raphanus). *The production of the fertile amphidiploid in this case from the* 2n = *18 hybrid occurred in an accidental fashion, but one similar to the method of producing tetraploids by using colchicine. (From A. M. Srb, R. D. Owen, and R. S. Edgar,* General Genetics, *2nd ed. Copyright © 1965, W. H. Freeman and Company. After Karpechenko,* z. Indukt. Abst. Vererb. *48:27, 1928.)*

However, one day a few seeds were in fact produced by this (almost!) sterile hybrid. On planting, these seeds produced fertile individuals with 36 chromosomes. These individuals were allopolyploids. They had apparently been derived from spontaneous accidental chromosome doubling in the sterile hybrid, presumably in tissue that eventually became germinal and underwent meiosis. Thus, in $2n_1 + 2n_2$ tissue, there is a pairing partner for each chromosome, and balanced gametes of the type $n_1 + n_2$ are produced. These fuse to give $2n_1 + 2n_2$ allopolyploid progeny, which are in turn fertile also. This kind of allopolyploid is sometimes called an **amphidiploid** (Figure 8-41). (Unfortunately for Karpechenko, his amphidiploid had the roots of a cabbage and the leaves of a radish.)

If the allopolyploid is crossed to either cabbage or radish, sterile offspring result. In the case of the cross to radish, these offspring would be $(2n_1 + n_2)$, constituted from an $(n_1 + n_2)$ gamete from the allopolyploid, and an n_1 gamete from the radish. You can see that the n_2 chromosomes will have no pairing partners, so sterility will result. Consequently, Karpechenko had effectively created a new species, with no possibility of gene exchange with its parents. He called his new species *Raphanobrassica.*

Nowadays, allopolyploids are routinely synthesized as a major tool in plant breeding. The goal obviously is to combine some of the worthwhile features of both parental species into one type. This kind of endeavor is very uncertain, as Karpechenko found out. In fact, only one amphidiploid has ever been produced that is of potentially widespread use. This is *Triticale,* an amphidiploid between wheat (*Triticum,* $2n = 6x = 42$) and rye (*Secale,* $2n = 2x = 14$). *Triticale* combines the high yields of wheat with the ruggedness of rye. A massive international *Triticale* testing program is now under way, and many

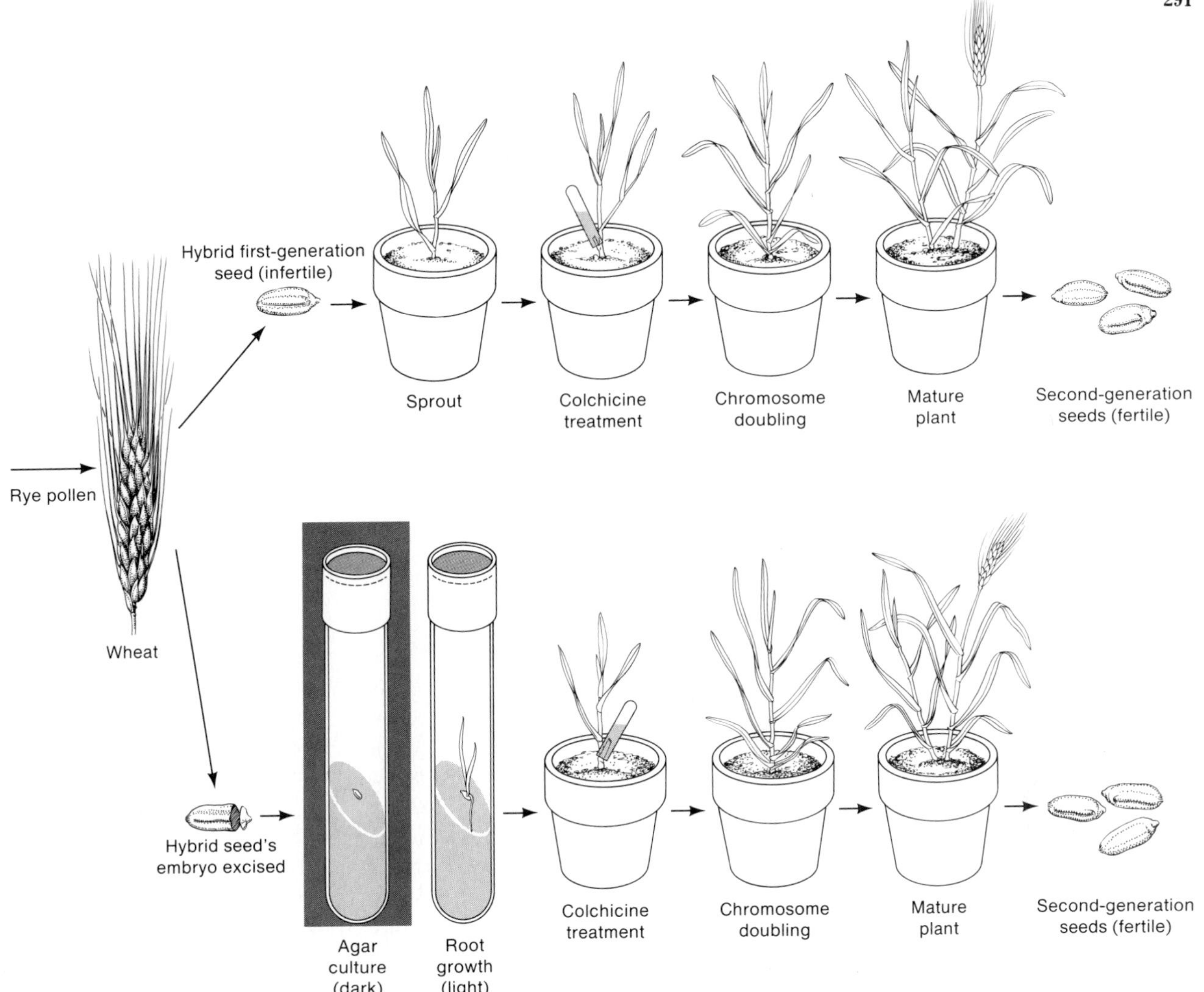

Figure 8-42. Techniques for the production of the amphidiploid Triticale. *If the hybrid seed does not germinate, tissue culture (below) may be used to obtain a hybrid plant. (From Joseph H. Hulse and David Spurgeon, "Triticale." Copyright © 1974 by Scientific American, Inc. All rights reserved.)*

breeders have great hopes for the future of this artificial amphidiploid. Figure 8-42 shows the procedure for synthesizing *Triticale.*

In nature, allopolyploidy seems to have been a major force in speciation of plants. There are many different examples. One particularly satisfying one is shown by the genus *Brassica* (Figure 8-43). Here three different parent species have been hybridized in all possible pair combinations to form new amphidiploid species. This has all taken place in nature, but *Brassica* amphidiploids also have been artificially synthesized (Figure 8-44).

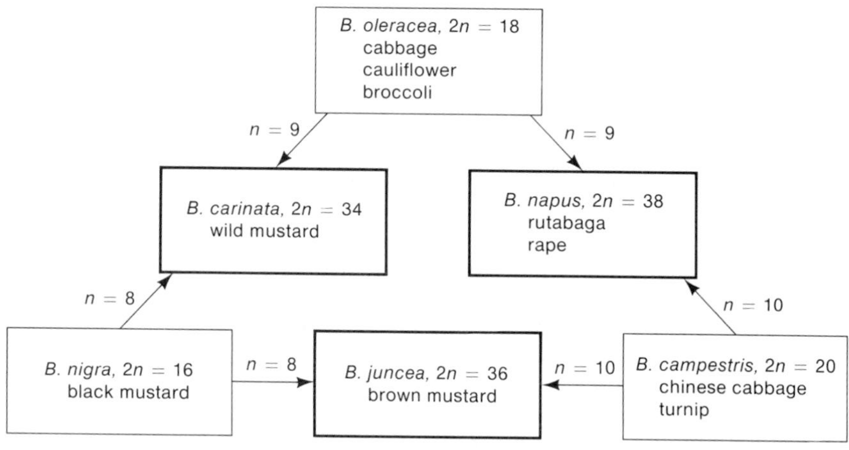

Figure 8-43. A species triangle, showing how amphidiploidy has been important in the production of new species of Brassica.

A particularly interesting natural allopolyploid is bread wheat, *Triticum aestivum* ($2n = 6x = 42$). By a study of various wild relatives, it has been possible to reconstruct a probable evolutionary history of bread wheat (Figure 8-45). In a wheat meiosis, there are always 21 pairs of chromosomes. Furthermore, it has been possible to establish that any given chromosome has only one specific pairing partner (homologous pairing) and not five other potential ones (which would be called **homoeologous pairing**). The suppression of homoeologous pairing (which would lead to much reduced stability of the species) is maintained by a gene *Ph* on the long arm of chromosome 5 of the B set. Thus *Ph* ensures a diploidlike genetics for this basically hexaploid species. Without *Ph,* bread wheat could probably never have arisen. It is interesting to speculate on whether civilization could have arisen or progressed without this species—in other words, without *Ph.*

Somatic Allopolyploids from Cell Hybridization. Another innovative approach to plant breeding is to try to make allopolyploidlike hybrids by asexual methods. If it were possible, this technique might permit combination of widely differing parental species. The technique does indeed work, but so far the only allopolyploids that have been produced are those that can also be made by the sexual methods we have studied already. The procedure is as follows. Cell suspensions of the two parental species are prepared and stripped of their cell walls by special enzyme treatments. The stripped cells are called **protoplasts.** The two suspensions (protoplast suspensions) are combined with polyethylene glycol, which enhances protoplast fusion. The parental cells and the fused cells will proliferate to form colonies (in much the same way as microbes) on agar medium. If these colonies (called **calluses**) are examined, a fair percentage of them are found to

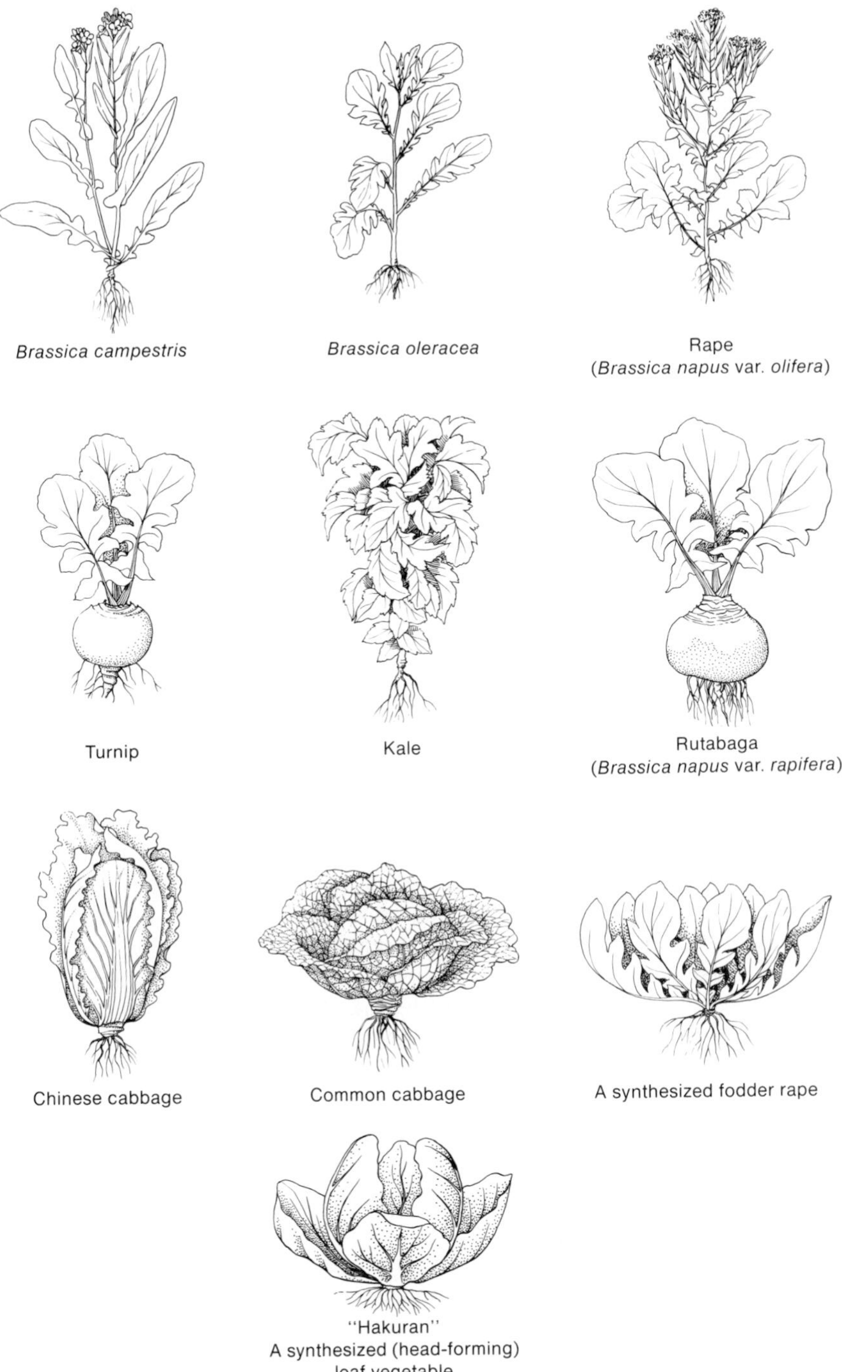

Figure 8-44. Some of the species from the species triangle of Brassica *and two man-made* Brassica *amphidiploids. (Courtesy of H. Kihara,* Seiken Ziho *20, 1968.)*

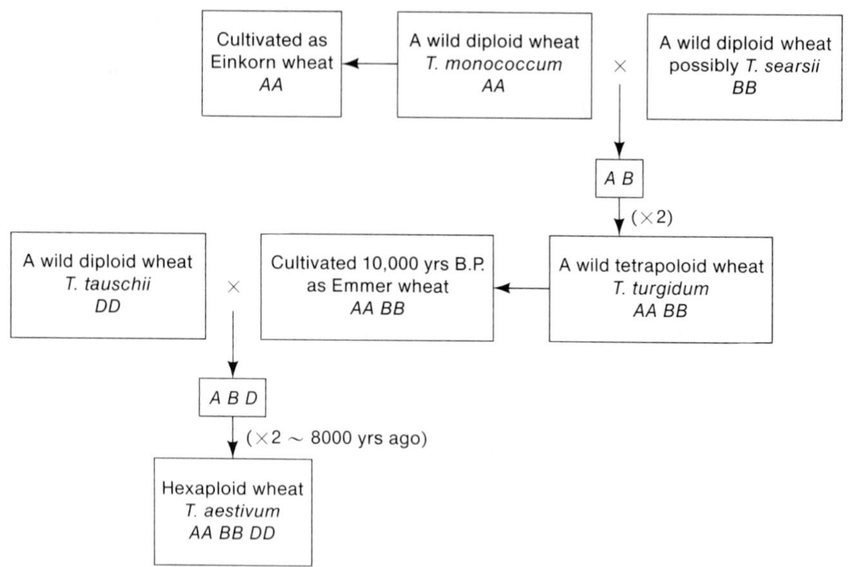

Figure 8-45. Diagram illustrating the proposed evolution of modern hexaploid wheat involving amphidiploid production at several points. A, B, *and* D *are different chromosome sets.*

be allopolyploidlike hybrids with chromosome number equal to the sum of the parental types. Thus, not only do the protoplast cell membranes fuse to form a kind of heterokaryon, but the nuclei fuse too.

The recovery of hybrids may be enhanced if a selective system is available. For example, two different monoploid lines of *Nicotiana tabacum* (tobacco) had light-sensitive yellowish and light-sensitive whitish leaves, respectively. The hybrid calluses (diploid in this example) proved to be green and light-resistant, as a result of recessiveness of the parental genotypes.

The calluses can be grown into plantlets, which then either are grafted onto a mature plant to develop or are themselves potted. A typical protocol is illustrated in Figure 8-46. This is another technique that offfers great promise for the future.

Polyploidy in Animals. You will have noticed that all our examples of polyploidy have been from plants. Polyploid animals do exist, mainly in such lower organisms as flatworms, leeches, and brine shrimp. In these organisms, reproduction is **parthenogenetic** (not involving a normal meiotic sexual cycle). Progeny are produced essentially by mitotic division of parental cells. Parthenogenetic (apomictic) plants, such as dandelions, also are commonly polyploid. Furthermore, **endopolyploidy** (polyploidy of a somatic sector) also is common in plants, animals, and tumor cells. The reason for the rarity of polyploidy in higher animals is not known, but the most widely held hypothesis is that their complex sex-determining mechanisms depend on a delicate balance of chromosome numbers. In humans, polyploids always abort in utero.

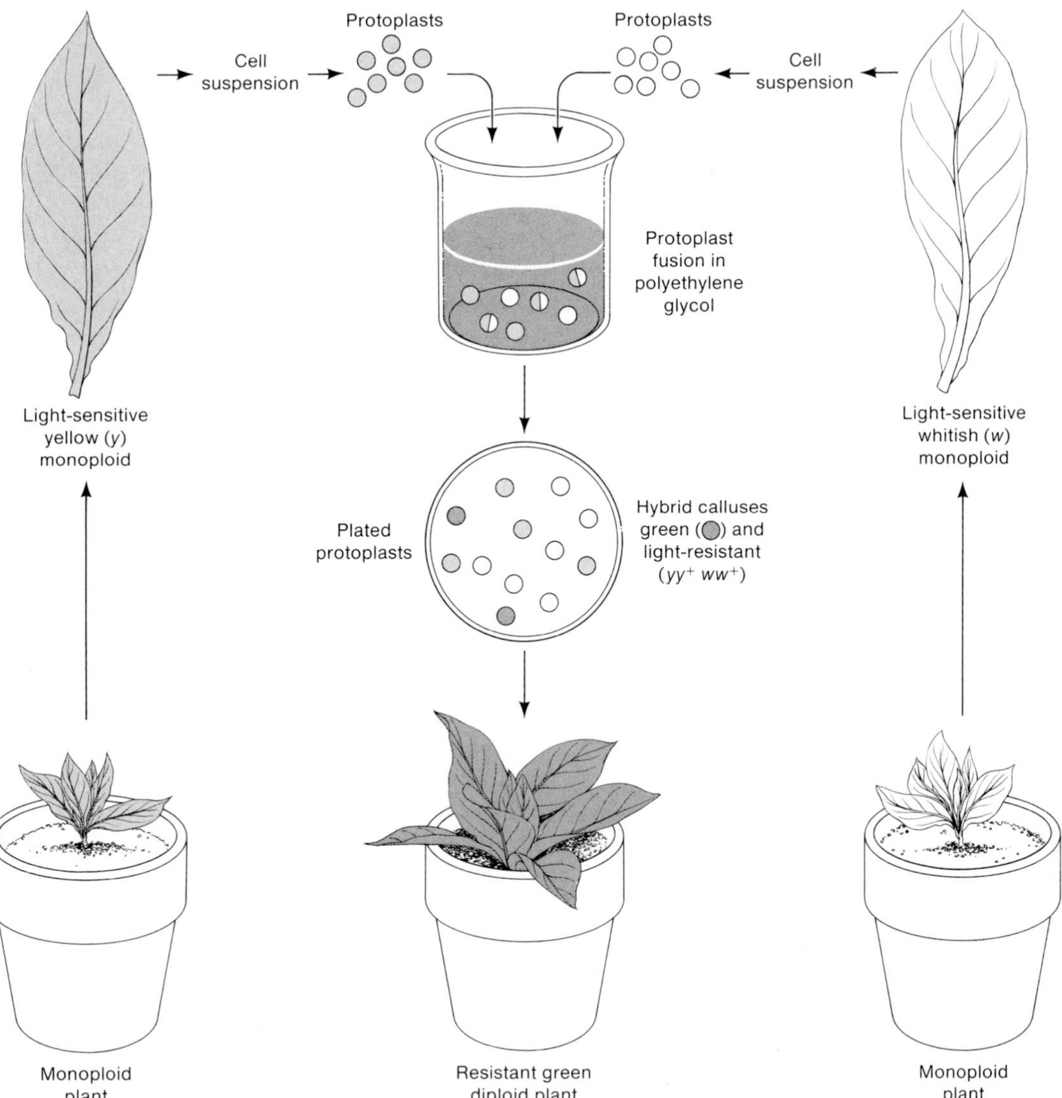

Figure 8-46. The use of cell hybridization in plant engineering of two strains of Nicotiana tabacum. *One strain has light-sensitive yellowish leaves, and the other has light-sensitive whitish leaves. Protoplasts are produced by enzymatically stripping the cell walls from the leaf cells of each strain. Fusion of the protoplasts can occur, as indicated, and those which fuse as hybrids can be grown into calluses that are light resistant, as a result of recessiveness of the parental genotypes. The calluses, under the appropriate hormone regime, can be grown into green diploid plants.*

Aneuploidy

Nullisomics (*2n* − *2*). Although nullisomy is a lethal condition in regular diploids, an organism like wheat (which "pretends" to be diploid but is fundamentally hexaploid) can tolerate nullisomy. In fact, all of the possible 21 wheat nullisomics have been produced (Figure 8-47). Their appearances are different from normal wheat; furthermore, most of them show less vigorous growth.

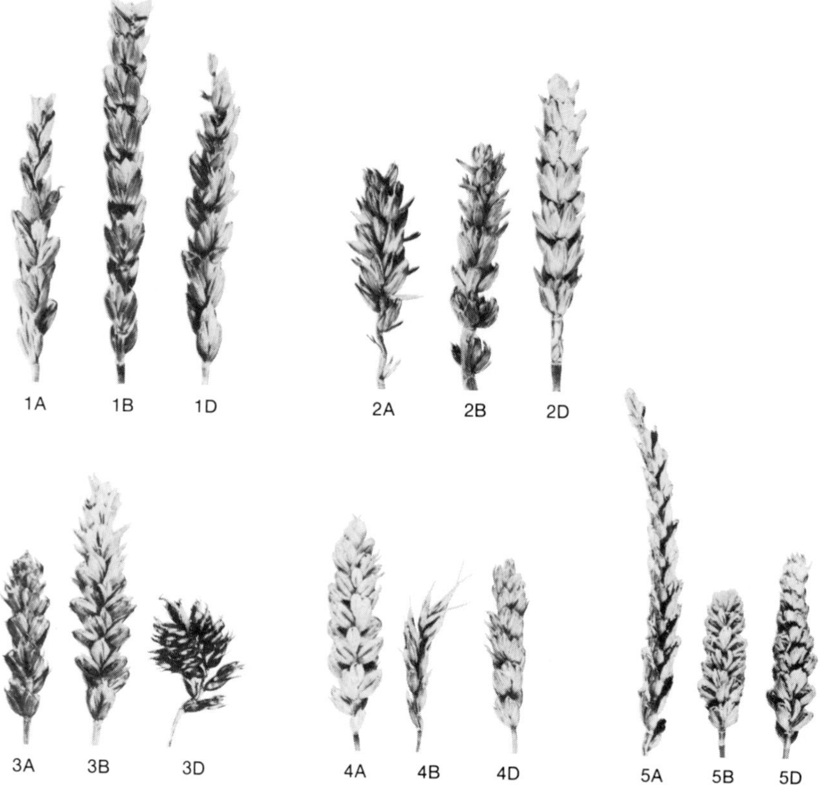

Figure 8-47. The nullisomics of wheat. Although nullisomics are usually lethal in regular diploids, organisms like wheat, which "pretends" to be diploid but is fundamentally hexaploid, can tolerate nullisomy. Nullisomics, however, are less vigorous growers. (Courtesy of E. R. Sears.)

Monosomics (2n − 1). Monosomic chromosome complements are generally deleterious for two main reasons. First, the balance of chromosomes that is necessary for a finely tuned cellular homeostasis, carefully put together during evolution, is grossly disturbed. For example, if a genome consisting of two of each of chromosomes a, b, and c becomes monosomic for c (that is, 2a + 2b + 1c), the ratio of these chromosomes is changed from 1c:1 (a + b) to 1c:2 (a + b). Second, any deleterious recessive on the single remaining chromosome becomes hemizygous and may be directly expressed phenotypically. (Note that these are the same effects as those produced by deletions.)

Monosomics, trisomics ($2n + 1$), and other chromosome aneuploids are probably produced by nondisjunction during mitosis or meiosis. In meiosis it can happen either at the first or second division (Figure 8-48). (You can ask yourself whether products of nondisjunction at these two times can be distinguished genetically.) If an $n - 1$ gamete is fertilized by an n gamete, a monosomic ($2n - 1$) zygote is produced; an $n + 1$ and an n gamete give a trisomic $2n + 1$; and an $n + 1$ and an $n + 1$ give a tetrasomic if the same chromosome is involved or a double trisomic if different chromosomes are involved, and so on.

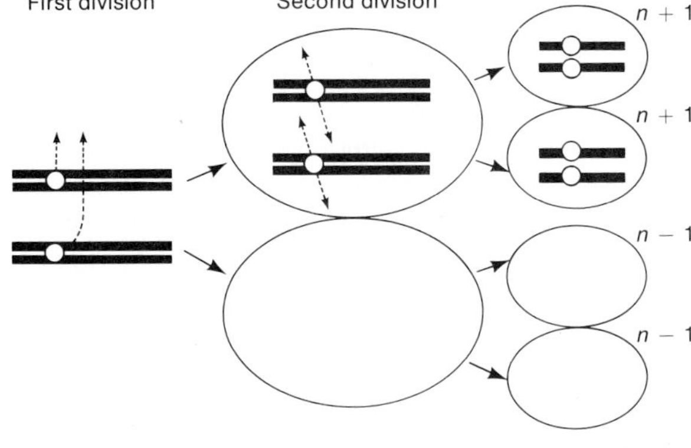

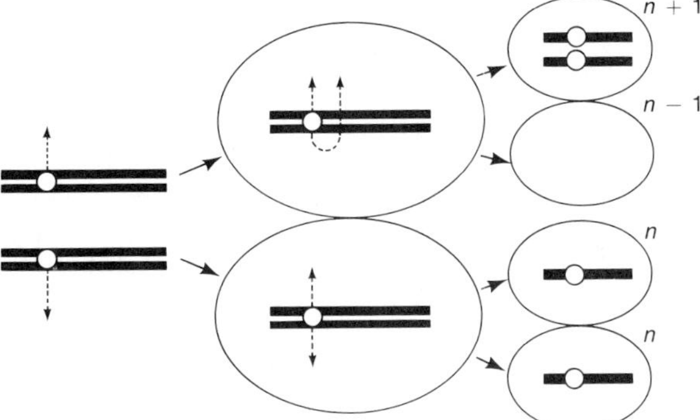

Figure 8-48. The origin of aneuploid gametes by nondisjunction at either the first or second meiotic division.

In *Neurospora* (a haploid), the $n - 1$ meiotic products abort and do not darken in the ascospore; so M_I and M_{II} nondisjunctions are detected as asci with 4:4 and 6:2 ratios of normal to aborted spores. (Diagram the chromosome content of the various spores to convince yourself of the relation of the spore pattern to nondisjunction.) For loci on the aneuploid chromosomes, what ascus genotypes are produced?

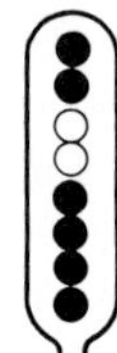

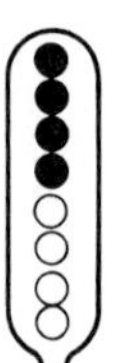

In humans, monosomics are well known, especially the sex chromosome monosomic (44 autosomes + 1 X), which produces a phenotype known as Turner's syndrome. People who have Turner's syndrome have a characteristic, easily recognizable phenotype: they are sterile females, are short in stature, and have webbed necks. Their intelligence varies from different degrees of mental retardation to normal. Monosomics for all autosomes except 13 and 18 die in utero.

If viable, nullisomics and monosomics are useful in locating newly found recessive genes on specific chromosomes in plants. In one such method, different monosomic lines lacking a different chromosome in each line are obtained. Homozygotes for the new gene are crossed with each monosomic line, and the progeny of each cross are inspected for expression of the recessive phenotype. The cross in which the phenotype appears identifies its chromosomal location. In nullisomics and monosomics, of course, $n - 1$ gametes are produced (see Figure 8-49 for monosomics). In general, these gametes tend to be more viable in a female parent than in a male. It is the union of these $n - 1$ gametes with n gametes, bearing the new mutation, that provides the crucial progeny types for the linkage test.

A similar trick can be used in humans. For example, two people whose vision

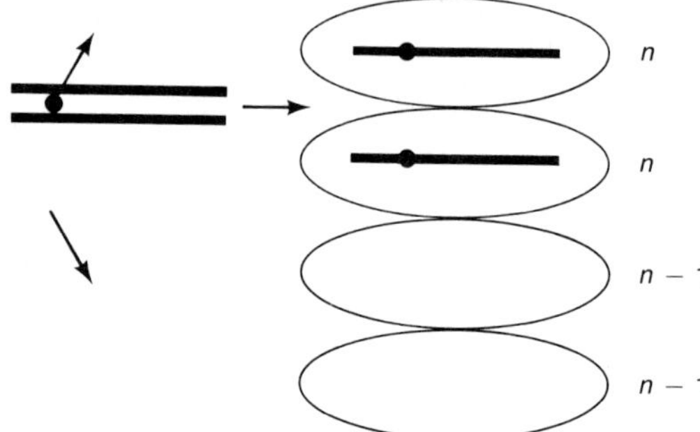

Figure 8-49. Behavior of a monosomic chromosome at meiosis. Two of the resulting haploid gametes will contain a normal set of chromosomes (n), *while two will contain a set missing the monosomic chromosome of the parent* (n - 1).

is normal may produce a daughter who has Turner's syndrome and is also red-green color blind. This shows that the allele for red-green color blindness is recessive, that it was on the X chromosome of the mother, and that nondisjunction must have occurred in the father. (Can you see why?)

Trisomics ($2n + 1$). In trisomics, trivalents are regularly seen (Figure 8-50). For genes that are tightly linked to the centromere of a trisomic chromosome set, the random segregations can be represented as shown in Figure 8-51 in a trisomic *Aaa*. All types occur equally frequently, and a gamete ratio of

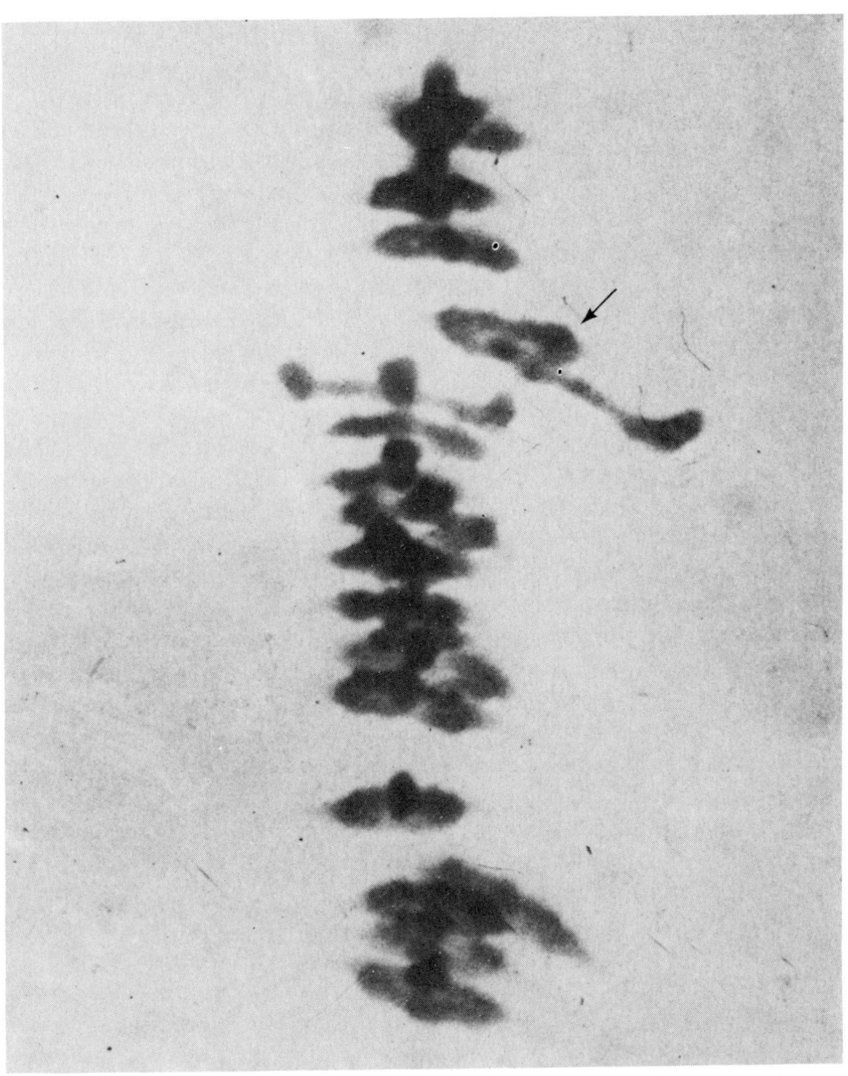

Figure 8-50. Chromosome pairing in a trisomic of wheat. The trivalent chromosome is shown by the arrow. (Courtesy of Clayton Person.)

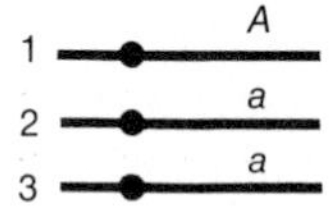

One pole	1 A 2 a	1 A 3 a	2 a 3 a
Other pole	3 a	2 a	1 A

Figure 8-51. Genotypes of meiotic products of a Aaa *trisomic. Three segregation patterns are equally likely.*

1*A*:2*Aa*:2*a*:1*aa* is produced. Trisomics are sometimes recognized by these ratios, which are also useful in locating genes on chromosomes. We have already observed a complete set of trisomic lines in *Datura* (Figure 3-4). Once again, note that chromosome imbalance produces highly chromosome-specific deviations from the normal appearance.

In humans there are several examples of trisomics: The combination XXY results in Klinefelter's syndrome, producing males that have lanky builds, are mentally retarded, and often are sterile. Another combination, XYY males, occurs in more than 1 in 3000 births. A lot of excitement was aroused when an attempt was made to link the XYY condition with a predisposition toward violence. This is still hotly debated, although it is now clear that an XYY condition in no way guarantees such behavior. Nevertheless, several enterprising lawyers have attempted to use the XYY genotype as grounds for acquittal or compassion in crimes of violence. The XYY males are usually fertile.

We have already looked at the generation of Down's syndrome through adjacent segregation in translocation heterozygotes. Down's syndrome also occurs as a result of nondisjunction during meiosis, and it is then called trisomy 21. In this form of Down's syndrome, there is generally no family history of the phenotype. However, the frequency of this form is dramatically higher among the children of older mothers (Figure 8-52). The average frequency of Down's syndrome births is about 0.15%. Down's syndrome is a severely incapacitating condition, causing much suffering. About one-third of all individuals die by the age of 10 years. Some hope of discovering the roots of Down's syndrome has come from recent advances in mapping the human genome. The identification of precisely those genes on the long arm of 21 that must be trisomic to produce Down's syndrome offers some hope of a more precise understanding of its chemical nature and possible therapy.

Chromosome mutation in general plays a prominent role in determining genetic ill health in humans. Figure 8-53 summarizes the surprisingly high levels of various chromosomal abnormalities at different developmental stages of the human organism. In fact, the incidence of chromosome mutations ranks close to that of gene mutations in human livebirths (Table 8-1). This is particularly surprising when one realizes that virtually all chromosome mutations arise

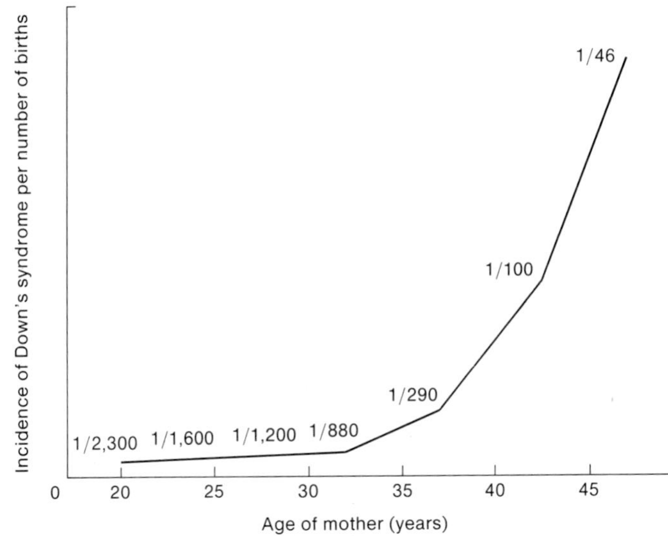

Figure 8-52. Maternal age and the production of Down's syndrome offspring. (From Down's Anomaly *by L. S. Penrose and G. F. Smith. Little, Brown and Company, 1966.)*

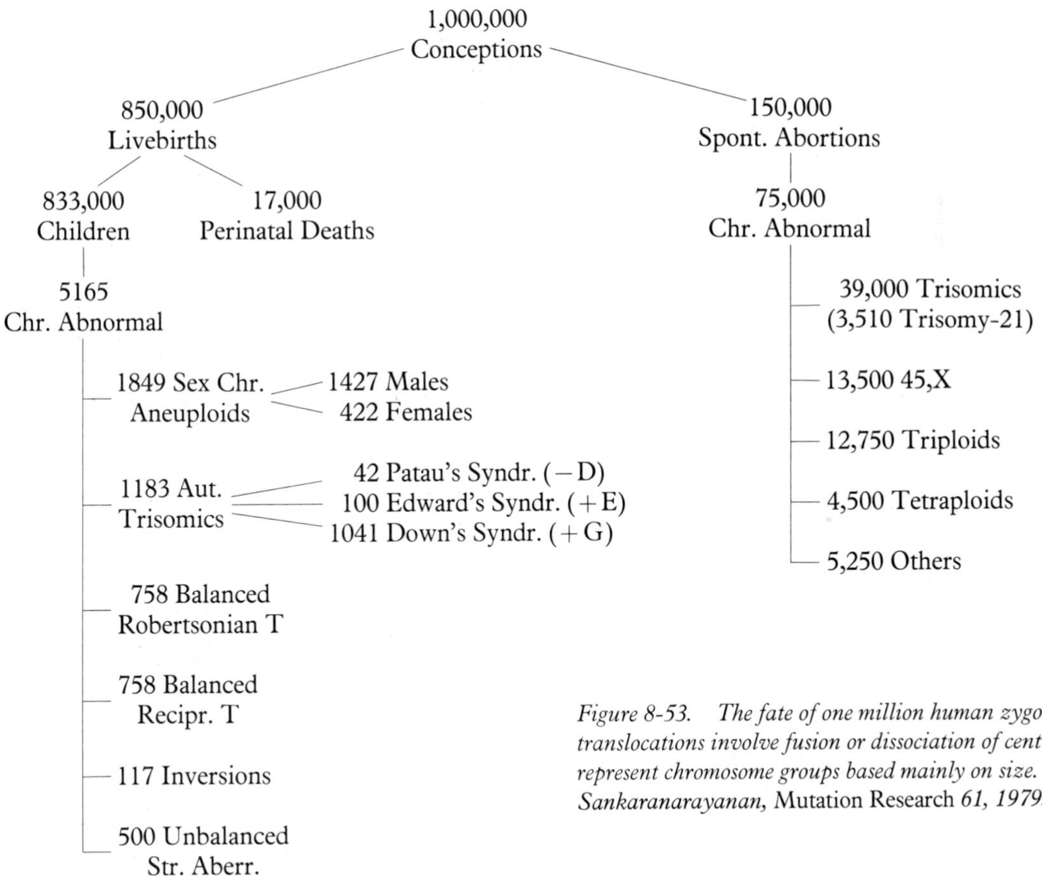

1,000,000 Conceptions

- 850,000 Livebirths
  - 833,000 Children
    - 5165 Chr. Abnormal
      - 1849 Sex Chr. Aneuploids
        - 1427 Males
        - 422 Females
      - 1183 Aut. Trisomics
        - 42 Patau's Syndr. (−D)
        - 100 Edward's Syndr. (+E)
        - 1041 Down's Syndr. (+G)
      - 758 Balanced Robertsonian T
      - 758 Balanced Recipr. T
      - 117 Inversions
      - 500 Unbalanced Str. Aberr.
  - 17,000 Perinatal Deaths
- 150,000 Spont. Abortions
  - 75,000 Chr. Abnormal
    - 39,000 Trisomics (3,510 Trisomy-21)
    - 13,500 45,X
    - 12,750 Triploids
    - 4,500 Tetraploids
    - 5,250 Others

Figure 8-53. The fate of one million human zygotes. Robertsonian translocations involve fusion or dissociation of centromeres. D, E, and G represent chromosome groups based mainly on size. (Courtesy of K. Sankaranarayanan, Mutation Research *61, 1979.)*

Table 8-1. Relative incidence of human ill health due to gene mutation and to chromosome mutation

Type of mutation	Percentage of live births
Gene mutation:	
Autosomal dominant	0.90
Autosomal recessive	0.25
X-linked	0.05
Total gene mutation	1.20
Chromosome mutation:	
Autosomal trisomies (mainly Down's syndrome)	0.14
Other unbalanced autosomal aberrations	0.06
Balanced autosomal aberrations	0.19
Sex chromosomes: XYY, XXY, and other ♂♂	0.17
XO, XXX, and other ♀♀	0.05
Total chromosome mutation	0.61

de novo each generation. This is in contrast to gene mutations, which (as we shall see in Chapter 19) owe their level of incidence to a complex interplay of mutation rates and environmental selection, acting over many generations of the history of the human species.

Chromosome Mechanics in Plant Breeding

Some of the material covered in this chapter no doubt seems esoteric to you. The purpose of this section is to convince you that the details we have covered are of immense significance in the genetic engineering that is so necessary to produce and maintain new crop types in our hungry world. The sole experiment to be described was performed by E. R. Sears in the 1950s. The experiment concerns the transfer of a gene for leaf-rust resistance from a wild grass, *Aegilops umbellulata*, to bread wheat, which was highly susceptible to this disease. It is a classic experiment of its kind.

The first problem was that these two species (Figure 8-54) are not interfertile, so one might think that the feat of gene transfer is impossible. Sears sidestepped this problem with a **bridging cross,** in which *A. umbellulata* was crossed to a wild relative of bread wheat called emmer, *Triticum dicoccoides.* (Follow the process in Figure 8-55.) *A. umbellulata* is a diploid, $2n = 2x = 14$. We will call its chromosome sets CC. *T. dicoccoides* is a tetraploid, $2n = 4x = 28$, with sets AA BB. From this cross, the resulting sterile hybrid ABC was doubled into a fertile amphidiploid AA BB CC with 42 chromosomes. This amphidiploid was fertile in crosses with wheat, which is represented as $2n = 6x = 42$, AA BB DD. The offspring were of course AA BB CD, and they were almost completely sterile because of pairing irregularities between the C and D sets. But crosses to wheat did produce a few rare seeds, some of which

Figure 8-54. Wheat and Aegilops umbellulata. *Whole plants are shown with seed heads inset. (Courtesy of E. R. Sears.)*

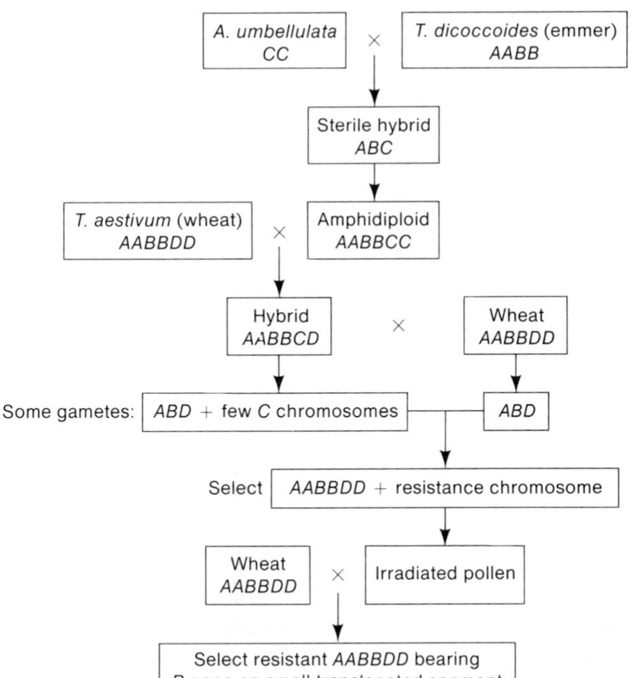

Figure 8-55. Summary of Sears' program for transferring rust resistance from Aegilops *to wheat. A, B, C, and D represent chromosome sets of diverse origin. R represents the genetic determinant (single gene?) of resistance.*

grew into resistant plants. Some of these were almost the desired types, having 43 chromosomes (42 of which were wheat, plus the one *Aegilops* chromosome bearing the resistance gene). Thus, in the AA BB CD hybrid, some aberrant form of chromosome assortment had produced a gamete with 22 chromosomes: ABD plus one from the C group.

Unfortunately, the extra chromosome carried just too many undesirable *Aegilops* genes along with the good one, and the plants were weedy and low producers. So the *Aegilops* gene linkage had to be broken. Sears accomplished this by irradiating pollen of these plants and using it to pollinate wheat. He was looking for translocations of parts of the *Aegilops* chromosome tacked onto the wheat chromosomes. These were quite common, but only one turned out to be ideal. This was a very small *insertional* unidirection translocation (Figure 8-56). When bred to homozygosity, the resistant plants were indistinguishable from wheat.

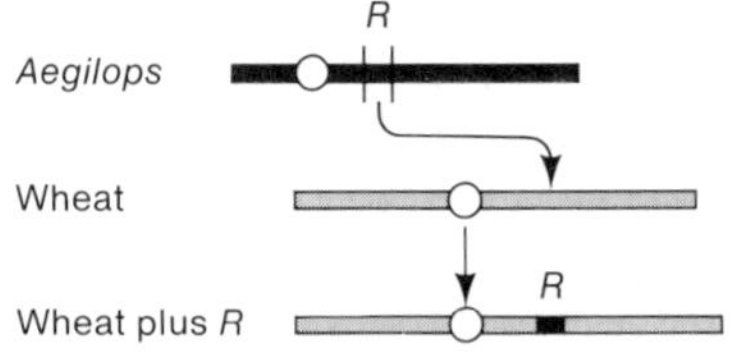

Figure 8-56. Translocation of Aegilops *R segment to wheat using radiation as a means of breaking the chromosomes.*

The study of chromosome mutation is obviously of great importance not only in pure biology, where there are ramifications in many areas (especially in evolution), but also in applied biology. The application of such studies is especially important in agriculture and medicine. Rather curiously, very little is known about the mechanisms of chromosome mutation, especially at the molecular level. More details are needed about the normal chemical architecture of chromosomes. We return to this subject in Chapter 14.

Summary

The morphology of chromosomes provides a useful means for identifying them within the nuclei of different organisms. The location of the centromere determines whether the chromosome is telocentric, acrocentric, or metacentric. Furthermore, chromosomes from certain tissues of some organisms, such as the Diptera, show distinctive blocks of heterochromatin and banding patterns, which aid the identification of the chromosome and cytological location of certain genes on the chromosome.

This chapter deals primarily with changes in the morphology of the chromosomes, or chromosomal mutation. Such mutations involve alterations in chromosomal structure or alterations in chromosome number. Four types of chromosomal structural changes are considered: deletions, duplications, inversions, and translocations.

Deletions remove sections of genetic material. If the gene removed is essential to life, a deletion homozygote will be lethal. Deletion heterozygotes can be viable or lethal and can express recessive genes uncovered by the deletion.

Duplications of genetic material can cause an imbalance in the number of genes, thereby causing a phenotypic effect on the organism. However, there is good evidence from a number of species, including humans, that duplications can lead to increased variety of gene functions. In other words, duplications can be a source of new material for evolution.

Inversions, a third type of alteration in chromosome structure, are caused by a 180° turn of a portion of a chromosome. In the homozygous state, these may cause little problem for an organism unless heterochromatin is involved and a position effect is thus exhibited. On the other hand, inversion heterozygotes often have difficulty in pairing at meiosis, and, if crossing over occurs within the inverted region, inviable products will result. The crossover products will be of different types for paracentric and pericentric inversions.

Translocations, or the movement of a portion of one chromosome to another chromosome, are the final type of structural change considered. When in a heterozygous state, translocation causes problems during anaphase I of meiosis. Aneuploid gametes can result. Both translocation and inversion heterozygotes can show reduced fertility.

Changes in chromosome number are apparently more easily tolerated in plants than in animals. In plants, entire sets of chromosomes may be added or removed, or single chromosomes may be added or deleted. The artificial method of increasing chromosome sets has been used by plant breeders and experimenters to increase size or productivity in commercially important species. Aneuploid karyotypes cause various disorders in humans, depending on which chromosome is involved.

Problems

1. The normal sequence of a certain *Drosophila* chromosome is 1 2 3 · 4 5 6 7 8, where the dot represents the centromere. Some chromosome aberrations were isolated with the following structures: (a) 1 2 3 · 4 7 6 5 8 9; (b) 1 2 3 · 4 6 7 8 9; (c) 1 6 5 4 · 3 2 7 8 9; (d) 1 2 3 · 4 5 6 6 7 8 9. Give the correct name for each type, and draw diagrams to show how each would pair with the normal chromosome.

2. A *Neurospora* heterokaryon is established between nuclei of the following genotypes in a common cytoplasm in which *leu, his, ad, nic,* and *met* are all recessive

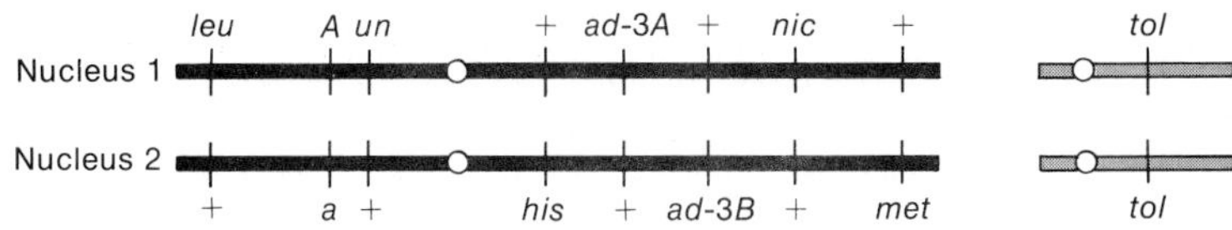

alleles causing specific nutritional requirements for growth. *A* and *a* are the mating-type alleles (for a cross to occur one parent must be *A* and the other *a*). Usually, (*A* plus *a*) heterokaryons are incompatible, but the recessive mutant *tol* suppresses this incompatibility and permits heterokaryotic growth on vegetative medium. The allele *un* is recessive, prevents the fungus from growing at 37°C (it is a temperature-sensitive allele), and cannot be corrected nutritionally. This heterokaryon grows well on minimal medium, as do most of the cells derived mitotically from it. Some rare cells are found, however, that show the following traits:

a. They will not grow on minimal medium unless it is supplemented with leucine.

b. When transferred to a crossing medium, a cross does not occur (they will not self).

c. They will not grow when moved into a 37°C temperature, *even* if supplied with leucine.

d. When haploid wild-type *a* cells are added to these aberrant cells, a cross occurs, but the addition of *A* does not cause a cross.

e. From the cross with wild-type *a*, progeny with the genotype of nucleus 1 are recovered, but no alleles from nucleus 2 ever emerge from the cross.

Formulate an explanation for the origin of these strange cells in the original heterokaryon, and account for the observations concerning them.

3. Certain mice execute bizarre steps—in contrast to the normal gait for mice—and are called "waltzers." The difference between normals and waltzers is genetic, with waltzing being a recessive characteristic. W. H. Gates crossed waltzers with homozygous normals and found among several hundred normal progeny a single waltzing mouse, a ♀. When crossed to a waltzing ♂, she produced all waltzing offspring. Crossed to a homozygous normal ♂, she produced all normal progeny. Some ♂♂ and ♀♀ of this normal progeny were intercrossed, and there were no waltzing offspring among their progeny. Painter examined the chromosomes of waltzing mice that were derived from some of Gates' crosses and that showed a breeding behavior similar to that of the original, unusual waltzing ♀. He found that these individuals had 40 chromosomes, just as in normal mice or the usual waltzing mice. In the unusual waltzers, however, one member of a chromosome pair was abnormally short. Interpret these observations, both genetic and cytological, as completely as possible.

4. In *Neurospora crassa*, mutants of the *ad-3B* gene are relatively easy to amass because they have a purple coloration as well as requiring adenine. One hundred spontaneous *ad-3B* mutants were obtained in haploid cultures, and cells from each were plated on medium containing no adenine to test for reversion. Thirteen cultures produced no colonies, even after extensive platings involving a wide array of mutagens. What is the probable nature of these mutants? Account for both the lack of revertability and the haploid viability of these strains.

5. In *Drosophila*, five recessive lethal mutations are all shown to map on chromosome 2. The chromosomes bearing the lethal mutations are then each paired with a chromosome 2 from a stock having six recessive mutations (*h*, *i*, *j*, *k*, *l*, and *m*), which are distributed throughout the chromosome in that order. The appearance of the resulting flies is shown in Table 8-2, where M stands for mutant for any particular phenotype, and W stands for wild.

Table 8-2.

Lethal mutation	Chromosome-2 marker					
	h	*i*	*j*	*k*	*l*	*m*
1	M	M	W	W	W	W
2	W	W	W	M	M	W
3	W	W	W	W	W	M
4	W	W	W	M	M	M
5	W	W	W	W	W	W

a. What is the probable nature of the lethal mutations 1 through 4?

b. What can you say about the nature of the lethal mutation 5?

6. Two pure lines of corn showed different recombinant frequencies in the region from *P1* to *sm* on chromosome 6. The normal strain (A) shows an RF of 26%, and the abnormal strain (B) shows an RF of 8%. The two lines are crossed, producing hybrids that are semisterile.

   a. Decide between an inversion and a deletion as possible causes of the low RF in B. List your reasons.

   b. Sketch the approximate relation of the chromosome aberration to the genetic markers.

   c. Why are the hybrids semisterile?

7. Duplication is important in evolution because (a) it holds favorable gene combinations together; (b) it inhibits recombination; (c) it uncovers recessive mutations; (d) it promotes sex and hence outbreeding; (e) it releases one copy for new function.

8. In the meiosis *P Bar Q* / *p Bar q*, the *P*/*p* and *Q*/*q* genes represent flanking markers very close to the left and right of a homozygous bar-eye mutation. In an appropriate testcross, some normal-eye and some double-bar types are recovered at low frequencies. These show the flanking marker combinations *Pq* or *pQ*. Explain with diagrams (a) the origin of the rare normal and double-bar types; (b) the association with the flanking marker genotypes.

9. In *Neurospora*, a nontandem duplication of the following type grows quite well as a haploid culture:

   A B ——●—————— *b a*

   From such a culture, rare cells of the following constitution are detected:

   A b ——●—————— *B a*

   a. What kinds of somatic pairing and crossover could produce such rare types?

   b. What other kinds of rare types might you expect to find if you looked hard enough?

10. In *Drosophila,* a pure line is developed, carrying a duplication of the X-chromosome segment that contains the vermillion-eye gene. The stock is

$$\frac{v^+ \quad\quad v^-}{v^+ \quad\quad v^-}$$

and has wild-type eye color. Females of this stock are crossed to nonduplicated vermillion males:

$$\text{Y} \;\frac{\quad}{\quad}\; \frac{\quad v^- \quad}{\quad}$$

The male offspring all have wild-type eye color, and the female offspring all have vermillion eyes. Explain why these are surprising results in regard to the theory of dominance. Explain the phenotype of (a) the female and male parents, and (b) the female and male progeny.

11. An aberrant corn plant gives the following results when testcrossed:

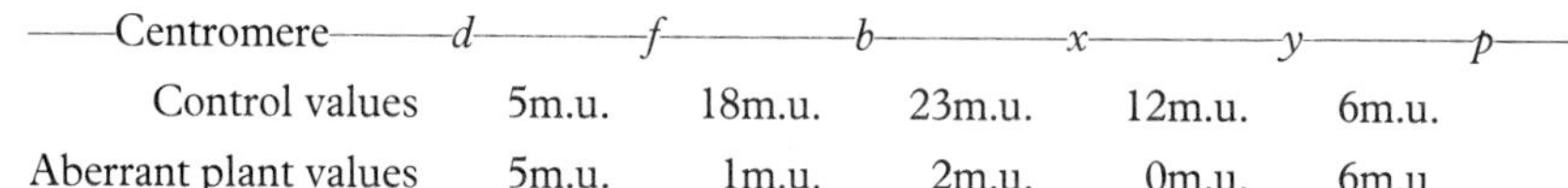

Centromere d f b x y p
Control values 5m.u. 18m.u. 23m.u. 12m.u. 6m.u.
Aberrant plant values 5m.u. 1m.u. 2m.u. 0m.u. 6m.u

The aberrant plant is a healthy plant, but it produces far fewer normal ovules and pollen than the control.

a. Propose a hypothesis to account for the abnormal recombination and the abnormal fertility.

b. Explain with diagrams the origin of the recombinants, according to your hypothesis.

12. In *Neurospora,* a cross is heterozygous for a paracentric inversion. The break points of the inversion are known to be very close to two loci that recombine with an RF of 10%.

a. Using the mapping function RF $= 1/2(1 - e^{-m})$, calculate the *mean* number of exchanges expected in the inversion loop per meiosis.

b. Use this mean frequency to calculate the frequency of meioses with
   i. no exchanges in the loop
   ii. one exchange in the loop
   iii. two exchanges in the loop

   The Poisson formula is $e^{-m}\left(\frac{1}{0!} + \frac{m}{1!} + \frac{m^2}{2!} + \cdots\right)$.

c. Remembering that ascospores bearing deficient chromosome complements do not darken in *Neurospora,* predict how many light and dark ascospores you would find in 8-spored asci resulting from meioses in which there had been
   i. no crossovers in the loop
   ii. one crossover in the loop
   iii. two crossovers in the loop
   (Remember that there are three kinds of double crossovers and so the progeny population of asci from part iii could be heterogeneous.)

d. Using your predicted frequencies of 0, 1, and 2 crossover meioses, what *overall* frequencies of

i. 8 dark:0 light
ii. 0 dark:8 light
iii. 4 dark:4 light

asci would you find from this cross? (NOTE: Your total from part b should be less than 100% because we ignored triple and higher exchanges, but it should be close to 100% because these events are rare. Simply make your total for this part equal your total for part b.)

13. A *Drosophila* geneticist has a strain of fruitflies that is true breeding and wild type. She crosses this strain with a multiply marked X-chromosome strain carrying the recessive genes *y* (yellow), *cv* (crossveinless), *v* (vermilion), *f* (forked), and *car* (carnation), which are equally distributed along the X chromosome from one end to the other. She collects the heterozygous F_1 female offspring and crosses them with *y cv v f B car* males. She gets the following classes among the male offspring:

1. *y cv v f* *car*
2. + + + + +
3. *y* + + + *car*
4. + *cv v f* +
5. *y cv* + *f* *car*
6. + + *v* + +
7. *y cv* + + *car*
8. + + *v f* +
9. *y* + + *f* *car*
10. + *cv v* + +
11. *y cv v f B car*

a. How would you account for the results in classes 1 through 10?

b. How can you account for class 11? You should be able to give two ways.

c. How would you test your hypotheses?

(Problem 13 courtesy of Tom Kaufmann.)

14. Suppose that you are given a *Drosophila* line from which you can get males or virgin females at any time. The line is homozygous for a second chromosome, which has an inversion to prevent crossing over, a dominant gene (*Cu*) for curled wings, and a recessive gene (*pr*, purple) for dark eyes. The chromosome can be drawn as

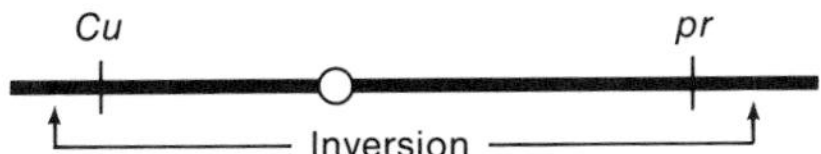

You have irradiated sperm in a wild-type male and wish to determine whether recessive lethal mutations have been induced in chromosome 2. How would you determine this? (HINT: Remember that each sperm carries a *different* irradiated second chromosome.) Indicate the kinds and number of flies used in each cross.

15. What chromosomal shapes do you predict would be produced at anaphase I of meiosis in a reciprocal translocation heterozygote undergoing (a) alternate segregations; (b) adjacent-1 segregations?

16. In *Neurospora,* the genes *a* and *b* are on separate chromosomes. In a cross of a standard *a b* strain with a wild type obtained from nature, the progeny are as follows: *a b* 45%, + + 45%, *a* + 5%, + *b* 5%. Interpret these results and explain the origin of all the progeny types under your hypothesis.

17. You discover a *Drosophila* male that is heterozygous for a reciprocal translocation between the second and third chromosomes, each break having occurred near the centromere (which for these chromosomes is near the center).
    a. Draw a diagram showing how these chromosomes would synapse at meiosis.
    b. You find that this fly has the recessive genes *bw* (brown eye) and *e* (ebony body) on the nontranslocated second and third chromosomes, respectively, and wild-type alleles on the translocated ones. It is crossed with a female having normal chromosomes that is homozygous for *bw* and *e.* What type of offspring would be expected and in what ratio? (Zygotes with an extra chromosome arm or deficient for one do not survive. There is no crossing over in *Drosophila* males.)

18. An *insertional* translocation consists of the insertion of a piece from the center of one chromosome into the middle of another (nonhomologous) chromosome. Thus

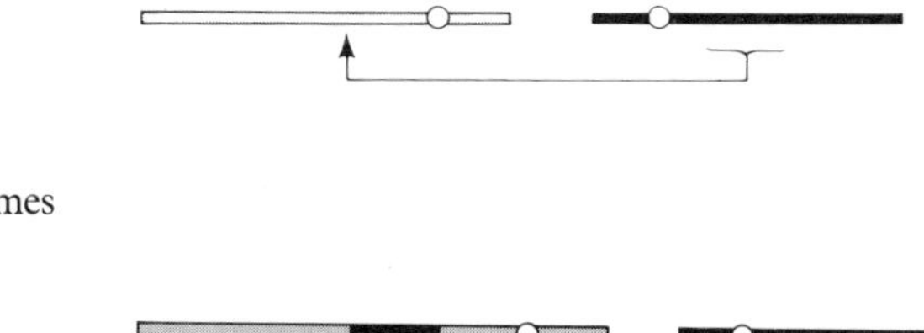

How will genomes heterozygous for such translocations pair at meiosis? In *Neurospora,* what spore abortion patterns will be produced and in what relative proportions in such translocation heterozygotes? (Remember, duplications survive and have dark spores, but deficiencies are light-spored.)

19. Assume that a corn plant is missing the knob at the end of one of its ninth chromosomes and has a knob at the end of one of its tenth chromosomes. Letting *K* stand for *knob,* the plant's chromosomal formula may be designated $\left(\frac{9K}{9}, \frac{10K}{10}\right)$. In corn, as in animals, the "male" tissues form four functional meiotic products, whereas in the "female" tissues three degenerate and only one is functional.
    a. Suppose that you examined cytologically the meiotic products in the male tissues of this exceptional plant. What proportion would you expect to find with both knobs (9*K*, 10*K*)? What other types would you expect to find, and in what proportions?
    b. Answer the questions in part a with regard to the chromosomal constitution of the functional products of meiosis in the female tissues.
    c. If this exceptional plant is self-fertilized, what proportion of the zygotes would you expect to be like the standard type of corn $\left(\frac{9K}{9K}, \frac{10}{10}\right)$?

d. Discuss the value of material like that in the preceding three questions with regard to the visible recognition of the independent assortment of chromosomes.

(Problems 19 and 20 are from A. M. Srb, R. D. Owen, and R. S. Edgar, *General Genetics,* 2nd ed. Copyright © 1965 by W. H. Freeman and Company.)

20. A corn plant *pr/pr* that has standard chromosomes is crossed with a plant homozygous for a reciprocal translocation between chromosomes 2 and 5 and for the *Pr* allele. The F_1 is semisterile and phenotypically Pr (a seed color). A backcross to the parent with standard chromosomes gives 764 semisterile Pr; 145 semisterile pr; 186 normal Pr; and 727 normal pr. What is the map distance of the *Pr/pr* locus from the translocation point?

21. In *Neurospora,* a reciprocal translocation of the following type is obtained

and the following cross is made:

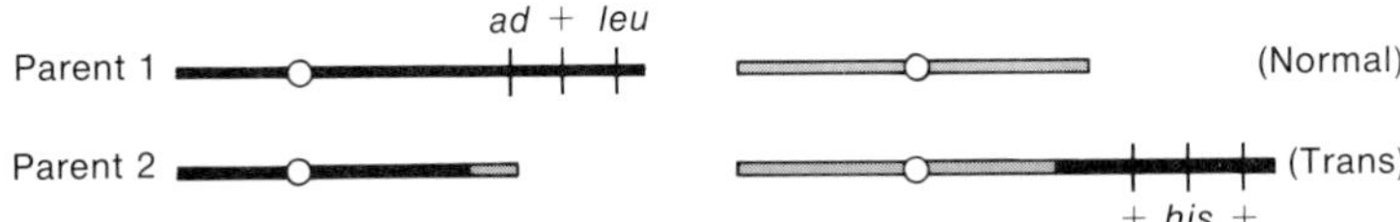

Assuming that the small, lightly shaded piece of the chromosome involved in the translocation does not carry any essential genes, how would you select products of meiosis that are duplicated for the translocated part of the solid chromosome?

22. Assume that, in a study of hybrid cells, three genes in humans have been assigned to chromosome 17. These genes are *a, b,* and *c* and are concerned with making the compounds a, b, and c—all of which are essential for growth. If $a^- b^- c^-$ mouse cells are fused with $a^+ b^- c^+$ human cells, assume that you find a hybrid in which the only human component is the right arm of chromosome 17 (17R), translocated by some unknown mechanism to a mouse chromosome. The hybrid can make the compounds a, b, and c. Treatment of cells with adenovirus causes chromosome breaks. Let's assume that you can isolate 200 lines in which bits of the translocated 17R have been clipped off. These lines are tested for ability to make a, b, and c and the results are:

Number	Can make
0	a only
0	b only
12	c only
0	a and b only
80	b and c only
0	a and c only
60	a, b, and c
48	nothing

a. How would these different types arise?

b. Are *a*, *b*, and *c* all on the right arm of 17?

c. If so, could you draw an approximate map indicating relative positions?

d. How would quinacrine dyes help you in this? (NOTE: This *kind* of approach has actually been used, although the details of this particular question are largely hypothetical.)

23. Complex translocations often are found in natural plant populations. The best example is in the evening primrose, *Oenothera* ($2n = 14$). In this genus, the centromeres tend to be more or less in the middle of the chromosomes. Furthermore, chromosome breakage in the production of translocations tends to be close to or at the centromere. The basic haploid chromosome set can be represented as follows, where a dot represents a centromere:

$$1^{L}\cdot 1^{R} \quad 2^{L}\cdot 2^{R} \quad 3^{L}\cdot 3^{R} \quad 4^{L}\cdot 4^{R} \quad 5^{L}\cdot 5^{R} \quad 6^{L}\cdot 6^{R} \quad 7^{L}\cdot 7^{R}$$

The species *O. lamarckiana* contains one basic haploid set as above, plus a haploid set bearing many translocations, as follows:

$$1^{L}\cdot 1^{R} \quad 3^{L}\cdot 5^{R} \quad 2^{L}\cdot 7^{L} \quad 6^{R}\cdot 5^{L} \quad 4^{L}\cdot 2^{R} \quad 7^{R}\cdot 3^{R} \quad 6^{L}\cdot 4^{R}$$

a. What patterns of chromosome pairing would you expect at meiosis if all homologous regions pair? (Draw a diagram.)

b. In *O. lamarckiana,* the segregation is always alternate. What are the cytological consequences of this?

c. *O. lamarckiana* always contains one "basic" plus one "translocated" set as above, never basic plus basic or translocated plus translocated. Can you think of a mechanism whereby the plant might maintain this situation?

d. Can you think of a reason why it is advantageous to the plant to maintain this situation?

24. Suppose that you are studying the cytogenetics of five closely related species of *Drosophila.* Figure 8-57 shows the gene orders (the letters indicate genes identical in all five species) and chromosome pairs in each species that you find. Show how these species probably evolved from each other, describing the changes that occurred at each step. (NOTE: Be sure to compare gene order carefully.)

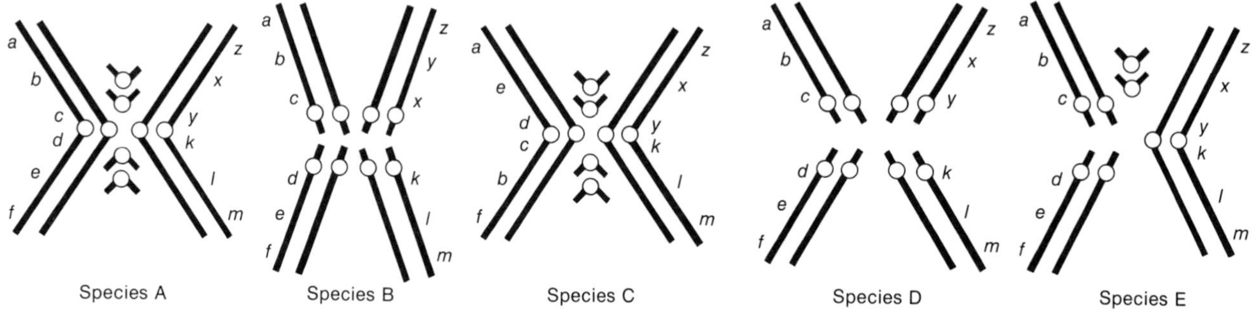

Figure 8-57.

25. H. Sharat Chandra recovered triploids of mealybugs, *Planococcus* ($n = 5$). He found that, in the gonads, at the end of meiosis I all cells had 15 chromosomes, and at the end of meiosis II cells had variable numbers of chromosomes ranging from 0 to 15. How do you interpret these results?

26. In corn, the part we eat (the kernel) is predominantly triploid tissue called endosperm. It is formed as follows. The haploid egg-cell nucleus divides to produce several identical nuclei, one of which acts as the gametic nucleus and two of which act as so-called polar nuclei. The pollen-cell nucleus also divides to form several identical haploid nuclei, one of which fuses with the gametic nucleus to produce the embryo, and one of which fuses with the two polar nuclei to form the endosperm. What are the constitutions of the endosperm types in the cross *Aa Bb* ♀ × *Aa Bb* ♂? (Assume independent assortment.)

27. Allopolyploids are (a) not fertile at all; (b) only fertile amongst themselves; (c) fertile with one parent only; (d) fertile with both parents only; (e) fertile with both parents and themselves.

28. Tetraploid yeast can be created by fusing two diploid cells. These tetraploids will undergo meiosis like any other tetraploid and will produce four diploid products of meiosis. Assuming that homologous chromosomes synapse randomly in pairs and that there is no crossing over in the gene–centromere interval, what nonlinear tetrads would be produced by a tetraploid of genotype *BBbb*? What would be the frequencies of the ascus types? (NOTE: This question involves tetrad analysis of tetraploid cells instead of the usual diploid cells.)

29. In a tetraploid *AAaa,* there is no pairing between chromosomes from the same parent. What phenotypic ratio will result from selfing? In another tetraploid *BBbb,* the only kind of pairing is between chromosomes from the same parent. What phenotypic ratio will result from selfing? (Assume that one parent carried the dominant allele and the other carried the recessive allele in each case.)

30. The New World cotton species *Gossypium hirsutum* has a $2n$ chromosome number of 52. The Old World species *G. thurberi* and *G. herbaceum* each have a $2n$ number of 26. Hybrids between these species show the following chromosome pairing arrangements at meiosis:

Hybrid	
G. hirsutum × *G. thurberi*	13 small pairs (bivalents) + 13 large univalents
G. hirsutum × *G. herbaceum*	13 large bivalents + 13 small univalents
G. thurberi × *G. herbaceum*	13 large univalents + 13 small univalents

Draw diagrams to interpret these observations phylogenetically, indicating clearly the relationships between the species. How would you go about proving that your interpretation is correct?

(Problem 30 is from A. M. Srb, R. D. Owen, and R. S. Edgar, *General Genetics*, 2nd ed. Copyright © 1965 by W. H. Freeman and Company.)

31. An autotetraploid is heterozygous for two gene loci, *FFff* and *GGgg*, each locus affecting a different character and located on a different set of homologous chromosomes very close to their respective centromeres.

a. What gametic genotypes will be produced by this individual, and in what proportions?

b. If the individual is self-fertilized, what proportion of the progeny will have the genotype *FFFf GGgg?* the genotype *ffff gggg?*

32. Which of the following is *not* caused by meiotic nondisjunction? (a) Turner's syndrome, (b) Down's syndrome, (c) Klinefelter's syndrome, (d) XYY syndrome, (e) achondroplastic dwarfism.

33. A patient with Turner's syndrome is found to be color-blind. Both her mother and father have normal vision. How can this be explained? Does this tell us whether nondisjunction occurred in the father or in the mother? If the color-blindness gene were close to the centromere (it is not in fact), would the clinical data tell us whether the nondisjunction occurred at the first or at the second meiotic division?

34. Individuals have been found who are color-blind in one eye but not in the other. What would this suggest (a) if these individuals were only or mostly females? (b) if they were only or mostly males? (Assume that this is an X-linked recessive trait.)

35. Down's syndrome men and women are able to mate with each other and have offspring, although this is rare. What chromosomal constitutions might be expected in the zygotes of such matings, and what would become of these zygotes?

36. When human sperms are treated with quinacrine dihydrochloride, about one-half of the sperms show a fluorescent spot thought to be the Y chromosome. About 1.2% of sperms, however, show two fluorescent spots. Some industrial workmen were exposed over about one year to the chemical dibromochloropropane. Their sperms were examined and the frequency of sperms with two spots was found to be on average 3.8%. Propose an explanation of these results, and explain how you would test it.

37. People with Down's syndrome have about a 15-fold higher risk of leukemia. In the progression of the disease leukemia, complex chromosome aneuploidies usually are seen in the cancer cells. Discuss the possible relationship between these two statements.

38. In humans, the only autosomal trisomics that survive until birth are those for chromosomes 13, 18, or 21. All three types are severely deformed. If you were a medical geneticist, how would you go about studying aneuploidy for the other chromosomes? Do you think aneuploids for the other chromosomes never occur, or are they very rare?

39. The incidence of trisomy 21 or Down's syndrome among children increases some 40-fold for mothers over the age of 35. What kinds of explanations for this increase in nondisjunction can you think of? (We don't know the answer yet, so if you come up with a super idea, let us know.)

40. Several kinds of sexual mosaics are well documented in humans, some examples of which are given below. Suggest how each may have arisen.

    a. XX/XO (that is, there are two cell types in the body, XX and XO)

    b. XX/XXYY

    c. XO/XXX

d. XX/XY

e. XO/XX/XXX

41. In a *Petunia* plant, the genes *A*, *B*, *C*, and *D* are very closely linked. A plant of genotype $a\,B\,c\,D/A\,b\,C\,d$ is irradiated with gamma rays and is then crossed to *aa bb cc dd.* In the progeny, plants of phenotype $A{-}\ B{-}\ C{-}\ D{-}$ are not rare. What are two possible modes of origin, and which is the most likely?

42. In *Neurospora,* a cross between the multiply marked chromosomes $a + c + e$ and $+ b + d +$ gave one product of meiosis that grew on minimal medium (assume *a*, *b*, *c*, *d*, and *e* are nutritional markers). The rare colony was grown up, and some *asexual* spores were $a + c + e$ in genotype, some were $+ b + d +$, and the rest grew on minimal medium. Explain the origin of the rare product of meiosis.

43. In yeast, a diploid was made of genotype

$$\frac{\textit{ade-3}\quad \textit{leu-1}\ \bullet\ \textit{cyh-2}\quad +}{+\quad +\ \bullet\ +\quad \textit{met-13}} \qquad \frac{\textit{ade-2}}{\textit{ade-2}}$$

where all mutant alleles are recessive. On supplemented medium, the colonies are red (due to an accumulation of red pigment at the block caused by *ade-2* in adenine synthesis). However, a few colonies are one-half red and one-half white. (White is the normal yeast color.) These white sectors are also leucine-requiring and cycloheximide-resistant (*cyh* is the resistant allele). The red sectors did not require methionine.

a. Given that *ade-3* is an earlier adenine block than *ade-2,* what mechanism may have given rise to the sectored colonies?

b. How would you test the hypothesis?

44. An *Aspergillus* diploid is

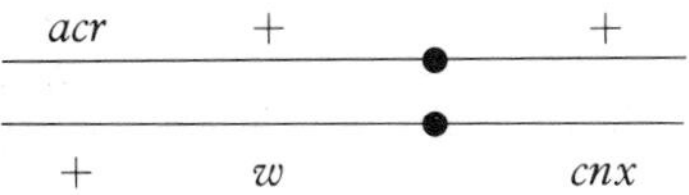

The diploid is green but produces rare white diploid sectors of three different genotypes:

$$\text{(A)}\ \frac{\textit{acr}\quad w\quad +}{+\quad w\quad \textit{cnx}} \qquad \text{(B)}\ \frac{+\quad w\quad +}{+\quad w\quad \textit{cnx}} \qquad \text{(C)}\ \frac{+\quad w\quad \textit{cnx}}{+\quad w\quad \textit{cnx}}$$

Which of these genotypes is most likely due (a) to mitotic crossing over; (b) to mutation; (c) to mitotic nondisjunction? Explain.

45. Design a test system for detecting agents in the human environment that are potentially capable of causing aneuploidy in eukaryotes.

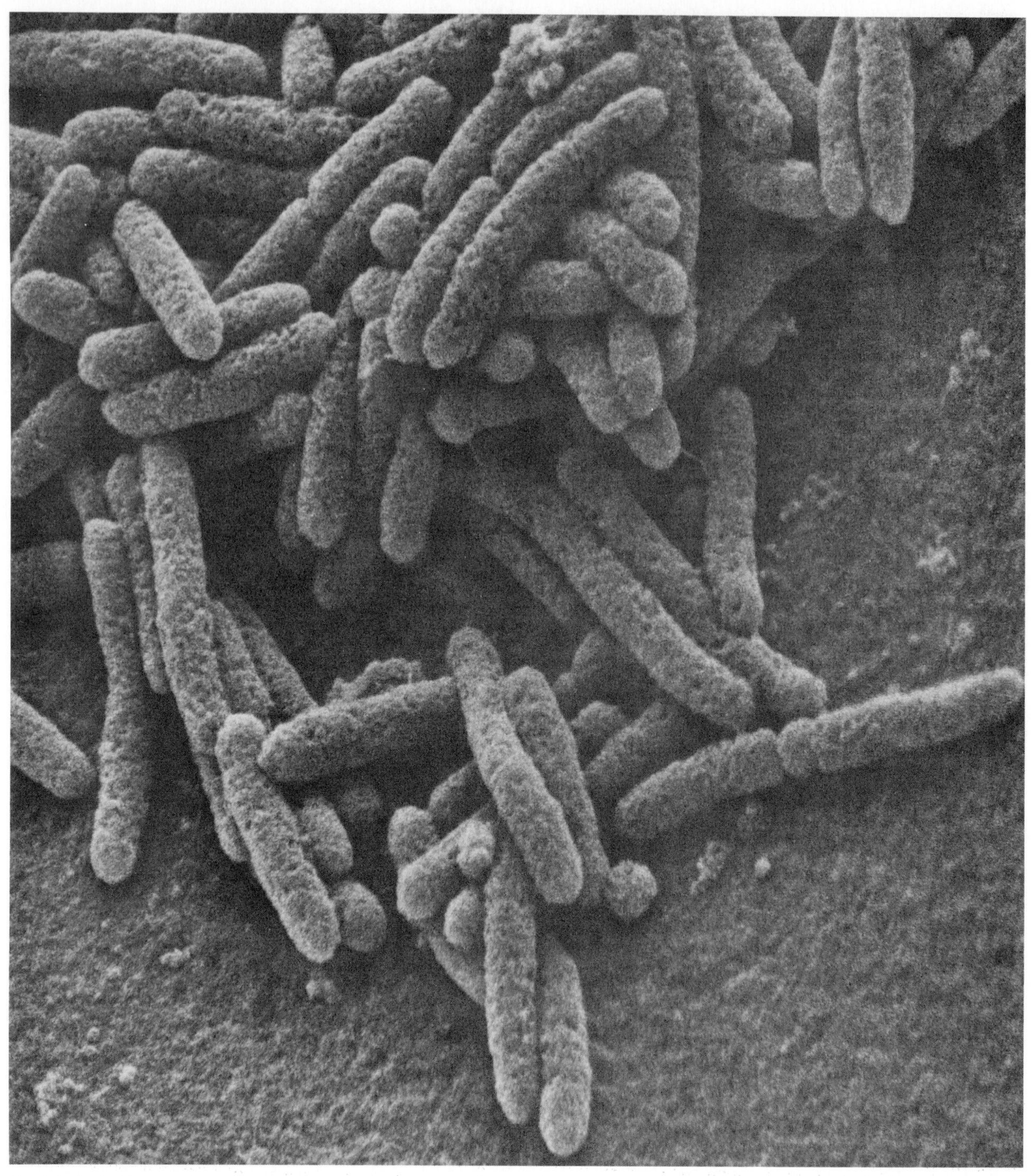

Cells of Escherichia coli *viewed under the scanning electron microscope (× 21,000). (Copyright © David Scharf/Peter Arnold, Inc.)*

9

Recombination in Bacteria and Their Viruses

Thus far, we have dealt almost exclusively with eukaryotic organisms—those whose genes are packed into chromosomes and enclosed within a nucleus. However, a very large part of the history of genetics and of current genetic analysis (particularly molecular genetics) is concerned with prokaryotic organisms and viruses. Viruses are a problem for biologists to classify. They aren't cells, and they cannot grow or multiply alone; they must parasitize living cells, using the cells' metabolic machinery to reproduce. Nevertheless, they do have hereditary properties and can be used for genetic analysis. Although viruses share some of the definitive properties of organisms, many biologists regard them as distinct entities that in some sense are not fully alive.

Prokaryotic organisms and viruses have very simple chromosomes (compared to those of eukaryotes) that are not contained within a nuclear membrane. Nor do these chromosomes undergo meiosis. However, although the ways in which they undergo reproduction may seem strange, there are stages that are analogous to meiosis. The approach to genetic analysis of recombination in these organisms is surprisingly similar to that for eukaryotes. Although we treat them in separate chapters here, there is no fundamental difference between the two systems.

The prokaryotes are bacteria and blue-green algae, but only the bacteria have been used extensively in genetic research. Among the viruses, the best studied

content of each arm was tested for cells able to grow on minimal medium, and none were found. In other words, *physical contact* between two strains was needed for wild-type cells to form. It looked as though some kind of "sex" is involved, and genetic **recombinants** are indeed produced.

Early Attempts to Identify Linkage in Bacteria

Having seen that bacteria can exchange genetic material, as can higher organisms, Lederberg and Tatum hypothesized that the bacterial genes might be organized into linkage groups, as are those of higher organisms. They recognized that two possibilities exist: either the bacterial genes are not organized into linkage groups (in which case something like Mendel's law of independent assortment would apply to all combinations of gene pairs) or, alternatively, the bacterial genes are organized into linkage groups (in which case departures from independent assortment could be detected). Furthermore, they recognized that a simple, two-step extension of their work with strain A and strain B could be used to test for independent assortment.

In step one, they crossed A with B, then plated onto minimal medium supplemented with biotin. Any surviving colonies *must* be met^+ thr^+ leu^+ thi^+ at these four loci but *could* be either bio^+ or bio^-, because either of these can grow when biotin is supplied. Thus, step one produced many colonies that could be designated as $bio^?$ met^+ thr^+ leu^+ thi^+. In step two, they reasoned that, if the loci are unlinked, then their assortment must be independent, and about one-half of the $bio^?$ colonies should be bio^+ and the other one-half bio^-. (Why? Because one-half of the "parental" alleles were bio^+ and one-half bio^-.) On the other hand, if the biotin locus is linked to any of the other four loci, then there should be a great departure from the 1:1 ratio. Specifically, if the biotin locus is linked to the methionine locus, they expected to see an excess of bio^+ over bio^- alleles (because they selected for the met^+ allele from strain B, and that would pull the linked bio^+ allele along). If the biotin locus is linked to any of the other three loci, they expected to see an excess of bio^- over bio^+ alleles (because the selection for the thr^+ leu^+ thi^+ alleles from strain A would pull the linked bio^- along). To determine how many of the $bio^?$ colonies were + and how many were −, they simply plated samples of them on minimal medium: if they grew, they were bio^+; if not, they were bio^-. Finally, to generate additional data, they repeated the whole procedure using each of the other supplements in turn. Their results are shown in Table 9-1.

The data clearly show linkage, because the allele frequencies at each of the unselected loci show great departures from a 1:1 ratio. By noting whether the departure from 1:1 is toward an excess of + alleles or an excess of − alleles, Lederberg and Tatum were able to determine the linkage arrangements among the loci. (As explained above, the excess of bio^+ over bio^- suggests that the biotin and the methionine loci are linked. To test your understanding of the experiment, you should try to work out the linkage arrangements among the other loci.)

You may be wondering why Lederberg and Tatum didn't try using methionine as a medium supplement. In fact, they did, but they had to throw those data

Table 9-1. Allele frequencies observed by Lederberg and Tatum

Medium supplement	Genotype of surviving colonies	Allele frequencies at the unselected (?) locus	
		+	−
Biotin	$met^+ bio^? thr^+ leu^+ thi^+$	60	10
Threonine	$met^+ bio^+ thr^? leu^+ thi^+$	37	9
Leucine	$met^+ bio^+ thr^+ leu^? thi^+$	51	5
Thiamine	$met^+ bio^+ thr^+ leu^+ thi^?$	8	79

away when they discovered that their methionine had been contaminated with biotin. We mention this to make two important points: first, even Nobel Prize winners sometimes make mistakes; second, it is still possible to obtain a great deal of information from a "damaged" experiment, provided care is taken to discard all of the tainted data.

Discovery of the Fertility Factor F

The experiments we have discussed represent a complicated way to do a genetic analysis. In fact, Lederberg found it increasingly difficult to interpret his data. In 1953, William Hayes made a remarkable discovery. He verified the results of Lederberg and Tatum, using a similar cross:

$$\begin{array}{ccc} \text{strain A} & & \text{strain B} \\ met^- \, thr^+ \, leu^+ \, thi^+ & \times & met^+ \, thr^- \, leu^- \, thi^- \end{array}$$

In a further experiment, Hayes treated one of the strains with the antibiotic streptomycin (which does not kill immediately but which prevents cell division) before mixing in the other strain. When he treated strain A with the streptomycin, mixed in strain B, and then plated the culture on minimal medium, he obtained the same frequency of colonies as he obtained in the control experiment with both strains untreated. However, when he treated strain B with streptomycin, mixed in strain A, and then plated the culture on minimal medium, he obtained no colonies.

How can these asymmetrical results be explained? It would seem that the ability of strain-B to divide is crucial to the production of the colonies on minimal medium, whereas it does not matter whether or not the strain-A cells can divide. One interpretation is that the genetic exchange is asymmetrical: cells of strain A donate genetic material to cells of strain B, which then can divide to produce colonies containing genes from both strains.

Message
The exchange of genetic material in E. coli *is not reciprocal. One cell acts as the* ***donor,*** *and the other cell acts as the* ***recipient.***

The donor can still transmit genes after exposure to streptomycin, but the recipient cannot divide to produce colonies if it has been exposed to streptomycin. This kind of unidirectional exchange of genes can be analogized to a sexual difference, with the donor being "male" and the recipient "female."

By accident, Hayes discovered a variant of his original strain A (male) that would not produce recombinants upon crossing with the B strain (female). Had the A males changed into females, or perhaps even into homosexuals? In his analysis of this sterile variant, Hayes discovered a central but surprising component of the fertility of *E. coli.* The original A male strain, and hence the sterile variant, were sensitive to streptomycin (str^s). By plating the sterile variant on medium containing streptomycin, Hayes isolated a streptomycin-resistant mutant (str^r), which proved to be very useful in solving the genetic jigsaw puzzle.

Hayes mixed the sterile A str^r cells with the fertile A male str^s cells and then plated on a medium containing streptomycin. He found that as many as one-third of the A str^r cells had become fertile and could cross with B females (Figure 9-4). Hayes explained these bizarre results by suggesting that maleness (or donor ability) is itself a hereditary state imposed by a **fertility factor,** or F. Females lack F and therefore are recipients. Thus, females can be designated F^-, and males F^+. The sterile A-strain variant must have lost F and become F^-, and the original pedigree must have been that shown in Figure 9-5. In other words, the variant strain (which was sterile as a male parent) had in fact become female, as which it was perfectly fertile.

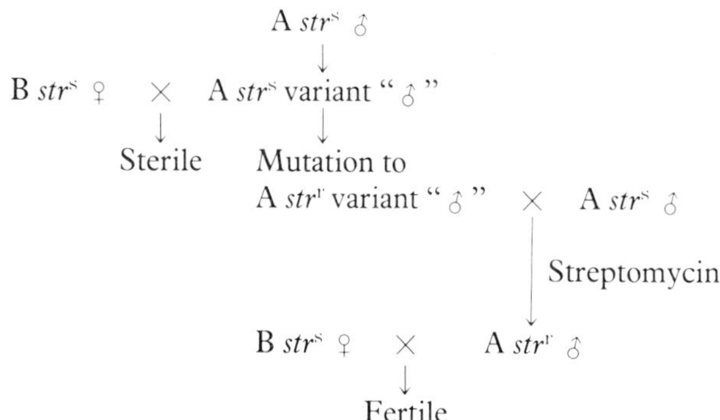

Figure 9-4. Pedigree involving an apparently sterile bacterial variant. The original cross of strain A (streptomycin-sensitive recipient) to strain B (streptomycin-sensitive donor) was sterile. However, a rare variant, or mutant, of strain A to streptomycin resistance allowed Hayes to cross this variant to the "normal" streptomycin-sensitive strain A. If this cross was plated on a medium containing streptomycin, only the streptomycin-resistant progeny would grow. These, he discovered, could then be crossed to strain B cells, yielding fertile offspring. From these and other data, Hayes concluded that maleness in these bacteria was determined by a factor called F, which could be passed to the female during conjugation, converting her to a male.

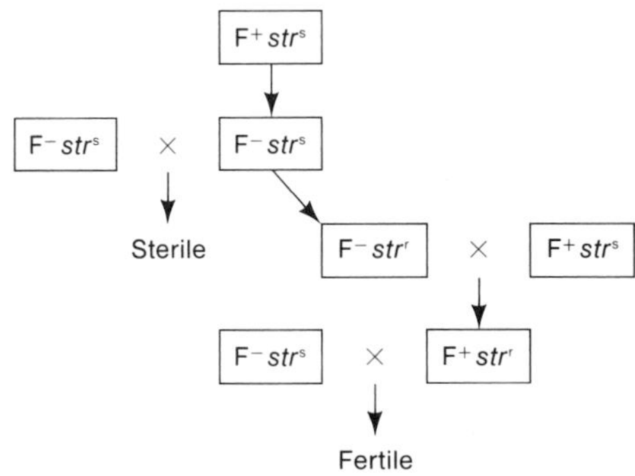

Figure 9-5. Symbolic explanation of pedigree in Figure 9-4.

Recombinant genotypes for marker genes are relatively rare in bacteria crosses, but the F factor apparently is transmitted very effectively at physical contact, or **conjugation.** Male fertile str^r cells were recovered very soon after mixing with F^+ str^s. There seems to be a kind of **infectious transfer** of the F factor that takes place far more quickly than the regular exchange of marker genes. The physical nature of the F factor was elucidated much later, but these early experiments showed clearly that it is some kind of errant particle not closely tied to the marker genes.

In $F^+ \times F^-$ crosses, Hayes found that the progeny selected as recombinants for specific marker genes generally tend to contain the remaining (unselected) alleles of the F^- parent. This is true regardless of the particular marker alleles used and regardless of their coupling or repulsion arrangement. These results suggest a nonreciprocal exchange of genetic material between the parents of a cross. Further substantiation came in a difficult experiment where F^+ and F^- cells were separated using a micromanipulator; only the F^- cells ever showed a recombinant phenotype of any kind. The process involved in these crosses obviously is quite different from any kind of meiosis. Is it possible that only *part* of the F^+ chromosome is being transferred into the F^- cell?

The big break in solving the puzzle came when Luca Cavalli-Sforza obtained a new kind of male from an F^+ strain. On crossing with F^- females, this new strain produced a thousand times more recombinants for marker genes than did its progenitor. Cavalli-Sforza called this derivative an **Hfr** strain (for *high frequency of recombination*). (It later became known as Hfr C, to distinguish it from a similar strain, Hfr H, found by Hayes.)

You will appreciate that, in the cells resulting from an $F^+ \times F^-$ cross, a large proportion of genotypes initially associated with the F^- parent are found to have been converted to F^+ by infectious transfer of the fertility particle. However, in the Hfr $\times$ F^- crosses, none of the basically F^- genotypes are converted to F^+ or to Hfr. Thus infectious transfer of the factor F does not seem to occur in these crosses, even though recombination of markers is manyfold more efficient than in $F^+ \times F^-$ crosses. Perhaps at this point you can understand why most geneticists until the late 1950s tried to ignore bacterial genetics!

Determining Linkage From Interrupted-Mating Experiments

The clues all began to fit together in 1957 when Ellie Wollman and François Jacob investigated the pattern of transmission of Hfr genes to F^- cells during a cross. They crossed Hfr $str^s\ a^+\ b^+\ c^+\ d^+$ with F^- $str^r\ a^-\ b^-\ c^-\ d^-$. At specific time intervals after mixing, they removed samples. Each of these samples was put into a kitchen blender for a few seconds to disrupt the mating cell pairs; it then was plated onto medium containing streptomycin to kill the Hfr donor cells. This is called an **interrupted-mating** procedure. The str^r cells then were tested for the presence of marker alleles from the donor. Those str^r cells that bear donor marker alleles must have been involved in conjugation; such cells are called **exconjugants.** Figure 9-6 shows a plot of the results; azi^r, ton^r, lac^+, and gal^+ correspond to the a^+, b^+, c^+, and d^+ mentioned in our generalized description of the experiment.

The most striking thing about these results is that each donor allele first appears in the F^- recipients at a specific time after mating begins. Furthermore,

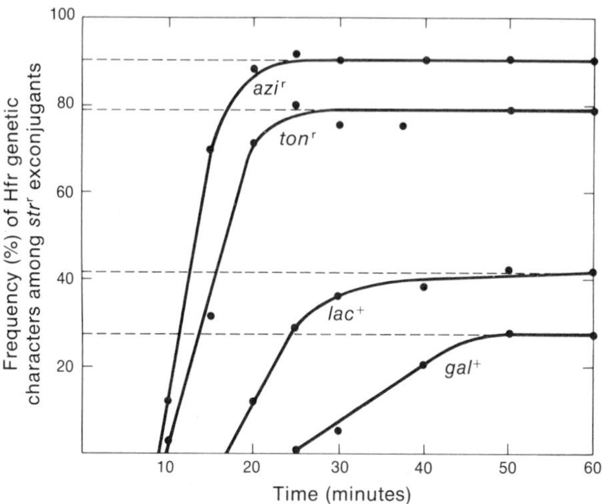

Figure 9-6. Interrupted mating/conjugation experiments with E. coli. *F^- cells that were* str^r *were crossed with Hfr cells that were* str^s. *The F^- cells had a number of mutants (indicated by* azi, ton, lac, gal*) that prevented them from carrying out specific metabolic steps. However, the Hfr cells were capable of carrying out all these steps. At different times after the cells were mixed, samples were withdrawn, disrupted in a blender in order to break conjugation between cells, and plated on media containing streptomycin. The antiobiotic killed the Hfr cells but allowed the F^- cells to grow and to be tested for their ability to carry out the four metabolic steps. Transfer of the donor allele of each of these steps is obviously dependent on the time that conjugation is allowed to continue. (From E. L. Wollman, F. Jacob, and W. Hayes,* Cold Spring Harbor Symposia on Quantitative Biology *21:141, 1950.)*

the donor alleles appear in a specific sequence. Finally, the maximal yield of cells containing a specific donor allele is smaller for the donor markers that enter later. Putting all these observations together, we obtain an interpretation of the results.

Message
The Hfr chromosome is transferred to the F^- cell in a linear fashion, beginning at a specific point (called the origin, or O). The farther a gene is from O, the later it is transferred to the F^-. For "later" genes, it is more likely that the transfer process will stop before they are transferred (hence the less steep slope and smaller maximal frequency for later genes).

Wollman and Jacob realized that it would be easy to construct linkage maps from the interrupted-mating results, using as a measure of "distance" the time at which the donor alleles first appear after mating. The units of distance in this case are minutes. Thus, if b^+ begins to enter the F^- cell 10 minutes after a^+ begins to enter, then a^+ and b^+ are 10 units apart (Figure 9-7). Like the maps based on crossover frequencies, these linkage maps are purely genetic constructions; at the time, they had no known physical basis.

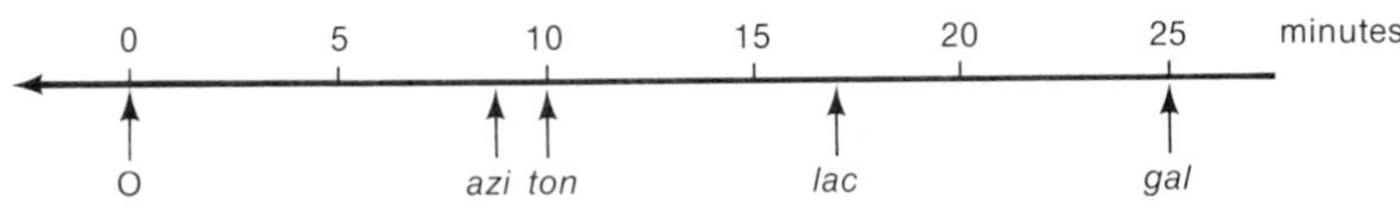

Figure 9-7. Chromosome map from Figure 9-6. A linkage map can be constructed for the E. coli *chromosome from interrupted-mating studies, using the time at which the donor alleles first appear after mating. The units of distance are given in minutes.*

A similar approach is used in any interrupted-mating experiment. The easily selectable alleles (prototrophic, or resistance alleles) are concentrated in the donor, and their counterpart alleles are concentrated in the recipient. One specific resistance allele (usually str^r) is used in the F^- cells to eliminate selectively the Hfr cells so that they won't obscure the transmission patterns. Thus the system permits study of genetic recombinants, but always in one specific coupling type of arrangement of the parental individuals. (The exconjugants which contain donor genes are of course recombinants.) We see, however, that the first bacterial maps were based not on recombinant frequency but on time of appearance of recombinants—a novel approach based on a system that was itself very much a novelty.

Chromosome Circularity

When Wollman and Jacob allowed Hfr × F^- crosses to continue for as long as 2 hours before blending, they found that some of the exconjugants were converted into Hfr. In other words, the fertility factor conferring maleness (or donor ability) is eventually transmitted, but at a very low efficiency and apparently as the last element of the linear "chromosome." We now have the following picture:

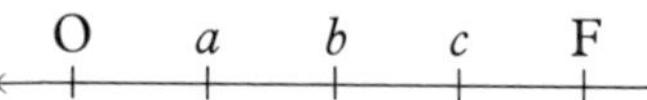

However, when several different Hfr linkage maps were derived by interrupted-mating and time-of-entry studies using different separately derived Hfr strains, the maps differed from strain to strain:

Hfr H	O *thr pro lac pur gal his gly thi* F
1	O *thr thi gly his gal pur lac pro* F
2	O *pro thr thi gly his gal pur lac* F
3	O *pur lac pro thr thi gly his gal* F
AB 312	O *thi thr pro lac pur gal his gly* F

At first glance, there seems to be a random reshuffling of genes. However, a pattern does exist; the genes are not thrown together at random in each strain. For example, note that in every case the *his* gene has *gal* on one side and *gly* on the other. Similar statements can be made about each gene, except when it appears at one end or the other of the linkage map. The order in which the genes are transferred is not constant. For example, in two Hfr strains the *his* gene is transferred before the *gly* gene (*his* is closer to O), but in three strains the *gly* gene is transferred before the *his* gene.

A startling hypothesis was proposed to account for these results. Suppose that, in an F^+ male, F is a small cytoplasmic element (and therefore easily transferred to an F^- cell on conjugation). If the "chromosome" of the F^+ male is a **ring,** then any of the linear Hfr chromosomes could be generated simply by inserting F at some particular place in the ring (Figure 9-8).

Chromosome circularity was a wildly implausible concept inferred solely from the genetic data; confirmation of its physical reality came only years later. Apparently, the insertion of F determines polarity, with the end opposite F being the origin. How might we explain F attachment? Wollman and Jacob suggested that some kind of crossover event between F and the chromosome might generate the Hfr chromosome. Alan Campbell then came up with a brilliant extension of that idea. He proposed that F, like the chromosome, is circular. Hence, a crossover between the two rings would produce a single larger ring with F inserted (Figure 9-9).

Now suppose that F consists of three different regions, as shown in Figure 9-10. The bacterial chromosome is pictured as having several regions of pairing

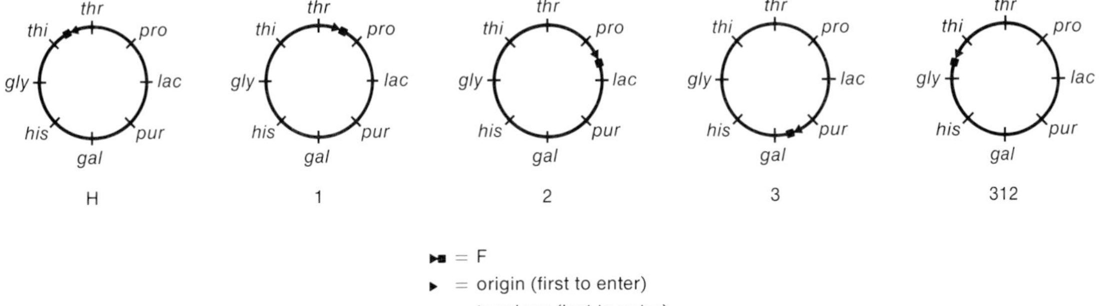

Figure 9-8. *Circularity of the* E. coli *chromosome. By using different Hfr strains (H, 1, 2, 3, 312), which have the fertility factor inserted into the chromosome at different points, interrupted-mating experiments indicate that the chromosome is circular. The mobilization points in the various strains are shown.*

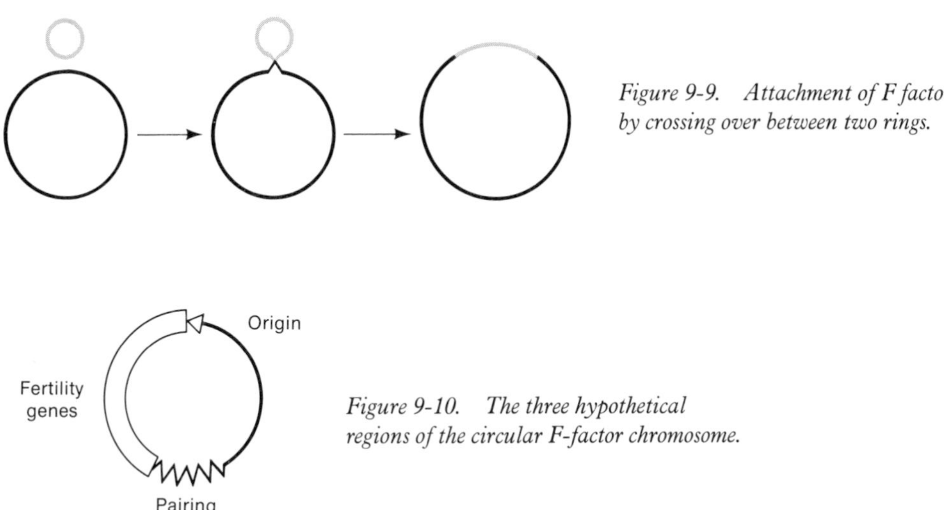

Figure 9-9. *Attachment of F factor by crossing over between two rings.*

Figure 9-10. *The three hypothetical regions of the circular F-factor chromosome.*

homology with the pairing region of F. Then different Hfr chromosomes would easily be generated by crossovers at these different sites (Figure 9-11). Furthermore, we now can suppose that the infrequent recombinants in an $F^+ \times F^-$ cross are the result of infrequent insertions of the F factor to produce a low frequency of Hfr cells.

The fertility factor exists in two states: as a free cytoplasmic element F that is easily transferred to F^- recipients and as an integrated part of a circular chromosome that is transmitted only very late in conjugation. A cell containing F in the

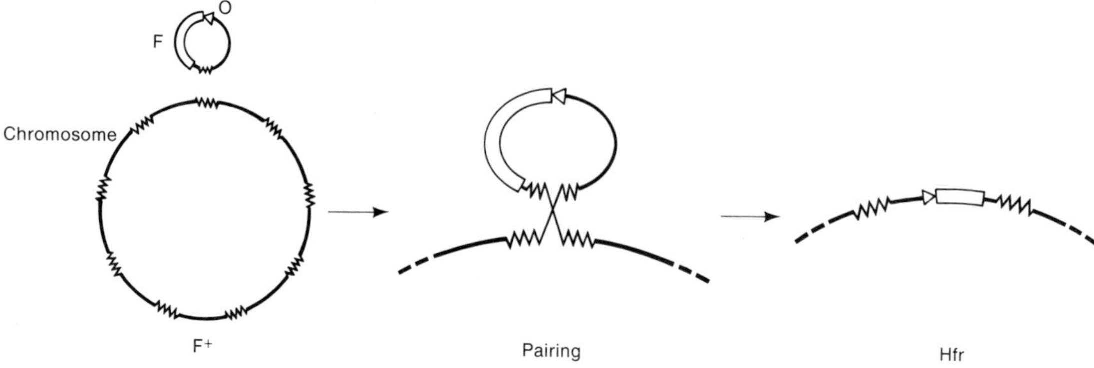

Figure 9-11. *Model for the insertion of F factor into the* E. coli *chromosome. If the F factor is considered to be circular, then an area of the fertility factor may have regions of pairing homology for different regions of the circular* E. coli *chromosome. Crossover between the F factor and the chromosome could thus insert the F factor into various points on the chromosome. (Jagged lines represent possible regions of homology.)*

first state is an F^+ cell; a cell containing F in the second state is an Hfr cell; and a cell lacking F is an F^- cell. The word *episome* was coined for genetic particles having such a pair of states.

Message
An ***episome*** *is a genetic factor in bacteria that can exist either as an element in the cytoplasm or as an integrated part of a chromosome. The F factor is an episome.*

Infectious elements other than F have been found in *E. coli* and other bacteria. Some are episomes that integrate into the ring chromosome as the F factor does. Others (called **plasmids**) do not integrate into the ring. Episomes and plasmids are of central importance in the mechanics of genetic engineering, and we shall discuss them in more detail later in this book.

Does an Hfr die after donating its chromosome to an F^- cell? The answer is no (unless the culture is treated with streptomycin). There is evidence that the Hfr chromosome replicates shortly before or during conjugation, thereby ensuring a complete chromosome for the donor after mating. Finally, we assume that the F^- chromosome also is circular, because F^- is readily converted into an F^+ from which Hfr can be derived.

The picture emerges of a circular Hfr unwinding a copy of its chromosome, which is then transferred in a linear fashion into the F^- cell. How is the transfer

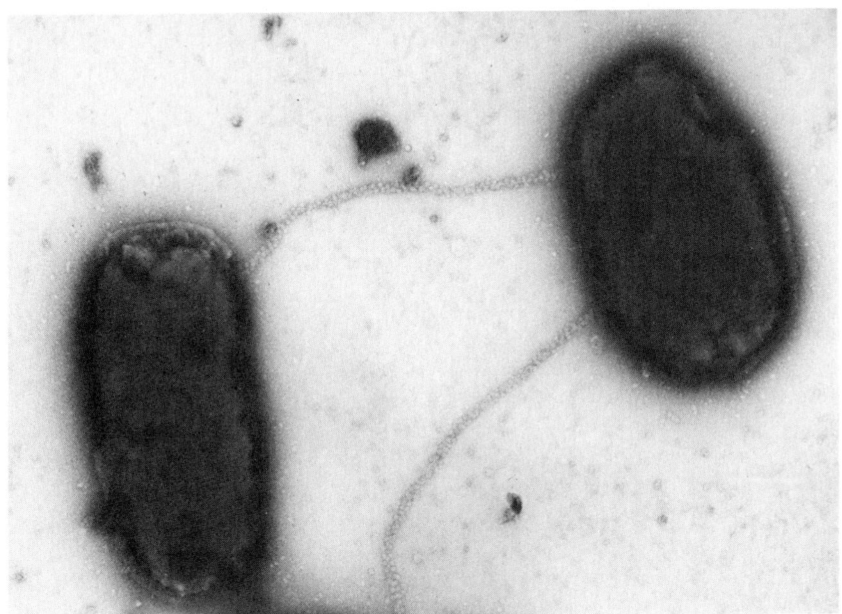

Figure 9-12. Electron micrograph of a cross between two E. coli *cells (× 34,300). The pili of the Hfr cell (not present in the F⁻ cell) have been visualized through the addition of viruses, which attach specifically to them. (Electron micrograph by David P. Allison, Biology Division, Oak Ridge National Laboratory.)*

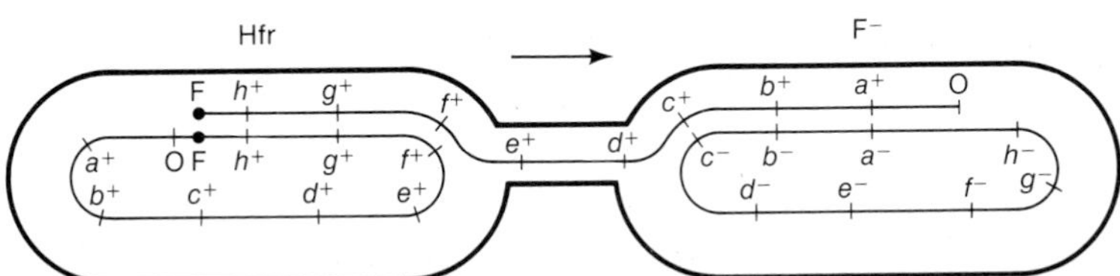

Figure 9-13. Diagrammatic representation of the sequential transfer of genes in a bacterial cross. (F = fertility factor; O = origin)

achieved? Electron-microscope studies show that Hfr and F^+ cells have **conjugation tubes** (or **sex pili;** singular, pilus) protruding from their cell walls. These tubes are hollow, and they almost certainly represent the path of chromosome transfer to the F^- cell, which has no pili. It is evidently the pili that are fractured during an interrupted-mating experiment, breaking the donor chromosome. Figure 9-12 shows conjugating bacteria, and Figure 9-13 is a diagrammatic representation of the passage of the chromosome.

We can now summarize the various aspects of the sex cycle in *E. coli* (Figure 9-14).

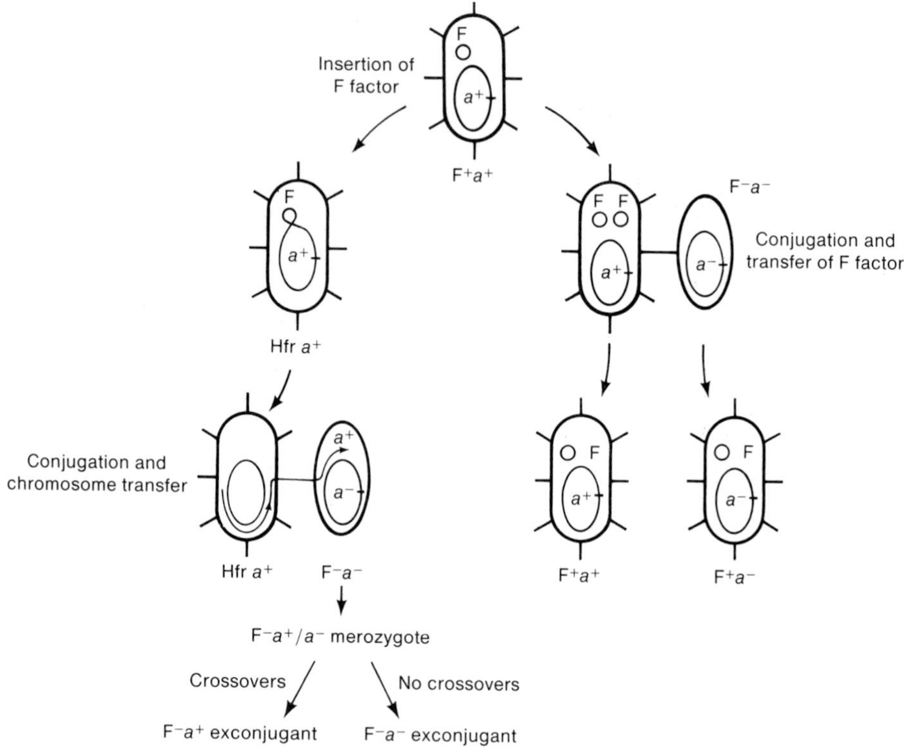

Figure 9-14. Summary of the various events that occur in the sexual cycle of E. coli.

Recombination Between Marker Genes After Transfer

Thus far we have discussed only the process of transfer of genetic information between individuals in a cross. This transfer was inferred from the existence of recombinants produced from the cross. However, before a stable recombinant can be produced, obviously the transferred genes must be "integrated" or incorporated into the host's genome by an exchange mechanism. We now consider some of the special properties of this exchange event. Genetic exchange in prokaryotes does not take place between two whole genomes (as it does in eukaryotes); rather, it takes place between one complete genome (called the F^- **endogenote**) and an incomplete one (called the donor **exogenote**). What we have in fact is a partial diploid, or **merozygote.** Bacterial genetics is merozygote genetics. Figure 9-15a is a diagram of a merozygote.

It is obvious that a single "crossover" would not be very useful in generating viable recombinants, because the ring is broken to produce a strange, partially diploid linear chromosome (Figure 9-15b). To keep the ring intact, there must be an even number of crossovers (Figure 9-15c). The fragment produced in such

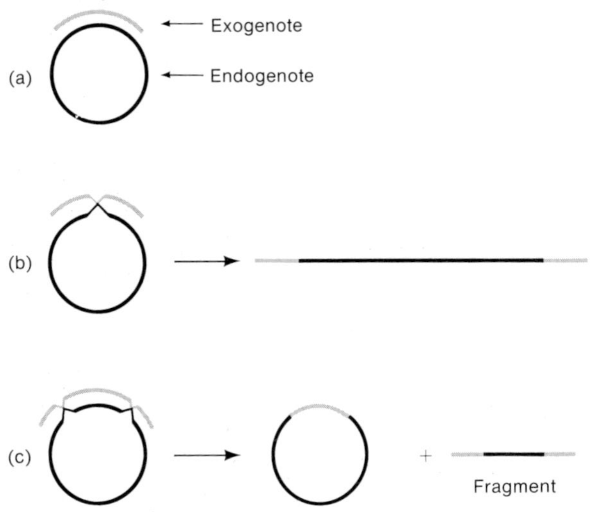

Figure 9-15. *Crossover between exogenote and endogenote in a merozygote. (a) The merozygote. (b) A single crossover leads to a partially diploid linear chromosome. (c) An even number of crossovers leads to a ring plus a linear fragment.*

a crossover is only a partial genome; in most cases, it is lost during subsequent cell growth. (We say "in most cases" because there are ways to maintain "stable partial diploids.") Hence, it is obvious that reciprocal products of recombination do not survive—only one does. A further unique property of bacterial exchange, then, is that we must forget about reciprocal exchange products.

Message

In the merozygote genetics of bacteria, we generally are concerned with double crossovers, and we do not expect reciprocal recombinants.

Higher-Resolution Mapping by Recombinant Frequency in Bacterial Crosses

Interrupted-mating experiments are ideal to obtain a rough overall set of gene locations over the entire map. In other words, this technique is useful for **low-resolution** mapping. Some other tricks are needed in order to obtain a higher resolution between marker loci that are quite close together. For "distances" of less than about two minutes, the interrupted-mating method is not reliable (it does not produce consistent results). Here we consider two approaches to this problem, both based on measuring the frequency of recombinants. We saw earlier that the first attempts to map in this way were failures. With the wisdom of hindsight, we can now see why they failed, and we can compensate for the problems.

The basic problem was the confusion due to the merozygotic nature of bacterial reproduction. In our RF calculations for eukaryotes, we assumed (quite rightly) an opportunity for recombination between two genes in a diploid cell. In merozygotes, however, only partial diploids exist. In the merozygote, some genes don't even get into the act! Our analysis must involve two steps in the process. We need methods that allow us to measure first the *potential* opportunity for recombination (resulting from gene transfer), and then the *actual* recombination that occurs.

Approach 1. In the interrupted-mating experiment considered earlier, the genes *ade* and *leu* are about 2 minutes apart, with *ade* entering after *leu*. Let's use these genes to illustrate the first approach. This method relies on selecting the Hfr marker that enters last, then testing these individuals for the presence or absence of the male marker that entered ahead of it. Because the unselected marker had to enter first, we can safely assume that it was part of the merozygote. In the present example, we select ade^+ exconjugants and see which of them are leu^+ and which leu^-. In the leu^- exconjugants, we know that one crossover has occurred between these gene pairs. We need not worry about the location of the other crossover of the double crossover; we are concentrating on the kind that breaks up the ade^+ leu^+ linkage. Figure 9-16 diagrams the two crossover patterns that lead to the two ade^+ types. Let (ade^+ leu^-) represent the frequency of this genotype; this is the frequency of types arising from a crossover between *ade* and *leu*. The total frequency of ade^+ types is (ade^+ leu^+) + (ade^+ leu^-). The fraction

$$\frac{(ade^+\ leu^-)}{(ade^+\ leu^+) + (ade^+\ leu^-)}$$

obviously will be related to the distance separating the two loci, and we can call this the recombinant frequency. In *E. coli,* 1 minute on the map equals about 20% recombinants calculated in this manner.

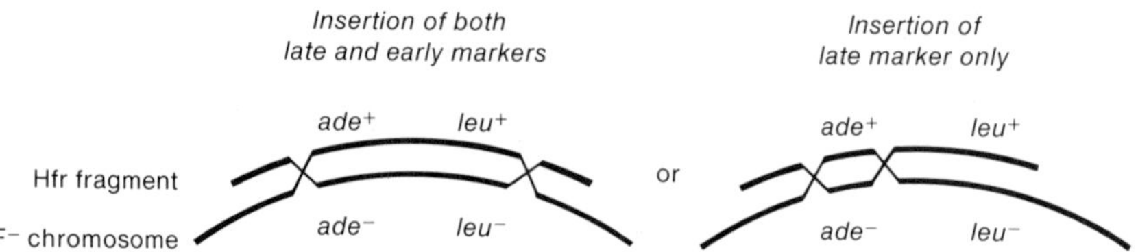

Figure 9-16. Incorporation of a later marker (ade^+) *into the* F^- E. coli *chromosome. The early marker* (leu^+) *may or may not be inserted. Whether or not it will be inserted depends on recombination between the Hfr fragment and the* F^- *chromosome at the appropriate place.*

Approach 2. In a repulsion cross, Hfr $a^- b^+ \times$ F$^-$ $a^+ b^-$, it is easy to select for $a^+ b^+$ recombinants by plating exconjugants on minimal medium. Obviously, the production of such recombinants will be related to the interlocus distance, but how can we obtain a frequency? In our calculation, what will be the denominator of the fraction? We want an expression of the form "recombinants per *Z*," but what entity is *Z*? We can't use total cells, because not all cells will have provided an opportunity for recombination.

There are at least two possible tricks that may be applied here. In the first, the denominator used is the number of prototrophic colonies in a cross Hfr $a^+ b^+ \times$ F$^-$ $a^+ b^-$. In the second, the denominator is the number of exconjugants that receive an Hfr allele at a completely different locus—say, *m* in a cross Hfr $m^+ a^- b^+ \times$ F$^-$ $m^- a^+ b^-$. What is the justification for such bizarre ratios? All that we are doing is comparing the specific "rare" exchange needed to generate $a^+ b^+$ with the opportunity for exchange, which in these techniques is represented as a general "nonspecific" kind of exchange needed to act on any one constant mutant allele (Figure 9-17). Of course, in these calculations, the values obtained are in no sense absolute. They can be compared with one another only if a constant "general" exchange locus is used for all the pairs studied.

These techniques can be used to study most of the merozygote systems in bacteria (and in phage transduction, as we shall see later). We hope you are beginning to appreciate the fact that bacteria do things quite differently from

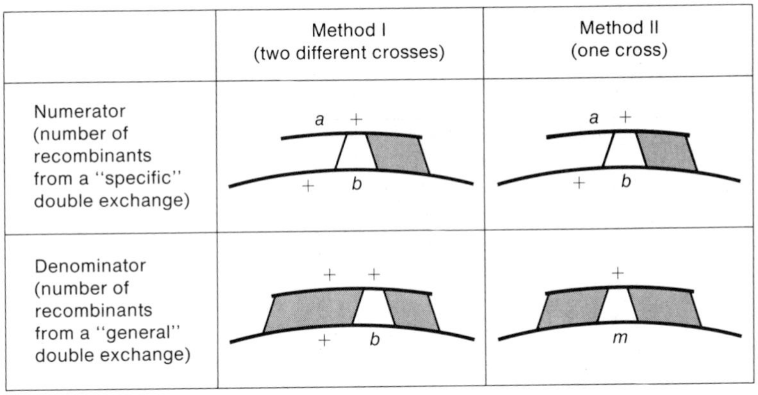

	Method I (two different crosses)	Method II (one cross)
Numerator (number of recombinants from a "specific" double exchange)	a + / + b	a + / + b
Denominator (number of recombinants from a "general" double exchange)	+ + / + b	+ / m

Figure 9-17. Two ways of approaching mapping of small genetic regions by selection of + + recombinants during conjugation. In Method I, the frequency of recombination (distance between two genes) is based on a denominator determined by using one of the two loci being mapped. In Method II, the denominator of the ratio is determined by using a third, unrelated marker. In neither of these crosses can the total number of cells be used, because not all will have been provided an opportunity for recombination. (Shading represents a region where an exchange may occur in any position.)

eukaryotes. But even though their analysis bears little superficial resemblance to that of eukaryotes, the principles of mapping are very similar. The underlying genetic approach is much the same, whether we are studying humans, peas, flies, or bacteria.

Deriving Gene Order

In many cases, loci are so close together that it becomes difficult to order them in relation to a third locus. For example, consider three loci linked in this way:

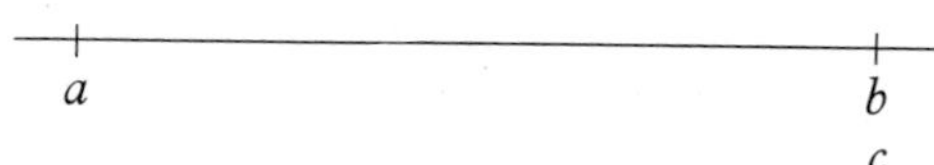

The RF between *a* and *b* under most experimental conditions will be more or less the same as that between *a* and *c*. Is the order *a–b–c* or *a–c–b*? Once again, we need to use a trick to answer the question. We need to make a pair of reciprocal crosses, using the same marker genotypes as both donor and recipient. Figure 9-18 shows the crossover events needed to generate prototrophs from the cross $a\,b\,c^+ \times a^+\,b^+\,c$, depending on the order of the loci.

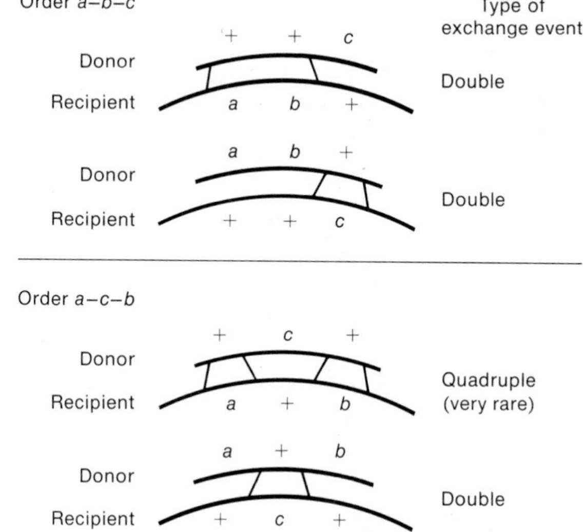

Figure 9-18. Inferring gene order from the relative frequencies of recombinants in reciprocal crosses involving parental genotypes a b + × + + c. *If the order is* a-b-c, *the frequency of* + + + *progeny from the reciprocal crosses will be approximately equal. However, if the order is* a-c-b, *the frequencies will be quite different.*

Thus, if the reciprocal crosses give dramatically different frequencies of wild-type survivors on minimal medium, then we know that the order is a–c–b. If we observe no difference in frequencies between the reciprocal crosses, then the order must be a–b–c. Once again, this same principle can be used in other bacterial and phage mapping systems.

Infectious Marker-Gene Transfer by Episomes

Now that we understand the F^-, F^+, and Hfr states, we can understand the initially confusing results of recombination work in *E. coli.* Knowing all this, Edward Adelberg began in 1959 to do recombination experiments with an Hfr strain. However, the particular Hfr strain he used kept producing F^+ cells, so the recombination frequencies were not very large. He called this particular fertility factor F′ to signify a difference from the normal F, for the following reasons:

1. The F′-bearing F^+ strain reverted back to an Hfr strain quite often.
2. F′ always integrated at the *same place* to give back the original Hfr chromosome (remember that randomly selected Hfr derivatives from F^+ males have origins at many different positions).

How could these properties of F′ be explained? The answer came from the recovery of a new F′ from an Hfr strain in which the *lac*$^+$ locus was near the end of the Hfr chromosome (that is, was transferred very late). Using this Hfr *lac*$^+$ strain, Jacob and Adelberg found an F^+ derivative that transferred *lac*$^+$ to F^- *lac*$^-$ recipients at a very high frequency. Furthermore, the recipients, which became F^+ *lac*$^+$ phenotypically, occasionally (at a frequency of 1×10^{-3}) produced F^- *lac*$^-$ daughter cells. Thus, the genotype of the recipients appears to be F′ *lac*$^+$/*lac*$^-$.

Now we have the clue: F′ is a cytoplasmic element that carries a part of the bacterial chromosome. Its origin and reintegration can be visualized as shown in Figure 9-19. This F′ is known as F-*lac*. Because F *lac*$^+$/*lac*$^-$ cells are *lac*$^+$ in phenotype, we know that *lac*$^+$ is dominant over *lac*$^-$. As we shall see later, the dominance–recessive relationship between alleles can be a very useful bit of information in interpretations of gene function. Partial diploidy (called **merodiploidy**) for specific segments of the genome can be made with an array of F′ derivatives from Hfr strains. The F′ cells can be selected by looking for infectious transfer of normally late genes in a specific Hfr strain.

The use of F′ elements to create partial diploids is called **sexduction,** or F′-duction. Some F′ strains can carry very large parts (up to one-quarter) of the bacterial chromosome; if appropriate markers are used, the merozygotes generated can be used for recombination studies.

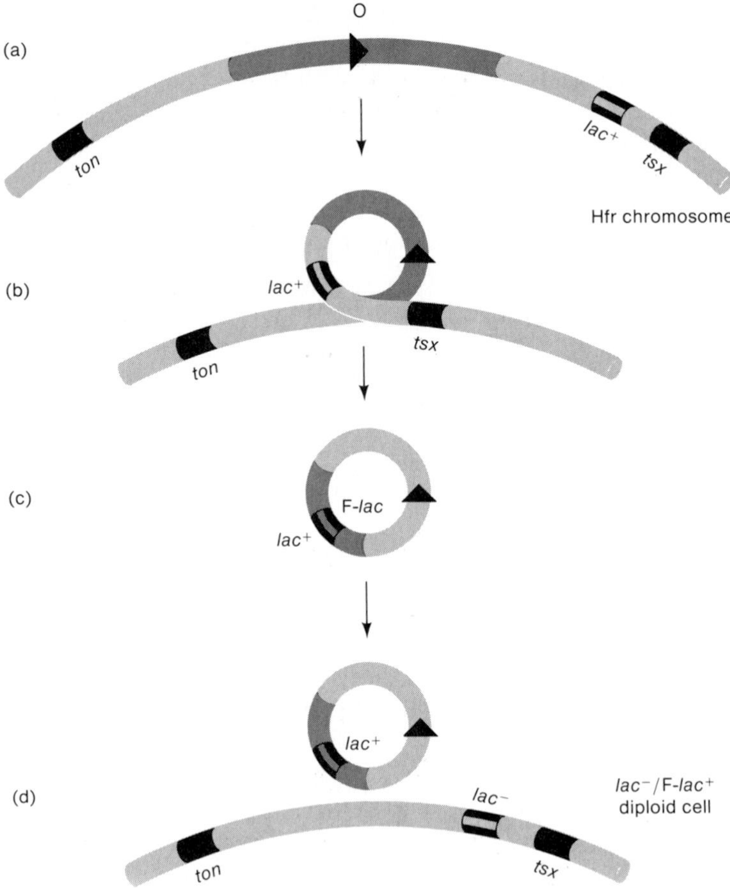

*Figure 9-19. Origin and reintegration of the F′ (or F-*lac*) factor. (a) F factor inserted in an Hfr strain between the* ton *and* lac^+ *alleles. (b) Abnormal "outlooping" (separation of F factor) to include* lac *locus. (c) The F-*lac^+ *particle (= F′). (d) Constitution of an F-*lac^+/lac^- *partial diploid produced by the transfer of the F-*lac^+ *particle into an* F^- lac^- *recipient. (From G. S. Stent and R. Calendar,* Molecular Genetics, *2nd ed. Copyright © 1978, W. H. Freeman and Company.)*

As mentioned earlier, F is by no means the only kind of episome. In *E. coli* and other intestinal bacterial species, a variety of inherited characters may be transferred in an infectious manner from cell to cell. Some of these characters are colicin production (colicin is a poison that kills other bacteria) and certain antibiotic resistances. In some cases, these episomes may be transferred between bacterial species. Furthermore, these episomes can "pick up" other bacterial genes and transfer them simultaneously in a manner resembling sexduction. These episomes and plasmids are important tools for genetic engineering, and we will encounter them again in Chapters 13 and 16.

Message
During conjugation between an Hfr donor and an F^- recipient, the genes of the donor are transmitted linearly, with the inserted fertility factor transferring last.

During conjugation between an F^+ donor carrying an F′ plasmid and an F^- recipient, a specific part of the donor genome may be transmitted infectiously to F^- cells, carried on the plasmid. This part was originally adjacent to the F locus in an Hfr strain from which the F^+ was derived.

Bacterial Transformation

The Discovery of Transformation

A puzzling observation was made by Frederick Griffith in the course of experiments on the bacterium *Streptococcus pneumoniae* in 1928. This bacterium causes pneumonia in humans and is normally lethal in mice. Griffith had two different strains of this bacterial species. One strain is a normal virulent type. The cells of this strain are enclosed in a polysaccharide capsule, giving colonies a smooth appearance; hence this strain is labeled S. The other strain is a mutant that has lost its virulence; it grows in mice but is not lethal. The polysaccharide coat is lost in this strain, giving colonies a rough appearance; this strain is called R.

Griffith killed some virulent cells by boiling them and injected the heat-killed cells into mice. The mice survived, showing that the carcasses of the cells do not cause death. However, if the mice were injected with a mixture of heat-killed virulent cells and live nonvirulent cells, then the mice died! Furthermore, live cells could be recovered from the dead mice; these cells gave smooth colonies and were virulent upon subsequent injection. Somehow the cell debris of the boiled S cells had converted the live R cells into live S cells. The process was called **transformation.** Griffith's experiment is summarized in Figure 9-20.

This same basic technique was also utilized to determine the nature of the **"transforming principle,"** the agent in the cell debris that is specifically responsible for transformation. In 1944 Oswald Avery, C. M. MacLeod, and M. McCarty separated the classes of molecules found in the debris of the dead S cells and tested them for transforming ability, one at a time. They showed first that the polysaccharides themselves do not transform the rough cells. Therefore the polysaccharides, although no doubt concerned with the pathogenic reaction, are only the phenotypic expression of virulence. In screening the different groups, they found that only one class of molecules, deoxyribonucleic acid (DNA), induces transformation of R cells (Figure 9-21). They deduced that DNA is the agent that determines the polysaccharide character and hence the pathogenic character. Furthermore, it seems that providing R cells with S DNA is tantamount to providing these cells with S genes!

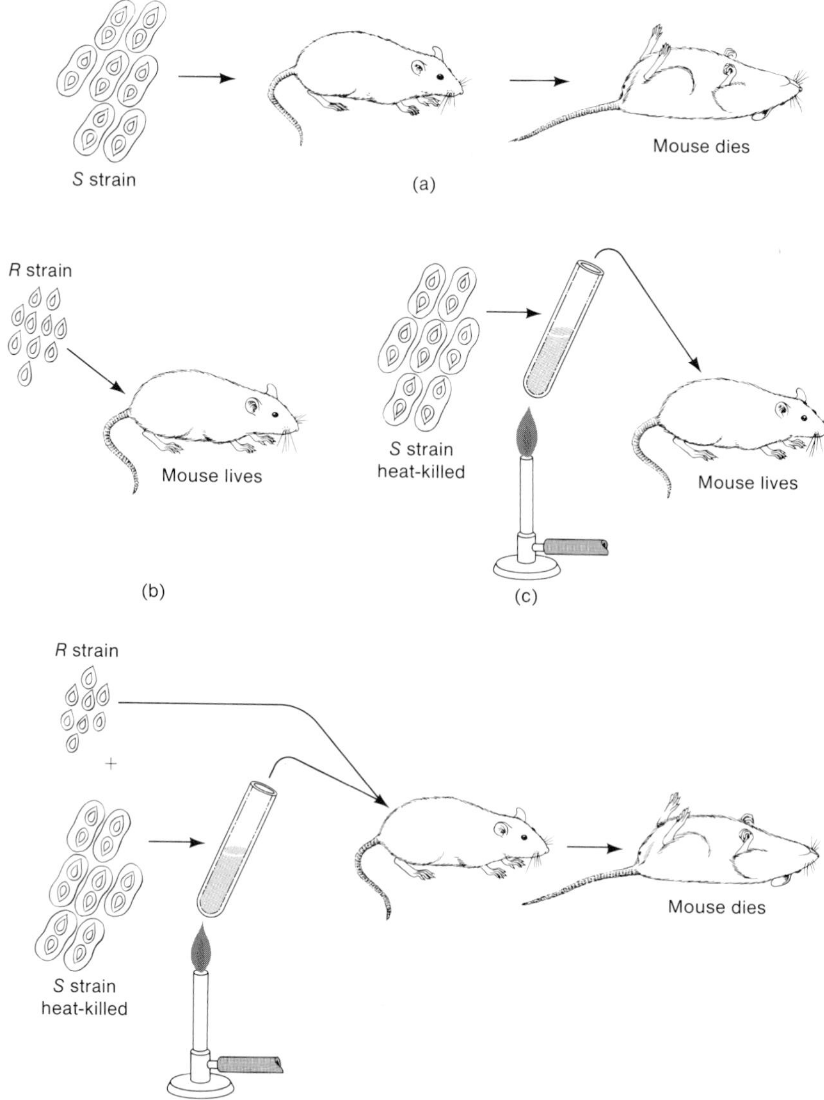

Figure 9-20. The first demonstration of bacterial transformation. (a) Mice die after injection with the virulent S strain. (b) Mice survive after injection with the R strain. (c) Mice survive after injection with heat-killed S strain. (d) Mice die after injection with a mixture of heat-killed S strain and live R strain. The heat-killed S strain somehow transforms the strain R to virulence. Parts a, b, and c act as control experiments for this demonstration. (From G. S. Stent and R. Calendar, Molecular Genetics, *2nd ed. Copyright 1978 by W. H. Freeman and Company. After R. Sager and F. J. Ryan,* Cell Heredity, *Wiley, 1961.)*

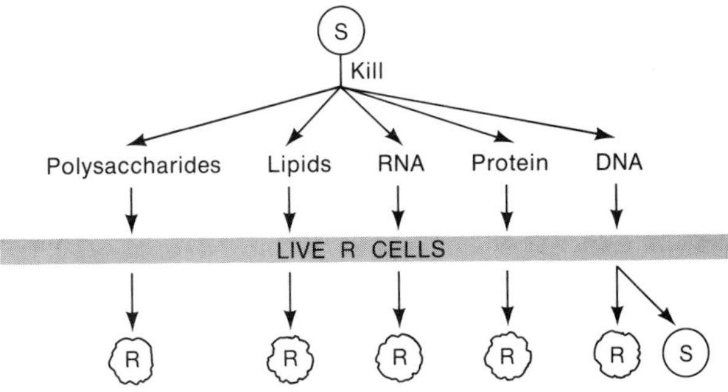

Figure 9-21. Demonstration that DNA is the transforming agent. DNA is the only agent that produces smooth (S) colonies when added to live R cells.

Message
The demonstration that DNA is the transforming principle was the first demonstration that genes are composed of DNA.

Transformation was subsequently demonstrated for other genes (such as drug resistance, Figure 9-22, and prototrophy) and in other bacterial species.

It seems that the transforming principle, or DNA, is actually physically incorporated into the bacterial chromosome by a physical breakage-and-insertion

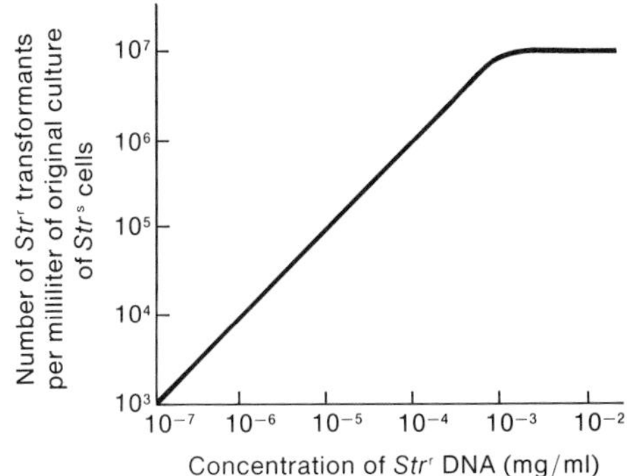

Figure 9-22. The genetic transfer of streptomycin resistance (Str^r) *into streptomycin-sensitive* (Str^s) *cells. The recovery of* Str^r *transformants among* Str^s *cells depends on the concentration of* Str^r *DNA. (From G. S. Stent and R. Calendar,* Molecular Genetics, *2nd ed. Copyright © 1978 by W. H. Freeman and Company.)*

process quite similar to crossing over. Thus, if radioactively labeled DNA from an arg^+ bacterial culture is added to unlabeled arg^- cells, the arg^+ transformants (selected by plating on minimal medium) can be shown to contain some of the radioactivity. We shall provide a molecular model for this after further discussion of DNA. For now, let's consider transformation simply as a genetic analytic tool.

Linkage Information from Transformation

Transformation has been a very handy tool in several areas of bacterial research. We shall learn later how it is used in some of the modern techniques of genetic engineering. Here we examine its usefulness in providing linkage information.

When DNA (the bacterial chromosome) is extracted for transformation experiments, some breakage into smaller pieces is inevitable. The closer together two donor genes are located, then the greater is the chance that they will be carried on the same piece of transforming DNA and hence will cause a **double transformation.** Conversely, if genes are widely separated on the chromosome, then they will be carried on separate transforming segments, and the frequency of double transformants will equal the product of the single-transformation frequencies. Thus, it should be possible to test for close linkage by testing for a departure from the product rule.

Unfortunately, the situation is made more complex by several factors, most important of which is that not all cells in a population of bacteria are competent to be transformed. Because single transformations are expressed as proportions, the success of the product rule obviously depends on the absolute size of these proportions. There are ways of calculating the proportion of competent cells, but we need not detour into that subject now. You can sharpen your skills in transformation analysis in one of the problems at the end of the chapter, which assumes 100% competence.

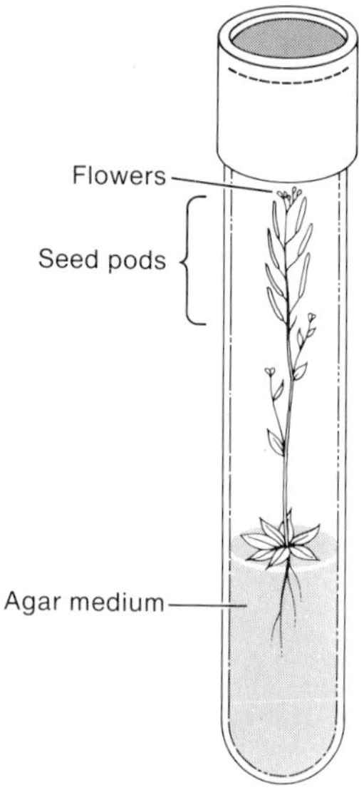

Figure 9-23. The tiny plant Arabidopsis thaliana *is suitable for some culturing techniques used with microbes. Furthermore, each plant produces thousands of seeds.*

Transformation in Higher Organisms

In recent years, there have been numerous attempts to transform higher organisms. This work has been done mainly in plants, where it offers prospects for another new weapon in the arsenal of modern crop-improvement techniques. Several encouraging results have been obtained. Although there is still disagreement over the significance of these results, enough positive cases are now known that the phenomenon should be given serious consideration.

One of the best-studied systems involves the tiny plant *Arabidopsis thaliana* (Figure 9-23). Recessive auxotrophic thiamine (thi^-) mutants are available in this plant, and these have been "transformed," not by using donor DNA from thi^+ *Arabidopsis,* but by using donor DNA from thi^+ bacterial or calf DNA! This is heterologous DNA with genes that must be of analogous function. A number of seeds from a thi^-/thi^- plant are sterilized and allowed to germinate and grow on thiamineless medium in the presence of donor DNA. Some "corrected" normal-looking plants result, which grow without thiamine.

When selfed, all the progeny remain transformed down through seven generations attempted. But when crossed to *thi*$^+$ or *thi*$^-$, complications arise, because in *both* kinds of crosses similar peculiar results occur. Typically, a great preponderance of normal plants, and small proportions of "leaky" (intermediate) and lethal types, are observed on thiamineless medium. More research is necessary before the significance of these results becomes clear.

Phage Genetics

Most bacteria are susceptible to attack by bacteriophages (a name that literally means "eaters of bacteria"). A phage consists of a nucleic acid "chromosome" (DNA or RNA) surrounded by a coat of protein molecules. The various strains of phage are identified as T1, T2, and so on. Figures 9-24 and 9-25 show the complicated structure of a phage belonging to the class called T-even phages (T2, 4, and so on).

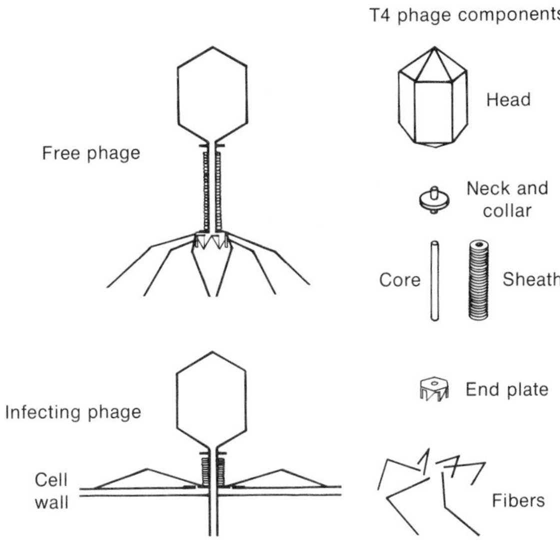

Figure 9-24. Phage T4, shown in its free state and in the process of infecting a cell of E. coli. *On the right, a phage has been diagrammatically exploded to show its highly ordered structure in three dimensions. (From "The Genetics of a Bacterial Virus" by R. S. Edgar and R. H. Epstein. Copyright © 1965 by Scientific American, Inc. All rights reserved.)*

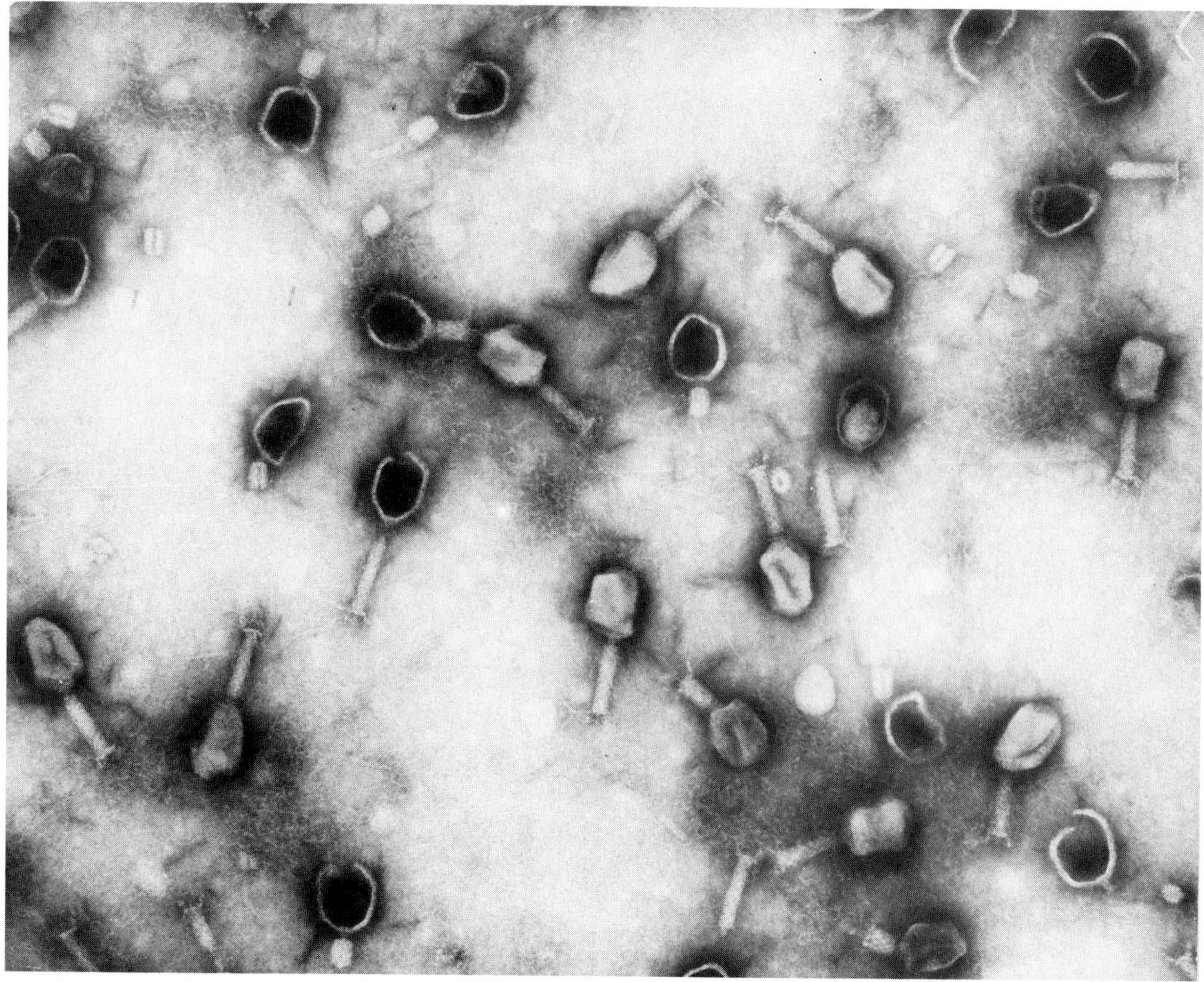

Figure 9-25. Mature particles of the E. coli *phage T4 (× 97,500). (From Grant Heilman.)*

The Phage Cross

A phage attaches to a bacterium and injects its genetic material into the bacterial cytoplasm (Figure 9-26a). The phage genetic information then takes over the machinery of the bacterial cell by turning off the synthesis of bacterial components and redirecting the bacterial synthetic material to make more phage components (Figure 9-26b). (The use of the word *information* is interesting in this connection; it literally means "to give form." And of course, that is precisely the role of the genetic material: to provide blueprints for the construction of form. In the present discussion, the form is the elegantly symmetrical structure

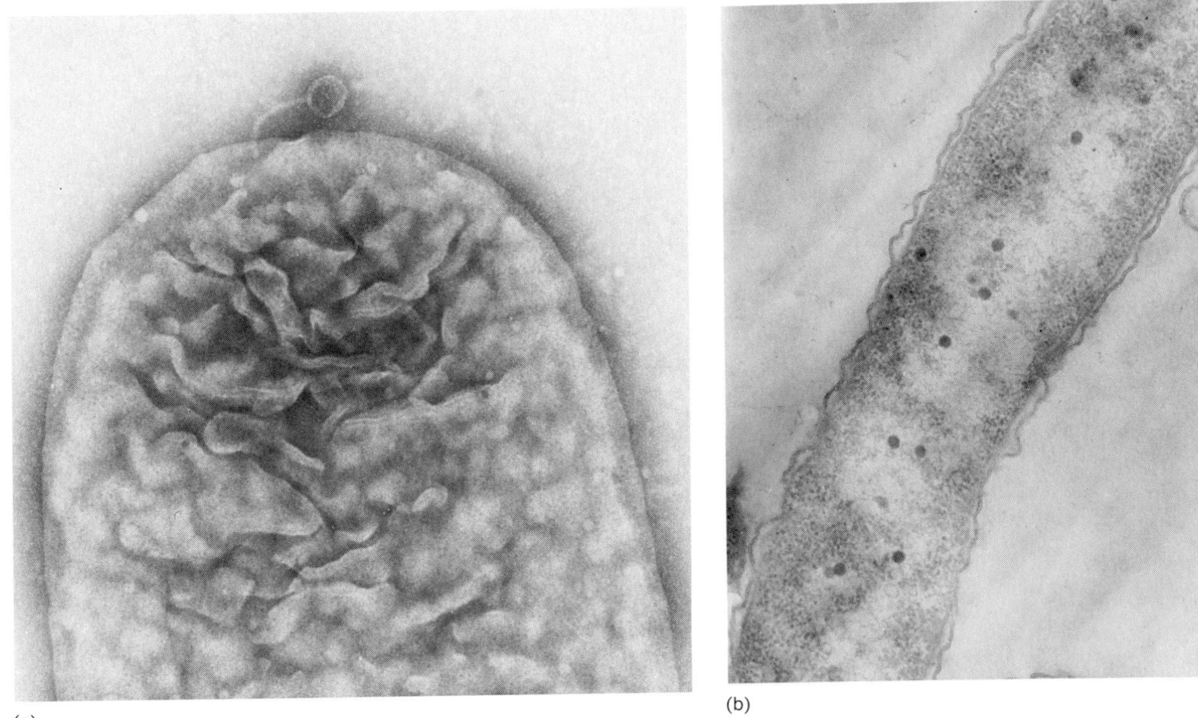

Figure 9-26. *(a) A bacteriophage (called λ) attached to an* E. coli *cell and injecting its genetic material. (b) Progeny particles of phage λ maturing inside an* E. coli *cell. (From Jack D. Griffith.)*

of the new phages.) Ultimately, many phage descendants are released when the bacterial cell wall breaks open. The process of breaking open is called **lysis.**

But how does one study inheritance in phages when they are so small that they are visible only with the electron microscope? In this case, we cannot produce a visible colony by plating. One phage character we can examine involves the effects of the phage on bacteria. When a phage lyses a bacterium, progeny phages infect neighboring bacteria, and these in turn lyse and infect other bacteria. This is an exponentially explosive phenomenon (an exponential increase in the number of lysed cells). Very soon after starting an experiment of this type (overnight), the effects are visible to the naked eye: a clear area, or **plaque,** is present on the opaque lawn of bacteria on the surface of a dish of solid medium (Figure 9-27). Depending on the phage genotype, such plaques can be large or small, fuzzy or sharp, and so forth. Thus **plaque morphology** is a phage character that can be analyzed.

Another phage phenotype that can be analyzed genetically is **host range.** Certain strains of bacteria are immune to adsorption (attachment) or injection by phages. Phages, in turn, may differ in the spectra of bacterial strains they can infect and lyse.

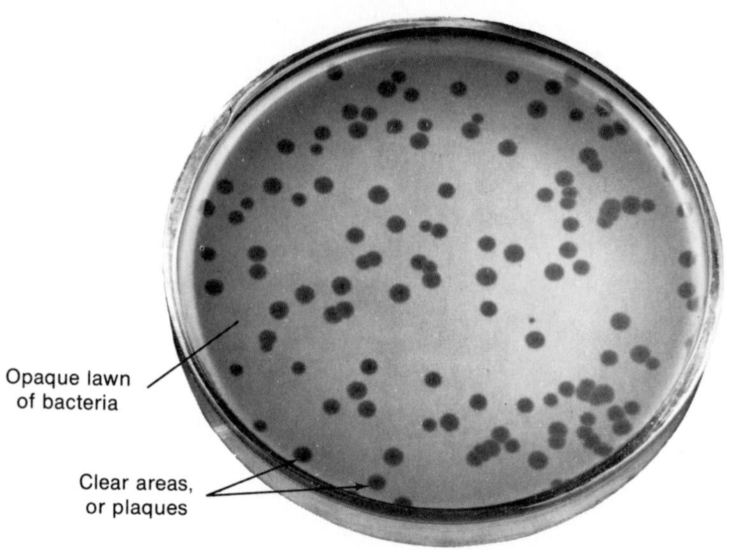

Figure 9-27. The appearance of phage plaques. Individual phage are spread on an agar medium that contains a fully grown "lawn" of E. coli. *Each phage infects one bacterial cell, producing 100 or more progeny phage that burst the* E. coli *cell and infect neighboring cells. They, in turn, are exploded with progeny, and the process continues until a clear area, or plaque, appears on the opaque lawn of bacterial cells. (From* Molecular Biology of Bacterial Viruses *by G. S. Stent. W. H. Freeman and Company. Copyright © 1963.)*

To illustrate a phage cross, we describe a cross of T2 phages originally studied by Alfred Hershey. These phages of *E. coli* are of a type called **virulent;** they are committed to a simple cycle of infection and lysis. The genotypes of the two parental strains of T2 phage in Hershey's cross were $h^- r^+ \times h^+ r^-$. The alleles are identified by the following characters: h^- can infect two different *E. coli* strains (which we call strains 1 and 2); h^+ can infect only strain 1; r^- rapidly lyses cells, thereby producing large plaques; and r^+ slowly lyses cells, thus producing small plaques.

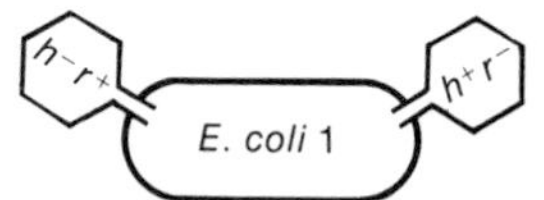

Figure 9-28. A double infection of E. coli *by phages.*

In the cross, strain 1 is infected with both parental T2 phage genotypes at a concentration (called **multiplicity of infection,** which is the ratio of phages to bacteria) sufficiently high to ensure a high proportion of cells that are simultaneously infected by both phage types. This kind of infection (Figure 9-28) is called a **mixed infection,** or **double infection.** The phage lysate (the progeny phage) is then analyzed by spreading it onto a bacterial lawn composed of a mixture of strains 1 and 2. Four plaque types are distinguishable (Figure 9-29 and Table 9-2). These four genotypes can easily be scored as parental (the first two in Table 9-2) and recombinant, and an RF can be calculated:

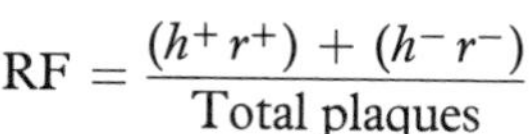

$$\text{RF} = \frac{(h^+ r^+) + (h^- r^-)}{\text{Total plaques}}$$

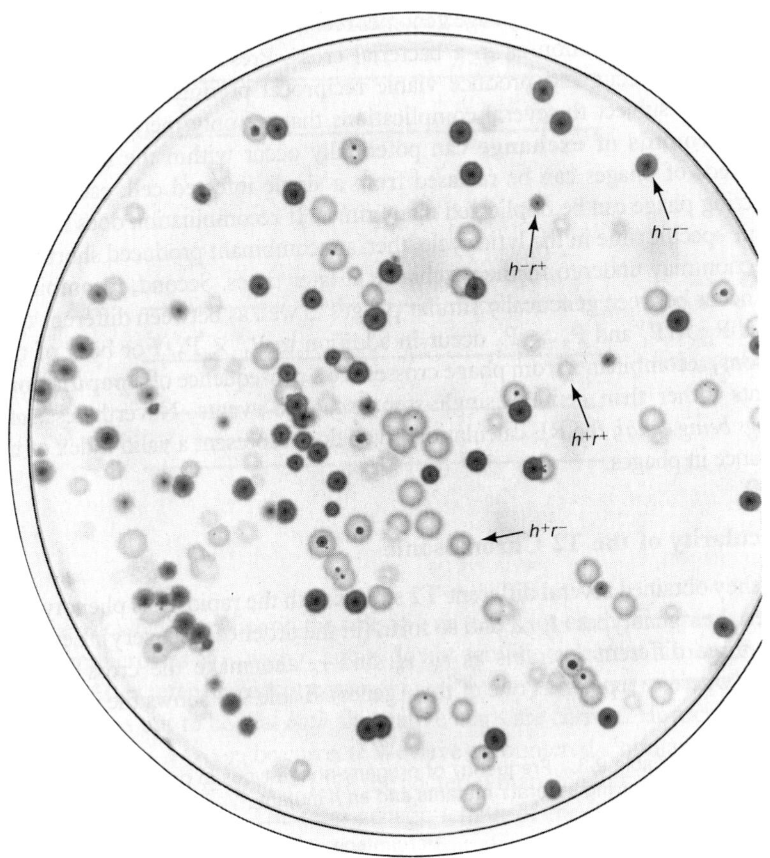

Figure 9-29. Plaque phenotypes produced by progeny of a cross $h^- r^+ \times h^+ r^-$*. Enough phage of each genotype are added to ensure that most bacterial cells are infected with at least one phage of each genotype. After lysis, the progeny phage are collected and added to an appropriate* E. coli *lawn. Four plaque phenotypes can be differentiated, representing two parental types and two recombinants. (From* Molecular Biology of Bacterial Viruses *by G. S. Stent. W. H. Freeman and Company. Copyright © 1963.)*

Table 9-2. Progeny-phage plaque types from cross $h^- r^+ \times h^+ r^-$

Phenotype	Inferred genotype
Clear and small	$h^- r^+$
Cloudy and large	$h^+ r^-$
Cloudy and small	$h^+ r^+$
Clear and large	$h^- r^-$

NOTE: Clearness is produced by the h^- allele, which allows infection of *both* bacterial strains in the lawn; cloudiness is produced by the h^+ allele, which limits infection to the cells of strain 1.

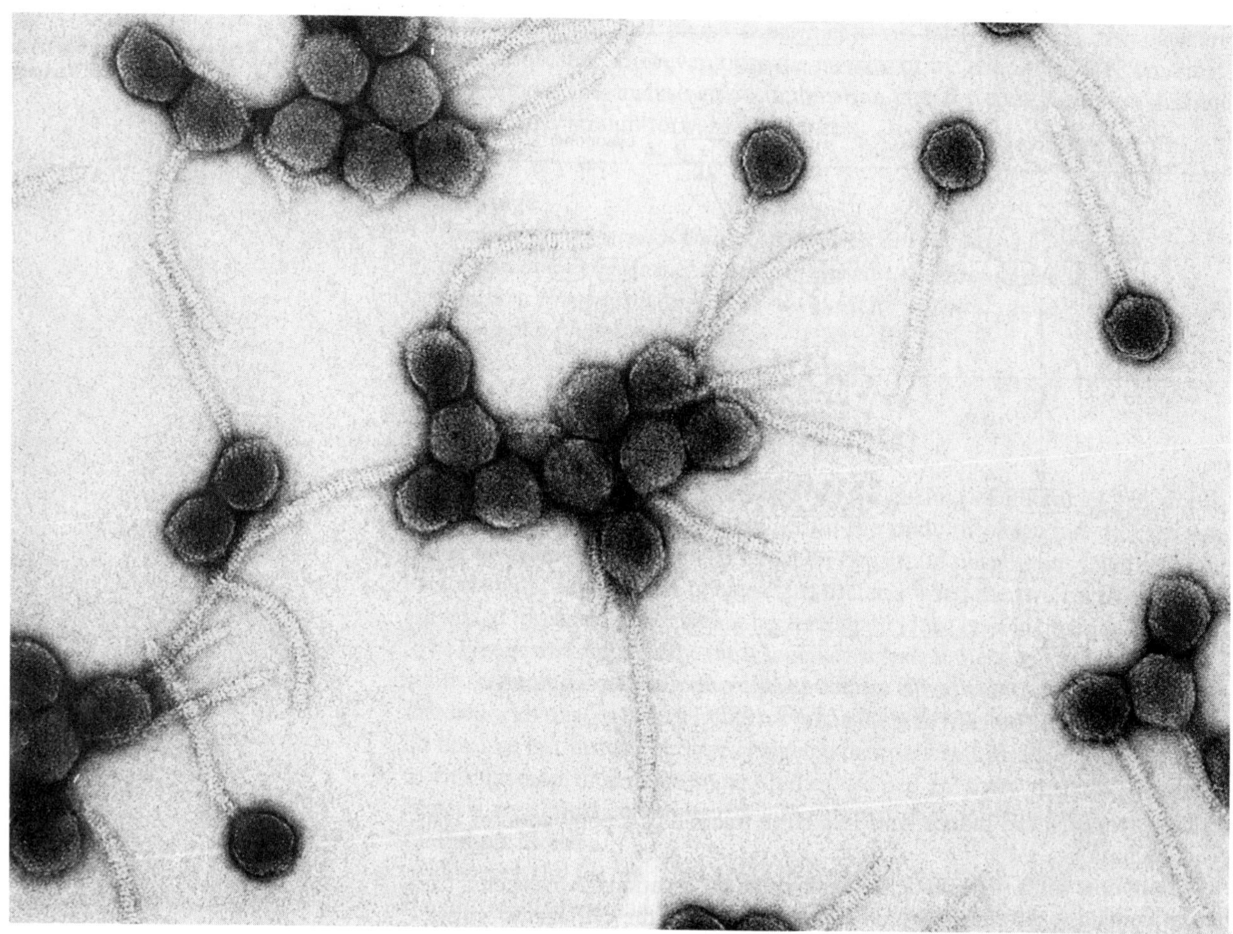

Figure 9-33. Electron micrograph of λ phages. (Photograph from Jack D. Griffith.)

terial cytoplasm, or is it somehow associated with the bacterial genome? By a fortuitous happenstance, the original strain of *E. coli* used by Lederberg and Tatum (page 319) proved to be lysogenic for a temperate phage called lambda (λ). Lambda has become the most intensively studied and best-characterized phage (Figure 9-33). Crosses between F^+ and F^- cells yielded interesting results. It turns out that $F^+ \times F^-$ (λ) crosses yield recombinant lysogenic recipients, whereas the reciprocal cross F^+ (λ) $\times$ F^- almost never gives lysogenic recombinants.

These results became understandable when Hfr strains were discovered. In the cross Hfr $\times$ F^- (λ), lysogenic F^- exconjugants with Hfr genes are readily recovered. However, in the reciprocal cross Hfr (λ) $\times$ F^-, the early genes from the Hfr are recovered among the exconjugants, but recombinants for late markers (those expected to transfer after a certain time in mating) are not recovered. Furthermore, lysogenic exconjugants are almost never recovered from this reciprocal cross. What is the explanation? The observations make sense if the λ prophage is behaving like a bacterial gene locus—that is, as part of the bacterial chromosome. In the cross of a lysogenic Hfr with a nonlysogenic F^- recipient, the entry of the λ prophage into the cytoplasm of a nonimmune cell triggers the prophage into a lytic cycle; this is called **zygotic induction.** In interrupted-mating experiments, the λ prophage proves always to enter the F^- cell at a specific time, closely linked to the *gal* locus. We can assign the λ prophage to a specific locus next to the *gal* region.

Entry of the λ prophage into an F^- cell immediately induces the lytic cycle. But in the cross Hfr (λ) × F^- (λ), any recombinants are readily recovered (that is, no induction of the prophage occurs). It would seem that the cytoplasm of the F^- cell must exist in two different states (depending on whether the cell contains a λ prophage), so that contact between an entering prophage and the cytoplasm of a nonimmune cell immediately induces the lytic cycle. Perhaps some cytoplasmic factor specified by the prophage somehow represses the multiplication of the virus. Entry of the prophage into a nonlysogenic environment immediately dilutes this repressing factor, and therefore the virus reproduces. But if the virus specifies the repressing factor, then why doesn't it shut itself off again? Perhaps it does, because a fraction of infected cells do become lysogenic. There may be a race between the λ gene signals for reproduction and those specifying a shutdown. The model of a phage-directed cytoplasmic repressor nicely explains immunity, because any superinfecting phage would immediately encounter a repressor and be inactivated.

How is the prophage attached to the bacterial genome? In the days before chromosome circularity was known, two models seemed possible (Figure 9-34).

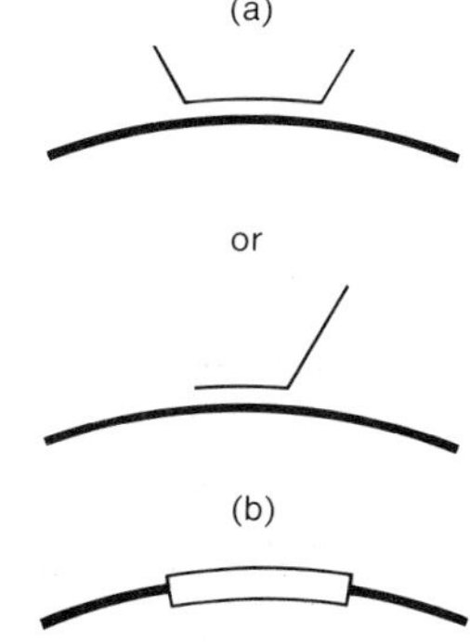

Figure 9-34. Two conceivable modes of prophage attachment. (a) Pairing association. (b) Physical incorporation. (Thick line = E. coli *chromosome; thin line = phage chromosome.)*

The clue for a choice between these models came from a mutant strain of λ. A large section of the chromosome was deleted in this strain, and it could not lysogenize, although it could reproduce. Perhaps the deleted region (which obviously does not contain the genes for λ production) is a region of homology with the bacterial chromosome (the Campbell model again). Crossing over between the λ and *E. coli* chromosomes at such a region could incorporate the entire λ genome at a specific point (and as a continuous part) of the *E. coli* chromosome (Figure 9-35).

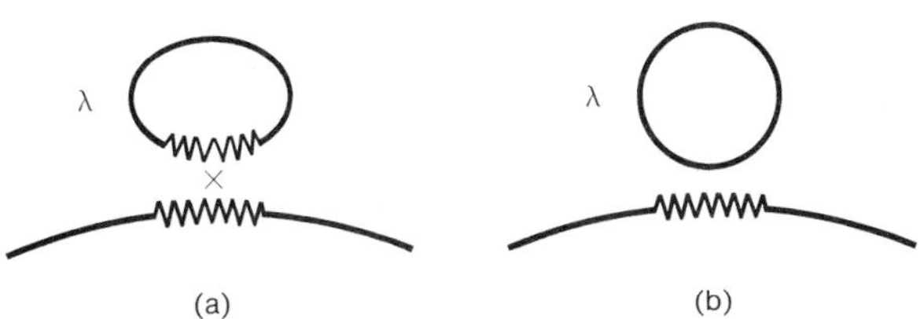

Figure 9-35. Models of λ affinity for the bacterial chromosome (a) in the normal phage; (b) in a phage mutant showing deletion.

Can this simple model of lysogeny be tested? The attraction of Campbell's proposal is that it does make testable predictions, and λ gives us a chance to test those predictions.

1. Physical integration of the prophage should increase the genetic distance between flanking bacterial markers (Figure 9-36). In fact, time-of-entry or recombination distances between the bacterial genes *are* increased by lysogeny.

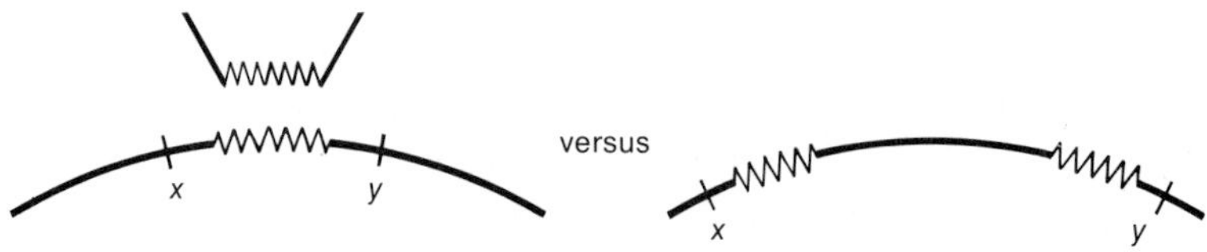

Figure 9-36. Insertion of λ *genome into the* E. coli *chromosome splits the attachment region, increasing the distance between markers* x *and* y.

2. Some deletions of bacterial segments adjacent to the prophage site should delete phage genes (Figure 9-37). Experimental studies do confirm this prediction.

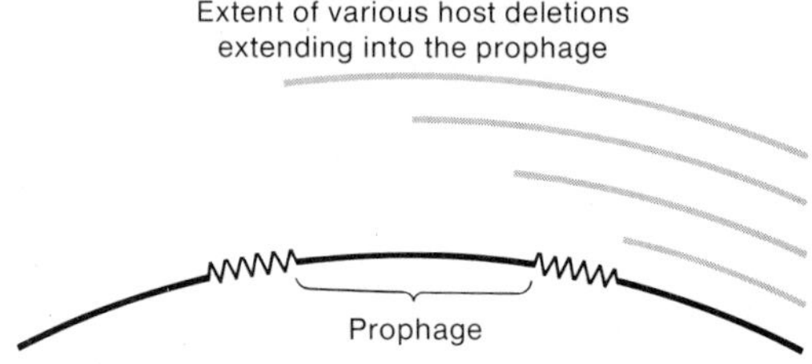

Figure 9-37. Host deletions may delete genes from an adjacent prophage.

The phenomenon of lysogeny is a very successful way for a temperate phage to avoid "eating itself out of house and home." Lysogenic cells can perpetuate and carry the phages around. Amazingly, the study of this biologically interesting but seemingly esoteric phenomenon has recently burst forth as a very relevant model for cancer. For example, polyma virus (a mouse virus) and simian virus 40 (SV40) (Figure 9-38) behave in many ways just like λ phage. Their properties may be studied in mammalian cells in culture. These cancer viruses have two possible destinies upon infecting a cell. First, they can lyse the cell much like phage lysis. Alternatively, they can enter a cell and become inserted into the host chromosome as a **provirus.** Provirus-containing cells are called

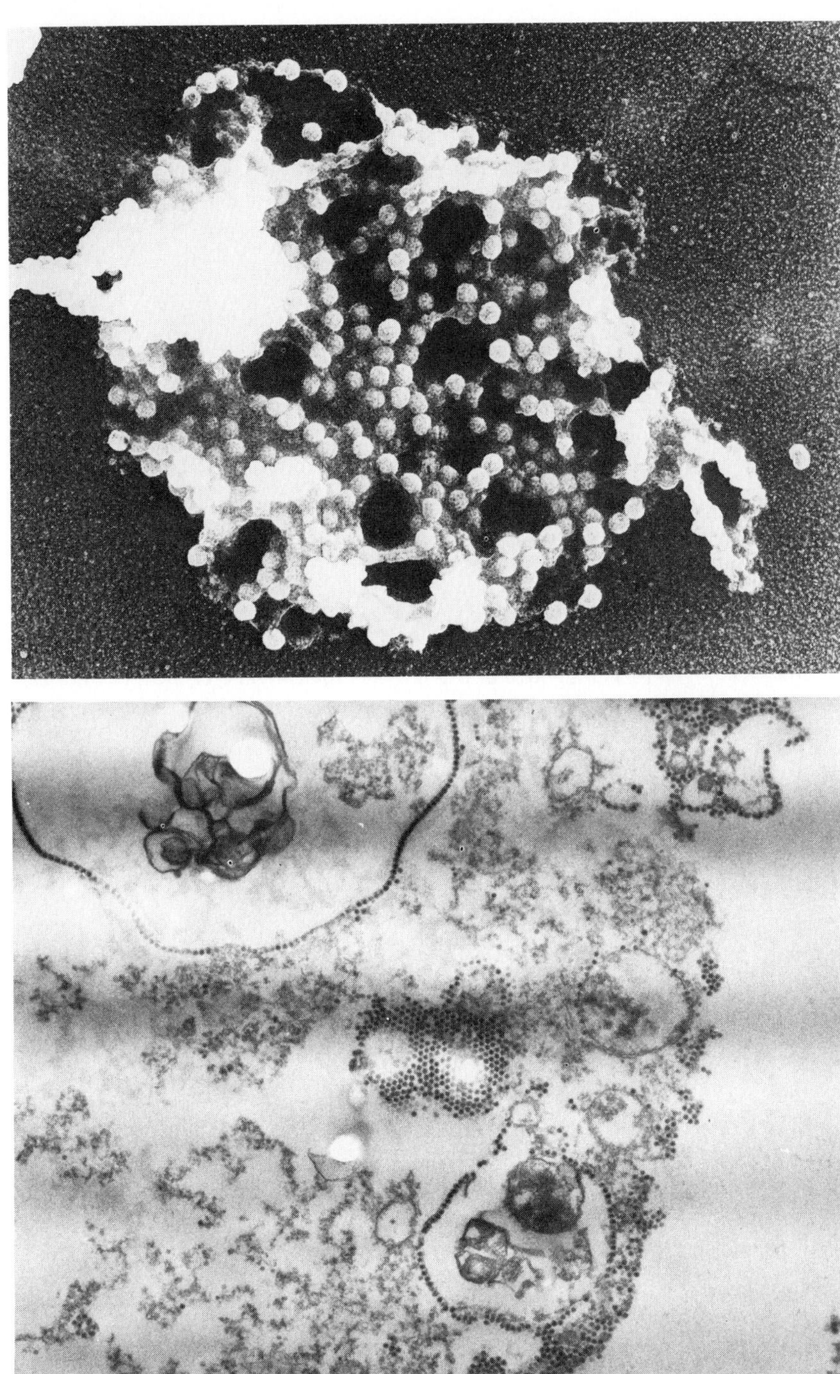

Figure 9-38. Simian virus 40 (SV40) in a monkey cell. (a) Virus particles on the nuclear membrane. (b) Packing of virus particles seen in section. (From Jack D. Griffith.)

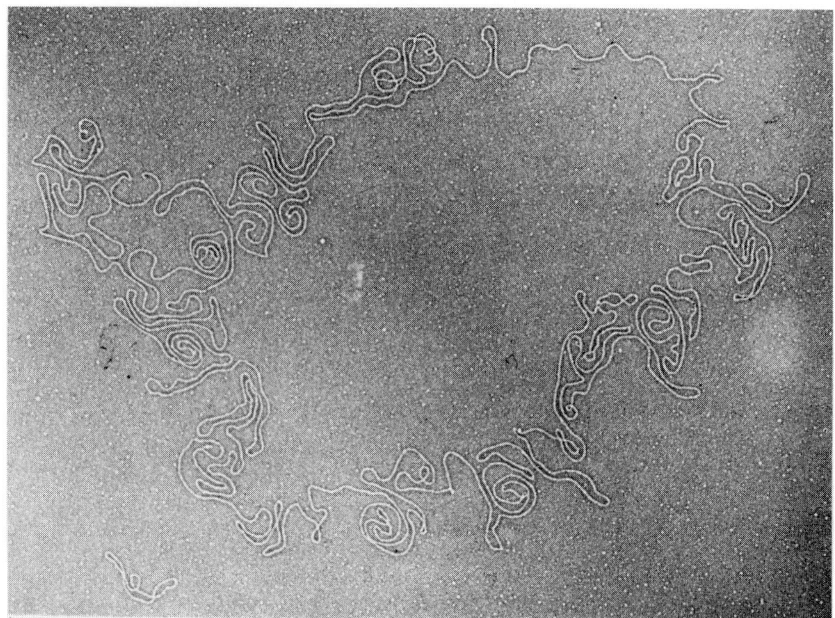

Figure 9-39. The very large circular DNA of the Epstein-Barr virus, which has been found in association with the human diseases mononucleosis, Burkitt's lymphoma, and nasal-pharyngeal carcinoma.

transformed cells (an unfortunate conflict of terms with bacterial genetics). In a transformed cell, the presence of the provirus causes many altered cell properties, most significant of which is that the transformed cells will form tumors when injected into suitable test animals. To date, however, no provirus insertions have been demonstrated in human cancers, although several viruses, such as the Epstein-Barr virus (Figure 9-39), have been found in association with certain human cancers. (Note that transformation can also occur spontaneously or be triggered by certain chemicals or radiation.)

Transduction

Phages are able to carry bacterial genes from one bacterial cell to another in a process called **transduction.** There are two kinds of transduction: generalized and specialized.

Generalized Transduction

In 1951, Joshua Lederberg and Norton Zinder were testing for recombination in the bacterium *Salmonella typhimurium,* using the techniques that had been successful with *E. coli.* They used two different strains: one was $phe^- \ trp^- \ tyr^-$ and the other was $met^- \ his^-$. (We won't worry about the nature of these markers except to note that the mutant alleles confer nutritional requirements.) When they plated either strain on minimal medium, they observed no wild-type cells.

However, after mixing the two strains, they found wild-type cells at a frequency of about 1 in 10^5. Thus far, the situation seems similar to that for recombination in *E. coli.*

However, in this case, the experimenters also recovered recombinants from a U-tube experiment, in which cell contact (conjugation) was prevented by a filter. Varying the size of the pores in the filter separating the two arms, they found that the agent responsible for recombination is about the size of the virus P22, a known temperate phage of *Salmonella.* Further studies supported the suggestion that the vector of recombination is indeed P22. The filterable agent and P22 are identical in properties of size, sensitivity to antiserum, and immunity to hydrolytic enzymes. Thus Lederberg and Zinder, instead of confirming conjugation in *Salmonella,* had discovered a new type of gene transfer mediated by a virus. They called this process transduction. Somehow, some virus particles during the lytic cycle pick up bacterial genes, which are then transferred to another host, where the virus inserts its contents. Transduction has subsequently been shown to be quite common among both temperate and virulent phages.

How are transducing phages produced? In 1965, K. Ikeda and J. Tomizawa threw light on this question in some experiments on the temperate *E. coli* phage P1. They found that, when a nonlysogenic donor cell is lysed by P1, the bacterial chromosome is broken up into small pieces. Occasionally, the forming phage particles mistakenly incorporate a length of *pure* bacterial DNA into a phage head. This is the origin of the transducing phage. A similar process can occur when the prophage of P1 is induced. Because it is the phage coat proteins that determine the phage's ability to attack a cell, such viruses or transducing particles can bind to a bacterial cell and inject their contents, which now happen to be donor bacterial genes. When the contents of a transducing phage are injected into a recipient cell, a merodiploid situation is created in which the transduced genes can be incorporated by recombination (Figure 9-40).

From such merozygotes, we can derive linkage information about bacterial genes. Transduction from an $a^+ b^+$ donor to an $a^- b^-$ recipient produces various transductants for a^+ and b^+. We can obtain linkage information from the ratio

$$\frac{\text{single-gene transductants}}{\text{total transductants}} = \frac{(a^+ b^-) + (a^- b^+)}{(a^+ b^-) + (a^- b^+) + (a^+ b^+)}$$

Presumably, the chance of either a^+ or b^+ being included individually in the transducing phage is proportional to the distance between them. If they are close together, they will usually be picked up and transduced by the phage together. Of course, the method gives linkage values only if the genes under test *are* close enough together for *both* to be included in the transducing phage and thus form an $a^+ b^+$ transductant (called a **cotransductant**). However, even the failure to find cotransductants provides some linkage information in a negative sense. Thus transduction joins the battery of modes of genetic transfer in bacteria—along with conjugation, infectious transfer of episomes, and transformation.

The phages P1 and P22 belong to a group that shows generalized transduction, that is, they transfer virtually any gene of the host chromosome. As prophages, P22 probably inserts into the host chromosome, and P1 remains free like a large plasmid. But both transduce by faulty headstuffing during lysis.

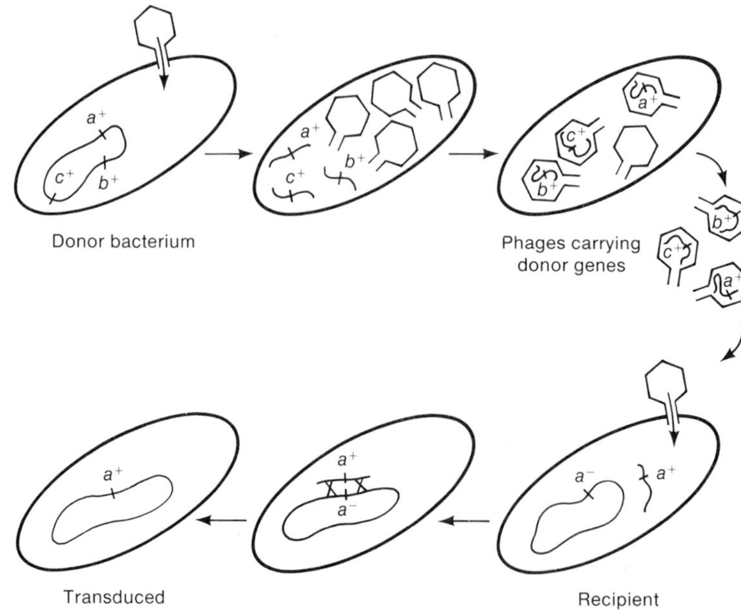

Figure 9-40. The mechanism of generalized transduction. In reality, only a minority of phage progeny would carry donor genes.

We can estimate the size of the piece of host chromosome that a phage can pick up from the following type of experiment using P1 phage:

$$\text{Donor } leu^+\ thr^+\ azi^{\text{r}} \rightarrow \text{Recipient } leu^-\ thr^-\ azi^{\text{s}}$$

We can select for one or more donor markers in the recipient and then (in true merozygote-genetics style) look for the presence of the other unselected markers (Table 9-4).

Experiment 1 tells us that *leu* is relatively close to *azi* and distant from *thr*, but we are left with two possibilities:

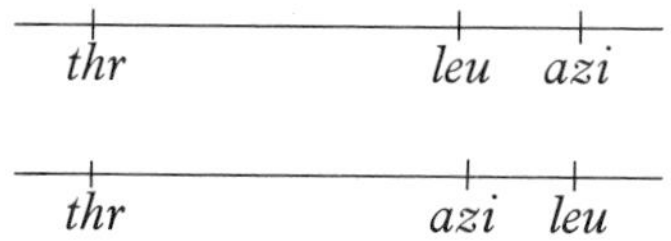

Experiment 2 tells us that *leu* is closer to *thr* than is *azi*, so the map must be

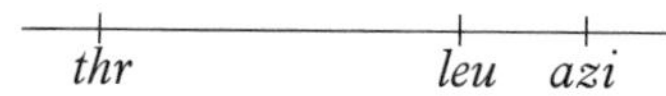

Table 9-4. Accompanying markers in specific P1 transductions

Experiment	Selected marker(s)	Unselected markers
1	leu^+	50% are azi^r; 2% are thr^+
2	thr^+	3% are leu^+; 0% are azi^r
3	leu^+ and thr^+	0% are azi^r

By selecting for *thr*$^+$ and *leu*$^+$ in the transducing phage in experiment 3, we see that the transduced piece of genetic material never includes the *azi* locus. If enough markers were studied to produce a more complete linkage map, we could estimate the size of a transduced segment.

Specialized Transduction

Another class called **specialized** (or **restricted**) transducing phages carry only restricted parts of the bacterial chromosome. We look now at this process of specialized transduction.

Lambda is a good example of a restricted transducer. You will recall that, as a prophage, λ always inserts next to the *gal* region of the *E. coli* host chromosome. In transduction experiments, λ can transduce only the *gal* gene and another closely linked gene, *bio.* Let's follow a typical experiment and try to elucidate the mechanism of λ transduction. *E. coli* strain K12(λ) is lysogenic for λ. This strain is *gal*$^+$. We can induce the lytic cycle with UV light and produce a lysate (progeny phage population). This lysate can be used to infect a *gal*$^-$ recipient culture which is nonlysogenic for λ. These infected cells are plated on minimal medium, and rare *gal*$^+$ transductants are the only ones to grow and produce colonies. About a third of these transductants are stable *gal*$^+$ types, all their descendant cells being *gal*$^+$, too. The remaining transductants are unstable. That is, they segregate out *gal*$^-$ progeny cells as well as *gal*$^+$ cells. If an unstable culture is lysed and used as a *gal*$^+$ donor in transduction, a very high frequency of transduction is obtained. This lysate is called a high-frequency transduction (HFT) lysate.

Putting all these facts together, we arrive at the model shown in Figure 9-41. We see that the initial transducing particles are produced by a rare faulty outlooping of the prophage. This particle is defective, in that some genes have been left behind in the host; consequently it is called λ *d gal* (λ defective gal). The λ *d gal* particle has a λ body and can infect bacteria. Once in, it has two possible fates, shown in Figure 9-41c. We see that the unstable types are in fact double lysogens. These can lyse because the normal λ prophage acts as a helper to supply the missing functions of the adjacent λ *d gal* prophage.

Message

Transduction occurs when bacteriophage pick up host genes prior to lysis. Generalized transduction is mediated by phage particles that have accidentally incorporated a piece of bacterial chromosome during phage packaging. This can occur in the lytic cycle of some virulent and temperate phages, or when lysogenic temperate phages are induced to lysis. Restricted (specialized) transduction is mediated by temperate phages, whose prophages always are inserted at one specific bacterial locus. The transducing phage in this case is produced by faulty separation of the prophage from the bacterial chromosome, so that the prophage includes both bacterial and phage genes.

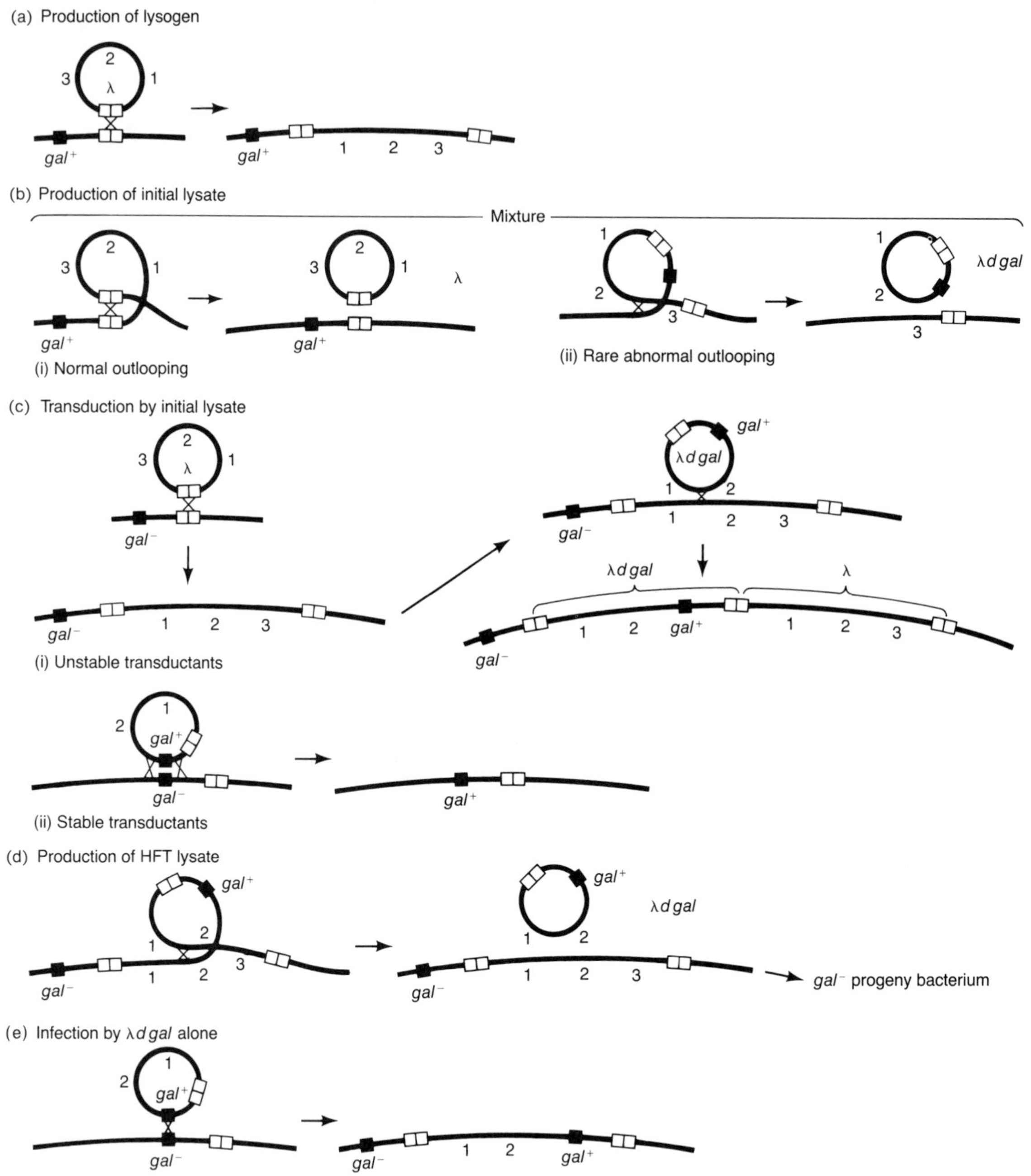

Figure 9-41. Transduction mechanisms in phage λ. (a) The production of a lysogenic bacterium by crossing over in a specialized region. (b) The lysogenic bacterial culture can produce normal λ, or, rarely, an abnormal particle, λ d gal, *which is the transducing particle. (c) Transduction by the mixed lysate can produce either stable transductants, by cross-overs flanking the* gal *gene, or unstable transductants, by the incorporation of* λ *and* λ d gal. *(d) The unstable double lysogen produces* λ d gal *at a very high frequency, which causes a high frequency of transduction (HFT). (e) At low multiplicities of infection,* λ d gal *can insert alone, where it generates a transductant which cannot lyse because of missing* λ *genes.*

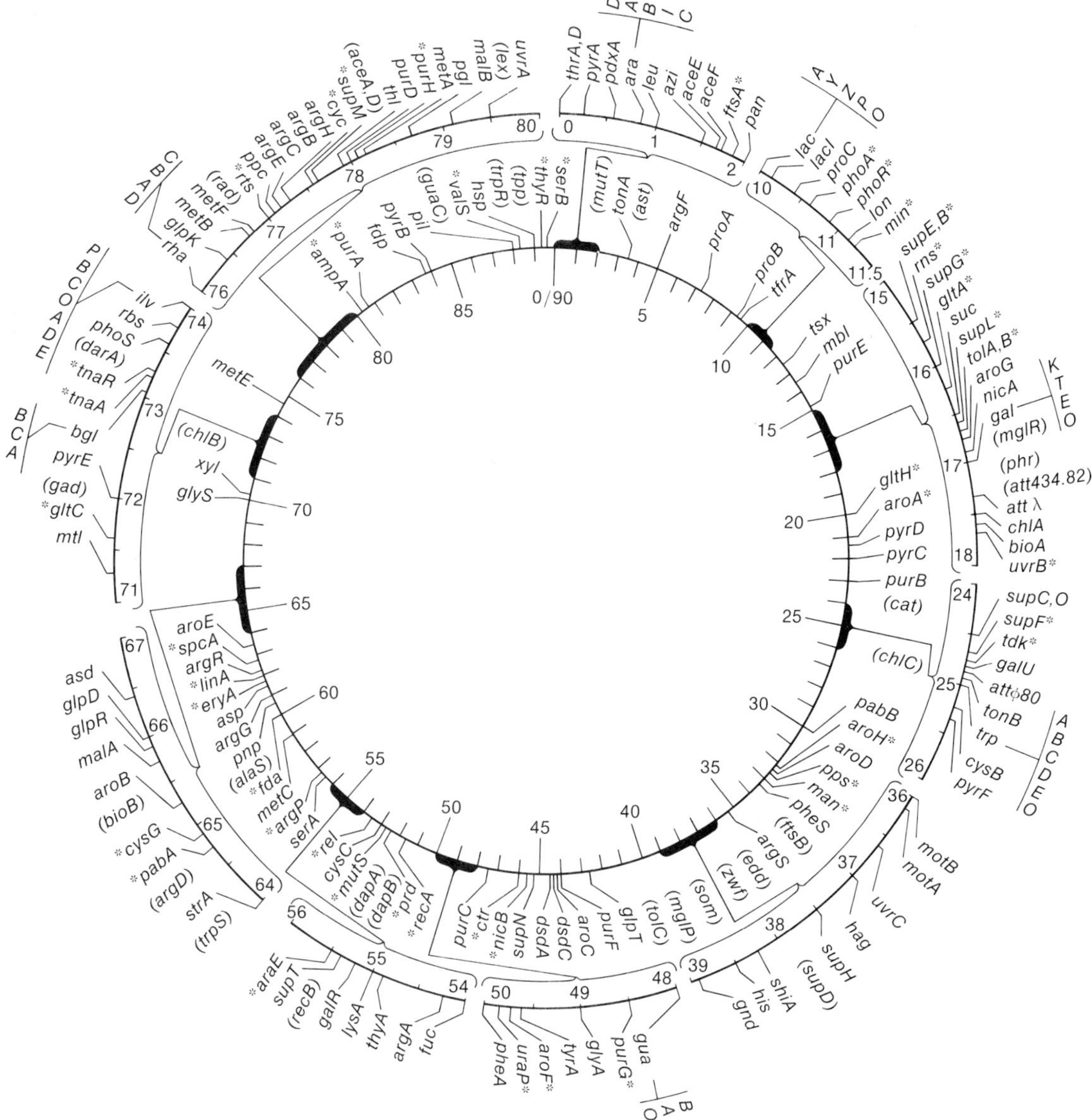

Figure 9-42. The genetic map of E. coli. *Units are minutes (timed from an arbitrarily located origin). (From* Molecular Biology of Bacterial Viruses *by G. S. Stent. W. H. Freeman and Company. Copyright © 1963.)*

When lysates are diluted to a multiplicity of infection of one, a *gal*$^+$ transductant type is detected which cannot lyse, even though it contains λ material as shown by its immunity to superinfection. This type represents λ *d gal* inserted with no helper phage.

Some very detailed chromosome maps for bacteria have been obtained by combining the mapping techniques of interrupted mating, recombination mapping, transformation, and transduction. The complexity of the *E. coli* map (Figure 9-42) illustrates the power and sophistication of genetic analysis at its best. More bacterial and phage genetics will be discussed in Chapter 16.

Summary

Advances in microbial genetics within the past four decades have provided the foundation for the recent advances in molecular biology (to be discussed in the next several chapters). Early in this period, it was discovered that gene transfer and recombination occur between different mating types in several bacteria. However, in bacteria, genetic material is passed in only one direction, from a donor cell (F^+ or Hfr) to a recipient cell (F^-). Donor ability is determined by the presence in the cell of a fertility (F) factor acting as an episome.

On occasion, the F factor present in a free state in F^+ cells can integrate into the *E. coli* chromosome and form an Hfr. When this occurs, gene transfer and subsequent recombination take place. Furthermore, since the F factor can insert at different places on the host chromosome, investigators were able to show that the *E. coli* chromosome is a single circle. Interruptions of the transfer at different times has provided geneticists with a new method for constructing a linkage map of the single chromosome of *E. coli* and other similar bacteria.

Genetic traits can also be transferred from one bacterial cell to another in the form of purified DNA. This process of transformation in bacterial cells was the first demonstration that DNA is the genetic material. For transformation to occur, DNA must be taken into a recipient cell, and then recombination between a recipient chromosome and the incorporated DNA must take place.

Bacteria also have viral diseases, caused by bacteriophage. Phage can affect bacteria in two ways. The phage chromosome may enter the bacterial cell and, using the bacterial metabolic machinery, produce progeny phage and burst the host bacteria. The new phage can then infect other cells. If two phage of different genotypes infect the same host, recombination between their chromosomes can take place during this lytic process. Mapping the genetic loci through these recombinational events has led to the discovery that phage chromosomes can also be circular.

In another infection method, lysogeny, the injected phage lies dormant in the bacterial cell. In many cases the prophage incorporates into the host chromosome and replicates with it. Either spontaneously or under appropriate stimulation, the dormant phage (prophage) can lyse the bacterial host cell.

Phages can carry bacterial genes from a donor to a recipient. In generalized transduction, pure host DNA is incorporated into the phage head during lysis. In restricted transduction, faulty outlooping of the prophage from a unique chromosomal locus results in the inclusion of some host DNA into the phage head.

Problems

1. A microbial geneticist isolates a new mutation in *E. coli* and wishes to map its chromosomal location. She uses interrupted-mating experiments with Hfr strains and generalized transduction experiments with phage P1. Explain why each technique by itself is insufficient for accurate mapping.

2. Four *E. coli* strains of genotype $a^+ b^-$ are labeled 1, 2, 3, and 4. Four strains of genotype $a^- b^+$ are labeled 5, 6, 7, and 8. The two genotypes are mixed in all possible combinations and (after incubation) plated to determine the frequency of $a^+ b^+$ recombinants. The following results were obtained, where M = many recombinants, L = low numbers of recombinants, and O = no recombinants.

	1	2	3	4
5	O	M	M	O
6	O	M	M	O
7	L	O	O	M
8	O	L	L	O

On the basis of these results, assign a sex type (either Hfr, F^+, or F^-) to each of the strains.

3. In *E. coli,* four Hfr strains donate the markers shown in the order given:

strain 1	Q	W	D	M	T
strain 2	A	X	P	T	M
strain 3	B	N	C	A	X
strain 4	B	Q	W	D	M

All these Hfr strains are derived from the same F^+ strain. What is the order of these markers on the circular chromosome of the original F^+?

4. An Hfr strain with genotype $a^+ b^+ c^+ d^- str^s$ is mated with a female strain of genotype $a^- b^- c^- d^+ str^s$. At various times, the culture is shaken violently to separate mating pairs, and the cells are plated on agar of the following three types (where nutrient A allows the growth of a^- cells, and so on; a plus indicates the presence of each nutrient, and a minus indicates its absence):

Agar type	str	A	B	C	D
1	+	+	+	−	+
2	+	−	+	+	+
3	+	+	−	+	+

a. What donor genes are being selected on each type of agar?

b. Table 9-5 shows the number of colonies on each type of agar for samples taken at various times after mixing of the strains. Using this information, determine the order of the genes *a*, *b*, and *c*.

Table 9-5.

Time of sampling (minutes)	Number of colonies on agar of type		
	1	2	3
0	0	0	0
5	0	0	0
7.5	100	0	0
10	200	0	0
12.5	300	0	75
15	400	0	150
17.5	400	50	225
20	400	100	250
25	400	100	250

c. One hundred colonies from each of the 25-minute plates are picked and transferred to a dish containing agar with all of the nutrients except D. The numbers of colonies that grow on this medium are 89 for the sample from agar type 1; 51 for the sample from agar type 2; and 8 for the sample from agar type 3. Using these data, fit gene *d* into the sequence of *a*, *b*, and *c*.

d. On agar containing C and streptomycin but no A or B, at what sampling time would you expect colonies to first appear?

(Problem 4 is from D. Freifelder, *Molecular Biology and Biochemistry*, © 1978 by W. H. Freeman and Company.)

5. You are given two strains of *E. coli.* One is an Hfr strain and is arg^+ ala^+ glu^+ pro^+ leu^+ T^s; the other is F^- and is arg^- ala^- glu^- pro^- leu^- T^r. The markers are all nutritional except *T*, which determines sensitivity or resistance to phage T1. The order of entry is that given, with arg^+ entering the recipient first and T^s last. You find that the F^- strain dies when exposed to penicillin (pen^s), but the Hfr does not (pen^r). How would you locate the locus for *pen* on the bacterial chromosome with respect to *arg*, *ala*, *glu*, *pro*, and *leu*? Formulate your answer in logical, well-explained steps, illustrated with explicit diagrams where possible.

6. A cross is made between Hfr met^+ thi^+ pur^+ × F^- met^- thi^- pur^-. Interrupted-mating studies show that met^+ enters the recipient last, so met^+ exconjugants are selected on medium containing thi and pur only. These exconjugants are tested for the presence of thi^+ and pur^+. The following numbers of individuals are found with each genotype:

met^+ thi^+ pur^+	280
met^+ thi^+ pur^-	0
met^+ thi^- pur^+	6
met^+ thi^- pur^-	52

a. Why was met left out of the selection medium?

b. What is the gene order?

c. What are the map distances in recombination units?

d. Why are there no individuals of the met^+ thi^+ pur^- genotype?

7. In the cross Hfr aro^+ arg^+ ery^r str^s × F^- aro^- arg^- ery^s str^r, the markers are transferred in the order given (with aro^+ entering first), but the first three genes are very close together. Exconjugants are plated on medium containing str (streptomycin, to contraselect Hfr cells), ery (erythromycin), arg (arginine), and aro (aromatic amino acids). Three hundred colonies from these plates are isolated and tested for growth on various media, with the following results. On ery only, 263 strains grow. On ery + arg, 264 strains grow. On ery + aro, 290 strains grow. On ery + arg + aro, 300 strains grow.

a. Draw up a list of genotypes and indicate the number of individuals of each.

b. Calculate the recombination frequencies.

c. Calculate the ratio of the size of the *arg*-to-*aro* region to the size of the *ery*-to-*arg* region.

8. You make the following *E. coli* cross. Hfr Z_1^- *ade*$^+$ *str*^{s} × F$^-$ Z_2^- *ade*$^-$ *str*^{r}, in which *str* determines resistance or sensitivity to streptomycin, *ade* determines adenine requirement for growth, and Z_1 and Z_2 are two very close sites whose Z^- alleles cause an inability to use lactose as an energy source. After about an hour, the mixture is plated on medium containing streptomycin, with glucose as the energy source. Many of the *ade*$^+$ colonies that grow are found to be capable of using lactose. However, hardly any of the *ade*$^+$ colonies from the reciprocal cross Hfr Z_2^- *ade*$^+$ *str*^{s} × F$^-$ Z_1^- *ade*$^-$ *str*^{r} are found to be capable of using lactose. What is the order of the Z_1 and Z_2 sites in relation to the *ade* locus? (Note that the *str* locus is terminal.)

9. Jacob selected eight closely linked *lac*$^-$ mutants (called *lac-1* through *lac-8*) and then attempted to order the mutants with respect to the outside markers *pro* (proline) and *ade* (adenine) by performing a pair of reciprocal crosses for each pair of *lac* mutants:

cross A: Hfr *pro*$^-$ *lac-x ade*$^+$ × F$^-$ *pro*$^+$ *lac-y ade*$^-$

cross B: Hfr *pro*$^-$ *lac-y ade*$^+$ × F$^-$ *pro*$^+$ *lac-x ade*$^-$

In all cases, prototrophs are selected by plating on minimal medium with lactose as the only carbon source. Table 9-6 shows the number of colonies in the two crosses for each pair of mutants. Determine the relative order of the mutants.

Table 9-6.

x	*y*	Cross A	Cross B	*x*	*y*	Cross A	Cross B
1	2	173	27	1	8	226	40
1	3	156	34	2	3	24	187
1	4	46	218	2	8	153	17
1	5	30	197	3	6	20	175
1	6	168	32	4	5	205	17
1	7	37	215	5	7	199	34

(Problem 9 is from Burton S. Guttman, *Biological Principles.* Copyright © 1971 by W. A. Benjamin, Inc., Menlo Park, California.)

10. Linkage maps in an Hfr bacterial strain are calculated in units of minutes, the number of minutes between genes indicating the length of time it takes for the second gene to follow the first after conjugation. In making such maps, microbial geneticists assume that the bacterial chromosome is transferred from Hfr to F$^-$ at a constant rate. Thus, two genes separated by 10 minutes near the origin end are assumed to be the same *physical* distance apart as two genes separated by 10 minutes near the F-attachment end. Suggest a critical experiment to test the validity of this assumption.

11. A particular Hfr strain normally transmits the *pro*$^+$ marker as the last one during conjugation. In a cross of this strain with an F$^-$ strain, some *pro*$^+$ recombinants are recovered early in the mating process. When these *pro*$^+$ cells are mixed with F$^-$ cells, the majority of the latter are converted to *pro*$^+$ cells that also carry the F factor. Explain these results.

12. F′ strains in *E. coli* are derived from Hfr strains. In some cases, these F′ strains show a high rate of integration back into the bacterial chromosome. Furthermore, the site of integration often is the same site that the sex factor occupied in the original Hfr strain (before production of the F′ strains). Explain these results.

13. You have two *E. coli* strains, F^- *str*^{r} *ala*$^-$ and Hfr *str*^{s} *ala*$^+$, in which the F factor is inserted close to *ala*$^+$. Devise a screening test to detect F′ *ala*$^+$ sexductants.

14. *Streptococcus pneumoniae* cells of genotype *str*^{s} *mtl*$^-$ are transformed by donor DNA of genotype *str*^{r} *mtl*$^+$ and (in a separate experiment) by a mixture of two DNAs with genotypes *str*^{r} *mtl*$^-$ and *str*^{s} *mtl*$^+$. Table 9-7 shows the results.

Table 9-7.

	Percentage of cells transformed to		
Transforming DNA	*str*^{r} *mtl*$^-$	*str*^{s} *mtl*$^+$	*str*^{r} *mtl*$^+$
str^r *mtl*$^+$	4.3	0.40	0.17
str^r *mtl*$^-$ + *str*^{s} *mtl*$^+$	2.8	0.85	0.0066

a. What does the first line of the table tell you? Why?

b. What does the second line of the table tell you? Why?

15. A transformation experiment is performed with a donor strain that is resistant to four drugs: A, B, C, and D. The recipient is sensitive to all four drugs. The treated recipient-cell population is divided up and plated on media containing various combinations of the drugs. Table 9-8 shows the results.

Table 9-8.

Drug(s) added	Number of colonies	Drug(s) added	Number of colonies
None	10,000	BC	51
A	1,156	BD	49
B	1,148	CD	786
C	1,161	ABC	30
D	1,139	ABD	42
AB	46	ACD	630
AC	640	BCD	36
AD	942	ABCD	30

a. One of the genes obviously is quite distant from the other three, which appear to be tightly (closely) linked. Which is the distant gene?

b. What is the probable order of the three tightly linked genes?

(Problem 15 is from Franklin Stahl, *The Mechanics of Inheritance,* 2nd. ed. Copyright © 1969 by Prentice-Hall, Englewood Cliffs, New Jersey. Reprinted by permission.

16. In the bacteriophage T4, gene a is 1.0 m.u. from gene b, which is 0.2 m.u. from gene c. The gene order is a–b–c. In a recombination experiment, you recover five double crossovers between a and c from 100,000 progeny viruses. Is it correct to conclude that interference is negative?

17. You have infected *E. coli* cells with two strains of T4 virus. One strain is minute (m), rapid lysis (r), and turbid (tu); the other is wild type for all three markers. The lytic products of this infection are plated and classified. Of 10,342 plaques, the following numbers are classified as each genotype

$m\ r\ tu$	3,467	$m\ +\ +$	520
$+\ +\ +$	3,729	$+\ r\ tu$	474
$m\ r\ +$	853	$+\ r\ +$	172
$m\ +\ tu$	162	$+\ +\ tu$	965

a. Determine the linkage distances between m and r, between r and tu, and between m and tu.

b. What linkage order would you suggest for the three genes?

c. What is the coefficient of coincidence in this cross, and what does it signify?

(Problem 17 is reprinted with the permission of Macmillan Publishing Co., Inc., from Monroe W. Strickberger, *Genetics.* Copyright © 1968 by Monroe W. Strickberger.)

18. Using P22 as a generalized transducing phage grown on a $pur^+\ pro^+\ his^+$ bacterial donor, a recipient strain of genotype $pur^-\ pro^-\ his^-$ is infected and incubated. Later, transductants for each donor gene are selected individually. In experiment I, pur^+ transductants are selected. In experiment II, pro^+ transductants are selected. In experiment III, his^+ transductants are selected.

a. What media are used for these selection experiments?

b. Next, the transductants are examined for the presence of unselected donor markers, with the following results:

I		II		III	
$pro^-\ his^-$	87%	$pur^-\ his^-$	43%	$pur^-\ pro^-$	21%
$pro^+\ his^-$	0%	$pur^+\ his^-$	0%	$pur^+\ pro^-$	15%
$pro^-\ his^+$	10%	$pur^-\ his^+$	55%	$pur^-\ pro^+$	60%
$pro^+\ his^+$	3%	$pur^+\ his^+$	2%	$pur^+\ pro^+$	4%

What is the order of the bacterial genes?

c. Which two genes are closest together?

d. On the basis of the order you have proposed, explain the relative proportions of genotypes observed in experiment II.

(Problem 18 courtesy of D. Freifelder, *Molecular Biology and Biochemistry,* © 1978 by W. H. Freeman and Company.)

19. Although most λ-mediated gal^+ transductants are unstable, a small percentage of these transductants in fact are completely stable. Control experiments show that these stable transductants are not produced by mutation. What is the likely origin of these types?

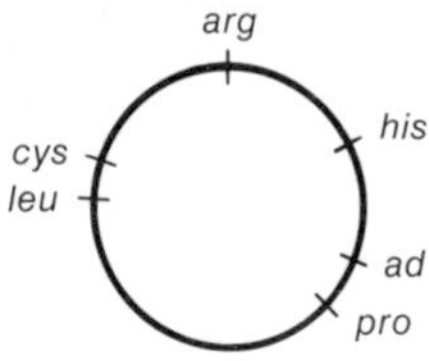

Figure 9-43.

20. An ade^+ arg^+ cys^+ his^+ leu^+ pro^+ bacterial strain is known to be lysogenic for a newly discovered phage, but the site of the prophage is not known. Figure 9-43 shows the bacterial map. The lysogenic strain is used as a source of the phage, and the phages are added to a bacterial strain of genotype ade^- arg^- cys^- his^- leu^- pro^-. After a short incubation, samples of these bacteria are plated on six different media, whose supplementation is indicated in Table 9-9. The table also shows whether or not colonies were observed on the various media.

Table 9-9.

Medium	Nutrient supplementation in medium						Presence of colonies
	ade	arg	cys	his	leu	pro	
1	−	+	+	+	+	+	N
2	+	−	+	+	+	+	N
3	+	+	−	+	+	+	C
4	+	+	+	−	+	+	N
5	+	+	+	+	−	+	C
6	+	+	+	+	+	−	N

NOTE: + indicates the presence of a nutrient supplement: N indicates no colonies; C indicates colonies present.

a. What genetic process is at work here?
b. What is the approximate locus of the prophage?

21. You have two strains of λ that can lysogenize *E. coli;* Figure 9-44 shows their linkage maps. The segment shown at the bottom of the chromosome and designated 1–2–3 is the region responsible for pairing and crossing over with the *E. coli* chromosome. (Keep the markers on all your drawings.)

a. Diagram the way in which strain X is inserted into the *E. coli* chromosome (so that the *E. coli* is lysogenized).
b. It is possible to superinfect the bacteria lysogenic for strain X by using strain Y. A certain percentage of these superinfected bacteria become "doubly" lysogenic (that is, lysogenic for both strains). Diagram how this will occur. (Don't worry about how double lysogens are detected.)
c. Diagram how the two λ prophages can pair.
d. It is possible to recover crossover products between the two prophages. Diagram a crossover event and the consequences.

Strain X

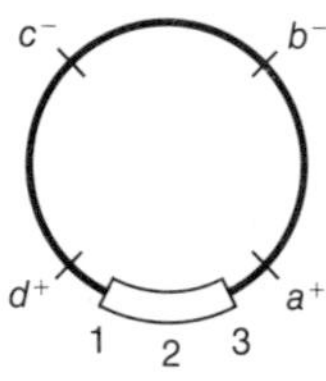

Strain Y

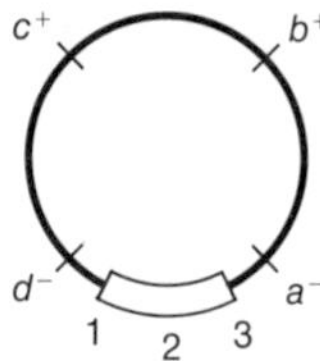

Figure 9-44.

22. You have three strains of *E. coli.* Strain A is F′ cys^+ *trp1*/cys^+ *trp1* (that is, both the F′ and the chromosome carry cys^+ and *trp1*, an allele for tryptophan requirement). Strain B is F^- cys^+ *trp2 Z* (this strain requires cysteine for growth and carries *trp2*, another allele causing a tryptophan requirement; the strain also is lysogenic for the generalized transducing phage Z). Strain C is F^- cys^+ *trp1* (it is an F^- derivative of strain A that has lost the F′).

a. How would you determine whether *trp1* and *trp2* are alleles of the same locus? (Describe the crosses and the results expected.)

b. Suppose that *trp1* and *trp2* are allelic and that the *cys* locus is cotransduced with the *trp* locus. Using phage Z to transduce genes from strain C to strain B, how would you determine the genetic order of *cys*, *trs1*, and *trp2*?

23. A generalized transducing phage is used to transduce an $a^- b^- c^- d^- e^-$ recipient strain of *E. coli* with an $a^+ b^+ c^+ d^+ e^+$ donor. The recipient culture is plated on various media with the results shown in Table 9-10. What can you conclude about the linkage and order of the genes?

Table 9-10.

Compounds added to minimal medium	Presence (+) or absence (−) of colonies
C D E	−
B D E	−
B C E	+
B C D	+
A D E	−
A C E	−
A C D	−
A B E	−
A B D	+
A B C	−

NOTE: The allele a^- determines a requirement for A as a nutrient, and so forth.

24. In a P1 phage transduction experiment involving three very closely linked *arg* loci (*arg-1*, *arg-2*, and *arg-3*) and another remote locus *pro*, the data shown in Table 9-11 are obtained. Draw a map of the loci represented by *arg-1*, *arg-2*, and *arg-3*. (HINT: Use the technique described for conjugation on page 335.)

Table 9-11.

	Recipient			
	pro⁻ arg-1		*pro⁻ arg-2*	
	Number of transductants for		Number of transductants for	
Donor	arg^+	pro^+	arg^+	pro^+
pro^+ *arg-1*	—	—	492	14,959
pro^+ *arg-2*	403	12,358	—	—
pro^+ *arg-3*	55	2,978	996	18,239

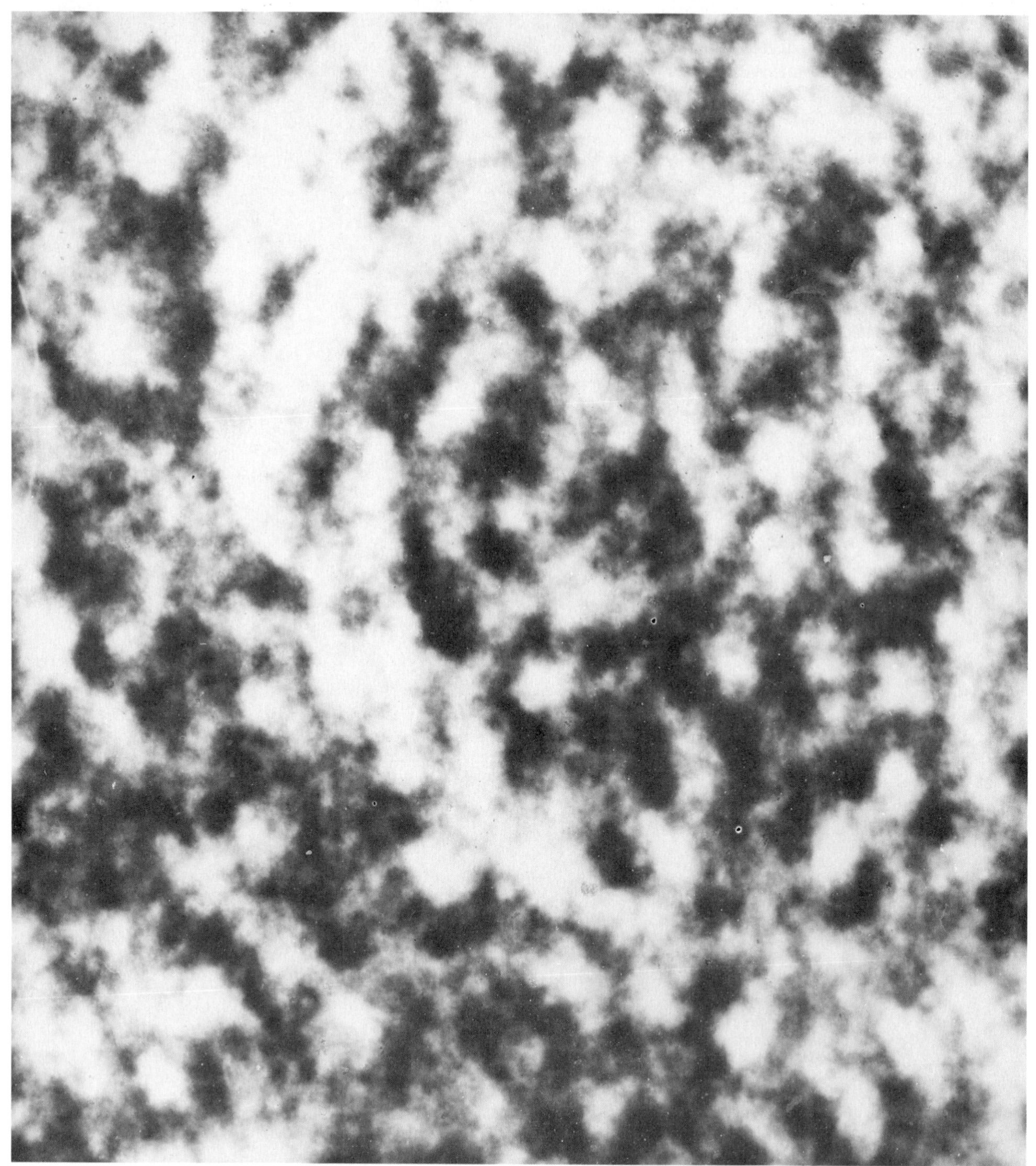

High magnification electron micrograph of a fragment of DNA, showing its helical structure. (Copyright © Jack D. Griffith.)

10

The Nature of the Gene

Our analyses thus far, both genetic and cytological, have led us to regard the chromosome as a linear (one-dimensional) array of genes, strung rather like beads on an unfastened necklace. Indeed, this model is sometimes called the **bead theory.** According to the bead theory, a gene is a unit of inheritance whose existence is recognized through its mutant alleles. All of these alleles affect a single phenotypic character, all map to one chromosomal locus, all give mutant phenotypes when paired, and all give Mendelian ratios when intercrossed. There are several points about the bead theory worth emphasizing:

1. The gene is viewed as a fundamental unit of *structure,* indivisible by crossing over. Crossing over occurs between genes (the beads in this model) but never within them.
2. The gene is viewed as a fundamental unit of *change,* or mutation. It changes from one allelic form to another, but there are no smaller components within it that can change.
3. The gene is viewed as the basic unit of *function.* Parts of a gene, if they exist, cannot function (although the precise function of the gene is not specified in this model).
4. The chromosome is viewed merely as a vector or transporter of genes. It exists simply to permit their orderly segregation and to shuffle them in recombination.

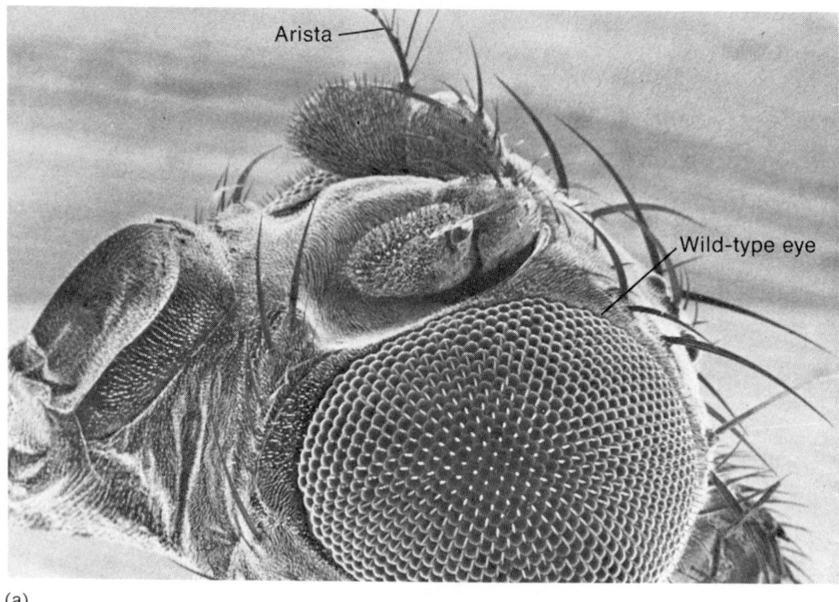

(a)

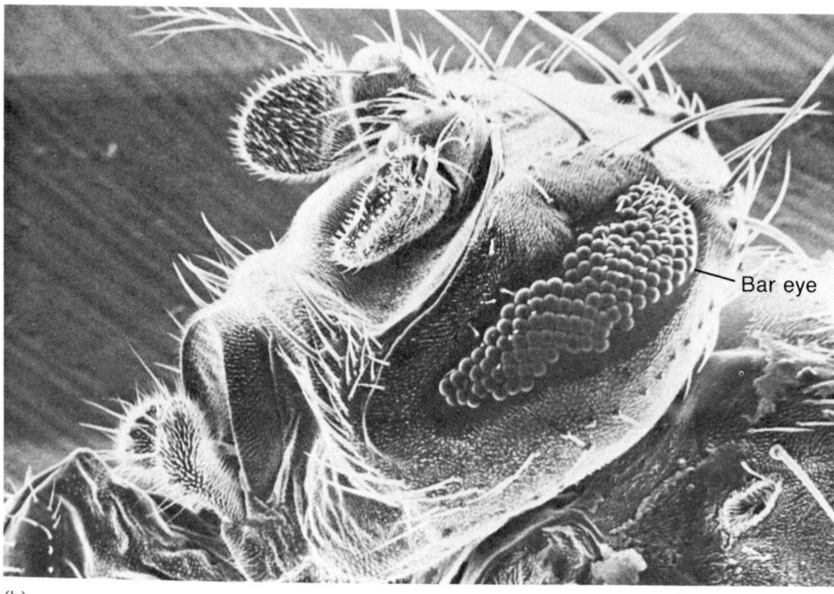

(b)

Figure 10-1. Scanning electron micrographs of Drosophila *eyes. (a) Wild type. (b) Bar eye.*

In this chapter, we see the demise of the bead theory on all four points. Although there will be more to say about point 4 in a later chapter, it is a convenient starting point for our attack on the bead theory.

Position Effect

Some of the earliest evidence for a significant role for the chromosome in the genetic mechanism came from the study of bar-eyed mutants in *Drosophila* (Figure 10-1). This bar mutation is sex-linked and dominant. The mutant allele *B* was shown to be the result of a tandem duplication of the wild-type region (allele +) for this gene. The double-bar phenotype (allele *BB*) proved to be a triplication of the wild-type region. Thus the various genotypes for this locus possess varying numbers of copies of the wild-type region.

A. H. Sturtevant obtained genotypes with various multiples of the wild-type region. For each genotype, he counted the number of facets (cells) of the compound eye as a measure of the phenotypic character of eye size. Table 10-1 shows his results. Note the significant effect on phenotype of a difference in the chromosomal position of the same number of multiple copies of the wild-type region (*B*/*B* versus *BB*/+). The arrangement of the copies on the chromosomes (and not just the number of copies) affects the phenotype; such an effect is called a **position effect.** Apparently, the immediate neighbors of a gene are important, and chromosomal location does have a functional significance. Point 4 in our outline of the bead theory is thereby thrown into question.

Table 10-1. Eye size for various bar-eye genotypes in *Drosophila*

Genotype	Number of copies of wild-type region	Eye size (number of facets)
+ / +	2	779
+ / *B*	3	358
B / *B*	4	68
BB / +	4	45
BB / *B*	5	36
BB / *BB*	6	25

Intragenic Recombination

One of the first setbacks to the concept of the indivisibility of the gene (point 1 on our list) also came from work on *Drosophila.* Edward B. Lewis studied a locus near the left end of chromosome 2 in *Drosophila.* He had two mutants. One was the dominant Star (mutant allele *S*), which produces an eye slightly smaller than wild type in *S*/+ flies. The other mutant, mapping in the same region, was the recessive asteroid (*ast*), which gives a much smaller eye in *ast*/*ast* homozygotes. However, *S*/*ast* heterozygotes have eyes that are smaller yet, thus indicating that *S* and *ast* must be functionally related. Together with their similar mapping location, this information indicates that *S* and *ast* are alleles. The order of eye size for the genotypes is $+/+ > S/+ > ast/ast > S/ast$. Lewis obtained a strain with markers *al* and *ho* closely flanking the *S* locus. He made the following cross: *al S ho*/+ *ast* + ♀ × *al ast ho*/*al ast ho* ♂.

In 57,000 progeny of this cross, Lewis found 16 flies with wild-type eyes! How could the + allele have appeared at the *S* locus? All these 16 flies were of the phenotype + + ho, clearly suggesting a crossover event somewhere between *al* and *ho.* How could such a crossover also generate a wild-type allele for eye size? Suppose that *S* and *ast* are actually two very tightly linked loci rather than alleles for the same locus. Then we could imagine a crossover between these two loci producing a wild-type phenotype:

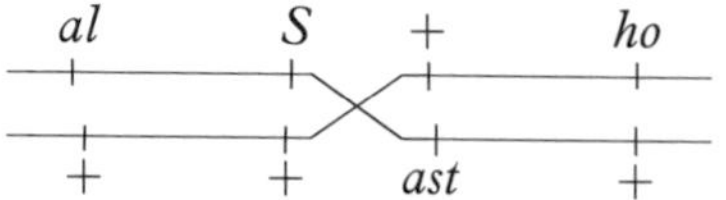

Following this model, Lewis decided that *S* and *ast* are *pseudoalleles.* (This term is of historical significance only; we shall see that it was soon dropped.) With this model, the true crossover genotype revealed by the wild-type flies is + + + *ho*. The reciprocal recombinant would be *al S ast* +. This is much harder to detect because it carries both *S* and *ast.* (How would you distinguish *S ast* from separate single alleles *S* and *ast* in genetic analysis?) The reciprocal recombinant was recovered, and again a kind of position effect was revealed. The phenotypes of *S* +/+ *ast* (very small eyes) and *S ast*/+ + (eyes only slightly smaller than wild type) are quite different, even though their total genetic content appears similar.

The existence of pseudoalleles was demonstrated at another locus, lozenge (*lz*), in *Drosophila* by Mel Green and Kathleen Green. In this case, mutant alleles produce eyes with a glossy, smooth surface. Green and Green had several independently isolated alleles of lozenge, all mapping at one chromosomal position. All heterozygotes carrying two different mutants were lozenge in phenotype, thereby showing that they are indeed functionally allelic. They intercrossed different *lz* mutants and testcrossed heterozygotes with different *lz* alleles (very much like *S*/*ast*), and they also found wild-type recombinants. The idea that these lozenge alleles are actually pseudoalleles (each with a slightly different chromosomal location) began to look very likely when Green and Green showed that wild-type recombinant frequency could be used to draw a linear genetic map of the pseudoallelic sites:

lz^{BS} lz^{k} lz^{l} lz^{g}

But are these pseudoalleles really any different from "true" alleles? Perhaps wild-type recombinants for alleles at other loci have not been found, simply because they are hard to detect or because no one carried out an intensive search for them. After all, recombination between pseudoalleles is a very rare event.

For example, in the coat-color allelic series in rabbits (say, between chinchilla and Himalayan), is it possible that study of tens of thousands of rabbit progeny would reveal recombinants (denoted by c^+, or C)?

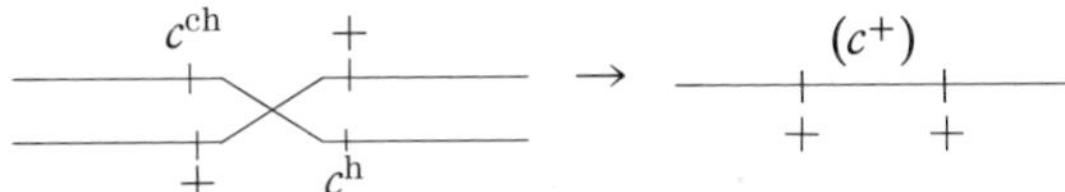

Pseudoalleles, of course, segregate in an almost perfect 1:1 Mendelian ratio because crossing over between them is so rare, so this hypothesis does seem tenable. Rearing and scoring 100,000 flies is quite a job—with rabbits, the equivalent experiment is almost unthinkable!

In the 1950s, Seymour Benzer performed some beautiful experiments that shed a great deal of light on the nature of alleles. Benzer capitalized on the fantastic **resolving power** made possible by using selective systems for rare events in phages. He worked with the virulent *E. coli* phage T4 (Figure 10-2) and a class of mutants that produces a phenotype called rapid lysis (*r*). Rapid-lysis

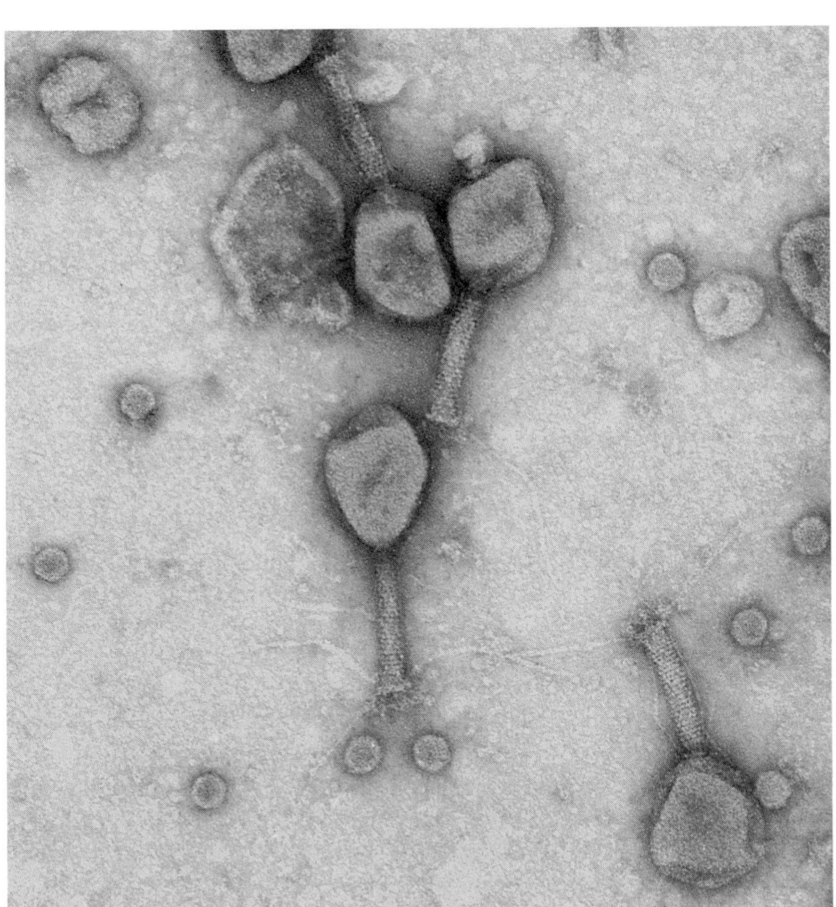

Figure 10-2. Enlargement of the E. coli *phage T4 showing details of structure: note head, tail, and tail fibers. This was the phage used by Benzer in his experiments on the nature of the* rII *(rapid lysis) gene. (From Jack D. Griffith.)*

strains produce large circular plaques (compared to the small and irregular plaques of normal phages). Genes at several different loci can mutate to cause rapid lysis. Benzer worked with the locus called *rII* for one important reason: all *rII* alleles are *conditional* lethals. That is, an *rII* mutant will grow and form large plaques of *E. coli* strain B, but it will not grow at all on the strain of *E. coli* that is lysogenic for λ called strain K(λ). (We need not worry now about the cause of the difference between these two *E. coli* strains.) In contrast, rII^+ (wild-type) phages will grow and form small ragged plaques on both strains. Thus we have the relationships shown in Table 10-2.

Table 10-2. Plaque phenotypes produced by different combinations of *E. coli* and phage strains

	E. coli strain	
T4 phage strain	B	K(λ)
rII	Large, round	No plaques
rII^+	Small, ragged	Small, ragged

This conditional growth of *rII* strains proved essential to the analysis because it allows the use of selective techniques to identify recombinants. Benzer started with an initial sample of eight independently derived *rII* mutant strains and set about crossing them in all possible combinations of pairs, simply by the double infection of *E. coli* B. He took the lysate (progeny phages) and plated onto a lawn of *E. coli* K(λ), on which the *rII* genotypes will not grow. But plaques *were* seen on K(λ), indicating that recombination had given rise to rII^+ genotypes (Figure 10-3). The frequency of plaques was too large for them to be due to back mutations. Furthermore, as with the lozenge locus, the alleles (or pseudoalleles, or whatever) could be mapped unambiguously to the right or left of each other to give what we now call a gene map. In this case, the map units are the frequency of rII^+ plaques.

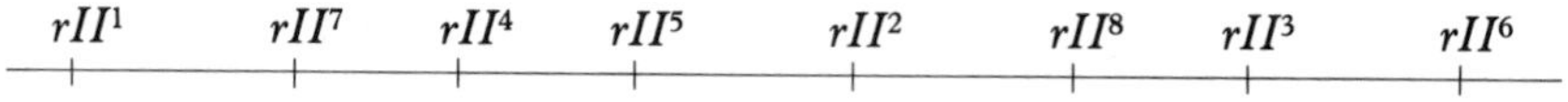

rII^1 — rII^7 — rII^4 — rII^5 — rII^2 — rII^8 — rII^3 — rII^6

At this point, it seems just as well to drop the notion of pseudoalleles. It was now apparent (and it became even more apparent when maps were made of many genes in microorganisms) that there is nothing pseudo about pseudoalleles. Recombination within a gene (**intragenic recombination**) seems to be the rule rather than the exception. It can virtually always be found at any locus if a suitable selection system is available to detect recombinants. In other words, a mutant allele can be pictured as a length of genetic material (the gene) that has a damaged or nonwild part (a **mutant site**) somewhere, and this partial damage is what causes the nonwild phenotype. Different alleles have different phenotypic effects because they involve damage to different parts (sites) of the wild-type allele.

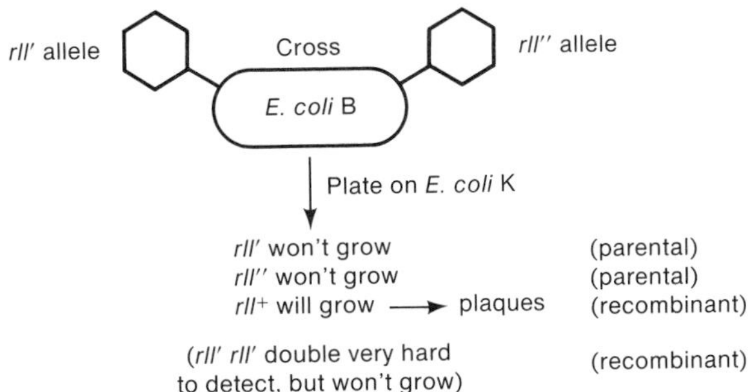

Figure 10-3. Selection of intragenic recombinants at the rII *locus of phage T4. Mutants within the gene* rII *cannot grow on* E. coli *K. When two different phage carrying different alleles of* rII *infect the same bacterial cell, some progeny phage can grow on* E. coli *K; in other words, some progeny have become* rII+. *This result indicates that recombination has occurred* within *a single gene and not just between genes.*

Thus, an allele a^1 can be represented as

+ + + + + * + + + + + + + + + + + + + + +

where the asterisk (*) represents the mutant site within an otherwise normal gene (denoted by the sites marked +). A cross between a^1 and another mutant allele a^2, in general, then, is something like this:

+ + + + + * + + + + + + + + + + + + + + +

×

+ + + + + + + + + + + + + + * + + + + + +

and it is easy to see how

+ +

could be generated by a simple crossover anywhere between the two mutant sites. Reconsider the star/asteroid, lozenge, and coat color examples discussed earlier in the light of this profound insight into the nature of the gene and recombination within it. You will see that all can be represented in terms of recombination between mutant sites of the same gene, and that the postulation of closely linked loci with "pseudoalleles" is not necessary.

Benzer extended his *rII* analysis to use many hundreds of alleles. He found that the frequency of rII^+ recombinants in a heteroallelic (different alleles) cross always at least 0.01%, although his analytical system was sufficiently fine to have allowed detection of frequencies as low as 0.0001% if they had occurred. He concluded that there is a minimal recombination distance within a gene. This led to the idea that the gene is composed of units called **recons.** A recon is a region within which no crossing over occurs. Crossing over can occur between recons, but never within a recon. Thus the recon replaced the gene as the basic unit that is indivisible by crossing over. Once again, a hypothetical entity (the recon) has been determined solely through genetic analysis—and once again it was soon to assume physical reality. The map of a gene is a completely linear sequence of recons.

Message
A gene is composed of units called recons that are not divisible by recombination.

Benzer also coined the term **muton** to describe the minimal element within a gene that can be altered to produce a mutant phenotype. This idea arose naturally from the concept of mutant sites.

Message
The muton is the smallest element within a gene that, if altered, can give rise to a nonwild (mutant) phenotype.

The concept of the muton is suggested by the fact that *part* of a gene can change (as we saw in our analysis of "pseudoalleles"). Thus we have discarded point 2 in our outline of the bead theory at the beginning of the chapter, which would have required the *entire* gene to act as the muton.

Theoretically, a muton need not necessarily be the same as a recon. Figure 10-4 illustrates a *hypothetical* gene composed of different recons and mutons. However, we shall see later that, in reality, a muton *is* the same as a recon.

Don't become impatient with this account of concepts that later required modification or replacement. These experiments are not merely of historical significance. They represent examples of recombinational and mutational genetic dissection at its best. Furthermore, these classical experiments established the formats for many subsequent investigations.

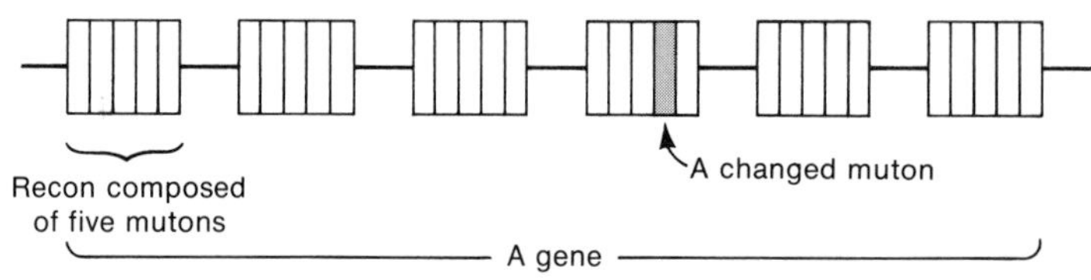

Figure 10-4. A hypothetical gene in which recons are not identical to mutons.

Message
The occurrence of intragenic recombination permits the construction of gene maps. The relative frequencies of intragenic recombinants in crosses between various mutant alleles reveals the order and relative positions of the mutant sites within a gene.

Deletion Mapping

We now digress somewhat to describe the way in which Benzer was able rapidly to locate new mutant sites in his *rII* gene map. He found some special *rII* mutants that would not give recombinants when crossed with any of several other mutants (which had been shown to be different recons) but would give recombinants when crossed to still others. He realized that such mutants behaved as if they were short deletions within the *rII* region, and this model was supported by their lack of reversion. They could be used for rapid location of mutant sites in newly obtained mutant alleles. For example, consider the following gene map showing 12 identifiable mutant sites:

1 2 3 4 5 6 7 8 9 10 11 12

One special mutant D_1 fails to give rII^+ recombinants when crossed with *1, 2, 3, 4, 5, 6, 7,* or *8*; therefore, D_1 behaves as if it involves a deletion of sites *1* through *8*:

9 10 11 12

Another special mutant D_2 fails to give rII^+ recombinants when crossed with *5, 6, 7, 8, 9, 10, 11,* or *12*; therefore, D_2 behaves as if it involves a deletion of sites *5* through *12*:

1 2 3 4

These overlapping deletions now define three areas of the gene—call them i, ii, and iii:

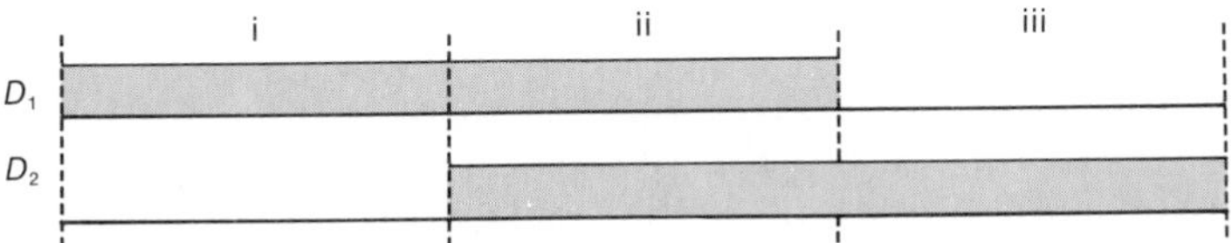

A new mutant that gives rII^+ recombinants when crossed with D_1 but not when crossed with D_2 must have its mutant site in area iii. One that gives

(Continued)

Deletion Mapping (*Continued*)

rII^+ recombinants with D_2 but not with D_1 must have its mutant site in area i. A new mutant that does not give rII^+ recombinants with either D_1 or D_2 must have its mutant site in area ii. For example, consider a mutant in area iii crossed with D_1:

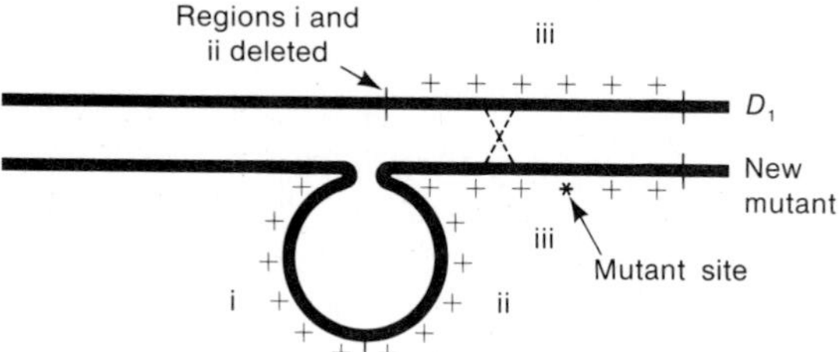

The more deletions there are in the tester set, the more areas can be uniquely designated, and the more rapidly new mutant sites can be located. Once assigned to a region, a mutant can be crossed with other alleles in the same region to obtain an accurate position. Figure 10-5 shows the complexity of Benzer's actual map.

Deletions themselves can be intercrossed and mapped just like point mutations. The deleted region is represented by a bar. If no wild-type recombinants are produced in a cross between different deletions, then the bars are shown as overlapping. A typical deletion map might be

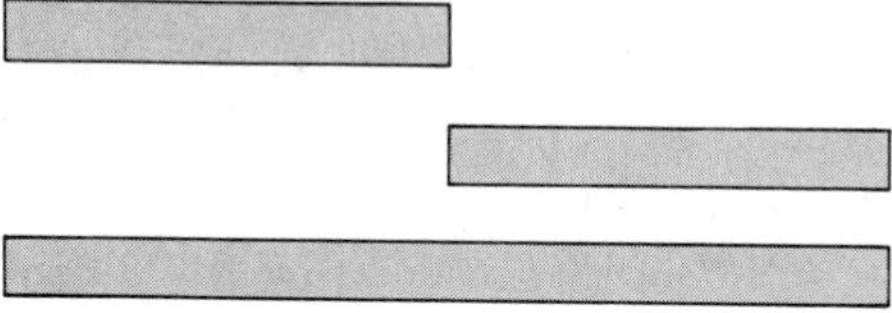

Such deletion maps are useful in delineating regions of the gene to which new point mutants can be assigned.

Intragenic Complementation

In another part of his studies, Benzer made a discovery that relates to the gene as the unit of function (point 3 in our list at the beginning of the chapter). Benzer found that, although *rII* mutants cannot individually infect *E. coli* K(λ) cells, some specific pairs of mutants can "help each other out" or **complement** each

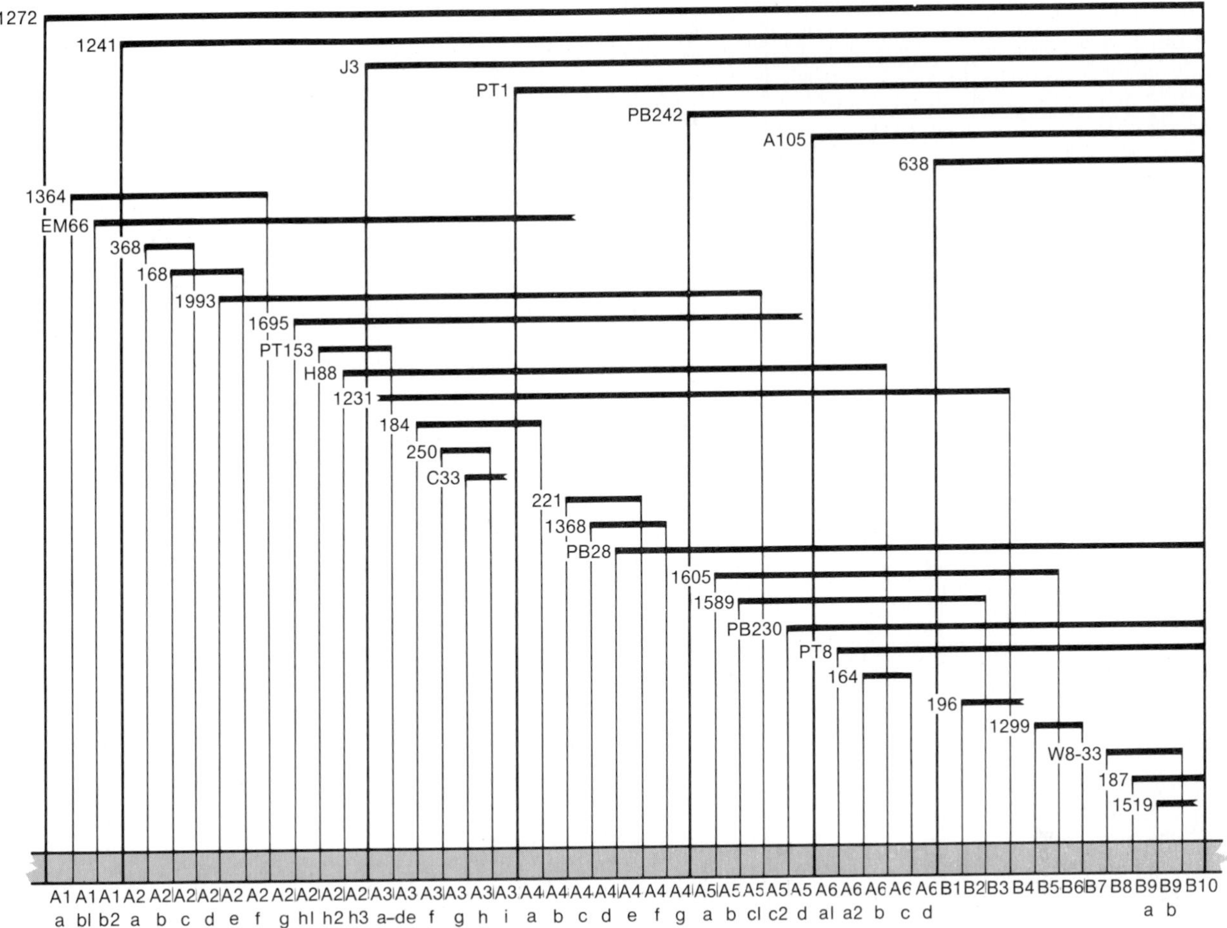

Figure 10-5. Detailed deletion map of the rII *gene. Each deletion (horizontal bar) has an identification number. Along the bottom are the arbitrary identification numbers of the regions defined by the deletions. Note that some deletions extend out of the* rII *gene. (From G. S. Stent and R. Calendar,* Molecular Genetics, *2nd ed. Copyright © 1978 by W. H. Freeman and Company. After S. Benzer,* Proc. Natl. Acad. Sci. USA *47:403, 1961).*

other in a double infection of K(λ), leading to lysis of the cells. In fact, he was able to divide all the *rII* mutants he tested into two groups, A and B. Any member of group A will complement any member of group B, but no complementation is possible between any pair of alleles in the same group. Thus it appears that mutants in the A group are defective in some function that mutants in the B group can supply, and vice versa. Very interestingly, all alleles in group

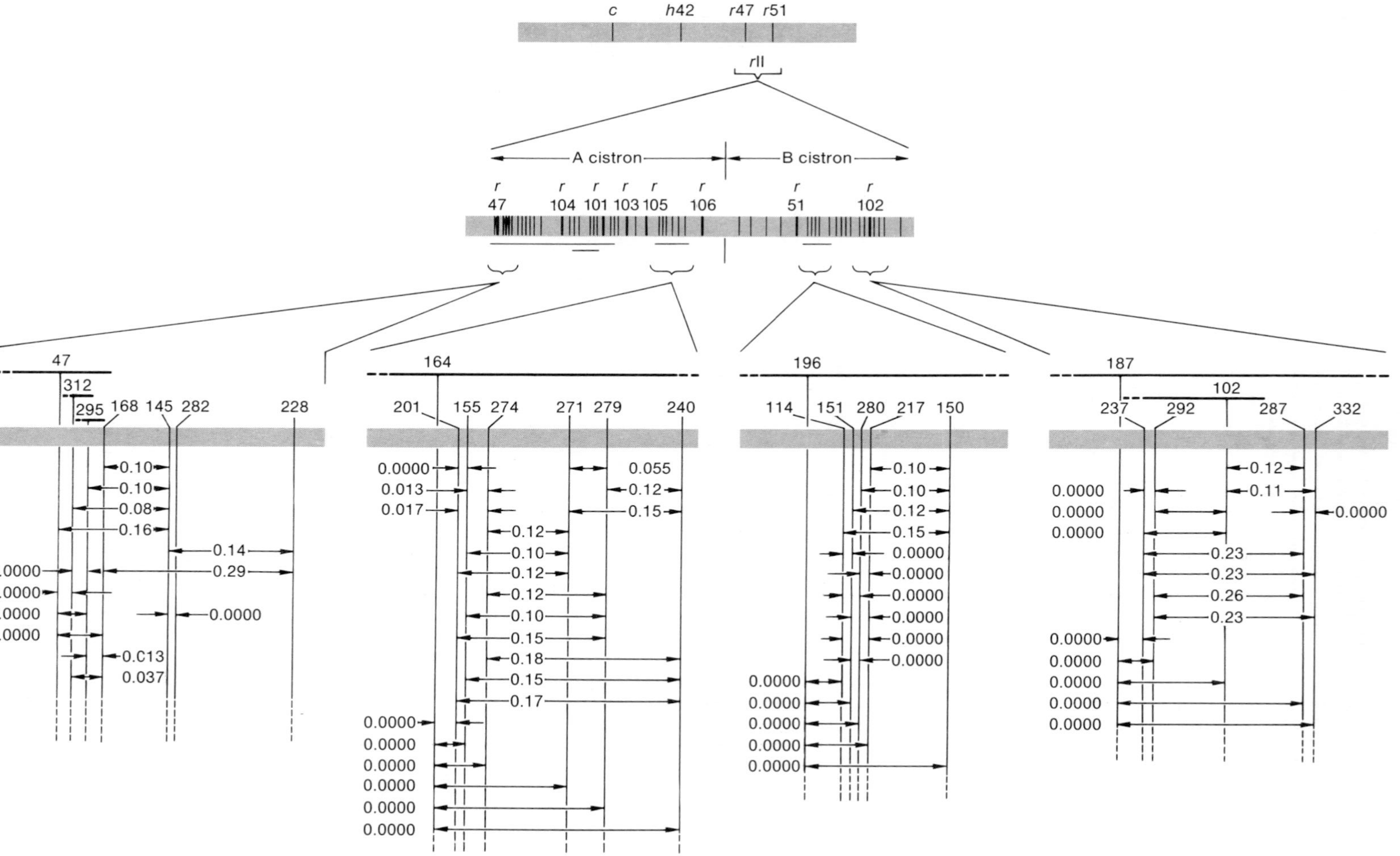

Figure 10-6. *Detailed recombination map of the* rII *region of the phage T4 chromosome. The map unit is the percentage of* rII^{+} *recombinants in crosses between the* rII *mutants. Typical regions are progressively enlarged. Numbers on the map represent mutant sites. Note that the two portions of the* rII *region, the A cistron and B cistron, are two different functional units of the region. (After S. Benzer,* Proc. Natl. Acad. Sci. USA *41:344, 1955. From G. S. Stent and R. Calendar,* Molecular Genetics, *2nd ed. Copyright © 1978 by W. H. Freeman and Company.)*

A mapped on one half of the *rII* locus, and all those in group B mapped in the other half:

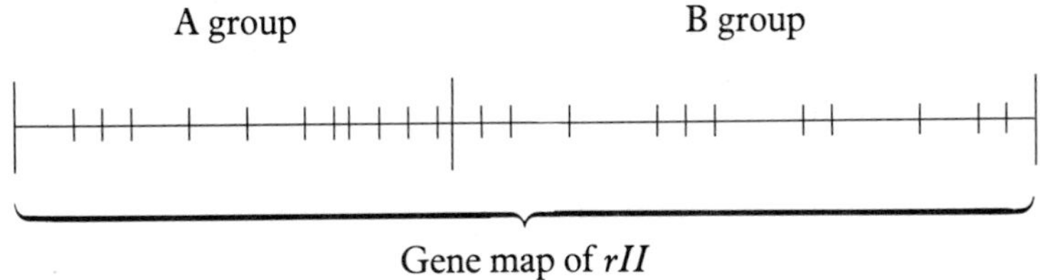

Figure 10-6 shows a complete map. Benzer called the A and B groups different **cistrons.**

Message
A cistron is a genetic region within which there is no complementation between mutations.

The cistron gets its name from the test that is performed to see whether two mutant sites are within the same cistron or are in different ones. In this **cis–trans test,** the complementation test is arranged with the mutant sites on the same chromosome (*cis*) or on opposite chromosomes (*trans*), as shown in Figure 10-7. The terms *cis* and *trans* used at the intragenic level are equivalent to the terms *coupling* and *repulsion* used at the intergenic level. The *trans* test is the

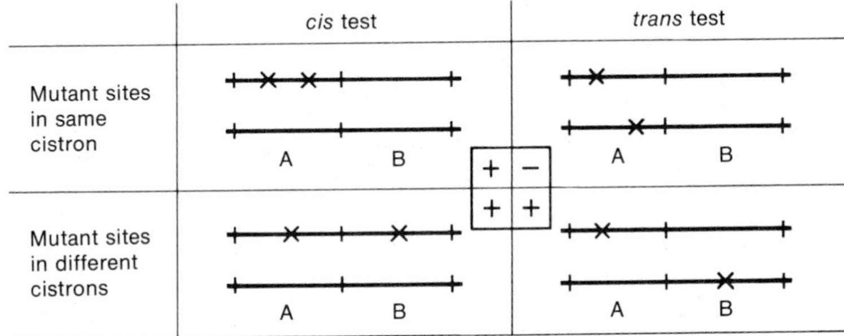

Figure 10-7. The cis–trans *test. A cistron is a genetic region within which there is no complementation between mutations. If two different* rII *mutant T4 phages infect the same bacterium (the* trans *configuration) and no growth of phage occurs, the mutants are said to be in the same cistron (upper right box of the figure). However, if complementation does occur (if progeny phage grow), the mutants are in different functional units or cistrons (lower right box of the figure). In the upper right box, only the B function can be expressed normally; in the lower right box, both A and B functions can be expressed normally. Plus represents a successful complementation; minus indicates no complementation; crosses represent mutant sites.*

critical one; the *cis* test serves mainly as a control. A complementation in a *trans* test means that the mutant sites are in different cistrons; a failure to complement in a *trans* test means that the sites are in the same cistron.

Of course, complementation was nothing new. When a diploid eukaryote of genotype $a^+a^-\ b^+b^-$ shows wild-type phenotype, this is nothing more than intergenic complementation between the normal a^+ and b^+ alleles. What was new in Benzer's work was the demonstration of **intragenic complementation.**

This proof of intragenic complementation demolishes point 3 in our bead-theory list at the beginning of the chapter, which holds that the gene is the basic unit of function. In the *rII* locus, obviously the cistron is the unit of function! Both the *rII*A and the *rII*B cistron must be undamaged (or unchanged) if the *rII* function is to be normal (that is, to produce normal slow lysis).

We now know that some genes consist of only one cistron, some (like *rII*) consist of two, and others consist of three or even more. This unforeseen complexity within the gene introduces some problems of terminology that we must consider.

A Functional Definition of Allelism

With intragenic complementation demonstrated as a reality in phages and in many other systems, we must take another long, hard look at the meaning of the term *allele.* Star (*S*) and asteroid (*ast*) still can be regarded as good alleles, because the combination *S*/*ast* in *trans* produces a mutant phenotype. However, another *Drosophila* gene poses a problem: the mutants miniature (*m*) and dusky (*dy*) both decrease wing size, and both map to the same locus on the X chromosome. With similar mutant phenotype and similar locus, they appear to be good alleles, but when combined in females in *trans* (*m*/*dy*), they produce a normal phenotype. Are these mutations allelic or not?

We intend to use the *trans* test as an operational definition of allelism. If two mutations do not complement in *trans,* then they are alleles. Thus the cistron replaces the gene as a unit of function, and alleles are defined in terms of function. Strictly speaking, then, we should restate our definition of an allele as "a form of cistron." Of course, if the gene in question has only one cistron, then the definition as "a form of a gene" is still tenable.

Returning to the examples from *Drosophila,* we make the following statements under our operational definition of allelism:

1. *S* and *ast* are alleles of the same cistron of a gene;
2. *m* and *dy* are not allelic (they represent mutations probably in separate cistrons of the same gene).

Message
An allele is a form of the genetic unit of function, the cistron. The trans *test provides a workable operational definition of allelism.*

How Genes Work

We have defined the cistron as the unit of function. But what is a cistron in chemical terms, and how does it function? The first clues to the nature of primary gene function came from studies of humans—unlikely subjects for genetic research. Early in this century, Archibald Garrod (a physician) noted that several hereditary human defects are produced by recessive mutations. Some of these defects can be traced directly to metabolic defects affecting the basic body chemistry; this observation led to the suggestion of "inborn errors" in metabolism. We know now, for example, that phenylketonuria, caused by an autosomal recessive allele, results from an inability to convert phenylalanine into tyrosine. Consequently, phenylalanine accumulates and is spontaneously converted into a toxic compound, phenylpyruvic acid. In a different character, it is the inability to convert tyrosine into the pigment melanin that produces an albino. In any case, the observations of Garrod focused attention on metabolic control by genes.

The One-Gene–One-Enzyme Hypothesis

Clarification of the actual function of genes came from research on *Neurospora* by George Beadle and Edward Tatum in the 1940s; they later received a Nobel Prize for this work. After irradiation to produce mutations, ascospores were grown and the resulting cultures tested for their ability to grow on minimal medium. Wild types can grow on minimal medium, but some cultures were unable to grow. However, the latter *could* grow when they were transferred to a medium containing certain specific chemical additives, showing that they were not dead. Some nongrowers responded to arginine, some to pyridoxine, and some to other specific chemicals. Each of these auxotrophic requirements was inherited as a single-gene mutation; each gave a 1:1 ratio when crossed with wild type (recall that *Neurospora* is haploid).

Beadle and Tatum concentrated on a group of independently isolated arginine-requiring auxotrophic mutants. They set out to map the chromosomal location of each mutation. We'll simplify the story somewhat by telling you that all of the arginine (*arg*) mutations eventually were mapped into three different locations on separate chromosomes—we'll call these loci the *arg-1, arg-2,* and *arg-3* genes. Beadle and Tatum found that the auxotrophs for each of the three loci differ in their response to the chemical compounds ornithine and citrulline, which are related to arginine (Figure 10-8). The *arg-1* mutants will grow if supplied with either ornithine *or* citrulline *or* arginine in addition to the minimal medium. The *arg-2* mutants will grow on either arginine or citrulline but not on ornithine. The *arg-3* mutants can grow only when arginine is supplied.

It was known that related compounds are interconverted in cells by biological catalysts called enzymes. Beadle and Tatum and their colleagues proposed a biochemical model for such conversions in *Neurospora:*

$$\text{Precursor} \xrightarrow[\text{Enzyme X}]{} \text{Ornithine} \xrightarrow[\text{Enzyme Y}]{} \text{Citrulline} \xrightarrow[\text{Enzyme Z}]{} \text{Arginine}$$

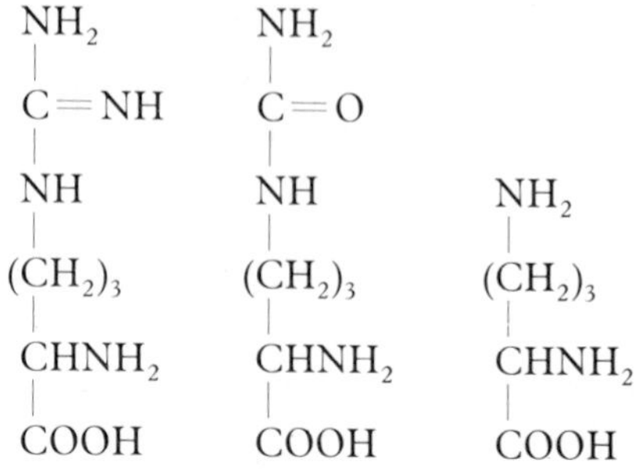

Arginine	Citrulline	Ornithine
NH_2	NH_2	
|	|	
$C{=}NH$	$C{=}O$	
|	|	
NH	NH	NH_2
|	|	|
$(CH_2)_3$	$(CH_2)_3$	$(CH_2)_3$
|	|	|
$CHNH_2$	$CHNH_2$	$CHNH_2$
|	|	|
$COOH$	$COOH$	$COOH$

Figure 10-8. Chemical structures of arginine and the related compounds citrulline and ornithine. In the work of Beadle and Tatum, different arg *auxotrophic mutants of* Neurospora *were found to grow when the medium was supplemented with citrulline or ornithine as an alternative to arginine.*

Now suppose that the *arg-1* mutants have defective enzyme X, so that they are unable to convert the precursor into ornithine as the first step in producing arginine. However, having normal enzymes Y and Z, they are perfectly able to produce arginine if supplied with either ornithine or citrulline. By similar reasoning, we conclude that the *arg-2* mutants lack enzyme Y and that the *arg-3* mutants lack enzyme Z. Thus, a mutation at a particular gene is assumed to interfere with production of a single enzyme. The defective enzyme, then, creates a block in some biosynthetic pathway. The block can be circumvented by feeding the cells any compound that normally comes *after* the block in the pathway. Note that this entire model was inferred from the properties of the mutant classes detected through genetic analysis; only later were the existence of the pathway and the presence of defective enzymes demonstrated through independent biochemical evidence.

This model became known as the **one-gene–one-enzyme hypothesis.** It provided the first exciting insight into the function of genes: genes somehow are responsible for the function of enzymes, and each gene apparently controls one specific enzyme. Other researchers obtained similar results for other biosynthetic pathways, and the hypothesis soon achieved general acceptance. It is one of the great unifying concepts in biology, because it provided a bridge to bring together the concepts and research techniques of genetics and chemistry.

Message
Genes control biochemical reactions by controlling the production of enzymes.

We should pause to let the significance of this discovery sink in. When we ask how genes function, we really are asking how a genotype produces a particular phenotype. For example, how is it that one allele of a gene can produce a wrinkled pea, whereas another allele of the gene produces a round smooth pea? The one-gene–one-enzyme hypothesis provides a model for this relationship, and it is worth spelling out this crucial breakthrough in our understanding of heredity.

1. The characteristic features of an organism are determined by the phenotypes of its parts.
2. The phenotype of a part (say, an organ) is determined by the phenotypes of the tissues that make up the part.
3. The tissue phenotype is determined by the phenotypes of the cells that compose the tissue.
4. The phenotype of a cell is determined by its internal chemistry.
5. Each cell is a reaction vessel in which metabolic reactions are catalyzed by enzymes, so the enzymes control the internal chemistry of the cell.
6. The enzymes present in a cell are determined by the genotype of the cell.

Enzymic Explanation of Genetic Ratios and Dominance

When the significance of the gene control of cellular chemistry became clear, a lot of other things fell into place. Many genetic generalizations now could be explained and tied together in a single conceptual model.

Most "classical" (Mendelian) gene-interaction ratios can be explained simply by the one-gene–one-enzyme concept. For example, recall the 9:7 F_2 dihybrid ratio for flower pigment

P	*AA BB* (purple) × *aa bb* (white)	
	↓	
F_1	*Aa Bb* (purple)	
	↓	
F_2	9 *A–B–*	(purple)
	3 *A–bb*	(white)
	3 *aa B–*	(white)
	1 *aa bb*	(white)

We can easily explain this result if we imagine a biosynthetic pathway leading ultimately to a purple petal pigment in which there are two colorless (white) precursors:

$$\text{White precursor 1} \xrightarrow[\text{Enzyme A}]{} \text{White precursor 2} \xrightarrow[\text{Enzyme B}]{} \text{Purple pigment}$$

Enzyme A ↑ *A* allele; Enzyme B ↑ *B* allele

(Try to work out your own models to explain such ratios as 9:3:4 and 13:3.)

The meaning of dominance and recessiveness also becomes a little clearer in light of the biochemical model. In most cases, dominance probably represents the presence of enzyme function, whereas recessiveness probably represents the lack of enzyme function. A heterozygote has one dominant allele that can produce the functional enzyme:

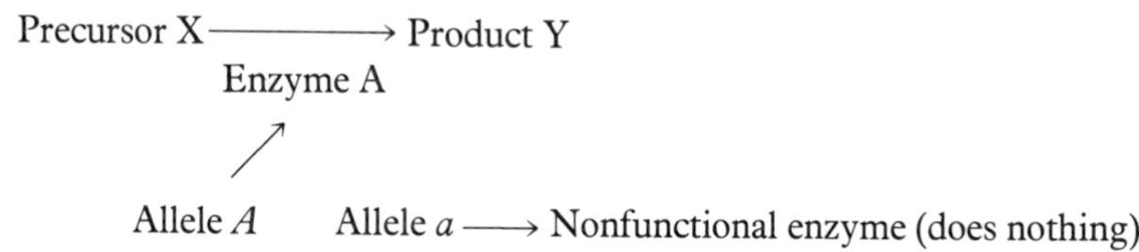

If phenotype Y is due to the presence of product Y, and if phenotype X is due to the absence of product Y, then it is clear that the heterozygote will show phenotype Y, and allele A will be dominant over allele a.

However, this is not the only possible model. We can build in the concept of a threshold. Suppose that phenotype Y is produced only when the concentration of product Y exceeds some threshold level. Suppose further that the homozygote AA produces more enzyme and hence more product than the heterozygote Aa. In this case, the phenotype of the heterozygote will depend upon the relationship between the threshold and the amount of product Y produced by the heterozygote. In Figure 10-9a, the heterozygote A_1A_2 does produce enough product Y to exceed the threshold, so the heterozygote has phenotype Y, and therefore A_2 is dominant over A_1. However, the situation could be like that shown for the alleles B_1 and B_2, where the heterozygote B_1B_2 does not exceed the threshold. In this case (which is less common), B_1 is dominant over B_2, and the dominant phenotype is the one involving a lack of product Y.

Figure 10-9b shows a situation in which no threshold exists. The heterozygote has an intermediate phenotype—exactly the situation observed in cases of incomplete dominance.

What determines whether a gene will behave like A, B, or C? Probably many interacting factors are involved—other genes, the chemical nature of the product, and (last but not least) the effect of the environment on that particular cell type. Finally, note that Figure 10-9 shows a simple linear relationship between enzyme concentration and the number of active alleles—this need not be the case, as we will see in Chapter 14.

The recessiveness of the *arg* mutants in *Neurospora* can be demonstrated by making heterokaryons between two *arg* mutants. If an *arg-1* mutant is placed on minimal medium with an *arg-2* mutant, the two cell types fuse and form a heterokaryon composed of both nuclear types in a common cytoplasm (Figure 10-10). The *arg-1* nuclei produce enzyme for one step, and the *arg-2* nuclei produce enzyme for the other step, so they combine their abilities to produce arginine in the shared cytoplasm.

The scale of involvement of the genes in controlling cellular metabolism is staggering. Most of us have boggled at the charts on laboratory walls showing

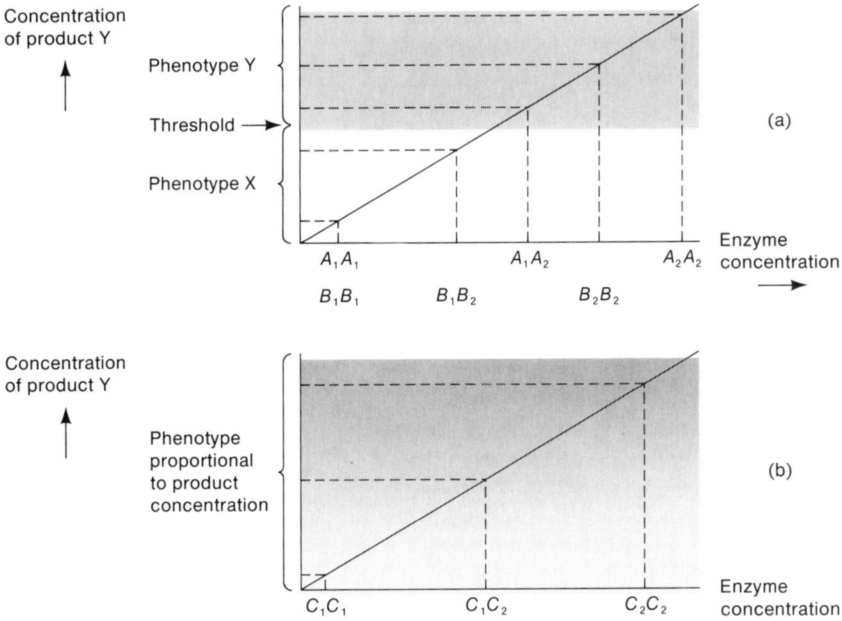

Figure 10-9. Hypothetical curves relating enzyme concentration to the amount of product. Two basic situations are possible. In (a), there is a threshold of the enzyme product above which a sharply contrasting phenotype, Y, is observed. Depending on where the levels in the three possible genotypes occur in relation to this threshold, then the heterozygote will show either phenotype X (for example, B_1B_2) or Y (for example, A_1A_2). If the subscript 2 alleles are the active ones, you can see that the active allele (A_2) can be dominant, which is normally the case. Less frequently, the inactive allele (B_1) can be dominant. In (b) there is no threshold, and the heterozygote has an intermediate phenotype. This situation explains incomplete dominance.

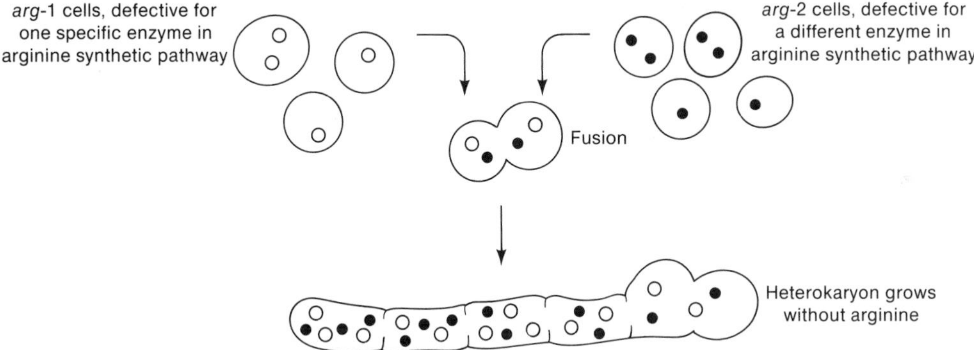

Figure 10-10. Formation of, and complementation in, a heterokaryon of Neurospora. *Vegetative cells of this fungus can fuse, allowing the nuclei from the two strains to intermingle within the same cytoplasm. If each cell is blocked at a different point in a pathway, as are* arg-1 *and* arg-2, *all functions are present in a heterokaryon and the* Neurospora *will grow; in other words, complementation takes place.*

the myriad interlocking, branched, and circular pathways along which the cell's chemical intermediates are shunted like parts on an assembly line. Bonds are broken, molecules cleaved, molecules united, groups added or removed, and so on. The key fact is that almost every step, represented by an arrow on the metabolic chart, is controlled (mediated) by an enzyme. And each of these enzymes is produced under the direction of a gene that specifies its function. Genes control the enzymes, and the enzymes control the chemical reactions that comprise metabolism. When we study a gene (or a few genes), we are focusing on one tiny corner of this vast network controlled by genes and their enzymes.

Humans provide some startling examples. Consider the list of specific enzyme-associated genetic diseases in Table 10-3, which suggests the magnitude of genetic involvement in human disease. Figure 10-11 shows a corner of the metabolic map to illustrate how a set of diseases, some of them common and familiar to us, can stem from blockage of adjacent steps in the pathways.

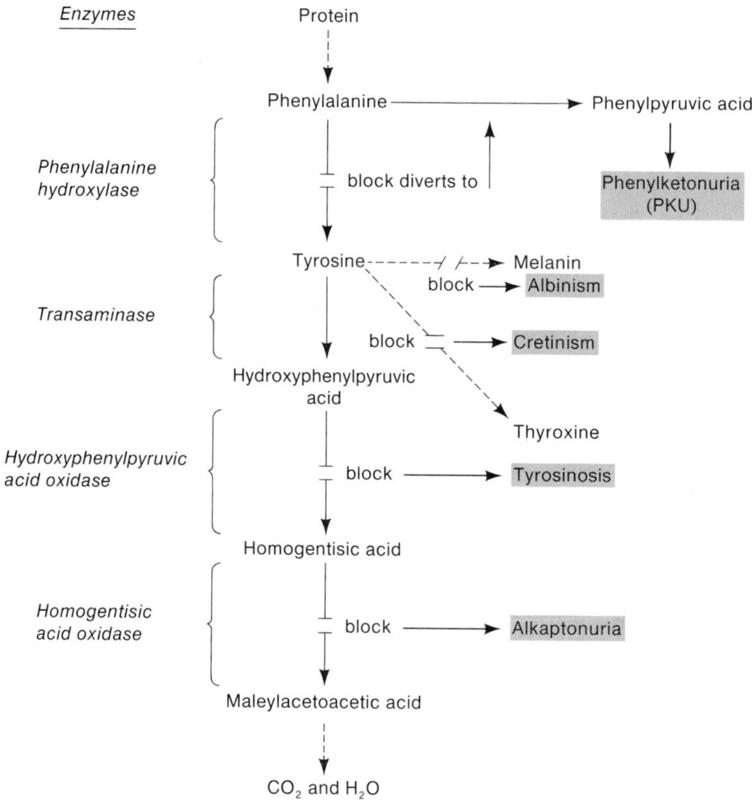

Figure 10-11. One small part of the human metabolic map, showing the consequences of various specific enzyme failures. (Disease phenotypes are shown in shaded boxes.) (After I. M. Lerner and W. J. Libby, Heredity, Evolution, and Society, *2nd ed. Copyright © 1976 by W. H. Freeman and Company.)*

Table 10-3. Enzymopathies: Inherited disorders in which altered activity (usually deficiency) of a specific enzyme has been demonstrated in humans

Condition	Enzyme with deficient activity[1]
Acatalasia	Catalase
Acid phosphatase deficiency	Acid phosphatase
Adrenal hyperplasia I	20,21-Desmolase*
Adrenal hyperplasia II	3-β-Hydroxysteroid dehydrogenase*
Adrenal hyperplasia III	21-Hydroxylase*
Adrenal hyperplasia IV	11-β-Hydroxylase*
Adrenal hyperplasia V	17-Hydroxylase*
Albinism	Tyrosinase
Aldosterone deficiency	18-OH-Dehydrogenase
Alkaptonuria	Homogentisic acid oxidase
Angiokeratoma, diffuse (Fabry disease)	Ceramide trihexosidase
Apnea, drug-induced	Pseudocholinesterase
Argininemia	Arginase
Argininosuccinic aciduria	Argininosuccinase
Aspartylglycosaminuria	Specific hydrolase (AADG-ase)
Ataxia, intermittent	Pyruvate decarboxylase
Carnosinemia	Carnosinase
Norum disease	Lecithin cholesterol acetyltransferase (LCAT)
Citrullinemia	Arginosuccinic acid synthetase
Crigler–Najjar syndrome	Glucuronyl transferase
Cystathioninuria	Cystathionase
Disaccharide intolerance I	Invertase
Disaccharide intolerance II	Invertase, maltase
Disaccharide intolerance III	Lactase
Ehlers–Danlos syndrome, type V	Lysyl oxidase
Ehlers–Danlos syndrome, type VI	Collagen lysyl hydroxylase
Ehlers–Danlos syndrome, type VII	Procollagen peptidase
Fanconi panmyelopathy	Exonuclease*
Farber lipogranulomatosis	Ceramidase
Formininotransferase deficiency	Formininotransferase*
Fructose intolerance	Fructose 1-phosphate aldolase
Fructosuria	Hepatic fructokinase
Fucosidosis	α-L-Fucosidase
Galactokinase deficiency	Galactokinase
Galactose epimerase deficiency	Galactose epimerase
Galactosemia	Galactose 1-phosphate uridyl transferase
Gangliosidosis, generalized, type I GM	β-Galactosidase A, B, C
Gangliosidosis, GM_1, type II or juvenile form	β-Galactosidase B, C

(*Continued*)

Table 10-3. (*Continued*)

Condition	Enzyme with deficient activity[1]
Gangliosidosis, GM(3)	Acetylgalactosaminyl transferase
Gaucher disease	Glucocerebrosidase
Glycogen storage disease I	Glucose 6-phosphatase
Glycogen storage disease II	α-1,4-Glucosidase
Glycogen storage disease III	Amylo-1,6-glucosidase
Glycogen storage disease IV	Amylo-(1,4 to 1,6)-transglucosidase
Glycogen storage disease V	Muscle phosphorylase
Glycogen storage disease VI	Liver phosphorylase*
Glycogen storage disease VII	Muscle phosphofructokinase
Glycogen storage disease VIII	Liver phosphorylase kinase
Gout	Hypoxanthine guanine phosphoribosyltransferase
Gout	PPRP synthetase (increased activity)
Granulomatous disease	NADPH oxidase
Hemolytic anemia	Adenosine triphosphatase
Hemolytic anemia	Adenylate kinase
Hemolytic anemia	Aldolase A
Hemolytic anemia	Diphosphoglycerate mutase
Hemolytic anemia	γ-Glutamylcysteine synthetase
Hemolytic anemia	Glucose 6-phosphate dehydrogenase
Hemolytic anemia	Glutathione peroxidase
Hemolytic anemia	Glutathione synthetase
Hemolytic anemia	Hexokinase
Hemolytic anemia	Hexosephosphate isomerase
Hemolytic anemia	Phosphoglycerate kinase
Hemolytic anemia	Pyrimidine 5′-nucleotidase
Hemolytic anemia	Pyruvate kinase
Hemolytic anemia	Trisephosphate isomerase
Histidinemia	Histidase
Homocystinuria I	Cystathionine synthetase
Homocystinuria II	N^5N^{10}-methylenetetrahydrofolate reductase
β-Hydroxyisovaleric aciduria and methylcrotonylglysinuria	β-Methylocrotonyl CoA carboxylase*
Hydroxyprolinemia	Hydroxyproline oxidase
Hyperammonemia I	Ornithine transcarbamylase
Hyperammonemia II	Carbamyl phosphate synthetase
Hyperglycinemia, ketotic form	Propionyl CoA carboxylase*
Hyperglycinemia, nonketotic form	Glycine formininotransferase
Hyperlipoproteinemia, type I	Lipoprotein lipase
Hyperlysinemia	Lysine-ketoglutarate reductase
Hyperprolinemia I	Proline oxidase

Table 10-3. (*Continued*)

Condition	Enzyme with deficient activity[1]
Hyperprolinemia II	δ-1-Pyrroline-5-carboxylate dehydrogenase*
Hypoglycemia and acidosis	Fructose 1,6-diphosphatase
Hypophosphatasia	Alkaline phosphatase
Immunodeficiency disease	Adenosine deaminase
Immunodeficiency disease	Uridine monophosphate kinase
Intestinal lactase deficiency (adult)	Lactase
Isovalericacidemia	Isovaleric acid CoA dehydrogenase
Ketoacidosis, infantile	Succinyl CoA: 3-ketoacid CoA-transferase
Krabbe disease	A β-Galactosidase
Lactosyl ceramidosis	Lactosyl ceramidase
Leigh necrotizing encephalomyelopathy	Pyruvate carboxylase
Lipase deficiency, congenital	Lipase (pancreatic)
Lysine intolerance	L-lysine: NAD-oxido-reductase
Male pseudohermaphroditism	testicular 17,20-desmolase
Male pseudohermaphroditism	testicular 17-ketosteroid dehydrogenase*
Male pseudohermaphroditism	α-Reductase*
Mannosidosis	α-Mannosidase
Maple sugar urine disease	Keto acid decarboxylase
Metachromatic leukodystrophy	Arylsulfatase A (sulfatide sulfatase)
Methemoglobinemia	NAD-methemoglobin reductase
Methylmalonicaciduria I (B12-unresponsive)	Methylmalonic CoA mutase
Methylmalonicaciduria II (B12-responsive)	Deoxyadenosyl transferase*
Methylmalonicaciduria III	Methylmalonyl-CoA racemase
Mucopolysaccharidosis I	α-L-Iduronidase
Mucopolysaccharidosis II	Sulfo-iduronide sulfatase
Mucopolysaccharidosis IIIA	Heparan sulfate sulfatase
Mucopolysaccharidosis IIIB	*N*-Acetyl-α-D-glucosaminidase
Mucopolysaccharidosis IV	6-Sulfatase*
Mucopolysaccharidosis VI	Arylsulfatase B
Mucopolysaccharidosis VII	β-Glucuronidase
Myeloperoxidase deficiency with disseminated candidiases	Myeloperoxidase (leukocyte)
Niemann–Pick disease	Sphingomyelinase
Ornithinemia	Ornithine ketoacid amino-transferase
Oroticaciduria I	Orotidylic pyrophosphorylase and orotidylic decarboxylase

(*Continued*)

Table 10-3. *(Continued)*

Condition	Enzyme with deficient activity[1]
Oroticaciduria II	Orotidylic decarboxylase
Oxalosis I (glycolic aciduria)	2-Oxo-glutarate-glyoxylase carboligase
Oxalosis II (glyceric aciduria)	D-glycerate dehydrogenase
Pentosuria	Xylitol dehydrogenase (L-xylulose reductase)
Phenylketonuria	Phenylalanine hydroxylase
Porphyria, acute intermittent	Uroporphyrinogen I synthetase
Porphyria, congenital	Uroporphyrinogen III cosynthetase
Pulmonary emphysema and/or cirrhosis	α-1-Antitrypsin
Pyridoxine-dependent infantile convulsions	Glutamic acid decarboxylase
Pyridoxine-responsive anemia	δ-Aminolevulinic acid synthetase*
Pyruvate carboxylase	Pyruvate carboxylase
Refsum disease	Phytanic acid oxidase
Renal tubular acidosis with deafness	Carbonic anhydrase B
Richner–Hanhart syndrome	Tyrosine aminotransferase
Rickets, vitamin-D-dependent	25-Hydroxycholecalciferol*
Sandhoff disease (GM_2-gangliosidosis, type II)	Hexosaminidase A, B
Sarcosinemia	Sarcosine dehydrogenase*
Sulfite oxidase deficiency	Sulfite oxidase
Tay–Sachs disease	Hexosaminidase A
Thyroid hormonogenesis, defect in, II	Peroxidase*
Thyroid hormonogenesis, defect in, IV	Iodotyrosine dehalogenase (deiodinase)
Trypsinogen deficiency	Trypsinogen
Tyrosinemia I	Para-hydroxyphenylpyruvate oxidase
Tyrosinemia II	Tyrosine transaminase
Valinemia	Valine transaminase
Wolman disease	Acid lipase
Xanthinuria	Xanthine oxidase
Xanthurenic aciduria	Kynureninase
Xeroderma pigmentosum	Ultraviolet specific endonuclease
Xylosidase deficiency	Xylosidase

SOURCE: Victor A. McKusick, *Mendelian Inheritance in Man,* 4th ed., (Johns Hopkins University Press, 1975).

NOTE: In some conditions marked * (as well as some not listed), deficiency of a particular enzyme is suspected but has not been proved by direct study of enzyme activity.

[1]The form of gout due to increased activity of PPRP is the only disorder listed here that is due to *increased* enzyme activity.

Gene–Protein Relations

The one-gene–one-enzyme hypothesis is an impressive step forward in our understanding of gene function, but just *how* do genes control the functioning of enzymes? Enzymes belong to a general class of molecules called **proteins,** and we must review the basic facts of protein structure in order to follow the next step in the study of gene function.

Protein Structure

In simple terms, a protein is a macromolecule composed of **amino acids** attached end-to-end in a linear string. The general formula for an amino acid is H_2N–CHR–COOH, in which the R group can be anything from a hydrogen atom (as in the amino acid glycine) to a complex ring (as in the amino acid tryptophan). There are 20 common amino acids in living organisms (Table 10-4), each having a different R group. Amino acids are linked together in proteins by covalent (chemical) bonds called **peptide bonds.** A peptide bond is formed through a condensation reaction that involves removal of a water molecule (Figure 10-12).

Several amino acids linked together by peptide bonds form a molecule called a **polypeptide;** proteins are large polypeptides of the kinds formed in living organisms. The linear arrangement of amino acids in a polypeptide chain is called the **primary structure** of the protein. Figure 10-13 shows the primary structures of beef insulin (a hormonal protein) and tryptophan synthetase (an

Table 10-4. The 20 amino acids common in living organisms

Amino acid	Three-letter abbreviation
Alanine	Ala
Arginine	Arg
Asparagine	Asn
Aspartic acid	Asp
Asparagine, or aspartic acid	Asx
Cysteine	Cys
Glutamine	Gln
Glutamic acid	Glu
Glutamine, or glutamic acid	Glx
Glycine	Gly
Histidine	His
Isoleucine	Ile
Leucine	Leu
Lysine	Lys
Methionine	Met
Phenylalanine	Phe
Proline	Pro
Serine	Ser
Threonine	Thr
Tryptophan	Trp
Tyrosine	Tyr
Valine	Val

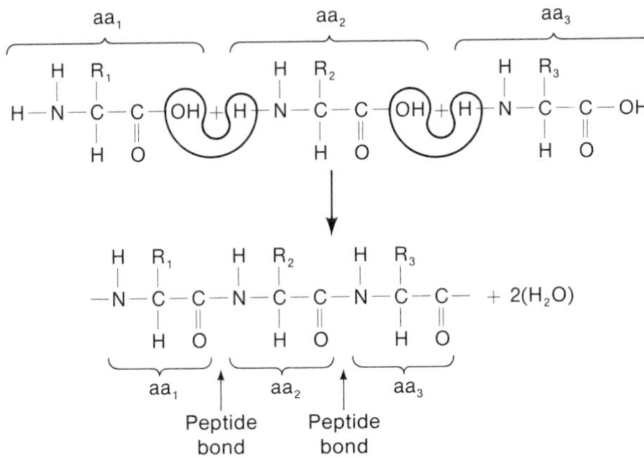

Figure 10-12. Formation of a polypeptide by the removal of water between amino acids to form peptide bonds. R_1, R_2, and R_3 represent side groups that differentiate the amino acids. R can be anything from a hydrogen atom (as in glycine) to a complex ring (as in tryptophan). Each aa indicates an amino acid.

Met-Glu-Arg-Tyr-Glu-Ser-Leu-Phe-Ala-Gln^{10}-Leu-Lys-Glu-Arg-Lys-Glu-Gly-Ala-Phe-Val^{20}-

Pro-Phe-Val-Thr-Leu-Gly-Asp-Pro-Gly-Ile^{30}-Glu-Gln-Ser-Leu-Lys-Ile-Ile-Asp-Thr-Leu^{40}-

Ile-Glu-Ala-Gly-Ala-Asp-Ala-Leu-Glu-Leu^{50}-Gly-Ile-Pro-Phe-Ser-Asp-Pro-Leu-Ala-Asp^{60}-

Gly-Pro-Thr-Ile-Gln-Asn-Ala-Thr-Leu-Arg^{70}-Ala-Phe-Ala-Ala-Gly-Val-Thr-Pro-Ala-Gln^{80}-

Cys-Phe-Glu-Met-Leu-Ala-Leu-Ile-Arg-Gln^{90}-Lys-His-Pro-Thr-Ile-Pro-Ile-Gly-Leu-Leu^{100}-

Met-Tyr-Ala-Asn-Leu-Val-Phe-Asn-Lys-Gly^{110}-Ile-Asp-Glu-Phe-Tyr-Ala-Gln-Cys-Glu-Lys^{120}-

Val-Gly-Val-Asp-Ser-Val-Leu-Val-Ala-Asp^{130}-Val-Pro-Val-Gln-Glu-Ser-Ala-Pro-Phe-Arg^{140}-

Gln-Ala-Ala-Leu-Arg-His-Asn-Val-Ala-Pro^{150}-Ile-Phe-Ile-Cys-Pro-Pro-Asn-Ala-Asp-Asp^{160}-

Asp-Leu-Leu-Arg-Gln-Ile-Ala-Ser-Tyr-Gly^{170}-Arg-Gly-Tyr-Thr-Tyr-Leu-Leu-Ser-Arg-Ala^{180}-

Gly-Val-Thr-Gly-Ala-Glu-Asn-Arg-Ala-Ala^{190}-Leu-Pro-Leu-Asn-His-Leu-Val-Ala-Lys-Leu^{200}-

Lys-Glu-Tyr-Asn-Ala-Ala-Pro-Pro-Leu-Gln^{210}-Gly-Phe-Gly-Ile-Ser-Ala-Pro-Asp-Gln-Val^{220}-

Lys-Ala-Ala-Ile-Asp-Ala-Gly-Ala-Ala-Gly^{230}-Ala-Ile-Ser-Gly-Ser-Ala-Ile-Val-Lys-Ile^{240}-

Ile-Glu-Gln-His-Asn-Ile-Glu-Pro-Glu-Lys^{250}-Met-Leu-Ala-Ala-Leu-Lys-Val-Phe-Val-Gln^{260}-

Pro-Met-Lys-Ala-Ala-Thr-Arg-Ser^{268}-

(a)

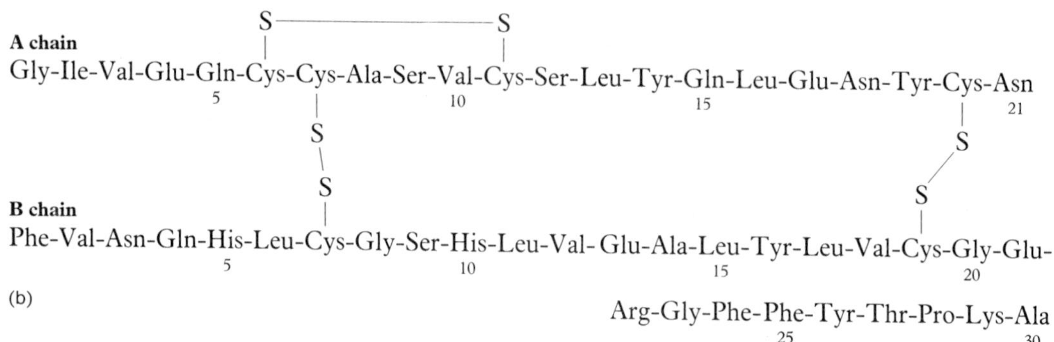

Figure 10-13. Primary sequences of the tryptophan synthetase A protein in E. coli *(a) and beef insulin protein (b). Note that the amino acid cysteine can form unique "sulfur bridges," since it contains sulfur.*

enzyme). Many of the side groups of amino acids attract or repel one another in the protein, resulting in a spiral or zigzag form called the **secondary structure** of the protein (Figure 10-14). The spiral may itself be folded to form a **tertiary structure** (Figure 10-15). In many cases, a number of folded structures can associate to form a **quaternary structure** that is **multimeric** (composed of several separate polypeptide monomers; see Figures 10-16 and 10-17). Many proteins (called **globular proteins**) are basically compact "blobs." Enzymes and antibodies are among the important globular proteins. Other proteins, unfolded (fibrous) proteins, are important components of such structures as hair and muscle.

Proteins act as biological catalysts, as hormones, and as structural elements in spindle fibers, hair, muscle, and so forth. Proteins play a very central role in living systems. If we purify a particular protein, we find that we can specify a particular ratio of the various amino acids for that specific protein. But the protein is not formed by a random hook-up of fixed amounts of the various amino acids—it also has a characteristic sequence. It is possible to take small polypeptides and determine their amino acid sequences by clipping off one amino acid at a time and identifying it. However, large polypeptides cannot be readily "sequenced" in this way.

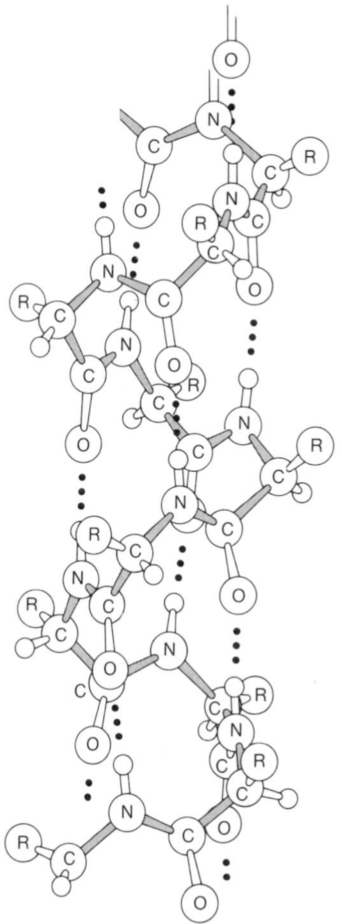

Figure 10-14. The helical formation that is a common basis of secondary protein structure. The backbone of the protein can be seen as the shaded lines. Each R is a specific side group on one amino acid. Dots are weak stabilizing bonds, which maintain the helical shape. (Reprinted from Linus Pauling, The Nature of the Chemical Bond. *Copyright 1939 and 1940 by Cornell University. Third edition © 1960 by Cornell University. Used by permission of Cornell University Press.)*

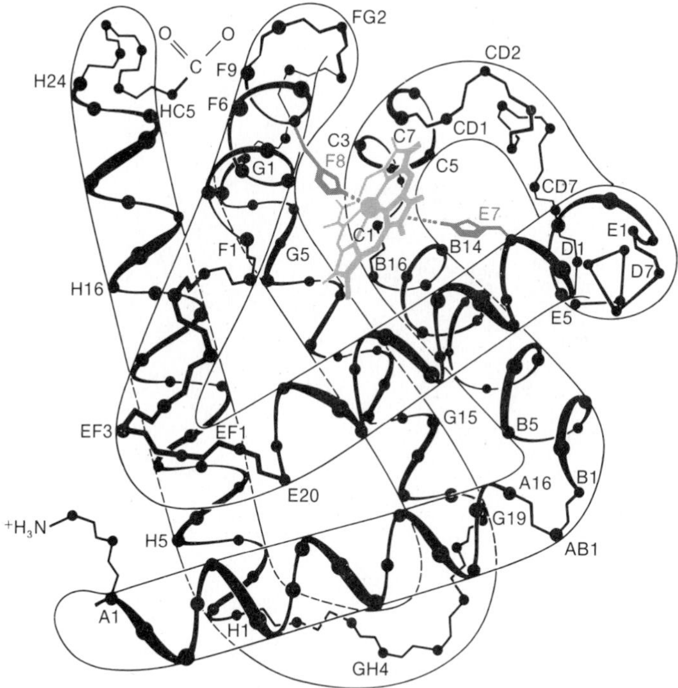

Figure 10-15. Folded tertiary structure of myoglobin, an oxygen-storage protein. Each dot represents an amino acid. (From L. Stryer, Biochemistry, *2nd ed. Copyright © 1981 by L. Stryer. Based on R. E. Dickerson. In* The Proteins, *H. Neurath, ed., 2nd ed., vol. 2. Academic Press, 1964.)*

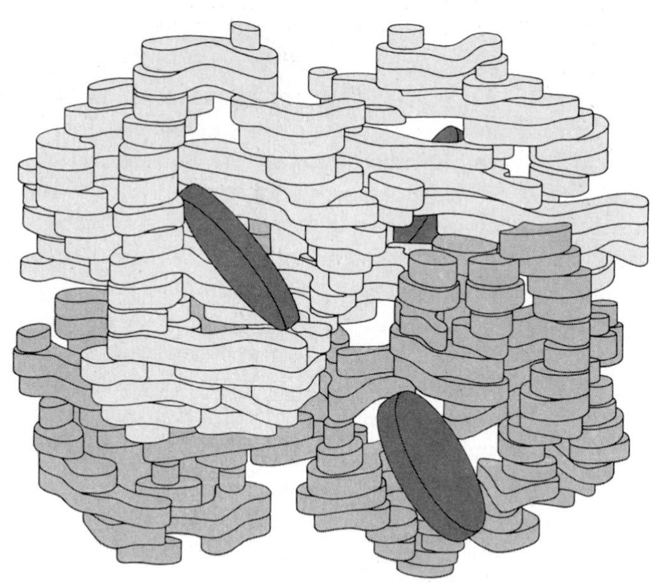

Figure 10-16. *A model of the hemoglobin molecule (shown as contoured layers to provide a visualization of the three-dimensional shape). The different shadings indicate the different polypeptide chains that combine to form the quaternary structure of the protein. The disks are heme groups, complex structures containing iron. (After M. F. Perutz, "The Hemoglobin Molecule." Copyright © 1964 by Scientific American, Inc. All rights reserved.)*

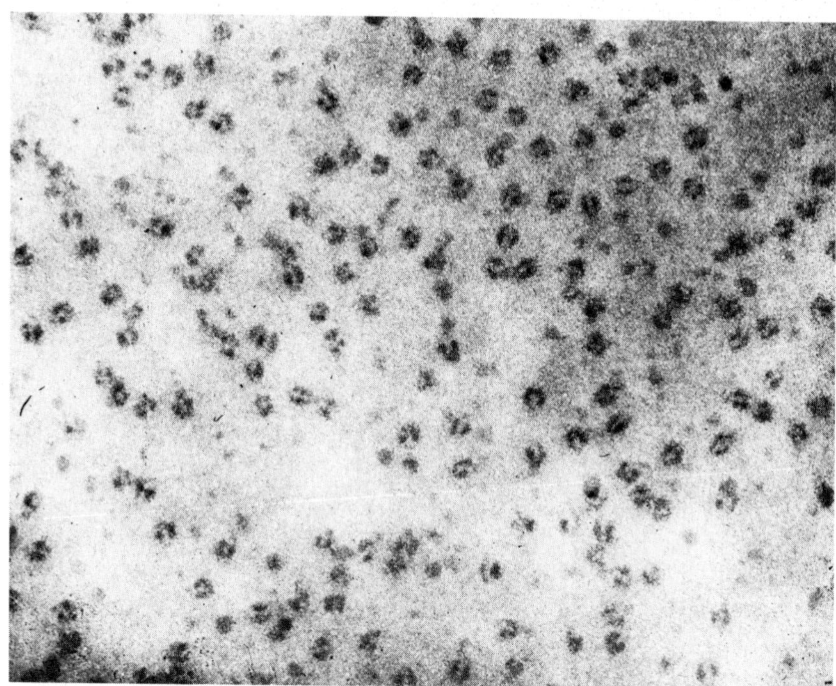

Figure 10-17. *Electron micrograph of the enzyme aspartate transcarbamylase. Each small "glob" is an enzyme molecule. Note the quaternary structure; the enzyme is composed of subunits. (From Jack D. Griffith.)*

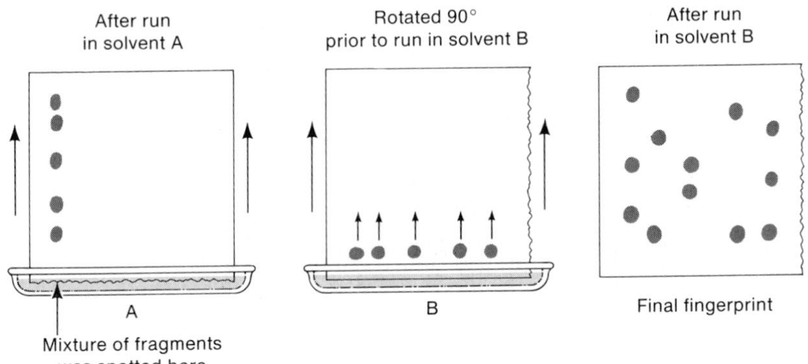

Figure 10-18. Two-dimensional chromatographic fingerprinting of a polypeptide fragment mixture. A protein is digested by a proteolytic enzyme into fragments that are only a few amino acids long. A piece of chromatographic filter paper is then spotted with this mixture and dipped into solvent A. As solvent A ascends the paper, some of the fragments are separated. The paper is then turned 90° and, as solvent B ascends, further resolution of the fragments is obtained.

Frederick Sanger worked out a brilliant method for deducing the sequence of large polypeptides. There are several different **proteolytic enzymes**—enzymes that can break peptide bonds only between specific amino acids in proteins. Thus a large protein can be broken by such enzymes into a number of smaller fragments. These fragments can be separated according to their migration speeds in a solvent on chromatographic paper. Because the speeds of mobility of different fragments may vary differently in various solvents, two-dimensional chromatography can be used to enhance the separation of the fragments (Figure 10-18). In this technique, the mixture is separated in one solvent; then the paper is turned 90° and another solvent used.

When the paper is stained, the polypeptides appear as spots in a characteristic chromatographic pattern called the **fingerprint** of the protein. Each of the spots can be cut out and the polypeptide fragments washed from the paper. Because each spot contains only small polypeptides, their amino acid sequences can easily be determined. Using different proteolytic enzymes to cleave the protein at different points, we can repeat the experiment to obtain other sets of fragments. The fragments from the different treatments will overlap (because the breaks have been made in different places with each treatment). The problem of solving the overall sequence then becomes one of fitting together the small-fragment sequences—almost like solving a tricky jigsaw or crossword puzzle (Figure 10-19).

Using this elegant technique, Sanger confirmed that the sequence of amino acids (as well as the amounts of the various amino acids) is specific to a particular protein. In other words, the amino acid sequence is what makes insulin insulin.

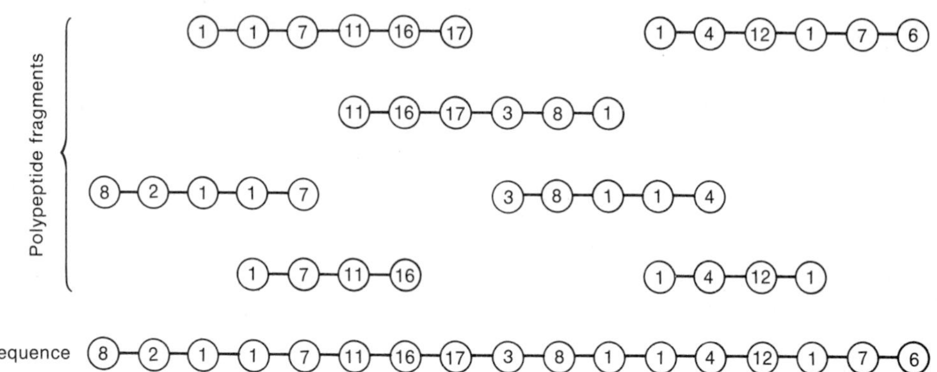

Figure 10-19. Alignment of polypeptide fragments to reconstruct an entire amino acid sequence. Different proteolytic enzymes can be used on the same protein to form different fingerprints. The amino acid sequence of each spot can be determined rather easily, and, because of overlap of amino acid sequences from different spots from different fingerprints, the entire amino acid sequence of the original protein can be determined. Using this procedure, it took Sanger about six years to determine the sequence of the insulin molecule, a relatively small protein.

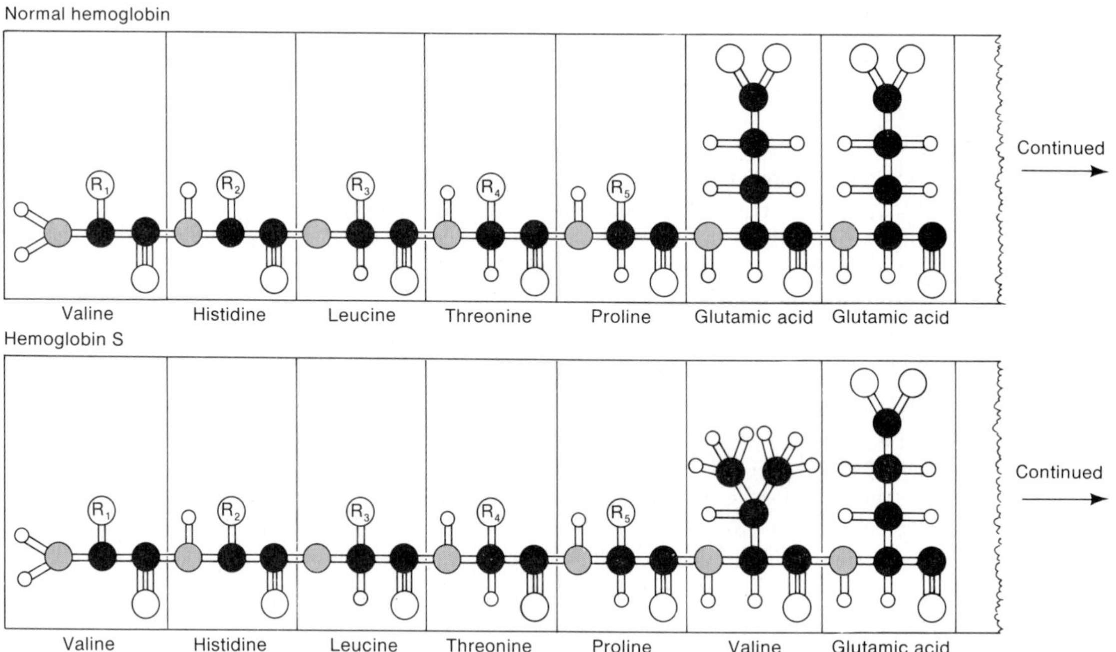

Figure 10-20. The difference at the molecular level between normalcy and sickle-cell disease. Shown are only the first seven amino acids; all the rest not shown are identical. (From "Cyanate and Sickle-Cell Disease" by Anthony Cerami and Charles M. Peterson. Copyright © 1975 by Scientific American, Inc. All rights reserved.)

We can now examine the results of some studies by Vernon Ingram (in 1957) on the globular protein hemoglobin—the molecule that transports oxygen in red blood cells. Ingram compared hemoglobin (HbA) from normal people with hemoglobin (HbS) from people homozygous for the mutant gene that causes sickle-cell anemia. Using Sanger's technique, he found that the fingerprint of HbS differs from that of HbA in only one spot. Sequencing that spot from the two kinds of hemoglobin, Ingram found that only one amino acid in the fragment differs in the two kinds. Apparently, of the 150 amino acids known to make up a hemoglobin molecule, a substitution of valine for glutamic acid at just one point in the molecule is all that is needed to produce the defective hemoglobin (Figure 10-20). Unless patients with HbS receive medical attention, this single error in an amino acid in one protein will cause their death. Figure 10-21 shows the sequence of events leading to the disease.

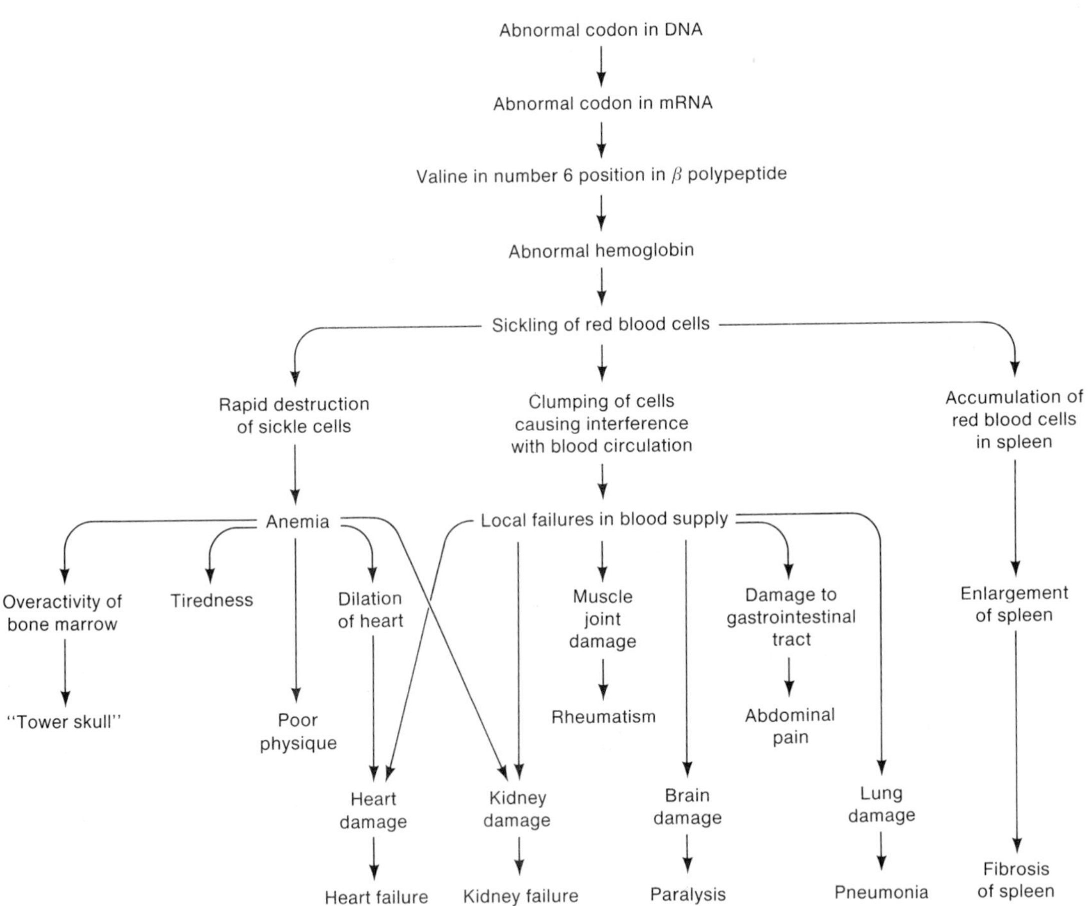

Figure 10-21. The compounded consequences of one amino acid substitution in hemoglobin to produce sickle-cell anemia. The terms codon *and* mRNA *will be explained later.*

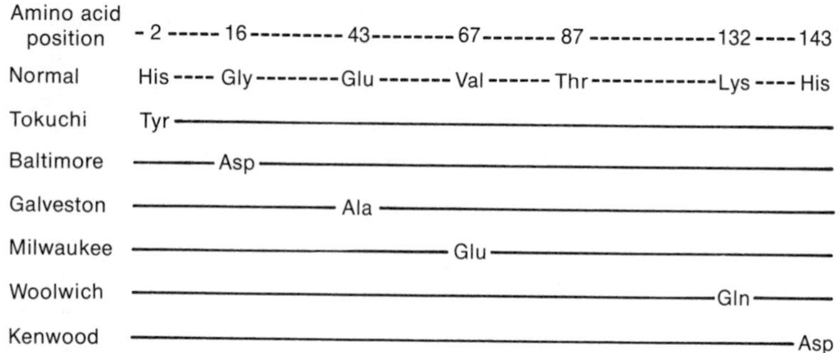

Figure 10-22. A variety of single amino acid substitutions in human hemoglobin. All amino acids except those indicated are normal. Each type of change causes disease. (Names indicate areas where cases were first identified.)

Notice what Ingram had accomplished. A gene mutation that is well established through genetic studies has been connected to an altered amino acid *sequence* in a protein. Subsequent studies have identified numerous changes in hemoglobin, and each is the consequence of a single amino acid difference in the entire chain of 150 amino acids. Figure 10-22 shows a few examples. We conclude that one mutation in a gene corresponds to a change of one amino acid in the sequence of a protein.

Message
Genes determine the primary sequences of amino acids in specific proteins.

Colinearity of Gene and Protein

Is there any relationship between the linear sequence of mutant sites in a gene (inferred from genetic analysis) and the linear sequence of amino acids in a protein (determined biochemically)? Let's consider the work done in the 1960s by Charles Yanofsky, who studied the enzyme tryptophan synthetase in *E. coli.* The enzyme catalyzes the conversion of indoleglycerol phosphate into tryptophan. The gene controlling the enzyme is called *trp*; it is composed of two cistrons, A and B. Each cistron controls a separate polypeptide; after the A and B polypeptides are produced, they combine to form the active enzyme (a multimeric protein). We concentrate here on the A cistron.

Yanofsky isolated many A mutants, all of which required tryptophan in order to grow. He carefully performed genetic mapping studies by intercrossing the mutants and measuring the frequency of trp^+ protrophs produced. Eventually he obtained a cistron map. He also examined the A polypeptide produced by each mutant. His results were similar to those of Ingram for hemoglobin: each

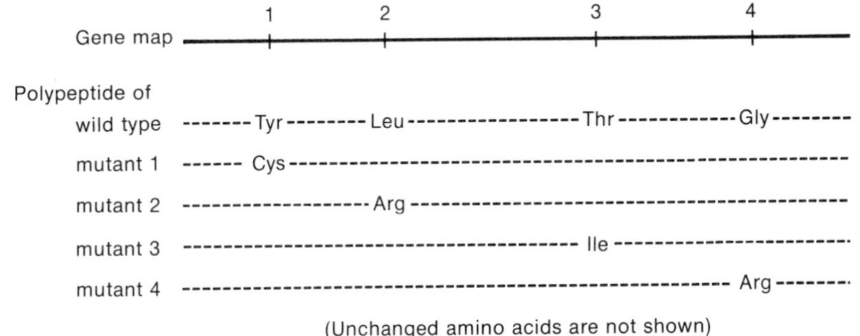

Figure 10-23. Simplified representation of colinearity of gene mutations. The genetic map of point mutations (determined by recombinational analysis) corresponds linearly to the changed amino acids in the different mutants (determined by fingerprint analysis).

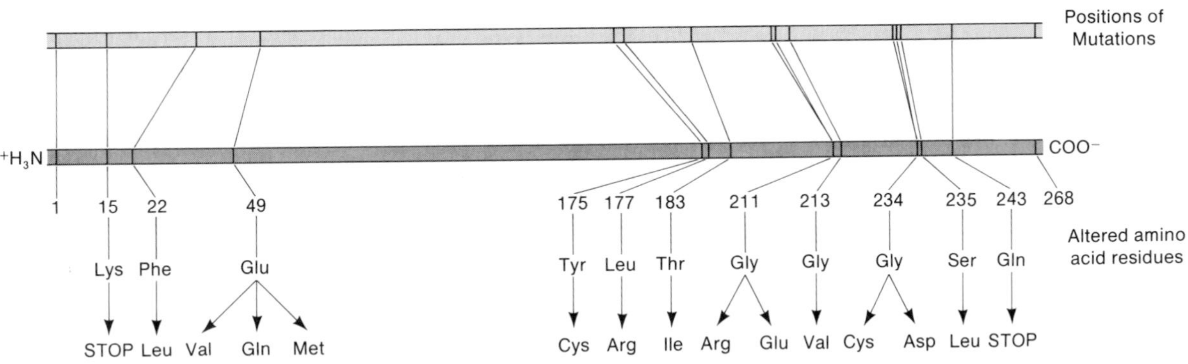

Figure 10-24. Actual colinearity shown in the A protein of tryptophan synthetase from E. coli. *There is a linear correlation between the position of mutations and the altered amino acid residues. (Based on C. Yanofsky, "Gene Structure and Protein Structure." Copyright © 1967 by Scientific American. All rights reserved.)*

mutation has a defective polypeptide associated with a specific amino acid substitution at a specific point. However, Yanofsky was able to show an exciting correlation that Ingram had not observed. There is an exact match between the sequence of the mutant sites in the gene map of the A cistron and the location of the corresponding altered amino acids in the A polypeptide chain. The farther apart two mutant sites are in map units, the more amino acids there are between the corresponding substitutions in the polypeptide (Figure 10-23). Thus, Yanofsky demonstrated **colinearity** between a gene and its corresponding polypeptide. Figure 10-24 shows the complete set of data.

Message
The primary structure of a protein is a direct reflection of the linear structure of the gene.

It seems, therefore, that each amino acid is somehow "coded" by a specific region of the gene. Let's call such regions **codons,** and we shall see how they correspond to the mutons or recons we have already defined.

In 1967, Yanofsky made further analyses that added to our understanding of the structure of the gene. He crossed two *trp-A*$^-$ mutants (A23 and A46). These two mutant sites map very close together on the genetic map, and each involves an amino acid substitution at position 210 of the polypeptide (Figure 10-25). He plated the progeny of this cross on a medium that does not contain tryptophan; he found that about 0.002% of the progeny grew, indicating that they were *trp-A*$^+$ recombinants. Yanofsky extracted enzyme from the recombinants and determined the amino acid present at position 210. As Figure 10-25 shows, the recombinants matched the original wild type.

This evidence demonstrates that the codon (the piece of gene coding for an amino acid) is composed of at least two recons—because we have just demonstrated recombination within the codon. Figure 10-26 shows a chromosomal or genic diagram of the cross that may make the concept a little clearer. The glycine codon is obtained from an arginine and a glutamic acid codon by crossing over within the codons. We also know that this codon corresponds to at least two mutons, because the two mutant sites studied here map within the codon and affect the same amino acid position in the polypeptide.

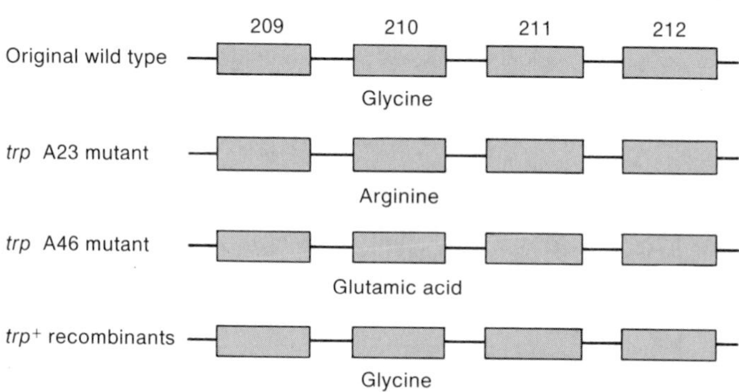

Figure 10-25. Intracodon recombination at codon 210. The mutants trp *A23 and* trp *A46 each have a different amino acid at position 210, and each protein is nonfunctional. When these two mutants are mated, however, an occasional* trp$^+$ *progeny individual occurs with functional tryptophan synthetase A protein. This occurrence indicates that genetic recombination can occur within the coding unit for one amino acid.*

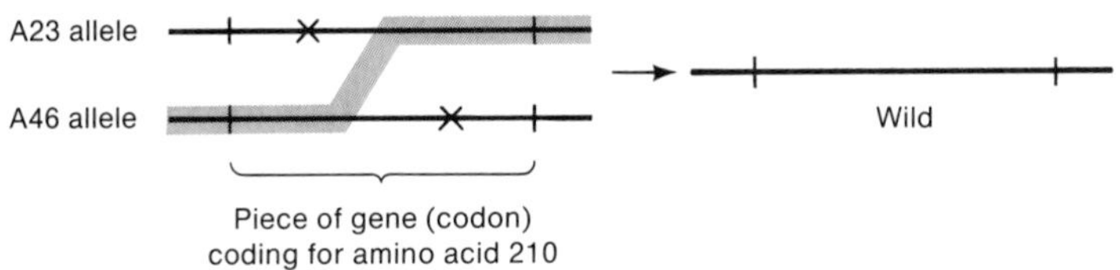

Figure 10-26. Intracodon recombination at codon 210. (Reciprocal not detected; × = mutant site.)

Message
Each codon is composed of at least two recons and at least two mutons.

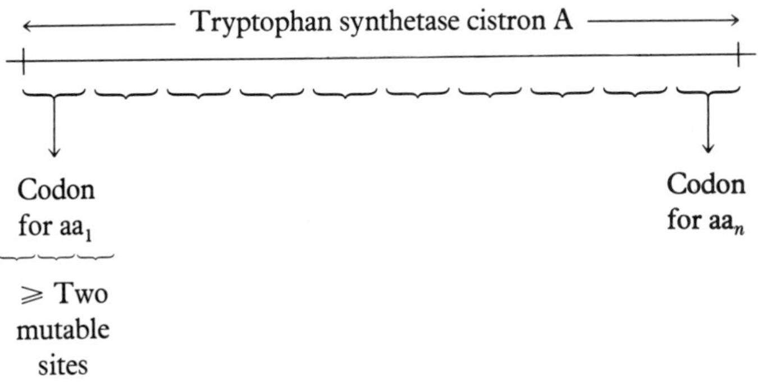

We now know that a cistron is a region of the genetic material that codes for one polypeptide chain. We could express the one-gene–one-enzyme hypothesis more precisely as the **one-cistron–one-polypeptide hypothesis,** thus also emphasizing that cistrons (genes) can code for proteins other than enzymes (as we shall see). Further clarification of the natures of codons, mutons, and recons will come in following chapters.

Protein Function

The genes truly are the master controllers of the cell. They not only dictate cell chemistry through the enzymes encoded by some genes, but they also dictate biological architecture through the structural proteins encoded by other genes. Furthermore, the blueprints of such important proteins as hormones and hemoglobin are encoded in the structure of genes.

How can a single amino acid substitution have such a profound effect on protein function? Take enzymes, for example. Enzymes are known to do their job of catalysis by physically grappling with their substrate molecules, twisting or bending the molecules to make or break chemical bonds. Figure 10-27

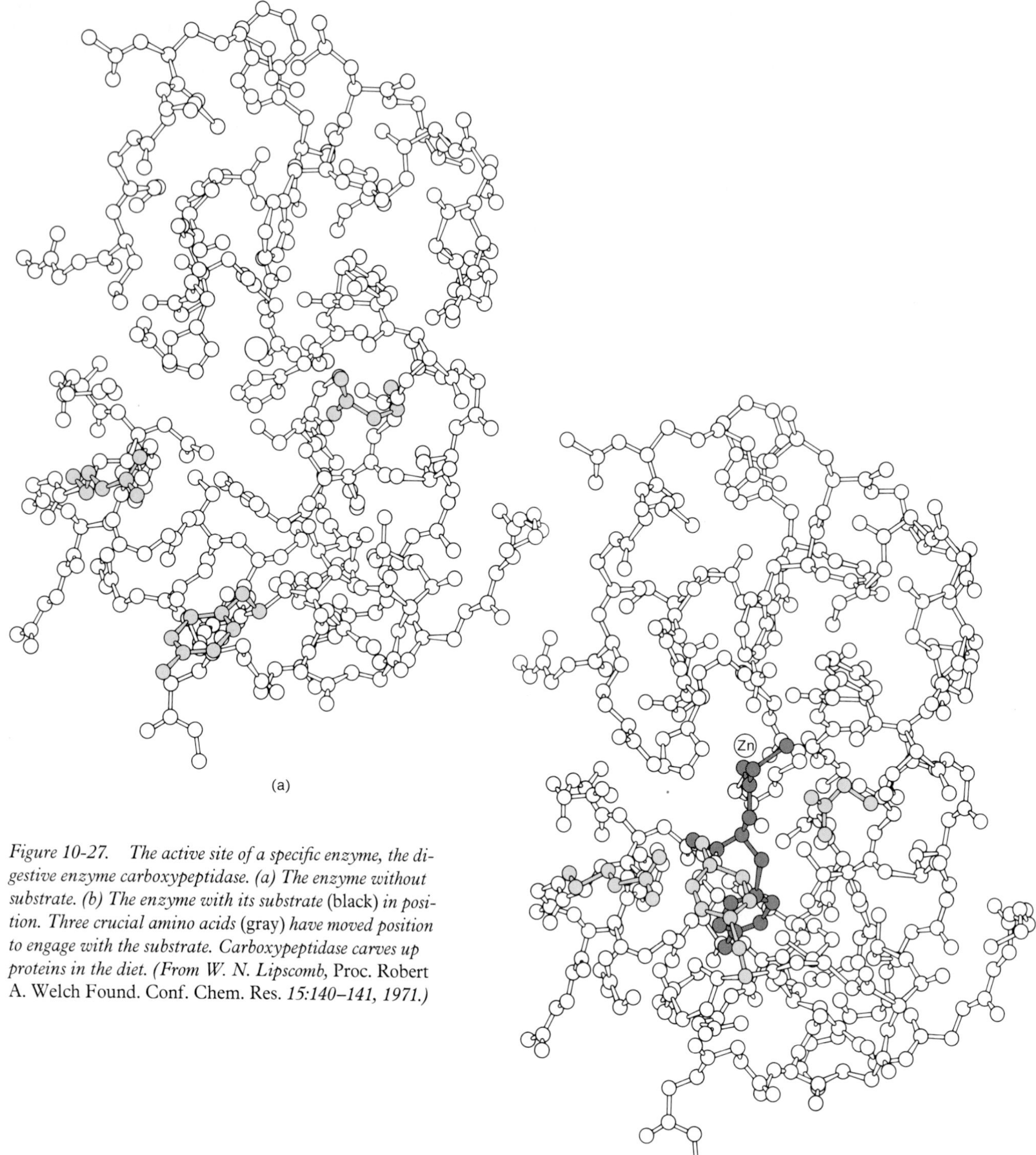

Figure 10-27. The active site of a specific enzyme, the digestive enzyme carboxypeptidase. (a) The enzyme without substrate. (b) The enzyme with its substrate (black) *in position. Three crucial amino acids* (gray) *have moved position to engage with the substrate. Carboxypeptidase carves up proteins in the diet. (From W. N. Lipscomb,* Proc. Robert A. Welch Found. Conf. Chem. Res. *15:140–141, 1971.)*

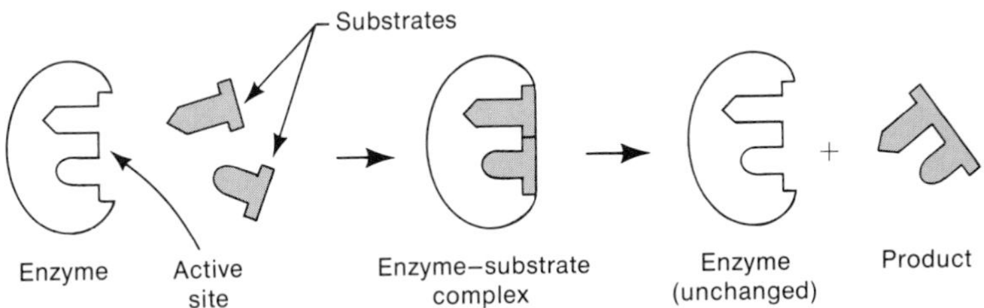

Figure 10-28. Diagrammatic representation of the action of a hypothetical enzyme in putting two substrate molecules together. The "key-in-lock" fit of the substrate into the enzyme's active site is very important in this model.

shows the gastric digestion enzyme carboxypeptidase in its relaxed position and after grappling with its substrate molecule, glycyltyrosine. The substrate molecule fits into a notch in the enzyme structure; this notch is called the **active site.** Figure 10-28 diagrams the general concept. (Note that here we have encountered the two basic types of reactions performed by enzymes—breakdown of substrate into simple products and synthesis of a complex product from simpler substrate(s).)

Much of the "blob" structure of an enzyme is nonreactive material that simply supports the active site. We might expect that amino acid substitutions through most of the structure would have little effect, but that very specific amino acids are required for the part of the enzyme molecule that gives the precise shape to the active site. Hence the possibility arises that a functional enzyme would not require a *unique* amino acid sequence for the entire polypeptide.

Yanofsky did an experiment that bears on this question. Starting with mutant A23 (arginine at position 210), he detected "back mutations" by plating many A23 cells on medium with no tryptophan. These back mutants (revertants) were examined to determine the amino acid at position 210 in their enzymes. Some did have the expected (wild-type) glycine at position 210, but others had such amino acids as threonine or serine. It is clear that position 210 can be filled by several alternative amino acids that are compatible with the enzyme function. At certain other positions in the polypeptide, only the wild-type amino acid will restore activity; most likely, these amino acids form critical parts of the active sites. In Figure 10-27, some of these critical amino acids are indicated by gray shading.

Message

Protein architecture is the key to gene function. A gene mutation typically results in substitution of a different amino acid into the polypeptide sequence of a protein. The new amino acid may have different chemical properties that are incompatible with the proper protein architecture at that particular position; in such a case, the mutation will lead to a nonfunctional protein.

Complex Complementation

The one-cistron–one-polypeptide hypothesis, as we have developed it in this chapter, was of paramount importance (and still is) in drawing attention to the fact that a gene can include several different functional subunits (cistrons). However, complementation tests sometimes give results that cannot be interpreted neatly in terms of the cistron concept.

Let's examine an example and see how the confusion arises. Two adjacent regions called *ad-3A* and *ad-3B* can be recognized in the *ad-3* (adenine-3) locus of *Neurospora*. All *ad-3A* mutants complement all *ad-3B* mutants in heterokaryons; *ad-3A* mutants map genetically at one end of the locus and *ad-3B* mutants at the other end; and *ad-3A* mutants affect one enzyme function whereas *ad-3B* mutants affect another (the enzymes represent two adjacent steps in the synthesis of adenine). No *ad-3A* mutant ever exhibits complementation with any other *ad-3A* mutant, so we seem to have a perfect pair of cistrons in the original Benzer definition. However, some (but not all) *ad-3B* mutants do complement *some* other *ad-3B* mutants!

Is it possible that *ad-3B* actually represents two or more cistrons? Let's look at some typical complementation data. The following grid shows the results of testing four *ad-3B* alleles in all possible combinations of heterokaryon pairing (+ indicates growth of the heterokaryon on minimal medium—that is, complementation):

	1	2	3	4
1	−	+	−	−
2	+	−	+	−
3	−	+	−	+
4	−	−	+	−

Can we make any sense of these data? There is no way to define separate cistrons that will account for them. The best we can do is to draw a **complementation map,** in which failure to complement is indicated by overlapping "bars":

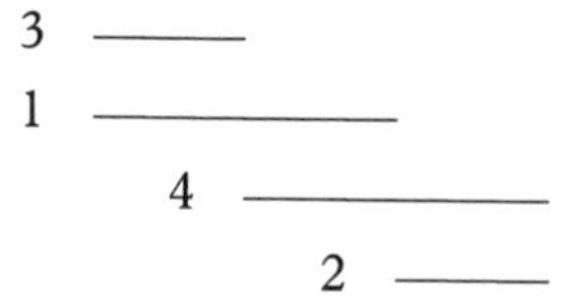

```
3 ———————
1 ——————————————
        4 ——————————————
                2 ———————
```

Mutants 3 and 1 do not complement each other, and mutants 2 and 4 do not complement each other. But we cannot define two separate cistrons here because 1 and 4 also fail to complement each other. The possibility that these mutants represent small deletions is ruled out by showing that the mutants are easily revertible. (The complementation map looks like a deletion map, but they should not be confused even though they are derived by the same principle.)

This form of behavior has been observed in many genetic regions of many organisms, so it is not a rare exception. What does the complementation map mean? The order of mutants in a complementation map (in our example,

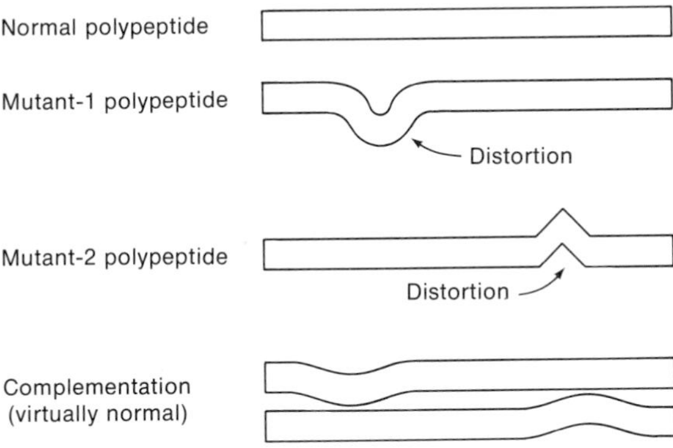

Figure 10-29. One possible mechanism of intracistronic complementation. Distorted portions of a polypeptide chain, caused by mutant substitutions of improper amino acids, may be mutually "propped up" if the functional enzyme is multimeric (has a quarternary structure). Such "propping up" could only occur to yield a functional enzyme if the distorted areas do not overlap.

3–1–4–2) is usually the same as the order found in the recombination-based allele map. The generally accepted explanation of this phenomenon is that a bar in a complementation map represents a region of a polypeptide where distortion has been created by substitution of an amino acid in the mutant. Two distorted polypeptides can come together, and, if the distortions do not overlap, they can complement by "propping up" each other's distorted region (Figure 10-29). As this hypothesis would predict, this kind of complementation is observed only in proteins that normally have a quaternary structure (multimeric proteins). Loci coding for monomeric proteins do behave as proper cistrons.

In some systems, complementation products have been obtained in vitro by mixing extracts from both mutants. Not only is protein (enzyme) function detected, but a hybrid molecule can be detected by using the appropriate techniques.

The term *cistron* cannot be applied to a genetic region showing a complex complementation map. What shall we call such regions? There is no official term, so we just use the word *region.*

Message
The correspondence between one cistron and one polypeptide is a useful general rule, but complementation between two mutations in the same polypeptide can occur in proteins made up of two or more polypeptide chains (multimeric proteins).

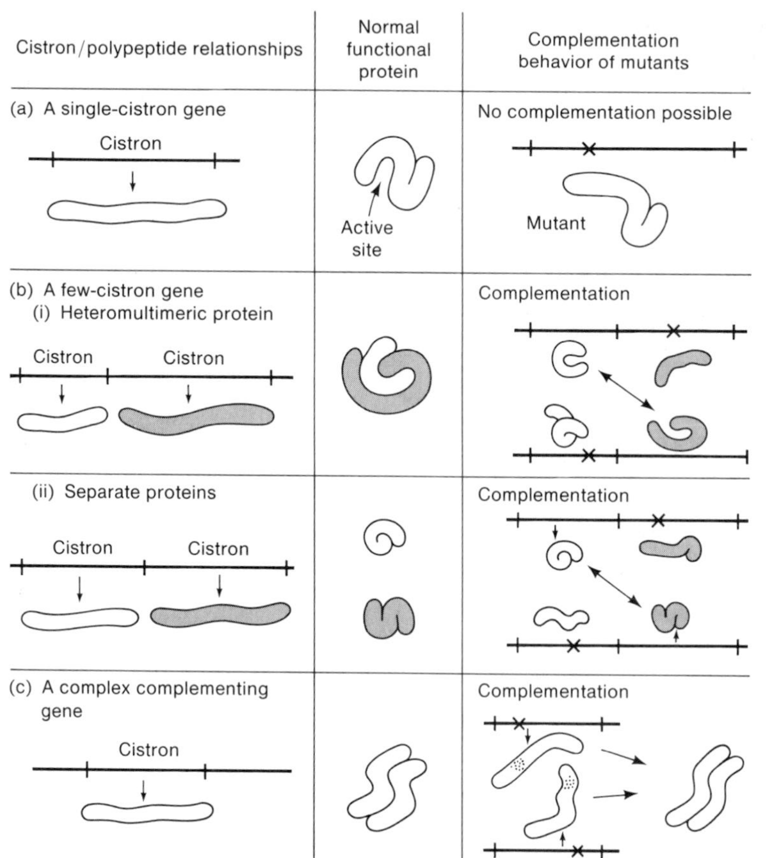

Figure 10-30. Complementation behavior in different types of genes.

The real world is so complex! We have now described three kinds of genes that show different patterns of intragenic complementation, but we have been able to explain each kind in terms of protein structure. The three gene types are

1. single-cistron genes (showing no complementation at all);
2. few-cistron genes (showing simple complementation patterns); and
3. genes showing complex complementation patterns.

Figure 10-30 summarizes the molecular explanations of these gene types.

Temperature-Sensitive Alleles

Recall that some mutants appear to be wild type at normal temperatures but can be detected as mutants at high or low temperature. We can now explain such mutations by assuming that a substitution of an amino acid produces a protein that is functional at normal (**permissive**) temperatures but is distorted and nonfunctional at high or low (**restrictive**) temperatures. Figure 10-31 shows a heat-sensitive example.

As we have seen, conditional mutations such as temperature-sensitive mutations can be very useful to geneticists. Stocks of the mutant culture can easily be maintained under permissive conditions, and the mutant phenotype can be studied intensively under restrictive conditions. Such mutants can be very useful in the genetic dissection of biological systems. For example, with a temperature-sensitive allele, the time at which a gene is acting can be determined by shifting to restrictive temperature at various times during development.

Normal protein

Mutant that is functionally normal at 25°C

Mutant at 37°C

Figure 10-31. Diagram of protein conformational distortion, probably the basis for temperature sensitivity in certain mutants. An amino acid substitution that has no significant effect at normal (permissive) temperatures may cause significant distortion at abnormal (restrictive) temperatures.

Genetic Dissection of Genes

This chapter provides a good indication of the power of genetic analysis. Using the delicate forceps and scalpel of selective systems, we have used fine-structure mapping to pick away at the bead theory of gene structure until it crumbled. Out of the results, we have constructed a much deeper understanding of the structure and function of the gene. Thus we have genetically dissected the gene; biochemical studies have confirmed many of the conclusions reached through genetic studies. In following chapters we shall use similar techniques to dissect other cell structures and functions.

Message
Genetic analysis can be used to dissect the processes of heredity itself, or it can be used to study any other cellular or organismal process.

Summary

The work of Beadle and Tatum in the 1940s showed that one gene codes for one protein. Further work by Benzer and others illustrated that the gene could be dissected into smaller and smaller pieces. A structural unit of mutation (called a muton) and one of recombination (called a recon) were identified. A cistron is defined at the phenotypic level as a genetic region within which there is no complementation between mutations. This is the unit that codes for the structure of a single functional polypeptide.

The failure of an enzyme to function normally because of a mutation yields a variant phenotype. These variant phenotypes are often the basis of genetic disease in any organism, including humans. In order to understand how abnormal enzymes can cause phenotypic change, we need to understand the structure of proteins. Composed of a specific linear sequence of amino acids connected through peptide bonds, the proteins assume specific three-dimensional shapes as a result of the interaction of the twenty amino acids that in different combinations constitute the polypeptide chain. Different areas of this folded chain are sites for the attachment and interaction of substrates. Furthermore, many functional enzymes and other proteins are built by combining, in multimeric form, several polypeptide chains.

Specific amino acid changes can be detected in a protein by the technique of fingerprinting and amino acid sequencing. Work of this type has demonstrated colinearity between mutant sites on the genetic map and the positions of altered amino acids in a protein. Finally, and of importance to work to be discussed in the succeeding chapters, it has been shown that recombination takes place within a codon (a coding unit for one amino acid.)

In contrast to the early days of genetics, when genes were represented as indivisible beads on a chain, we have now arrived at a far more complex picture of the gene. Multiple mutable sites exist, and recombination may occur anywhere within a gene. In addition, a closer connection between genotype and phenotype was realized when it was established that one cistron is responsible for the synthesis of one polypeptide.

Problems

1. A common weed, Saint-John's-wort, is toxic to albino animals. It also causes blisters on animals that have white areas of fur. Suggest a possible genetic basis for this reaction.

2. In humans, the disease galactosemia causes mental retardation at an early age because lactose in milk cannot be broken down, and this failure affects brain function. How would you provide a secondary cure for galactosemia? Would you expect this phenotype to be dominant or recessive?

3. Amniocentesis is a technique in which a hypodermic needle is inserted through the abdominal wall of a pregnant woman into the amnion, the sac that surrounds the developing embryo, to withdraw a small amount of amniotic fluid. This fluid contains cells that come from the embryo (not from the woman). The cells can be cultured; they will divide and grow to form a population of cells on which enzyme analyses and karyotype analyses can be performed. Of what use would this technique be to a genetic counselor? Name at least three specific conditions under which amniocentesis might be useful. (NOTE: This technique involves a small but real risk for the health of both woman and embryo; take this fact into account in your answer.)

4. Table 10-5 shows the ranges of enzyme activity (in units we need not worry about) observed for enzymes involved in two recessive metabolic diseases of humans. Similar information is available for many metabolic genetic diseases.

Table 10-5.

Disease	Enzyme involved	Range of enzyme activity: Patients	Parents of patients	Normal individuals
Acatalasemia	Catalase	0	1.2–2.7	4.3–6.2
Galactosemia	Gal-1-P uridyl transferase	0–6	9–30	25–40

a. Of what use is such information to a genetic counselor?

b. Indicate any possible sources of ambiguity in interpreting studies of an individual patient.

c. Reevaluate the concept of dominance in the light of such data.

5. Two albinos marry and have a normal child. How is this possible? Suggest at least two ways. (This question appeared first in Chapter 4. Reconsider it now in the light of biochemical pathways.)

6. In humans, PKU (phenylketonuria) is a disease caused by an enzyme inefficiency at step A in the following simplified reaction sequence, and AKU (alkaptonuria) is due to an enzyme inefficiency in one of the steps summarized as step B here:

$$\text{Phenylalanine} \xrightarrow{A} \text{Tyrosine} \xrightarrow{B} CO_2 + H_2O$$

A person with PKU marries a person with AKU. What phenotypes do you expect for their children? (a) All normal; (b) All having PKU only; (c) All having AKU only; (d) All having both PKU and AKU; (e) Some having AKU and some having PKU.

7. Three independently isolated tryptophan-requiring strains of yeast are called trpB, trpD, and trpE. Cell suspensions of each are streaked on a plate supplemented with just enough tryptophan to permit weak growth for a *trp*⁻ strain. The streaks are arranged in a triangular pattern so that they do not touch one another. Luxuriant growth is noted at both ends of the trpE streak and at one end of the trpD streak (Figure 10-32).

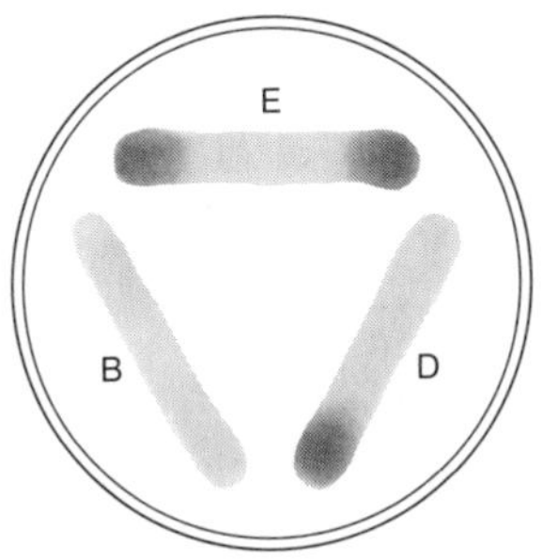

Figure 10-32.

a. Do you think complementation is involved?

b. Explain briefly the patterns of luxuriant growth.

c. In what order in the tryptophan-synthesizing pathway are the enzymic steps defective in trpB, trpD, and trpE?

d. Why was it necessary to add a small amount of tryptophan to the medium in order to demonstrate such a growth pattern?

8. In *Drosophila* pupae, certain structures called imaginal disks can be detected as thickenings of the skin; after metamorphosis, these imaginal disks develop into specific organs of the adult fly. George Beadle and Boris Ephrussi devised a means of transplanting eye imaginal disks from one larva into another larval host. When the host metamorphoses into an adult, the transplant can be found as a colored eye located in its abdomen. They took two strains of flies that were phenotypically identical in having bright scarlet eyes: one because of the sex-linked mutant vermillion (*v*); the other because of cinnabar (*cn*) on chromosome 2. If *v* disks are transplanted into *v* hosts or *cn* disks into *cn* hosts, then the transplants develop as mutant scarlet eyes. Transplanted *cn* or *v* disks in wild-type hosts develop wild-type eye colors. A *cn* disk in a *v* host develops a mutant eye color, but a *v* disk in a *cn* host develops wild-type eye color. Explain these results and outline the experiments you would propose to test your explanation.

9. In *Drosophila,* the autosomal recessive *bw* causes a dark brown eye, and the unlinked autosomal recessive *st* causes a bright scarlet eye. A homozygote for both genes has a white eye. Thus we have the following correspondences between genotypes and phenotypes:

$+/+ \quad +/+ =$ red eye (wild type)
$+/+ \quad bw/bw =$ brown eye
$st/st \quad +/+ =$ scarlet eye
$st/st \quad bw/bw =$ white eye

Construct a hypothetical biochemical pathway showing how the gene products interact and why the different mutant combinations have different phenotypes.

10. Several mutants are isolated, all of which require compound G for growth. The compounds (A through E) in the biosynthetic pathway are known, and each compound is tested for its ability to support the growth of each mutant (1 through 5). In the following table, + indicates growth and – indicates no growth:

		A	B	C	D	E	G
Mutant	1	–	–	–	+	–	+
	2	–	+	–	+	–	+
	3	–	–	–	–	–	+
	4	–	+	+	+	–	+
	5	+	+	+	+	–	+

a. What is the order of compounds A through E and G in the pathway?

b. At which point in the pathway is each mutant blocked?

c. Would a heterokaryon composed of double mutant 1,3 plus double mutant 2,4 grow on minimal medium? 1,3 plus 3,4? 1,2 plus 2,4 plus 1,4?

11. In *Neurospora* (a haploid), assume that two genes participate in the synthesis of valine. Their mutant alleles are called *val-1* and *val-2,* and their wild-type alleles

are called *val-1*$^+$ and *val-2*$^+$. These two genes are linked on the same chromosome, and a crossover occurs between them on the average in one of every two meioses.

a. In what proportion of meioses are there no crossovers between the genes?

b. Use the map function to determine the recombinant frequency between these two genes.

c. Progeny from the cross *val-1 val-2*$^+$ $\times$ *val-1*$^+$ *val-2* are plated on medium containing no valine. What proportion of the progeny will grow?

d. The *val-1 val-2*$^+$ strains accumulate intermediate compound B, and the *val-1*$^+$ *val-2* strains accumulate intermediate A. The *val-1 val-2*$^+$ strains will grow on valine or A, but the *val-1*$^+$ *val-2* strains grow only on valine and not on B. Show the pathway order of A and B in relation to valine, and indicate which gene controls each conversion.

12. In a certain plant, the flower petals are normally purple. Two recessive mutations arise in separate plants and are found to be on different chromosomes. Mutation 1 (m_1) gives blue petals when homozygous (m_1m_1). Mutation 2 (m_2) gives red petals when homozygous (m_2m_2). Biochemists working on the synthesis of flower pigments in this species have already described the following pathway:

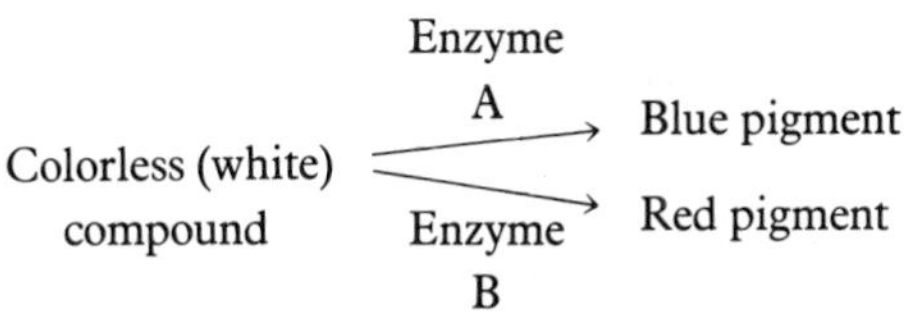

a. Which mutant would you expect to be deficient in enzyme-A activity?

b. A plant has genotype $M_1m_1\ M_2m_2$. What do you expect its phenotype to be?

c. If the plant of part b is selfed, what colors of progeny are expected, and in what proportions?

d. Why are these mutants recessive?

13. In sweet peas, the synthesis of purple anthocyanin pigment in the petals is controlled by two genes, *B* and *D*. The pathway is

White intermediate —Gene-B enzyme→ Blue intermediate —Gene-D enzyme→ Anthocyanin (purple)

a. What color petals would you expect in a pure-breeding plant unable to catalyze the first reaction?

b. What color petals would you expect in a pure-breeding plant unable to catalyze the second reaction?

c. If the plants of parts a and b are crossed, what color petals would the F_1 plants have?

d. What ratio of purple:blue:white plants would you expect in the F_2?

14. Various pairs of *rII* mutants of phage T4 are tested in *E. coli* in both the *cis* and *trans* positions. Comparisons are made of the average number of phage particles produced per bacterium (a measure called the "burst size"). Table 10-6 shows a hypothetical set of results for six different *r* mutants: *rU*, *rV*, *rW*, *rX*, *rY*, and *rZ*.

If we assign rV to the A cistron, what are the locations of the other five rII mutations with respect to the A and B cistrons.

Table 10-6.

Cis genotypes	Burst size	*Trans* genotypes	Burst size
rU rV / + +	250	rU + / + rV	258
rWrX / + +	255	rW + / + rX	252
rY rZ / + +	245	rY + / + rZ	0
rU rW / + +	260	rU + / + rW	250
rU rX / + +	270	rU + / + rX	0
rU rY / + +	253	rU + / + rY	0
rU rZ / + +	250	rU + / + rZ	0
rV rW / + +	270	rV + / + rW	0
rV rX / + +	263	rV + / + rX	270
rV rY / + +	240	rV + / + rY	250
rV rZ / + +	274	rV + / + rZ	260
rWrY / + +	260	rW + / + rY	240
rWrZ / + +	250	rW + / + rZ	255

(Problem 14 is from M. Strickberger, *Genetics.* Copyright © 1968 by Monroe W. Strickberger. Reprinted with permission of Macmillan Publishing Co., Inc.)

15. Mutagenic treatment of a haploid organism produces ten strains showing mutation in a hypothetical gene *Q*, which is involved in the synthesis of the substance Q. Each of the ten mutant strains requires Q, and the position of the mutations can be located in the *Q* gene by mapping analysis. Luckily, a means for detecting complementation is available, so all mutants are tested against each other pairwise in a *trans* arrangement to see if the pairs can complement to synthesize their own Q substance. Table 10-7 shows the results (with + indicating successful complementation).

Table 10-7.

	1	2	3	4	5	6	7	8	9	10
1	–	+	+	+	+	+	+	+	+	+
2	+	–	+	–	+	–	–	+	+	–
3	+	+	–	+	–	+	+	–	–	+
4	+	–	+	–	+	–	–	+	+	–
5	+	+	–	+	–	+	+	–	–	+
6	+	–	+	–	+	–	–	+	+	–
7	+	–	+	–	+	–	–	+	+	–
8	+	+	–	+	–	+	+	–	–	+
9	+	+	–	+	–	+	+	–	–	+
10	+	–	+	–	+	–	–	+	+	–

a. How many cistrons are there in the Q gene? (Assume here that complementation indicates separate cistrons.)

b. Which mutations belong in which cistron?

c. How would you order the cistrons in the gene map? Can you do it with the available data? If not, what procedure would be needed?

16. All *pur* alleles result in defective enzyme P and map at one genetic locus. A complementation test among six mutant *pur* strains produces the following results (where a plus indicates complementation):

	1	2	3	4	5	6
1	−	−	−	−	+	−
2	−	−	−	−	+	+
3	−	−	−	−	−	−
4	−	−	−	−	−	+
5	+	+	−	−	−	+
6	−	+	−	+	+	−

a. Draw a complementation map.

b. What kind of mutant might mutant 3 be?

c. What can you say about the structure of enzyme P?

17. Some complementing mutants are *polar complementing* (showing intracistronic complementation only with mutations to one side of the mutant site). Speculate on the biochemical nature of the polar mutations.

18. There is evidence that occasionally during meiosis either one or both homologous centromeres will divide and segregate precociously at the first division rather than at the second division (as is the normal situation). In *Neurospora, pan2* alleles produce a pale ascospore, aborted ascospores are completely colorless, and normal ascospores are black. In a cross between two complementing alleles *pan2x* × *pan2y*, what ratios of black:pale:colorless would you expect in asci resulting from precocious division (a) of one centromere? (b) of both centromeres? (Assume that *pan2* is near the centromere.)

19. *Protozoon mirabilis* is a hypothetical single-celled haploid green alga. It orients to light by means of a red "eye-spot." Fourteen "white-eye-spot" mutants (eye^-) are isolated after mutation by selecting cells that do not move toward the light. It is possible to fuse haploid cells to make diploid individuals. The 14 eye^- mutants are paired in all combinations, and the color of the eye-spot is scored in each. Table 10-8 shows the results, where + indicates a red eye-spot and − indicates a white eye-spot.

a. Mutant 14 obviously is different from the rest. Why might this be?

b. Excluding mutant 14, how many complementation groups are there, and which mutants are in which group?

c. Three crosses are made with the results shown in Table 10-9. Explain these genetic ratios with symbols.

Table 10-8.

	1	2	3	4	5	6	7	8	9	10	11	12	13	14
1	−	+	+	+	−	+	+	−	−	+	+	+	+	−
2	+	−	−	−	+	+	+	+	+	+	+	−	+	−
3	+	−	−	−	+	+	+	+	+	+	+	−	+	−
4	+	−	−	−	+	+	+	+	+	+	+	−	+	−
5	−	+	+	+	−	+	+	−	−	+	+	+	+	−
6	+	+	+	+	+	−	−	+	+	−	−	+	−	−
7	+	+	+	+	+	−	−	+	+	−	−	+	−	−
8	−	+	+	+	−	+	+	−	−	+	+	+	+	−
9	−	+	+	+	−	+	+	−	−	+	+	+	+	−
10	+	+	+	+	+	−	−	+	+	−	−	+	−	−
11	+	+	+	+	+	−	−	+	+	−	−	+	−	−
12	+	−	−	−	+	+	+	+	+	+	+	−	+	−
13	+	+	+	+	+	−	−	+	+	−	−	+	−	−
14	−	−	−	−	−	−	−	−	−	−	−	−	−	−

Table 10-9.

	Number of progeny		
Mutants crossed	*eye*$^+$	*eye*$^-$	Total
1 × 2	31	89	120
2 × 6	5	113	118
1 × 14	0	97	97

d. How many genetic loci are involved altogether, and which of the 14 mutants are at each locus?

e. What is the linkage arrangement of the loci? (Draw a map.)

20. The Martian bacterium *Martibacillus novellus* is green and can make phlizic acid. A mutation (*wht*) produces white bacteria. A number of phlizic acid auxotrophs (*phl*$^-$) are selected; *wht* and *phl* are cotransducible by the phage pm-22, and we arbitrarily place *wht* on the right. A series of transductions is performed using donors that are *wht*$^+$ *phl-x* and recipients that are *wht phl-y*. The transductants are plated on minimal medium to select *phl* prototrophs, but no selection is made for color. Table 10-10 gives two kinds of data: first, the ratio of green/white colonies among the transductants; and second, the presence (+) or absence (−) of microcolonies from abortive transductions. (Abortive transduction is the failure of a transducing DNA segment to be incorporated into the recipient chromosome. The segment is functional but does not divide, so only one daughter cell receives it, and a microcolony is produced.)

a. Place the six markers on a map in their proper order.

Table 10-10.

	Donors					
Recipients	*phl-1*	*phl-2*	*phl-3*	*phl-4*	*phl-5*	*phl-6*
phl-1 wht	—	0.1 +	0.2 +	0.1 +	0.2 +	0.1 +
phl-2 wht	3.0 +	—	2.7 +	0.1 −	2.6 +	2.4 −
phl-3 wht	2.4 +	0.1 +	—	0.1 +	0.2 −	0.1 +
phl-4 wht	3.1 +	3.4 −	3.0 +	—	2.8 +	2.6 −
phl-5 wht	2.7 +	0.2 +	2.4 −	0.1 +	—	0.1 +
phl-6 wht	2.7 +	0.2 −	2.5 +	0.1 −	2.4 +	—

b. Divide them into complementation groups, and mark the lines between cistrons on the map.

(Problem 20 is from Burton S. Guttman, *Biological Principles.* Copyright © 1971 by W. A. Benjamin, Inc., Menlo Park, California.)

21. You have the following map of the *rII* locus:

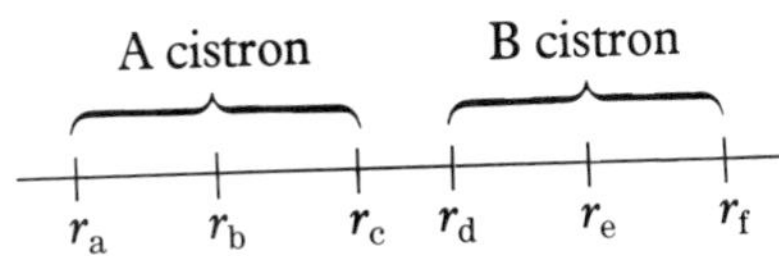

You detect a new mutation r_x, and you find that it does not complement any of the mutants in the A or B cistrons. You find that wild-type recombinants are obtained in crosses with r_a, r_b, r_e, and r_f, but not with r_c or r_d. Suggest possible explanations for these results. Describe tests you would use to choose between the explanations.

22. In *Salmonella,* four mutations (A through D) affecting one enzyme involved in the histidine biosynthetic pathway are obtained. The mutants can be tested in all combinations of pairs by transduction. The results are the following (with + indicating recovery of wild-type transductants):

	A	B	C	D
A	−	+	−	+
B	+	−	−	−
C	−	−	−	+
D	+	−	+	−

a. Considering that some or all of these mutations may be deletions, draw a possible gene map of this region.

b. If a point mutation produced wild-type recombinants with all these mutations except C, at which position on this map is C most likely located?

23. The deletion map at the left shows four deletions (1 through 4) involving the *rIIA* cistron of phage T4. Five point mutations (*a* through *e*) in *rIIA* are tested against

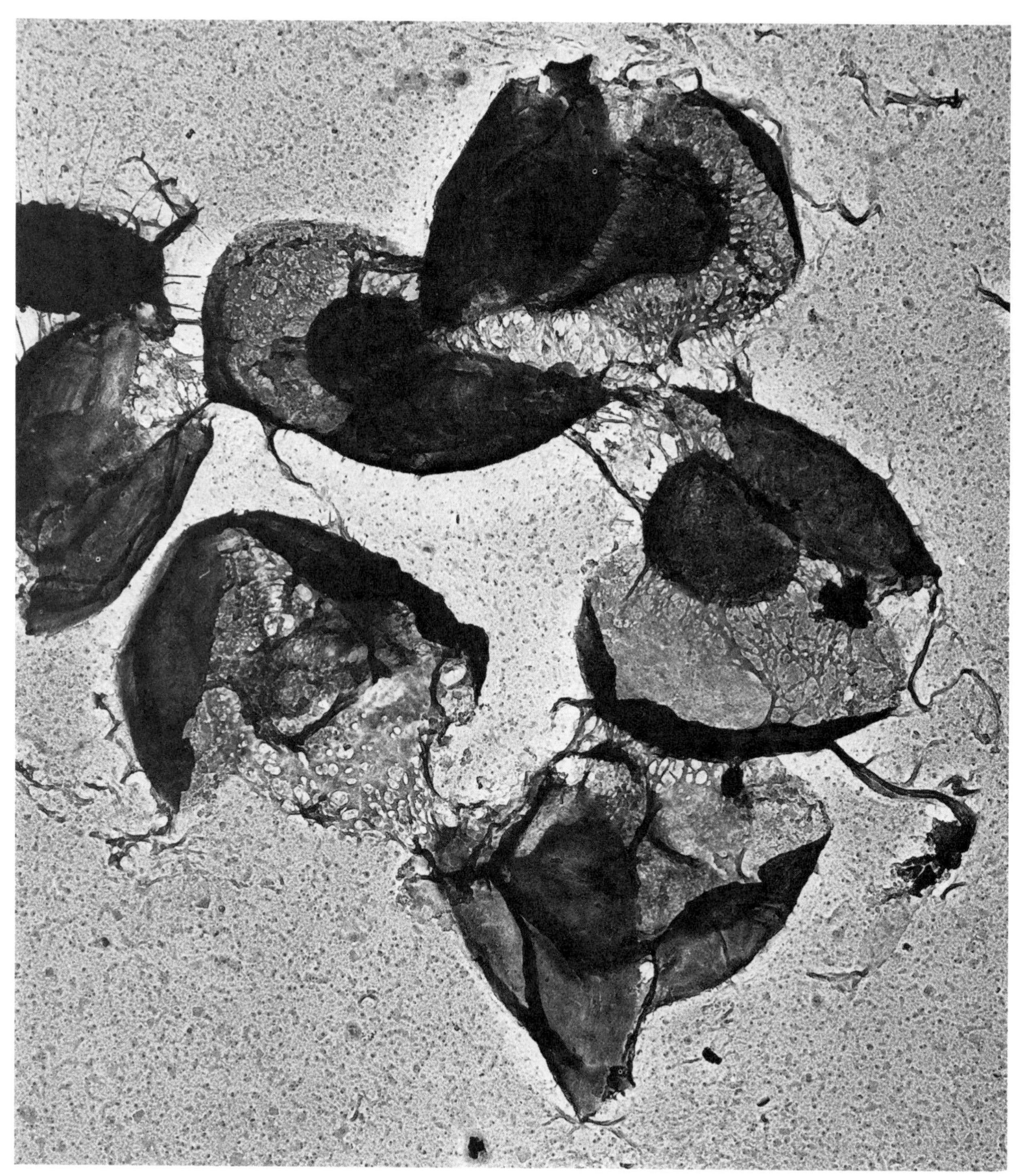

Germinating spores of the bacterium Bacillus subtilis. *(Copyright © Jack D. Griffith.)*

11

DNA Structure

We know now that genes are linear arrays of codons, which are colinear with the primary sequence of amino acids in proteins. We assume that the gene has some molecular counterpart that embodies its linearity. Can we determine the molecular species that corresponds to genes?

We have seen (Chapter 9) that Griffith's observations on transformation were explained by Avery's demonstration that DNA is the basis of these induced heritable changes. Avery's experiments were definitive, but many scientists were very reluctant to accept the conclusion that DNA (rather than proteins) is the genetic material. The final clincher was provided in 1952 by Alfred Hershey and Martha Chase using the phage T2. Phage infection obviously must involve the introduction into the bacterium of the specific information that dictates viral reproduction. The phage is relatively simple in molecular constitution. Most of its structure is protein, with DNA contained inside the protein sheath of its "head."

Phosphorus is not found in proteins but is an integral part of DNA; conversely, sulfur is present in proteins but never in DNA. Hershey and Chase incorporated the radioisotope of phosphorus (^{32}P) into phage DNA, and that of sulfur (^{35}S) into the proteins of a separate phage culture. They then used each phage culture independently to multiply infect *E. coli*. After sufficient time for injection to occur, they sheared the empty phage carcasses (called "ghosts") off the bacterial cells by agitating in a kitchen blender. They used centrifugation to separate the bacterial cells from the phage ghosts, and then they measured the radioactivity in the two fractions. When the ^{32}P-labeled

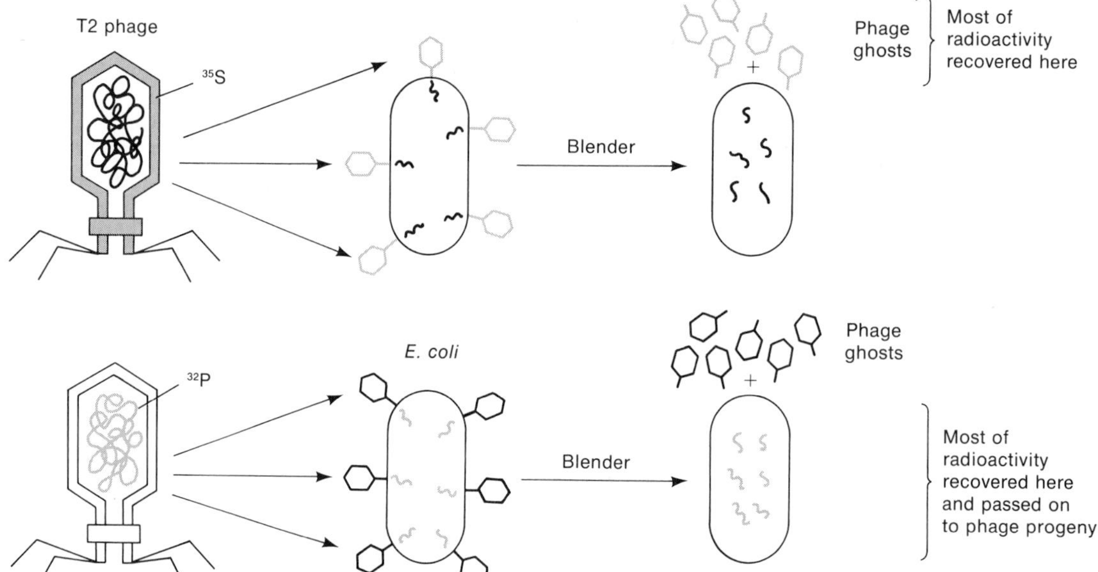

Figure 11-1. The Hershey–Chase experiment demonstrated that the genetic material of phage is DNA, not protein. The experiment uses two sets of T2 bacteriophage. In one set, the protein coat is labeled with radioactive sulfur (^{35}S) not found in DNA. In the other set, the DNA is labeled with radioactive phosphorus (^{32}P) not found in protein. Only the ^{32}P is injected into the E. coli, *indicating that DNA is the agent necessary for the production of new phage.*

phage were used, most of the radioactivity ended up inside the bacterial cells, indicating that the phage DNA entered the cells. When the ^{35}S-labeled phage were used, most of the radioactive material ended up in the phage ghosts, indicating that the phage protein never enters the bacterial cell (Figure 11-1). The conclusion is inescapable: DNA is the hereditary material, and the phage proteins are mere structural packaging that is discarded after delivering the vital DNA to the bacterial cell.

Why such reluctance to accept this conclusion? DNA was known to be a rather simple chemical. How could all the information about the wonderously variable protein structures (with their sequences of the 20 amino acids) be stored in such a simple molecule? What is DNA like?

The Structure of DNA

Although the DNA structure was not known, the basic building blocks of DNA had been known for many years. The basic elements of DNA had been isolated and determined by partly breaking up purified DNA. These studies showed that DNA is composed of only four basic molecules called **nucleotides,** which are identical except that each contains a different nitrogen base. Each nucleotide contains phosphate, sugar (of the deoxyribose type), and one of the four bases (Figure 11-2). The four bases are **adenine** (A), **guanine** (G), **cytosine** (C), and **thymine** (T). The full chemical names of the nucleotides are deoxyadenosine 5′-monophosphate (or deoxyadenylate, or dAMP), deoxyguanosine 5′-monophosphate (or deoxyguanylate, or dGMP), deoxycytidine 5′-monophosphate (or deoxycytidylate, or dCMP), and deoxythymidine 5′-

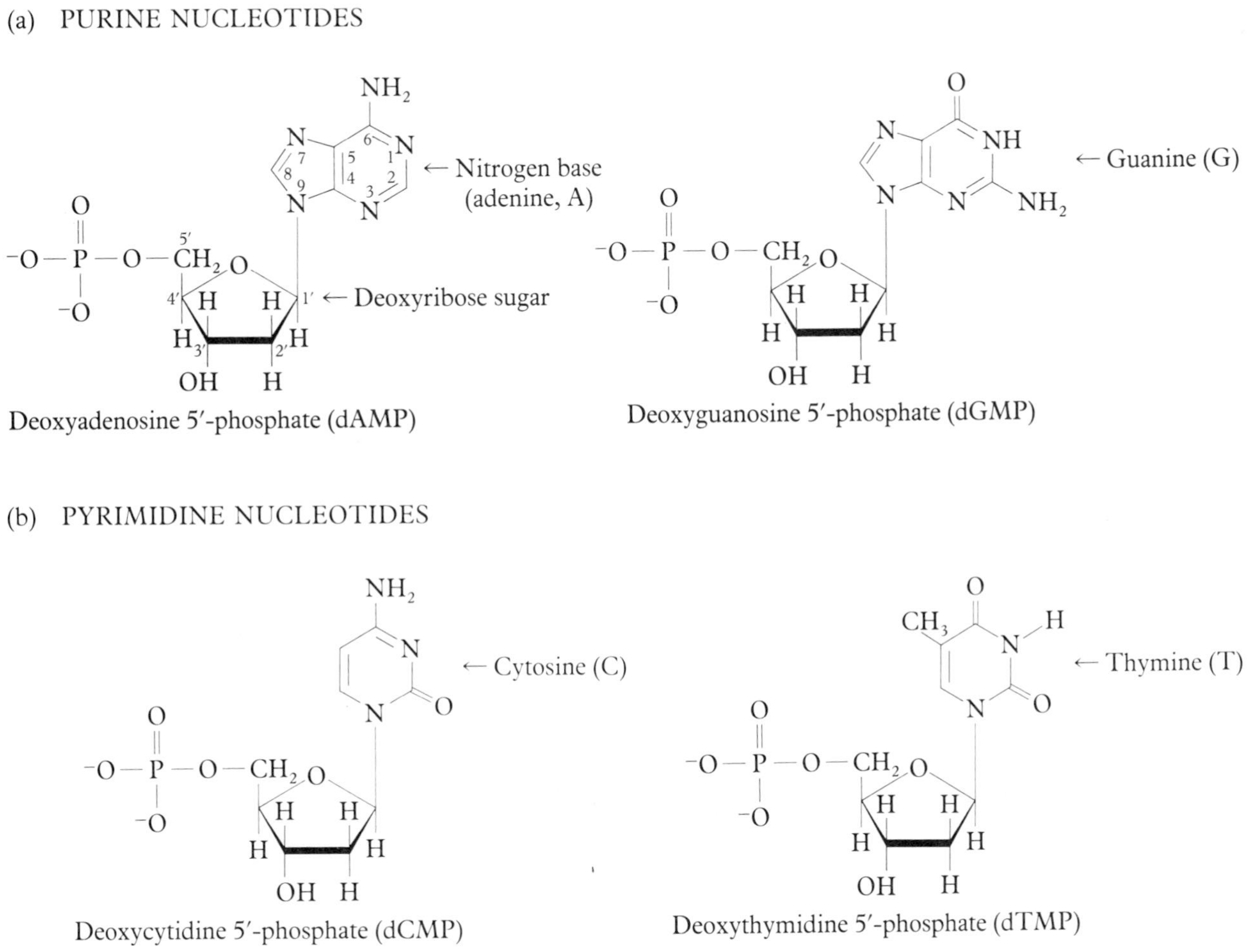

(a) PURINE NUCLEOTIDES

Deoxyadenosine 5′-phosphate (dAMP)

Deoxyguanosine 5′-phosphate (dGMP)

(b) PYRIMIDINE NUCLEOTIDES

Deoxycytidine 5′-phosphate (dCMP)

Deoxythymidine 5′-phosphate (dTMP)

Figure 11-2. Chemical structure of the four nucleotides (two with purine bases and two with pyrimidine bases) that are the fundamental building blocks of DNA. The sugar is called deoxyribose because it is a variation of a common sugar, ribose, that has one more oxygen atom.

monophosphate (or deoxythmidylate, or dTMP). However, for our purposes, it will be quite sufficient to just refer to each nucleotide by the abbreviation of its base. Two of the bases, adenine and quanine, are similar in structure and are called **purines.** The other two bases, cytosine and thymine, also are similar and are called **pyrimidines.**

After the central role of DNA in heredity became clear, many scientists set out to determine the exact structure of DNA. How can a molecule with such a limited range of different components possibly store the vast range of information about all the protein primary structures of the living organism? The first to succeed in finding a reasonable DNA structure were James Watson and Francis Crick in 1953. They worked from two kinds of clues. First, other researchers had amassed a lot of X-ray crystallographic data on DNA structure. In such experiments, X rays are fired at DNA crystals, and the scatter of the rays from the crystal is observed by catching them on photographic film, where the X rays produce spots. The angle of scatter represented by each spot on the film gives information about the position of an atom or certain groups of atoms in the DNA molecule. This procedure is not simple to carry out (or to explain), and the interpretation of the spot patterns is very difficult. The available data suggested that DNA is long and skinny and that it has two similar parts that are parallel to one another and run along the length of the molecule. Other regularities were present in the spot patterns, but no one had yet thought of a three-dimensional structure that could account for just those spot patterns.

The second set of clues available to Watson and Crick came from work done several years earlier by Erwin Chargaff. Studying a large selection of DNAs from different organisms, Chargaff found certain empirical rules about the amounts of each component of DNA.

1. The total amount of pyrimidine nucleotides (T + C) always equals the total amount of purine nucleotides (A + G).
2. The amount of T always equals the amount of A, and C always equals G. But the amount of A + T is not necessarily equal to the amount of G + C.

The structure that Watson and Crick derived from these clues is now well known—a double helix looking rather like two interlocked bedsprings. Each bedspring (helix) is a chain of nucleotides held together by phosphodiester bonds, and the two bedsprings (helices) are held together by hydrogen bonds between the bases. Figure 11-3 shows a part of this structure with the helices uncoiled. Figure 11-4 shows a simplified picture of the coiling, with each of the base pairs represented by a "stick" between the "ribbons" that represent the so-called "sugar–phosphate backbones" of the chains. In Figure 11-3, note that the two backbones run in opposite directions; they are said to be **antiparallel,** and (for reasons apparent in the figure) one is called $5' \rightarrow 3'$ and the other $3' \rightarrow 5'$.

In three dimensions, the bases actually form rather flat structures (more like

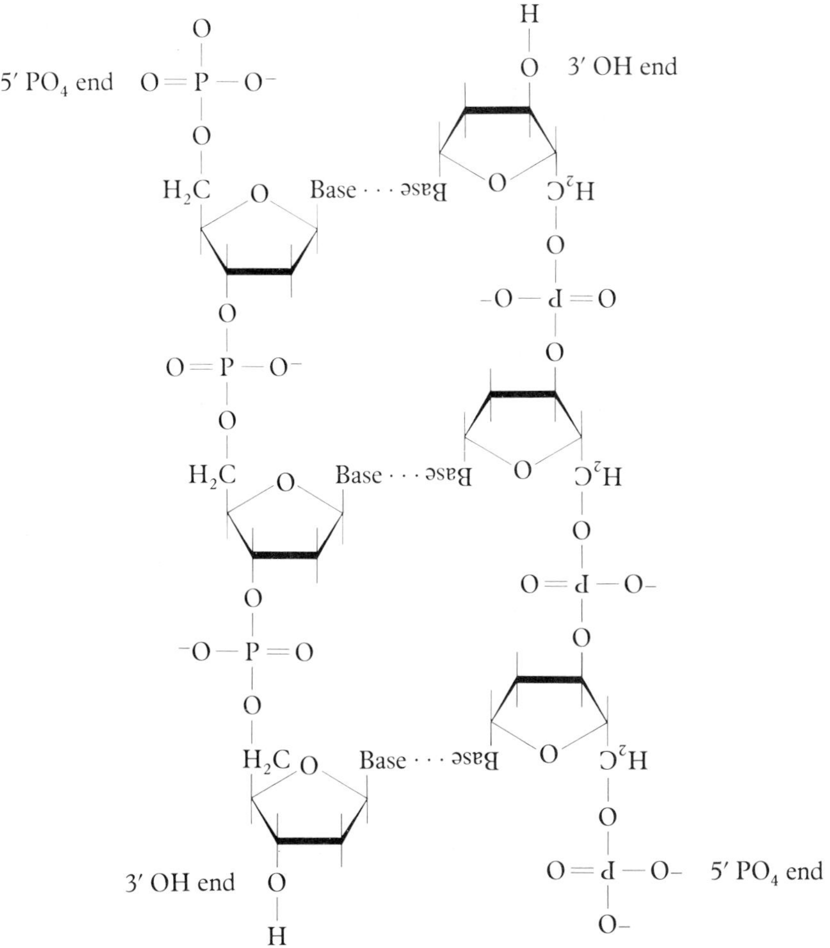

Figure 11-3. Antiparallel nature of the two sugar-phosphate backbones of DNA. Dots between bases represent hydrogen bonds that help hold the two antiparallel strands together.

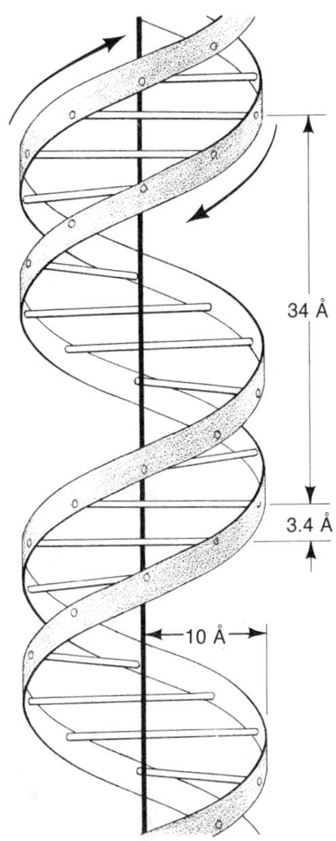

Figure 11-4. A simplified model showing the helical structure of DNA. The sticks represent base pairs, and the ribbons represent the sugar–phosphate backbones of the two antiparallel chains. In fact, the base pairs are more like flat "steps" than round sticks as shown here.

steps in a ladder than like the sticks shown in Figure 11-4), and these flat bases stack on top of one another in the twisted structure of the double helix. This stacking of bases adds tremendously to the stability of the molecule by excluding water molecules from the spaces between the base pairs. (This phenomenon is very much like the stabilizing force that you can feel when you squeeze two plates of glass together under water and then try to separate them.) Figure 11-5 summarizes the three kinds of forces hooking the nucleotides together, and Figure 11-6 shows how regularly the bases stack within the double helix.

The double helix accounted nicely for the X-ray data, and it also tied in very nicely with Chargaff's data. Studying models they made of the structure,

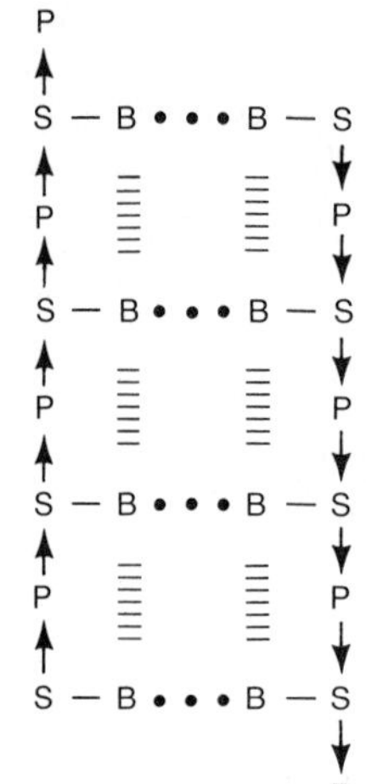

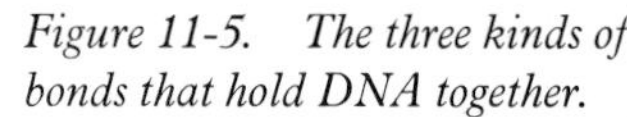

Figure 11-5. The three kinds of bonds that hold DNA together.

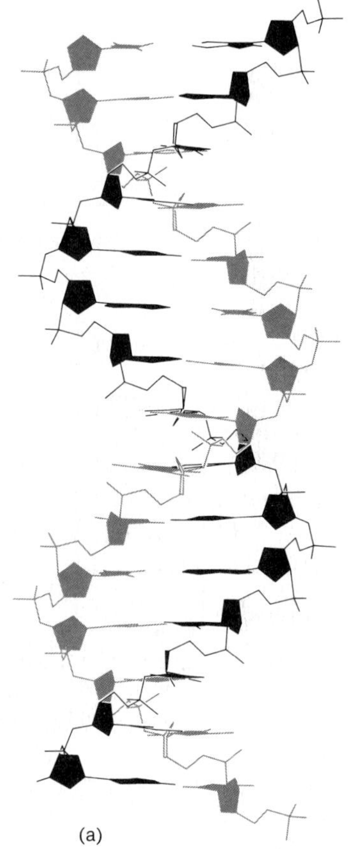

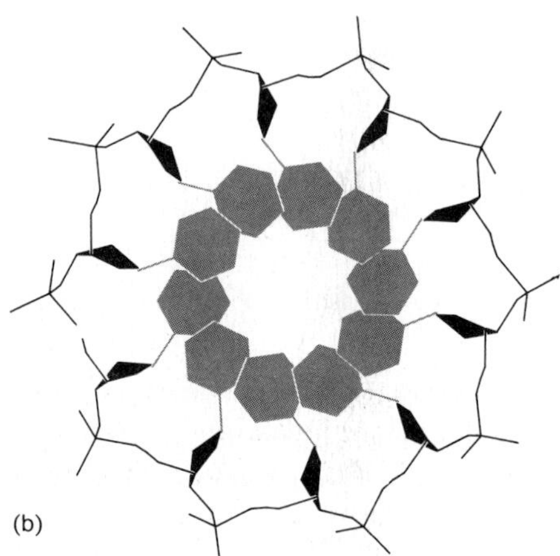

Figure 11-6. An accurate model of the DNA structure. (a) A side view of the helix. One strand is drawn in black, the other in gray. (b) The helix viewed from one end; note the regular stacking of the base rings. The sugars are drawn in black, and the bases in gray. (From L. Stryer, Biochemistry, *2nd ed. W. H. Freeman and Company. Copyright © 1981.)*

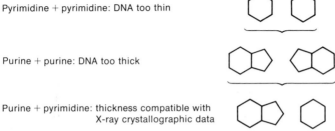

Pyrimidine + pyrimidine: DNA too thin

Purine + purine: DNA too thick

Purine + pyrimidine: thickness compatible with X-ray crystallographic data

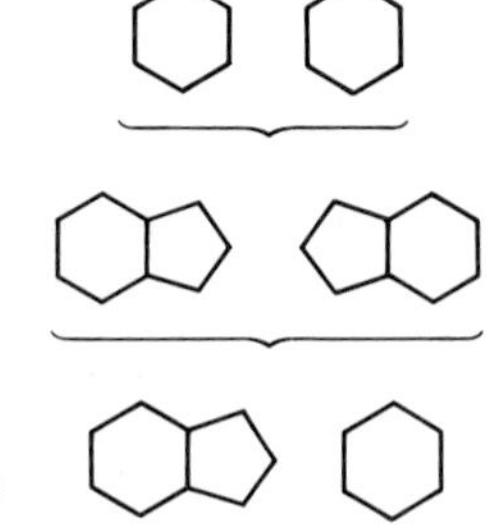

Figure 11-7. The pairing of purines with pyrimidines accounts exactly for the diameter of the DNA double helix as determined from X-ray crystallographic data.

Watson and Crick realized that the observed radius of the double helix (known from the X-ray data) would be explained if a purine base always pairs (by hydrogen bonding) with a pyrimidine base (Figure 11-7). Such pairing would account for the (A + G) = (T + C) regularity observed by Chargaff, but it would predict four possible pairings: T···A, T···G, C···A, and C···G. Chargaff's data, however, indicate that T pairs only with A and C pairs only with G.

Watson and Crick showed that only these two pairings have the necessary complementary "lock-and-key" shapes to permit efficient hydrogen bonding. Hydrogen bonds occur between hydrogen atoms with a small positive charge and acceptor atoms with a small negative charge. For example,

$\delta+$ H   H $\delta+$:O: $\delta-$
N

Each hydrogen atom in the NH_2 group is slightly positive ($\delta+$) because the nitrogen atom tends to "hog" the electrons involved in the N–H bond, thereby leaving the hydrogen atom slightly short of electrons. The oxygen atom has six unbonded electrons in its outer shell that form an electron cloud around it, making it slightly negative ($\delta-$). A hydrogen bond forms between one H and the O. Hydrogen bonds are quite weak (only about 3% of the strength of a covalent chemical bond), but this weakness (as we shall see) plays an important role in the hereditary function of the DNA molecule. One further important chemical fact: the hydrogen bond is much stronger if the participating atoms are "pointing at each other" in the ideal orientations.

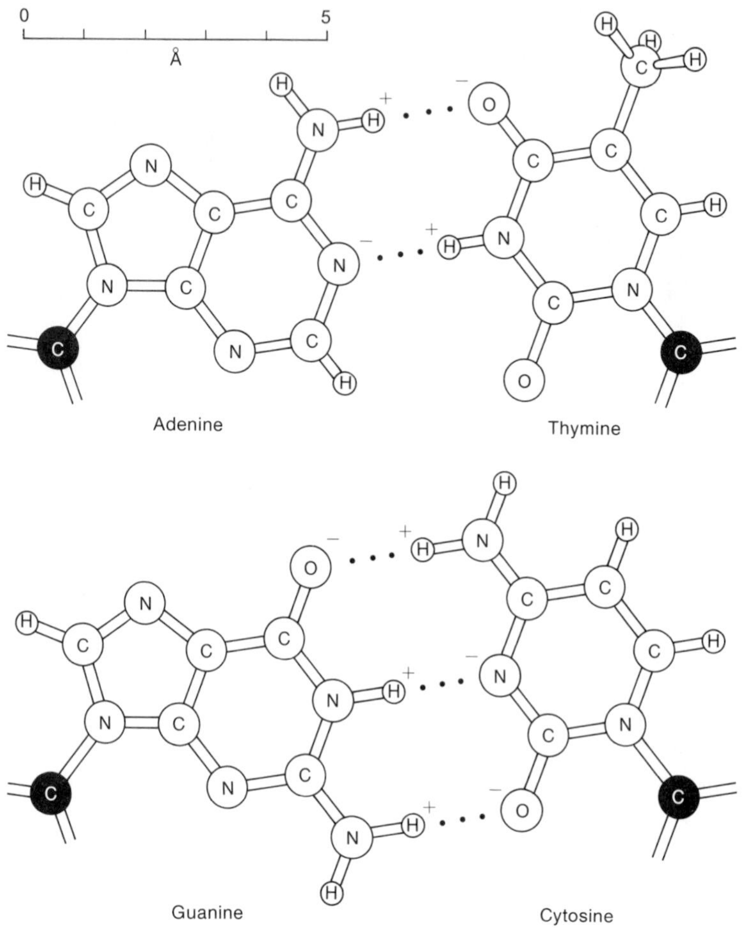

Figure 11-8. The lock-and-key hydrogen bonding between A and T and between G and C. (From G. S. Stent, Molecular Biology of Bacterial Viruses. *Coypright © 1963 by W. H. Freeman and Company.)*

Looking at the hydrogen-bonding potential between the various purine–pyrimidine pairs, we find that only two pairs have the necessary arrangement of $\delta+$ hydrogen atoms and $\delta-$ acceptor atoms. These two are the T–A pair and the C–G pair, both of which show beautiful lock-and-key fit (Figure 11-8), providing just the proper "width" of the base pair to explain the known radius of the DNA double helix.

Note that the C–G pair has three hydrogen bonds, whereas the T–A pair has only two. One would predict that DNA containing many G–C pairs would be more stable than DNA containing many A–T pairs. In fact, this prediction is confirmed. We now have a neat explanation for the data of Chargaff in terms of DNA structure (Figure 11-9). We also have a structure that is consistent with the X-ray data.

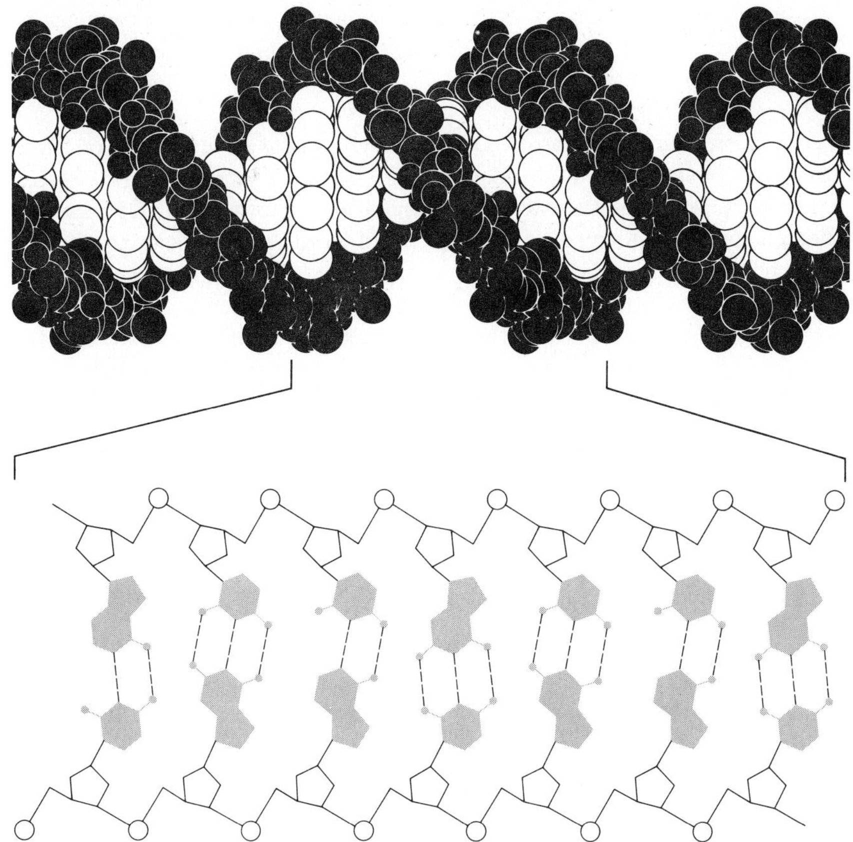

Figure 11-9. At the top is a space-filling model of the DNA double helix. At the bottom is an unwound representation of a short stretch of nucleotide pairs, showing how A–T and G–C pairing produces the Chargaff ratios. (Space-filling model from C. Yanofsky, "Gene Structure and Protein Structure." Copyright © 1967 by Scientific American, Inc. All rights reserved. Unwound structure based on A. Kornberg, "The Synthesis of DNA." Copyright © 1968 by Scientific American, Inc. All rights reserved.)

Elucidation of the DNA structure caused a lot of excitement in genetics (and in all areas of biology) for two basic reasons:

1. The structure suggests an obvious way in which the molecule could be **duplicated** (or **replicated**). This essential property of a genetic molecule had been a mystery up until then.
2. The structure suggests a possible reason for the colinearity of gene and polypeptide: perhaps the *sequence* of nucleotide pairs in DNA is dictating the sequence of amino acids in the protein organized by that gene. In other words, there is some sort of **genetic code** whereby

information in DNA is written as a sequence of nucleotide pairs and then translated into a different language of amino acid sequences in protein.

This basic information about DNA is now familiar to almost anyone who has read a biology text in elementary or high school, or even magazines and newspapers. It may seem trite and obvious. But try to put yourself back into the scene in 1953 and imagine the excitement! Until then, the evidence that the uninteresting DNA is the genetic molecule seemed disappointing and discouraging. But the Watson–Crick structure of DNA suddenly opened up the possibility of explaining two of the biggest "secrets" of life. James Watson has told the story of this discovery (from his own point of view, strongly questioned by others involved) in a fascinating book called *The Double Helix;* it reveals the intricate interplay of personality clashes, clever insights, hard work, and simple luck in such important scientific advances.

Like every good model, the Watson–Crick model was ideally suited for further testing, and we now look at some of its predictions and their testing.

Replication of DNA

Figure 11-10 diagrams the possible mechanism for DNA replication as proposed by Watson and Crick. Here the sugar–phosphate backbones are represented by lines, and the sequence of base pairs is random. Let's imagine that the double helix is like a zipper that unzips starting at one end (the top in this figure). You can see that, if this zipper analogy is valid, the unwinding of the two strands will expose single bases of either strand. Because the pairing requirements imposed by the DNA structure are strict, each exposed base will

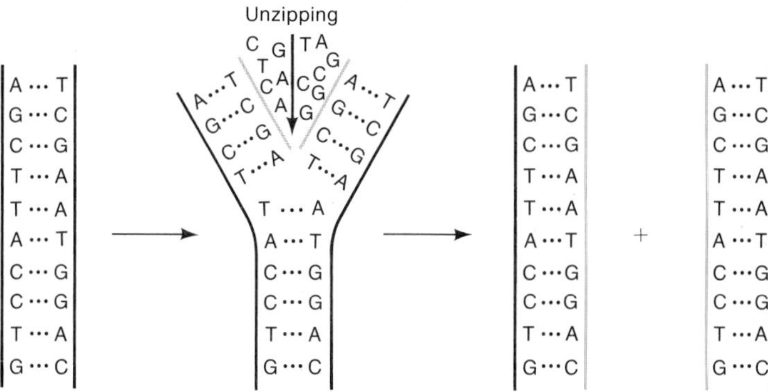

Figure 11-10. The model of DNA replication proposed by Watson and Crick is based on the hydrogen-bonding specificity of the base pairs. The lines represent the chain backbones, and the lighter lines indicate newly synthesized strands.

pair only with its *complementary* base. Because of this base complementarity, each of the two single strands will act as a **template** or mold and will begin to reform a double helix identical to the one from which it was unzipped. The newly added nucleotides are assumed to come from a pool of free nucleotides that must be present in the cell.

If this model is correct, four predictions should be testable:

1. The daughter molecules should have nucleotide compositions identical with that of the original parent molecule.
2. Each daughter molecule should contain one parental nucleotide chain (black line in Figure 11-10) and one newly synthesized nucleotide chain (gray line).
3. A replication "fork" should be visible during replication.
4. During replication, the genes at the starting point of replication should be present in two copies more often than those at the other end.

The following experiments not only test these four predictions, they also illustrate many of the techniques and procedures used in molecular genetics.

Testing Prediction 1

The first prediction states that daughter molecules have nucleotide compositions identical to the original parent molecule. This prediction can be tested by actually replicating some DNA in a chemical (cell-free) system in a test tube (in vitro rather than in vivo). In the late 1950s, Arthur Kornberg succeeded in identifying and purifying an enzyme, DNA polymerase (Figure 11-11), that catalyzes the replication reaction:

$$\text{Primer (parental) DNA} + \begin{matrix}\text{dATP}\\+\\\text{dGTP}\\+\\\text{dCTP}\\+\\\text{dTTP}\end{matrix} \xrightarrow{\text{DNA polymerase}} \text{Progeny DNA}$$

This reaction works only with the triphosphate forms of the nucleotides (such as deoxyadenosine triphosphate, or dATP). The total amount of DNA at the end of the reaction can be as much as 20 times the amount of original primer DNA, so most of the DNA present at the end must be progeny DNA. Thus, an analysis of this final DNA mixture can be regarded as largely indicative of the nature of the progeny DNA.

As the Watson–Crick model predicts, the ratio of (G + C)/(A + T) in the progeny DNA is identical to the ratio in the primer DNA, no matter what ratio of dATP:dGTP:dCTP:dTTP is put into the reaction pot. This (G + C)/

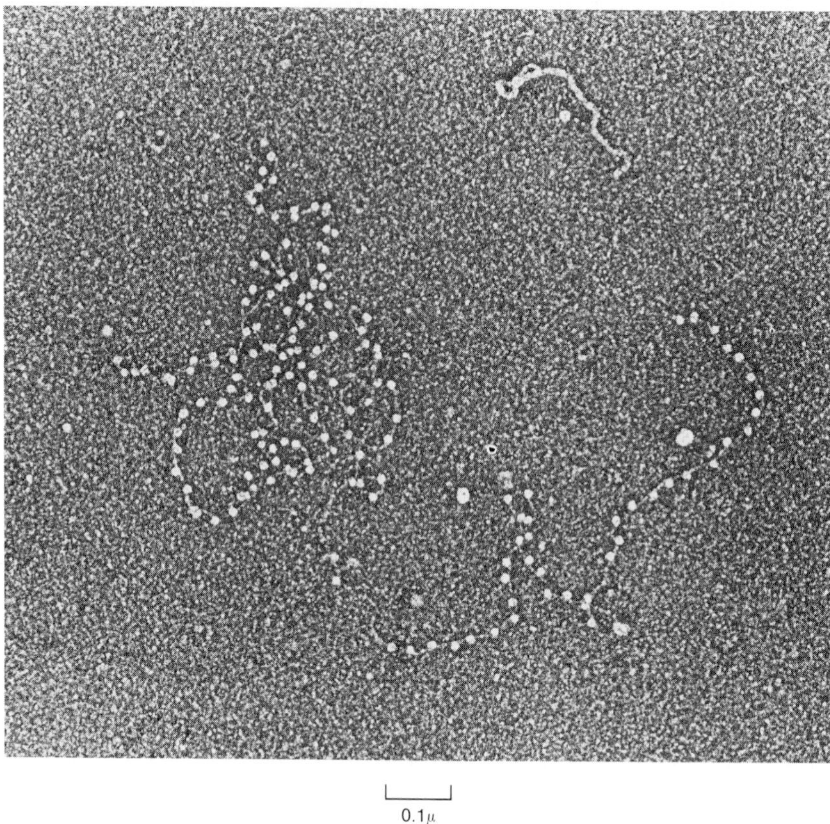

Figure 11-11. Electron micrograph of molecules of the enzyme DNA polymerase attached to sections of DNA. No progeny DNA is visible in the photograph. (From Jack D. Griffith.)

(A + T) ratio is interesting because it directly reflects the linear abundance of G–C and A–T base pairs along the DNA molecule. A primer DNA with a (G + C)/(A + T) ratio of 2 should produce daughter molecules with a ratio of 2 if the model is correct, and this duplication of the parental ratio is confirmed for many primers with different ratios.

Testing Prediction 2

The second prediction states that a daughter molecule will contain one parental nucleotide chain and one newly synthesized chain. This prediction has been tested in both prokaryotes and eukaryotes. A little thought shows that there are at least three different ways in which a parental DNA molecule might possibly be related to the daughter molecules. These hypothetical modes are called semiconservative (the Watson–Crick model), conservative, and dispersive (Figure 11-12).

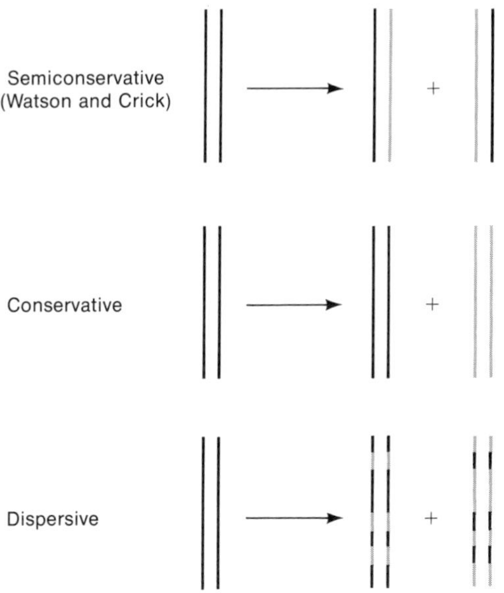

Figure 11-12. Three alternative patterns for DNA replication; the Watson–Crick model would produce the first (semiconservative) pattern. Lighter lines represent the newly synthesized strands.

In 1958, Matthew Meselson and Franklin Stahl set out to distinguish among these possibilities in an experiment with *E. coli.* They grew *E. coli* cells in a medium containing the heavy isotype of nitrogen (^{15}N) rather than the normal light (^{14}N) form. This isotope is inserted into the nitrogen bases, which then are incorporated into newly synthesized DNA strands. After many cell divisions in ^{15}N, the DNA of the cells is well labeled with the heavy isotope. The cells were then removed from the ^{15}N medium and put into ^{14}N medium; after one and two cell divisions, samples were taken. DNA was extracted from the cells in each of these samples and put into a solution of cesium chloride (CsCl) in an ultracentrifuge.

If cesium chloride is spun in a centrifuge at tremendously high speeds (50,000 rpm) for many hours, the salt ions tend to be pushed by centrifugal force toward the bottom of the tube. Ultimately, a **gradient** of Cs^+ and Cl^- ions is established in the tube, with the highest ion concentration at the bottom. Molecules of DNA in the solution also are pushed toward the bottom by centrifugal force. But, as they travel down the tube, they encounter the increasing salt concentration, which tends to push them back up because of DNA's buoyancy (or tendency to float). Thus, the DNA finally "settles" at some point in the tube where the centrifugal forces just balance the buoyancy of the molecules in the cesium chloride gradient. The buoyancy of DNA depends on its density (which in turn reflects the ratio of G–C to A–T base pairs). The presence of

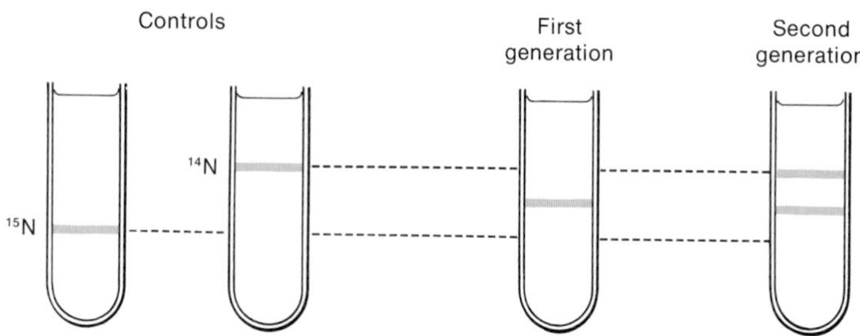

Figure 11-13. Centrifugation of DNA in a cesium chloride gradient. Cultures grown for many generations in ^{15}N and ^{14}N media provide control positions for "heavy" and "light" DNA bands, respectively. When the cells grown in ^{15}N are transferred to a ^{14}N medium, the first generation produces an intermediate DNA band, and the second generation produces two bands, one intermediate and the other light.

the heavier isotope of nitrogen changes the buoyant density of DNA. The DNA extracted from cells grown for several generations on ^{15}N medium can readily be distinguished from the DNA of cells grown on ^{14}N medium by the equilibrium position reached in a cesium chloride gradient. Such samples are commonly called "heavy" and "light" DNA, respectively.

Meselson and Stahl found that, one generation after the "heavy" cells were moved to ^{14}N medium, the DNA formed a single band with density intermediate between the densities of the heavy and light controls. After two generations in ^{14}N medium, the DNA formed two bands, one at the intermediate position and the other at the light position (Figure 11-13). This result would be expected from the semiconservative mode of replication, and in fact the result is compatible *only* with this mode *if* one begins with chromosomes composed of individual double helices (Figure 11-14).

The Meselson–Stahl experiment on *E. coli* was essentially duplicated in 1958 by Herbert Taylor on the chromosomes of bean root-tip cells, using a cytological technique. He put root cells into a solution containing *tritiated* (^{3}H-labeled) thymidine—that is, thymine nucleotide containing a radioactive hydrogen isotope called tritium. He allowed the cells to undergo mitosis in this solution, so that the [^{3}H]-thymidine could be incorporated into DNA. He then washed the tips and transferred them to a solution containing nonradioactive thymidine. Addition of colchicine to such a preparation inhibits the spindle apparatus so that chromosomes in metaphase fail to separate, and sister chromatids remain "tied together" by the centromere.

The cellular location of ^{3}H can be determined by **autoradiography.** As ^{3}H decays, it emits a beta particle (an energetic electron). If a layer of photographic emulsion is spread over a cell that contains ^{3}H, a chemical reaction takes place wherever a beta particle strikes the emulsion. The emulsion can then be devel-

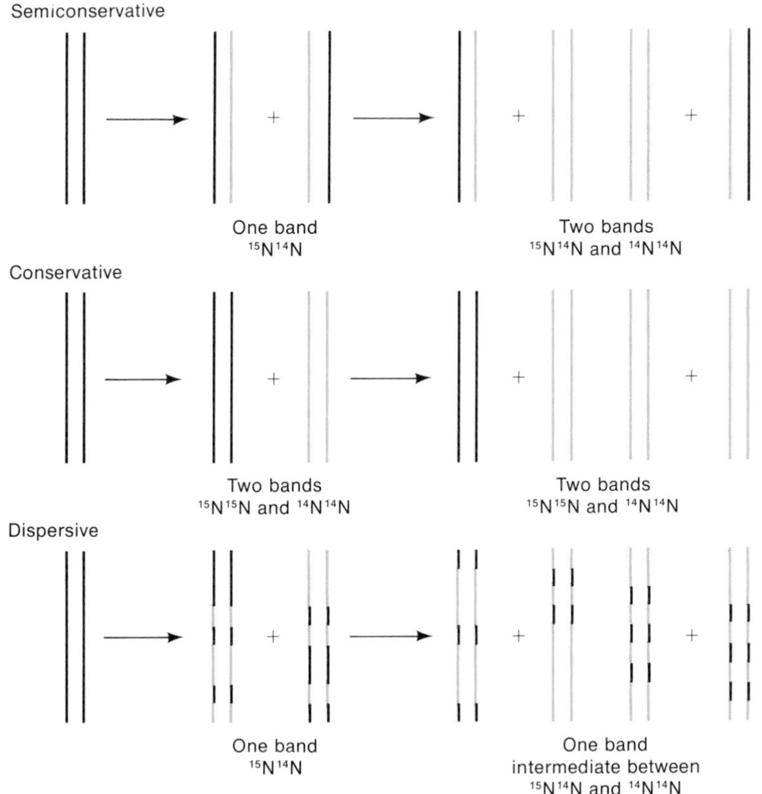

Figure 11-14. Only the semiconservative model of DNA replication predicts results like those shown in Figure 11-13: that is, a single intermediate band in the first generation and one intermediate and one light band in the second generation. (See Figure 11-12 for explanation of symbols.)

oped like a photographic print, so that the emission track of the beta particle appears as a black spot or grain. The cell can also be stained, so that the structure of the cell is visible, to identify the location of the radioactivity. In effect, autoradiography is a process in which radioactive cell structures "take their own pictures."

Figure 11-15 shows the results observed when colchicine is added during the division in [^{3}H]-thymidine or during the subsequent mitotic division. It is possible to interpret these results by representing each chromatid as a single DNA molecule that replicates semiconservatively (Figure 11-16).

Using a more modern staining technique, it is now possible to visualize the semiconservative replication of chromosomes at mitosis without the aid of autoradiography. In this procedure, the chromosomes are allowed to go through two rounds of replication in bromodeoxyuridine. The chromosomes

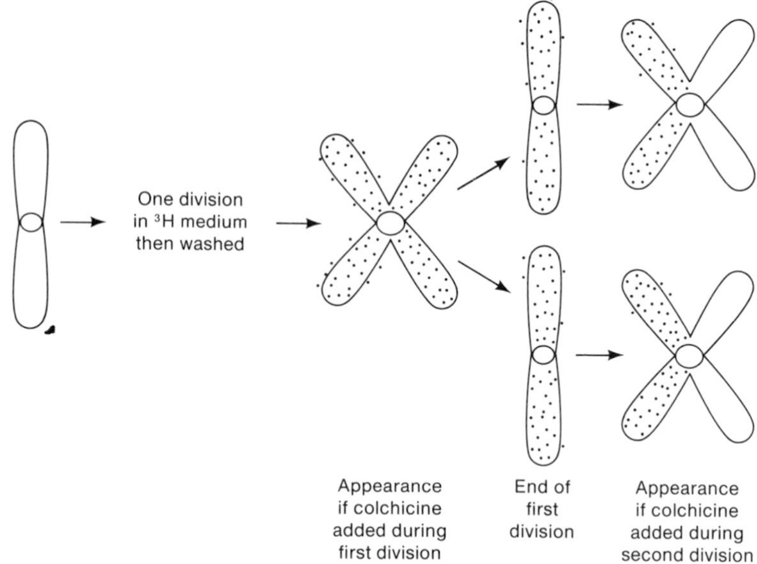

Figure 11-15. *Diagrammatic representation of autoradiography of chromosomes from cells grown for one cell division in the presence of the radioactive hydrogen isotope ^3H (tritium). Each dot represents the track of a particle of radioactivity.*

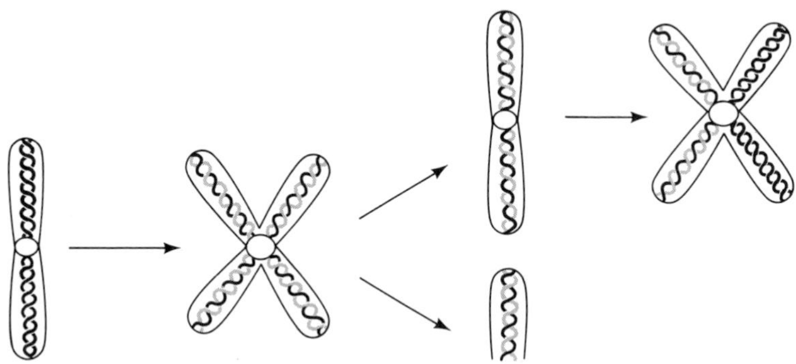

Figure 11-16. *An explanation of Figure 11-15 at the DNA level. Lighter lines represent radioactive strands.*

are then stained with fluorescent dye and giemsa stain, and this process produces so-called **harlequin chromosomes** (Figure 11-17). The DNA strands that are newly synthesized in bromodeoxyuridine stain differently from the "original" DNA strands. The basis of this pattern is exactly identical to that of Figure 11-16. (Note, in passing, that harlequin chromosomes are particularly

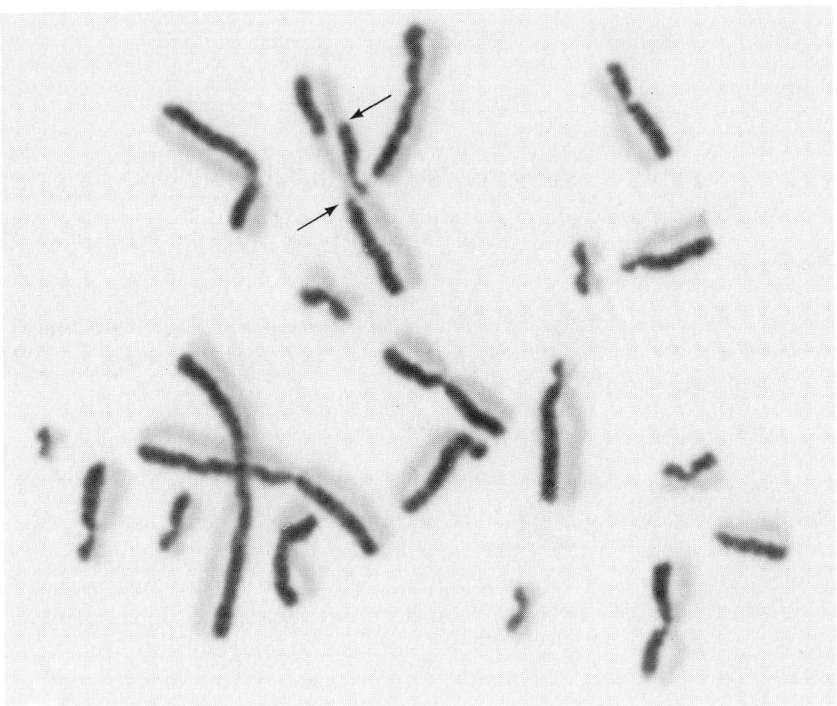

Figure 11-17. *Harlequin chromosomes in a Chinese hamster ovary (CHO) cell. The procedure involves letting the chromosomes go through two rounds of replication in the presence of bromodeoxyuridine (BUdR), which replaces thymidine in the newly synthesized DNA. The chromosomes are then stained with a fluorescent dye and geimsa stain, producing the figure above. The DNA strands that are newly replicated in BUdR stain differentially from the "original" DNA strands. A chromosome at the top has two sister chromatid exchanges. (Photo courtesy of Sheldon Wolff and Judy Bodycote.)*

favorable for the detection of sister-chromatid exchange at mitosis; two examples are seen in Figure 11-17.)

Using similar techniques, Taylor also showed that chromosome replication at meiosis also is semiconservative. This result drove another nail in the coffin of the copy-choice theory of crossing over (Chapter 5), which would require *conservative* chromosome replication at meiosis.

Figures 11-15 and 11-16 bring up one of the remaining great unsolved questions of genetics: Is a eukaryotic chromosome basically a single DNA molecule surrounded by a protein matrix? Two things strongly suggest that this is, in fact, the case. First, if there were many DNA molecules in the chromosome—whether side-by-side, or end-to-end, or randomly oriented—it would be almost impossible for the chromosome to replicate semiconservatively (with all of the label going into one chromatid as in Taylor's results). Look at Figure 11-18

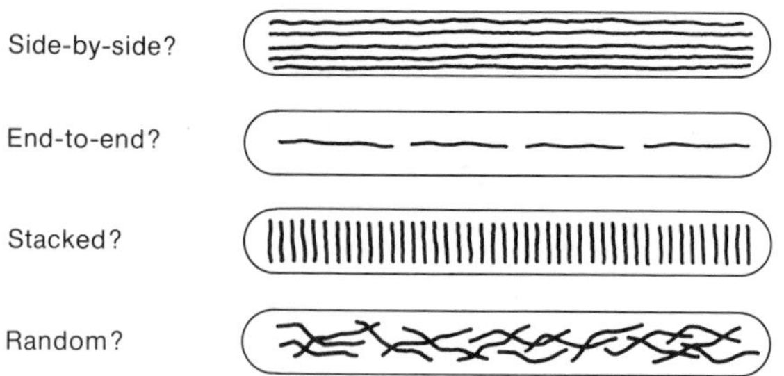

Figure 11-18. Some theoretical alternative packing arrangements of DNA in a eukaryotic chromosome. Any of these models is hard to reconcile with the data supporting a semiconservative model of DNA replication.

and try to figure out how it could be done. Recent studies on isolated chromosomes and long DNA molecules are consistent with the suggestion that *each chromatid is a single molecule of DNA.* That makes a very long molecule. There is enough DNA in a single human chromosome, for example, to stretch out to several inches in length. (That raises another interesting problem: How is this long molecule packed into the chromosome to permit easy replication?) The second fact supporting a single-molecule hypothesis is that DNA and genes behave as though they are attached end-to-end in a single string or thread that we call a linkage group. All genetic linkage data tell us that we need nothing more than a single linear array of genes per chromosome to explain the genetic facts.

As we mentioned parenthetically, there is far too much DNA in a chromosome for it to extend linearly along the chromosome. It must be packed very efficiently into the chromosome. Current thinking (supported by good microscopic evidence) tends toward a process of coiling and supercoiling of the DNA. Twist a rubber band with your fingers and notice the way it coils; chromosomes may be like this. We return to these questions in Chapter 14.

Testing Prediction 3

The third prediction of the Watson–Crick model of DNA replication is that a fork will be found in the DNA molecule during replication. In 1963, John Cairns tested this prediction by allowing replicating DNA in bacterial cells to incorporate tritiated thymidine. Theoretically, each newly synthesized daughter molecule should then contain one radioactive ("hot") strand and another nonradioactive ("cold") strand. After varying intervals and varying numbers of replication cycles in hot medium, the DNA was extracted from the cells, put on a slide, and autoradiographed for examination under the electron microscope. After one replication cycle in [^{3}H]-thymidine, rings of dots were seen in the autoradiogram, and these were interpreted as shown in Figure 11-19.

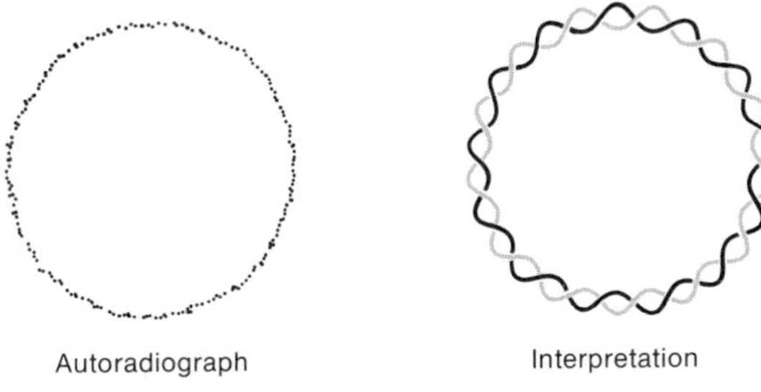

Figure 11-19. Autoradiograph of bacterial chromosome after one replication in tritiated thymidine. According to the semiconservative model of replication, one of the two strands should be radioactive. The interpretation of the autoradiograph is at the right.

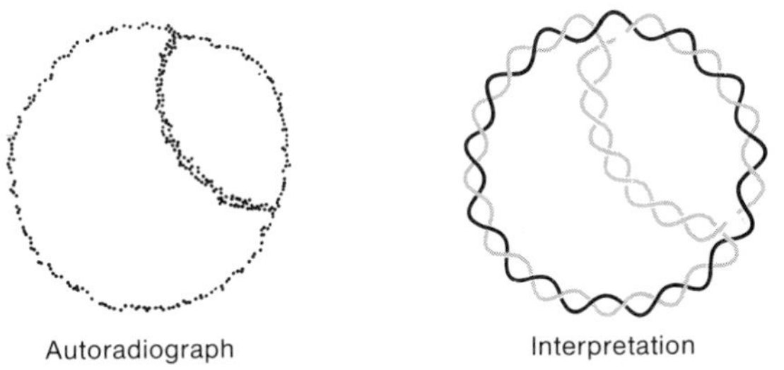

Figure 11-20. Autoradiograph of bacterial chromosome during the second round of replication in tritiated thymidine. In this theta *structure, the newly replicated double helix that crosses the circle could show both strands as radioactive. The double thickness of the radioactive tracing on the autoradiogram appears to confirm this.*

During the second replication cycle, the forks predicted by the model were indeed seen. Furthermore, the density of grains in the three segments was such that the interpretation shown in Figure 11-20 could be made. All sizes of these moon-shaped autoradiographic patterns were seen, corresponding to the progressive movement of the replication zipper, or fork, around the ring. Structures of the sort shown in Figure 11-20 are called theta (θ) structures.

This process of replication looks easy in Figure 11-20, but certain problems exist:

1. The double helix must rotate in the process of replication, because the two strands are intertwined (or interlocked). Some theorists have proposed molecular "swivels" along the chain to circumvent this problem in the bacterial ring chromosome.

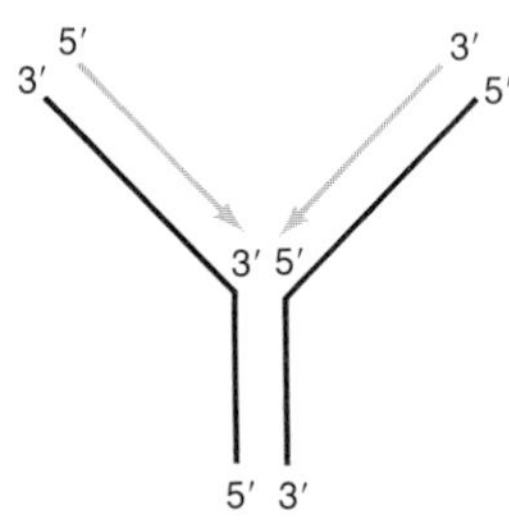

Figure 11-21. Model requiring polymerization of nucleotides at the replication fork from both the 5' and 3' ends of the DNA strands. Replication experiments in vitro strongly suggest that this does not occur. All growth of DNA uses the 5' end as the beginning and adds to the 3' end.

2. It is still unclear *how* the fork is generated. The simple notion of an unzipping molecule is inadequate. All of the known DNA polymerases initiate synthesis of a matching strand only at the 3′ end of the parental strand. In the zipper model, growth of the new strands would have to begin at the 3′ end on one strand and at the 5′ end on the other strand (Figure 11-21). Several fascinating alternatives have been proposed, including one in which new chains are synthesized in short segments and then later hooked together by enzymes called ligases (Figure 11-22); such enzymes are known to exist.

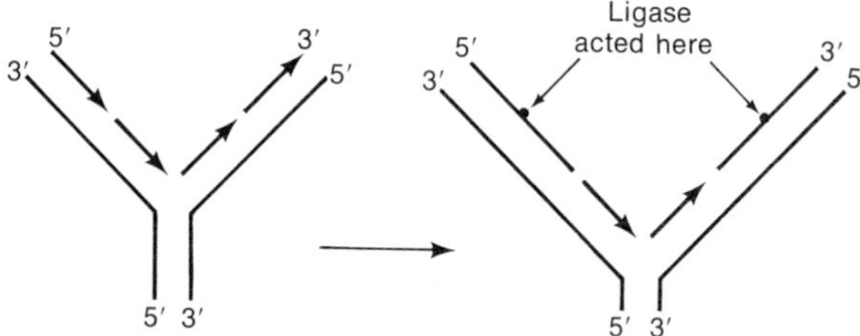

Figure 11-22. Replication model involving only 3' growth. In this model, short lengths of DNA are replicated and then joined together by enzymes called DNA ligases.

3. The original DNA polymerase that Kornberg discovered does not seem to be the main enzyme involved in DNA replication. It assists the main enzyme by completing the short fragments; it also acts to repair DNA after radiation damage. This information came from the discovery of a bacterial mutant that lacks activity of the Kornberg enzyme but still is capable of replicating its DNA.

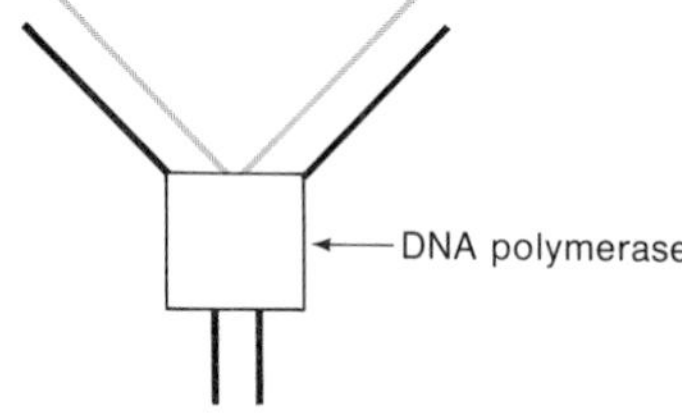

Figure 11-23. The actual molecular action of DNA polymerase is still cloaked in mystery. We shall use the general idea of the action of DNA polymerase as a simplified summary of whatever complex processes really go on.

For the purposes of our testing of the Watson–Crick model, the prediction of a fork structure has been adequately confirmed. Biochemists are still working on the problem of the actual molecular mechanism that occurs in the crotch of the replicating fork. The reactions must surely be complex. Several enzymes are known to be involved: two kinds of polymerases, and an unwinding enzyme, and ligase, and initiating enzymes. Furthermore, short strips of RNA (ribonucleic acid) are known to be involved as primers to get the polymerization process started. It is best to just leave the intimate details of the crotch cloaked in a loin cloth that we'll call the "DNA polymerase" for now (Figure 11-23).

Testing Prediction 4

The fourth prediction of the Watson–Crick model of replication is that, if replication is sequential and begins from a constant origin, then the DNA from a culture of replicating *E. coli* should contain more copies of the genes near the origin than of genes near the terminal end. (Note that failure to confirm this prediction might simply mean that there is no constant origin for replication, but confirmation of the prediction would support the model.)

This prediction was tested in 1963 by Toshio Nagata, who used *E. coli* lysogenic for two different phages, λ and 424. These two phages insert at different sites of the bacterial chromosome. Nagata synchronized the replication of his cells by filtering them through 18 layers of filter paper. Those that get through are all about the same size and therefore are all at the same stage in the cell cycle. At different times, he took a sample of the synchronized cells and irradiated them with ultraviolet light (which induces the prophages into a lytic cycle). He then harvested the phages released and measured the relative proportions of the two kinds. If we assume that the number of phages released is in direct proportion to the number of original prophages present on the bacterial chromosome, then this ratio should vary with the time of replication (Figure 11-24). The observations are consistent with a replication that begins in a set direction from a set initiation point in the bacterial chromosome.

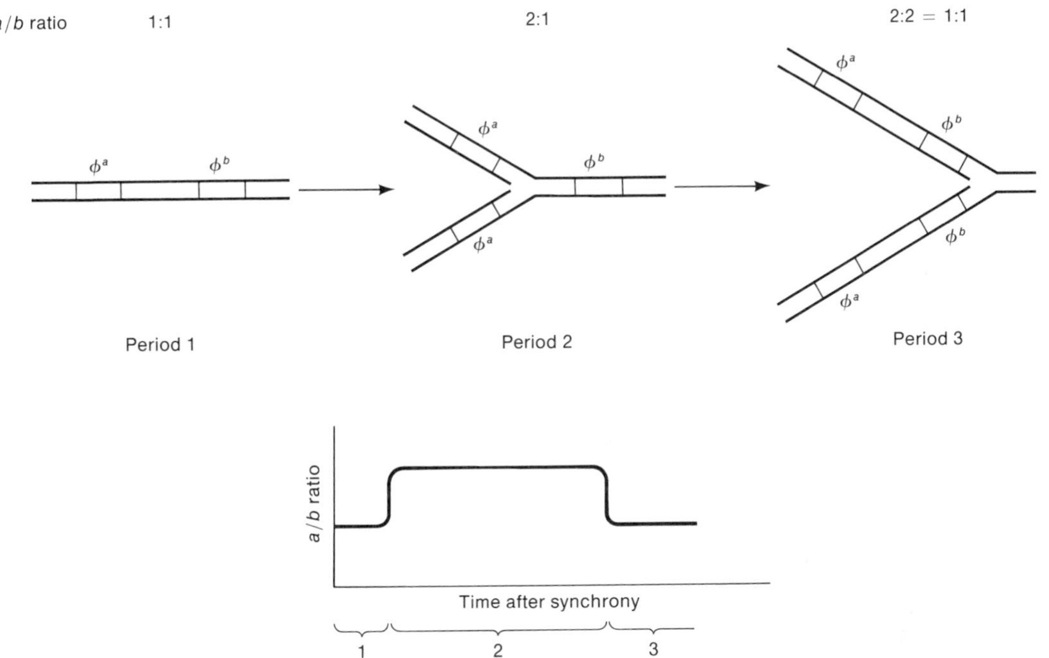

Figure 11-24. *"Doses" of prophage found at various times during replication of* E. coli. *($\phi^a = \lambda$; $\phi^b = 424$.) The* E. coli *host used in these experiments was synchronized in its growth, so replication of DNA should begin at approximately the same time in all cells. The doses of the two phage were determined by inducing the lytic cycle at different time periods and counting the numbers of λ and 424 phage present in each cell. These data are consistent with a polarized replication from a set initiation point in the bacterial chromosome.*

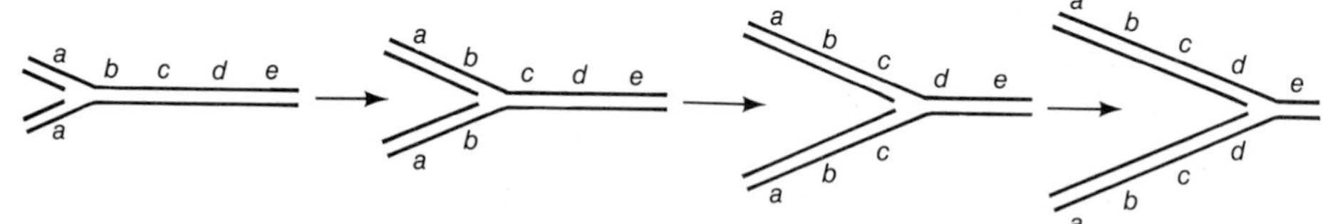

Figure 11-25. "Doses" of genes assayed at various times during replication in B. subtilis. *Measurement is in terms of transforming activity. It is reasoned that even in a nonsynchronized culture there should be a larger number of copies of genes near the initiation point of replication than distant from the initiation point. The data obtained from this experiment are consistent with this reasoning. On the average, of the samples shown here, the ratio of* a/e *would be 2:1, and the ratio of* d/e *would be 5:4.*

In 1963, Hiroshi Yoshikawa and Noboru Sueoka used a different method to test for gene dosage with *Bacillus subtilis,* which has no sexual cycle. They extracted DNA from a nonsynchronized population of cells and used the frequency of transformation for different markers as a measure of the relative proportions of each gene. They reasoned that, if replication does start at a unique point, then (even in a nonsynchronized population) the relative frequencies of genes will always be highest at the initiation end and will decrease in direct proportion to distance from that end (Figure 11-25).

The transforming ability of different markers may vary, so Yoshikawa and Sueoka used the number of transformants per amount of DNA extracted from nonreplicating spores as their basic control measurement. Transformants per amount of DNA from dividing cells could then be standardized against this control, so that they could compare the ratio of transforming activity for each of the markers. You can see from Figure 11-25 that the maximal ratio possible would be 2.0 (for a/e), and the minimal ratio would be 1.0 (for two closely linked sites). Experimental results confirmed the predictions. Furthermore, it is now possible to construct an independent linkage map showing the *relative positions of the loci on the replicating chromosome* as determined by transforming ability.

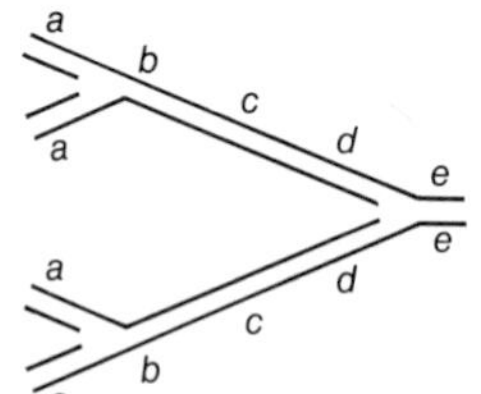

Figure 11-26. A second round of replication initiated before the end of the first, as determined for bacterial cells.

By growing cells from spores, Sueoka and his associates later obtained synchronized cultures. They extracted DNA at different times and assayed for gene dosage by transformation. The gene order they obtained was the same as that obtained in the earlier experiments, but they now obtained a maximal ratio of $a/e = 4$. Apparently, even before one round of replication is completed, a second round begins at the initiation end (Figure 11-26). In any case, these experiments do confirm a directional replication beginning from a fixed point on the chromosome.

Suppose that a eukaryotic cell is briefly exposed to [^{3}H]-thymidine (called a **pulse** exosure) and then provided an excess of cold thymidine (called the **chase**); the DNA is then extracted, and autoradiographs are made. Figure 11-27 shows the results of such a procedure, with what appear to be distinct

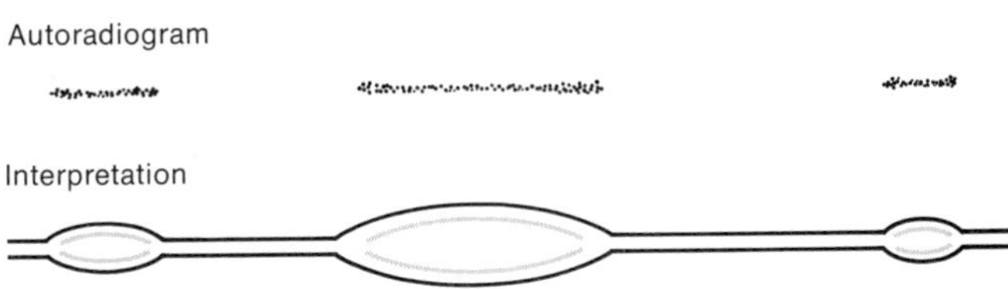

Figure 11-27. A replication pattern in DNA revealed by autoradiography. A cell was briefly exposed to ³H-thymidine (pulse) and then provided an excess of nonradioactive (cold) thymidine (chase). DNA is spread on a slide and autoradiographed. The interpretation shown in this figure is that there are several initiation points for replication within one double helix of DNA.

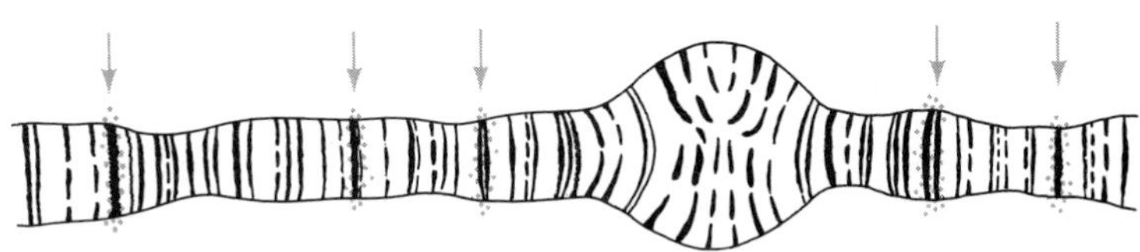

Figure 11-28. Replication pattern in a Drosophila *chromosome revealed by autoradiography. Several points of replication are seen without a single chromosome, as indicated by the arrows.*

simultaneously replicating regions along the DNA molecule. Replication appears to begin at several different sites on these eukaryotic chromosomes. Similarly, a pulse-and-chase study of DNA replication in polytene (giant) chromosomes of *Drosophila* by autoradiography reveals many replication regions within single chromosome arms (Figure 11-28). As yet there is no firm proof that these regions are indeed different start points on a single DNA molecule; they could also be interpreted as evidence that the chromosome is made up of many separate DNA molecules. The structure of the eukaryotic chromosome still remains as one of the most exciting unresolved problems in genetics (see Chapter 14).

All of the evidence we have reviewed thus far is consistent with a unidirectional (polarized) replication of DNA. However, autoradiography does provide a clue that something is happening that is inconsistent with that idea. When autoradiographs of the labeled bacterial DNA loops are examined, the distribution of grains is not found to be uniform. In fact, the concentration of grains is lowest in the center of the loop and increases toward the ends. This is beautifully demonstrated in DNA from *B. subtilis* (Figure 11-29). How can we explain these observations? If the lightest area of grains represents the earliest-replicating DNA (because the [^{3}H]-thymidine had not yet reached a high con-

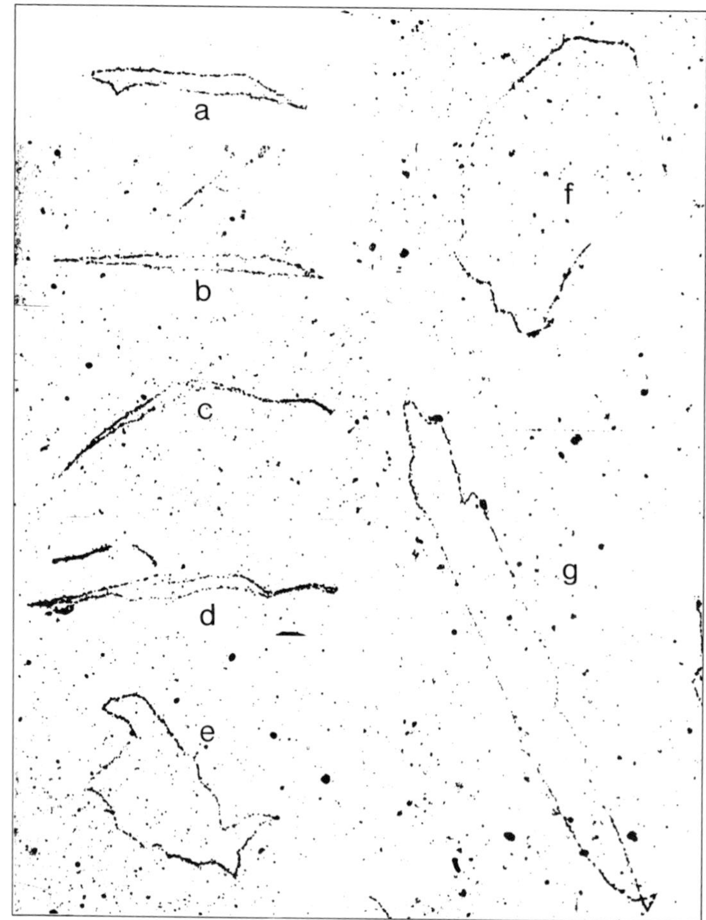

Figure 11-29. Replication loops (labeled a through g) in B. subtilis *DNA molecules as visualized through autoradiography. Note that there are fewer grains in the center of each of the strands of the loop. This phenomenon indicates that replication is occurring bidirectionally. (From E. B. Gyurasitis and R. G. Wake,* Journal of Molecular Biology *73:55, 1973.)*

centration inside the cell), then it appears that the DNA is replicating **bidirectionally** (with moving forks at both ends of the replicating piece).

Can we trust the grain patterns? In 1966, R. B. Inman devised another way of testing bidirectionality. He showed that if λ DNA is treated with alkali (high pH) for short periods, then the DNA strands in the regions rich in A–T pairs tend to separate before regions containing more G–C. This is called **partial denaturation,** or **partial melting.** (You will recall that an A–T pair shares two hydrogen bonds, whereas a G–C pair shares three.) By preventing the two complementary strands from hooking back together (through use of formaldehyde), Inman observed a highly reproducible pattern of "bubbles" along the length of the molecule. This pattern is called a denaturation map (Figure 11-30).

Inman isolated θ structures of λ DNA and made denaturation maps of them. (Incidentally, he always found that the maps of the two segments of equal length are identical, thus proving that newly synthesized strands are identical with each other.) If we take a specific bubble as a reference point in the molecule, then the unidirectional and bidirectional models of DNA replication lead to very different predictions (Figure 11-31). The unidirectional model predicts that the bubble should remain a constant distance from the origin of replication (the fixed fork), whereas the bidirectional model predicts that the distance from both forks will change. The results supported the bidirectional model.

Then what about the *B. subtilis* transformation experiment we discussed earlier? The transformation polarity obtained there suggests unidirectional replication. However, these results could be obtained from bidirectional replication if all of the markers are on the same side of the starting point for replication.

Bidirectional replication probably occurs in both linear and circular DNA molecules, and it has now been reported in phage, bacteria, yeast, and *Drosophila* DNA. Presumably the situation is something like that shown in Figure 11-32.

In 1972, Lucien Caro's research group measured the relative number of copies of different chromosome regions during replication. They began with F^- cells lysogenic for λ, and they then selected from among these cells the

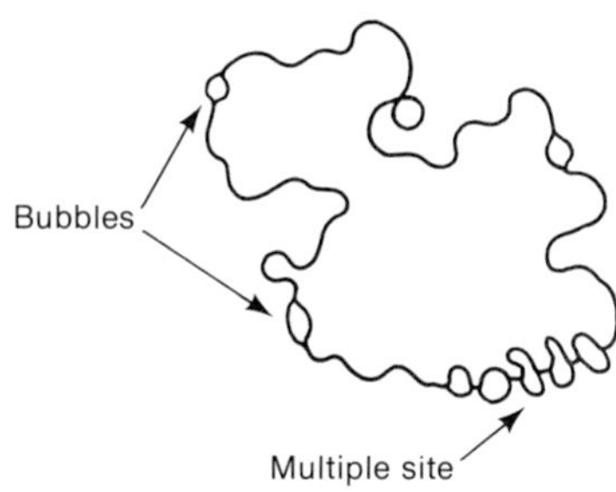

Figure 11-30. Partly denatured DNA as seen under the electron microscope. Several regions rich in A–T base pairs are apparent, including one multiple site. A–T rich regions denature into separate strands more easily when treated for short periods with alkali than do G–C rich regions.

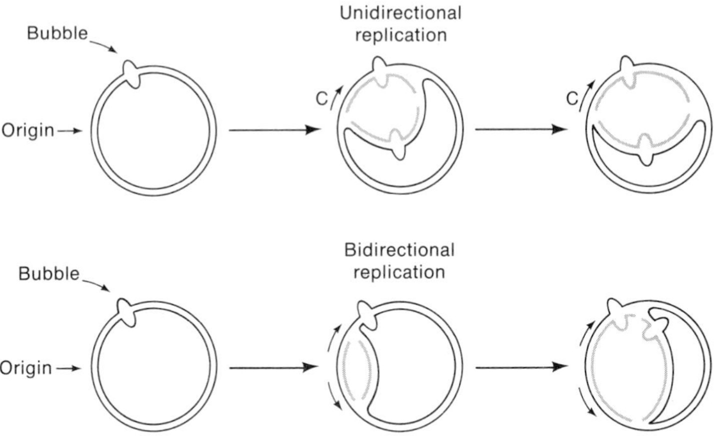

Figure 11-31. Testing unidirectional versus bidirectional replication by means of a marker denaturation bubble. For undirectional replication, the distance C from the replication fork to the bubble should remain constant. However, that distance should increase for bidirectional replication. (From A. Kornberg, DNA Synthesis. *Copyright © 1974, W. H. Freeman and Company.)*

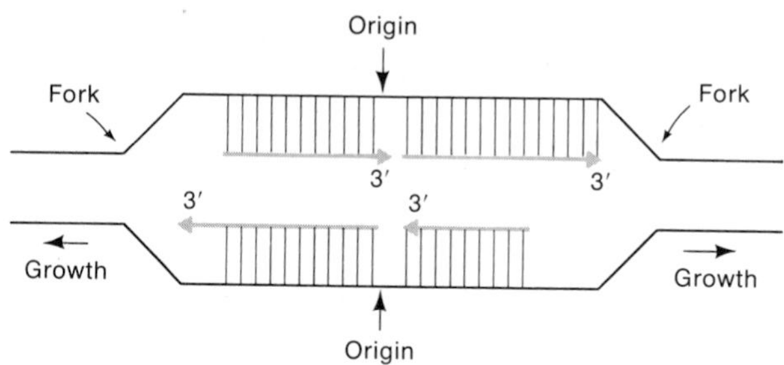

Figure 11-32. *Diagrammatic representation of 3′ polymerization in bidirectional replication of DNA. (From A. Kornberg,* DNA Synthesis. *Copyright © 1974, W. H. Freeman and Company.)*

ones that were also lysogenic for another phage called Mu-1. Mu-1 is very unusual in that it can insert itself anywhere in the *E. coli* chromosome. If it inserts within a cistron, then the function of that gene is lost. Thus, Mu-1 lysogens in specific loci can be recovered by selecting for auxotrophy for the corresponding compounds. Caro's group selected several strains with Mu-1 inserted at different positions in the bacterial chromosome. The closer Mu-1 is to the origin of replication, the more copies of Mu-1 there will be relative to the λ prophage, which is inserted at a fixed site. Thus the ratio of the two prophages should indicate the direction of replication.

To measure this ratio, it is necessary to use an analytical tool that has had enormous impact on molecular biology. In 1960, Paul Doty and Julius Marmur observed that, when DNA is heated to 100°C, all of the hydrogen bonds between the complementary strands are destroyed, and the DNA becomes single-stranded (this is called *melting,* or *denaturation*—we have already seen partial denaturation in experiments described earlier). If the solution is cooled slowly, some double-stranded DNA is formed that is biologically normal (for example, it may have transforming ability). Presumably, this **reannealing** (or **renaturation**) process occurs when two single strands happen to collide in such a way that the complementing base sequences can align and reconstitute the original double helix (Figure 11-33). This reannealing is very specific and precise, making it a very powerful tool because stretches of complementary base sequences in *different* DNAs also will anneal after melting and mixing. Thus the efficiency of annealing provides a measure of similarity between two different DNAs.

Message
The similarity between DNA molecules can be measured by melting them together and examining the amount of intermolecular hybridization that occurs upon slow cooling. This technique is called ***DNA/DNA hybridization.***

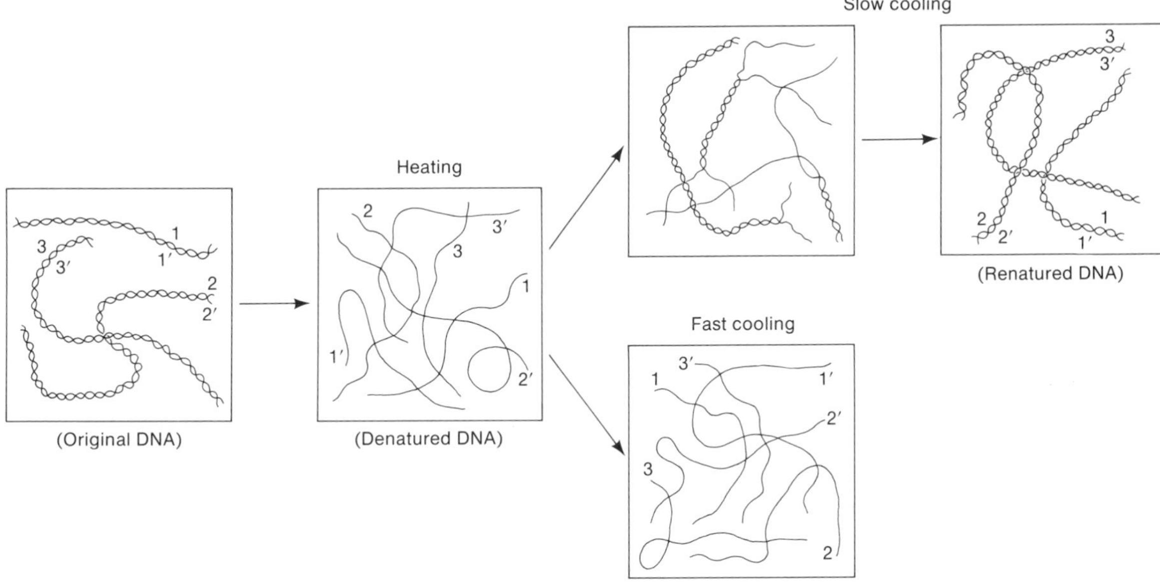

Figure 11-33. *DNA denaturation (melting) by heat and subsequent reannealing by slow cooling. Fast cooling does not allow reannealing and produces only single strands.*

Caro and his associates made use of DNA/DNA hybridization by extracting DNA from mature λ and Mu-1 phages, denaturing it, and fixing it to separate filters. The average number of copies of λ and Mu-1 prophage DNA were then measured by adding denatured labeled DNA extracted from the double lysogen. The amount of label binding to each filter is a measure of the number of copies of that prophage in the replicating bacterial DNA. Their results show that the ratio of Mu-1 to λ prophages is lowest when Mu-1 is inserted at *trp*, and the ratio increases in either direction from *trp* to a maximum at *ilv* (Figure 11-34). The interpretation of this result is that replication begins at a fixed point near *ilv* and then proceeds bidirectionally around the loop. Each replication fork must proceed at the same rate, because the replication seems to terminate from both directions around *trp*, which is exactly 180° from *ilv* on the genetic map.

Message
All of the genetic, biochemical, and cytological evidence remains consistent with the semiconservative replication of a duplex DNA structure, with replication beginning at a fixed initiation point and proceeding at a steady rate in both directions.

You may encounter the term **replicon,** which is simply a unit of replication (from the initiation point to the termination point).

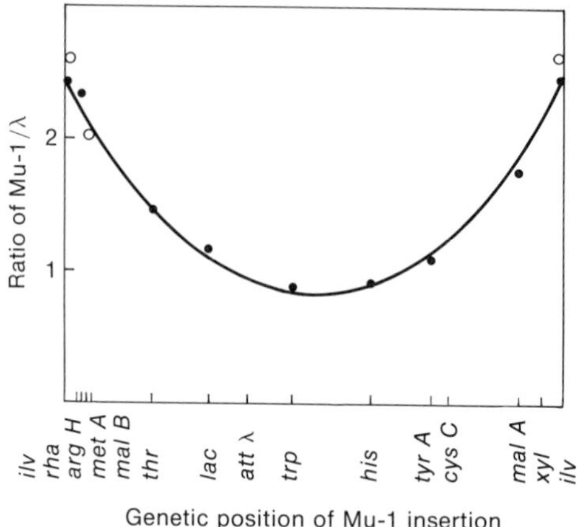

Figure 11-34. Genetic proof of bidirectional replication by use of DNA reannealing techniques. DNA was extracted from mature λ and Mu-1 phages, denatured, and fixed to separate filters. The average number of copies of λ and Mu-1 prophage DNA was then measured by adding denatured labeled DNA extracted from the double lysogen. The amount of label binding to the filter is a measure of the number of prophage gene copies in the replicating bacterial DNA. Mu-1 prophage can be inserted at various sites in the bacterial chromosome. The results show that the ratio of Mu-1 to λ prophages is lowest when Mu-1 is inserted at trp *and increases in either direction from* trp. *(From R. D. Bird and L. Caro,* Journal of Molecular Biology *70:557, 1972.)*

DNA and the Gene

We have seen in Chapter 10 that the gene is a linear sequence of codons that is colinear with the sequence of amino acids in a protein. Now we learn that DNA is the genetic material and consists of a linear sequence of nucleotide pairs. The obvious conclusion is that the allele maps represent a genetic equivalent of the nucleotide-pair sequences in DNA. We can validate this assumption if we can show that *the genetic maps are congruent with DNA maps.* This step has been taken, using some elegant genetic and biochemical tricks.

The DNA of the λ phage turns out to be a linear stretch of DNA that can be circularized because each 5′ end of the two strands has an extra terminal extension of 12 bases that is complementary to the other 5′ end (Figure 11-35). Because they are complementary, these ends can pair to join the DNA into a circle; they are known as "cohesive," or "sticky," ends.

When a linear object such as a DNA molecule is subjected to shear stress (say, by pipetting or stirring), the mechanics of the stress cause breaks to occur principally in the middle of the molecule. When DNA from the λ phage is sheared in half, the two halves happen to differ in G–C ratio, which means that

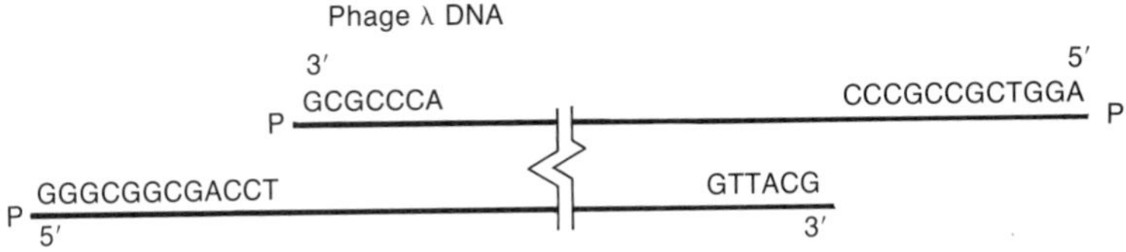

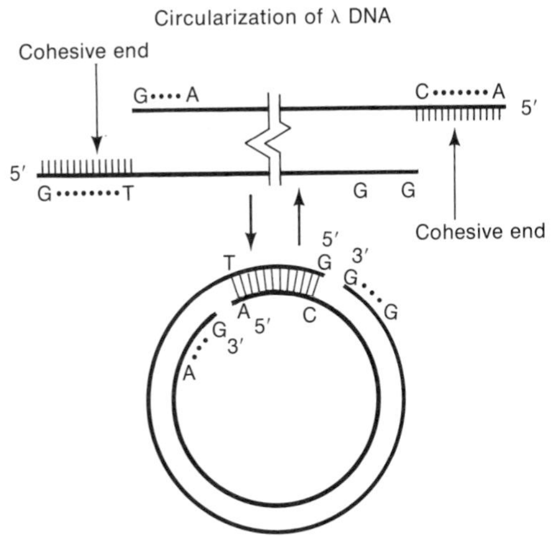

Figure 11-35. The λ DNA is linear in the phage, but once in a host cell it circularizes as a prelude to insertion or replication. Circularization is achieved by joining the complementary ("sticky") single-stranded ends. (From A. Kornberg, DNA Synthesis. *Copyright © 1974, W. H. Freeman and Company.)*

their buoyant densities differ and they can be separated by centrifugation in CsCl. When the two half-molecules are separated, their genetic content can be assayed by introducing the DNA into bacteria in the following way. The DNA is introduced into the bacteria simultaneously upon infection with mutant phage λ. Recombination can occur between the phage DNA and the fragment as shown in Figure 11-36, in which *a* through *f* are phage genes. The introduced DNA can be incorporated into the normal λ DNA and "rescued" by inducing lysis. You can see that, by using different strains, this technique can lead to analysis of the marker content of the DNA fractions. This is called a **marker-rescue experiment.** Such experiments demonstrated that one particular half of the DNA molecule carries the information of one particular half of the linkage map.

Dale Kaiser and his associates separated one of the DNA halves carrying a cohesive end and attached the other end to chromatographic material over which DNA could be passed (Figure 11-37). The single-stranded ends dangle free like fish hooks. If the other half of the DNA is sheared into smaller and

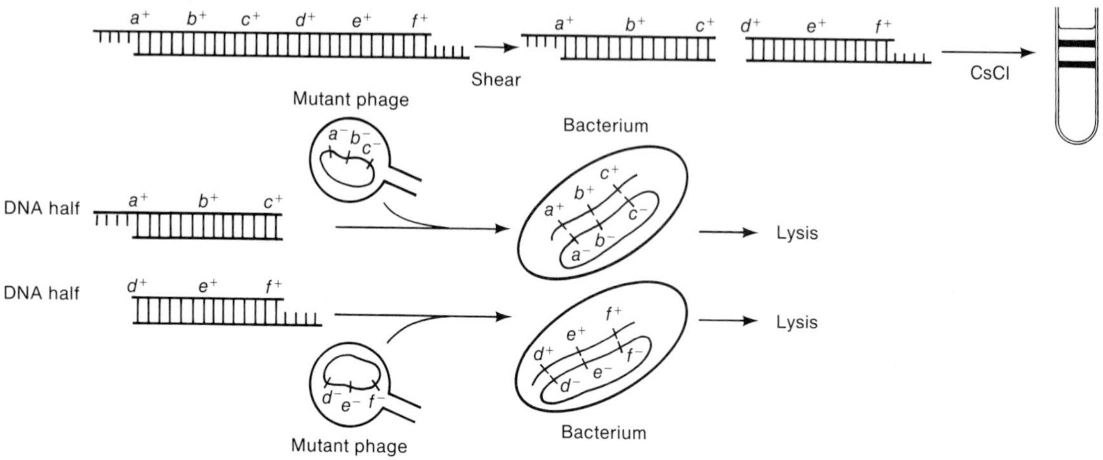

Figure 11-36. *When λ DNA is broken, it forms roughly equal halves that happen to have different buoyant densities. Each half can be "rescued" by a multiply mutant phage during simultaneous transformation and infection. From these experiments, it was shown that only genes from one specific half of the genetic map could be rescued from one specific half of the DNA. (Note that the bacterial chromosome has been omitted from this diagram for simplicity.)*

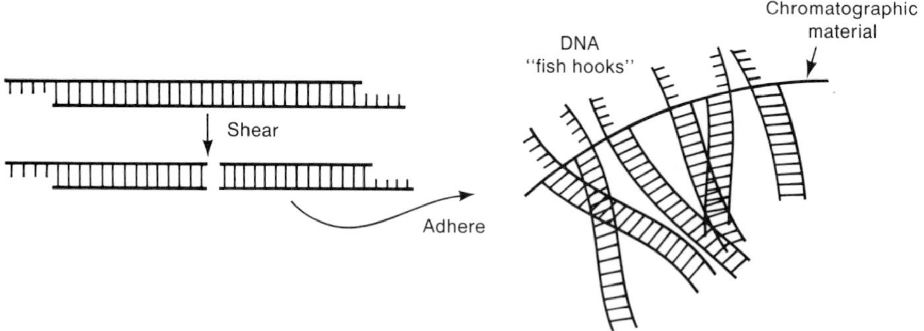

Figure 11-37. *One specific sticky end can be used as a "fish hook" for the other sticky end plus any genes that may be attached to it. One-half of the λ phage DNA is attached to chromatographic material so that the cohesive or sticky ends are exposed. And DNA passed over this chromatographic material that has single-stranded base sequences complementary to the "fish hook" cohesive ends will stick. Other DNA will pass through. Later, the "stuck" DNA can be removed from the chromatographic material by heating it to break the hydrogen bonds.*

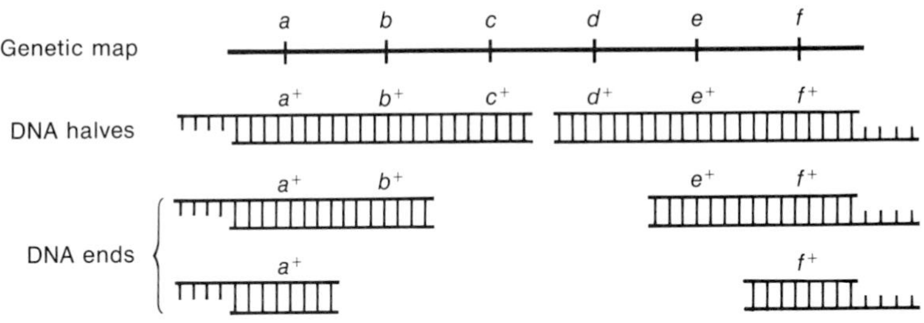

Figure 11-38. As the λ DNA is sheared progressively smaller and smaller, genes are lost to the fish hooks (Figure 11-37) in a progressive order, the same as their order on the genetic map.

smaller molecules and then passed over the sticky ends, the complementary sequences will stick. These stuck pieces can be easily detached by raising the temperature to break the hydrogen bonds in the sticky ends. In this way, a series of fractions (of varying size) of one end of the λ DNA can be separated and tested for content by marker rescue. Kaiser and his associates showed that an unambiguous arrangement of genes on the DNA can be determined, and this sequence is completely congruent with the genetic map (Figure 11-38).

Message
We are now justified in concluding that the sequence of bases in the DNA is indeed congruent with the gene map.

The Genetic Code

If genes are segments of DNA, and if DNA is just a string of nucleotide pairs, then how does the sequence of nucleotide pairs dictate the sequence of amino acids in protein? The analogy to a code springs to mind at once. The cracking of the genetic code is the story of the rest of this chapter and much of the following one. The experimentation was sophisticated and swift, and it did not take long for the code to be deciphered once its existence was strongly indicated.

Armchair logic tells us that if nucleotide pairs are the *letters* in a code, then a combination of letters could form words representing different amino acids. Then we must ask how many letters make up a word (a codon) and which specific codon or codons represent each specific amino acid.

Reading a DNA molecule from one particular end, one nucleotide pair at a time, only four nucleotide pairs can be encountered:

$$\begin{matrix} -\text{A}- \\ -\text{T}- \end{matrix} \quad \begin{matrix} -\text{T}- \\ -\text{A}- \end{matrix} \quad \begin{matrix} -\text{G}- \\ -\text{C}- \end{matrix} \quad \text{and} \quad \begin{matrix} -\text{C}- \\ -\text{G}- \end{matrix}$$

Thus, if the words are one letter long, then only four words are possible. This cannot be the genetic code because we must have a word for each of the 20 amino acids commonly found in cellular proteins. If the words are two letters long, then $4^2 = 16$ words are possible; for example,

–AT– or –CT– or –CC–
–TA–   –GA–   –GG–

where the length of the DNA molecule runs across the page (with one strand shown at the top and the other at the bottom). This vocabulary still is not large enough.

If the words are three letters long, then $4^3 = 64$ words are possible; for example,

–ATT– or –GCG– or –TGC–
–TAA–   –CGC–   –ACG–

This code would provide more than enough words to describe the amino acids. We can conclude that the code word must consist of at least three nucleotide pairs. However, if all words are "triplets," we have a considerable excess of possible words over the 20 needed to name the common amino acids.

Convincing proof that a codon is, in fact, three letters long (and no more than three) came from beautiful genetic experiments first reported in 1961 and later extended by Crick and his group, using mutants in the *rII* locus of T4 phage. Mutations causing the rII phenotype were induced using a chemical called proflavin, which was thought to act by the addition or deletion of single nucleotide pairs in DNA. (This assumption is based on experimental evidence that we shall not discuss.) The following examples illustrate the action of proflavin:

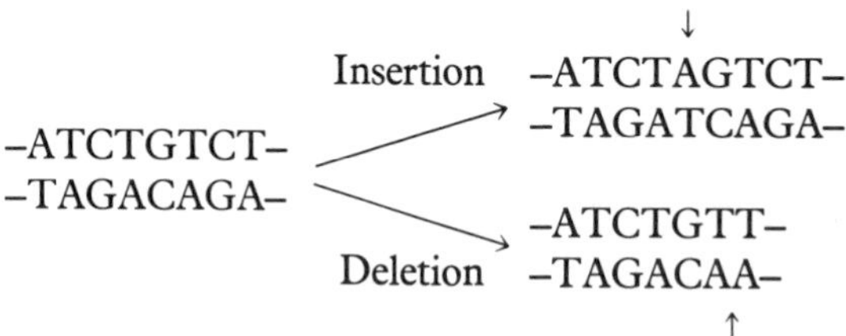

Then, starting with one particular proflavin-induced mutation called FCO, they again used proflavin to induce "reversions" that were detected by their wild-type plaques on *E. coli* strain K(λ). Genetic analysis of these plaques revealed that the "revertants" were not identical to true wild types, thereby suggesting that the back mutation was not an exact reversal of the original forward mutation. In fact, the reversion was found to be caused by the presence of a *second mutation* at a different site from—but in the same cistron as—that of FCO; this second mutation suppressed mutant expression of the original FCO. More surprisingly, the **suppressor mutation** could be separated from

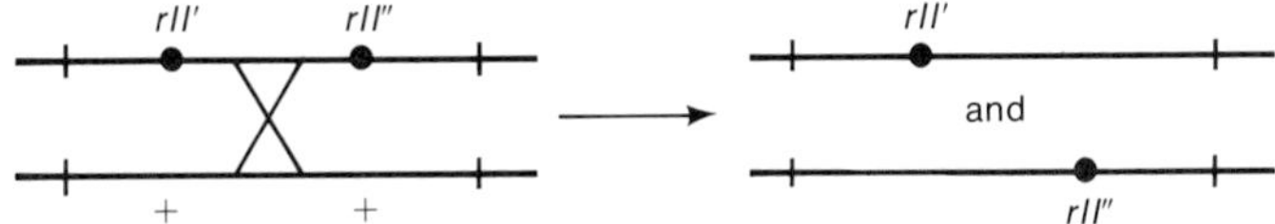

Figure 11-39. The supressor of an initial rII *mutation is shown to be an* rII *mutation itself after separation by crossing over. The original mutant, FCO, was induced by proflavin. Later, when the FCO strain was treated with proflavin again, a revertant was found, which on first appearance seemed to be wild type. However, it was found that a second mutation within the* rII *region had been induced, and the double mutant,* rII' rII''*, was shown not to be quite identical to the original wild type.*

the original forward mutation by recombination. When this was done, the suppressor was shown to be an *rII* mutation by itself (Figure 11-39).

How can we explain these results? *If* the cistron is "read" from both ends, then the two proflavin-induced mutations should still give a mutant phenotype when combined:

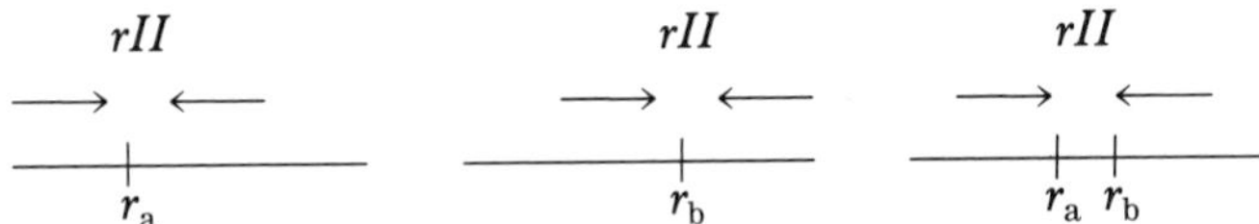

However, if reading is polarized—that is, if the cistron is read from one end only—then the original proflavin-induced addition or deletion could be mutant because it interrupts a normal reading mechanism that establishes the groups of bases to be read as words. For example, if each three nucleotide pairs make a word, then the "reading frame" might be established by taking the first three pairs from the end as the first word, the next three pairs as the second word, and so on. In that case, a proflavin-induced addition or deletion of a single pair would shift the reading frame from that point on, causing all following words to be misread. Such a **frame-shift mutation** could reduce most of the genetic message to gibberish. However, the proper reading frame could be restored by a compensatory insertion or deletion somewhere else, giving only a short stretch of gibberish between the two. Consider the following example that uses three-letter English words to represent the codons:

	THE FAT CAT ATE THE BIG RAT
Delete C:	THE FAT ATA TET HEB IGR AT
Insert A:	THE FAT ATA ATE THE BIG RAT

The insertion suppresses the effect of the deletion by restoring most of the sense of the sentence. By itself, however, the insertion also disrupts the sentence:

THE FAT CAT AAT ETH EBI GRA T

If we assume that the FCO mutant is caused by an addition, then the second (suppressor) mutant would have to be a deletion, because this would restore the reading frame of the gene as we have seen (a second insertion would not correct the frame). In the following diagrams, we use a hypothetical nucleotide chain to represent DNA for simplicity, realizing that the complementary chain is automatically dictated by the specificity of the base pairing. We also assume that the code words are three letters long and that the words are read in one direction (left to right in our diagrams).

	CAT	CAT	CAT	CAT	CAT		Wild-type message
Addition ↓ (between CAT and ACA)	CAT	ACA	TCA	TCA	TCA	T	*rII′* message: distal words changed (x) by frame-shift mutation (words marked ✓ are unaffected)
	✓	x	x	x	x		
Deletion ↓ (in TCT)	CAT	ACA	TCT	CAT	CAT		*rII′ rII″* message: few words wrong, but reading frame restored for later words
	✓	x	x	✓	✓		

The few wrong words in the suppressed genotype could account for the fact that the "revertants" (suppressed phenotypes) that Crick and his associates recovered did not look exactly like the true wild types phenotypically.

We have assumed here that the original frame-shift mutation was an addition, but the explanation works just as well if we assume that the original FCO mutation is a deletion and the suppressor is an addition. If the FCO is defined as plus, then suppressor mutations are automatically minus; hence, proflavin-induced mutations are also called sign mutations. Experiments confirmed that a plus cannot suppress a plus, nor can a minus suppress a minus. In other words, two mutations of the same sign never act as suppressors of each other. However, very interestingly, it was found that combinations of *three* pluses or *three* minuses can act together to restore a wild-type phenotype. This observation provided the first experimental confirmation that a word in the genetic code consists of three successive nucleotide pairs, or a **triplet.** This is because three additions or three deletions within a gene would automatically restore the reading frame if the words are triplets. For example,

	Deletions					
	↓	↓	↓			
CAT	CAT	CAT	CAT	CAT	CAT	CAT
CAT	ACA	TAT	CAT	CAT	CAT	
√	×	×	√	√	√	

Crick's work also suggested that the code is **degenerate.** That expression is not a moral indictment! It simply means that each of the 64 triplets must have some meaning within the code, so that at least some amino acids must be specified by two or more different triplets. If only 20 triplets were used (with the other 44 being nonsense), then most frame-shift mutations would be expected to produce nonsense words, which presumably would stop the protein-building process. However, if all triplets specify some amino acid, then the changed words would simply result in the insertion of incorrect ("gibberish") amino acids into the protein. Thus Crick reasoned that many or all amino acids must have several different names in the base-pair code; this hypothesis was later confirmed biochemically.

Proof that the genetic deductions about proflavin were correct came from an analysis of proflavin-induced mutants in a gene whose protein product could be analyzed. George Streisinger worked with the gene that controls the enzyme lysozyme, whose amino acid sequence is known. He induced a mutation in the gene with proflavin and selected for proflavin-induced "revertants," which were shown genetically to be double mutants (with mutations of opposite sign). When the protein of the double mutant was analyzed, a stretch of "gibberish" amino acids lay between two wild-type ends, just as predicted:

Wild type	–Thr–Lys–Ser–Pro–Ser–Leu–Asn–Ala–
"Revertant" type	–Thr–Lys–Val–His–His–Leu–Met–Ala–

Thinking back to Yanofsky's work (Chapter 10), it now looks as though we have pinned down the nature of the recon. Recall that Yanofsky showed that the piece of gene coding for one amino acid (a codon) is divisible by recombination. We have now seen that this piece of gene is three nucleotide pairs long, so surely a nucleotide pair must be the recon, the piece of a gene that is indivisible by recombination. Perhaps what occurred in Yanofsky's experiment was something like the situation diagrammed in Figure 11-40.

In talking about one nucleotide pair being "changed" (we have not specified how), it looks as though we have also identified the muton—that piece of a gene that *is* the unit of change. In other words, the nucleotide pair is both the recon and the muton. The codon contains three recons (three mutons). We consider the mechanism whereby a nucleotide pair can change in Chapter 16.

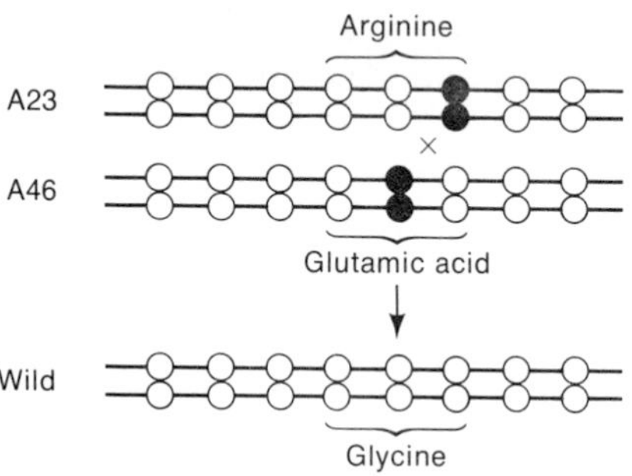

Figure 11-40. Intracodon recombination can regenerate a wild codon from two codons containing different nucleotide-pair substitutions.

Message
A codon is three nucleotide pairs. A muton and a recon are each one nucleotide pair.

We can use a diagram to summarize our stage of understanding in this book so far:

Codon sequence in DNA ⟶ Amino acid sequence in protein

Replication

At this stage, the story is essentially over as far as geneticists are concerned. The remaining details were for the most part worked out by people with expertise in chemistry rather than in genetic analysis. We shall summarize the findings because they are very interesting to geneticists, but geneticists can lay little claim to their specific elucidation. However, we should point out that many questions have been resolved that geneticists thought would never be resolved within their lifetimes.

1. What is the nature of the gene? It is a strip of DNA that can be viewed as a continuous sequence of triplets of nucleotide pairs, each triplet being a codon.

2. How do genes control phenotypes? Each codon stands for an amino acid in a polypeptide sequence. Polypeptides (proteins) catalyze most

metabolic reactions and form most of the important body structures; thus, changes in polypeptides produce changes in phenotypic expressions.

3. What is the molecular nature of mutations? We shall see that there are several kinds, but thus far we have seen two kinds: nucleotide-pair change (or substitution) and nucleotide-pair addition or deletion.

Alternative DNA Models

The Watson–Crick model for DNA structure has been the central pivotal point in biology over the past two decades. The impact and importance of the model has been immense, catapulting our understanding of cellular and evolutionary processes onto an entirely new and lofty level. In fact, so great has been the success of the intertwined double helix that only recently, after the dust has settled somewhat, have physical scientists begun to reexamine X-ray data (of the kind Watson and Crick interpreted) and suggest some alternative conformations for DNA.

The basic structure remains unchallenged—two nucleotide chains held together by hydrogen bonds between specifically paired bases—but the way in which the two strands are associated has been challenged in one particularly interesting alternative structure. In this structure, the strands are viewed as being not interlocked (Watson and Crick), but rather *side-by-side.* Figure 11-41 shows a highly simplified representation of the basic differences involved.

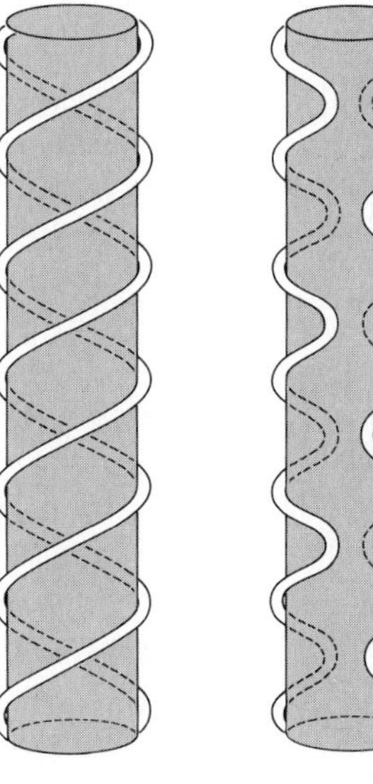

Figure 11-41. Alternative models for the double-helical structure of DNA. These are simplified diagrams to show the difference between the interlocked double helix and the side-by-side double helix. (From G. A. Rodley et al., "A Possible Conformation for Double Stranded Polynucleotides." PNAS *73, 1976.)*

Both versions are compatible with X-ray scattering and other physical data. The side-by-side model does have one highly appealing feature from the genetic point of view. There have always been nagging doubts about how strand separation and the associated unscrewing of the Watson–Crick interlocked helix could be achieved during replication. The side-by-side model, of course, makes such separation much easier to achieve, at least at the conceptual level. Whereas it is relatively easy to imagine interlocked helices swiveling and unwinding in a phage or bacterium, a eukaryotic chromosome represents a monstrous logistics problem for DNA swivels. Obviously, all the evidence is not yet available for a reliable decision between these models. However, the new alternatives do illustrate how science can proceed by using models that are not necessarily 100% correct.

Summary

Experimental work on the molecular nature of hereditary material has conclusively demonstrated that DNA (and not protein, RNA, or some other substance) is indeed the genetic material. Using data supplied by others, Watson and Crick created a double helical model with two DNA strands wound around each other running in antiparallel fashion. Specificity of binding the two strands together is based on the fit of adenine (A) to thymine (T) and guanine (G) to cytosine (C), the former pair held by two hydrogen bonds and the latter by three.

The Watson–Crick model shows how DNA can be replicated in an orderly fashion, a prime requirement for genetic material. Replication is accomplished semiconservatively; that is, one double helix is replicated into two identical helices, each with identical linear orders of nucleotides, and each of the two new double helices is composed of one old and one newly polymerized strand of DNA. This semiconservative replication occurs in both prokaryotes and eukaryotes.

It has been shown in several organisms that replication of DNA originates at an initiation point and proceeds in both directions, adding nucleotides to the 3′ ends of the growing chains. Eukaryotes may differ from prokaryotes by having many initiation points within one chromosome rather than a single initiation point.

The similarity between DNA molecules can be measured by melting them and examining the amount of intermolecular hybridization. Data obtained in this fashion and in other ways led to the conclusion that the sequence of bases in the DNA is congruent with the gene map.

By studying frame-shift mutations, geneticists concluded that a codon consists of three nucleotide pairs. A muton and a recon are each one nucleotide pair; that is, the nucleotide pair is the unit of mutation and of recombination.

Recently, new models have been presented for the double helical structure of DNA. These models accommodate the data known about DNA and offer some tantalizing solutions to problems raised by the Watson–Crick model.

Problems

1. Assume that thymine makes up 15% of the bases in a specific DNA molecule. What percentage of the bases are cytosine?

2. Draw a graph of DNA content against time in a cell that undergoes mitosis and then meiosis.

3. Consider an *E. coli* chromosome in which every nitrogen atom is labeled—that is, every nitrogen atom is the heavy isotope ^{15}N instead of the normal isotope ^{14}N. The chromosomes then are allowed to replicate in an environment in which all the nitrogen is ^{14}N. Using a solid line to represent a heavy polynucleotide chain and a dotted line for a light chain, sketch the following:
   a. The heavy parental chromosome and the products of the first replication after transfer to ^{14}N medium, assuming that the chromosome is one DNA double helix and that replication is semiconservative.
   b. Repeat part a, but assume conservative replication.
   c. Repeat part a, but assume that the chromosome is in fact two side-by-side double helices, each of which replicates semiconservatively.
   d. Repeat part c, but assume that each side-by-side double helix replicates conservatively and that the overall *chromosome* replication is semiconservative.
   e. Repeat part d, but assume that the overall chromosome replication is conservative.
   f. If the daughter chromosomes from the first division in ^{14}N are spun in a cesium chloride density gradient and a single band is obtained, which of possibilities a through e can be ruled out? Reconsider the Meselson–Stahl experiment: What does it *prove?*

4. R. Okazaki found that the immediate products of DNA replication in *E. coli* include single-stranded DNA fragments approximately 1000 nucleotides in length after the newly synthesized DNA is extracted and denatured. When he allowed DNA replication to proceed for a longer period of time, he found a lower frequency of these short fragments, and he found long single-stranded DNA chains after extraction and denaturation. Explain how this result might be related to the fact that all known DNA polymerases synthesize DNA only in a $5' \rightarrow 3'$ direction.

5. When plant and animal cells are given pulses of [^{3}H]-thymidine at different times in the cell cycle, it is found that heterochromatic regions on chromosomes are invariably "late replicating." Can you suggest any biological significance this observation might have?

6. On the planet of Rama, the DNA is built of six nucleotide types: A, B, C, D, E, and F. A and B are called marzines; C and D are orsines; and E and F are pirines. The following rules are valid in all Raman DNAs:
   1. Total marzines = total orsines = total pirines
   2. A = C = E
   3. B = D = F

   a. Prepare a model for the structure of Raman DNA.

   b. On Rama, mitosis produces three daughter cells. Bearing this fact in mind, propose a replication scheme for your DNA model.

   c. Proflavin studies reveal that the code is (inevitably) a triplet code. How many different codons are possible in this system?

   d. Consider the process of meiosis on Rama. What comments or conclusions can you suggest?

7. If you extract the DNA of the coliphage ϕX174, you will find that its composition is 25% A, 33% T, 24% G, and 18% C. Does this make sense in terms of Chargaff's rules? How would you interpret this result? How might such a phage replicate its DNA?

8. The temperature at which a DNA sample denatures can be used to estimate the proportion of its nucleotide pairs that are G–C. What would be the basis for this determination, and what would a high denaturation temperature for a DNA sample indicate?

9. You extract DNA from a small virus, denature it, and allow it to reanneal with DNA taken from other strains that carry a deletion, an inversion, or a duplication. What would you expect to see on inspection with an electron microscope?

10. DNA extracted from a mammal is heat-denatured and then slowly cooled to allow reannealing. Figure 11-42 shows the results obtained. There are two "shoulders" in the curve. The first shoulder indicates the presence of a very rapidly annealing part of the DNA—so rapid in fact, that it occurs before strand interactions take place.

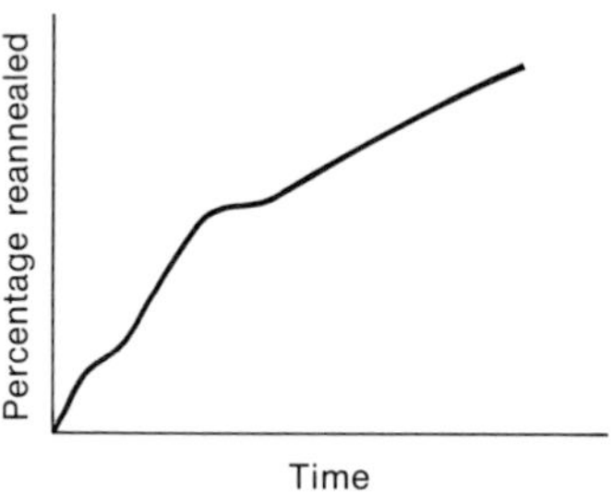

Figure 11-42.

a. What could this part of the DNA be?

b. The second shoulder is a rapidly reannealing part as well. What does this evidence suggest?

11. Design tests to determine the physical relationship between highly repetitive and unique DNA sequences in chromosomes. (HINT: It is possible to vary the size of DNA molecules by the amount of shearing they are subjected to.)

12. In mice, there are viruses that are known to cause cancer. You have a pure preparation of virus DNA, a pure preparation of DNA from the chromosomes of mouse cancer cells, and pure DNA from chromosomes of normal mouse cells. Virus DNA will hybridize with cancer-cell DNA, but not with normal-cell DNA. Explore the possible genetic significance of this observation, its significance at the molecular level, and its medical significance.

13. Ruth Kavenaugh and Bruno Zimm have devised an elegant technique to measure the maximal length of the longest DNA molecules in solution. They studied DNA samples from the three *Drosophila* karyotypes shown in Figure 11-43. They found the longest molecules in karyotypes a and b to be of similar length and about twice the length of the longest in c. How would you interpret these results?

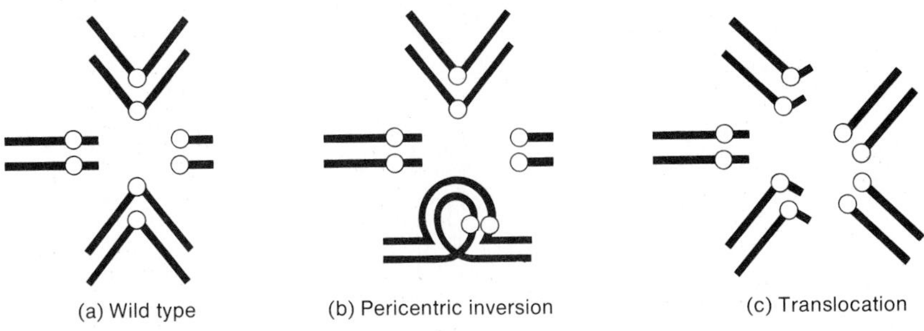

Figure 11-43.

14. In the harlequin-chromosome technique, you allow *three* rounds of replication in bromodeoxyuridine and then stain the chromosomes. What result do you expect to obtain?

15. Several proflavin-induced plus and minus mutations are obtained in the *rII* region of phage T4. They are assumed to all be in a region that is not essential to the normal function of the polypeptide but that is flanked by essential regions. For each of the following combinations, predict whether the phenotype will be wild or mutant. (a) +; (b) + −; (c) + +; (d) − −; (e) + − +; (f) + + +; (g) − − +; (h) − − −; (i) + + + +; (j) + − + + +.

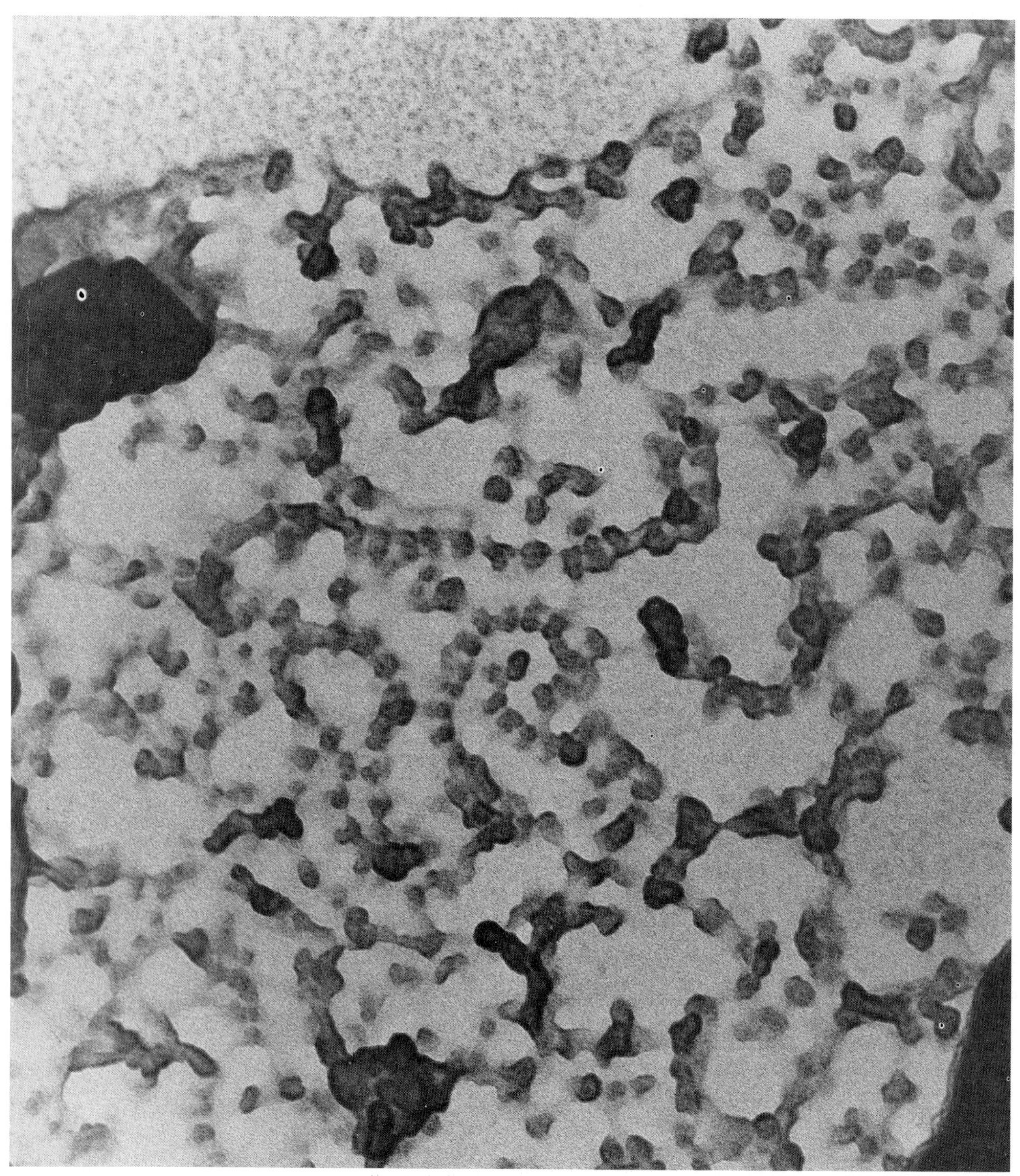

Human ribosomes embedded in the membranes of the endoplasmic reticulum of the cell. (Copyright © Jack D. Griffith.)

12

DNA Function

The genetic information embodied in DNA can either be copied into more DNA during replication or be translated into protein. These are the processes of information transfer that constitute DNA function. We considered replication in some detail in Chapter 11. It is now time to explore the way that genetic information is turned into protein. Information really is an appropriate word here. Literally, information means "that which is necessary to give form" (to something), which is precisely what DNA does as it influences phenotype. It gives form to organisms. The form itself is to a large extent embodied in protein.

Very little of this part of our story was revealed through purely genetic analysis. Although mutants have been useful, they have been used as tools to shortcut a lot of the biochemical work. Nevertheless, DNA function is important to genetics, and the kind of reasoning used in investigating it illustrates an analytical approach also characteristic of genetics.

Much of the work discussed in this chapter has been done in the past two decades, and the many areas of uncertainty reflect the problems just now being posed. New discoveries have been made with spectacular rapidity in molecular biology, bringing us within striking range of genetic engineering with all its dazzling promise and horrendous possible consequences.

Transcription

It soon became evident that information is not transferred directly from DNA into protein. Another nucleic acid, ribonucleic acid (RNA), is necessary as an intermediary. There are good reasons for thinking this. For one thing, DNA is found in the nucleus (of eukaryotic cells), whereas protein is known to be synthesized in the cytoplasm. If cells are fed radioactive RNA precursors, the labeled RNA shows up first of all in the nucleus, indicating that the RNA is synthesized there. In a pulse–chase experiment, a brief "pulse" of labeling is followed by a "chase" of nonlabeled RNA precursors. In samples taken after the chase, the labeled RNA is found in the cytoplasm (Figure 12-1). Apparently, the RNA is synthesized in the nucleus and then moves into the cytoplasm. Thus it is a good candidate as an information-transfer intermediary between DNA and protein.

In 1957, Elliot Volkin and Lawrence Astrachan made a significant observation. They found that one of the most striking molecular changes when *E. coli* is infected with the phage T2 is a rapid burst of RNA synthesis. Furthermore, this phage-induced RNA "turns over" rapidly, as shown in the following experiment. The infected bacteria are first pulsed with radioactive uracil (a specific precursor of RNA). The bacteria are then chased with cold uracil; the RNA recovered shortly after the pulse is labeled, but that recovered somewhat later after the chase is unlabeled, indicating that the RNA has a very short lifetime. Finally, when the nucleotide contents of *E. coli* and T2 DNA are compared to the nucleotide content of the induced RNA, the RNA is found to be very similar to the phage DNA.

The tentative conclusion is that RNA is synthesized from DNA and that it passes into the cytoplasm where it is somehow used to synthesize protein. We can outline three stages of information transfer: **replication, transcription,** and **translation** (Figure 12-2).

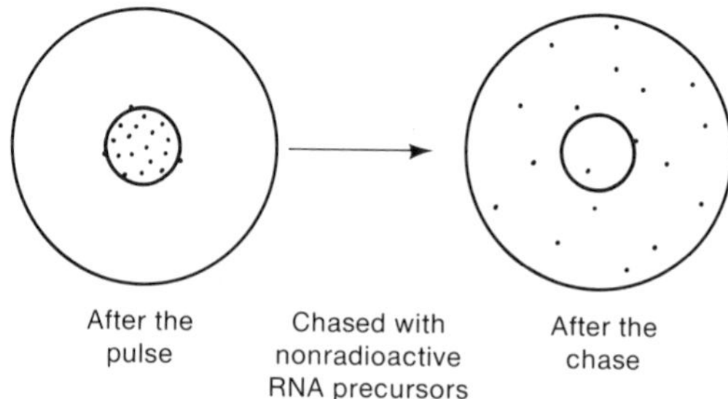

Figure 12-1. RNA synthesized during one short time period is labeled by feeding the cell a brief "pulse" of radioactive RNA precursors, followed by a "chase" of nonradioactive precursors. In an autoradiograph, the labeled RNA appears as dark grains, showing that the RNA moves from the nucleus to the cytoplasm. Apparently, the RNA is synthesized in the nucleus (small circle) and then moves out into the cytoplasm.

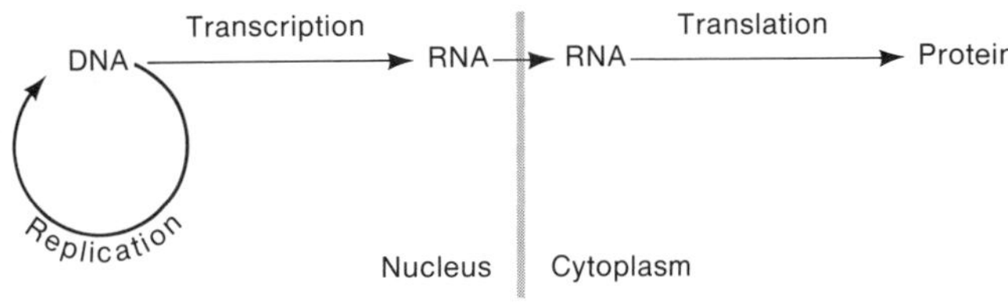

Figure 12-2. The three processes of information transfer: replication, transcription, and translation.

Although RNA is a long-chain macromolecule of nucleic acid (as is DNA), it has very different properties. First, RNA is single stranded, not a double helix. Second, RNA has ribose sugar rather than deoxyribose in its nucleotides (hence its name):

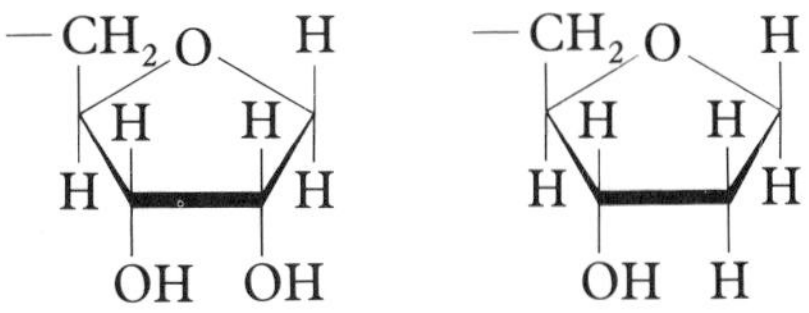

ribose

deoxyribose

Third, RNA has the pyrimidine base **uracil** (abbreviated U) instead of thymine. However, uracil does form hydrogen bonds with adenine just as thymine does.

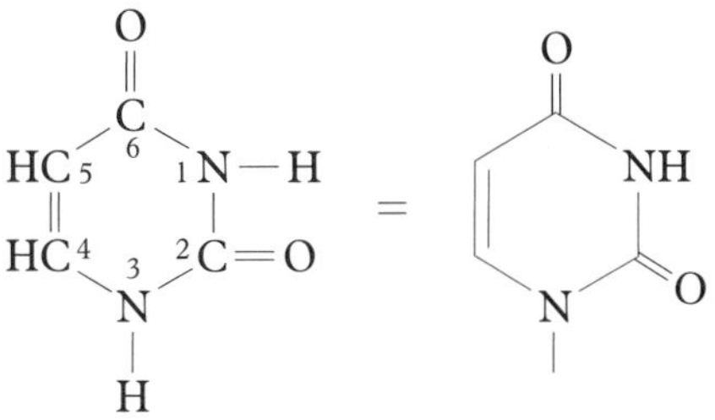

uracil

No one is absolutely sure why RNA has uracil instead of thymine or why it has ribose instead of deoxyribose. The most important aspect of RNA is that it is single stranded, but otherwise it is very similar in structure to DNA. This suggests that transcription may be based on the complementarity of bases, which was also the key to DNA replication. A transcription enzyme, RNA polymerase, could perform the transcription in a fashion quite similar to replication (Figure 12-3).

In fact, this model of transcription is confirmed cytologically (Figure 12-4). The fact that RNA can be synthesized with DNA acting as a template is demonstrated by synthesis in vitro of RNA from nucleotides in the presence

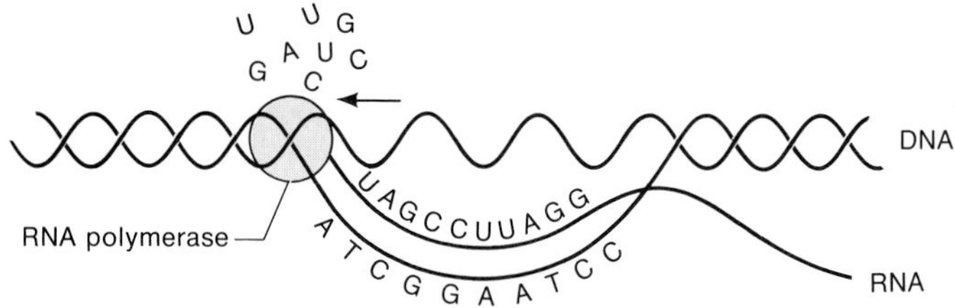

Figure 12-3. Synthesis of RNA on a single-stranded DNA template using free nucleotides. The process is catalyzed by RNA polymerase. Uracil (U) pairs with adenine (A).

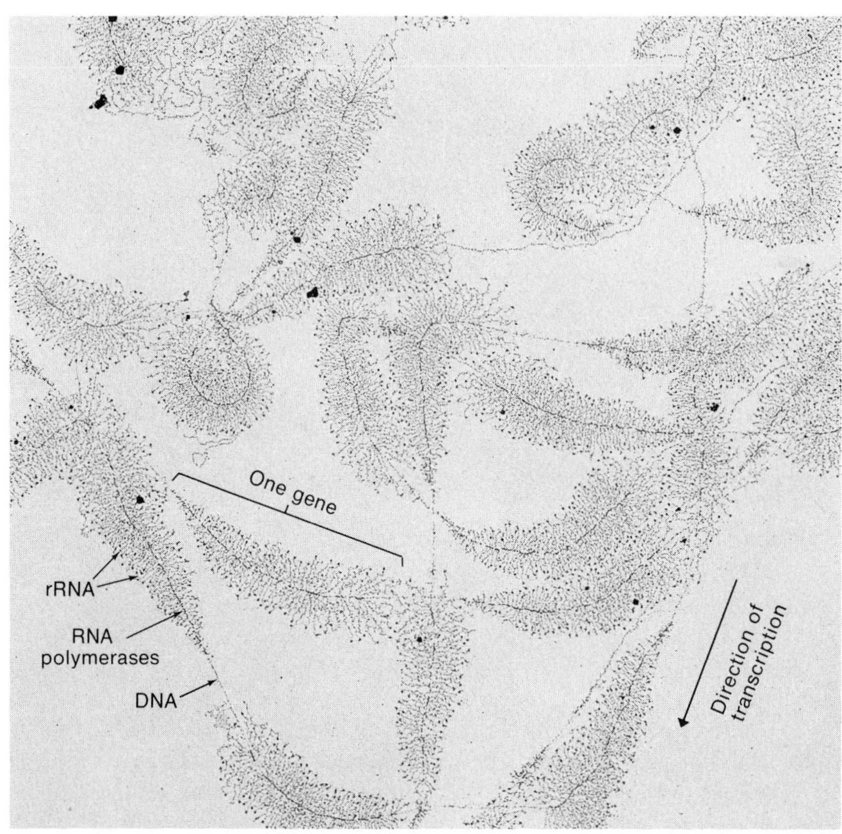

Figure 12-4. Tandemly repeated ribosomal RNA genes being transcribed in the nucleolus of Triturus viridiscens *(an amphibian). Along each gene, many RNA polymerase molecules are attached and transcribing in one direction. The growing RNA molecules appear as threads extending out from the DNA backbone. The shorter RNA molecules are nearer the beginning of transcription; the longer ones have almost been completed, hence the "Christmas tree" appearance. (Photograph by O. L. Miller, Jr., and Barbara A. Hamkalo.)*

Table 12-1. Nucleotide ratios in various DNAs and in their transcripts (in vitro)

DNA source	$\frac{(A + T)}{(G + C)}$ of DNA	$\frac{(A + U)}{(G + C)}$ of RNA synthesized
T2 phage	1.84	1.86
Cow	1.35	1.40
Micrococcus (bacterium)	0.39	0.49

of DNA, using an extractable RNA polymerase. Whatever the source of DNA used, the RNA synthesized has an (A + U)/(G + C) ratio similar to the (A + T)/(G + C) ratio of the DNA (Table 12-1). This experiment does not indicate whether the RNA is synthesized from both DNA strands or just from one, but it does indicate that the linear frequency of the A–T pairs (in comparison with the G–C pairs) in DNA is precisely mirrored in the relative abundance of (A + U) in the RNA. (These points are difficult to grasp without drawing some diagrams; Problem 2 at the end of this chapter provides some opportunities to clarify these notions.)

To test the complementarity of DNA with RNA, one can apply the specificity and precision of nucleic acid hybridization. DNA can be denatured and mixed with RNA formed from it. On slow cooling, some of the RNA strands anneal with complementary DNA to form a DNA:RNA hybrid. The DNA:RNA hybrid differs in density from the DNA:DNA duplex, so its presence can be detected by ultracentrifugation in cesium chloride. Nucleic aids will anneal in this way only if there are stretches of base-sequence complementarity, so the experiment does prove that the RNA transcript is complementary in base sequence to the parent DNA.

Can we determine whether RNA is synthesized from only one or from both of the DNA strands? It seems reasonable that only one strand would be used, because transcription of RNA from both strands would produce two complementary RNA strands from the same stretch of DNA, and these presumably would produce two different kinds of protein (with different amino acid sequences). In fact, a great deal of chemical evidence confirms that transcription takes place on only one of the DNA strands (though not necessarily the same strand throughout the entire chromosome).

The hybridization experiment can be extended to explore this problem. If the two strands of DNA have distinctly different purine:pyrimidine ratios, they can be purified separately because they have different densities in cesium chloride. The RNA made from a stretch of DNA can be purified and annealed separately to each of the strands to see whether it is complementary to only one. J. Marmur and his colleagues were able to separate the strands of DNA from the *B. subtilis* phage SP8. They denatured the DNA, cooled it rapidly to prevent reannealing of the strands, and then separated the strands in cesium chloride. They showed that the SP8 RNA hybridizes to only one of the two strands, thus proving that transcription is **asymmetrical**—that is, it occurs only on one DNA strand.

*nu*1 A W B C *nu*3 D E FI FII Z U V G T H M L K I J *b*2 *int xis red* γ *c*III N *rex c*I *cro c*II O P Q S R

Figure 12-5. Map of λ *phage DNA, with arrows indicating direction of transcription into mRNA. The upper arrows indicate transcription from one DNA strand, and the lower arrows indicate transcription from the other strand. The DNA is circular during transcription, so the lower two arrows are actually one stretch of transcription. (From G. S. Stent and R. Calendar,* Molecular Genetics, *2nd ed. Copyright © 1978, W. H. Freeman and Company.)*

RNA is transcribed from a single strand of DNA at a time. However, the same strand is not necessarily transcribed throughout the entire chromosome or through all stages of the life cycle. The RNA produced at different stages in the cycle of a phage hybridizes to different parts of the chromosome, showing the different genes that are activated at each stage. In λ phage, each of the two DNA strands is partially transcribed at a different stage (Figure 12-5). In phage T7, however, the same strand is transcribed for both early-acting and late-acting genes (Figure 12-6). In any case, the RNA always is synthesized in the $5' \rightarrow 3'$ direction (Figure 12-7).

Translation

The information-bearing RNA is appropriately called **messenger RNA** (mRNA). It acts as a copy of information from the DNA in the nucleus, sent out to direct protein synthesis in the cytoplasm (rather like a copy of a blueprint sent from the executive office to the production department). The details of the translation process (in which mRNA information is translated into protein) are still somewhat unclear, but we do have a very good general picture. If you mix mRNA and all 20 amino acids in a test tube and hope to make protein, you will be disappointed. Other components are needed; the discovery of the nature of these components provided the key to understanding the mechanism of translation.

Another important experimental tool is involved in this part of our story: sucrose density-gradient centrifugation. The cesium chloride gradient discussed earlier is created by ultracentrifugation of a uniform solution. However, the sucrose gradient is created in a test tube by layering successively lower concentrations of sucrose solution, one on top of the other. The material to be studied is carefully placed on top. When the solution is centrifuged in a machine that allows the test tube to swivel freely, the sedimenting material travels through the gradient at different rates that are related to the sizes and shapes of the molecules. Larger molecules migrate farther in a given period of time than do smaller molecules. The separated molecules can be collected individually by collecting sequential drops from a small opening in the bottom

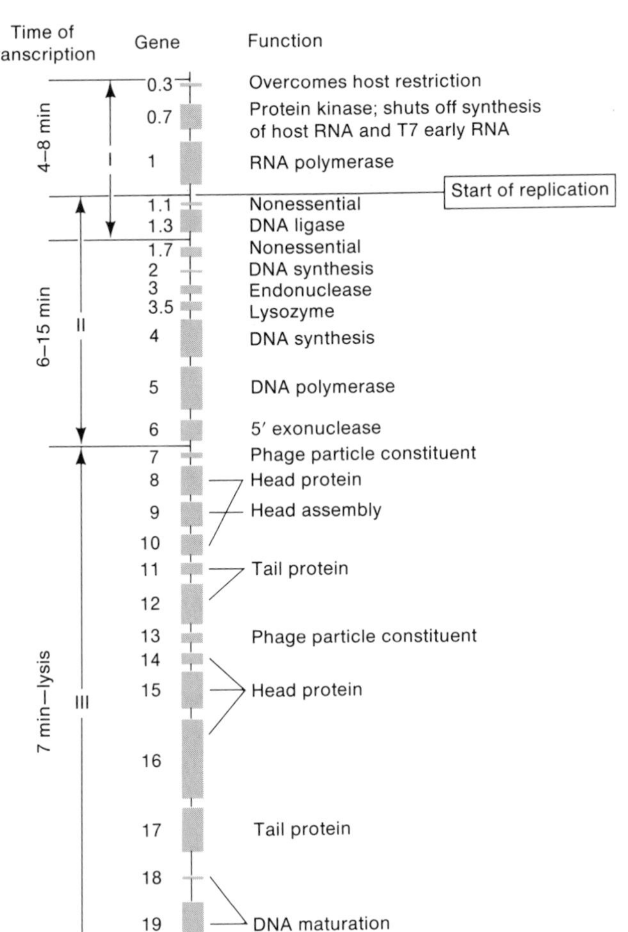

Figure 12-6. The various parts of the T7-phage genome are transcribed at various times during the phage growth cycle. (After G. S. Stent and R. Calendar, Molecular Genetics, *2nd ed. Copyright © 1978, W. H. Freeman and Company.)*

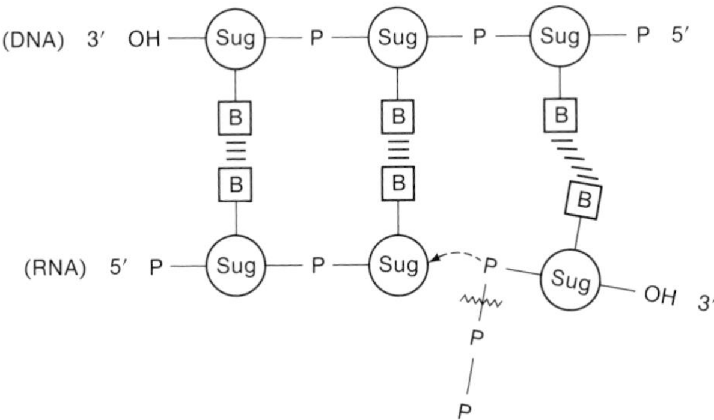

Figure 12-7. Chain elongation during transcription. A nucleoside triphosphate is aligning at the 3′ growing point. Two phosphate groups (P) will be lost as the phosphodiester bond is synthesized. (B = base; Sug = sugar.)

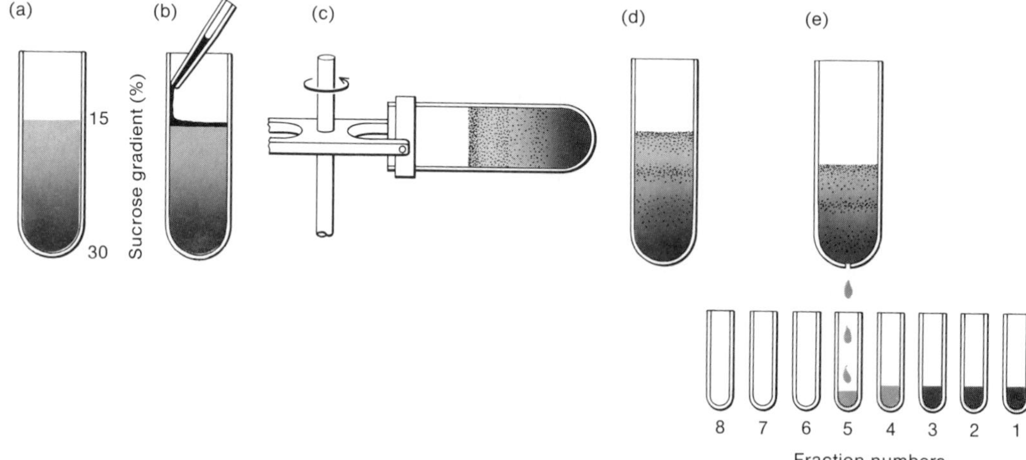

Figure 12-8. The sucrose-gradient technique. (a) A sucrose density gradient created in a centrifuge tube by layering of solutions of differing density. (b) The sample to be tested is placed on top of the gradient. (c) Centrifugation causes the various components (fractions) of the sample to sediment differentially. (d) The different fractions appear as bands in the centrifuged gradient. (e) The different bands can be collected separately by collecting samples from the bottom of the tube at fixed time intervals. The S value for the fraction is based on its position in the gradient, which is determined by the time at which it drips from the bottom of the tube. (From A. Rich, "Polyribosomes." Copyright © 1963 by Scientific American, Inc. All rights reserved.)

of the tube (Figure 12-8). The time that a fraction takes to move the fixed distance to the tube bottom indicates its position or sedimentation (S) value, which is a measure of the size of the molecules in the fraction.

Using the separatory powers of the sucrose-gradient ingredient technique, it is possible to identify several macromolecules and macromolecule aggregates in a typical protein-synthesizing system. The main components are **transfer RNA** (tRNA), **ribosomes,** and messenger RNA (mRNA). Transfer RNA is a class of small (4S) RNA molecules of rather similar type and function. In fact, complete nucleotide sequences have been determined for some tRNA molecules; they all appear to have some hydrogen-bonded regions and some single-stranded regions, and they form variations of a cloverleaf structure (Figure 12-9).

Ribosomes, on the other hand, are cellular organelles. They are composed of very complex aggregations of ribosomal proteins and **ribosomal RNA** (rRNA) components. At least three separate RNA molecules can be distinguished by size in ribosomes: 23S, 16S, and 5S. The precise function of ribosomes is still the subject of much research; the story of how we have learned what we do know is a fascinating one that lies outside the scope of this book.

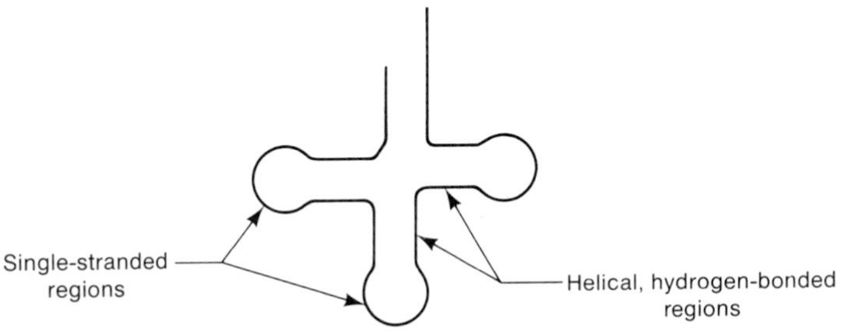

Figure 12-9. The cloverleaf structure of tRNA.

We do not know precisely how rRNA is bound up with the protein components or how each component functions. Under the electron microscope, a ribosome appears as a "blob." On chemical treatment, the blob splits into two main sub-blobs (50S and 30S), and we represent it by the simplified symbol shown in Figure 12-10.

Figure 12-10. The simplified "two-blob" representation of the ribosome.

Both rRNA and tRNA molecules will form RNA:DNA hybrids in vitro, indicating that they are transcribed from the DNA. Figure 12-11 diagrams an exploded view of the components of an *E. coli* ribosome, giving some idea of their relative sizes. Table 12-2 summarizes the main types of RNA that can be found in a typical protein-synthesizing system.

Other components are needed to make protein synthesis work in vitro. These include several enzymes (aminoacyl-tRNA synthetases and peptidyl transferase), several mysterious protein "factors" whose role is probably enzymic, and a chemical source of energy. The energy donor is needed because an orderly structure is being created out of a mess of components—a process that requires energy because the system must lose entropy (a measure of disorder or randomness).

We can regard protein synthesis as a chemical reaction, and we shall take this approach at first. Then we shall take a three-dimensional look at the physical interactions of the major components.

1. Each amino acid (aa) is attached to a tRNA molecule that is specific to that amino acid by a high-energy bond derived from GTP. The process is catalyzed by a specific enzyme (the tRNA is said to be "charged" when the amino acid is attached):

$$\text{aa}_1 + \text{tRNA}_1 + \text{GTP} \xrightarrow{\text{synthetase}_1} \text{aa}_1\text{–tRNA}_1 + \text{GDP}$$

2. The energy of the charged tRNA is converted into a peptide bond linking the amino acid to another on the ribosome:

$$\text{aa}_1\text{–tRNA}_1 + \text{aa}_2\text{–tRNA}_2 \xrightarrow[\substack{\text{peptidyl}\\\text{transferase}\\\text{on a}\\\text{ribosome}}]{} \underbrace{\text{aa}_1\text{–aa}_2}_{\substack{\text{small}\\\text{polypeptide}}}\text{–tRNA}_2 + \underset{\text{released}}{\text{tRNA}_1}$$

Table 12-2. RNA molecules in *E. coli*

Type	Relative amount in cell	Sedimentation coefficient (S)	Molecular weight	Number of nucleotides
Ribosomal RNA (rRNA)	80%	23	1.2×10^6	3700
		16	0.55×10^6	1700
		5	3.6×10^4	120
Transfer RNA (tRNA)	15%	4	2.5×10^4	75
Messenger RNA (mRNA)	5%		Heterogeneous	

SOURCE: From L. Stryer, *Biochemistry,* 2nd ed. Copyright © 1981 by W. H. Freeman and Company.

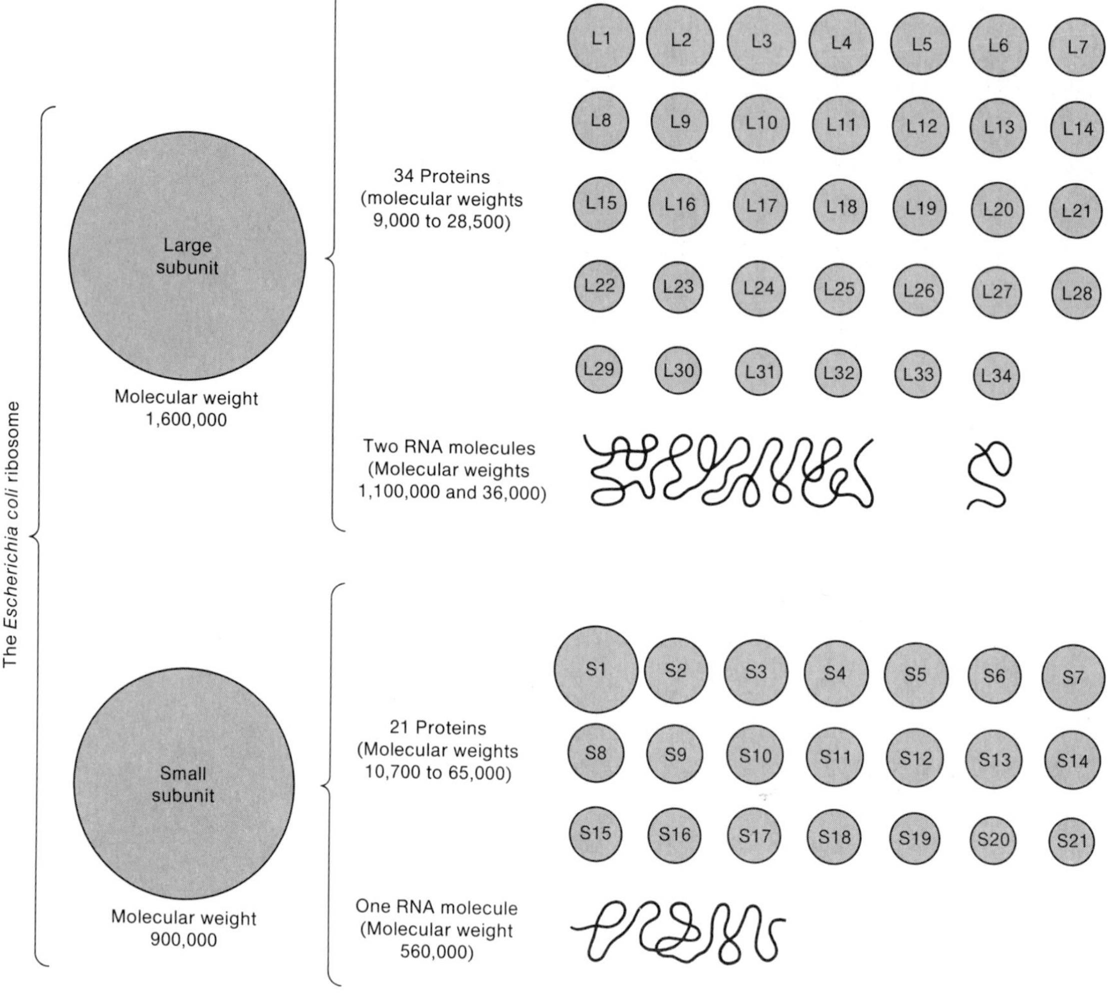

Figure 12-11. The parts of a ribosome of E. coli, *showing the complexity of what is the single most important entity in all biological synthesis. The precise roles of the parts are not understood. Peptidyl transferase, the enzyme activity responsible for the formation of peptide bonds between incoming amino acids and the growing polypeptide chain, appears to be built into the ribosome. (From "Neutron-scattering Studies of the Ribosome" by Donald M. Engelman and Peter B. Moore. Copyright © 1976 by Scientific American, Inc. All rights reserved.)*

3. New amino acids are linked by means of a peptide bond to the growing chain:

$$\mathrm{aa_3{-}tRNA_3 + aa_1{-}aa_2{-}tRNA_2 \rightarrow \underbrace{aa_1{-}aa_2{-}aa_3}_{\text{larger polypeptide}}{-}tRNA_3 + \underbrace{tRNA_2}_{\text{released}}}$$

4. This process continues until aa_n (the final amino acid) is added. Of course, the whole thing works only in the presence of mRNA, ribosomes, several additional protein factors, enzymes, and inorganic ions.

To visualize this incredible process, we must recognize the main interacting sites of the components. We begin with tRNA. Is it the tRNA or the amino acid that recognizes the codon on mRNA? A very convincing experiment answered this question. In the experiment, cysteinyl tRNA ($tRNA_{Cys}$, the tRNA specific for cysteine) charged with cysteine was treated with nickel hydride, which converted the cysteine (while still bound to $tRNA_{Cys}$) into another amino acid, alanine, without affecting the tRNA:

$$\text{cysteine—}tRNA_{Cys} \xrightarrow{\text{nickel hydride}} \text{alanine—}tRNA_{Cys}$$

Protein synthesized with this hybrid species had alanine wherever you would expect cysteine. Thus the experiment demonstrated that the amino acids are illiterate; they are inserted at the proper position because the tRNA "adaptors" recognize the mRNA codons and insert their attached amino acids appropriately. We expect then to find some site on the tRNA that recognizes the mRNA codon, presumably by complementary base pairing.

Figure 12-12a shows several sites of tRNA in terms of function. The site that recognizes an mRNA codon is called the **anticodon;** its bases are complementary to the bases of the codon. Another operationally identifiable site is the amino acid-binding region. The other arms probably assist in binding the tRNA to the ribosome. Figure 12-12b shows a specific tRNA (yeast alanine tRNA). The "flattened" cloverleafs shown in these diagrams are not the normal conformation of tRNA molecules; tRNA normally exists as an L-shaped folded cloverleaf (Figure 12-12c). These diagrams are supported by very sophisticated chemical analysis of tRNA nucleotide sequences and by X-ray crystallographic data on the overall shape of the molecule.

Where does tRNA come from? If radioactive tRNA is put into a cell nucleus in which the DNA has been partially denatured by heating, the radioactivity appears (by autoradiography) in localized regions of the chromosomes. These regions probably reflect the location of **tRNA genes;** they are regions of DNA that produce tRNA rather than producing mRNA that will produce a protein. The labeled tRNA hybridizes to these sites because of the complementarity of base sequences between the tRNA and its parent gene. A similar situation holds for rRNA. Thus we see that even the one-gene–one-polypeptide idea is not completely valid. Some genes do not code for protein; rather, they specify RNA components of the translational apparatus.

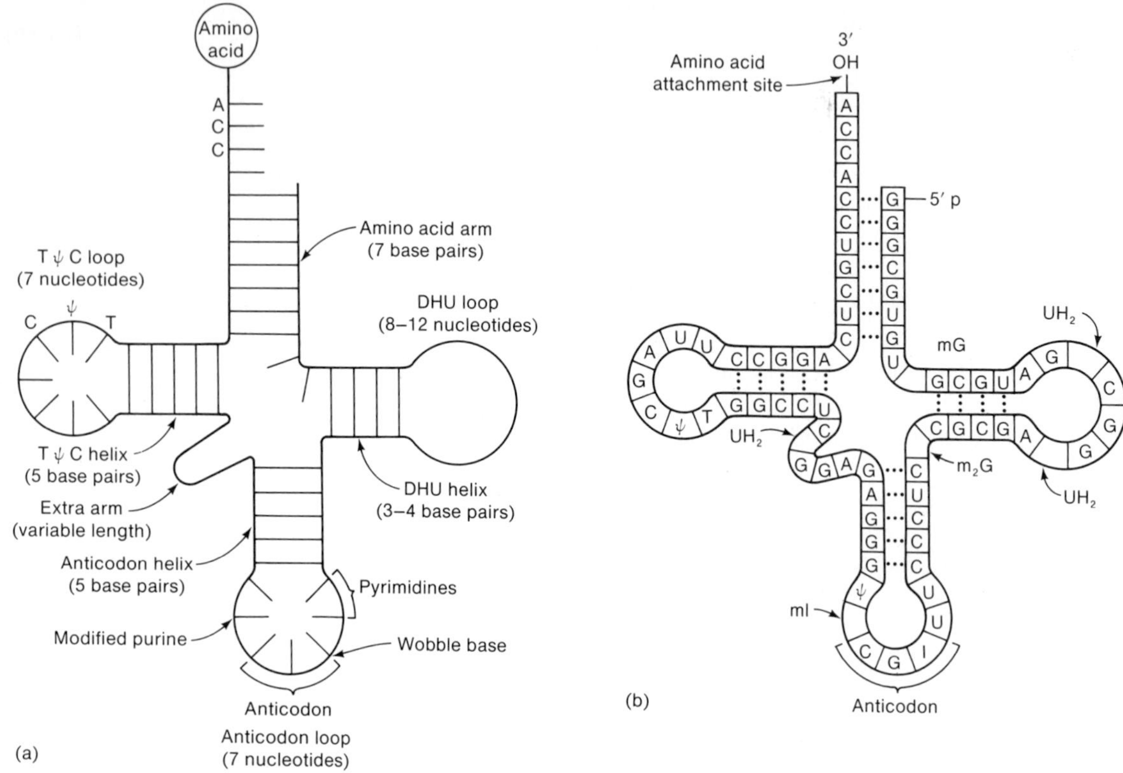

Figure 12-12. Structure of transfer RNA. (a) The functional areas of a generalized tRNA molecule. (b) The specific sequence of yeast alanine tRNA. Arrows indicate several kinds of rare modified bases. (c) The actual three-dimensional structure of yeast phenylalanine tRNA. (Part a from S. Arnott, "The Structure of Transfer RNA," Progress in Biophysics and Molecular Biology 22 (1971): 186; parts b,c from L. Stryer, Biochemistry, 2nd ed. W. H. Freeman and Company, copyright © 1981; part c based on a drawing by Dr. Sung-Hou Kim.)

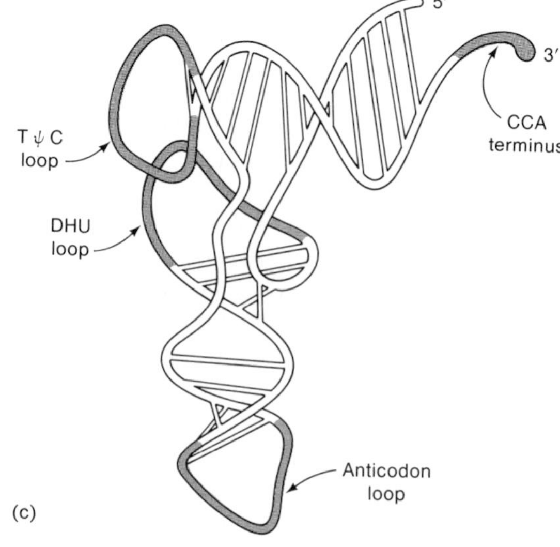

Message
Some genes code for proteins; other genes have RNA (tRNA or rRNA) as their final product.

How does tRNA get its fancy shape? It probably folds up spontaneously into a conformation that produces maximal stability. Transfer RNA contains many "odd" or modified bases (such as pseudouracil, ψ) in its nucleotides, and these might play a role in folding. These bases have also been implicated in other tRNA functions.

Now we turn to the ribosomes, which have several sites that we can predict from our model of translation, including binding sites for mRNA and tRNA. By adding radioactive tRNA to a solution of ribosomes and measuring how much of it becomes bound to the ribosomes, one can calculate that there are two tRNA binding sites per ribosome. There also is an mRNA binding site, and the enzyme peptidyl transferase is built up into the ribosome (Figure 12-13).

We can now assemble all of the components in a diagram (Figure 12-14) that shows an amino acid being added to a growing polypeptide. The ribosome "moves along" the mRNA strand. As each new codon enters the ribosome, conditions are right for the insertion of the appropriate aa–tRNA complex in the amino acid binding site (A site) and for the formation of a bond between the amino acid in the A site and the polypeptide being held to a tRNA in the polypeptide binding site (P site). When the peptide bond is formed, the tRNA in the P site is liberated, and the new tRNA–polypeptide complex moves into the unoccupied P site as the ribosome moves farther along the mRNA.

Much of the model shown in Figure 12-14 was constructed by inference based on genetic and chemical analyses. Its validity was gratifyingly demonstrated recently by electron micrographs of the process of protein synthesis

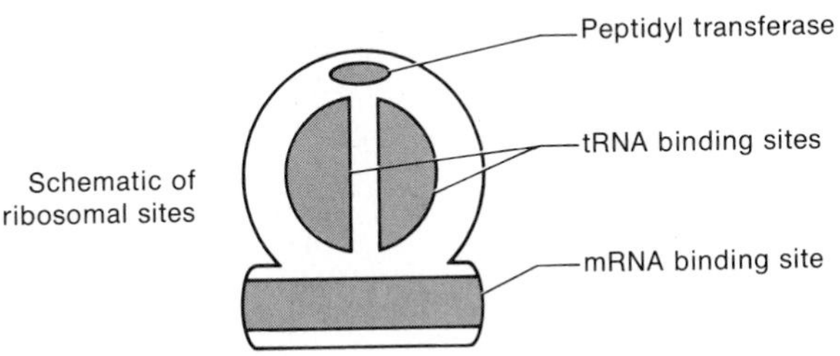

Figure 12-13. A diagrammatic representation of some important functional regions of the ribosome.

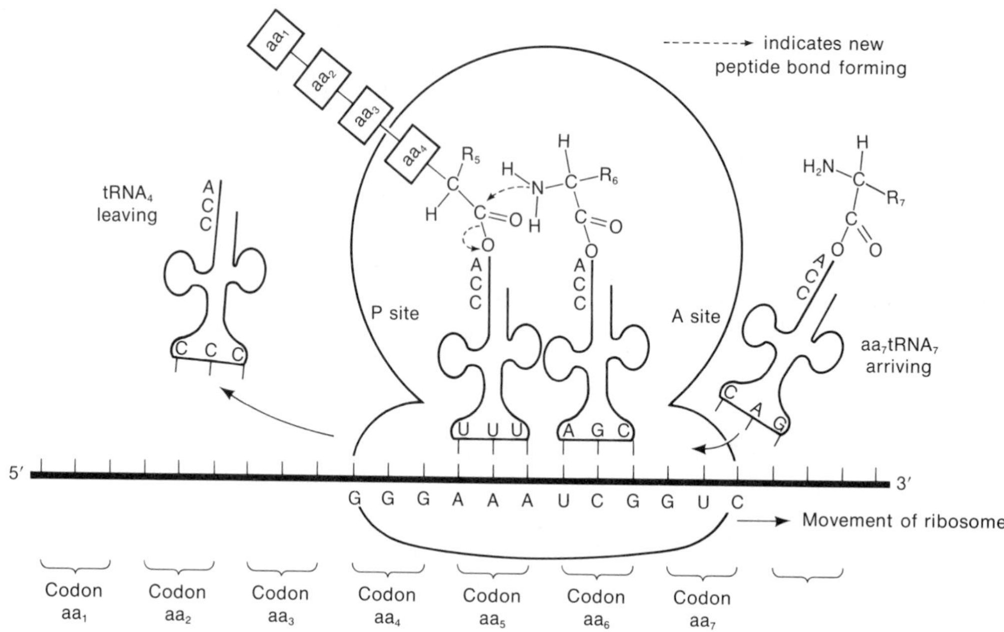

Figure 12-14. Mechanism of the addition of a single amino acid to the growing polypeptide chain during translation of mRNA. Amino acid 6 is being added; amino acids 1 through 5 are already part of the growing polypeptide. The A site is the amino acid binding site; the P site is the polypeptide binding site.

in an *E. coli* gene (Figure 12-15). As the figure shows, protein synthesis in this prokaryotic organism does not wait for the completion of mRNA synthesis. As soon as part of the mRNA is synthesized, ribosomes attach to it and begin translation! Protein synthesis in eukaryotes is different, because mRNA is made in the nucleus and then shipped to the cytoplasm for translation; thus transcription and translation in the eukaryote are separated in both time and space.

In Figure 12-14, a few imaginary codons have been inserted as illustrations of the process. Of course, this was the general notion in the minds of the investigators at the time also, but the knowledge of *which* specific codons represented *which* specific amino acids had to await another round of sophisticated investigations. This work came to be known as "cracking the genetic code."

Cracking the Code

The actual deciphering of the genetic code—determining the amino acid specified by each triplet—is one of the most exciting breakthroughs of the past two decades. Once the necessary experimental techniques became available, the genetic code was broken in a rush.

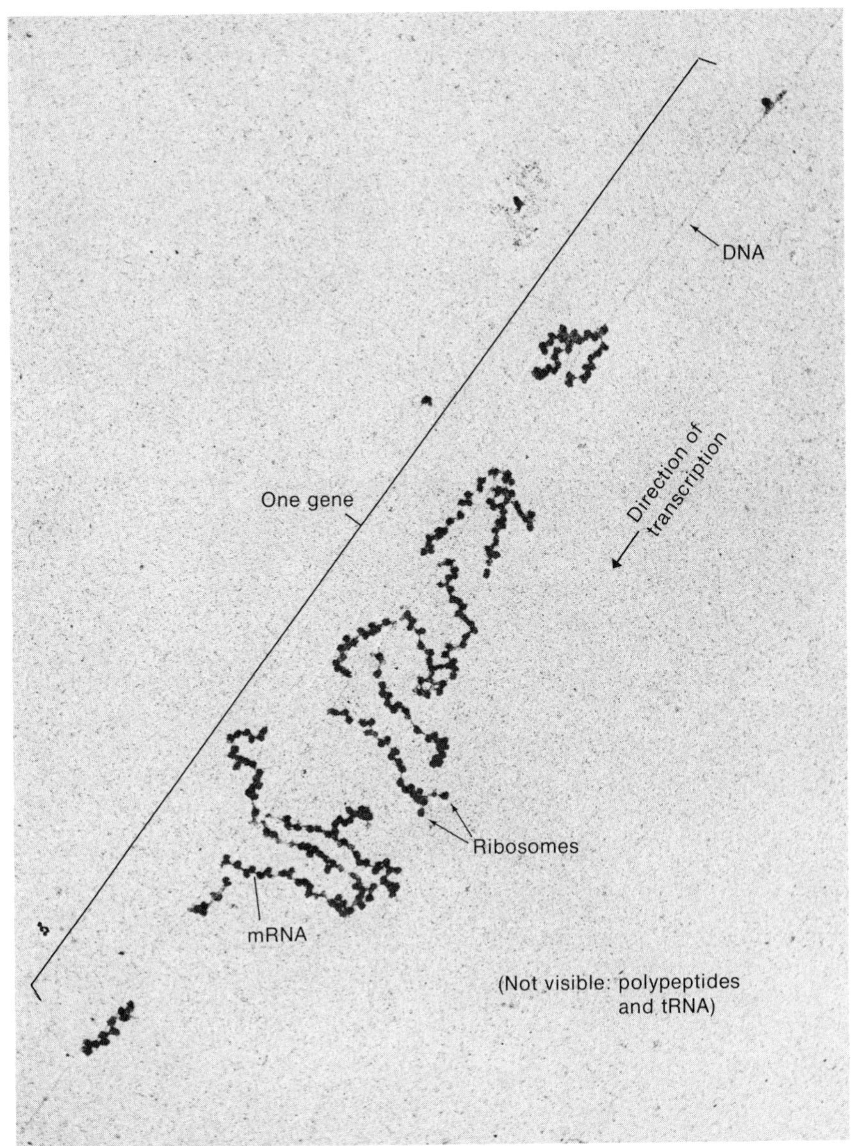

Figure 12-15. A gene of E. coli *being simultaneously transcribed and translated. (Electron micrograph by O. L. Miller, Jr. and Barbara A. Hamkalo.)*

The first breakthrough was the discovery of how to make synthetic mRNA. If the nucleotides of RNA are mixed with a special enzyme (polynucleotide phosphorylase), a single-stranded RNA is formed in the reaction. No DNA is needed for this synthesis, and so the nucleotides are incorporated at random. The ability to synthesize mRNA offered the exciting prospect of creating specific RNA sequences and then seeing what proteins they would create when acting as mRNA. The first synthetic messenger obtained was poly-U (· · ·–U–U–U–U–· · ·), made by reacting only uracil nucleotides with the RNA-synthesizing enzyme. In 1961, Marshall Nirenberg and Heinrich Mathaei mixed poly-U with the protein-synthesizing machinery of *E. coli* (ribosomes, aminoacyl tRNAs, a source of chemical energy, several enzymes, and a few other things) in vitro and *observed the formation of a protein!* Of course, the main excitement was the question of the amino acid sequence of this protein. It proved to be polyphenylalanine—that is, a string of phenylalanine molecules attached to form a polypeptide. Thus, the triplet U–U–U must code for phenylalanine:

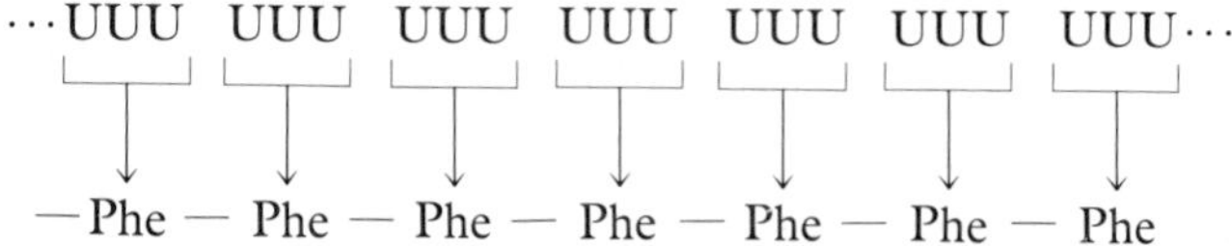

This type of analysis was extended by mixing nucleotides in a known fixed proportion when making synthetic mRNA. In one experiment, the nucleotides uracil and guanine were mixed in a ratio of 3:1. If they are incorporated at random into synthetic mRNA, then we can calculate the relative frequency at which each triplet will appear in the sequence (Table 12-3). The amino acids produced by this RNA in the protein-synthesizing system in vitro should reflect the same distribution of probabilities. In fact, the ratios of amino acids in the protein produced were those shown in Table 12-4. From this evidence, we deduce that codons consisting of one guanine and two uracils (G + 2U) code for valine, leucine, and cysteine, although we cannot distinguish the specific sequence for each of these amino acids. Similarly, one uracil and two guanines (U + 2G) must code for tryptophan, glycine, and perhaps one other. It looks as though the Watson–Crick model is correct in predicting the importance of the precise sequence (not just the ratios of bases). Many provisional assignments (such as those we have just outlined for G and U) were soon obtained, primarily by groups working with Nirenberg or with Severo Ochoa.

Table 12-3. Expected frequencies of various codons in synthetic mRNA composed of 3/4 uracil and 1/4 guanine

Codon	Probability	Ratio
UUU	$p(\text{UUU}) = 3/4 \times 3/4 \times 3/4 = 27/64$	1.00
UUG	$p(\text{UUG}) = 3/4 \times 3/4 \times 1/4 = 9/64$	0.33
UGU	$p(\text{UGU}) = 3/4 \times 1/4 \times 3/4 = 9/64$	0.33
GUU	$p(\text{GUU}) = 1/4 \times 3/4 \times 3/4 = 9/64$	0.33
UGG	$p(\text{UGG}) = 3/4 \times 1/4 \times 1/4 = 3/64$	0.11
GGU	$p(\text{GGU}) = 1/4 \times 1/4 \times 3/4 = 3/64$	0.11
GUG	$p(\text{GUG}) = 1/4 \times 3/4 \times 1/4 = 3/64$	0.11
GGG	$p(\text{GGG}) = 1/4 \times 1/4 \times 1/4 = 1/64$	0.03

Specific code words were finally deciphered through two kinds of experiments. The first involved making "mini mRNAs," each only three nucleotides in length. These, of course, are too short to promote translation into protein, but they do stimulate the binding of aminoacyl tRNAs (aa–tRNAs) to ribosomes in a kind of abortive attempt at translation. One can make a specific mini mRNA and ask *which* aminoacyl tRNA it will bind to ribosomes.

For example, the G + 2U problem discussed above can be resolved by using the following mini mRNAs:

GUU stimulates binding of valyl $tRNA_{Val}$

UUG stimulates binding of leucyl $tRNA_{Leu}$

UGU stimulates binding of cysteinyl $tRNA_{Cys}$

Analogous mini RNAs provided a virtually complete cracking of all the $4^3 = 64$ possible codons.

The second kind of experiment that was useful in cracking the code is described in Problem 13 at the end of this chapter. In solving it, you can pretend that you are H. Gobind Khorana, the worker who received a Nobel Prize for directing the experiments.

Figure 12-16 gives the code dictionary of 64 words. You should inspect this figure carefully for long periods of time in which you ponder the miracle of molecular genetics. Such an inspection should reveal several points that require further explanation.

Table 12-4. Observed frequencies of various amino acids in protein translated from mRNA composed of 3/4 uracil and 1/4 guanine

Amino acid	Ratio
Phenylalanine	1.00
Leucine	0.37
Valine	0.36
Cysteine	0.35
Tryptophan	0.14
Glycine	0.12

Multiple Codons for a Single Amino Acid

First, you will note that the number of codons for a single amino acid varies from one (tryptophan = UGG) to as many as six (serine = UCU or UCA or

First letter	Second letter: U	Second letter: C	Second letter: A	Second letter: G	Third letter
U	UUU Phe	UCU Ser	UAU Tyr	UGU Cys	U
	UUC Phe	UCC Ser	UAC Tyr	UGC Cys	C
	UUA Leu	UCA Ser	UAA Stop	UGA Stop	A
	UUG Leu	UCG Ser	UAG Stop	UGG Trp	G
C	CUU Leu	CCU Pro	CAU His	CGU Arg	U
	CUC Leu	CCC Pro	CAC His	CGC Arg	C
	CUA Leu	CCA Pro	CAA Gln	CGA Arg	A
	CUG Leu	CCG Pro	CAG Gln	CGG Arg	G
A	AUU Ile	ACU Thr	AAU Asn	AGU Ser	U
	AUC Ile	ACC Thr	AAC Asn	AGC Ser	C
	AUA Ile	ACA Thr	AAA Lys	AGA Arg	A
	AUG Met	ACG Thr	AAG Lys	AGG Arg	G
G	GUU Val	GCU Ala	GAU Asp	GGU Gly	U
	GUC Val	GCC Ala	GAC Asp	GGC Gly	C
	GUA Val	GCA Ala	GAA Glu	GGA Gly	A
	GUG Val	GCG Ala	GAG Glu	GGG Gly	G

Figure 12-16. The genetic code.

UCG or UCC or AGU or AGC). Why? The answer is complex but not difficult, and it can be divided into two parts.

1. Certain amino acids can be brought to the ribosome by several *alternative* tRNA types (species) having different anticodons, whereas certain other amino acids are brought to the ribosome by only one tRNA.

2. Certain tRNA species can bring their specific amino acids in response to several codons, not just one, through a loose kind of base pairing at one end of the codon and anticodon. This sloppy pairing is called **wobble.**

Table 12-5. Codon–anticodon pairings allowed by the wobble rules

5′ end of anticodon	3′ end of codon
G	U or C
C	G only
A	U only
U	A or G
I	U, C, or A

We had better discuss wobble first, and it will lead us into a discussion of the various species of tRNA. Wobble is caused by a third nucleotide of the anticodon (at the 5′ end) that is not quite aligned. This out-of-line nucleotide sometimes can form hydrogen bonds not only with its normal complementary nucleotide in the third position of the codon but also with a different nucleotide in that position. Crick established certain "wobble rules" that dictate which nucleotides can and which cannot form new hydrogen-bonded associations through wobble (Table 12-5). In Table 12-5, inosine (I) is one of the rare bases found in tRNA, often in the anticodon.

Figure 12-17 shows the possible codons that one tRNA serine species can recognize. As the wobble rules indicate, G can pair with U or with C. Table 12-6 lists all the codons for serine and shows how different tRNAs can service these codons. This is a good example of the effects of wobble in the genetic code.

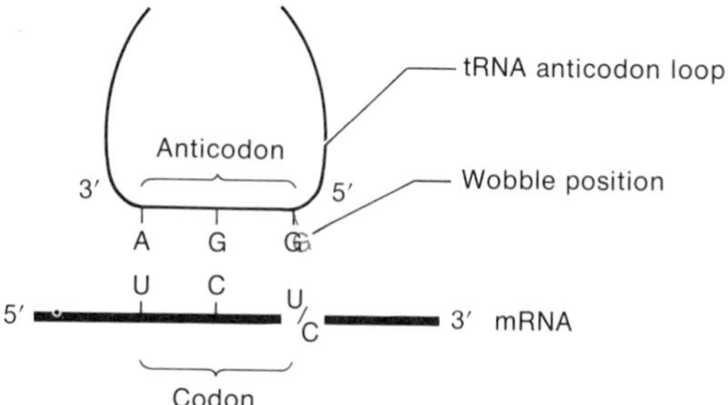

Figure 12-17. In the third site (5′ end) of the anticodon, G can take either of two wobble positions, thus being able to pair with either U or C. This means that a single tRNA species carrying an amino acid (in this case, serine) can recognize two codons in the mRNA, UCU and UCC.

Sometimes there can be an additional tRNA species that we represent as $tRNA_{Ser_4}$; it has an anticodon identical with any one of the three anticodons shown in Table 12-6, but it differs in its nucleotide sequence elsewhere in the tRNA molecule. These four tRNAs are called **isoaccepting tRNAs** because they accept the same amino acid, but they are probably all transcribed from different tRNA genes.

Table 12-6. Different tRNAs that can service codons for serine

Codon	tRNA	Anticodon
UCU UCC	$tRNA_{Ser_1}$	AGG + wobble
UCA UCG	$tRNA_{Ser_2}$	AGU + wobble
AGU AGC	$tRNA_{Ser_3}$.	UCG + wobble

Stop Codons

The second point you may have noticed in Figure 12-16 is that some codons do not specify an amino acid at all. These codons are labeled as **stop codons.** They can be regarded as something like periods or commas punctuating the message encoded in the DNA.

One of the first indications of the existence of stop codons came in 1965 from work by Sidney Brenner with the T4 phage. Brenner analyzed certain mutations (m_1 through m_6) in a single cistron that controls the head protein of the phage. These mutants had two things in common. First, the head protein of each mutant was a shorter polypeptide chain than that of the wild type. Second, the presence of a suppressor mutation (*su*) at a separate chromosomal location would cause the phage to develop a head protein of normal (wild-type) chain length despite the presence of the *m* mutation (Figure 12-18).

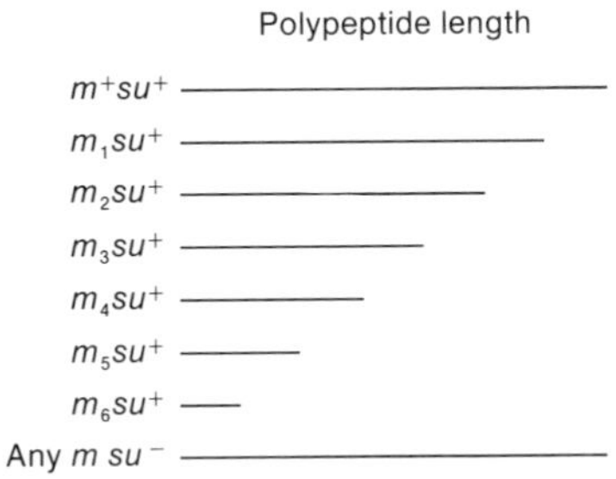

Polypeptide length

m^+su^+
m_1su^+
m_2su^+
m_3su^+
m_4su^+
m_5su^+
m_6su^+
Any m su^-

Figure 12-18. Polypeptide-chain lengths of phage-T4 head protein in wild type (top) *and various amber mutants* (m). *An amber suppressor* (su) *leads to phenotypic development of the wild-type chain.*

Brenner examined the ends of the shortened proteins and compared them with wild-type protein, recording for each mutant the next amino acid that *would* have been inserted to continue the wild-type chain. These amino acids for the six mutations were glutamine, lysine, glutamic acid, tyrosine, tryptophan, and serine. There is no immediately obvious pattern to these results, but Brenner brilliantly deduced that certain codons for each of these amino acids are similar in that each of them can mutate to the codon UAG by a single change in a DNA nucleotide pair. He therefore postulated that UAG is a stop (or termination) codon, a signal to the translation mechanism that the protein is now complete.

UAG was the first stop codon deciphered, and it is called the **amber codon.** Mutants that are defective because of the presence of an abnormal amber codon are called amber mutants, and their suppressors are amber suppressors. UGA (**opal**) and UAA (**ocher**) are also stop codons and also have suppressors. Stop codons often are called **nonsense codons** because they designate no amino acid, but their sense is real enough; it is unfortunate that this misnomer persists. (**Missense mutations** are those that lead to a change in a single amino acid in the translated polypeptide.) Not surprisingly, stop codons do not act as mini mRNAs in binding aa–tRNA to ribosomes in vitro.

Message

A missense mutation changes a codon so that it stands for a different amino acid. A nonsense mutation changes a codon so that it means "stop" to the translation system.

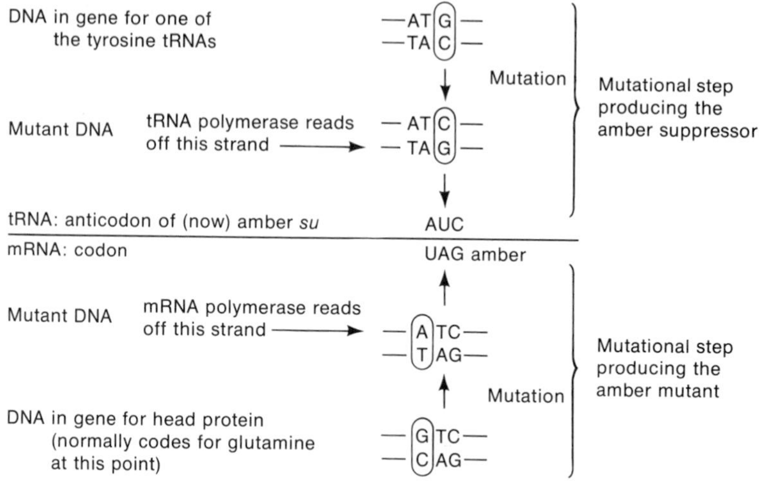

Figure 12-19. Mutation can produce an amber suppressor (top) *that counteracts the effect of the amber mutant codon* (bottom). *Amber mutations occur when the base sequence in the DNA is changed from a codon representing an amino acid to one representing a stop signal. Suppressor mutations are mutations in genes coding for specific tRNA molecules. One nucleotide in the anticodon of the tRNA is changed so that the tRNA molecule will now recognize the mutant stop signal and insert an amino acid at that point. Translation can thus continue through the entire gene.*

It is interesting to consider the nonsense suppressor mutations, which now are known to be mutations in the anticodon loop of specific tRNAs that allow recognition of a nonsense codon in mRNA. Thus an amino acid is inserted in response to the stop codon, and translation continues past that triplet. Figure 12-19 shows an example for m_1 in the T4 head-protein analysis. The amber mutation replaces the codon for glutamine by a stop codon, which by itself would prematurely cut off the protein at this position. The suppressor mutation produces a $tRNA_{Tyr}$ with an anticodon that complements the mutant stop codon. This mutant tRNA inserts tyrosine in the protein, so that translation continues past the m_1 stop codon. The suppressed m_1 mutant thus does not have a fully wild-type protein, because one glutamine has been replaced by a tyrosine. Obviously, the insertion of tyrosine is not the critical factor that restores protein activity (probably any of a variety of amino acids would suffice). The important thing is that some amino acid *is* inserted, and the translation apparatus trundles by and completes the rest of the protein.

Two questions often are asked at this point:

1. What is doing the job of putting tyrosine where tyrosine should be when the gene for tyrosine tRNA becomes a nonsense suppressor? Remember that there typically are several isoaccepting tRNA forms (and hence several $tRNA_{Tyr}$ genes) and that two of these may have

the same anticodon loop. The "other" (one or more) wild-type $tRNA_{Tyr}$ gene takes over when a suppressor locus is created from the first.

2. What happens to normal stop signals (indicating the ends of normal proteins) when nonsense suppressors are present? There is evidence that the normal end of a protein is signaled by two different consecutive stop codons, so that a single suppressor will not bypass the stop message. It is not yet known whether this situation exists for most proteins.

Start Codon

The third point regarding Figure 12-16 is that a single codon (AUG) acts as a **start codon**—rather like the capital letter indicating the start of a sentence. This codon also acts as the codon for the amino acid methionine. In *E. coli* and in some other organisms, the first amino acid in any newly synthesized polypeptide always is one called N-formyl methionine. It is not inserted by $tRNA_{Met}$, however, but by an initiator tRNA called $tRNA_{N\text{-f-Met}}$. This initiator tRNA has the normal methionine anticodon, but it inserts *N*-formylmethionine rather than methionine in the polypeptide chain (Figure 12-20). Either the formyl group or a sequence in the tRNA apparently mimics a polypeptide chain; at least the ribosome recognizes some specific signal and shifts one codon to amino acid position 2. Methionine won't cause this shift; if methionyl $tRNA_{Met}$ inserts, nothing happens. The system then must wait until the Met–$tRNA_{Met}$ diffuses away and an *N*-f-Met–$tRNA_{N\text{-f-Met}}$ chances along to get things going. Similarly, when AUG occurs in the middle of a protein, *N*-f-Met–$tRNA_{N\text{-f-Met}}$ cannot form a peptide bond with the growing chain, so the system must wait for Met–$tRNA_{Met}$.

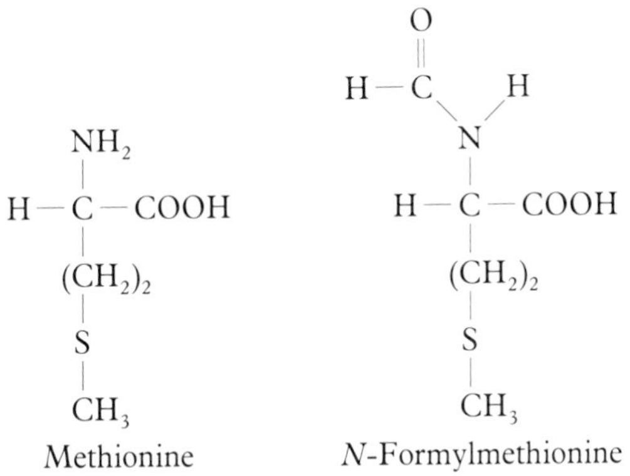

Methionine: NH_2 — H—C—COOH — $(CH_2)_2$ — S — CH_3

N-Formylmethionine: O=C(H)—N(H) — H—C—COOH — $(CH_2)_2$ — S — CH_3

Figure 12-20. The structures of methionine (Met) and N*-formylmethionine (*N*-f-Met). A tRNA bearing* N*-f-Met can initiate a polypeptide chain but cannot be inserted in a growing chain; a tRNA bearing Met can be inserted in a growing chain but will not initiate a new chain. Both of these tRNAs bear the same anticodon complementing the codon AUG.*

Unsolved Problems

The ribosome is an immensely challenging structure. Very little is known about the precise arrangement of the ribosomal protein and rRNA subunits, or about how they unite to grapple physically with the mRNA, the tRNA, the amino acid, and the peptide chain. The role of rRNA has always been a mystery. The role of the proteins can be visualized much like a complex enzyme, but the occurrence of what is normally an informational molecule in this structure is baffling. There is evidence now suggesting that the rRNA is crucial in the binding of various translational components (such as tRNA) to the ribosome. This role must be nonspecific with respect to the tRNA species because a ribosome is basically a nonspecific translational apparatus. Ribosomal RNA is the most abundant RNA in a cell and obviously serves a crucial cell function. It is known to be synthesized in the nucleolus in a repetitive tandem array of rRNA genes, budded off the chromosomal region known as the **nucleolar organizer.**

We have followed a highly simplified version of protein synthesis. There are several other components or "factors" required for the system to work. Figure 12-21 summarizes the action of some of them.

There are many other chemical mysteries surrounding protein synthesis. In eukaryotic cells, for example, mRNA is transcribed in the nucleus as **heterogeneous nuclear RNA** (HnRNA), which is much longer than the final messenger and also acquires a 3′ tail of about 150 to 200 adenine nucleotides (a **poly A tail**). The mRNA is enzymatically chopped out of the HnRNA, although the poly A tail remains. Certainly there is still much work to do in a field that has moved at lightning speed in the past two decades.

Our discussion has been limited chiefly to microbes, but the amazing fact is that the information-transfer, coding, and translation processes are virtually identical in all organisms that have been studied. For example, all of the different amino acid substitutions known to occur in human hemoglobin can in theory result from single nucleotide-pair substitutions based on the genetic code derived from *E. coli* (Table 12-7). Such observations suggest that the

Table 12-7. Mutations in human hemoglobin, in *E. coli* tryptophan synthetase, and in tobacco-mosaic-virus (TMV) coat protein

Protein	Amino acid substitution	Inferred codon change
Hemoglobin	Glu → Val	GAA → GUA
Hemoglobin	Glu → Lys	GAA → AAA
Hemoglobin	Glu → Gly	GAA → GGA
Tryptophan synthetase	Gly → Arg	GGA → AGA
Tryptophan synthetase	Gly → Glu	GGA → GAA
Tryptophan synthetase	Glu → Ala	GAA → GCA
TMV coat protein	Leu → Phe	CUU → UUU
TMV coat protein	Glu → Gly	GAA → GGA
TMV coat protein	Pro → Ser	CCC → UCC

SOURCE: From L. Stryer, *Biochemistry*, 2nd ed. Copyright © 1981 by W. H. Freeman and Company.

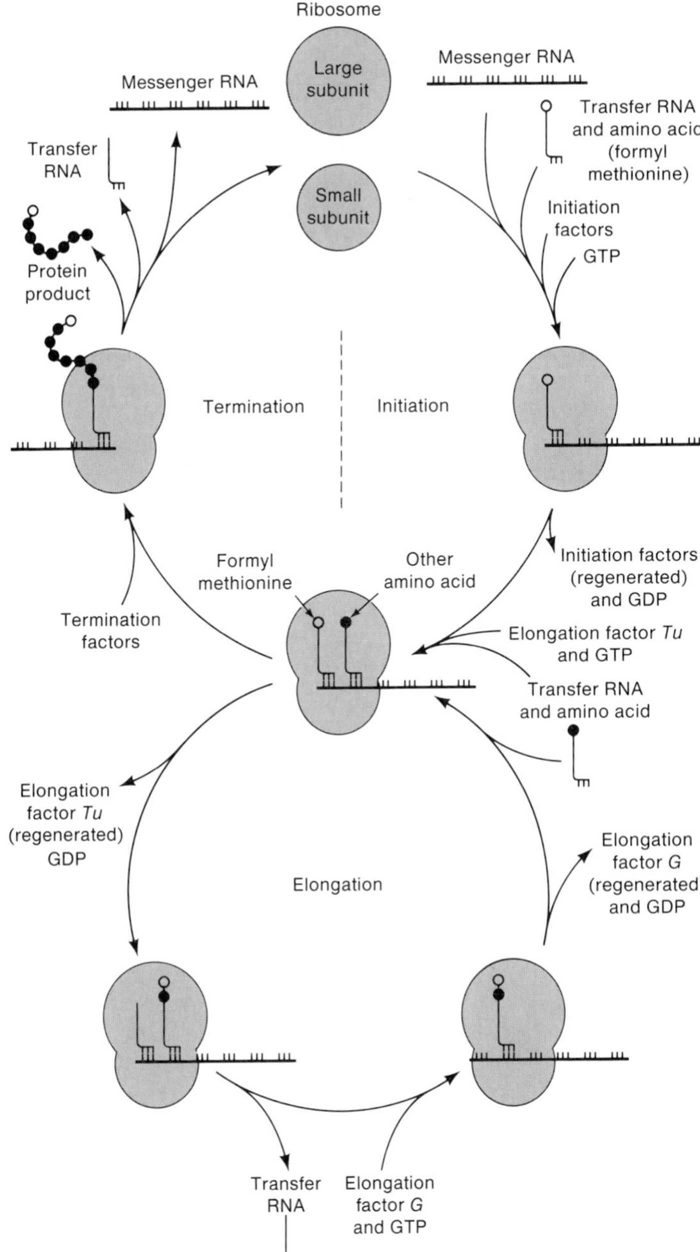

Figure 12-21. *The transactions of the ribosome. At initiation, the ribosome recognizes the starting point in a segment of mRNA and binds a molecule of tRNA bearing a single amino acid. In all bacterial proteins, this first amino acid is* N-*formylmethionine. In elongation, a second amino acid is linked to the first one. The ribosome then shifts its position on the mRNA molecule, and the elongation cycle is repeated. When the stop codon is reached, the chain of amino acids folds spontaneously to form a protein. Subsequently, the ribosome splits into its two subunits, which rejoin before a new segment of mRNA is translated. Protein synthesis is facilitated by a number of catalytic proteins (initiation, elongation, and termination factors) and by guanosine triphosphate (GTP), a small molecule that releases energy when it is converted into guanosine diphosphate (GDP). (From "Neutron-scattering Studies of the Ribosome" by Donald M. Engelman and Peter B. Moore. Copyright © 1976 by Scientific American, Inc. All rights reserved.)*

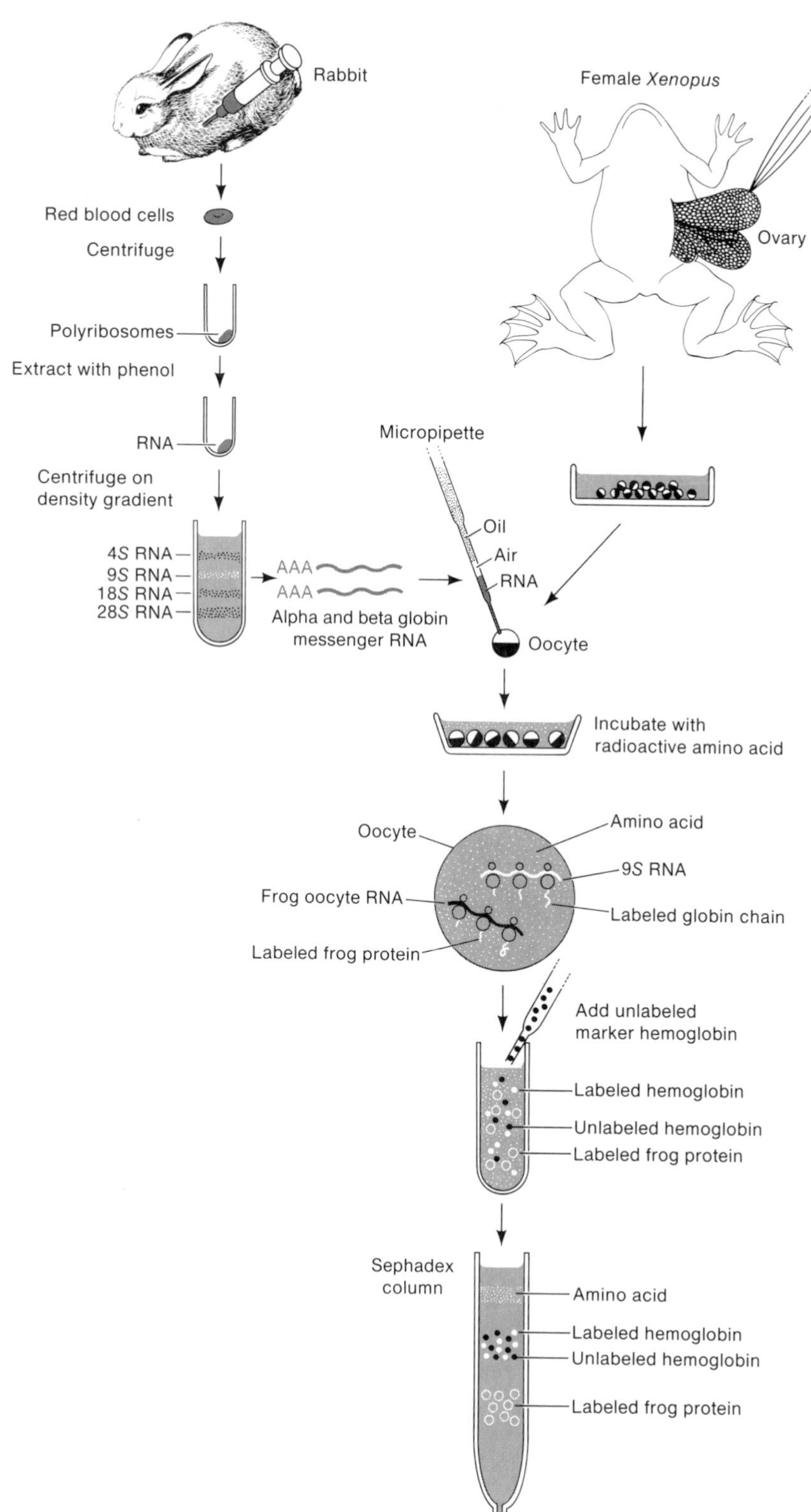
Rabbit
Female *Xenopus*
Ovary
Red blood cells
Centrifuge
Polyribosomes
Extract with phenol
RNA
Centrifuge on density gradient
4*S* RNA
9*S* RNA
18*S* RNA
28*S* RNA
AAA
AAA
Alpha and beta globin messenger RNA
Micropipette
Oil
Air
RNA
Oocyte
Incubate with radioactive amino acid
Oocyte
Amino acid
9*S* RNA
Frog oocyte RNA
Labeled globin chain
Labeled frog protein
Add unlabeled marker hemoglobin
Labeled hemoglobin
Unlabeled hemoglobin
Labeled frog protein
Sephadex column
Amino acid
Labeled hemoglobin
Unlabeled hemoglobin
Labeled frog protein

genetic code is shared by all organisms. Furthermore, an information-bearing molecule, such as rabbit red-blood-cell mRNA, which is predominantly hemoglobin gene transcript, will be translated in an alien environment (such as a frog egg) into rabbit hemoglobin (Figure 12-22). Apparently, the translation apparatus is functionally the same in a wide range of different organisms. Finally, as we shall see from genetic engineering experiments in Chapter 13, DNA is DNA no matter what its origin. The nature and message of the DNA represent a universal language of life on earth.

Does this interspecific equivalence of parts in the genetic apparatus indicate a common evolutionary ancestry for all life forms on earth? Or does it simply reflect the fact that this is the only workable biochemical option in the earth environment (biochemical predestination)? Whatever the answer, the wonderful uniformity of the molecular basis of life on earth is firmly established. Minor variations do exist, but they do not detract from the central uniformity of the mechanism that we have described.

Message

The processes of information storage, replication, transcription, and translation are fundamentally similar in all living systems. In demonstrating this fact, molecular genetics has provided a powerful unifying force in biology. We now know the tricks that life uses to achieve persistent order in a randomizing universe.

Chapters 11 and 12 have described the development of the central theory of molecular genetics. A linear sequence of nucleotides in DNA is transcribed into a comparable linear sequence of nucleotides in mRNA. This mRNA sequence then is translated into an amino acid sequence in protein by a complex translational apparatus. The protein thus made has importance to the organism either as a structural component (such as hair, muscle, or skin protein) or as a regulator of the body chemistry (such as enzymes or hemoglobin). Figure 12-23 summarizes the structural relationship between DNA and protein.

Figure 12-22. Rabbit hemoglobin mRNA is translated into rabbit hemoglobin in a frog (Xenopus) *egg. This translation occurs with the* Xenopus *translation apparatus. This experiment is one of many that point toward the uniformity of the genetic molecular mechanism in all forms of life. (Sephadex is a separatory material used in chromatography.) (From C. Lane "Rabbit Hemoglobin from Frog Eggs." Copyright © 1976 by Scientific American, Inc. All rights reserved.)*

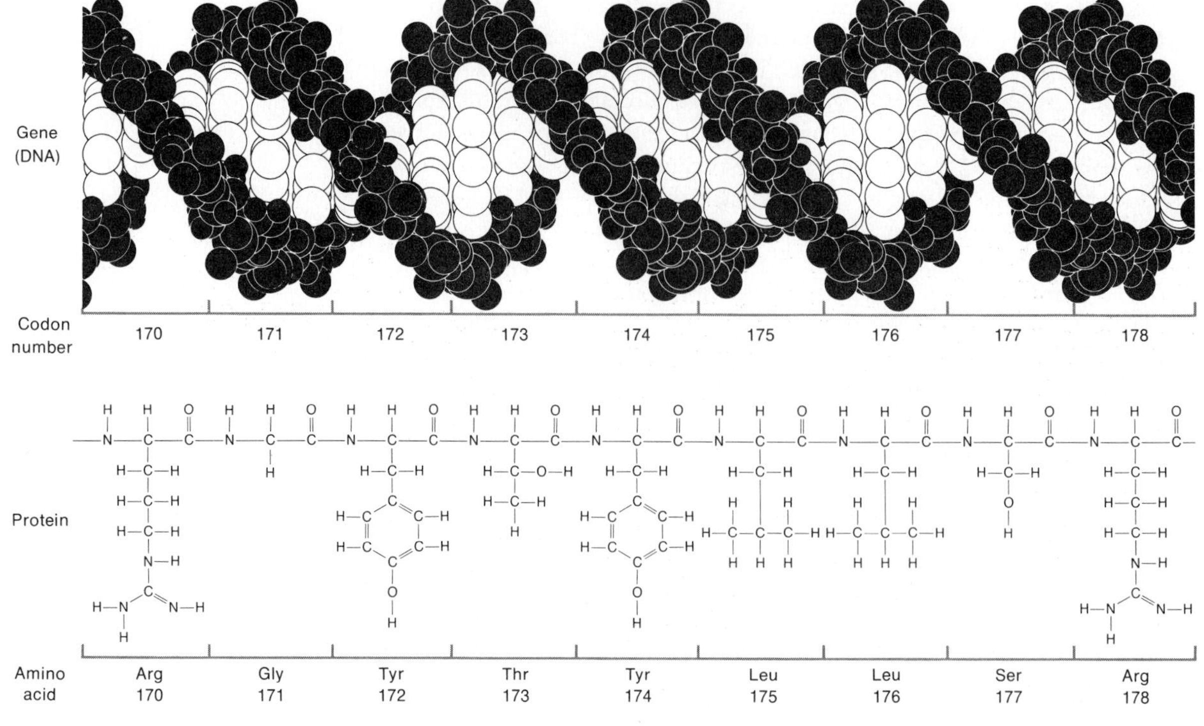

Summary

We have discovered in earlier chapters that DNA is the genetic material and is responsible for directing the synthesis of proteins. But just how does DNA accomplish this feat? The first clue came from eukaryotes, when it was shown that RNA is synthesized in the nucleus and then transferred to the cytoplasm. However, most of the details of this transfer of information from DNA to protein were worked out with experiments in bacteria and phage.

RNA is synthesized from only one strand of a double-stranded DNA helix. This transcription is catalyzed by an enzyme, RNA polymerase, and follows rules similar to those followed in replication: A complements with T, G with C, and U (uracil) with A. Ribose is the sugar used in RNA, and uracil replaces thymine. Extraction of RNA from a cell yields three basic varieties: ribosomal, transfer, and messenger RNA. The three sizes of ribosomal RNA (rRNA) combine with an array of specific proteins to form ribosomes that are the machines used for protein synthesis (translation). Transfer RNAs (tRNA) are a group of rather small RNA molecules, each with specificity for a particular amino acid; they carry the amino acids to the ribosome, where they can be attached to a growing polypeptide.

13

Manipulation of DNA

We have seen the power of genetic analysis in revealing the principles of heredity. Such analysis depends completely on the ability to mix hereditary materials through some kind of genetic cross, for the different genetic combinations are created and disrupted only through crosses. Even in cases where we do not know the molecular or cellular basis for phenotypic differences, we can use those differences as "markers" to be mixed in the progeny of a cross. By following such phenotypic markers through successive generations of crosses, we have deduced the mechanisms of genetic exchange, the circularity of some chromosomes, transduction, lysogeny, recombination, the fine structure of the gene, and transformation in microorganisms. In eukaryotes, crosses revealed Mendel's laws, linkage, chromosome aberrations, and mutation. Even at the molecular level, the basic elements of replication and the genetic code were inferred on purely genetic grounds. However, as hinted in Chapter 12, the purely genetic analysis is not sufficient to solve all the mysteries at the molecular level. Point mutations and crosses are useful tools for the molecular biologist, but biochemical techniques and experiments in vitro have become the central methodology of molecular genetics.

Yet many of the operational criteria used in molecular biology are identical to those used by geneticists working at the level of organisms:

1. Recognizable differences (comparable to phenotypes) must exist so that the genetic molecules under study can be distinguished from one another.
2. There must be techniques that permit the study of individual molecules or groups of molecules over time—that is, ways to follow the fate of individuals or particular groups.
3. There must be quantitative measures of the molecules involved (comparable to progeny counts or frequencies).
4. There must be ways to mix and even to hybridize the molecules in different combinations (equivalent to crosses in organisms).
5. There must be controls for comparison and standardization of results.

In this chapter, we discuss experiments that do not require genetic crosses but that nevertheless involve similar processes of manipulation.

Manipulations of DNA in vitro have become incredibly sophisticated, employing a wide array of techniques. The enormous variation in base sequence seemed at first quite intractable to chemical identification, but rapid and routine laboratory procedures now provide such identifications. Advances in this field are so numerous and rapid that the details of any discussion of DNA manipulation are certain to be obsolete before publication. However, there are key elements that make such work possible, and these form the topic of our discussion in this chapter.

The Power of Base Complementarity

The manipulation of DNA depends on two properties of DNA: the extreme precision of complementary base-sequence recognition and the numbers of hydrogen bonds that link the complementary base pairs. We look first at some of the experimental techniques and the information they have provided. In the second part of this chapter, we turn to the controversial topic of genetic engineering.

Denaturation of DNA by Heat

The hydrogen bonds that link complementary base pairs are disrupted by high temperature. Because a G–C pair has three hydrogen bonds, whereas an A–T pair has only two, the **melting point** (defined as the temperature at which one-half of the DNA sample is no longer hydrogen bonded) is directly proportional to the G–C content (Figure 13-1). Thus, any experimental technique that allows separation of single-stranded from double-stranded DNA will permit separation of molecules with different melting points and hence with different G–C contents.

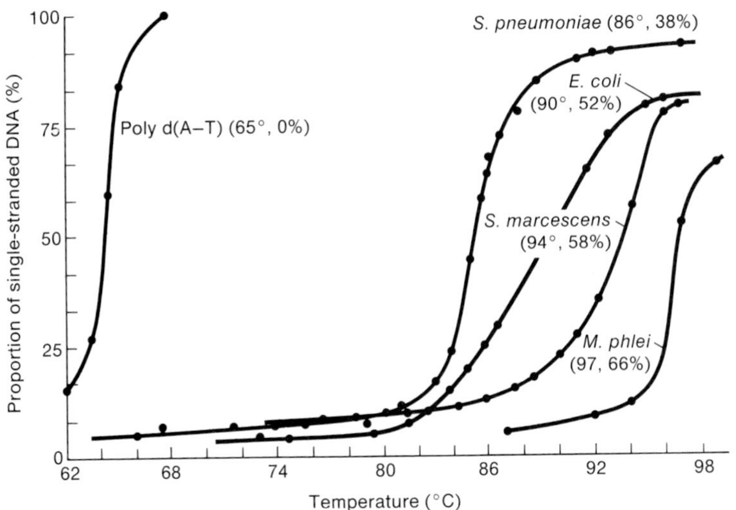

Figure 13-1. Denaturation of DNA from different sources at various temperatures. In parentheses for each sample, the first value is the melting temperature (at which one-half of the DNA molecules have been denatured), and the second value is the percentage of base pairs in the sample that are G–C pairs. There is an obvious relationship between G–C content and melting temperature. (From J. Marmur and P. Doty, Nature *183:1427, 1959.)*

Message

The relative G–C content of DNA can be inferred from its melting temperature. DNA with a relatively higher content of A–T is denatured (melted) at a lower temperature than is DNA with a greater amount of G–C.

Within a DNA molecule, the proportion of G–C versus A–T pairs may vary along the length of the molecule. We have seen that partial denaturation can be produced by a hydrogen-bond-breaking agent, such as high temperature, and that the denatured regions within the molecule can then be mapped (Chapter 11). The positions and sizes of the melted regions can be reproducibly identified in a population of identical molecules, thus indicating that such maps do reflect the base composition within the molecules.

In 1976, David Wolstenholme used the heat denaturation of AT-rich regions to map the mitochrondrial DNAs of different *Drosophila* species (Figure 13-2). Using the AT-rich segment in mitochondrial DNA of *Drosophila melanogaster* as a marker, Wolstenholme then inspected a series of replicating molecules. He could map the initiation point for replication, and he showed that replication is unidirectional and begins in the center of the AT-rich segment.

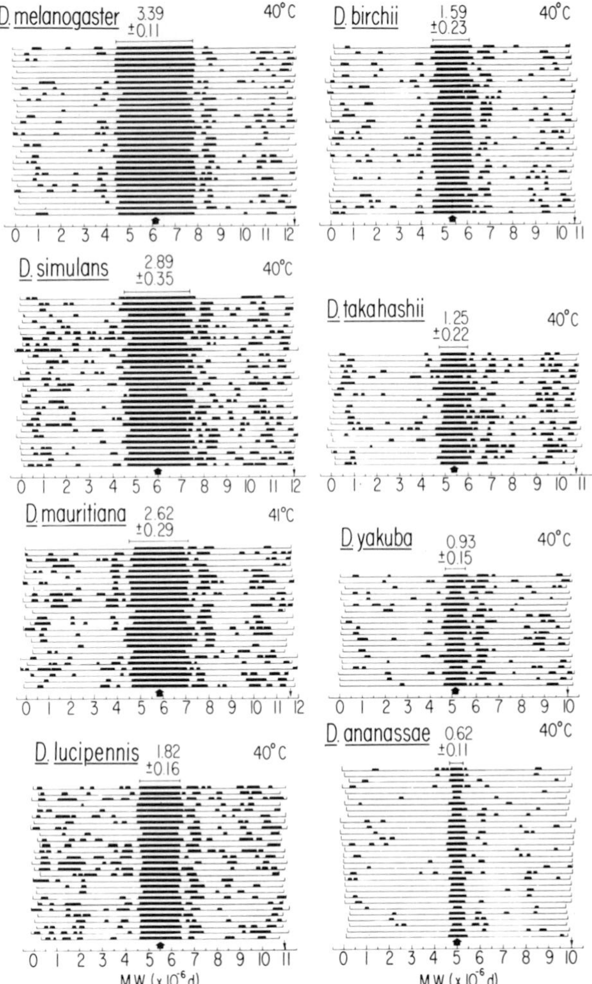

Figure 13-2. Denaturation maps of open circular mitochondrial DNA molecules of different Drosophila *species. Each horizontal line represents a single DNA molecule examined; the dark portions of the lines indicate AT-rich regions that denature at 40°C to 41°C. The molecules are aligned by the positions of the major AT-rich regions. Numbers above each group of molecular maps for a species represent the average length of the AT-rich region. (From C. Fauron and D. Wolstenholme,* Proc. Natl. Acad. Sci. USA *73:3626, 1976).*

The histone genes provide another example of differing melting points. Histones are the basic proteins found in association with nuclear DNA. They are rich in arginine and lysine, whose codons are relatively rich in G–C pairs, so the histone genes should have a greater density than other parts of the

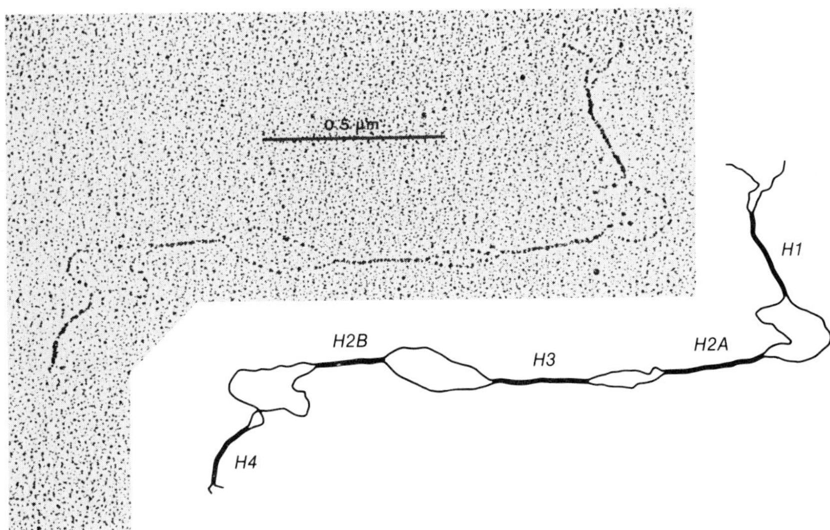

Figure 13-3. Electron micrograph of a partially denatured (at 61°C) DNA molecule from the histone genes of the sea urchin, with a drawing showing the sequence of the histone genes (H), *separated from one another by the denatured AT-rich spacers. (From R. Portmann and M. Birnstiel,* Nature *246:31, 1976.)*

chromosome. This density difference permits isolation of the histone genes. Furthermore, the histone genes should resist heat denaturation. Indeed, the segments of DNA that code for five histones in sea urchins occur in a single region, with each gene separated from the next by an AT-rich "spacer" that denatures at lower temperatures (Figure 13-3).

Message
Regions within DNA that are relatively rich in A–T pairs denature at a lower temperature than the surrounding regions. Denaturation maps show the effects of these intramolecular differences in melting temperature.

Reassociation of Complementary Single Strands

In completely denatured DNA, all strands are single. These single strands can reassociate by random collisions that permit matching between complementary sequences. The specific matching of strands after reannealing can be demonstrated by testing the biological activity of reformed duplexes in bacterial transformation.

Message
Strand matching by base complementarity is the property that makes DNA accessible to manipulation.

The specificity of such reannealing is demonstrated by the elegant experiments of Jonathan Beckwith and his group in 1969. They intensively studied the *lac* locus in *E. coli* using the special episome called F*lac*, which carries lac^+ linked to the fertility factor. They recovered a temperature-sensitive mutant, $F_{ts}lac$, in which the episome is lost at high temperature because of its inability to replicate autonomously. If $F_{ts}lac$ is transferred to F^-lac^- cells that are then grown at high temperatures, some lac^+ colonies grow. They are found to have $F_{ts}lac$ integrated into the genome at the *lac* region. If $F_{ts}lac$ is transferred to an F^- mutant in which the *lac* locus is deleted, then lac^+ colonies again are found after plating at high temperatures, but such colonies appear at a lower frequency than in the first experiment. In this case, $F_{ts}lac$ is inserted at a number of places throughout the genome.

Using this technique, Beckwith and his colleagues recovered two different *lac*-insertion strains: one inserts adjacent to the insertion site for λ phage, and the other inserts adjacent to the insertion site for φ80 phage. The phages λ and φ80 are related, and they are quite similar in structure and DNA base sequence. Beckwith's group recovered specialized transducing phages of λ and φ80 that carry lac^+. If we represent each strand of DNA by a line, the phage genes by letters, and the 5′ end of a strand by an arrowhead, we can diagram the two phages as in Figure 13-4. The two strands differ in density, and the heavy strand is indicated by the primed symbols. You can see that the *lac* locus in the two strains is inverted in its sequence relative to the phage marker genes.

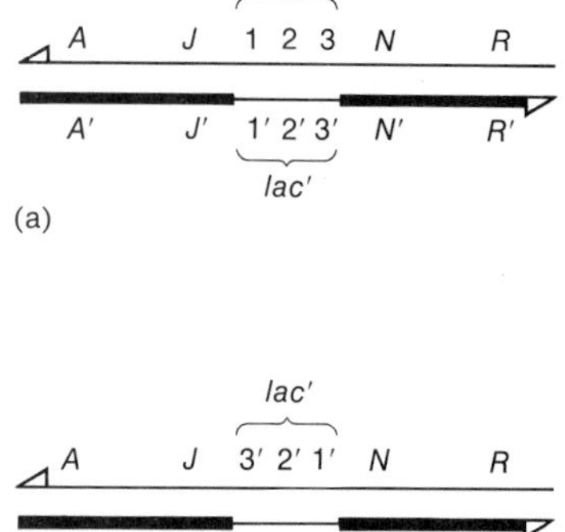

Figure 13-4. DNA of lac*-transducing phages λ (a) and φ80 (b). Each double-stranded DNA is only a portion of the phage genome with* lac *integrated into it. There is considerable homology between the genes of λ and φ80, as we see with the labels,* A, J, N, *and* R. *Arrows represent the 5′ end of the DNA strand. The two strands differ in density; the "heavy" strand is indicated by the prime symbols. The* lac *locus in the two strains is inverted relative to the phage markers.*

In each strain, the two strands of DNA were separated by heat denaturation followed by spinning the DNA in a cesium chloride gradient to separate the light and heavy strands by their buoyant densities. Figure 13-5 shows the two heavy strands. You can see that these two strands have only one region, *lac*, in which there is the complementarity needed for duplex formation (Figure 13-6).

The actual complex formed can be seen with the aid of an electron microscope (Figure 13-7). Using an enzyme that cleaves only single-stranded DNA,

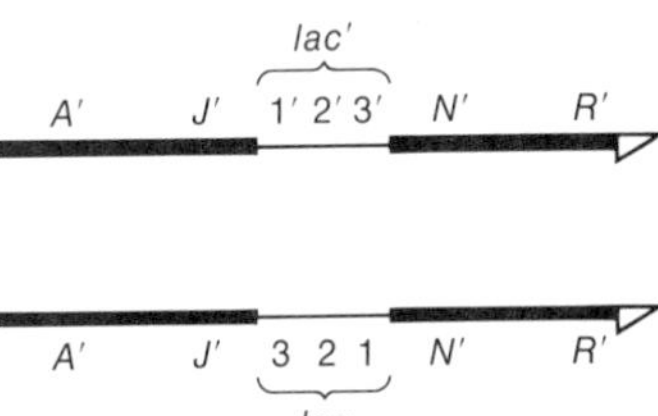

Figure 13-5. The heavy strands of the ϕ80 and λ phages that transduce lac^{+}.

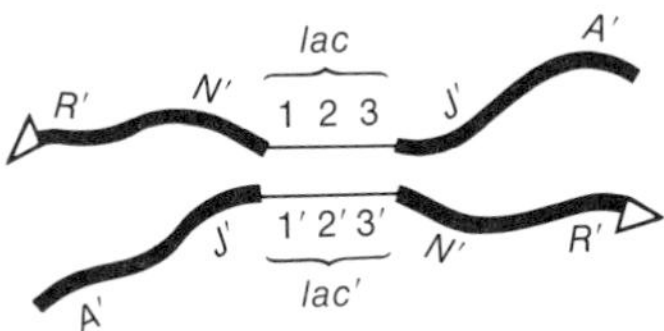

Figure 13-6. Duplex formation between the heavy strands of λ and ϕ80, each containing an inserted lac *region. Since the* lac *insertion in the two phages is in reverse direction, they will reanneal only at the* lac *regions of homology.*

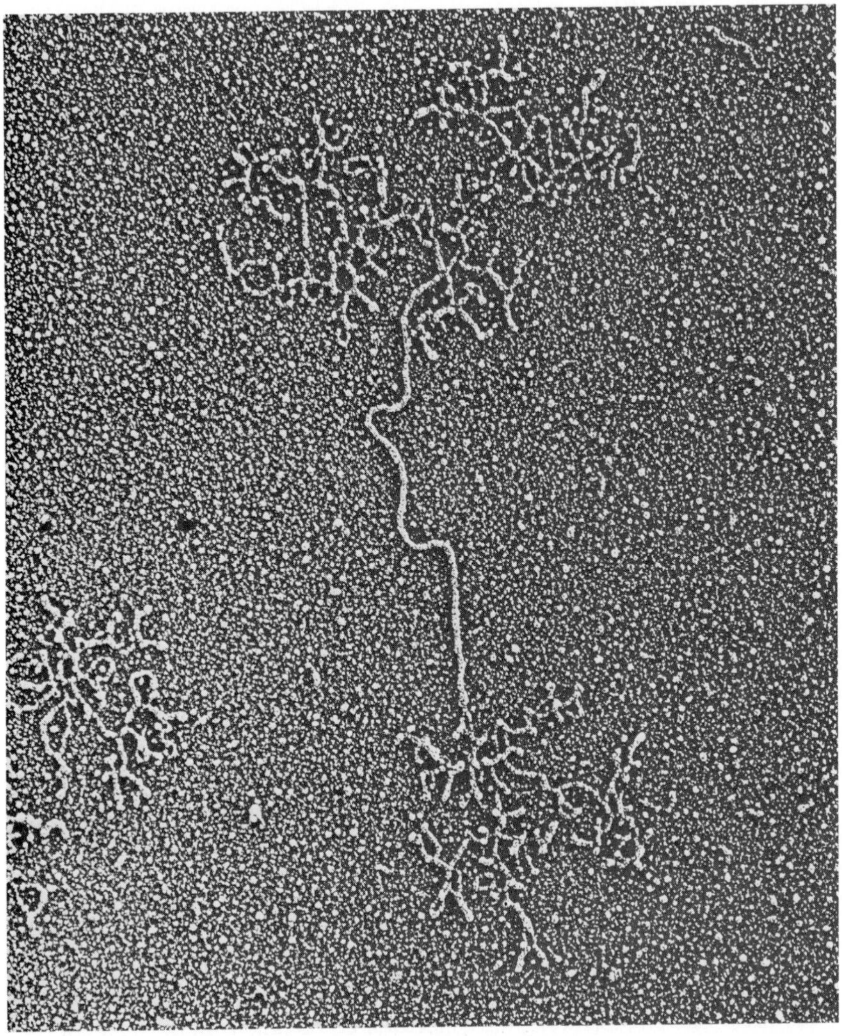

Figure 13-7. Electron micrograph showing the paired strands of the lac *region* (center) *and unpaired "whiskers" of the phage strands. This configuration is derived from reannealing the heavy strands from λ and ϕ80, each containing an insertion of the* lac *region. (From J. Beckwith,* Nature *224:722, 1969.)*

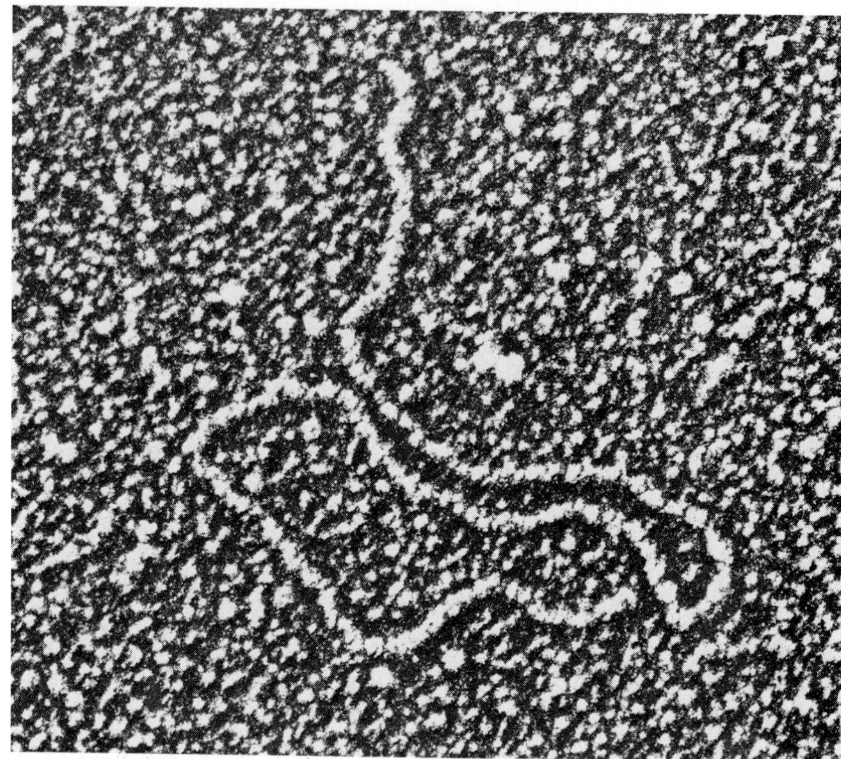

Figure 13-8. Electron micrograph of the lac *gene. After reannealing the heavy strands from λ and φ80, each containing the* lac *insertion, the "whiskers," or the nonannealed portions of the phage genomes, have been digested away with a specific enzyme. This leaves only the* lac *gene intact as a double-stranded piece of DNA. (From J. Beckwith,* Nature *224:772, 1969.)*

Beckwith's group was able to clip away the "whiskers" of unpaired strands on either side of *lac*, leaving a pure gene (Figure 13-8). Using this technique, therefore, it is possible to prepare a sizeable sample of one kind of gene that is freed from the rest of the genome—a very handy capability for the genetic engineer. However, as we shall see, techniques exist that make this task much easier.

Kinetics of Reassociation

The rate of reassociation (reannealing) of DNA strands is strongly affected by a number of factors, including the concentration of cations in the medium (the cations neutralize the negatively charged single strands that would normally repel each other), the temperature of incubation, the initial concentration of DNA, and the length of the fragments (varying from 3×10^2 to 10^5 base pairs). Generally, cation concentrations and fragment length are standardized for optimal reassociation.

The kinetics of DNA reassociation (as well as those of DNA:RNA hybridization) follow a **second-order curve** (rather than a straight line or linear curve) because the interaction involves two components, the two single strands that must rejoin. The rate at which the concentration (C) of single strands decreases by annealing is $-dC/dt = kC^2$ (where t is the time in seconds and k is a constant that is affected by the four parameters we have mentioned). The equation $-dC/dt = kC^2$ can be rearranged as $-dC/C^2 = kdt$ and integrated. Using the boundary condition $C = C_0$ (initial concentration) at $t = 0$, we obtain

$$\frac{1}{C} - \frac{1}{C_0} = kt$$

or

$$\frac{C}{C_0} = \frac{1}{1 + kC_0t}$$

We see that the proportion of single-stranded DNA remaining in a reaction mixture is a function of C_0t. We can illustrate the rate of reassociation by a so-called "Cot" curve, with the proportion of single-stranded DNA plotted against C_0t (Figure 13-9).

Single-stranded DNA can be distinguished from DNA duplexes in a number of ways. For example, dissociated DNA absorbs more ultraviolet (UV) light

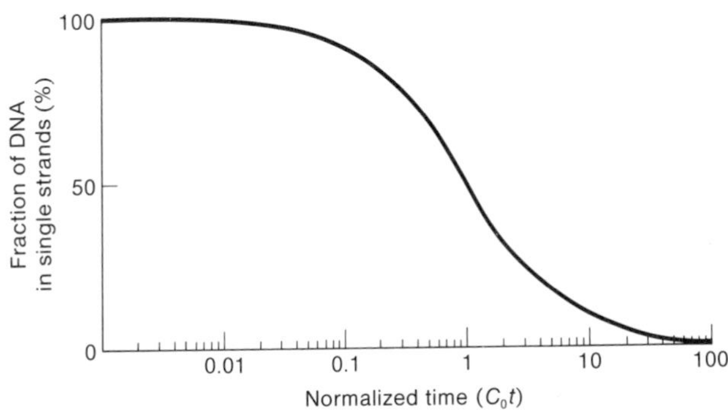

Figure 13-9. Ideal time course for DNA reassociation as seen in a Cot curve. The midpoint represents the reassociation of half of the strands. Single-stranded DNA is mixed under standard conditions of temperature, cation concentration, and length of DNA fragments. The single strands will reanneal, following a second order *reaction curve, indicating that it is a bimolecular reaction. The higher the initial concentration* (C_0) *of complementary strands, the shorter will be the time* (t) *required for their collision and the formation of double-stranded structures; the lower the initial concentration, the longer will be the time required for the appropriate molecular collision and consequent reannealing. The function, initial concentration × time (sec), or Cot, is the abcissa of the graph. (From R. J. Britten and D. E. Kohne, "Repeated Sequences in DNA,"* Science *161:529–540, 1968. Copyright 1968 by the American Association for the Advancement of Science.)*

than does the more organized double-stranded DNA. Hence UV absorption is directly proportional to the concentration of single strands (and inversely proportional to the concentration of double strands). Single-stranded DNA can be immobilized on a nitrocellulose filter or an agar matrix. When radioactively labeled DNA is passed over this immobilized DNA, the number of duplexes formed can be estimated by measuring the radioactivity remaining on the filter or gel. Another useful technique involves passing DNA through a column of calcium phosphate hydroxide (hydroxyapatite). Double-stranded DNA is retained by ionic binding in such a column, while single-stranded DNA passes through. Thus there are several ways to assess the extent of renaturation.

The **normalized time** (C_0t) required for reassociation of one-half of the DNA single strands provides a useful measure of the rate of reassociation; let's call this the **half-reassociation time.** Measurements on many DNA samples show that the half-reassociation time is largest for DNA from mammals; the half-reassociation time is much shorter for *E. coli* DNA, and shorter yet for phage T4 DNA. The rate of reassociation is slower for more complex organisms. We would expect the more complex organisms to have a larger genome with more nucleotide pairs in the DNA. But why should this reduce the rate of renaturation?

Let's take a simplified example. Take two viruses, one having three genes and the other having ten. We fragment the DNA from each source into gene-sized pieces and denature it (Figure 13-10). If the concentration of DNA pieces is the same in each case, it is obvious that a given gene from the smaller genome will be more likely to run into its complementary partner than will a gene from the larger genome. To put it another way, the probability of a collision with another piece of DNA depends on the concentration of such pieces, but the probability that a collision (once it occurs) will lead to reassociation depends on the proportion of appropriate complementary strands, and that is greater for the

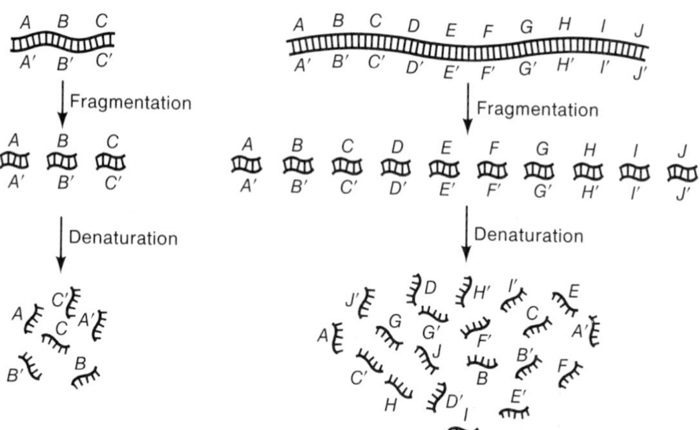

Figure 13-10. The DNAs of two hypothetical viruses are fragmented into gene-sized pieces and then denatured.

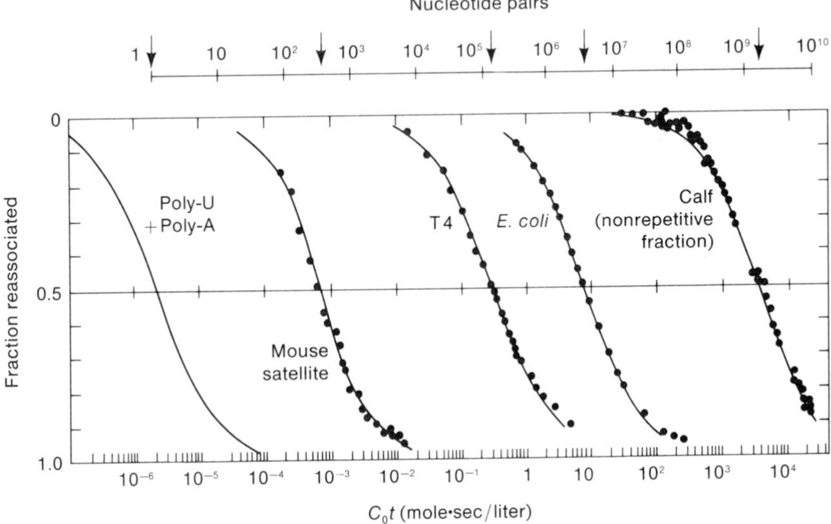

Figure 13-11. The rate of reassociation is strongly affected by the complexity of the DNA. For each of the DNA samples tested, the number of base pairs in the genome is indicated by an arrow on the logarithmic scale at the top of the graph. The poly-U + poly-A sample is a double helix of RNA, with one strand containing only A and the other strand only U. The mouse satellite DNA is a fraction of nuclear DNA in mouse cells that differs in its physical properties from the bulk of the DNA. The calf DNA represents only those sequences that are present in single copies per haploid genome. The denatured DNA samples were fragmented by mechanical shear to chain lengths of about 400 nucleotides and incubated at a temperature near 60° C. The fraction reassociated was measured by the decrease in UV absorption as double strands formed. (After R. J. Britten and D. E. Kohne, "Repeated Sequences in DNA," Science *161:529, 1968. Copyright 1968 by the American Association for the Advancement of Science.)*

smaller genome. The same reasoning can be applied to larger strands, assuming that reannealment must begin when some complementary portions of the two strands come into contact. Increasing DNA diversity effectively dilutes the concentration of the sequences appropriate for pairing with a given sequence and hence reduces the probability of a collision that can initiate reassociation.

In the late 1960s, Roy Britten used Cot curves to measure the relative complexity of the DNA sequences carried by different organisms (Figure 13-11). You can see that the normalized time required for half-reassociation is proportional to the number of nucleotide base pairs in the genome of the DNA. This number is a reflection of the sequence complexity of the genome.

Message
Single-stranded DNA can find its complementary chain and reanneal at a rate that reflects the extent of sequence diversity in the genome.

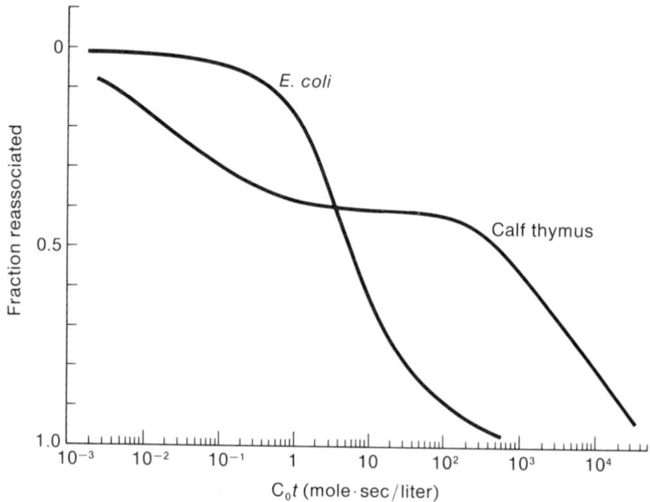

Figure 13-12. The Cot curve for the DNA from calf thymus shows an initial quick reassociation of a sizeable part of the denatured DNA. This early reassociation is much more rapid than that of the E. coli *DNA. The early-reassociating fraction of the calf-thymus DNA represents a part of the genome with a very large number of copies of a relatively small number of genes. The second drop in the curve (representing reassociation that is slower than that of the* E. coli *DNA) is due to the reassociation of the unique sequences in the calf genome, which are far more complex than the* E. coli *genome. (After R. J. Britten and D. E. Kohne, "Repeated Sequences in DNA,"* Science *161:529, 1968. Copyright 1968 by the American Association for the Advancement of Science.)*

Upon close examination of his results for the DNA from calf thymus, Britten observed that there are two parts to the Cot curve. The first part represents a fraction with a shorter half-reassociation time than that of *E. coli* DNA, producing the first sharp dip in the calf-thymus Cot curve (Figure 13-12). The second fraction of the calf-thymus DNA reassociates more slowly than the *E. coli* DNA; this slow fraction is shown separately in Figure 13-11. Why is a portion of this mammalian DNA able to reanneal so rapidly, when most mammalian DNA is quite slow in reassociation? Two explanations are possible. First, some sequences in the DNA might be repeated many times, so that this long DNA is very simple in sequence complexity, acting more like the simple artificial DNA at the far left in Figure 13-11. Second, there might be complementary sequences within a single strand that could form a duplex when the strand simply folds back on itself to form a "hairpin" without waiting for collision with another strand. In fact, both explanations are valid.

About 40% of the DNA from calf-thymus nuclei represents sequences that are **redundant**—up to 10^5 or 10^6 copies of each sequence are present in a

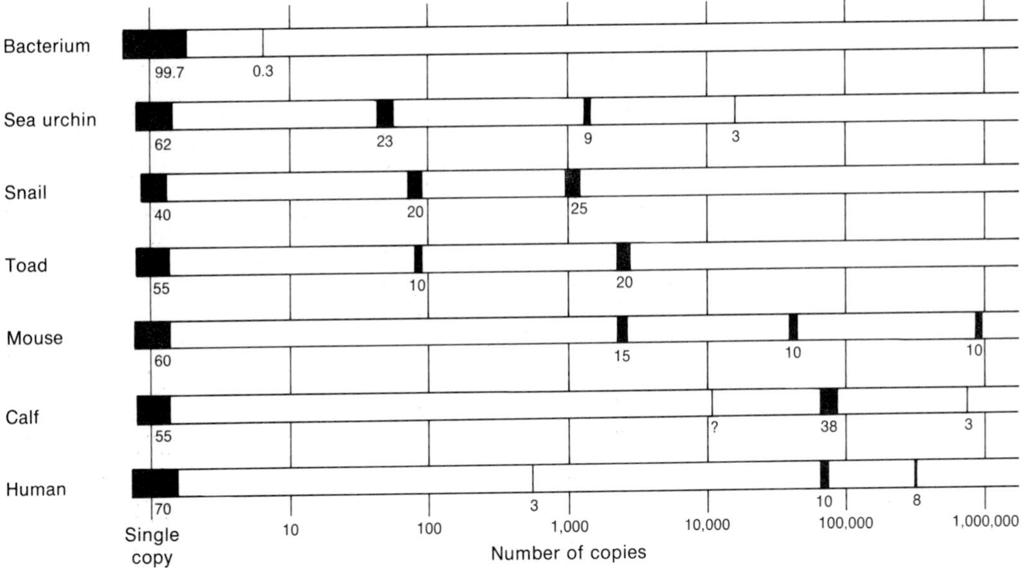

Figure 13-13. Abundance of repeated sequences in the DNA of seven organisms. The width of each band represents the percentage (also given below the band) of the total DNA that appears with a given degree of repetition. (From R. J. Britten and D. E. Kohne, "Repeated Segments of DNA." Copyright © 1970 by Scientific American, Inc. All rights reserved.)

single genome. As we shall see, a small part of this redundancy is due to a low number of multiple copies of such important loci as histone, ribosomal RNA, and transfer RNA genes. The highly repetitive DNA is found clustered in heterochromatin adjacent to centromeres and at the tips of chromosomes. In contrast, the middle repetitive sequences are dispersed throughout the genome and separated from each other by unique sequences of varying length. Figure 13-13 shows the degree and biological distribution of levels of redundancy. To date, the biological significance of the redundant regions remains largely a mystery.

Message
Rates of renaturation provide information on the relative abundance of unique sequences. The proportion of repetitive sequences rises with increasing complexity of organisms.

Making a close examination of renaturation kinetics in 1974, Charles Thomas observed that a small fraction of duplexes can be detected almost at time zero, *before* any collisions between molecules in solution can occur. This

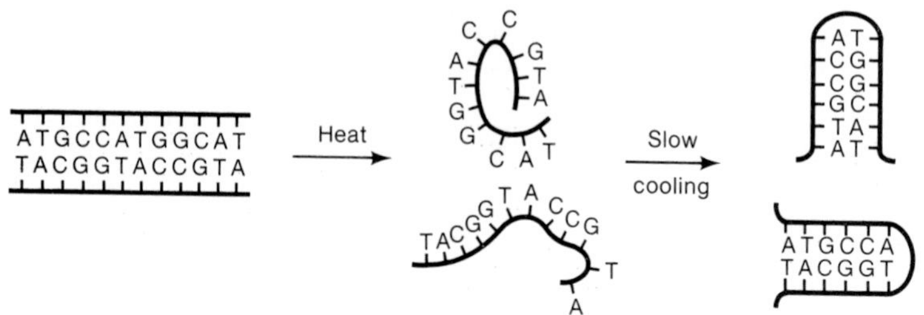

Figure 13-14. Formation of a section of double helix by pairing of complementary sequences within a single strand. Such sequences are called palindromes. This "snap-back" double-strand formation within one chain of DNA occurs very rapidly under reannealing conditions.

can be explained if internal complementary sequences occur in single strands (Figure 13-14). Comparing the sequences in the two strands in the figure (recalling that each is read in the opposite direction), you can see that the sequences are the same. In semantics, a sentence reading the same forward or backward (such as ABLE WAS I ERE I SAW ELBA) is called a **palindrome,** and this name is used to describe DNA sequences such as the one in Figure 13-14. As we shall see, palindromes are important in genetic manipulations, but their biological role remains obscure. As Thomas points out, palindromes give an alternative second dimension to a DNA duplex (Figure 13-15), which he calls a **cruciform configuration** because of its crosslike shape. It could very well be that such cruciform configurations of varying sizes and shapes serve as recognition structures for enzymes.

Figure 13-15. A cruciform (crosslike) structure formed from a palindrome within a DNA molecule. The pairing of a strand with itself in the palindromic stretches will form these cruciform configurations of varying sizes, depending on the lengths of the palindromic stretches.

Verification of Genetic Predictions

Because of the remarkable specificity of duplex formation, a variety of questions can be answered by direct observation of the reassociated molecules. For example, on purely genetic grounds, geneticists had suggested that the

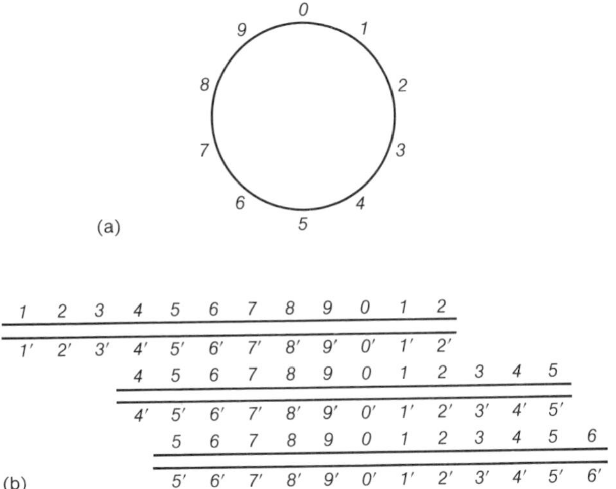

Figure 13-16. Location of genes in the phage T4 chromosome. (a) Circular T4 genetic map. (b) A model for the T4 genome with a number of linear DNA molecules whose ends occur at different positions of the circular genetic map. At any point around the chromosome, there are a number of copies of the DNA due to overlapping ends of the linear molecules. (After L. A. MacHattie, D. A. Ritchie, and C. A. Thomas, Jr., Journal of Molecular Biology *23:355, 1967. Copyright by Academic Press, Inc., London, Ltd.)*

phage T4 chromosome could have ends at different points around a circular map, with the ends overlapping. Thus the genetic map is circular (Figure 13-16a), but the phage DNA is suggested to be a population of linear rods with ends at different spots around the circle and overlapping the neighboring rods at each end (Figure 13-16b). In 1967, Lorne McHattie and Charles Thomas verified this prediction by denaturing a population of T4 DNA molecules, allowing the mixture to reanneal, and then inspecting the resulting duplexes. They found circles with single-stranded tails representing the alignment of strands from different DNA molecules that have ends in inappropriate places (Figure 13-17). Such ends could not exist if all the DNA molecules terminated at the same place.

Another use of reassociation is the physical mapping of deletions or of regions of nucleotide differences in heteroduplexes formed by DNA from different sources. For example, the position and size of a deletion can be determined by annealing DNA from a deletion mutant with DNA from wild type (Figure 13-18). Figure 13-19 shows an electron micrograph of such a heteroduplex. In the same way, hybridization of DNA sequences from sources that are related evolutionarily provides a visual indication of the extent of divergence in the nucleotide sequences.

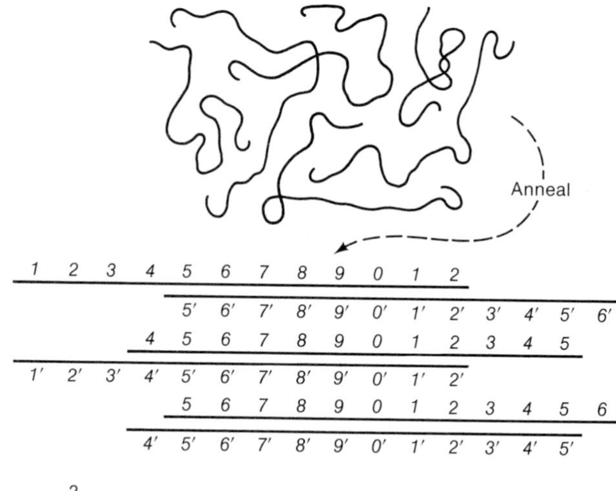

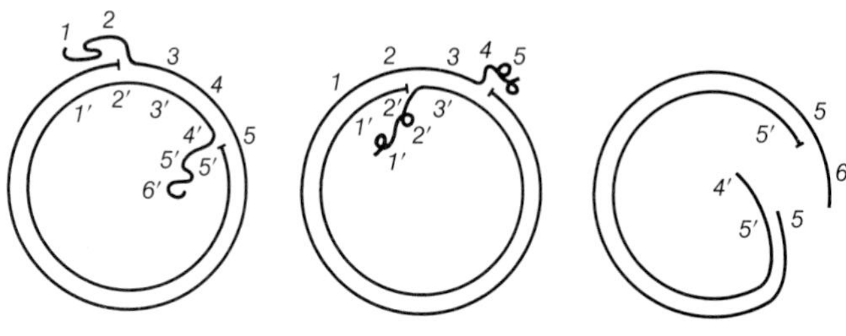

Figure 13-17. The molecules of Figure 13-16b are denatured and then reannealed. The various molecules had nonidentical sequences, so pairing is likely to be imperfect (with single-stranded ends left over). The single strands can anneal to form the ring chromosome again. Because the duplicated portions on different strands are not the same, one end per strand will remain unpaired. (After L. A. MacHattie, D. A. Ritchie, and C. A. Thomas, Jr., Journal of Molecular Biology *23:355, 1967. Copyright by Academic Press, Inc., London, Ltd.)*

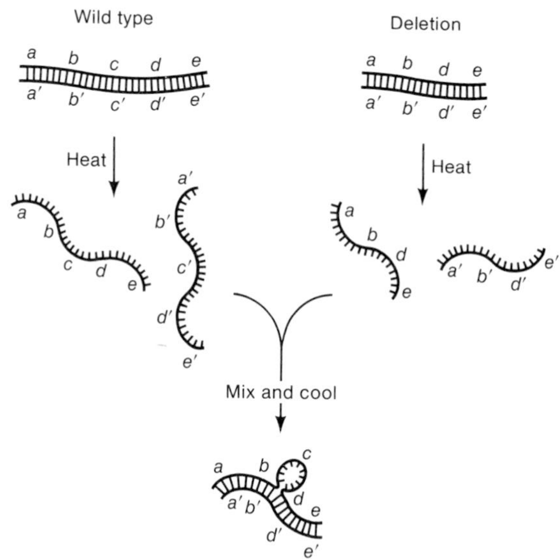

Figure 13-18. Formation of a heteroduplex with deletion and normal DNA.

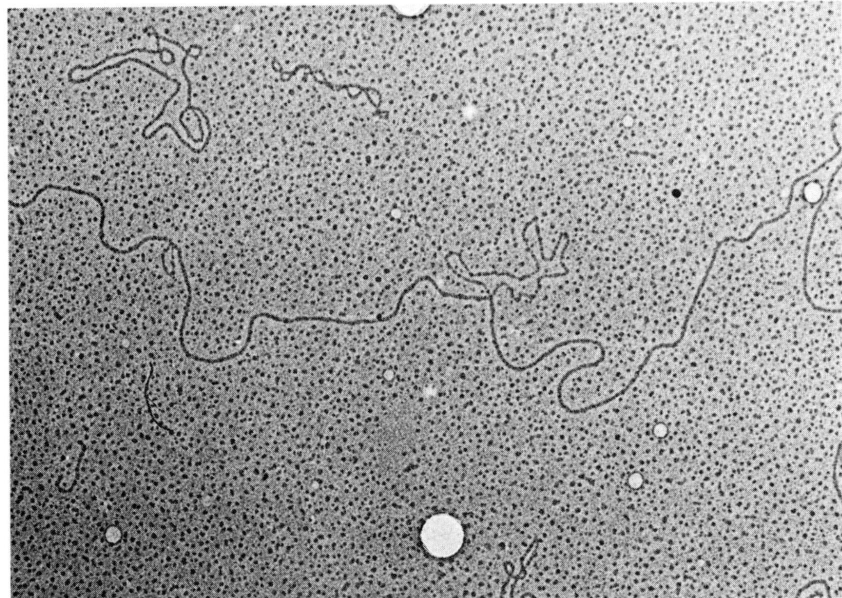

Figure 13-19. A heteroduplex of deletion and wild-type DNA of phage T4. Two different deletion loops are visible. (From T. Homyk.)

Message
The remarkable specificity of complementary base-sequence recognition provides a powerful tool for actually comparing base sequences in heteroduplexes.

Locating DNA Sequences on Chromosomes

Can the base sequences of a purified nucleic acid be used to locate those sequences of DNA within a chromosome that are complementary to them? In the late 1960s, Mary Lou Pardue and Joseph Gall made use of the giant chromosomes of *Drosophila* salivary glands to answer this question. Because the DNA is duplicated several hundredfold in the polytene chromosome, specific nucleic acids have many potential pairing partners at a complementary site. Suppose that we wish to locate the genes coding for an RNA.

The chromosomes are initially squashed and fixed on the slides. Then the DNA is denatured with a mild alkaline treatment that breaks hydrogen bonds without liberating the DNA from the chromosome. The single strands are prevented from renaturing by treatment with formamide, which combines with

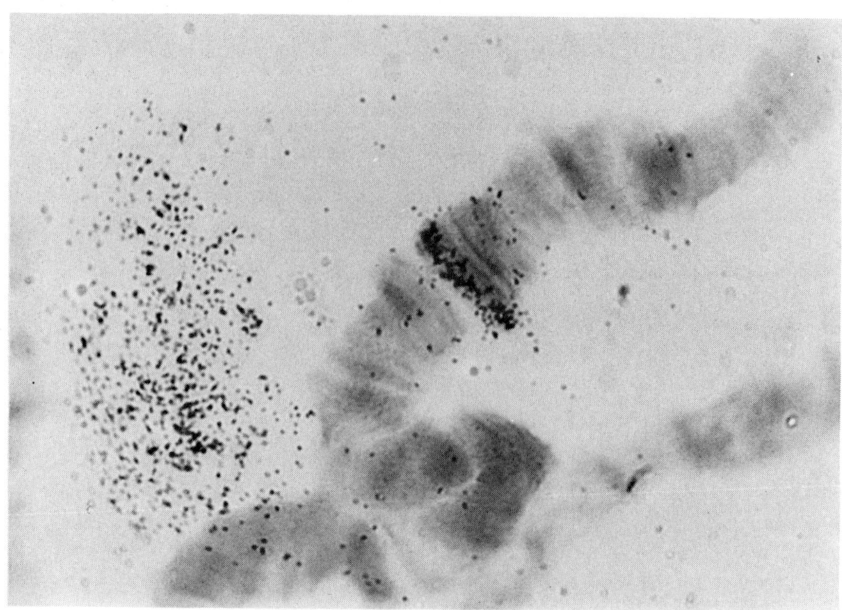

Figure 13-20. Autoradiograph showing ribosomal RNA binding to the chromocenter and at 56EF in the right arm of chromosome 2. (From T. Grigliatti et al., Cold Spring Harbor Symposia on Quantitative Biology *38:461, 1974.)*

free amino groups and inhibits duplex reformation. Now radioactive RNA is incubated on the chromosomes. After sufficient time for hybridization, the excess label is washed off, single-stranded RNA is removed by treatment with ribonuclease (which breaks single-stranded RNA into nucleotides while leaving untouched any RNA that is double-stranded or paired to DNA), and a photographic emulsion is placed onto the chromosomes to locate DNA–RNA hybrids in an autoradiograph.

In this way, it was possible to locate the DNA that codes for 18S and 28S ribosomal RNA in the chromocenter (Figure 13-20). The locus for 5S ribosomal RNA is at 56EF in the right arm of chromosome 2. Several sites for tRNA genes have now been identified by localization on chromosomes in situ. RNA made from highly repetitive sequences of DNA (by a procedure we shall discuss later in this chapter) has been used to show the centromere position of this redundant DNA in mammalian chromosomes (Figure 13-21).

Message

Chromosome positions of the DNA complementary to specific RNAs can be identified by hybridization to chromosomes in situ.

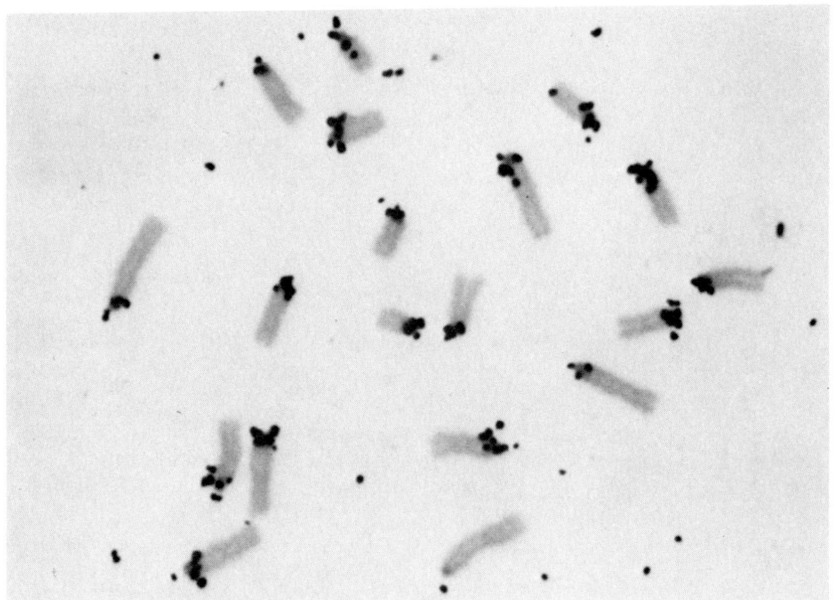

Figure 13-21. Autoradiograph of mouse metaphase chromosomes hybridized in situ with radioactive RNA made from highly redundant mouse DNA. The RNA sequences seem to be present in all chromosomes and concentrated in the region immediately adjacent to the centromere. (From M. L. Pardue and J. Gall, Chromosomes Today *3:47, 1971.)*

Isolation of Specific DNA Sequences

In order to probe the properties of DNA in relation to specific functions, we must separate specific classes or segments of DNA from the rest. Several techniques are now available to accomplish this task.

1. One of the earliest techniques was the separation of the rapidly annealing portion of mammalian DNA that represents highly redundant sequences from the more slowly annealing unique segments that are present only in single copies per genome. This separation is accomplished on nitrocellulose filters or columns of hydroxyapatite. The amount of highly repetitive sequences varies greatly from 0% in many prokaryotes to 30% or 40% in some eukaryotes. The highly repetitive sequences involve repetitions of a short basic unit. For example, in the guinea pig, the repeating unit is

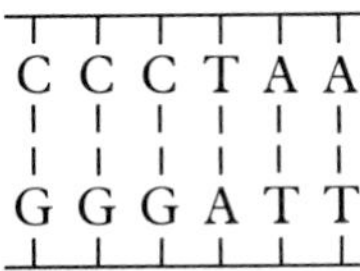

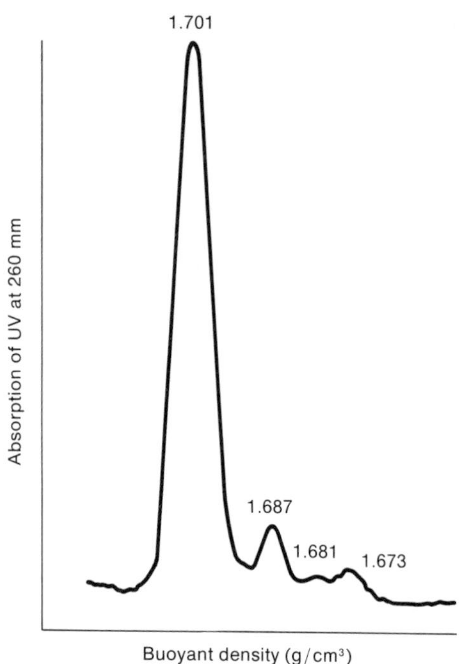

Figure 13-22. Distribution of Drosophila melanogaster *DNA in a cesium chloride density gradient that separates DNA molecules of differing G–C content. The bulk of the DNA appears at 1.701 g/cm³ in the gradient, but there are several satellite bands with a lower buoyant density. (From S. A. Endow, M. L. Polan, and J. G. Gall,* Journal of Molecular Biology *96:670, 1975. Copyright by Academic Press, Inc., London, Ltd.)*

2. Because the highly redundant sequence is a repetition of a short unit, the ratio of GC to AT in such segments may deviate considerably from the ratio for the bulk of the organism's DNA. Because the equilibrium position of DNA in a cesium chloride gradient is determined by its GC content, DNA fragments with different buoyant densities can be identified as **satellites** of the main DNA band in chromatography (Figure 13-22). The satellite DNA may be more or less dense than the main band.

The first two methods provide DNA fractions whose physical properties distinguish them from the main DNA component. The challenge then is to try to find biological functions for these fractions. But are there ways of purifying DNA sequences whose function is already known—sequences that therefore are of specific interest? Such methodologies do exist, and they can provide the most useful kind of information.

3. A very effective method for separating small plasmids (episomes) from the bacterial chromosomes was developed in 1967 by Jerome Vinograd. The molecule ethidium (Figure 13-23) can insert itself between bases of a DNA molecule if the molecule is sufficiently flexible to untwist a bit, thereby providing the space for ethidium to squeeze in. When DNA has ethidium bound to it, this complex has a density less than that of the pure DNA. Vinograd realized that a bacterial chromosome is so long that mechanical stresses will break the DNA when it is isolated, thus producing linear DNA molecules that do allow ethidium intercalation. In contrast, smaller circular molecules, such as the fertility factor, mitochondrial DNA, or chloroplast DNA, remain intact on isolation and have a restricted ability to untwist. Hence, when DNA is isolated from bacterial or eukaryotic cells, the large linear molecules bind much more ethidium than the small circular molecules, which thus remain more dense. In a cesium chloride gradient, the circular molecules are readily purified and collected (Figure 13-24). As we shall see, episomes have played a key role in manipulation of genes, so the purification of such molecules is an important technique.

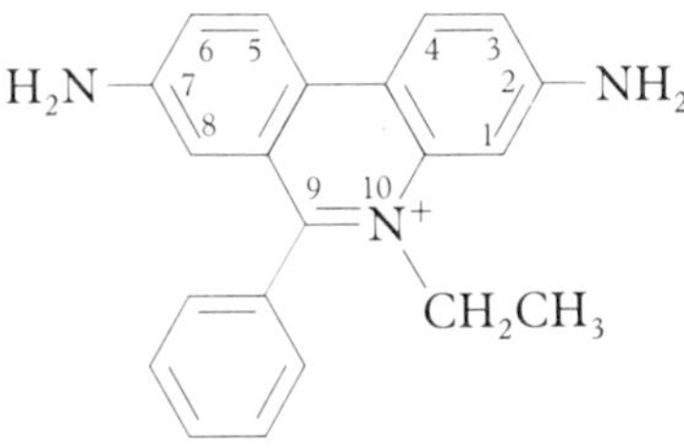

Figure 13-23. The structure of ethidium. This chemical can be used to separate DNA of E. coli *from DNA of its episomes.*

4. Another method of separating specific DNA sequences utilizes the ability of some proteins to recognize such sequences. For example, the enzyme RNA polymerase that is responsible for transcription must initiate the process by attachment to specific regions at the beginnings of genes. After a preparation of DNA is exposed to RNA polymerases, the mixture can be digested with a nuclease that breaks down the DNA

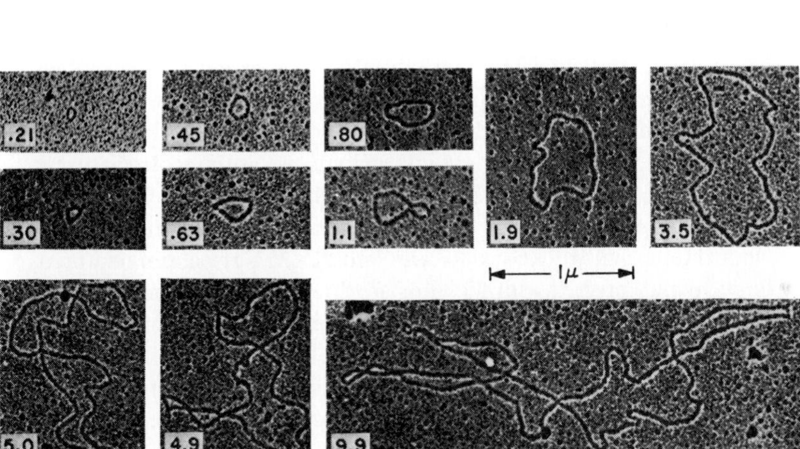

Figure 13-24. Electron micrographs of circular DNA from HeLa cells. The number in each micrograph is the length of the molecule in microns. These may be similar to episomes isolated from bacterial cells. (From R. Radloff, W. Bauer, and J. Vinograd, Proc. Natl. Acad. Sci. USA *57:1514, 1967.)*

5′ TGGAATTGTGAGCGGATAACAATT 3′
3′ ACCTTAACACTCGCCTATTGTTAA 5′

Figure 13-25. Base sequence of the control region of the lac *gene. This DNA sequence was isolated, purified, and sequenced because it has the ability to bind specifically to a protein called the* lac *repressor. All other DNA in the preparation can be destroyed by DNase treatment, but the bound protein protects these 24 base pairs.*

chains. Those sequences to which the enzyme is bound are physically sheltered from the nucleolytic action, so they can be recovered intact after the nuclease treatment. In principle, any DNA sequence that is bound by a specific protein can be recovered in this manner. This is the principle that permitted the purification of the beginning of the lactose gene, which is regulated by a protein called a repressor (see Chapter 14). The DNA segment protected by a repressor is 24 base pairs in length, with the sequence shown in Figure 13-25. Note that the regions indicated by the lines are palindromic (with a few intervening base pairs that do not fit the palindromic pattern). As the amino acid sequences and three-dimensional structures of proteins are determined, along with the base sequences of the DNA sites they recognize, we shall learn about the factors responsible for the specificity of the DNA–protein interactions that appear to be the key to regulation of gene expression.

5. An important tool for the isolation of specific DNA sequences resulted from a study of tumor-causing viruses. Thus far, we have assumed a unidirectional relationship between DNA, RNA, and protein: DNA directs the synthesis of RNA, which in turn specifies protein production (shown as DNA → RNA → protein). The ability of RNA to function both as the hereditary information and as the molecule to be translated is shown by the existence of viruses that contain only RNA. An exception to the DNA → RNA → protein rule was discovered with animal tumor viruses known to carry only RNA. Normal cells cultured in vitro and infected with the virus are "transformed" phenotypically into tumorlike cells. However, such transformed cells normally do not produce infective viruses. It was suggested that the genetic material of the virus is integrated like a prophage into the DNA of the transformed cells. But, if the virus genome is RNA, how can it integrate into chromosomal DNA?

   In the early 1960s, Howard Temin proposed a mechanism by suggesting that the information in the viral RNA is "transcribed" into DNA that is then inserted into the host cell's genome. In 1970, Temin and David Baltimore independently discovered an enzyme that catalyzes this step; it is called **reverse transcriptase.** The existence of reverse transcriptase requires an amendment of the central dogma to DNA $\rightleftarrows$ RNA $\rightarrow$ protein, and it provides a tool for making DNA that complements any RNA that can be purified. Cell biologists have long recognized that cells that are specialized to carry out a specific function typically contain large quantities of a single gene product. For example, the silk-producing cells in the glands of a silkworm produce large quantities of silk protein, fibroin; cells in the mammalian pancreas excrete insulin; red blood cells (erythrocytes) are rich in hemoglobin; cells in chick oviducts make a great deal of ovalbumin; and so on. Isolation of RNA from such specialized cells provides a sample that is enriched for a specific mRNA. Thus, pure hemoglobin mRNA can readily be prepared from blood cells. With reverse transcriptase, cDNA (DNA complementary to the mRNA) is formed in vitro from the mRNA. The cDNA can be subjected to sequence analysis or used to recover complementary sequences from total cellular DNA or from preparations of RNA by the formation of hybrids. When the cDNA is hybridized with RNA, DNA:RNA hybrids reveal the existence of large RNA transcripts, which are precursors of the active mRNA.

Message

There are several techniques for isolating specific classes of DNA from a heterogeneous population of molecules. This isolation allows us to focus on a restricted part of the genome, just as geneticists used various techniques to isolate specific loci for genetic analysis.

Genes in Eukaryotes

We have constantly stressed the underlying similarities between diverse organisms at the genetic level. Molecular genetics has shown us the key role of nucleic acids for all organisms, the universality of the genetic code, and the apparent similarity of the translational apparati for producing gene products in different organisms. In studying eukaryotes, the challenge has been to find ways of analyzing phenomena that may not exist in prokaryotes. We have assumed that all of the mechanisms operative in a bacterium may also be found in an elephant, but that the elephant may have mechanisms that do not exist in the bacterium. Obviously, multicellularity, differentiation, and coordination of diverse tissues pose problems not faced by bacteria or viruses, so these multicellular organisms are likely to possess novel mechanisms not encountered in the simpler forms.

To explore such phenomena, many scientists have focused their attention on highly specialized cells whose protein contents reflect their functions. As we have mentioned, silk-producing cells in moths, hemoglobin-producing blood cells, and immunoglobulin-producing tumor cells provide useful study material. An important model for the study of differentiation and its associated genetic activity is the oviduct in chickens. The oviduct transports the egg toward the area for shell production while surrounding the yolk with egg-white proteins. One of the major constituents of the egg white is the protein ovalbumin, which has been used as an indicator of oviduct differentiation. In newly hatched chicks, there is no recognizable oviduct tissue and no detectable ovalbumin. This is hardly surprising in a sexually immature animal. However, injection of estrogen into a young chick stimulates a massive growth of oviduct tissue and synthesis of ovalbumin. Removal of estrogen is followed by regression of the oviduct. So, we have here a system of genetic induction that can readily be controlled by the experimenter.

From chick oviducts, mRNA coding for ovalbumin was isolated and then paired with complementary DNA sequences. Figure 13-26 shows the DNA–RNA hybrid that was recovered. Apparently, all of the bases in the mRNA have their complements in the DNA, but not all of the DNA bases lying between the ends specified by the message are found in the transcript. In other words, although the ovalbumin messenger is found in one continuous RNA molecule, the gene coding for it contains several sequences that are not found in the messenger; these sequences intervene between segments that do appear in the messenger. Similar nontranslated intervening sequences have been found in a number of other eukaryotic genes, and they are called **introns.** Those DNA segments that are expressed in the mRNA are called **exons.** You can see that the introns can be very large relative to the size of the extrons. It is also known that there can be long nontranslated sequences at the beginning (leader) and end (tail) of the mRNA.

But how does the cell produce the mRNA (lacking the introns) from the complete DNA segment? Figure 13-27 summarizes the possibilities. First, the

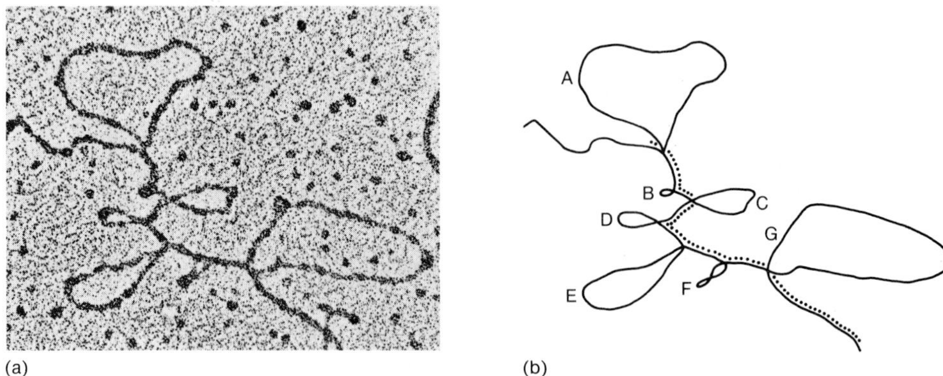

Figure 13-26. Ovalbumin mRNA hybridized to ovalbumin DNA. (a) Electron micrograph of the RNA–DNA hybrid. (b) An interpretation of the structure in the electron micrograph. The dotted line represents the mRNA, and the solid line represents the DNA. (From Dugaiczyk et al., Proc. Natl. Acad. Sci. USA *76:2256, 1979. Courtesy of B. W. O'Malley.)*

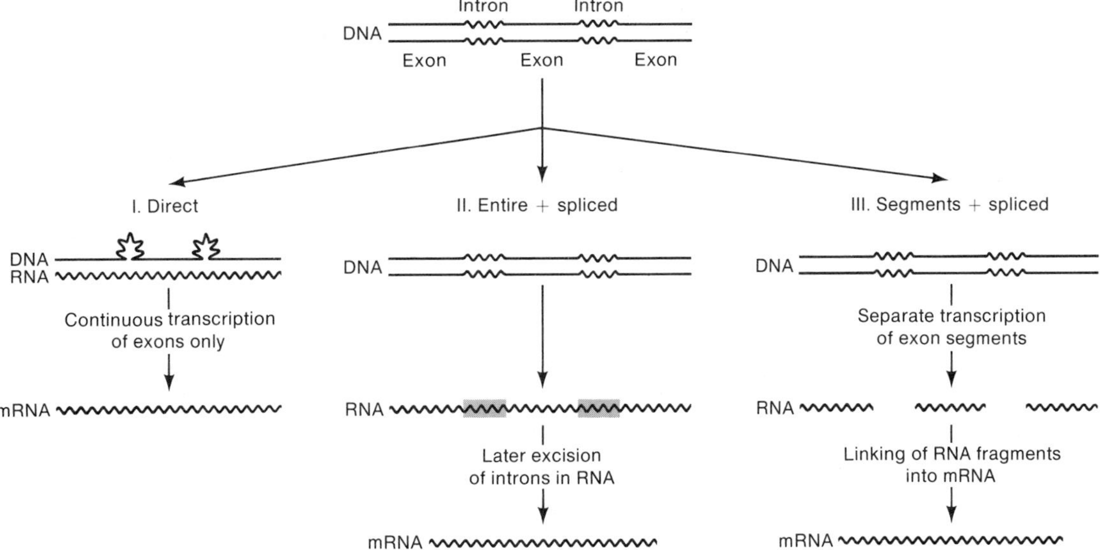

Figure 13-27. Possible mechanisms for production of mRNA from a eukaryotic gene. The mRNA contains all of the extrons but none of the introns from the gene. All processing of the RNA to mRNA must occur within the nucleus since only mature mRNA is found in the cytoplasm.

mRNA might be directly transcribed only from the exons as a single continuous molecule, with the transcription process skipping over the introns. Second, everything might be transcribed, with the introns being excised from the RNA later to produce functional mRNA. Third, the exons may be transcribed as individual pieces that then are linked together to yield functional mRNA. Any of these mechanisms would require a transcription apparatus that has yet to be elucidated.

The mRNAs are extracted from the cytoplasm, but it has long been known that RNA molecules found in the nucleus are larger than the cytoplasmic messengers and contain many sequences that are not found in mature mRNA. This "heterogeneous nuclear" RNA is often called HnRNA. When HnRNA is separated on gels, the molecules corresponding to the ovalbumin gene can be identified by their ability to bind to ovalbumin DNA. Studies show that an entire transcript of the ovalbumin gene is present in HnRNA; this result seems to support possibility II in Figure 13-27.

Message

Eukaryotic genes differ from prokaryotic genes in that they contain sequences that are transcribed but that neither code for the protein nor are found in the mature RNA.

Genetic Engineering

New experimental techniques and ingenious ideas constantly add to the arsenal of sophisticated methods available for identifying specific classes of DNA segments. As we have seen in the case of reverse transcriptase, new techniques can often arise in unexpected—indeed, unpredictable—ways. In the past decade, such unexpected discoveries have produced a startling result: we now are on the verge of being able to manipulate DNA at will.

Restriction Enzymes

The Phenomenon of Host Restriction. We have already seen that a bacterial genotype determines the cell's susceptibility to infection by various phages. Similarly, a phage's genotype determines the range of bacterial cells that it can infect. Clearly, host and parasite have each evolved genetic strategies for coping with the other.

Let's consider the infective capacity of phages grown in bacteria of different genotypes. For example, suppose that a hypothetical phage X is found in two bacterial strains, A and B. We'll label phage from these two sources as X.A and X.B, respectively. Suppose that we use X.A to infect strain B, and we use X.B to infect strain A. We find that X.B phages do poorly at infecting A cells, whereas X.A phages infect both strains equally well (Figure 13-28). When the

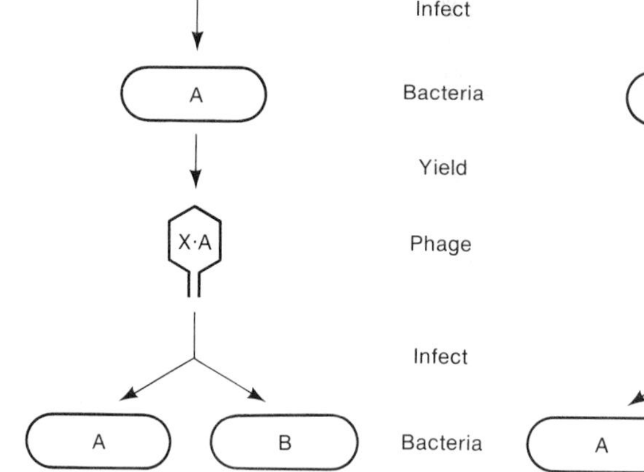

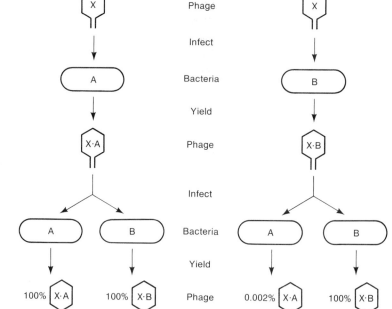

Figure 13-28. Infectivity of hypothetical phage X from two different bacterial genotypes, A and B. Phages derived from strain B are poorly infective on strain A, whereas phages derived from strain A infect both strains equally well. The percentages indicate the efficiency of plaque formation on each host.

few X.B phages that are recovered from A are used to infect strain A, they are found to be *fully infective!* Perhaps some A-infecting mutants already existed in the X.B population, and selection has now isolated them. However, when the phages are allowed to infect strain B again, and then returned to strain A, we find that they are once again poorly infective in strain A. We have not isolated an A-infecting pure-breeding strain of X.B phages.

Apparently, the host range of the phage depends on the bacterial strain in which it matured, not on the phage's genotype. In other words, the host genotype apparently modifies the phage particle without altering the DNA sequence in the phage. It would seem that the small number of phages released in the restricting cross are produced in a few cells whose physiological state happens to be tolerant of the infective particle.

DNA Modification. But what restricts phage X.B from multiplying in strain A? The mature phage is able to attach to the bacterial cell wall and to inject its DNA into the cell. The factor that prevents it from reproducing is the nucleolytic activity of a host enzyme (nuclease) that breaks the phage DNA into a number of noninfective fragments. The host cell seems to have a defense system that destroys unfamiliar DNA! But what protects the host DNA and the infective X.A phage DNA from this nuclease? The answer to this question revealed the nature of the phage-modification process.

Enzymes exist that modify specific bases at specific sequences *without altering the coding properties of those bases.* For example, in certain bacterial strains, cytosine and adenine are altered by the addition of a methyl group to the forms shown in Figure 13-29. In the 1960s, Werner Arbor was able to show that the alteration of cytosine and adenine by addition of a methyl group (a process called methylation) is responsible for protection of the DNA from nuclease activity. He showed that mutations involving replacement of a normally methylating base by a base that cannot be methylated results in a loss of protection from the nuclease action. The total number of bases that are methylated is

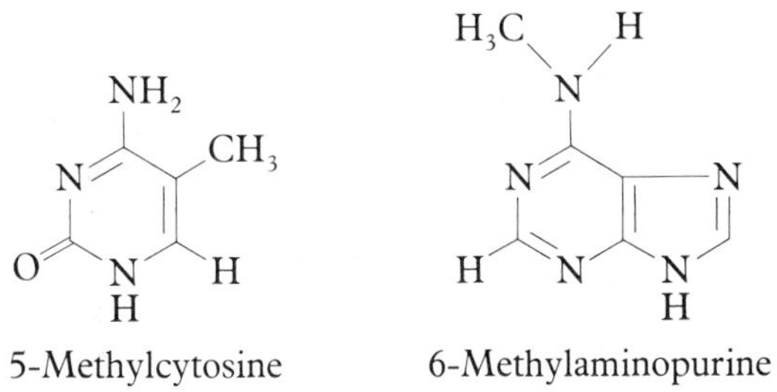

NH_2, CH_3, N, O, N, H, H

5-Methylcytosine

H_3C, H, N, N, N, H, N, N, H

6-Methylaminopurine

Figure 13-29. Forms of cytosine and adenine that have been modified by the addition of a methyl group (methylation). Such methylation does not change the coding capacity of the DNA but does protect the DNA from attack by some highly specific nucleases.

very small, so these particular bases must occupy very specific regions that are attacked by the nucleases. In our example, bacterium A evidently has the capacity to modify DNA whereas B does not.

The phenomenon of host restriction is due to the action of nucleases that degrade any DNA not specifically modified for protection from the nucleases of a given host cell.

Message
There are two interrelated processes involved in the phenomenon of host restriction: the ***restriction*** *of phage multiplication by enzymic destruction of invading viral DNA, and the* ***modification*** *of phage DNA to render it immune to effects of the restriction enzymes.*

Specificity of Restriction Enzymes. The next step in understanding restriction phenomena did not come until 1970, when Hamilton Smith made great progress in his studies of the restriction enzyme from *Haemophilus influenzae.* This enzyme (called *Hin*dII) cuts T7 phage DNA into 40 specific fragments. What do the cleavage sites have in common? To find out, Smith took the mixture of fragments and marked the 5′ ends by attaching the radioisotope ^{32}P. He then used hydrolyzing enzymes to break the labeled fragments into still smaller pieces (Figure 13-30). He was able to separate the labeled end fragments in small segments only a few nucleotides long, so that he could determine the base sequence of each fragment.

Smith found that the label was always attached to an adenine or a guanine (the purines). Fragments with labeled adenine were always A–A or A–A–C; fragments with labeled guanine were always G–A or G–A–C. Smith suggested that such fragments could be produced if the enzyme cleaves only at the points indicated by the arrows in the following specific sequence (where Py represents a pyrimidine and Pu represents a purine):

```
                 ↓
5′  –G–T–Py–Pu–A–C–  3′
                 •
3′  –C–A–Pu–Py–T–G–  5′
                 ↑
```

Note the rotational symmetry of this sequence: a rotation of 180° around the dot in the center leaves the sequence unchanged. Verify that cleavage at the arrows will produce just the 5′-end fragments that Smith identified: Pu–A– and Pu–A–C–. *Hin*dII apparently is very specific in identifying the sequence at which it will cleave. Furthermore, Smith showed that the host-induced methylation that confers protection from *Hin*dII activity occurs at this same cleavage site: protection from cleavage by *Hin*dII is obtained when the adenines one base away from the cleavage site are methylated (m):

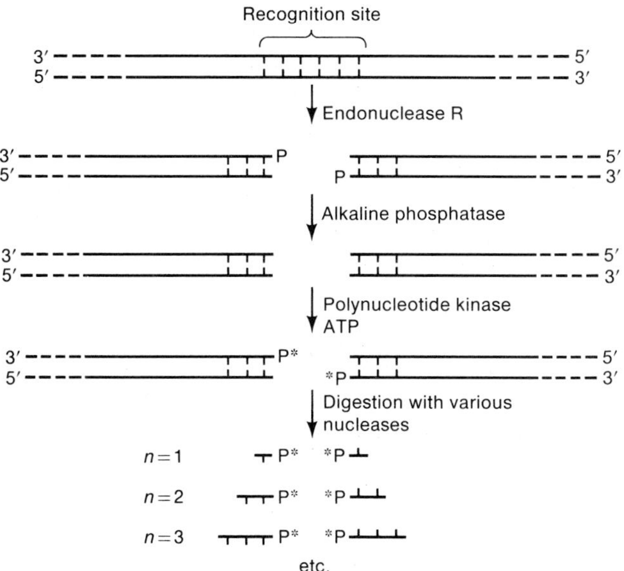

Figure 13-30 Smith's method for identifying the bases at the HindII *recognition site. The T7 DNA is cleaved by* HindII *(endonuclease R), which leaves a phosphate on the 5′ end of the cleaved strands. This phosphate is removed with alkaline phosphatase, and then polynucleotide kinase is used to catalyze the attachment to the 5′ end of a radioactive phosphorus atom in a phosphate group (*P) from labeled ATP. Various nucleases are then used to cleave the fragments into smaller fragments of varying lengths. The labeled fragments are separated, and their base sequences are identified. (From J. T. Kelly, Jr. and H. O. Smith,* Journal of Molecular Biology *51:397, 1970. Copyright by Academic Press, Inc., London, Ltd.)*

```
                m
5′  –G–T–Py–Pu–A–C–  3′
            •
3′  –C–A–Pu–Py–T–G–  5′
       m
```

Studies of other restriction enzymes produced similar results. For example, the enzyme *Eco*RI is produced by a gene on an R plasmid in *E. coli*. *Eco*RI cleaves the circular DNA of SV40 (a small mammalian virus) at only one site, thus converting the DNA ring into a linear molecule that has sticky ends like those of linear λ DNA. The sequence at the cleavage site for *Eco*RI is

```
            m  ↓
3′  –C–T–T–A–A–G–  5′
          •
5′  –G–A–A–T–T–C–  3′
     ↑   m
```

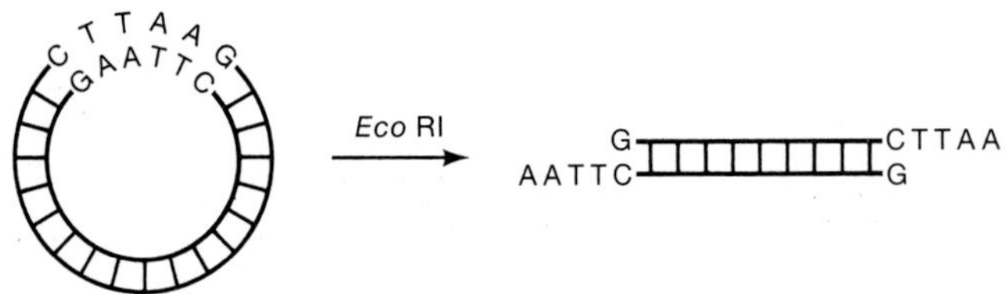

Figure 13-31. Cleavage of SV40 DNA by the restriction enzyme Eco*RI produces a linear DNA with sticky ends.*

Again the sequence is rotationally symmetrical, but in this case the cuts are staggered and therefore produce sticky ends (Figure 13-31). Once again, methylation of adenines near the cleavage site provides protection against the cleaving action (the methylated bases are indicated in the sequence, although, of course, *either* methylation *or* cleavage would occur at such a sequence, *not* both). Note that the *Eco*RI site is a palindrome.

These studies have opened an explosively expanding area. Dozens of restriction enzymes with different sequence specificities have now been identified, some of which are shown in Table 13-1. They are powerful tools for analysis, because they can locate specific base sequences and cut the DNA in a very specific way. Now we are ready to see how we can approach the task of genetic engineering.

Message
Restriction enzymes provide a way of cleaving DNA from any source at a specific sequence, thereby producing a heterogeneous population of fragments with identical ends.

Restriction-Enzyme Mapping. The restriction-enzyme target sites can be used as markers for DNA, just as the bubbles of AT-rich melt regions are used. The DNA from a specific source is subjected to successive digestion by different restriction enzymes. When the fragments are separated electrophoretically on polyacrylamide gels, the "map" of the restriction sites can be deduced, as shown below.

For example, consider a hypothetical DNA molecule having the distribution of restriction sites shown in Figure 13-32a. The 3′ ends can be labeled with ^{32}P before the DNA is cleaved by each enzyme in separate samples (Figure 13-32b). The resulting fragments are separated on gels. The recovery of three fragments after treatment with each restriction enzyme shows that there are two recognition sites for each enzyme. The absence of radioactivity in fragments B and E shows that these are the center pieces.

Table 13-1. Recognition, cleavage, and modification sites for various restriction enzymes

Enzyme	Source organism	Restriction site	Number of cleavage sites in DNA from		
			ϕX174	λ	SV40
*Eco*RI	*Escherichia coli*	m ↓ –C–T–T–A–A–G– 5′ ● 5′ –G–A–A–T–T–C– ↑ m	0	5	1
*Eco*RII	*E. coli*	m ↓ –C–G–G–A–C–C–G– 5′ ● 5′ –G–C–C–T–G–G–C– ↑ m	2	>35	16
*Hin*dII	*Hemophilus influenza*	m ↓ –C–A–Pu–Py–T–G– 5′ ● 5′ –G–T–Py–Pu–A–C– ↑ m	13	34	7
*Hin*dIII	*H. influenza*	↓m –T–T–C–G–A–A– 5′ ● 5′ –A–A–G–C–T–T– m↑	0	6	6
*Hae*III	*H. aegyptius*	↓ –C–C–G–G– 5′ ● 5′ –G–G–C–C– ↑	11	>50	19
*Hpa*II	*H. parainfluenzae*	↓ –G–G–C–C– 5′ ● 5′ –C–C–G–G– ↑	5	>50	1
*Pst*I	*Providencia stuartii*	↓ –G–A–C–G–T–C– 5′ ● 5′ –C–T–G–C–A–G– ↑	1	18	2
*Sma*I	*Serratia marcescens*	↓ –G–G–G–C–C–C– 5′ ● 5′ –C–C–C–G–G–G– ↑	0	3	0
*Bam*I	*Bacillus amyloliquefaciens*	↓ –C–C–T–A–G–G– 5′ ● 5′ –G–G–A–T–C–C– ↑	0	5	1
*Bgl*II	*B. globiggi*	↓ –T–C–T–A–G–A– 5′ ● 5′ –A–G–A–T–C–T– ↑	0	5	0

NOTE: An asterisk (*) is commonly used to indicate methylation sites rather than the m we have used here (to avoid confusion with radioactive labeling).

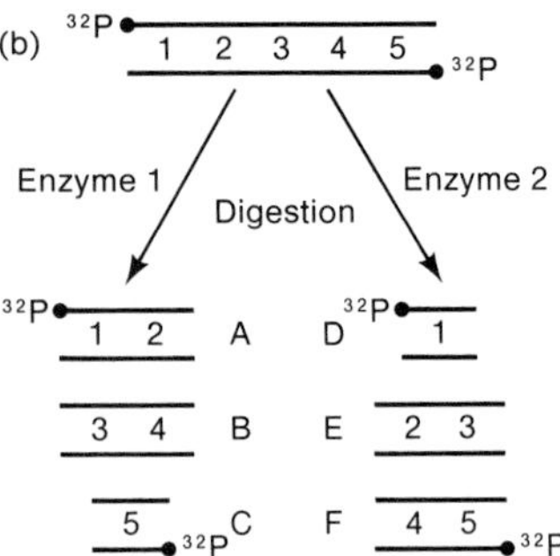

Figure 13-32. (a) A hypothetical DNA with recognition sites for two different restriction enzymes. (b) After labeling of the 3′ ends, digestion of the DNA by enzyme 1 produces fragments A, B, and C, whereas digestion of the DNA by enzyme 2 produces fragments D, E, and F. The fragments can be separated electrophoretically, and the end fragments can be identified by the radioactive labeling.

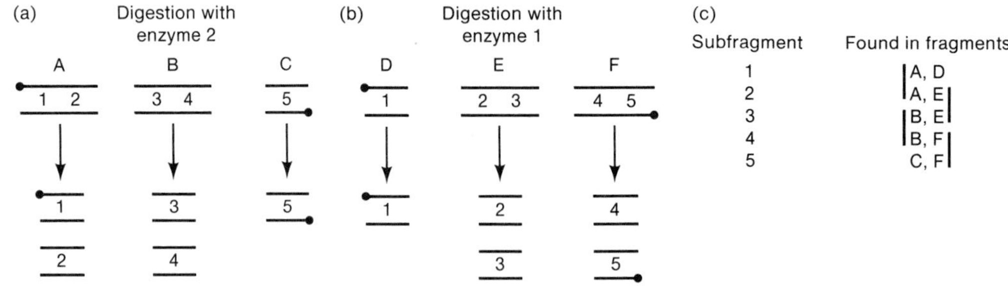

Figure 13-33. (a) Fragments A, B, and C (obtained from the original DNA by digestion with enzyme 1) are now digested with enzyme 2, producing subfragments 1 through 5. (b) The fragments obtained by digestion with enzyme 2 are now digested with enzyme 1 to produce subfragments 1 through 5. (c) Comparison of the subfragments (identified by position on the electrophoretic gel) obtained from each fragment indicates which fragments must overlap (indicated here by vertical lines).

The relative positions of the recognition sites can be determined by taking the fragments from one enzyme treatment and treating these fragments with the other enzyme (Figure 13-33). If smaller subfragments are produced by the second treatment, we know that the original fragments contained sites for the second enzyme. The number of subfragments produced by this double treat-

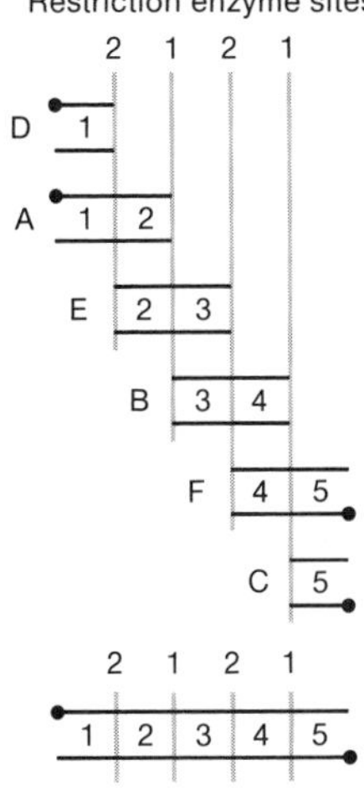

Figure 13-34. Arranging the fragments in the order indicated by their overlapping subfragments, we obtain the map of the restriction-enzyme sites along the DNA (bottom). *Compare this map to Figure 13-32a (which, of course, would not be known at the beginning of such an analysis).*

ment is always one more than the total number of restriction sites (for both enzymes). We now compare the subfragments and determine overlaps. For example, the subfragment 2 is obtained by further breakdown of both fragments A and E, so we know that A and E must overlap. (The similarity of subfragments in the two experiments is indicated by similar positions on the electrophoretic gel.) With this kind of reasoning (Figure 13-33c), we conclude that the overlapping order of fragments in the original DNA must have been DAEBFC.

We know that fragments D and A are both end fragments (because of the radioactive labeling). They share subfragment 1, so we can place subfragment 1 at the end of our map. Fragments A and E share subfragment 2, so we place subfragment 2 next in the map. Proceeding in this way, we map the subfragments on the original DNA (Figure 13-34). We know that the original treatment with enzyme 2 separated fragments D and E, so we conclude that a site for enzyme 2 lies betwen subfragments 1 and 2. With similar reasoning, we can locate the other restriction-enzyme sites on the map. The final map at the bottom of Figure 13-34 is, of course, identical to the one shown in Figure 13-32a (which would not have been known in a real experiment). You may find it useful to review this reasoning to be sure you see how the map could be constructed in a real experiment where the order of subfragments is not known initially.

In 1976, Smith and Birnstiel developed a similar method for locating the restriction-enzyme sites. This method utilizes the ability to distinguish between DNA molecules of differing length. Suppose that we have a circular DNA molecule with a single *Eco*RI site, so that the molecule can be converted into a linear rod by treatment with *Eco*RI (Figure 13-35). Several samples are each partially digested (using concentrations such that around 1 of every 50 sites is actually recognized and cleaved) by one of a series of test enzymes, producing

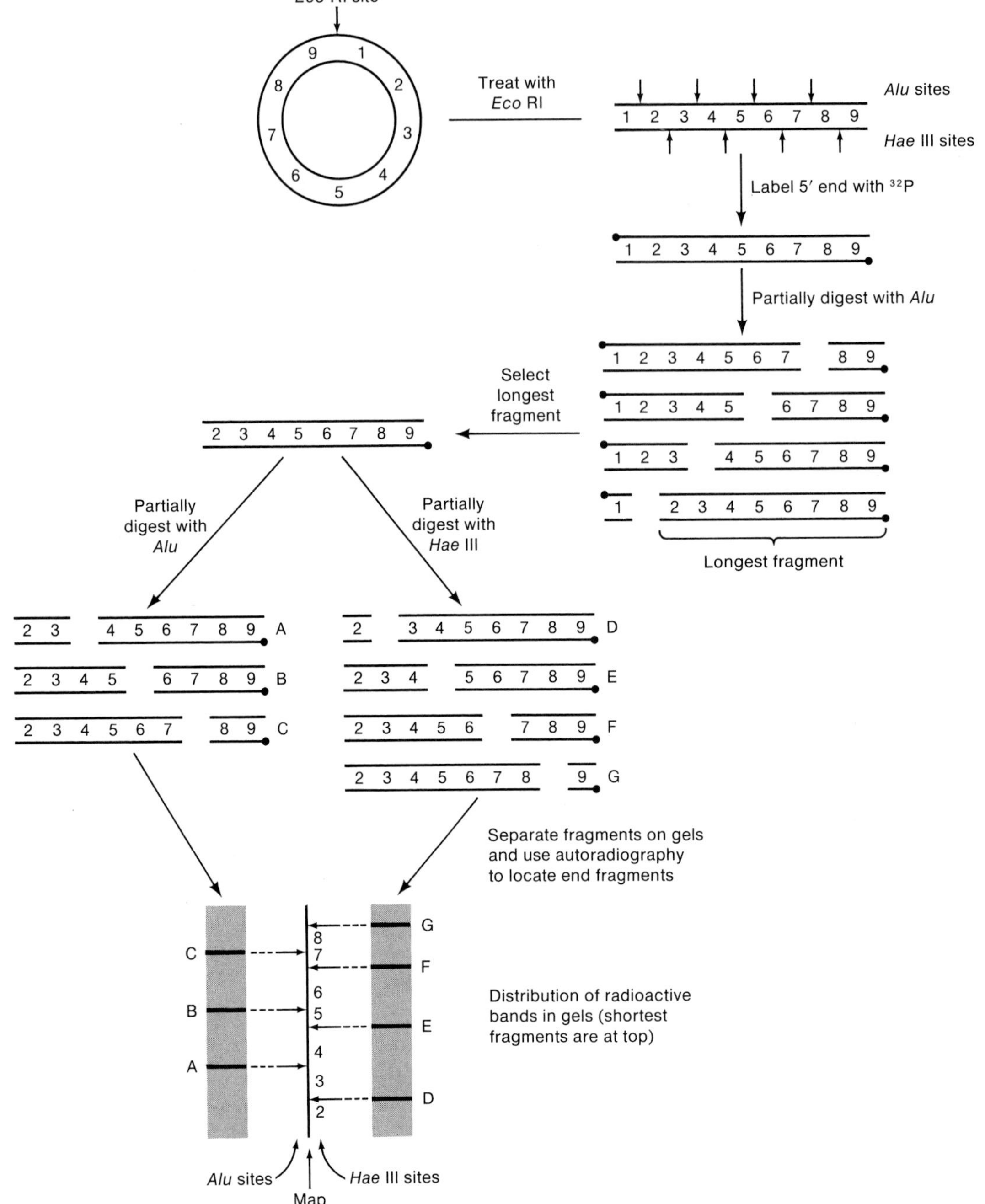

Figure 13-35. Identification of the sequence of restriction-enzyme sites in a circular DNA molecule. In this example, the ring of DNA with a single Eco*RI site is opened into a linear molecule by cleavage with* Eco*RI. The molecule is labeled on its 5′ ends with* ^{32}*P and then partially digested with* Alu *(it could just as easily have been with* Hae*III) to produce fragments of varying length. Each fragment now carries the radioactive label at only one end. The longest fragment is selected (after testing on a gel to compare lengths), and it is partially digested with* Alu *or* Hae*III. The new sets of fragments are separated on gels. The location of the labeled fragments can be determined by autoradiography of gels. The sequence of restriction sites can now be read in order as the two gels are compared.*

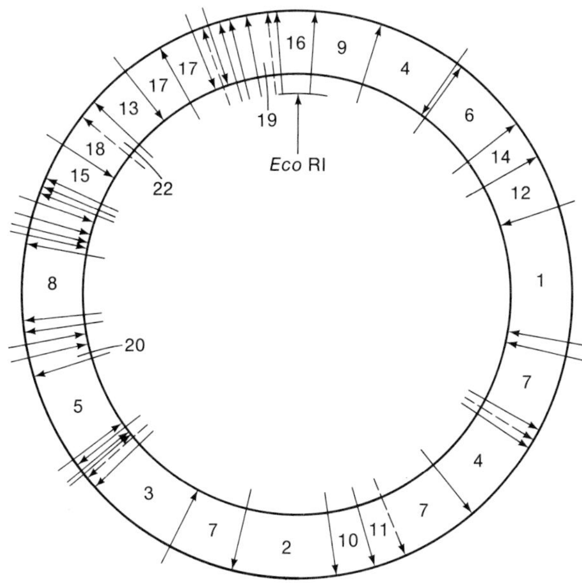

Figure 13-36. Restriction-enzyme cleavage map of SV40 phage DNA. Arrows pointing outward indicate sites for Alu *(restriction enzyme from* Arthrobacter luteus*), and arrows pointing inward indicate sites for* Hae*III (from* Hemophilus aegyptius*). (From Yang et al.,* European Journal of Biochemistry *61:119, 1976.)*

different populations of DNA fragments of varying lengths. These fragments can be separated on gels, where they migrate in order of increasing length. By reading across the gels from the bottom up, we can obtain the order of cleavage sites. A map of enzyme sites can now be prepared. Furthermore, the distances that the various fragments migrate in the gels provide a measurement of their sizes, so that the relative distances between cleavage sites can also be determined. Restriction enzyme maps prepared in this way can be very detailed (Figure 13-36). Using such maps, specific segments can be identified and separated for further analysis.

Message

The relative order of restriction-enzyme cleavage sites on a DNA molecule can be determined, providing a new kind of chromosome map. Furthermore, the number of nucleotides between any two sites can be estimated from the size of the fragments obtained in the analysis.

Formation of Recombinant DNA. Using restriction enzymes, we can cleave DNA from any source at a specific sequence. The sticky ends exposed will anneal to the same sticky ends of other DNA molecules, regardless of their source. For example, the plasmid SC101 carries one *Eco*RI site, so digestion with *Eco*RI converts the circular plasmid DNA to a linear molecule with single stranded sticky ends. DNA from any other source (say, *Drosophila*) can be treated with *Eco*RI to produce a population of fragments carrying

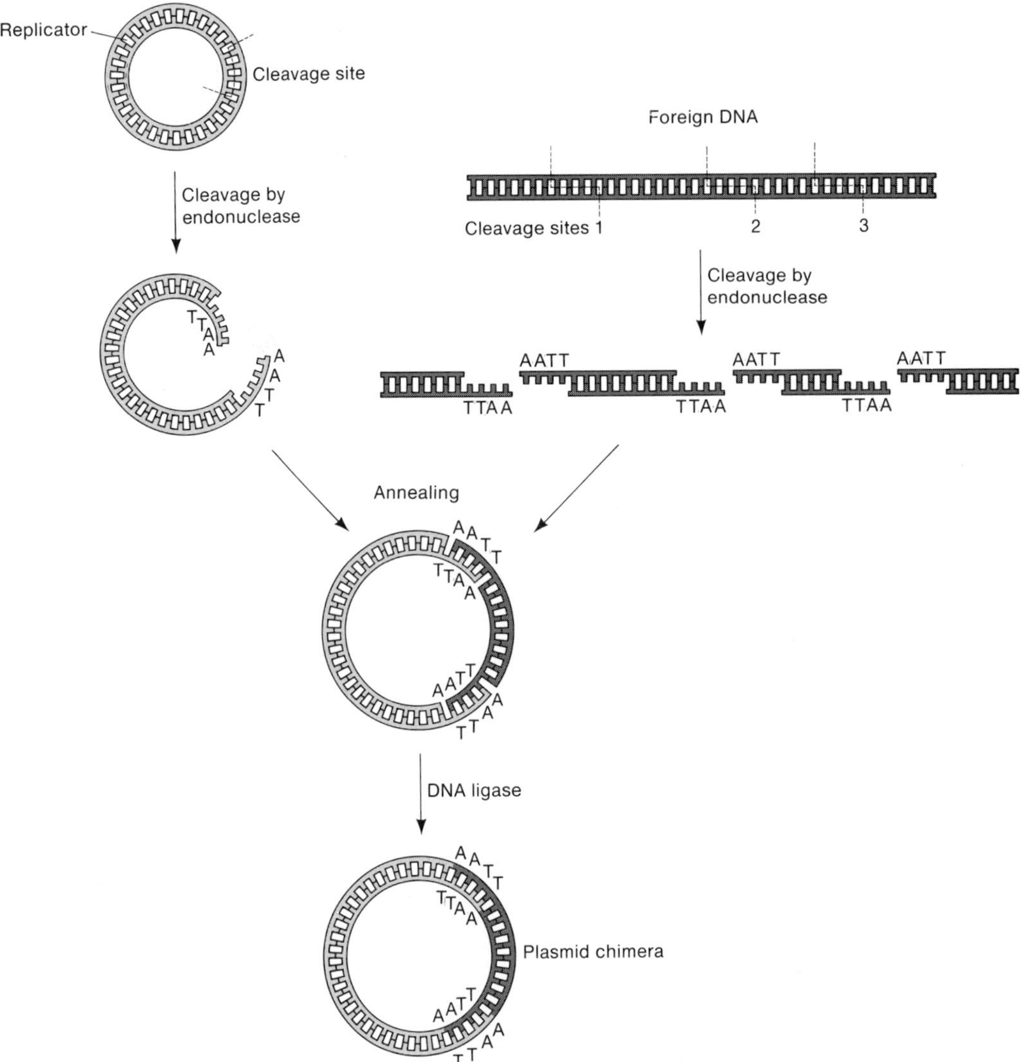

Figure 13-37. Method for generating a chimeric DNA plasmid containing genes derived from foreign DNA. (From S. N. Cohen, "The Manipulation of Genes." Copyright © 1975 by Scientific American, Inc. All rights reserved.)

the same sticky sequence at the fragment ends. When the two populations are mixed, the DNAs from the two sources can combine as duplexes form between their sticky ends (Figure 13-37). The new hybrid (chimeric) molecules represent new combinations of nonhomologous DNA; they are called **recombinant DNA.**

Bacteria can be made permeable to DNA by treatment with calcium chloride, so recombinant DNA may be taken up by living cells. But will such introduced

Message
Recombinant DNA molecules can be made by joining nonhomologous DNAs from virtually any sources. This technique offers the possibility of bypassing all biological restraints to genetic exchange and mixing—even permitting combination of genes from widely differing species.

DNA persist if taken up, and how can we recognize its presence if it does persist? The problem of persistence is overcome by using bacterial plasmids that are autonomous cytoplasmic entities carrying all of the information needed for their own replication. The problem of recognition is solved by using plasmids that carry genes for drug resistance. If a plasmid enters a bacterium and expresses its own drug-resistance genes, it confers drug resistance on the host cell, so the persistence of the plasmid can readily be detected by plating on a drug-containing medium.

The method of producing recombinant DNA outlined thus far is not very effective. Because all the molecules involved in the process carry the same sticky ends, a DNA is more likely to recombine with itself or with other molecules from the same source than it is to form a recombinant DNA with the foreign source. Can we produce recombinant DNAs more efficiently? This was made possible by tagging one set of molecules with a string of As, while attaching a string of Ts to the end of the other molecules (Figure 13-38).

Cloning DNA. By using a bacterial recipient and a plasmid carrier, DNA from virtually any other source can be inserted into a host cell. If a fragment of DNA from a eukaryotic cell is isolated and recombined into a plasmid, it can be inserted into bacterial host cells, where it will be replicated. The copies of the eukaryotic DNA fragment can be distinguished in a bacterial host far more readily than in their original eukaryotic cells. This procedure of replicating a DNA segment in bacterial cells is known as **cloning** that DNA segment.

One of the early methods of DNA cloning was the "shotgun" approach applied by David Hogness in the early 1970s. Hogness fragmented *Drosophila* DNA, made random recombinants with a bacterial plasmid, and selected hundreds of different bacteria using the phenotype of drug resistance. In this way, he recovered many different fragments of DNA derived from different regions throughout the *Drosophila* genome. He could identify the chromosomal segment for any clone by hybridization in situ with salivary gland chromosomes.

The entire *Drosophila* genome has now been cloned in fragments using the shotgun approach. A number of interesting observations have resulted. For example, several clones that have each been isolated from a single chromosomal region nevertheless hybridize with many other parts of the *Drosophila* genome, suggesting that redundant sequences may be scattered around the genome and not necessarily closely linked as tandem repeats. The ability to clone gene segments from such organisms as frogs and mammals permits the study of their genes without carrying out genetic crosses. With the Hogness shotgun tech-

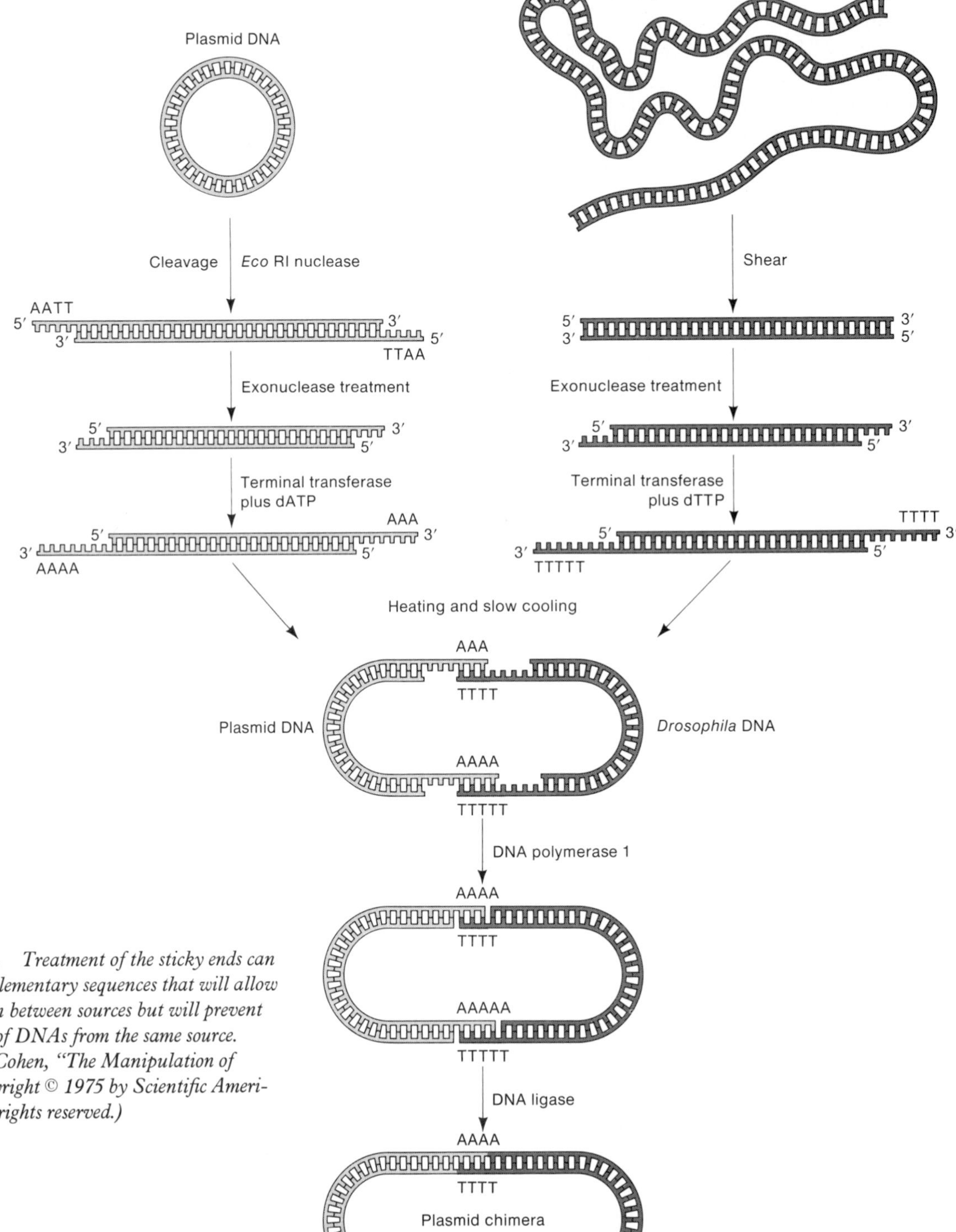

Figure 13-38. Treatment of the sticky ends can produce complementary sequences that will allow recombination between sources but will prevent combination of DNAs from the same source. (After S. N. Cohen, "The Manipulation of Genes." Copyright © 1975 by Scientific American, Inc. All rights reserved.)

nique, it is possible to search for DNA clones derived from specific chromosome regions and to correlate their molecular and genetic properties. Is it possible to *select* a specific clone from the clone library produced by the shotgun technique? Specific function is often used; for example, the *leu3*$^+$ gene from yeast can be cloned by using a *leu*$^-$ *E. coli* strain lacking the equivalent enzyme. Where a specific transcript is known, this or its cDNA can be used as a hybridization **probe** to identify the desired clone. The availability of specific probes is often the key to successful cloning.

Another fruitful use of recombinant DNA involves the direct cloning of segments whose RNA or protein product is known. One of the first eukaryotic DNA segments cloned in *E. coli* contained 18S and 28S ribosomal genes from the toad *Xenopus*. In 1974, John Morrow and his colleagues inserted the genes into a plasmid that was then taken up by *E. coli*. They were able to recover *Xenopus* rRNA from the bacteria, thereby showing that bacteria can correctly "read" eukaryotic genes. Where mRNAs have been purified, the use of reverse transcriptase provides the DNA coding for them; that DNA in turn can then be cloned by use of a plasmid vector.

DNA Sequence Determination

The technology for DNA manipulation that we have discussed is extensive; combined with a technique for identifying the bases in isolated fragments, it opens the way for genetic engineering. A mere decade ago, the ability to determine base sequences easily seemed a long way in the future. Today, however, "DNA sequencing" is performed as routinely as genetic crosses.

We do know what sequence is recognized by each restriction enzyme, so an extensive restriction-enzyme map such as that in Figure 13-36 represents the distribution of known short sequences along the DNA. Ideally, we would like to take each DNA fragment, tag the end with a radioactive label, and then clip it off with an enzyme for identification. However, repeating this process over and over to identify several hundred nucleotides in a typical eukaryotic DNA fragment is far too laborious and time-consuming. Using such a method, it took Gilbert two years to determine the order of the 24-base segment that controls expression of the *lac* gene (Figure 13-25). So long as only this approach was available, the sequencing of a small piece of DNA was a major task, and genetic manipulation as a routine procedure was impossible.

The art of DNA sequencing advanced explosively after 1975, when new techniques were developed that made the task astonishingly simple. The crucial technique involves placing single-stranded DNA on a special gel material (acrylamide or agarose) and subjecting it to an electric current to separate strands on the basis of their lengths. (We have already mentioned the use of this technique in our discussion of restriction-enzyme mapping). The mobility of a strand is inversely proportional to the logarithm of its length. This technique is so sensitive that fragments differing in length by only a single nucleotide can be separated.

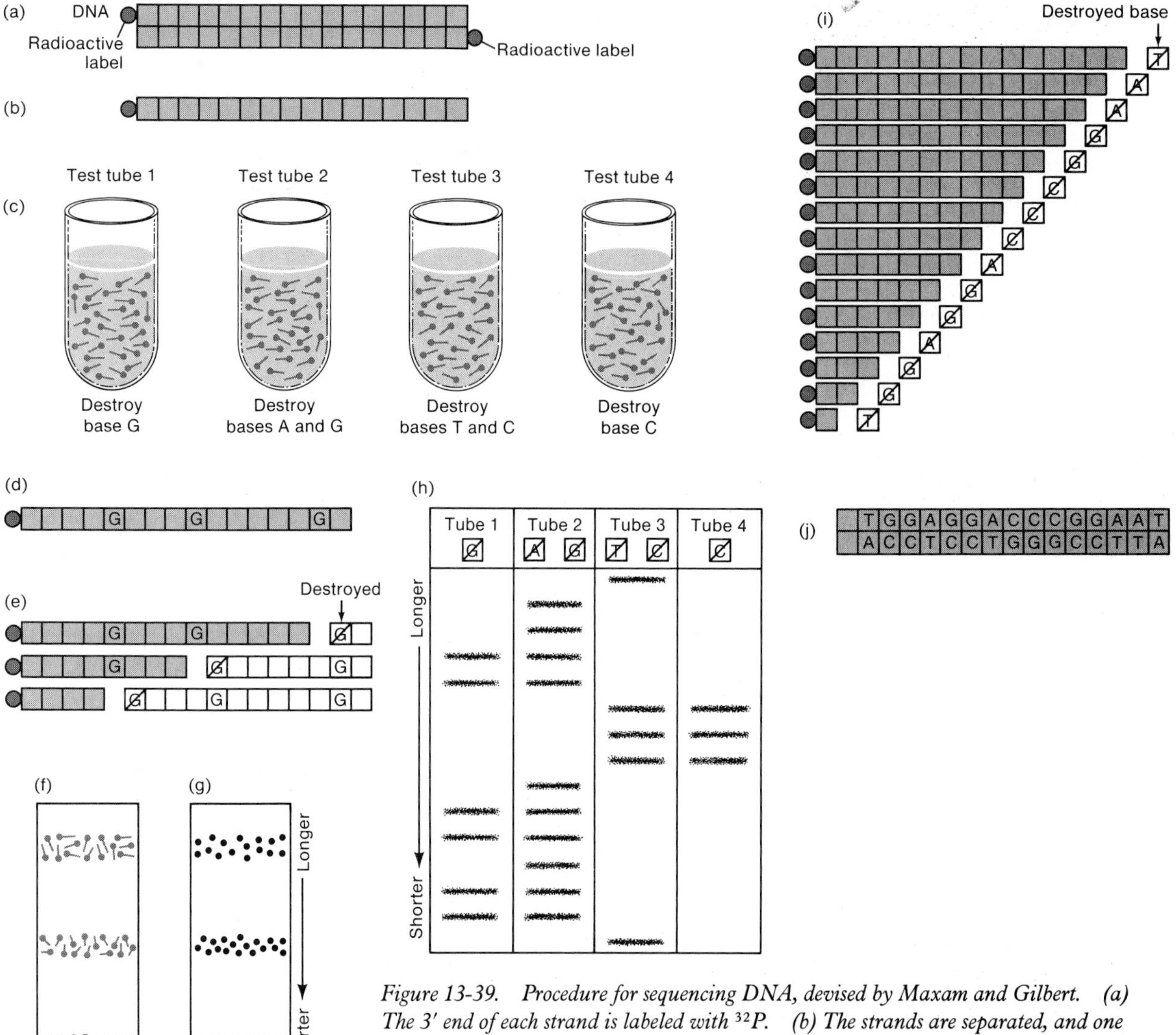

Figure 13-39. *Procedure for sequencing DNA, devised by Maxam and Gilbert.* *(a) The 3′ end of each strand is labeled with* ^{32}P. *(b) The strands are separated, and one specific strand is retained.* *(c) The strands are separated into four equivalent fractions and placed in four test tubes. Each tube is treated with a different reagent that selectively destroys one or two of the four bases. The concentration is adjusted so that only a small proportion of the target bases is attacked, thereby generating a population of fragments of different lengths.* *(d) A hypothetical example: a strand containing three G bases.* *(e) After treatment in test tube 1, this strand yields a population of labeled fragments of three different lengths.* *(f) The fragments are separated on a gel.* *(g) The bands of labeled fragments are identified by autoradiography.* *(h) Comparison of the bands produced by labeled fragments from the four different treatments provides a display of successively shorter fragments when read from top to bottom on the gels.* *(i) The appearance of the band on one or two particular gels indicates which base was destroyed to yield the fragment.* *(j) The sequence of base pairs in the original DNA can then be inferred.* *(From W. Gilbert and L. Villa-Komaroff, "Useful Proteins from Recombinant Bacteria." Copyright © 1980 by Scientific American, Inc. All rights reserved.)*

The most widely used method is that advanced by Maxam and Gilbert. They label the 3′ ends of DNA with ^{32}P and then separate one strand from the other to yield a population of identical strands labeled on one end. They then divide the mixture into four samples, each of which is subjected to a different chemical reagent that destroys one or two specific bases (Figure 13-39). The four reagents destroy (1) only G, (2) only C, (3) A and G, or (4) T and C. The loss of a base makes the sugar–phosphate backbone more likely to break at that point. The reagent concentration is adjusted so that only about 1 in 50 of the target bases is destroyed. The procedure is similar to that outlined for restriction-enzyme mapping in Figure 13-35. It results in a mixture of different-sized pieces carrying the ^{32}P label. When these are separated on gels, they can be arranged in order of length, and the base destroyed at each site can be determined by noting on which gel or gels the band appears. Thus the sequence of bases in the strand can quite simply be read from the pattern of bands on the gels. Figure 13-40 shows an actual set of gels used to determine a 25-nucleotide sequence from the Kilham rat virus.

Message

It is now possible to carry out rapid determination of nucleotide sequences in DNA. This procedure has become so routine that many hundreds of sequences are known, and the number of known sequences increases daily.

Fred Sanger developed a different sequencing method, with which he and his associates set out to determine the complete nucleotide sequence of ϕX174 DNA, which codes for nine proteins in this virus. They completed this *tour de*

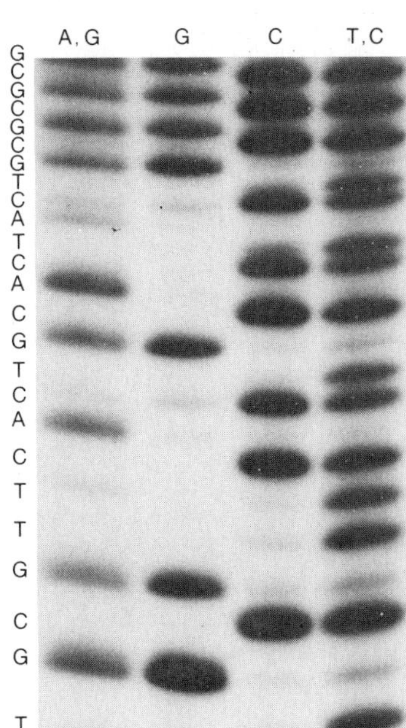

Figure 13-40. *Actual autoradiographs of DNA fragments on gels. These fragments were produced by the Maxam–Gilbert method from DNA of the Kilham rat virus. The sequence in this case is read from the bottom up. (From M. Smith.)*

force in 1977. The molecule contains just under 5400 nucleotides! This investigation was not undertaken solely to achieve the sequencing of a remarkably long DNA molecule; it also sought to resolve an interesting paradox. From the molecular weight of the nine proteins encoded in the ϕX174 DNA, one can estimate the number of nucleotides required for the coding. That estimate is significantly higher than the number of nucleotides indicated by physical properties of the DNA. It was hoped that knowledge of the complete sequence would lead to an explanation of this puzzle.

Different teams within Sanger's laboratory group worked on the sequencing of different fragments of the ϕX174 chromosome. The paradox of the "missing nucleotides" was quickly resolved in a surprising way. Within the coding sequence for one protein, a second protein code proved to exist with a different reading frame! Another pair of such "overlapping" genes was discovered by another team. The *genetic* map of ϕX174 has thus been completed by the evidence from the molecular study (Figure 13-41). Figure 13-42 shows the nature of the overlapping codes for genes *D* and *E*.

Message

A single stretch of nucleotides can be read in more than one way by initiating reading of the sequence in two different reading frames that are offset from each other.

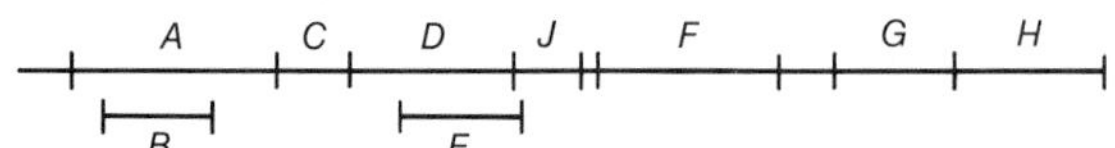

Figure 13-41. The genetic map of the virus ϕX174, including overlapping genes.

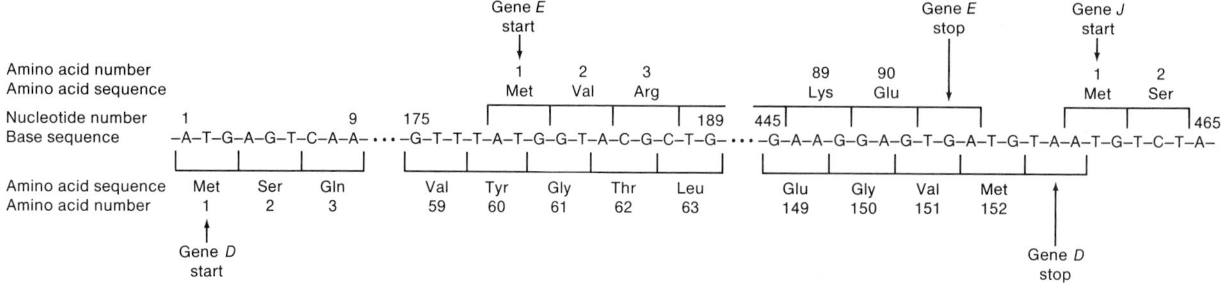

Figure 13-42. The beginnings and ends of genes D *and* E *(and of their corresponding proteins) in the DNA of ϕX174. The nucleotide sequence is numbered from the start triplet of gene* D. *The* E *gene begins at nucleotide 179, with its triplet sequence offset from the reading frame of gene* D. *Gene* E *is completely contained within gene* D *and codes for a protein about 60% as large as the protein of gene* D. *The final base in the stop triplet for gene* D *is used as the first base of the start triplet for gene* J.

The concept of overlapping genes with offset reading frames had earlier been discarded because it raises problems about the simultaneous evolution of the two proteins. However, once undeniable evidence of such overlapping genes was available for ϕX174, old observations were reassessed and new studies undertaken, and there is now evidence indicating the existence of overlapping genes in other organisms, including *E. coli.* Note that a single point mutation within the region of overlap will produce amino acid substitutions in two different proteins; this creates an interesting problem for those studying the evolutionary effects of various mutations. (What happens if a certain mutation is favorable for one protein but unfavorable for the other?)

Because of the ease with which DNA can be sequenced, it now is often convenient to infer the amino acid sequence of a protein by determining the sequence of codons in the gene that encodes it.

Gene Synthesis

We now have methods to determine the sequence of nucleotides in a DNA. Is it possible to synthesize a DNA of some desired sequence from scratch? (We might determine a desired DNA sequence by finding the amino acid sequence of a desired protein or by determining the nucleotide sequence of an RNA that codes for a desired protein.) Suppose we wish to construct a DNA that has the sequence

```
T–A–G–C–C–T–C–C–A–G–T–A–A–T
⋮ ⋮ ⋮ ⋮ ⋮ ⋮ ⋮ ⋮ ⋮ ⋮ ⋮ ⋮ ⋮ ⋮
A–T–C–G–G–A–G–G–T–C–A–T–T–A
```

We might set out to construct the bottom strand, with the intention of later synthesizing the complementary strand through normal replication reactions. We add the nucleotides of A and T to an appropriate reaction mix to obtain the A–T dinucleotide, but we must stop the reaction very quickly before A–T–T or A–T–A is formed. Therefore, the yield of the A–T polymer is only a fraction of the amount of A and T supplied. We then add the A–T and C to an appropriate reaction mix, and again must stop the reaction quickly to avoid forming polymers longer than the desired A–T–C. Thus the yield of A–T–C represents only a percentage of the A–T. At each step, we have a low yield, and this obviously imposes harsh limits on the length of strands that can be constructed in vitro. Such stepwise synthesis of long-chain polynucleotides is impossible.

In the mid-1960s, Gobind Khorana attempted to synthesize the DNA coding for an alanine tRNA molecule whose sequence had been determined by Robert Holley. Figure 13-43a shows the sequence of bases in the tRNA, which must be dictated by the DNA sequence shown in Figure 13-43b. Khorana developed a method that bypassed the need to synthesize the entire sequence by

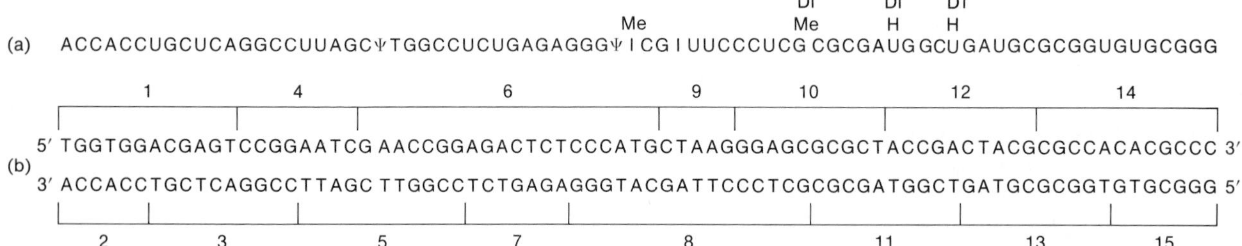

Figure 13-43. (a) The base sequence of the major alanine tRNA from yeast. (b) The base sequence of the DNA coding for the alanine tRNA. The segments used by Khorana in synthesizing this DNA molecule are indicated by the numbers. The segments were then annealed and connected with a ligase.

the stepwise approach we have discussed. He synthesized short polynucleotides of each strand by adding one base at a time. The short fragments were selected to have overlapping complementary sequences. For example, he first synthesized the fragment G–G–T–G–G–A–C–G–A–G–T, and then the fragments C–C–A–C–C and T–G–C–T–C–A–G–G–C–C. When the first two fragments are mixed, the complementary sequences anneal to form a double helix with a single-stranded end:

```
G–G–T–G–G–A–C–G–A–G–T
⋮ ⋮ ⋮ ⋮ ⋮
C–C–A–C–C
```

When the third fragment is added, it anneals to the sticky end and produces a new sticky end:

```
G–G–T–G–G–A–C–G–A–G–T
⋮ ⋮ ⋮ ⋮ ⋮ ⋮ ⋮ ⋮ ⋮ ⋮ ⋮
C–C–A–C–C T–G–C–T–C–A–G–G–C–C
```

A fourth fragment can then be synthesized to overlap this sticky end and extend beyond. The entire sequence can be "stitched together" in this fashion, obtaining a double helix composed of short fragments. The ends of the fragments then can be linked by the enzyme ligase (Figure 13-44).

As Khorana developed this methodology, it became clear that the gene specifying the tRNA extends beyond the limits indicated by the length of the mature

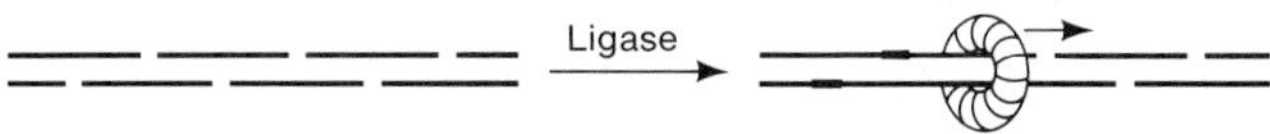

Figure 13-44. Treatment with the enzyme ligase leads to formation of an uninterrupted DNA molecule. In effect, the ligase works its way along the molecule and bonds together any consecutive nucleotides that are not joined in one strand or the other.

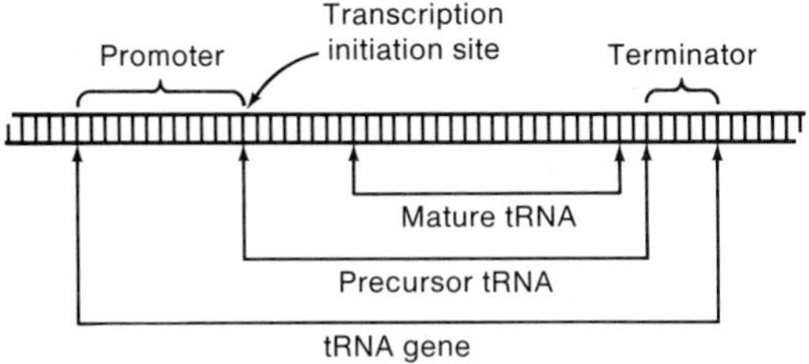

Figure 13-45. The DNA regions included in a complete tRNA gene. The promoter site is the region where the RNA polymerase attaches to begin transcription. Transcription begins at the initiation site and ends at a termination sequence. The precursor tRNA that is transcribed is larger than the active tRNA, which is produced by removing segments from the beginning and end of the precursor molecule.

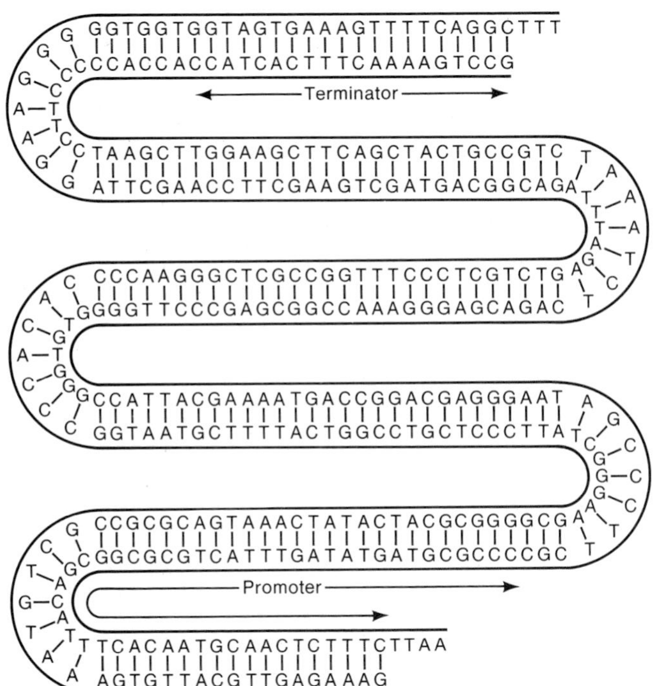

Figure 13-46. The complete sequence of the alanine tRNA gene constructed by Khorana's group. Note the sticky ends that were added to permit insertion in the λ DNA after that DNA was cleaved by EcoR*I. (From Graham Chedd, "The Making of a Gene." This first appeared in* New Society, *London, the weekly review of the social sciences, 30 September 1976. Reprinted with permission of New Science Publications.)*

tRNA molecule. He did eventually construct the entire tRNA gene (Figure 13-45). Appended to each end of his artificial gene was a cohesive terminus of –T–T–A–A complementary to *Eco*RI–induced ends so that he could insert the complete gene into λ DNA (Figure 13-46). When the gene was inserted into *E. coli* by the λ vector, the bacterial cell did synthesize the tRNA. This verified that the start and termination signals inferred to exist did, in fact, work.

Message
By construction of short nucleotide sequences which overlap with complementary sequences, large DNA duplexes can be generated.

Applications of Recombinant DNA

The Khorana method no longer is needed because there are easier methods to recover specific pieces of DNA. However, it remains an elegant and sophisticated technique, and it was applied by Herbert Boyer's group to make the gene coding for a small human growth-regulating hormone, somatostatin. The hormone is a short polypeptide with the sequence shown in Figure 13-47. Boyer's group synthesized the gene using overlapping fragments, and they added a triplet specifying methionine and an *Eco*RI cleavage site on the amino end. On the other end, they placed two consecutive stop triplets and a *Bam*HI site (Figure 13-48). The entire gene was inserted into a plasmid carrying the bacterial gene β-galactosidase, within which there is an *Eco*RI site. The other end of the somatostatin gene hybridized with a *Bam*HI site elsewhere in the plasmid. The *E. coli* selected by their possession of the plasmid were found to

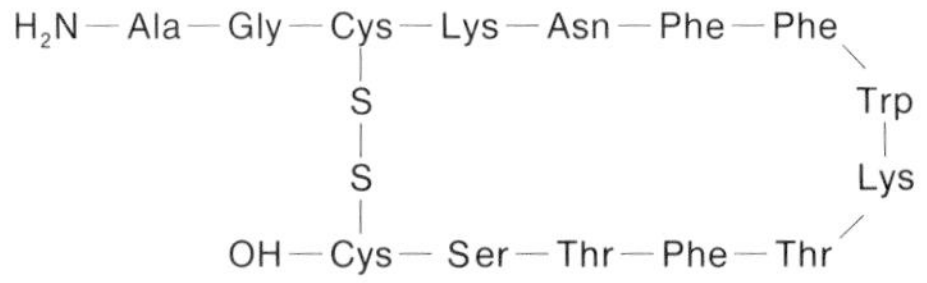

H_2N—Ala—Gly—Cys—Lys—Asn—Phe—Phe—Trp—Lys—Thr—Phe—Thr—Ser—Cys—OH (Cys—S—S—Cys disulfide bridge)

Figure 13-47. The amino acid sequence of the hormone somatostatin.

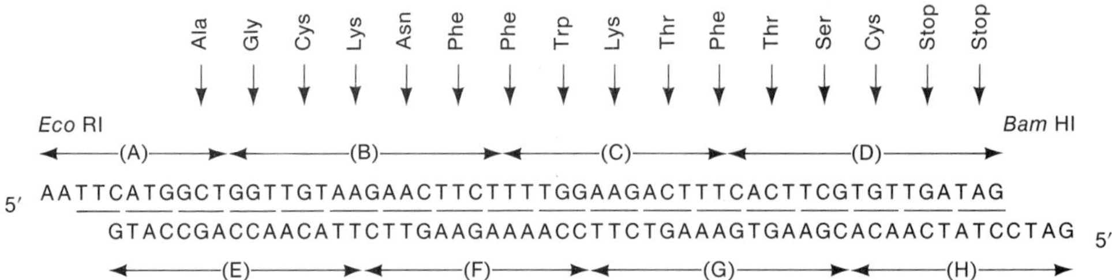

Figure 13-48. The overlapping complementary sequences synthesized to produce the somatostatin gene. A triplet specifying methionine was added to the 5′ end of the somatostatin coding region (inferred from the amino acid sequence), and, adjacent to this, an Eco*RI restriction sequence was added. A* Bam*HI restriction sequence was added at the other end of the "artificial" gene. (From K. Itakura et al., "Expression in* Escherichia coli *of a Chemically Synthesized Gene for the Hormone Somatostatin."* Science *198:1056–1063, 1977. Copyright 1977 by the American Association for the Advancement of Science.)*

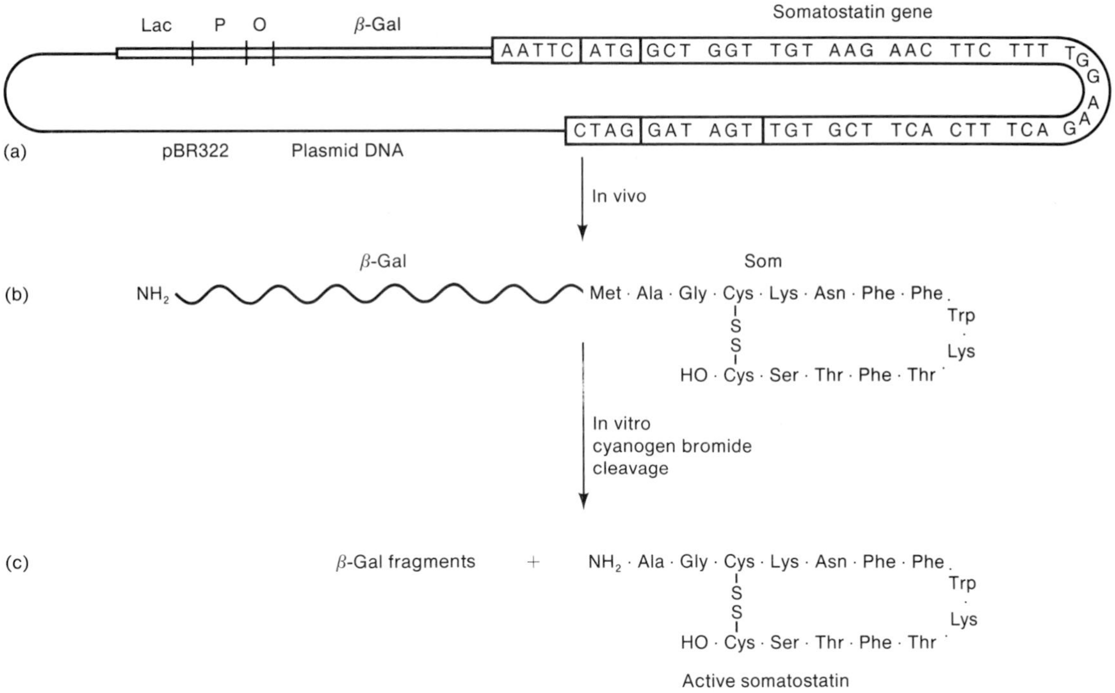

Lac P O β-Gal Somatostatin gene

AATTC ATG GCT GGT TGT AAG AAC TTC TTT TGGAA

CTAG GAT AGT TGT GCT TCA CTT TCA G

(a) pBR322 Plasmid DNA

In vivo

(b) β-Gal Som

NH_2 ∿∿∿ Met · Ala · Gly · Cys · Lys · Asn · Phe · Phe · Trp · Lys · Thr · Phe · Thr · Ser · Cys · HO (Cys–S–S–Cys)

In vitro cyanogen bromide cleavage

(c) β-Gal fragments + NH_2 · Ala · Gly · Cys · Lys · Asn · Phe · Phe · Trp · Lys · Thr · Phe · Thr · Ser · Cys · HO (Cys–S–S–Cys)

Active somatostatin

Figure 13-49. Production of somatostatin by E. coli. *(a) The plasmid carrying the synthetic DNA sequence is added to the bacterial cell. (b) A chimeric polypeptide is produced. (c) The desired somatostatin is liberated by treatment with cyanogen bromide. (From K. Itakura et al., "Expression in* Escherichia coli *of a Chemically Synthesized Gene for the Hormone Somatostatin."* Science *198:1056–1063, 1977. Copyright 1977 by the American Association for the Advancement of Science.)*

produce a protein chimera containing part of β-galactosidase fused to somatostatin via a methionine residue. Methionine is cleaved by cyanogen bromide, so the active hormone could be liberated by such treatment (Figure 13-49).

Techniques such as this already have been used to produce recombinant plasmids bearing DNA sequences for human insulin, growth hormone, and interferon. Commercially profitable quantities of such human proteins as blood-clotting factors may soon be obtained from bacterial cultures in which the synthetic genes have been inserted. Already bacteria producing human insulin have gone into commercial production.

Message

Recombinant DNA technology is sufficiently advanced to promise in the near future the economically profitable production of human proteins through direct engineering of the genetic content of microorganisms.

Recombinant DNA and Social Responsibility

Recombinant DNA techniques have revolutionized biology with their revelations about gene and chromosome organization, and they promise enormous potential benefits for humanity. However, as scientists began to exploit these techniques in the early 1970s, some scientists (and some other people) began to express concern about the possible hazards of manipulating gene segments. For example, SV40 is a mammalian virus known to cause cancer in monkeys. *E. coli* is a bacterium that normally lives in the human digestive tract. When DNA from SV40 is inserted into *E. coli*, is it possible that a carcinogenic bacterium might be produced, escape, and thrive as a parasite in humans? Others have wondered whether the combination of genes from eukaryotes with prokaryotic cells might generate new types of pathogenic organisms against which humans would have no natural defenses. Of course, the question of whether such fears are reasonable could be settled definitively only by carrying out the experiments to explore the possible results.

In an unprecedented step, eleven eminent molecular biologists published a letter in 1974 pointing out some of their concerns about potential biohazards of work with recombinant DNA. They called for the development of guidelines to regulate such research. They asked scientists to observe a moratorium on certain kinds of experiments deemed particularly hazardous (cloning genes for toxins or cancer-causing agents). For the first time in history, a group of scientists publicly declared certain areas of scientific inquiry to be "off limits" and called for possible restrictions on such research.

The call for a moratorium attracted widespread public notice and caused many people to conclude that recombinant DNA research is dangerous. Under considerable public pressure, the National Institutes of Health (NIH) in the United States set out to establish categories of biohazards and guidelines for conducting experiments in each category. Eventually (on June 23, 1976), the NIH announced categories for experiments based on their potential hazards and defined four categories of physical conditions to contain the experiments. These conditions range from the P1 requirements of standard sterile techniques and common-sense precautions to the P4 facilities with the most extreme precautions against the escape of any organisms. Standards were set as well for biological restrictions that would minimize the chances of escape. For example, special strains of genetically enfeebled *E. coli* were constructed to use as recipients of recombinant DNA. These bacteria would not survive except in special laboratory conditions.

Interestingly, on the day the NIH guidelines were released, the city of Cambridge (Massachusetts) held a public meeting to discuss the application of Harvard University's Mark Ptashne to build a special containing facility for research with recombinant DNA. The result of the meeting was the establishment of a citizen's review board to decide whether Ptashne's proposed experiments posed a health hazard to the citizens of Cambridge. This step established a new precedent of involvement by nonscientists from the general public in controlling the research policy of scientists. The review board did approve Ptashne's application.

Over the decade of the 1970s, evidence accumulated to indicate that the biohazards of recombinant DNA research are not as serious as some had feared. Meanwhile, increasing impatience built up in the scientific community about the delays in scientific progress and the regulation of scientific research by those with no training in the field. The NIH guidelines have now been relaxed significantly, and the use of recombinant DNA techniques has become routine laboratory practice.

Nonetheless, the turmoil about recombinant DNA did raise important social issues. For example, what is the social responsibility of scientists who are developing powerful new technologies? Should the people who are doing the experiments be the ones who set the guidelines? At what point should the public have an input? Who should be legally liable for any accidental damage that results from scientific research? Should limits be placed on the freedom of scientists to design and conduct research projects? Should a scientist attempt to foresee possible adverse effects from future use of discoveries and refuse to advance knowledge in certain directions that might have unfortunate applications?

Other kinds of questions are raised by the controversy over recombinant DNA. Can we predict with confidence the properties of an organism modified by inserting DNA from a totally unrelated source? Can there be deleterious effects that will not be detected until large populations have been exposed for years (as was the case with oral contraceptives)? In the long run, will increasing sophistication in DNA manipulation inevitably lead to genetic manipulation of human beings? If so, who will decide the conditions?

Although much of the worry about dangers of recombinant DNA research has been laid to rest, the issue has served to raise far more profound questions about the relationship between science and society. These questions have not been answered satisfactorily, and they are likely to persist and become even more important in the coming years.

Summary

The highly specific pairing of complementary bases (to form A–T and G–C pairs) permits the experimental manipulation of DNA. A DNA strand can recognize a complementary strand and anneal with it to renature a double helix or to produce DNA:RNA duplexes.

Because even a short polynucleotide has a sequence that is relatively unlikely to occur by chance, base pairing provides a highly specific method of matching strands. (If all four bases have an equal probability of occurring at any position, then a specific sequence n bases in length should occur with a frequency of $1/4^n$. For example, a particular sequence 5 bases long should occur by chance only about once in 1000 such segments.)

By annealing single strands into double helices, it is possible to identify the sites of chromosomal DNA that are complementary to RNA, to compare DNAs from different sources, and to isolate the DNA that codes for specific RNAs. Furthermore, the rate at which a population of single strands reassociates is a function of their initial concentration (C_0) and time (t), so that the value C_0t (normalized time) provides a means of distinguishing sequence redundancy in a genome (because redundant sequences have a high concentration in the genome). Studies using DNA:DNA and DNA:RNA hybrids have revealed the existence in eukaryotes of highly redundant short sequences of DNA and of large sections of DNA that do not code for amino acids but are inserted into genes between regions that do code for amino acids.

Synthetic genes have been constructed by producing short polynucleotides with regions of overlapping complementarity. In this way, a longer duplex can be formed by annealing several short overlapping strands and then using ligase to seal the gaps between adjacent ends.

Restriction enzymes recognize specific nucleotide sequences and cleave the DNA molecule at such sites; they provide a powerful tool for fragmenting DNA in a controlled fashion. Coupled with electrophoretic gels that permit separation of strands varying in length (by as little as a single base), restriction enzymes have made genetic engineering simple. Large DNA molecules can be cut into small fragments; the fragments can be separated, and their base sequences determined. Such a study determined the entire base sequence for the DNA of the phage ϕX174 and revealed the surprising fact that some genes are contained within other genes (but using offset reading frames).

Recombinant DNA is produced by linking DNA fragments with sticky ends. That is, two molecules are prepared with complementary single-stranded ends, and these ends are then annealed. A gene from a eukaryote can be isolated or constructed and then inserted into the DNA of a bacterial plasmid. This recombinant DNA can then be inserted into bacteria, where the recombinant plasmid can persist as a self-replicating cytoplasmic entity. DNA coding for human proteins has been constructed or isolated and inserted into bacteria, where the human protein is produced in significant quantity.

Recombinant DNA technology provides powerful insights into the structure and regulation of genes. It also offers promise as a way to produce modified organisms that will have great benefits for humans. However, like any powerful new technique, genetic engineering also involves potential hazards for society that must be assessed carefully.

Problems

1. The bacteriophage ϕX174 has in its head a single strand of DNA as its genetic material. Upon infection of a bacterial cell, the phage forms a complementary strand on the infective strand to yield a double-stranded replicative form (RF).

Design an experiment using ϕX174 to determine whether or not transcription occurs on both strands of the RF double helix.

2. After irradiation of wild-type T4 phages, an *rII* mutation is recovered that fails to complement with mutants in either the A or the B cistron. DNAs from the mutant and the wild-type strains are mixed, heat denatured, and cooled slowly to allow reannealing. What hybrid molecules would be seen if the *rII* mutation is (a) a double point mutation? (b) a deletion? (c) an inversion? (d) a transposition? (e) a tandem duplication?

3. Suppose that the actual function of the *rII* locus is not known. How would you go about determining its primary gene products? Assume that some techniques not presently available may become available in the future.

4. Noboru Sueoka showed that some species of crabs contain DNA, of which 30% is dAT, a polymer of alternating sequences of adenine and thymine. Suppose that you want to study the cell bology of crab dAT. How would you show (a) where it is located in the cell? (b) if it is nuclear, in which chromosomes it is located? (c) whether there are other DNA sequences linked to the dAT?

5. In 1975, Norman Davidson and his colleagues isolated *Drosophila* DNA and sheared it into pieces. The DNA was then denatured into single strands and allowed to renature for a very short period. The DNA was then filtered through a hydroxyapatite column, which retains double-helical DNA while allowing single strands to pass through. About once in every 40 to 80 thousand bases (kilobases, or kb), they recovered a double-helical structure ranging in size from very short to more than 15 kb. In many cases the structure had the appearance shown in Figure 13-50, and they found around 2000 to 4000 such structures per genome. What is the explanation of these structures?

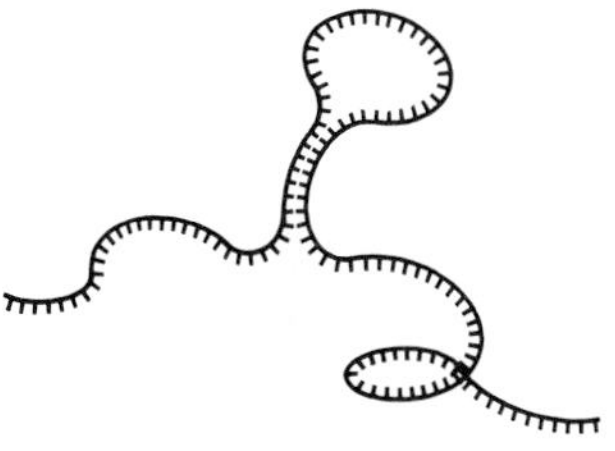

Figure 13-50.

6. In 1973, Eric Davidson and his associates took DNA from *Xenopus* and broke it into pieces a few kb in length. The DNA was then denatured and allowed to anneal under conditions in which only redundant sequences anneal. The double-stranded sequences were retained on hydroxyapatite columns and then inspected by electron microscope. They obtained such duplexes as those in Figure 13-51.

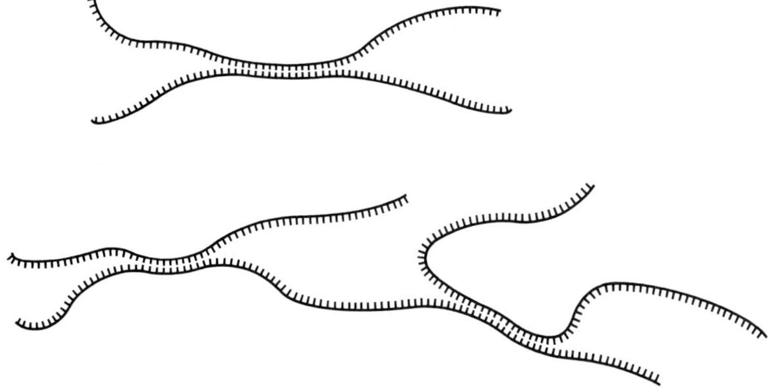

Figure 13-51.

On the average, the duplexes were 0.3 kb in length, and the separation between duplexes was around 0.8 kb. Interpret these results.

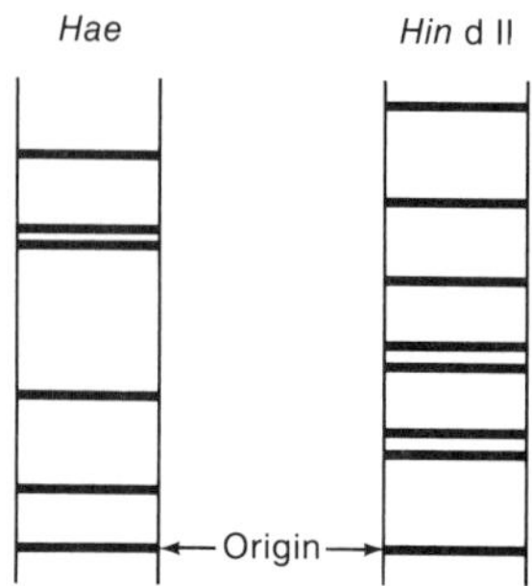

Figure 13-52.

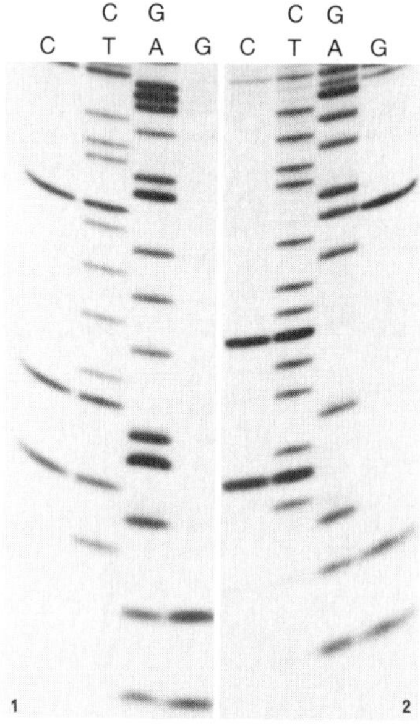

Figure 13-53.
(From Caroline R. Astell.)

7. You have a purified DNA molecule, and you wish to map restriction-enzyme sites along its length. After digestion with *Eco*RI, you obtain four fragments: 1, 2, 3, and 4. After digestion of each of these fragments with *Hin*dII, you find that fragment 3 yields two subfragments (3_1 and 3_2), and fragment 2 yields three (2_1, 2_2, and 2_3). After digestion of the entire DNA molecule with *Hin*dII, you recover four pieces: A, B, C, and D. When these pieces are treated with *Eco*RI, piece D yields fragments 1 and 3_1, A yields 3_2 and 2_1, and B yields 2_3 and 4. The C piece is identical to 2_2. Draw a restriction map of this DNA.

8. After treating *Drosophila* DNA with restriction enzyme, the fragments are attached to plasmids and selected as clones in *E. coli.* Using this "shotgun" technique, David Hogness has recovered every DNA sequence of *Drosophila* in a cloned line.

   a. How would you go about identifying the clone that contains DNA from a particular chromosome region of interest to you?

   b. How would you identify a clone coding for a specific tRNA?

9. You have isolated and cloned a segment of DNA that is known to be a unique sequence in the genome. It maps near the tip of the X chromosome. It is about 10 kb in length. You label the 5′ ends with ^{32}P and cleave the molecule with *Eco*RI. You obtain two fragments, one that is 8.5 kb, and the other, 1.5 kb. You separate the 8.5 kb fragments into two fractions, partially digesting one with *Hae* and the other with *Hin*dII. You then separate each sample on an agarose gel. Figure 13-52 shows the autoradiographs you obtain of the gels. (The origin is at the bottom.) Draw a restriction-enzyme map of the complete (10 kb) molecule.

10. Figure 13-53 is an autoradiograph of two DNA fragments from the *a* mating-type locus in yeast. The Maxam–Gilbert technique has been used for sequencing each one. The bases above each gel are the ones attacked by the reagent. What is the base sequence of these two DNA fragments?

11. Suppose that a specific tRNA is encoded by six tandemly duplicated tRNA genes. You don't know what phenotype a tRNA mutation would yield. Design experiments to determine whether there are spacers between the tRNA genes and how the cluster of genes is transcribed. How could you detect mutations in these genes?

12. What result would you expect when adenine is substituted for guanine at nucleotide position 181 of the *D* gene in ϕX174? (See Figure 13-42).

13. Design an experiment to allow the purification of DNA sequences in the Y chromosome.

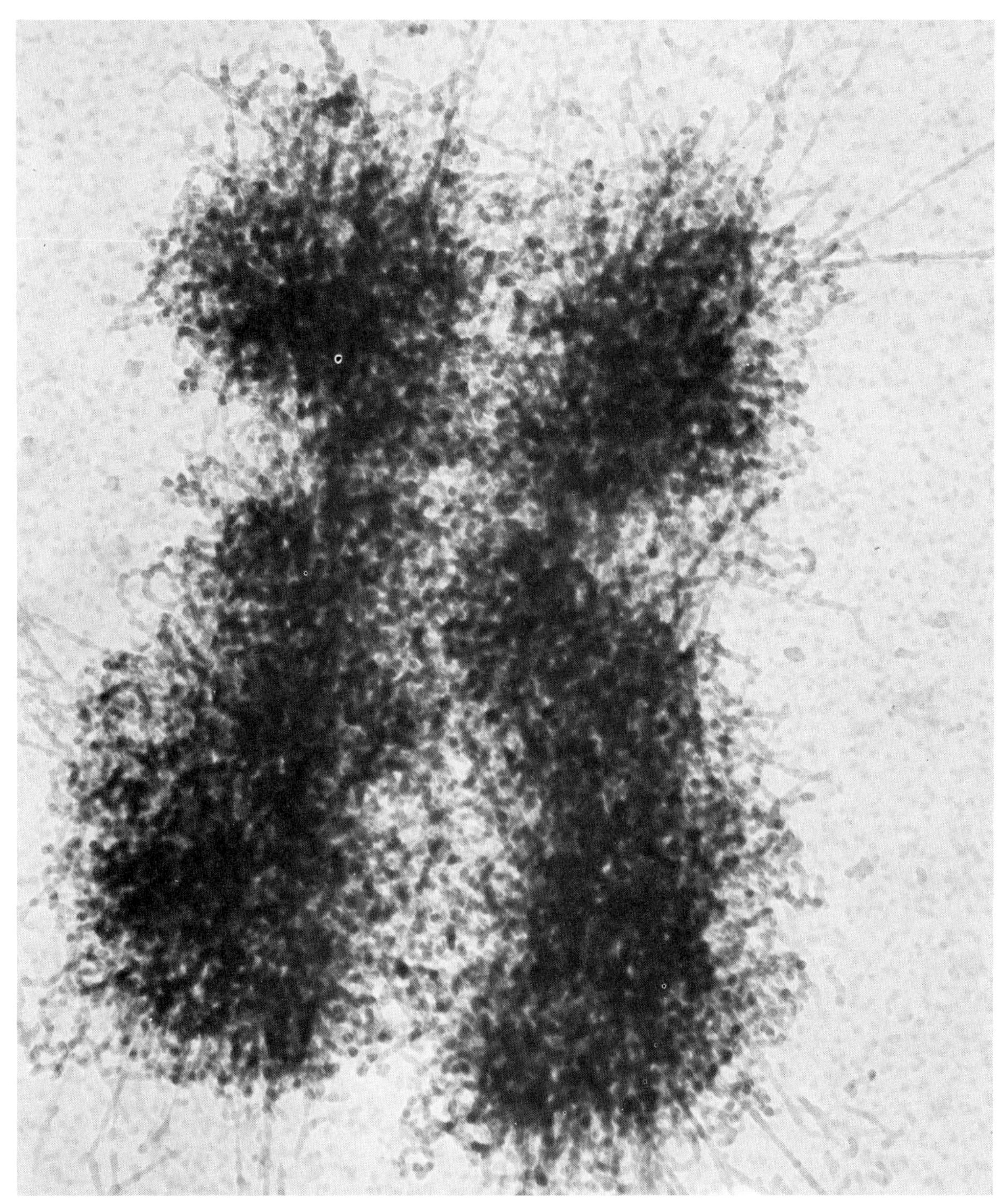

Human chromosome (sister chomatid pair) disrupted to show strands composed of DNA and associated histone protein (× 30,000). (Armed Forces Institute of Pathology.)

14

The Structure and Function of Chromosomes

With the extensive repertoire of modern techniques for manipulating DNA, we can now look again at the physical nature of the genes in their locations upon the chromosomes. Is the chromosome simply a structural framework that assembles the genes for proper partition during cell division? Or does the structure of the chromosome play some role in determining the functions of the genes within it? In Chapter 10, we saw that there is a position effect for complex loci such as *Bar* (or for pseudoalleles) where the chromosomal arrangement of the genetic elements determines their phenotypic expression. But more extensive effects of chromosomal location have been detected, and these effects offer tantalizing problems for the molecular biologist. In this chapter we encounter many phenomena and observations for which we have as yet no explanation in terms of molecular biology; these pose the problems to be tackled with the growing arsenal of techniques and ideas. We begin with classical cytological and genetic observations that await reexamination in molecular terms.

By cytological examination, we can distinguish eukaryotic chromosome segments of varying staining intensity. Heterochromatic elements (heterochromatin) are densely stained; these regions generally are assumed to reflect a state of genetic inactivity. Euchromatic regions (euchromatin) are less densely stained and typically less compact than the heterochromatin; these regions are

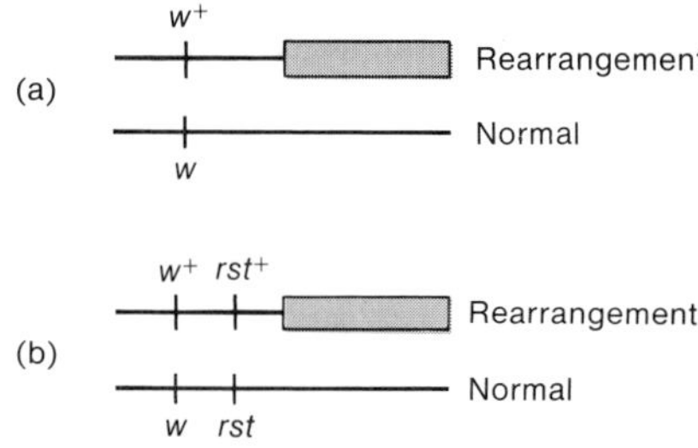

Figure 14-4. Chromosome rearrangements that lead to variegated position effects. (a) In the rearranged chromosome, the allele w$^+$ *is moved to a position near the heterochromatin* (dark block). *A heterozygote carrying this* R(w$^+$) *chromosome and a structurally normal chromosome with the* w *allele may exhibit the variegated phenotype. (b) The* R(w$^+$ rst$^+$) *rearrangement shown at the top can also produce a variegated phenotype when heterozygous with a normal chromosome carrying the* w rst *alleles.*

$R(w^+)$ to represent a chromosome rearrangement (an inversion or translocation) that moves the w^+ allele to a position near the proximal heterochromatin (Figure 14-4a). In some cases, an $R(w^+)/w$ heterozygote shows the expected wild-type phenotype, but many such individuals are **variegated** (with eyes that are mosaics of wild-type and white patches). Some kind of position effect must be involved. Apparently, the heterochromatin causes the inactivation of the wild-type allele in some (but not in all) somatic cells.

How far does this effect of the heterochromatin extend into the euchromatin? Let's include a second locus near *white* in our study. The *roughest* (*rst*) locus affects the surface texture of eye facets. Now we consider the heterozygote $R(w^+\ rst^+)/w\ rst$ (Figure 14-4b). Many such heterozygotes do have variegated eyes, but the mosaic patches in the eyes do not exhibit all of the possible phenotypes. Some patches are smooth and red (wild type), some are roughest and red, and some are roughest and white, but no sectors are ever found to be smooth and white. It would appear that the heterochromatin has a **spreading effect** that moves outward progressively across the adjacent euchromatin. The effect cannot inactivate the w^+ allele without first inactivating the rst^+ allele. The effect can extend quite far from the heterochromatin—to genes that are as much as 60 bands away from the heterochromatin in maps made from the giant (polytene) salivary chromosomes.

Message
Position-effect variegation results when heterochromatin inactivates adjacent euchromatic loci in some somatic cells. This inactivation is a spreading effect that moves linearly outward from the heterochromatin through the sequence of genes in the adjacent euchromatin.

The mutant tissues in the variegated phenotype appear in patches rather than in a "salt-and-pepper" mixture of individual cells with differing phenotypes. Apparently, all of the cells in a given patch are related by some common event. Could the inactivating effect of the heterochromatin occur at some early stage in development, so that it affects all daughter cells derived from the affected cell?

Some information about development in the *Drosophila* eye was obtained in 1957 by Hans Becker. Becker used larvae heterozygous for w^{co} *and* w (different alleles of the *white* locus). He chose these genotypes because the phenotypes of w^{co}/w^{co}, w/w, and w^{co}/w can be distinguished from one another. Becker irradiated young larvae to induce mitotic crossovers. Such crossovers in the w^{co}/w genotype lead to twin spots of homozygous w/w and w^{co}/w^{co} tissue (Figure 14-5). If the mitotic crossover is induced at an early stage of eye development, a large segment of the adult eye will derive from the daughter cells of the mitosis where the crossover occurred, so the twin spot will cover a large part of the eye. If the crossover is induced late in eye development, then the twin spot will involve relatively few facets of the adult eye. Studying many

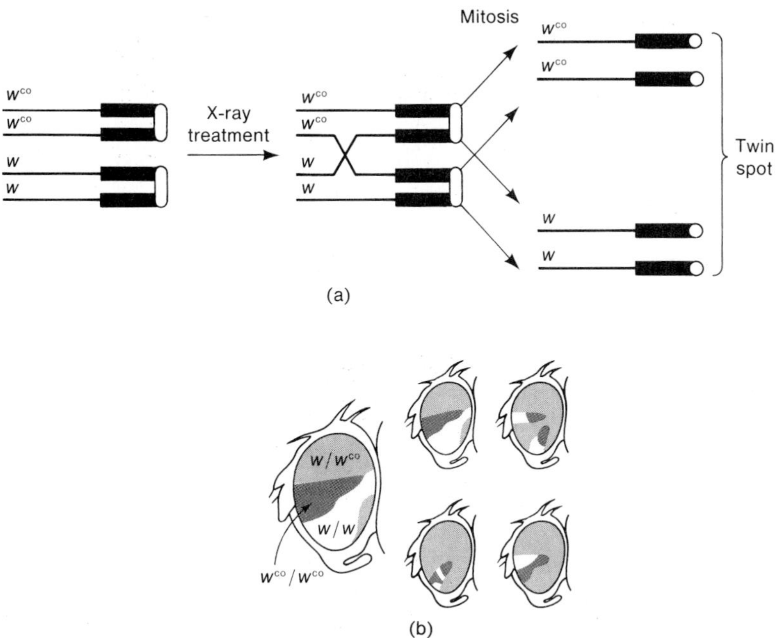

Figure 14-5. Production of twin spots of mutant tissue in a background of cells of a different phenotype. (a) Larvae heterozygous for the sex-linked alleles w *and* w^{co} *are irradiated to induce a mitotic crossover in prospective eye cells. This crossover produces cells homozygous for* w *and* w^{co} *that are phenotypically distinct from the surrounding* w/w^{co} *tissue. (b) Some representative mosaic eyes. (Part b is after H. J. Becker,* Verhandl. deutsch. zool. Ges. *1956.)*

such crossovers and mapping the twin spots, Becker was able to trace the **cell lineage** of the lower half of the eye to eight larval cells. The daughter cells of each larval cell occupy one of the eight sectors shown in Figure 14-6.

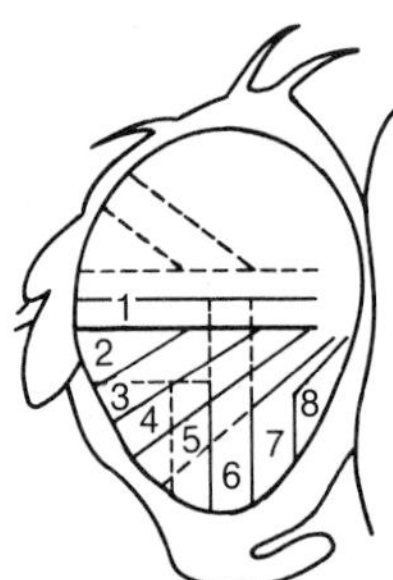

Figure 14-6. All of the cells in each sector of the lower eye of Drosophila *are derived from one of eight original larval cells. (After H. J. Becker,* Verhandl. deutsch. zool. Ges. *1956.)*

In 1963, William Baker mapped the areas of the mutant patches in the variegated eye. He found that these patches occupy the same regions as the clonal sectors that Becker mapped. Apparently, each mutant patch does represent a clone of cells that are daughters of one larval cell in which the determinative event must have occurred. This indicates that it occurs early in larval development, several days before any pigment actually is synthesized. The determinative event then must involve a *potential* to produce pigment, a potential that is not actually realized until days later. The result also implies that, once the determinative event occurs, all of the daughter cells "remember" that decision. In each generation, the variegating chromosome begins with the potential to variegate. Within the development of a single individual, however, the functional state of the locus (whether the wild-type allele will or will not be expressed) is determined in an individual somatic cell at some stage of development, and all daughter cells then inherit the chromosome in its fixed functional state.

What effect does the heterochromatin have on the wild-type allele? What determines the stage of development when this effect will be exercised? Why is the effect "permanent" through the divisional cycles of somatic cells, although no fixing of the activity of the allele occurs from generation to generation? These fascinating puzzles about constitutive heterochromatin await molecular explanations.

Message

A rearranged chromosome has the potential in each generation to express or not to express its variegating loci. In the somatic cells of an individual, at a specific developmental stage, the potential activity of the loci is determined. After that determination, all daughter cells inherit the loci with their fixed functional states, even if the actual expression of the genetic activity does not occur until days later.

Facultative Heterochromatin

The problems posed by facultative heterochromatin are just as intriguing as those posed by the constitutive heterochromatin. We consider here examples drawn from two types of organisms: mealybugs and mammals.

Mealybugs

The insects classified as "true bugs" provide a classic example of the effects of facultative heterochromatin. It has long been known that there is a striking sex difference in chromosome behavior in the scale insects called coccids, or mealybugs. In many coccid species, the diploid ($2n$) number is 10 chromosomes. In females, the chromosomes behave normally, disappearing during interphase and condensing for cell division. In contrast, early in the embryonic development of males, one chromosome of each pair becomes heterochromatic, and these chromosomes remain visible as a clumped chromocentral mass through interphase (Figure 14-7). After that point in the developmental process, the same chromosomes appear to remain heterochromatic throughout subsequent mitoses.

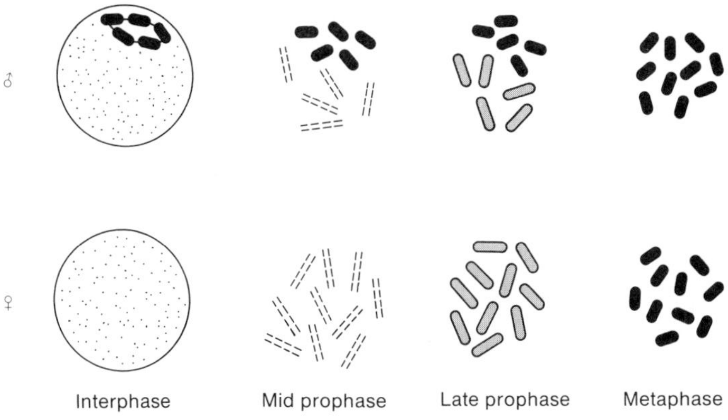

Figure 14-7. Mitosis in male and female mealybugs (coccids) with a diploid chromosome number of $2n = 10$. *In males, five of the chromosomes remain visible during interphase as densely staining heterochromatic elements. During prophase, the other five chromosomes appear less densely stained at first but eventually become heterochromatic. In contrast, none of the chromosomes in the female is visible during interphase; all ten chromosomes in the female behave like the euchromatic set in males. (From S. W. Brown and U. Nur, "Heterochromatic Chromosomes in the Coccids."* Science *145:130–136, 1964. Copyright 1964 by the American Association for the Advancement of Science.)*

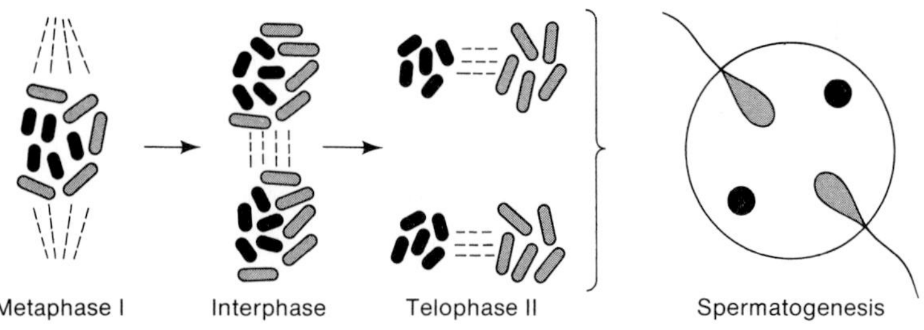

Figure 14-8. Spermatogenesis in the mealybug (coccid). The first division is equational and the second is reductional in the male, the reverse of the usual meiotic process. In the second meiotic division, all heterochromatic chromosomes go to one pole. Only the euchromatic products form sperm; the heterochromatic products appear as deep-staining residues that slowly degenerate. (From S. W. Brown and U. Nur, "Heterochromatic Chromosomes in the Coccids." Science *145:130, 1964. Copyright 1964 by the American Association for the Advancement of Science.)*

Chromosome behavior in meiosis also differs between the two sexes. In female coccids, meiosis is normal. However, in spermatogenesis, the first division is the equational one in which the centromere splits to allow sister chromatids to separate, and the second division is reductional with the homologous chromosomes separating (Figure 14-8). This pattern of meiosis in the mealybug males is, of course, the opposite of the normal meiotic sequence of divisions. At the second meiotic division in the males, the movement of chromosomes is nonrandom—all of the heterochromatic chromosomes go to one pole, and all of the euchromatic ones go to the other pole. Only the nuclei containing euchromatic elements form functional sperm. What determines which chromosomes become heterochromatic?

In the 1920s, Franz Schrader and Sally Hughes-Schrader described this strange behavior of chromosomes in coccids and suggested that the heterochromatic chromosomes (1) are those coming from the male parent and (2) are genetically inert. Both hypotheses were confirmed in 1957 by Spencer Brown and Walter Nelson-Rees through irradiation studies. In the male offspring of X-irradiated males, radiation-induced chromosome aberrations were found only in the heterochromatic chromosomes. In the male offspring of X-irradiated females, radiation-induced chromosome aberrations were found only in the euchromatic chromosomes. This observation confirmed the paternal origin of the heterochromatic chromosomes.

If the heterochromatic chromosomes are inert, then males are functionally haploid, because only the euchromatic chromosomes are genetically active. Brown and Nelson-Rees applied increasing doses of X rays to male coccids and studied the survival of their offspring. As the dosage was increased, the survival of daughters declined because of the induction of dominant lethal mutations. However, the survival of the sons remained constant, thus supporting the notion that any mutations induced in the father are not expressed in the active chromosomes of the sons.

This unusual genetic system poses some interesting puzzles. The differing functional states of the chromosomes cannot be determined by genetic differences between chromosomes. A male's euchromatic chromosomes came from his mother, but they in turn will become the heterochromatic chromosomes of his sons. What then controls the heterochromatization of certain chromosomes? Once the male or female origin of chromosomes (and therefore their functional fate in males) is set, each chromosome retains that imprint through subsequent cell generations in the somatic cells of the individual. What event establishes this functional state, and how is it retained through somatic cell divisions? Once again, a well-documented phenomenon awaits a molecular explanation.

Message

In the male coccid, paternally derived chromosomes remain inactive and heterochromatic in the somatic cells and segregate together into nonfunctional nuclei during spermatogenesis. No molecular explanation is yet known for this pattern of facultative heterochromatin.

Mammals

Another striking example of facultative heterochromatin is found in mammals. In many of the cells from a mammalian female, the nucleus is characterized by a densely staining heterochromatic element called a **Barr body** (named after its discoverer, Murray Barr). The Barr body contains DNA and is not found in males, so it was suggested that it may represent an X chromosome. Indeed, the number of Barr bodies is always one less than the number of X chromosomes in the genome (Figure 14-9). Another name commonly applied to the Barr body is **sex chromatin.**

In coccids, we found that the heterochromatic elements are genetically inert. Does the presence of the Barr body indicate that all but one of the female's X chromosomes are inactivated in somatic cells? Such a mechanism would provide an answer to the puzzling problem of how mammals adjust to the presence of twice as many X-linked genes in females as in males, because the X chromosome is known to carry many loci that are necessary for viability. Measurements of enzymes produced by sex-linked genes (glucose 6-phosphate dehydrogenase, for example) show little quantitative difference in enzyme production between the sexes. There must be some **dosage compensation mechanism** to overcome the differences in gene numbers between sexes for X-linked genes, and inactivation of all but one of the female's X chromosomes would provide just such a mechanism.

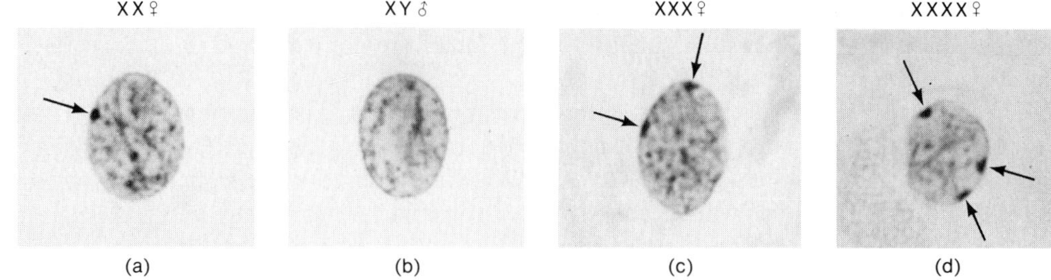

Figure 14-9. Nuclei obtained from cells in the mucous membrane of the human mouth. (a) Nucleus from a female, showing one Barr body (arrow). *(b) Nucleus from a male, with no Barr body. (c) Nucleus from an XXX female, showing two Barr bodies. (d) Nucleus from an XXXX female, showing three Barr bodies. (Parts a and b from M. M. Grumbach and M. L. Barr,* Rec. Progr. Hormone Res. *14:26, 1958. Parts c and d courtesy of Dr. M. L. Barr.)*

Variegation Due to Dosage Compensation. In the late 1950s, Liane Russell obtained a genetic clue to the mode of dosage compensation in mice through studies on mutations after radiation. She irradiated wild-type male mice and crossed them with females homozygous for several recessive autosomal coat-color mutations. Any F_1 individual exhibiting a mutant coat color would be presumed to carry a radiation-induced mutant allele of one of the loci. Among the F_1 progeny, she recovered several females that exhibited a variegated phenotype, with patches of mutant and wild-type fur. She testcrossed these variegated females and recovered two types of male progeny: completely mutant or completely wild type. Testcrossing the wild-type male progeny, she obtained completely mutant males and variegated females. Figure 14-10 outlines the crosses and their outcomes. The last testcross shows that the phenotype is sex-linked in males. What is going on?

Russell and Jean Bangham found that variegation results from a translocation between the chromosome carrying the wild-type color-coat allele and the X chromosome. Apparently, in some cells of females heterozygous for the translocation, the translocated wild-type allele does not function in the production of pigment in fur (so that the cells are mutant). In other cells, the X–autosome part does produce wild-type gene product. In males, on the other hand, the translocated wild-type allele always functions to produce normal fur color in all cells (Figure 14-11). Again it appears that some mechanism is inactivating genes on one X chromosome in the female. The variegation suggests that inactivation occurs during development and that it may involve random selection of one of the X chromosomes, so that different patches show inactivation of different X chromosomes.

X-Chromosome Inactivation. In 1961, Liane Russell and Mary Lyon independently noted that, in addition to variegating chromosome rearrangements, many sex-linked point mutations in mice and humans exhibit a variegated

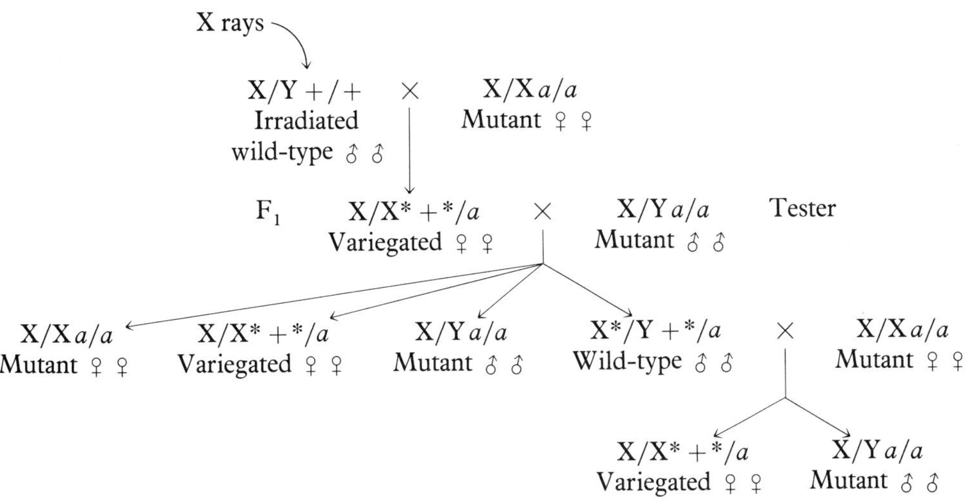

Figure 14-10. Induction of variegating chromosomes in mice. After irradiation, wild-type males are crossed to females homozygous for a recessive autosomal marker (a). *Some F_1 females carrying the irradiated chromosomes (indicated by asterisks) are variegated for the autosomal recessive. Such females are testcrossed, and the offspring show segregation for the mutant and variegating phenotypes in females and for the mutant and wild-type phenotypes in males. Testcrosses of the F_2 wild-type males show that they still transmit the variegating gene, which behaves like a sex-linked locus. These experiments indicate that the radiation must have induced a translocation linking the X chromosome to the autosomal locus* a.

phenotype in heterozygous females. They therefore suggested that dosage compensation in mammals may occur by the inactivation of one of the female's two X chromosomes—thus producing a functional equivalence of X chromosome genes between males and females.

This hypothesis can be tested in humans using females heterozygous for two alleles of the *glucose 6-phosphate dehydrogenase* (*G-6-PD*) locus that produce electrophoretically distinct enzymes (F and S). When isolated cells from a skin biopsy of a heterozygote are cloned, each clone contains either the F or the S form of G-6-PD, but never both forms. This observation tells us a number of things:

1. The *G-6-PD* locus and, by inference, most or all of the loci on *one* of the female's X chromosomes are inactive.

2. Either the X chromosome received from the mother or that from the father can be the one inactivated.

3. Because skin samples taken from a small area contain cells of both phenotypes, the inactivation must occur at a time when there are many prospective skin cells. If the inactivation occurred at an early stage of development, very large homogeneous patches would have resulted.

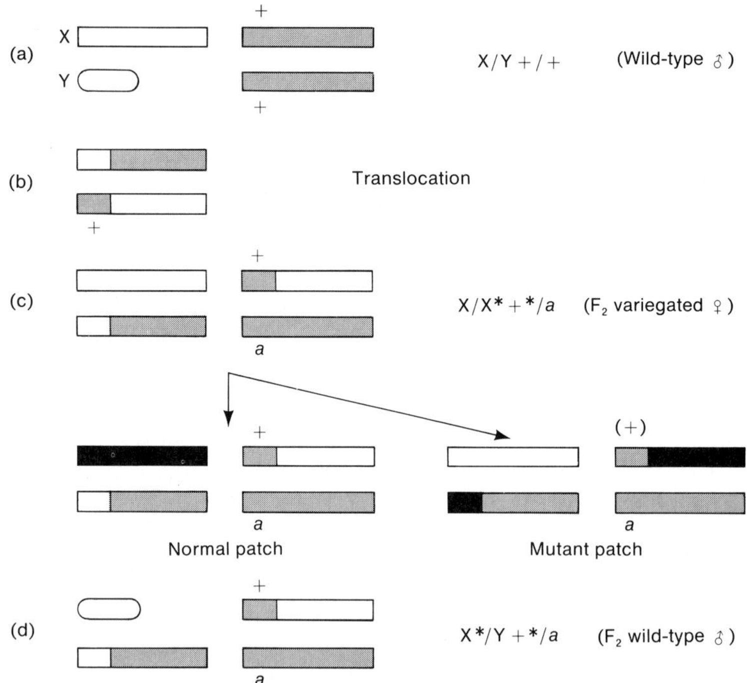

Figure 14-11. A model to explain variegation through an X–autosome translocation. (a) The X/Y +/+ genotype of the parental wild-type males before irradiation. (b) The translocation between X and the autosome induced by irradiation. (c) The X/X +*/*a *genotype in some F_1 females can produce variegation if one or the other X chromosome is inactivated randomly in various somatic cell lines during development. In cells where the X chromosome is inactivated* (dark shading), *the daughter cells will produce a normal patch because the +/*a *genotype yields a wild phenotype. In cells where the X* translocation is inactivated, the + allele may be inactivated because of its proximity to the inactivated X heterochromatin; the daughter cells of this cell will have an* a *genotype and will exhibit the mutant phenotype. If the inactivation occurs once at some stage of development and then is inherited through all following somatic cell divisions, variegated patches will develop. (d) In the X*/Y +*/*a *males of the F_2 generation, there is no X inactivation, and so the +/*a *genotype leads to a wild phenotype.*

4. Once the inactivation occurs, the state is inherited somatically, because a clone of cells is uniform in the expression of the same allele.

Message
In a mammalian female at some stage in development, either of the X chromosomes in a somatic cell is inactivated by heterochromatization. After that, all subsequent daughter cells inherit the same inactive X chromosome.

Another very dramatic demonstration of X inactivation is seen in human females who are heterozygous for X-linked anhidrotic ectodermal dysplasia. The mutant allele causes the absence of sweat glands, and mutant sectors can be detected by altered electrical resistance of the skin or by effects of various sprays. Figure 14-12 shows the phenotypic effects in three generations of women.

Sex Determination. What activates the dosage compensation in mammalian females? Is it the femaleness (two X chromosomes) or the lack of maleness (absence of a Y chromosome)? To answer this question, we must first understand how sex is determined.

In 1959, William Welshons and William Russell reported a study of a dominant X-linked mutation, *Tabby* (*Ta*), that produces a dark coat color in mice. The *Ta*/*Ta* females have the same dark-furred Ta phenotype as the

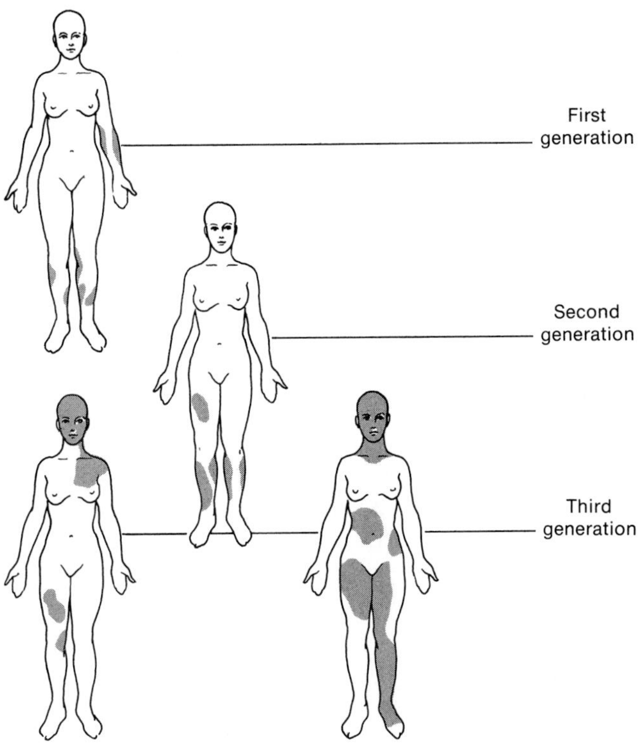

Figure 14-12. Somatic mosaicism in three generations of females heterozygous for sex-linked anhidrotic dysplasia. This abnormality in sweat-gland secretion can be demonstrated with a harmless dye. The location of the mutant tissue is determined by chance, but each female does exhibit the characteristic mosaic expression of a single X chromosome.

Ta/Y males, but the *Ta*/+ females have patches of light and dark fur. Wild-type +/+ females and +/Y males are a uniform light color. Crossing *Ta*/+ females with +/Y males, Welshons and Russell recovered female progeny that were phenotypically similar to *Ta*/*Ta* females. (You can see that, if disjunction is normal, all females should be *Ta*/+ or +/+.) If sex determination is like that in *Drosophila,* these exceptions probably are *Ta*/*Ta*/Y nondisjunctional females. Welshons and Russell then crossed these Ta exceptions with +/Y males. The progeny phenotypes were wild-type females, patched females, and Ta males in a 1:1:1 ratio. If the females were indeed *Ta*/*Ta*/Y, this cross should have produced the progeny genotypes shown in Table 14-2.

Table 14-2. Progeny expected if the exceptional Ta females are XXY

			Sperm	
			+	Y
Eggs	Normal	*Ta* *Ta*/Y	*Ta*/+ ♀ Ta/+/Y ♀	*Ta*/Y ♂ *Ta*/Y/Y ♂
	Nondisjunctional	*Ta*/*Ta* Y	? +/Y ♂	*Ta*/*Ta*/Y ♀ Y/Y lethal

The presence of wild-type females and the absence of Ta females among the progeny obviously show that the assumption of *Ta*/*Ta*/Y for the parental female is wrong. Welshons suggested that the exceptional Ta female might have only a single X chromosome and no Y; we write the genotype as *Ta*/O. Such a female would produce *Ta* and O eggs, which would result in *Ta*/+ and +/O female offspring in the cross. Welshons concluded that, in mice, the Y chromosome determines maleness, and its absence determines femaleness. This hypothesis was confirmed cytologically by the finding that the exceptional Ta females had only a single X chromosome, and it was confirmed genetically by the discovery of X/X/Y males.

In humans also, XO and XXY individuals are now recognized. Females who are XO are not completely normal, indicating that some activity of the second X is required at some time or in some tissues during female development. The XO phenotype is called **Turner's syndrome** and includes underdeveloped sex organs, webbing of the neck, and short stature (Figure 14-13). Cells from females with Turner's syndrome lack the Barr body. An abnormal male genotype is XXY, which produces a recognizable phenotype called **Klinefelter's syndrome** (Figure 14-14), including long legs, small testes, and development of breasts. A Barr body is present in cells from males with Klinefelter's syndrome. Obviously, phenotypic sex is not the factor that determines the formation of a heterochromatic X chromosome (a Barr body). Rather, it is the *number*

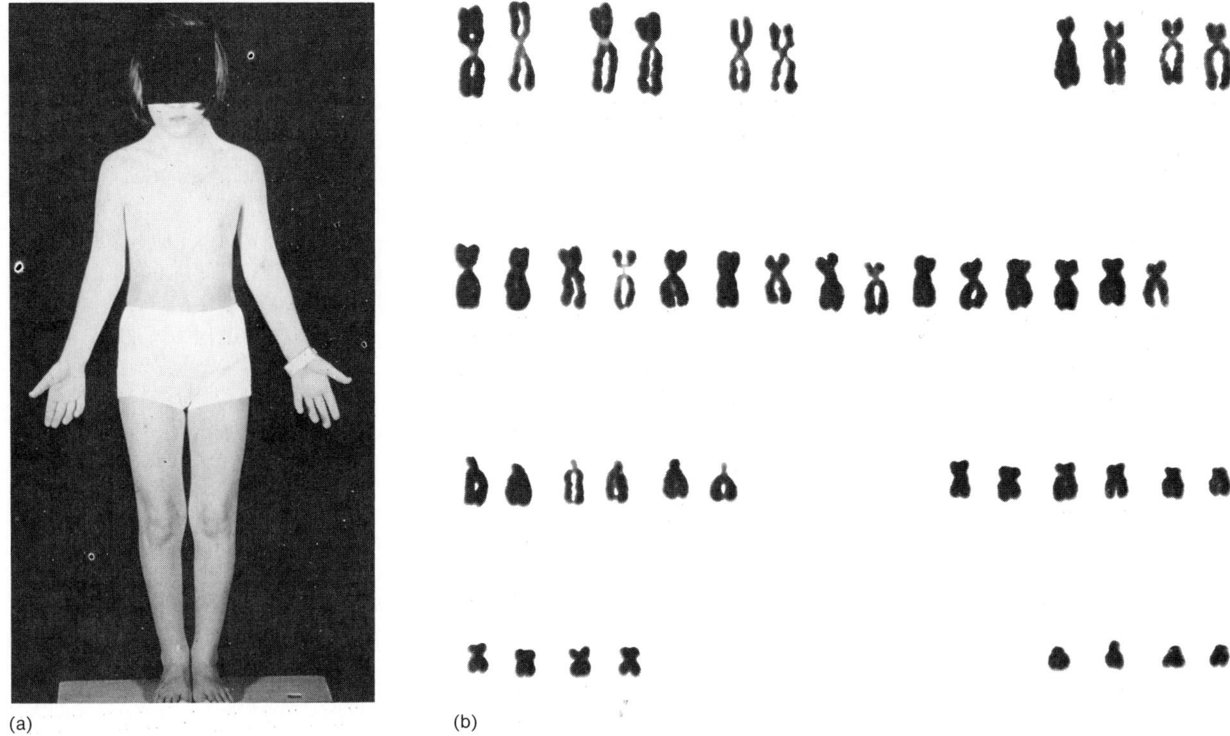

(a) (b)

Figure 14-13. An XO female. (a) The typical XO phenotype known as Turner's syndrome. (b) The karyotype of her chromosomes. (Courtesy of P. A. Baird.)

of X chromosomes that determines whether facultative heterochromatization will occur. This is confirmed through the recognition of individuals carrying several sex chromosomes (many of these individuals are mosaics for different numbers of sex chromosomes), as shown in Table 14-3.

Table 14-3. Correlation between numbers of sex chromosomes and Barr bodies.

Sex-chromosome constitution		Number of
Males	Females	Barr bodies
XY, XYY	XO	0
XXY, XXYY, XXYYY	XX	1
XXXY	XXX	2
XXXXY	XXXX	3
XXXXXY	XXXXX	4

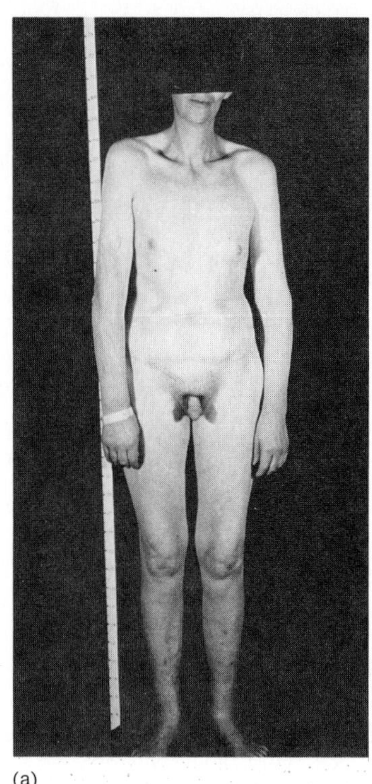

(a)

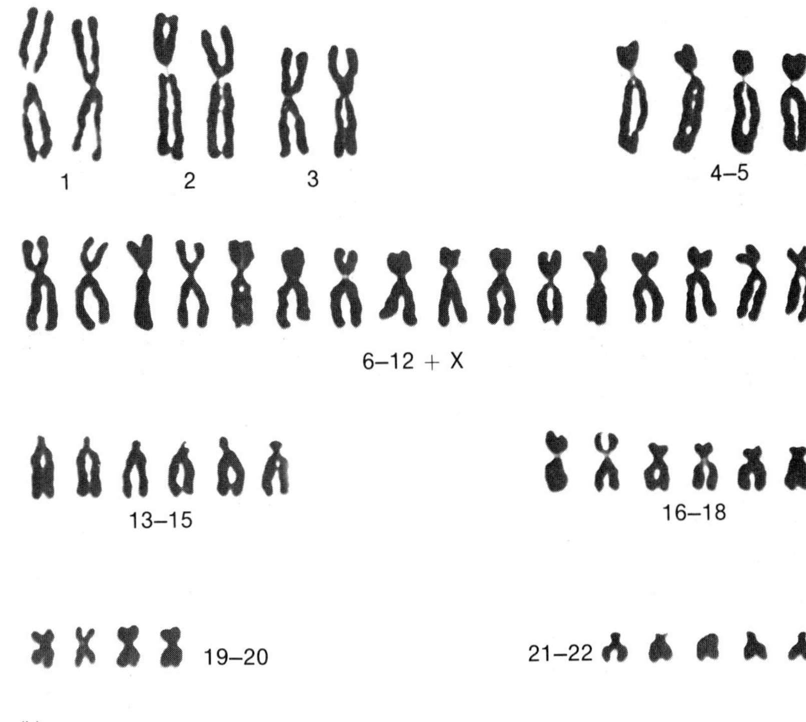

(b)

Figure 14-14. An XXY male. (a) The typical XXY phenotype known as Klinefelter's syndrome. (b) The karyotype of his chromosomes. (Courtesy of P. A. Baird.)

Message
X chromosome inactivation in mammals is determined by the number of X chromosomes per nucleus, not by the phenotypic sex of the individual. Phenotypic sex is determined by the presence or absence of a Y chromosome.

Here is another fascinating phenomenon for which we have no molecular explanation as yet: How does a cell recognize the presence of multiple X chromosomes and ensure that only one X functions?

Eukaryote Gene Function and Organization

Many fascinating phenomena and problems exist in the study of the way that genes function in eukaryotes. How are the genes organized so that their functions are coordinated in the developmental and biochemical cycles of the organism? How are specific genes activated under certain conditions or at certain times in the life cycle of the organism? What is the physical nature of the chromosome and of the gene placement along it? We look now at some of these problems.

Dosage Compensation in *Drosophila*

Heterochromatization inactivates a large genetic segment, but this is a rather crude method for control of gene expression. Are there modulations of gene expression within a more restricted portion of a chromosome? In fact, dosage compensation was originally defined in regard to the observation in *Drosophila* that some alleles of the *white* locus produce the same eye phenotype in males and females, whereas other alleles produce more eye pigment in females than in males. The former alleles are said to be dosage-compensated; the excess dosage of X-linked genes in the female is not reflected in the phenotype. The deletion of the *Notch* gene on the X chromosome produces a mutant phenotype of nicked wings in females heterozygous for the deletion and a normal chromosome, thus proving that both X chromosomes do function in a normal *Drosophila* female.

In 1965, Ed Grell demonstrated the compensatory capabilities of sex-linked loci, using two loci that affect the enzymes xanthine dehydrogenase (XDH). He constructed strains carrying different numbers of the wild-type alleles of the sex-linked locus *maroon-like* (*ma-l*) and the chromosome-3 gene *rosy* (*ry*). Upon measuring the relative amounts of enzyme activity in these *Drosophila* individuals, he obtained the results summarized in Table 14-4. Obviously, there is a dosage-compensation for the activity of $ma\text{-}l^+$, whereas the number of ry^+ genes is directly reflected in enzyme activity.

In a single cell, is only one allele active or do both alleles function? In *Drosophila,* various alleles of a single gene produce different electrophoretic mobilities of the enzyme 6-phosphogluconate dehydrogenase (6-PGD). Electrophoresis of the enzyme from heterozygous females reveals a hybrid band between the

Table 14-4. Relative enzyme activities and numbers of gene copies

	Relative enzyme activity for	
Number of gene copies	$ma\text{-}l^+$	ry^+
1	1.0	0.5
2	1.0	1.0
3	1.0	1.5

two parental bands. The hybrid band can be explained by assuming simultaneous activity of both alleles, with dimer formation by polypeptides from the two loci. Thus, unlike mammals, in *Drosophila* females, both X chromosomes function in all cells.

Measurement of enzymes controlled by X-linked loci shows that each dosage-compensated locus in a male produces twice as much product as each locus in a female. In different X–autosome translocations, the loci lying in the X portions remain compensated, whereas the autosomal loci are not compensated. These results suggest that no single region or small number of regions control X activity. Rather, each locus or small region apparently has information about the number of gene copies present and is able to control its own activity accordingly. No satisfactory molecular explanation of this method of gene regulation yet exists.

Number of Genes in *Drosophila*

Measurements of DNA show that a haploid nucleus from *Drosophila* contains about 1.5×10^8 base pairs, one genome of *E. coli* contains about 4.56×10^6 base pairs, and a human haploid nucleus contains about 3.1×10^9 base pairs. In *Drosophila,* unique sequences make up about 70% of the total nuclear DNA, with the remaining 30% consisting of intermediate or highly repetitive sequences. A typical *E. coli* protein is encoded by about 1000 base pairs, so the unique sequences of *Drosophila* could code for around 100,000 such proteins. Looking at the arrangement of DNA in the giant salivary gland chromosomes, we see that most of the DNA is located in about 5000 to 6000 densely staining bands (chromomeres). Cytogeneticists have long assumed that each chromomere represents the cytological position of a gene locus. In the haploid nucleus, each chromomere is equivalent to from 5000 to 100,000 base pairs, with an average of 30,000. How can this 30-fold excess over the size of a typical *E. coli* gene be explained?

Perhaps the initial assumption of one gene per band is erroneous, and each chromomere in fact represents many genes. Another possibility is that the average *Drosophila* gene is indeed 30 times larger than the average *E. coli* gene, and certainly the existence of large introns can account for a lot of the excess. Most of the basic metabolic processes exist both in *E. coli* and in multicellular eukaryotes, so the major differences in the size of the genome could represent codings involved in regulation of gene activity rather than codings for new gene functions. Much of the eukaryotic DNA may be involved in regulating the gene activity in relation to developmental time, the nature of the tissue, and the extent of activity required. In order to distinguish between these alternative explanations, Burke Judd and Thomas Kaufman set out in the late 1960s to detect all of the genes within a limited region of the *Drosophila* X chromosome by inducing and recovering enough mutations to ensure that every locus was represented by at least one mutant.

They chose the interval from *zeste* to *white,* a span of 10 or 12 bands, for

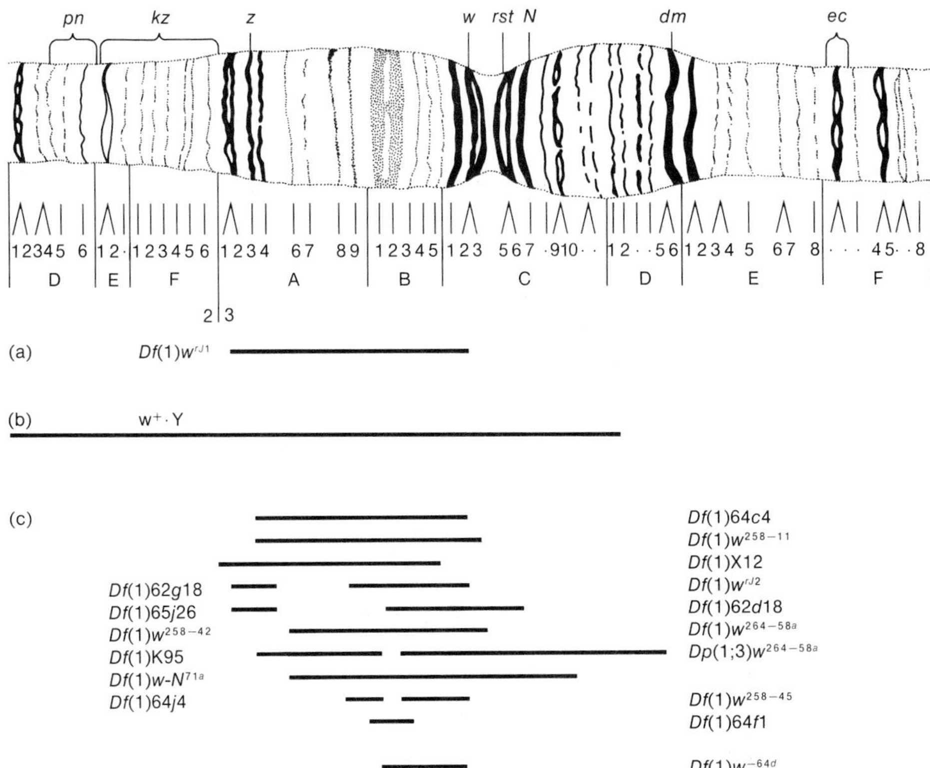

Figure 14-15. The segment of the polytene X chromosome within which the z–w *region lies. The known positions of mutations are shown above the chromosome, and the numbering nomenclature is indicated underneath (the* z–w *region extends from 3A3 to 3C3). The deficiencies and duplications covering this region are indicated at the bottom, with the extent of each deficiency or duplication shown by the horizontal line. Mutations were detected by their mutant phenotype when heterozygous with* Df(1)w^{rJ1} *(a), maintained in males with* w^+·*Y (b), and localized by their phenotypes when heterozygous with different deficiencies (c). (From B. Judd, M. Shen, and T. Kaufman,* Genetics *71:139, 1972.)*

which several overlapping duplications and deficiencies exist with which to map new mutations (Figure 14-15). They selected newly induced mutations in the *z–w* interval using the method for detection and recovery shown in Figure 14-16. Each mutagen-treated X chromosome was initially recovered in an F_1 female and then tested for its survival or phenotype when heterozygous with a *z–w* deficiency. All mutant chromosomes were "captured" in males carrying a duplication. Each mutant was tested with the duplications and deficiencies to localize it, and all mutations falling within the same segment were tested for complementation and map position by crossing over. In 1972, Judd and

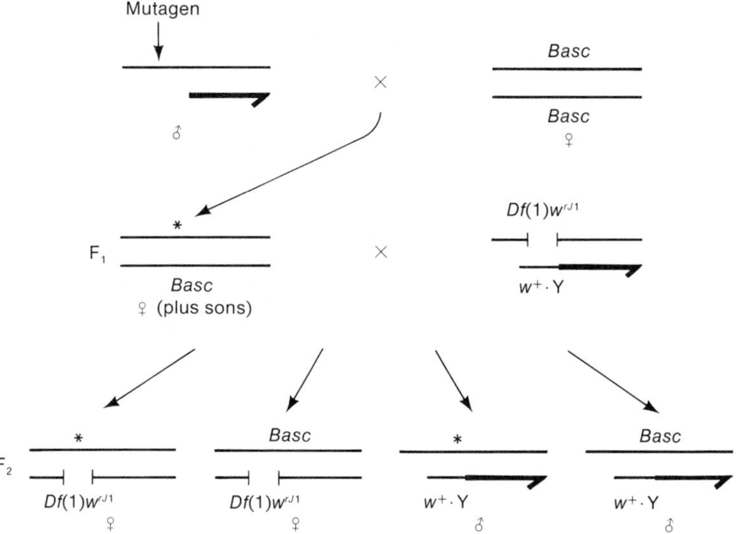

Figure 14-16. The mating scheme used to recover mutations in the z–w *region. Wild-type males were treated with a mutagen and then crossed to females carrying Basc, a multiply inverted chromosome marked with* B *and* w^a*. Then* F_1 *females carrying the mutagen-treated chromosome (indicated by an asterisk) were selected and individually mated to males carrying a deficiency on the X chromosome,* Df*(1)*w^{rJ1}*, and a duplication for this region on the Y chromosome,* w^+*•Y. A mutation in the* z–w *region is indicated by the absence (if lethal) or the mutant phenotype of the non-Bar-eyed females. The chromosome is not lost because the mutation is covered by the duplication in non-Basc males. (From B. Judd, M. Shen, and T. Kaufman,* Genetics *71:139, 1972.)*

Kaufman reported that 121 point mutations had been assigned to 16 complementation groups that could be mapped by crossing over (Figure 14-17). The position of each complementation group could be assigned to a single band, thus leading to the conclusion that *each gene corresponds to one chromomere.* We now realize that there exist nontranslated spacers between genes, large introns that are excised from mRNA, and sequences that are removed during mRNA maturation. Whether these factors account for the 30-fold excess of DNA per chromomere has not yet been determined.

Message

In Drosophila, *each polytene chromosome chromomere apparently corresponds to one gene detectable by mutation and complementation. This means that there are 5000 to 6000 genes in* Drosophila.

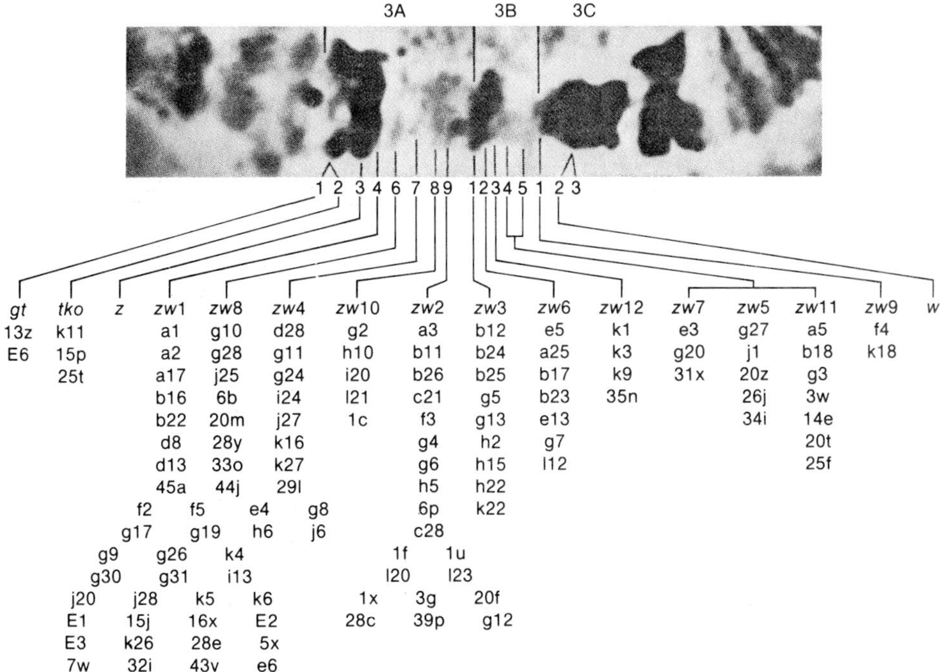

Figure 14-17. A photograph of the 3A–3C region of the X chromosome of a Drosophila melanogaster *female. The complementation groups, with the point mutations in each, are shown below the photograph. The cytological position of each complementation group is determined by deletion mapping. (From B. Judd, M. Shen, and T. Kaufman,* Genetics *71:139, 1972.)*

DNA Molecules in Chromosomes of Eukaryotes

Now we turn to the analysis of chromosomes, which may eventually provide explanations for the genetic and cytological phenomena described in the preceding subsections. By staining chromosomes with dyes that bind to specific macromolecules, we can demonstrate that the visible chromosome contains DNA, RNA, and protein. But how is that material organized during cell division? When cells are disrupted mechanically (by squeezing them under high pressure) or osmotically (by exploding them in hypotonic solutions), their contents can be mounted for electron microscopic examination. Such study reveals that the chromosome resembles a mass of spaghettilike fibrils with diameters of about 230 Å, each composed of DNA with associated protein (Figure 14-18). Ernest DuPraw showed that few if any ends protrude from the fibrillar mass, as if there is a single long fiber that is coiled up in a compact mass.

Could it be that each chromosome contains a small number of DNA molecules? In 1973, Ruth Kavenoff and Bruno Zimm resolved this question using

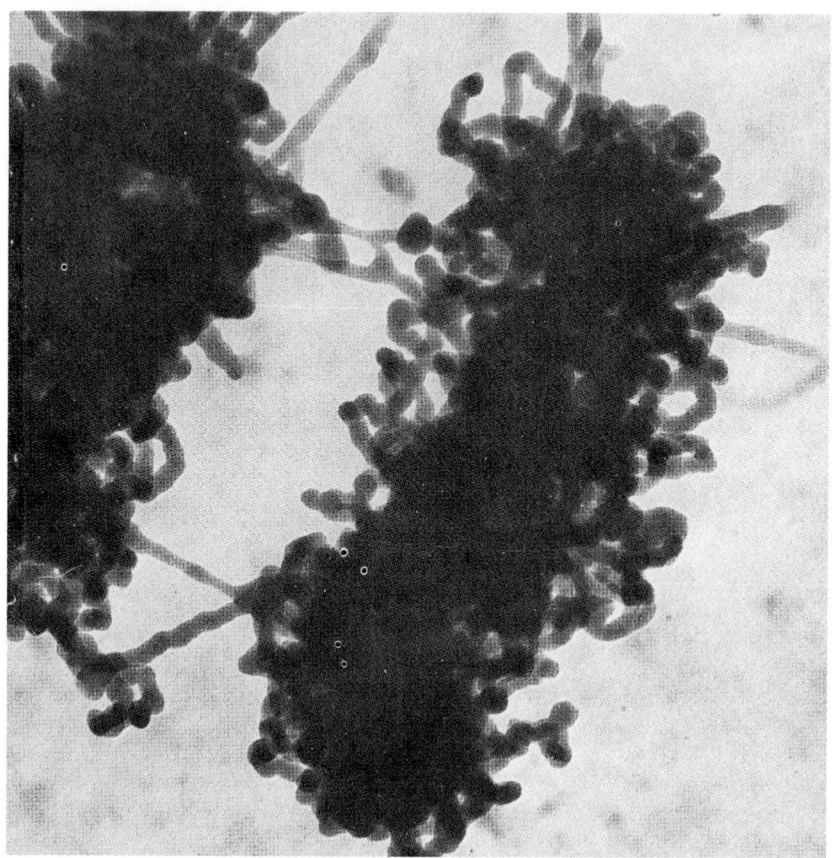

Figure 14-18. Electron micrograph of a metaphase chromatid from an embryonic honeybee cell. It appears to be a tangle of one continuous strand made up of a DNA core complexed with nuclear protein. (From E. J. DuPraw, Cell and Molecular Biology, *p. 531, fig. 18-4, 1968, Academic Press, Inc.)*

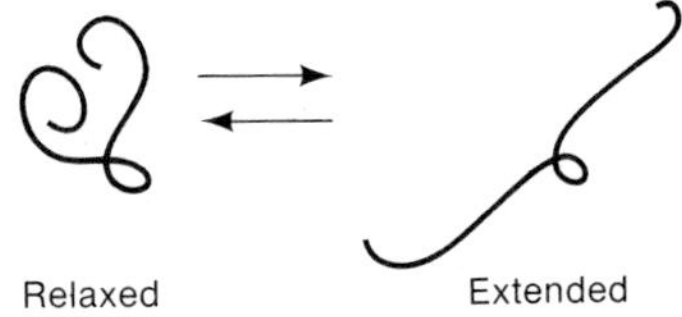

Figure 14-19. The relaxed and extended states of DNA.

a **viscoelastic technique** that essentially measures the size of the largest DNA molecules by their elastic properties in solution. Very simply, if DNA is stretched to an extended state (for example, by spinning a paddle in a DNA solution) and then allowed to recoil towards a random relaxed state (Figure 14-19), the recoil requires a time that is proportional to molecular size. This technique provides a sensitive indicator of the *largest molecules in the solution,* even when they represent a small fraction of the total number of molecules.

Kavenoff and Zimm studied the DNA molecules of *Drosophila melanogaster* (which has four pairs of chromosomes; Figure 14-1). They extracted DNA gently (to avoid shear) from wild-type nuclei and obtained a value of 41×10^9 ($\pm 3 \times 10^9$) from the viscoelastic measurements for the largest DNA molecule in solution. This value is remarkably close to George Rudkin's measurement

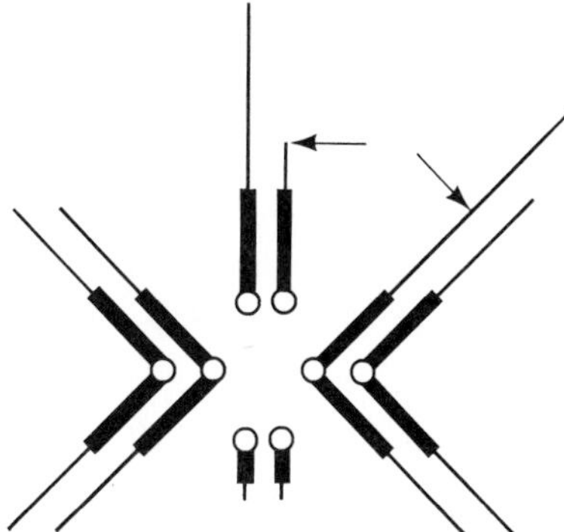

Figure 14-20. Karyotype of a Drosophila *female heterozygous for an X–autosome translocation. Breakpoints are indicated by arrows. About 60% of the mitotic chromosome length of the X is attached to the tip of chromosome 3, thereby increasing the length of chromosome 3 by 37%.*

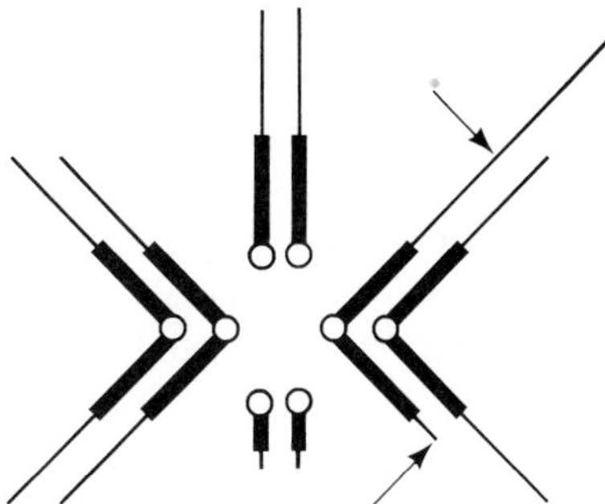

Figure 14-21. Karyotype of a Drosophila *female heterozygous for a pericentric inversion of chromosome 3. Breakpoints are indicated by arrows. The inversion changes the ratio of the lengths of the two arms of chromosome 3 from 1:1 to about 7:1, but it causes no change in the total length of the chromosome.*

Using DNA molecules from a mutant having an X-autosome translocation that increases the length of the autosomal portion by one-third (Figure 14-20), Kavenoff and Zimm obtained a viscoelastic value of 58×10^9 for the largest DNA molecule. Rudkin's measurement for the translocated autosome was 59×10^9.

Finally, they studied DNA molecules from mutants with a pericentric inversion (Figure 14-21) that increased the length of one chromosome arm without affecting the total DNA content of the chromosome. In this case, they obtained a value of 42×10^9 ($\pm 4 \times 10^9$), not significantly different from the wild-type value. In studies of other *Drosophila* species, Kavenoff and Zimm obtained values for the largest DNA molecules that were proportional to the cytological lengths observed for the largest chromosomes. Therefore, they concluded that each chromosome is composed of a single DNA molecule that extends from one end of the chromosome through the centromere to the other end.

Message
The eukaryotic chromosome is a single continuous molecule of DNA.

DNA Packaging in Chromosomes of Eukaryotes

Each chromosome that becomes visible under the light microscope at division contains one remarkably long molecule of DNA. The typical bacterial or viral chromosome also is a single DNA molecule, but it is a relatively short molecule with a single site at which bidirectional replication begins. As we have seen in earlier chapters, the eukaryotic DNA contains multiple sites for initiation of replication.

But how are such long molecules packed into the compact entities we recognize as chromosomes? In terms of end-to-end lengths, the DNA molecule contained in a chromosome may be more than 100,000 times longer than the chromosome itself. How can this compact packaging be achieved in a way that permits orderly replication?

An important insight into the packaging problem came with the recovery from nuclease-digested chromatin of a histone–DNA complex called a **nucleosome,** or **nu body.** About 140 base pairs of DNA are wrapped around a core of pairs of four histones called H2A, H2B, H3, and H4. Viewed under the electron microscope after special treatment, chromatin appears to be composed of a chain of granules (nucleosomes) that is about 100 Å in diameter (Figure 14-22). The DNA appears to be coiled around the protein octamer in two twists (Figure 14-23).

There is evidence that the nucleosomes in turn are thrown into a coil that has been called a **solenoid structure** (resembling a solenoid coil in an automobile engine). The solenoid structure has a diameter of 200 Å to 300 Å (Figure 14-24). Obviously, the DNA can be compacted extensively by superimposing coils on coiled structures during cell division (Figure 14-25). An incredible cycle of coiling and supercoiling at the level visible in the light microscope appears during mitosis of gut parasites of termites (Figure 14-26). How is this process of nucleosome formation and superimposed layers of coiling controlled? That problem remains unsolved.

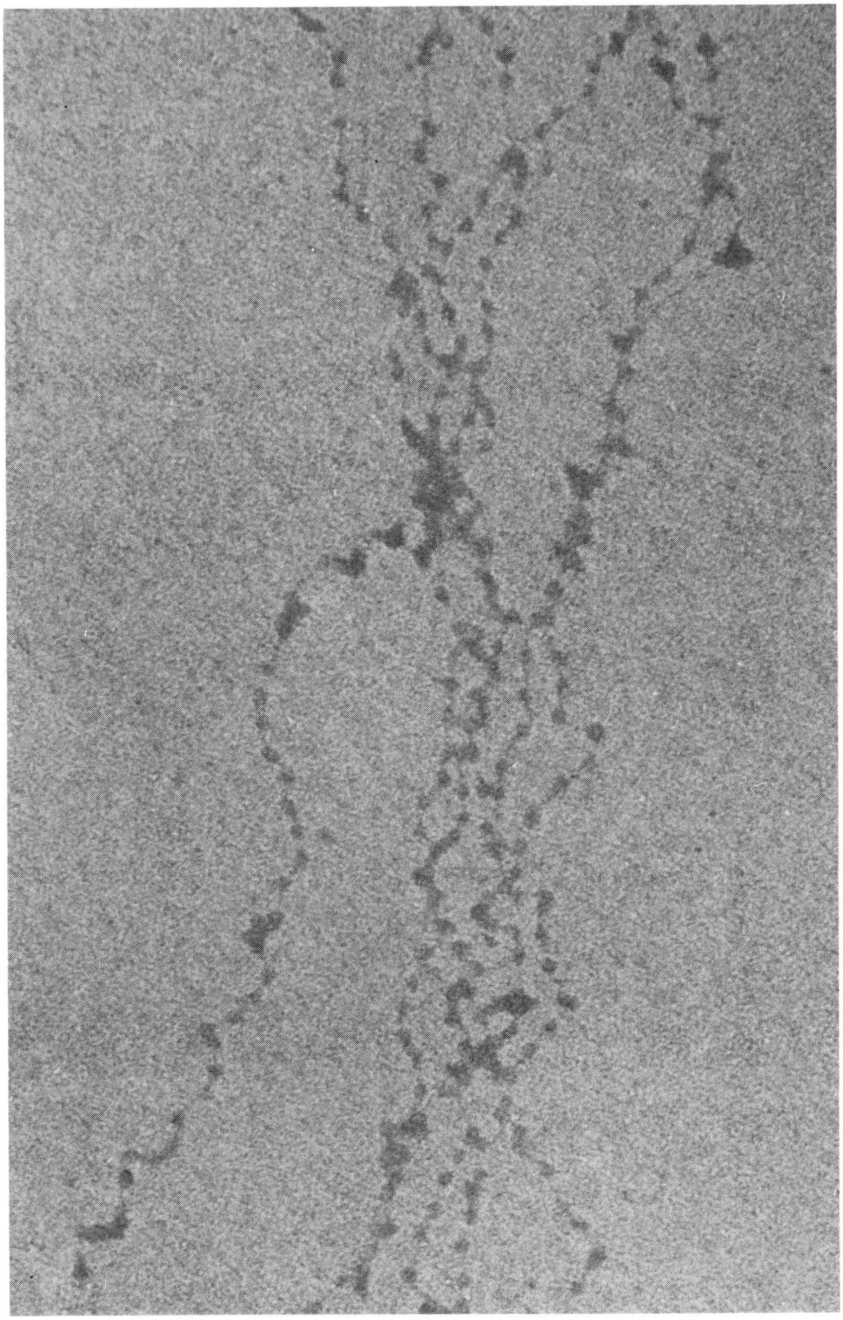

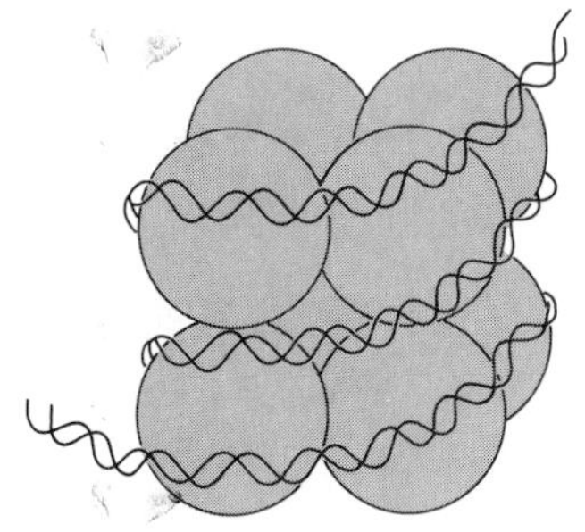

Figure 14-23. A diagrammatic representation of the four parts of histone molecules around which DNA is wound to form a nucleosome.

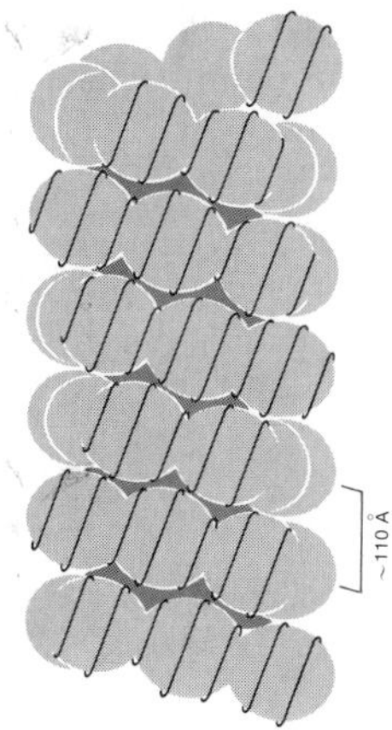

Figure 14-24. A higher level of packing is achieved in DNA when nucleosomes are wound into a solenoid structure. (From J. T. Finch and A. Klug, Proc. Natl. Acad. Sci. USA *71:1897, 1976.)*

Figure 14-22. Electron micrograph of the central core of chromatin isolated from the eukaryote Physarum. *The individual granules that form the chains presumably are nucleosomes.*

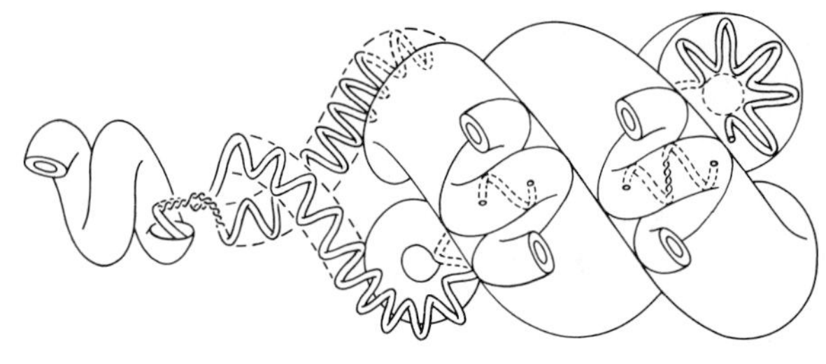

Figure 14-25. A hypothetical scheme for superimposition of coiled structures upon coiled structures for packaging DNA into chromatin. (From C. Person and D. T. Suzuki, Canadian J. Genet. Cytol. *10, 3, 1968. Reprinted with permission of the Genetics Society of Canada.)*

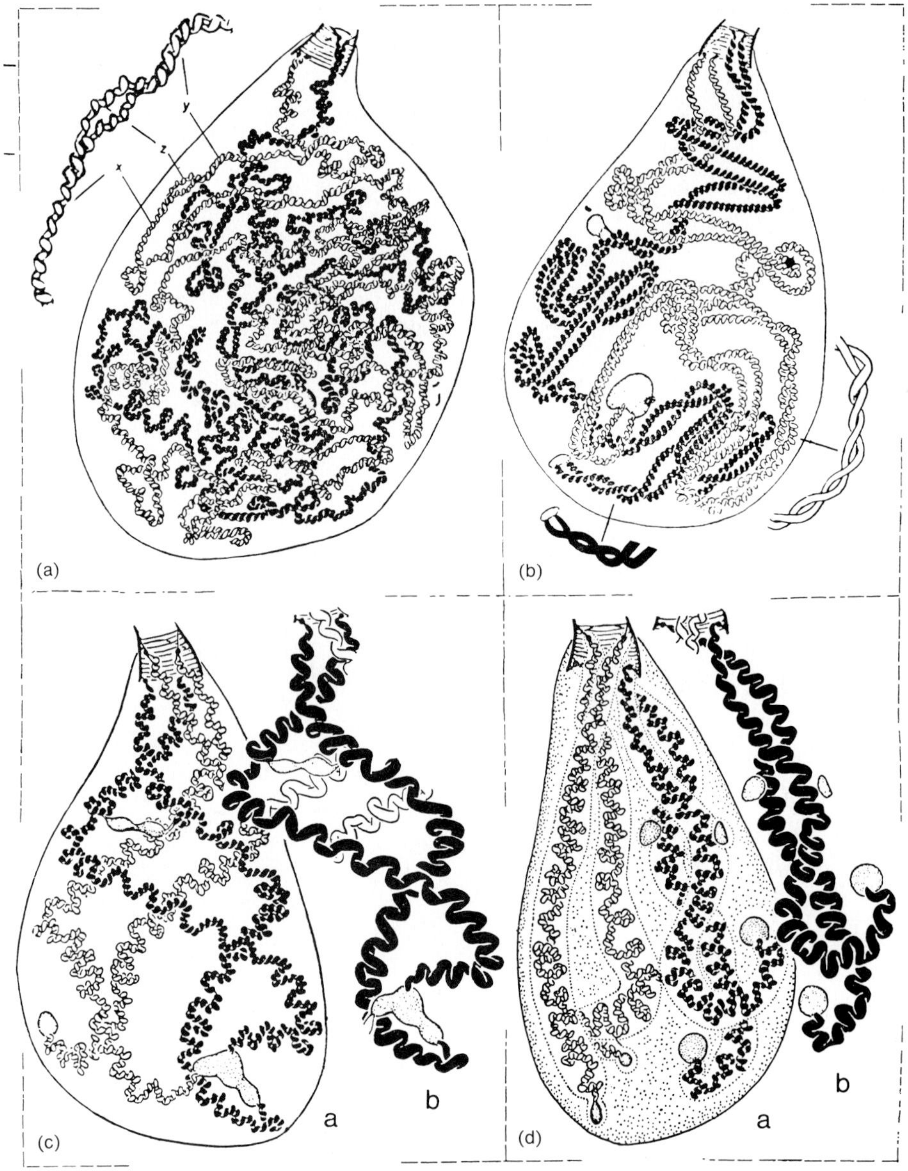

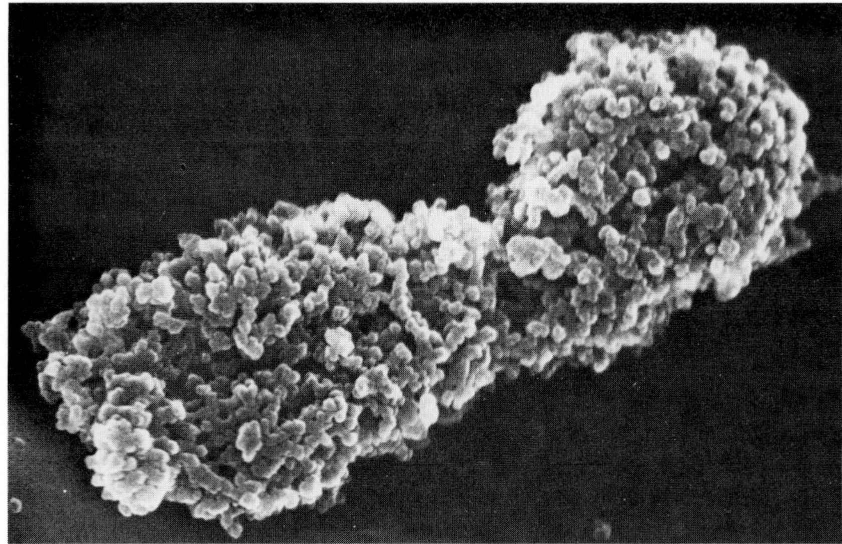

Figure 14-27. A scanning electron micrograph of a mitotic chromosome, showing the compact folding of the strand. (Wayne Wray.)

Message
DNA in the eukaryotic chromosome is shortened more than 100,000-fold by superimposing different levels of coiling. At the core of this structure, the DNA is wrapped around a cluster of histone proteins; the fundamental histone–DNA complex thus formed is called the nucleosome.

Inspection of the chromosome with a scanning electron microscope fails to reveal how the tightly coiled strand is held together, but such study does show a mat of fibers that do not appear to have a free end (Figure 14-27). Ulrich Laemmli and his associates developed a method for gentle removal of histones from chromosomes, which can then be inspected under the electron microscope. This inspection reveals that sister chromatids remain paired and that each has a central structure surrounded by a halo of DNA. The chromosome appears to have a central "scaffold" of histone to which the DNA is anchored. Close inspection reveals the incredible density of DNA packaging (Figure 14-28).

Figure 14-26. Drawings of chromosomes in meiotic prophase in a protozoan, demonstrating different degrees of coiling and supercoiling visible with the light microscope. Two large chromosomes are shown, one black and the other white; (a) through (d) are a progression. (a) Coiling is seen though duplication becomes apparent. (b) Duplication is well advanced. (c) Supercoiling is beginning. (d) Supercoiling is well advanced. (From L. R. Cleveland, "The Whole Life Cycle of Chromosomes and Their Coiling Systems." Transactions of the American Philosophical Society *39, 1, 1949.)*

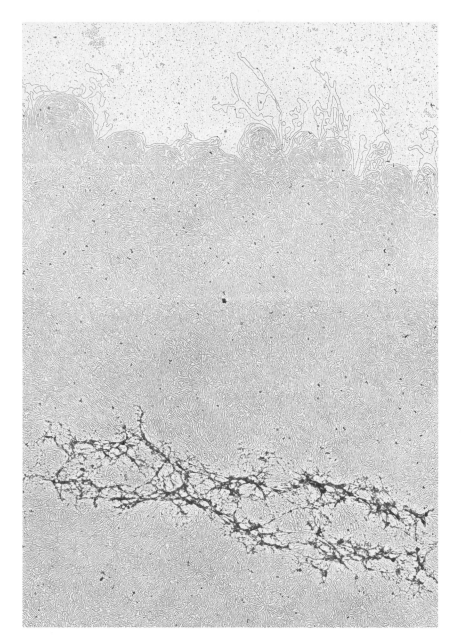

Figure 14-28. Electron micrograph of a metaphase chromosome from a cultured human cell. Note the central core, or scaffold, from which the DNA strands extend outward. No free ends are visible at the outer edge. At even higher magnification, it is clear that each loop begins and ends near the same region of the scaffold. (From Baumbach and Adolph, CSHSQB, *Cold Spring Harbor Laboratory, 1977.)*

The ends of each loop of DNA that extend from the scaffold are attached near the same point, suggesting that the packaging process is regular and precise.

Message
The DNA in the eukaryotic chromosome is organized about a central core from which segments may extend to form loops.

Genes with Related Functions

At this point, we must ask whether the chromosome is simply a vehicle to ensure the proper physical distribution of genes during cell division or whether the structure itself has some role in determining the expression of genes. The genetic and cytological properties of heterochromatin and euchromatin clearly indicate that these regions are controlled differently. The Y chromosome appears to be a specialized chromosome that has evolved a special control mechanism. But are there more subtle patterns of chromosome regulation at the level of groups of genes?

Metabolic Pathways

A linkage study of all the loci affecting a particular phenotypic feature (say, eye color in *Drosophila*) reveals no apparent pattern; these loci seem to be scattered randomly through the genome. However, when Milislav Demerec studied the distribution of loci affecting a common biosynthetic pathway, he found that the genes controlling steps in the synthesis of the amino acid tryptophan in *Salmonella typhimurium* are clustered together in a restricted part of the genome. In 1964, Demerec then looked at the distribution of genes involved in a number of different metabolic pathways. Analyzing auxotrophic mutations representing 87 different cistrons, he found that 63 could be located in 17 functionally similar clusters. A cluster is defined as two or more loci that control related functions, where the loci are carried on a single transducing fragment and are not separated by an unrelated gene.

Similar clusters of genes involved in a metabolic pathway have been found in a number of eukaryotes, including *Neurospora*, yeast, and *Drosophila*, so the phenomenon of tight linkage of metabolically related genes is widespread. Furthermore, in cases where the sequence of catalytic activity is known, there is a remarkable congruence between the sequence of genes on the chromosome and the sequence in which their products act in the metabolic pathway. This congruence is strikingly illustrated by the histidine cluster in *Salmonella* that was extensively studied in the early 1960s by Philip Hartman and Bruce Ames (Figure 14-29).

Morphogenetic Sequences

Some point mutations have phenotypes that are conditionally expressed—that is, the phenotype is mutant under *restrictive* conditions but wild-type under

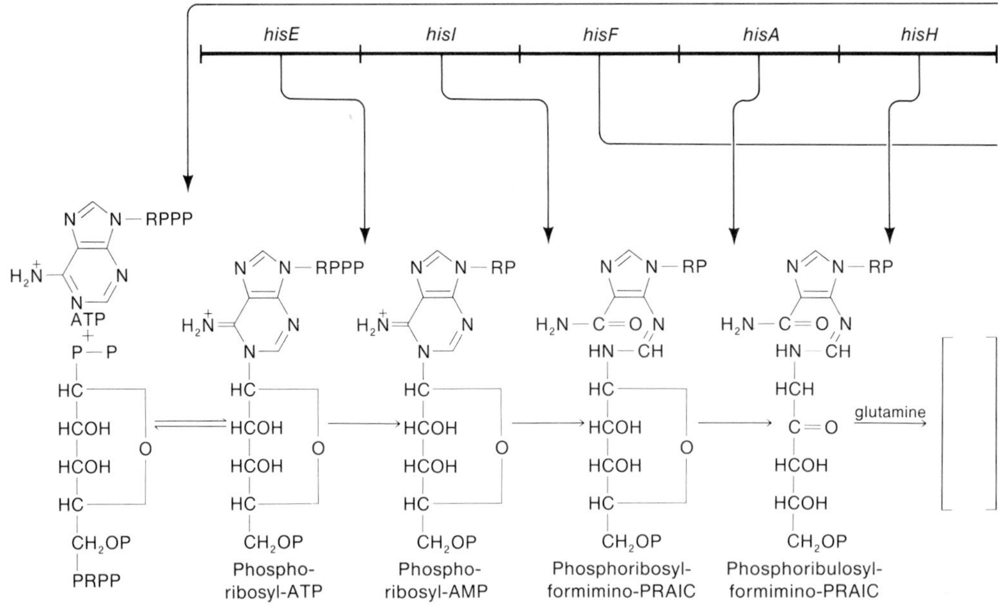

Figure 14-29. *The histidine* (his) *gene cluster and the metabolic pathway that it controls. Note that the sequence of genes in the cluster generally corresponds to the sequence of steps that each gene catalyzes in the pathway*

permissive conditions. We have seen that such mutations provide a means of detecting variants in virtually any essential gene. Two commonly used classes of conditional mutants are temperature-sensitive (*ts*) and amber (*am*).

Robert Edgar and Richard Epstein recovered a large number of *ts* and *am* mutants in T4 phages. Each mutation could be mapped genetically and tested for allelism with closely linked mutations by mixed infections under restrictive conditions. Finally, by examining the hosts infected with mutants under restrictive conditions, Edgar and Epstein were able to identify different phenotypes in which phage formation stopped at different points (Figure 14-30). We can see immediately that genes affecting the same morphogenetic pathway (such as formation of the tail, or synthesis of DNA) are clustered together.

In addition, one can determine the time at which a gene product is produced by shifting *ts* strains from permissive to restrictive temperatures at different times after the start of infection and noting whether the loss of the gene activity interferes with normal functioning (Figure 14-31). The genetic map of the phage T7 provides a striking example of the clustering of genes with related functions and of those that act at the same time (Figure 12-6). There are three

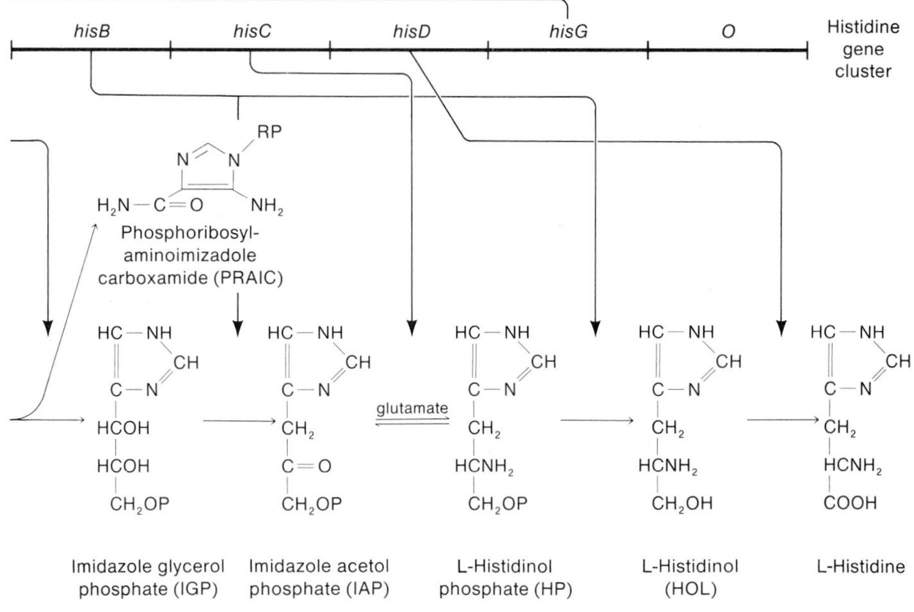

for synthesis of histidine. The fact that the final gene in the sequence (hisG) *catalyzes the first step in the reaction sequence is probably not a coincidence; this pattern is commonly found.*

groups of genes that are transcribed as blocks. The genes expressed early in development are involved in altering the host environment to make it favorable for phage multiplication. The genes for DNA replication are activated next. Finally, the late genes concerned with the structural elements of the virus are activated.

Message
Genes involved in the same metabolic or morphogenetic pathway are tightly clustered on the chromosome, often in the same sequence as the reactions that they control. Furthermore, the genes within a cluster often are expressed at the same time.

Taken together, these data suggest that the chromosome is more than a structural framework for the proper assortment of genes at division. The position of genes relative to each other is a key to their expression. For example,

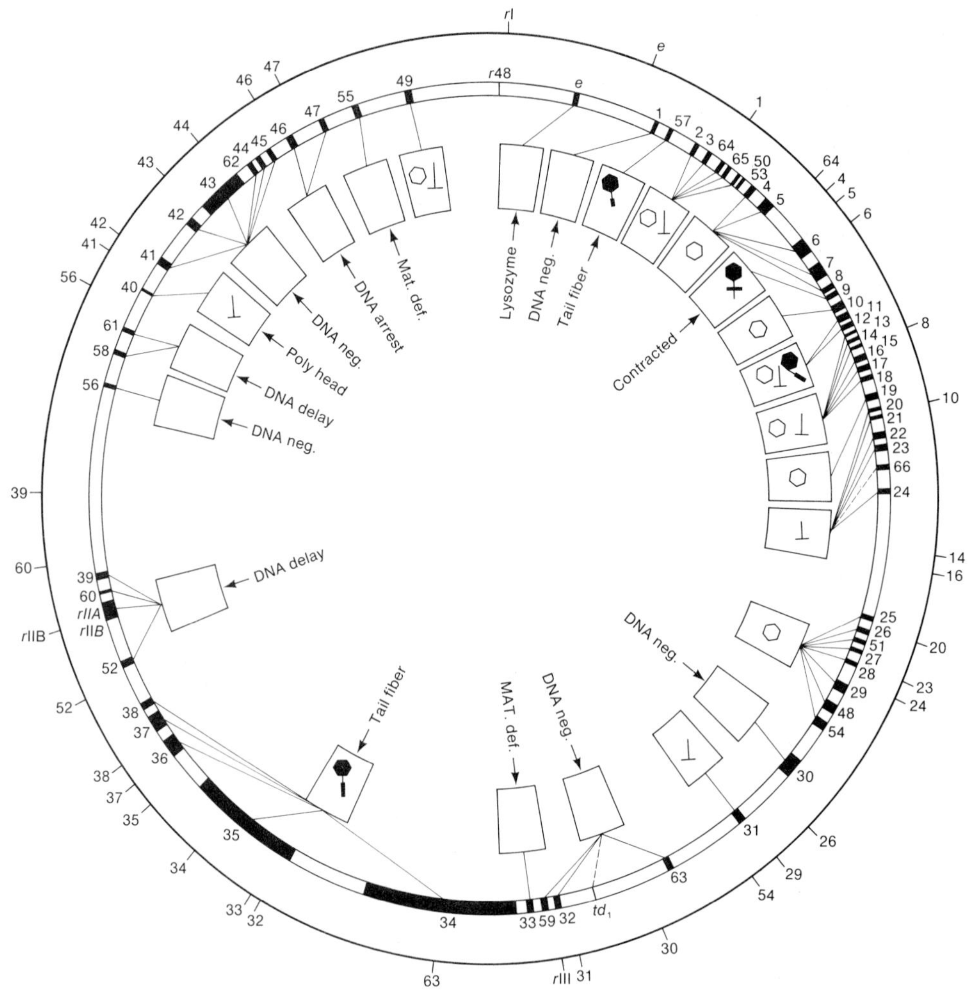

Figure 14-30. The genetic map of conditionally lethal mutations in phage T4, showing the structures that are present when mutants are grown under restrictive conditions. It is clear that genes involved in related parts of phage morphogenesis are clustered together. (From G. Mosig, Adv. Genetics *15:1, 1970.)*

the temperature-shift studies of Edgar and Epstein suggest that sets of genes function at different times. Where different gene products are required at different stages of the life cycle, it seems eminently reasonable that such genes should be clustered in groups that could be activated at the appropriate times by some mechanism. Similarly, a group of genes involved in the synthesis of a particular substance can be activated together when that substance is needed.

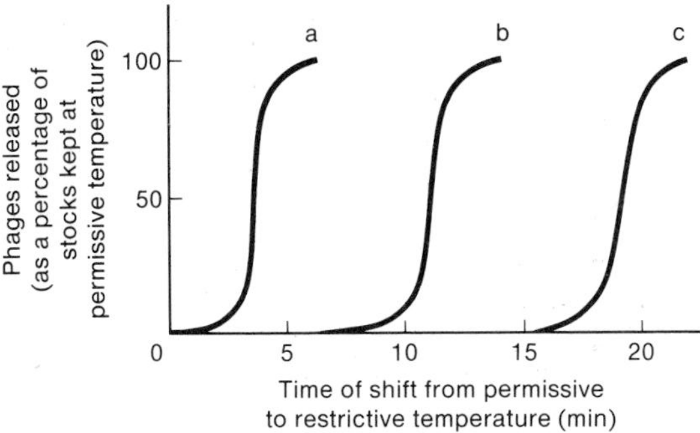

Figure 14-31. Phage formation in different ts *lethal stocks (a, b, c) when shifted from permissive to restrictive temperatures at different times after infection. For each stock, there is a specific* ts *period after which the restrictive temperature has no effect. Apparently, the loss of expression of the mutant gene is unimportant to the organism after a certain point in its development.*

In higher organisms, some mechanism is needed to activate specific genes only in certain cells. For example, the loci involved with specifying the translational apparatus and process will be required to function in all cells, but the gene responsible for silk production in the glands of silkworms should be activated only in a few specialized cells. Again, clustering of related genes would provide an appropriate situation for an activation mechanism. Even in a bacterial cell, there seems to be some form of controlled gene activation because different polypeptides range in abundance from 10 to 500,000 molecules per cell.

Does a mechanism exist to activate a cluster of genes? If so, how is this mechanism controlled, and how does it activate the genes?

The *Lactose* Genes: Positive Control

If lactose is present in the medium on which *E. coli* is grown, the cells contain the enzyme β-galactosidase for which lactose is a substrate (Figure 14-32). If the lactose in the medium is replaced with glucose, the enzyme disappears after a time. If lactose is again added to the medium, there is a rapid increase in β-galactosidase activity within the cell. In his studies on lactose utilization, Jacques Monod showed that any wild-type cell possesses the *capacity* to produce β-galactosidase and that the enzyme is synthesized de novo in response to

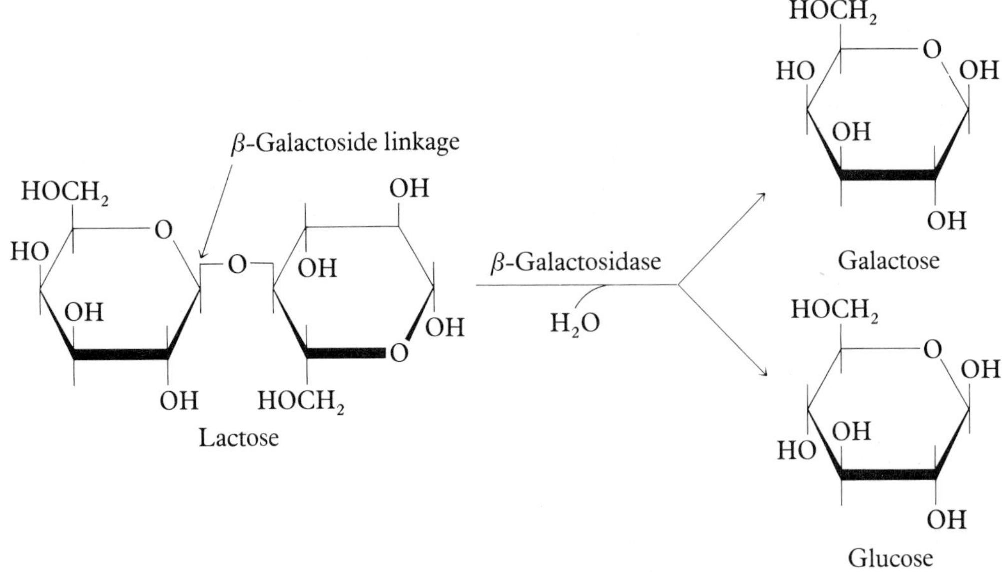

Figure 14-32. The metabolism of lactose. The enzyme β-galactosidase catalyzes a reaction in which water is added to the β-galactoside linkage to break lactose into separate molecules of galactose and glucose.

the presence of lactose. In this case, the substrate acts to *induce* production of its enzyme. We have here an example of genetic activation and inactivation regulated by a known molecule.

In the late 1950s, François Jacob and Monod were able to select missense mutants that produced a protein lacking β-galactosidase catalytic activity but still recognizable as a form of the enzyme because of its reaction with anti-β-galactosidase serum. They mapped such mutations within the *z* cistron, lying betwen the *ara* (*arabinose*) and *trp* (*tryptophan*) loci. They also detected other mutants that produced wild-type β-galactosidase in the presence of lactose but produced an *amount* of enzyme directly proportional to the external concentration of lactose, the inducer molecule. These mutants were found to lack galactoside permease, a membrane-pump protein whose function is to concentrate lactose by transporting the molecules into the cell. These mutations were mapped in the *y* cistron adjacent to *z*. Finally, they obtained mutations affecting the third enzyme involved in lactose metabolism, thiogalactoside transacetylase, and mapped these adjacent to *y* in the *a* cistron. Thus, the gene sequence is *ara–z–y–a–trp*.

This cluster of cistrons represents the familiar pattern of a sequence of genes specifying proteins that participate in various aspects of a specific metabolism. What is most interesting is that all three proteins appear *simultaneously* when lactose is added to the medium of wild-type cells. That is, all three gene products are **coinduced** by the same molecule, lactose. Because F can insert near this region, it is possible to create sexduced partial diploids (called **merodiploids**) for this region. The $z^+ y^+ a^+/z^- y^- a^-$ heterozygotes are wild type in phenotype—that is, inducible.

The genetic clustering and the coinduction of the related cistrons demand further investigation. What is the mechanism that turns the three tightly linked cistrons on and off?

Regulators of Structural Genes. Jacob and Monod found two other kinds of mutations that mapped in the vicinity of z, y, and a. These are **regulatory mutations** that do not affect the structures of β-galactosidase, permease, and acetylase; instead, these mutations affect the process by which lactose induces synthesis of the enzymes.

The first group of regulatory mutants showed **constitutive enzyme production**—that is, they synthesized the three enzymes in the absence of the inducing lactose. The site for this mutation is called the operator (O), and the constitutive mutation is called O^c; it maps between *ara* and z, very close to z. Most interestingly, the O^c mutation affects only the z, y, and a cistrons that are adjacent to it on the same chromosome (in a *cis* arrangement). Let's consider an example. (We'll refer only to z, but the same comments apply to all three loci.) We can construct a merodiploid cell by transfer of a plasmid carrying genes of the bacterial chromosome. In an $O^c z^+/O^+ z^-$ merodiploid, the cells produce β-galactosidase constitutively, so O^c appears to be dominant over the wild-type O^+ that confers the inducible state. However, when the same alleles are organized differently in the merodiploid $O^c z^-/O^+ z^+$, the enzyme is inducible rather than constitutive. The allele O^c causes constitutive synthesis only in genes which are linked in *cis* arrangement to it, so it is said to be *cis*-dominant. The O–z–y–a region is a coordinated unit of function called the **lactose** (*lac*) **operon.**

The second group contained regulatory mutations that were not limited in their effects to the *cis* arrangement with z, y, and a. These are called **repressor mutations,** and their locus i maps between *ara* and O. One mutant allele i^S is dominant and exhibits **superrepression,** completely preventing induction of enzyme synthesis. For example, enzyme synthesis cannot be induced in the merodiploid $i^S O^+ z^-/i^+ O^+ z^+$. On the other hand, another mutant allele i^c is recessive to i^+ and causes constitutive enzyme production in $i^c O^+ z^+$, although $i^c O^+ z^+/i^+ O^+ z^+$ cells are inducible.

A further class called **polar mutations** is of interest. These were found in the region from O to the a gene and had the effect of knocking out all functions to the right. For example, a polar mutation mapping in z would also knock out y and a functions. Since these mutations were revertible, they could not be deletions.

Jacob and Monod brilliantly recognized that the *i* and *O* loci interact in the control of the "structural" cistrons *z*, *y*, and *a* that code for the structures of the *lac* proteins. The operator locus affects only the adjacent cistrons and is located at one end of the sequence of genes that are transcribed together. These observations suggest that the operator may be the site for initiation of transcription. In the wild-type operator, transcription is initiated only when lactose is present. In the mutant O^c, transcription is initiated whether or not lactose is present. The repressor gene, on the other hand, must act by way of the cytoplasm because some of its alleles exhibit *trans*-dominance. Perhaps it produces a product that controls the initiation of transcription at *O*. Jacob and Monod suggested the following model of gene regulation. In the absence of lactose, a **repressor molecule** produced by the gene *i* interacts physically with the operator site to prevent transcription. When the inducer is present, a complex forms between the inducer and the repressor, altering the repressor in such a way that it no longer blocks the operator site (Figure 14-33), thus permitting transcription of the structural genes as one piece of mRNA. What about the polar mutations? These were shown to be the result of stop codons which arrested production of all protein past that point. The existence of such polar mutations is a good indication of the transcriptional unit in an operon.

Later studies revealed further complications. Some deletions of the *O* region that extend from *z* to different positions to the left of *O* do not express y^+ and a^+ activity, even though there is no site for repressor binding. Apparently, still another site, now called the **promoter** (*p*), must be present if the *lac* genes are to be expressed. Indeed, mutations mapping between *i* and *O* that tend to reduce *lac* gene activity but are still under *i* gene control do appear to be *p* mutations. It is now fairly certain that *p* represents the site to which RNA polymerase, the enzyme for transcription, binds to the DNA; transcription then begins when the repressor is released from *O*. Figure 14-34 shows a photograph of the enzyme actually bound to the promoter region. Thus, the entire map of the *lac* region is *ara–i–p–O–z–y–a–trp*. Figure 14-35 summarizes the control of the *lac* operon as it is now understood.

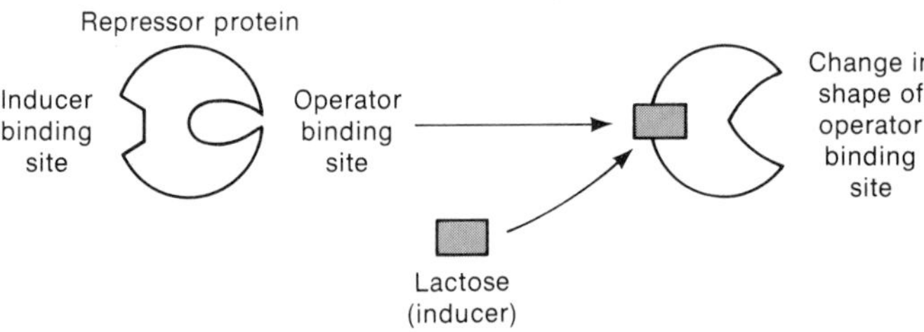

Figure 14-33. When lactose combines with the repressor protein, it alters the shape of the repressor so that it will no longer bind to the operator.

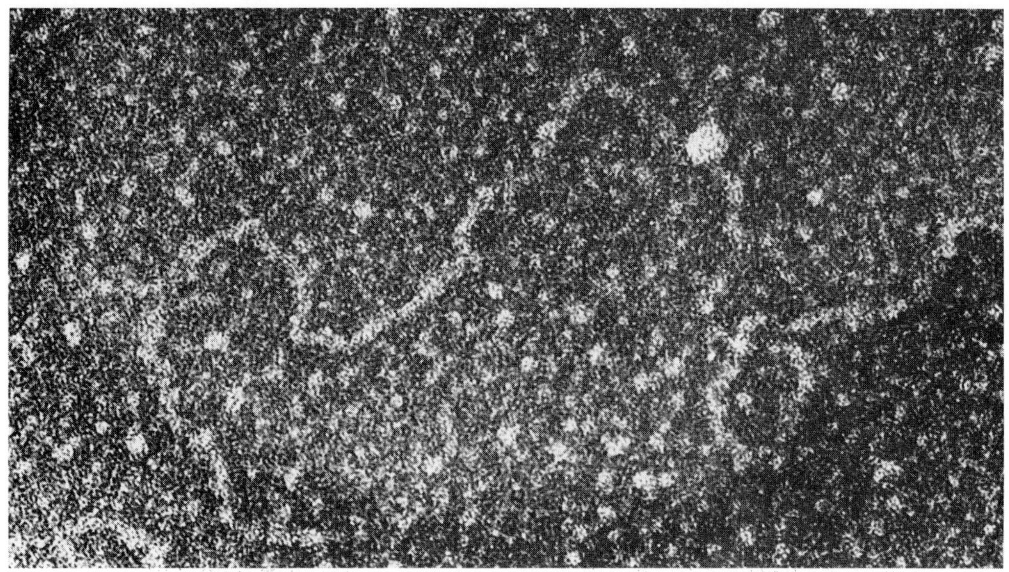

Figure 14-34. *The* lac *repressor* (large whitish sphere) *bound to DNA at the promoter region of the operon. (From Jack D. Griffith.)*

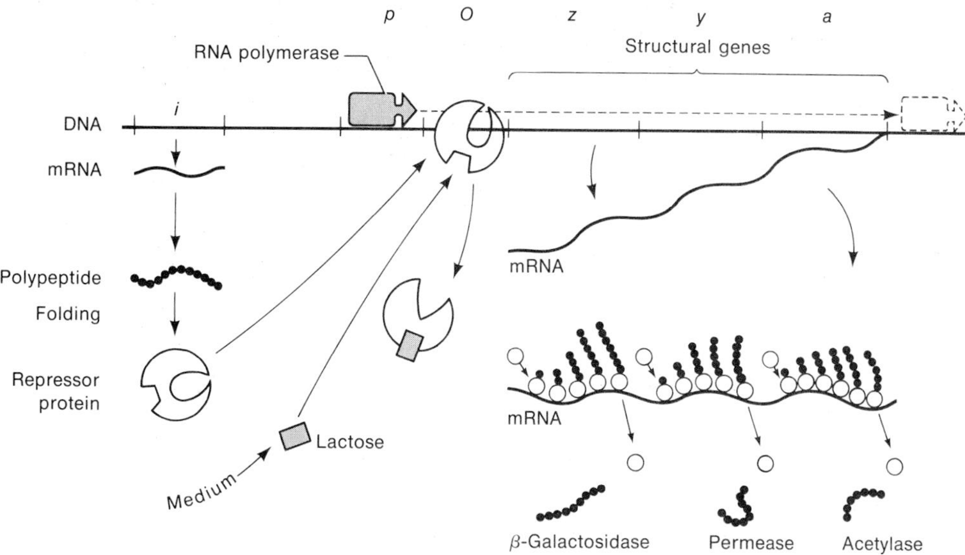

Figure 14-35. *Regulation of the* lactose *operon. The* i *gene (which is not considered a part of the operon and is somewhat separated from the operon) continually makes repressor. The repressor binds to the* O *(operator) region, blocking the RNA polymerase bound to* p *from transcribing the adjacent structural genes. When lactose is present, it binds to the repressor and changes its shape so that the repressor no longer binds to* O. *The RNA polymerase is then able to transcribe the* z, y, *and* a *structural genes, so the three enzymes are produced.*

Message

The lac *operon is a cluster of structural genes that specify enzymes involved in lactose metabolism. These genes are controlled by the coordinated actions of* cis-*dominant promoter and operator regions. The activity of these regions is in turn determined by a repressor molecule that is specified by a separate regulator gene.*

This model explains the various regulatory mutants. The O^c mutants have altered DNA that does not permit repressor binding, so transcription of the structural genes occurs continuously. The i^S mutants produce altered repressor proteins that will not bind to inducer and hence cannot be released from the operator site, so transcription of the structural genes is permanently blocked. The i^c mutants produce repressor protein that is altered in a different way so that it cannot bind to the operator region, thus allowing continuous transcription of the structural genes. Note that this model does explain why the O^c mutation affects only the adjacent structural genes (and not those on a different chromosome in a merodiploid), whereas the i^c mutation affects any *lac* structural genes in the same cell.

Verification of the operon model became possible in 1966 when Walter Gilbert and Benno Müller-Hill isolated and purified the repressor molecule. Studies in vitro showed that the repressor protein binds to DNA that contains the *lac* operon but does not bind when the DNA carries a deletion of the *lac* genes. Repressor mixed with inducer does not bind to *lac*-containing DNA, and repressor alone does not bind to DNA from an O^c mutant stock. Gilbert used the enzyme DNase to break apart DNA bound to repressor, and he was able to recover short DNA strands shielded from the enzyme activity by the repressor molecule and hence presumed to represent the operator sequence. This sequence was determined, and each operator mutation was shown to involve a change in the sequence (Figure 14-36). These results not only confirm the identity of the operator sequence, they also show the incredible speci-

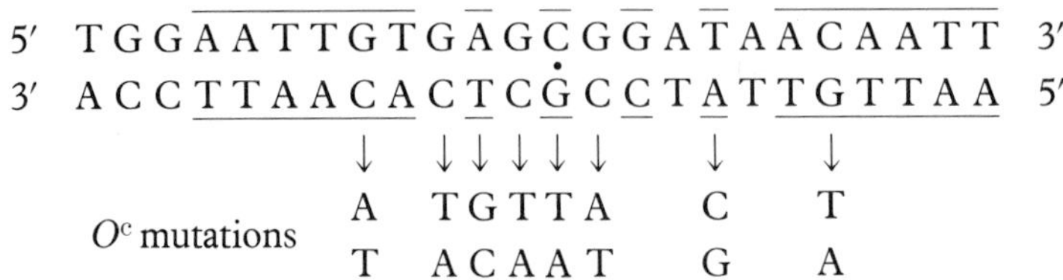

Figure 14-36. The DNA base sequence of the lactose operator, and the base changes associated with eight O^c *mutations. Regions of twofold rotational symmetry are indicated by horizontal lines above and below the symmetrical base pairs and by a dot at their axis of symmetry. (From W. Gilbert, A. Maxam, and A. Mirzabekov, in N. O. Kjeldgaard and O. Malløe, eds.,* Control of Ribosome Synthesis. *Academic Press, © 1976. By permission of Munksgaard International Publishers Ltd., Copenhagen.)*

ficity of the repressor–operator recognition, which is disrupted by a single base substitution.

When the sequence of bases in the *lac* mRNA (transcribed from the *lac* operon) was determined, the first 21 bases on the 5′ initiation end proved to be complementary to the operator sequence Gilbert had determined. All of these pieces fit together to confirm the general model. The repressor protein binds to the operator, preventing initiation of transcription by the RNA polymerase bound at the promoter site.

Catabolite Repression of the lac *Operon.* Thus far we have discussed the induction of β-galactosidase when cells are transferred from a medium containing glucose to one containing lactose. If both lactose *and* glucose are present, synthesis of β-galactosidase is not induced until all of the glucose has been utilized. Thus the cell conserves its metabolic machinery by utilizing any existing glucose before going through the steps of creating new machinery to exploit the lactose. The operon model we have outlined will not account for the suppression of induction by glucose, so we must modify it.

Studies indicate that in fact it is some catabolic breakdown product of glucose (whose exact identity is not yet known) that prevents activation of the *lac* operon by lactose, so this effect is called **catabolite repression.** The effect of the glucose catabolite apparently is exerted on an important cellular constituent called cyclic adenosine monophosphate (cAMP). When glucose is present at high concentration, the cAMP concentration is low; as the glucose concentration decreases, the concentration of cAMP correspondingly increases. The high concentration of cAMP is necessary for induction of the *lac* operon. Mutants that cannot convert ATP to cAMP cannot be induced to produce β-galactosidase because the concentration of cAMP is not great enough to activate the *lac* operon. In addition, there are other mutants that do make cAMP but cannot be induced because they lack yet another protein called CAP (catabolite activator protein) made by the *crp* gene. The CAP forms a complex with cAMP, and it is this complex that induces the *lac* operon (Figure 14-37).

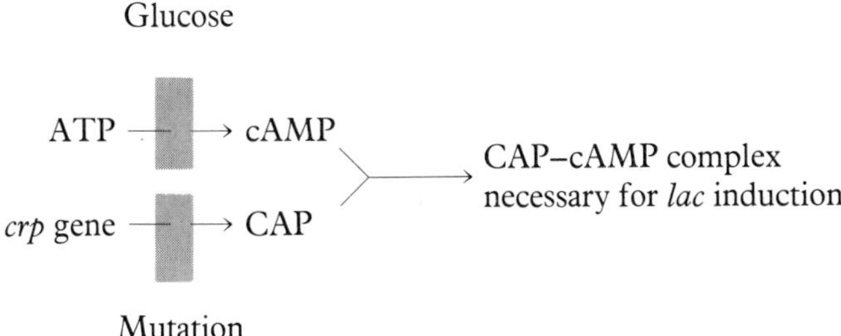

Figure 14-37. Catabolite control of the lac *operon. The operon is inducible by lactose when cAMP and CAP form a complex. The* lac *operon cannot be induced if formation of cAMP is blocked by excess glucose, or if formation of CAP is blocked by a mutation of the* crp *gene. (CAP = catabolic activator protein; cAMP = cyclic adenosine monophosphate;* crp *= structural gene responsible for synthesizing CAP.)*

How does catabolite repression fit into our model for the structure and regulation of the *lac* operon? Recall the technique that Gilbert used to identify the operator base sequence. In a similar experiment, the CAP–cAMP complex was added to DNA and the DNA then subjected to digestion by the enzyme DNase. The surviving strands are presumably those shielded from digestion by an attached CAP–cAMP complex, and the sequence of these strands is that shown in Figure 14-38. This sequence clearly is different from the operator sequence (Figure 14-36), but it also has a rotational twofold symmetry.

G T G A G T T A G C T C A C
C A C T C A A T C G A G T G

Figure 14-38. The DNA base sequence to which the CAP–cAMP complex binds. Regions of twofold rotational symmetry are indicated by horizontal lines above and below the symmetrical base pairs and by a dot at their axis of symmetry.

The entire *lac* operon can be inserted into λ phage in such a way that the initiation of transcription is prompted by the phage gene adjacent to *lac*. In this case, the transcribed product carries a complementary copy of the base sequence from the *lac* control regions (sequences not transcribed in the *lac* mRNA). We already know the amino acid sequences for the repressor and β-galactosidase, so these sequences can be identified, and the remaining sequences can be assigned to the control regions (Figure 14-39). We can also fit the known repressor and CAP–cAMP binding sites into the detailed model. It now appears that the control mechanism for this one operon is active at the level of the DNA base sequence.

The knowledge about the *lac* operon provides insight into the elegance of gene regulation. The *E. coli* cell normally processes glucose as a source of energy and carbon. It possesses an "emergency" capability to process lactose, but it does not waste energy or materials in preparing the mechanism for that processing so long as glucose is available, even if lactose is also present. This control is accomplished because a glucose-breakdown product inhibits formation of the CAP–cAMP complex that is required for attachment of RNA polymerase at the *lac* promoter site. Even when there is a shortage of glucose

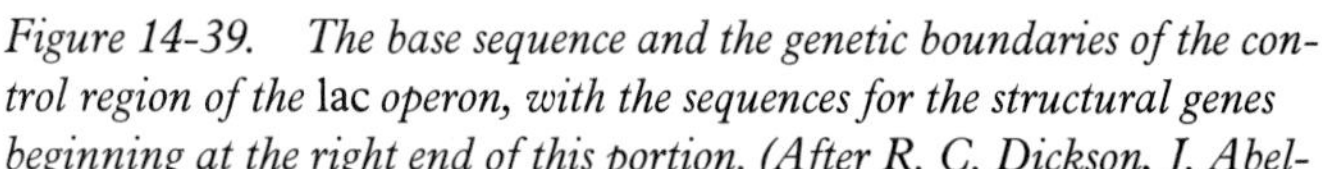
Figure 14-39. The base sequence and the genetic boundaries of the control region of the lac *operon, with the sequences for the structural genes beginning at the right end of this portion. (After R. C. Dickson, J. Abel-*

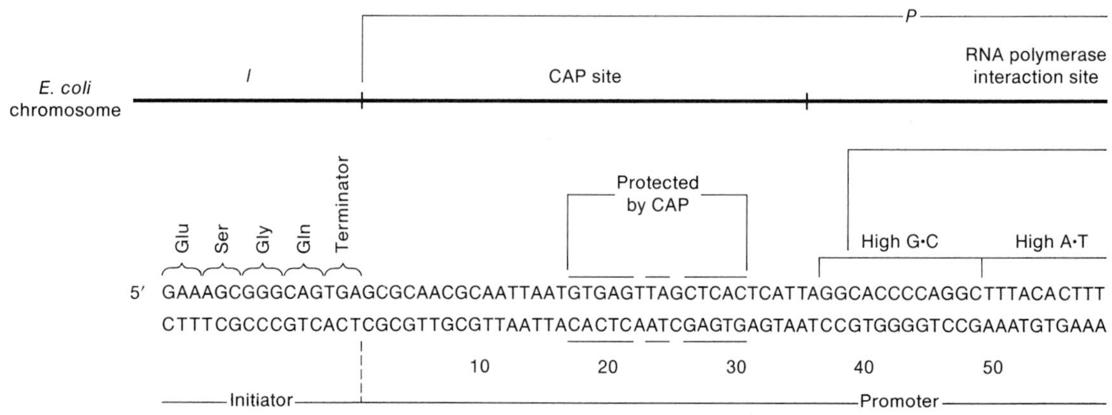

catabolites and CAP–cAMP forms, the mechanism for lactose processing will be created only if lactose is present. This control is accomplished because lactose must bind to the repressor protein to remove it from the operator site and permit transcription of the *lac* operon. Thus the cell conserves its energy and resources by producing the lactose-processing enzymes only when they are both needed and useful.

Message
The lac *operon has an added level of control so that the operon remains inactive in the presence of glucose even if lactose is also present. A high concentration of glucose catabolites produces low concentrations of cyclic AMP, which must form a complex with CAP to permit induction of the* lac *operon.*

The *Tryptophan* Genes: Negative Control

The *lac* operon is an example of **positive control,** in the sense that synthesis of an enzyme is induced by the presence of its substrate. **Negative control** systems also exist, in which an excess of product leads to a shutdown of the production of enzymes involved in synthesizing that product. Such a control system has been identified for a cluster of genes controlling enzymes in the pathway for tryptophan production. Synthesis of tryptophan is shut off when there is an excess of tryptophan in the medium. Jacob and Monod suggested

son, W. M. Barnes, and W. S. Reznikoff, "Genetic Regulation: The Lac *Control Region."* Science *187:27, 1975. Copyright 1975 by the American Association for the Advancement of Science.)*

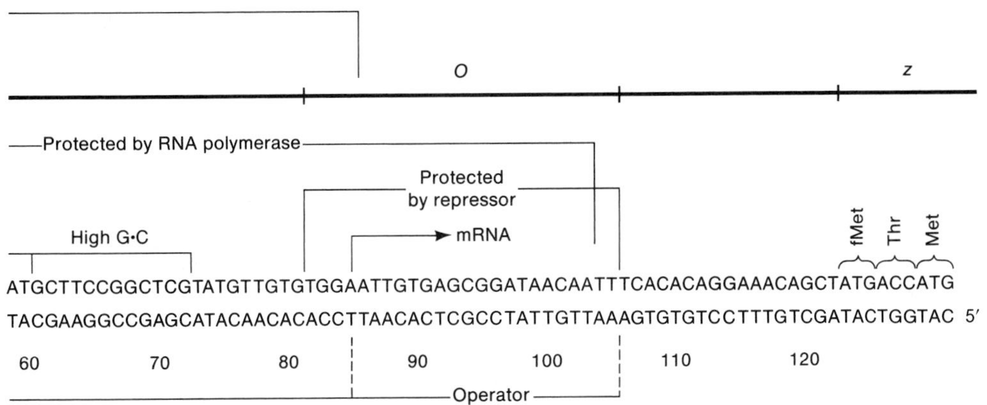

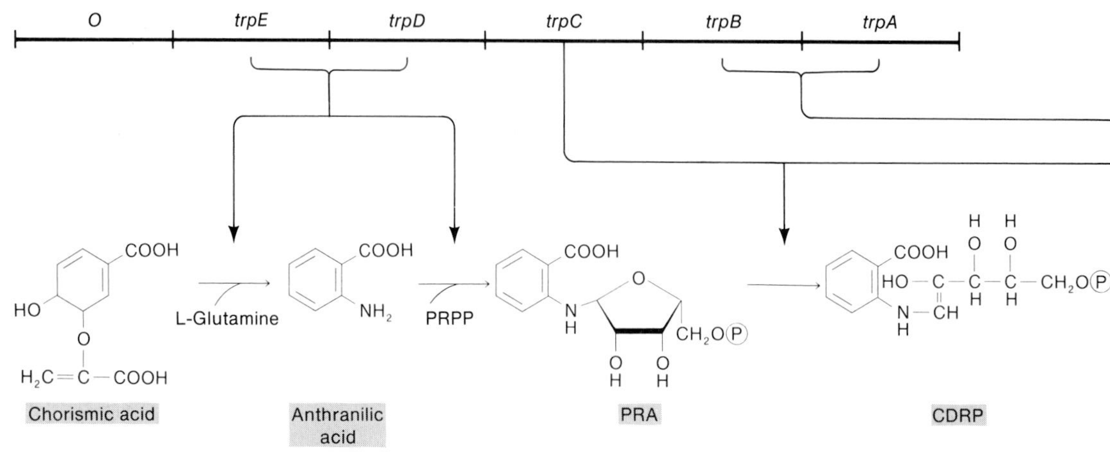

Figure 14-40. The genetic sequence of cistrons in the trp *operon of* E. coli, *and the sequence of reactions catalyzed by the enzyme products of the* trp *structural genes. The products of genes* trpD *and* trpE *form*

that the cluster of five *trp* cistrons in *E. coli* forms another operon, differing from the *lac* operon in that the tryptophan repressor will bind to the *trp* operator only when it *is* bound to tryptophan (Figure 14-40). (Recall that the *lac* repressor binds to the operator except when it is bound to lactose.)

As with the *lac* operon, further analysis of the *trp* operon revealed another level of control superimposed on the basic repressor–operator mechanism. Charles Yanofsky was studying constitutive mutant strains (carrying a mutation in *trpR*, the repressor locus) that continue to produce *trp* mRNA in the presence of tryptophan. Yanofsky found that removal of tryptophan from the medium leads to almost a tenfold increase in *trp* mRNA production in these constitutive mutant strains. Furthermore, Yanofsky isolated a constitutive mutant strain that produces *trp* mRNA at the maximal level even in the presence of tryptophan, and he showed that this mutation has a deletion located between the operator and the *trpE* cistron.

Yanofsky was able to isolate the polycistronic *trp* operon mRNA. Upon sequencing it, he found a long sequence of 160 bases at the 5′ end before the first triplet in the *trpE* gene. The deletion mutant that always produces *trp* mRNA at maximal levels has a deletion extending from base 130 to base 160 (Figure 14-41). Yanofsky called this the **attenuator** region, because its presence apparently leads to a reduction of the rate of mRNA transcription when tryptophan is present. But what is the role of the leader sequence from bases 1 through 130? A surprising observation provided the key to solving this problem.

When studying the mRNAs transcribed from the *trp* operon, Yanofsky dis-

InGP

L-Serine

Indole

L-Tryptophan

a complex that catalyzes specific steps, as do the products of genes trpB *and* trpA. *(After S. Tanemura and R. H. Bauerle,* Genetics, *in press.)*

covered that the original constitutive mutant strains continue to produce the first 141 bases of the mRNA at maximal rate, even in the presence of tryptophan. In other words, the segment from base 141 through base 160 apparently normally acts as a chain terminator to halt transcription of about 9 in every 10 mRNAs if tryptophan is present. When tryptophan is absent, every transcription is carried through this attentuator region and on to completion. In the deletion mutants, the attenuator region is missing, so transcription is carried through in every case regardless of the presence or absence of tryptophan.

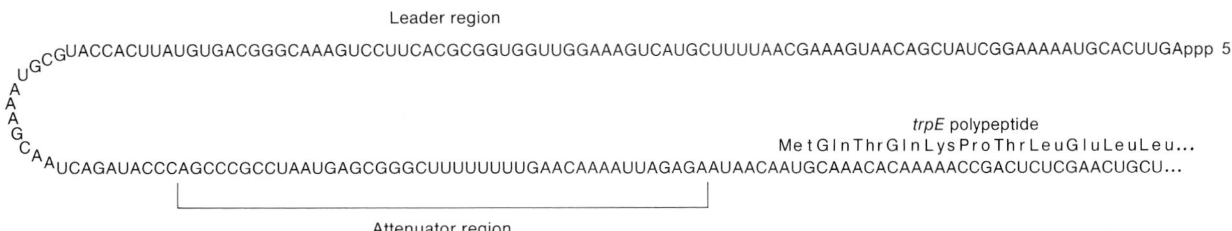

Figure 14-41. The leader sequence and attenuator sequence of the trp *operon, with the beginning of the* trpE *structural sequence (showing the amino acid sequence of the* trpE *polypeptide). (From G. S. Stent and R. Calendar,* Molecular Genetics, *2nd ed. Copyright © 1978, W. H. Freeman and Company. Based on unpublished data provided by C. Yanofsky.)*

Message

The trp *operon is regulated by a negative repressor–operator control system that represses synthesis of tryptophan enzymes when tryptophan is present in the medium. A second level of control involves an attenuator region where termination of transcription is induced by the presence of tryptophan.*

What causes the antitermination at the attenuator in the absence of tryptophan? Two other loci are involved, because mutations at either *trpS* or *rho* result in a high level of mRNA transcription in the presence of tryptophan. The exact mechanism of the attenuator control system is not yet known.

The λ Phage: A Complex of Operons

At the time they proposed the operon model, Jacob and Monod suggested that the genetic activity of temperate phages might be controlled by a system analogous to the *lac* operon. In the lysogenic state, the prophage genome is inactive—that is, repressed. In the lytic phase, the phage genes for reproduction are active—that is, induced. Since Jacob and Monod proposed the idea of operon control for phages, the λ phage has become one of the organisms whose genetic system is best understood. This phage does indeed have an operon-type system controlling its two functional states. By now, you should not be surprised to learn that this system proved to be more complex than initially suggested.

Alan Campbell induced and mapped many conditionally lethal mutations in the λ phage (Figure 14-42). There is clear evidence for the clustering of genes

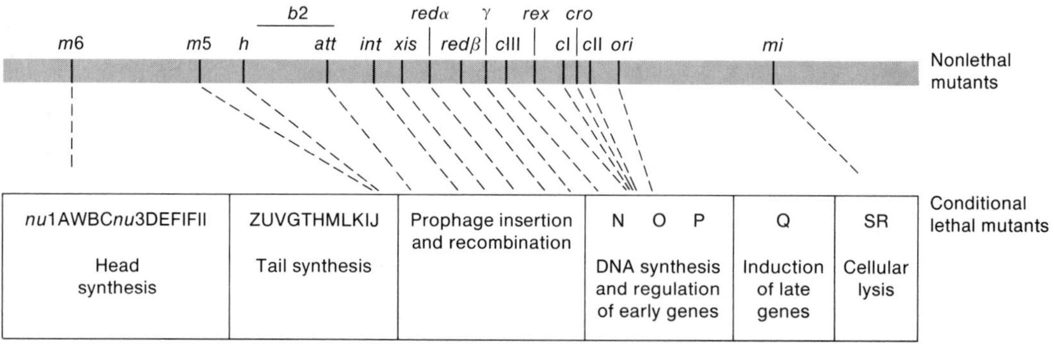

Figure 14-42. The genetic map of the λ phage. The positions of nonlethal and conditionally lethal mutations are indicated, and the characteristic clusters of genes with related functions are shown. (From A. Campbell, The Episomes. *Harper & Row, 1969.)*

with related functions. Furthermore, mutations in *N*, *O*, or *P* prevent most of the genome from being expressed after phage infection, with only those loci lying betwen *N* and *O* being active. We shall soon see the significance of this observation.

When normal bacteria are infected by wild-type λ phage, two possible sequences may follow: (1) a phage may be integrated into the bacterial chromosome as an inert prophage (thus lysogenizing the bacterial cell), or (2) the phage may produce the necessary enzymes to guide production of products needed for phage maturation and cell lysis. When wild-type phage particles are placed on a lawn of sensitive bacteria, clearings (plaques) appear where bacterial cells are infected and lysed, but these plaques are turbid because lysogenized bacteria (which are resistant to phage superinfection) grow within the plaques.

Mutant phages that form clear plaques can be selected as a source of phages that are unable to lysogenize cells. Such *clear* (*c*) mutants prove to be analogous to *i* and *O* mutants in *E. coli.* For example, conditional mutants for a site called *cI* are unable to establish a lysogenic state under restrictive conditions, but they fail to induce lysis in a cell that has been lysogenized by a wild-type prophage. Apparently, the *cI* mutation produces a defective repressor in the phage control system.

Virulent mutants were isolated that do not lysogenize cells but will grow in a lysogenized cell, thus providing defects that are insensitive to the λ repressor. Genetic mapping revealed *two* operators designated O_L and O_R located on the left and right of *cI*, respectively. Furthermore, each operator has a promoter; the promoters are called P_L and P_R. Figure 14-43 shows the genetic map of the phage control regions. Because transcription extends from the *O* regions away from *cI*, the transcripts must be read from different strands of the DNA.

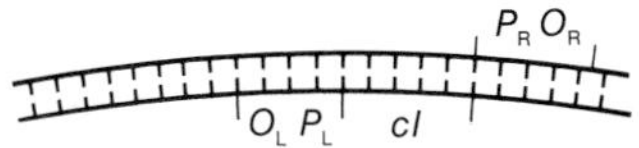

Figure 14-43. Genetic map of the control elements of λ reproduction on the double-stranded phage DNA. Each symbol is placed next to the strand that is transcribed for the corresponding gene.

Mark Ptashne was able to purify the λ repressor. He showed that each λ genome binds six repressor molecules, three at each operator. Furthermore, the repressor-binding sites of O_R overlap with the promoter for the *cI* gene itself.

When a phage infects a cell, there is no repressor present. A host-cell enzyme ("*transcriptase*") initiates transcription, beginning at O_L and extending through the *N* gene, and beginning at O_R and extending through the *cro* gene (Figure 14-44). Now we see why the *N* gene is important: its product interacts with the host transcriptase to modify it in such a way that transcription will not terminate in an attenuator region at the ends of *N* and *cro*. Instead, transcription proceeds through to the *int* and *P* genes on the left and right, respectively (Figure 14-45). These genes are involved in phage DNA replication, recombination, and lysogeny. The name of the *cro* gene is derived from the expression "*c*ontrol of *r*epressor and *o*ther things." The *cro* gene product acts to inhibit production of the repressor by binding to a promoter region adjacent to the *cI* locus. Thus, the *cro* product appears to be a repressor of a repressor, and it must act during the lytic phage of λ growth. On the other hand, λ repressor binds to O_R, thus shutting off transcription of the *cro* gene. Thus *cro* and *cI* activity must be mutually exclusive, with *cro* activity required for lysis and *cI* activity required for lysogeny.

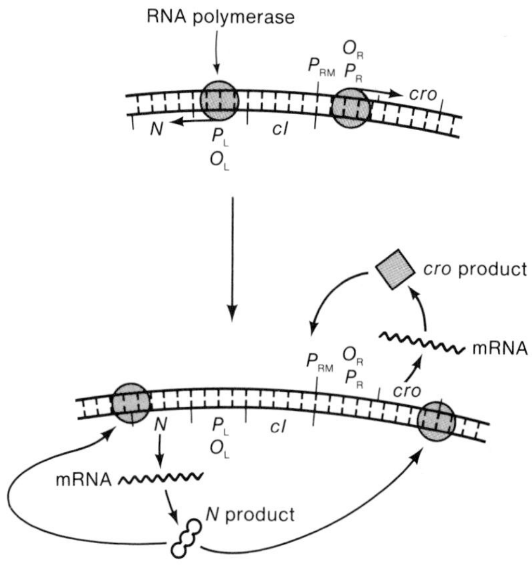

Figure 14-44. Initial transcriptional events immediately after λ infection. Bacterial RNA polymerase initiates transcription to the left and right of cI *at* P_L *and* P_R. *The product of the* N *gene attaches to the RNA polymerase, thereby modifying it so that it no longer terminates transcription at the ends of* N *and* cro. *The* cro *product binds at* P_{RM}, *the promoter for* cI, *and therefore shuts off* cI.

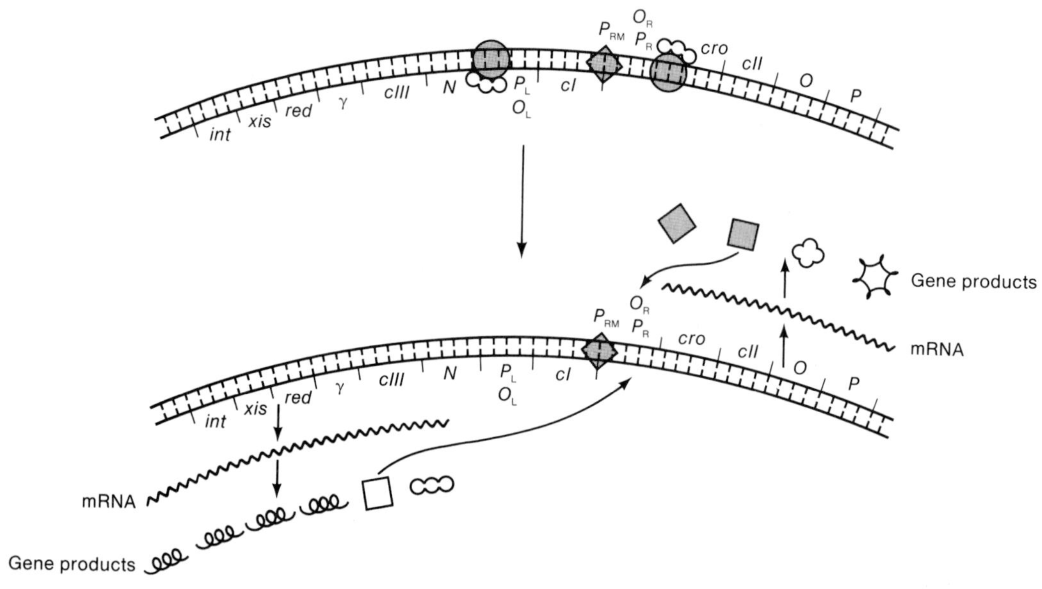

Figure 14-45. Second stage of transcription following phage infection. The cI *gene is shut off because of the* cro *product binding to it. Transcription proceeds beyond* cro *and* N *through genes concerned with DNA replication, recombination, and lysogeny. The* cII *and* cIII *gene products act to initiate transcription of* cI *at a position closer to* cro *than* P_{RM}.

The complexity of the λ cycle only increases with further study. With the aid of the *N* gene product, genes *cII* and *cIII* are activated. The products of these *c* genes enable RNA polymerase to initiate *cI* transcription from a different promoter, P_{RE}. This site is much more effective at producing repressor than is the P_{RM} site used initially. But now we have the *cro* product trying to repress *cI* while the *cII* and *cIII* products act to increase *cI* activity. It is still not clear what determines whether the phage will enter a lytic or a lysogenic phase, but for our purposes we can regard the decision as the outcome of a "race" between the *cI* and the *cro* genes.

Let's assume that the *cro* gene wins the race and shuts off *cI*. Then the *O*, *P*, and *Q* products act to initiate transcription of the rest of the genome, starting at P_{LATE} (the promoter of late genes). The description we have given here is an abbreviated version of a well-characterized process that provides further modifications of the operon model.

Message

The regulation of the lytic and lysogenic states of the λ phage provides a model for interacting control systems that may be useful for interpreting gene regulation in eukaryotes.

Functionally Related Genes in Eukaryotes

The operon model and the close linkage of functionally related genes in prokaryotes indicate the importance of gene position within a linkage group in those organisms. But do the same relationships exist in eukaryotes? It is clear that, at the level of chromosomes, groupings such as the Y and satellite chromosomes and the heterochromatic segments are special and exceptional. In euchromatin, there are few obvious blocks of genes related by function or by time of action. For example, in *Drosophila,* we do not find all the genes affecting a wing or all those acting in a certain stage of embryonic development clustered together in one part of the genome. Although the loci for tRNAs also appear to be widely distributed through the *Drosophila* genome, duplicate genes for ribosomal RNA and histone and those for purine synthesis are clustered. The best-studied systems are in yeast and *Neurospora.*

To date, there has been no conclusive demonstration of the existence of an operon in any eukaryote. Most attempts to find operons have been carried out on lower eukaryotes, particularly on fungi. We can summarize such studies very briefly.

In some cases, functionally related genes are found clustered next to each other. For example, the *ad-3A* and *ad-3B* loci (which are structural genes for two successive steps of purine synthesis in *Neurospora*) are found next to each other in the genome. Also, the five genes that control five sequential steps in the synthesis of aromatic amino acids in *Neurospora* are found next to each other in the *arom* cluster and are called *arom-2* through *arom-6*. At first glance, these two examples seem to represent possible operons. But read on.

By far the most common eukaryotic situation is that in which functionally related genes are not adjacent. For example, in the bacterium *Salmonella,* the nine genes that control nine adjacent steps in histidine synthesis are linked together in an operon, but those same genes are scattered throughout the various linkage groups of *Neurospora.*

Even when genes are grouped, the requirements for an operon may not be fulfilled. For one thing, coordinate control by means of an operator has not been demonstrated in any eukaryotic case. However, many examples have been demonstrated of mutations that seem similar to the bacterial polar mutations (which inactivate all adjacent cistrons on one side of the mutation)—for example, in the *arom* cluster. However, as explained in the next paragraph, even these observations are subject to an alternative explanation. In some cases (such as *ad-3A* and *ad-3B*), there is *no* evidence of polar mutants to indicate transcriptional unity for the loci.

The interpretation of eukaryotic gene control is further complicated by the demonstrated occurrence of large superproteins that are encoded by a single region and yet have more than one enzyme function. This is the situation at the *arom* region. Such a region can mimic the properties of an operon because a point mutation (a stop mutation or a missense mutation causing a major conformational distortion) in one function can affect the other functions. Thus, a single cistron can easily be misinterpreted as a linked group of cistrons coding for several polypeptides by way of a single mRNA molecule. For example it is not yet known whether the *arom* cluster is an operon or a single giant gene; all of the enzyme activities are in fact controlled by a single sequence of DNA, but either alternative explanation can be invoked to explain the observations currently available.

Perhaps the *arom* cluster simply represents a way of ensuring coordinate protein function but without an operon control system, or perhaps it is a single gene producing a giant protein that channels metabolic intermediates from one enzymic site to the next like a bucket brigade, or perhaps the cluster actually *is* an operon (Figure 14-46).

There are good examples of regulatory genes in eukaryotes that can be separated by recombination from their structural loci. The great majority of such regulatory genes act in such a way that the mutated regulatory gene can no longer activate its structural gene (or genes).

Message

No operon control system of regulated gene activity has yet been conclusively demonstrated in eukaryotes, but there are numerous candidates that may represent such systems.

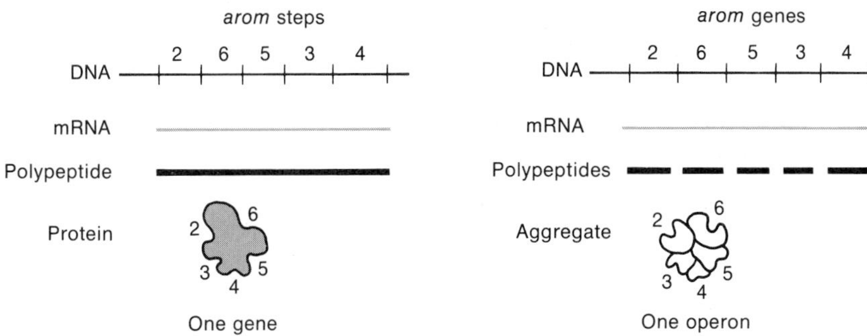

Figure 14-46. Alternative explanations for the activity of the arom *gene complex in* Neurospora. *The locus may be a single gene producing a large polypeptide with various enzymic sites that catalyze the sequence of reactions. Thus, a polar mutation in one of the catalytic sites would prevent that step and all following steps in the sequence. On the other hand, the* arom *locus could be a genuine operon whose polypeptide products aggregate to form a single complex that catalyzes the various reaction steps.*

Control of Ubiquitous Molecules in Eukaryotic Cells

Several classes of molecules are found in abundance in every eukaryotic cell: histones, the translational apparatus, membrane components, and so on. Several strategies have evolved to maintain a sufficient amount of these products in the eukaryotic cell. These strategies include continuous and repeated transcription throughout the cell cycle, repetition of genes, and extrachromosomal amplification of specific gene sequences.

Genetic Redundancy. Genes coding for rRNA and tRNA have been extensively analyzed because the availability of purified gene products provides a probe with which to recover complementary DNA. Simple DNA/RNA hybridization shows that each *Xenopus* chromosome carrying a nucleolus organizer has 450 copies of the DNA coding for 18S and 28S rRNA. In contrast, there are 20,000 copies in each nucleus of the genes coding for 5S rRNA, and these genes are not located in the NO region. Donald Brown and his collaborators during the 1970s have analyzed the NO DNA extensively and have shown that it carries tandem duplications of a large (40S) transcription unit separated from its neighbors by nontranscribed **spacers.** The initial transcript is then processed to release smaller rRNAs finally found in the ribosome.

As we saw in Chapter 13, Max Birnstiel and his associates have shown that the five histone genes in sea urchins also are found as a tightly linked unit (Figure 13-3). There are several hundred copies of a repeating unit with the genes in the sequence *H4–H2B–H3–H2A–H1*. Transcription begins at *H4* and ends at *H1*; each gene is transcribed as a single message. As we have seen in earlier chapters, tandem duplications can pair asymmetrically between homologous or within the same chromosome; exchanges then increase or reduce the number of copies. It remains to be seen how (or whether) the number of tandem repeats is kept constant in a particular species.

Gene Amplification. Specific gene amplification has been studied extensively in the case of the rRNA genes in amphibian oocytes. As we have noted, each chromosome of *Xenopus* carries about 450 copies of the DNA coding for 18S and 28S rRNA. However, the oocyte contains up to 1000 times this number of copies of these genes. The oocyte is a very specialized cell that is loaded with the nutritive material needed to maintain the embryo until the tadpole stage without any ingestion of food. The cytoplasm also is prepared with the translational apparatus needed to carry out the complex program of differentiation in the many cells of the developing embryo.

During oogenesis, the maturing oocyte increases greatly in size as material is poured in from adjacent nurse cells. (Amazingly, so many ribosomes are present in the egg at fertilization that *an/an* homozygotes carrying no genes for 18S and 28S rRNA will develop and differentiate to the twitching tadpole stage before death.) The oocyte nucleus also contributes material as the cell proceeds to the first meiotic prophase, where further develoment is arrested. The amphibian meiotic chromosomes are enormous in size, with numerous lateral loops representing regions of intense genetic activity (Figure 14-47). In the oocyte nucleus, there are hundreds of extrachromosomal nucleoli of varying size. Each nucleolus contains a ring of rDNA of different size that has been replicated and released from the chromosome. The steps involved in this process of DNA amplification are completely unknown. The DNA rings

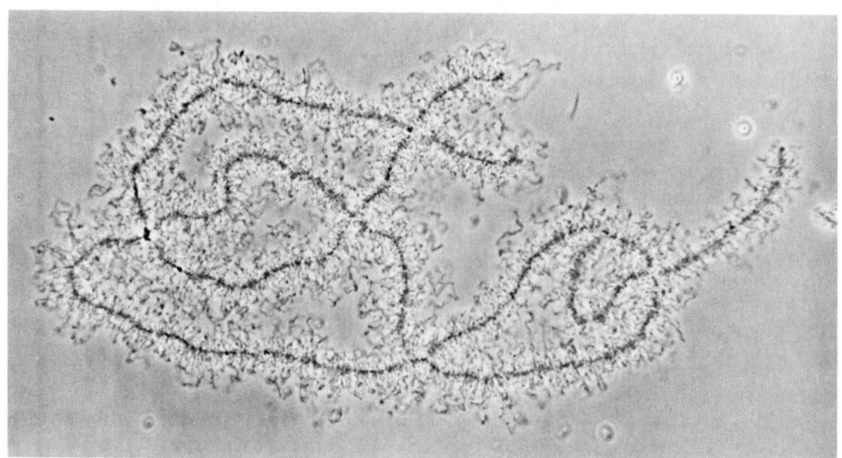

Figure 14-47. The "lampbrush" chromosomes of an amphibian oocyte. (Photograph courtesy of J. Gall.)

actively produce rRNA that is assembled into ribosomes, which are stored in the nucleolus until their release during meiosis.

Other examples of genetic amplification are known. Specific puffs of the polytene chromosomes of Diptera (such as the giant chromosomes from *Drosophila* salivary glands) are found to produce excess DNA rather than mRNA. Indeed, George Rudkin showed in the early 1960s that the polytene chromosomes themselves may represent specific amplification of euchromatic DNA about a thousandfold, while the proximal heterochromatic regions may be replicated only a few times. Thus, in addition to the mechanism for replicating chromosomal DNA normally for cell division, there exists means for amplifying specific loci and not others.

Message
Cellular mechanisms exist to ensure an adequate supply of gene products that are vital to all cells. These mechanisms include increasing the number of of gene copies per chromosome (redundancy) and increasing the number gene copies in the cell (amplification).

Finally, we must remember that any overall scheme for chromosome structure must take into account the large amounts of repetitive DNA of unknown function, some interspersed between the unique DNA, and some in highly repetitive clusters.

Summary

Each eukaryotic chromosome apparently represents a single continuous DNA molecule extending from one end through the centromere to the other end. The genome of *Drosophila* includes 5000 to 6000 genes detectable by mutations, but it has 25 to 30 times as much unique DNA as would be needed for this number of *E. coli*-sized genes. This discrepancy may be explained by large nontranscribed spacers between genes, nontranslated sections on either end of transcripts, and large intervening sequences within genes.

The molecule of DNA may be 10^5 times as long as the chromosome itself, and it is apparently packaged in a very regular and complex fashion within the chromosome. The packaging begins with the nucleosome, an aggregate of four pairs of histones around which the DNA is coiled twice. Chains of nucleosomes are then coiled, the coils in turn are coiled, and so on.

In eukaryotic chromosomes, differences in gene activity can be recognized by the degree of compaction (which is indicated by staining intensity). Densely staining heterochromatin represents nonactive material. Within a chromosome, heterochromatic regions are typically found at the tips, around the centromere, and around the nucleolus organizer. Such constitutive heterochromatin is generally highly redundant and relatively devoid of sequences for which RNA transcripts may be found, and it contains sequences of satellite DNA.

In some cases, a chromosome may exist in one of two states: functionally active or functionally inactive. Constitutive heterochromatin represents entire chromosomes or specific segments which are predictably heterochromatized and inactive during the cell cycle. In other cases, a chromosome or segment may or may not become heterochromatin—when it does, it is referred to as facultative heterochromatin. Once the functional state of the chromosome is determined, that state is replicated from cell to cell in further reproduction. Early in the development of a mammalian female, one of the two X chromosomes in each cell is inactivated (becoming a heterochromatic Barr body), thereby rendering the cell functionally equivalent to a male cell (which contains only one X chromosome). In contrast, organisms like *Drosophila* compensate for the double dose of sex-linked genes in females (compared to males) by a 50% reduction in gene activity on the female X chromosomes.

The regulation of gene activity is related to the position of a gene in the chromosome. In phages, this position effect is graphically illustrated by the clustering of genes that are related by their times of action and/or their functions. In bacteria, genes coding for enzymes in the same metabolic pathway typically are tightly linked and often appear in the same sequence on the chromosome as the sequence of reactions their products catalyze in the metabolic pathway.

The operon model explains how prokaryotic genes are controlled through a mechanism that coordinates the activity or inactivity of a number of related genes. The initiation of transcription is controlled at the operator by a repressor whose binding affinities may be altered by inducer molecules. Superimposed on this simple operator-repressor control system are more subtle backup controls: catabolite repression (in the *lac* operon), modulation of transcription by an attenuator sequence (in the *trp* operon), multiple binding sites for repressors (in the λ phage), and repressors of repressors (also in λ). No operons have yet been proven to exist in eukaryotes.

In eukaryotes, specific DNA segments may be present as tandem duplications within the chromosome or as amplifications that represent extra copies transcribed and then released from the chromosome.

Problems

1. Which of the following is *not* an advantage from having genes arranged in chromosomes rather than floating free?
   a. Favorable combinations of genes can be inherited more or less as a unit.
   b. Functionally related genes can be controlled in a coordinated fashion.
   c. DNA can pass through the nuclear pore into the cytoplasm.
   d. Orderly separation of genetic information can occur at cell division.
   e. Replication of DNA can occur in orderly fashion.
2. The mealybug, *Planococcus citri,* has a diploid number of 10. The meiotic divisions in *P. citri* are thought to be reversed—that is, an equational division is

followed by a reductional division. In the tetrads formed at meiosis I, it is difficult to distinguish sister chromatids from their paired homologs because they form a tightly paired unit. Sharat Chandra obtained a triploid strain of *P. citri.* What would you expect to observe at the end of meiosis I and II in this strain (a) if the division sequence is conventional? (b) if the division sequence is reversed?

3. Considering the length of each DNA molecule in a eukaryote and the levels of complexity in its packaged form in a chromosome, the mechanism of pairing between homologous sequences becomes staggering. How could pairing between chromosomes possibly take place so that specific base pairs are properly aligned?

4. In 1917, Alfred Sturtevant noted that crossing over in any chromosome pair in a *Drosophila* female is increased if the female is heterozygous for an inversion in some different pair. This is called the "interchromosomal effect" on crossing over. It has since been shown that an inversion of one chromosome always increases crossing over in the others. Devise a hypothesis to explain this effect. Design an experimental test of your hypothesis.

5. In 1956, R. Alexander Brink noted some aberrant results from crosses with corn involving alleles of the *R* locus, which affects the color pattern in seeds. The alleles studied (and their corresponding phenotypes) were r^{r} (colorless seed), R^{r} (dark purple seed), and R^{st} (stippled seed, having irregular spots on a light background). In the cross $R^{r}r^{r} \times r^{r}r^{r}$, Brink obtained 50% dark and 50% colorless seeds. In the cross $R^{st}r^{r} \times r^{r}r^{r}$, he obtained 50% stippled and 50% colorless seeds. In the cross $R^{r}R^{st} \times r^{r}r^{r}$, he obtained 50% stippled and 50% "weakly colored" seeds. Apparently, in the $R^{r}R^{st}$ heterozygote, the R^{r} alleles change to a new and stable allelic state labeled R'. This effect is called **paramutation.** How does paramutation differ from mutation? Devise a hypothesis to explain the results. Design an experimental test of your hypothesis.

6. In embryos of the fungus gnat, *Sciara,* after the seventh cell division, the paternally inherited X chromosomes in all somatic cells migrate through the nuclear membrane and disintegrate in the cytoplasm. Helen Crouse studied the properties of a series of X–autosome translocations with the following X-chromosome breakpoints:

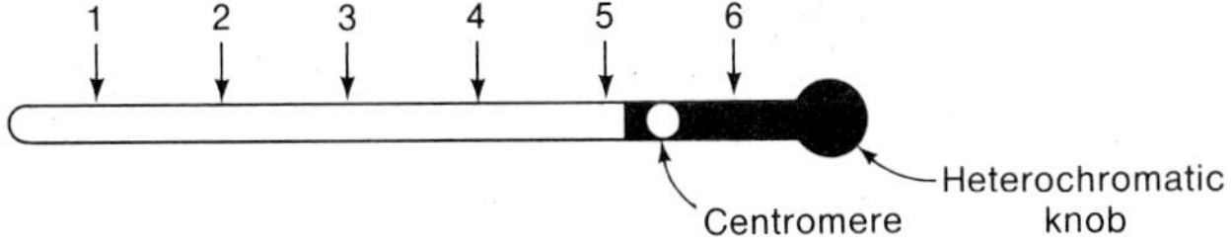

She found that, when males contributed the translocations to the offspring, the translocated portion carrying the X centromere in rearrangements 1 through 5 was lost during embryonic development. However, with translocation 6, the autosomal centromere carrying the right arm of the X was lost. Suggest an interpretation of these results.

7. Erich Wolff has studied the polytene chromosomes of two strains of a European midge, *Phryne cincta.* One strain (Berlin) has no heterochromatic elements on the X chromosome, whereas another strain (alpine) has two heterochromatic sites at which large amounts of DNA are produced and released into a nucleoluslike structure.

a. Design experiments to study the nature of the DNA in the heterochromatic sites. (NOTE: The strains will mate with each other, and fertile F_1 progeny are obtained.)

b. Wolff also has observed that some *Phryne* strains carry up to seven supernumary chromosomes that are not present in the normal genome. How would you study the DNA in these supernumary chromosomes?

8. a. When DNA from *Drosophila* is isolated from salivary-gland nuclei and from testes and spun in a CsCl gradient, the distributions obtained are those shown in Figure 14-48a. What does this result suggest?

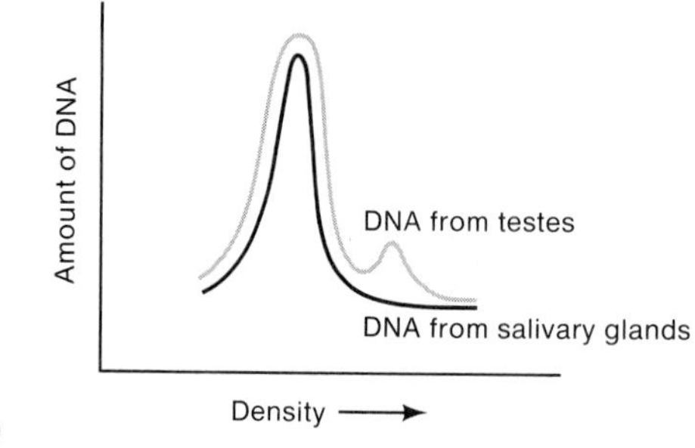

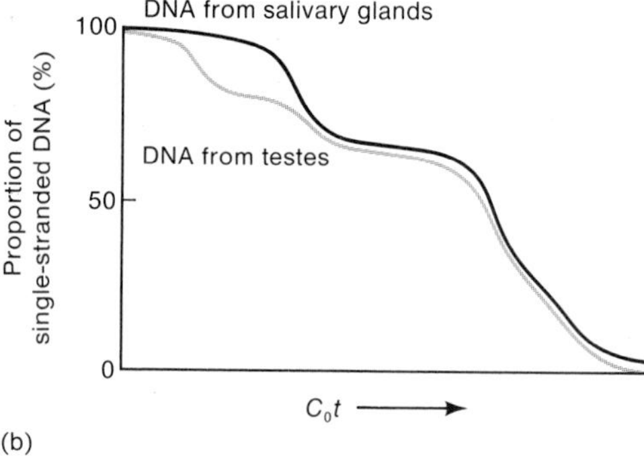

Figure 14-48.

b. When the DNA from the two tissues is heat denatured and then allowed to reanneal, the Cot curves obtained are those shown in Figure 14-48b. What does this result suggest?

9. Joseph Gall and Mary Lou Pardue recognized that polytene chromosomes are like columns of DNA laid out in proper sequence. They developed a method for denaturing the DNA in these chromosomes without disrupting the spatial organization of the chromosomes. By incubating radioactive RNA on the denatured chromosomes and then removing the unbound RNA, they were able to localize the RNA/DNA hybrids by autoradiography. (This technique was described in Chapter 13.)

a. What parameters would affect the resolving power of this technique—that is, the accuracy with which two adjacent sites can be distinguished?
b. How would you demonstrate that such RNA binding does indeed reflect complementary sequences of DNA?
c. What do you infer when a certain RNA binds to a certain locus?

10. Highly redundant DNA sequences in *Drosophila* are concentrated in heterochromatin around all of the centromeres. However, in relation to their DNA content, these same regions have a disproportionately small number of loci that can be detected genetically. For that reason, Painter and Müller suggested that this proximal heterochromatin is "inert." Furthermore, very little crossing over occurs within the heterochromatin. What biological function might be filled by these redundant sequences? Your hypothesis should explain the absence of detectable loci and the nonrandom distribution of the heterochromatin. (NOTE: This is an unsolved problem. No one knows the correct answer, although there are many hypotheses.).

11. When tandemly duplicated DNA is sequenced, the duplicate segments are identical. Suggest a mechanism that might preserve the exact duplication of these segments over time against the effects of random mutations.

12. When lampbrush chromosomes (Figure 14-47) are treated with restriction enzymes, a small number of the lateral loops are each cut into several pieces. What do these results suggest?

13. You are a molecular biologist on a field trip in the jungles of Surinam, where you discover a purple tree frog with yellow strips. Struck by its beauty, you capture several dozen males and females and smuggle them back to your laboratory. They breed easily and profusely, and you decide to use them in an investigation of the organization of the ribosomal RNA genes. Design experiments to carry out this investigation without the need for genetic crosses.

14. The genes shown in Table 14-5 are from the *lac* operon sysem of *E. coli.* The symbols *a*, *b*, and *c* represent the repressor (*i*) gene, the operator (*O*) region, and the structural gene (*z*) for β-galactosidase, although not necessarily in that order.

Table 14-5.

Genotype	Activity (+) or inactivity (−) of *z* gene: Inducer absent	Inducer present
$a^-b^+c^+$	+	+
$a^+b^+c^-$	+	+
$a^+b^-c^-$	−	−
$a^+b^-c^+/a^-b^+c^-$	+	+
$a^+b^+c^+/a^-b^-c^-$	−	+
$a^+b^+c^-/a^-b^-c^+$	−	+
$a^-b^+c^+/a^+b^-c^-$	+	+

Furthermore, the order in which the symbols are written in the genotypes is not necessarily the actual sequence in the *lac* operon.

a. State which symbol (a, b, or c) represents each of the *lac* genes i, O, and z.

b. In Table 14-5, a superscript minus on a gene symbol merely indicates a mutant, but you know there are some special mutant behaviors in this system that are given special mutant designations. For each of the genotypes in the table, indicate the genotype using the conventional gene symbols for the *lac* operon.

(Problem 14 is from J. Kuspira and G. W. Walker, *Genetics: Questions and Problems,* McGraw-Hill, 1973.)

15. The map of the *lac* operon is

$$i \qquad\qquad p \quad O \quad z \quad y$$

The promoter (p) region is thought to be the site of initiation of transcription through the binding of the RNA polymerase molecule before actual mRNA production. Mutants of the promoter (p^-) apparently cannot bind the RNA polymerase molecule. Certain predictions can be made about the effect of p^- mutations. Use your predictions and your knowledge of the lactose system to complete Table 14-6. Insert a plus where enzyme is produced and a minus where no enzyme is produced. (NOTE: O^o is a polar mutation now known to be in z, close to the O region.)

Table 14-6.

		β-Galactosidase		Permease	
Part	Genotype	No lactose	Lactose	No lactose	Lactose
Example	$i^+p^+O^+z^+y^+/i^+p^+O^+z^+y^+$	−	+	−	+
(a)	$i^cp^-O^cz^+y^+/i^+p^+O^+z^-y^-$				
(b)	$i^+p^-O^+z^+y^+/i^cp^+O^+z^+y^-$				
(c)	$i^+p^+O^oz^-y^+/i^+p^-O^+z^+y^-$				
(d)	$i^sp^+O^+z^+y^+/i^cp^+O^+z^+y^+$				
(e)	$i^cp^+O^oz^+y^+/i^cp^+O^+z^-y^+$				
(f)	$i^cp^-O^+z^+y^+/i^cp^+O^cz^+y^-$				
(g)	$i^+p^+O^+z^-y^+/i^cp^+O^+z^+y^-$				

16. In a haploid eukaryotic organism, you are studying two enzymes that perform sequential conversions of a nutrient A supplied in the medium:

$$A \xrightarrow[E_1]{} B \xrightarrow[E_2]{} C$$

Treatment of cells with mutagen produces three different mutant types with respect to these functions. Mutants of type 1 show no E_1 function; all type 1 mutations map to a single locus on linkage group II. Mutants of type 2 show no E_2 function; al type 2 mutations map to a single locus on linkage group VIII. Mutants of type 3 show no E_1 or E_2 function; all type 3 mutants map to a single locus on linkage group I.

a. Compare this system with the *lac* operon of *E. coli,* pointing out the similarities and the differences. (Be sure to account for each mutant type at the molecular level.)

b. If you were to intensify the mutant hunt, would there be any other mutant types you would expect to find on the basis of your model? Explain.

17. In *Neurospora,* all mutants affecting the enzymes carbamyl phosphate synthetase or aspartate transcarbamylase map at the *pyr-3* locus. If you induce *pyr-3* mutations in ICR-170 (a chemical mutagen), you find that either both enzyme functions are lacking or only the transcarbamylase function is lacking; in no case is the synthetase activity lacking when the transcarbamylase activity is present. (ICR-170 is assumed to induce frame shifts.) Interpret these results in terms of a possible operon.

18. In *Drosophila,* an autosomal locus Adh^+ codes for alcohol dehydrogenase, an enzyme that converts alcohols into aldehydes. Wild-type flies, when fed pentenol, convert that alcohol into a toxic compound and die, whereas Adh^- mutants survive. Conversely, ethanol-fed flies die if they are Adh^- but survive if they are Adh^+ because they can convert the ethanol to a harmless aldehyde. Describe methods for the selection of mutants regulating the production and the amino acid sequence of alcohol dehydrogenase. How would you distinguish operationally among various mutant types?

19. There is a class of mutations in *Drosophila* that includes loci on all four chromosomes. These mutants are called *Minute,* and they are characterized by similar properties: recessive lethality, a dominant effect on bristles (which become slender and short) and development time (which is delayed by several days), and a *Minute* phenotype in the presence of a deficiency for any of the loci. There are at least 50 to 60 *Minute* loci. Speculate on the possible biological function of *Minute* genes. How would you study a *Minute* gene from a genetic point of view and from a molecular point of view?

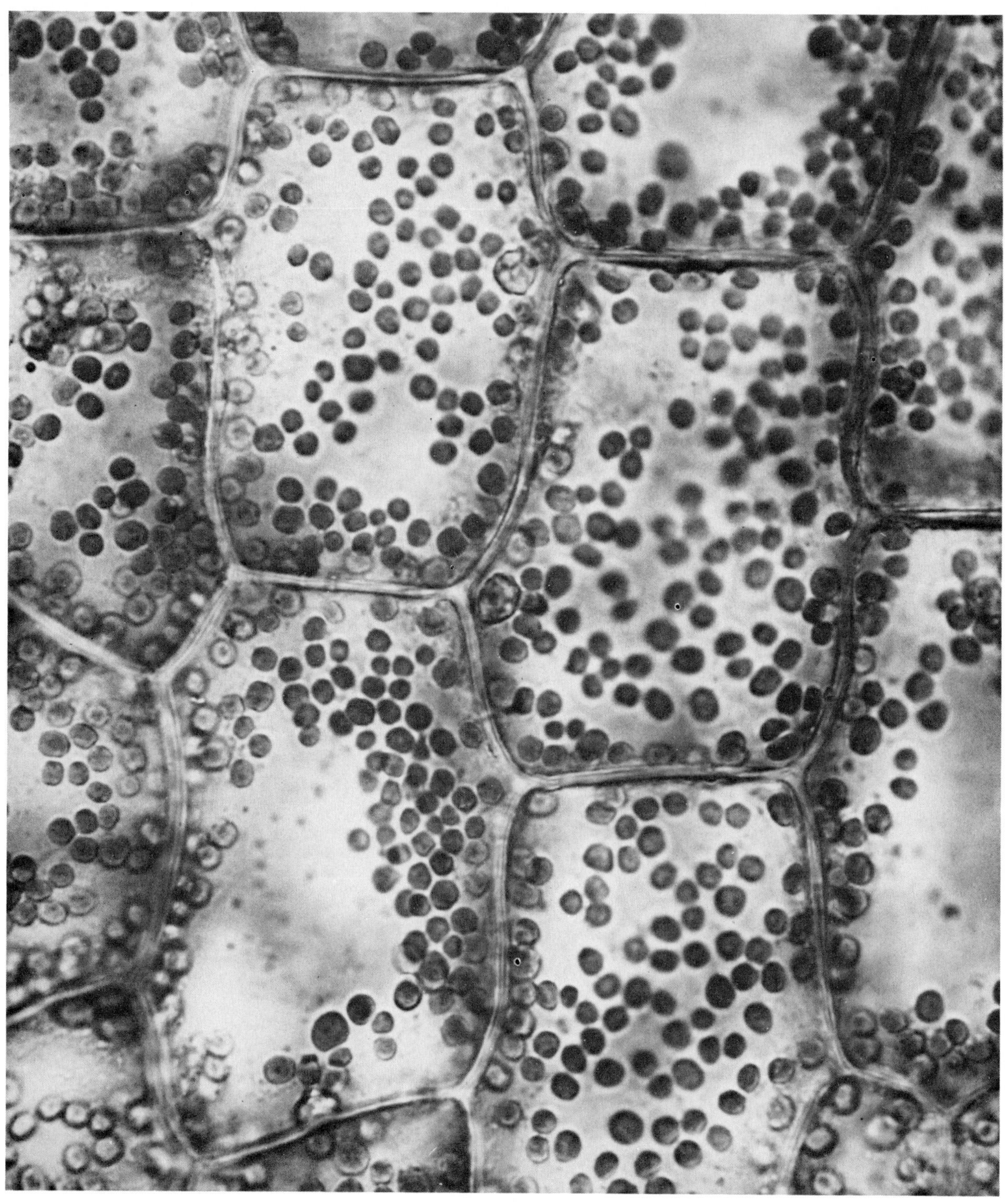

Leaf cells of the plant Elodea, *showing the large number of chloroplasts per cell (×500). (Copyright ©* Grant Heilman.)

15

Organelle Genes

In the analysis of eukaryotic organisms, the notion that genes reside in the nucleus has become a firm part of the general dogma of genetics. Indeed, there is no need to question the fundamental truth of this notion. The mechanics of the nuclear processes of meiosis and mitosis, together with the understanding that genes are located on nuclear chromosomes, provide the basic set of operational rules for genetics. In fact, the great successes of genetics in this century have been in large part due to the success of these operational rules in predicting inheritance patterns – from simple eukaryotes to human beings.

However, like most fundamental truths, this one does have exceptions, and these exceptions are the subject of this chapter. In eukaryotes, the special inheritance patterns of some genes reveal that these genes must be located outside the nucleus. Such **extranuclear genes** ("extra-" means "outside") have been found in a variety of eukaryotes. At first, the very idea of such genes seemed heretical, but their existence has gradually been accepted over the years as more and more results have turned up to confirm their reality. The inheritance patterns shown by extranuclear genes are less commonly encountered than those of nuclear genes, and indeed there are fewer extranuclear genes than normal chromosomal genes. However, although they are exceptional in this sense, it is becoming increasingly clear that extranuclear genes play normal (but highly specialized) roles in the control of eukaryote phenotypes.

In the past decade or so, intense research activity has been directed toward study of extranuclear inheritance. This activity resulted from the discovery (and subsequent general availability) of new genetic and technological approaches. The great success of these analyses has led to a vastly improved understanding of

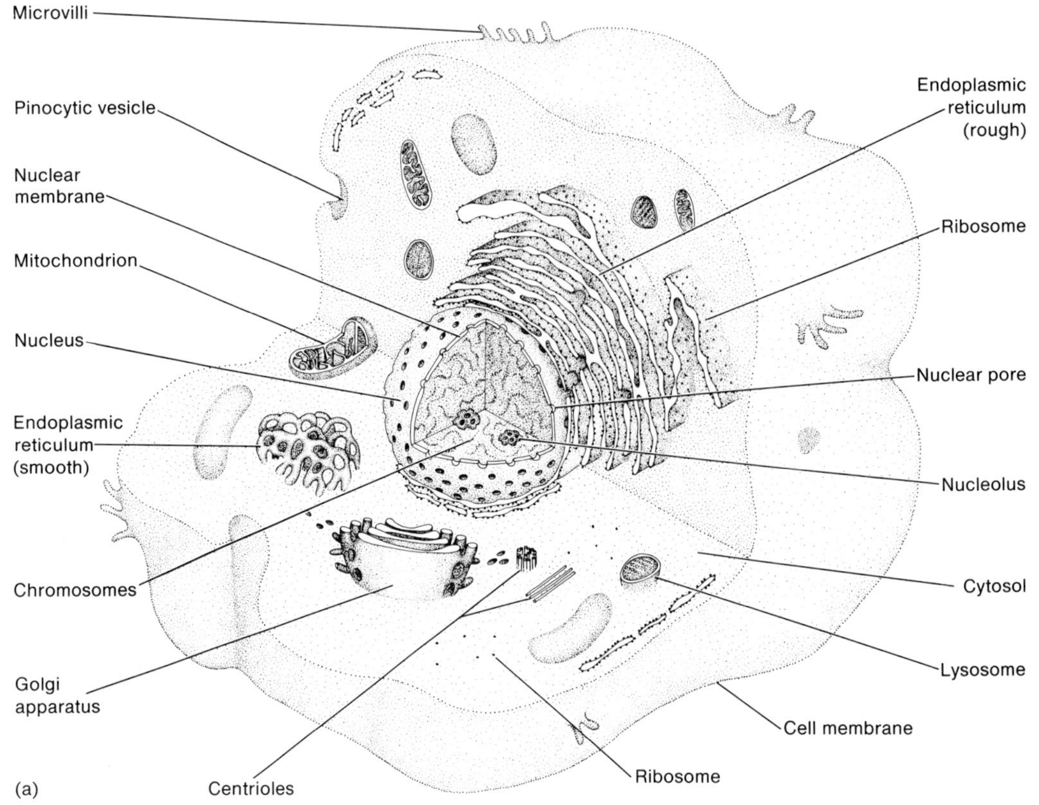
Microvilli
Pinocytic vesicle
Nuclear membrane
Mitochondrion
Nucleus
Endoplasmic reticulum (smooth)
Chromosomes
Golgi apparatus
(a)
Centrioles
Endoplasmic reticulum (rough)
Ribosome
Nuclear pore
Nucleolus
Cytosol
Lysosome
Cell membrane
Ribosome

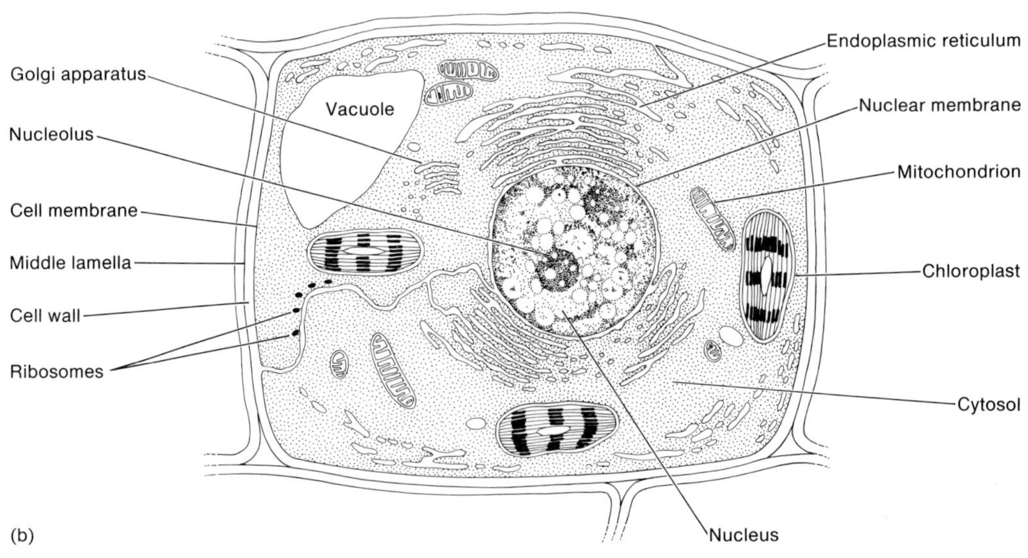
Golgi apparatus
Vacuole
Nucleolus
Cell membrane
Middle lamella
Cell wall
Ribosomes
(b)
Endoplasmic reticulum
Nuclear membrane
Mitochondrion
Chloroplast
Cytosol
Nucleus

Figure 15-1. The heterogeneity of the cytoplasm. The area called the cytoplasm (everything between the nuclear membrane and the cell membrane) embraces the cytosol (liquid phase) plus a rich assortment of organelles, membranes, and other structures. Extranuclear genes are found in the mitochondria and in the chloroplasts of green plants. (a) An animal cell. (b) A plant cell. (Part a from S. Singer and H. R. Hilgard, The Biology of People. *Copyright © 1978, W. H. Freeman and Company; part b from J. Janick et al.,* Plant Science. *Copyright © 1974, W. H. Freeman and Company.)*

the nature and function of extranuclear genes. We see in this chapter that the operational approaches in such research are often quite different from those used in studying nuclear genes (although there are similarities also). These unique analytical methods alone would merit special treatment, but their results also have revealed fascinating genetic systems that are novel and distinct enough in themselves to warrant major coverage in such a text as this.

A word of caution is in order here. A large proportion of the material presented in this chapter is based on frontier research. Although the experimental approaches to extranuclear inheritance are now well established, and a wealth of indisputable information has been accumulated, there are still many unanswered questions. For one thing, only a rather small number of species has proved suitable for the specialized techniques thus far developed. Even in the case of the well-studied organisms, we must await further research in cell biology to fill in many details not yet understood.

Conventionally, the eukaryotic cell is divided into two domains: the nucleus and the cytoplasm. The cytoplasm itself is a highly organized heterogeneous array of organelles, membranes, and various molecules in solution (Figure 15-1). Extranuclear genes are found in the mitochondria and in the chloroplasts of green plants. These organelles serve the major functions of photosynthesis and of ATP synthesis, respectively. Each organelle contains its own set of unique and autonomous genes, linked together in an organellar chromosome. Extranuclear genes often are called extrachromosomal genes—a term that is potentially confusing because the genes found in the mitochondria and chloroplasts do in fact comprise extranuclear chromosomes. Nevertheless, we must emphasize that the organellar chromosomes show inheritance patterns and organization quite different from those of the nuclear chromosomes.

A great deal is now known about the genetics and biochemistry of organelle genes. This chapter outlines some of the major steps in the evolution of our present understanding. It is a beautiful example of the power of genetic analysis when combined with chemical technology.

Message

The genes that have been called cytoplasmic genes, extrachromosomal genes, or extranuclear genes are in fact located on a unique kind of chromosome inside cytoplasmic organelles.

Variegation in Leaves of Higher Plants

One of the first convincing examples of extranuclear inheritance was found in higher plants. In 1909, Carl Correns reported some surprising results from his studies on four-o'clock plants (*Mirabilis jalapa*). The blotchy leaves of these variegated plants showed patches of green and white tissue, but some branches carried only green leaves and others carried only white leaves (Figure 15-2).

Flowers appear on all types of branches. They may be intercrossed in a variety of different combinations by transferring pollen from one flower to another. Table 15-1 shows the results of such crosses. Two features of these results are surprising. First, there is a difference between reciprocal crosses: for example, white ♀ × green ♂ gives a different result from green ♀ × white ♂. Recall that Mendel never observed differences between reciprocal crosses. In fact, in conventional genetics, differences between reciprocal crosses are normally encountered only in the case of sex-linked genes. However, the results of the four-o'clock crosses cannot be explained by sex linkage.

The second surprising feature is that the phenotype of the maternal parent is solely responsible for determining the phenotype of all progeny. The pheno-

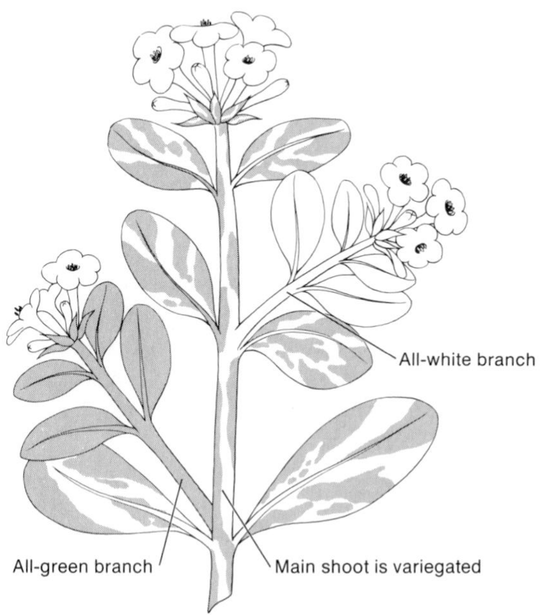

Figure 15-2. Leaf variegation in Mirabilis jalapa, *the four-o'clock plant. Flowers may form on any branch (variegated, green, or white), and these flowers may be used in crosses.*

Table 15-1. Results of crosses of variegated four-o'clock plants

Phenotype of branch bearing egg parent (♀)	Phenotype of branch bearing pollen parent (♂)	Phenotype of progeny
White	White	White
White	Green	White
White	Variegated	White
Green	White	Green
Green	Green	Green
Green	Variegated	Green
Variegated	White	Variegated, green, or white
Variegated	Green	Variegated, green, or white
Variegated	Variegated	Variegated, green, or white

type of the male parent appears to be irrelevant, and its contribution to the progeny appears to be zero! This phenomenon is known as **maternal inheritance.** Although the white progeny lines do not live for long because they lack chlorophyll, the other progeny types do survive and can be used in further generations of crosses. In these subsequent generations, maternal inheritance always appears in the same patterns as those observed in the original crosses.

How can such curious results be explained? The differences in leaf color were known to be due to the presence of either green or colorless chloroplasts. The inheritance patterns might be explained if these cytoplasmic organelles are somehow genetically autonomous and are never transmitted via the pollen parent. For an organelle to be genetically autonomous, it must have its own genetic determinants that are responsible for its phenotype. In this case, the chloroplasts would carry their own genetic determinants responsible for chloroplast color. Thus, the suggestion arises that this organelle has its own genome. The failure to be transmitted via the pollen parent is reasonable because the bulk of the cytoplasm of the zygote is known to come from the maternal parent via the cytoplasm of the egg. Figure 15-3 diagrams a model that formally accounts for all the inheritance patterns in Table 15-1.

The variegated branches apparently produce three kinds of eggs: some contain only white chloroplasts, some contain only green chloroplasts, and some contain both kinds of chloroplasts. The egg type containing both green and white chloroplasts produces a zygote that also contains both kinds of chloroplasts. In the subsequent mitotic divisions, some form of cytoplasmic segregation occurs that segregates the chloroplast types into pure cell lines, thus producing the variegated phenotype in that progeny individual (Figure 15-4).

This process of sorting might be described as "mitotic segregation." However, we wish to emphasize that this is an extranuclear phenomenon, not to be confused with mitotic segregation of nuclear genes, so we shall invent a new term.

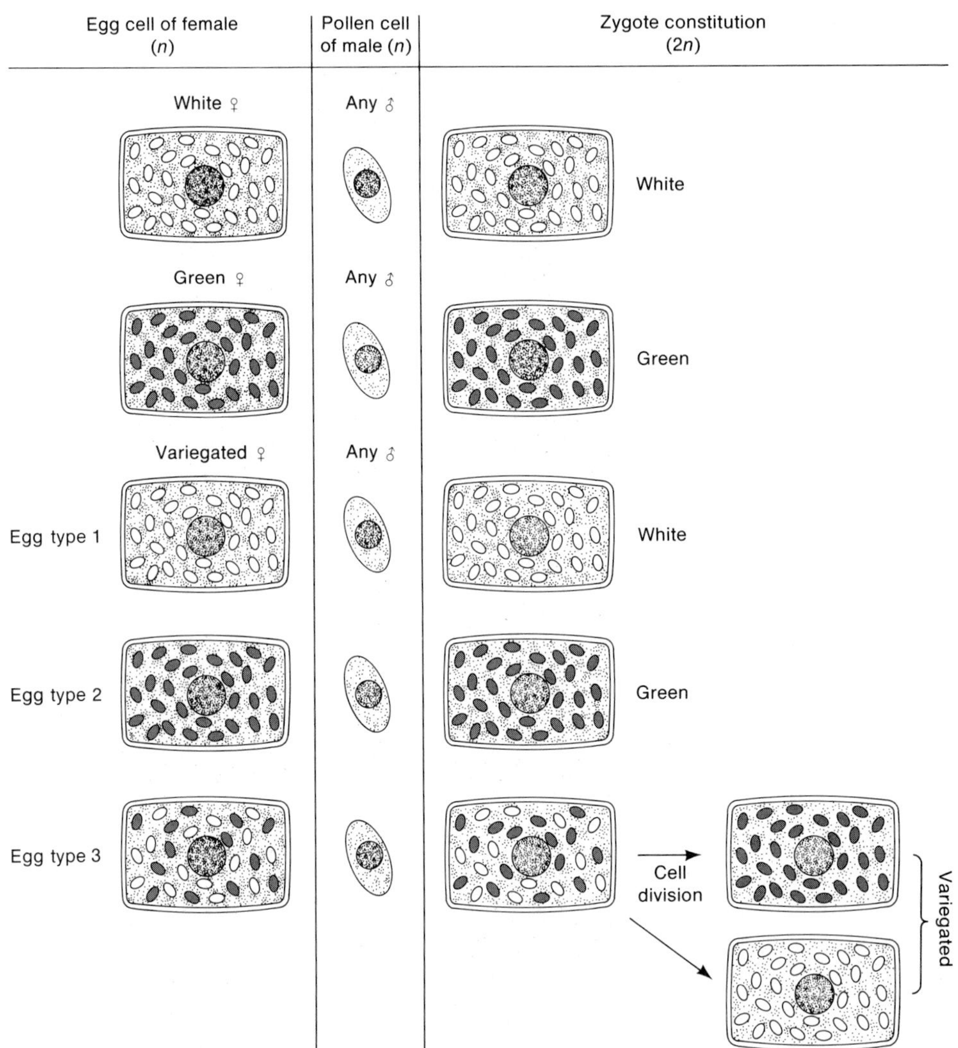

Figure 15-3. A model explaining the results of the Mirabilis jalapa *crosses in terms of autonomous chloroplast inheritance. The large dark spheres are nuclei. The smaller bodies are chloroplasts, either green* (shaded) *or white* (unshaded). *Each egg cell is assumed to contain many chloroplasts, and each pollen cell is assumed to contain no chloroplasts. The first two crosses exhibit strict maternal inheritance. If the maternal branch is variegated, three types of zygotes can result, depending on whether the egg cell contains only white, only green, or both green and white chloroplasts. In the latter case, the resulting zygote can produce both green and white tissue, so a variegated plant results.*

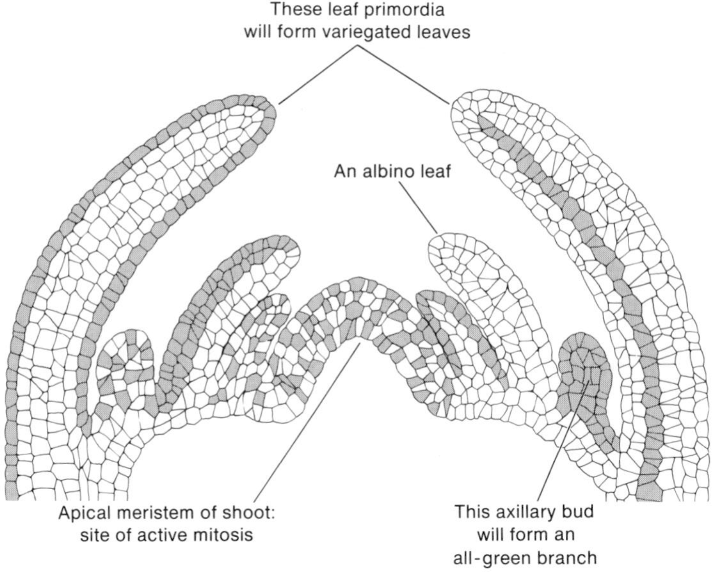

Figure 15-4. In a plant containing both green and white chloroplasts, different green or white cell lines are established in the meristem (growing point). The egg from which such a plant is derived contains both types of chloroplasts; after fertilization and subsequent mitotic divisions, cytoplasmic segregation occurs, segregating the chloroplast types into pure cell lines. The color of any part of the plant thus depends on the particular cell types that happened to form the relevant leaf or shoot primordia.

This term represents a hypothetical "black box" in which the segregation and (as we shall see) recombination of organelle genotypes occur. We call it **cytoplasmic segregation and recombination** and apply the convenient acronym CSAR. Throughout this chapter, CSAR crops up quite often; it appears to be a common behavior of extranuclear genomes. In one sense, CSAR is the cytoplasmic equivalent of meiosis, because meiosis is the process whereby segregation and recombination of nuclear genes regularly occurs. However, CSAR is purely hypothetical—no physical counterpart is known for this process.

You might suspect that random chloroplast assortment could explain the green/white segregation in variegated plants, but this hypothesis has not been proved. Furthermore, the large numbers of chloroplasts in cells make this an unlikely possibility. In a cell with, let's say, 40 chloroplasts, consisting of 20 green and 20 colorless chloroplasts, the production by random assortment of one daughter cell containing only green and one daughter cell containing only colorless chloroplasts would be at best a very rare event. To be on the safe side, we shall treat the CSAR process as hypothetical. This hypothetical CSAR process is diagrammed in Figure 15-5.

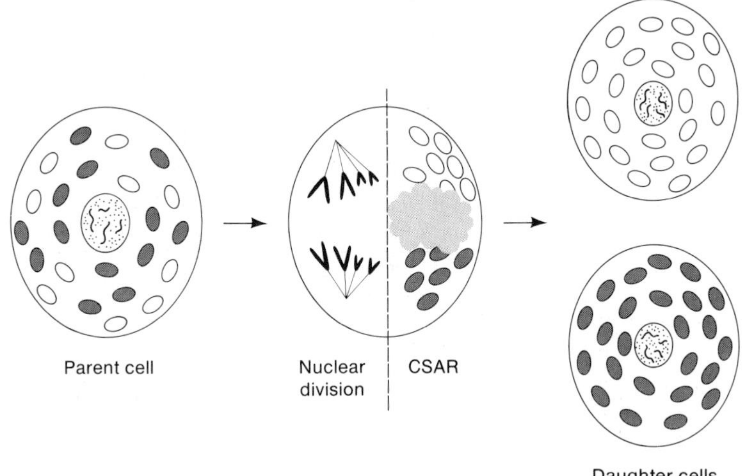

Figure 15-5. *Two processes may be distinguished at cell division. Nuclear division (mitosis) is an observed physical process that segregates the genes on the nuclear chromosomes. The hypothetical CSAR process in the cytoplasm (for which no physical counterpart has been observed) would account for segregation and recombination of organelle genes, producing the new organelle sets of the daughter cells. The diagrams show how CSAR can generate two different organelle sets from a mixed parental cell.*

Poky *Neurospora*

In 1952, Mary Mitchell isolated a mutant strain of *Neurospora* that she called poky. This mutant differs from the wild type in a number of ways: it is slow growing, it shows maternal inheritance, and it has abnormal amounts of cytochromes. (Cytochromes are electron-transport proteins necessary for the proper oxidation of foodstuffs to generate ATP energy. There are three main types: cytochromes *a*, *b*, and *c*. In poky, there is no cytochrome *a* or *b*, and there is an excess of cytochrome *c*.)

How can maternal inheritance be expressed in a haploid organism? It is possible to cross some fungi in such a way that one parent contributes the bulk of cytoplasm to the progeny, and this cytoplasm-contributing parent is called the female parent, even though no true sex is involved. Maternal inheritance for the poky phenotype was demonstrated in the following crosses:

poky ♀ × wild-type ♂ → progeny all poky
wild-type ♀ × poky ♂ → progeny all wild-type

However, in such crosses, any nuclear genes that differ between the parental strains are observed to segregate in the normal Mendelian manner and to pro-

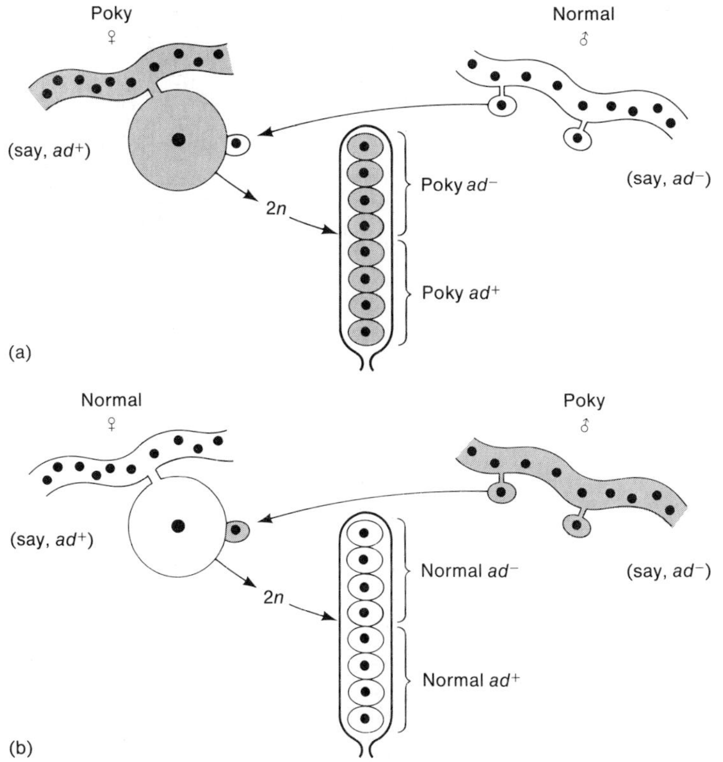

Figure 15-6. Explanation of the different results from reciprocal crosses of poky and normal Neurospora. *Most of the cytoplasm of the progeny cells comes from the parent called female. Shading represents cytoplasm with the poky determinants. A nuclear gene* ad^+/ad^- *is used to illustrate the segregation of the nuclear genes in the expected 1:1 Mendelian ratio.*

duce 1:1 ratios in the progeny (Figure 15-6). All poky progeny behave like the original poky strain, tansmitting the poky phenotype down through many generations when crossed as females. Because poky does not behave like a nuclear mutation, it was termed an **extranuclear mutation.** To hammer the point home, it also has been called a **cytoplasmic mutation** because it seems to be carried in the cytoplasm of the female parent.

But where in the cytoplasm is the mutation carried? The important clue is that several aspects of the mutant phenotype seem to involve mitochondria. For instance, the cells' slow growth suggests lack of ATP energy, which is normally produced by mitochondria. There are abnormal amounts of cytochromes in the mutants, and cytochromes are known to be located in the mitochondrial membranes. These clues led to the conclusion that mitochondria are involved and inspired several interesting experiments designed to investigate mitochondrial autonomy.

David Luck labeled mitochondria with radioactive choline, a membrane component, and he then followed their division autoradiographically in unlabeled medium. He found that even after several doublings of mass, the mitochondria were still evenly labeled. Luck concluded that mitochondria can grow and divide in a fashion that seems to be autonomous rather than nucleus directed. (If mitochondria were synthesized anew, or de novo, some of the resulting population of mitochondria should be unlabeled, whereas the original ones would remain heavily labeled.)

Another research group extracted mitochondria from a mutant similar to poky, a mutant called *abnormal.* Using an ultrafine needle and syringe, they injected the mitochondria into wild-type recipient cells, using appropriate controls. The recipient cells were cultured. After several of these subcultures, the abnormal phenotype appeared! In transferring the mitochondria, the experimenters had transferred the hereditary determinants of the abnormal phenotype. Presumably, then, the extranuclear genes involved in this phenotype are located in the mitochondria. These inferences gained credibility with the discovery of DNA in the mitochondria of *Neurospora* and other species (but more of that story later).

The Heterokaryon Test

Maternal inheritance is one criterion for recognizing inheritance through organelles. Another test has been applied in filamentous fungi such as *Neurospora* and *Aspergillus.* In principle, this test could be applied to other systems involving heterokaryons. A heterokaryon is made between the prospective extranuclear mutant (say, a slow-growth mutant) and a strain carrying a known nuclear mutant. If the phenotype of the nuclear mutation can be recovered from the heterokaryon together with the phenotype of the slow-growth mutant being tested, then the growth mutant is a good candidate for an organelle-based mutation. This is because no diploidy occurs in a heterokaryon, so that there is no genetic exchange between nuclei. The slow-growth phenotype must have been transferred solely by cytoplasmic contact. The segregation of the pure extranuclear type from the mixed cytoplasm of the heterokaryon presumably involves the CSAR process (Figure 15-7).

Message
Extranuclear inheritance can be recognized by uniparental (usually maternal) transmission through a cross or by transmission via cytoplasmic contact.

Shell Coiling in Snails: A Red Herring

Does maternal inheritance always indicate extranuclear inheritance? Usually, but not always. It is possible for a maternal-inheritance pattern of reciprocal crosses to be generated by nuclear genes. Alfred Sturtevant (who studied crossing over in *Drosophila*) found a good example in the water snail *Limnaea.* Sturtevant crossed snails that differed in the direction of their shell coiling. Looking into

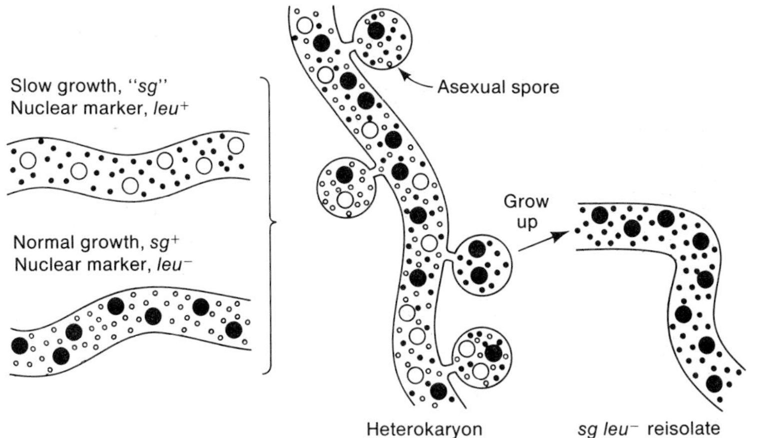

Figure 15-7. The heterokaryon test is used to detect extranuclear inheritance in filamentous (threadlike) fungi. A possible extranuclear mutation (here, sg$^-$, *causing slow growth) is combined with a nuclear mutation* (leu$^-$) *to form a heterokaryon. Cultures of* leu$^-$ *can be derived from the heterokaryon. If some of these cultures also are sg$^-$ in phenotype, then* sg *is very likely an extranuclear gene, borne in an organelle. Since no nuclear recombination occurs in a heterokaryon, the sg$^-$ phenotype must have been acquired by cytoplasmic contact.*

the opening of the shell, you can see that some snails coil to the right (dextral) and others to the left (sinistral). All of the F_1 progeny of the cross dextral ♀ × sinistral ♂ are dextral, but all of the F_1 progeny of the cross sinistral ♀ × dextral ♂ are sinistral. Thus far, the situation seems similar to that observed with poky or with chloroplast inheritance. However, Sturtevant found that the F_2 generation is all dextral from both pedigrees!

The F_3 generation in this case revealed that the inheritance of coiling direction involves nuclear genes, not extranuclear genes. The F_3 is produced by individually selfing the F_2 snails. (This is possible because snails are hermaphroditic.) Three-fourths of the F_3 snails were found to be dextral, and one-fourth were found to be sinistral (Figure 15-8). This ratio reveals a Mendelian segregation in the F_2 generation. Apparently, dextral ($+^s$) is dominant to sinistral (*s*), but, strangely enough, the shell-coiling phenotype of any individual animal is determined by the genotype (not the phenotype) of its mother! We now know that this happens because the genotype of the mother's body determines the initial cleavage pattern of the developing embryo. Note that the segregation ratios shown in Figure 15-8 would never appear in the phenotypes of true organelle genes. This example shows that one generation of crosses is not enough to provide certain evidence that maternal inheritance is due to organelle-based inheritance. Let us use the term **maternal effect** for cases like the shell-coiling example in order to distinguish them from organelle-based inheritance.

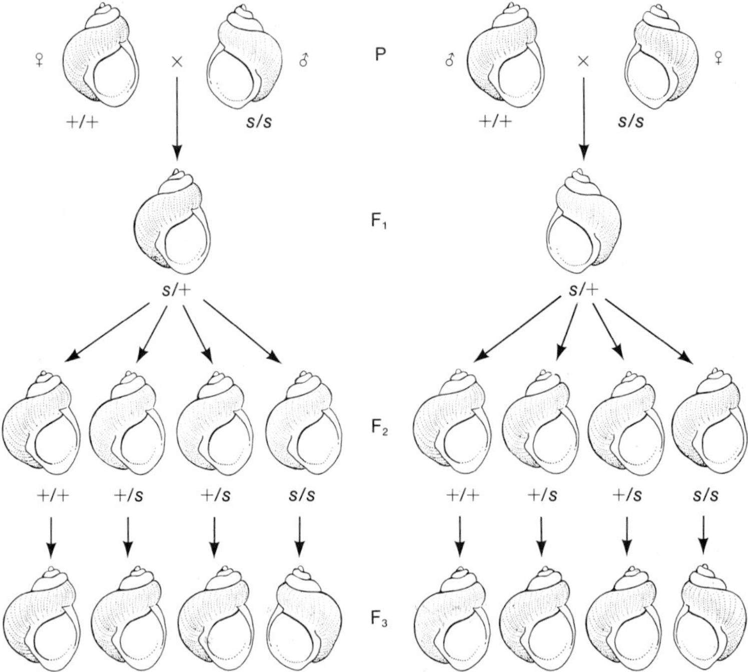

Figure 15-8. The inheritance of dextral (+) and sinistral (s) *nuclear genes for shell coiling in a species of water snail. The direction of coiling is determined by the nuclear genotype* of the mother, *not the genotype of the individual involved. This explanation accounts for the initial difference between reciprocal crosses as well as the phenotypes of later generations. No organelle inheritance is involved (such a hypothesis would not explain the phenotypes of later generations).*

Extranuclear Genes in *Chlamydomonas*

If one were to design an ideal experimental organism with a simple life cycle, the result would be something like the unicellular freshwater alga *Chlamydomonas*. Like fungi, algae rarely have different sexes, but they do have mating types. In many algal and fungal species, there are two mating types that are determined by alleles at one locus. A cross can occur only if the parents are of different mating types. The mating types are physically identical but physiologically different. Such species are called **heterothallic** (literally, "different-bodied"). In *Chlamydomonas,* the mating-type alleles are called mt^+ and mt^- (in Neurospora they are *A* and *a*; in yeast *a* and α). Figure 15-9 diagrams the Chlamydomonas life cycle. As you might guess, tetrad analysis (see Chapter 6) is possible and in fact is routinely performed.

In 1954, Ruth Sager isolated a streptomycin-sensitive mutant of *Chlamydomonas* with a peculiar inheritance pattern. In the following crosses, *sm-r* and

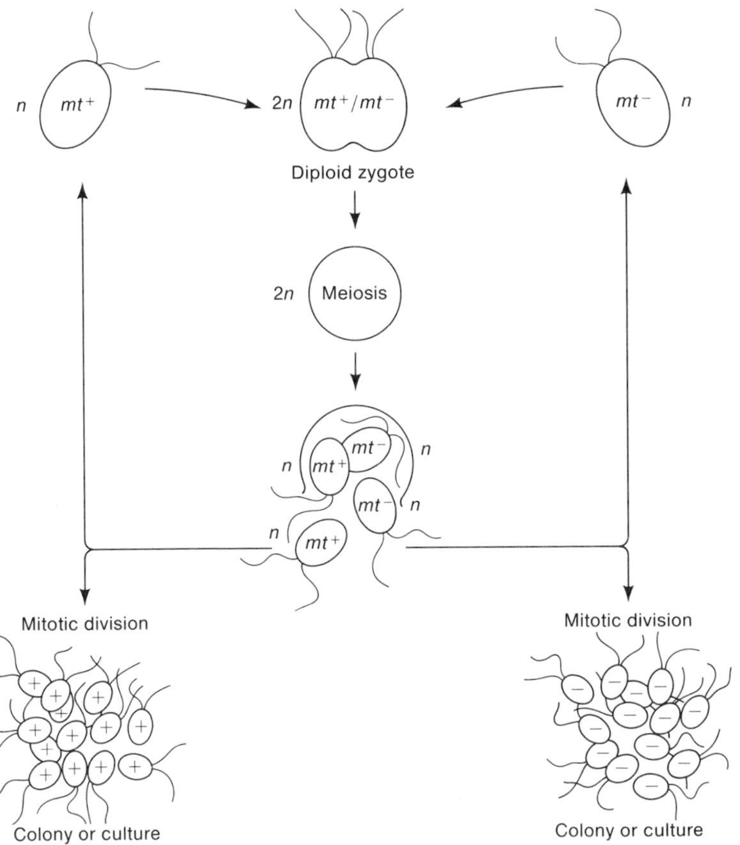

Figure 15-9. Diagrammatic representation of the life cycle of Chlamydomonas, *a unicellular green alga. This organism has played a central role in research on organelle genetics. The nuclear mating-type alleles* mt^+ *and* mt^- *must be heterozygous in order for the sexual cycle to occur.*

sm-s indicate streptomycin resistance and sensitivity, and *mt* is the mating-type gene discussed earlier:

$$sm\text{-}r\ mt^+ \times sm\text{-}s\ mt^- \longrightarrow \text{progeny all } sm\text{-}r$$
$$sm\text{-}s\ mt^+ \times sm\text{-}r\ mt^- \longrightarrow \text{progeny all } sm\text{-}s$$

Here we see a difference in reciprocal crosses; all progeny cells show the streptomycin phenotype of the mt^+ parent. Like the maternal inheritance we discussed earlier, this is a case of **uniparental inheritance.** In fact, Sager now refers to the mt^+ mating type as the female, using this analogy. However, there is no observable physical distinction between the mating types as there would be if true sex were involved. In these crosses, the conventional nuclear marker genes (such as *mt* itself) all behave normally and give 1:1 progeny ratios. For example, one-half of the progeny of the cross *sm-r* mt^+ × *sm-s* mt^- are *sm-r* mt^+, and one-half are *sm-r* mt^-.

To determine the sm-s or sm-r phenotypes of progeny, the cells are plated onto medium containing streptomycin. While performing these experiments, Sager observed that the drug streptomycin itself acts as a mutagen. If a pure strain of *sm-s* cells is plated onto streptomycin, quite a significant proportion of streptomycin-resistant colonies will appear. In fact, treatment with streptomycin will produce a whole crop of different kinds of mutant phenotypes, all of them showing uniparental inheritance (and hence presumably extranuclear). In fact, streptomycin treatment does not produce nuclear mutants. Most of the mutants produced by streptomycin treatment show either (1) resistance to one of several different drugs or (2) defective photosynthesis. The photosynthetic mutants are unable to make use of CO_2 from the air as a source of carbon, so they survive only if soluble carbon (in some form such as acetate) is added to their medium. (How would you go about selecting such mutants?) Table 15-2 lists several different extranuclear mutant types produced by the streptomycin treatment.

These experiments reveal the existence of a mysterious "uniparental genome" in *Chlamydomonas*—that is, a group of genes that all show uniparental transmission in crosses. Where is this uniparental genome located? What is the mechanism of the uniparental transmission? When there seems to be no physical difference between mating types, why are these genes transmitted only by the mt^+ parent?

There is evidence to suggest that the uniparental genome in this case is in fact chloroplast DNA (cpDNA). About 15% of the DNA of a *Chlamydomonas* cell in

Table 15-2. Some of the *Chlamydomonas* mutants showing uniparental inheritance

Gene	Mutant phenotype
ac1 through *ac4*	Requires acetate
tm1, *tm3* through *tm9*	Cannot grow at 35°C
tm2	Conditional: grows at 35°C only in presence of streptomycin
ti1 through *ti5*	Forms tiny colonies on all media
ery1	Resistant to erythromycin at concentration of 50 μg/ml
kan1	Resistant to kanamycin at concentration of 100 μg/ml
spc1	Resistant to spectinomycin at concentration of 50 μg/ml
spi1 through *spi5*	Resistant to spiramycin at concentration of 100 μg/ml
ole1 through *ole3*	Resistant to oleandomycin at concentration of 50 μg/ml
car1	Resistant to carbamycin at concentration of 50 μg/ml
ele1	Resistant to eleosine at concentration of 50 μg/ml
ery3 and *ery11*	Resistant to erythromycin, carbamycin, oleandomycin, and spiramycin (at concentrations listed above for individual drugs)
sm2 and *sm5*	Resistant to streptomycin at concentration of 500 μg/ml
sm3	Resistant to streptomycin at concentration of 50 μg/ml
sm4	Requires streptomycin for survival

NOTE: All of these mutants were produced by treatment with streptomycin except for the *ti* mutants, which were produced by treatment with nitrosoguanidine.

SOURCE: R. Sager, *Cytoplasmic Genes and Organelles,* Academic Press.

rapidly growing culture is found in the single chloroplast of the cell. This DNA forms a band in a CsC1 gradient that is distinct from the band of nuclear DNA (Figure 15-10). Furthermore, the precise position of the cpDNA band can be altered by adding the heavy isotope of nitrogen, ^{15}N, to the growth medium. The position of the band can be calibrated in terms of the buoyant density of the cpDNA. With this technique, the cpDNA of the two parents in a cross can be labeled differently, one light (^{14}N) and one heavy (^{15}N). The buoyant densities of the cpDNA from these parental cells are 1.69 and 1.70, respectively. Although this difference seems small, it provides appreciably different band positions in a CsC1 gradient. With differently labeled parents, the zygote DNA can be examined to see how it compares with the parents (Table 15-3). The results seem to indicate that the cpDNA of the mt^- parent is in fact lost, inactivated, or destroyed in some way. This loss of cpDNA from the mt^- parent, of course, parallels the loss of uniparental genes (such as the *sm* genes) borne by the mt^- parent.

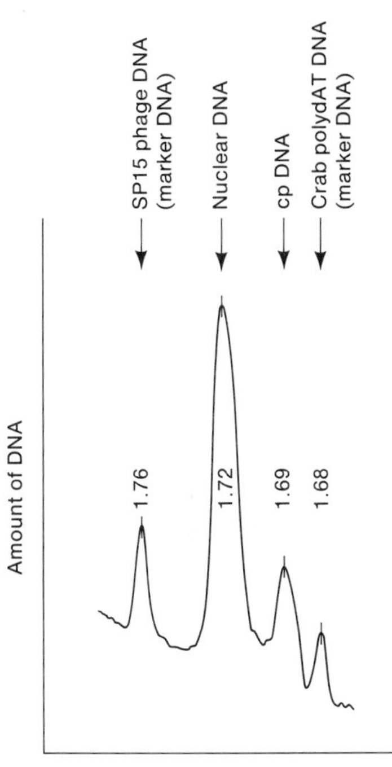

Figure 15-10. The chloroplast DNA (cpDNA) of Chlamydomonas *can be detected in a CsCl density gradient as a band of DNA distinct from the nuclear DNA. In this particular experiment, two other DNA types were sedimented along with the* Chlamydomonas *DNA to act as known reference points in the gradient. The numbers represent buoyant densities (g/cm³) of the various DNA types. (After R. Sager,* Cytoplasmic Genes and Organelles, *Academic Press.)*

Table 15-3. Buoyant densities of cpDNA in progeny from various *Chlamydomonas* crosses

Cross	Buoyant density of zygote cpDNA
$^{14}N\,mt^+ \times {}^{14}N\,mt^-$	1.69
$^{15}N\,mt^+ \times {}^{15}N\,mt^-$	1.70
$^{15}N\,mt^+ \times {}^{14}N\,mt^-$	1.70
$^{14}N\,mt^+ \times {}^{15}N\,mt^-$	1.69

SOURCE: R. Sager, *Cytoplasmic Genes and Organelles,* Academic Press.

Other experiments have been performed using similar logic. For example, crosses can be made between *Chlamydomonas* strains having cpDNAs with distinctly different patterns of digestion (breakdown) by restriction enzymes. Again, the cpDNA passed on to the progeny is that of the mt^+ parent only.

Message

The cpDNA of Chlamydomonas *is inherited uniparentally. Because the behavior of this DNA parallels the behavior of the uniparentally transmitted genes, it appears that these genes are located in the cpDNA.*

Mapping Chloroplast Genes in *Chlamydomonas*

We have seen (Table 15-2) that *Chlamydomonas* has a large number of uniparentally inherited genes. Is there linkage among these genes? Are they all in one linkage group (chromosome), or are they arranged in several linkage groups? Or does each gene assort independently as though on its own separate piece of

DNA? Of course, the way to approach this question in the true tradition of classical genetics is to perform a recombination analysis. But here we run into a problem. Both sets of parental DNA must be present if there is to be an opportunity for recombination, and we have seen that the *mt*$^-$ parent's cpDNA is eliminated in the zygote cell. Luckily, there is a way out of this dilemma.

In crosses of *mt*$^+$ *sm-r* × *mt*$^-$ *sm-s*, about 0.1% of the progeny zygotes were found to contain both *sm-r* and *sm-s*. The presence of both alleles was inferred from the fact that the products of meiosis of such cells show both *sm-r* and *sm-s* phenotypes among their number. Such segregation of the two alleles presumably arises from a CSAR process in the zygote or at a later stage of division. These zygotes were called **biparental zygotes,** for obvious reasons, and their genetic condition is described as a **cytohet** (standing for "*cyto*plasmically *het*erozygous"). It appears as though the inactivation of the *mt*$^-$ parent's cpDNA fails in these rare zygotes. However, whether this is what happens or not, the cytohets provide just the opportunity we need to study recombination —they are cells containing *both* sets of parental cpDNA. The rarity of cytohet zygotes does pose a problem for research, but their frequency can be increased by treatment of the *mt*$^+$ parent with ultraviolet light before mating. After such treatment, from 40% to 100% of the progeny zygotes are cytohets. (Note in passing that this observation implies a normal role of the *mt*$^+$ cell in actively eliminating the *mt*$^-$ cell's cpDNA. The ultraviolet irradiation of the *mt*$^+$ cell must inactivate such a function.)

Message

In Chlamydomonas, *biparental zygotes (or cytohets) must be the starting point for all studies on the segregation and recombination of chloroplast genes.*

A good map of the cpDNA has been obtained using the cytohets. We shall examine two mapping methods: mapping in relation to a hypothetical attachment point and cosegregation mapping.

Mapping in Relation to a Hypothetical Attachment Point. The first mapping technique obtains mapping data from the analysis of cytohets that are heterozygous for only one gene pair. This approach is not as implausible as it may seem at first. Recall that a single nuclear gene pair can be mapped in relation to the centromere in fungi (Chapter 6). The present analysis is similar in some ways, although the analogy cannot be pressed too far.

The four products of a single meiosis are plated onto an agar surface, where the cells then undergo mitotic division. The daughter cells are gently separated, and a record is kept of which cells are sisters. After the first mitotic division, there are eight cells. After a second mitotic division, there are 16 cells, and these form the standard set for mapping analysis. (Using the 16-cell stage as the standard set is somewhat arbitrary.) Figure 15-11 shows the pedigree for this analysis.

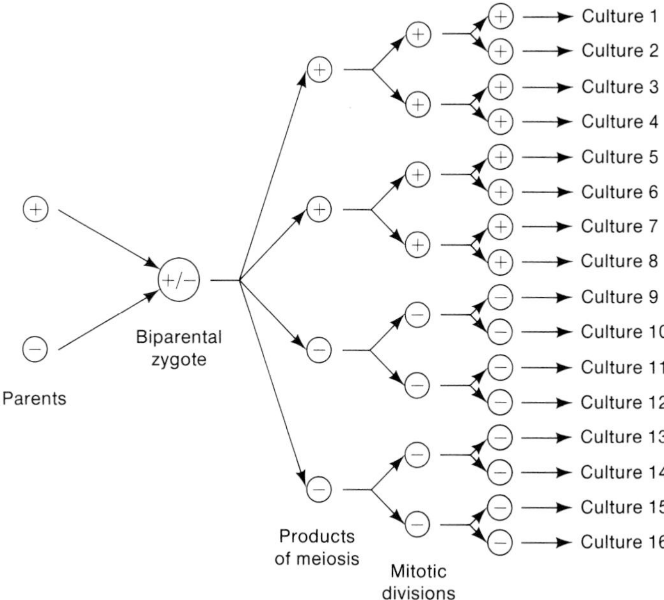

Figure 15-11. Derivation of the standard set of 16 cultures used for mapping of organelle genes in Chlamydomonas. *The four products of meiosis are carefully followed through two mitotic division cycles. The resulting 16 cells are isolated to form 16 cultures. The + and − symbols here represent the mating types,* mt^+ *and* mt^-.

The 16 cells are isolated and allowed to grow into 16 separate cultures, which then are subjected to tests. For example, in the cross mt^+ *sm-r* × mt^- *sm-s*, the 16 cultures derived from cytohet zygotes would be tested for two characteristics.

1. Test for streptomycin sensitivity or resistance.
2. Test for ability to undergo further segregation. If the cells of a culture will segregate, they are still cytohets. If they will not segregate further, then the culture is genotypically of one pure parental type (P1) or the other (P2).

The first column of Table 15-4 shows an illustrative set of 16 cultures. This set was chosen to show all possible patterns; a typical set is less diversified than this one. In the table, P1 indicates pure-breeding *sm-r*, and P2 indicates pure-breeding *sm-s*. From any one cross, a researcher would obtain as many sets as possible and would study them all.

Because the order of the cultures reflects the origin of each of the 16 cells in the pedigree, we can infer the genotypes of the cells at the 8-cell stage and (extrapolating further) at the 4-cell stage (the meiotic products). In the case

Table 15-4. An illustrative set of 16 *Chlamydomonas* cultures derived from a single biparental zygote (as shown in Figure 15-11)

Culture	Phenotype	Type of second mitotic division	Inferred genotypes of cells at 8-cell stage	Type of first mitotic division	Inferred genotypes of 4 products of meiosis
1	P1	IV	P1	II	Cytohet
2	P1				
3	Cytohet	II	Cytohet		
4	P2				
5	P1	IV	P1	III	Cytohet
6	P1				
7	P2	IV	P2		
8	P2				
9	P1	III	Cytohet	I	Cytohet
10	P2				
11	P1	II	Cytohet		
12	Cytohet				
13	Cytohet	I	Cytohet	II	Cytohet
14	Cytohet				
15	P2	IV	P2		
16	P2				

NOTE: The first column represents the experimental observations; the information in the other columns is inferred by inspection of the results. P1 and P2 are pure parental phenotypes. Types I through IV are defined on the basis of the two daughter cells produced: I = Het → (Het + Het); II = Het → (Het + P1) or (Het + P2); III = Het → P1 + P2; IV = Pure breeding, P1 → (P1 + P1) or P2 → (P2 + P2).

of Table 15-4, it appears that all four of the meiotic products were cytohets. However, cells of either parental genotype began to appear in the mitotic divisions, presumably as a result of a CSAR process. (Sometimes CSAR does occur during meiosis, so that a product of meiosis can be P1 or P2 in phenotype.)

Each cell division (represented by a bracket in the table) can be classified as one of four types:

Type I: Cytohet → (Cytohet + Cytohet)
Type II: Cytohet → (Cytohet + P1) or (Cytohet + P2)
Type III: Cytohet → (P1 + P2)
Type IV: Pure breeding; P1 → (P1 + P1), or P2 → (P2 + P2)

Divisions of types I, II, and III are the ones that we must consider for mapping, because each starts with a cytohet cell and an opportunity to detect recombination. Ruth Sager has interpreted the three types as follows (we discuss them in reverse order).

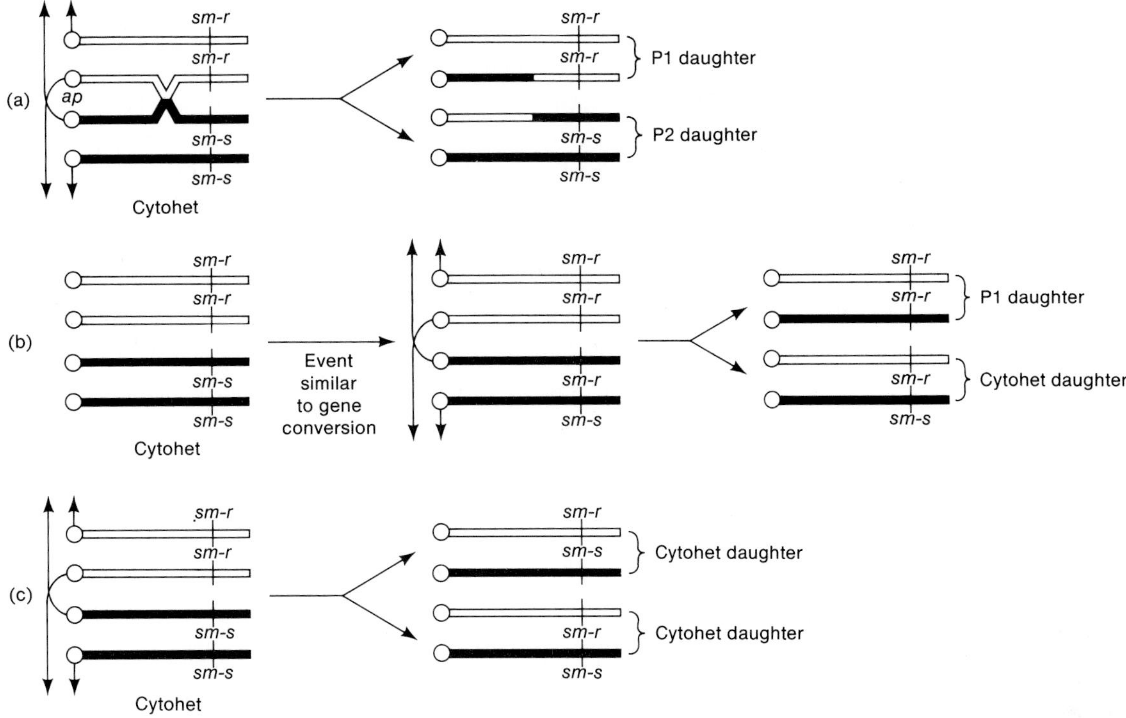

Figure 15-12. Models to explain the three types of cytohet cell division. (a) In a division of type III, a crossover between ap *and* sm *produces P1 and P2 daughter cells. (b) In a division of type II, a gene-conversionlike event produces one parental cell and one cytohet cell without crossing over. (c) In a division of type I, segregation without crossing over produces two cytohet daughter cells.*

1. A division of type III represents a reciprocal crossover event between the *sm* locus and some hypothetical centromerelike point called the attachment point (*ap*). The crossover must occur at some four-stranded stage of the paired parental cpDNA molecules (Figure 15-12a). Note that this explanation assumes a "diploidlike" state of the cpDNA in most normal cells. A transient "tetraploidlike" state, where CSAR occurs as shown in Figure 15-12a, would be found in zygotes and prior to cell division. There is some cytological evidence to support these assumptions.

2. A division of type II represents a "gene-conversionlike" event (see Chapters 5 and 16). After one allele has been converted, the observed products will be produced without crossovers (Figure 15-12b). A similar event will explain the production of P2 + Cytohet from a cytohet cell.

3. A division of type I represents no exchange (Figure 15-12c).

The frequency of type II divisions is relatively constant for all heterozygous chloroplast gene pairs. However, the frequency of type III varies and is highly specific to the particular gene pair involved. In fact, the frequency of type III divisions can be used to map gene pairs from the hypothetical attachment point. For example, if the frequency of type III divisions for *sm-r/sm-s* were twice as great as the frequency of type III divisions for ac^+/ac^-, then the *sm* locus would be mapped twice as far from the attachment point as the *ac* locus.

In an extension of this type of analysis, cytohets may be grown in liquid culture and sampled at various time intervals. The cells sampled are tested and scored according to phenotype as P1, P2, or Cytohet. We might expect genes close to the hypothetical attachment point to remain heterozygous for several cell generations. In fact, the *rate* of disappearance of cytohets from the culture does prove to be highly specific to the particular gene pair under study. This rate can be used to generate another map, using rates of disappearance as the map units. The sequence and relative distances in such a map prove to be very similar to those in the map based on the type III frequencies. For example, the map for four illustrative genes (by either mapping method) is the following:

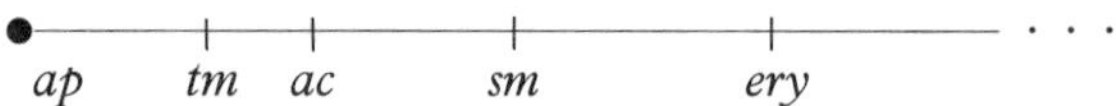

We must stress that no physical counterpart of the attachment point (*ap*) has been observed; its presence is inferred only to make sense of the type III frequencies.

Mapping by Cosegregation. In a cross that is heterozygous for several different pairs of organelle genes, the products of the CSAR process can be examined for evidence of linkage between the genes. In this approach, of course, the gene pairs are studied in relation to each other and not in relation to any hypothetical attachment point. Thus *ap* cannot be located with this method. The approach involves examining type III and type II divisions for the cosegregation of parental gene combinations in the CSAR process. The closer together two loci are in the chloroplast genome, the more frequently the parental combinations should be carried together by the shuffling mechanisms of CSAR. Let's look at an example.

The following cross is carried out:

$$mt^+\ sm\text{-}s\ ac^+ \times mt^-\ sm\text{-}r\ ac^-$$

Biparental zygotes are obtained. Division is followed past meiosis to the 16-cell stage, and the resulting 16 cultures are tested. Several kinds of type III events are possible in this case (Figure 15-13). Presumably, the type III events for *ac* only or for *sm* only arise from crossovers between the loci, so the distance between these loci should be proportional to the relative frequency of such events.

The total frequency of type III cosegregation events (Figure 15-13a) can be used to develop a map. A high frequency of cosegregation, of course, indicates

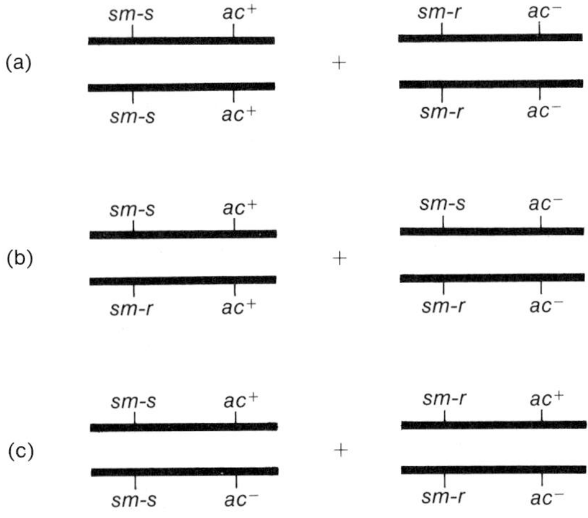

Figure 15-13. Daughter genotypes arising from three different kinds of type III divisions in study of two gene pairs. (a) Cosegregation of the pairs shows type III results for both ac *and* sm. *(b) Crossover between the pairs can produce type III results for* ac *only. (c) Or crossover between the pairs can produce type III results for* sm *only.*

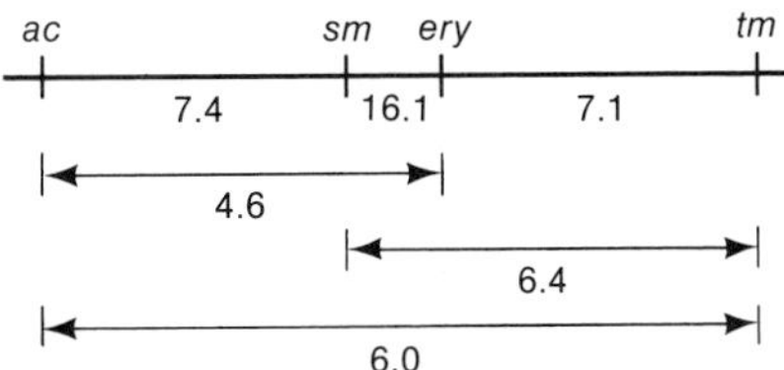

Figure 15-14. Map of the Chlamydomonas *chloroplast genes based upon cosegregation studies. Note that this map is consistent with the one derived from single-gene studies only if the map is circular. The map distances shown are frequencies of cosegregation, which are inversely proportional to distances between loci.*

tight linkage and a short map distance, whereas a low frequency of cosegregation indicates weak linkage and a long map distance. Figure 15-14 shows the cosegregation map for the same genes discussed earlier, where the map distances represent the percentage of type III cosegregations.

The cosegregation map agrees with the map derived earlier for the positions of *ac*, *sm*, and *ery*. However, the position of the *tm* locus will agree between the two maps only if the map is circular! Extensive studies with other markers have

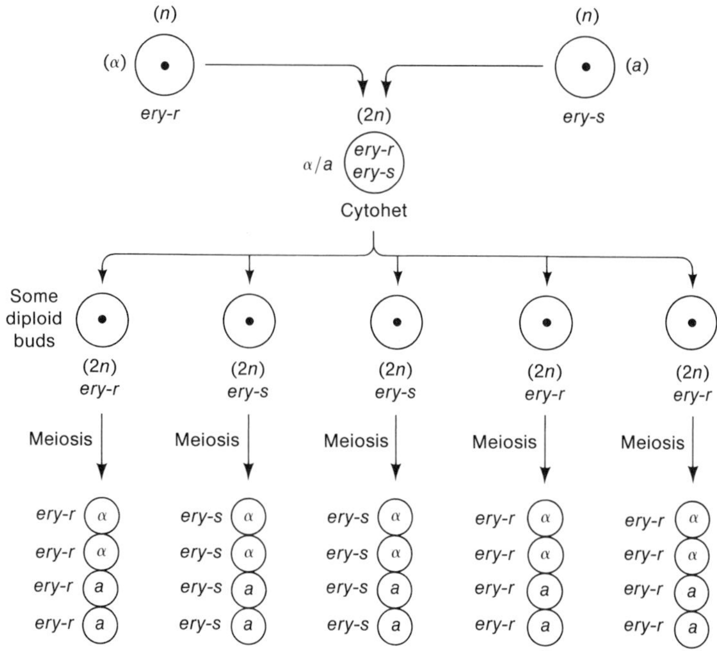

Figure 15-17. The special inheritance pattern shown by certain drug-resistant phenotypes in yeast. When diploid buds are induced to sporulate (undergo meiosis), the products of each meiosis show uniparental inheritance. (Only a representative sample of the diploid buds is shown here.) Note that the nuclear genes, represented here by the mating-type alleles a *and* α*, segregate in a strictly Mendelian pattern. The alleles* ery-r *and* ery-s *determine erythromycin resistance or sensitivity, respectively.*

important differences, as we shall see. It is clear that uniparental inheritance is occurring at the level of a single meiosis. This kind of inheritance looks like a very good candidate for mitochondrial inheritance.

The inheritance patterns observed here are reminiscent of those for the cytoplasmic petites. However, recall that the CSAR process in the case of the petites seems consistently to favor one genome to the complete exclusion of the other. In the present case, the ratio of *ery-r* to *ery-s* diploid buds can deviate moderately from 1:1. This **transmission ratio** is different for different markers and different strains. For a suppressive petite, however, the transmission ratio would approach infinity. (In the present discussion, we need not worry about why the transmission ratios vary slightly from 1:1. We wish simply to note that the situation in the case of these mutants is different from that for the petite mutants.)

We can set out to look for recombinants by using a cross between parents differing in two extranuclear gene pairs (a kind of "dihybrid cross"). For example, we can carry out the cross *ery-r spi-r* × *ery-s spi-s*, allow the resulting diploid cell to bud through several generations, and then induce the resulting cells to sporulate. We can then identify the genotype of each bud cell by observing the phenotype common to all its ascospores. (Of course, we could observe the phenotype of the bud directly without resorting to sporulation. However, the testing of the four products of meiosis with identical extranuclear genotypes provides a control against experimental errors.)

Four genotypes could result from such a cross (Figure 15-18), and all were in fact observed. An early cross yielded the following results:

ery-r spi-r	63 tetrads
ery-s spi-s	48 tetrads
ery-s spi-r	7 tetrads
ery-r spi-s	1 tetrad

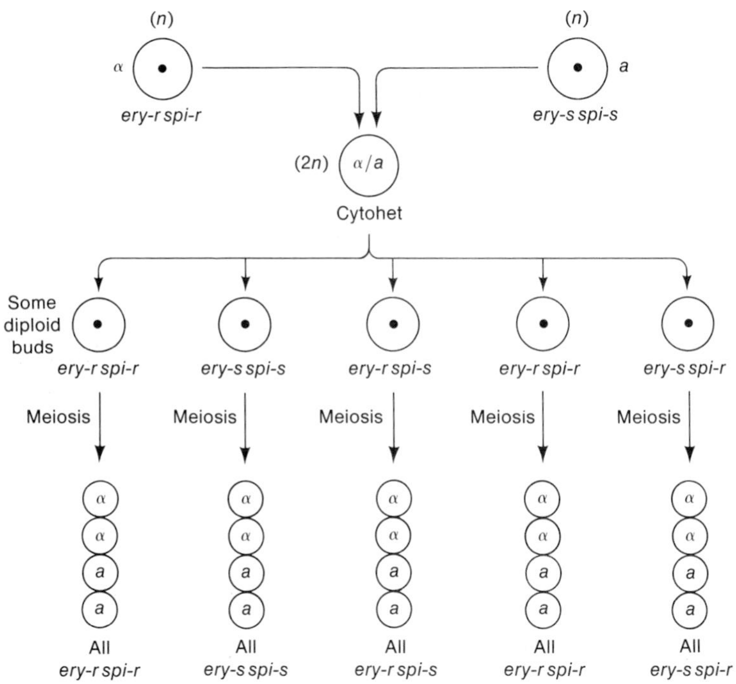

*Figure 15-18. The study of inheritance in a cross between yeast cells differing with respect to two different drug-resistance alleles (*ery-r *= erythromycin resistance;* spi-r *= spiramycin resistance). Each diploid bud can be classified as parental or recombinant on the basis of these results. Note that the identity of the extranuclear genotype for all four products of meiosis confirms that the CSAR process must occur during the production of the diploid buds.*

The genotypes *ery-s spi-r* and *ery-r spi-s*, of course, represent recombinants that have been produced in the CSAR process during formation of diploid buds. (Note that these recombinants cannot have been produced by meiosis because the products of any given meiosis are identical with respect to the drug-resistance phenotypes.)

Message
In yeast, extranuclear segregation and recombination (CSAR) are achieved during bud formation in diploid cytohets. The segregation and/or recombination may be detected directly in cultures of the diploid buds or may be detected by observing the products of meiosis that result when the buds are induced to sporulate.

We now seem to be just a short step away from the development of a complete map of all the drug-resistance genes in yeast. Unfortunately, this did not prove to be such an easy task. Several problems arose that prevented the production of a consistent map.

The first problem involved the existence of a **mitochondrial sex locus** called *omega* (ω). A cell can be either ω^+ or ω^-, and the ω gene itself is inherited in a pattern similar to that observed for the drug-resistance genes. A mitochondrial cross in yeast can be described as either **heterosexual** ($\omega^+ \times \omega^-$) or **homosexual** ($\omega^+ \times \omega^+$ or $\omega^- \times \omega^-$). The results of recombination analysis proved to be very different for these two kinds of crosses (Table 15-5). Note that the ω constitution of the cross affects the values of the recombinant frequency (RF), the transmission ratio, and the polarity (in this case defined as the ratio of one recombinant class to its reciprocal). Also note that all these effects are independent of the *nuclear* mating types *a* and α. Particularly interesting is the observation that only certain genes seem to be affected by ω. In the example shown in Table 15-5, only the *cap* gene is affected. Some models have been proposed to explain the action of ω, but the actual mechanism is not yet known. What is

Table 15-5. Effect of the locus ω on results of a yeast cross
cap-r oli-r par-r × *cap-s oli-s par-s*

Variable observed		Value for a heterosexual cross ($\omega^+ + \omega^-$)	Value for a homosexual cross ($\omega^+ + \omega^+$)
Recombinant frequency:	*cap* to *oli*	47.3	24.0
	cap to *par*	42.1	16.0
	oli to *par*	20.9	22.4
Polarity:	*cap* to *oli*	77.0	1.2
	cap to *par*	68.5	0.9
	oli to *par*	0.6	0.8
Transmission ratio:	*cap-r*/*cap-s*	11.2	0.9
	oli-r/*oli-s*	0.8	0.8

known is that ω affects only about 5% of the mitochondrial genome and that ω^+ is in fact a short DNA insertion of about 1000 nucleotide pairs.

A second problem that delayed recombination mapping was the failure to find a minority class in three-point ("trihybrid") crosses. As you should recall from Chapter 5, this class normally provides the chief clue for determining gene sequence. The reason for this lack of a minority class is not known.

Despite such problems, however, a great body of data was obtained for recombination mapping. These data became useful when viewed in the light of subsequent developments in the mapping of mtDNA—developments that came through less conventional genetic analyses.

Mapping of Yeast mtDNA by Petite Analysis. The petite mutations and the drug-resistance mutations are both apparently inherited on the mitochondrial genome of yeast. Some very effective techniques for mapping that genome have been developed through the combined study of both classes of mutants. Most of these approaches are based on the fact that petites probably represent deletions of the mtDNA. This fact opens up a new and different kind of genetic analysis that has been combined with new techniques of DNA manipulation to produce a rather complete genetic map of the mtDNA.

The pivotal observation came in studies where drug-resistant grande strains (such as *ery-r*) were used as the starting material for induction of petite mutants. We have seen that petite mutants form spontaneously, but they can also be induced by use of various specific mutagens, notably ethidium bromide. When petite cultures were obtained from grande *ery-r* strains, it was of interest to determine their drug resistance. Are they still *ery-r*, or has the petite mutation caused them to become *ery-s*? Unfortunately, the answer cannot be determined directly because drug resistance cannot be tested in petite cells!

However, genetic trickery comes to the rescue. The petites are combined with grande *ery-s* cells (which, of course, lack resistance to erythromycin) to form diploids, and the diploids are allowed to form diploid buds. Depending on the type of petite used, varying amounts of petite and grande diploid buds are formed. It is the grande diploid buds that are of interest here because they *can* be tested for drug resistance. If the original petites retain the drug resistance, then some of these diploid grande cells might be expected to have acquired that resistance through recombination during the CSAR process. In fact, for some petites, the derived grande diploid buds did prove to be of two phenotypes: *ery-s* (derived from the grande *ery-s* haploid) and *ery-r* (which must have been derived from the petite haploid). This result indicates that only the genetic determinants for the grande phenotype are lost in the original mutation event that generates these particular petite mutants; the genetic determinants for the drug resistance were not affected by the petite mutation.

Of particular interest, however, were those petites whose derived grande diploids were all *ery-s*. In this case, the original mutation event apparently inactivated or destroyed both the determinant for the grande phenotype *and* the determinant for drug resistance (Figure 15-19). The coincidental loss of several

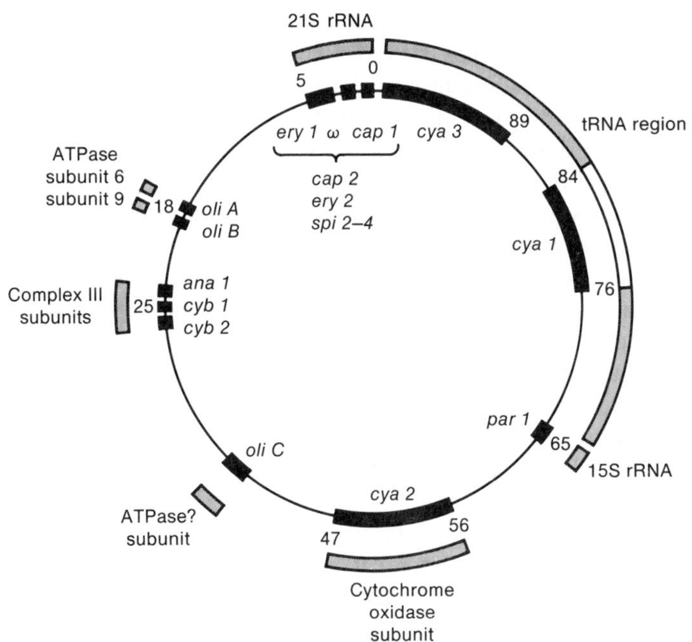

Figure 15-25. Genetic map of yeast mtDNA, showing mutant loci (black) *and gene functions* (shaded). *Positions are indicated on an arbitrary scale of 100 units reading counterclockwise from the top of the map. Mutants are labeled inside the circle. (After A. Linnane, "Mitochondrial Genetics in Perspective."* Plasmid *1, 1978.)*

The map in Figure 15-25 shows the location of the familiar drug-resistance genes as well as the genes for the mitochondrial rRNA and tRNA. (Positions are indicated in a scale of 100 arbitrary units starting from an arbitrary point.) Mutants labeled *cya* and *cyb* also are shown; these are examples of the so-called *mit* mutants. The *mit* mutants have phenotypes similar to the petite mutants, but the *mit* mutants retain mitochondrial protein synthesis. They can be regarded as "point-mutation" petites. The designations *cya* and *cyb* are based on biochemical studies of a kind that we now describe briefly.

Several lines of biochemical research built up the view of gene function indicated in Figure 15-25. Protein synthesis in vitro using grande mtDNA showed that the protein products are small polypeptides of average molecular weight from 12,000 to 20,000. These small proteins were identified immunologically as subunits of the electron-transport-chain proteins such as cytochrome *a*, cytochrome *c*, and ATPase. Further studies showed that certain mutants lack these subunits. Notably, the *oli-r* mutants have defects in either of two subunits of mitochondrial ATPase, and some *mit* mutants lack a subunit of cytochrome *a* (these are the *cya* mutants) and others lack a subunit of cytochrome *b* (the *cyb* mutants).

Restriction-enzyme techniques also have been used to prove that cytoplasmic recombination occurs through the production of recombinant DNA molecules. Parental cells carrying different genetic markers and having mtDNAs that can be distinguished by restriction-enzyme treatment are used to produce genetic recombinants. These recombinant cells contain only one type of mtDNA, which can be treated with restriction enzymes. Some of the resulting fragments resemble those from one parent, and some resemble fragments from the other parent. This result confirms that the new genetically recombinant mitochondrial genome is physically composed of recombinant DNA.

A Summary of the Yeast Mitochondrial System

The studies we have reviewed yield a picture of a mtDNA concerned with the production of subunits for electron-transport proteins and of the RNA and protein components of a unique mitochondrial protein-synthesizing apparatus. Many of the mutations produced in the protein-synthesizing apparatus confer resistance to various drugs, presumably because of altered protein conformations. The normal conformations of these proteins are sensitive to damage by the drugs. Figure 15-26 summarizes the information about the yeast mitochondrial system.

Why should this mtDNA system exist at all? One answer, although not an adequate one, is that the mtDNA encodes hydrophobic mitochondrial membrane components that could not pass into the mitochondrion from the cytosol and so must be synthesized by a special system from the inside of the mitochondrion. Although plausible, this explanation cannot provide the entire story

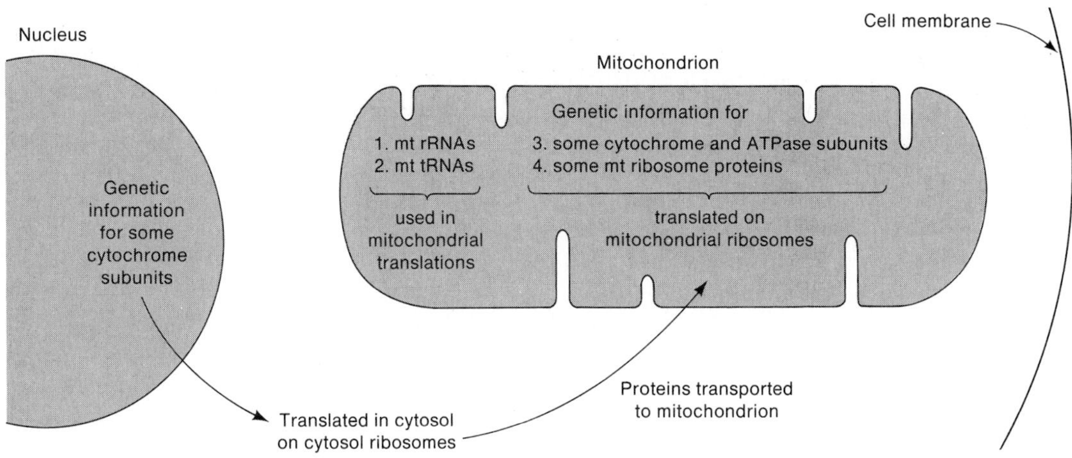

Figure 15-26. *A summary of the functions of the yeast mitochondrial system and its interrelation with the nuclear genome. The cytochrome subunits (and the ATPase subunits) are proteins needed for a functional electron transport system, which is essential to normal aerobic energy production.*

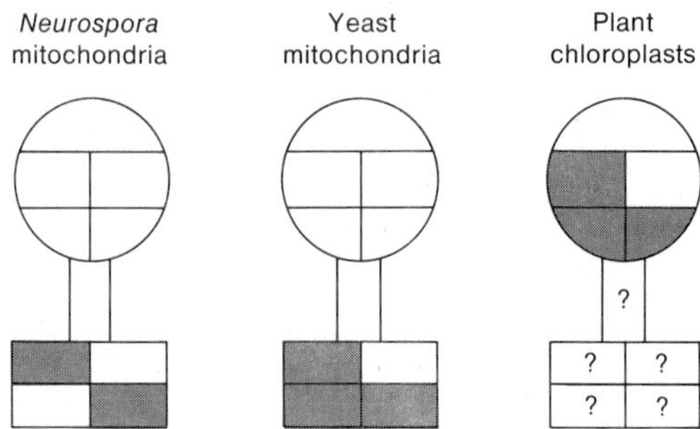

Figure 15-27. The source of the subunits of the enzyme ATPase, which is found in mitochondria and chloroplasts as part of the ATP-generating electron-transport system. Subunits synthesized on organelle ribosomes are shaded; subunits synthesized on cytosol ribosomes are unshaded. The origins of some subunits of the ATPase in chloroplasts are not yet known. (After P. Borst and A. Grivell, "The Mitochondrial Genome in Yeast." Cell *15:705, 1978. Copyright © 1978 M.I.T. Press.)*

because some of the components of mitochondrial ATPase (encoded by the mtDNA in yeast) are encoded by nuclear DNA in *Neurospora* (Figure 15-27). As suggested in Figure 15-27, an analogous story is emerging about the organization of the chloroplast genome, although that story is not yet as complete.

Most researchers in this field feel that the mutant hunt is over in yeast; few new mitochondrial mutant types are likely to be found in the future. This possibility raises a puzzling problem. The total DNA involved in the known genes makes up only about 15% of the mtDNA! What function does the other 85% of the mtDNA serve? Apparently, this large proportion of the mtDNA is "silent" or **spacer** DNA. The spacer DNA stretches are known to be rich in A + T (only 5% G + C), whereas the active parts of the DNA are much richer in G + C (32%). The spacer DNA also is highly repetitive.

Although no function is known for this spacer DNA, it does provide a neat explanation for the highly variable G + C content observed in petite mutants. If the region retained and amplified in a particular petite (Figure 15-23) includes an abnormally large or small proportion of spacer DNA, then the atypical G + C content of the retained fragment will be amplified and will cause an atypical G + C content for the entire petite mtDNA.

Message
The mtDNA has genes for mitochondrial translation components and for some electron-transport protein subunits; it also includes spacer regions thought to be genetically inactive.

As a footnote to this section, it is worth noting that studies on the anticodons of mitochondrial tRNA have shown that the genetic code in mitochondrial protein synthesis is slightly different from that at work in the cytosol. For example, in *Neurospora,* yeast, and humans, one normal tRNA reads the UGA codon as Trp, not as a stop codon.

How Many Copies?

For the genetic and the biochemical approaches to the study of organelle genomes that we have discussed, it does not matter much how many copies of organelle genome are present per cell. (Even in the case of the nuclear genome, the number of genomes present per chromosome was in doubt until very recently, and many uncertainties still exist in that area.) However, it is interesting to ask how many copies of the organelle genome do exist, and an answer can be given.

The answer turns out to vary among species. More surprisingly, it also can vary within a single species. The leaf cells of the garden beet have about 40 chloroplasts per cell. The chloroplasts themselves contain specific areas that stain heavily with DNA stains; these areas are called **nucleoids,** and they are a feature commonly found in many organelles. Each beet chloroplast contains from 4 to 18 nucleoids, and each nucleoid can contain from 4 to 8 cpDNA molecules. Thus, cells of a single beet leaf can contain as many as $40 \times 18 \times 8 = 5760$ copies of the chloroplast genome! Although *Chlamydomonas* has only one chloroplast per cell, the chloroplast contains from 500 to 1500 cpDNA molecules, commonly observed to be packed in nucleoids.

A "typical" haploid yeast cell can contain from 1 to 45 mitochondria, each having 10 to 30 nucleoids, with 4 or 5 molecules in each nucleoid.

How does this genome duplication relate to the CSAR process? How many genomes are present at the beginning of CSAR? Do all copies of the genome actually become involved in the CSAR process? Does CSAR occur when organelles fuse, and must the nucleoids fuse also? It seems that the number of organelles in a cell generally is too large to permit cytoplasmic segregation by random assortment of organelles. How can the CSAR process segregate the many copies of the genome scattered through the cell? Few answers to such questions are available at present. The situation is rather like that of early geneticists who knew about genes and linkage groups but knew nothing about their relation to the process of meiosis. As we have seen, such problems pose few limitations to progress in genetic research—they are problems for the cell biologist.

Other Examples of Extranuclear Inheritance

We have concentrated in this discussion on a few extranuclear systems that are relatively well understood. However, quite a few other examples of extranuclear inheritance have been encountered in the history of genetic analysis. They are detected experimentally as differences between reciprocal crosses, and many diverse phenotypes are affected.

One such phenotype of great importance in agriculture is that of male sterility in plants. Sterile male plants produce no functional pollen, but the trait of male sterility itself is inherited (rather obtusely) only via the female parent. This situation has been put to profitable use in the production of hybrid corn. The

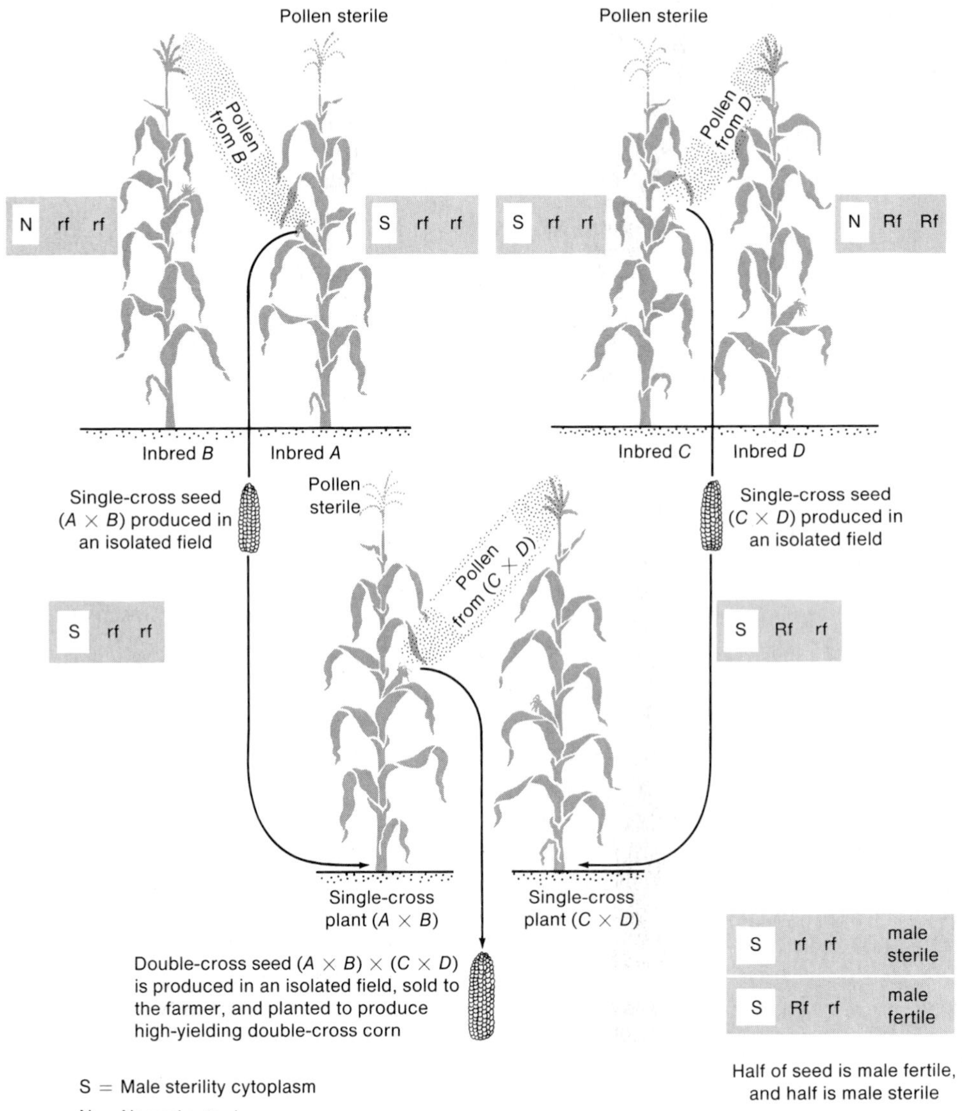

Figure 15-28. The use of cytoplasmic male sterility to facilitate the production of hybrid corn. In this scheme, the hybrid corn is generated from four pure parental lines, A, B, C, *and* D. *Such hybrids are called double-cross hybrids. At each step, selfing is prevented by appropriate combinations of cytoplasmic genes and nuclear restorer genes to ensure that the female parents will be pollen sterile (male sterile). (From J. Janick et al.,* Plant Science. *Copyright © 1974, W. H. Freeman and Company.)*

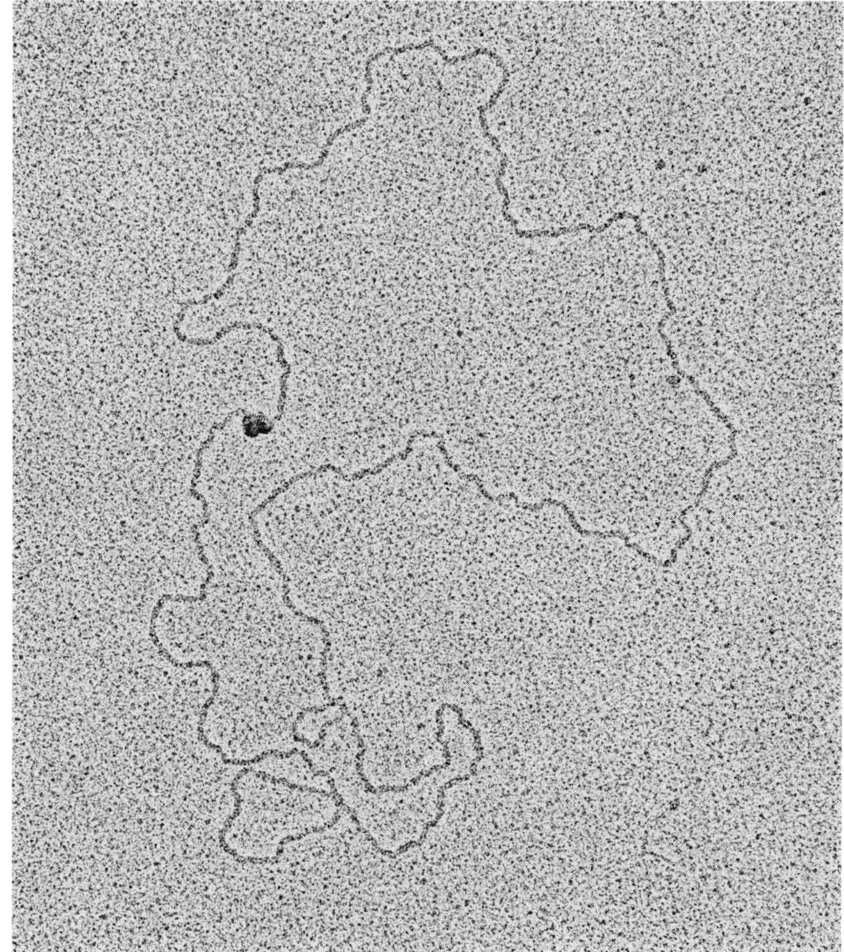

Figure 15-29. Circular mitochondrial DNA molecule from a human cell. The dark blob is a protein involved in anchoring the DNA to the mitochondrial membrane. (Photo by Jack D. Griffith.)

hybrid corn seeds that one buys from a seed supplier are produced from crosses between two specific parental strains. These parents, however, must be prevented from selfing (which would not produce hybrid kernels). The selfing is conveniently prevented by incorporating cytoplasmic male sterility into one parent (Figure 15-28). Surprisingly, examples of cytoplasmic male sterility have been found in natural populations. There is no reason why many of the elegant approaches to organelle genetics developed in yeast should not be applied to other organisms, even to man (Figure 15-29).

Summary

The extranuclear genome supplements the nuclear genome of eukaryotic organisms discussed in earlier chapters. The existence of the extranuclear genome is recognized at the genetic level chiefly by uniparental transmission of the relevant mutant phenotypes. At the cellular level, a combination of genetic and biochemical techniques has demonstrated that the extracellular genome is organelle DNA—either mitochondrial (mtDNA) or chloroplast (cpDNA).

The mtDNA of yeast is the best-understood organelle DNA. This DNA codes for unique mitochondrial translational components, and it also codes for some components of the respiratory enzymes normally found in the mitochondrial membranes. Mutations in the translational-component genes typically produce drug-resistant phenotypes; mutations in the respiratory-enzyme genes typically produce phenotypes involving respiratory insufficiency.

The chloroplast DNA is larger (about 2.5 times greater in diameter) and more complex than the yeast mtDNA. Mutations in cpDNA typically produce photosynthetic defects or drug resistance.

The organelle genes have been fully investigated in only a few organisms, but these studies have produced general analytical techniques that should have widespread application in the study of extranuclear genomes in many other organisms.

Problems

1. In the genus *Antirrhinum*, a yellowish leaf phenotype called prazinizans (pr) is inherited as follows:

   $$\text{normal} \times \text{pr} \rightarrow 41{,}203 \text{ normal} + 13 \text{ variegated}$$
   $$\text{pr} \times \text{normal} \rightarrow 42{,}235 \text{ pr} + 8 \text{ variegated}$$

   Explain these results with a hypothesis involving cytoplasmic inheritance. (Explain both the majority *and* the minority classes of progeny.)

2. You are studying a plant whose tissue includes green sectors and white sectors. You wish to decide whether this phenomenon is due to (1) a chloroplast mutation of the type discussed in this chapter, or (2) a dominant nuclear mutation that inhibits chlorophyll production and is present only in certain tissue layers of the plant as a mosaic. Outline the experimental approach you would use to resolve this problem.

3. A dwarf variant of tomato appears in a research line. The dwarf is crossed as female to normal plants, and all the F_1 progeny are dwarf. These F_1 individuals are selfed, and the F_2 progeny are all normal. Each of the F_2 individuals is selfed, and the resulting F_3 generation is 3/4 normal and 1/4 dwarf. Can these results be explained (a) by cytoplasmic inheritance? (b) by cytoplasmic inheritance plus nuclear suppressor gene(s)? (c) by maternal influence on the zygotes? Explain your answers.

4. Assume that diploid plant A has a cytoplasm genetically different from that of plant B. To study nuclear–cytoplasmic relations, you wish to obtain a plant with the cytoplasm of plant A and the nuclear genome predominantly of plant B. How would you go about producing such a plant?

5. Two species of *Epilobium* (fireweed) are intercrossed reciprocally as follows:

   ♀ *E. luteum* × ♂ *E. hirsutum* → progeny all very tall
   ♀ *E. hirsutum* × ♂ *E. luteum* → progeny all very short

   The progeny from the first cross were backcrossed as females to *E. hirsutum* for 24 successive generations. At the end of this crossing program, the progeny still were all tall, like the initial hybrids. (a) Interpret the reciprocal crosses. (b) Explain why the program of backcrosses was performed.

6. One form of male sterility in maize is maternally transmitted. Plants of a male-sterile line crossed with normal pollen give male-sterile plants. In addition, some lines of maize are known to carry a dominant nuclear restorer gene (*Rf*) that restores pollen fertility in male-sterile lines.

   a. Research shows that introduction of restorer genes into male-sterile lines does not alter or affect the maintenance of the cytoplasmic factors for male sterility. What kind of research results would lead to such a conclusion?

   b. A male-sterile plant is crossed with pollen from a plant homozygous for gene *Rf*. What is the genotype of the F_1? What is its phenotype?

   c. The F_1 plants from part b are used as females in a testcross with pollen from a normal plant (*rf rf*). What would be the result of this testcross? Give genotypes and phenotypes, and designate the kind of cytoplasm.

   d. The restorer gene already described can be called *Rf-1*. Another dominant restorer, *Rf-2*, has been found; it is located in a different chromosome from that of *Rf-1*. Either or both of the restorer alleles will give pollen fertility. Using a male-sterile plant as a tester, what would be the result of a cross where the male parent was

   (i) heterozygous at both restorer loci? (ii) homozygous dominant at one restorer locus and homozygous recessive at the other? (iii) heterozygous at one restorer locus and homozygous recessive at the other? (iv) heterozygous at one restorer locus and homozygous dominant at the other?

7. In the flax cultivar called Stormont Cirrus, the following experiments were performed. Plants grown on soil with a high ratio of nitrogen to phosphorus (N/P) were large, and they produced large progeny in the next generation (G_2). In all generations following G_2, large plants were produced *even when not grown* in soil of high N/P ratio. When the original cultivar was grown in soil of low N/P ratio, small plants were produced, and this phenotype showed similar persistence through later generations. (The large and small variants are called **genotrophs.**) When a large plant is crossed to a small plant, the progeny are all of the original cultivar type. The large genotroph has 60% more DNA and 60% more ribosomal genes than the normal types. Can you explain these results? (Few people can.)

8. In *Chlamydomonas,* application of nitrosoguanidine when the nucleus is preparing for division leads to many nuclear mutations but no uniparentally inherited mutations. If the application of nitrosoguanidine is made when the chloroplast is prepar-

ing for division, then there is an increase in uniparentally inherited mutations. What is the relevance of this observation?

9. Treatment with streptomycin induces the formation of streptomycin-resistant mutant cells in *Chlamydomonas*. In the course of subsequent mitotic divisions, some of the daughter cells produced from some of these mutant cells show the normal phenotype. Suggest a possible explanation of this phenomenon.

10. In *Chlamydomonas,* chloroplast genes may be mapped in relation to the hypothetical attachment point (*ap*) by growing zygotes in liquid culture and tracing the rate of disappearance of cytohets with time. (Meiotic products appear and go through subsequent mitotic divisions.) Figure 15-30 summarizes the results of such an experiment. The vertical axis of the graph is measured as the proportion of cells that are cytohets for a given marker at time *t*, divided by the proportion that were cytohets at time 0. The *ant* symbols represent various different antibiotic-resistance markers.

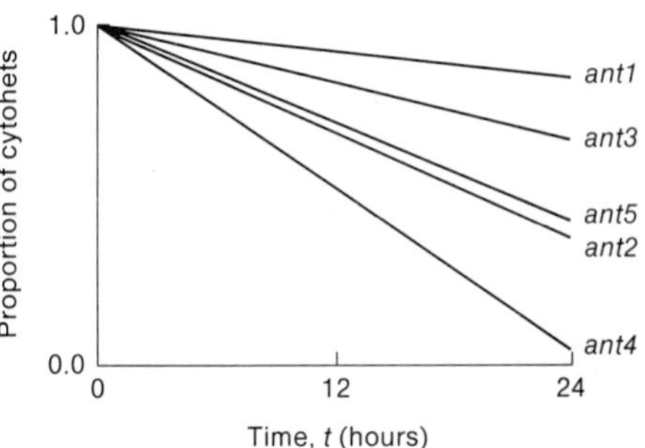

Figure 15-30.

a. What is the order of the drug-resistance markers in relation to *ap*?

b. Of what use might the slopes of these curves be in genetic mapping?

11. Cosegregation mapping is performed in *Chlamydomonas* on four chloroplast markers, *m1*, *m2*, *m3*, and *m4*. The markers are considered pairwise in heterozygous condition, and cosegregation frequencies are obtained as indicated in Table 15-6. (For example, *m1* and *m2* cosegregate in 29% of the cell divisions followed.) Draw a rough genetic map based on these results.

Table 15-6.

	m1	*m2*	*m3*	*m4*
m1	—	29.0	18.0	18.4
m2		—	10.9	26.2
m3			—	8.8
m4				—

12. In *Aspergillus,* a "red" mycelium arises in a haploid strain. You make a heterokaryon with a nonred haploid that requires *para*-aminobenzoic acid (PABA). From this heterokaryon, you obtain some PABA-requiring progeny cultures that are red, along with several other phenotypes. What does this information tell you about the gene determining the red phenotype?

13. On page 624, an experiment is described in which abnormal mitochondria are injected into normal *Neurospora.* The text mentions that "appropriate controls" are used. What controls would you use?

14. Adrian Srb crossed two closely related species, *Neurospora crassa* and *N. sitophila.* In the progeny of some of these crosses, there appeared a phenotype called aconidial (ac) that involves a lack of conidia (asexual spores). The observed inheritance was

♀ *N. sitophila* × ♂ *N. crassa* → progeny 1/2 ac and 1/2 normal
♀ *N. crassa* × ♂ *N. sitophila* → progeny all normal

a. What is the explanation of this result? Explain all components of your model with symbols.

b. From which parent(s) did the genetic determinants for the ac phenotype originate?

c. Why were neither of the parental types ac?

15. The following crosses were made in *Neurospora*. Explain the results of each cross, and assign genetic symbols for each of the strains involved.

a. Nonpoky strain B ♀ × poky strain A ♂ → progeny all nonpoky

b. Nonpoky strain C ♀ × poky strain A ♂ → progeny all nonpoky

c. Poky strain A ♀ × nonpoky strain B ♂ → progeny all poky

d. Poky strain A ♀ × nonpoky strain C ♂ → progeny 1/2 poky, all identical, e.g., strain D, and 1/2 nonpoky, all identical, e.g., strain E

e. Nonpoky strain E ♀ × nonpoky strain C ♂ → progeny all nonpoky

f. Nonpoky strain E ♀ × nonpoky strain B ♂ → progeny 1/2 poky, 1/2 nonpoky

g. Poky strain D behaves just like poky strain A in all crosses.

16. In yeast, an antibiotic-resistant haploid strain ant^r arises spontaneously. It is combined with a normal ant^s strain of opposite mating type to form a diploid culture that is then allowed to go through meiosis. Three tetrads are isolated:

tetrad 1	tetrad 2	tetrad 3
$\alpha\ ant^r$	$\alpha\ ant^r$	$a\ ant^s$
$\alpha\ ant^r$	$a\ ant^r$	$a\ ant^s$
$a\ ant^r$	$a\ ant^r$	$\alpha\ ant^s$
$a\ ant^r$	$\alpha\ ant^r$	$\alpha\ ant^s$

a. Interpret these results.

b. Explain the origin of each ascus.

c. If an ant^r grande strain were used to generate petites, would you expect some of the petites to be ant^s? Explain.

17. In yeast, two haploid strains are obtained that are both defective in their cytochromes; the mutants are named *cyt1* and *cyt2*. The following crosses are made:

$cyt1^- \times cyt1^+$
$cyt2^- \times cyt2^+$

One tetrad is isolated from each cross; they are the following:

$cyt1^-$	$cyt2^-$
$cyt1^-$	$cyt2^-$
$cyt1^+$	$cyt2^-$
$cyt1^+$	$cyt2^-$

a. Explain the difference in the natures of these two mutants.

b. What other ascus types might be expected from each cross?

c. How might the two genes involved in these mutants interact at the functional level?

18. In a marker-retention analysis in yeast, a multiply resistant strain apt^r bar^r cob^r is used to induce 500 petites. Table 15-7 shows, for different pairs of markers, the number of petites in which only one marker of the pair was lost. (For example, 87 petites were apt^s bar^r, and 120 were apt^r bar^s.) Use these results to draw a rough map of these mitochondrial genes.

Table 15-7.

Gene pair	Petites in which first marker is lost	Petites in which second marker is lost
apt bar	87	120
apt cob	27	18
bar cob	48	69

19. A grande yeast culture of genotype cap^r ery^r oli^r par^r is used to obtain petites by treatment with ethidium bromide. The petites are tested for (1) their drug resistance and (2) the ability of their mtDNA to hybridize with various specific mtRNA types: $rRNA_{small}$, $rRNA_{large}$, $tRNA_1$, $tRNA_2$, $tRNA_3$, $tRNA_4$, and $tRNA_5$. In all, 12 petites are tested, and Table 15-8 shows the results. Using this information, plot a map of the mtDNA showing the sequence of the 11 genetic loci involved in these phenotypes. Be sure to state your assumptions and draw a complete map.

Table 15-8.

	Resistant (r) or sensitive (s) to				Petite mtDNA is able (+) or unable (−) to hybridize with mitochondrial						
Petite culture	cap	ery	oli	par	$rRNA_{large}$	$rRNA_{small}$	$tRNA_1$	$tRNA_2$	$tRNA_3$	$tRNA_4$	$tRNA_5$
1	r	s	s	s	+	−	−	−	−	−	−
2	s	s	s	s	−	−	−	−	−	+	−
3	s	s	s	s	−	−	−	+	−	−	+
4	s	s	r	s	−	−	+	−	−	−	−
5	s	r	s	s	+	+	−	−	−	−	−
6	r	s	s	s	−	−	−	−	+	−	−
7	s	s	r	r	−	−	−	+	−	−	−
8	s	s	s	s	−	+	−	−	−	+	+
9	s	r	s	s	+	−	−	−	−	−	−
10	s	s	s	s	−	−	+	−	+	−	−
11	r	s	r	r	−	+	+	+	+	+	+
12	s	r	s	s	+	+	−	−	−	−	−

20. The mtDNA is compared from two haploid strains of bakers' yeast. Strain 1 (mating type α) is from North America, and strain 2 (mating type a) is from Europe. A single restriction enzyme is used to fragment the DNAs, and the fragments are separated on an electrophoretic gel. The sample from strain 1 produces two bands, corresponding to one very large and one very small fragment. Strain 2 also produces two bands, but they are of more intermediate sizes. If a standard diploid budding analysis is performed, what results do you expect to observe in the resulting cells and in the tetrads derived from them? In other words, what kinds of restriction-fragment patterns do you expect?

21. In yeast, some strains are found to have in their cytoplasm circular DNA molecules that are 2 microns (μ) in circumference. In some strains, this 2 μ DNA has a single *Eco* RI restriction site; in other strains, there are two such sites. A strain with one site is mated to a strain with two sites. All of the resulting diploid buds are found to contain both kinds of 2 μ DNA.

    a. Is the 2 μ DNA inherited in the same fashion as mtDNA?

    b. What do you think the 2 μ DNA is likely to represent functionally?

22. Using restriction-enzyme techniques, how would you demonstrate that mitochondrial recombination is due to a process similar to crossing over between two mitochondrial genomes (and not to assortment of separate particles)?

Kernels of corn, showing mutational instability of the a_1 *mutation under the influence of the controlling element* Dt. *Each dot represents a reversion event. (Kernels provided by Derek Styles.)*

16

Mechanisms of Genetic Change

Genetic change is of fundamental interest to the genetic analyst because the processes of genetic change generate the raw material that geneticists study. This raw material, of course, is an ample supply of hereditary variants. Such variants are easily obtainable from plants and animals in natural populations, or they may be induced by laboratory manipulation. Many geneticists pursue their studies of these variants unmindful of their precise modes of origin, using them merely as tools for their genetic dissections. Such dissections comprise much of the content of our discussion in the preceding chapters. However, genetic change in itself is a cellular process of central significance in biology and in many aspects of human affairs. Thus, the mechanisms of genetic change are of great relevance to the geneticist, the biologist, the physician, the politician, and everyone else.

In this chapter, we come to grips with the molecular mechanisms of genetic change. Although genetic change has been a part of genetic research from its very beginnings, only recently have we begun to achieve some glimmer of understanding of the molecular mechanisms involved. The models discussed in this chapter represent the current level of understanding; they should not be regarded as final mechanistic truths. What we have is a rough framework on which to hang future discoveries. Many areas of uncertainty remain in this field of research.

Genetic change is not a single process, but rather it is a highly complex heterogeneous collection of processes that can produce change. A complete coverage is impossible in a text such as this, but we divide the subject into four major topics. If you as an analyst are studying an individual organism that represents a variant from your established control population, what mechanism could have produced the change that created this variant individual? The following are four possibilities:

1. chromosomal mutation
2. gene mutation
3. recombination
4. transposable genetic elements

All of these headings should be familiar to you, except perhaps the last, which is in fact a relative newcomer to the list of possibilities. In this chapter, we shall have the least to say about the first possibility, chromosomal mutation, because relatively little is known about possible mechanisms for such changes. We begin with gene mutation.

Gene Mutation

To understand mechanisms of gene mutation, we must clearly understand the nature of mechanisms at the level of DNA and protein molecules. Much of this model has been described in preceding chapters. The box draws together this information to provide an explanation of the nature of gene mutations at the molecular level.

Gene mutations can arise spontaneously, or they can be induced. Spontaneous mutations occur at quite low but constant rates in individuals. Induced mutations are produced when an organism is treated with a mutagenic agent, or **mutagen;** such mutations typically are recovered at much higher frequencies than the spontaneous mutations.

Types of Gene Mutation

I. Forward mutation
   A. Single nucleotide-pair (base-pair) substitutions
      1. At the DNA level
         a. **Transitions** (purine replaced by a different purine, or pyrimidine replaced by a different pyrimidine):

            AT → GC   GC → AT   CG → TA   TA → CG

b. **Transversions** (purine replaced by a pyrimidine, or pyrimidine replaced by a purine):

AT → CG   AT → TA   GC → TA   GC → CG
TA → GC   TA → AT   CG → AT   CG → GC

2. At the protein level
   a. **Neutral mutation**
      i. To a triplet coding for the same amino acid, such as AGG → CGG (both triplets code for Arg)
      ii. To a triplet coding for a different but chemically equivalent amino acid, such as CAC → CGC (changing the basic His to the basic Arg)
   b. **Missense mutation** (to a triplet coding for a different and nonfunctional amino acid)
   c. **Nonsense mutation** (to a chain-termination triplet, such as CAG → UAG, changing from a codon for Gln to the amber termination codon)

B. Single nucleotide-pair addition or deletion (a **frameshift mutation**)

C. Intragenic deletion of several nucleotide pairs

II. Reverse mutation

A. Exact reversions, such as

$$\underset{\text{wild type}}{\text{AAA (Lys)}} \xrightarrow{\text{forward}} \underset{\text{mutant}}{\text{GAA (Glu)}} \xrightarrow{\text{reverse}} \underset{\text{wild type}}{\text{AAA (Lys)}}$$

B. Equivalent reversions, such as

$$\underset{\text{wild type}}{\text{UCC (Ser)}} \xrightarrow{\text{forward}} \underset{\text{mutant}}{\text{UGC (Cys)}} \xrightarrow{\text{reverse}} \underset{\text{wild type}}{\text{AGC (Ser)}}$$

$$\underset{\text{wild type}}{\text{CGC (Arg, basic)}} \xrightarrow{\text{forward}} \underset{\text{mutant}}{\text{CCC (Pro, not basic)}} \xrightarrow{\text{reverse}} \underset{\text{pseudo-wild type}}{\text{CAC (His, basic)}}$$

C. **Suppressor mutations**

1. Frameshift of opposite sign at a second site, such as

```
CATCATCATCATCATCAT
   (+)  (−)
    ↓    ↓
CATXCATATCATCATCAT
 √   ×   ×   √   √   √
```

(*Continued*)

Types of Gene Mutation (*Continued*)

2. Second-site missense mutation. This type is still not fully understood at the level of protein function, but it is very common. It is explained in terms of a second distortion that restores a more-or-less wild-type protein conformation after a primary distortion. This mutation type has not been mentioned before in this text; Figure 16-1 shows a highly diagrammatic visualization of a possible mode of action for such a mutation. The existence of such suppression mutations is revealed by amino-acid-sequencing analyses.

3. Extragenic suppressors

   a. **Nonsense suppressors.** A gene (say, for tyrosine tRNA) undergoes a mutational event in its anticodon region that enables it to recognize and align with a mutant nonsense codon (say, the amber UAG) to insert an amino acid (tyrosine, in this case) and permit completion of the translation.

   b. **Missense suppressors.** A heterogeneous set of mutations whose molecular mechanisms are not fully understood. In *E. coli*, there is a missense suppressor that is an abnormal tRNA that carries glycine but inserts it in response to arginine codons. Such a mutation might be expected to be very dangerous because *normal* arginine codons would also be mistranslated. However, the observed mutations are not lethal, probably because the efficiency of the abnormal substitution is very low.

   c. **Frameshift suppressors.** Very few examples of these mutations have been found. In one, there is a four-nucleotide anticodon in a single tRNA that can "read" a four-letter codon caused by a single nucleotide-pair insertion.

   d. **Physiological suppressors.** A defect in one chemical pathway is circumvented by another mutation—for example, one that opens up another chemical pathway to the same result, or perhaps one that permits more efficient transport of a compound produced in small quantities because of the original mutation. As you can see, these mutations act as one form of missense suppressors. This group is very heterogeneous.

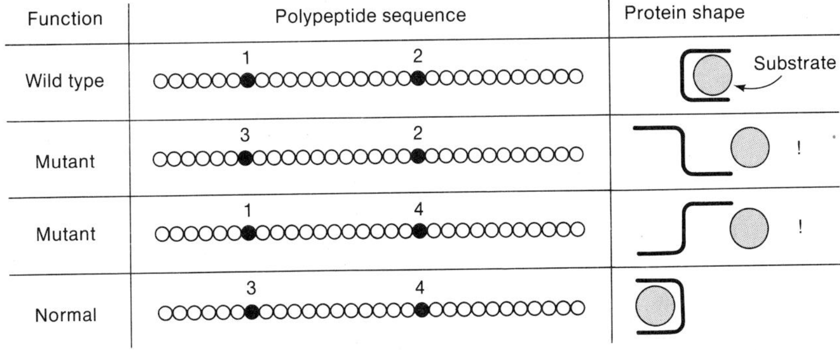

Figure 16-1. A second-site mutation within the same gene as the first mutation can act as a suppressor of the mutant phenotype. In this highly simplified representation, the diagrams at the right assume that a C-shaped protein is necessary for normal function. The first mutation event changes 1 to 3, distorting the protein to a nonfunctional shape. A second-site mutation changing 2 to 4 might restore the protein function (typically the restoration is only partial, as in the 34 polypeptide), even though that mutation by itself would also produce a nonfunctional protein.

Spontaneous Mutation

Spontaneous mutation was once attributed to the effects of natural background radiation and chemical mutagens in the environment. However, studies revealed that spontaneous mutations occur at a rate too high to be entirely attributable to these natural mutagens (Figure 16-2). Thus it is now thought that most spontaneous mutations are due to mistakes made by normal cellular enzymes. To see how such a mistake can be made, consider an example. Suppose that a polymerase on occasion allows an illegitimate nucleotide pair (say, A–C) to form during DNA replication, thereby generating an altered codon. Recombination enzymes or recombination itself by unequal exchange also can create heritable changes.

Several examples exist of mutant alleles that increase the frequency of spontaneous mutation elsewhere in the genome. The existence of these **mutator alleles** obviously is consistent with the idea that enzymes or other proteins (possibly those concerned with nucleic-acid metabolism) play a major role in producing spontaneous mutations.

If the frequency of spontaneous mutation is controlled by the genome itself, we can draw the interesting conclusion that this frequency can be (and presumably has been) affected by natural selection. The spontaneous rates existing in organisms today probably represent a balance between (1) the input of advantageous new alleles needed to retain evolutionary flexibility and (2) the input of disadvantageous new alleles, which ultimately must be eliminated from the population in dead or infertile individuals (genetic death).

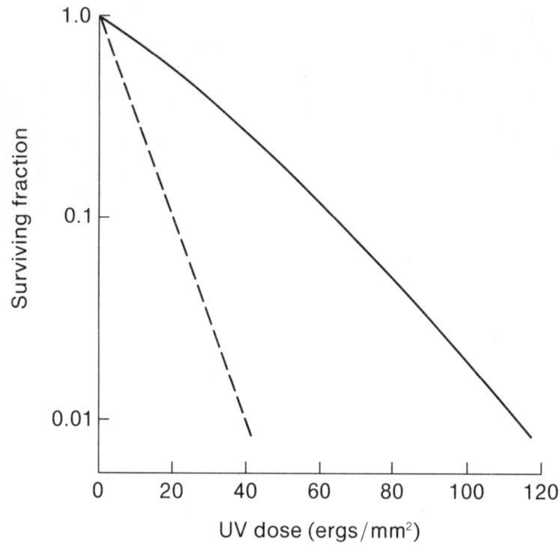

Figure 16-7. Survival curves for human skin cells in tissue culture after exposure to various dosages of UV irradiation. The solid line shows the curve for normal cells; the dashed line shows the survival curve for cells from an individual with xeroderma pigmentosum. (After J. E. Cleaver, Advances in Radiation Biology, *vol. 4, 1974.)*

dark-repair systems. The existence of repair systems based on *recA* is now well established.

Interestingly, the *polA* mutants survive quite well in a normal non-UV environment, replicating and undergoing cell division in a normal fashion. Apparently, the pol I DNA polymerase is used only during excision repair and not in normal DNA replication. In fact, we now know that other important DNA polymerases carry out the major polymerase functions during DNA replication.

The most interesting fact about the *recA* mutants is that very few UV-induced mutations appear in *recA* strains treated with UV. The lack of normal *recA* function somehow eliminates the ability to produce nonlethal UV-induced mutations. We conclude that *recA* function is the major source of UV-induced mutation. The *recA* gene seems in fact to be involved in two important repair systems, **recombination repair** and **SOS repair,** both acting during replication. (Another gene product, *lexA,* is also needed for SOS repair.) In recombination repair, the DNA replication system, which would stall at a dimer, is allowed to skip over it, leaving a single-strand gap; this gap is patched by DNA cut from the sister molecule (see Figure 16-8). This process seems to lead to few errors. SOS repair is, however, highly error prone. Here the replication system is helped across the dimer, accepting *illegitimate* nucleotides for new strand synthesis. Hence many random changes are incurred. These systems probably act on other kinds of gross structural damage to DNA, permitting replication and survival, but at the cost of mutation.

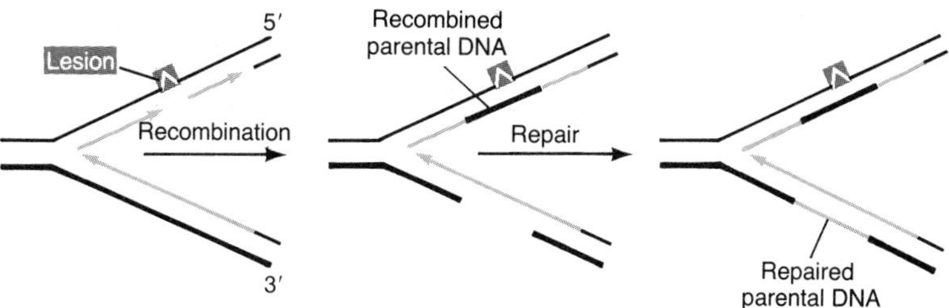

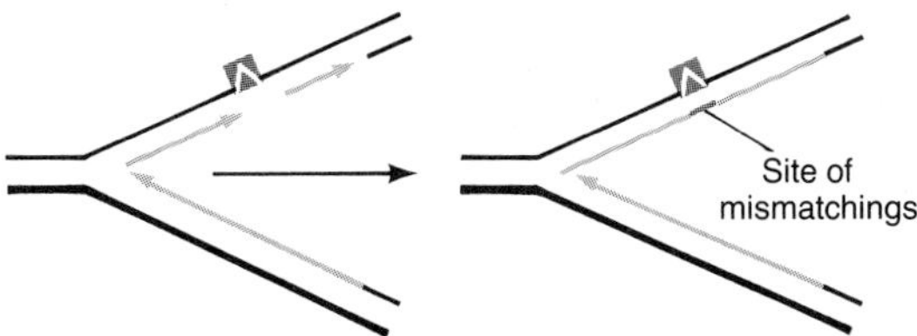

Figure 16-8. Schemes for repair, bypassing removal of the lesion, by recombination or by error-prone replication of the sequence containing the lesion. (From A. Kornberg, DNA Replication. *Copyright © 1980 by W. H. Freeman and Company.)*

Message
UV-induced nonlethal mutations are produced not by the thymine dimers that the radiation creates but rather by errors incurred during the repair of these dimers in the dark.

Can we generalize about the kind of mutation produced by UV treatment? As we shall see, a generalization of this sort is rarely possible in the study of mutagenesis.

1. Typically, a single mutagen produces a *spectrum* of mutational types, perhaps with an excess of one type or the absence of one type.
2. The precise mutational spectrum obtained depends on the organism studied.
3. Even within a single species, the mutational spectrum obtained depends very much on the specific genotype of the individual organism. (Once again, we see that cellular processes play a very important role in induction of mutations.)

In the case of UV-induced mutations, transitions are certainly common in all organisms studied, and transversions are relatively rare in most cases. In some organisms (such as *Neurospora*), a large proportion of the UV-induced mutations are frameshifts. Even duplications and deletions have been induced by UV. Evidently, the ways in which dimer repair can go awry are many and varied.

Message

A typical mutagen produces an array of mutant types. The precise nature of the mutational spectrum observed depends on the species and even on the individual genotype.

This brings us to an important problem. How can we determine the molecular nature of a specific mutation? What tests can we perform? Ideally, we would like to obtain complete DNA sequences for the wild-type and mutant genes. Modern sequencing techniques (Chapter 13) make this approach feasible, and several such studies have been performed. The next-best approach is to compare the amino-acid sequences of the wild-type and mutant protein products. This study has been performed for several different proteins, including tryptophan synthetase in *E. coli,* cytochrome *c* in yeast, and hemoglobin in humans. From the amino-acid substitutions, it often is possible to use the known genetic code to infer specific transitions, transversions, or frameshifts.

Purely genetic tests also can be used. Suppressors can be identified by linkage arrangement. Suppressors that act on several different mutations (called **super-suppressors**) are immediately suspected to be nonsense suppressors. However, the test that probably has been most useful in the widest range of studies is the reversion test. For example, a mutation that can be reversed with proflavin treatment is probably a frameshift mutation. We shall extend this reasoning as we discuss more specific kinds of mutagens.

X Rays. Although X rays were among the first agents shown to be mutagenic, the mode of their action remains obscure. Undoubtedly, the rays form free radicals and ions that are involved in damaging the DNA, but the biological effects of X rays are so complex that this probably is not the entire story. X rays are renowned for their ability to produce chromosome breaks that lead to gross structural rearrangements. However, they do also induce point mutations (such as nucleotide-pair substitutions) at high frequencies.

We would expect the frequency (m_1) of point mutations to be proportional to the dose (d), whereas the frequency (m_2) of two-break rearrangements (for example, deletions) should be proportional to the square of the dose. Using k_1 and k_2 as proportionality constants, we can write

$$m_1 = k_1 d$$

$$m_2 = k_2 d^2$$

Logarithmically,

$$\log m_1 = \log k_1 + \log d$$

$$\log m_2 = \log k_2 + 2 \log d$$

Thus, if we plot m against d on a logarithmic scale, a slope of 1 for the curve should indicate a point mutation, and a slope of 2 should indicate a two-break event. A convincing demonstration of this has been made in the case of X-ray-induced *ad-3* mutants of *Neurospora.* Genetic analysis shows that some of these mutations are point mutations within the *ad-3* region, whereas others are deletions of part or all of the *ad-3* region. The plotted curves show the expected slopes for these two cases (Figure 16-9). (On a nonlogarithmic plot, of course, the two-break case appears as a curve rather than a straight line; see Figure 16-10.) In other cases, two-break rearrangements yield exponents (slopes)

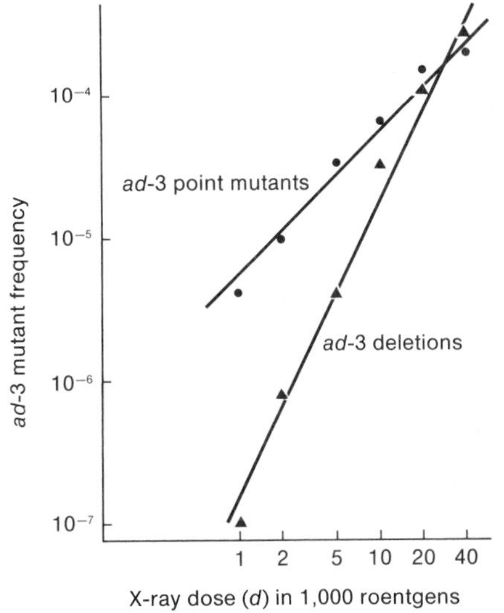

Figure 16-9. X-ray induction of ad-3 *mutations in* Neurospora. *The logarithm of the mutant frequency is plotted against the logarithm of the X-ray dose. The induction of point mutations should require only single hits; this expectation is confirmed by the slope of 1 for the induction of point mutations. On the other hand, deletions should require two hits; in fact, the data show a curve with a slope of 2. (From F. J. deSerres and H. V. Malling,* Japanese J. Genet. *44 (Suppl. 1):107, 1969.)*

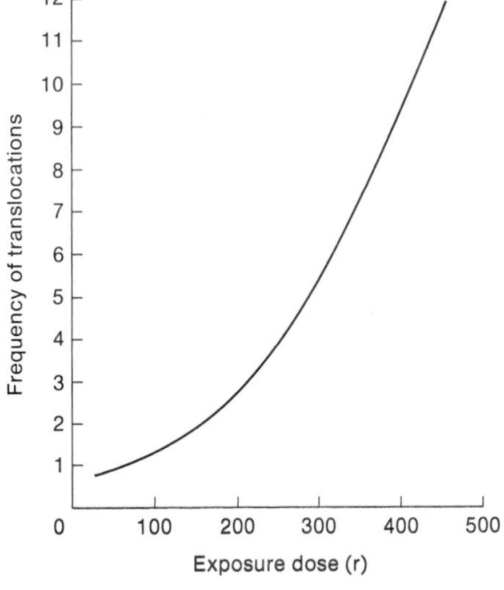

Figure 16-10. On a nonlogarithmic plot, two-break mutations produce exponential curves. The production of translocations should require two X-ray-induced breaks. Thus, the frequency of translocations should be proportional to the square of the dose, and in fact it is. (From W. F. Bodmer and L. L. Cavalli-Sforza, Genetics, Evolution, and Man. *Copyright © 1976, W. H. Freeman and Company.)*

between 1 and 2—for example, 1.5. This result is due to the fact that some broken ends rejoin to reconstitute the original chromosomal configurations, so that not all two-break events produce mutations. In a genetically untested situation, an exponent (slope) of 1.5 could also indicate a mixture of one-break and two-break events.

There is good evidence that some of the point mutations induced by X rays are base substitutions, not only in *Neurospora* but also in other organisms. For example, some temperature-sensitive mutants are induced by X-ray treatment of *Drosophila.* Most such mutations that have been fully studied have proved to involve substitution of a single amino acid due to a base substitution in the DNA. Under permissive conditions, the slightly altered protein can function normally. Hence the X-ray-induced temperature-sensitive mutants on *Drosophila* are probably examples of X-ray-induced base-substitution mutations.

Whatever the nature of the X-ray damage to DNA (whether bond breakage of various kinds or perhaps cross-linking between DNA strands), the nature of the mutation event undoubtedly depends on the cell's reaction to the initial damage (lesion). In other words, the types of mutations ultimately detected are probably a result of the actions of the cellular repair systems.

Base Analogs. Some chemical compounds are sufficiently similar to the normal nitrogen bases of DNA that they occasionally are incorporated into DNA in place of normal bases; such compounds are called **base analogs.** Once in place, these bases have pairing properties unlike those of the bases they replace, and thus they can produce mutations by causing insertions of incorrect nucleotides opposite them during replication. The original base analog exists in only a single strand, but it can cause a nucleotide-pair substitution that is replicated in all DNA copies descended from the original strand.

For example, 5-bromouracil (5BU) is an analog of thymine that has bromine at the C-5 position in place of the CH_3 group found in thymine. This change does not involve the atoms that take part in hydrogen bonding during base pairing, but the presence of the bromine significantly alters the distribution of electrons in the base. The normal structure (*keto* form) of the 5-bromouracil undergoes a relatively frequent spontaneous change to the *enol* form (Figure 16-11). (Such a spontaneous change is called a **tautomeric shift.**) The enol form has hydrogen-bonding properties almost identical to those of cytosine! Thus the nature of the pair formed during replication will depend on the form of the 5-bromouracil at the moment of pairing (Figure 16-12). The compound 2-aminopurine acts in a similar manner. It is an analog of adenine that sometimes bonds with cytosine, thus causing transitions.

DNA Modifiers. **DNA modifiers** are agents that chemically react with DNA and change the chemical properties of the bases. For example, nitrous acid (NA) reacts with the C-6 amino groups of cytosine and adenine, replacing the amino groups (positive in hydrogen bonding) with oxygen (negative in hydrogen bonding). Thus the hydrogen-bonding properties of the modified base are changed: the former cytosine now bonds like thymine, and the former adenine

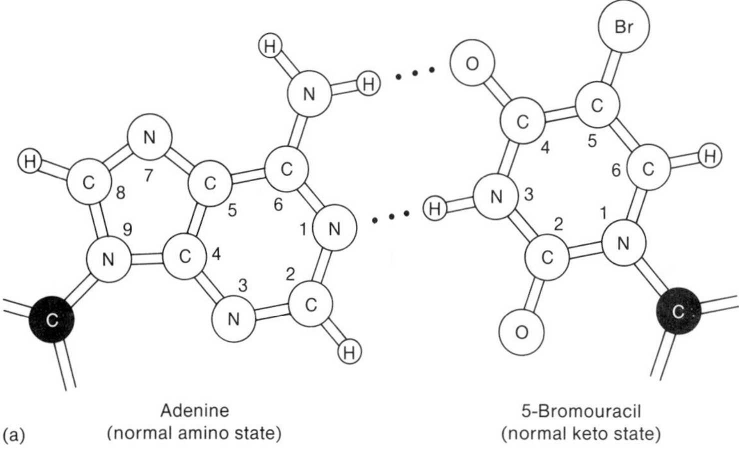

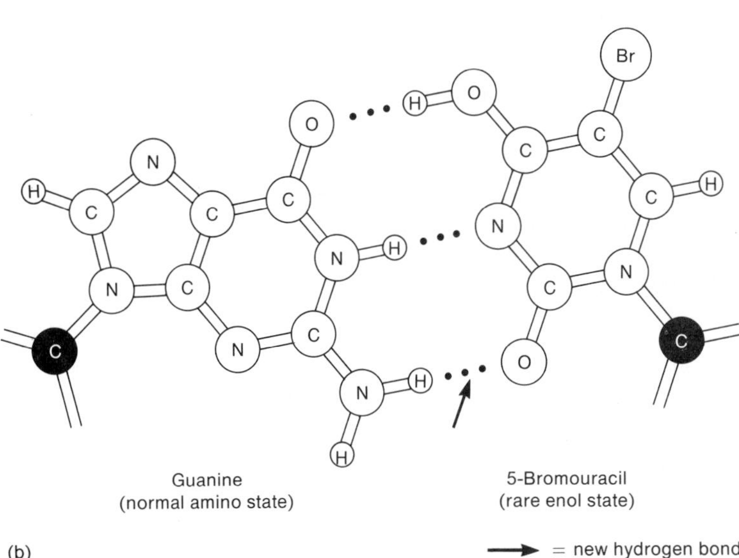

Figure 16-11. The alternative pairing possibilities for 5-bromouracil (5BU). 5BU is an analog of thymine that can be mistakenly incorporated into DNA as a base. It has a bromine atom in place of the methyl group. (a) In its normal keto state, 5BU mimics the pairing behavior of the thymine it replaces, pairing with adenine. (b) The presence of the bromine atom, however, causes a relatively frequent redistribution of electrons so that 5BU can spend part of its existence in the rare enol form. In this state, it pairs with guanine, mimicking the behavior of cytosine, and can thus induce mutations during replication.

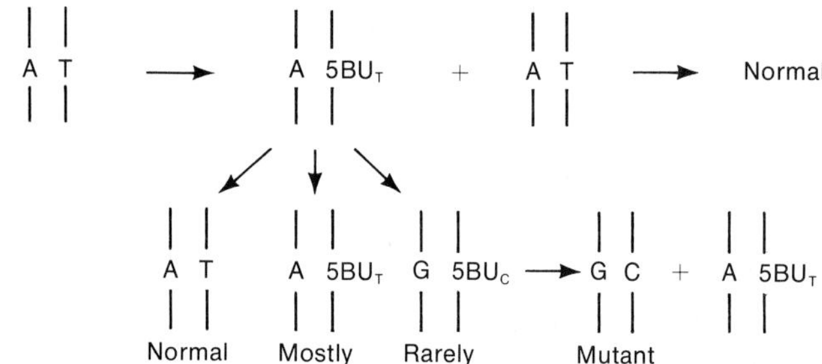

(b) Transition 2 GC → AT

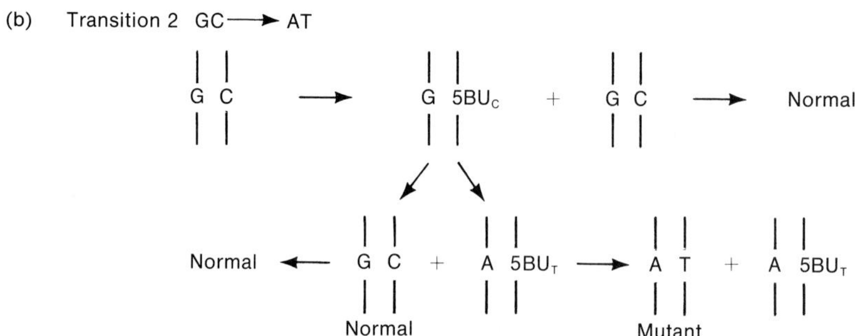

Figure 16-12. The mechanism of 5BU mutagenesis. (a) In its keto state, 5BU pairs like thymine (this state indicated as $5BU_T$). (b) In its enol state, 5BU pairs like cytosine (this state indicated as $5BU_C$).

now bonds like guanine. From these chemical considerations, we predict the transitions GC → AT and AT → GC (Figure 16-13).

The **alkylating agents** form a major class of mutagenic chemicals that react with DNA in a variety of ways. These compounds have reactive alkyl groups, and they can bond to a variety of positions on the DNA. Some alkylating agents (called **polyfunctional**) have more than one reactive alkyl group and hence can enter into several reactions simultaneously, creating various kinds of cross-linkages within the DNA.

Two of the best-known alkylating agents are ethyl methane sulfonate (EMS) and nitrogen mustard (HN2):

$$C_2H_5-O-\overset{\overset{\displaystyle O}{\|}}{\underset{\underset{\displaystyle O}{\|}}{S}}-CH_3$$

EMS (monofunctional)

$$HN\begin{matrix} \diagup CH_2CH_2Cl \\ \diagdown CH_2CH_2Cl \end{matrix}$$

HN2 (bifunctional)

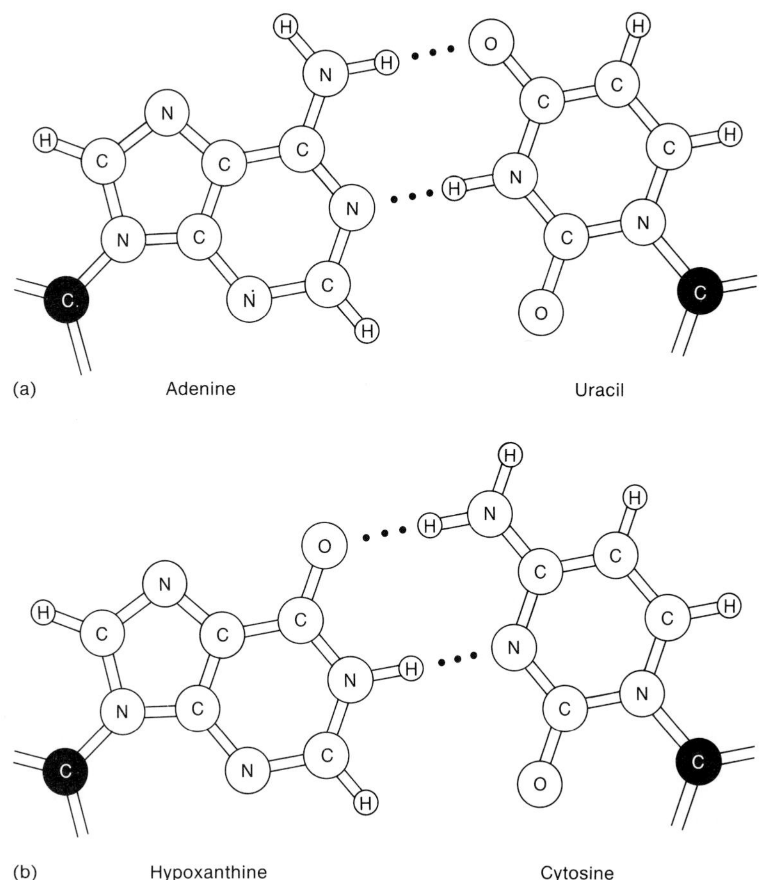

Figure 16-13. Nitrous acid (NA) mutagenesis. (a) NA deaminates cytosine to form uracil, which bonds like thymine. (b) NA deaminates adenine to form hypoxanthine, which bonds like guanine. These altered bonding patterns can lead to mutations. For example, AT may become GC, or GC may become AT. (From E. Freese, in Structure and Function of Genetic Elements, *Brookhaven Symposia in Biology, No. 12, Brookhaven National Laboratory, Upton, N.Y., 1959.)*

EMS is used routinely to generate point mutations in a wide variety of organisms; its action is not yet fully understood. It is known to react mainly at the N-7 and O-6 atoms in guanine. At both these positions, the added ethyl group can shift the electron distribution within the molecule so that guanine will pair with illegitimate partners. In T4 phage, *E. coli,* and yeast, the most common substitution observed is GC → AT. In *Neurospora,* however, AT → GC transitions appear to be favored (Figure 16-14).

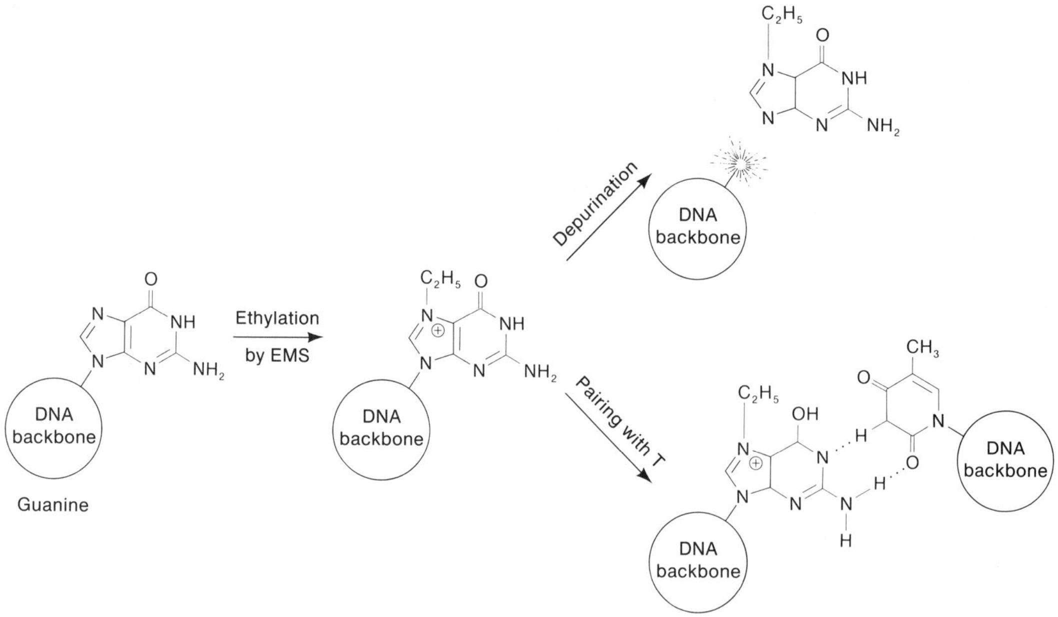

Figure 16-14. Alkylation of guanine by ethyl methane sulfate (EMS). There are at least two possible consequences that will result in point mutations. Depurination (upper right) *is the complete loss of the modified guanine from the DNA backbone and can result in the insertion of any base into a newly synthesized opposite strand of DNA. A second possible change is the pairing of the modified guanine with thymine* (lower right) *in a newly synthesized opposite strand of DNA. This change will result in a transition from GC to AT.*

As shown in Figure 16-14, alkylation also can lead to depurination. The apurinic gap left by the removal of guanine, for example, may lead to the insertion of any nucleotide in the opposite strand upon replication. This effect could produce virtually any transition, transversion, or frameshift. Another possible consequence of an apurinic gap is breakage of the DNA backbone.

The ***N*-nitroso compounds** form another major class of DNA modifiers. These compounds are highly carcinogenic, and some of them are used routinely as agents for the experimental induction of mutations in laboratory genetics. One such compound is nitrosoguanidine (NG):

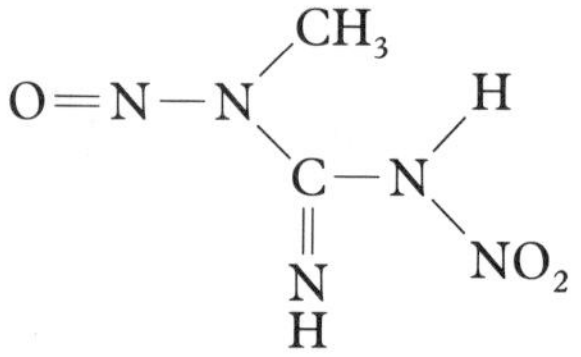

$$O{=}N{-}N(CH_3){-}C({=}NH){-}N(H)(NO_2)$$

Although NG is an alkylating agent at pH values above 7, it has its strongest mutagenic effect at acidic pH values. Thus its mutagenic action probably does not involve alkylation. In some organisms, NG seems relatively specific for GC → AT transitions, but other kinds of nucleotide-pair substitutions are commonly observed in other organisms. No precise reaction mechanism has been generally accepted.

Hydroxylamine (HA) is an interesting mutagen because, although it is not powerful, it has been advanced as the most specific inducer of GC → AT transitions. This specific effect is observed particularly in phage and *Neurospora;* the effects of HA are less specific in *E. coli.* Its structure is

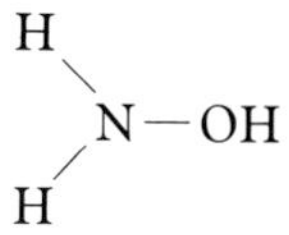

The relative specificity of HA is probably due to the fact that it preferentially hydroxylates the amino nitrogen bound to C-6 of cytosine, causing the cytosine to bind like thymine (Figure 16-15).

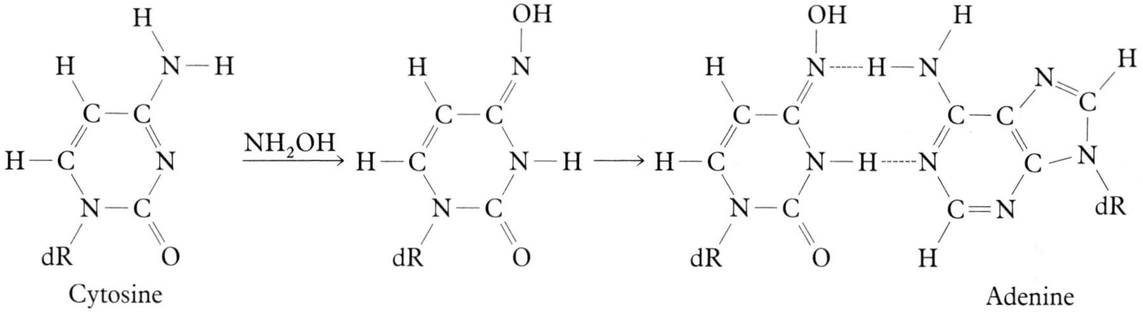

Figure 16-15. A possible explanation for the GC → AT specificity of HA in some organisms. Cytosine is modified to pair like thymine, resulting in a GC to AT transition.

The **intercalating agents** form another important class of DNA modifiers. This class includes proflavin, acridine orange, and ICR-170. Chemical studies show that these agents are able to slip themselves in between the stacked nitrogen bases at the core of the DNA double helix in a process called intercalation (Figure 16-16). In this intercalated position, the agent can cause single nucleotide-pair insertions or deletions. Again, the precise mechanism has not been ascertained, but reasonable models are available (such as the one in Figure 16-17).

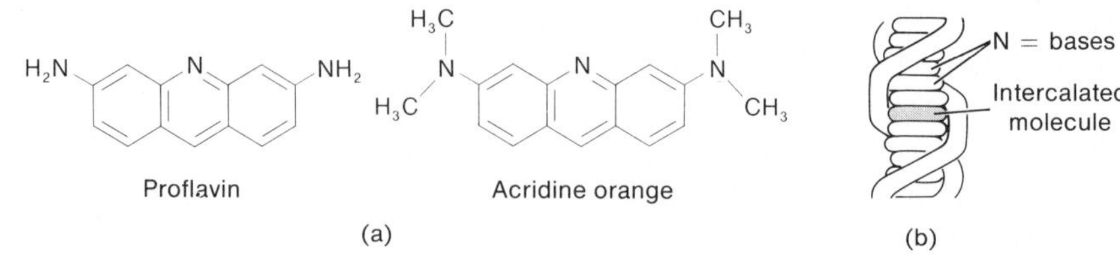

Figure 16-16. Intercalating agents. (a) Structures of the common agents proflavin and acridine orange. (b) An intercalating agent slips between the stack of bases at the center of the DNA molecule. This occurrence can lead to single nucleotide pair insertions and deletions. (From L. S. Lerman, Proc. Natl. Acad. Sci. USA *49:94, 1963.)*

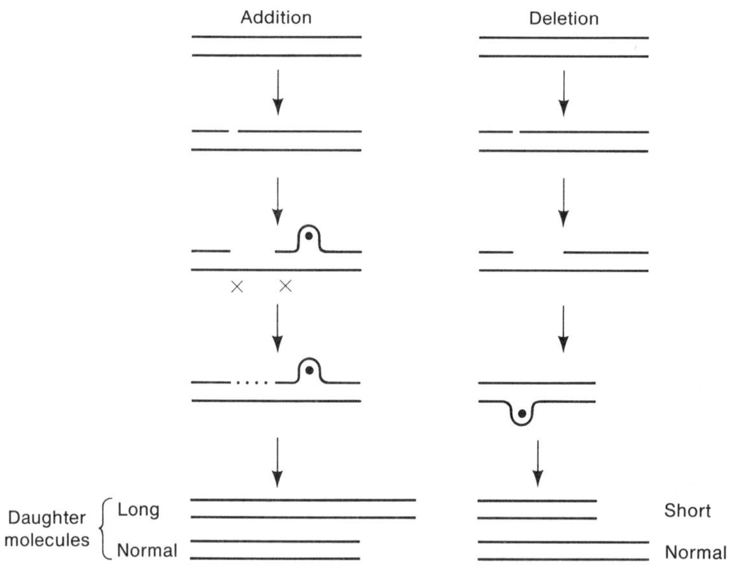

Figure 16-17. Model for the action of intercalating agents to cause short deletions or insertions. In this model, we assume that the agents are active only during DNA processing (repair or recombination), when they insert into a single-strand loop and stabilize it. For an addition to occur, the loop can form only if there is a short repeat length of complementary sequence (indicated by ×).

The modes of action for many mutagens are still very speculative, but it is evident that they act in many different ways to cause a primary lesion in the DNA (Figure 16-18). In some cases (such as the base analogs), little further action is needed to cause mutation. In most cases, however, some cellular function is necessary to generate a mutation from the original DNA lesion. These

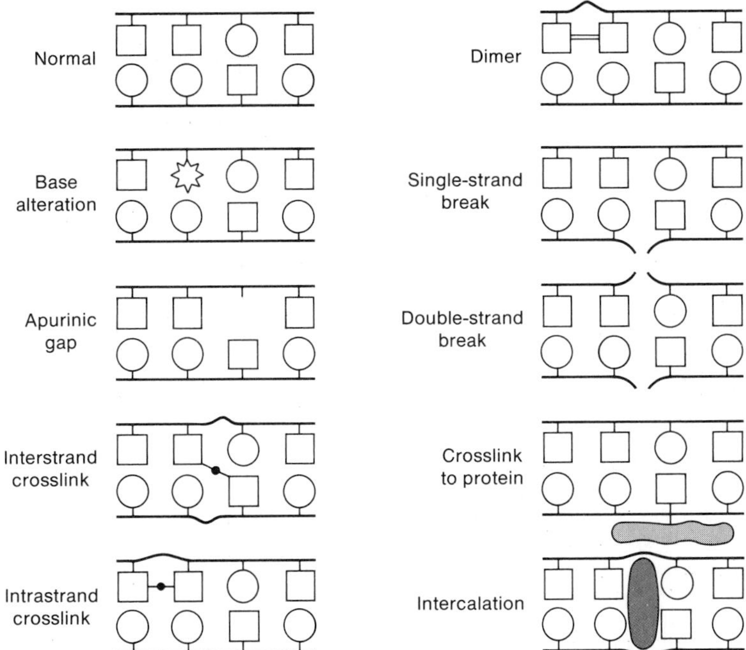

Figure 16-18. Various ways of modifying DNA that can lead to mutation. (From H. J. Evans, in Progress in Genetic Toxicology, *D. Scott et al., eds. Elsevier/North-Holland Biomedical Press.)*

cellular functions include replication, nuclease action, polymerase action, and ligase action. Different systems may be necessary for different kinds of mutagens, but it seems likely that the repair systems known to exist for UV-induced lesions also will act on other lesions:

$$\text{DNA lesion} \longrightarrow \text{Error-prone repair} \longrightarrow \text{Mutation}$$

Reversion Analysis. We can now begin to see how reversion analysis can tell us something about the nature of a mutation or something about the action of a mutagen. For example, if a mutation cannot be reverted by action of the mutagen that induced it, then the mutagen must have some relatively specific unilateral action. In the case of HA-induced mutation, it would be reasonable to expect that the original mutation is GC → AT, which of course cannot be reverted by another specific GC → AT event. Similarly, mutations that can be reverted by proflavin are very likely frameshift mutations, but NA-induced mutations (which are transitions) should not be revertible by proflavin. Transversions are not as easy to detect by reversion, but they are known definitely to be common among spontaneous mutations, as shown by studies of DNA and protein sequencing. Thus, in the reversion test, if a mutation does revert spontaneously but does not revert in response to a transition mutagen or a frameshift mutagen, then by elimination it is probably a transversion. Note that the kinds of logic employed in the reversion test rely heavily on the assumption that the reversion events are not due to suppressors; such suppressors would make inference from reversion more difficult. Table 16-1 summarizes some reversion expectations based on simple assumptions.

Table 16-1. Different types of point mutations can in theory be distinguished by their reversion behaviors in response to a battery of specific mutagens

	Reversion mutagen			
Mutation	NA	HA or EMS	Proflavin	Spontaneous
Transition, GC → AT	+	−	−	+
Transition, AT → GC	+	+	−	+
Transversion	−	−	−	+
Frameshift	−	−	+	+

NOTE: A plus (+) indicates a measurable rate of reversion due to a given mutagen. NA = nitrous acid; HA = hydroxylamine; EMS = ethyl methane sulfonate.

The system outlined in Table 16-1 is intended merely to illustrate the kinds of inferences possible from reversion analysis. Recall that mutagen specificities depend on the organism, the genotype, the gene studied, and perhaps even the *region* of the gene studied. The existence of **mutational hotspots** within genes has been known since the first mapping of the *rII* region; furthermore, the locations of the hotspots vary for different mutagens (Figure 16-19). DNA-sequencing studies of the *lac* repressor gene in *E. coli* have shown that mutational hotspots for spontaneous mutation are due to methylation of a cytosine molecule at those sites.

Usefulness of Mutagens in Genetic Analysis. The mutagenic agents are potent tools for the geneticist engaged in mutational dissection. Let's consider a few examples of the great usefulness of mutagens in such research.

First, Table 16-2 shows the relative frequencies of *ad-3* forward mutants in *Neurospora* after various mutagenic treatments. Note that recovery of a single mutant requires the testing of some 2.5 million cells without mutagenic treatment, whereas treatment with ICR-70 produces about 1 mutant in each 450 surviving cells tested. Table 16-3 provides another example of the great increase in available mutants obtained with an appropriate mutagen. The new term **supermutagen** has been introduced to describe some of these highly potent agents.

For a third example, we turn to *Drosophila.* Until the mid-1960s, most researchers working with *Drosophila* used radiation to induce mutations (Figure 16-20). Doses of 4000 roentgens produce lethal mutations in perhaps 5% to 6% of all X chromosomes of irradiated males. At higher doses, however, there

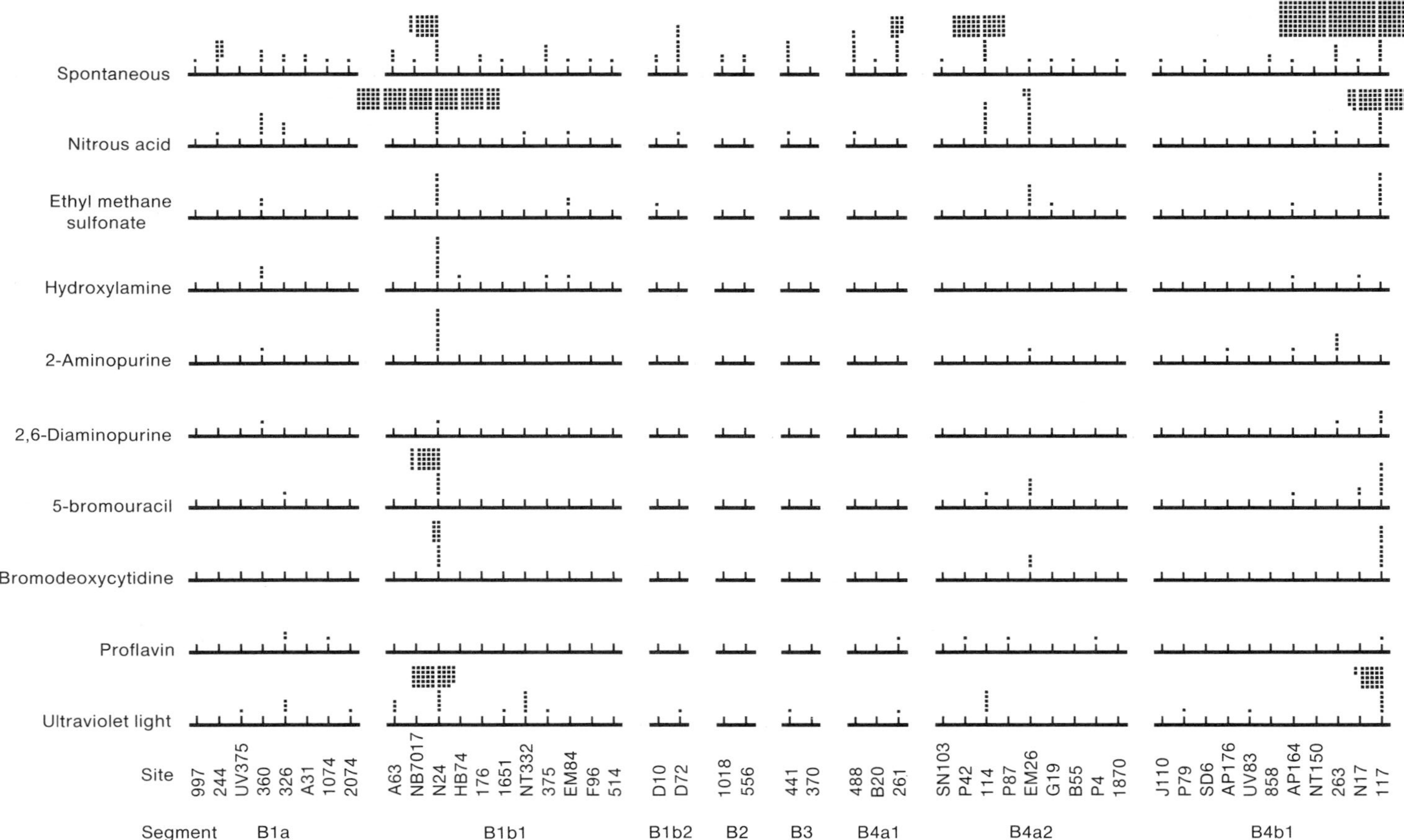

Figure 16-19. *Mutational hotspots within the* rIIB *cistron exist in varying positions for different mutagens. Each square represents a mutation obtained at that site. Different mutagens yield hotspots at different sites within the gene. (From S. Benzer, "The Fine Structure of the Gene." Copyright © 1961 by Scientific American Inc. All rights reserved.)*

Table 16-2. Forward mutation frequencies obtained with various mutagens in *Neurospora*

Mutagenic treatment	Exposure time (minutes)	Survival (%)	Number of *ad-3* mutants per 10^6 survivors
No treatment (spontaneous rate)	—	100	~ 0.4
Amino purine (1 to 5 mg/ml)	During growth	100	3
Ethyl methane sulfonate (1%)	90	56	25
Nitrous acid (0.05 M)	160	23	128
X rays (2000 r/min)	18	16	259
Methyl methane sulfonate (20 mM)	300	26	350
UV (600 erg/mm² per min)	6	18	375
Nitrosoguanidine (25 μM)	240	65	1500
ICR-170 acridine mustard (5 μg/ml)	480	28	2287

NOTE: The assay measures the frequency of purple *ad-3* colonies among the white colonies produced by the wild-type *ad-3*$^+$.

is greater infertility of the treated flies. Furthermore, many of the mutations induced by X rays involve chromosome rearrangements, and these can greatly complicate the genetic analysis.

In contrast, the alkylating agent EMS induces mutations that are almost exclusively point mutations. This mutagen is very easily administered simply by placing adult flies on a filter pad saturated with a mixture of sugar and EMS. Simple ingestion of the EMS produces very large numbers of mutations. For example, males fed 0.025 M EMS produce sperm carrying lethals on more than 70% of all X chromosomes and on virtually every chromosome 2 and chromosome 3. At these levels of mutation induction, it becomes quite feasible to screen for many mutations at specific loci or for defects with unusual phenotypes.

Table 16-3. The potency of various chemical mutagens for *his*$^-$ reversion in *Salmonella*

Mutagen	Revertants/nanomole	Ratio
1,2-Epoxybutane	0.006	1
Benzyl chloride	0.02	3
Methyl methane sulfonate	0.63	105
2-Naphthylamine	8.5	1,400
2-Acetylaminofluorene	108	18,000
Aflatoxin B_1	7,057	1,200,000
Furylfuramide (AF-2)	20,800	3,500,000

SOURCE: From J. McCann and B. N. Ames, in *Advances in Modern Toxicology*, vol. 5, W. G. Flamm and M. A. Mehlman, eds., Hemisphere Publishing Corp.

Figure 16-20. Radiation-induced rudimentary winged mutants in Drosophila melanogaster. *(Brookhaven National Laboratory.)*

It is possible to screen mutagenized chromosomes immediately in the F_1 generation if the region of interest is hemizygous. The X chromosome can be tested in this way by crossing EMS-fed males with females carrying attached-X chromosomes and a Y chromosome ($\widehat{X\,X}/Y$). All F_1 males carry a mutagenized paternal X chromosome, and each fly represents a different treated sperm. Thus, if individual F_1 males are crossed with $\widehat{X\,X}/Y$ females, each culture will represent a single cloned X. All F_1 zygotes carrying a sex-linked lethal will die, but any newly induced visible mutation will be expressed (Figure 16-21). In this way, it has been possible to select a wide range of behavioral and visible mutants involving a particular region of the genome (an approach known as **saturating** the region). The F_1 flies can be reared at 22°C, and then clones of each individual can be established at 22°C, 17°C, and 29°C to permit ready detection of heat-sensitive or cold-sensitive lethals. These temperature-sensitive (ts) lethals are found to represent about 10% to 12% of all EMS-induced lethals. We have already discussed the utility of such conditional mutants in genetic analysis. Many laboratories now routinely test for ts mutants in any *Drosophila* screening tests. Again, the approach is possible because large numbers of mutants are so easily obtained.

$\widehat{XX}/Y$ ♀ × X/Y ♂ (EMS-treated)

		Sperm	
		X*	Y*
Eggs	$\widehat{XX}$	$\widehat{XX}$ X* (dies)	$\widehat{XX}$/Y* ♀
	Y	X*/Y ♂	YY* (dies)

Figure 16-21. The use of attached-X chromosomes (XX) in Drosophila *to facilitate the recovery of X-linked mutations. Sperm treated with EMS or another mutagen will fertilize eggs containing either the attached-X chromosome or a Y chromosome. The treated X chromosome from the male will show up as the hemizygous X of the sons, revealing phenotypically recessive mutations. (Recall that the ability to carry out this test is dependent on the mechanism of sex determination in* Drosophila; *see Chapters 3 and 17). The asterisk (*) denotes the chromosome exposed to the mutagen.*

The recovery of ts lethals has a side benefit in simplifying laboratory procedures. One of the bothersome tasks involved in any large-scale *Drosophila* experiment is the need to separate all females from males within 12 hours after emergence from the pupae to prevent undesired mating. (Newly emerged males and females do not mate for 12 to 14 hours after emergence.) The tiresome procedure of collecting virgin females can be eliminated by using ts lethals (represented as l^{ts}) to produce unisexual cultures at will. For example, the cross $\widehat{XX}/Y$ ♀ × l^{ts}/Y ♂ produces progeny of both sexes at permissive temperature. If the culture is shifted to restrictive temperature, the l^{ts}/Y males die, and only the females hatch. Similarly, homozygosis of an l^{ts} in an $\widehat{XX}$ chromosome will produce only wild-type males at restrictive temperatures. (As an exercise, you might try to devise procedures for incorporating ts recessive and dominant lethals into inversions or translocations to construct autosomal unisexual stocks. That's chromosome mechanics.)

Message
The mutagen EMS has revolutionized the genetic versatility of Drosophila *by providing a potent method for the recovery of a wide range of point mutants.*

The Ames Test. There is increasing awareness of a correlation between mutagenicity and carcinogenicity. One study showed that 157 of 175 known carcinogens are also mutagens (a ratio of 90%). Some theories hold that these

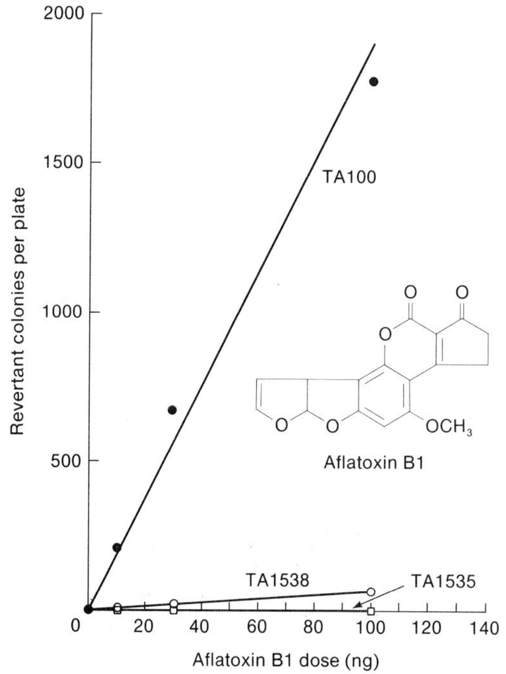

Figure 16-22. The mutagenicity of aflatoxin B1, which is also a potent carcinogen. The Salmonella *strain TA100 is a* his *mutant strain that is highly sensitive to reversion through base-pair substitution. The strains TA1535 and TA1538 are* his *mutant strains sensitive to reversion through frameshift mutation. The results of this test show that aflatoxin B1 is a potent mutagen causing base-pair substitutions but not frameshifts. This is the Ames test used to detect mutagens. (From J. McCann and B. N. Ames, in* Advances in Modern Toxicology, *vol. 5, W. G. Flamm and M. A. Mehlman, eds., Hemisphere Publishing Corp.)*

agents cause cancer by inducing mutation of somatic cells. This means that mutagenesis is of great relevance to our society. The modern environment exposes each individual to a wide variety of chemicals in drugs, cosmetics, food preservatives, pesticides, compounds used in industry, pollutants, and so on. Many of these compounds have been shown to be carcinogenic and mutagenic (Figure 16-22). Examples include the food preservative AF-2, the food fumigant ethylene dibromide, the antischistosome drug hycanthone, several hair-dye additives, and the industrial compound vinyl chloride; all are potent, and some have subsequently been subjected to government control. However, hundreds of new chemicals and products appear in the market each week. How can such vast numbers of new agents be tested for carcinogenicity before much of the population has been exposed to them?

Many test systems have been devised to screen for carcinogenicity. These are time-consuming tests, typically involving laborious research with small mammals. More rapid tests do exist that make use of microbes (such as fungi or bacteria) and test for mutagenicity rather than carcinogenicity. The most widely used test was developed in the 1970s by Bruce Ames, using *Salmonella typhimurium.* This **Ames test** uses two auxotrophic histidine mutations. One is reverted by nucleotide-pair substitution, and the other by frameshift mutation. Further properties were genetically engineered into these strains to make them suitable for mutagen detection. First, they carry a mutation that inactivates the excision-repair system. Second, they carry a mutation eliminating the protective lipopolysaccharide coating of wild-type *Salmonella,* so that any escaping bacteria will be unable to survive in such natural (but chemically hostile) environments as sewers or intestines.

Bacteria are evolutionarily a long way removed from humans. Can the results of such a test in bacteria have any real significance in detecting chemicals dangerous for humans? First, we have seen that the genetic and chemical nature of DNA is identical in all organisms, so that a compound acting as a mutagen in one organism is likely to have some mutagenic effects in other organisms. Second, Ames devised a way to simulate the human metabolism in the bacterial system. Much of the important processing of ingested chemicals in mammals occurs in the liver, where externally derived compounds normally are detoxified or broken down. In some cases, the action of liver enzymes can create a toxic or mutagenic compound from a substance that was not originally dangerous. Ames set out to incorporate the mammalian liver enzymes in his bacterial test system.

Rat livers are normally used for this purpose. First, the liver enzymes are mobilized by injection of Arochlor, a polychlorinated biphenyl (PCB), into the rats. The rats are then killed, and their livers are homogenized and centrifuged to remove cell debris. The supernatant, called S-9 mix, contains the solubilized batteries of rat liver enzymes. The S-9 mix is added to a suspension of auxotrophic bacteria in a solution of the suspected carcinogen being tested. The solution is then plated on medium containing no histidine, and later the plates are scored for colonies of revertants (Figure 16-23).

The Ames test has detected potential mutagens among a wide variety of heterogeneous types of chemicals. Of course, chemicals detected by this test can be regarded not only as potential carcinogens (sources of somatic mutations), but also as possible causes of mutations in germinal cells. Because the test system is so simple and inexpensive, many laboratories throughout the world now routinely test large numbers of potentially hazardous compounds for mutagenicity and potential carcinogenicity.

Recombination

A normal crossover (a reciprocal intrachromosomal recombination event) is really a miraculous process. Somehow the genetic material from one parental chromosome and the genetic material from the other parental chromosome are "cut up and pasted together" during each meiosis, and this is done with complete reciprocity. In other words, neither chromosome gains or loses any genes in the process. In fact, it is probably correct to say that neither chromosome gains or loses even one nucleotide in the exchange. How is this remarkable precision attained? We do not know for sure, but many interesting phenomena provide important clues about the nature of the answer.

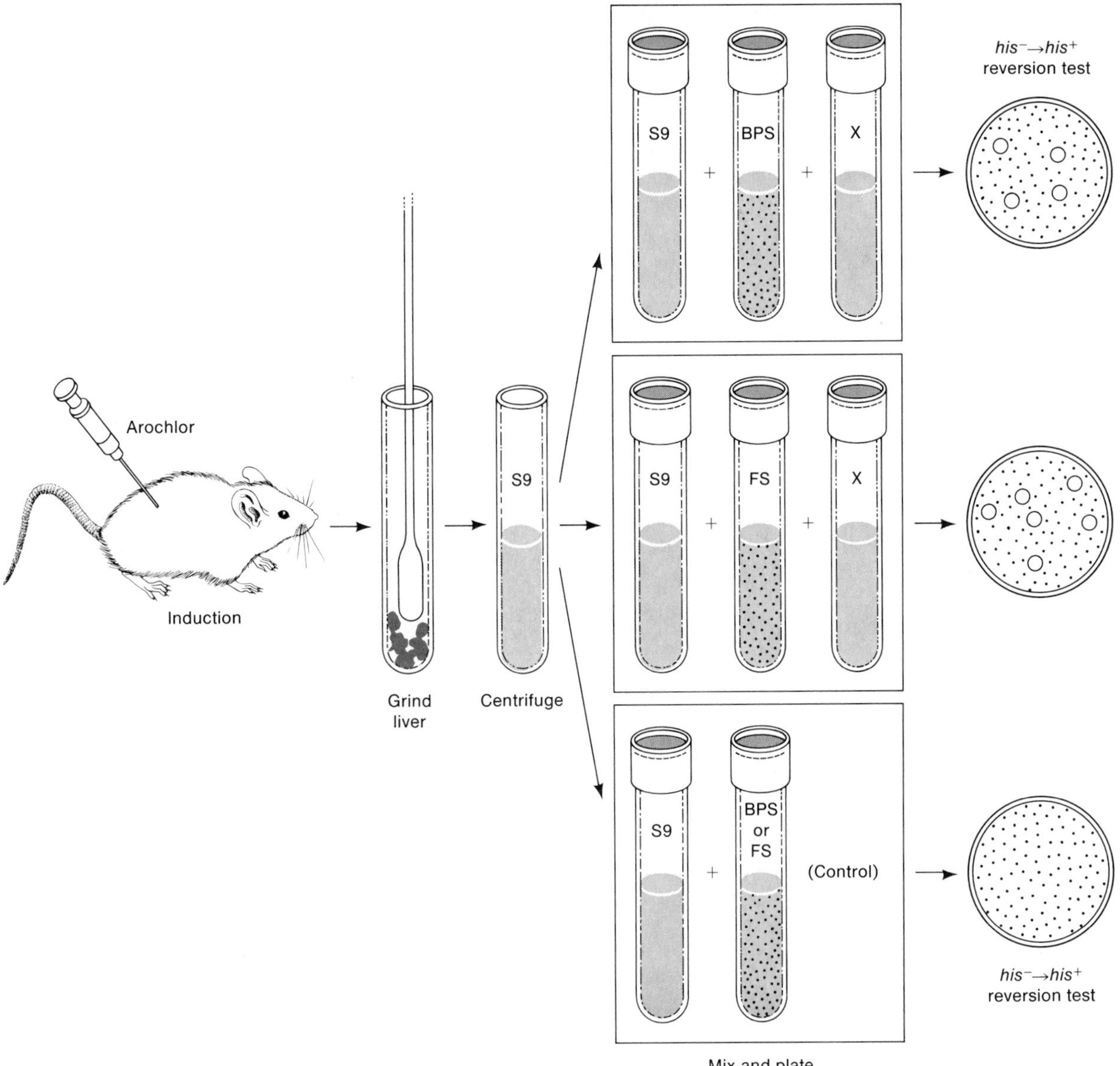

Figure 16-23. Summary of the procedure used for the Ames test. First, rat liver enzymes are mobilized by injecting the animals with aroclor. (Enzymes from the liver are used because it carries out the processes of detoxifying and toxifying bodily chemicals.) The rat liver is then homogenized, and the supernatant of solubilized rat liver enzymes (S-9) is added to a suspension of auxotrophic bacteria in a solution of the potential carcinogen (X). This mixture is plated on a medium containing no histidine, and revertants of a mutant strain containing a base-pair substitution (BPS) or a strain containing a frameshift mutation (FS) are looked for. A control experiment containing no potential carcinogen is always run simultaneously. The presence of revertants indicates that the chemical is a mutagen and possibly a carcinogen as well.

Breakage and Reunion of DNA Molecules

Throughout our analysis of linkage, we implicitly assumed that crossing over occurs by some process of breakage and reunion of chromatids. The evidence against the copy-choice hypothesis (Chapter 5) provides good *indirect* evidence in favor of breakage and reunion. Furthermore, there is good genetic and cytological evidence that crossing over occurs during prophase of meiosis rather than during interphase, when chromosomal DNA is replicating. A small amount of DNA synthesis does occur during prophase, but certainly chromosome replication is not associated with crossing over. One of the first direct proofs that chromosomes (albeit viral chromosomes) can break and rejoin came from experiments on λ phage done in 1961 by Matthew Meselson and Jean Weigle.

They multiply infected *E. coli* with two strains of λ. One strain had the genetic markers *c* and *mi* at one end of the chromosome, and this chromosome was "heavy" because the phages were produced from cells growing in heavy isotopes of carbon (^{13}C) and nitrogen (^{15}N). The other strain was + + for the markers and had "light" DNA because it was harvested from cells grown on the normal light isotopes ^{12}C and ^{14}N. The two DNAs (chromosomes) can be represented as shown in Figure 16-24a.

The progeny phage released from the cells were spun in a cesium chloride density gradient. A wide band was obtained, indicating that the virus DNAs ranged in density from the heavy parental value to the light parental value, with a great many intermediate densities (Figure 16-24b). Interestingly, some recombinant phages were recovered with density values very close to the heavy

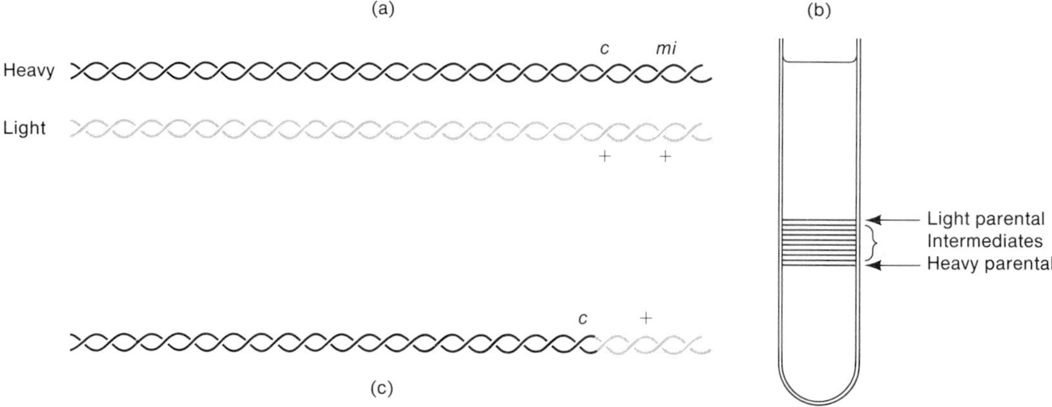

Figure 16-24. Evidence for chromosome breakage and reunion in λ phage. (a) The chromosomes of the two λ strains used to multiply infect E. coli. *(b) Band produced when progeny phage are spun in a cesium chloride density gradient. The fact that intermediate densities are obtained indicates a range of chromosome compositions with partly light and partly heavy components. (c) The chromosome of the heavy* c + *progeny resulting from crossover between the two markers. The density of this crossover product confirms that the crossover involved a physical breakage and reunion of the DNA.*

parental value. They were of genotype *c* +, and they must have arisen through an exchange event between the two markers (Figure 16-24c). The heavy density of the chromosome would be expected because only the small tip of the chromosome carrying the mi^+ allele would come from the "light" parental chromosome. When heavy + + phage were crossed with light *c mi*, the heavy recombinants were found to be + *mi*, and the light recombinants were found to be *c* +, as expected. These results can be explained in only one way: the recombination event must have occurred through the physical breakage and reunion of DNA. Of course, it is unwise to extrapolate from viral to eukaryotic chromosomes. However, this evidence shows that the breakage and reunion of DNA strands is a chemical possibility.

Chiasmata Are Crossover Points

In Chapter 5, we made the simple assumption that chiasmata are the actual sites of crossovers. Mapping analysis gives indirect support for this idea: since an average of one crossover per meiosis produces 50 genetic map units, there should be correlation between the size of the genetic map of a chromosome and the observed mean number of chiasmata per meiosis. This correlation has been made in well-mapped organisms.

However, the harlequin chromosome staining technique has made it possible to test the idea directly. In 1978, C. Tease and G. H. Jones prepared harlequin chromosomes in meioses of the locust. You will remember that the harlequin technique produces sister chromatids, one of which is dark and the other light. When a crossover occurs, it can involve two dark, two light, or a dark and a light nonsister chromatid. This latter situation is crucial because mixed (part dark and part light) crossover chromatids are produced. Tease and Jones found that the dark/light transition occurred right at the chiasma, proving beyond reasonable doubt that these are the sites of crossing over, and settling a question that had been unresolved since the early part of the century (Figure 16-25).

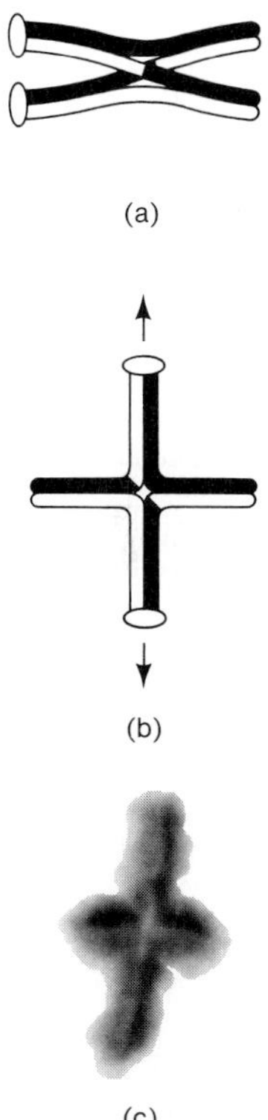

Figure 16-25. Crossing over between dark and light-stained nonsister chromatids in a meiosis in the locust. (a) Representation of the chiasma. (b) The best stage for observing is when the centromeres have pulled apart slightly, forming a cross-shaped structure with the chiasma at the center. (c) Photograph of the stage shown in b. (Photo courtesy of C. Tease and G. H. Jones, Chromosoma *69:163–178, 1978.)*

A Crossover Model

Much of the available information on the mechanism of intrachromosomal recombination, especially at the chemical level, has come from study of bacteria and phage. However, we shall concentrate on the information that is available about eukaryotes. This is evidence of a different kind. It is based more on genetic analysis and less on chemistry. We shall be studying what can be called the genetics of genetics! Several clues about recombination mechanisms have emerged in study of eukaryotes.

Clue 1: Chromatid Conversion and Half-Chromatid Conversion. Much information about recombination in eukaryotes has come from studies of fungi. There is one good reason for this: the ascus. All products of a single meiosis can be recovered and examined, so that records of each meiosis can be kept

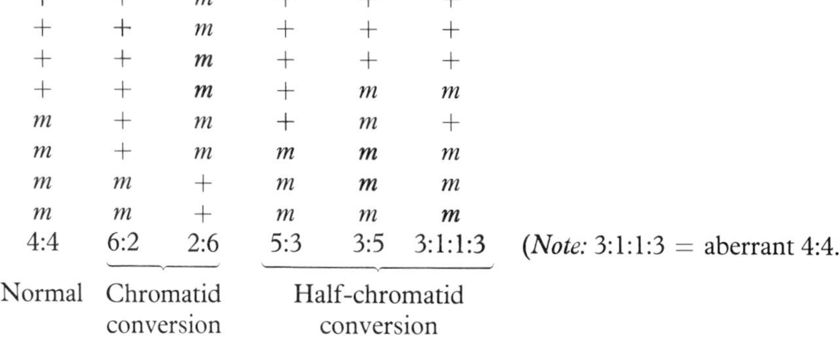

+ + m + + +
+ + m + + +
+ + m + + +
+ + m + m m
m + m + m +
m + m m m m
m m + m m m
m m + m m m
4:4 6:2 2:6 5:3 3:5 3:1:1:3
(Note: 3:1:1:3 = aberrant 4:4.)
Normal
Chromatid conversion
Half-chromatid conversion

Figure 16-26. Rare aberrant allele ratios observed in a cross of type + × m in fungi. (Ascus genotypes are represented here.) When the Mendelian ratio of 4:4 is not obtained, some of the alleles in the cross have been converted to the opposite allele. In some asci, it appears that the entire chromatid has been converted (6:2 or 2:6 ratios). In others, it appears that only half-chromatids have been converted (5:3, 3:5, or 3:1:1:3 ratios).

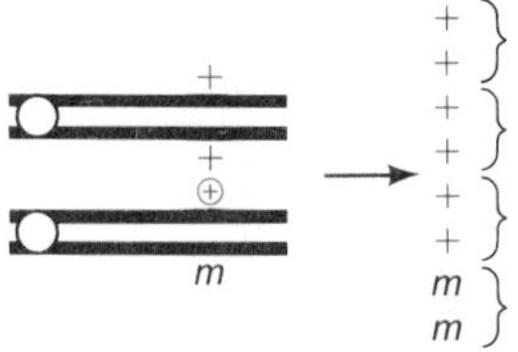

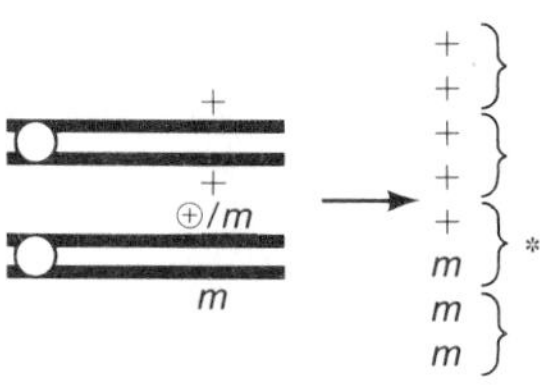

Figure 16-27. Gene conversions are inferred from the patterns of alleles observed in asci. (a) In a chromatid conversion, the allele on one chromatid seems somehow to have been converted to an allele like those on the other chromatid pair. The converted allele is shown by the symbol ⊕. One spore pair is of the opposite genotype from that expected in Mendelian segregation. (b) In a half-chromatid conversion, one sport pair (*) has nonidentical alleles. Somehow, one chromatid seems to be "half-converted," giving rise to one spore of the original genotype and one spore converted to the other allele.

quite precisely in terms of the total genetic information being lost or gained. Because we can recover all four products of a specific meiosis, we can make inferences about events occurring during that meiosis with a high degree of confidence. The ascus represents a tight system of internally self-consistent controls.

Non-Mendelian allele ratios are detectable in some asci. Mendel would have predicted 4:4 segregations for all monohybrid crosses, but other ratios are obtained very rarely (in 1% to 0.1% of all asci, depending on the fungus species). Figure 16-26 gives the most common aberrant ratios obtained. It appears as though some genes in the cross have been "converted" to the opposite allele (Figure 16-27). The process therefore became known as **gene conversion;** it can occur only where there is heterozygosity for two different alleles of a gene. In some asci, the entire chromatid in meiosis seems to have converted (**chromatid conversion**). In other asci, only half of the chromatid seems to have converted (**half-chromatid conversion**). In half-chromatid conversions, different members of a spore pair have different genotypes. Recall that each spore pair is produced by mitosis from a single product of meiosis.

We should emphasize that alleles heterozygous at other loci in the same cross typically segregate 4:4, so the aberrant ratios are not the results of accidents of isolation. The process of conversion cannot be mutation because the process is directional. The allele that is converted always changes to the other specific allele involved in the cross. This specificity has been confirmed in molecular studies on the gene products of converted alleles.

Clue 2: Polarity. In genes for which accurate allele maps are available, we can compare the conversion frequencies of alleles at various positions within the gene. In almost every case, the sites closer to one end show higher frequencies than do those farther away from that end. In other words, there is a gradient, or **polarity,** of conversion frequencies along the gene.

Clue 3: Conversion and Crossing Over. In heteroallelic crosses where the locus under study is closely flanked by other genetically marked loci, the conversion event is very often accompanied by an exchange in one of the flanking regions. (To keep the argument simple, we will say that the exchange is observed in 50% of the conversions). This exchange nearly always occurs on the side nearest the allele that has converted. Furthermore, it almost always involves the chromatid in which conversion has occurred.

For example, consider the chromatids diagrammed in Figure 16-28. Suppose the polarity is such that alleles toward the left end of the chromatid convert more often than those toward the right end. The cross diagrammed here is between $a^+ m_2 b^+$ and $a\, m_1\, b$, where m_1 and m_2 are different alleles of the m locus, and a and b represent closely linked flanking markers. If we look at asci in which conversion has occurred at the m_1 site (the most frequent kind of conversion in this locus), we will find that one-half of these asci will also have a crossover in region I, and one-half will have no crossover. In the smaller number of asci showing gene conversion at the m_2 site, one-half will also have a crossover in region II, and one-half will have no crossover. Such events are detected in ascus genotypes like that shown in Figure 16-29, which can be inter-

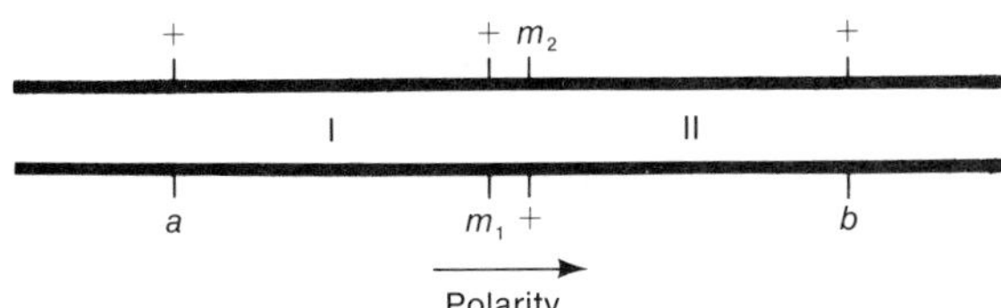

Figure 16-28. Diagram of chromatids involved in a cross. The arrow indicates the polarity of gene conversion in the m *locus, pointing toward the end with lower conversion frequency.*

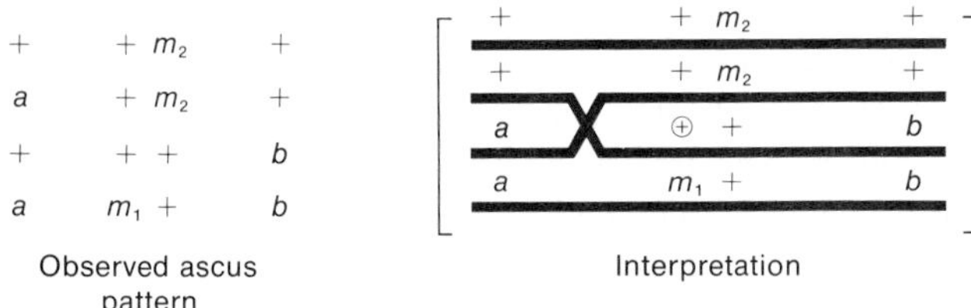

Figure 16-29. A specific ascus pattern can be explained by both a crossover and a chromatid conversion. In this case, a conversion of $m_1 \rightarrow +$ *is accompanied by a crossover in the region between* a *and* m_1.

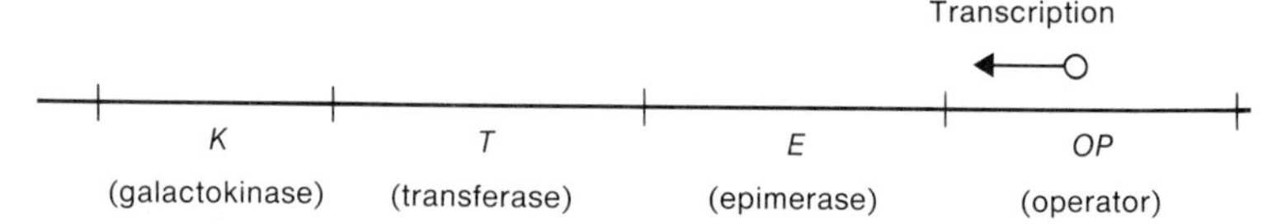

Figure 16-38. The structure of the galactose (gal) *operon in* E. coli. *Galactokinase, transferase, and epimerase are structural genes of the galactose operon. The operator (OP) gene is a regulatory element for the operon.*

sugar galactose. The operon consists of a positively controlled operator and three adjacent structural genes (epimerase, transferase, and kinase), as shown in Figure 16-38.

The IS elements were found in mutants that had been selected by their deficiency in kinase activity. These mutations were found to map at various sites throughout the operon. In any particular mutant, the mutant site might map somewhere in the kinase itself, or in the transferase, or in the epimerase, or even in the operator. Each mutation obviously affects all structural gene functions transcriptionally "downstream" in the operon. Such **polar mutations** were not new (see Chapter 14), but this particular kind was new. The mutants were observed to be capable of spontaneous reversion to wild type (showing that the mutation is not a deletion), but the reversion rate is not increased by the effect of any mutagen. Thus they do not seem to be frameshifts or any form of point mutation, either! If not deletions or frameshifts or point mutations, what are they?

Recall that the phage λ inserts next to the *gal* operon and that it is a simple matter to obtain λ*dgal* phage particles that have picked up the *gal* region. (Review the material on this topic in Chapter 9.) Let us represent one of the polar mutations by the symbol gal^{m}. The $\lambda dgal^{\mathrm{m}}$ phages were isolated from the polar mutants, and their DNA was used to synthesize radioactive RNA in vitro. Certain fragments of this RNA were found consistently to hybridize with the mutant DNA but not with wild-type DNA. This must mean that the mutant contains an extra piece of DNA. These particular $\lambda dgal^{\mathrm{m}}$ RNA fragments would also hybridize to DNA from other polar mutants, showing that the same bit of DNA has been inserted in different places in the different polar mutants.

By hybridizing denatured $\lambda dgal^{\mathrm{m}}$ DNA to denatured $\lambda dgal^{+}$ DNA, the extra piece of DNA can actually be located under the electron microscope. Some of the DNA molecules that form in the mixture are heteroduplexes between one mutant and one wild-type strand. Normally, such heteroduplexes under the electron microscope are indistinguishable from parental DNA molecules; however, in the case of the DNA involving the polar mutants, each heteroduplex showed a single-stranded buckle, or loop (Figure 16-39). This single-stranded buckle confirmed the presence of an inserted sequence in the $\lambda dgal^{\mathrm{m}}$ DNA that has no complementary sequence in the $\lambda dgal^{+}$ DNA. The length of this single-stranded loop can be calibrated by including standard-sized marker DNA in the preparation. It proved to be 800 nucleotides in length.

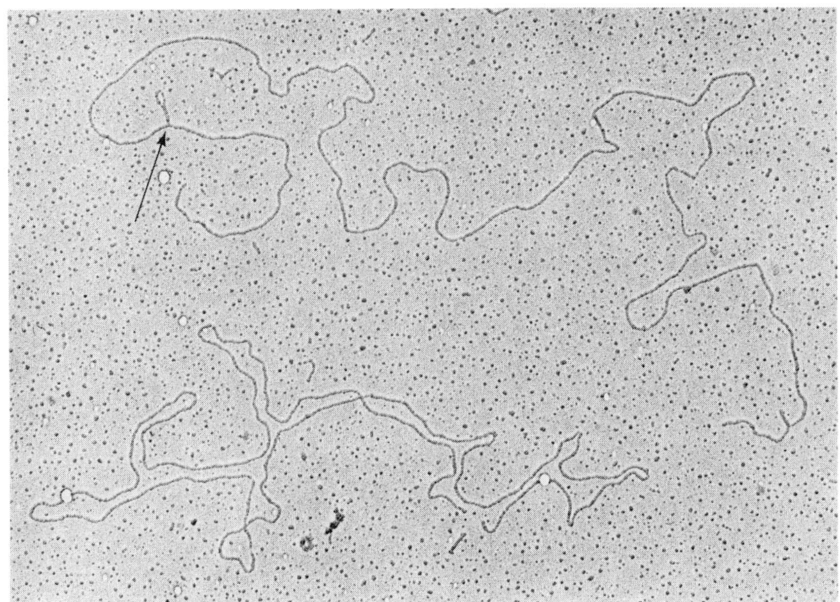

Figure 16-39. Electron micrograph of a λdgal $^{+}$/λdgal^{m} *DNA heteroduplex. The single-stranded loop* (arrow) *is caused by the presence of an insertion sequence in* λdgal^{m}. *(From A. Ahmed and D. Scraba,* Mol. Gen. Genet. *136:233, 1975.)*

We now have a model indicating that the polar mutations of the *gal* operon involve the insertion of an extra bit of DNA (800 base pairs long) anywhere within the operon. This insertion sequence was later called IS1 after different insertion sequences were identified in other mutants. For example, IS2 is 1350 base pairs long.

Because of the base sequence, the two strands of λ*dgal*$^{+}$ DNA happen to have different buoyant densities. After DNA denaturation, they can be recovered separately in the ultracentrifuge. In some cases, the *same* strands from two different IS1 mutants would form an unexpected hybrid with each other. Under the electron microscope, these hybrids had a peculiar appearance—a double-stranded region with four single-stranded tails (Figure 16-40). This observation was explained by assuming that the IS1 elements are inserted in opposite directions in the two mutants (Figure 16-41).

We now know that the genome of the standard wild-type *E. coli* is rich in IS elements: it contains eight copies of IS1, five copies of IS2, and copies of other less-well-studied IS types as well. We should emphasize that the sudden appearance of an insertion sequence at any given locus under study means that these elements are truly mobile, with a capability for transposition throughout the genome. Presumably, they produce a mutation or some other detectable alteration of normal cell function only when they happen to end up in an "abnormal" position, such as the middle of a structural gene.

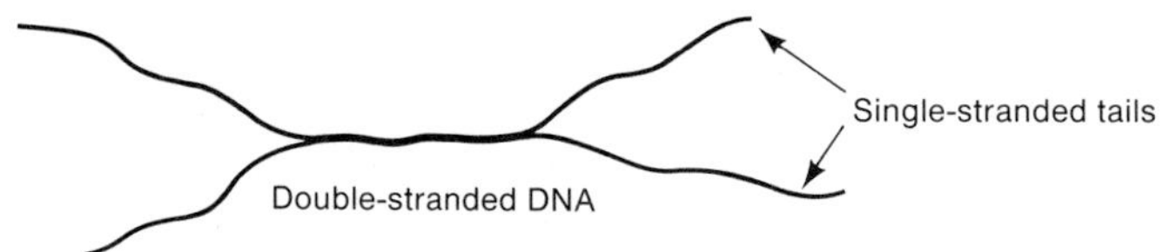

Figure 16-40. Diagrammatic representation of the appearance of a hybrid DNA molecule formed by annealing corresponding strands of λdgal^{m} *DNA from two specific mutants caused by insertion sequences. This unexpected hybridization between corresponding strands (normally having the same rather than complementary base sequences) can be explained by the model shown in Figure 16-41.*

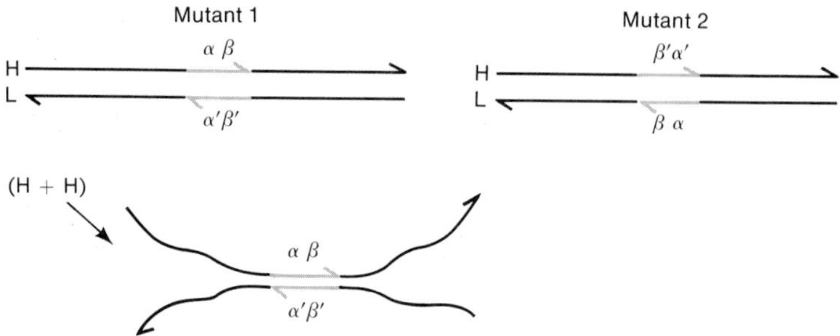

Figure 16-41. A model to explain the hybrid DNA structure of Figure 16-40. IS1 is represented by the shaded lines; α and β represent the ends of the IS1 molecule; H and L represent the strands of λdgal *DNA with high and low buoyant density, respectively. The hybrid can be explained by assuming that the IS1 sequence is inserted in opposite directions in the two mutants.*

In this connection, it is interesting to note a difference in the effects of IS1 and IS2. As we have seen, IS1 reduces the expression of structural genes "downstream" from its site of insertion in the *gal* operon, and it does this regardless of its orientation of insertion. However, IS2 reduces expression downstream only in one orientation of insertion. If it is inserted with the opposite orientation, it causes a higher level of production of the downstream enzymes! Apparently, IS2 contains a promoter sequence that enhances RNA polymerase attachment at a point where this enzyme would not normally attach. The effects of the IS elements in reducing downstream production are probably due to the presence of a terminator sequence that interrupts the process of transcription by the RNA polymerase.

The mode of insertion of the IS elements remains largely a mystery. We do know that they will insert in the absence of the normal bacterial recombination enzyme system (that is, in *recA* bacteria). Hence they either use a different set of enzymes or possess some intrinsic insertion ability of their own.

We should caution that the insertion sequences just described are not to be confused with the intervening sequences (introns) mentioned in Chapter 13.

Transposons and Plasmids. A frightening ability of pathogenic bacteria was discovered in Japanese hospitals in the 1950s. Bacterial dysentery is caused by bacteria of the genus *Shigella*. This bacterium proved sensitive to a wide array of antibiotics, which were used originally to control the disease. In the Japanese hospitals, however, *Shigella* isolated from patients with dysentery proved to be simultaneously resistant to many of these drugs, including penicillin, tetracycline, sulfanilamide, streptomycin, and chloramphenicol. This multiple drug-resistance phenotype was inherited as a single genetic package, and it could be transmitted in an infectious manner—not only to other sensitive *Shigella* strains, but also to other related species of bacteria! This talent is an extraordinarily useful one for the pathogenic bacterium, and its implications for medical science were terrifying. From the point of view of the geneticist, however, the situation is very interesting. The vector carrying these resistances from one cell to another proved to be a plasmid similar to the F particle. These **R plasmids** are transferred rapidly upon cell conjugation, much like the F particle in *E. coli* (Chapter 9).

In fact, these R plasmids proved to be just the first of many similar F-like plasmids to be discovered. These plasmids have been found to carry many different kinds of genes in bacteria. Table 16-5 shows just some of the characteristics that can be borne by plasmids. What is the mode of action of these plasmids? How do they acquire their new genetic abilities? How do they carry them from cell to cell? Some answers to these questions can now be given.

Table 16-5. Genetic determinants borne by plasmids

Characteristic	Plasmid examples
Fertility	F, R1, Col
Bacteriocin production	Col E1
Heavy-metal resistance	R6
Enterotoxin production	Ent
Metabolism of camphor	Cam
Tumorigenicity in plants	T1 (in *Agrobacterium tumifaciens*)

If the DNA of a plasmid (carrying the genes for kanamycin resistance, for example) is denatured to single-stranded forms and then allowed to renature slowly, some of the strands form an unusual shape under the electron microscope—a large circular DNA ring is attached to a "lollipop-shaped" structure (Figure 16-42). The "stick" of the lollipop is double-stranded DNA, which has formed through the annealing of two **inverted repeat (IR) sequences** in

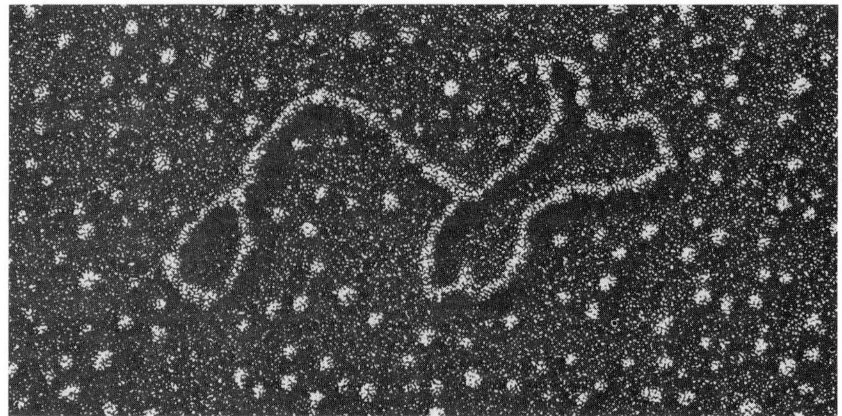

Figure 16-42. Peculiar structure formed when denatured plasmid DNA is reannealed. The double-stranded IR region separates a large circular loop from the small "lollipop" loop. (From S. N. Cohen.)

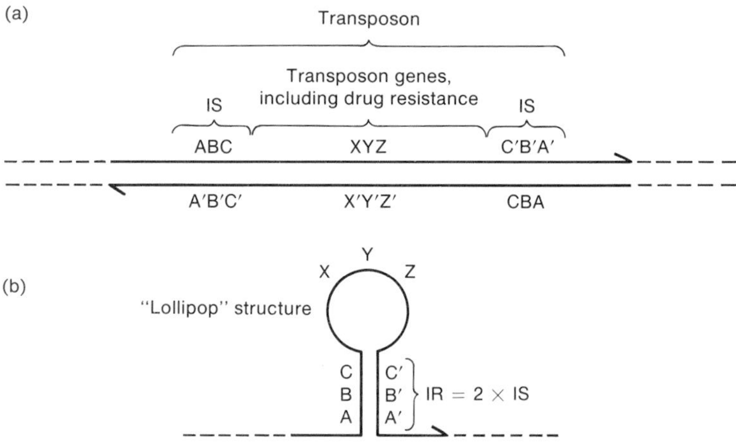

Figure 16-43. An explanation at the nucleotide level for the "lollipop" structure seen in Figure 16-42. The structure is called a transposon. (a) The transposon in its double-stranded form before denaturing. Note the presence of oppositely oriented copies of an insertion sequence (IS). (b) The lollipop structure formed by intrastrand annealing after denaturing. The two IS regions anneal to form the IR of the transposon; the transposon genes are carried in the lollipop loop.

the plasmid (Figure 16-43). Subsequent studies have shown that the IR sequences in fact are a pair of IS elements. The IS1 element is involved in the case of plasmids carrying kanamycin resistance, and IS3 is involved in the case of plasmids carrying tetracycline resistance (Figure 16-44).

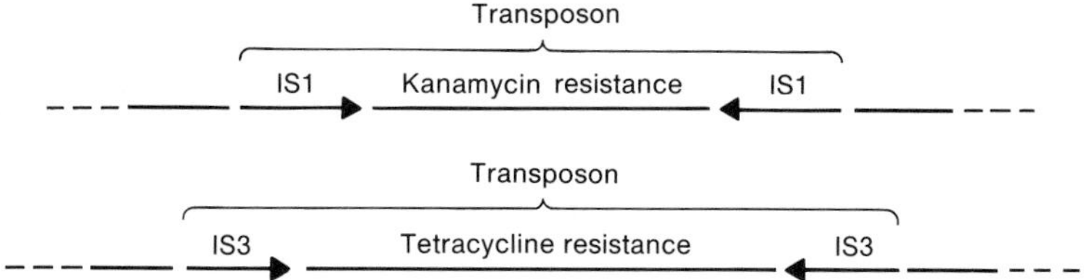

Figure 16-44. Two different transposons having different IR regions and carrying different drug-resistance genes.

The genes for drug resistance or other genetic abilities carried by the plasmid are located between the IR sequences (on the "lollipop head"). The IR sequences together with their contained genes have been called a **transposon** (Tn). The remainder of the plasmid, bearing the genes coding for resistance-transfer functions (RTF), is called the **RTF region** (Figure 16-45).

A transposon can jump from one plasmid to another plasmid or from a plasmid to a bacterial chromosome. Let us consider an example of this transposition. A transposon (Tn3) containing an ampicillin-resistance gene (Ap^{r}) is carried in a large *E. coli* plasmid R64-1. This Tn3 was transferred to a small plasmid RSF1010, which carried the genes for sulfonamide resistance (Su^{R}) and streptomycin resistance (Sm^{R}) in another transposon (Tn4):

$$\text{R64-1 } (Ap^{R} \text{ in Tn3}) \longrightarrow \text{RSF1010 } (Su^{R}\ Sm^{R} \text{ in Tn4})$$

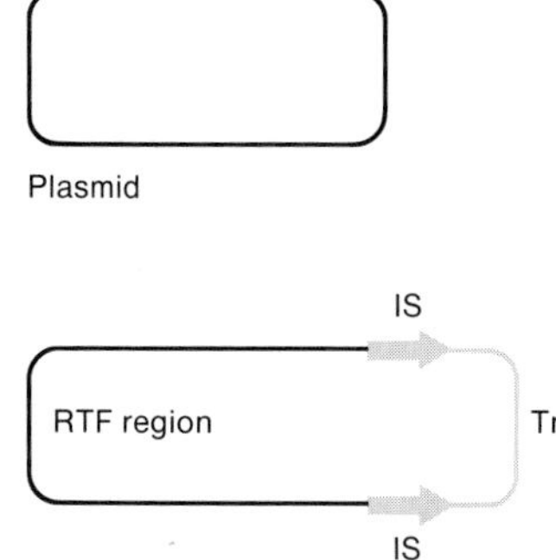

Figure 16-45. The insertion of a transposon (Tn) into a plasmid. RTF = the resistance-transfer functional genes of the plasmid. The Tn includes both the IS elements and the drug-resistance genes.

The following procedure was used to carry out this transposition. First, R64-1 DNA was isolated and added as transforming DNA to RSF1010 that had been treated with $CaCl_2$ (a treatment that enhances the probability of uptake of the donor DNA). The Ap^{R} transformants were selected by plating on ampicillin medium. These transformants proved to be of several genotypes:

$$Ap^{R}\ Su^{R}\ Sm^{R} \qquad Ap^{R}\ Su^{R}\ Sm^{S} \qquad Ap^{R}\ Su^{S}\ Sm^{R} \qquad Ap^{R}\ Su^{S}\ Sm^{S}$$

Apparently, the Ap^{R} transposon, once it had entered the recipient cell, could insert itself into the recipient's genome either outside the $Su^{R}\ Sm^{R}$ transposon or within it. If it inserted within the recipient's original Tn4 transposon, then it knocked out either one or both of the resistance functions originally possessed, depending on the precise point of insertion. New lollipop structures were observed in the transformed genotypes, corresponding to the acquisition of the new transposon. Figure 16-46 shows a composite diagram of an R plasmid, indicating the various places where transposons can be located, including the transposons just described. Techniques similar to those used in this experiment can also be used to demonstrate transposition from plasmid to bacterial DNA.

The diagram in Figure 16-46 also shows some direct repeat (DR) sequences (or noninverted sequences). Also, a single IS sequence can sometimes be found inserted in the resistance genes of a Tn, where it inactivates the drug-resistance function (Figure 16-47). Insertion sequences also are commonly observed in the F particle. Figure 16-48 shows an example in an F *lac* plasmid.

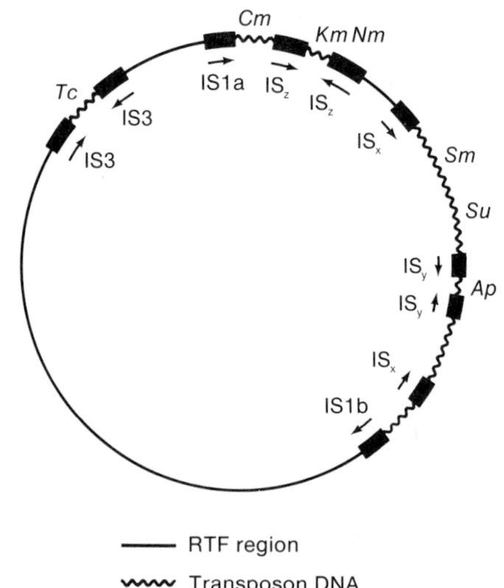

Figure 16-46. Genetic map of an R plasmid, showing the positions where various transposons have been found. The drug-resistance genes carried in the various transposons are indicated, and the IS regions forming the IRs of the transposons are also identified. A number of IS types (such as ISx, ISy, and ISz) are indicated. Note that IS1a and IS1b are direct repeats (DR). (After S. N. Cohen.)

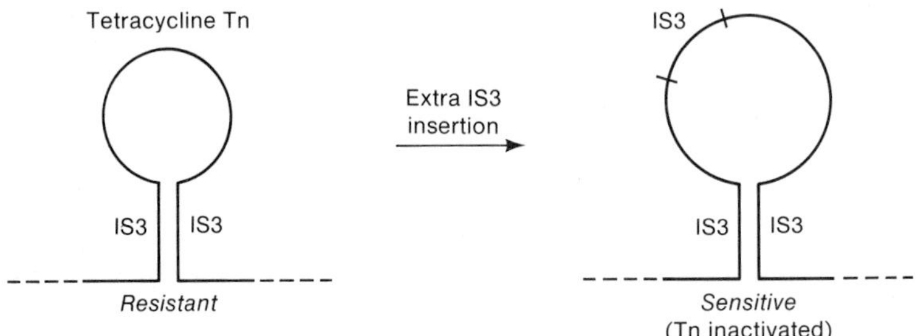

Figure 16-47. A single (unpaired) insertion sequence can inactivate the drug-resistance gene of a Tn.

The model emerges of the IS regions as genetic "snap fasteners" that (by some unknown molecular mechanism) can unhook themselves from the DNA and reinsert elsewhere. Not only can they mobilize themselves in this way, but they can carry with them the genes conferring various kinds of bacterial talents, usually with the IS elements acting as flanking snap fasteners. Thus multiply drug-resistant plasmids are generated. Even the F plasmid, judging from the strategically located pair of inverted IS elements, seems to owe its mobilization powers to this same process. It is now known that even a single IS element itself has a pair of short terminal inverted repeats about 50 base pairs long.

IS elements turn up in many other places. For example, in the adenovirus (which causes certain respiratory diseases), the linear genome is enclosed by an IR. Herpes viruses (which have been indirectly implicated in some human cancers) are composed of two tandemly arranged adenovirus genomes with their IRs.

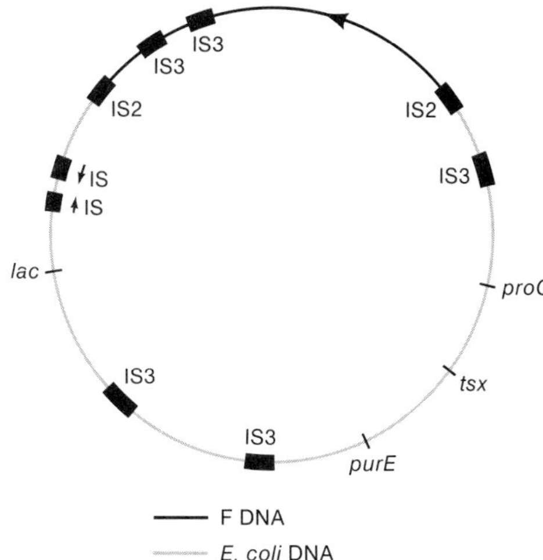

Figure 16-48. A map of an F lac *plasmid showing positions of IS elements;* proC, tsx, purE, *and* lac *are bacterial markers. (From H. Ohtsubo and E. Ohtsubo, in* DNA Insertion Elements, Plasmids, and Episomes, *A. I. Bukhari et al., Cold Spring Harbor Laboratory.)*

Undoubtedly, the IR sequences are crucial to the mobilization of a transposon. However, there is evidence that there may be additional components to the transposon's transposition mechanism. For example, Tn3 (which bears the gene for ampicillin resistance) also has a region that has been shown by mutational studies to be essential for transposition. Both this "transposase" region and the ampicillin-resistance gene are part of the region between the pair of IR sequences.

Phage Mu. Phage *mu* is a normal-appearing phage. We discuss it here because, although it is a true virus, it has many features in common with IS elements. The DNA double helix of this phage is 36,000 nucleotides long—much larger than an IS element. However, it does appear to be able to insert itself anywhere in a bacterial (or plasmid) genome, in either orientation. Once inserted (again, like an IS), it causes mutation at the locus of insertion. (The phage was named for this ability: *mu* stands for *mutator*.) Normally, these mutations cannot be reverted, but reversion can be produced by certain kinds of genetic trickery. When this reversion is produced, the phages that can be recovered show no deletion, proving that excision is exact and that the insertion of the phage therefore does not involve any loss of phage material either.

Each mature phage particle has a piece of flanking host DNA on each end (Figure 16-49). However, this DNA is not inserted anew into the next host. Its function is unclear. Phage *mu* also has an IR pair, but they are not terminal.

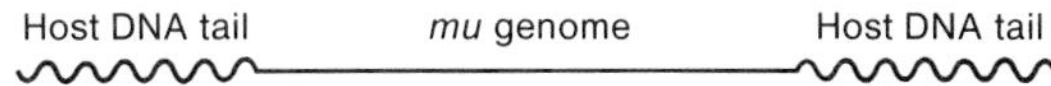

Figure 16-49. The DNA of a free mu *phage has tails derived from its previous host.*

Mu can also act like a genetic snap fastener, mobilizing any kind of DNA and transposing it anywhere in the genome. For example, it can perform this trick on another phage (such as λ) or on the F particle. In such situations, the inserted DNA is flanked by two *mu* genomes (Figure 16-50). It can also transfer bacterial markers onto a plasmid; here again, the transferred region is flanked by a pair of *mu* genomes (Figure 16-51). Finally, the phage *mu* can mediate various kinds of structural chromosome rearrangements (Figure 16-52).

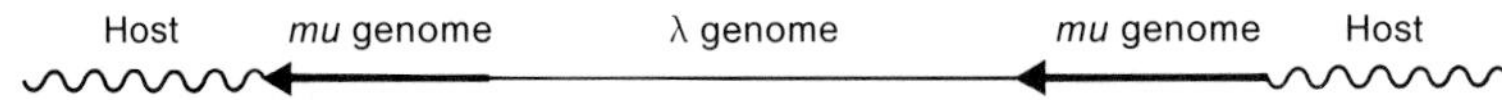

Figure 16-50. Phage mu *can mediate the insertion of phage λ into the bacterial chromosome, resulting in a structure like the one shown here.*

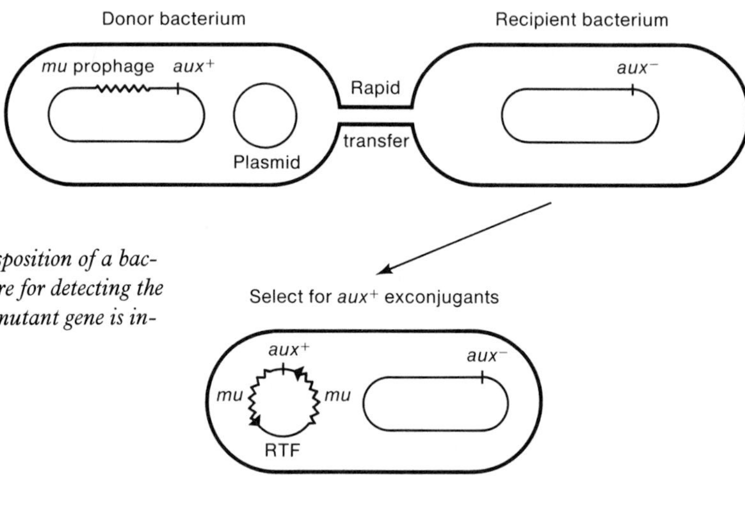

Figure 16-51. Phage mu *can mediate the transposition of a bacterial gene into a plasmid. The selection procedure for detecting the transposition is indicated here. An auxotrophic mutant gene is indicated by* aux$^-$.

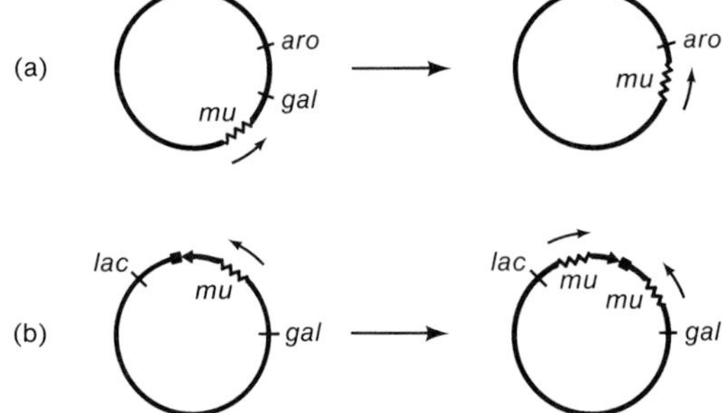

Figure 16-52. Phage mu *can cause deletion or inversion of adjacent bacterial segments. (a) The* gal *region is deleted by transposition of phage* mu. *(b) The F-factor region of an* Hfr *strain is inverted by transposition of phage* mu.

Message
Some DNA sequences in bacteria and phage seem to act as genetic snap fasteners. They are capable of joining different pieces of DNA and thus are capable of splicing DNA fragments into or out of the middle of a DNA molecule.

Transposable Genetic Elements in Eukaryotes

We turn now to two examples of transposable genetic elements in eukaryotic organisms—one in corn and the other in yeast. We shall see to what extent our understanding of the prokaryotic elements will provide models to explain the elements in eukaryotes.

Controlling Elements in Maize. In 1938, Marcus Rhoades analyzed an ear of Mexican black corn. The ear came from a selfing of a pure-breeding pigmented genotype, but it surprisingly showed a modified Mendelian dihybrid segregation ratio of 12:3:1 among pigmented, dotted, and colorless kernels. Analysis showed that two simultaneous events had occurred at unlinked loci. At one locus, a pigment gene A_1 had mutated to a_1; at another locus, a dominant allele *Dt* (*Dotted*) had appeared. The effect of *Dt* was to produce pigmented dots in the otherwise colorless phenotype of a_1a_1 (Figure 16-53). That is, the original line must have

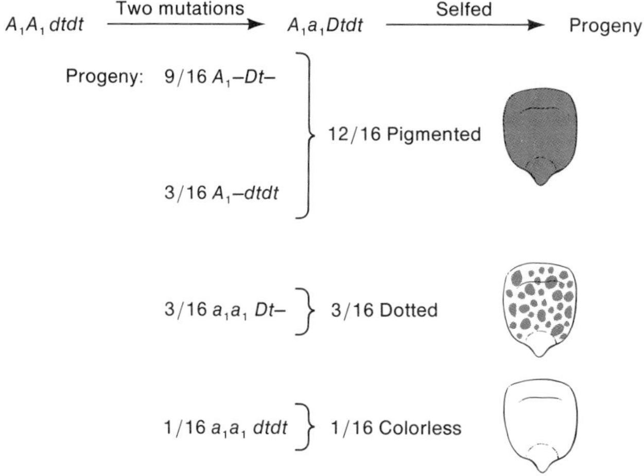

Figure 16-53. A formal genetic explanation of the appearance of the Dotted phenotype in corn. The A_1a_1 Dtdt *genotype is created by simultaneous mutations of* $A_1 \rightarrow a_1$ *and* dt → Dt. *Upon selfing, this genotype yields the observed 12:3:1 ratio of kernel phenotypes.*

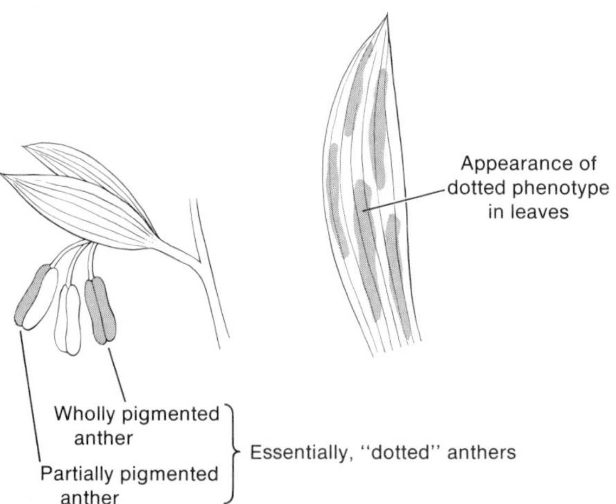

Figure 16-54. Rhoades used special stocks of a_1a_1 Dt– *corn plants carrying special genes that allow pigmented sectors to be detected in tissue other than the kernels. (After M. M. Rhoades,* Genetics *23:382, 1938.)*

been A_1A_1 *dtdt*, and the simultaneous mutations generated an A_1a_1 *Dtdt* plant, which upon selfing gave the observed ratio of progeny.

What was causing the dotted phenotype? A reverse mutation of $a_1 \rightarrow A_1$ in somatic cells would be an obvious possibility, but the large numbers of dots in the *Dotted* kernels would require extremely high reversion rates. Using special stocks, Rhoades was able to find anthers in the flowers of a_1a_1 *Dt*–plants that showed patches of pigment (Figure 16-54). He reasoned that these anthers might contain pollen grains bearing the reverted pigment genotype, and he used the pollen from these anthers to fertilize a_1a_1 tester females. Sure enough, some of the progeny were completely pigmented, showing that each dot in the parental plants was in fact the phenotypic manifestation of a genetic reversion event. Thus a_1 was the first known example of an **unstable mutant allele**—an allele for which reverse mutation occurs at a very high rate. The instability, however, is dependent on the presence of the unlinked *Dt* gene. Once the reverse mutations have occurred, they are stable; the *Dt* gene can be crossed out of the line with no loss of the A_1 character.

In the 1950s, Barbara McClintock demonstrated an analogous situation in another study of corn. She found a genetic factor *Ds* that causes a high tendency toward chromosome breakage at the location where it appears. These breaks could be located either cytologically (Figure 16-55a) or by the uncovering of recessive genes (Figure 16-55b). As you will appreciate, this action of *Ds* is another kind of instability. Once again, this instability proved to be dependent on the presence of an unlinked gene *Ac* (*Activator*), in the same way that the instability of a_1 is dependent on *Dt*.

McClintock tried to map *Ac*, but she found it impossible to map! In some plants it mapped to one position, and in other plants of the same line it mapped to different positions. As if this were not enough of a curiosity, the *Ds* locus itself was constantly changing position on the chromosome arm, as indicated by the differing phenotypes of the variegated sections (as different recessive

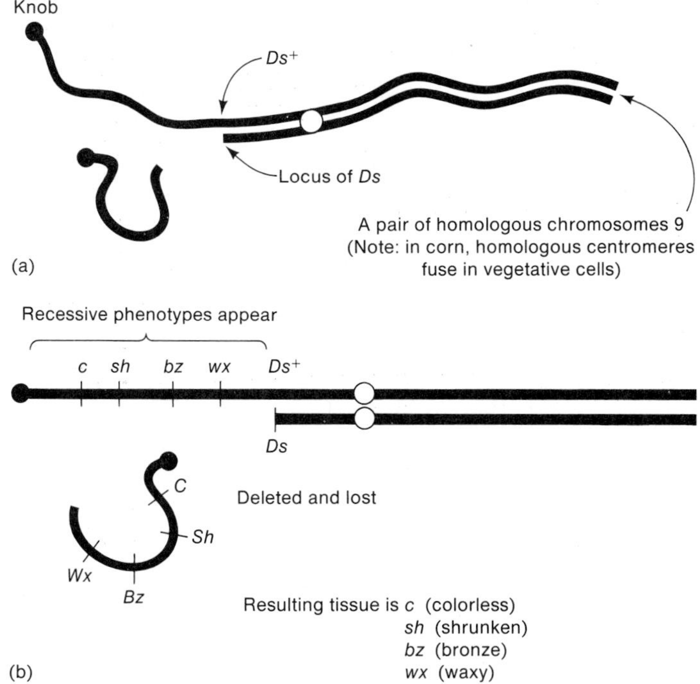

Figure 16-55. Detection of chromosomal breakage (instability) due to action of Ds *in corn. (a) Cytological detection of the breakage. (*Ds^+ *indicates a lack of* Ds.*) (b) Genetic detection of the breakage.*

gene combinations were uncovered in a system such as that illustrated in Figure 16-55b).

The wanderings of the *Ds* element take on new meaning for us in the context of this chapter when we consider the results of the following cross:

$$\male\ CC\,DsDs\,AcAc^+ \times \female\ cc\,Ds^+Ds^+\,Ac^+Ac^+$$

where C allows color expression and Ds^+ indicates the lack of a *Ds* element. Most of the kernels from this cross were of the expected types (Figure 16-56), but one exceptional kernel was very interesting. In this individual, the *Ds* element seems to have wandered into the middle of the C gene and inactivated it, producing the colorless phenotype. Although the presence of *Ac* is still necessary to cause this instability, it now seems to take on the function of rendering the c mutation unstable (c^u), and patches of revertant color appear in the kernel. When *Ac* is crossed out of this line, the c^u becomes a stable mutant.

The analogy of this system with the a_1 *Dt* system is obvious. Perhaps the earlier situation also was due to the insertion of a *Ds*-like element into the A_1 gene. It is natural to ask whether a_1 will respond to *Ac*, or c^u to *Dt*. The answer is no; there

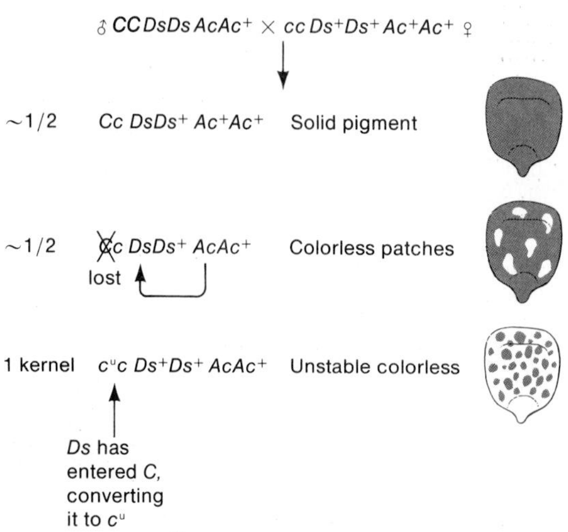

Figure 16-56. Results that indicate the transposition of Ds *into the* C *gene in corn. (*C *allows color expression;* c *does not.* Ac = *activator.) The action of* Ds *is dependent on the presence of the unlinked gene,* Ac.

is some kind of specificity involved that prevents this cross-activation of mutational instability.

The *Ds* element can wander not only into the middle of the *C* gene but also into other genes, rendering them also unstable mutants dependent on *Ac*. One such locus, *wx* (*waxy*), has been the subject of an intense study on the effects of the *Ds* element. Oliver Nelson has paired many unstable *waxy* alleles in the absence of the *Ac* mutation. In such $wx^{m\text{-}1}/wx^{m\text{-}2}$ heterozygotes, he has looked for rare wild-type *Wx* recombinants by staining the pollen with KI-I_2 reagent, which stains *Wx* pollen black and *wx* pollen red. By counting the frequency of *Wx* pollen grains in each kind of heterozygote, Nelson was able to do fine-structure recombination mapping of the *waxy* gene. He showed that the different "mutable *waxy*" mutant alleles are in fact due to the insertion of the *Ds* element in different positions of the gene. In a subsequent experiment, he allowed *Wx*-bearing pollen to fertilize *wx* plants and produce rare *waxy* kernels, which could be detected by staining sliced-off slivers. The *waxy* kernels were then raised into plants, and Nelson was able to show that the *Wx* pollen grains arose from chromosome exchange, which also involved exchange of flanking markers.

Several systems like a_1 *Dt* and *Ds Ac* have now been found in corn. Each shows similar action, having a **target gene** that is inactivated, presumably by the insertion of some **receptor element** into it, and a distant **regulator gene** that maintains the mutational instability of the locus, presumably through its ability to "unhook" the receptor element out of the target locus and thus return the locus to normal function. The receptor and the regulator are called **controlling elements.**

In the examples discussed thus far, the receptor is **nonautonomous.** It can render the target gene unstable (mutable) only in the presence of the regulator. Sometimes, however, a system such as *Ds Ac* can produce a mutable allele that is **autonomous.** Such mutants are recognized because they show single-locus Mendelian ratios (such as 3:1 for pigmented to dotted) apparently independent of any external element. Such alleles can subsequently regenerate a nonautonomous system. In such cases, the autonomy is thought to result from the integration of the receptor *and* regulator elements in the target gene.

Figure 16-57 summarizes the overall behavior of controlling elements in corn. Note that the apparently simultaneous mutation events that gave rise to Rhoades' original ratio are nicely explained by this model. Only *one* event is necessary—the splitting of the regulator from the receptor, thus generating both a_1 and *Dt*.

Controlling elements have been found in several other higher plants, including snapdragons (where the mutational instability can be manifested as various kinds of pigment variegation of the petals). Although such examples are not common, they are regularly encountered. It is beginning to appear as if controlling elements may play a role in the normal functioning of higher-plant genomes.

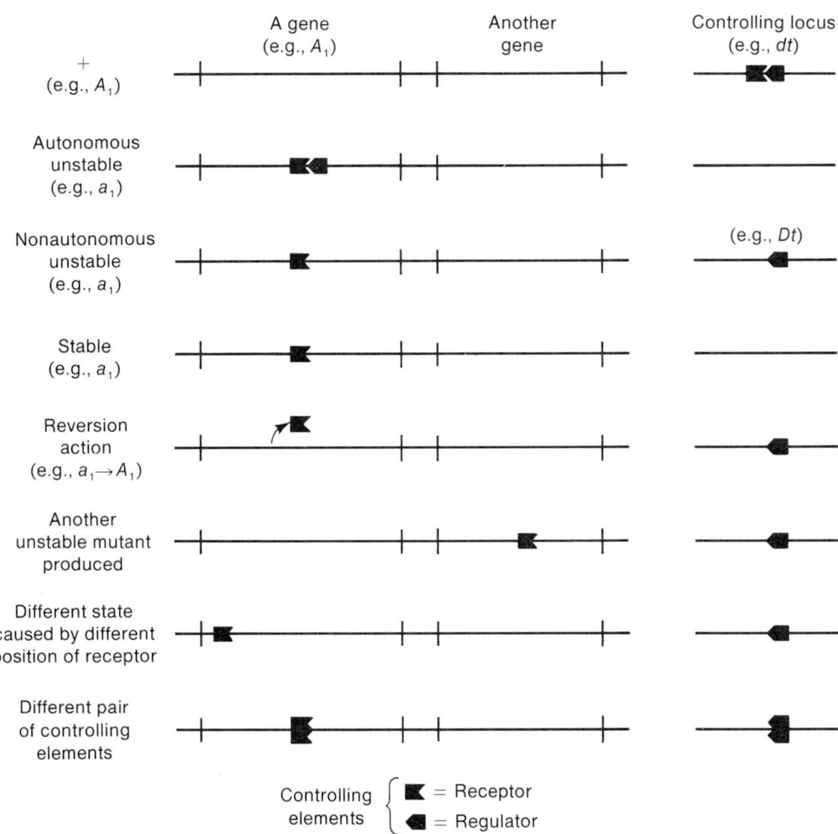

Figure 16-57. A model to explain the behavior of controlling elements in corn.

Unstable mutations are also reported (and have been thoroughly studied) in *Drosophila.*

The analogies between these controlling-element systems and the IS systems in bacteria (and phage *mu*) are striking, with transposition and inactivation being the main areas of similarity. Although evolutionary homology does not seem likely, these comparisons may point to some common role that has arisen independently to serve some common function in several different kinds of genomes.

Message
Transposable genetic elements in higher plants have been shown to be of two kinds. Receptors inactivate a target gene by inserting into it, and regulators control the excision of the receptor. Sometimes the receptor and the regulator are inserted together into the target gene. No effects of these genes are known when the receptor is not inserted into the specific target gene affected by the system.

Cassettes in Yeast. In yeast, transposable genetic elements have been identified directly in studies on the genetics of homothallism. The terms *homothallic* and *heterothallic* are used to distinguish between two types of sexual cycles in some lower eukaryotes. **Homothallic** species are those that can undergo a complete sexual cycle starting from one product of meiosis; **heterothallic** species need two individuals of opposite **mating type** to accomplish sexual reproduction. In the yeast *Saccharomyces,* some species are homothallic and some are heterothallic. Geneticists working with the homothallic species *S. chevalieri* were anxious to determine the genetic basis of this homothallism. They set out to cross the genetic determinant of homothallism into the heterothallic species *S. cerevisiae.* There are many good reasons for trying this approach, mostly stemming from the fact that a vast fund of refined genetic and chemical knowledge has been accumulated for *S. cerevisiae,* thus making it far more amenable to genetic analysis. (At the level of molecular genetics, *S. cerevisiae* is probably the best-understood eukaryote.)

When transferred into *S. cerevisiae,* the *S. chevalieri* homothallism gene also produced homothallism. The homothallism was found to act by causing the mating-type allele in some cells to change to the opposite type within the first few mitotic division. (These studies were performed by carefully separating and testing the daughter cells in the early divisions of an ascospore.) Thus, cells of the two mating types come to coexist in a single culture, and the sexual cycle can begin (Figure 16-58).

The genes involved are found at four loci. The *HO* gene is the main switching gene that controls the change of mating type. However, the *HO* gene can function only if another gene is present. This other gene can be a specific allele at either of two separate loci, which we call HM_1 and HM_2. Either the hm_1 gene or the nonallelic HM_2 gene must be present in order for *HO* to switch *a* to α; the *HO* gene cannot carry out this switch in the genotype $HM_1\,hm_2$. The reverse switch of α to *a* requires the presence of either hm_2 or HM_1; this switch

Initial genotype	Change induced by homothallism genes at one of first cell divisions	Typical situation in culture after several cell divisions
a	$a \rightarrow \alpha$	Sex cycle
α	$\alpha \rightarrow$ a	Sex cycle

Figure 16-58. The seeming absence of mating types in a homothallic yeast culture is an illusion. In fact, conversion of some cells to the opposite mating type during early mitotic divisions provides populations of both mating types.

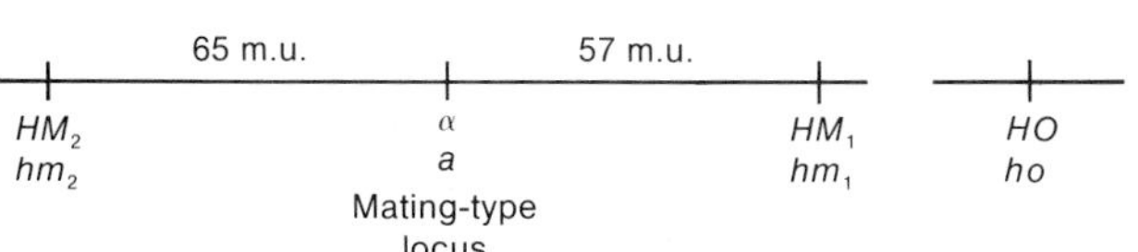

Figure 16-59. Linkage relationships of the genes involved in yeast homothallism. The hm_1 *or the nonallelic* HM_2 *gene is needed for* HO *to switch* a *to* α. *The* hm_2 *or* HM_1 *gene is needed for* HO *to switch* α *to* a.

cannot be accomplished in the genotype $hm_1 HM_2$. Figure 16-59 shows the linkage arrangement of these four interacting loci.

The situation became clearer through studies on a mutant α allele that we call α^m; the mutant is sterile and cannot carry out sexual reproduction with an *a* partner. In the presence of *HO* and either HM_1 HM_2 or hm_1 hm_2, the α^m mutant can change to a fertile *a* genotype (due to the HM_1 or hm_2 allele) and then back to an α genotype (due to the HM_2 or hm_1 allele). Surprisingly, this α genotype is fertile! The flip-flop of mating types has somehow "cleaned up" the mutation.

This situation has been explained by an exciting model in which the HM_1/hm_1 and HM_2/hm_2 loci are pictured as depots of "silent" blocks of mating-type information, rather whimsically called **cassettes.** Only at the mating-type locus itself are these cassettes "played" by proximity to a promoter gene of some sort. The mating-type changes are explained by the transposition and insertion of a cassette copy into the mating-type locus, replacing an old cassette that presumably is ejected and discarded (Figure 16-60).

The *alpha* mating-type gene of yeast has now been cloned. Yeast has its own plasmids. These can be joined to *E. coli* plasmids, such as pBR322, to form hybrid plasmids which can replicate in and transform either *E. coli* or yeast! One of these, YEp13, was used as a cloning vector for the *alpha* gene. *Bam*HI restriction fragments from *alpha* were cloned into YEp13 and then grown in *E. coli.* These plasmids were then used to try to transform *a* yeast cells to *alpha.* Once an *alpha* clone was identified, it was used as a hybridization probe to demonstrate that the *HM* genes did in fact contain homologous DNA, a highly satisfying confirmation of the cassette model derived from purely genetic analysis.

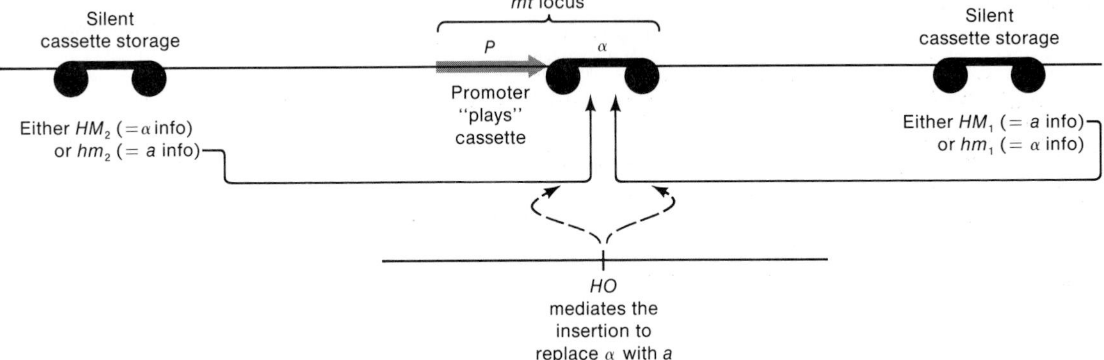

Figure 16-60. The cassette model to explain α → a conversion in homothallic strains of yeast. Mating-type-a cassette copies may be used from either storage locus, depending on which alleles are found there in the particular strain. A similar explanation accounts for a → alpha switches.

Message
In yeast, the regular change of one mating-type allele to another appears to be caused not by conventional mutation but by the insertion and activation of copies of silent mating-type genes from elsewhere in the genome.

One of the attractions of the cassette model is that it implicates the involvement of transposable genetic elements in a relatively normal cell function, homothallism. As more complex regulatory processes in higher organisms are further dissected genetically and chemically, more examples of such normal roles for transposition will surely appear. Although the transposable genetic elements appear at first glance to be agents of genetic chaos, they may turn out to be integral aspects of the regulation and organization of higher genomes. The appearance of chaos in the known examples probably results simply from the fact that geneticists can identify such elements only in cases where mutation has caused a disruption of the normal functioning. As a final word of caution, it should be pointed out that in no case has transposition been shown to be due to free particles. It seems much more likely that transposable elements are moved during juxtaposition of nonhomologous chromosome parts, by a crossover or conversionlike event.

Summary

Nature has devised many different ways of changing the genetic architecture of organisms. We are now beginning to understand the molecular processes behind some of these processes. Gene mutation, recombination between homologous chromosomes, and transposition can all be reasonably explained at the DNA level. Far from merely producing genetic waste, these processes undoubtedly all have important roles in evolution. This idea is strengthened through the knowledge that the processes themselves are to a large extent under genetic control: there are genes that affect the efficiency of mutation, recombination, and transposition.

Gene mutation works through the transversion, transition, addition, and deletion of small numbers of nucleotide pairs. This kind of change can be regarded as the basic raw material of evolution. Recombination and transposition act upon this raw material to produce new combinations of DNA sequences that can be matched against various environmental situations, so that their adaptive worth can be evaluated by natural selection.

Recombination and transposition work by quite different mechanisms. Whereas what we have called recombination works essentially by exchanging homologous chromosome parts, transposition works by realigning nonhomologous segments. But both processes produce new gene combinations. Transposable genetic elements can produce quite large chromosome rearrangements. It would be tempting to conclude that these elements are a major source of chromosome mutation in general, but the truth is that the precise mechanism behind most major chromosome rearrangements is not known at present. It is for this reason that chromosome rearrangements were not given specific coverage in this chapter.

Finally, let us note that transposition produces recombinant DNA in the same sense that experimental genetic engineering produces recombinant DNA—that is, between nonhomologous fragments. Nature has probably been using such recombinant DNA methods for billions of years.

Problems

1. Proflavin is a highly effective mutagen in the T phage of *E. coli,* but it does not cause mutations in *E. coli*. Suggest a possible explanation, and outline the experiments you would use to test your hypothesis.

2. Many mutagens increase the frequency of sister-chromatid exchange. Discuss possible explanations for this observation.

3. Normal ("tight") auxotrophic mutants will not grow at all in the absence of the appropriate supplement to the medium. However, in mutant hunts for auxotrophic mutants, it is common to find some mutants (called "leaky") that grow very slowly in the absence of the appropriate supplement but normally in the presence of the supplement. Propose an explanation for the molecular nature of the leaky mutants.

4. Strain A of *Neurospora* contains an *ad-3* mutation that reverts spontaneously at a rate of 10^{-6}. Strain A is crossed with a newly acquired wild-type isolate, and *ad-3*

strains are recovered from the progeny. When 28 different *ad-3* progeny strains are examined, 13 lines are found to revert at the rate of 10^{-6}, but the remaining 15 lines revert at the rate of 10^{-3}. Formulate a hypothesis to account for these findings, and outline an experimental program to test your hypothesis.

5. In *E. coli,* how would you expect each of the following double mutants to behave with respect to UV sensitivity and UV mutability in the light and in the dark? (a) *uvrA recA*; (b) *uvrA polA*; (c) *polA recA*.

6. a. Why is it not possible to induce nonsense mutations (represented at the mRNA level by the triplets UAG, UAA, and UGA) by treating wild-type strains with mutagens that cause only AT → GC or TA → CG transitions in DNA?
   b. Hydroxylamine causes only GC → AT or CG → TA transitions in DNA. Will hydroxylamine produce nonsense mutations in wild-type strains?
   c. Will hydroxylamine treatment revert nonsense mutations?

7. Several auxotrophic point mutants in *Neurospora* are treated with various agents to see if reversion will occur. Table 16-6 shows the results (with + indicating reversion).

Table 16-6.

	Mutagen			
Mutant	5BU	Hydroxylamine	Proflavin	Spontaneous
1	−	−	−	−
2	−	−	+	+
3	+	−	−	+
4	−	−	−	+
5	+	+	−	+

   a. For each of the five mutants, describe the nature of the original mutation event (not the reversion) at the molecular level. Be as specific as possible.
   b. For each of the five mutants, name a possible mutagen that could have caused the original mutation event (spontaneous mutation is not an acceptable answer).
   c. In the reversion experiment for mutant 5, a particularly interesting prototrophic derivative is obtained. When this type is crossed to a standard wild-type strain, the progeny consists of 90% prototrophs and 10% auxotrophs. Provide a full explanation, including a precise reason for the frequencies observed.

8. You are using nitrous acid to "revert" mutant *nic-2* alleles in *Neurospora.* You treat cells and plate them on medium without nicotinamide and look for prototrophic colonies. You obtain the following results for two mutant alleles. Explain these results at the molecular level, and indicate how you would test your hypotheses.
   a. With *nic-2* allele 1, you obtain no prototrophs at all.
   b. With *nic-2* allele 2, you obtain three prototrophic colonies, and you cross each separately with a wild-type strain. From the cross prototroph A × wild type, you obtain 100 progeny and all are prototrophic. From the cross prototroph B × wild type, you obtain 100 progeny, of which 78 are prototrophic and 22 are

nicotinamide requiring. From the cross prototroph C × wild type, you obtain 1000 progeny, of which 996 are prototrophic and 4 are nicotinamide requiring.

9. Mutations in locus 46 of the Ascomycete fungus *Ascobolus* produce light-colored ascospores; we'll call them *a* mutants. In the following crosses between different *a* mutants, asci are observed for the appearance of dark wild-type spores. In each cross, all such asci were of the genotypes indicated:

$a_1 \times a_2$	$a_1 \times a_3$	$a_2 \times a_3$
↓	↓	↓
a_1 +	a_1 +	a_2 +
+ +	a_1 +	a_2 +
+ a_2	+ +	+ +
+ a_2	+ a_3	+ a_3

Interpret these results in light of the models discussed in this chapter.

10. In the cross $A\, m_1\, m_2\, B \times a\, m_3\, b$, the order of the mutant sites m_1, m_2, and m_3 is unknown in relation to each other and to A/a and B/b. One nonlinear conversion ascus is obtained:

$A\, m_1\, m_2\, B$

$A\, m_1 \quad b$

$a\, m_1\, m_2\, B$

$a\, m_3 \quad b$

Interpret this result in the light of the hybrid DNA theory, and derive as much information as possible about the order of the sites.

11. In the cross $a_1 \times a_2$ (alleles of one locus), the following acus is obtained. Deduce what events may have produced it (at the molecular level).

a_1 +

$a_1\ a_2$

$a_1\ a_2$

+ a_2

12. G. Leblon and J.-L. Rossignol have made the following observations in *Ascobolus*. Single nucleotide-pair insertion or deletion mutations show gene conversions of the 6:2 or 2:6 type, rarely 5:3, 3:5, or 3:1:1:3. Base-pair transition mutations show gene conversion of the 3:5, 5:3, or 3:1:1:3 type, rarely 6:2 or 2:6.
    a. In terms of the hybrid DNA model, propose an explanation for these observations.
    b. Furthermore, they have shown that there are far less 6:2 than 2:6 for insertions and far more 6:2 than 2:6 for deletions (where the ratios are $+:m$). Explain these results in terms of hybrid DNA. (You might also perhaps think about excision of thymine dimers.)
    c. Finally, they have shown that, when a frameshift mutation is combined in a meiosis with a transition mutation at the same locus in a *cis* configuration, the asci showing *joint* conversion are all 6:2 or 2:6 for *both* sites—that is, the frameshift conversion pattern seems to have "imposed its will" on the transition site. Propose an explanation for this.

13. At the *grey* locus in the Ascomycete fungus *Sordaria*, the cross $+ \times g_1$ is made. In this cross, hybrid DNA sometimes extends across the site of heterozygosity, and two hybrid DNA molecules are formed (as we have discussed in this chapter). However, correction of hybrid DNA is not 100% efficient. In fact, 30% of all hybrid molecules are not corrected at all, whereas 50% are corrected to +, and 20% are corrected to g_1. What proportion of aberrant-ratio asci will be (a) 6:2? (b) 2:6? (c) 3:1:1:3? (d) 5:3? (e) 3:5?

14. Noreen Murray crossed α and β, two alleles of the *me-2* locus in *Neurospora*. Included in the cross were two markers, *trp* and *pan*, which flank *me-2* each at a distance of 5 m.u. The ascospores were plated onto a medium containing tryptophan, pantothenate, and no methionine. The methionine prototrophs that grew were isolated and scored for the flanking markers, yielding the results shown in Table 16-7. Interpret these results in light of the models presented in this chapter. Be sure to account for the asymmetries in the classes.

Table 16-7.

	Genotype of *me-2*$^+$ prototrophs			
Cross	*trp* +	+ *pan*	*trp pan*	+ +
trp α + × + β *pan*	26	59	16	56
trp β + × + α *pan*	84	23	87	15

15. In *Neurospora*, the cross $A\,x \times a\,y$ is made, in which x and y are alleles of the *his-1* locus, and A and a are mating-type alleles. The recombinant frequency between the *his-1* alleles is measured by prototroph frequency when ascospores are plated on medium lacking histidine; it is measured as 10^{-5}. Progeny of parental genotype are backcrossed to the parents, with the following results. All $a\,y$ progeny backcrossed to the $A\,x$ parent show prototroph frequencies of 10^{-5}. When $A\,x$ progeny are backcrossed to the $a\,y$ parent, two prototroph frequencies are obtained: one-half of the crosses show 10^{-5}, but the other one-half show the much higher frequency of 10^{-2}. Propose an explanation for these results, and describe a research program to test your hypothesis. (NOTE: Intragenic recombination is a *meiotic* function that occurs in a diploid cell. Thus, even though this is a haploid organism, dominance and recessiveness could be involved in this question.)

16. When Rhoades took pollen from wholly pigmented anthers on plants of genotype a_1a_1 *DtDt* and used this pollen to pollinate a_1a_1 *dtdt* tester females, he found wholly pigmented kernels and, in addition, some dotted kernels. Explain the origin of *both* these phenotypes.

17. In yeast, the *his4* gene has three cistrons, A, B, and C, in that order, each mediating an enzymatic step of histidine synthesis. A certain spontaneous mutation is mapped in the cistron A, but these mutants are defective for all three functions (A, B, and C). The mutation is not suppressible by nonsense or frameshift suppressors. Spontaneous reversion occurs quite frequently, but this rate is not enhanced by any mutagen. Discuss the possible nature of the mutation.

18. In *Drosophila*, M. Green found a *singed* allele *sn77-27* with some unusual characteristics. Females homozygous for this X-linked allele have singed bristles, but they

have numerous patches of sn^+ (wild-type) bristles on their heads, thoraxes, and abdomens. When these flies are bred to *sn* males, some females give only singed progeny, but others give both singed and wild-type progeny, in variable proportions. Explain these results.

19. An alternative model to explain mating-type interconversion in *Saccharomyces* yeast is called the "promoter flip-flop" model. Given just this descriptive name and your knowledge of the system (and of the cassette model), speculate on what the flip-flop model is.

20. You have two separate homothallic cultures of *Saccharomyces cerevisiae* with genotypes α *HO* HM_1 hm_2 and α *HO* HM_1 HM_2. Each culture is allowed to sporulate, and one tetrad is removed from each. What mating behavior is expected in the ascospores? For each tetrad, indicate the mating behavior of the four spores as α, *a*, or homothallic (the order within each tetrad is unimportant).

21. A stable disomic (with chromosome number $n + 1$) of yeast has been used to detect recessive mutations blocking intragenic recombination. The duplicated chromosome carries the genotypes *leu2x a*/*leu2y* α, where *leu2x* and *leu2y* are noncomplementing mutations of the *leu2* gene, and *a* and α are the mating-type alleles. When this strain is placed on sporulation medium, meiosis occurs because all the non-aneuploid chromosomes produce identical copies of themselves.

    a. How would you use this strain to detect recessive mutations anywhere on the nonaneuploid parts of the genome that affect recombination within the *leu2* gene?

    b. Why must the mating-type gene be carried on the aneuploid chromosome?

22. Devise very clever screening procedures for detecting

    a. nerve mutants in *Drosophila;*

    b. mutants lacking flagella in a haploid unicellular alga;

    c. supercolossal-sized mutants in bacteria;

    d. mutants that overproduce the black compound melanine in haploid fungus cultures (which normally are white);

    e. individual humans (in large populations) whose eyes polarize incoming light;

    f. negatively phototropic *Drosophila* or unicellular algae;

    g. UV-sensitive mutants in haploid yeast.

23. Crown-gall tumors are found in many dicotyledonous plants infected by the bacterium *Agrobacterium tumifaciens.* The tumors are caused by the insertion of DNA into the plant DNA from a large plasmid carried by the bacterium. A tobacco plant of type A (there are many types of tobacco plants) is infected, and it produces tumors. You remove tumor tissue and grow it on synthetic medium. Some of these tumor cultures produce aerial "shoots," and you graft one of these shoots onto a normal tobacco plant of type B. The graft grows to an apparently normal A-type shoot and flowers.

    a. You remove cells from the graft and place them in synthetic medium, where they grow like tumor cells. Explain why the graft appears normal.

    b. When seeds are produced by the graft, the resulting progeny are normal A-type plants. No trace of the inserted plasmid DNA remains. Propose a possible explanation for this "reversal."

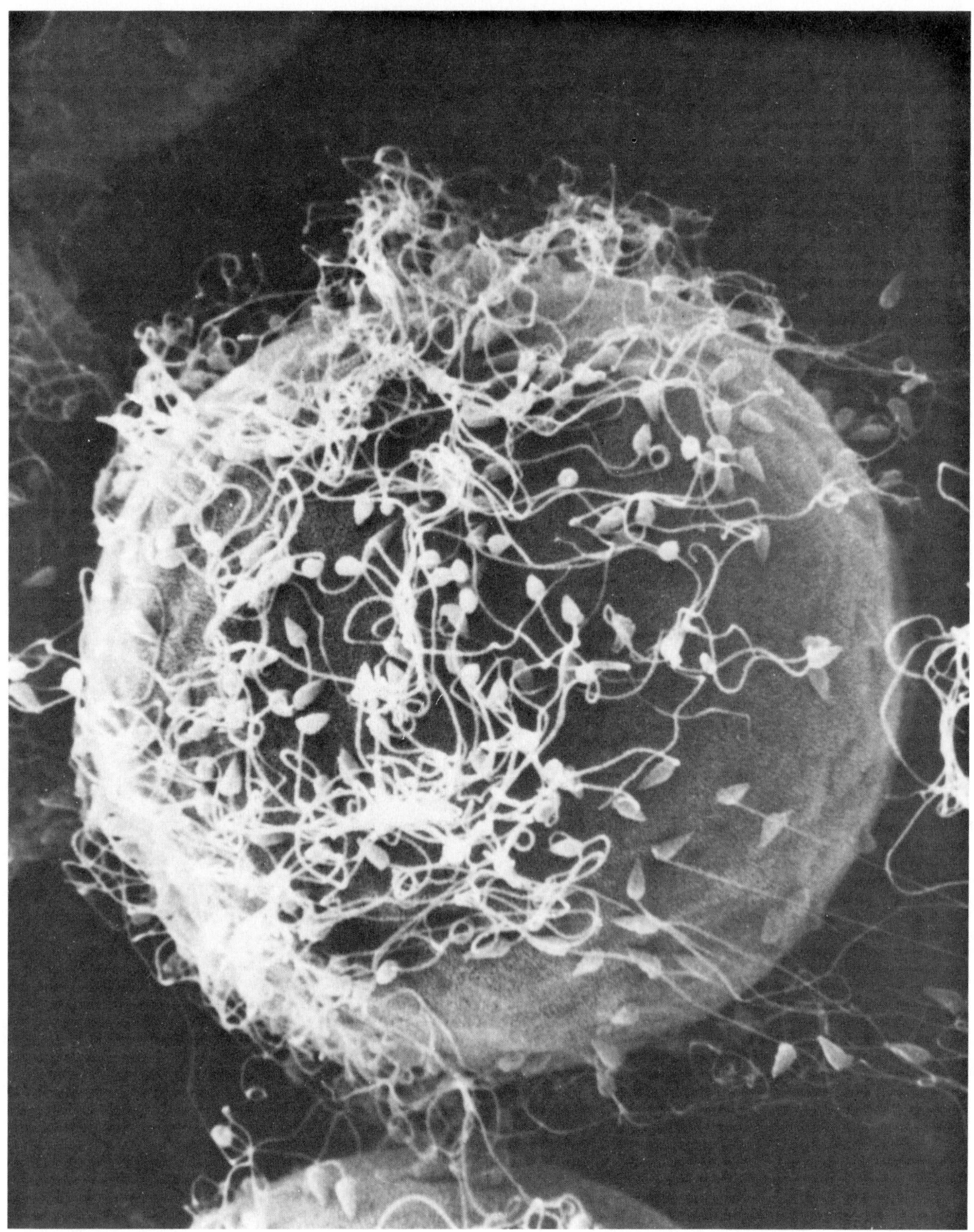

Sea urchin egg and sperm (× 1300). (Mia Tegner, Scripps Institution of Oceanography.)

17

Developmental Genetics

The developmental biologist faces an intriguing challenge in trying to understand the dramatic transformation of a fertilized egg into a complex, multicellular organism containing differentiated cells and tissues that are wonderfully integrated and coordinated. In humans, a single sperm (from perhaps a billion released on ejaculation) combines with a single egg to form a zygote containing all the cellular information to be passed on from parents to offspring. That tiny mass of protoplasm (weighing 5×10^{-8} oz and measuring only 0.14 mm in diameter) develops into an organism of some 6×10^{13} cells at adulthood. The understanding of this developmental process would provide major advances in medicine—and perhaps would open the way for the deliberate engineering of organisms for desired characteristics. Although many phenomena have been observed and studied, biologists have not yet achieved a detailed and sweeping understanding of the mechanisms of development. However, many different lines of research are now beginning to yield results that may ultimately contribute to such an understanding. This chapter explores some of those results.

Development as a Regulated Process

It is clear that the DNA in the zygote must carry all of the genetic information needed for complete development of the adult organism. Many questions arise, however, in trying to construct a model for the genetic regulation of development. Does every cell of the adult organism contain a complete copy of the original genome? If so, is each cell of the adult potentially capable of acting as a zygote and developing into a complete adult? What role does the cytoplasm (and the cytoplasmic genome) play in the regulation of development? We look now at the available evidence bearing on these questions.

Is the Entire DNA Complement Present in All Cells?

It is clear that genes play a central role in determining any aspect of the adult phenotype of an organism—the presence or absence of hair on the knuckles, the color of the eyes, the texture of the hair, the shape of ear lobes, and so forth. The numerous mutant phenotypes (dwarfs, multiple fingers, certain kinds of mental retardation, and so on) demonstrate that genes can interfere with the normal developmental process. Can we conclude that the developmental process is completely under the control of the genes? If so, how does it happen that two cells derived from the same ancestry will develop in different ways in the organism (for example, one becoming a skin cell and another becoming a nerve cell)?

During the embryological development (**embryogenesis**) of most organisms, mitotic division appears to provide each cell with an identical complement of chromosomes. But does each cell of the organism in fact inherit a complete set of DNA, or could the genome be parceled out in some way during embryogenesis, so that different cell lines acquire different parts of the genome? Such a division of the genome could provide an explanation of the differing fates of the cell lines during further development.

DNA hybridization is a useful technique for studying similarities among DNA samples. In the most informative experiment, radioactively labeled DNA from cultures of mouse-embryo cells is hybridized with unlabeled DNA from similar cultures. Samples of other DNA can be added to measure the effectiveness with which the DNA being tested will compete with the labeled DNA in hybridization. Samples from various differentiated mouse tissues compete equally successfully with the labeled mouse-embryo DNA (Figure 17-1). However, this experiment reflects mainly the hybridization of repetitive sequences; it does not test the similarities or differences between the unique sequences. Other studies to compare unique DNA sequences from various tissues have found no detectable differences from embryo DNA, but the margin of error is such that a difference in 5% to 10% of the unique DNA between the various cultures would not be detected. We conclude that each cell does include most (and possibly all) of the DNA complement from the original zygote.

Are Differentiated Nuclei Totipotent?

We could demonstrate that each cell does retain a complete set of DNA after differentiation if we could show that each cell is **totipotent**—that is, capable of developing into a complete individual. There are cases of multiple births in which "identical" siblings are derived from a single fertilized human egg (Figure 17-2). The Dionne quintuplets were genetically identical. Thus we conclude

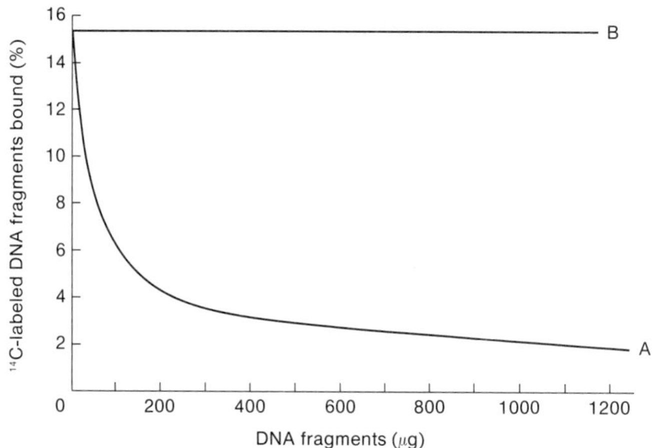

Figure 17-1. Competition for binding to mouse-embryo DNA between radioactively labeled mouse-embryo DNA and unlabeled DNA from other sources. If the unlabeled DNA comes from a cell culture obtained from differentiated mouse tissue (brain, kidney, thymus, spleen, or liver), the results are the same as those obtained if the unlabeled DNA comes from a mouse embryo (curve A). Curve B is obtained when the unlabeled DNA comes from Bacillis subtilis *(a control experiment). (From B. J. McCarthy and B. H. Hoyer,* Proc. Natl. Acad. Sci. USA *52:915, 1964.)*

Figure 17-2. Identical quadruplets. (Photograph courtesy of Erika Stone, © Peter Arnold, Inc.)

Figure 17-3. Two "arms" (rays) of this starfish (Asterias) *were severed. New rays have begun to form, and they will eventually grow into normal-sized rays. (Photograph courtesy of Grant Heilman.)*

that the genetic information is faithfully reproduced through at least the first three cleavages after fertilization (two cleavages produce only four cells). But what of the nuclei in cells resulting from later divisions? Do they remain totipotent?

Many highly differentiated organisms can regenerate new organs and tissues. For example, a starfish can regenerate a lost "arm" (Figure 17-3), a reptile can regenerate a lost tail, and a human can regenerate a damaged liver. However, such regeneration is possible only in certain tissues. Regeneration of a complete organism from a single somatic cell is not observed among animals in nature.

In the 1950s, Frederick Steward demonstrated that highly differentiated phloem cells in the root of a carrot plant are totipotent. He was able to obtain an entire carrot plant from a single phloem cell (Figure 17-4). Using this method, he obtained a number of genetically identical plants from the somatic cells of a single plant. This asexual method of reproduction is now called **cloning** (in fact, the word *clone* is derived from a Greek word referring to a plant cutting). This

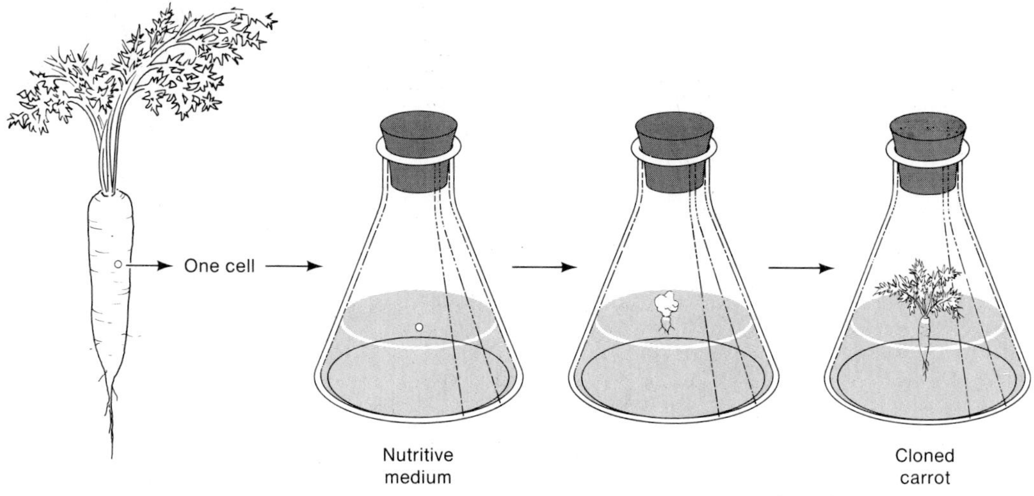

Figure 17-4. Cloning of a carrot plant from a single cell taken from the phloem tissue of a mature root. The nutritive media in the different flasks must contain appropriate plant hormones.

work has important implications because of the economic potential for obtaining duplicates of any individual plant with a particularly useful phenotype. One of the problems in breeding any organism that reproduces sexually is the constant disruption of useful gene arrangements by random assortment and crossing over. Cloning retains intact genetic combinations, and this process is now used commercially to maintain lines of trees. Of course, gardeners and farmers have long used asexual reproduction in the grafting of a branch from one genotype onto another or in the generation of new plants from cuttings.

Message
Cloning of plants shows that differentiated cells remain totipotent.

Individual cells of multicellular organisms can be treated and analyzed like microorganisms when grown in cell cultures. Mutations can be induced and selected from these cells for such qualities as increased content of certain amino acids. If such mutant cells can be stimulated to differentiate into mature plants, then the process of plant breeding is tremendously simplified and accelerated. Such a procedure seems quite possible in plants, because the totipotency of differentiated plant cells has been demonstrated. However, are differentiated animal cells similarly totipotent?

In the early 1950s, Robert Briggs and Thomas King developed techniques for manipulating nuclei in amphibian cells. They created **enucleated** frog eggs (eggs lacking nuclei) by inserting fine glass pipettes into unfertilized eggs and

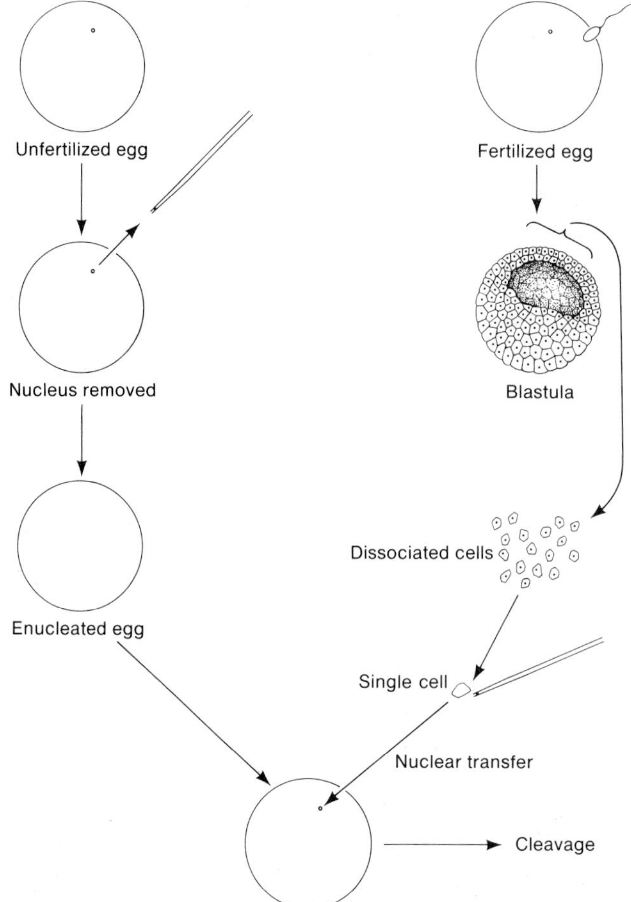

Figure 17-5. Transplantation of a nucleus from an embryonic amphibian cell into an enucleated egg. The nucleus of an unfertilized egg is removed mechanically by micropipette (left). *Cells of a fertilized egg that has developed to the blastula stage are dissociated* (right) *so that the nucleus of a single cell can be removed and inserted into the enucleated egg* (bottom).

sucking the nuclei out. Such manipulation makes possible the study of totipotency in animal cells. For example, Briggs and King were able to remove nuclei from differentiated cells and inject them into enucleated eggs (Figure 17-5). Thus they could test the developmental capacity of the nucleus from a differentiated cell within the cytoplasmic environment of an unfertilized egg. They found that an enucleated egg will divide when provided with a somatic nucleus. Furthermore, they found that a nucleus removed from a cell of the **blastula** (a stage when the embryo is still a hollow ball of cells) is totipotent.

When the nucleus from a blastula cell is inserted into an enucleated egg, the egg divides normally and develops into an adult.

However, different results were obtained when the nucleus was taken from a cell at the **gastrula** stage of development (when infolding of the layers of cells begins). A nucleus from a gastrula cell will not support normal development when inserted into an enucleated egg. It appears that some process of nuclear differentiation begins at gastrulation, destroying the totipotency of the nucleus from a frog somatic cell.

Even if each cell does retain a complete complement of DNA, it is obvious that some regulation process must activate and inactivate specific parts of the genome as development occurs. Perhaps it takes a certain amount of time for such activation and inactivation to take place. Then, the renucleated egg may divide before the transplanted nucleus can return to a condition appropriate to the first division. Such an effect could lead to abnormal development even though the transplanted nucleus was in theory totipotent. To eliminate this possibility, Briggs and King performed another group of experiments. They kept nuclei in the cytoplasmic environment that encourages totipotency (cytoplasm from pregastrulation cells) by serially transplanting nuclei from gastrula embryos into enucleated eggs (Figure 17-6). Even after many serial transplantations, the nuclei still remained incapable of supporting normal development.

These experiments show that somatic nuclei of the frog *Rana* differentiate irreversibly after gastrulation—at least, under the conditions of the nucleus-transplantation experiments. However, John Gurdon repeated these experiments in 1964 with a different amphibian, the African clawed toad, *Xenopus laevis.* He was able to take nuclei from highly differentiated cells of the gut of tadpoles and insert them into eggs whose nuclei had been destroyed by a beam of ultraviolet light (Figure 17-7). Although many of the eggs failed to develop or produced abnormal growth, a significant number of eggs did develop normally into adults whose genetic markers showed them to be identical clones of the nucleus-donor tadpole (not of the toad from which the unfertilized egg was taken). The result clearly differs from those obtained with the frog *Rana.* However, the result with *Xenopus* does indicate that a nucleus can remain totipotent at a highly differentiated stage in at least one vertebrate organism.

We do not yet know why the nuclei of the frog *Rana* cease to be totipotent at gastrulation whereas those of the toad *Xenopus* can remain totipotent in the tadpole stage. Therefore, we must be very cautious about extrapolating from one species to another in this kind of research. Of course, the implications of techniques for cloning humans have received extensive popular discussion. Aldous Huxley's *Brave New World* develops one classic scenario based on this possibility. But is there any reason to believe that nuclei remain totipotent in animals other than *Xenopus?*

Karl Illmensee has obtained evidence to support the idea of totipotency in differentiated cells from both mice and *Drosophila.* In mice, Ilmensee takes cells from a teratocarcinoma—a type of tumor of the gonads, in which are found such differentiated cells from both mice and *Drosophila.* In mice, Illmensee takes cells cells retain their phenotypes to the extent that injection of a single cell into the

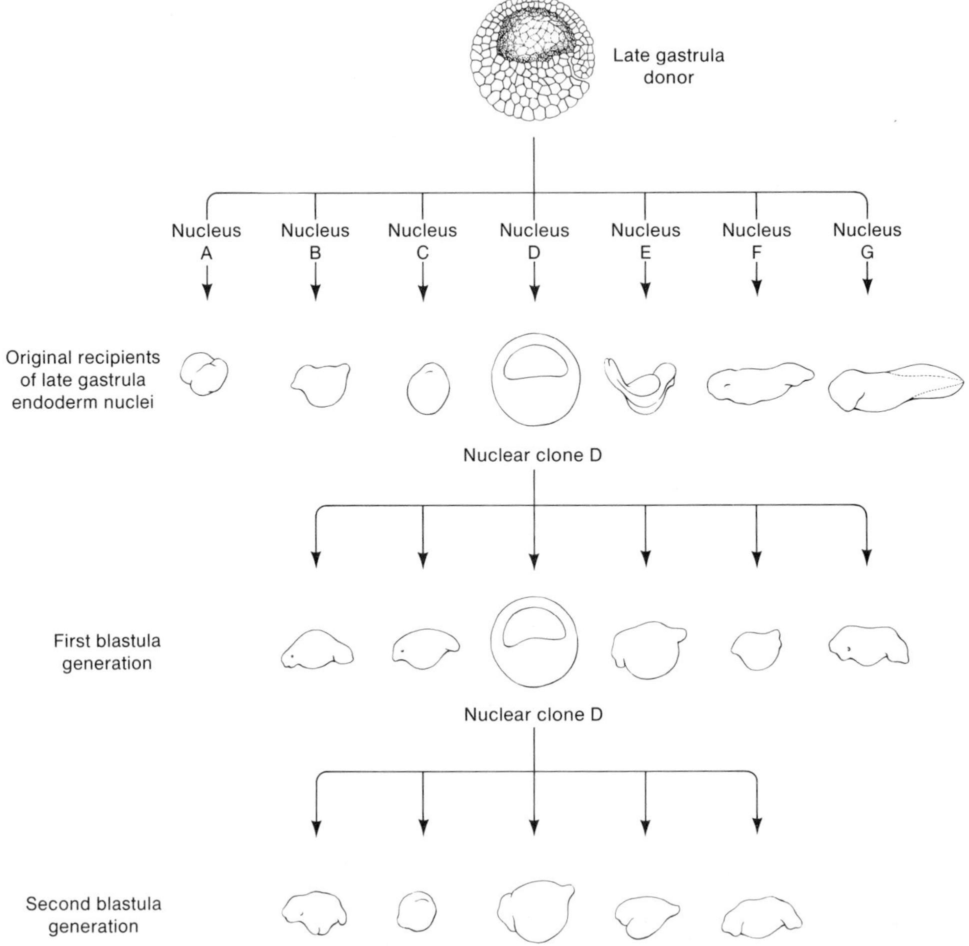

Figure 17-6. Serial transplants of gastrula nuclei in the frog Rana. *All eggs except D are allowed to proceed to the end of their development (all develop abnormally). The D clone is allowed to proceed to the blastula stage. Then, nuclei from the blastula cells are transplanted to enucleated eggs, and the same procedure is repeated. Normal development of any renucleated cells was never observed, even after many repetitions of this procedure. (From R. Briggs,* Cold Spring Harbor Symp. Quant. Biol. *21:277, 1956.)*

peritoneal cavity of a normal mouse results in the growth of a teratocarcinoma. Single cells from one genetically marked line are injected into an embryo from another line (Figure 17-8). The embryo is then implanted into the uterus of a "pseudopregnant" female—a female that has been mated to a sterile male and hence has a uterus receptive to implantation of the embryo (Figure 17-9). The embryo develops normally. The mouse that is born proves to be a chimera (or mosaic) containing a clone of cells derived from the injected tumor cell. This

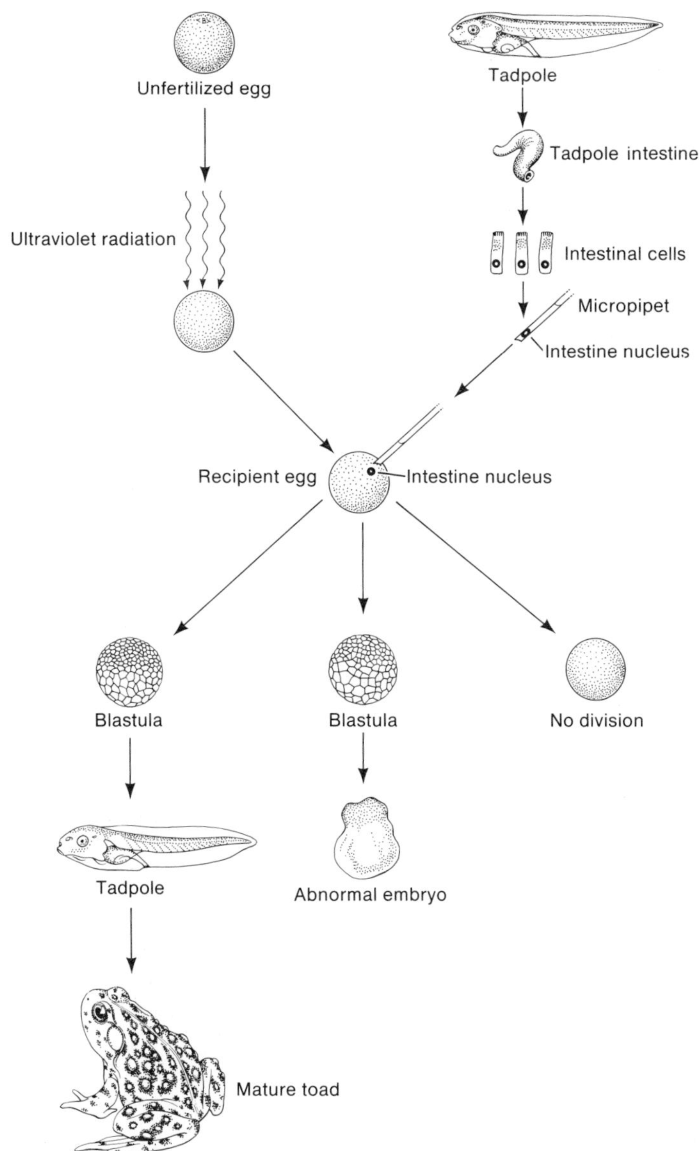

Figure 17-7. Cloning of the toad Xenopus *by nuclear transplantation. The nucleus of an unfertilized egg is destroyed by a beam of ultraviolet light* (upper left), *which specifically damages DNA. A nucleus taken from a gut cell of a tadpole* (upper right) *is transplanted into the enucleated egg* (center). *Many of the resulting zygotes develop abnormally or fail to divide, but a few of the zygotes develop into mature individuals* (bottom left). *(From J. B. Gurdon, "Transplanted Nuclei and Cell Differentiation." Copyright © 1968 by Scientific American, Inc. All rights reserved.)*

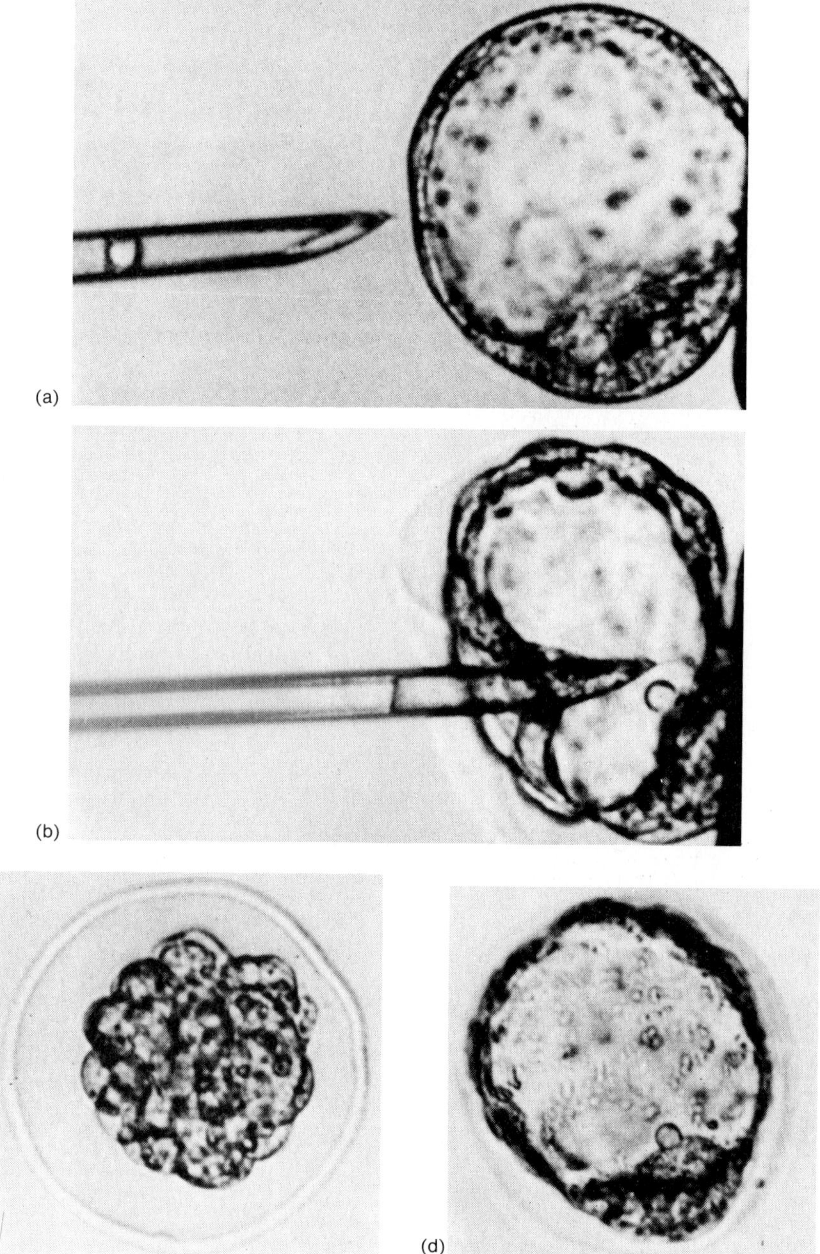

Figure 17-8. Formation of a mosaic embryo. (a) A mouse embryo is held on the end of a pipette while a teratocarcinoma cell is brought up to it. (b) The teratocarcinoma cell is injected into the center of the embryo. (c) The teratocarcinoma cell becomes integrated into the embryo. (d) All of the cells of the embryo (including that from the teratocarcinoma) multiply as development proceeds. The descendants of the injected cell will be integrated into tissues of the embryo—which tissues will depend on the location of the injection. (From K. Illmensee, in Genetic Mosaics and Chimeras in Mammals, *Liane B. Russell, ed., Plenum Publishing Corp., 1978.)*

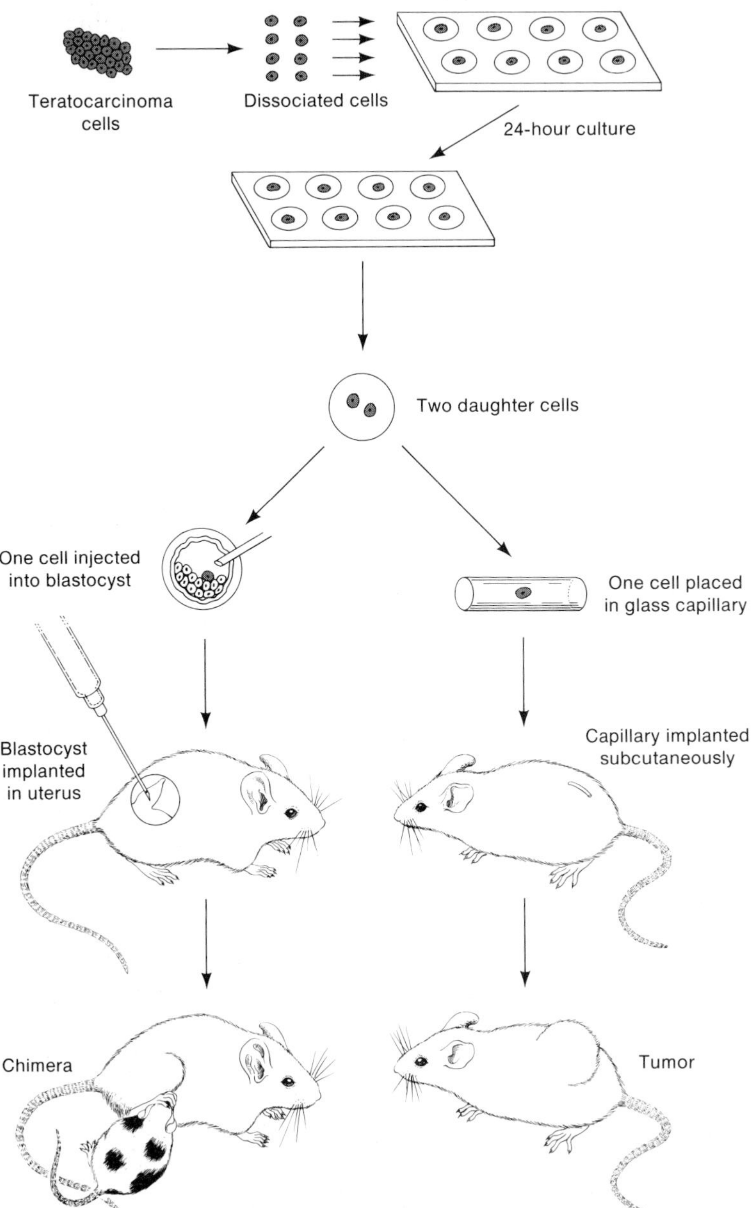

Figure 17-9. Experimental analysis of teratocarcinoma cells. Individual cells are allowed to divide once, and one of the daughter cells is injected into an embryo of the blastocyst stage (left). *The resulting mosaic embryo is then inserted into the uterus of a recipient female, where it develops normally. The other cell is placed in a glass capillary tube that is placed under the skin of another mouse, where it develops into a tumor* (right). *Material from the tumor can be used for comparison with cells from the chimera. (From K. Illmensee and L. C. Stevens, "Teratomas and Chimeras." Copyright © 1979 by Scientific American, Inc. All rights reserved.)*

clone is completely integrated into the organ or organs in which it is found. If the clone of injected cells is included in the gonadal region, the animals can be mated and normal offspring derived from gametes formed by the clone. Hence, the nuclei from the teratocarcinoma are indistinguishable from normal cells in their totipotency. Using a similar microinjection technique, Illmensee has introduced single marked cells of *Drosophila* embryos into unfertilized eggs, which then develop into adult flies. Again, the differentiated embryonic cells prove to be totipotent.

From these results, we conclude that differentiated cells in many different organisms can be totipotent under certain experimental conditions. It seems reasonable to conclude that every cell in a highly differentiated organism does contain a complete copy of the DNA from the original zygote. In many such cells, the expression of parts of the genome is blocked in such a way that the nucleus from the differentiated cell cannot direct normal development when transplanted to an unfertilized egg. The nature of this blockage then becomes the subject for further research.

Message
Mitosis accurately duplicates and distributes DNA sequences among the daughter cells at each division during development. Hence each somatic cell contains the same genetic information as the newly fertilized egg from which the organism developed.

Is Development Controlled by the Nucleus or by the Cytoplasm?

For decades, embryologists have argued that the nucleus plays a relatively passive role in directing the primary events that determine the developmental fates of cells and tissues. Different cells develop in different ways within the embryo, but each cell seems to inherit an identical set of nuclear genetic material. On the other hand, asymmetries are known to exist in the distribution of cytoplasmic material during cell divisions. Thus it seems reasonable that the varying developmental patterns in different cells are under cytoplasmic rather than nuclear control.

Mitochondria and yolk, for example, are not evenly distributed through the egg. The planes of division in early cleavages after fertilization result in the inclusion of differing amounts of yolk and mitochondria in different daughter cells (Figure 17-10). Within a few divisions, differences in division rates and cell size become apparent between cells in the embryo. These differences between parts of the embryo mean that environmental factors (oxygen, waste materials, temperature, and so on) impinge on the different parts in different ways.

The importance of the cytoplasm can be demonstrated experimentally. A part of a cell cortex (outer layer) without nuclei is removed from one embryo and

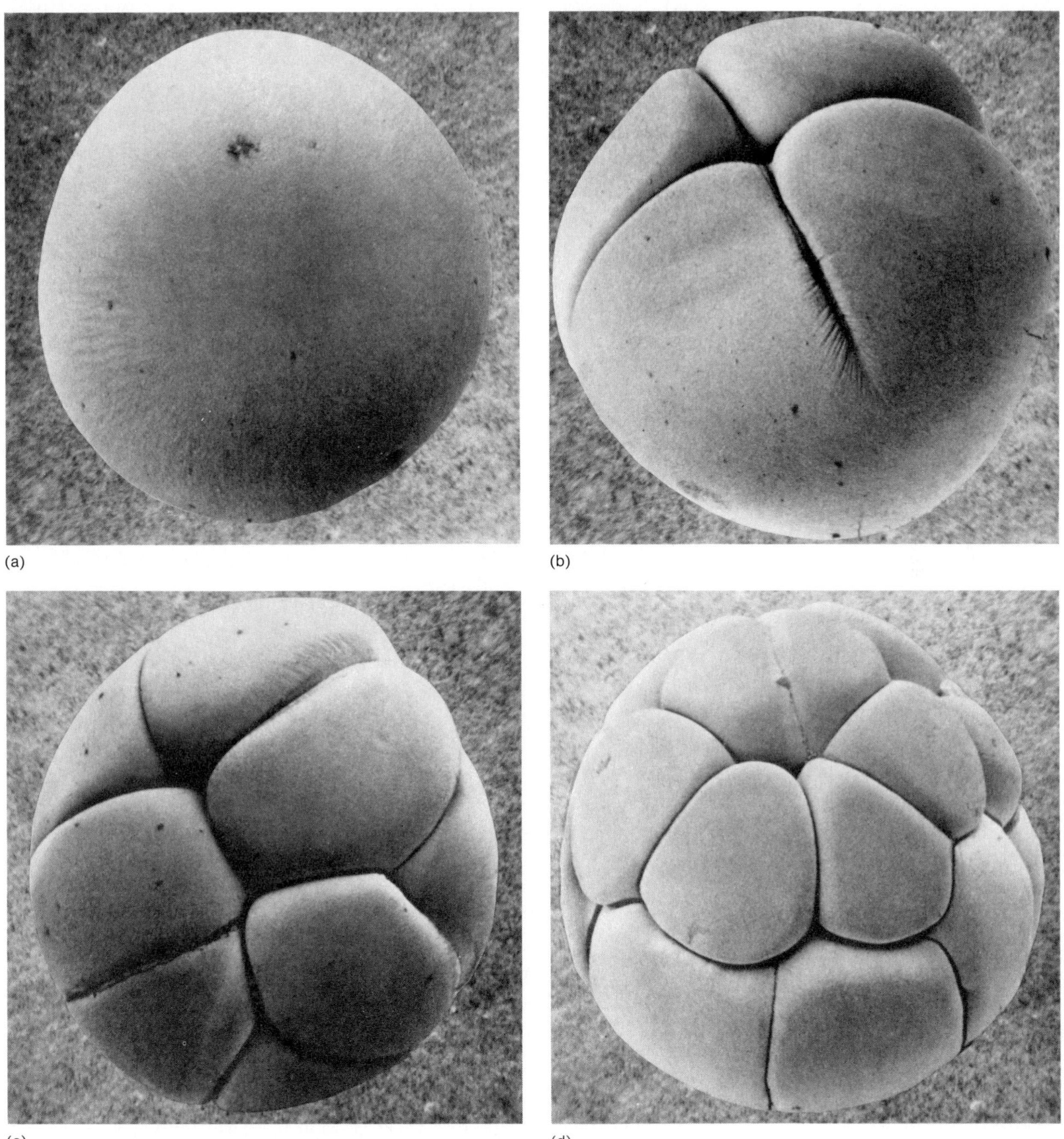

Figure 17-10. *Early cleavages of an amphibian egg. It can be seen that asymmetries soon develop due to the presence of yolk at the lower (vegetal) pole of the zygote cell. (a) Undivided egg. (b) Second cleavage. (c) 8-cell stage. (d) 16-cell stage. (Lloyd M. Beidler, Florida State University.)*

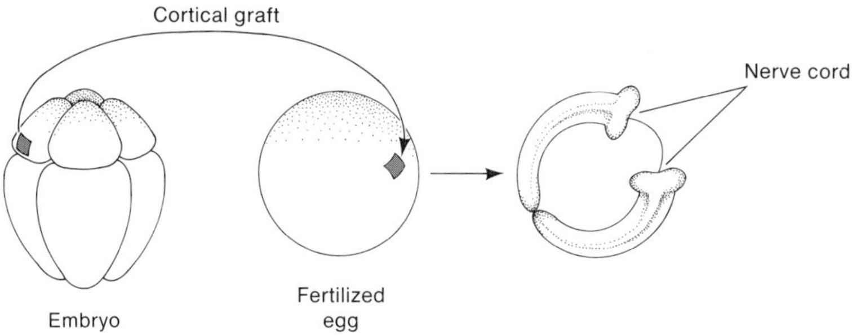

Figure 17-11. Experimental demonstration of extranuclear control of development. A graft of cell membrane and underlying cytoplasm is taken from the smaller cells of an embryo at the 8-cell stage and transplanted to the equatorial region of a fertilized egg. A second spinal cord develops in the resulting embryo. (From A. S. G. Curtis, J. Embr. Exp. Morphol. *10:416, 1972.)*

grafted onto another (Figure 17-11). The resulting embryo may develop an extra set of some structure (twinned spinal cords in the example illustrated), showing that the cytoplasm (without the nucleus) can exert a major influence on development.

Environmental effects also are important in determining the course of development. In the 1940s, Richard Goldschmidt showed that a variety of environmental factors can produce phenotypic abnormalities in *Drosophila* that resemble the abnormal phenotypes of known genetic mutations. This phenomenon is called **phenocopy.** Phenocopy has subsequently been demonstrated in a number of organisms, produced by a variety of agents, including temperature shock, radiation, and an array of chemical compounds. Goldschmidt showed that the agent must be applied at a specific critical period in development in order to obtain each particular phenocopy. In 1945, Walter Landauer demonstrated such **phenocritical periods** in the embryonic response of chickens to such compounds as insulin, boric acid, and pilocarpin. These studies revealed the existence of a nongenetic process that can produce abnormalities—a process called **teratogenesis.** We now know that a variety of agents (German measles and various drugs) can produce abnormalities in the human fetus.

Message

The egg cytoplasm is not homogeneous, and regional differences in the cytoplasm play a role in development. Furthermore, environmental agents can provoke specific defects during development.

While embryologists were searching in the cytoplasm for the key factors controlling development, geneticists were seeking evidence of nuclear control. There is no doubt that the egg cytoplasm does regulate the early events in embryonic development—for example, an enucleated frog egg will still go through the early cleavages. However, the architecture and molecular composition of the oocyte cytoplasm are themselves under rigid control by the nuclear genome of the mother! We have already discussed an example: the *anucleolate* (*an*) mutation in *Xenopus*. The cross *an*/+ × *an*/+ forms some *an*/*an* homozygotes that are incapable of forming ribosomes. Nonetheless, these embryos develop to twitching tadpoles before dying. Their survival is a result of the prefertilization amplification of ribosomal genes and the consequent store of ribosomes (which must reflect the maternal genotype). The most striking illustration of maternal genotypic influences on the phenotype of the offspring is seen in the coiling of snail shells (Chapter 15).

Another striking illustration of maternal genotype on the oocyte is found in *Drosophila*. In the *Drosophila* egg, the anterior end can be distinguished from the posterior end, where a special region of the cortex can be seen to have densely staining **polar granules.** The granules define a region containing pole cytoplasm; the cytoplasm of all gonadal cells is derived from the pole cytoplasm. An autosomal recessive mutation discovered in *D. pseudoobscura* is called *grandchildless* (*gs*) for reasons that will become apparent. In the cross *gs*/+ × *gs*/+, some *gs*/*gs* homozygotes are formed. These homozygotes are fertile, but all eggs produced by *gs*/*gs* females lack polar granules. Regardless of the genotype of their mates, all of the offspring of *gs*/*gs* females are born without gonads (and thus are sterile and fail to produce grandchildren of the *gs*/*gs* females). Obviously, in this case the maternal genotype plays a major role in determining the cytoplasmic makeup of eggs, thus affecting the phenotype of the offspring.

Message
Maternal genotype affects the architecture and composition of the egg cytoplasm and thus can affect the phenotype of the offspring.

Is the cytoplasm or the nucleus the site for control of the developmental process? It should be obvious by now that this is rather like asking whether the chicken or the egg came first! In the context of development, the nucleus in isolation from the cytoplasm is as useless as a cell without its nucleus. (Enucleate red blood cells are very useful, but they do not play a major role in development.) Much of the immediate control of developmental processes seems to come from the cytoplasm, but nuclear genetic factors play a major role in determining the nature of the cytoplasm.

Differentiation and Variable Gene Activity

We have seen that cell divisions just after fertilization are relatively rapid and are controlled by maternally derived information in the cytoplasm, with little genetic control from the zygotic nuclei. At what point in development do the embryonic nuclei begin to play a major role? (We know that they must play such a role because the zygote genotype does have a major effect on the phenotype.)

Nineteenth-century embryologists devised ingenious experimental approaches to the study of this question. For example, Theodor Boveri studied species of sea urchin in which the embryos differ in morphology. He enucleated an egg from one species and fertilized it with sperm from another species. The sperm nucleus can promote development in such situations. He showed that development of the embryo follows the pattern dictated by the genotype of the sperm, and he determined the point in the developmental sequence when this paternal influence becomes obvious. A similar approach can now be used to measure the time of appearance of some gene product (either RNA or protein) encoded by the paternal genes. Drugs such as actinomycin D that inhibit RNA synthesis can be tested to determine the time in development when they have lethal effects—thus showing when the products of the embryonic nuclei are required for survival. Such studies show that zygotic genes probably begin to act around the time of gastrulation in amphibians. However, lethal effects of actinomycin D on mouse embryos begin to appear around the 8-cell or 16-cell stage of development.

Visible Gene Activity

We have seen that each cell carries a complete nuclear genome, but we know that certain genes have their effects only at certain stages of development. A hypothesis involving differential activation of different genes at different stages of development seems very attractive. Can such differential gene activation be demonstrated?

Once again, we turn to the giant polytene chromosomes of Diptera. Certain stains show that these chromosomes contain regions that are rich in RNA, suggesting that these regions may be cytologically visible sites of RNA synthesis. The suggestion can be verified by incubating polytene chromosomes in ^{3}H-labeled uracil (which is incorporated into RNA, thereby labeling it) and then preparing autoradiographs of the chromosomes. These autoradiographs reveal that certain parts of the chromosome do indeed incorporate the RNA precursor, especially in the swollen **puffs** and **Balbiani rings** (Figure 17-12). The polytene chromosomes are so large that they can be removed from cells with forceps and cut into specific pieces. When Balbiani rings are clipped from the giant salivary gland chromosomes of the midge *Chironomus*, they prove to differ in their base ratios of RNA. Morever, the base ratio of RNA in a particular Balbiani ring is similar to the base ratio of the DNA in the corresponding chromosome segment.

Message
The RNA present in polytene chromosomes apparently represents transcripts of specific gene sequences. The RNA-rich regions (which can be identified visually as puffs and Balbiani rings) thus apparently represent regions of genetic activity of the DNA.

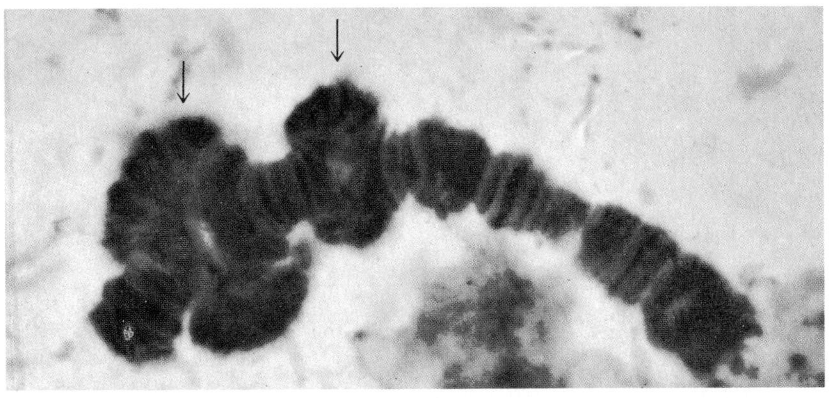

(a)

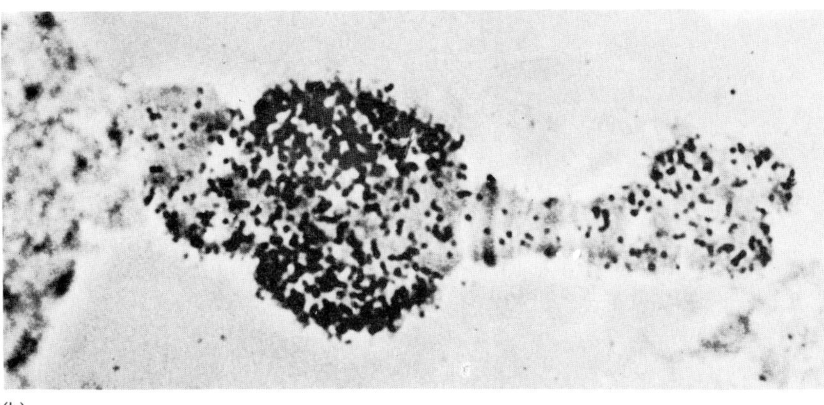

(b)

Figure 17-12. Balbiani rings in salivary gland chromosomes from the midge Chironomus. *(a) Electron micrograph showing visible Balbiani rings* (arrows). *(b) Autoradiograph after incubation in medium containing* ^{3}*H-labeled uracil. The incorporation of the* ^{3}*H-labeled uridine shows that heavy RNA synthesis is occurring in the Balbiani ring. (Part a by Ulrich Clever, part b by Claus Pelling, Max Planck Institute for Biology.)*

As we have mentioned before, polytene chromosomes are found in several different tissues of flies. If puffs really do reflect genetic activity, and if differentiation of tissues is a result of differential gene activity, then the patterns of puffing on polytene chromosomes should vary from tissue to tissue. In the 1950s, Wolfgang Beermann compared the puffing in polytene chromosomes from different tissues of the same chironomid fly. He was able to identify each chromosome and its bands, and he found that the pattern of puffs present in a cell is specific to the tissue from which the cell was taken. Some puffs are unique to one or two tissues, whereas other puffs appear in all tissues (Figure 17-13). If the visible size of a puff or ring reflects the amount of gene activity, then changes in gene activity during development are indicated by progressive swelling and regression of puffed regions during development (Figure 17-14). These studies

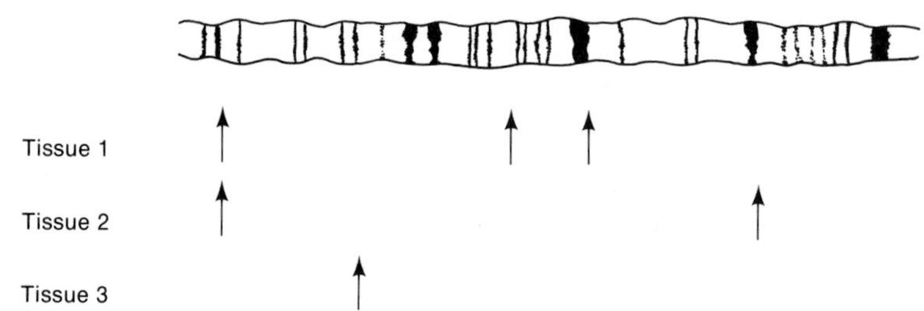

Figure 17-13. Locations of puffs (arrows) *in polytene chromosomes from three different tissues of a chironomid fly. This demonstrates tissue specificity for puffing patterns, and presumably specificity for RNA synthesis.*

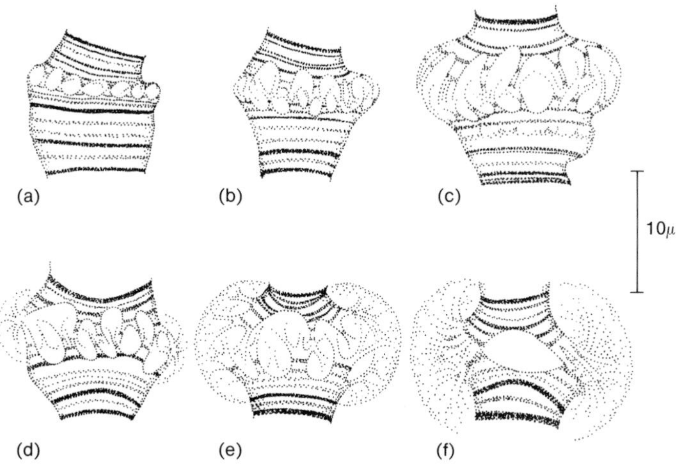

Figure 17-14. Sequence of formation of a Balbiani ring in the same tissue at successively later stages of larval development. (From W. Beermann, Chromosoma *5:173, 1952.)*

provide no evidence on whether the different patterns of puff formation are the cause or the result of tissue differentiation. However, they do provide a very nice cytological demonstration of genes being turned on and off.

Rhynchosciara is a fly that lays batches of eggs that then develop synchronously. This characteristic is handy for the experimenter because puffing patterns can be studied at different stages of development simply by sampling a few larvae from a single egg batch at different times after the batch was laid. Crowaldo Pavan was able to demonstrate that puffing patterns in a given tissue from *Rhynchosciara* do change in a regular and ordered sequence during development. These observations were subsequently verified by similar observations on salivary gland chromosomes from *Drosophila.* Not only are puffing patterns

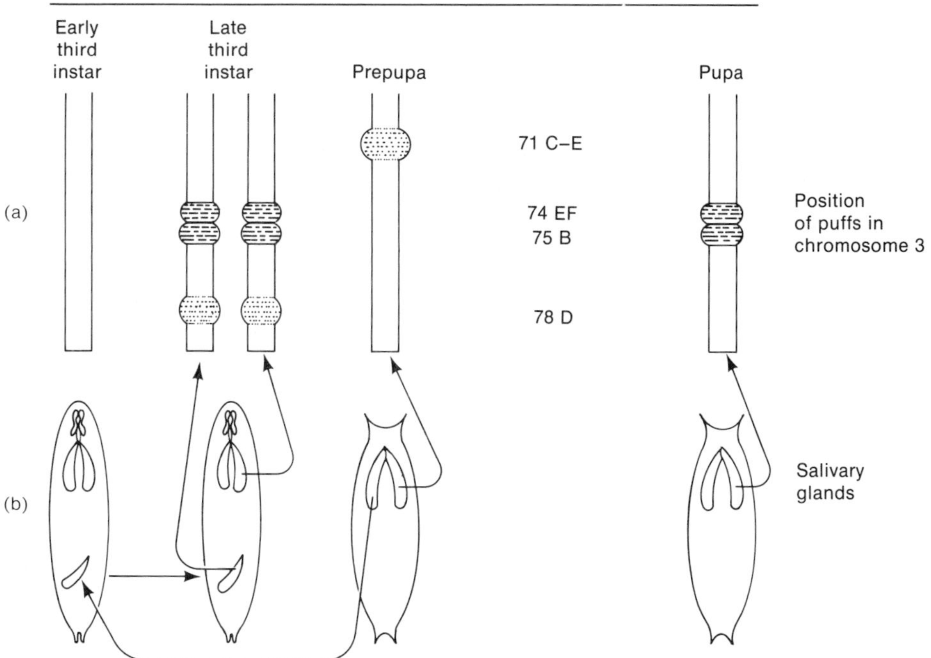

Figure 17-15. Reversal of puffing patterns in transplanted salivary chromosomes in Drosophila. *(a) The positions of the puffs present in chromosome 3 in glands at different stages of development. (b) A gland from a prepupa larva is transplanted to a larva in the early third instar stage of development. The larva is allowed to develop to the late third instar stage, and the puffs in the transplanted chromosome are then examined. The puff pattern is that normal for the late third instar stage. (From H. J. Becker,* Chromosoma *13:363, 1962.)*

specific to the particular developmental stage of a particular tissue, but they can even be reversed by transplanting a tissue from a more advanced embryo into a younger embryo. When salivary glands from a *Drosophila* larva in the prepupa stage were transplanted into a younger larva, the puffs observed in the transplanted glands were similar to those normal for the host's stage of development (Figure 17-15). This experiment demonstrates the ability of the cellular environment to modify gene activity.

The major changes in puffing pattern that occur at pupation seem to be primarily controlled by an increasing concentration of the molting hormone ecdysone, which is secreted by ring glands near the head. A larva can be pinched in two with a fine thread so that the salivary glands protrude from both sides of the constriction (Figure 17-16). The part of the larva on the anterior (head) side of the ligature pupates in response to ecdysone secretion, but the part on the posterior side does not pupate. Hans Becker showed that the chromosomes in

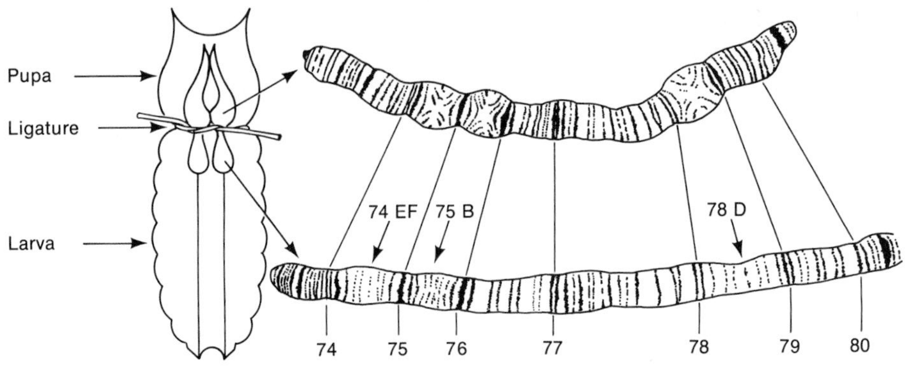

Figure 17-16. Effect of ligation of a larva on puffing patterns in polytene chromosomes of the salivary glands in Drosophila. *The ligature prevents diffusion of ecdysone from the ring glands in the head region into the posterior area. Hence the anterior part pupates while the posterior part remains larval. Puffs that appear in chromosome 3 in the pupal region* (top) *are absent in the larval region* (bottom). *(From H. J. Becker,* Chromosoma *13:372, 1962.)*

the salivary gland on the anterior side puff in a pupal fashion, whereas those in the larval part retain the larval puff pattern. Further confirmation has come through injection of ecdysone into young larvae, whose polytene chromosomes then prematurely develop pupal puff patterns. In this case, it appears that a developmental change in one tissue (secretion of ecdysone by the ring glands) controls developmental changes in other tissues (a switch to the pupal patterns of gene activation).

The puffing patterns of the polytene chromosomes are characteristic and specific for different tissues and developmental stages, and they seem to provide a visible indication of differential gene activity. But can we find any evidence to implicate a particular puff directly in some particular developmental change that is involved in cell differentiation? In the salivary gland of *Chironomus pallidivittatus,* four cells situated near the opening of the gland have dark cytoplasmic granules, whereas none of the other cells of the gland have such granules. A Balbiani ring at the end of chromosome 4 is found only in the cells with granules, whereas the puff patterns in these cells are identical in all other respects to the patterns of other cells in the gland (Figure 17-17). In a related species, *Chironomus tentans,* there are no such cytoplasmic granules, nor is there a puff at the end of chromosome 4. Hybrids between the two species have the ring on the *C. pallidivittatus* homolog of chromosome 4, and they have dark granules in the four cells (indicating that the gene or genes for the granules show dominance). Crossing over can occur between the two homologs of chromosome 4, and Beermann showed that the puff and the granules appear only in those progeny carrying the *C. pallidivittatus* chromosome-4 tip. Thus, the granules are a direct reflection of the presence of a Balbiani ring at the tip of chromosome

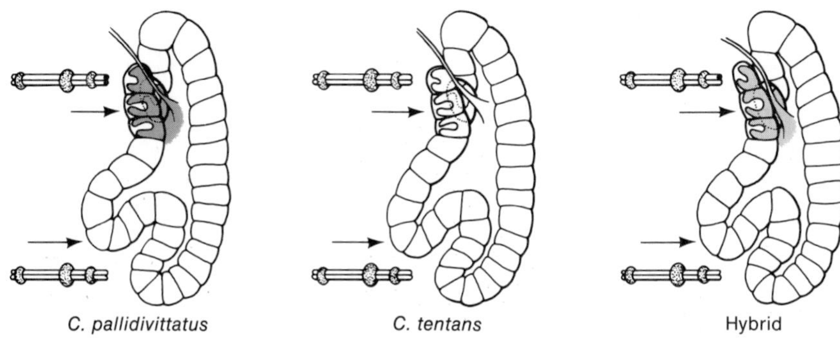

Figure 17-17. Puffs present in the synapsed chromosome 4 of Chironomus *in different cells of the salivary glands. The puff at the tip (represented as a black dot at the right end of the chromosome) is found in* C. pallidivittatus *cells with granules* (left) *and in hybrids with* C. tentans (right), *but only in the* C. pallidivittatus *homolog in the hybrids. (From W. Beermann and U. Clever, "Chromosome Puffs." Copyright © 1964 by Scientific American, Inc. All rights reserved.)*

4, and we conclude that genetic activity in this region almost certainly is responsible for the production of the granules.

Message
In each tissue that contains polytene chromosomes, puffing patterns can be shown to be specific to the tissue and its developmental stage. The pattern is at least partially under the control of the cytoplasmic environment and of such factors as hormones. There is some evidence that a particular puff can be responsible for the development of a particular phenotypic cell feature.

The appearance and regression of puffs appears to be a cytological analog to the induction and repression of such loci as *lac* in microorganisms. In 1962, F. M. Ritossa made an observation that permitted similar analysis of puffs. He noted the induction of a unique set of puffs when a short heat shock is administered to *Drosophila* larvae (Figure 17-18). A 40-minute exposure to 37° C elicits nine specific puffs on chromosome arms 2L (puff 33B), 3L (puffs 63C, 64F, 67B, and 70A), and 3R (puffs 87A, 87C, 93D, and 95D). Figure 17-19 shows two of these "heat-shock puffs."

The effects of the heat shock are almost immediate, and several things happen:

1. Puffing is detectable within a minute after the heat shock begins.
2. Puffs present when the heat shock begins show regression during the heat shock.

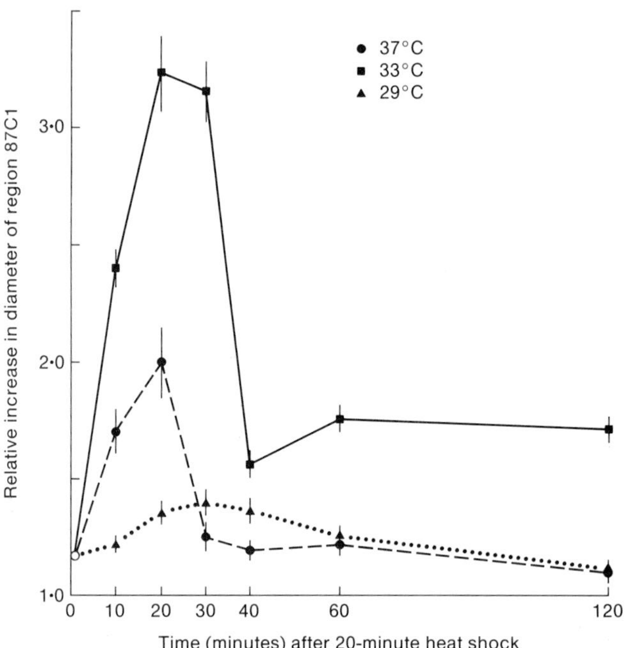

Figure 17-18. The size of the puff at 87C1 on the right arm of Drosophila *chromosome 3 (3R) at different times after a heat shock. (The puff size is measured as the ratio of the diameter of region 87C1 to an unpuffed reference band 87D1,2). (From M. Lewis et al.,* Proc. Natl. Acad. Sci. USA *72:3604, 1975.)*

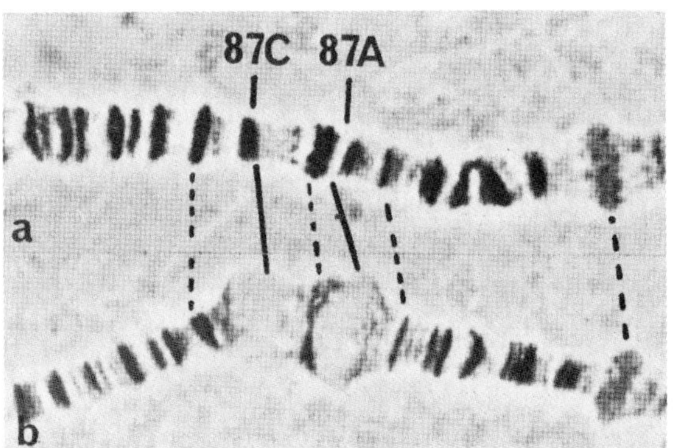

Figure 17-19. Induction of puffs in chromosome 3 of Drosophila *by a 40-minute exposure to 37°C. (a) Control experiment with no heat shock. (b) Puffs formed after heat shock. (From Ashburner and Bonner,* Cell *17:241, 1979. Copyright M.I.T. Press.)*

3. There is a halt in all RNA synthesis that was under way at the start of the heat shock.

4. Specific heat-shock RNA synthesis is induced in *all* tissues (including cells in culture). This RNA hybridizes in situ to the heat-shock puff regions.

5. Most normal protein synthesis ceases, and synthesis of a small number of specific heat-shock proteins is induced.

We know nothing about the biological role of this response to heat shock. Nevertheless, this system has made possible the isolation of inducible puff mRNA molecules and the complementary DNA sequences. This provides an ideal system for the genetic and molecular dissection of the elements that control the inducible loci.

Message
The induction of puffs by heat shocks provides an inducible system for analysis of the gene-activity control system in eukaryotes.

Biochemical Evidence for Gene Activation

Changes in puff patterns reflect changes in gene activity during development, but the fact that different RNAs are made in different cells can be verified directly by DNA:RNA hybridization tests. Denatured DNA can be fixed to filters made of nitrocellulose. When radioactive RNA is then passed through the filters, any RNA complementary to the DNA can anneal to it, forming hydrogen bonds to create a hybrid DNA:RNA duplex.

The RNA that is hybridized to DNA is resistant to the destructive effects of ribonuclease (RNase), an enzyme that specifically degrades single-stranded RNA. Thus, after a period of incubation sufficient for maximal duplex formation, the filters can be treated with RNase to remove unhybridized RNA and leave only the hybridized RNA on the filter. The amount of hybridized RNA can be measured by the radioactivity (in disintegrations per minute) after all small nucleotide fragments are washed off. The amount of DNA fixed to the filter is known, and the amount of RNA binding to it has been measured, so we can compute the proportion of the DNA that hybridizes to the RNA under test. Thus we can measure the proportion of the genome that codes for the RNA found in a specific tissue (Figure 17-20).

How much of the RNA present in two different tissues of the same organism is similar, and how much is unique to each tissue? Competition studies are used to answer this question. In a **competition experiment,** constant amounts of single-stranded DNA and complementary radioactive RNA are incubated together with varying amounts of unlabeled RNA from a specific source. The

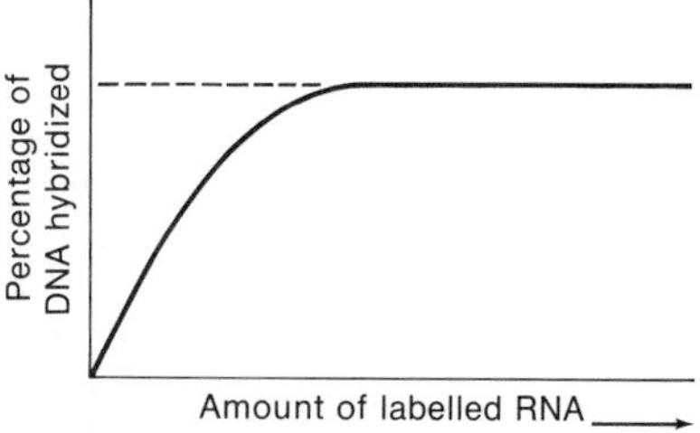

Figure 17-20. Measurement of genome activity by hybridization of RNA transcripts to a fixed amount of DNA. The RNA extracted from a specific cell type or tissue represents transcripts of the active gene. When all complementary DNA sequences are hybridized to RNA, then further increases in RNA will not increase the hybridization. The percentage of DNA that binds to RNA at "saturation" represents the amount of genetic activity in this sample of DNA.

extent to which the unlabeled RNA competes with the labeled RNA (preventing it from hybridizing) provides a measure of the similarity between the two kinds of RNA.

For example, we can ask what proportion of the RNA found in liver cells is also found in lung cells, and what proportion is found only in the liver cells. By the hybridization of liver RNA to liver DNA, we determine the conditions for maximal formation of DNA:RNA hybrids. Now we fix a known amount of liver DNA to a filter and add enough labeled (hot) liver RNA to give maximal hybridization. To different samples, we add increasing amounts of nonradioactive (cold) RNA from either lung or liver (to compete with the hot RNA for pairing with the DNA). If DNA:RNA hybrids are formed by the chance alignment of complementary sequences, then an increase in the number of cold RNA molecules should lead them to bind more frequently to a suitable DNA region rather than a hot RNA molecule. Thus, as the concentration of cold competitors increases, the binding of hot RNA decreases *except for those RNA sequences in the hot sample that are not present in the cold sample* (Figure 17-21). The experimental results show that a certain proportion of the RNA found in liver cells is also found in lung cells (hence the addition of cold lung RNA decreases the binding of labeled liver RNA). However, even a great excess of cold lung RNA cannot totally eliminate the binding of the labeled liver RNA, so we conclude that this residual binding represents sequences not present in the lung RNA (this is liver-specific RNA).

Message
The mRNA sequences present in cells vary from tissue to tissue. We infer therefore that differentiation proceeds through differential gene activity.

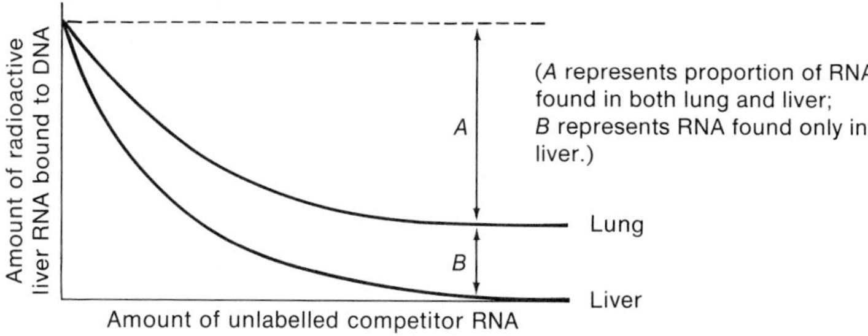

Figure 17-21. Measurement of competition for hybridization with liver DNA between labeled liver RNA and unlabeled RNA samples from either lung or liver. The ratio A:B represents the ratio between (a) DNA segments that code for RNAs common to the two tissues and (b) DNA segments that code for RNAs unique to the liver tissue.

Self-Assembly of Complex Structures

Cellular differentiation involves the formation of many subcellular structures in the differentiated cells. Many of these structures are assemblages of a number of different molecules. How are these molecules assembled into the proper structure? There are many proteins that serve to facilitate such assembly steps as folding, unwinding, and transportation. However, many complex proteins are capable of spontaneous self-assembly in the proper sequence. A classic example is the formation of the tobacco mosaic virus (TMV), which is a single RNA molecule (6400 bases in length) around which are packed 2130 identical polypeptide subunits (Figure 17-22).

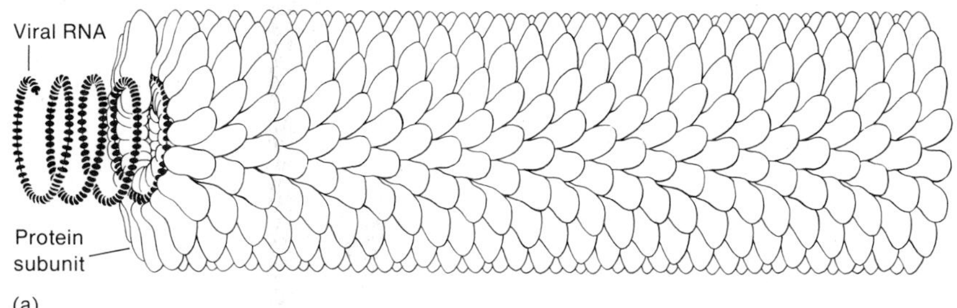

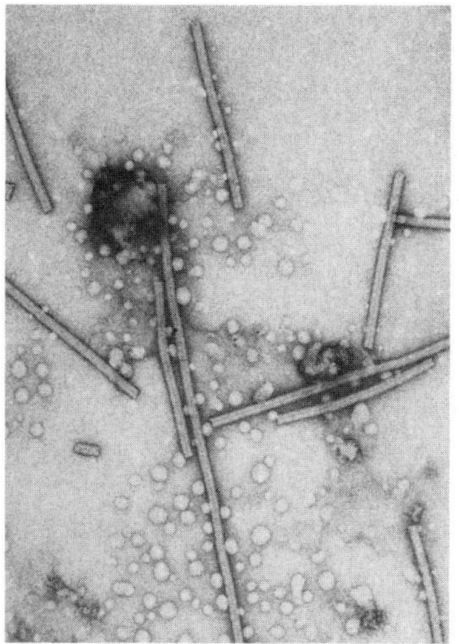

Figure 17-22. The tobacco mosaic virus (TMV). (a) The basic structure of the virus is a shell of 2130 identical polypeptide subunits surrounding a single coiled RNA molecule. (b) Electron micrograph of TMV particles. (Part a from H. Fraenkel-Conrat, "The Genetic Code of a Virus." Copyright © 1964 by Scientific American, Inc. All rights reserved. Part b courtesy of Jack D. Griffith.)

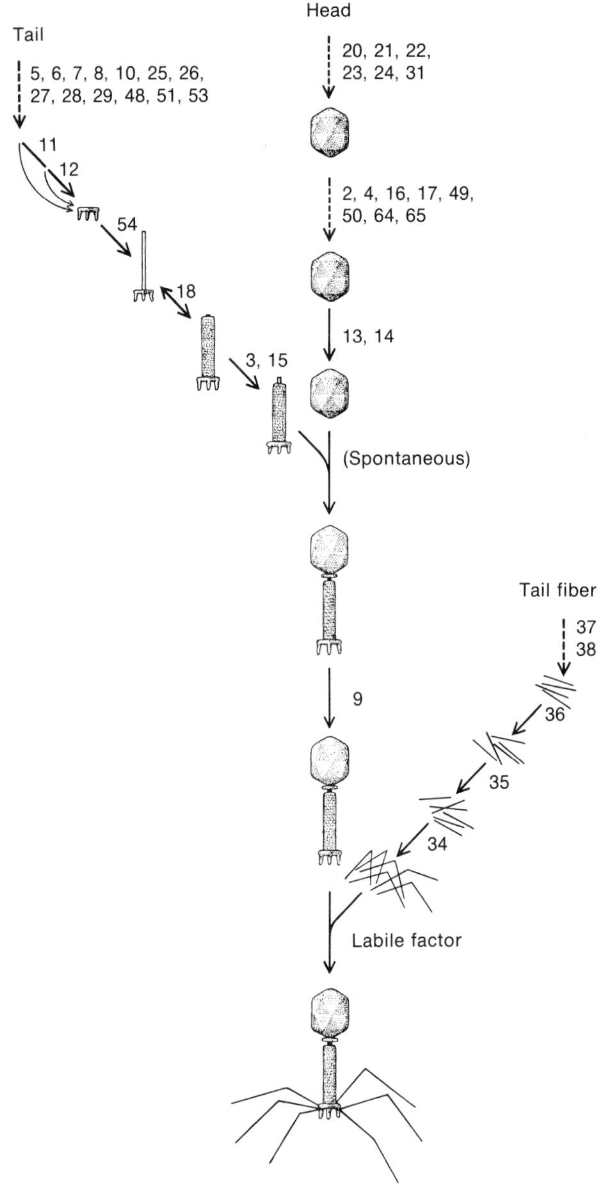

Figure 17-23. The sequence of events in formation of a mature T4 phage particle, as inferred from studies of mutant phages. Each number represents a specific gene in which a mutation blocks any further development. For example, after infection by mutant 54, no complete phages are formed, but there are separate end plates, tail fibers, and heads. Mutant 20 produces separate tails and tail fibers, but no heads. Mutant 37 produces complete phages lacking tail fibers. If lysates from two different mutants are mixed, these components can assemble spontaneously to form complete phages. (From W. B. Wood and R. S. Edgar, "Building a Bacterial Virus," Copyright © 1967 by Scientific American, Inc. All rights reserved.)

In 1955, Heinz Fraenkel-Conrat and Roblyn Williams were able to form infective virus particles by mixing purified TMV RNA and purified TMV coat protein in vitro. Apparently, formation of the complex phage particle occurs spontaneously when both components are present. This remarkable property of self-assembly has been demonstrated for much more complex aggregates of macromolecules.

In the phages T4 and λ, a genome of only about 100 genes controls the complex processes of infection, phage maturation, and lysis. In the late 1960s, an astonishing discovery was made in the study of various phage mutants. One mutant is blocked in the formation of a component of the tail fibers, and another mutant is blocked in the formation of a head component. When researchers mixed together lysates from bacteria infected by these two different phage mutants, they obtained infective phage particles fully capable of carrying out the complete lysis! Apparently, components from the two phage mutants can combine to form fully infective phage. By mixing lysates of different mutants, William Wood and his colleagues have been able to show which phage components are capable of self-assembly, and they determined the temporal sequence of the assembly process (Figure 17-23). Figure 17-24 shows a peculiar developmental "mistake" in self-assembly.

One of the most important cellular organelles is the ribosome, the site for translation of mRNA. The constituents of the ribosome are three RNA molecules and 55 distinctive proteins (see Figure 12-11). In the late 1960s, Masayasu

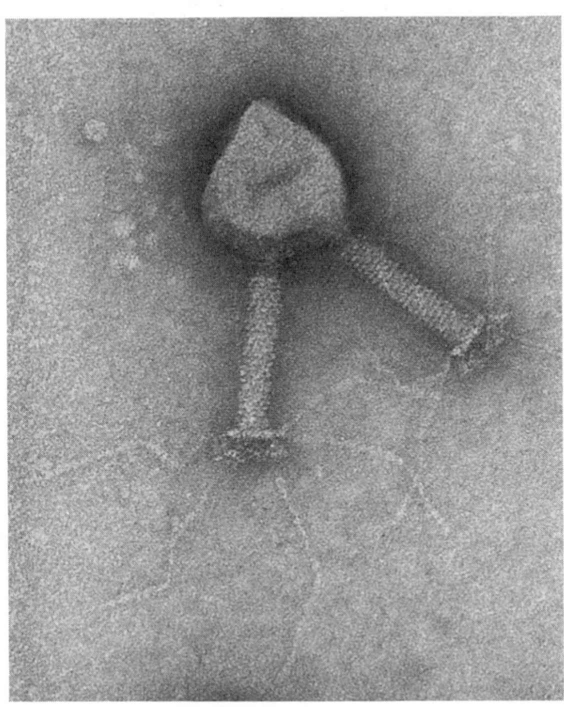

Figure 17-24. A rare developmental abnormality of phage T4 in which two phage heads have been assembled together. This is not a mutation (it is not inherited) but is rather an error in the self-assembly process. (Courtesy of Jack D. Griffith.)

Nomura recovered functional 30S subunits from a mixture of the 21 proteins and the 16S RNA of that subunit. The self-assembly of the larger 50S subunit has also been demonstrated.

The genetic analysis of ribosomes is made possible by the recovery of mutations affecting the ribosomal proteins. A number of antibiotics (including streptomycin) act by attacking the translational apparatus, so selection for resistance to such antibiotics provides a screen for ribosome-specific alterations. Indeed, selection of streptomycin-resistant mutants yields mutations only in a single locus (*str*) that codes for the S12 protein of the small subunit. Furthermore, Nomura found that assembly of the 30S components in vitro is strongly temperature dependent and requires a high energy of activation. He reasoned that selection of cold-sensitive (*cs*) mutants should permit preferential recovery of mutations affecting ribosome assembly. This screening technique did lead to recovery of mutants with defects in ribosome-subunit assembly that survive at 40°C but die at 20°C. By growing *cs* mutants at the restrictive temperature and analyzing the ribosomal structures formed, Nomura was able to show that assembly is sequential and that various mutants form various incomplete intermediates. He showed that the requirement for the 16S RNA is absolute—that is, proteins will not assemble in its absence.

By adding the different proteins to 16S RNA one at a time and in different combinations, Nomura worked out the sequence of assembly. Any one of seven proteins will bind directly to the RNA in the absence of other proteins. The binding of these proteins in turn permits the addition of other proteins. Figure 17-25 summarizes the entire sequence of interactions.

Message

Many complex structures composed of different macromolecules are formed by the spontaneous and sequential assembly of the molecules. This self-assembly must be controlled by inherent properties of the aggregating molecules.

Sex Determination as a Developmental Phenotype

The differentiation of individual organisms of a species into two different sexes is a remarkable example of the role of genes in development, and it merits special mention here. An extreme example of the importance of genes in this process is provided by the occurrence of XY human individuals who appear to be well-developed females. This phenotype is called **testicular feminization.** It apparently is produced by a mutation on the X chromosome that acts only in XY zygotes. In most higher organisms, the potential sex of a fertilized cell is determined at the time of gametic fusion by the combination of sex chromosomes or sex-determining genes in the zygote. Nevertheless, the actual differentiation of distinct sexual characteristics typically takes place much later in development and may even be altered by various factors from the sex that is chromosomally determined.

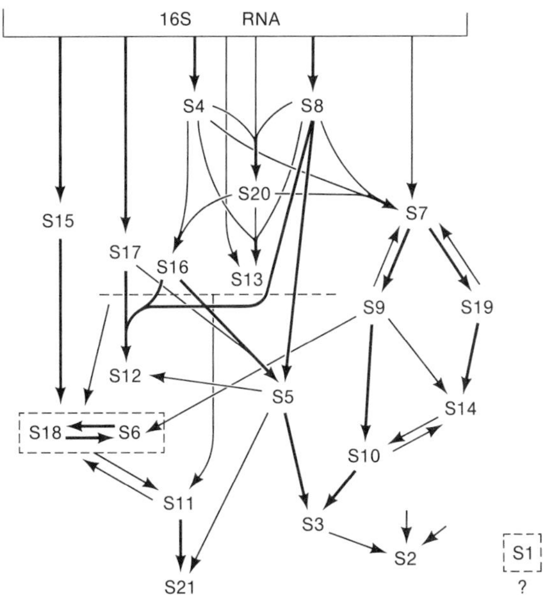

Figure 17-25. Map of the sequence of association of RNA and protein constituents of the 30S subunit of the ribosome. The arrows indicate the facilitating effect of one protein on the binding of another, with the thickness of the arrow proportional to the intensity of the effect. For example, the thick arrow from 16S RNA to S4 (a protein) indicates that S4 binds strongly to the RNA in the absence of all other proteins, whereas the thin arrow from 16S RNA to S7 indicates that binding of S7 to the RNA occurs only weakly in the absence of other proteins. Seven proteins (S15, S17, S4, S13, S20, S8, and S7) will bind independently to the 16S RNA in the absence of other proteins. Further binding to the complex of other components is dependent on the specific components already bound. For example, absence of S7 prevents binding of S9 and S19, and this in turn prevents attachment of S6, S10, and S14, which in turn must be present for the binding of S18, S11, S21, S3, and S2. (From M. Nomura and W. Held, Ribosomes, *Cold Spring Harbor Laboratory.)*

In many species of plants and animals, a single individual may function as both male and female. Such individuals are called **hermaphroditic** in animal species. Hermaphroditic plants have both ♂ and ♀ sexual organs in the same flower, whereas monoecious plants (such as corn) have flowers of separate sexes on the same plant (see Chapter 3). Earthworms and snails, which have both testes and ovaries in each individual, are examples of hermaphrodites in which the two kinds of sexual organs develop from the same genotype. In such cases, some developmental mechanism must regulate the expression of specific sets of genes in the two regions of the organism where the different sexual organs develop.

In still other organisms, a single genotype can produce an individual of either sex. In such organisms, environmental and cytoplasmic elements play an important role in determining the sex of the individual. A few striking examples illustrate this **phenotypic sex determination.** The marine annelid *Ophryotrocha* differentiates into a sperm-producing male as a young animal and then changes into an egg-laying female when it gets older. If part of an older female is amputated, the worm reverts to the male form, indicating that size rather than age is the important factor controlling the sex of the individual. In the marine archiannelid *Dinophilus,* on the other hand, sex appears to depend solely on the size of the egg produced by a female. Small eggs always produce males, whereas eggs 27 times as large always develop into females. Recent research indicates that the sex of a turtle may depend upon the temperature at which the individual develops. The mechanisms involved in environmental control of gene activation for development of sexual organs obviously is analogous to that involved in control of gene activation during differentiation of a zygote.

In many organisms, the sex of an individual can be correlated with a specific genotype. The X–Y and W–Z sex-determining mechanisms were clearly established by the 1920s. In these cases, the heterogametic sexes produce two types of gametes, and the sex of a zygote is determined on fertilization by the genotypes of the fusing gametes. Among individuals from such species, some mosaic individuals have been detected in which some parts of the organism are male and other parts are female. In *Drosophila,* for example, a cross *w m*/*w m* ♀ × + +/Y ♂ can occasionally produce a sexual mosaic (called a **gynandromorph**) that is phenotypically w m and male on one side and phenotypically + + and female on the other side (Figure 17-26). In this case, somatic mutation cannot be the cause of the mosaic because of the simultaneous involvement of two recessive markers. Instead, the gynandromorphism is explained by the loss

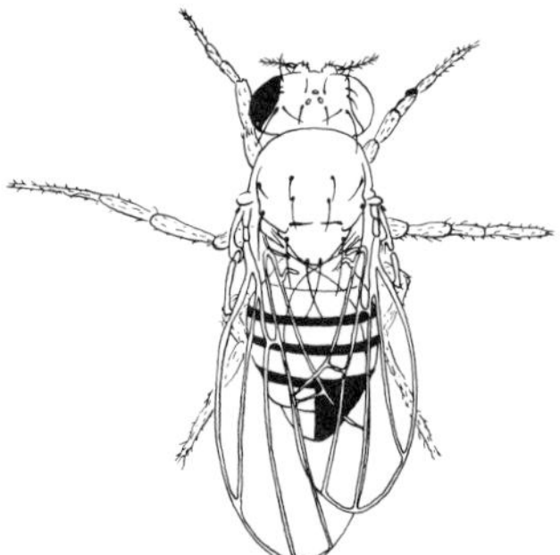

Figure 17-26. A gynandromorph that is bilaterally symmetrical. The left half of the fly is wild type and female, whereas the right side is white eyed, minature winged, and male. This arose from a cross of a w m/ w m *female* × + +/*Y male. The original zygote was probably a female of* w m/+ + *genotype. However, at an early mitotic division of the zygote, the X chromosome bearing the + alleles was lost in one cell, leaving one-half of the fly as an XO male, showing mutant phenotypes for eye color and wing length. (From A. M. Srb, R. D. Owen, and R. S. Edgar,* General Genetics, *2nd ed. Copyright © 1965, W. H. Freeman and Company.)*

of the wild-type X chromosome in some cells, thus producing a mosaic of *w m*/+ + female cells and *w m*/O male cells. We shall discuss gynandromorphs later, because they are useful in working out cell-lineage relationships among different cell and tissue types.

In many cases, hormones are important factors in the control of sex determination. Cases where hormones can counteract chromosomal sex determination have been strikingly demonstrated in fish and frogs. Genetic markers can be used to distinguish male and female eggs of the Japanese killifish called medaka. A male egg treated with female hormone (estrogen) will develop into an individual with a male (XY) genotype and a female phenotype. These XY females can be mated with XY males to produce YY males. Treatment of a YY egg with estrogen also leads to development of an individual with female phenotype. A cross of YY females with YY males obviously must produce only male progeny. It is clear that the X and Y chromosomes in this species must be very similar in their genetic content; otherwise, the YY genotype would produce an abnormal phenotype because of a lack of genes carried on the X chromosome. Analogous results have been obtained in frogs, which have a W–Z sex-determining mechanism. A WZ female that is fed male hormones (androgens) will develop into a male phenotype. A cross of WZ males with WZ females produces some WW females, and these also can be modified by hormones to produce a male phenotype.

In hymenopteran species (bees, wasps, and so on), sex is typically determined by the ploidy of the egg, which is controlled by the mother. For example, a honeybee queen (whose diploid number is 32) can lay two types of eggs. By controlling the sphincter of her sperm receptacle (which holds sperm previously obtained in matings with males), she can produce a fertilized egg (a zygote having 32 chromosomes and developing into a female) or an unfertilized egg (a zygote having 16 chromosomes and developing into a male). The diploid (female) zygotes can differentiate into either workers or queens, depending on the diet they consume during development. This is a striking example of chromosomal control of basic sexual constitution with environmental factors controlling subsequent differentiation.

Recall C. B. Bridges' studies of nondisjunction in *Drosophila* (Chapter 3). He showed that XXY flies form phenotypically normal females and that XO flies form sterile males. Obviously, the Y chromosome is necessary for male fertility but not for development of the male phenotype. What does determine sex in *Drosophila?* Triploid flies carrying three X chromosomes and three sets of autosomes are normal females. We shall represent a complete set of autosomes as A, so we can write this genotype as 3X:3A. When triploid females are crossed with normal males, some offspring having different combinations of X chromosomes and autosomes can survive. Bridges found that flies carrying two X chromosomes and three sets of autosomes (2X:3A) are **intersexes,** having phenotypic characteristics intermediate between those of the two sexes. He concluded that sex-determining genes are present on both the X chromosome and the autosomes and that the *balance* between these two kinds of chromosomes determines the phenotypic sex.

For example, suppose that the male-determining genes are on the autosomes and that the female-determining genes are on the X chromosome. In the normal flies, the XX genotype (with an X:A ratio of 2X:2A = 1.0) is female, and the XY genotype (1X:2A = 0.5) is male. If we assume that the X:A ratio determines the sex, we are not surprised to find that 3X:3A flies (ratio = 1.0) are female and that XO flies (1X:2A = 0.5) are male. The 2X:3A flies have an intermediate ratio of 0.67, and the corresponding phenotype is intermediate. This model is confirmed by the observation of 1X:3A (= 0.33) individuals that have an extreme male phenotype and 3X:2A (= 1.5) individuals that have an extreme female phenotype. (Both of these abnormal genotypes produce very weak individuals that would have poor chances of survival in normal environments.) This **balance model** for sex determination in *Drosophila* seems to explain the observations very well.

Message
In Drosophila, *sex is determined by the balance between male-determining genes on the autosomes and female-determining genes on the X chromosome.*

Cell Lineage During Development

If you have ever watched time-lapse movies of embryogenesis, you know that cell and tissue movements are extremely complex. In the dynamic process of embryogenesis, folding or migration can bring a group of cells to an area of the embryo very far from its original position in the blastula. How can we trace these cell movements during embryogenesis? The classical approach was to label physically specific cells of amphibian embryos with visible carbon particles, so that movements of the marked cells (or their descendants) could be traced during development. With this marking technique, researchers constructed **fate maps** showing the destinies of the descendants of particular cells in early stages of embryogenesis.

An alternative method is to mark cells *genetically* so that they and their descendants can be distinguished phenotypically. This approach is possible if we have genetic mosaics, individuals composed of several genetically distinct cell populations. Such mosaics do occur naturally. For example, the exchange of blood cells in the placentas of twin cattle can produce chimeras that carry two blood types. As discussed in Chapter 14, somatic nondisjunction can produce a person whose cells carry different numbers of X and Y chromosomes. Furthermore, every female mammal is a mosaic in terms of the functional states of her two X chromosomes. We have also mentioned the sexual mosaics called gynandromorphs.

Recall Illmensee's experimental generation of mosaics by injection of single cells into embryos of *Drosophila* and mice (Figure 17-8). Beatrice Mintz has developed a very elegant technique for fusing the developing embryos of two different mouse genotypes (Figure 17-27). When implanted in a host female, such a **tetraparental** embryo develops as a single mosaic individual.

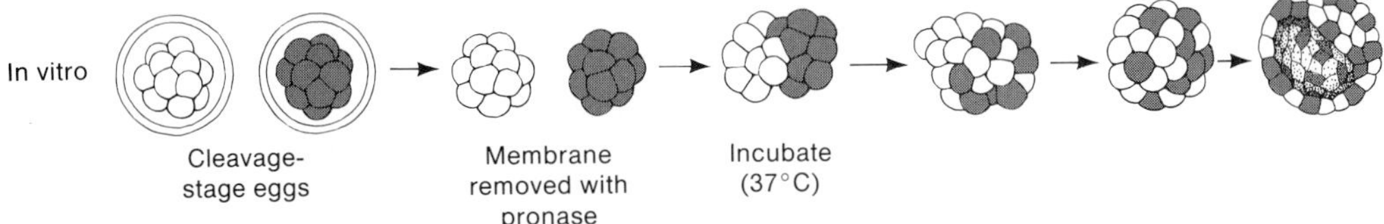

Figure 17-27. Formation of a genetic chimera by fusion of two mouse embryos of different genotypes. The embryos (at the 8-cell or 16-cell stage) are stripped of their surrounding membranes by treatment with the enzyme pronase, which breaks down proteins. The two cell clusters are placed together and incubated at 37° C. The cells adhere and mix together to form a single mosaic embryo. (From B. Mintz, Proc. Natl. Acad. Sci. USA *58:345, 1967.)*

The sex of the embryos being fused is not known at the time of fusion, so by chance about one-half of the fusions will be between a male embryo and a female embryo. Under the influence of the male hormones, most of these sexual chimeras differentiate as males even though the presence of the XX cells can be demonstrated. All the daughter cells of those introduced at the time of embryo fusion represent a genetic clone, so the recovery of a genotypically identical cluster of cells indicates their common origin through division. Mintz fused embryos from a strain having black fur color with embryos from a stock having white fur. If cells from either genotype can become precursors of skin, then the pattern of fur color should reflect the clonal origins of the skin cells. She found that the fur pattern of such mice always involves bands of black or white fur that circle the body to form a stripe from the stomach (ventral surface) to the back (dorsal surface) on each side (Figure 17-28). This information indicates that the cells whose descendants will produce fur pigment line up

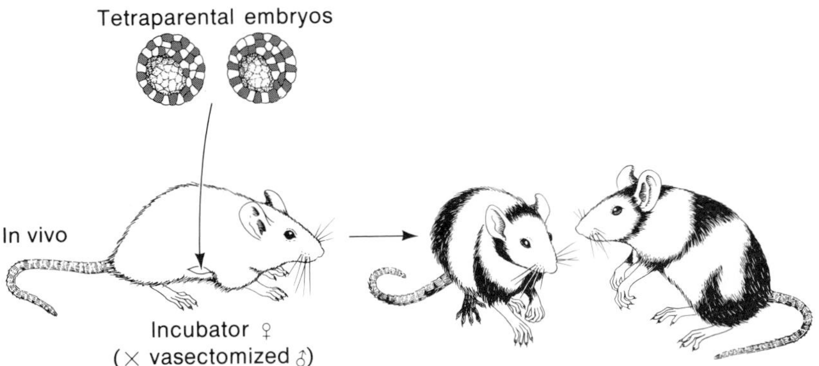

Figure 17-28. Tetraparental mouse embryos are transplanted into a female whose womb is receptive for implantation after mating with a sterile male. The resulting offspring are mosaics for the fur genotypes of the two strains from which they are derived. The patterning of the fur phenotypes in dorsal-to-ventral stripes indicates the regions of skin that have each derived from a single embryonic cell. (From B. Mintz, Proc. Natl. Acad. Sci. USA *58:345, 1967.)*

randomly in pairs along the midline of the embryo and then divide to form clonal sheets that meet on its other side. You can see how useful genetic markers can be in tracing cell lineage during development. As we shall see later in this chapter, such genetic dissection of development has been honed to a fine edge in *Drosophila.*

Using tetraparental mice, Mintz set out to obtain a definitive solution to the problem of the origin of the multiple nuclei in a striated muscle cell. A muscle cell differentiates from precursor cells called **myoblasts,** each of which has a single nucleus. Does a muscle cell arise through successive fusion of different myoblasts, or does a mononucleate myoblast undergo a series of nuclear divisions without division of the cytoplasm? To distinguish between these two models, Mintz produced tetraparental mice from two lines that differ in the electrophoretic mobility of the enzyme isocitrate dehydrogenase-1. Electrophoretically distinct enzymes are produced by $Id\text{-}1^a$ and $Id\text{-}1^b$ homozygotes, and the $Id\text{-}1^a/Id\text{-}1^b$ heterozygotes also exhibit an intermediate hybrid band indicative of a dimer containing one polypeptide unit from each phenotype (Figure 17-29). Formation of the hybrid dimer can occur only if both polypeptides are synthesized in the same cell, as shown by the fact that only the parental enzyme types are derived from the livers of mice that are chimeric for cells of both homozygous lines.

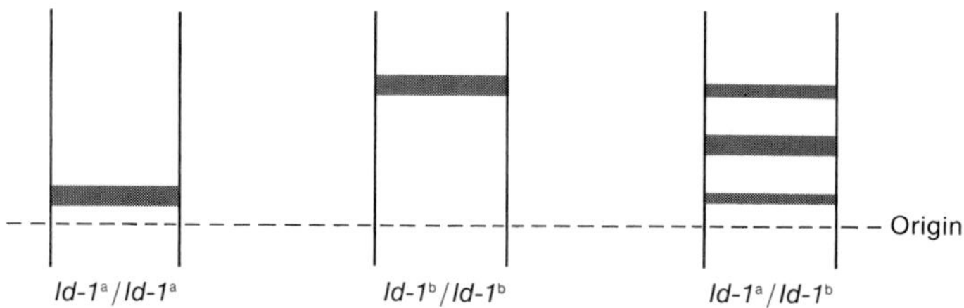

Figure 17-29. Electrophoretic mobility of isocitrate dehydrogenase-1 of different genotypes. Each homozygote shows a single isocitrate dehydrogenase-1 band with a specific mobility in the electrical field. However, the heterozygote ($Id\text{-}1^a/Id\text{-}1^b$) *shows three bands, one each with the mobility of the parental strain and a band of intermediate mobility. The intermediate band stains approximately twice as intensely as either parental band. This pattern indicates that isocitrate dehydrogenase-1 is a dimer, and the intermediate band contains one polypeptide from each of the parental alleles.*

The pattern of enzyme phenotypes in tetraparental mice permits a test between the two hypotheses for the origin of the muscle-cell nuclei (Figure 17-30). The division model predicts that a single muscle cell will contain nuclei of only a single genotype, so enzyme derived from muscle cells of a tetraparental mouse should show only the parental patterns. The fusion model predicts that a single cell in a tetraparental mouse may contain nuclei of both genotypes, so that the

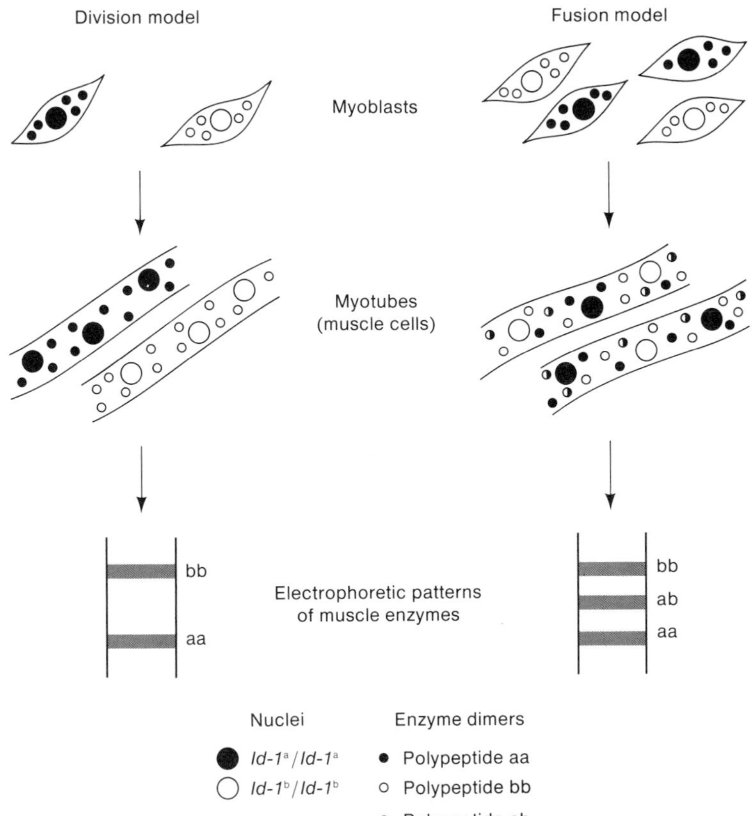

Figure 17-30. Predictions of differing results for electrophoresis of enzymes based on two different models for the origin of multinucleate muscle cells (myotubes). If a myotube is formed by successive divisions of the nucleus in a single myoblast, the electrophoretic pattern should show only the parental enzyme bands. If a myotube is formed by fusion of numerous myoblasts, a third band should appear as a result of the formation of hybrid enzyme dimers within the cell that has nuclei of both genotypes. The experimental results support the fusion model.

hybrid dimer can be formed. Therefore, Mintz's results (the formation of the hybrid dimer in muscle cells of tetraparental mice) provided definitive proof that a striated muscle cell is formed by myoblast fusion.

Tetraparental mice have been used to provide definitive solutions to other problems in developmental biology. For example, the sex-linked condition called Duchenne's muscular dystrophy involves a muscular degeneration that develops progressively in boys. It is difficult to determine biologically whether this degeneration of muscle tissues is a direct result of the mutation or is a secondary effect due to genetically controlled changes in other tissues (such as nerve cells). A comparable hereditary condition exists in mice. Alan Peterson has created tetraparental mice from normal lines and lines carrying the muscular dystrophy. These two lines also produce distinct forms of an enzyme. Peterson used the enzyme as a marker to determine the genetic origin of muscle cells in tetraparental mice showing the dystrophic phenotype. He found that such a mouse may have muscle cells derived from the normal line but that the nerve cells enervating the deteriorating muscles are always derived from the mutant line. Thus it appears that the mutation has its direct effect on the nerve tissue, with the muscular deterioration appearing as a secondary effect due to the changes in nerve cells.

Message
Genetic mosaics can be generated in a variety of ways. Such mosaics provide an extremely useful tool for following the destinies of cells and their daughters through development.

Immunogenetics: The Immune System as a Developmental Model

The immune system of a vertebrate organism provides the body's main line of defense against invasion by such disease-causing organisms as bacteria, viruses, and fungi. The system also attacks cancer cells produced by the organism itself. This fascinating biological system provides evidence of a variety of novel genetic mechanisms. The developmental biologist is interested in understanding the differentiation processes that produce the many components of the immune system. As we have seen, genetic analysis is a powerful tool for such studies. Furthermore, the molecular model that explains antibody specificity and diversity seems to require genetic mechanisms quite different from the conventional ones that have been discussed thus far in this text.

The Components of the System

One part of the immune system is the process by which a foreign molecule called an **antigen** (usually carried on the surface of a cell) is recognized and disabled by a host-produced **antibody** specific to the antigen. This is called the **humoral system** because antibodies are in solution in the blood humor. Its main job is to combat bacterial and viral infections. After an initial exposure to a pathogen, this system can rapidly produce antibodies to counteract future exposures, thus providing immunity against future infections by the same pathogen. A second part of the immune system causes rejection of transplanted tissue or organs. It is also believed to play an important role in the recognition and destruction of cancer cells. This is called the **cell-mediated system,** as it works mainly by cell–cell interaction, possibly through antibodylike molecules in cell membranes. The occasional malfunctioning of the humoral system leads to antibody attacks on molecules of the body and is responsible for the autoimmune diseases. Well-characterized autoimmune diseases include Hashimoto's thyroiditis (in which antibodies attack the thyroid gland) and Addison's disease (in which antibodies attack the adrenal glands). Such inflammatory joint diseases as rheumatoid arthritis are now considered to be due to autoimmune defects.

What characteristics of a cell lead to its recognition by the cell-mediated system as foreign? An investigation of the genetic basis for such recognition in the case of transplants can be made through a study of the fate of skin grafts between individuals of varying degrees of relatedness. If a graft from one individual is not rejected by the recipient individual, then the two individuals are said to be **histocompatible.** Skin grafts between unrelated individuals are almost invariably rejected (the transplanted skin deteriorates and sloughs off). However, grafts between genetically similar individuals (such as identical twins or animals from inbred lines) are almost invariably accepted (the transplanted skin or organ is incorporated into the structure of the recipient individual). We introduced this topic in Chapter 4.

By noting the acceptance or rejection of skin grafts between different inbred lines of mice, researchers have defined genetic loci that code for **histocompatibility antigens**—that is, for substances within the tissue that are recognized and rejected by nonhistocompatible lines. These loci can be grouped into two classes: the loci of the **major histocompatibility complex (MHC)** produce antigens that cause rapid graft rejection in cases of noncompatibility, and the minor histocompatibility loci produce antigens that cause much slower rejection effects. The recipient's cell-mediated system will act against antigens produced by certain alleles of these genes and will accept antigens produced by other alleles. In mice, there are at least 30 minor histocompatibility loci. We now know that the loci of the MHC are clustered in a single region of the genome. In mice, this is the *H-2* region, which behaves as a segregating unit in genetic crosses (Figure 17-31). In the *H-2* cluster are "strong" antigen-producing genes *H-2K* and *H-2D*. Between these loci are other loci that control the manufacture of other antigens (*H-2I*) and a molecule that aids antibodies (*H-2S*). In addition, loci with related functions (*t* and *Tla*) are located nearby.

In humans, the equivalent major histocompatibility complex is the *HLA* region on chromosome 6. The loci *HLA-D, HLA-B, HLA-C,* and *HLA-A* are tightly linked in the *HLA* region. There are numerous alleles of the genes in the *HLA* region, so two unrelated individuals will very seldom happen to have identical genotypes for this region. (We have discussed *HLA-A* and *HLA-B* in Chapter 4.) In most human matings, then, each parent will be heterozygous for a different combination of *HLA* alleles. Because a child receives one chromosome from each parent, grafts between progeny and either parent typically are rejected (Figure 17-32). Four different genotypes are possible among siblings (ignoring the rare cases of crossing over within the tightly linked *HLA* complex), so there is a probability of 0.25 that grafts will be accepted between a pair of siblings.

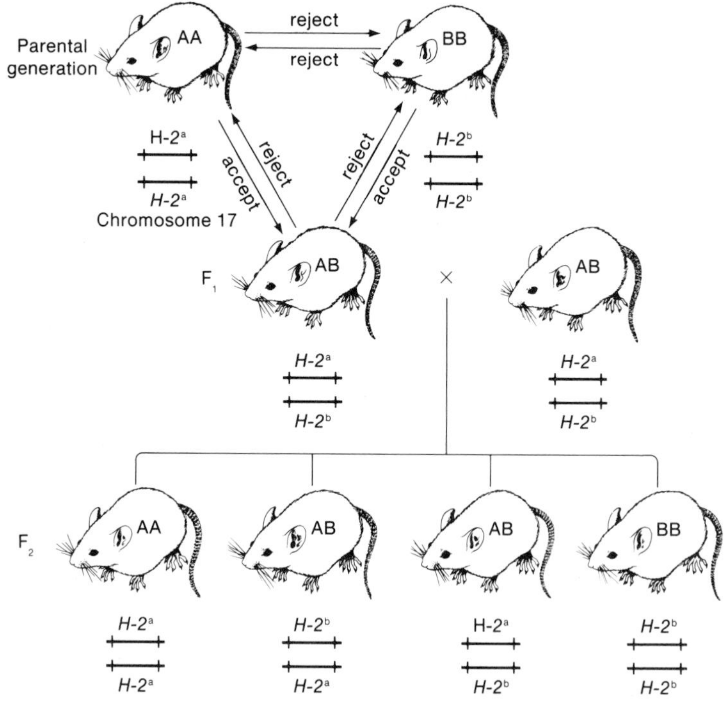

Figure 17-31. The genetics of histocompatibility antigens in inbred lines of mice. Each inbred line is homozygous for the H-2 *complex (either for* H-2ª *or for* H-2ᵇ*). Both sets of products are expressed in the heterozygous* F_1 *hybrids, so tissue from an* F_1 *mouse is rejected by either parental type (the direction of each arrow indicates transplantation toward the recipient). In the* F_2*, the alleles segregate according to Mendelian expectations. The symbols AA, BB, and AB are used to identify the histocompatibility phenotypes. (From L. E. Hood, I. L. Weissman, and W. B. Wood,* Immunology, *1978, Benjamin/Cummings.)*

The Immune Response

How does the body attack an antigen? The key lies in the embryonic cells that will differentiate to form parts of the blood and the lymph-producing tissues. Two important components belong to a family of blood cells called **lymphocytes.** The **T cells** are processed in the thymus gland, whereas the **B cells** are processed in the bone marrow and fetal liver tissue. Prior to any exposure to an antigen, these virgin lymphocytes become immunologically competent—that is, their surfaces acquire the ability to respond to antigens. The T cells play the role of "roving policemen." When a T cell detects foreign antigens on foreign tissue, it is activated to produce a line of "killer" cells that destroy the tissue. When a B cell is stimulated by a foreign antigen, it differentiates into plasma cells that make massive amounts of antibodies. These antibodies are released to

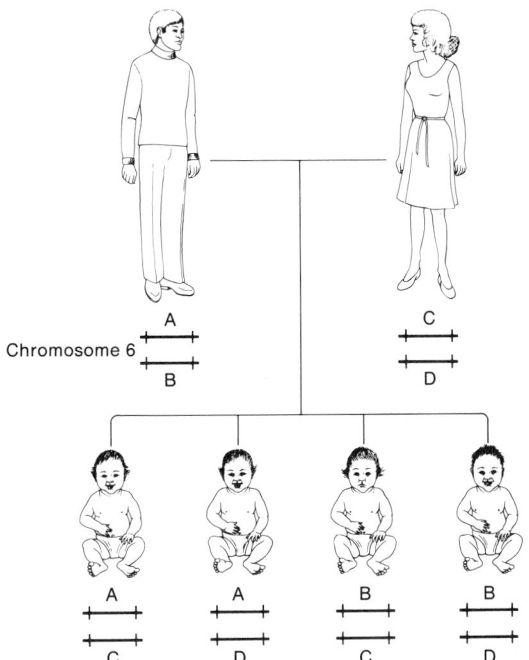

Figure 17-32. The genetics of histocompatibility antigens in humans. Different allelic combinations in the HLA *complex are indicated here by single letters (arbitrary histocompatibility types). Four different* HLA *genotypes are expected among siblings with equal frequencies. With the large complex of alleles present in this system, grafts between progeny and either parent will usually be rejected. (From L. E. Hood, I. L. Weissman, and W. B. Wood,* Immunology, *1978, Benjamin/Cummings.)*

circulate through the bloodstream, where they will attack particles carrying the same foreign antigen. A single plasma cell can produce from 3000 to 30,000 antibody molecules per second! Thus T cells are the main agents of the cell-mediated response, and B cells are the main agents of the humoral response.

Is each lymphocyte capable of responding to a variety of different antigens, or does a particular lymphocyte recognize only one antigen? In the 1950s, Niels Jerne and MacFarlane Burnet proposed a model of the immune system based on the second assumption. This **clonal selection model** seemed unlikely to many biologists at the time of its statement, but subsequent research has verified its predictions, and it is now generally accepted. According to the clonal selection model, each cell in a population of mature lymphocytes is capable of recognizing only one specific antigen. This capacity exists *before* any exposure of the cell to the stimulating antigen.

The population of lymphocytes in a mammal is enormous, ranging from 10^8 to 10^{12} cells. This population consists of some 10^5 to 10^8 clones of cells. The cells of each clone recognize a particular antigen. Thus the organism has a predetermined "library" of cells capable of recognizing and attacking many thousands of different antigens. When an antigen is introduced into the body for the first time, there is a delay before the immune response begins. This is because nothing happens until the antigen encounters a lymphocyte from the particular clone that recognizes that particular antigen. The lymphocyte then proliferates, and the daughter cells differentiate into killer or plasma cells. The concentration in the bloodstream of a specific antibody appropriate to an antigen increases gradually as the population of this particular lymphocyte clone increases. When an antigen-bearing cell has been destroyed, the antibody level decreases, but it does not return to the near-zero level that existed before the challenge by the antigen. The population of the specific lymphocyte clone remains much larger than it was before the challenge.

If the same antigen is introduced to the body at a later time, it encounters an appropriate lymphocyte much more rapidly. The immune response begins almost immediately after such a secondary challenge by a particular antigen. Because the clone population is large initially, the antibody level increases very rapidly and reaches a value much higher than that produced in response to the primary challenge (Figure 17-33). This secondary response is very specific to

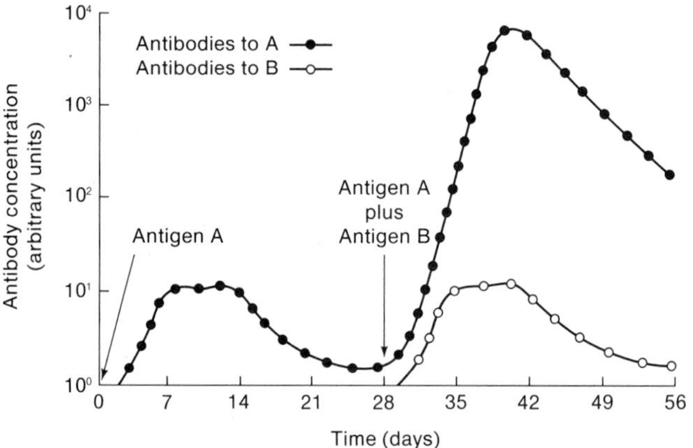

Figure 17-33. Changes in antibody concentration after the body is challenged by an antigen. When antigen A is introduced, the concentration of the corresponding antibody A in the blood is undetectably small. After a day or two, the level of antibody A begins to increase, reaching a peak after about a week and then declining considerably after a few weeks. A second exposure to antigen A produces an immediate response that leads to antibody concentrations several orders of magnitude greater than those produced in response to the first challenge. This secondary response is very specific; the response to antigen B (injected at the same time as the second injection of antigen A) follows the first-exposure pattern. (From L. E. Hood, I. L. Weissman, and W. B. Wood, Immunology, *1978, Benjamin/Cummings.)*

the particular antigen involved in the primary response—if a second antigen is introduced for the first time along with the secondary challenge, production of antibodies to the second antigen will follow the course of a normal primary response. We see that there is an **immunological memory**—the immune system "remembers" and responds more rapidly and effectively to antigens that it has encountered before.

Antibodies work by binding to and clumping antigenic cells, rendering them more vulnerable to attack by a family of membrane-puncturing proteins and by amoeboid-engulfing cells. Killer cells work more directly on foreign tissue, causing cell death.

This very simplified account of the immune response and the clonal selection explanation of it should give you some idea of the challenge that the immune system provides to developmental geneticists. We shall look at a few aspects of the system in more detail.

Structure of the Antibody

Each lymphocyte clone produces an antibody that very specifically recognizes a particular antigen molecule. What is the nature of the antibody, and how does it recognize the antigen? In the 1930s, antibodies were shown to be proteins of high molecular weight. This means that the structures of the antibody proteins must be specified by genes. But how can genes specify the many thousands of different antibodies?

The antibodies belong to a class of proteins called **immunoglobulins,** which Gerald Edelman (in 1959) showed to be complexes of polypeptides held together by disulfide bridges (Figure 17-34). Each immunoglobulin molecule is composed of a pair of identical large molecules (called heavy, or H) and a pair of smaller (light, or L) chains. A carbohydrate portion is attached to each of the H chains. (Because the molecule includes polypeptide and carbohydrate portions, it is called a glycoprotein.)

As we might expect from the number of different antibodies, the immunoglobulins exist in thousands of different varieties. The problem of purifying one specific antibody is a formidable one because the variation in the physical and chemical properties of the different antibodies is too subtle to allow their separation. The solution to this problem involved a surprising research tool—a type of tumor called a **myeloma,** which occurs in lymphoid cells. The myeloma tumor apparently begins with the uncontrolled proliferation of a single immunoglobulin-producing cell to form a clone that then releases a large amount of a homogeneous immunoglobulin product. It was first reported by Henry Bence-Jones in 1847 that patients with multiple myeloma tumors excrete proteins in their urine. Edelman and his collaborators showed that these "Bence-Jones proteins" are homogeneous populations of light chains released by the tumor. With this convenient supply of naturally purified immunoglobulins, it was possible to analyze the amino-acid sequences of the proteins from the urine of different

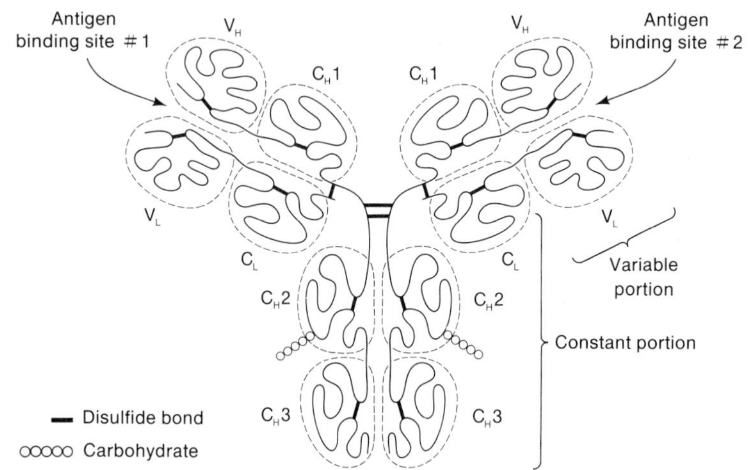

Figure 17-34. Human immunoglobulin C is a tetramer of a pair of identical heavy (H) polypeptides and a pair of identical light (L) polypeptides. Disulfide bonds connect parts of each strand, connect each L chain to an H chain, and connect the two H chains. Each chain has a region (V) whose amino-acid sequence varies considerably between antibodies to different antigens, and a region (C) that is similar in all antibodies. (From J. A. Gally and G. M. Edelman, Ann. Rev. Genetics *6:1, 1972. Copyright © 1972 by Annual Reviews Inc.)*

patients. This analysis revealed the source of the great variability of antibodies and of their specificity.

The amino-acid sequences of the carboxy-terminal portions (called the C_L region) of the light chains proved to be identical in all patients. The sequences of the amino-terminal portions (called the V_L region) varied from patient to patient. Studies of heavy chains obtained from these patients revealed that there is a variable region (V_H) occupying about one-fourth of the chain at the amino terminus. The remaining constant portion seems to be composed of three closely related repeat sequences called C_H1, C_H2, and C_H3 (see Figure 17-34). Figure 17-35 illustrates the kinds of differences found among the variable regions of heavy chains from nine different patients. Certain segments within the variable region (called hypervariable segments) show even more extreme variation than that found elsewhere in the region (Figure 17-36).

We see that the variability among antibodies might indeed be controlled by a few loci that code for these variable regions in the proteins. Furthermore the variable regions undoubtedly represent the basis of antibody–antigen binding specificity. But how can the few alleles present at each locus in one individual code for the many thousands of different antibodies present in that individual? Much research has been devoted to the problem, and clues to the answer have come from many sources.

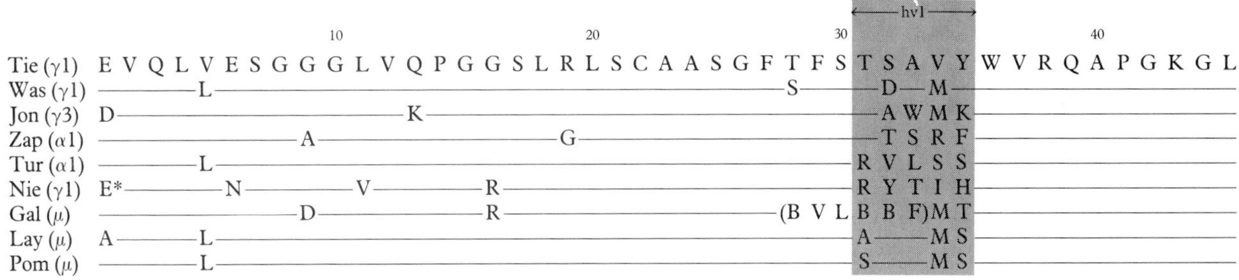


```
                    hv2                                                           hv3
         50              60                 70              80                   90
Tie   E W V G W R Y E G S S L T H Y A V S V Q G R F T I S R N D S K N T L Y L Q M L S L E P Z B
Was   ——————A—K—Q E A—N S—F—D T—M————————————————————————————————————N R——A——
Jon   ——————V——V—Q V V E L A F—N——N—————————————————————————————————I—V T———
Zap   ——————E F—V Q——A I S———D————A——————————————————————————————N T G—A——
Tur   ——————S G—L N A——N L—F———————A———————————————————————————————Q A——
Nie   ——————A V M S Y B G B B K———D——N————————————————————————N——N——R———
Gal   ——————A N I K Z B G—Z Z B—V D——K————————D N A———S————————N——R V——
Lay   ——————A—K——M G M D L———D——N—————————————————————————————N G—A A—V
Pom   ——————A—K——B G N D K———D——N————————————————————————L——N G—Q A——
```

```
                          hv4
                    100                 110             117
Tie   T A V Y Y C A R V T P A A A S L T F S A V W G Q G T L V
Was   ————————————————F R Q—F V Q[– –]—F D—F————————
Jon   ————————————————V V S T [– – – –] S M D————————P—
Zap   ————————————————T R—G G Y [– – –] D———————————
Tur   ————————————————L S V T—V [– – –] A F D——————————
Nie   ————————————————I R D T—M [– – –]———H————————K—
Gal   ————————————————G W G [– – – – – –] G G D Y——————
Lay   S—I—————————————D A G P Y V—P—F——H——————————
Pom   ——L—————————————D A G P Y V—P—F——H Y—————————
```

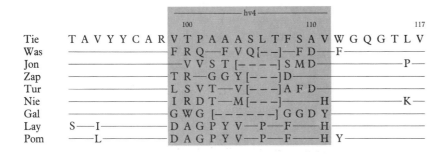

Figure 17-35. Amino-acid sequences of the variable portion of the heavy chain from human immunoglobulins. The numbers represent the amino-acid positions, counting from the amino terminus. The letters signify different amino acids. Unbroken lines represent sequences identical to that of the reference protein Tie. Dashes in brackets represent deletions of sequences present in the protein Tie. The sequence variations tend to cluster in the hypervariable (hv) segments (shaded). *(From J. D. Capra and J. M. Kehoe,* Proc. Natl. Acad. Sci. USA *71:4032, 1974.)*

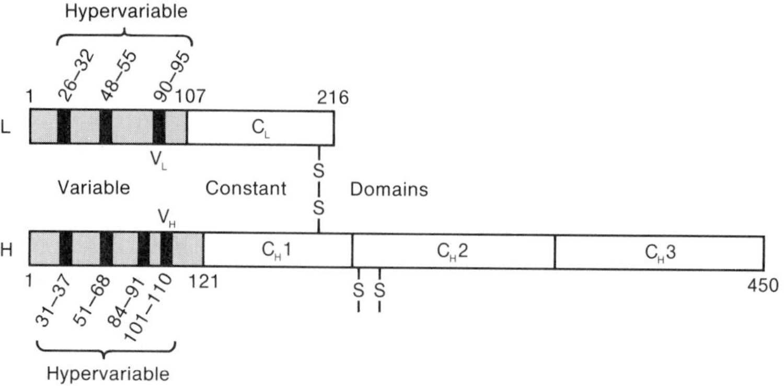

Figure 17-36. Light and heavy chains of human immunoglobulin, showing the variable and constant regions and the hypervariable (hv) segments. The hv segments are responsible for the binding specificity of the antibody. (From A. Williamson, Ann. Rev. Biochem. *45:467, 1976. Copyright © 1976 by Annual Reviews, Inc.)*

Genetics of the Immunoglobulins

The detailed analysis of the immunoglobulins showed that they can be divided into five distinct classes, chiefly on the basis of the size and amino-acid sequence of the heavy polypeptide chain (the classes also differ in the size of the associated carbohydrate portions). The five classes (immunoglobulins A, G, D, E, and M) have five different kinds of heavy chains (called α, γ, δ, ϵ, and μ, respectively). Within the *constant* region, each class of heavy chain differs from the other classes in at least 60% of the amino-acid sequence. The light chains can be divided into two classes (λ and κ) whose constant regions differ in more than 60% of the amino acids. Each class of immunoglobulin can occur with either kind of light chain (Table 17-1).

Even within a single class, there is some variation of sequence within the

Table 17-1. Classes of immunoglobulins (Ig)

Class	Concentration in blood serum (mg/ml)	Molecular weight	Type of heavy chains	Type of light chains	Chain structure
IgA	3	180,000 to 500,000	α	κ or λ	$(\kappa_2\alpha_2)_n$ or $(\lambda_2\alpha_2)_n$
IgD	0.1	175,000	δ	κ or λ	$\kappa_2\delta_2$ or $\lambda_2\delta_2$
IgE	0.001	200,000	ϵ	κ or λ	$\kappa_2\epsilon_2$ or $\lambda_2\epsilon_2$
IgG	12	150,000	γ	κ or λ	$\kappa_2\gamma_2$ or $\lambda_2\gamma_2$
IgM	1	950,000	μ	κ or λ	$(\kappa_2\mu_2)_5$ or $(\lambda_2\mu_2)_5$

NOTE: n = 1, 2, or 3.
SOURCE: After L. Stryer, *Biochemistry,* 2nd ed., W. H. Freeman and Company. Copyright © 1981.

constant regions. For example, three different sequences are found in the constant region of human κ light chains. These sequences differ in the amino acids at positions 151 and 191 (Figure 17-37). The classes of heavy chains can be divided into subclasses that are similarly limited to variations in one or two positions. It seems possible that each subclass might represent a set of alleles for a single cistron encoding the polypeptide chain.

Amino acid position

	151		191		
——	Val	——	Leu	——	COO^-
——	Ala	——	Leu	——	COO^-
——	Ala	——	Val	——	COO^-

Figure 17-37. Amino-acid positions where variation is found among the human kappa light chains in the constant region. The differences in amino acids at positions 151 and 191 define three different alleles of the kappa chain, and they segregate in a Mendelian fashion. COO^- indicates the carboxy-terminus of the polypeptide.

Genetically, we can identify alleles by studying the segregation of phenotypes involving the various amino-acid sequences. After mapping the sequences in immunoglobulin samples from various members of a pedigree, we can look for Mendelian inheritance patterns. For example, the segregation of the κ sequences shown in Figure 17-37 indicates that three different alleles of a single gene produce these three sequences.

Geneticists have long used the specificity of antibodies as a phenotypic indicator of the molecular structure of the protein product of a gene. The protein is injected into an animal, and the antibody produced may be tested for cross reaction with, say, the protein from a mutant allele, or compared with the antibody produced by a mutant protein. In a similar way, the genetics of the variable region can be analyzed by injecting a purified antibody from one animal into another animal, which in turn will produce antibodies to the injected antibody. Various phenotypes can be defined in terms of the amino-acid sequences of the V regions in the antibodies produced in response to standard test injections. Marker phenotypes can be defined and their segregation studied in animal experiments or human pedigrees. With such experiments, for example, the V region of the human κ light chain has been shown to be encoded by multiple alleles of at least three separate genes. In general, the genetic studies show that the immunoglobulin polypeptides are encoded in three unlinked clusters of autosomal genes. One cluster (*H*) codes for all classes of heavy chains, another cluster (κ) codes for the κ light chains, and a third cluster (λ) codes for the λ light chains.

One interesting observation from these experiments is that the cells of a lymphocyte clone express only one of each allele pair in an individual heterozygous for markers on the light and heavy chains. This **allele exclusion** is reminiscent of the X-chromosome inactivation in mammalian females.

The Basis of Antibody Variability

A very early model for the antigen–antibody interaction suggested that the antibody structure is molded by its interaction with the antigen, so that the antibody adapts itself to bind the antigen that it encounters. However, this model was ruled out by evidence showing that the antibody specificity is a function of its amino-acid sequence, which in turn dictates its three-dimensional structure (including the shape of the binding site) before it ever encounters an antigen. How then are the thousands of different antibodies produced with different variable regions suited to thousands of possible antigens?

The suggestion was made some time ago that different cistrons encode the variable and constant regions of any given chain. A striking corroboration of this hypothesis became possible in 1978, when S. Tonegawa and his colleagues were able to clone the DNA that encodes the variable portion of the λ light chain and determined its base sequence (Figure 17-38). The results indicate that a single

leader sequence
Met Ala Trp Thr Ser Leu Ile Leu Ser Leu Leu Ala Leu Cys Ser Gly
ATG GCC TGG ACT TCA CTT ATA CTC TCT CTC CTG GCT CTC TGC TCA GGT CAG CAG CCT TTC TAC ACT GCA GTG GGT ATG CAA CAA
TAO CGG ACC TGA AGT GAA TAT GAG AGA GAG GAC CGA GAG ACG AGT CCA GTC GTC GGA AAG ATG TGA CGT CAC CCA TAC GTT GTT

leader sequence
1
UGA Pne Ala Thr Asp Asp Trp Ile Ser Tyr Leu Phe Ala Gly Ala Ser Ser Glp Ala Val Val Thr Gin Glu Ser
TAC ACA TCT TGT CTC TGA TTT GCT ACT GAT GAC TGG ATT TCT TAC CTG TTT GCA GGA GCC AGT TCC CAG GCT GTT GTG ACT CAG GAA
ATG TGT AGA ACA GAG ACT AAA CGA TGA CTA CTG ACC TAA AGA ATG GAC AAA CGT CCT CGG TCA AGG GTC CGA CAA CAC TGA GTC CTT

10 20 hv1 30
Ala Leu Thr Thr Ser Pro Gly Gly Thr Val Ile Leu Thr Cys Arg Ser Ser Thr Gly Ala Val Thr Thr Ser Asn Tyr Ala Asn
TCT GCA CTC ACC ACA TCA CCT GGT GGA ACA GTC ATA CTC ACT TGT CGC TCA AGT ACT GGG GCT GTT ACA ACT AGT AAC TAT GCC
AGA CGT GAG TGG TGT AGT GGA CCA CCT TGT CAG TAT GAG TGA ACA GCG AGT TCA TGA CCC CGA CAA TGT TGA TCA TTG ATA CGG

40 50 hv2 60
Trp Val Gin Glu Lys Pro Asp His Leu Phe Thr Gly Leu Ile Gly Gly Thr Ser Asp Arg Ala Pro Gly Val Pro Val Arg Phe
AAC TGG GTT CAA GAA AAA CCA GAT CAT TTA TTC ACT GGT CTA ATA GGT GGT ACC AGC AAC CGA GCT CCA GGT GTT CCT GTC AGA
TTG ACC CAA GTT CTT TTT GGT CTA GTA AAT AAG TGA CCA GAT TAT CCA CCA TGG TCG TTG GCT CGA GGT CCA CAA GGA CAG TCT

70 80 90
Ser Gly Ser Leu Ile Gly Asp Lys Ala Ala Leu Thr Ile Thr Gly Ala Gin Thr Glu Asp Asp Ala Met Tyr Phe Cys Ala Leu
TTC TCA GGC TCC CTG ATT GGA GAC AAG GCT GCC CTC ACC ATC ACA GGG GCA CAG ACT GAG GAT GAT GCA ATG TAT TTC TGT GCT
AAG AGT CCG AGG GAC TAA CCT CTG TTC CGA CGG GAG TGG TAG TGT CCC CGT GTC TGA CTC CTA CTA CGT TAC ATA AAG ACA CGA

hv3 100 110 120
Trp Tyr Ser Thr His Phe | His Asn Asp Met Cys Arg Trp Gly Ser Arg Thr Arg Thr Leu Trp | Tyr Ser Leu Thr Thr Ile Phe
CTA TGG TAC AGC ACC CAT|TTC CAC AAT GAC ATG TGT AGA TGG GGA AGT AGA ACA AGA ACA CTC|TGG TAC AGT CTC ACT ACC ATC
GAT ACC ATG TCG TGG GTA|AAG GTG TTA CTG TAC ACA TCT ACC CCT TCA TCT TGT TCT TGT GAG|ACC ATG TCA GAG TGA TGG TAG
V|C(1) V|C(2)

130
Leu Thr Gly Gly Tyr Met Ser Leu Val Cys Ser Leu Leu Leu UAG
TTC TTA ACA GGT GGC TAC ATG TCC CTA GTC TGT TCT CTT TTA CTA TAG AGA AAT TTA TAA AAG CTG TTG TCT CGA GCA ACA AAA
AAG AAT TGT CCA CCG ATG TAC AGG GAT CAG ACA AGA GAA AAG GAT ATC TCT TTA AAT ATT TTC GAC AAC AGA GCT CGT TGT TTT

AGT TTT ATT CAA CAA ATT GTA TAA TAA TTA TGC CTT GAT GAC AAG CTT TGT TTA TCA ACT TGG CAG AAC ATA GAA TC
TCA AAA TAA GTT GTT TAA CAT ATT ATT AAT ACG GAA CTA CTG TTC GAA ACA AAT AGT TGA ACC GTC TTG TAT CTT AG

Figure 17-38. The complete nucleotide sequence of a gene (λe) *coding for the V region of the* λ *light chain of immunoglobulin from mouse embryo. The protein sequence encoded by the gene is also indicated. VC(1) indicates the point at which the* λe *sequence ceases to match the amino-acid sequence of the* λ *protein. The sequence of 21 amino acids beginning at VC(2) matches the sequence at the end of the constant region of the* λ *protein. (From S. Tonegawa et al.,* Proc. Natl. Acad. Sci. USA *75:1485, 1978.)*

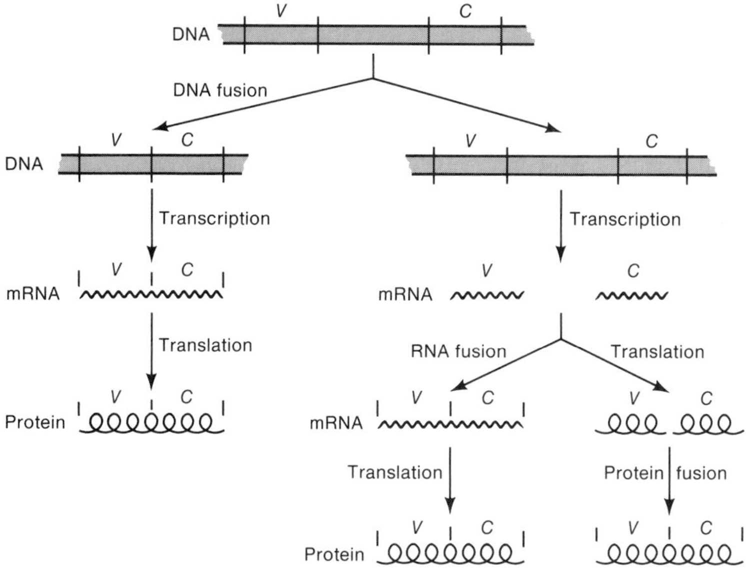

Figure 17-39. Possible ways to produce a single polypeptide chain from two different cistrons coding for the variable (V) and constant (C) regions. At the DNA level, the two cistrons could be fused into a single transcriptional unit. At the RNA level, the two separate transcripts could be fused into a single mRNA. At the protein level, the two polypeptides translated separately could be fused into a single chain.

gene encodes the V region and only 21 amino acids of the constant region. Presumably, the remainder of the C region is encoded in a different gene.

If the V and C regions of an immunoglobulin are encoded in different genes, then how are the separate sequences joined to form a single polypeptide chain? The fusion could take place at any of three different levels: DNA, RNA, or protein (Figure 17-39). Myeloma tumor cells of mice can be cultured in vitro. Single mRNA molecules isolated from such cultures contain sequences for both the C and V regions. This evidence rules out the possibility that the polypeptide fragments are translated from different transcripts and then fused into a single chain. Tonegawa and his group demonstrated that DNA fusion of the V and C regions occurs. They fragmented DNA of embryos with restriction enzymes and then hybridized these DNA fragments with labeled mRNA of the type known to contain sequences for both regions. As expected, the mRNA hybridized to two different DNA fragments, corresponding to the separate *V* and *C* genes (Figure 17-40). However, when they used DNA fragments derived from antibody-producing myeloma tumors, in this hybridization experiment they found that the labeled mRNA hybridized to a single DNA fragment (Figure 17-41). They concluded that the *V* and *C* genes are somehow brought into proximity in the myeloma cells, so that the restriction enzymes do not separate them. This evidence leads to the current assumption that *V* and *C* genes are joined

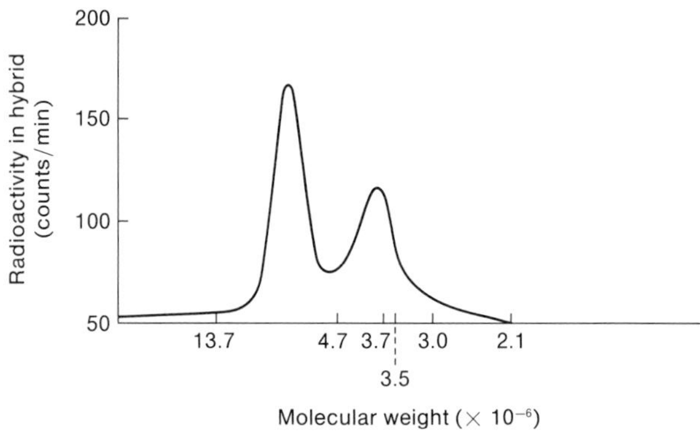

Figure 17-40. Hybridization of radioactive immunoglobulin mRNA to fragments of embryonic DNA. The curve represents the size distribution of the DNA fragments hybridized to the RNA. It is clear that the RNA hybridizes chiefly to fragments of two different sizes. (From L. E. Hood, I. L. Weissman, and W. B. Wood, Immunology, *1978, Benjamin/Cummings.)*

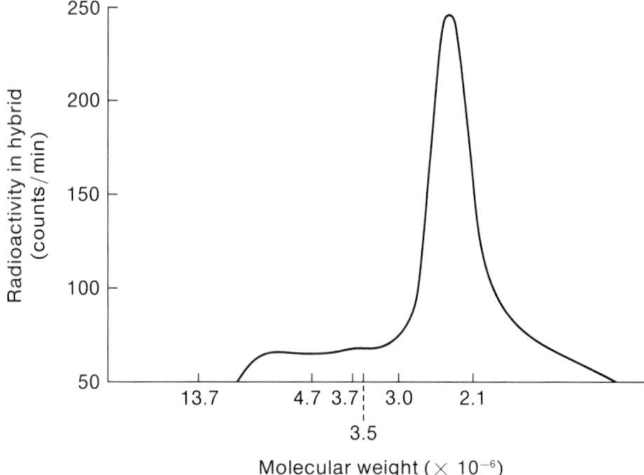

Figure 17-41. Hybridization of radioactive immunoglobulin mRNA to fragments of myeloma-cell DNA. The DNA fragments that hybridize in this case are chiefly of a single size. (From L. E. Hood, I. L. Weissman, and W. B. Wood, Immunology, *1978, Benjamin/Cummings.)*

(by some unknown mechanism) into a single transcription unit in antibody-producing cells.

You can see that the immune system has evolved mechanisms that lie outside the realm of conventional genetic behavior. The final genetic problem that must be resolved is the source of the fantastic array of different antibodies (from 10^5 to 10^8 of them) in a single individual. Two classes of models have been proposed.

1. **Germ-line theories.** These theories postulate that a separate gene codes for each variable part of the antibody. Multiple copies of the numerous *V* genes have accumulated in the genome through evolution.

2. **Somatic variation theories.** These theories postulate that a limited number of *V* genes are shuffled by some mechanism (mutation or recombination or mobile genetic elements) to generate a large number of different sequences in different somatic cells.

At present, the evidence does not permit us to rule out either group of theories. However, it is clear that genetic study of the immune system will provide insights into many novel genetic processes.

Message
The genetic mechanisms that have evolved in the immune system are quite different from those that have been studied in other parts of the organism. We still do not have a complete understanding of the process by which antibodies are produced, but geneticists have learned much about novel genetic processes in their study of the immune system.

Developmental Genetics of *Drosophila*

The small fly *Drosophila melanogaster* played a central role in the studies that revealed the basic mechanisms of chromosome behavior. For a period after the elucidation of the principles of transmission genetics, further research with *Drosophila* seemed to be of minor importance in genetics. However, *Drosophila* has once again become the organism of choice for studies of development and behavior. In large part, the resurgence of interest in *Drosophila* is due to the extensive array of genetic manipulations that have been perfected by generations of researchers working with this insect (such researchers are often called Drosophilists). In previous chapters, we have regarded the flies as convenient systems for the study of chromosomes. Now let's examine them again as complex aggregates of cells and tissues that are somehow coordinated to function as complete animals.

The Sequence of Development

After fertilization, the zygotic nucleus undergoes a rapid series of divisions without separation into cells. Thus, the newly fertilized egg develops as a **syncitium** (a single multinucleated cell). In the posterior part of the egg is a region characterized by cytoplasmic occlusions called polar granules; as we have seen, these are maternally deposited elements that determine differentiation of the gonadal tissue. The nuclei divide synchronously. After the ninth division, when about 512 nuclei are present, they migrate to the periphery of the egg cytoplasm. After three or four more synchronous divisions, when there are 4000 to 8000 nuclei, cell membranes enclose them to form the mononucleated cells of the **blastoderm,** which is essentially a cell monolayer enclosing the yolk (Figure 17-42). All of these events occur within 3 hours after fertilization. An important feature of blastoderm formation is that the pattern of nuclear migration is very ordered. Those nuclei most recently related by nuclear division remain nearer to one another than do more distantly related nuclei.

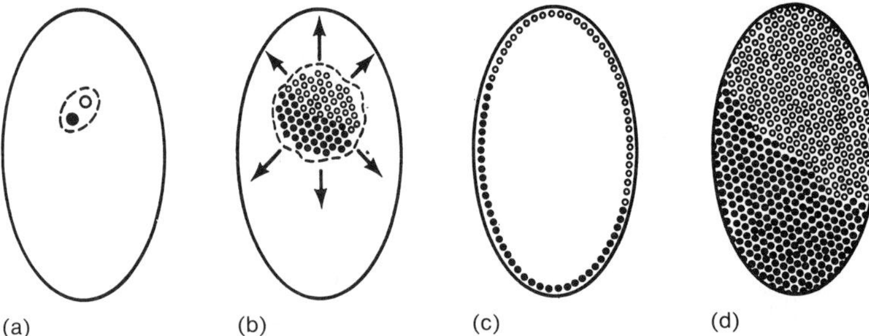

Figure 17-42. Early divisions leading to blastoderm formation in Drosophila. *(a) The first division of the nucleus occurs shortly after fertilization. No division of the cytoplasm occurs. (b) After nine synchronous nuclear divisions without cell cleavage, the embryo is a syncitium with about 500 nuclei. The nuclei now begin to migrate to the periphery of the egg cytoplasm. (c) After a few more nuclear divisions, the cell has some 4000 to 8000 nuclei in a layer near its surface. (d) Cell membranes form to enclose each nucleus in a separate cell, thus forming the blastoderm stage of the embryo. (From S. Benzer, "Genetic Dissection of Behavior." Copyright © 1973 by Scientific American, Inc. All rights reserved.)*

After the blastoderm stage, cell movements and folding of sheets of cells create the cell layers from which tissues will differentiate. Up to the blastoderm stage, the nuclei are totipotent, as Illmensee demonstrated by recovering adults after transplantation of blastoderm nuclei into enucleated eggs. However, after the blastoderm stage, each nucleus is restricted to a limited potential fate. As the nuclei differentiate, they fall into two classes with very different prospects. One set of cells must develop into a larva that tunnels into its food and, although lacking vision, responds to temperature, gravity, light and odors while crawling, eating, and excreting. It is a very sophisticated organism with a central nervous system that coordinates its behavior in response to environmental stimuli. The second set of cells will form a second organism that will emerge from the larval carcass as an adult fly bearing no resemblance to its larval predecessor and capable of seeing, walking, flying, and mating. This remarkable transformation is anticipated in the larva by the presence of packets of cells programmed to differentiate into adult tissue upon exposure to the proper hormonal cues. These packets of cells are called **imaginal disks** (often known simply as disks). They can be recognized by their sizes, shapes, and locations in the larva (Figure 17-43).

How do we know that the disks are already programmed (determined) for an adult fate? An eye disk, for example, can be removed and implanted in another larval host that then pupates and emerges as an adult. This adult will have somewhere within it extra adult eye structures derived from the implanted disk! Similar results are obtained with each of the other disk types. The stimulus for the differentiation of the already determined disk is the molting hormone ecdysone, which is released in the late third instar stage of the larva. This effect can be demonstrated by isolating disks from late third instar larvae and exposing them to ecdysone; they will then begin to differentiate.

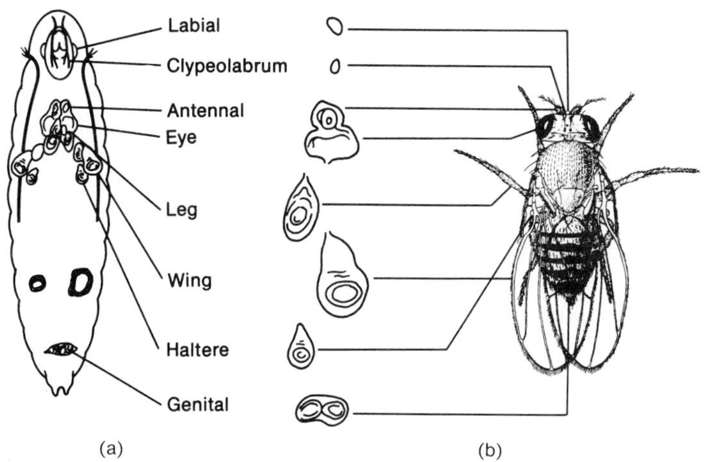

Figure 17-43. The imaginal disks of Drosophila. *(a) The larva in the third instar stage of development, showing the positions of the disks. The larva is significantly larger than the adult. (b) The adult fly, showing the part that is derived from each disk. Note that the disks can be identified by their differing sizes and shapes as well as by their positions in the larva. (From J. W. Fristrom, in* Problems in Biology, *University of Utah Press, 1969.)*

On pupation, the larval carcass begins to break down. Its residues act as a thick medium nourishing and embedding each of the disks. The disks now begin to differentiate into their specific adult structures, so that each part of the adult forms as a separate element. The separate elements then fuse with the correct neighbors to form a complete adult.

The imaginal disks play no functional role in the larva. A series of mutants has been recovered lacking all imaginal disks or specific sets of disks. These mutant larvae are completely viable until after pupation, when they die. The incredibly intricate and precise program whereby the disks are activated and differentiate into adult structures is acted out each time a fly is "born." You can see that mutations blocking various parts of this complex developmental process provide a great deal of information about the process itself. They also provide evidence about the genetic system controlling the process.

Early Determination

Let's now go back and examine a few parts of the *Drosophila* development process in more detail. At what point in development are the nuclei of the embryo committed to become parts of specific imaginal disks? The answer was provided by using techniques for culturing cells in vivo. Larval cells can be transplanted into an adult host (by injection). The cells escape exposure to ecysone, which triggers metamorphosis, so they simply proliferate as larval cells in the adult host. The clone of larval cells can then be removed from the adult host and used for experimentation. For example, the cell mass can be implanted in a larval host about to undergo metamorphosis, so that the differentiation of the cells upon ecdysone stimulation can be studied. We have mentioned Ilmensee's experiments, in which blastoderm nuclei were shown to be totipotent because they developed normally when transplanted into enucleated eggs.

In 1971, Lilian Chan and Walter Gehring performed an experiment with a *Drosophila* genotype distinct from the wild type. They took a blastoderm and dissected it at the equator into anterior and posterior halves. They then dissociated the cells from each half, compacted them into pellets by centrifugation, and injected the pellets into adult hosts for culturing (Figure 17-44). They then removed the cultures of larval cells from the adult host and injected them into wild-type larval hosts, where their differentiation could be observed after metamorphosis. (The distinctive genotype of the experimental cells distinguished them from the cells of the host.) They found that the cells from the

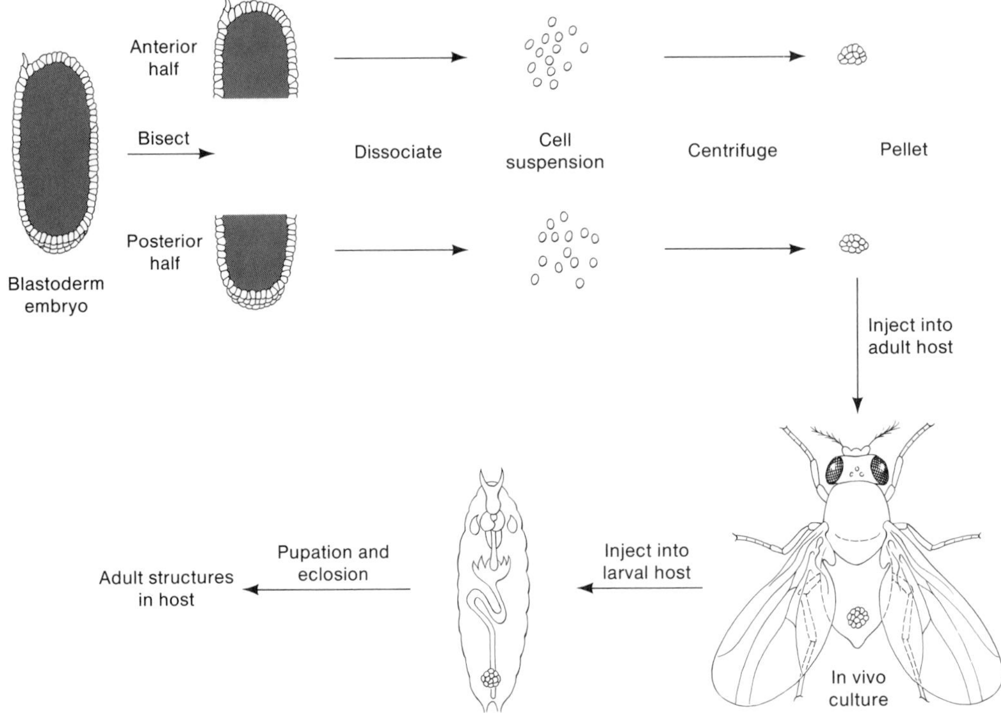

Figure 17-44. Experimental technique for testing the developmental fate of blastoderm cells. The cells multiply by mitotic division in the adult host but do not undergo differentiation. After a larger mass of cells is obtained from in vivo culture in the adult, their determination state is tested by implanting them into a larval host. These cells will now undergo differentiation just as will the imaginal structures of the host larva. When an adult fly emerges, it can be dissected, and the adult fate of the donor cells can be observed.

anterior half of the blastoderm developed into anterior adult structures (head and thorax), whereas the cells from the posterior half of the blastoderm developed into posterior adult structures (thorax and abdomen). These results indicate that the developmental fate of the blastoderm cells was already determined and that the cells retained their determined status even through repeated rounds of division in the adult host.

Further evidence of early determination comes from studies on damaged embryos. If the blastoderm is punctured, burned, or subjected to ultraviolet irradiation in specific regions of the embryo, then the adult fly that develops from the embryo shows damage in specific structures corresponding to the damaged part of the blastoderm. From Ilmensee's results, we know that the cell nuclei remain totipotent until after the blastoderm stage. Therefore, we conclude that the early determination is an effect of the cytoplasm. Further experiments indicate that this early determination occurs at roughly the time when the nuclei are enclosed by membranes to form the blastoderm cells.

As we have seen in the example of snail-shell coiling (Chapter 15), the egg cytoplasm contains developmental cues dictated by the genotype of the mother. Therefore, we would expect to find mutations that are expressed in females as an abnormal phenotype *in their offspring.* The *grandchildless* mutation of *Drosophila* mentioned earlier in this chapter is an example of such a mutation.

Message
The pattern of determination of embryonic cells is apparently established in the cytoplasm of the multinucleate egg, and it is under the control of the maternal genotype.

Disk Determination

In the embryo, the imaginal disk originates as a small number of cells that are programmed to form specific parts of adult structures. The number of founding cells of a disk can be estimated by inducing genetic mosaics by chromosome loss or mitotic crossing over. Suppose that we obtain XO/XX gynandromorphs formed by a chromosome loss. If a disk originates from a single determined cell, an adult derivative that is entirely male or entirely female should be recoverable in a mosaic. On the other hand, if two cells begin a disk, then the adult structure they form could only be a mosaic of 50% male and 50% female cells. You can see that the largest proportion occupied by mutant tissue in a mosaic structure provides an estimate of the number of founding cells, and they range from disk to disk from eight to forty. Determination must occur progressively as the cell number increases, since there are far more structures derived from a mature disk than there are cells in the founding group.

It should be obvious that we could derive a **fate map** of the cells within a disk by cutting the disk into different fragments and implanting the fragments into

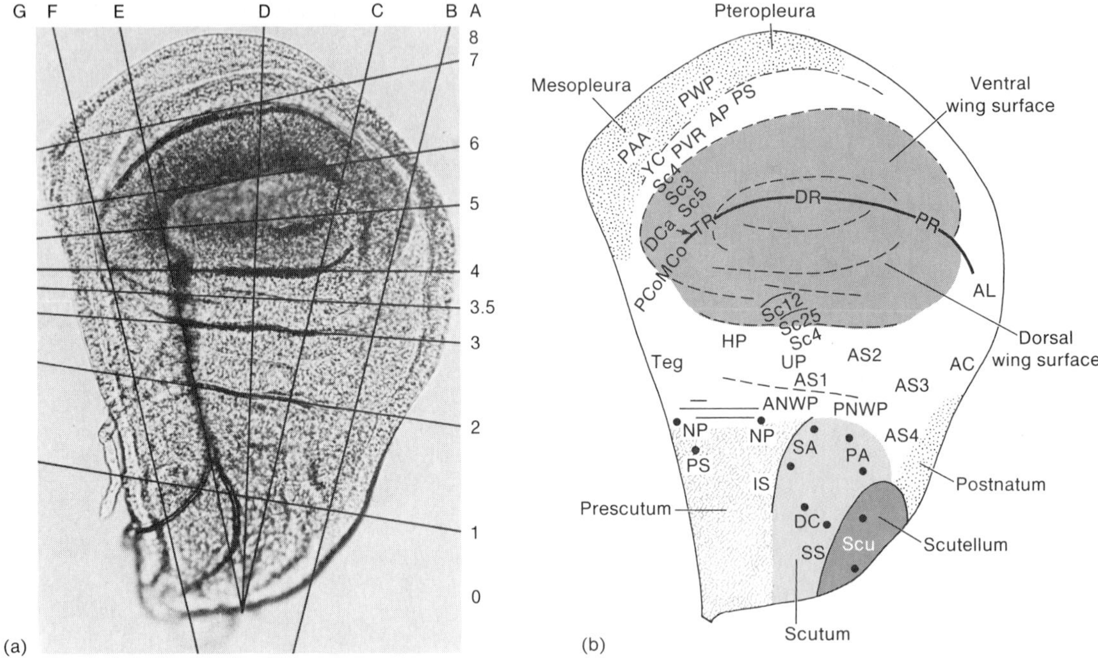

Figure 17-45. The fate map for a Drosophila *wing disk. (a) A freshly dissected wing disk, showing the planes of cutting to obtain fragments for the mapping procedure. (b) The fate map of the wing disk, showing the adult structures that develop from cells in each part of the disk. The symbols represent specific bristles and other small structures on the adult wing (see Figure 17-46). (From P. J. Bryant,* Cell Patterning, *Elsevier/North-Holland Biomedical Press, 1975.)*

larval hosts to see what adult structures they form (Figures 17-45 and 17-46). However, an interesting result is obtained when the fragments are first cultured in an adult host to increase the population of cells and then tested by implantation in a larva. When a disk is bisected, one fragment regenerates the missing part, but the other fragment forms a mirror-image duplicate of itself to replace the missing part. For example, if a leg disk is bisected along various planes, the upper, medial, or proximal portions will regenerate complete disks, whereas the lower, lateral, or distal portions will only create duplicates of themselves. Such duplication can also occur after radiation treatment of an embryo, presumably as a result of the destruction of those disk cells capable of regeneration (Figure 17-47). This phenomenon of different regenerative potentials for different body parts is also observed in amphibians and in annelid worms. We do not yet know the explanation.

Homeotic Mutations. Two exceptions have been encountered to the normal process of disk development. One exception occurs in a group of mutations

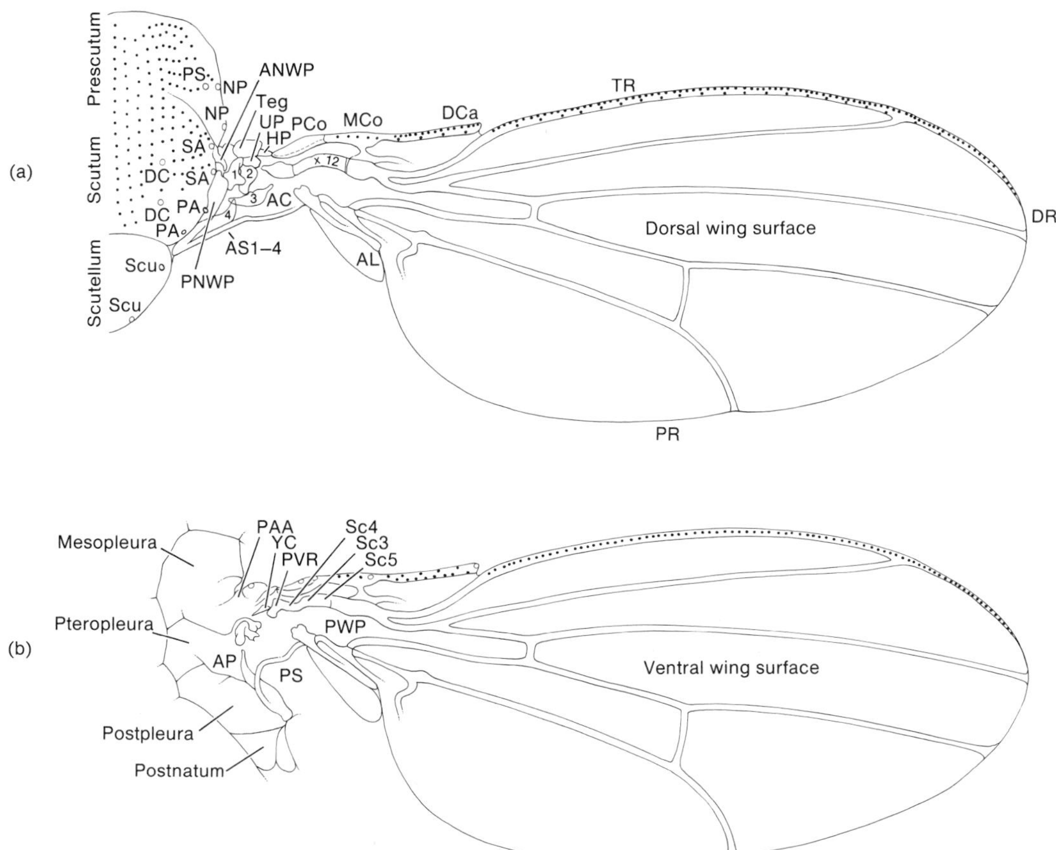

Figure 17-46. The adult structures derived from the wing disk (Figure 17-45). The dots indicate bristles and other small structures called bracts. (a) Dorsal view of the wing. (b) Ventral view. (From P. J. Bryant, Cell Patterning, *Elsevier/North-Holland Biomedical Press, 1975.)*

called **homeotic,** in which the imaginal disks have fates other than their normal ones. For example, the mutation *ophthalmoptera* causes wing structures to develop from an eye disk, and the mutation *aristapedia* causes the feathery arista of the antennal complex to develop as a leg instead (Figure 17-48). In the 1940s, many of the homeotic mutations were found to be temperature sensitive. Temperature-shift experiments were performed to determine the developmental stage at which the homeotic transitions occur. Almost all of these studies indicate that the temperature-sensitive periods occur during the third instar stage of larval development. After that stage, the fate of each disk seems to be firmly determined in a way that cannot be affected by these mutations.

In 1969, John Postlethwait and Howard Schneiderman studied the dominant mutation *Antennapedia,* which converts part of the antenna into leg structures.

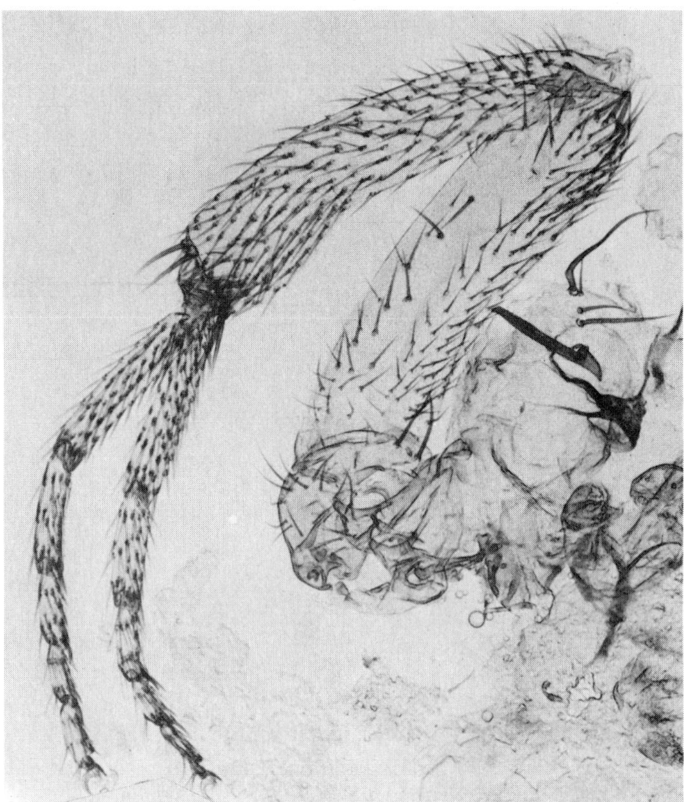

Figure 17-47. A partially duplicated Drosophila *leg that developed from an embryo treated with X rays 24 hours after fertilization. (Courtesy of John H. Postlethwait.)*

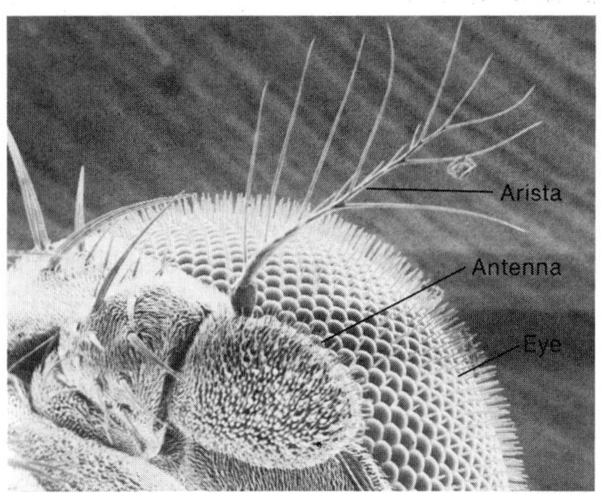

(a)

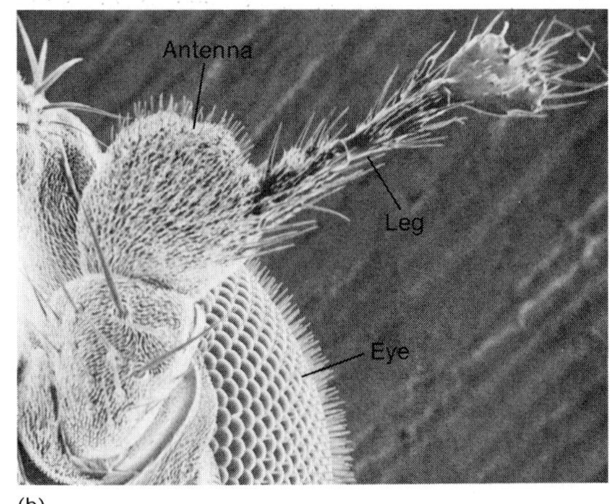

(b)

Figure 17-48. Scanning electron micrographs of the antennal complex of Drosophila, *showing the effects of a homeotic mutation. (a) A wild-type fly. (b) A mutant fly in which the feather arista is replaced by the distal part of a leg. This mutation is called* aristapedia.

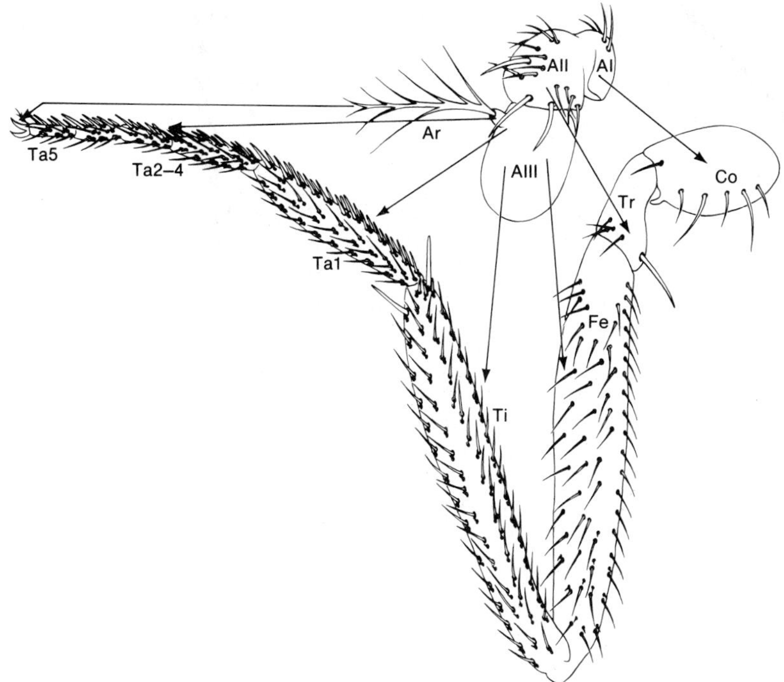

Figure 17-49. The correspondence between antennal and leg structures in Drosophila, *based on position-specific transformations in homeotic antennae of* Antennapedia. *The symbols identify various corresponding parts of the structures. (From Postlethwait and Schneiderman,* Developmental Biology *25:606, 1971.)*

By inducing mitotic crossovers in a heterozygote, they were able to generate a mosaic of wild-type cells in an *Antennapedia* background. They showed that the replacement of parts is position specific—that is, a given antennal segment is always replaced by a specific part of the leg (Figure 17-49). These observations suggest that there is some overall developmental plan on which the specific details of leg or antennal development are overlaid.

Transdetermination. Ernst Hadorn and his students investigated various aspects of disk development, using the techniques of adult culture and larval implantation that he had perfected. In the 1960s, he made an interesting observation in the course of these supposedly routine studies. One of the standard procedures used in his laboratory involves injection of a disk or disk fragment into an adult host to increase the cell number in the research sample. Serial transplantation through a series of successive adult hosts can provide a larger sample of material (Figure 17-50). Samples were tested after each transplantation to verify that the cells were unchanged by the culturing procedure. In some cases, an unexpected result was obtained. On successive transplants through

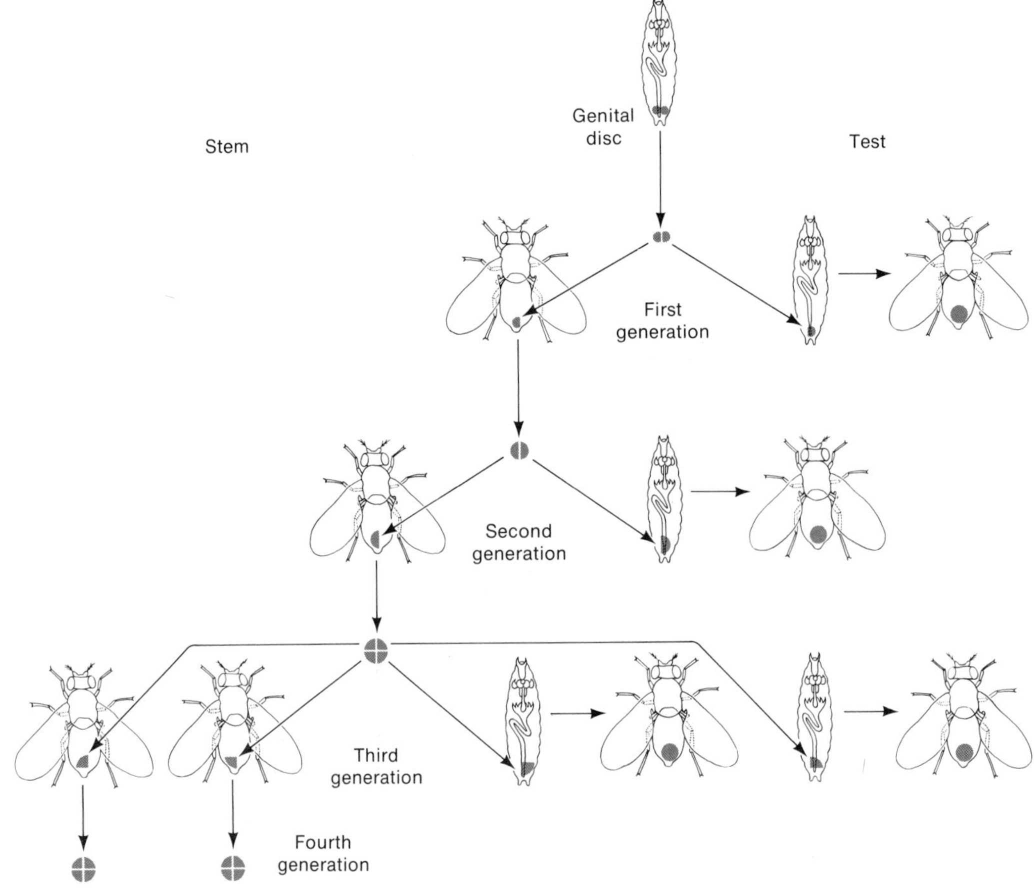

Figure 17-50. Serial transplants of a genital disk through adult Drosophila. *At each generation, the population of disk cells is subdivided. Some of the cells are implanted in another adult for further growth, and the other cells are implanted into a larva where they will develop into adult structures. (From E. Hadorn, "Transdetermination in Cells." Copyright © 1968 by Scientific American, Inc. All rights reserved.)*

adult hosts, the determined state of the disk sometimes changes. For example, a genital disk might eventually change so that it forms leg structures.

Hadorn called this change of state a **transdetermination.** The nature of observed transdeterminations is not random. For example, a genital disk can become a leg disk, but it never changes directly to an eye disk (Figure 17-51). We do not yet understand the molecular basis of transdetermination, but further study of the phenomenon should provide clues about the nature of the determination process. It is interesting to note that homeotic mutations cause alterations only in the same directions as the known transdetermination pathways.

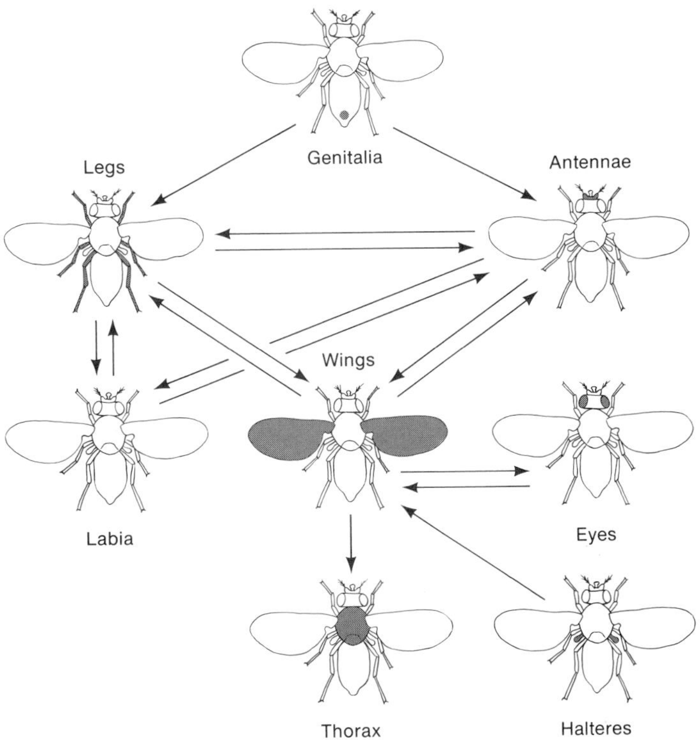

Figure 17-51. Types of transdetermination observed for disk cells. In each case, the shaded areas indicate the parts of the adult that develop from the serially transplanted disk material. The arrows indicate the observed changes of fate. Note that cells of genital disks do transdetermine to form leg or antennal structures, but the reverse transdeterminations have not been observed. However, some transdeterminations do occur in either direction. (From E. Hadorn, "Transdetermination in Cells." Copyright © 1968 by Scientific American, Inc. All rights reserved.)

A great deal of research has been done on imaginal disks, but they remain a fertile source of new information about developmental mechanisms and processes.

Compartmentalization

In the early 1970s, Antonio Garcia-Bellido made some interesting discoveries about disk development by inducing mitotic crossovers in the embryonic stage to produce a mosaic adult in which homozygous clones of cells are scattered in a heterozygous background. The clones are quite visible if the original embryo is a heterozygote for some mutation such as *multiple wing hairs* (*mwh*), which causes several hairs to appear where each single hair normally grows on the

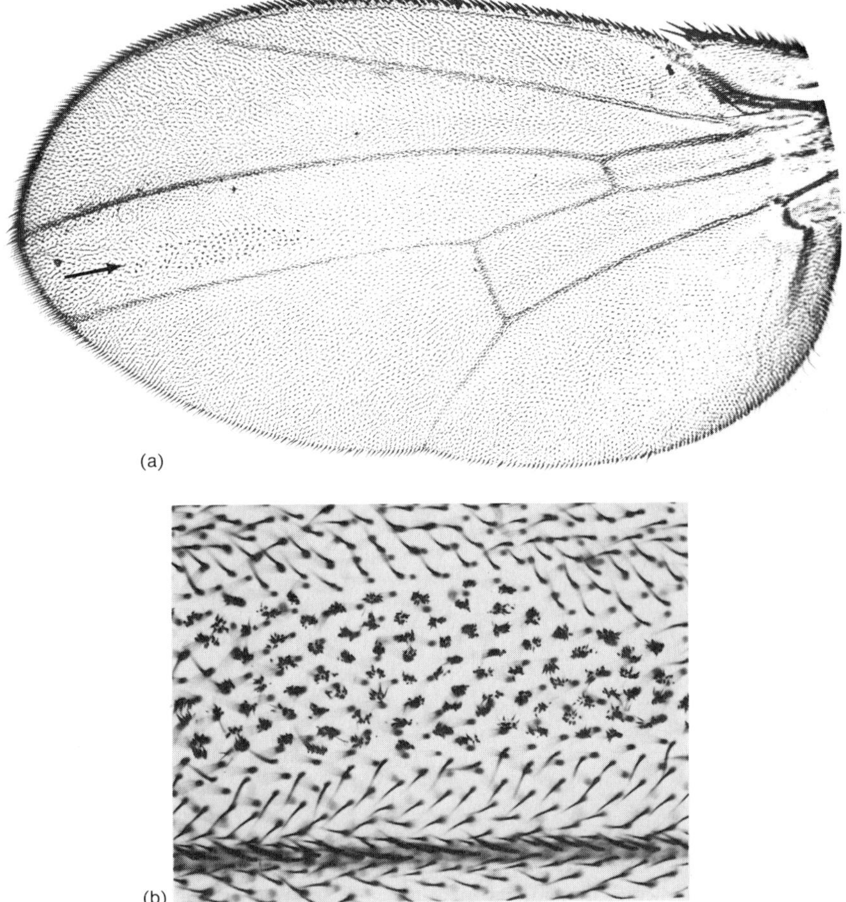

Figure 17-52. The wing of a Drosophila *individual with clones of* mwh/mwh *cells among a background of wild-phenotype* mwh/+ *cells. (a) The clones* (arrow) *are readily visible. (b) Higher magnification clearly shows the nature of the* mwh *phenotype. Such clones are produced by inducing somatic recombination with X-radiation. (From P. A. Lawrence, Medical Research Council, Cambridge, England.)*

wing. A clone of *mwh*/*mwh* derivatives is readily visible against the background of the wild phenotype (Figure 17-52). In different mosaic flies, these clones appear at positions all over the wing, and there is overlap of the boundaries of clones from different flies (Figure 17-53a).

Garcia-Bellido also produced mosaics with *Minute* (*M*) mutations, which retard development when heterozygous and are lethal when homozygous. Thus, when crossovers are induced in an *M*/+ embryo, the *M*/*M* cells die, but the reciprocal +/+ cells divide much more quickly than the surrounding *M*/+ cells. The rates of division are so different that a single +/+ cell in an early

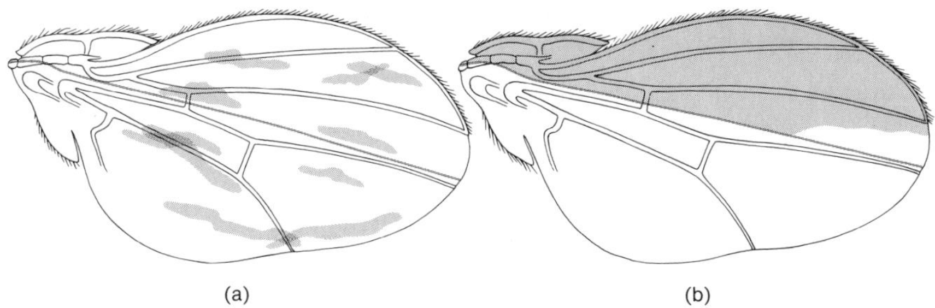

Figure 17-53. The fate of cells and their daughters in Drosophila *disks is not rigidly fixed. (a) A plot of the positions of several mosaic patches* (shaded area) *induced in different flies shows that they overlap and occupy many different parts of the wing. However, no patch crosses the boundary between the third and fourth longitudinal wing veins. (b) Even when +/+ cells grow much more rapidly than the surrounding* M/+ *cells, the clone of +/+ cells never occupies an area in the adult greater than the compartment delineated by the boundary between the third and fourth veins. (From A. Garcia-Bellido, P. A. Lawrence, and G. Morata, "Compartments in Animal Development." Copyright © 1979 by Scientific American, Inc. All rights reserved.)*

stage of an imaginal disk would be predicted to produce most of the cell population of the final disk. In fact, this does not happen; there seem to be rigid zones beyond which a clone will not expand. In the wing, the **boundary** lies between the third and fourth wing veins (Figure 17-53b). The areas separated by such boundaries are called **compartments.**

By noting the distribution and the boundaries of mutant tissue in mosaics induced at different times in development, Garcia-Bellido and his colleagues were able to map changes in cell determination during development. For example, consider the disk that forms the wing and the notum, the bit of thorax to which the wing is attached (Figure 17-54). If mutant clones are induced by irradiation of an embryo, this disk has two compartments that produce the anterior and posterior parts of the wing complex. If the mutant spots are induced in the early larval stage, the disk is divided into four compartments, and a further compartmentalization between the wing and the notum occurs late in the larval stage. These results provide a clear picture of the successively more detailed determination of the cells in the wing disk through the process of development.

Behavior

Thus far, we have dealt only with physical and biochemical aspects of the phenotype. However, as a functional entity, the adult *Drosophila* must control the movements of its legs and wings in order to walk and fly; it responds to sexual stimuli by courting and mating; it detects light, gravity, odor, and movement. All of this activity is coordinated through its central nervous system, which therefore can be probed by the use of mutants exhibiting abnormal behavioral

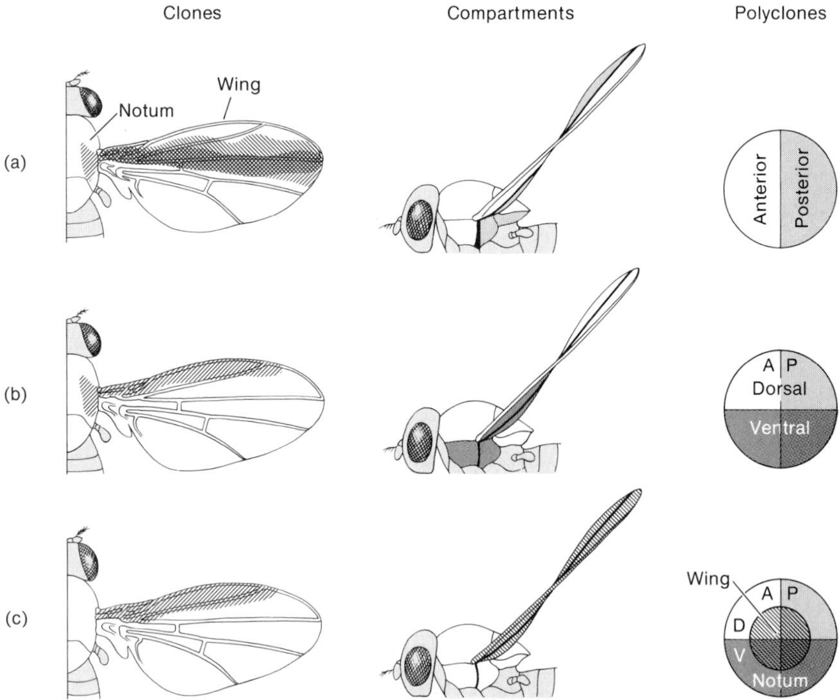

Figure 17-54. Changes in the compartment boundaries of the wing disk observed when genetically marked clones are induced at different developmental stages. (a) When the induction occurs in the embryo stage, a boundary separates the anterior from the posterior parts of the wing and notum. (b) When the induction occurs in young larvae, a second boundary separates the dorsal and ventral parts of the wing. (c) When the induction occurs in older larvae, a third boundary exists between wing and notum. (From A. Garcia-Bellido, P. A. Lawrence, and G. Morata, "Compartments in Animal Development." Copyright © 1979 by Scientific American, Inc. All rights reserved.)

phenotypes. Some care is needed in distinguishing behavioral from physical mutations. Some obvious cases cause no problems, such as the inability to fly due to a vestigial-wing mutation, or the blindness of an eyeless mutant. However, more subtle physical mutations can also mimic the effects of behavioral mutations. Nonetheless, a number of studies of mutant behavioral phenotypes do show that genetic study of the neurobiology of *Drosophila* is possible. Such behavioral phenotypes as hyperactivity, paralysis, blindness, flightlessness, shock sensitivity, odor detection, and even learning have been traced to mutant effects on the central nervous system.

Alterations in *Drosophila* behavior are readily detected because the behavior of the adult fly follows a rigid pattern that is genetically determined. Wild-type flies are positively phototactic—that is, they are attracted toward light. Thus, mutant nonphototactic flies can easily be separated from wild type by placing a population of flies into a vial with a light at one end (Figure 17-55). Of course,

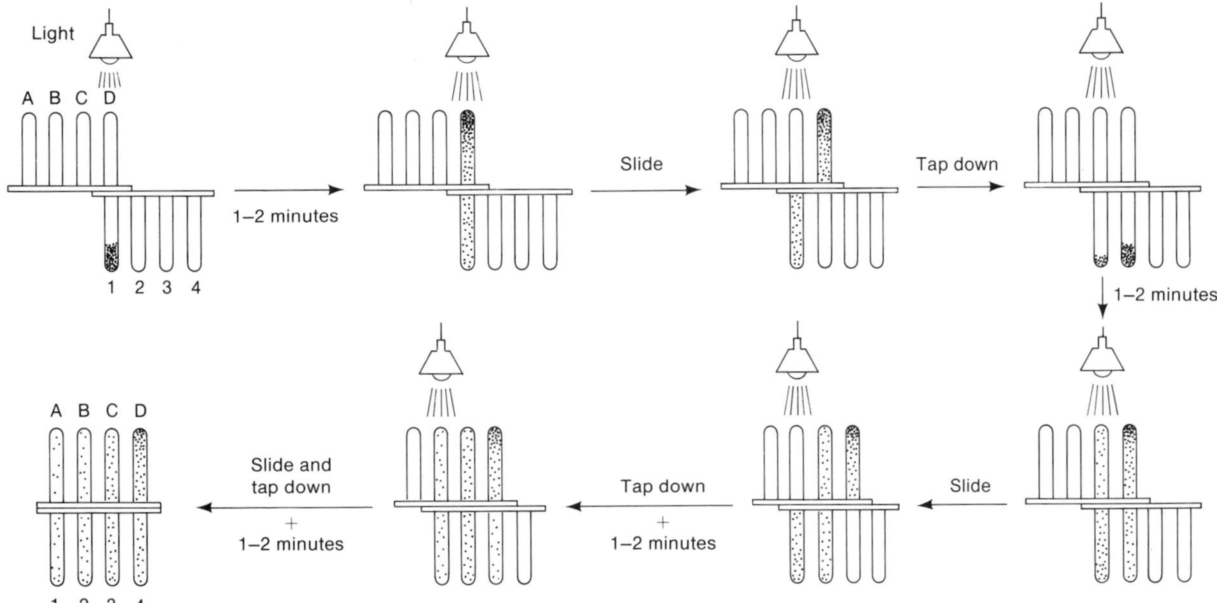

Figure 17-55. A top view of the apparatus used to separate nonphototactic mutants from the positively phototactic wild-type flies. Vials D and 1 are aligned, and a fly population is introduced at the end of vial 1 opposite to the light source. The wild-type flies move toward the light and cluster at the end of vial D. After a minute or two, the racks of vials are moved to align vial 1 with C and 2 with D. The apparatus is tilted to knock the flies to the ends of vials 1 and 2 and then returned to the horizontal position. Any phototactic flies remaining in vial 1 now have another chance to move toward the light. After a few minutes, the vials are realigned again. By the time vials 1 and A are aligned and then separated, vial 1 will contain a population of nonphototactic flies that have failed to respond to four successive light exposures. This procedure eliminates flies that might end up in the wrong vial simply by the chance of where they happened to be at the moment of the separation of two vials.

the population of nonphototactic flies will include mutations such as sluggishness, legless, eyeless, and so forth. The flies that are blind because of defects in the eye or nervous system must also be eliminated. Various tests can be used to eliminate these other mutants and obtain a population that is abnormal only in its failure to respond by moving toward light.

Another interesting behavior phenotype involves an optomotor orientation response. A fly is placed inside a rotating cylinder whose inner wall is marked with alternating vertical strips of black and white. If the wings of the fly have been removed so that it cannot fly, it will turn in the direction of the rotating stripes. Mutants lacking this behavior have been isolated, and a test for this behavior can be used to separate blind flies from those selected for lack of phototaxis.

Many of the stimuli for behavior affect the flies' feet. If a wild-type fly is suspended by a wire glued to its thorax, it will immediately begin to beat its wings in a flight pattern. If pressure is applied to its feet, it stops beating its wings. If the signal for wing beating is the absence of pressure against the feet, then how does a fly take off in the first place? Careful observation shows that the fly first leaps into the air by contracting its middle pair of legs, and the wings begin to beat only after the feet leave the ground. The legs then are tucked under the body as the wings beat rapidly enough to lift the weight of the body. You can see that many different screening systems could be used to obtain behavioral mutants with defects in this complex sequence.

Stimuli affecting feeding responses also are detected by the feet. When a solution of sugar is applied to the feet of a tethered fly, its proboscis automatically drops down to a position that would permit it to suck up material from a surface on which it was walking. Tests can be made to detect the range of substances and concentrations to which the fly will respond, and such tests also provide screens for mutants.

Walking is another complex behavioral pattern in which various mutations can be detected. This pattern is invariant in the wild type. The fly moves the front and hind legs on one side and the middle leg on the other side in a simultaneous stepping movement, and then it steps with the other three legs. This pattern leaves the fly always balanced on a stable tripod.

Courtship, another complex process, includes the male's recognition of a female fly and his dogged pursuit of her. He bumps her and flicks one wing in a specific beat frequency that apparently creates a "love song," to which the female may respond by raising her wings and lifting her abdomen while the male mounts and curls his abdomen underneath himself to mate.

The flies must also detect airborne molecules, including pheromones, repellants, and food, with appropriate inherited behavioral patterns for response to each. They also respond to temperature, gravity, and vibration. Mutations that affect such behaviors provide information about the neural control of the behaviors. Such research may eventually provide a complete model of the fly's central nervous system.

In many cases, newly detected mutations provide unexpected pleiotropic effects. For example, a screening program to detect muscle defects isolated a class of mutations selected by the mutant phenotype of reversible temperature-sensitive adult paralysis. One such mutation, *shibire*^{ts1} (*shi*^{ts1}), recovered by Thomas Grigliatti in 1971, is wild type at 22°C and completely paralyzed when shifted to 29°C. Upon its return to 22°C, the paralyzed mutant quickly recovers mobility. Further study showed that the *shi*^{ts1} mutants also have the temperature-sensitive mutant phenotypes of larval paralysis and homozygotic lethality in any developmental stage, as well as a range of visible abnormalities (Figure 17-56). One of the striking phenotypes is a vertical band in the eye in which the facets are destroyed. The position of this strip varies from the posterior to the anterior part of the eye as the heat shock is administered closer and closer to the time of pupation.

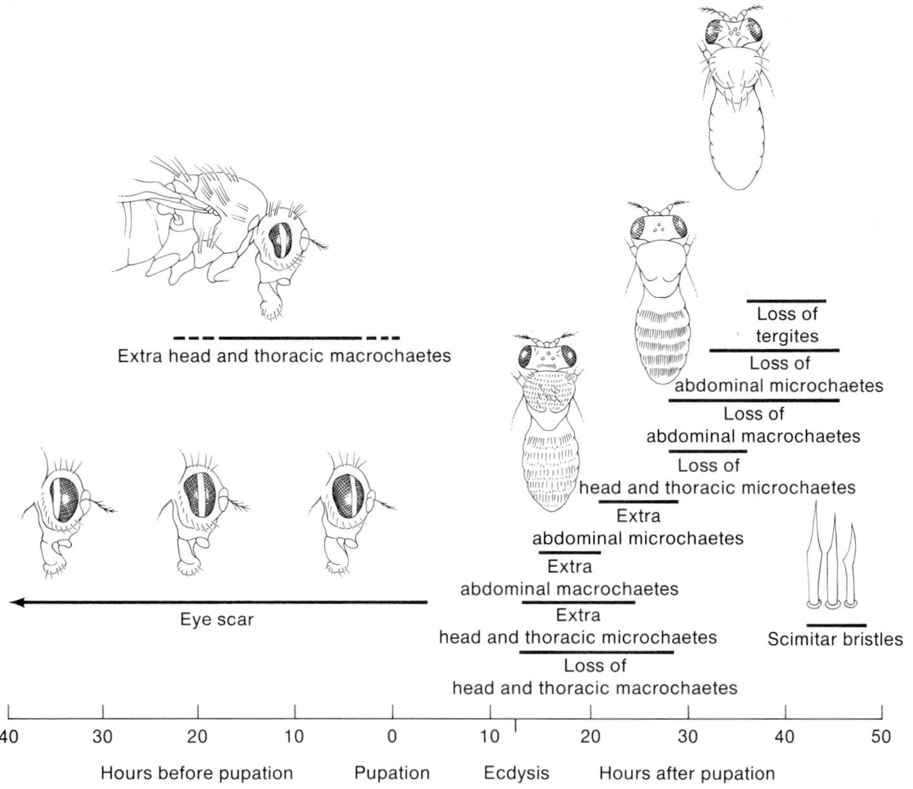

Figure 17-56. Temporal relationships of the temperature-sensitive periods for the various visible abnormalities of shi^{ts1} *flies, with diagrammatic representations of some of these abnormalities. The horizontal lines indicate the time during development when a heat pulse can produce the various abnormalities. Note the change in position of the eye scar with the changing time of the heat pulse (29° C). (From Poodry et al.,* Devel. Biol. *32:381, 1973.)*

These multiple temperature-sensitive abnormalities in the *shi*^{ts1} mutants reveal cellular relationships hitherto unsuspected. The effects apparently are due to a cell-surface defect, probably at the nerve junctions.

Message

The variety of behavioral mutants that can be recovered in Drosophila *seems to be limited only by the ingenuity of the researcher in designing screening procedures. These mutants are useful research tools for the dissection of both developmental processes and the neurobiology of the fly.*

Mosaic Analysis

We have already seen the value of genetic mosaicism as an analytical tool. However, the chance detection or tedious experimental production of such mosaics is a procedural difficulty that makes any systematic use of this approach very difficult in many organisms. In *Drosophila,* a variety of procedures have been developed to make the production of numerous mosaics a relatively simple matter. Two general mechanisms are involved in the formation of the mosaics—mitotic crossing over and chromosome loss.

In 1971, Antonio Garcia-Bellido and John Merriam used somatic crossing over in an attempt to determine the time when a particular gene ceases to be active. They induced crossovers by irradiating different developmental stages of females with the genotype *y Hw* + / + + sn^3. (*Hw* is a dominant mutation that causes the growth of extra bristles along wing veins and is closely linked to *y,* the mutation for yellow body color; sn^3 produces short, gnarled bristles and is more distantly linked.) Induced crossovers between the sn^3 locus and the centromere produce twin spots, with one spot of the genotype *y Hw* + / *y Hw* + and the other + + sn^3 / + + sn^3 (Figure 17-57). Garcia-Bellido and Merriam found that y Hw and sn^3 mutant patches can be induced by irradiation at any time during the third instar larval stage or during the first 24 hours of the pupal stage. However, if the irradiation occurs during the last 12 hours of the third instar stage or the first 24 hours of the pupal stage, the sn^3 patches also show the Hw phenotype. In other words, one spot of the twin spot is y Hw and the other spot is Hw sn^3. No crossover pattern can explain this observation; the *Hw* alleles cannot be present in the sn^3 spot cells. Garcia-Bellido and Merriam concluded that the *Hw* allele somehow acts before the last 12 hours of larval life to imprint its phenotype irreversibly on the cell of the wing disk.

In 1929, Sturtevant observed that eggs from females of *Drosophila simulans* homozygous for the autosomal recessive mutation ca^{nd} (claret–nondisjunction) commonly lose a chromosome during the first and second divisions after fertilization. If the lost chromosome was an X, then a mosaic of XX (♀) and XO (♂) cells is formed. If the original fertilized egg was heterozygous (say, + + / *w m*),

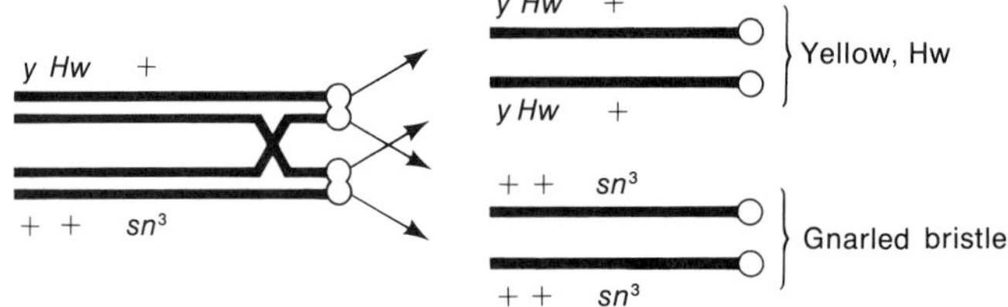

Figure 17-57. Production of twin spots in a y Hw + / + + sn^3 *fly by a mitotic crossing over induced with X-radiation at different developmental stages.* Hw *is a dominant mutation that causes the growth of extra bristles along wing veins and is closely linked to* y *(yellow body color);* sn^3 *produces short, gnarled bristles.*

then loss of the wild-type chromosome permits expression of the recessive phenotype in the hemizygous tissue (Figure 17-58). The resulting mosaic flies do *not* show a "salt-and-pepper" phenotype with male and female cells randomly arranged. The mosaic patterns observed indicate that nuclei more closely related by cell division tend to stay together. This observation led Sturtevant to the realization that the distribution of mutant and nonmutant tissue in a mosaic can provide information about the spatial relationships *in the embryo* of prospective adult nuclei.

Since Sturtevant's time, several different mutants that cause chromosome loss have been studied. The most useful was described by Claude Hinton in 1955. It is a ring X chromosome called $In(1)w^{vC}$ (we'll call it w^{vC} for short) that has the completely mysterious property of great instability in the newly formed zygote nucleus. Thus, w^{vC} is lost at a very high frequency and, if the egg was initially w^{vC}/m, gynandromorphs of w^{vC}/m (wild-type ♀) and m/O (mutant ♂) cells are created. This is an extremely useful aberration.

Using data originally obtained by Sturtevant in the 1930s, Garcia-Bellido and Merriam in 1969 set out to determine whether the pattern of mosaicism in adult flies can be used to infer relationships between cells in the embryo. Such an analysis should be possible if the related nuclei are arranged in close proximity in the blastoderm. It appears that the orientation of the division plane for the first nuclear division is tilted randomly with respect to the anterio-posterio and dorso-ventral axes of the egg, because all kinds of mosaic patterns are formed in the adult (Figure 17-59). We can also assume that prospective cells for imaginal disks and for larval structures are distributed throughout the surface of the blastoderm.

If cells destined to form two imaginal disks are located close to each other within the blastoderm sphere (and hence are probably closely related), then the chance that they are genetically *different* in a mosaic is small. (This is because

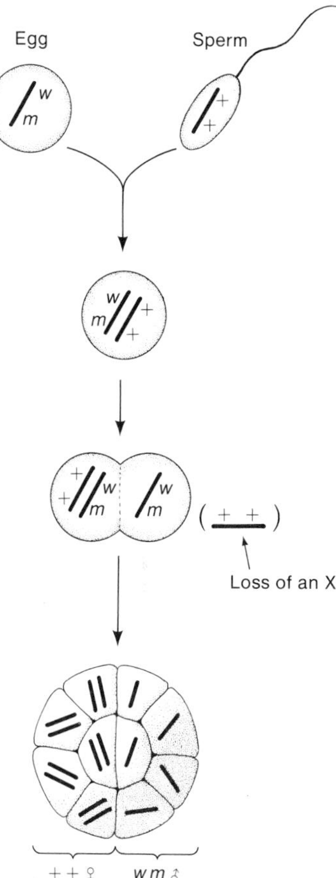

Figure 17-58. Production of a mosaic by the loss of an X chromosome shortly after fertilization in Drosophila. *If the XO cells of the gynandromorph carry recessive markers on the X, then the mutant phenotype is visible. Loss of the X can be induced in a number of ways, including the presence of certain mutations or by irradiation. (From A. M. Srb, R. D. Owen, and R. S. Edgar,* General Genetics, *2nd ed. Copyright © 1965 by W. H. Freeman and Company.)*

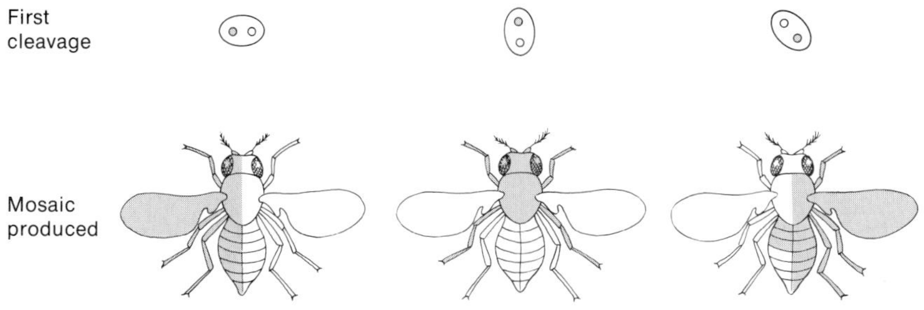

Figure 17-59. Orientation of the plane of the first nuclear division after fertilization. The spindle fibers can be tilted within the egg in any orientation with respect to the anterior-posterior pole of the cytoplasm. Because daughter nuclei tend to remain together and migrate to nearby locations at the egg surface, the orientation of the division plane determines the mosaic pattern observed in the adult (if the two nuclei produced from the first division are genetically different). (From S. Benzer, "Genetic Dissection of Behavior." Copyright © 1973 by Scientific American, Inc. All rights reserved.)

the dividing line between mutant and nonmutant cells determined at first cleavage is randomly distributed on the blastoderm.) The farther apart the positions of the cells in the blastoderm, the greater is the probability that the dividing line will fall between them so that they will develop into genetically different adult structures. Thus, the "distance" between blastoderm cells destined to become different adult structures can be quantified as the percentage of mosaics in which the two adult structures are genotypically different:

$$\text{“Distance between blastoderm cells”} = \frac{\text{Number of mosaics in which structures differ} \times 100}{\text{Total number of mosaics scored}}$$

We have no idea what these "distances" mean in real physical terms, but the same situation applied when linkage maps were first constructed. In this case, we develop a two-dimensional map because the spatial distribution is over the surface of a hollow spheroid. We can standardize the surface as that seen when the spheroid is cut along the axis of symmetry (Figure 17-60). Suppose that we map structures A and B as 10 units apart. We can introduce a second dimension by measuring their distances from a third structure, C. If A is 8 units from C, and B is 4 units from C, then we obtain the map shown in Figure 17-61. This point to which an adult structure maps on the blastoderm surface can be regarded as a **focus** from which the growth and movement of the structure proceeds during development and differentiation.

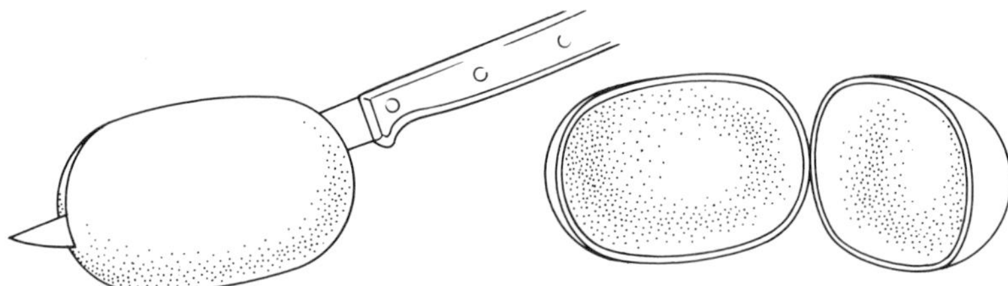

Figure 17-60. A hypothetical cut through the blastoderm to separate the right and left halves. The blastoderm fate maps are representations of the surface of one such half.

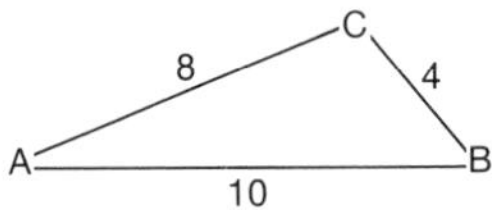

Figure 17-61. Embryonic map of cells destined to form adult structures A, B, and C. The map positions presumably represent the relative positions in the blastoderm of the cells from which these adult structures will eventually develop. The "distance" between blastoderm cells determined for different adult structures can be quantified as the percentage of mosaics in which two adult structures are genotypically different:

$$\frac{\textit{Number of mosaics in which structures are different}}{\textit{Total number of mosaics recorded}} \times 100$$

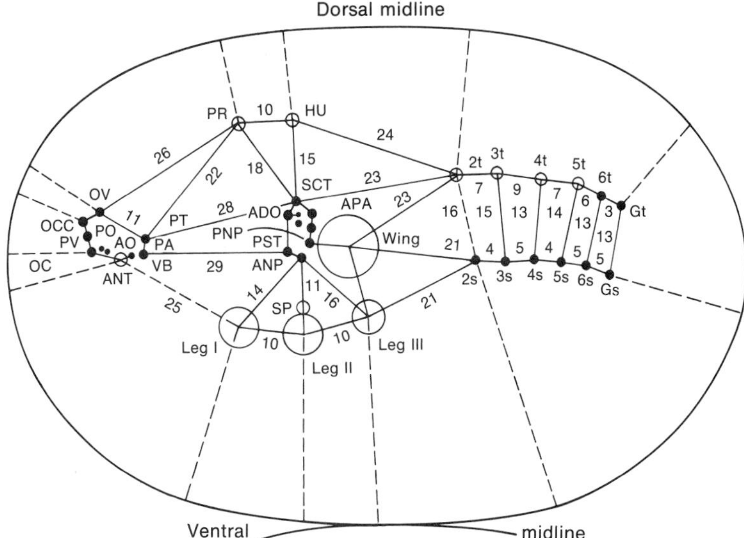

Figure 17-62. The embryonic fate map of adult structures. The map shows the foci for various structures and individual bristles. The distances between them are indicated in map units called **sturts.** *(From S. Benzer, "Genetic Dissection of Behavior." Copyright © 1973 by Scientific American, Inc. All rights reserved.)*

Using this map distance between embryonic foci for adult structures, Garcia-Bellido and Merriam were able to construct an embryonic "fate map" like that shown in Figure 17-62. Such maps can be made as refined as we desire by studying ever more detailed features of the adult structures. The same procedure can be used to construct detailed fate maps for each *disk* in its early form in the blastoderm. We assume that these maps have some congruence with the actual spatial distribution of determined cells in the blastoderm.

In 1972, Yasuo Hotta and Seymour Benzer took this analysis one step further by analyzing the foci pertinent to a mutant *behavioral* phenotype. If we have a recessive mutation that causes the legs of a fly to twitch when the fly is anesthetized, we can map the cells responsible for the twitch in relation to the external markers. Let's call the behavioral mutant *kic* (for *kicker*) and make a cross to generate $y\,kic/+\ +$ flies. A female that has no yellow tissue but exhibits leg kicking must have internal cells that are $y\,kic/\mathrm{O}$. We can calculate the distance from the focus for *kic* to the focus for the right front leg, for example, as

$$\frac{\text{Number of } y \text{ nonkicking legs} + \text{Number of } y^{+} \text{ kicking legs}}{\text{Total mosaics}} \times 100$$

where only the right front legs of mosaic flies are counted. Again, by triangulation with another focus (such as that for a thoracic bristle), we can map the focus for kicking.

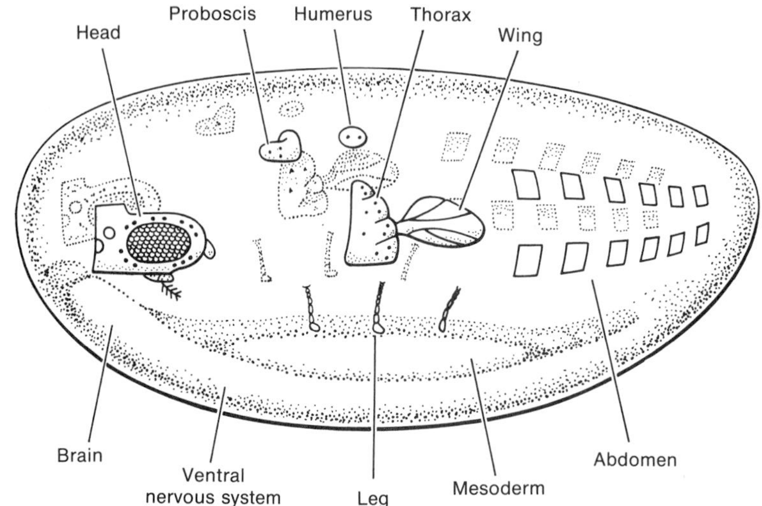

Figure 17-63. A redrawn version of the fate map in Figure 17-62, showing the adult fate of the cells that occupy specific regions of the embryo. Note that the anterior–posterior arrangement of adult structures is approximately retained in the arrangement of blastoderm cells. (From S. Benzer, "Genetic Dissection of Behavior." Copyright © 1973 by Scientific American, Inc. All rights reserved.)

Hotta and Benzer found that the mutants with presumed defects in leg muscles and nerves mapped near each other and near the leg foci of the structural fate map. Their positions correspond very well with cytological studies of the neural and muscular tissues developing from blastoderm cells. Such cytological studies provide the reference points for orientation of the fate map to the poles of the blastoderm. With all of this evidence, we can construct a hypothetical map of the locations in the blastoderm of the ancestors of the imaginal disks (Figure 17-63). Of course, the mapping procedure becomes much more complex in the cases of the many neurological mutants that involve simultaneous defects in several anatomical positions.

Message
The locations of cells destined to form particular adult structures can be mapped in relation to one another in the embryo by the use of mosaics.

Our brief discussion cannot do justice to the complexity and intricate beauty of developmental events that emerge in the study of embryology. We have seen, however, how the tools of genetic dissection can be combined with biochemical and cytological methods to further our understanding of the regulatory mechanisms for this complex developmental process.

Summary

Developmental geneticists seek an understanding of the mechanisms whereby a fertilized egg differentiates into the many cell types of a multicellular organism. Each somatic cell receives a complete complement of DNA and therefore should be totipotent. However, only a small fraction of the genome is transcriptionally active in any given cell, and differentiation is the consequence of transcription from different specific loci. Puff growth and regression in polytene chromosomes is a cytologically visible indicator of gene activity. Some of the complex specialized structures of differentiated cells assemble spontaneously from polypeptide products of gene translation. The cytoplasm of the oocyte is not homogeneous; as early cleavage divisions parcel out its components to daughter cells, cell differences arise in different parts of the embryo. The maternal genotype determines much of the molecular architecture of the oocyte cytoplasm and thus controls many aspects of the early events in development.

Sex determination and differentiation provide an illustration of the interaction between genes and environment during development. Hermaphroditic animals and plants have the sexual organs of both sexes in a single organism. In some organisms, sex is determined by factors other than genotypic differences. Even in organisms where the sex of an individual is determined primarily by sex chromosomes or alleles present at fertilization, the final phenotype realized can be altered by such factors as diet, hormones, or changes in the number of male-determining and female-determining loci.

The genetics of the immune system is another example of a highly specialized developmental process. The control of this system involves a number of genetic phenomena not yet observed in the study of other biological events. The acceptance or rejection of grafts between individuals is determined by a number of histocompatibility loci. If the alleles at these loci are similar in two individuals (as in identical twins or highly inbred animals), then grafts are accepted between them. The major histocompatibility genes that control rapid graft rejection are located at the *H-2* locus in mice and at *HLA* in humans.

The immune response involves recognition of a foreign antigen, synthesis of a specific antibody to the antigen, and the inactivation and removal of the foreign substance. The body contains thousands of clones of B lymphocytes, with each clone having an innate ability to recognize a particular antigen and synthesize the antibody to that antigen. When a B cell encounters the antigen to which it is sensitive, it differentiates into plasma cells that generate a massive amount of the corresponding antibody. When a T lymphocyte encounters the antigen to which it is sensitive, it attacks and destroys the foreign particle.

An antibody is an immunoglobulin, a glycoprotein whose major constituents are pairs of identical light (L) and heavy (H) polypeptide chains held together by disulfide bonds. Certain regions of these chains vary greatly in amino-acid sequence from one antibody to another, and these regions provide the highly specific binding sites for antigens. The variable (V) and constant (C) regions of the L and H chains are encoded in separate loci, but they are somehow joined in lymphocytes to form continuous transcription units. We do not yet know many of the details of the genetic mechanisms involved in producing variable regions and linking the V and C regions.

Drosophila has become a favorite organism for the study of development because of the extensive repertoire of techniques for its genetic manipulation. The two life stages of the fly (larva and adult) are anticipated in the embryo, where the developmental fates of cells are determined. Prospective larval cells begin to differentiate in the blastoderm, while cells destined to form adult structures proliferate as discrete imaginal disks that are already programmed to form specific adult parts when triggered by the hormone ecdysone. Serial transplantation of disks in adult hosts or homeotic mutations can alter the developmental potential of disk cells. Fate maps of disk cells can be constructed because pieces of disks differentiate into adult structures when injected into a host.

The induction of a mitotic crossover or chromosome loss generates a marked cell whose lineage can be traced as a clone differing in phenotype from the surrounding tissue. The later in development this mutation is generated, the smaller is the clone of marked tissue. Such studies show that the developmental fate of cells is progressively restricted within boundaries that delineate developmental compartments. As development progresses, new boundaries are added to cut existing compartments into smaller compartments whose fate is more specifically determined.

Descendants of genetically different nuclei created in the first or second cleavage tend to remain together at blastoderm formation. Thus, if the boundary between genetically different nuclei in a mosaic is randomly oriented in different embryos, then the "distance" between blastoderm cells destined to form different adult structures can be calculated. The closer the locations of two cells in the blastoderm, the smaller is the probability that they will be genetically different in any given mosaic. Hence the percentage of mosaic flies in which two adult structures are phenotypically different is a measure of the distance between the ancestral cells (foci) of these structures in the blastoderm. This method can be extended to map the embryonic foci of cells involved in production of mutant behavioral phenotypes. The resulting fate maps supplement those derived from cultures in vivo of disk fragments.

Problems

1. Review preceding chapters and prepare a list of all the ways in which variegation (sectoring) of biological tissue can occur.
2. a. Of what significance are phenocopies in medical genetics?
   b. How could phenocopies be used to study gene action?

c. How can an investigator determine whether an altered phenotype is due to a phenocopy or to a mutation?

3. Throughout history and mythology, there are stories of the birth of children to virgin mothers. Such parthenogenesis definitely does occur in many organisms. In theory, it could occur in mammals, including humans. Describe the mechanism that might be involved if any of the claims of virgin birth of a son are true.

4. Cells of organisms that are phylogenetically distantly related can be fused to form single mononucleate hybrids. For example, human–mouse, human–fish, human–bird, human–insect, and human–plant hybrid cells have been created. These hybrids even grow and divide.

   a. What conclusions can be drawn from these observations?

   b. When a chicken erythrocyte (which has a nucleus and actively synthesizes hemoglobin) is fused to a mouse fibroblast, some of the inactive chicken DNA becomes transcriptionally active, but hemoglobin synthesis is shut off. What conclusions can you draw from this observation?

5. In his book *In His Image: The Cloning of a Man*, David Rorvik claimed that a baby had been cloned from the cells of an elderly man. Few (if any) geneticists believe this story, but many scientists do feel that human cloning is a definite possibility in the future. Naturally occurring human clones (identical twins) do exist. How would the ability to produce human clones in the laboratory be useful in studying the relative importance of heredity and environment in determining the final phenotype of a human individual?

6. Figure 17-64 illustrates an experiment performed by Hämmerling in 1943 on two species of the unicellular alga *Acetabularia*. *A. mediterranea* has an intact, umbrella-like "hat," whereas the hat of *A. crenulata* is deeply indented. The nucleus is embedded in the base of the single cell in both species, and the hat is borne on a long cytoplasmic stem.

   a. The stem of an *A. mediterranea* individual is cut off just at the base, and the hat is removed by a cut at the top of the stem. The cut stem is grafted to the base of an *A. crenulata* individual, and a new hat forms (regenerates) on the top of the stem (right side of Figure 17-64). A reciprocal transplant of an *A. crenulata* stem to an *A. mediterranea* base also leads to regeneration of a new hat (left side of Figure 17-64). In each case, the phenotype of the new hat is consistent with that normal for the base to which the stem is transplanted rather than that normal for the stem from which the hat grows. On the basis of these results, discuss the cytoplasmic versus nuclear control of regeneration in *Acetabularia.*

   b. If the nucleus of an *Acetabularia* cell is removed and the hat is then cut off, the enucleated cell immediately regenerates a new hat. However, if the upper part of the stem is removed along with the hat, the enucleated cell does not regenerate a new hat. Assume that some "hat-forming substance" is responsible for regeneration. Discuss the origin, qualitative control, and distribution of this substance in the intact cell.

   c. If a stem with intact hat is transplanted from one species of *Acetabularia* to the base of another species, the hat retains its original character. How does this observation modify your answer to part b?

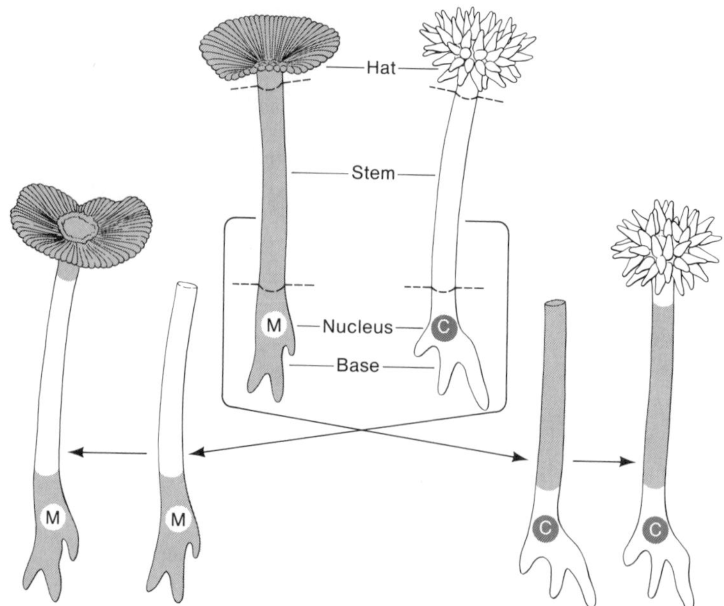

Figure 17-64. Grafting experiments in Acetabularia *have provided important information about the control of patterns of regeneration (see Problem 6). When a cut stem from* A. crenulata *(C) is grafted to a base from* A. mediterranea *(M), the stem regenerates a hat of the* A. mediterranea *type* (left). *In the reciprocal experiment, the regenerated hat again matches the phenotype of the base* (right). *(After Hämmerling,* Z. Abstg. Vererb. *81:114, 1943.)*

7. Boris Ephrussi and George Beadle have developed (and applied to good advantage) a technique for the study of genic effects on hormonelike materials in *Drosophila.* Figure 17-65 diagrams this technique and some of the results of its application. A piece of the larval disk that would later produce an adult eye is transplanted to a genetically different larva. The developmental interactions between host and transplant are then observed, particularly in terms of the color of pigment developed by the transplanted eye disk as the host larva matures.

   a. The reciprocal transplants shown in Figure 17-65a are typical of a large majority of *Drosophila* transplantation experiments. The larval disk from a wild-type fly develops the pigment color characteristic of its own genotype, even when its differentiation and pigment development occur in a white-eyed host. Similarly, an eye disk from a white-eyed larva develops according to its own genotype and is not influenced in any noticeable fashion by a wild-type host environment. Would you conclude that the white-eyed gene affects production of a circulating (hormonelike) material or that it acts directly in the developing eye tissue itself?

   b. The experiment shown in Figure 17-65b represents the kind of exception from which a good deal of information has been derived. Again, when wild-type eye disks are transplanted to either vermilion (*v*) or cinnabar (*cn*) host larvae, the

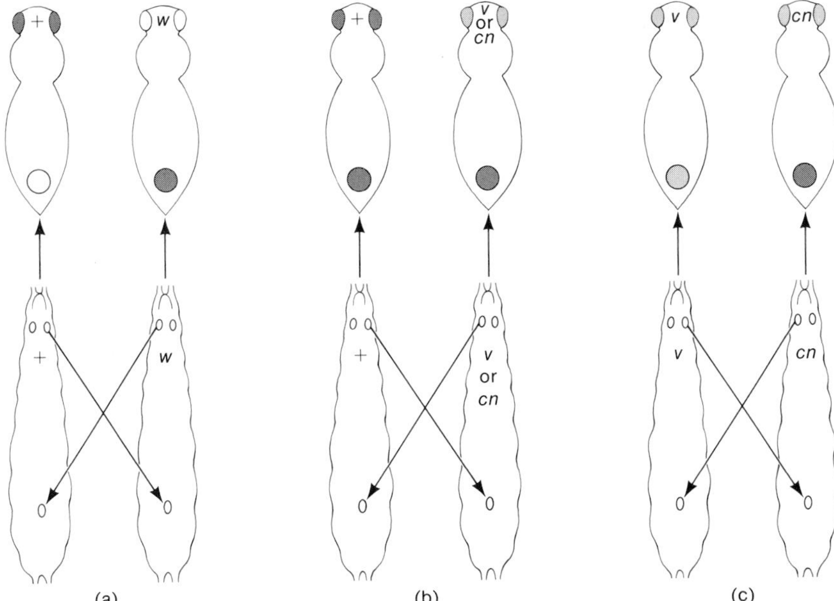

Figure 17-65. Transplants of larval tissue have elucidated some aspects of gene action in Drosophila *(see Problem 7). (After B. Ephrussi,* Quart. Rev. Biol. *17:329, 1942.)*

transplants develop autonomously into wild-type eyes. However, when vermilion or cinnabar disks are transplanted to wild-type larvae, the disks do not develop according to their own genotypes but rather produce wild-type eyes! Provide a detailed explanation of these experimental results, assuming that the body of the wild-type host is capable of providing the developing eye tissue with circulating or diffusing materials that compensate for the genetic blocks in vermilion and cinnabar flies.

c. Figure 17-65c shows the results of reciprocal transplants between vermilion and cinnabar larvae. Cinnabar disks developing in vermilion hosts maintain their cinnabar phenotype, but vermilion disks develop as wild type in cinnabar hosts. These genes are now known to be involved in control of sequential steps in a biochemical synthesis of hormonelike materials directly involved in production of eye pigment:

$$\text{tryptophane} \xrightarrow{v^+} \underset{\text{("}v^+\text{ substance)}}{\text{kynurenine}} \xrightarrow{cn^+} \underset{\text{("}cn^+\text{ substance)}}{\text{hydroxykynurenine}} \rightarrow \rightarrow \text{pigment}$$

Using this information, give a detailed explanation of *all* the transplantation results shown in Figure 17-65.

8. One theory of aging (senescence) assumes that randomly occurring somatic mutations accumulate over time in different cells within an individual, gradually disrupting the normal cellular processes necessary for life. Another theory suggests

that lifespan is a genetically determined character and that aging and death are simply the final stages of development and differentiation. These hypotheses are sufficiently specific to permit design of direct tests. Design experiments to test each of these theories. (NOTE: Choose your organism and the kinds of tests carefully.)

9. You observe that, during larval development of *Drosophila*, an enzyme specifically found in the salivary glands appears, increases, and disappears in exactly the same pattern as a specific puff on the polytene X chromosome.

   a. Does this observation prove that the enzyme is specified by the puff? Explain.

   b. Design experiments to test the hypothesis that the puff specifies the enzyme.

10. Figure 17-66 shows a hypothetical operon circuit in which the structural genes (*G*) adjacent to the operators (*O*) of two operons specify enzymes (E) that act by converting substrates (S) into products (P). Each product interacts with the repressor produced by the repressor gene (RG) to shut off the operator of the *other* operon. What would be the reaction of this system to variations in supplies of the substrates S_1 and S_2?

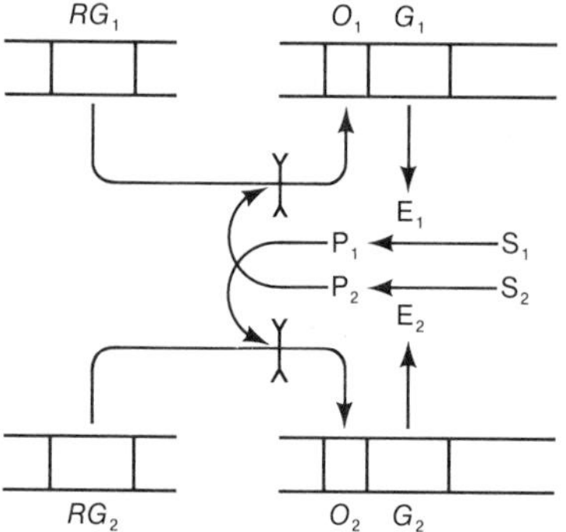

Figure 17-66.

11. As we have seen, the molecules in many complex structures (such as ribosomes and phage particles) can assemble spontaneously in a specific sequence. Design experiments to determine whether nucleosomal elements possess this capacity and whether the histones distinguish between DNAs.

12. The X-linked allele for glucose 6-phosphate deficiency (Gd^-) also confers resistance to malaria. You make a microscopic examination of the blood of a woman suffering from malaria. You observe that about one-half of the cells contain parasites, whereas one-half of the cells are unaffected. Explain this result.

13. Women who are XO have a distinctive phenotype that includes short stature, webbed necks, and sterility. Men who are XXY can also be recognized phenotypically by reduced axillary hair, enlarged breasts, and sterility. However, XYY males seem to be fertile and phenotypically normal. What does this observation indicate about sex-chromosome function?

14. Through some aberrant circumstance, chromosome doubling occurs in certain somatic cells in a triploid *Drosophila* female. Will these clones of hexaploid cells differentiate as "male" or as "female" tissue?

15. The addition of duplication fragments of X chromosomes into nuclei from *Drosophila* intersexes (2X:3A) shifts the balance of sexuality. Will this shift be in the direction of maleness or femaleness?

16. In *Drosophila*, an autosomal recessive *tra* acts only in females and transforms them into sterile males. What ratio of males and females would you expect among the progeny of a $+/tra$ ♀ × $+/tra$ ♂ cross?

17. There is a sex-linked gene in mice that causes muscular dystrophy. Alan Peterson fused embryos from the muscular dystrophic strain with wild-type embryos. Because the two strains also differ in enzyme patterns, he could determine the parental origin of particular cells in the tetraparental mice. He found that animals with muscle cells from the wild type and enervated by nerve cells from the dystrophic strain always have muscular dystrophy. However, in parabiosis studies (in which two animals are surgically bound together), attaching nerves from a dystrophic animal to the muscles of a wild-type mouse did not induce dystrophy in the muscles. Explain the significance of these results.

18. Jonathan Jarvik and David Botstein developed the following genetic method for determining the order of activity of different genes. Suppose that you obtain two different phage mutations (call them *A* and *B*) with an identical phenotype of incomplete head assembly. You recover a heat-sensitive mutation of *A* (the mutant dies at 40°C) and a cold-sensitive mutation of *B* (the mutant dies at 25°C); both mutants survive at 35°C. You construct a phage carrying both mutations ($A^{hs}\,B^{cs}$) and infect bacteria with it. You keep one culture of infected bacteria at 25°C for 10 minutes and then shift it to 40°C (call this the lo→hi culture). You keep another culture (hi→lo) at 40°C for 10 minutes and then shift it to 25°C. Interpret each of the following possible results of this experiment.

    a. Phages are released only in the hi→lo culture.

    b. Phages are released only in the lo→hi culture.

    c. No phages are released in either culture.

    d. Some phages are released in both cultures.

19. Bruce Ames found that over 85% of all known carcinogens can be detected by their action as mutagens. (It does not necessarily follow that a mutagen can confidently be predicted to be a carcinogen.) What does this result indicate about the basis for cancer?

20. In many forms of cancer, immunologically detectable changes occur at the surface of cancer cells. In many cases, the new antigens made by a cancer cell are identical to those found on normal embryonic cells. Propose an explanation for these observations.

21. Methylcholanthrene is a potent tumor-causing chemical in mice. It is also a potent mutagen. If a methylcholanthrene-induced tumor T_A is removed from mouse A, the mouse is "cured." The T_A cells can be kept alive in another mouse. If the T_A cells are reintroduced into the cured mouse, the mouse is seen to be immune (no tumors develop), although tumors can be induced by injection of T_A cells into mouse B, which has not previously been exposed to T_A cells. However, the mice surgically cured of tumor T_A are not immune to T_C cells removed from a methylcholanthrene-induced tumor in mouse C; such an injection induces a T_C-type tumor in mouse A. Propose an explanation for these observations in terms of gene-protein

relations, bearing in mind that cell-surface proteins can act as immunological antigens.

22. A degenerative kidney disease develops in a boy, and he requires a kidney transplant. He and his parents are analyzed to determine the genotypes of the *B* and *D* distrons in the *HLA* locus. The boy is *D-1 D-2 B-7 B-6*; his mother is *D-1 D-1 B-7 B-9*; and his father is *D-2 D-8 B-6 B-7*. What is the probability that the boy's brother can act as an organ donor (considering only these two histocompatibility loci)?

23. Assume that the number of genes coding for the variable portions of immunoglobulins is small. Propose a hypothesis to explain how the enormous range of antibody specificities could be generated. Outline a test for your hypothesis.

24. In *Drosophila,* heterozygotes for dominant temperature-sensitive lethal mutations are viable and fertile at the permissive temperature, but they die at the restrictive temperature. What could be the molecular basis for such dominant lethality?

25. Let us assign numbers to each leg of *Drosophila* as follows: legs 1, 2, and 3 are the front, mid, and hind legs on the left side, respectively, and legs 4, 5, 6 are the front, mid, and hind legs on the right side. A fly normally walks by moving legs 1, 3, and 5 together and then moving legs 2, 4, and 6 together. A mutation called *wobbly* causes the fly to get its mid legs tangled with its front or its hind legs. What could be wrong with the mutant, and how would you study it?

26. You are a molecular biologist studying the properties of muscle proteins in *Drosophila.* Design an experimental procedure for the detection and recovery of appropriate mutants. (This experiment has not been attempted, so you may be able to propose a clever scheme that is worth trying. NOTE: What would be the phenotype of a muscle mutant? Wouldn't it be lethal?)

27. The sex-linked dominant mutation *Hyperkinetic-1* (Hk^1) causes shaking of a fly's legs while the fly is etherized. Seymour Benzer and Yasuo Hotta generated mosaic flies with the tissue gentoypes $y\ Hk^1/+\ +$ and $y\ Hk^1/O$. They scored 600 fly sides for leg shaking and mutant cuticle tissue. Table 17-2 summarizes their results. Draw a fate map of the foci for Hk^1-caused shaking, the three legs, the antenna, and the humeral bristle.

Table 17-2.

Structure	Color of cuticle tissue	Shaking of leg 1	
		Normal	Mutant
Leg 1	Wild type	277	50.5
	Yellow	33	223.5
Leg 2	Wild type	261.5	69
	Yellow	44	215.5
Leg 3	Wild type	250.5	82.5
	Yellow	46.5	215.5
Antenna	Wild type	253	82
	Yellow	84.5	179.5
Humeral bristle	Wild type	241	94
	Yellow	66.5	197.5

28. Which of the following mRNA types will compete most effectively with the hybridization of radioactive liver mRNA to total cell DNA? (a) liver mRNA; (b) lung mRNA; (c) kidney mRNA; (d) brain mRNA; (e) muscle mRNA.

29. Eukaryotic cells can be lysed without breaking down the nuclear membrane. This technique permits separation of cytoplasmic from nuclear material. In competition studies, unlabeled nuclear RNA is found to compete with the binding of labeled cytoplasmic RNA to the DNA. However, unlabeled cytoplasmic RNA fails to interfere with the binding of a considerable amount of labeled nuclear RNA to the DNA. What do these studies show? Can you suggest a possible biological interpretation of these observations?

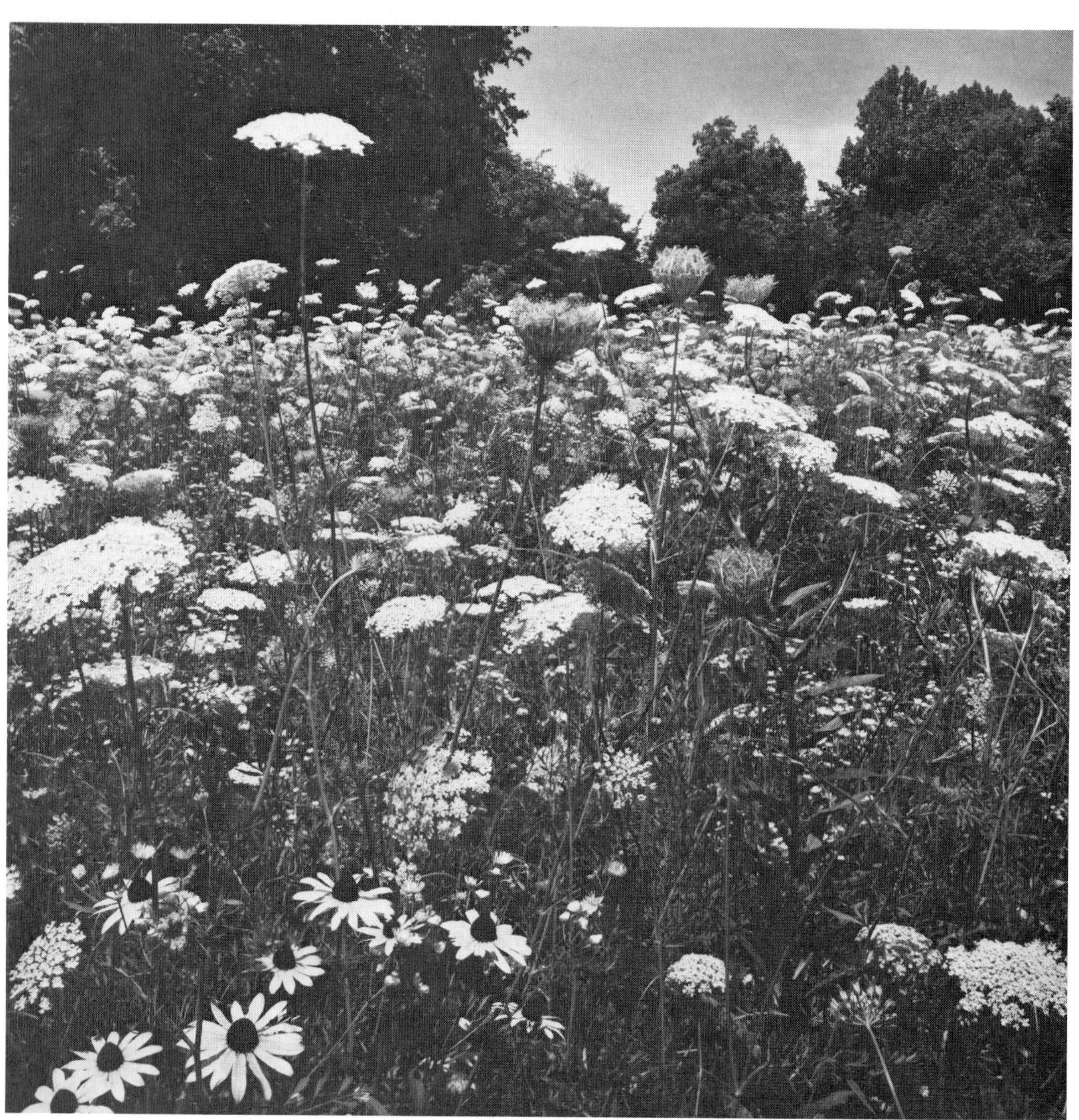

*A population of Queen Anne's Lace (*Daucus carota*), illustrating quantitative variation in nature. (Copyright © Grant Heilman.)*

18

Quantitative Genetics

The possibility of working out the laws of transmission genetics and of the molecular genetics underlying them has depended critically on the existence of a special kind of phenotypic trait. These are the traits that allow classification of individuals into discrete and unambiguous phenotypic classes. They may be the tall and short plants or red and white flowers of Mendel's peas, or the lactose fermenters and nonfermenters of modern microbial genetics. In all such cases, the phenotypic effect of an allelic substitution at some locus is sufficiently drastic that the substitution produces a clear effect. There is a uniquely distinguishable phenotype for each such genotype, and there is only a single genotype for each phenotype. At most, because of dominance, two genotypes may produce the same phenotype, but this confusion can be sorted out by a simple genetic cross. Such a one-to-one relation between genotype and phenotype is necessary for basic genetic experiments, because usually we can identify (and therefore study) the genotype only through the phenotype. For the most part, then, the study of genetics presented in the previous chapters is the study of allelic substitutions causing *qualitative* differences in phenotype.

However, most of the actual variation among organisms is not qualitative but quantitative. Wheat plants in a cultivated field or wild asters at the side of the road do not sort themselves neatly into two categories, "tall" and "short," any more than humans are neatly sorted into "black" and "white." Height, weight, shape, color, metabolic activity, reproductive rate, and behavior are characteristics that vary more or less continuously over a range. Even when the character is intrinsically countable (such as eye-facet or bristle number in *Drosophila*), the number of countable classes may be so large that the variation is nearly continuous. Even if we consider extreme individuals—say, a corn plant 8 feet tall and one 3 feet tall—a cross between them will not reproduce Mendel's result. Such a cross will produce plants about 6 feet tall, with some clear variation among sibs. The F_2 from selfing the F_1 will not fall into two or three discrete height classes in ratios 3:1 or 1:2:1. Instead, the F_2 will be continuously distributed in height from one parental extreme to the other. It is important to realize that, when we depart from the level of primary gene products, this behavior of crosses is not an exception but is the rule for most characters in most species. In general, size, shape, color, physiological activity, and behavior do not behave in a simple way in crosses.

The fact that most phenotypic characters vary continuously does not mean that their variation is the result of some genetic mechanisms different from the Mendelian genes we have been dealing with. The continuity of phenotype is a result of two phenomena. First, each genotype does not have a single phenotypic expression but a norm of reaction (see Chapter 1) that covers a wide phenotypic range. As a result, the phenotypic differences between genotypic classes become blurred, and we are not able to assign a particular phenotype unambiguously to a particular genotype. Second, there may be many segregating loci whose alleles make a difference to the phenotype being observed. Suppose, for example, that there are five equally important loci affecting some trait and that each has two alleles—call them + and −. For simplicity, suppose that there is no dominance and that a + allele adds 1 unit to the trait whereas a − allele adds nothing. There are $3^5 = 243$ different possible genotypes, ranging from $\frac{+}{+}\frac{+}{+}\frac{+}{+}\frac{+}{+}\frac{+}{+}$ through $\frac{+}{-}\frac{+}{-}\frac{+}{-}\frac{+}{-}\frac{+}{-}$ to $\frac{-}{-}\frac{-}{-}\frac{-}{-}\frac{-}{-}\frac{-}{-}$, but there will only be 11 phenotypic classes (10, 9, 8, ..., 0), because many of the genotypes will have the same numbers of + and − alleles. For example, although there is only one genotype with 10 + alleles and therefore an average phenotypic value of 10, there are 51 different genotypes with 5 + alleles and 5 − alleles. These include a single pentuple heterozygote $\frac{+}{-}\frac{+}{-}\frac{+}{-}\frac{+}{-}\frac{+}{-}$, 20 different triple heterozygotes, such as $\frac{+}{-}\frac{+}{-}\frac{+}{-}\frac{+}{+}\frac{-}{-}$ and $\frac{+}{+}\frac{+}{-}\frac{-}{-}\frac{+}{-}\frac{+}{-}$, and 30 different single heterozygotes, such as $\frac{+}{+}\frac{+}{+}\frac{-}{-}\frac{+}{-}\frac{-}{-}$ and $\frac{+}{-}\frac{+}{+}\frac{+}{+}\frac{-}{-}\frac{-}{-}$. Thus, many different genotypes may have the same average phenotype. At the same time, because of environmental variation, two individuals of the same genotype may not have the same phenotype. This lack of a one-to-one correspondence between genotype and phenotype obscures the underlying Mendelian mechanism.

If we cannot directly study the behavior of the Mendelian factors controlling such traits, what then can we learn about their genetics? The following questions that can be asked about the genetics of quantitative variation are those for which some kind of experimental answers can be given with current experimental techniques.

1. Is the observed variation in the character influenced *at all* by genetic variation? Are there alleles segregating in the population that produce some differential effect on the character, or is all the variation simply the result of environmental variation and developmental noise (see Chapter 1)?
2. If there is genetic variation, what are the norms of reaction of the various genotypes?
3. How important is genetic variation as a source of total phenotypic variation? Are the norms of reaction and the environments such that nearly all the variation is a consequence of environmental difference and developmental instabilities, or does genetic variation predominate?
4. Are many loci (or only a few) varying with respect to the character? How are they distributed over the genome?
5. How do the different loci interact with each other in influencing the character? Is there dominance or epistasis?
6. Is there any nonnuclear inheritance—for example, any maternal effect?

The precision with which these questions can be framed and answered varies greatly. On the one hand, it is relatively simple in experimental organisms to determine whether there is any genetic influence at all, but extremely laborious experiments are required to localize the genes, even approximately. On the other hand, in humans it is extremely difficult to answer even the question of the presence of genetic influence for most traits because of the near impossibility of separating environmental from genetic effects in an organism that cannot be manipulated experimentally. As a consequence, we know a relatively large amount about the genetics of bristle number in *Drosophila* but essentially nothing about the genetics of any complex human trait, except that some (such as skin color) are influenced by genes whereas others (such as the specific language spoken) are not.

Some Basic Statistical Notions

In order to discuss the answers to these questions about the most common kinds of genetic variation, we must introduce a number of statistical tools that are essential in the study of quantitative genetics.

Distributions

The outcome of a cross for a Mendelian character is described in terms of the proportion of the offspring falling in each of several distinct phenotypic classes, or often simply by the presence or absence of a class. For example, a cross between a red-flowered and a white-flowered plant might be expected to yield all red-flowered plants, or, if it were a backcross, 1/2 red-flowered and 1/2 white-flowered plants. However, we require a different mode of description for quantitative characters. The basic concept is that of the **statistical distribution.** If the heights of a large number of male undergraduates are measured to the nearest 5 cm, they will vary (say, between 145 cm and 195 cm), but many more individuals will fall in the middle categories (say, between 170 and 180 cm) than at the extremes.

Representing each measurement class as a bar, with its height proportional to the number of individuals in each class, we can graph the result as in Figure 18-1a. Such a graph of numbers of individuals observed against measurement class is a **frequency histogram.** Suppose now that the number of individuals measured is multiplied by five and that each individual is measured to the nearest centimeter. The classes in Figure 18-1a are now subdivided to produce a histogram like that in Figure 18-1b. If we continue this process, refining the measurement but proportionately increasing the number of individuals measured, the histogram eventually takes on the continuous appearance of Figure 18-1c, which is the **distribution function** of heights in the population.

Of course, this continuous curve is an idealization because no measurement can be taken with infinite accuracy or on an unlimited number of individuals. Moreover, the measured variate itself may be intrinsically discontinuous because it is the count of some number of discrete objects, such as eye facets or bristles. It is sometimes convenient, however, to develop concepts using this slightly idealized picture as a shorthand for the more cumbersome observed frequency histogram of Figure 18-1a. One should not forget, however, that the distribution function is indeed an idealization.

The Mode. Most distributions of phenotypes look roughly like that of Figure 18-1. There is a single most frequent class, the **mode,** near the middle of the distribution, with frequencies decreasing on either side. There are exceptions to this pattern, however. Figure 18-2a shows the very asymmetric distribution of seed weights in the plant *Crinum longifolium.* Figure 18-2b shows a **bimodal** (two-mode) distribution of larval survival probabilities for different second-chromosome homozygotes in *Drosophila willistoni.*

A bimodal distribution may indicate that the population being studied could better be considered as a mixture of two populations, each with its own mode. For example, suppose we sample 100 undergraduates and measure their heights. These 100 individuals will include some males and some females. Thus, our sample is really a mixture of a sample from the population of male undergraduates and one from the population of female undergraduates, even though we

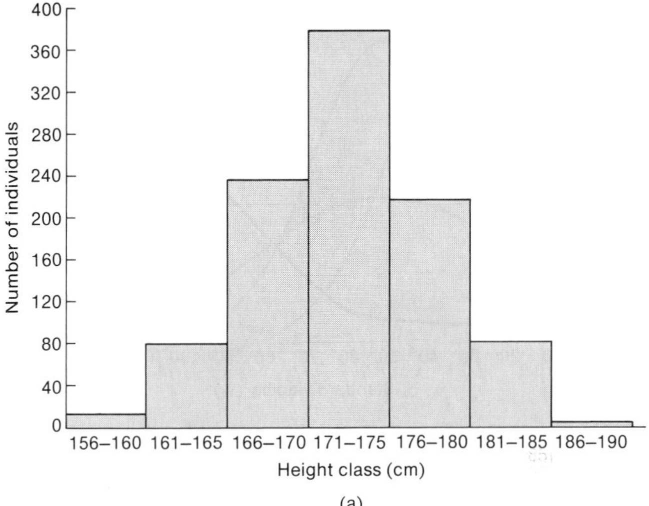

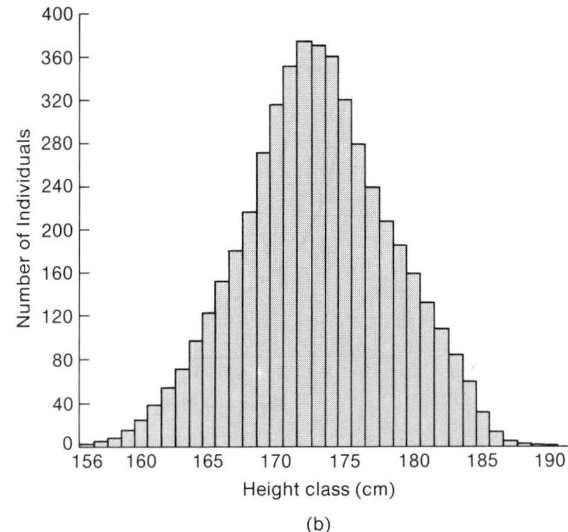

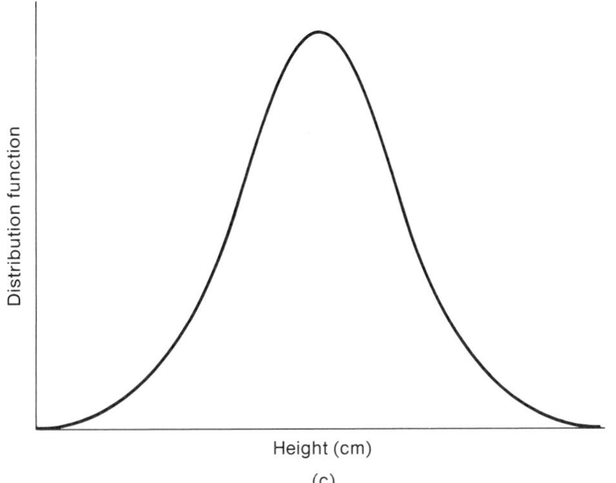

Figure 18-1. Frequency distributions for height of males. (a) A histogram with 5 cm class intervals. (b) A histogram with 1 cm class intervals. (c) The limiting continuous distribution.

have chosen the group in a single operation of sampling. In Figure 18-2b, the lefthand mode probably represents single severe point mutations that are extremely deleterious when homozygous, whereas the righthand mode is part of the distribution of "normal" viability. If the heights of both male and female undergraduates had been plotted in Figure 18-1, the distribution would have been bimodal, because the heights of females are distributed around a mode at a value considerably smaller than the mode of the males' heights.

The Mean. Complete information about the distribution of a phenotype in a population can be given only by specifying the frequency of each measured

Because the variance has squared units (square centimeters, for example), it is common to take the square root of variance, which then has the same units as the measurement itself. This square-root measure of variation is called the **standard deviation** of the distribution:

$$\text{standard deviation} = s = \sqrt{\text{variance}} = \sqrt{s^2}$$

Figure 18-3 shows two distributions having the same mean but different standard deviations (curves A and B) and two distributions having the same standard deviation but different means (curves A and C).

The mean and variance of a distribution do not describe it completely, of course. They will not distinguish a symmetrical distribution from an asymmetrical one, for example. One can even construct symmetrical distributions that have the same mean and variance but still have somewhat different shapes. Nevertheless, for the purposes of dealing with most quantitative genetic problems, the mean and variance suffice to characterize a distribution.

Correlation

Another statistical notion that we need is that of association or **correlation** between variables. Because of complex paths of causation, many variables in nature vary together but in an imperfect or approximate way. Figure 18-4a is an example, showing the lengths of two particular teeth in several individual specimens of a fossil mammal, *Phenacodus primaevis.* The longer an individual's first lower molar is, the longer its second molar is, but the relation is imprecise. Figure 18-4b shows that the total length and tail length of individual snakes (*Lampropeltis polyzona*) are quite closely related to each other, whereas Figure 18-4c shows that length and the number of caudal (tail) scales seem to have no relation at all.

The usual measure of the precision of a relationship between two variables x and y is the **correlation coefficient,** r_{xy}. It is calculated from the product of the deviation of each observation of x from the mean of the x values and the deviation of each observation of y from the mean of the y values, a quantity called the **covariance** of x and y:

$$\text{cov } xy = \frac{(x_1 - \bar{x})(y_1 - \bar{y}) + (x_2 - \bar{x})(y_2 - \bar{y}) + \cdots + (x_N - \bar{x})(y_N - \bar{y})}{N}$$

$$= \frac{1}{N} \sum (x_i - \bar{x})(y_i - \bar{y})$$

$$\text{correlation} = r_{xy} = \frac{\text{cov } xy}{s_x \times s_y}$$

This formula for the covariance is rather awkward for computation because it requires subtracting every value of x and y from the respective means, $\bar{x}$ and

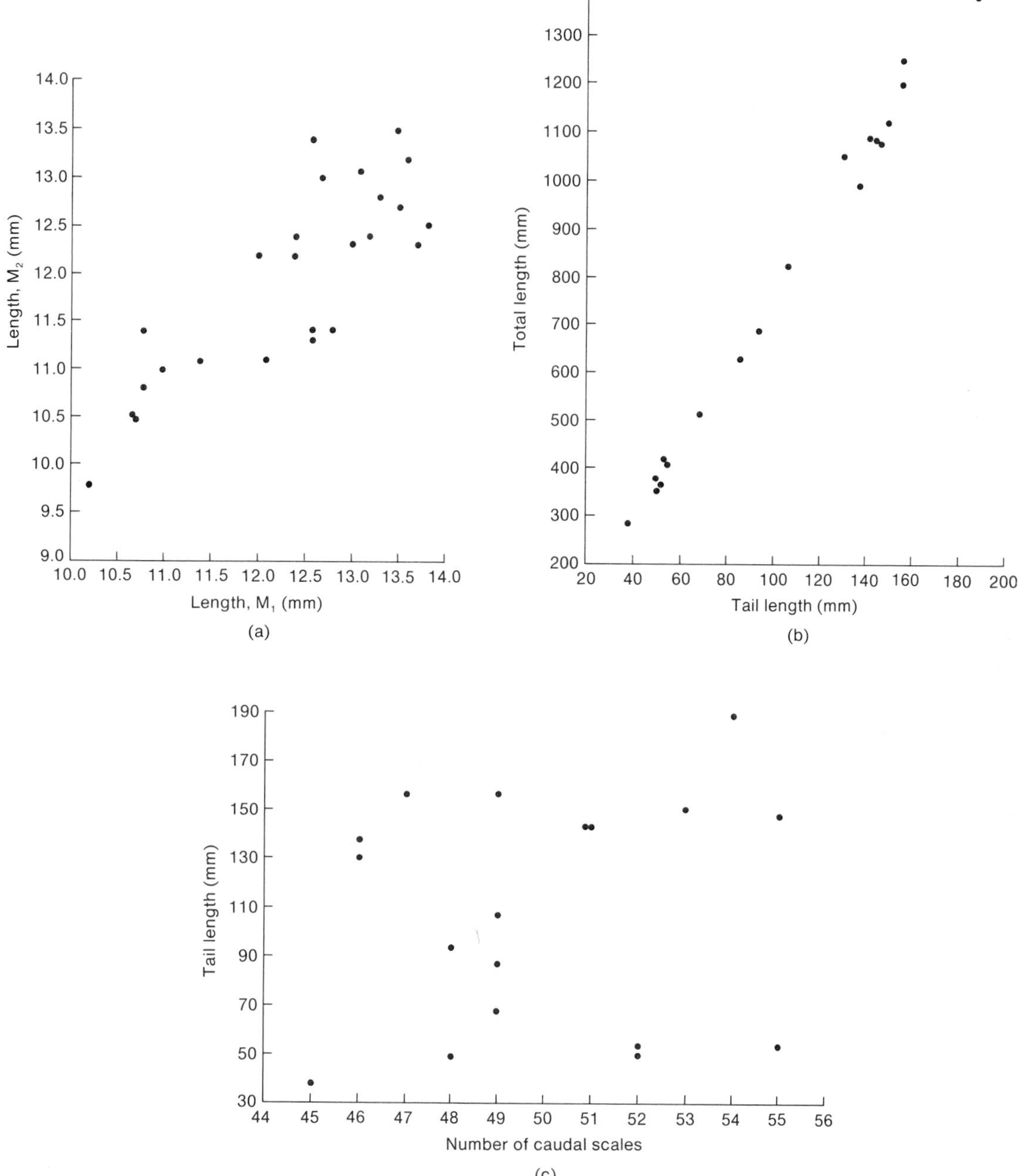

Figure 18-4. Scatter diagrams of relationships between pairs of variables. (a) Relationship between the lengths of the first and second lower molars, M_1 and M_2, in an extinct mammal Phenacodus. *Each point gives the two measurements, M_1 and M_2, for one individual. (b) Tail length and body length of 19 individuals of a snake* Lampropeltis polyzona. *(c) Number of caudal scales and tail length related in the same 19 snakes.*

$\bar{y}$. A formula that is exactly algebraically equivalent, but makes computation easier, is

$$\text{cov}\ xy = \frac{1}{N}\sum x_i y_i - N\overline{xy}$$

In the formula for correlation, the products of the deviations are divided by the product of the standard deviations of x and y (s_x and s_y). This normalization by the standard deviations has the effect of making r_{xy} a dimensionless number, independent of the units in which x and y are measured. So defined, r_{xy} will vary from -1, signifying a perfectly linear negative relation between x and y, to $+1$ for a perfectly linear positive relation. If $r_{xy} = 0$, there is no linear relation between the variables. It is important to notice, however, that sometimes when there is no *linear* relation between two variables, if there is a regular *nonlinear* relationship between them, one variable may be perfectly predicted from the other. Consider, for example, the parabola shown in Figure 18-5. The values of y are perfectly predictable from the values of x, yet $r_{xy} = 0$, because, on the average over the whole range of x values, larger x values are not associated with either larger or smaller y values. The data in Figures 18-4a, b, and c have r_{xy} values of 0.82, 0.99, and 0.10, respectively.

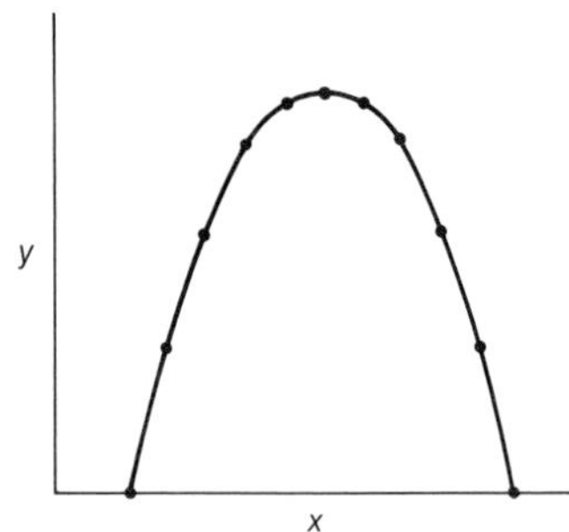

Figure 18-5. A parabola. Each value of y *is perfectly predictable from the value of* x, *but there is no linear correlation.*

In these examples, we have described correlations as the relation between a pair of measurements taken on the same individual, but a single measurement taken on pairs of individuals is also a subject for correlation analysis. Thus we can determine the correlation between the height of a parent (x) and that of an offspring (y), or the heights of an older sib (x) and a younger sib (y). This use of correlation is directly relevant to the problems of quantitative genetics. Figure 18-6 is an example showing the relation between the wing length of an offspring and the average wing length of its two parents (midparent value) in *Drosophila.*

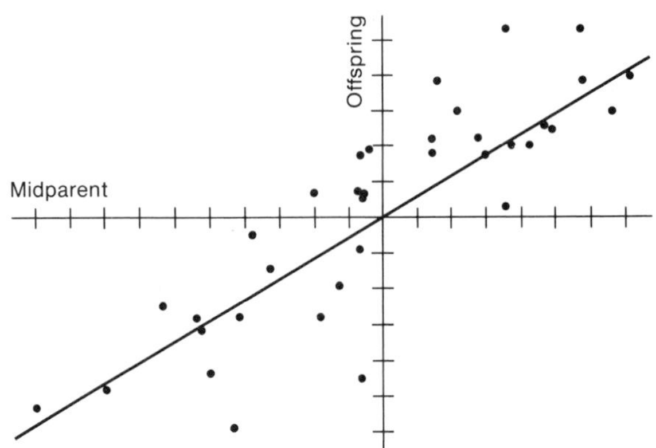

Figure 18-6. Relation between the wing lengths of individual Drosophila *and the mean wing lengths of their two parents. (From D. Falconer,* Quantitative Genetics, *Longman Group Limited. Copyright © 1981.)*

Correlation and Identity. It is important to notice that correlation is not the same thing as identity. Values can be perfectly correlated without being equal. The variables x and y in the pairs

x	y
1	22
2	24
3	26

are perfectly correlated ($r = +1.0$), although each value of y is 20 units greater than the corresponding value of x. Two variables are perfectly correlated if, for a unit to increase in one, there is a constant increase in the other (or decrease if r is negative). The importance of the difference between correlation and identity arises when we consider the effect of environment on heritable characters. Parents and offspring could be perfectly correlated in some trait, yet, because of an environmental difference between generations, every child could be larger than its parents. This phenomenon appears in adoption studies, where children may be correlated with their biological parents, yet, because of a change in social situation, may be on the average quite different from the parents.

Samples and Populations

We have described the distributions and some statistics of particular assemblages of individuals that have been collected in some experiments or sets of observations. For some purposes, however, we are not really interested in the particular 100 undergraduates or 27 snakes that have been measured. We are interested rather in a wider world of phenomena, of which those particular individuals are representative. Thus, we might want to know the average height of undergraduates *in general,* or the average seed weight *in general* of plants of the species *Crinum longifolium.* That is, we are interested in the characteristics of a *universe,* of which our small collection of observations is only a *sample.* The characteristics of any particular sample are, of course, not identical with those of the universe but will vary from sample to sample. Two samples of undergraduates drawn from the universe of all undergraduates will not have exactly the same mean, $\bar{x}$, or variance, s^2, nor will these values for a sample typically be exactly equal to the mean and variance of the universe. We distinguish between statistics of a sample and the values in the universe by using Roman letters, such as $\bar{x}$, s^2, and r, for sample values and Greek letters μ (for the mean), σ^2 (for the variance), and ρ (for the correlation) for the values in the universe.

The sample mean $\bar{x}$ is an approximation to the true mean μ of the universe. It is a statistical *estimate* of that true mean. So, too, s^2, s, and r are estimates of σ^2, σ, and ρ in the universe from which a sample has been taken.

If we are interested in a particular collection of individuals not for its own sake but as a way of getting information about a universe, then we want the sample statistics such as $\bar{x}$ and s^2 to be good estimates of the true values of μ, σ^2, and so on. There are many criteria of what a "good" estimate is, but one that seems clearly desirable is that, if we take a very large number of samples, then the average value of the estimate over these samples should be the true value in the universe. That is, the estimate should be *unbiased*. It turns out that $\bar{x}$ is indeed an unbiased estimate of μ. If a very large number of samples are taken and $\bar{x}$ is calculated in each one, then the average of these $\bar{x}$ values will be μ. Unfortunately, s^2, as we have defined it, is not an unbiased estimate of σ^2. It tends to be a little too small, so that the average of many s^2 values is less than σ^2. The amount of bias is precisely related to the size N of each sample, and it can be shown that $[N/(N-1)]s^2$ is an unbiased estimate of σ^2. Thus, whenever one is interested in the variance of a set of measurements, not as a characteristic of the particular collection but as an estimate of a universe from which the sample has been taken, then the appropriate quantity to use is $[N/(N-1)]s^2$ rather than s^2 itself. You may note that this new quantity is equivalent to dividing the sum of squared deviations by $N-1$ instead of N in the first place. That is,

$$\left(\frac{N}{N-1}\right)s^2 = \left(\frac{N}{N-1}\right)\frac{1}{N}\sum\left(x_i - \bar{x}\right)^2 = \frac{1}{N-1}\sum\left(x_i - \bar{x}\right)^2$$

It is this latter quantity that is often called the "sample variance," but that term is really not correct. The sample variance is s^2, whereas $[N/(N-1)]s^2$ is an adjustment to the sample variance to make it an unbiased estimate of σ^2. Which of the two quantities is to be used depends on whether one is interested primarily in the sample or in the universe.

All of these considerations about bias also apply to the sample covariance. In the formula for the correlation coefficient, however, the factor $N/(N-1)$ appears in both the numerator and denominator and cancels out, so we can ignore it for purposes of computation.

Genotypes and Phenotypic Distribution

Using the concepts of distribution, mean, and variance, we can understand the difference between quantitative and Mendelian genetic traits. Suppose that a population of plants contains three genotypes, each of which has some differential effect on growth rate. Further, suppose that there is some environmental variation from plant to plant because of inhomogeneity in the soil in which the population is growing and that there is some developmental noise (see Chapter 1). For each genotype separately, there will be a distribution of phenotypes whose mean and standard deviation depend on the genotype and the set of environments. We may suppose that they look like the three height distributions in Figure 18-7a. Suppose, finally, that the population consists of a mixture of the three genotypes but in unequal proportions, 1:2:3. Then the phenotypic distribution of individuals in the population as a whole will look like the solid line in Figure 18-7b, which is the result of summing the three underlying separate genotypic distributions, weighted by their frequencies in the popula-

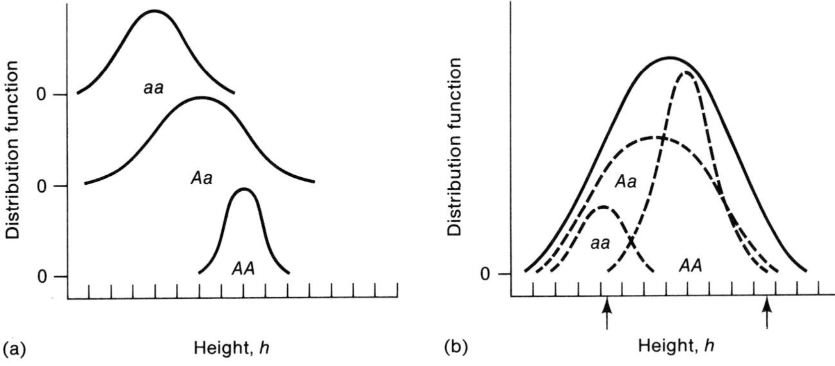

Figure 18-7. (a) Phenotypic distributions of three genotypes. (b) A population phenotypic distribution results from mixing individuals of the three genotypes in proportions 1:2:3.

tion. The mean of this total distribution is the average of the three genotypic means, weighted by the frequencies of the genotypes in the population. The variance of the total distribution is produced partly by the environmental variation within each genotype and partly by the slightly different means of the three genotypes.

Two features of the total distribution are noteworthy. First, there is only a single mode. Despite the existence of three separate genotypic distributions underlying it, the population distribution as a whole does not reveal the separate modes. Second, any individual whose height lies between the two arrows could have come from any one of the three genotypes, because they overlap so much. The result is that we cannot carry out any simple Mendelian analysis. For example, suppose the three genotypes are in fact the two homozygotes and the heterozygote for a pair of alleles at a locus. Let *aa* be the short homozygote and *AA* the tall one, with the heterozygote being intermediate in height. Because there is so much overlap of the phenotypic distribution, we cannot know to which genotype a given individual belongs. Conversely, if we cross two individuals that happen to be a homozygote *aa* and a heterozygote *Aa*, the offspring will not fall into two discrete classes in a 1:1 ratio but will cover almost the entire range of phenotypes smoothly. Thus, we could not know that the cross had in fact been *aa* × *Aa* and not *aa* × *AA* or *Aa* × *Aa*.

If the hypothetical plants of Figure 18-7 are grown in a stress environment, but care is taken that each plant has the same environment, then the picture changes. The phenotypic variance of each separate genotype is reduced because all the plants are grown under identical conditions, but at the same time the differences between genotypes may become greater because the small differences in physiology may be exaggerated under stress. The result (Figure 18-8) is a separation of the population as a whole into three nonoverlapping phenotypic distributions, each characteristic of one genotype. We could now carry out a

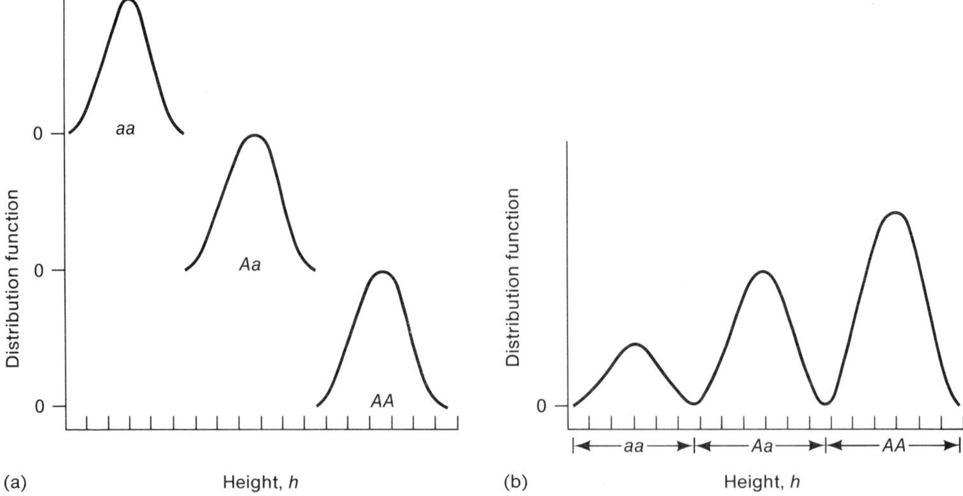

Figure 18-8. When the same genotypes as those in Figure 18-7 are grown in carefully controlled stress environments, the result is a smaller phenotypic variation in each genotype and greater difference between genotypes.

perfectly conventional Mendelian analysis of plant height. A "quantitative" character has been converted into a "qualitative" one! This conversion has been accomplished by making the differences between the means of the genotypes large as compared with the variation within genotypes.

Message
A quantitative character is one for which the average phenotypic differences between genotypes are small as compared with the variation between individuals within genotypes.

It is sometimes said that continuous variation in a character is a consequence of a large number of segregating genes that influence the measurement, so that continuous variation is to be taken as *prima facie* evidence for multigene control. But, as we have just shown, that is not necessarily true. If the difference between genotype means is small compared to the environmental variance, then the simple one gene–two allele case results in continuous variation.

Of course, if the range of a character is limited, and if there are many segregating loci influencing it, then we expect the character to show continuous variation, because each allelic substitution must account for only a small difference in the trait. This **multiple-factor hypothesis** (that large numbers of genes, each with a small effect, are segregating to produce quantitative variation) has long been the basic model of quantitative genetics. A special name, **polygenes,** has been coined for these factors of small-but-equal effect as opposed to the genes of simple Mendelian analysis.

It is important to remember, however, that the *number* of segregating loci that influence a trait is not what separates quantitative and qualitative characters. Even in the absence of large environmental variation, a very few genetically varying loci will produce variation that is indistinguishable from the effect of many loci of small effect. As an example, we can consider one of the earliest experiments in quantitative genetics, that of Wilhelm Johannsen on pure lines. Johannsen produced, by inbreeding, 19 homozygous lines of bean plants from an originally genetically heterogeneous population. Each line had a characteristic average seed weight ranging from 0.64g per seed for the heaviest line down to 0.35g per seed for the lightest. It is by no means clear that all these lines were genetically different. For example, five of the lines had seed weights of 0.450, 0.453, 0.454, 0.454, and 0.455 grams, but let us take the most extreme position—that they *were* all different. If there are 5 loci, each with 3 alleles, in the original population, then there would be $3^5 = 243$ different possible homozygous lines that could be created, and the chance that 19 chosen at random would be all different is about 50%.

Message
The critical difference between Mendelian and quantitative traits is not the number of segregating loci but the size of phenotypic differences between genotypes as compared with the individual variation within genotypic classes.

Norm of Reaction and Phenotypic Distribution

The phenotypic distribution of a trait, as we have seen, is a function of the average differences between genotypes and of the variation between genotypically identical individuals. But both of these are in turn functions of the environments in which the organisms develop and live. For a given genotype, each environment will result in a given phenotype (for the moment, ignoring developmental noise). Then a *distribution of environments* will be reflected biologically as a *distribution of phenotypes.* The way in which the environmental distribution is transformed into the phenotypic distribution is determined by the norm of reaction (Figure 18-9). The horizontal axis is environment (say, temperature) and the vertical axis is phenotype. The distribution on the horizontal axis represents the relative frequency of different environments. For example, most of these individuals developed at a temperature of 20°C, fewer at 18°C and 22°C, and so on. The norm of reaction for the genotype falls steeply at lower temperatures but flattens at high temperatures. Because each temperature produces a corresponding phenotype, as shown by the dotted lines, the distribution of temperatures results in the phenotypic distribution on the vertical axis. The norm of reaction is like a distorting mirror that reflects the environmental distribution onto the phenotypic axis.

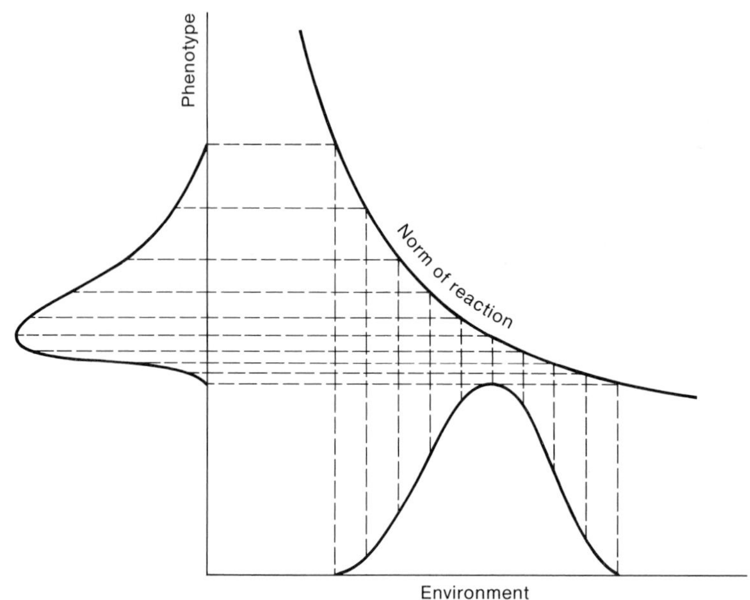

Figure 18-9. The distribution of environments on the horizontal axis is converted to the distribution of phenotypes on the vertical axis by the norm of reaction of a genotype.

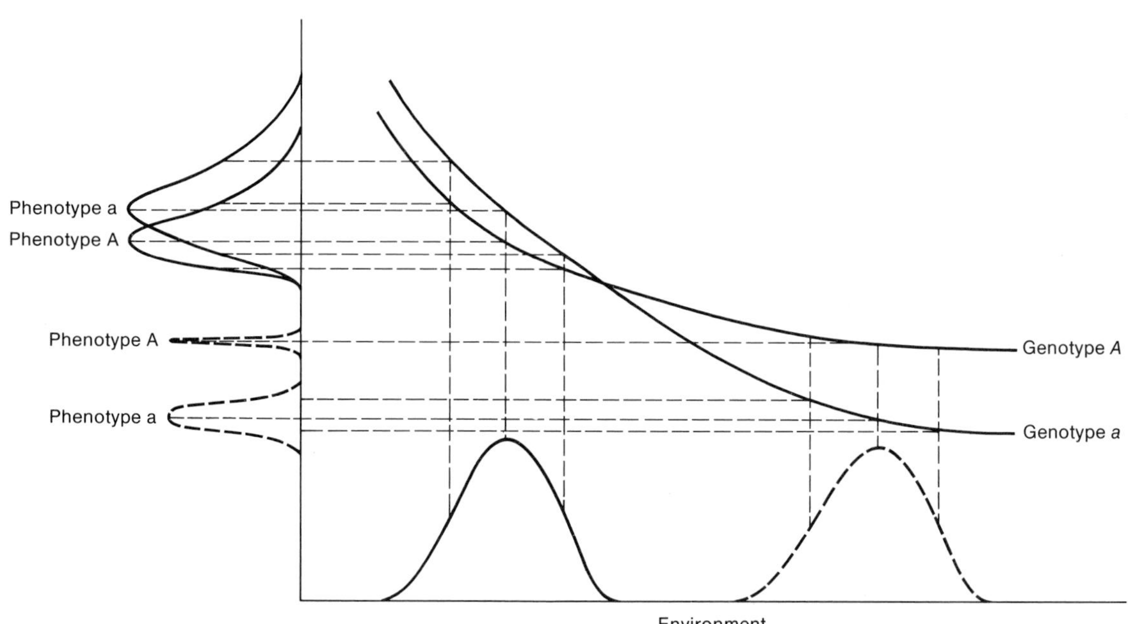

Figure 18-10. Two different environmental distributions are converted into different phenotype distributions by two different genotypes.

Using the same analysis, Figure 18-10 shows how a population consisting of two genotypes with different norms of reaction has a phenotype distribution that depends on the distribution of environments. If the environments are distributed as in the solid curve, then the resulting population of plants will have a unimodal distribution, because the difference between genotypes is very small in this range of environments as compared with the sensitivity of the norms of reaction to small changes in temperature. If the distribution of environments is shifted to the right, however, as shown by the dashed curve, a bimodal distribution of phenotypes results, because the norms of reaction are nearly flat in this environmental range but very different from each other.

Message
A distribution of environments is reflected biologically as a distribution of phenotypes. The transformation of environmental distribution into phenotypic distribution is determined by the norm of reaction.

The Heritability of a Trait

The most basic question to be asked about a quantitative trait is whether the observed variation in the character is influenced by genes at all. It is important to note that this is not the same question as whether genes play any role in the character's development. Gene-mediated developmental processes lie at the base of every character, but *variation* from individual to individual is not necessarily the result of *genetic variation.* Thus, the possibility of speaking any language at all depends critically on the structure of the central nervous system as well as those of vocal cords, tongue, mouth, and ears, which in turn depend on the nature of the human genome. There is no environment in which cows will speak. But although the particular language that is spoken by humans varies from nation to nation, that variation is totally nongenetic.

Message
The question of whether a trait is heritable is a question about the role that differences in genes play in the phenotypic differences *between individuals or groups.*

Familiality and Heritability

In principle, it is easy to determine whether there is any genetic variation. If genes are involved, then (on the average) biological relatives should resemble each other more than unrelated individuals do. This resemblance would be reflected as a positive correlation between parents and offspring or between sibs (offspring of the same parents). Parents larger than the average should have offspring larger than the average; the more seeds a plant produces, the more seeds its sib should produce. Such correlations between relatives, however, are evidence for genetic variation *only if the relatives do not share common environments more than do nonrelatives.* It is absolutely fundamental to distinguish *familiality* from *heritability.*

Message
Traits are ***familial*** *if members of the same family share them, for whatever reason. Traits are* **heritable,** *however, only if the similarity arises from shared genotypes.*

In experimental organisms, there is no problem in separating environmental from genetic similarities. The offspring of a cow producing milk at a high rate and the offspring of a cow producing milk at a low rate can be raised together in the same environment to see whether, despite the environmental similarity, each resembles its own parent. In natural populations, and especially in humans, the problem is difficult. Because of the nature of human societies, members of the same family not only share genes but also have similar environments. Thus, the observation of simple familiality of a trait is uninterpretable genetically. In general, people who speak Hungarian have Hungarian-speaking parents, whereas Japanese speakers have Japanese-speaking parents. Yet the massive experience of immigration to North America has demonstrated that these linguistic differences, although familial, are nongenetic. The highest correlation between parents and offspring for any social traits in the United States are those for political party and religious sect. But they are not heritable. The distinction between familiality and heredity is not always so obvious. The Public Health Commission that originally studied the vitamin-deficiency disease pellagra in the southern United States came to the conclusion that it was genetic because it ran in families!

To determine whether a trait is heritable in human populations, we must use adoption studies to avoid the usual environmental similarity between biological relatives. The ideal experimental subject is the case of identical twins raised apart, because they are genetically identical but environmentally different. Such adoption studies must be so contrived that there is no correlation between the social environment of the adopting family and that of the biological family. These requirements are exceedingly difficult to meet, so that in practice we know very little about whether human quantitative traits that are familial are also heritable. Skin color is clearly heritable, as is adult height—but even here we must be very careful. We know that skin color is affected by genes because of the result of crossracial adoptions and because the offspring of black African slaves were black even when born and raised in Canada. But are the differences in height between Japanese and Europeans affected by genes? The children of Japanese immigrants to North America, born and raised in North America, are taller than their parents but shorter than the North American average, so we might conclude that there is some influence of genetic difference. However, the second generation of Japanese-Americans are even taller than their American-born parents! It appears that some environmental-cultural influence, or perhaps a maternal effect, was still felt in the first generation of births in North America. We cannot yet say whether genetic differences in height distinguish North Americans of Japanese and Swedish ancestry.

Personality traits, temperament, and cognitive performance (such as IQ scores) are all familial, but there are no well-designed adoption studies to show whether they are heritable. The only large-sample studies of human IQ performance that claimed to involve randomized environments (Cyril Burt's reports on identical twins raised apart) have recently been shown to be completely fraudulent.

Figure 18-11 summarizes the usual method for testing heritability in experimental organisms. Individuals from both extremes of the distribution are mated with their own kind, and the offspring are raised in a common controlled environment. If there is an average difference between the two offspring groups, the trait is heritable. Most morphological traits in *Drosophila,* for example, turn out to be heritable—but not all. If flies whose right wings are slightly longer than their left wings are mated together, their offspring have no greater tendency to be "rightwinged" than do offspring of "left-winged" flies.

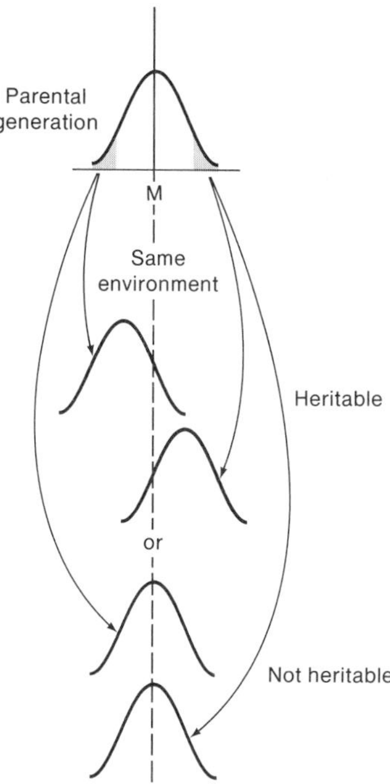

Figure 18-11. *Standard method for testing heritability in experimental organisms. Crosses are performed within two populations of individuals selected from the extremes of the phenotypic distribution in the parental generation. If the phenotypic distributions of the two groups of offspring are significantly different from each other, then the trait is heritable. If the offspring distributions both resemble the distribution for the parental generation, then the trait is not heritable.*

Message
In experimental organisms, environmental similarity may be readily distinguished from genetic similarity (or heritability). In humans, however, it is very difficult to determine whether a particular trait is heritable.

Determining Norms of Reaction

Remarkably little is known about the norms of reaction for any quantitative traits in any species. This is partly because it is difficult, in most sexually reproducing species, to replicate a genotype so that it can be tested in different environments. It is for this reason, for example, that we do not have a norm of reaction for any genotype for any human quantitative trait.

A few norm-of-reaction studies have been carried out with plants that can be clonally propagated. The results of one of these experiments are discussed in Chapter 1. It is possible to replicate genotypes in sexually reproducing organisms by the technique of mating of close relatives, or **inbreeding.** By selfing (where that is possible) or by mating brother and sister repeatedly generation after generation, an originally **segregating line** (that is, one that contains both homozygotes and heterozygotes at a locus) can be made homozygous.

In corn (which can be self-pollinated), for example, a single individual is chosen and self-pollinated. Then, in the next generation, a single one of its offspring is chosen and self-pollinated. In the third generation, a single one of *its* offspring is chosen and self-pollinated, and so on. Suppose the original individual in the first generation was already a homozygote at some locus. Then all of its offspring from self-pollination will also be homozygous and identical at the locus. Future generations of self-pollination will simply preserve the homozygosity. If, on the other hand, the original individual was a heterozygote, then the selfing $Aa \times Aa$ will produce $1/4$ AA homozygotes and $1/4$ aa homozygotes. If a single offspring is chosen in this subsequent generation to propagate the line, there is then a 50% chance that it is now a homozygote. If, by bad luck, the chosen individual should be still a heterozygote, there is another 50% chance of the selected individual in the third generation being homozygous, and so on. Considering the ensemble of all heterozygous loci, then, after one generation of selfing, only 1/2 will still be heterozygous; after two generations, 1/4; after three, 1/8. In the nth generation,

$$\text{Het}_n = \frac{1}{2^n}\,\text{Het}_0$$

where Het_n is the proportion of heterozygous loci in the nth generation and Het_0 is the proportion in the 0th generation. When selfing is not possible, brother–sister mating will accomplish the same end, although more slowly. Table 18-1 is a comparison of the amounts of heterozygosity left after n generations of selfing and brother–sister mating.

Table 18-1. Heterozygosity remaining after various generations of inbreeding for two systems of mating

Generation	Remaining heterozygosity	
	Selfing	Brother–sister mating
0	1.000	1.000
1	0.500	0.750
2	0.250	0.625
3	0.125	0.500
4	0.0625	0.406
5	0.03125	0.338
10	0.000977	0.114
20	1.05×10^{-6}	0.014
n	$\text{Het}_n = \frac{1}{2}\text{Het}_{n-1}$	$\text{Het}_n = \frac{1}{2}\text{Het}_{n-1} + \frac{1}{4}\text{Het}_{n-2}$

To carry out a norm-of-reaction study of a natural population, a large number of lines are sampled from the population and inbred for a sufficient number of generations to guarantee that each line is virtually homozygous at all its loci. Each line is then homozygous at each locus for a randomly selected allele present in the original population. Each line can then be crossed to each other line to produce heterozygotes that reconstitute the original population, and an arbitrary number of individuals from each cross can be produced. If inbred line 1 has the genetic constitution *AA BB cc dd EE* · · · and line 2 is *aa BB CC dd ee* · · ·, then a cross between them would produce a large number of offspring, all of whom are identically *Aa BB Cc dd Ee* · · ·, and these can be raised in different environments.

Inbreeding by mating of close relatives results in total homozygosity for the entire genome. In genetically well-marked species, particularly *Drosophila,* it is possible to produce lines that are homozygous for only a single chromosome, rather than the whole set, by an application of Müller's ClB technique (Chapter 7). The scheme as applied to an autosome is shown in Figure 18-12. A single male from the population to be sampled is crossed to a female carrying a chromosome with a crossover suppressor C (usually a complex inversion), a recessive lethal l, and a dominant visible marker M_1 heterozygous with a second dominant visible M_2. In the F_1 a *single* male carrying the ClM_2 chromosome is chosen. This male, which is also carrying a wild-type chromosome from the population, is again crossed to the marker stock. In the F_2 all flies showing the M_1 trait but not M_2 are necessarily all heterozygotes for copies of the original wild-type chromosome because $\text{ClM}_1/\text{ClM}_1$ is lethal, and no crossovers have taken place. In the F_3 all wild-type flies are identically homozygous for the wild-type chromosome and are now available to make a stock for norm-of-reaction studies and for crosses.

Figure 18-13 shows the norms of reaction of abdominal bristle number as a function of temperature for second-chromosome homozygotes of *Drosophila*

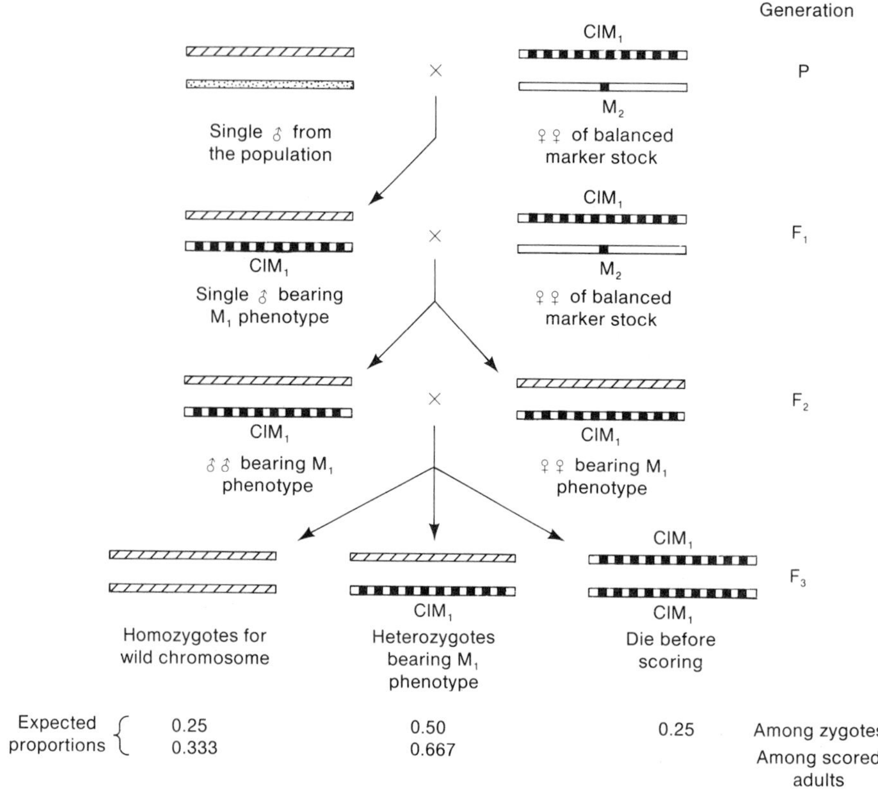

Figure 18-12. Adaptation of Müller's ClB method for autosomes to make homozygotes.

pseudoobscura. Like the growth-rate norms of *Achillea* (Chapter 1), the norms cross each other, with different genotypes having different temperature maxima. The heterozygotes, (which are the natural genotypes) are more similar to each other than are the chromosomal homozygotes (which seldom if ever occur in a natural population).

Results of Norm-of-Reaction Studies

Very few norm-of-reaction studies have been carried out for quantitative characters for the normally heterozygous genotypes found in natural populations. Only easily inbred species, such as maize or *Drosophila,* and clonal species, such as strawberries, have been studied to any degree. Whenever such studies have been made, the outcomes resemble Figure 18-13. That is, no genotype is consistently above or below other genotypes; there are rather small differences among genotypes, and the direction of these differences is not consistent over a wide range of environments.

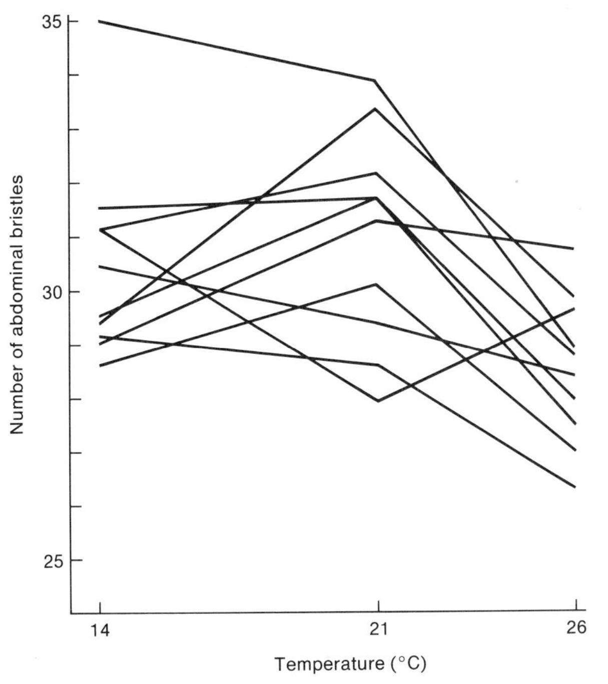

Figure 18-13. The number of abdominal bristles in different homozygous genotypes of Drosophila pseudoobscura *at three different temperatures. (Data courtesy of A. P. Gupta.)*

These facts have two important consequences. First, the selection of "superior" genotypes in domesticated animals and cultivated plants will result in very specifically adapted varieties that may not show their superior properties in other environments. To some extent, this problem is overcome by deliberately testing genotypes in a range of environments—for example, over several years and in several locations. It would be even better, however, if plant breeders could test their selections in a variety of controlled environments in which different environmental factors could be separately manipulated. The consequences of actual plant-breeding practices can be seen in Figure 18-14. The yields of two varieties of corn are shown as a function of different farm environments. Variety 1 is an older variety of hybrid corn, whereas variety 2 is a later "improved" hybrid. These performances are compared at a low planting density, which prevailed when variety 1 was developed, and at a high density characteristic of farming practice when hybrid 2 was selected. At the high density, the new variety is clearly superior to the old variety in all environments (Figure 18-14a). At the low density, however, the situation is quite different. First, note that the new variety is less sensitive to environment than the older hybrid, as evidenced by its flatter norm of reaction. Second, the new "improved" variety is actually poorer under the best farm conditions. Third, the yield improvement of the new variety is not apparent under the low densities characteristic of the earlier agricultural practice.

The second consequence of the nature of reaction norms is that, in the human species, even if it should turn out that there is genetic variation for various

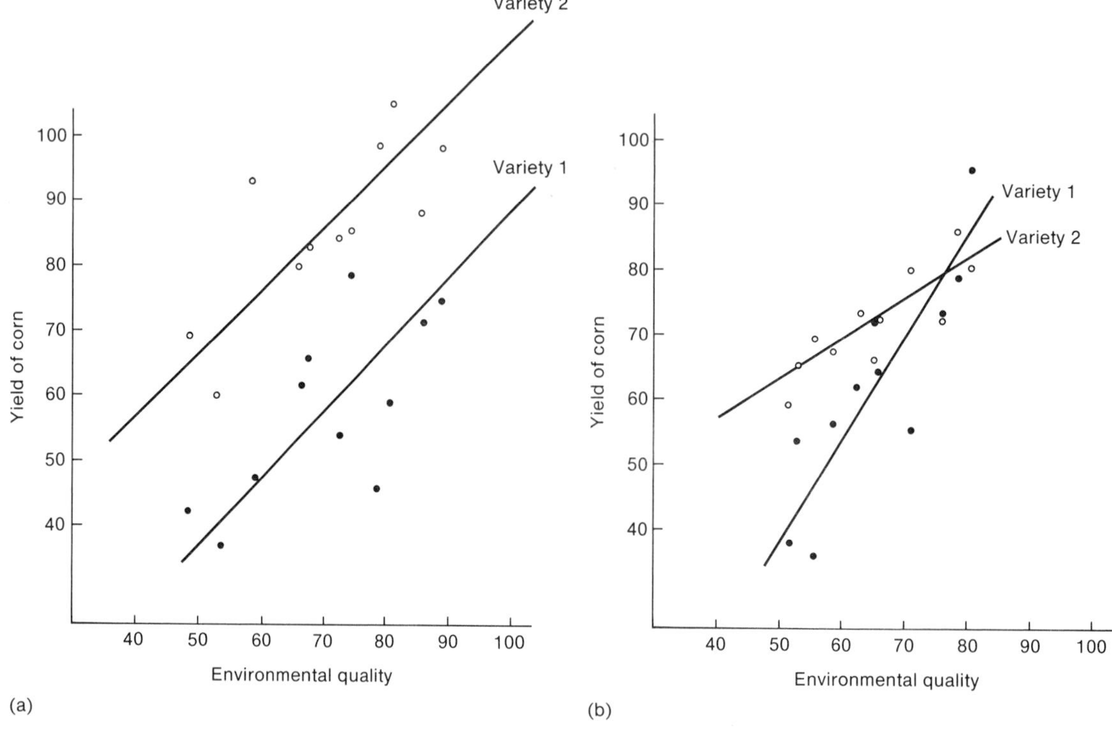

Figure 18-14. Yields of grain of two varieties of corn in different environments. (a) At a high planting density. (b) At a low planting density. (Data of W. A. Russell, Proc. 29th Annual Corn and Sorghum Research Conference, *1974.)*

mental and emotional traits (which is by no means clear), this variation is unlikely to favor one genotype over another across a range of environments. One must beware of hypothetical norms of reaction for human cognitive traits that show one genotype unconditionally superior to another. Even putting aside all questions of moral and political judgment, there is simply no basis for describing different human genotypes as "better" or "worse" on any scale, unless one is able to make a very exact specification of environment.

Message

Norm-of-reaction studies show only small differences among natural genotypes, and these differences are not consistent over a wide range of environments. Thus, "superior" genotypes in domesticated animals and cultivated plants may be superior only in certain environments. If it should turn out that humans exhibit genetic variation for various mental and emotional traits, this variation is unlikely to favor one genotype over another across a range of environments.

Quantifying Heritability

If a trait is shown to have some heritability in a population, then it is possible to quantify the degree of heritability. In Figure 18-7, we see that the variation among phenotypes in a population arises from two sources. First, there are average differences between the genotypes, and second, each genotype has phenotypic variance because of environmental variation. The total phenotypic variance of the population s_p^2 can then be broken into two portions, the variance among genotypic means (s_g^2) and the remaining variance (s_e^2). The former is called the **genetic variance** and the latter is called the **environmental variance,** although, as we shall see, these names are quite misleading. The degree of heritability can then be defined as the proportion of the total variance that is due to the genetic variance:

$$H^2 = \frac{s_g^2}{s_p^2} = \frac{s_g^2}{s_g^2 + s_e^2}$$

H^2, so defined, is called the **broad heritability** of the character.

It must be stressed that this measure of "genetic influence" tells what proportion of the population's *variation* in phenotype is assignable to *variation* in genotype. It does not tell what proportion of an *individual's* phenotype can be ascribed to its heredity and what proportion to its environment. This latter distinction is not a reasonable one. An individual's phenotype is a consequence of the interaction between its genes and its sequence of environments. It clearly would be silly to say that I owe 60 inches of my height to genes and 10 inches to environment. All measures of the "importance" of genes are framed in terms of the proportion of variance ascribable to their variation. This approach is a special application of the more general technique of the **analysis of variance** for apportioning relative weight to contributing causes. The method was, in fact, invented originally to deal with experiments in which different environmental and genetic factors were influencing the growth of plants. (For a sophisticated but accessible treatment of the analysis of variance written for biologists, see R. Sokal and J. Rohlf, *Biometry,* 2nd ed., W.H. Freeman and Company, 1980.)

Methods of Estimating H^2

The genetic variance and the heritability can be estimated in several ways. Most directly, one could obtain an estimate of s_e^2 by making a number of homozygous lines from the population, crossing them in pairs to reconstitute individual heterozygotes, and measuring the phenotypic variance *within* each heterozygous genotype. Because there is no genetic variance within a genotypic class, these variances would (when averaged) provide an estimate of s_e^2. This value can then be subtracted from the value of s_p^2 in the original population to give s_g^2.

Other estimates of genetic variance can be obtained by considering the genetic similarity between relatives. Using simple Mendelian principles, we can see that full sibs will (on the average) be identical for one-half of their genes. For identification purposes, we can label the alleles at a locus carried by the parents differently—so that they are, say, A_1A_2 and A_3A_4. Now the older sib has a probability of 1/2 of getting A_1 from its father, and so does the younger sib, so they have a chance $1/2 \times 1/2 = 1/4$ of both carrying A_1. But they also have a 1/4 chance of carrying A_2. Thus, the chance is $1/4 + 1/4 = 1/2$ that they got the same allele from their father. Exactly the same reasoning applies to the allele they got from their mother. Thus, the *genetic* correlation between full sibs is 1/2. If the same reasoning is carried out for half-sibs, who have only one common parent, their *genetic* correlation can be seen to be 1/4, so the difference in the *genetic* correlation between full sibs and half sibs is $1/2 - 1/4 = 1/4$.

If we now consider the difference between the *phenotypic* correlations of half-sibs and full sibs, it should be one-fourth of that proportion of all the phenotypic variance that is genetic variance. That is,

$$(\textit{phenotypic}\text{ correlation of full sibs}) - (\textit{phenotypic}\text{ correlation of half sibs}) = \tfrac{1}{4}H^2$$

so an estimate of H^2 is

$$H^2 = 4\,[(\text{correlation of full sibs}) - (\text{correlation of half-sibs})]$$

where the correlation here is the *phenotypic* correlation.

We can use similar arguments about genetic similarity between parents and offspring and between twins to obtain two other estimates of H^2:

$$H^2 = 4\,(\text{correlation of full sibs}) - 2\,(\text{parent–offspring correlation})$$

$$H^2 = 2\,[(\text{correlation of monozygotic twins})] - (\text{correlation of dizygotic twins})$$

These formulas come from considering the genetic similarity between relatives. They are only approximate and depend on assumptions about the way genes act. The first two formulas, for example, assume that genes at different loci add together in their effect on the character. The last formula is particularly inaccurate, because it also assumes that the alleles at each locus show no dominance. (See the discussion of components of variance on page 840.) If we ignore these problems of gene interaction, the genetic correlation between full sibs is 1/2 and between half sibs is 1/4. Substituting the values in the first formula gives $H^2 = 1$, which would be the case if all the variation were genetic. If there is nongenetic variation, it will be common to both half sibs and full sibs, and this will make their correlations more similar and reduce the value of H^2. A similar argument applies, for example, to the twin formula, since monozygotic twins have a genetic correlation of 1, while dizygotic twins are just full sibs and have a genetic correlation of 1/2.

All of these estimates, as well as others based on correlations between relatives, depend critically on the assumption that there are no environmental correlations between relatives. If such environmental correlations exist, then the estimates of heritability are biased. It is reasonable that most environmental correlations between relatives are positive, in which case the heritabilities would be overestimated. Negative environmental correlations can also exist. For example, if the members of a litter must compete for food in short supply, negative correlations in growth rate between sibs could occur.

In general, the presence of environmental correlation between relatives makes heritability estimates uninterpretable. It is for this reason that we have no legitimate estimates of heritability for human quantitative traits. Despite their widespread use in human genetics, parent–offspring correlations are estimates of *familiality,* but not of heritability. Even the difference between identical and fraternal twin correlations will not do, because there is the implicit assumption that identical twins are treated no more alike than fraternal twins—an assumption that is unwarranted by the facts. Volumes have been written on the heritability of human IQ, for example, and many modern textbooks of genetics treat the numerical estimates of the heritability of IQ seriously (0.8 is the usual figure given). The absence of studies that treat the environmental correlations between relatives, however, make all the numbers meaningless.

The Meaning of H^2

Attention to the problems of estimating broad heritability distracts from the deeper questions of the meaning of the ratio when it can be estimated. Despite its widespread use as a measure of how "important" genes are in influencing a trait, H^2 actually has a special and limited meaning.

First, H^2 is not a fixed characteristic of a trait but depends on the population in which it is measured and the set of environments in which the population has developed. In one population, alleles segregating at many loci may influence a trait. In another population, these loci may be homozygous. In such a homozygous population, the trait will show no heritability because $s_g^2 = 0$. That value does not mean that genes have no role in influencing the trait's development, but only that none of the variation between individuals within that population can be ascribed to genetic variation. Similarly, a population developing in a very homogeneous environment will have a smaller s_e^2 for a trait than one in a varying environment. The lack of environmental heterogeneity will result in a high value of H^2, but that value does not mean that the trait is insensitive to all environments.

Message
In general, the heritability of a trait is different in each population and each set of environments, and it cannot be extrapolated from one population and set of environments to another.

Second, the separation of variance into genetic and environmental components, s_g^2 and s_e^2, does not really separate the genetic and environmental causes of variation. Consider Figure 18-15, which shows two genotypes in a population, with their two norms of reaction. In Figure 18-15a, the population is assumed to consist of the two genotypes in equal frequency, and the distribution of environments (shown on the horizontal axis) is centered toward the right. The phenotypic distribution (shown on the vertical axis) is composed of two underlying distributions that are very different in their means and have rather little environmental variation. H^2 has a high value. In Figure 18-15b, the environmental distribution has been shifted to the left. As a result, the average difference between genotypes is very much less. The important point is that the so-called *genetic variance* has been changed by shifting the *environment.* In Figure 18-15c, we suppose that the environments are the same as in Figure 18-15b, but now genotype I has become extremely common in the population and genotype II is rare. Then the phenotypic distribution is almost entirely a reflection of norm of reaction I. The population has a larger environmental variance, s_e^2, than does the population of Figure 18-15b because it has become enriched for the less developmentally stable genotype. In this case, the *environmental variance* has been changed by a change in *genotypes.* In general, genetic variance depends on the environments to which the population is exposed, and environmental variance depends upon the frequencies of the genotypes.

Message
Because genotype and environment interact to produce phenotype, no partition of variation can actually separate causes of variation.

Third, as a consequence of the argument just given, knowledge of the heritability of a trait does not permit prediction of how the distribution of the trait will change if either genotypic frequencies or environment are changed markedly.

Message
A high heritability does not mean that a trait is unchangeable by environment.

Compare Figures 18-15a and 18-15b, for example. All that high heritability means is that, for the particular population developing in the particular distribution of environments in which the heritability was measured, average differences between genotypes are large compared to environmental variation within genotypes. If the environment is changed, immense differences in phenotype may occur.

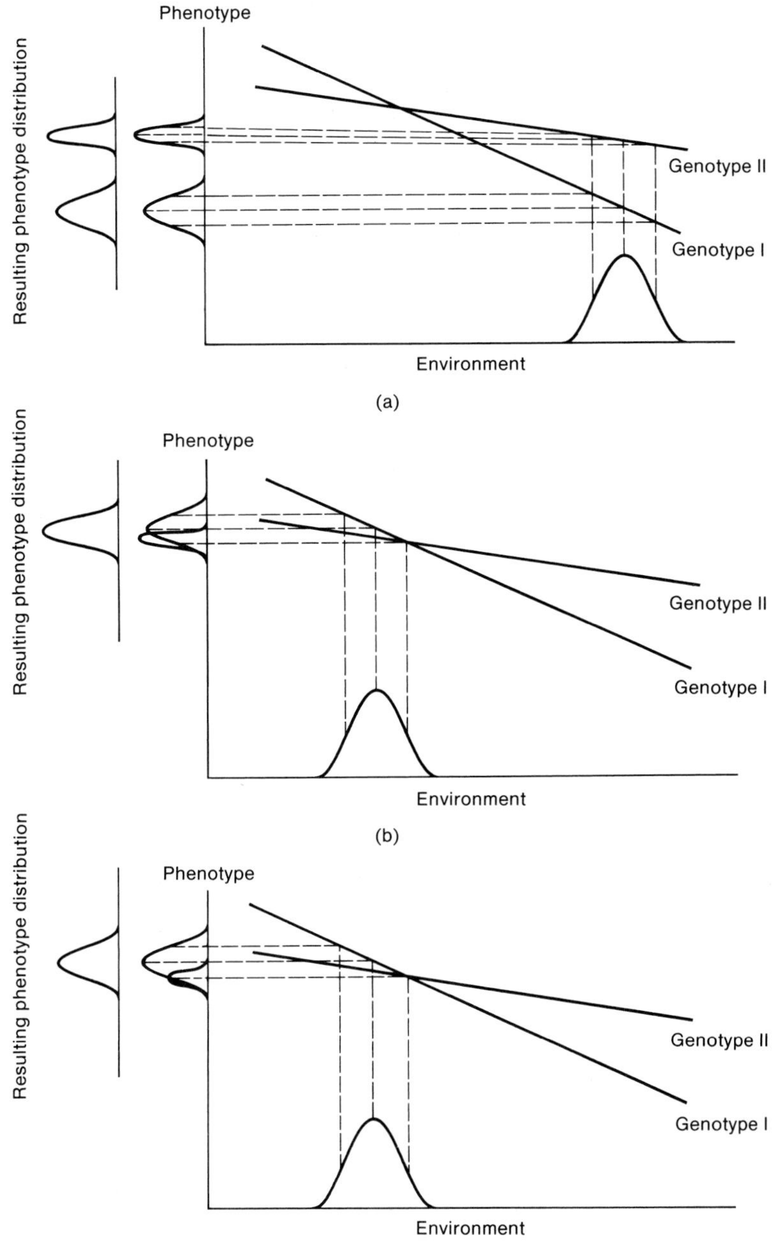

Figure 18-15. *Genetic variance can be changed by changing the environment, and environmental variance can be changed by changing the relative frequencies of genotypes.*

How Useful Is H^2?

If we cannot predict the changeability of a trait due to environmental or genetic manipulation from knowing the trait's broad heritability, of what use is a measure of H^2? We have introduced the concept as a *pedagogical step* toward understanding a related concept (narrow heritability) that is of considerable importance in plant and animal breeding. In fact, plant and animal breeders never use H^2, and the British statistician and geneticist R. A. Fisher, who invented the analysis of variance into genetic and environmental fractions, did not use the concept at all. Unfortunately, the actual history of teaching and research in genetics has led to a widespread but erroneous use of H^2 in characterizing traits. Instead of remaining an intermediate step in pedagogy, H^2 has taken on a life of its own. As important as it is to understand what can be done by the genetic analysis of quantitative variation, it is equally important to understand its limitations, especially when it is widely misused.

Despite its limited meaning, H^2 has been used over and over again, especially by human geneticists and psychologists. They have done so in the erroneous belief that they can estimate H^2 without strict adoption studies and that, if they could estimate H^2, it would tell them something important about clinical or social policy. Students should not suppose that a practice is justified just because many professionals adhere to it. Sometimes scientists simply do not know what else to do, so they continue to pursue useless and even incorrect lines. Sometimes they literally do not understand the basic structure of assumptions that underly this practice. Sometimes they use incorrect methods and concepts in an attempt to justify an ideological or social prejudice (as is clearly the case for some of the studies of the heritability of human temperamental and cognitive functions, such as IQ performance).

Since early in the nineteenth century, there has been a form of biological determinism that has attempted to explain human social differentiation as the direct consequence of biological differences. For example, in 1905, E. L. Thorndike (the most influential founder of modern American psychology) wrote that "in the actual race of life, which is not to get ahead, but to get ahead of somebody, the chief determining factor is heredity." This dictum, which appeared in a paper in a scientific journal, could only have been pure prejudice in 1905—only five years after the rediscovery of Mendel's paper, but five years *before* the chromosome theory of inheritance, ten years *before* the development of the theory of correlation, and thirteen years *before* Fisher's foundations of the theory of quantitative genetics. Much of the history of human psychological genetics since that time describes the vain attempt to justify Thorndike's claim.

If one were seriously interested in knowing how the genes might constrain or influence the course of development of any trait in any organism, it would be necessary to study directly the norms of reaction of the various genotypes in the population over the range of projected environments. No less detailed information will do. Summary measures such as H^2 are not first steps toward a more complete analysis and therefore are not valuable in themselves.

Message

Because H^2 *characterizes present populations in present environments only, it is fundamentally flawed as a predictive device.*

Counting and Locating the Genes

It is not possible with purely genetic techniques to identify all genes that influence the development of a given trait. That is true even for simple qualitative traits—for example, the genes involved in determining the total antigenic configuration of the membrane of the human red blood cell. About 40 loci determining human blood groups are known at present, each having been discovered by finding at least one person with an immunological specificity different from those of other people. There may be many other loci determining red-cell membrane structure that remain undiscovered because all individuals studied are genetically identical. That is, *genetic analysis* detects genes only when there is some allelic variation. In contrast, of course, *molecular analysis,* by dealing directly with DNA and its translated information, can identify genes even when they do not vary—provided one can identify the gene products.

Moreover, the power of classical genetic analysis to detect loci depends on the amount of genetic variation. Even in the case of qualitative human blood groups, only five were discovered in the first 40 years of research on the subject, these five being genetically variable in human populations. Then, during World War II, vast numbers of individuals were blood-typed in connection with the treatment of burns and other war-related injuries, so that loci for which most individuals are identically homozygous were discovered from the rare genetic variants.

The Method of Artificial Selection

The standard method of finding the loci that are segregating with respect to a given quantitative trait in a genetically well-marked species such as *Drosophila melanogaster* is to begin with two populations that are very divergent for the character. These may be two natural populations, but more frequently they are two subpopulations that have been created by artificial selection. For example, in one subpopulation the largest individuals are chosen as parents each generation, while in the other subpopulation the smallest individuals are chosen. As generations pass, the upwardly selected line will become enriched for alleles that lead to larger size, while in the downwardly selected line the alternative alleles will accumulate. If selection is carried on for a long enough time, the two populations will become virtually homozygous for "high" and "low" alleles. Alternatively, many inbred lines could have been made from the original population, and the most divergent lines chosen as the "high" and "low" populations. Whichever method is used, the two divergent populations will probably differ only for those loci that were somewhat heterozygous in the original population.

Once the divergent lines have been established, each chromosome in one line can be separately substituted into the other line using dominant marker stocks with crossover suppressors. An idealization of the method is shown in Figure 18-16, where A_{I}, A_{II}, A_{III}, B_{I}, . . . are dominant marker systems for chromosomes I, II, and III. Lines homozygous and heterozygous for various

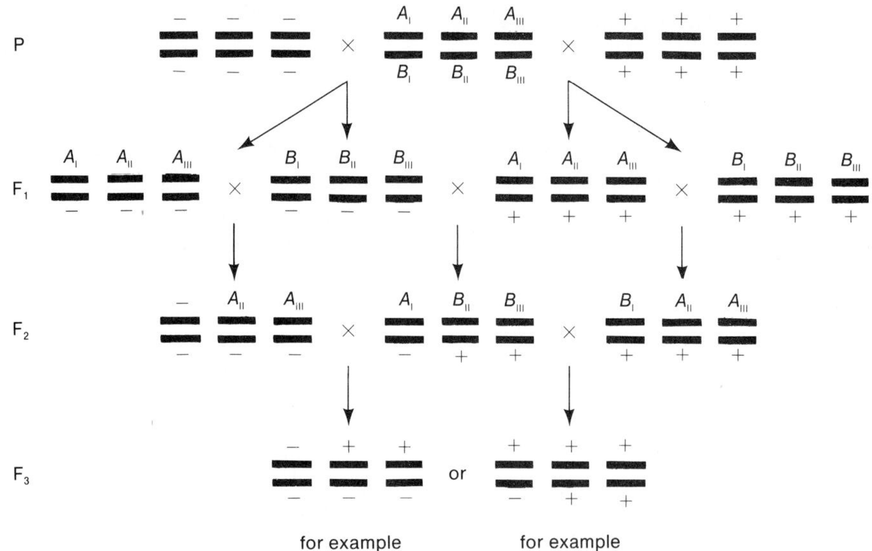

Figure 18-16. Scheme of mating to produce individuals with different combinations of chromsomes from a high selection line (+) and a low selection line (−). A_I, B_I, . . . *are dominant marker crossover suppressors.*

combinations of "+" and "−" chromosomes are measured for the trait, and in this way the contribution of each chromosome to the genetic differences between the lines is determined. An example of the application of this technique by J. Crow to DDT resistance in *Drosophila melanogaster* is shown in Figure 18-17. Crow did not manufacture lines that were homozygous for introduced chromosomes, so we see only the effect of chromosomes when heterozygous. The figure shows that every chromosome seems to have some genes that differ between the resistant and the susceptible line.

Linkage Analysis

The experiment of Figure 18-16 must be elaborated to find individual genes. A more detailed analysis can be carried out with recessive marker stocks that allow recombination. A cross between a multiply marked susceptible chromosome and a resistant chromosome, followed by a backcross to the marker stock, will result in a large number of recombinant types of varying resistance. In the simplest situation, all recombinants carrying a given short region will be resistant, while those not carrying the region will be susceptible. This would be strong evidence for a single locus of major effect in the region. In the worst case, all the recombinants will show partial resistance, so that no localization has been accomplished.

Where localization experiments have been carried out, as for example for the sterno-pleural bristles on the side of the thorax of *Drosophila melanogaster*, most

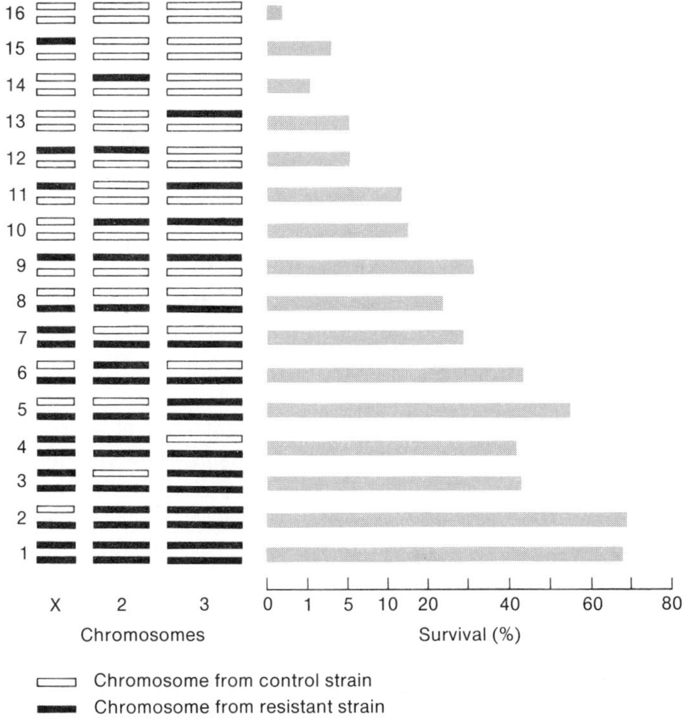

Figure 18-17. Survival of Drosophila melanogaster *genotypes with different combinations of chromosomes from a selected line* (dark chromosomes) *and a nonselected line* (light chromosomes). *Dark chromosomes are from strains that have been selected for resistance to DDT. (From J. Crow,* Ann. Rev. Entomology *2:228, 1957.)*

of the difference between selected lines has been localized to two or three genes with major effects, plus a residue of "genes with small effects."

The possibility exists that many quantitative characters (perhaps even most) will turn out to vary not as a consequence of the segregation of alleles at large numbers of loci with small effects (as the multiple-genes hypothesis supposes), but rather as a consequence of segregation of very few loci with major effects. This study of the "genetic architecture" of quantitative traits is a leading problem of quantitative genetics. It can be solved by straightforward but excessively tedious experiments in well-marked species, such as *Drosophila melanogaster.* In *Homo sapiens,* such experiments cannot be done, and we have no information on the number or localization of loci contributing to quantitative traits. Even for a character with obvious genetic influence and with a relatively simple biochemical and developmental basis (such as skin color), we do not know the number of loci (except to say that there must be several) or their distribution over the human chromosome set.

Gene Action

The question of gene action is closely tied to the problem of detecting the loci. The methods described in the preceding section are biased toward detection of loci where allelic differences are of large effect. The process of selection by which the contrasting populations are established is relatively efficient only for allelic substitutions of large effect.

Even without being able to localize the genes, we can obtain some information about their actions and interactions. We can judge average dominance at the chromosomal level by comparing homozygotes and heterozygotes in chromosomal substitution experiments. In Figure 18-17, a comparison of lines 1 and 2 shows that the X chromosomes do not differ, whereas lines 15 and 16 show that they do. An explanation of this discrepancy would be the complete dominance of resistance alleles over susceptibility alleles. A comparison of lines 7, 9, and 10 supports the hypothesis of dominance because 7 and 9 are equally resistant, whereas 10 is much less so.

There is also evidence of specific epistatic interactions. Line 5 has both the substitutions of lines 2 and 3 and so should be at least as susceptible as line 3, yet it is intermediate between 2 and 3. It would seem that heterozygosity for a susceptible X chromosome actually decreases the susceptibility when in combination with a susceptible second-chromosome heterozygote. Such evidences of dominance and epistasis of whole chromosomes throw only indirect light on individual gene action because a single gene of major effect with dominance or strong epistasis would hide the effects of several other genes of small effect.

Evidence for nonnuclear effects can also be deduced from the chromosomal substitution experiments. For example, in Figure 18-17, lines 8 and 9 have the same chromosomal constitution but are the result of reciprocal crosses and so have different cytoplasm. There clearly is a maternal effect.

More on Analyzing Variance

Knowledge of the broad heritability (H^2) of a trait in a population is not very useful in itself, but a finer subdivision of phenotypic variance can provide important information for plant and animal breeders. The genetic variation and the environmental variation can themselves be subdivided in a way that can provide information about gene action and the possibility of shaping the genetic composition of a population.

Additive and Dominance Variance

Our previous consideration of gene action suggests that the phenotypes of homozygotes and heterozygotes ought to have a simple relation. If one of the alleles codes for a less active gene product or one with no activity at all, and if one unit of gene product is sufficient to allow full physiological activity of the organism, then we expect complete dominance of one allele over the other, as Mendel observed for flower color in peas. If, on the other hand, physiological activity is proportional to the amount of active gene product, we would expect the heterozygote's phenotype to be exactly intermediate between the homozygotes (no dominance). For many quantitative traits, however, neither of these simple cases is the rule. In general, heterozygotes are not exactly intermediate between the two homozygotes but are closer to one of the homozygotes (partial dominance). The complexity of biochemical and developmental pathways is such that intermediate activity of primary gene product is not exactly scaled as intermediate phenotype. Indeed, in some cases the heterozygote may lie outside the phenotypic range of the homozygotes altogether (**overdominance**). For example, newborn babies who are intermediate in size have a higher chance of survival than very large or very small newborns. Thus, if survival were the phenotype of interest, heterozygotes for genes influencing growth rate would show overdominance.

Suppose there are two alleles (*a* and *A*) segregating at a locus influencing height. In the environments encountered by the population, the mean phenotypes (heights) and frequencies of the three genotypes are:

	aa	*Aa*	*AA*
Phenotype	10	18	20
Frequency	0.36	0.48	0.16

There is genetic variance in the population; the phenotypic means of the three genotypic classes are different. Some of the variance arises because there is an average effect on phenotype of substituting an allele *A* for an allele *a*. That is, the average height of all individuals with *A* alleles is greater than that of all individuals with *a* alleles. This average effect is easily calculated by simply counting the *a* and *A* alleles and multiplying them by the height of the individuals in which they appear. So, 0.36 of all the individuals are homozygous *aa*, each *aa* individual has two *a* alleles, and the average height of *aa* individuals is 10 cm. Heterozygotes make up 0.48 of the population, each has only one *a* allele, and the average phenotypic measurement of *Aa* individuals is 18 cm. The total "number" of *a* alleles is 2(0.36) + 1(0.48). Thus, the average effect of all the *a* alleles is

$$\bar{a} = \text{average effect of } a = \frac{2(0.36)(10) + 1(0.48)(18)}{2(0.36) + 1(0.48)} = 13.20 \text{ cm}$$

and, by a similar argument,

$$\bar{A} = \text{average effect of } A = \frac{2(0.16)(20) + 1(0.48)(18)}{2(0.16) + 1(0.48)} = 18.80 \text{ cm}$$

This average difference in effect between *A* and *a* alleles of 5.60 cm accounts for some of the variance in phenotype, but not for all, because the heterozygote is not exactly intermediate between the homozygotes. There is some dominance. The total genetic variance associated with this locus can then be partitioned between the **additive genetic variance** (s_a^2) associated with the average effect of substituting *A* for *a*, and the **dominance variance** (s_d^2) resulting from the partial dominance of *A* over *a* in heterozygotes. Thus, $s_g^2 = s_a^2 + s_d^2$.

The components of variance in this example can be calculated using the definitions of mean and variance that were developed earlier in this chapter. The mean phenotype is

$$\bar{x} = \Sigma f_i x_i = (0.36)(10) + (0.48)(18) + (0.16)(20) = 15.44 \text{ cm}$$

The total genetic variance that arises from the variation among the mean phenotypes of the three genotypes is

$$s_{\mathrm{g}}^2 = \Sigma f_i (x_i - \bar{x})^2 = (0.36)(10 - 15.44)^2 + (0.48)(18 - 15.44)^2 + (0.16)(20 - 15.44)^2 = 17.13 \text{ cm}^2$$

The additive variance is calculated from the squared deviation of the average a effect from the mean, weighted by its frequency, plus the squared deviation of the average A effect from the mean, weighted by its frequency. The frequency of the a allele is

$$f_a = \frac{2(0.36) + 1(0.48)}{2} = 0.60$$

and the frequency of the A allele is

$$f_A = \frac{2(0.16) + 1(0.48)}{2} = 0.40$$

Thus,

$$s_{\mathrm{a}}^2 = 2[f_a(\bar{a} - \bar{x})^2 + f_A(\bar{A} - \bar{x})^2] = 2[(0.60)(13.20 - 15.44)^2 + (0.40)(18.80 - 15.44)^2] = 15.05 \text{ cm}^2$$

and

$$s_{\mathrm{d}}^2 = s_{\mathrm{g}}^2 - s_{\mathrm{a}}^2 = 17.13 - 15.05 = 2.08 \text{ cm}^2$$

The usefulness of this subdivision of genetic variance is in the prediction that can be made about the effect of selective breeding. This use becomes clear in an extreme case. Suppose that there is overdominance and the phenotypic means and frequencies of three genotypes are

	AA	*Aa*	*aa*
Phenotype	10	12	10
Frequency	0.25	0.50	0.25

It is apparent (and a calculation like the preceding one will confirm it) that there is no average difference between a and A alleles, because each has an effect of 11 units. So there is no *additive genetic variance,* although there is dominance variance. The largest individuals are heterozygotes. If one attempts to increase height in this population by selective breeding, mating these heterozygotes together will simply reconstitute the original population. Selection will be totally ineffective. This illustrates the general law that the effect of selection depends on the additive genetic variance and *not* on genetic variance in general.

The total phenotypic variance can now be written as:

$$s_{\mathrm{p}}^2 = s_{\mathrm{g}}^2 + s_{\mathrm{e}}^2 = s_{\mathrm{a}}^2 + s_{\mathrm{d}}^2 + s_{\mathrm{e}}^2$$

We define a heritability, h^2, the **heritability in the narrow sense,** as

$$h^2 = \frac{s_a^2}{s_p^2} = \frac{s_a^2}{s_a^2 + s_d^2 + s_e^2}$$

It is this heritability, not to be confused with H^2, that is useful in determining whether a program of selective breeding will succeed in changing the population. The greater the h^2, the more of the difference between selected parents and the population as a whole will be preserved in the next generation.

What we have described as the "dominance" variance is really more complicated. It is all the genetic variation that cannot be explained by the average effect of substituting A for a. If there is more than one locus affecting the character, then any epistatic interactions between loci will appear as variance not associated with the average effect of substituting alleles at the A locus. In principle one can separate this interaction variance s_i^2 from the dominant variance s_d^2, but in practice this cannot be done with any semblance of accuracy, so all the nonadditive variance appears as "dominance" variance.

Estimating Genetic Variance Components

Genetic components of variance are estimated from covariance between relatives. Ignoring the epistatic contributions to variance that are contained to some degree in all covariances between relatives, Table 18-2 shows some of the components of variance estimated from relatives. These relations can be used, together with the total phenotypic variance, to estimate h^2. For example, we see from Table 18-2 that the covariance between parent and offspring contains half the additive variance. That is,

$$\text{cov (parent–offspring)} = s_a^2/2$$

Therefore,

$$2 \text{ cov (parent–offspring)} = s_a^2$$

and, by the definition of correlation,

$$\frac{2 \text{ cov (parent–offspring)}}{s_p^2} = 2 \text{ correlation (parent–offspring)} = \frac{s_a^2}{s_p^2} = h^2.$$

Table 18.2. Proportion of the additive variance (s_a^2) and the dominance variance (s_d^2) contained in the genetic covariance between various related individuals

Relatives	Estimated proportion of s_a^2	Estimated proportion of s_d^2
Cov (identical twins)	1	1
Cov (parent–offspring)	1/2	0
Cov (half-sibs)	1/4	0
Cov (full sibs)	1/2	1/4

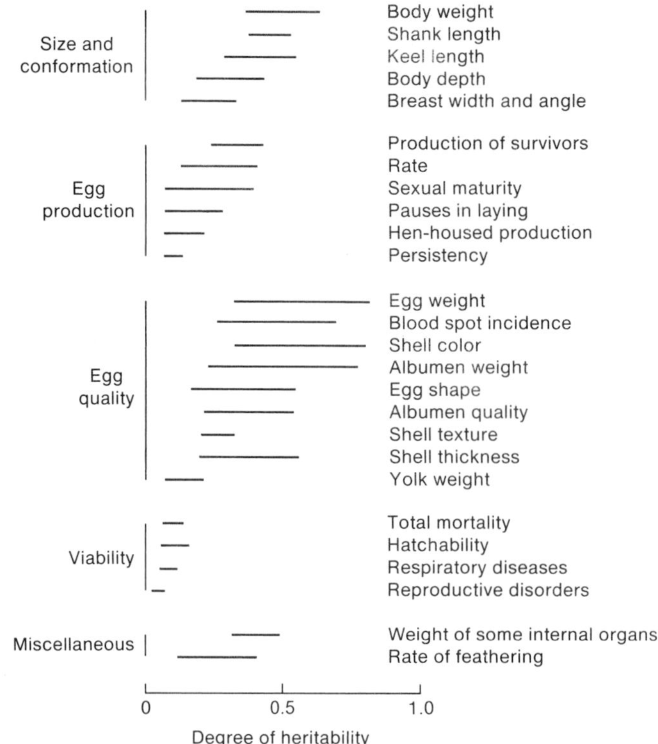

Figure 18-18. Ranges of heritabilities (h^2) *reported for a variety of characters in chickens. (From I. M. Lerner.)*

Thus, twice the parent–offspring correlation is an estimate of h^2.

Remember that any estimate of h^2, like H^2, depends on the assumption of no environmental correlation between relatives. Moreover, h^2 in one population in one set of environments will not be the same as that in a different population at a different time. Figure 18-18 shows the range of heritabilities reported for a number of traits in chickens. The very small ranges are generally close to zero. For most traits for which a substantial heritability has been reported in some population, there are big differences from study to study.

Partitioning Environmental Variance

Environmental variance, like genetic variance, can also be further subdivided. In particular, developmental noise is usually confounded with environmental variance, but when a character can be measured on left and right sides of an organism, or over repeated body segments, it is possible to separate noise from environment. Table 18-3 shows the complete partitioning of variation for two characters in a population of *Drosophila melanogaster* raised in standard laboratory conditions. For each character there is a substantial h^2, so we might expect that selective breeding could increase or decrease bristle number and ovary

Table 18.3. Partition of the total phenotypic variance for two characters in a population of *Drosophila melanogaster*

Source of variation			Percentage of variance: Number of abdominal bristles	Percentage of variance: Ovary size
Additive genetic	s_a^2	s_g^2	52	30
Dominance + epistatic variance	s_d^2		9	40
Environmental variance	s_e^2	s_e^2	1	3
Developmental noise	s_n^2		38	27
Total	s_p^2		100	100

SOURCE: From D. Falconer, *Quantitative Genetics*, Longman Group Limited. Copyright © 1981.

size. Moreover, nearly all the nongenetic variation is due to developmental noise. These values, however, are a consequence of the relatively rigorously controlled environment of the laboratory. Presumably s_e^2 in nature would be considerably larger, with a consequent diminution in the relative size of h^2 and of s_n^2. As always, such studies of variation are applicable only to a particular population in a given distribution of environments.

The Use of h^2 in Breeding

Even though h^2 is a number that applies to a particular population and a given set of environments, it is still of great practical importance to breeders (Figure 18-19). A poultry geneticist interested in increasing, say, growth rate, is not

Figure 18-19. Quantitative genetic theory has been extensively applied to poultry breeding. (Photograph courtesy of Welp, Inc., Bancroft, Iowa.)

concerned with the genetic variance over all possible flocks and all environmental distributions. The question is, given a particular flock (or a choice between a few particular flocks) under the environmental conditions approximating present husbandry practice, can a selection scheme be used to increase growth rate and, if so, how fast? If one flock has a lot of genetic variance and another only a little, the breeder will choose the former to carry out selection. If the heritability in the best flock is very high, then the mean of the population will respond quickly to the selection imposed, because most of the superiority of the selected parents will appear in the offspring. The higher the h^2, the higher is the parent–offspring correlation. If, on the other hand, h^2 is low, then only a small fraction of the increased growth rate of the selected parents will be reflected in the next generation.

If h^2 is very low, some alternative breeding scheme may be needed. It is this case where H^2 together with h^2 can be of use to the breeder. Suppose that h^2 and H^2 are both low. This means that there is a lot of environmental variance as compared to genetic variance. Some scheme of reducing s_e^2 must be used. One method is to change the husbandry conditions so that environmental variance is lowered. Another is to use **family selection.** Rather than choosing the best individuals, the breeder allows pairs to produce several progeny, and the *mating* is selected on the basis of the average performance of the progeny. By averaging over progeny, uncontrolled environmental and developmental-noise variation is cancelled out, and a better estimate of the genotypic difference between pairs can be made.

If, on the other hand, h^2 is low but H^2 is high, then there is not much environmental variance. The low h^2 is the result of a small amount of additive genetic variance as compared to dominance and interaction variance. Such a situation calls for special breeding schemes that make use of nonadditive variance. One widely used scheme is the hybrid–inbred method used almost universally for maize. A large number of inbred lines are created by selfing. These are then crossed in many different combinations (all possible combinations, if that is economically feasible), and the cross that gives the best hybrid is chosen. Then new inbred lines are developed from this best hybrid, and again crosses are made to find the best second-cycle hybrid (Figure 18-20). This scheme selects for dominance effects, because it takes the best heterozygotes, and it has been responsible for major genetic advances in hybrid maize yield in North America since 1930. However, it is still debatable whether this technique *ultimately* produces higher-yielding varieties than those that would have resulted from years of simple selection techniques based on additive variance.

Message

The subdivision of genetic variation and environmental variation provides important information about gene action that can be used in plant and animal breeding.

Figure 18-20. *Collecting corn pollen to be used in a cross. Quantitative genetic theory plays a central role in practical plant breeding. (Photograph by John Colwell, from Grant Heilman.)*

Summary

Many—perhaps most—of the phenotypic traits we observe in organisms vary continuously. In many cases, the variation of the trait is determined by more than a single segregating locus. Each of these loci may contribute equally to a particular phenotype, but it is more likely that they contribute unequally. Measurement of these phenotypes and the determination of the contribution of specific alleles to the distribution must be done, in these cases, on a statistical basis. Some of these variations of phenotype, such as height in some plants, may show a normal distribution around a mean value; others, such as seed weight in some plants, will illustrate a skewed distribution around a mean value.

In other characters, the variation of one phenotype may be correlated with the variation in another. A correlation coefficient may be calculated for these two variables.

A quantitative character is one for which the average phenotypic differences between genotypes are small as compared with the variation between individuals within genotypes. This situation may be true even for characters that are determined by alleles at one locus. The distribution of environments is reflected biologically as a distribution of phenotypes. The transformation of environmental distribution into phenotypic distribution is determined by the norm of reaction. Norms of reaction can be characterized in organisms where large numbers of genetically identical individuals can be produced.

By the use of genetically marked chromosomes, it is possible to determine the relative contributions of different chromosomes to variation in a quantitative trait, to observe dominance and epistasis from whole chromosomes, and, in some cases, to map genes that are segregating for a trait.

Traits are *familial* if members of the same family share them, for whatever reason. Traits are *heritable,* however, only if the similarity arises from shared genotypes. In experimental organisms, environmental similarities may be readily distinguished from genetic similarities, or heritability. In humans, however, it is very difficult to determine whether a particular trait is heritable. Norm of reaction studies show only small differences among genotypes, and these differences are not consistent over a wide range of environments. Thus, "superior" genotypes in domesticated animals and cultivated plants may be superior only in certain environments. If it should turn out that humans exhibit genetic variation for various mental and emotional traits, this variation is unlikely to favor one genotype over another across a range of environments.

The attempt to quantify the influence of genes on a particular trait has led to the determination of heritability in the broad sense (H^2). In general, the heritability of a trait is different in each population and each set of environments and cannot be extrapolated from one population and set of environments to another. Because H^2 characterizes present populations in present environments only, it is fundamentally flawed as a predictive device. Heritability in the narrow sense, h^2, measures the proportion of phenotypic variation that results from substituting one allele for another. This quantity, if large, predicts that selection for a trait will succeed rapidly. If h^2 is small, special forms of selection are required.

Problems

1. Suppose that two triple heterozygotes *Aa Bb Cc* are crossed. Assume that the three loci are in different chromosomes.
   a. What proportions of the offspring are homozygous at 1, 2, and 3 loci, respectively?
   b. What proportions of the offspring carry 0, 1, 2, 3, 4, 5 and 6 alleles represented by capital letters, respectively?
2. In Problem 1, suppose that at the *A* locus the average phenotypic effect of the three genotypes is $AA = 4, Aa = 3, aa = 1$, and similar effects exist for the *B* and *C* loci.

Moreover, suppose that the effects of loci add to each other. Calculate and graph the distribution of phenotypes in the population (assuming no environmental variance).

3. In Problem 2, suppose that there is a threshold in the phenotypic character so that when the phenotypic value is above 9, the individual has three bristles; when it is between 5 and 9, the individual has two bristles; and when the value is 4 or less, the individual has one bristle. Discuss the outcome of crosses within and between bristle classes. Given the result, could you infer the underlying genetic situation?

4. Suppose the general form of a distribution of a trait for a given genotype is

$$f = 1 - \frac{(x - \bar{x})^2}{s_e^2}$$

over the range of x where f is positive.

a. Plot on the same scale the distributions for three genotypes whose means and environmental variances are as follows:

Genotype	$\bar{x}$	s_e^2	*Approximate range of phenotype*
1	0.20	0.3	$x = 0.03$ to $x = 0.37$
2	0.22	0.1	$x = 0.12$ to $x = 0.24$
3	0.24	0.2	$x = 0.10$ to $x = 0.38$

b. Plot the phenotypic distribution that would result if the three genotypes were equally frequent in a population. Can you see distinct modes?

5. Table 18-4 shows a distribution of bristle number in *Drosophila.* Calculate the mean, variance, and standard deviation of the distribution.

Table 18-4.

Bristle number	Number of individuals
1	1
2	4
3	7
4	31
5	56
6	17
7	4

6. The following sets of hypothetical data represent paired observations on two variables (x, y). Plot each set of data pairs as a scatter diagram. Looking at the plot of the points, make an intuitive guess at the correlation between x and y. Then calculate the correlation coefficient for each set of data pairs and compare this value with your estimate.

a. (1, 1); (2, 2); (3, 3); (4, 4); (5, 5); (6, 6).

b. (1, 2); (2, 1); (3, 4); (4, 3); (5, 6); (6, 5).

c. (1, 3); (2, 1); (3, 2); (4, 6); (5, 4); (6, 5).

d. (1, 5); (2, 3); (3, 1); (4, 6); (5, 4); (6, 2).

7. A recent book on the problem of heritability of IQ makes the following three statements. Discuss the validity of each statement and its implications about the *authors' understanding of* h^2 and H^2.

a. "The interesting question then is . . . 'how heritable'? The answer [0.01] has a very different theoretical and practical application from the answer [0.99]." [The authors are talking about H^2.]

b. "As a rule of thumb, when education is at issue H^2 is usually the more relevant coefficient, and when eugenics and dysgenics are being discussed, h^2 is ordinarily what is called for."

c. "But whether the different ability patterns derive from differences in genes . . . is not relevant to assessing discrimination in hiring. Where it could be relevant is in deciding what, in the long run, might be done to change the situation."

8. Using the concepts of norms of reaction, environmental distribution, genotypic distribution, and phenotypic distribution, try to restate in terms that are more exact the statement that "80% of the difference in IQ performance between the two groups is genetic." What would it mean to talk about the heritability of a difference between two groups?

9. Describe an experimental protocol involving studies of relatives that could estimate the broad heritability of alcoholism. Remember that you must make an adequate observational definition of the trait itself!

10. A line selected for high bristle number in *Drosophila* has a mean of 25 sternopleural bristles, whereas a low selected line has a mean of only 2. Marker stocks involving the two large autosomes II and III were used to create stocks with various mixtures of chromosomes from the high (h) and low (l) lines. The mean number of bristles for each chromosomal combination is as follows:

$\frac{h\ h}{h\ h}$	25.1	$\frac{h\ h}{l\ h}$	22.2	$\frac{l\ h}{l\ h}$	19.0
$\frac{h\ h}{h\ l}$	23.0	$\frac{h\ h}{l\ l}$	19.9	$\frac{l\ h}{l\ l}$	14.7
$\frac{h\ l}{h\ l}$	11.8	$\frac{h\ l}{l\ l}$	9.1	$\frac{l\ l}{l\ l}$	2.3

What conclusions can you reach about the distribution of genetic factors and their action from these data?

11. Suppose that number of eye facets is measured in a population of *Drosophila* under various conditions of temperature. Further suppose that it is possible to estimate total genetic variance s_g^2 as well as the phenotypic distribution. Finally, suppose that there are only two genotypes in the population. Draw pairs of reaction norms that would lead to the following results:

a. An increase in mean temperature decreases the phenotypic variance.

b. An increase in mean temperature increases H^2.

c. An increase in mean temperature increases s_g^2 but decreases H^2.

d. An increase in temperature *variance* changes a unimodal into a bimodal phenotype distribution (one reaction norm is sufficient here).

12. The following variances and covariances between relatives have been found for egg weight in poultry: $s_p^2 = 14.8$; cov (mother–daughter) $= 1.7$; cov (sisters) $= 2.7$; cov (half-sisters) $= 0.8$. Calculate the broad and narrow heritabilities, H^2 and h^2.

13. Francis Galton compared the heights of male undergraduates to the heights of their fathers, with the results shown in Figure 18-21. The average height of all fathers is the same as the average of all sons, but the individual height classes are not equal across generations. The very tallest fathers had sons somewhat shorter than the fathers, whereas very short fathers had sons somewhat taller than the fathers. As a result, the best line that can be drawn through the points on the scatter diagram has a slope of about 0.67 (*solid line*) rather than 1.00 (*dashed line*). Galton used the term *regression* to describe this tendency for the phenotype of the sons to be closer to the population mean than was the phenotype of their fathers.

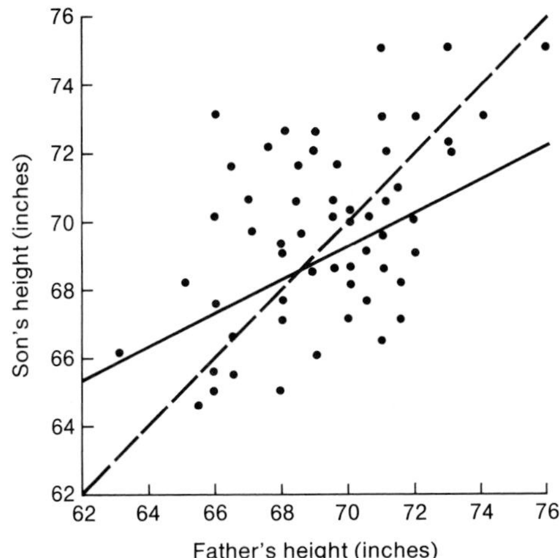

Figure 18-21. Scatter diagram for height data in father–son pairs (see Problem 13). The best line that can be drawn through the scatter points (solid line) *has a slope less than the line of identity* (dashed line). *(After W. F. Bodmer and L. L. Cavalli-Sforza,* Genetics, Evolution, and Man. *Copyright © 1976, W. H. Freeman and Company.)*

a. Propose an explanation for regression.

b. How are regression and heritability related?

Shell banding and shell color morphs in the snail Cepaea nemoralis. *All were collected from one population.*

19

Population Genetics

Mendel's investigations of heredity—and indeed all of the interest in heredity in the nineteenth century—arose from two related problems, how to breed improved crops and how to understand the nature and origin of species. What is common to these problems (and what differentiates them from the problems of transmission and gene action) is that they are concerned with *populations* rather than with *individuals*. Studies of gene replication, of protein synthesis, of development, and of chromosome movement are all concerned with processes that go on within the cells of individual organisms. But the transformation of a species, either in the natural course of evolution or by the deliberate intervention of human beings, is a change in the properties of a collectivity—of an entire population or a set of populations.

Message
The problem of population genetics is to relate the heritable changes in populations of organisms to the underlying individual processes of inheritance and development.

Darwin's Revolution

Many people think that the concept of organic evolution was first proposed by Charles Darwin, but that is certainly not the case. Most scholars had abandoned the notion of fixed species (unchanged since their origin in a grand creation of life) long before publication of Darwin's *The Origin of Species* in 1859. Most biologists agreed that new species arise through some process of evolution from older ones; the problem was to explain *how* this evolution can occur. The theories preceding Darwin's were *transformational theories,* which postulated the same transformation of quality for each individual organism within a species, much as every individual in its lifetime changes from an infant to an adult. Jean-Baptiste Lamarck, for example, claimed that each individual changes slightly because of an inner striving or will to adapt itself to its environment. These small individual changes are passed on to the offspring, who in turn continue the process of change by their own strivings to adapt.

Darwin broke fundamentally with these transformational hypotheses by creating a *variational theory.* In this theory, variation exists among organisms within a species. Individuals of one generation are qualitatively different from one another. Evolution of the species as a whole results from the differential rates of reproduction of the various types, so that the relative frequencies of the types change over time. Evolution, in this view, is a sorting process rather than a transformational one. For Lamarck, evolution was the sum of individual changes, with each individual changing itself slightly and passing that change along to its offspring. For Darwin, evolution was the sum of population changes, with each generation having offspring that survived in a slightly altered proportion of various types.

Message

Darwin proposed a new explanation to account for the accepted phenomenon of evolution. He argued that the population of a given species at a given time includes individuals of varying characteristics. The population of the next generation will contain a higher frequency of those types that most successfully survive and reproduce under the existing environmental conditions. Thus the frequencies of various types within the species will change over time.

There is an obvious similarity between the process of evolution as Darwin described it and the process by which the plant or animal breeder improves a domestic stock. The plant breeder selects the highest-yielding plants from the current population and (as far as possible) uses them as the parents of the next generation. If the characteristics causing the higher yield are heritable, then the next generation should produce a higher yield. It was no accident that Darwin chose the term **natural selection** to describe his model of evolution through differential rates of reproduction of different variants in the population. He had in mind as a model for this evolutionary process the selection that the breeder exercises on successive generations of domestic plants and animals.

We can summarize Darwin's theory of evolution through natural selection in three principles.

1. **The principle of variation.** Among individuals within any population, there is variation in morphology, physiology, and behavior.
2. **The principle of heredity.** Offspring resemble their parents more than they resemble unrelated individuals.
3. **The principle of selection.** Some forms are more successful at surviving and reproducing than are others in a given environment.

Clearly, a selective process can produce change in the population composition only if there are some variations to select among. If all individuals are identical, no amount of differential reproduction of individuals can affect the composition of the population. Furthermore, the variation must be in some part heritable if differential reproduction is to alter the population's genetic composition. If large animals have more offspring than small ones, but if their offspring are no larger on the average than those of small animals, no change from one generation to another in population composition can occur. Finally, if all variant types leave, on the average, the same number of offspring, then we expect the population to remain unchanged.

Message
Darwin's principles of variation, heredity, and selection must hold true if there is to be evolution by a variational mechanism.

Variation and Its Modulation

Population genetics is the translation of Darwin's three principles into precise genetic terms. Thus it deals with the following problems.

1. The description of the genetic variation within and between populations.
2. The study of the introduction of new variation into populations by mutation, recombination, and the migration of individuals.
3. The study of the patterns of differential reproduction of genotypes as a result of variation in mating patterns, fertility, and survival of individuals of different genotypes.

4. The creation of a formal machinery of deduction (theoretical population genetics) that allows prediction of the effects of the introduction of variation and of the differential rate of reproduction of variants on the genetic composition of a population.

5. The observation of actual changes in the composition of populations over time, either in nature or in controlled culture, and the comparison of these changes with those predicted from the theory, in order to check the adequacy of the entire theoretical structure.

6. The application of the theory to the controlled evolution of domesticated plants and animals, pests, pathogens, and other organisms relevant to human welfare.

Message
Population genetics is the study of inherited variation and its modulation in time and space.

Observations of Variation

Population genetics necessarily deals with genotypic variation, but only phenotypic variation can be observed. Some of this phenotypic variation has a simple one-to-one correspondence to genotype—for example, the phenotypic variation in amino-acid sequence of a protein, coded by allelic substitutions at a structural-gene locus. For reasons of experimental convenience, most population genetics of nondomesticated species has concentrated on such phenotypes. For plant and animal breeding, however, the characters of interest (such as yield, growth rate, rate, or body shape) typically are quantitative and have a complex relation to genotype. As a consequence, the study of such variation is a two-step process involving (1) the description of the phenotypic variation, and (2) a genetic analysis of that variation by the methods of Chapter 18.

The simplest description of Mendelian variation is the frequency distribution of genotypes in a population. Table 19-1 shows the frequency distribution of the three genotypes at the MN blood-group locus in several human populations. There is variation both within and between populations. More typically, instead of the frequencies of the diploid genotypes, the frequencies of the alternative alleles are used. If f_{AA}, f_{Aa}, and f_{aa} are the proportions of the three genotypes at a locus with two alleles, then the frequencies $p(A)$ and $q(a)$ of the alleles are obtained by counting alleles:

$$p = f_{AA} + \frac{1}{2} f_{Aa} = \text{frequency of } A$$

$$q = f_{aa} + \frac{1}{2} f_{Aa} = \text{frequency of } a$$

$$p + q = f_{AA} + f_{aa} + f_{Aa} = 1.00$$

Table 19-1. Frequencies of genotypes for alleles at the MN blood-group locus in various human populations

Population	Genotype			Allele frequencies	
	MM	*MN*	*NN*	*p*(*M*)	*q*(*N*)
Eskimos	0.835	0.156	0.009	0.913	0.087
Australian aborigines	0.024	0.304	0.672	0.176	0.824
Egyptians	0.278	0.489	0.233	0.523	0.477
Germans	0.297	0.507	0.196	0.550	0.450
Chinese	0.332	0.486	0.182	0.575	0.425
Nigerians	0.301	0.495	0.204	0.548	0.452

SOURCE: W. C. Boyd, *Genetics and The Races of Man.* Boston: D. C. Heath, 1950.

If there are multiple alleles, then the frequency for each allele is simply the frequency of its homozygote plus one-half the sum of the frequencies for all the heterozygotes in which it appears. Table 19-1 shows the values of p and q for each of the populations.

As an extension of p (the **gene frequency** or **allele frequency**), one can describe variation at more than one locus simultaneously by the **gametic frequencies.** There is a locus *S* (the secretor factor) closely linked to the MN locus in humans. Table 19-2 shows the gametic frequencies of the four gametic types. (*MS*, *Ms*, *NS*, and *Ns*) in various populations.

A *measure* of genetic variation (as opposed to its *description* by gene frequencies) is the amount of **heterozygosity** in a population, which is given by the total frequency of heterozygotes at a locus. If one allele is in very high frequency and all others are near zero, then there will be very little heterozygosity because, by necessity, most individuals will be homozygous for the common allele. We expect heterozygosity to be greatest when there are many alleles at a locus, all at equal frequency. In Table 19-1, the heterozygosity is simply equal to the frequency of the MN genotype in each population. When more than one locus

Table 19-2. Frequencies of gametic types for the MNS system in various human populations

Population	Gametic type				Heterozygosity (*H*)	
	MS	*Ms*	*NS*	*Ns*	From gametes	From alleles
Ainus	0.024	0.381	0.247	0.348	0.672	0.438
Ugandans	0.134	0.357	0.071	0.438	0.658	0.412
Pakistanis	0.177	0.405	0.127	0.291	0.704	0.455
English	0.247	0.283	0.080	0.390	0.700	0.469
Navahos	0.185	0.702	0.062	0.051	0.467	0.286

SOURCE: A. E. Mourant, *The Distribution of the Human Blood Groups,* Blackwell Scientific Pub., 1954.

is considered, there are two possible ways of calculating heterozygosity. First, one may average the frequency of heterozygotes at each locus separately. Alternatively, one may take the gametic frequencies, as in Table 19-2, and calculate the proportion of all individuals who carry two different gametic forms. So, for example, an individual MS/Ms is a heterozygote, even though it is heterozygous only at one of the two loci. These two methods do not, in general, give the same value for heterozygosity. In Table 19-2, the results of both calculations are given.

Simple Mendelian variation can be observed within and between populations of any species at various levels of phenotype, from external morphology down to the amino-acid sequence of enzymes and other proteins. Indeed, with the new methods of DNA sequencing, variations in DNA sequence (such as third-position variants that are not differentially coded in amino-acid sequences, and even variations in nontranslated intervening sequences) have been observed. Every species of organism ever examined has revealed considerable genetic variation, or **polymorphism,** that is reflected at one or more levels of phenotype, either within populations or between populations or both. Genetic variation that might be the basis for evolutionary change is ubiquitous. The tasks for population genetics are to describe that ubiquitous variation quantitatively in terms that allow predictions of evolution and to build a theory of evolutionary change that can use the observations in prediction.

It is quite impossible in this book to provide an adequate picture of the immense richness of genetic variation that exists in species. We can only give examples of the different kinds of Mendelian variation to provide a superficial sense of the genetic diversity within species. Each of these examples can be multiplied many times over in other species and with other traits.

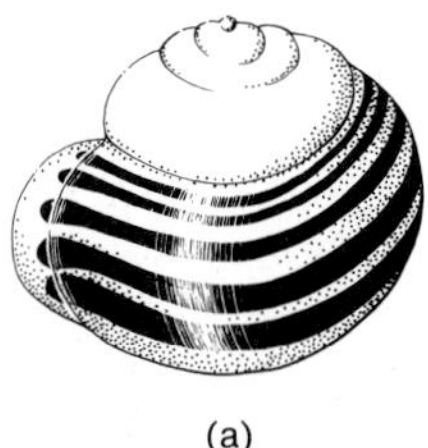

(a)

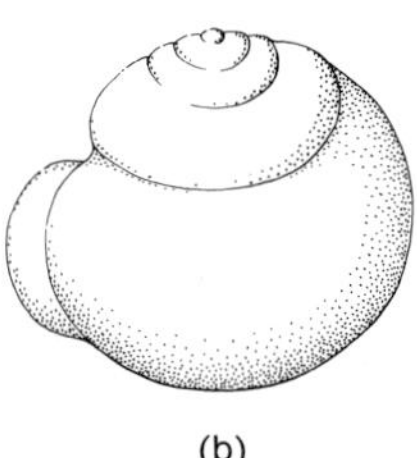

(b)

Figure 19-1. Shell patterns of the snail Cepaea nemoralis. *(a) Banded. (b) Unbanded.*

Morphological Variation. The shell of the land snail *Cepaea nemoralis* may be pink or yellow, depending on two alleles at a single locus, with pink dominant to yellow. Also, the shell may be banded or unbanded (Figure 19-1) as a result of segregation at a second (unlinked) locus, with unbanded dominant to banded. Table 19-3 shows the variation of these two loci in several European colonies of the snail. The populations also show polymorphism for the number of bands and the height of the shells, but these characters have a complex genetic basis.

Table 19-3. Frequencies of snails (*Cepaea nemoralis*) with different shell colors and banding patterns in three French populations

	Yellow		Pink	
Population	Banded	Unbanded	Banded	Unbanded
Guyancourt	0.440	0.040	0.337	0.183
Lonchez	0.196	0.145	0.564	0.095
Peyresourde	0.175	0.662	0.100	0.062

SOURCE: Lamotte, *Bulletin Biologique de France et Belgique,* supplement 35: 1-238, 1951.

Figure 19-2. Dimorphism in the most common species of mussel found on the west coast of North America. Two morphological forms are found wherever the species occur: the blue form (right) *and the brown form* (left). *Typically, the blue form is more frequent. The phenotypic difference is caused by an allelic difference at a single locus:* B = *blue and* b = *brown.*

Figure 19-2 shows another example of a naturally occurring morphological genetic polymorphism (in this case, a dimorphism). The natural population of this mussel species includes two forms, one blue and the other brown. The difference is caused by an allelic difference at one locus.

Examples of naturally occurring morphological variation within plant species were discussed in Chapter 1 (*Plectritis*), Chapter 2 (*Collinsia*), and Chapter 4 (clover).

Chromosomal Polymorphism. Although the karyotype is often regarded as a distinctive characteristic of a species, in fact numerous species are polymorphic for chromosome number and morphology. Extra chromosomes (supernumeraries), reciprocal translocations, and inversions segregate in many populations of plants, insects, and even mammals. Table 19-4 gives the frequencies of various inversions in some populations of *Drosophila pseudoobscura* from western North America. Although one chromosomal arrangement is called "Standard," it is in fact impossible to say which of these types is "normal" and which a derived inversion. Indeed, in Mexico, there are no "Standard" chromosomes.

allelic classes I^A, I^B, and i of the ABO blood group. Each point represents the allelic composition of a population, where the three allelic frequencies can be read by taking the lengths of the perpendiculars from each side to the point. The diagram shows that all human populations are bunched together in the region of high i, intermediate I^A, and low I^B frequencies. Moreover, neighboring points (enclosed by dashed lines) do not correspond to geographic races, so that geographic races are not distinguished from each other by characteristic allele frequencies for this gene. The study of polymorphic blood groups and enzyme loci in a variety of human populations has shown that about 85% of total human genetic diversity is found within local populations, about 8% between local populations within major geographical "races," and the remaining 7% between the major "races." Clearly the genes influencing skin color, hair form, and facial form that are well differentiated between races are not a random sample of structural gene loci.

Message
In general, the genetic difference between individuals within human races is much greater than the average difference between races.

Quantitative Variation

The variation in quantitative characters cannot be described in terms of allelic frequencies because individual loci and their alleles cannot be identified. Such variation can be characterized, however, by the amount of genetic variance (or the heritability of the trait) in the population. Figure 18-18 shows that many morphological and physiological traits in poultry have genetic variance of different amounts in different populations. A simple technique for estimating the additive genetic variance (Chapter 18) of a character is to choose two groups of parents that are extremely different and then measure the difference between their offspring groups. The difference between offspring groups divided by the difference between parental groups is a measure of the heritability, h^2. Where this technique has been applied to morphological variation in *Drosophila,* for example, virtually every variable trait is found to have some genetic variance, so evolution of the trait can occur. Indeed, the method of estimating heritability just described is itself a kind of one-generation artificial-selection experiment.

It should not be supposed that all variable traits are heritable, however. Certain metabolic traits (such as resistance to high salt concentrations in *Drosophila*) show individual variation but no heritability. Left–right asymmetry is also nonheritable: "left-winged" flies (those with left wings slightly longer than their right wings) have no more left-winged offspring than do their right-winged companions. In general, behavioral traits have lower heritabilities than morphological ones, especially in organisms with more complex nervous systems, where there is immense individual flexibility in central nervous states. Before any judgment can be made about the evolution of a particular quantitative trait, it is essential to determine whether there is genetic variance for it in the population whose evolution is to be predicted. Thus, suggestions that, in the human species, such traits as performance on IQ tests, temperament, or social organization are in the process of evolving or have evolved at particular epochs in human history, depend critically on evidence about genetic variation for these traits. No such evidence is presently available.

One of the most important findings in evolutionary genetics has been the discovery of substantial genetic variation underlying characters that show no morphological variation! These are called **canalized characters,** because the final outcome of their development is held within narrow bounds despite disturbing forces. Development is such that, for canalized characters, all the different genotypes have the same constant phenotype over the range of environments that is usual for the species. The genetic differences are revealed if the organisms are put in a stress environment or if a severe mutation stresses the developmental system. For example, all wild-type *Drosophila* have exactly four scutellar bristles (Figure 19-6). If the recessive mutant *scute* is present, the number of bristles is reduced, but, in addition, there is variation from fly to fly. This variation is heritable, and one can breed lines with 0 or 1 bristles and lines with 3 or 4 bristles in the presence of the *scute* mutation. When the mutation is removed, these lines now have 2 and 6 bristles, respectively. Similar experiments have been performed using environmental shock in place of mutants.

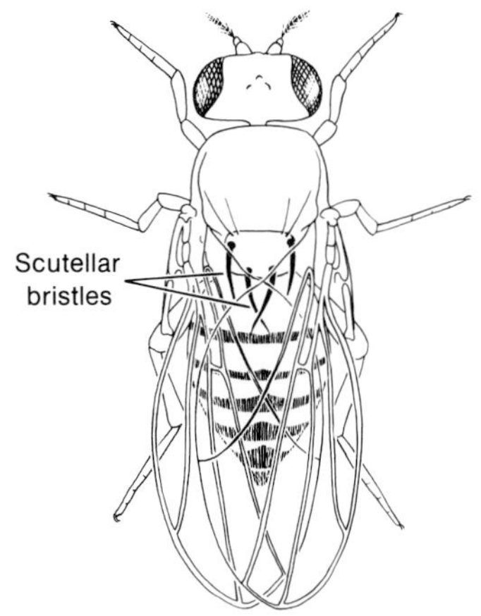

Figure 19-6. The scutellar bristles of the adult Drosophila. *This is an example of a canalized character—all wild-type* Drosophila *will have four scutellar bristles in a very wide range of environments.*

Message
Even characters with no apparent phenotypic variance may evolve when developmental conditions are changed drastically.

The Sources of Variation

The variational theory of evolution has a peculiar self-defeating property. If evolution occurs by the differential reproduction of different variants, we expect that the variant with the highest rate of reproduction will eventually take over the population, and all other genotypes will disappear. But then there is no longer any variation for further evolution. Genetic variation is the fuel for the evolutionary process, but differential reproduction consumes that fuel and so destroys the very condition necessary for further evolution. The possibility of continued evolution is critically dependent on renewed variation.

For a given population, there are three sources of such variation—mutation, recombination, and immigration of genes. However, recombination by itself does not produce variation unless there are already alleles segregating at different loci, because otherwise there is nothing to recombine, and immigration cannot provide variation if the entire species is homozygous for the same allele. Ultimately, the source of all variation must be mutation.

Variation from Mutations

Mutations are the *source* of variation, but the *process* of mutation does not itself drive evolution. The rate of gene-frequency change from the mutation process is very low because spontaneous mutation rates are low (Table 19-10). Let μ be the **mutation rate** from allele A to allele a (the probability that a gene copy A will become a during meiosis). If p_t is the frequency of the A allele in generation t, and $q_t = 1 - p_t$ is the frequency of the a allele, then the change in allelic frequency in one generation is

Table 19-10. Some point-mutation rates in different organisms

Organism	Gene	Mutation rate per generation
Bacteriophage	Host range	2.5×10^{-9}
Escherichia coli	Phage resistance	2×10^{-8}
Zea mays (corn)	*R* (color factor)	2.9×10^{-4}
Zea mays	*Y* (yellow seeds)	2×10^{-6}
Drosophila melanogaster	Average lethal	2.6×10^{-5}

SOURCE: Th. Dobzhansky, *Genetics and the Origin of Species,* 3rd ed., revised. New York: Columbia University Press, 1951.

$$\Delta p = p_t - p_{t-1} = -\mu p_{t-1}$$

That is, the frequency of A decreases (and the frequency of a increases) by an amount that is proportional to the mutation rate μ and to the proportion p of all the genes that are still available to mutate. Thus, Δp gets smaller as the frequency of p itself decreases, because there are fewer and fewer A alleles to turn into a alleles.

If we rearrange the formula for a change in allele frequency, we find that

$$p_t = p_{t-1} - \mu p_{t-1} = (1 - \mu)\, p_{t-1}$$

That is, the value of p in the next generation is simply $(1 - \mu)$ times the value in the previous generation. Then, after yet another generation,

$$p_{t+1} = (1 - \mu)p_t = (1 - \mu)^2\, p_{t-1}$$

and so on for as many generations as we please. After n generations of mutation, then, the value of p is

$$p_n = (1 - \mu)^n\, p_0$$

When a variable x is very small, the value of $(1 - x)^n$ is very close to e^{-nx}, where e is the base of the natural logarithms. So, we can make the approximation

$$p_n \cong p_0 e^{-n\mu}$$

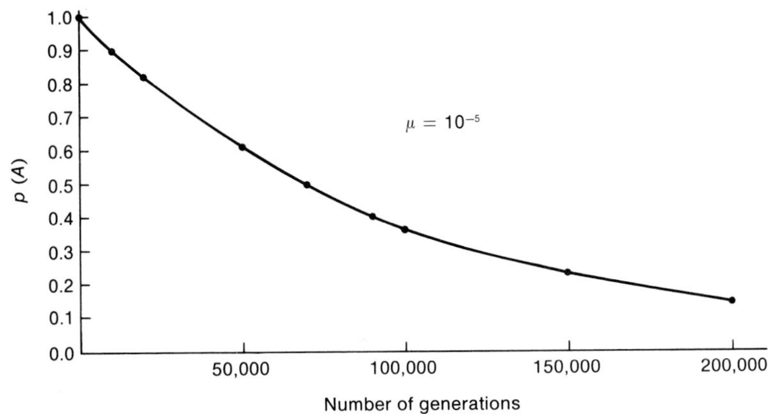

Figure 19-7. The change over generations in the frequency of a gene A *because of mutation from* A *to* a *at a constant mutation rate* (μ) *of* 10^{-5}.

which is shown in Figure 19-7. If μ is, for example, 10^{-5} (a rather high rate for mutations), then after 10,000 generations

$$p = p_0 e^{-(10^4)(10^{-5})} = p_0 e^{-0.1} = 0.904 p_0.$$

So, if the population starts with only A alleles ($p_0 = 1.0$), it would still have only 10% a alleles after 10,000 generations, and it would require 60,000 additional generations to reduce p to 0.5. Even if mutation rates were doubled (say, by environmental mutagens), the rate of evolution would be very slow. Radiation of intensity sufficient to double the mutation rate over the reproductive lifetime of an individual human would be more than the amount necessary to cause death, so rapid genetic change in the species would not be one of the effects of increased radiation. Although we have many things to fear from environmental radiation pollution, turning into a species of monsters is not one of them.

If we look at the mutation process from the standpoint of the increase of a particular new allele rather than the decrease of the old form, the process is even slower. Most mutation rates that have been determined are the sum of all mutations of A to any mutant form with a detectable effect. Any *specific* base substitution is likely to be at least two orders of magnitude lower in frequency than the sum of all changes. So, precise back-mutations to the original allele A are unlikely, although many mutations may produce alleles that are *phenotypically* similar to the original.

It is not possible to measure locus-specific mutation rates for continuously varying characters, but the rate of accumulation of genetic variance can be determined. Beginning with a completely homozygous line of *Drosophila* derived from a natural population, between 1/1000 and 1/500 of the genetic variance for bristle number in the original population is restored each generation by spontaneous mutation. On the other hand, for genetic variance in probability of survival, about 1/60 of the genetic variance present in a natural population is regenerated each generation in an inbred line.

Variation from Recombination

The creation of genetic variance by recombination can be a great deal faster than that due to mutation. When two chromosomes of *Drosophila* with "normal" survival are allowed to recombine for a single generation, they produce an array of chromosomes that have between 25% and 75% as much genetic variance in survival as does the original wild population from which the parent chromosomes were sampled. This is simply a consequence of the fact that a single homologous pair of chromosomes that is heterozygous at n loci (taking into account only single and double crossovers) can produce $n(n-1)/2$ new unique gametic types from one generation of recombination. If the heterozygous loci are well spread on the chromosomes, these new gametic types will be frequent, and a considerable variance will be generated.

Variation from Migration

A further source of variation is migration into a population from other populations with different gene frequencies. If p_t is the frequency of an allele in the recipient population in generation t, and P is the average allele frequency over all the donor populations, and m is the proportion of migrants, then in the next generation the gene frequency is the result of mixing $(1-m)$ genes from the population with m genes from the donor populations. Thus,

$$p_{t+1} = (1-m)p_t + mP = p_t + m(P - p_t)$$

and

$$\Delta p = p_{t+1} - p_t = m(P - p_t)$$

The change in gene frequency is proportional to the difference in frequency between the recipient population and the average of the donor populations. Unlike the mutation rate, the migration rate m can be large, so the change in frequency may be substantial.

We must understand *migration* as meaning any form of introduction of genes from one population into another. So, for example, genes from Europeans have "migrated" into the population of African origin in North America steadily since the Africans were introduced as slaves. One can use the frequency-change formula to determine the amount of this migration by looking at the frequency of an allele that is found only in Europeans and not in Africans and comparing the frequency among blacks in North America.

We can use the formula for the change in gene frequency from migration if we modify it slightly to take account of the fact that several generations of admixture have taken place. If the rate of admixture has not been too great, then (to a close order of approximation) the sum of the single-generation migration rates over several generations (let us call it M) will be related to the total change in the recipient population after these several generations by the same expression as that for changes due to migration. If, as before, P is the allele frequency in the donor population and p_0 the original frequency among the recipients,

$$\Delta p_{\text{Total}} = M(P - p_0)$$

so

$$M = \frac{\Delta p_{\text{Total}}}{P - p_0}$$

For example, the Duffy blood-group allele Fy^a is absent in Africa but has a frequency of 0.42 in whites from the state of Georgia. Among blacks from Georgia, the Fy^a frequency is 0.046. Therefore,

$$M = \frac{\Delta p_{\text{Total}}}{P - p} = \frac{0.046 - 0}{0.42 - 0} = 0.1095$$

When the same analysis is carried out on American blacks from Detroit or Oakland, M comes out to be 0.26 and 0.22, respectively, showing either (1) greater admixture rates in these cities than in Georgia or (2) differential movement into these cities by American blacks with more European ancestry. In any case the genetic variation at the *Fy* locus has been increased by this admixture.

The Origin of New Functions

Point mutations or chromosome rearrangements are themselves a limited source of variation for evolution because they can only alter a function or change one kind of function into another. To add quite new functions requires expansion in the total repertoire of genes through duplication and polyploidy, followed by a divergence between the duplicated genes, presumably by the usual process of mutation. Expansion of the genome by polyploidy has clearly been a frequent process, at least in plants. Figure 19-8 shows the frequency distribution of *haploid* chromosome numbers among dicotyledonous plant species. Even numbers are much more common than odd numbers, a consequence of frequent polyploidy.

Once an expansion in total DNA of the genome has occurred, it may require only a few base substitutions in a gene to provide it with a new function. For example, B. Hall has experimentally produced a gene with a new function in *E. coli.* In addition to the *lac Z* genes specifying the usual lactose-fermenting β-galactosidase activity in *E. coli,* there is another structural gene locus *ebg* that specifies another β-galactosidase that does not ferment lactose although it is induced by lactose. The natural function of this second enzyme is unknown. Hall was able to alter this gene into one specifying an enzyme that will ferment lactobionate. To do so, it was necessary to alter the regulatory element to a constitutive state and to produce three successive structural-gene mutations.

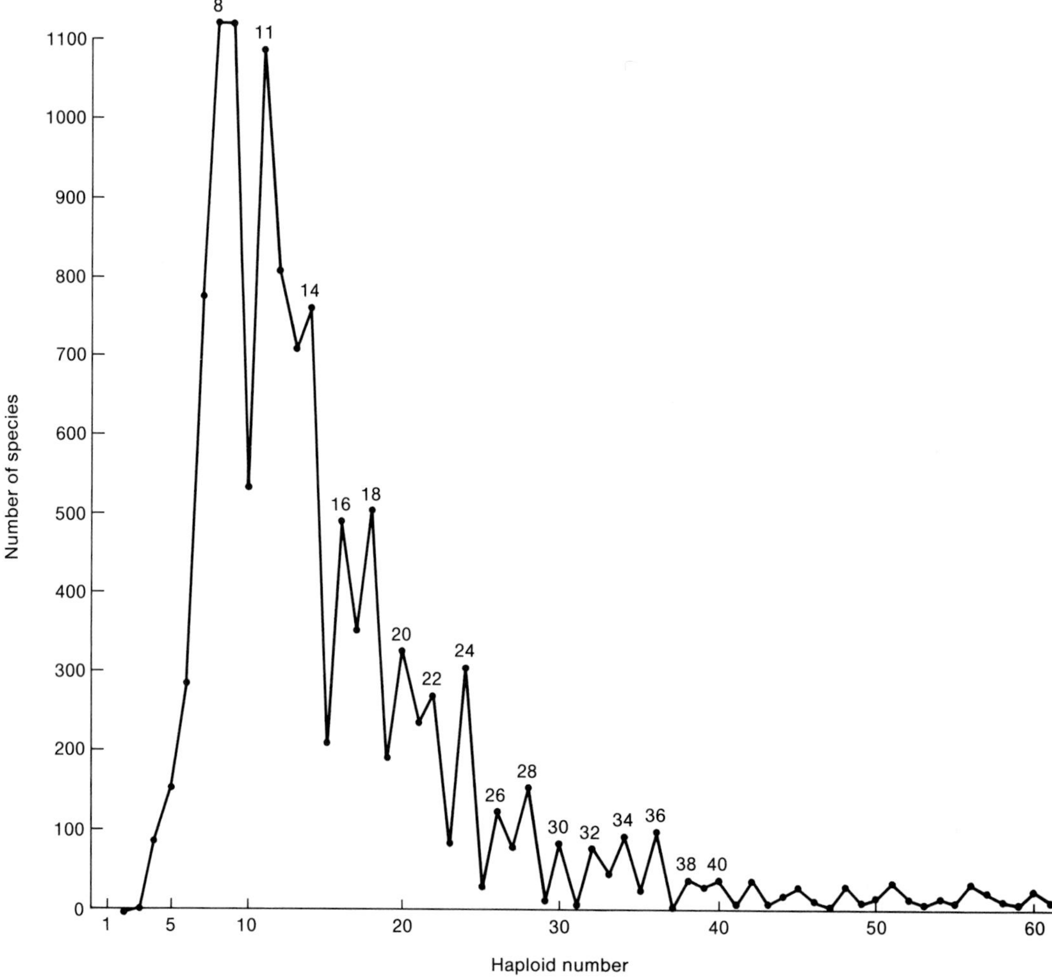

Figure 19-8. Frequency distribution of haploid chromosome numbers in dicotyledonous plants. (From Verne Grant, The Origin of Adaptations, *Columbia University Press, 1963.)*

Message
Evolution would come to a stop by running out of variation if new genetic variation were not added to populations by mutation, recombination, and migration. Ultimately, all new variation derives from gene and chromosome mutations.

The Effect of Sexual Reproduction on Variation

For the evolutionary theorists of the nineteenth century, there was a fundamental difficulty in Darwin's theory of evolution through natural selection. The possibility of continued evolution by natural selection is limited by the amount of genetic variation. But biologists of the nineteenth century, including Darwin, believed in one form or another of **blending inheritance,** a model postulating that the characteristics of each offspring are some intermediate mixture of the parental characters. Such a model of inheritance has fatal implications for a theory of evolution that depends on variation.

Suppose that some trait (say, height) has a distribution in the population and that individuals mate more or less at random. If intermediate individuals mate with each other, they will produce only intermediate offspring according to a blending model. But the mating of a tall with a short individual will also produce only intermediate offspring. Only the mating of tall with tall and short with short will preserve extreme types. The net result of all matings will be an increase in intermediate types and a decrease in the extremes. The variance of the distribution will shrink, simply as a result of sexual reproduction. In fact, it can be shown that the variance is *cut in half* each generation, so that before very many generations have passed the population will be essentially uniformly intermediate in height. There will then be no variation on which natural selection can operate. This was a very serious problem for the early Darwinists, and it was necessary for Darwin to assume that new variation is generated at a very rapid rate by the inheritance of characters acquired by individuals during their lifetimes.

The rediscovery of Mendelism changed the picture completely. The discrete nature of the Mendelian genes and the segregation of alleles at meiosis have the result that a cross of intermediate with intermediate individuals does *not* result in all intermediate offspring. On the contrary, extreme types (homozygotes) segregate out of the cross. To see the consequence of Mendelian inheritance for genetic variation, consider a population in which males and females mate with each other at random with respect to some gene locus *A,a*. That is, individuals do not choose their mates preferentially with respect to the partial genotype at the locus. Such random mating is equivalent to mixing all the sperm in the population together and all the eggs, and then matching randomly drawn sperm with randomly drawn eggs.

If the frequency of the allele A is p in both sperm and eggs, and the frequency of a is $q = 1 - p$, then the consequences of random unions of sperm and eggs are those shown in Figure 19-9. The probability that both sperm and egg carry A is $p \times p = p^2$, so this will be the frequency of AA homozygotes in the next generation. In like manner, the chance of heterozygotes Aa will be $(p \times q) + (q \times p) = 2pq$, and that of homozygotes aa will be $q \times q = q^2$. The three genotypes, after a generation of random mating, will be in the frequencies $p^2{:}2pq{:}q^2$. As the figure shows, the allele frequency of A has not changed and is still p. Therefore, in the second generation, the frequencies of the three genotypes will again be $p^2{:}2pq{:}q^2$, and so on forever.

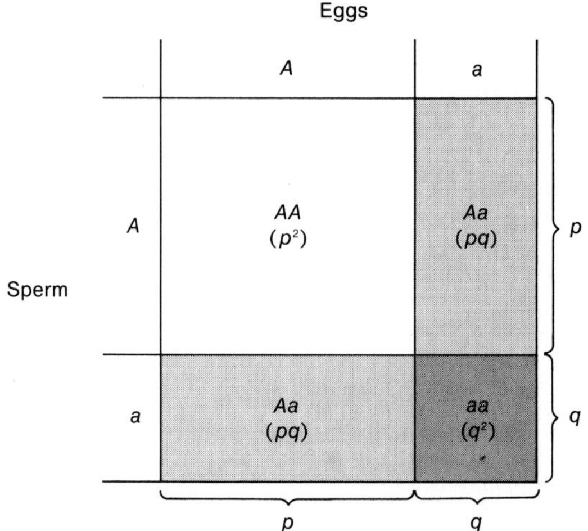

Figure 19-9. The Hardy–Weinberg equilibrium frequencies that result from random mating. The frequencies of A *and* a *among both eggs and sperms are* p *and* q *(= 1 −* p*), respectively. The total frequencies of the zygote genotypes are* p^2 *for* AA, $2pq$ *for* Aa, *and* q^2 *for* aa. *The frequency of the allele* A *in the zygotes is the frequency of* AA *plus one-half the frequency of* Aa, *or* $p^2 + pq = p(p + q) = p$.

Message
Mendelian segregation has the property that random mating results, after only one generation, in an equilibrium distribution of genotypes.

The equilibrium distribution

AA	*Aa*	*aa*
p^2	$2pq$	q^2

is called the **Hardy–Weinberg** (or Hardy–Weinberg–Tschetverikov) **equilibrium** after those who independently discovered it.

The Hardy–Weinberg equilibrium means that sexual reproduction does not cause a constant reduction in genetic variation each generation; on the contrary, the amount of variation remains constant generation after generation, in the absence of other disturbing forces. The equilibrium is the direct consequence of the segregation of alleles at meiosis in heterozygotes.

Numerically, the equilibrium shows that, irrespective of the particular mixture of genotypes in the parental generation, the genotypic distribution after one round of mating is completely specified by the allelic frequency p. For example, consider three hypothetical populations all having the same frequency of A ($p = 0.3$):

	AA	Aa	aa
I	0.3	0.0	0.7
II	0.2	0.2	0.6
III	0.1	0.4	0.5

After one generation of random mating, each of the three populations will have the same genotypic frequencies,

AA	Aa	aa
$(0.3)^2 = 0.09$	$2(0.3)(0.7) = 0.42$	$(0.7)^2 = 0.49$

and they will remain so indefinitely.

In our derivation of the equilibrium, we assumed that the allele frequency p is the same in sperm and eggs. If, in the initial generation, p for males is not equal to p for females, then it takes one generation to equalize the frequencies between the sexes and then a *second* generation to reach the Hardy–Weinberg equilibrium. As an extension of this effect, the Hardy–Weinberg equilibrium theorem does not apply to sex-linked genes even after two generations if males and females start with unequal gene frequencies (see Problem 3).

Random mating with respect to particular genes is quite common. Human beings, for example, do not choose mates with respect to their blood groups. Table 19-11 shows the result of sampling MN blood groups in various populations. Clearly, the populations are mating at random with respect to this locus.

Inbreeding and Assortative Mating

Random mating with respect to a locus is common, but it is not universal. There are two kinds of deviation from random mating that must be distinguished. First, individuals may mate with each other nonrandomly because of their degree of common ancestry or degree of genetic relationship. If mating between relatives occurs more commonly than would occur by pure chance, the population is

Table 19-11. Comparison between observed frequencies of genotypes for the MN blood-group locus and the frequencies expected from random mating

	Observed			Expected		
Population	*MM*	*MN*	*NN*	*MM*	*MN*	*NN*
Eskimos	0.835	0.156	0.009	0.834	0.159	0.008
Egyptians	0.278	0.489	0.233	0.274	0.499	0.228
Chinese	0.332	0.486	0.182	0.331	0.488	0.181
Australian aborigines	0.024	0.304	0.672	0.031	0.290	0.679

NOTE: The expected frequencies are computed according to the Hardy–Weinberg equilibrium, using the values of p and q computed from the observed frequencies.

inbreeding. If mating between relatives is less common than chance, the population is undergoing **enforced outbreeding,** or negative inbreeding.

Second, individuals may choose each other as mates, not because of their degree of genetic relationship but because of their resemblance to each other at some locus. If like mates with like, it is called **positive assortative mating.** Mating with unlike partners is called **negative assortative mating.**

Inbreeding levels in natural populations are a consequence of geographical distribution, of the mechanism of reproduction, and of behavioral characteristics. If close relatives occupy adjacent areas, then simple proximity may result in inbreeding. The seeds of many plants, for example, fall very close to the parental source, and pollen is not widely spread, so there will be a high frequency of sib mating. Some plants (such as corn) can be self-pollinated as well as cross-pollinated, so that wind pollination results in some very close inbreeding. Yet other plants, like the peanut, are obligatorily selfed. Many small mammals (such as house mice) live and mate in restricted family groups that persist generation after generation. Humans, on the other hand, generally have complex mating taboos and proscriptions that reduce inbreeding.

Assortative mating for some traits is common. In humans, there is positive assortative mating for skin color and height, for example. In plants and in many insects with one generation per year, there is positive assortative mating for time of development to sexual maturity. An important difference between assortative mating and inbreeding is that the former is specific to a trait whereas the latter applies to the entire genome. Individuals may mate assortatively with respect to height but at random with respect to blood group. Cousins, on the other hand, resemble each other genetically on the average to the same degree at all loci.

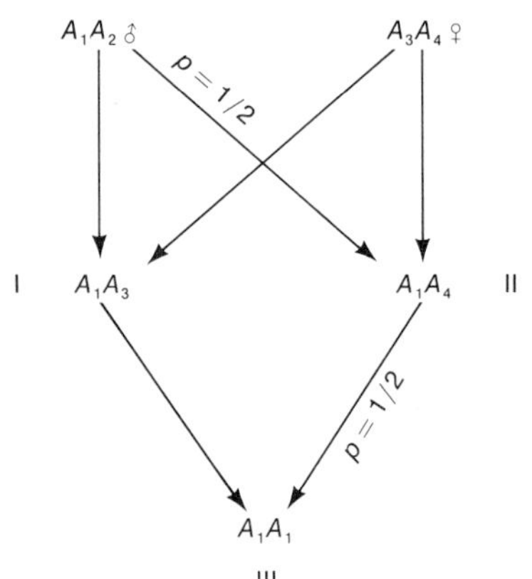

Figure 19-10. Calculation of homozygosity by descent for an offspring (III) of a brother–sister (I–II) mating. The probability that II will receive A_1 *from its father is 1/2, and, if it does, the probability that II will pass* A_1 *to the generation producing III is 1/2. Thus, the probability that III receives an* A_1 *from II is* $1/2 \times 1/2 = 1/4$.

For both positive assortative mating and inbreeding, the consequence to population structure is the same: There is an increase in homozygosity above the level predicted by the Hardy–Weinberg equilibrium. If two individuals are related, they have at least one common ancestor. Thus there is some chance that an allele carried by one of them and an allele carried by the other are both descended from the identical DNA molecule. The result is that there is an extra chance of **homozygosity by descent,** to be added to the chance of homozygosity ($p^2 + q^2$) that arises from the random mating of unrelated individuals. The probability of homozygosity by descent is called the **inbreeding coefficient,** F. Figure 19-10 illustrates the calculation of this probability of homozygosity by descent. Individuals I and II are full sibs because they share both parents. We label each allele in the parents uniquely to keep track of them. Individuals I and II mate to produce individual III. If individual I is A_1A_3 and the gamete that it contributes to III contains the allele A_1, then we would like to calculate the probability that the gamete produced by II is also A_1. The chance is 1/2 that II will receive A_1 from its father, and if it does so, the chance is 1/2 that II will pass A_1 to the gamete in question. Thus the probability that III receives an A_1 from II is $1/2 \times 1/2 = 1/4$, and this is the chance that III, the product of a full-sib mating, will be homozygous by descent.

To see one of the deleterious consequences of such close inbreeding, consider a rare deleterious allele a, which, when homozygous, causes a metabolic disorder. If the frequency of the allele in the population is p, then the probability of a random couple producing a homozygous offspring is only p^2. Thus if p is, say, 1/1000, then the frequency of homozygotes will be 1 in 1,000,000. But suppose now that the couple are brother and sister. If one of their common parents was a heterozygote for the disease, they may both receive it and both pass it on to the offspring they produce. The probability of a homozygous aa offspring is

$$\begin{aligned}&(\text{Prob one or the other grandparent is } Aa) \times (\text{Prob } a \text{ is passed to male sib}) \times \\ &(\text{Prob } a \text{ is passed to female sib}) \times (\text{Prob of a homozygous } aa \text{ offspring} \\ &\qquad \text{from } Aa \times Aa \text{ mating}) = (2pq + 2pq) \times 1/2 \times 1/2 \times 1/4 = pq/4\end{aligned}$$

We assume that the chance that both grandparents are Aa is negligible. If p is very small, then q is nearly 1.0, and the chance of an affected offspring is close to $p/4$. For $p = 1/1000$, there is 1 chance in 4000 of an affected child as compared to the one-in-a-million chance from a random mating. In general, for full sibs, the ratio of risks will be

$$\frac{\frac{p}{4}}{p^2} = \frac{1}{4p}$$

so the rarer the gene, the worse the relative risk of a defective offspring from inbreeding. For more distant relatives the chance of homozygosity by descent is, of course, less but still substantial. For first cousins, for example, the relative risk is $1/16p$ as compared to random mating.

The population consequences of inbreeding depend on its intensity and form. We next consider some examples.

Systematic Inbreeding. In experimental genetics (especially in plant and animal breeding), generation after generation of systematic selfing, full-sib, parent–offspring, or other form of mating between relatives may be used to increase homozygosity. Such systematic mating schemes, when they are between close relatives, lead eventually to complete homozygosity of the population, but at different rates. Table 18-1 shows the amount of heterozygosity still left within lines after various numbers of generations of inbreeding. Which allele is fixed within a line is a matter of chance. If, in the original population from which the inbred lines are taken, allele A has frequency p and a has frequency $q = 1 - \text{p}$, then a proportion p of the homozygous lines established by inbreeding will be homozygous AA, and q of the lines will be aa. What inbreeding does is to take the genetic variation present *within* the original population and to convert it into variation *between* homozygous inbred lines sampled from the population (Figure 19-11).

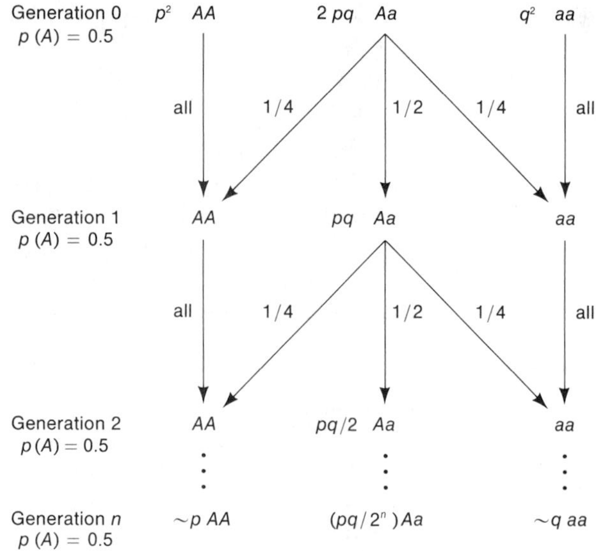

Figure 19-11. Repeated generations of self-fertilization (or inbreeding) will eventually split a heterozygous population into a series of completely homozygous lines. The frequency of AA *lines among the homozygous lines will be equal to the frequency of the allele* A *in the original heterozygous population.*

Random Inbreeding. In a natural population, there will be some fraction of mating between relatives (or even selfing if that is possible) because of spatial proximity. However, there is no continuity of inbreeding within any specific family. If some proportion of wind-pollinated plants are selfed in a particular generation, these are not necessarily the progeny of selfed plants in the previous generation, but they are distributed at random over selfed and outcrossed progeny. A consequence of such random inbreeding is that there is an equilibrium frequency of homozygotes and heterozygotes similar to the Hardy–Weinberg equilibrium but with more homozygotes. Thus, genetic variation is still preserved, in contrast to the result of systematic experimental inbreeding.

Small Populations. Suppose that a population is founded by some small number of individuals who mate at random to produce the next generation but that no further immigration into the population ever occurs again. (For example, the rabbits now in Australia probably descend from a single introduction of a few animals in the nineteenth century.) Then, in later generations, everyone is related to everyone else, because if their family trees are traced back they will be found to have common ancestors here and there in their pedigrees. Such a population is then inbred, in the sense that there is some probability of a gene being homozygous by descent. As generations go on, F increases and finally reaches 1.00, so that the population is totally homozygous. The rate of loss of heterozygosity in such a closed, finite, randomly breeding population is inversely proportional to the total number ($2N$) of haploid genomes, where N is the number of diploid individuals in the population. In each generation $1/2N$ of the remaining heterozygosity is lost so that

$$H_t = H_0\left(1 - \frac{1}{2N}\right)^t \cong H_0 e^{-t/2N}$$

where H_t and H_0 are the proportions of heterozygotes in the tth and original generation, respectively. As the number t of generations becomes very large, H_t approaches zero.

Which of the original alleles becomes fixed at each locus is a chance matter. If allele A_i has frequency p_i in the original population, then the probability is p_i that eventually the population will become homozygous A_iA_i. Suppose that a number of isolated island populations are founded from some large, heterozygous mainland population. Eventually, if these islands are completely isolated from each other, each will become homozygous for one of the alleles at each locus. Some will be homozygous A_1A_1, some A_2A_2, and so on. Thus the result of this form of inbreeding is to cause genetic differentiation between populations.

Message

Once again we see that inbreeding is a process that converts genetic variation within a population into differences between populations by making each separate population homozygous for a randomly chosen allele.

Within each population, there is a change in allele frequency from the original p_i to either 1 or 0, depending on which allele is fixed, but the average allele frequency over all such populations remains p_i. Figure 19-12 shows the distribution of allelic frequencies among islands in successive generations, where $p(A_1) = 0.5$. In generation zero, all populations are identical. As time goes on, the gene frequencies among the populations diverge and some become fixed. After about $2N$ generations, every allelic frequency except the fixed classes ($p = 0$ and $p = 1$) is equally likely, and about one-half of the populations are totally homozygous. By the time $4N$ generations have gone by, 80% of the populations are fixed, one-half of them being homozygous AA and one-half homozygous aa.

The process of differentiation by inbreeding in island populations is slow, but not on an evolutionary or geological time scale. If an island can support, say, 10,000 individuals of a rodent species, then after 20,000 generations (about 7000 years, assuming 3 generations per year), the population will be homozygous for about one-half of all the loci that were initially at the maximum of heterozygosity. Moreover, the island will be differentiated from other similar islands in two ways. For the loci that are fixed, many of the other islands will

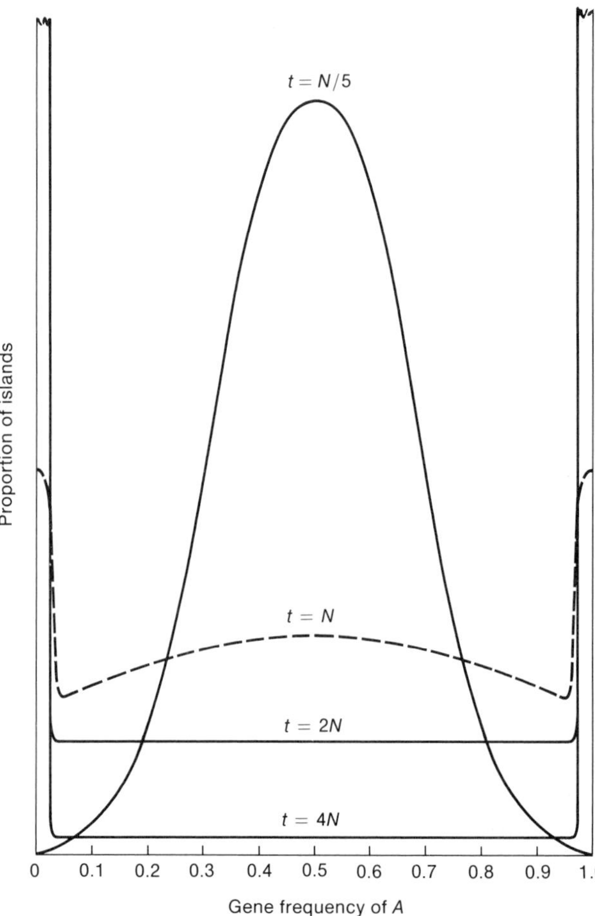

Figure 19-12. *Distribution of gene frequencies among island populations after various numbers of generations of isolation.*

still be segregating, whereas others will be fixed at a different allele. For the loci that are still segregating in all the islands, there will be a large variance in gene frequency from island to island, as shown in Figure 19-12.

The Balance Between Inbreeding and New Variation

Any population of any species is finite in size, so all populations should eventually become homozygous and differentiated from one another because of inbreeding. Evolution would then cease. In nature, however, new variation is always being introduced into populations by mutation and by some migration between localities. The actual variation available for natural selection thus is a balance between the introduction of new variation and its loss through local inbreeding. The rate of loss of heterozygosity in a closed population is $1/2N$, so any effective differentiation between populations will be negated if the rate of introduction of new variation is at this rate or higher. If m is the migration rate into a given population and μ is the rate of mutation to new alleles, then roughly (to an order of magnitude) a population will retain most of its heterozygosity and will not differentiate much from other populations by local inbreeding if

$$m \geq \frac{1}{N} \text{ or } \mu \geq \frac{1}{N}$$

or if

$$Nm \geq 1 \text{ or } N\mu \geq 1$$

For populations of intermediate and even fairly large size, it is unlikely that $N\mu \geq 1$. For example, if the population size is 100,000, then the mutation rate must exceed 10^{-5}, which is somewhat on the high side for known mutation rates, although not unknown. On the other hand, a migration rate of 10^{-5} per generation is not unreasonably large. In fact,

$$m = \frac{\text{Number of migrants}}{\text{Total population size}} = \frac{\text{Number of migrants}}{N}$$

Thus the requirement that $Nm \geq 1$ is equivalent to the requirement that

$$Nm = N \times \frac{\text{Number of migrants}}{N} \geq 1$$

or

$$\text{Number of migrant individuals} \geq 1$$

irrespective of population size! For many populations, more than a single migrant individual per generation is quite likely. Human populations (even isolated tribal populations) have more migration than this minimal value and, as a result, show remarkably little gene-frequency differentiation among populations. There is, for example, no locus known in humans for which one allele is fixed in some populations and an alternative allele in others (see Table 19-9).

Selection

Fitness and the Struggle for Existence

Darwin recognized that evolution consists of two processes, both of which must be explained. One is the origin of the *diversity* of organisms, and the second is the origin of the *adaptation* of these same organisms. Evolution is not simply the origin and extinction of different organic forms, but it is also a process that creates some kind of match between the phenotypes of species and the environments in which they live. Darwin regarded "organs of extreme perfection" (such as the eye) as tests of his theory. His explanation of such organs was that there is a constant *struggle for existence.* Organisms with phenotypes better suited to the environment have a greater probability of surviving the struggle and will leave more offspring. Presumably, the better an organism can see, the better is its chance to find food, defend itself, find mates, and so on. The relative probability of survival and rate of reproduction of a phenotype or genotype is now called its **Darwinian fitness.**

Although one sometimes speaks loosely of the fitness of an individual, the concept of fitness really applies to classes of individuals and is a statement about the average survival and reproduction of the individuals in that class. Because of chance events in the life histories of individuals, even two organisms with identical genotypes and identical environments will differ in their survival and reproduction. No evolutionary prediction can be made from the unique life history of a single organism. It is the fitness of a genotype on the average over all its possessors that matters.

Fitness is a consequence of the relationship between the phenotype of the organism and the environment in which the organism lives, so the *same genotype will have different fitnesses in different environments.* In part this is because different environments during development will result in different phenotypes for the same genotypes. But even if the phenotype is the same, the success of the organism depends on the environment. Having webbed feet is fine for paddling in water but a positive disadvantage for walking on land, as a few moments of observation of a duck will reveal. No genotype is unconditionally superior in fitness to all others in all environments.

Furthermore, the environment is not a fixed situation that is experienced passively by the organism. The environment of an organism is defined by the activities of the organism itself. Dry grass is part of the environment of a junco, and so juncos who are more efficient at gathering it may waste less energy in nest building and thus have a higher reproductive fitness. But dry grass is part of a junco's environment *because juncos gather it to make nests.* The rocks among which the grass grows are not part of the junco's environment, although the rocks are physically present there. However, the rocks are part of the environment of thrushes, who use them to break snails against. Moreover, the environment that is defined by the life activities of an organism evolves as a result of those activities. The structure of the soil that is in part determinative of the kinds of plants that will grow is altered by the growth of those very plants. Organisms define and alter the environment. Thus, as they evolve in response to the present environment, they find themselves in new environments that are direct consequences of their own evolution. Environment is both the cause and the result of evolution of organisms. Organisms are both the cause and result of changes in the environment. The human hand is at the same time the organ of human labor and the evolutionary product of that labor.

Darwinian or reproductive fitness is not to be confused with "physical fitness" in the everyday sense, although they may be related. No matter how strong, healthy, and mentally alert the possessor of a genotype may be, that genotype has a fitness of zero if, for some reason, the possessor is sterile. Thus, such statements as "the unfit are outreproducing the fit so the species may become extinct" are meaningless. By definition, the unfit cannot outreproduce the fit, although some aspect of the phenotype of the more fit may be disadvantageous for some purpose. The fitness of a genotype is a consequence of all the phenotypic effects of the genes involved. Thus, an allele that doubles the fecundity of its carriers while at the same time reducing the average lifetime of its possessors by 10% will be more fit than its alternatives, despite its life-shortening property. The most common example is parental care. An adult bird that expends a great deal of its energy gathering food for its young will have a lower probability of survival than one that keeps all the food for itself. But a totally selfish bird will leave no offspring because its young cannot fend for themselves. As a consequence, parental care is favored by natural selection.

Two Forms of the Struggle for Existence

Darwin saw the "struggle for existence" as having two quite different forms, with different consequences for fitness. In one form, the organism "struggles" with the environment directly. Darwin's example was the plant that is struggling for water at the edge of a desert. The fitness of a genotype in such a case does not depend on whether it is frequent or rare in the population, because fitness is not mediated through interactions of individuals but is a direct consequence of the physical relation to the external environment. Fitness is then **frequency independent.** Other examples are the differential probability of a seedling surviving freezing temperatures, or the differential ability of ground squirrels to dig burrows for nesting.

The other form of struggle is between organisms competing for a resource in short supply or otherwise interacting so that their relative abundances determine fitness. If prey are in short supply but relatively easy to catch, then the faster of two predators will have the higher fitness. Suppose that the faster lion, F, always wins out when it competes directly with the slower lion, S. Then, as F types become more numerous, the S type will have to compete directly with more F lions and so will have fewer and fewer chances to acquire prey. Thus its fitness will decrease as compared with that of F. A more complex example is mimicry in butterflies. Some species of butterflies (such as the brightly colored orange and black monarchs) are distasteful to birds, who learn, after a few trials, to avoid attacking them (Figure 19-13). It is then advantageous for a palatable species (such as the viceroy butterfly) to evolve to look like the distasteful one, because birds will avoid the tasty mimics as well as the distasteful models. But as the frequency of the mimics increases, birds will increasingly have the experience that butterflies with this morphology are, in fact, good to eat. They will no longer avoid them, and the mimics will lose their fitness advantage. These are examples of **frequency-dependent fitness.**

For reasons of mathematical convenience, most models to explain mechanisms of natural selection have been constructed with frequency-independent fitness. In actual fact, however, a very large number of selective processes (perhaps most) are frequency dependent. The kinetics of the evolutionary process depend on the exact form of frequency dependence, and, for that reason alone, it is difficult to make any generalizations. The result of positive frequency dependence (such as the competing predators, where fitness increases with

Figure 19-13. A blue jay eating a monarch butterfly, which induces vomiting in the jay. Because of this experience, the jay later will refuse to eat a viceroy butterfly that is similar in appearance to the monarch, although jays that have never tried monarchs will eat the viceroys with no ill effects. (Photographs courtesy of Lincoln Brower.)

increasing frequency) is quite different from the case of negative frequency dependence (such as the butterfly mimics, where fitness of a genotype declines with increasing frequency). For the sake of simplicity, and to illustrate the main qualitative features of selection, we deal only with models of frequency-independent selection in this chapter, but convenience should not be confused with reality.

Measuring Fitness Differences

For the most part, the differential fitness of different genotypes can be most easily measured when the genotypes differ at many loci. In very few cases (except for laboratory mutants, horticultural varieties, and major metabolic disorders) does the effect of an allelic substitution at a single locus make enough difference to the phenotype to be reflected in measurable fitness differences. Figure 19-14 shows the probability of survival from egg to adult (the **viability**) of a number of second-chromosome homozygotes of *Drosophila pseudoobscura* at three different temperatures. As is generally the case, the fitness (in this case, a component of the total fitness, viability) is different in different environments. A few homozygotes are lethal or nearly so at all three temperatures, whereas a few have consistently high viability. Most genotypes, however, are not con-

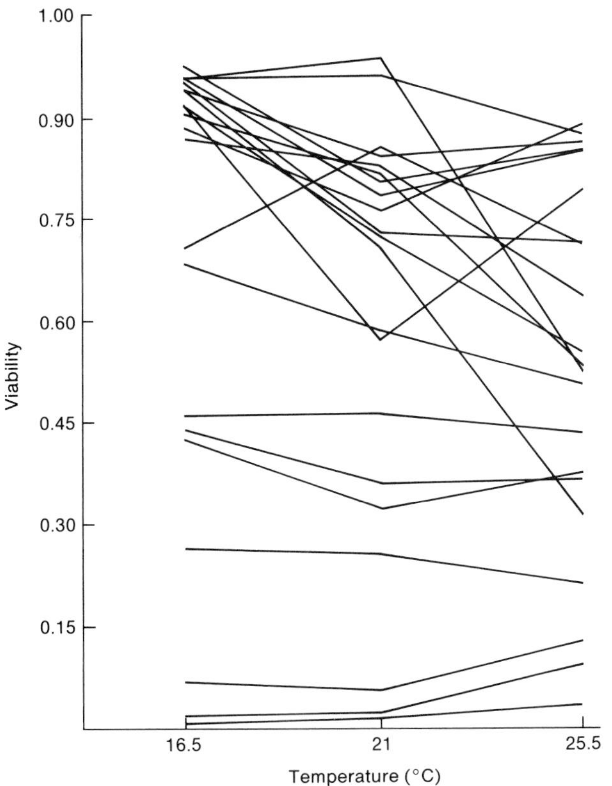

Figure 19-14. Viabilities of various chromosomal homozygotes of Drosophila pseudoobscura *at three different temperatures.*

sistent in viability between temperatures, and no genotype is unconditionally the most fit at all temperatures. The fitnesses of these chromosomal homozygotes were not measured in competition with each other, but all are measured against a common standard, so we do not know whether they are frequency dependent. An example of frequency-dependent fitnesses is shown in the estimates for inversion homozygotes and heterozygotes of *Drosophila pseudoobscura* in Table 19-12.

Table 19-12. Comparison of fitnesses for inversion homozygotes and heterozygotes in laboratory populations of *Drosophila pseudoobscura* when measured in different competitive combinations

	Homozygotes			Heterozygotes		
Experiment	*ST/ST*	*AR/AR*	*CH/CH*	*ST/AR*	*ST/CH*	*AR/CH*
ST and *AR* alone	0.8	0.5	—	1.0	—	—
ST and *CH* alone	0.8	—	0.4	—	1.0	—
AR and *CH* alone	—	0.86	0.48	—	—	1.0
ST, AR, and *CH* together	0.83	0.15	0.36	1.0	0.77	0.62

Examples of clearcut fitness differences associated with single gene substitutions are the many "inborn errors of metabolism," where a recessive allele interferes with a metabolic pathway and causes lethality of the homozygotes. Two examples in humans are phenylketonuria (where tissue degeneration is the result of the accumulation of a toxic intermediate in the pathway of tyrosine metabolism) and Wilson's disease (where death results from copper poisoning because the pathway of copper detoxification is blocked). A case that illustrates the relation of fitness to environment is sickle-cell anemia. An allelic substitution at the structural-gene locus for the β chain of hemoglobin results in a substitution of valine for the normal glutamic acid at chain position 6. The abnormal hemoglobin crystallizes at low oxygen pressure, and the red cells deform and hemolyze. Homozygotes $Hb^{S}Hb^{S}$ have a severe anemia, and survivorship is low. Heterozygotes have a mild anemia and under ordinary circumstances have the same or only slightly lower fitness than normal homozygotes $Hb^{A}Hb^{A}$. However, in regions of Africa with a high incidence of falciparum malaria, heterozygotes ($Hb^{A}Hb^{S}$) have a *higher* fitness than normal homozygotes because the presence of some sickling hemoglobin apparently protects them from the malaria. Where malaria is absent, as in North America, the fitness advantage of heterozygosity is lost.

In contrast to chromosomal homozygotes and metabolic diseases, it has not been possible to measure fitness differences for most single-locus polymorphisms. The evidence for differential net fitness for different ABO or MN blood types is shaky at best. The extensive enzyme polymorphism present in all sexually reproducing species is for the most part unconnected with measurable fitness differences, although in *Drosophila* clearcut differences in fitness of different genotypes have been demonstrated in the laboratory for a few loci, such as α-amylase and alcohol dehydrogenase.

How Selection Works

Suppose that a population is mating at random with respect to a given locus with two alleles and that the population is so large that (for the moment) we can ignore inbreeding. Just after eggs have been fertilized, the eggs will be in Hardy–Weinberg equilibrium:

Genotype	AA	Aa	aa
Frequency	p^2	$2pq$	q^2

and $p^2 + 2pq + q^2 = (p + q)^2 = 1.0$, where p is the frequency of A.

Further suppose that the three genotypes have probabilities of survival to adulthood (viabilities) of $W_{AA} : W_{Aa} : W_{aa}$. Then, among the progeny when they have reached adulthood, the frequencies will be

Genotype	AA	Aa	aa
Frequency	$p^2\ W_{AA}$	$2pq\ W_{Aa}$	$q^2\ W_{aa}$

These adjusted frequencies do not add up to unity. However, we can readjust them so that they do, without changing their relation to each other, by dividing each frequency by the sum of the frequencies after selection, $\overline{W}$:

$$\overline{W} = p^2\, W_{AA} + 2pq\, W_{Aa} + q^2\, W_{aa}$$

So defined, $\overline{W}$ is called the **mean fitness** of the population because it is, indeed, the mean of the fitnesses of all individuals in the population. After this adjustment, we have

Genotype	AA	Aa	aa
Frequency	$p^2 \dfrac{W_{AA}}{\overline{W}}$	$2pq \dfrac{W_{Aa}}{\overline{W}}$	$q^2 \dfrac{W_{aa}}{\overline{W}}$

We can now determine the frequency p' of the allele A in the next generation by counting up genes:

$$p' = AA + (1/2)Aa = p^2 \frac{W_{AA}}{\overline{W}} + \frac{pq\,W_{Aa}}{\overline{W}} = p\,\frac{p\,W_{AA} + q\,W_{Aa}}{\overline{W}}$$

Finally, we note that the expression $p\,W_{AA} + q\,W_{Aa}$ is the mean fitness of A alleles, because A alleles occur with frequency p in homozygotes with another A and in that condition have a fitness of W_{AA}, whereas they occur with frequency q in heterozygotes with a and have a fitness of W_{Aa}. Using $\overline{W}_A$ to denote $p\,W_{AA} + q\,W_{Aa}$, we can give the final answer for the new gene frequency:

$$p' = p\,\frac{\overline{W}_A}{\overline{W}}$$

In other words, after one generation of selection, the new value of the frequency of A is equal to the old value (p) multiplied by the ratio of the average fitness of A alleles to the fitness of the whole population. If the fitness of A alleles is greater than the average fitness of all alleles, then $\overline{W}_A / \overline{W}$ is greater than unity, and p' is larger than p. The allele A increases in the population. Conversely, if $\overline{W}_A / \overline{W}$ is less than unity, A decreases. But the mean fitness of the population, $\overline{W}$, is the average fitness of the A alleles and of the a alleles. So if $\overline{W}_A$ is greater than the mean fitness of the population, it must be greater than $\overline{W}_a$, the mean fitness of a alleles.

Message
The allele with the higher average fitness increases in the population.

It should be noted that the fitnesses W_{AA}, W_{Aa}, and W_{aa} may be expressed as absolute probabilities of survival and absolute reproduction rates, or they may all be rescaled relative to one of the fitnesses, which is given the standard value 1.0. This rescaling has absolutely no effect on the formula for p' because it cancels out in numerator and denominator.

Message
The course of selection depends only on relative fitnesses.

An increase in the allele with the higher fitness means that the average fitness of the population as a whole increases, so that selection can also be described as a process that *increases mean fitness.* This rule is strictly true only for frequency-independent genotypic fitnesses, but it is close enough to a general rule to be used as a fruitful generalization. This maximization of fitness does not necessarily lead to any optimal property for the species as a whole because fitnesses are only defined relative to each other within a population. It is relative and not absolute fitness that is increased by selection. The population does not necessarily become larger, faster growing, or less likely to become extinct.

The Rate of Change of Gene Frequency

An alternative way to look at the process of selection is to solve for the *change* in allele frequency in one generation:

$$\Delta p = p' - p = \frac{p\overline{W}_A}{\overline{W}} - p = \frac{p(\overline{W}_A - \overline{W})}{\overline{W}}$$

But $\overline{W}$, the mean fitness of the population, is the average of the allelic fitnesses, $\overline{W}_A$ and $\overline{W}_a$. That is,

$$\overline{W} = p\overline{W}_A + q\overline{W}_a$$

Substituting this expression for $\overline{W}$ in the formula for Δp, we obtain (after some algebraic manipulation)

$$\Delta p = \frac{pq(\overline{W}_A - \overline{W}_a)}{\overline{W}}$$

which is the general expression for a change in allele frequency as a result of selection. This general expression is particularly illuminating. It says that Δp will be positive (A will increase) if the mean fitness of A alleles is greater than that of a alleles, as we saw before. But it also shows that the speed of the change depends not only on the difference in fitness between the alleles but also on the factor pq, which is proportional to the frequency of heterozygotes ($2pq$). For a given difference in fitness of alleles, gene frequency will change most rapidly when the alleles A and a are in intermediate frequency so that pq is large. If p is near zero or 1 (that is, if A or a is nearly fixed), then pq is nearly zero and selection will proceed very slowly. Figure 19-15 shows the S-shaped curve that represents the course of selection of a new favorable allele A that has recently entered a population of homozygotes aa. At first, the change in frequency is very small because p is still close to zero. Then it accelerates as A becomes more frequent,

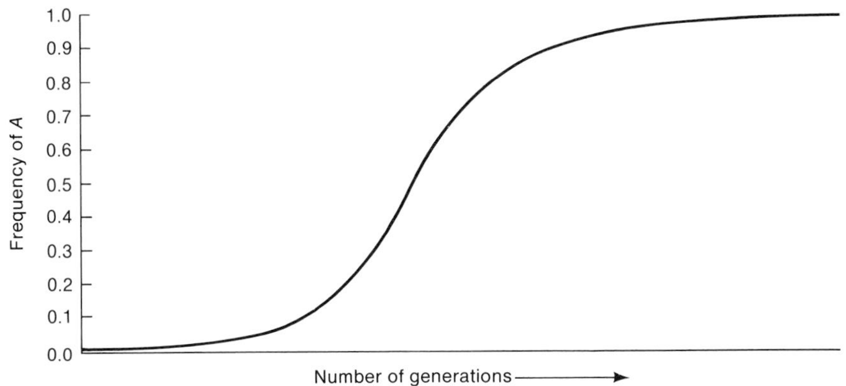

Figure 19-15. The time pattern of increasing frequency of a new favorable allele A *that has entered a population of* aa *homozygotes.*

but it slows down again as A takes over and a becomes very rare. This is precisely what is expected from a *selection* process. When most of the population is of one type, there is nothing to select. For evolution by natural selection, there must be genetic variance, and the more variance, the faster the process.

An important consequence of the dependence of selection on genetic variance can be seen in a case of artificial selection. In the early part of this century, it became fashionable to advocate a program of **negative eugenics.** It was proposed that individuals with certain undesirable genetic traits (say, metabolic or nervous disorders) should be prevented from having any offspring. By this means, it was thought, the frequency of the trait in the population would be lowered, and the trait could eventually be eradicated. Suppose that such a program is completely efficient, so that every homozygote aa is prevented from reproducing. The fitnesses of the genotypes then are

Genotype	AA	Aa	aa
Fitness	1.0	1.0	0

If p is the frequency of A, and q is the frequency of a, then

$$\begin{aligned}\overline{W}_A &= pW_{AA} + qW_{Aa} = p(1) + q(1) = 1.0\\ \overline{W}_a &= pW_{Aa} + qW_{aa} = p(1) + q(0) = p\\ \overline{W} &= p^2W_{AA} + 2pqW_{Aa} + q^2W_{aa} = p\overline{W}_A + q\overline{W}_a\\ &= p(1) + q(p) = p(1+q)\end{aligned}$$

so

$$q' = q\frac{\overline{W}_a}{\overline{W}} = q\frac{p}{p(1+q)} = \frac{q}{1+q}$$

If we iterate this formula over generations, we get

$$q'' = \frac{q'}{1 + q'} = \frac{\dfrac{q}{1+q}}{1 + \dfrac{q}{1+q}} = \frac{q}{1 + 2q}$$

$$q''' = \frac{q}{1 + 3q}$$

$$q^{(n)} = \frac{q}{1 + nq} = \frac{1}{n + \dfrac{1}{q}}$$

From this sequence we can see what would be the fate of a negative eugenics program. A deleterious gene will already be rare in a population. Suppose that $q = 1/100$, for example. Then after one generation the frequency would be 1/101, after two generations 1/102, and so on. It would take 100 generations to reduce the frequency to 1/200, and then another 200 generations again to cut it in half to 1/400. But a human generation is 25 years, so it would require 2500 years (the time since the founding of the Roman republic) with perfectly efficient selection against the recessive just to reduce the frequency from 1/100 to 1/200. The negative eugenics plan clearly is impractical.

Of course, if one could detect the *heterozygote* for the deleterious genes (as, for example, in sickle-cell anemia), then in a single generation all copies of the gene could be removed from the population if the heterozygotes were all prevented from having offspring. The only trouble with this suggestion is that every human being is heterozygous for several different deleterious recessive genes, so no one would be allowed to breed. Negative eugenics is no longer seriously proposed by geneticists.

When alternative alleles are not rare, selection can cause quite rapid changes in allele frequency. Figure 19-16 shows the course of elimination of a malic dehydrogenase allele in a laboratory population of *Drosophila melanogaster*. The fitnesses in this case are

$$W_{AA} = 1.0 \qquad W_{Aa} = 0.75 \qquad W_{aa} = 0.40$$

Of course, the frequency of *a* was not reduced to zero, and further reduction in frequency will require longer and longer times, as we showed in the negative eugenics case.

Message

Unless alternative alleles are present in intermediate frequencies, selection (especially against recessives) is quite slow. Selection depends on genetic variation.

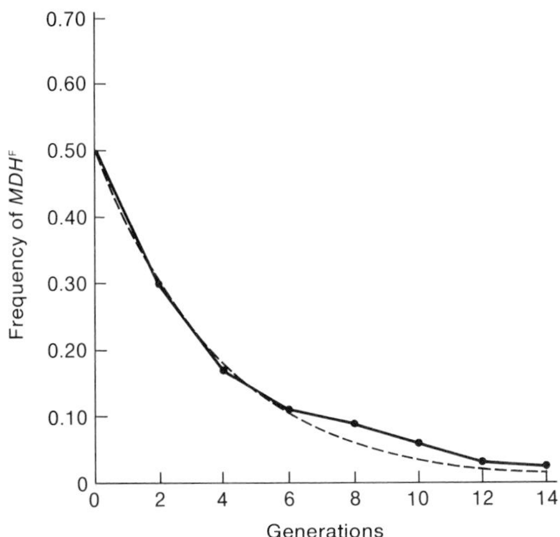

Figure 19-16. *The loss of an allele of the malic dehydrogenase locus,* MDH^F, *due to selection in a laboratory population of* Drosophila melanogaster. *The dashed line shows the theoretical curve of change computed for the fitnesses* $W_{AA} = 1.0$, $W_{Aa} = 0.75$, *and* $W_{aa} = 0.4$. *(From R. C. Lewontin,* The Genetic Basis of Evolutionary Change, *Columbia University Press, 1974.)*

Balanced Polymorphism

Examine the general formula for allele frequency change:

$$\Delta p = pq \frac{(\overline{W}_A - \overline{W}_a)}{\overline{W}}$$

Under what conditions will the process stop? When is $\Delta p = 0$? Two immediately obvious answers are when $p = 0$, or when $q = 0$—that is, when either allele A or allele a has been eliminated from the population. One of these events will eventually occur if $\overline{W}_A - \overline{W}_a$ is consistently positive or negative so that Δp is always positive or negative irrespective of the value of p. The condition for such unidirectional selection is that the heterozygote fitness be somewhere between the fitnesses of the two homozygotes:

$(\overline{W}_A - \overline{W}_a)$ positive: $W_{AA} \geq W_{Aa} \geq W_{aa}$ so A is favored
$(\overline{W}_A - \overline{W}_a)$ negative: $W_{AA} \leq W_{Aa} \leq W_{aa}$ so a is favored

But there is another possibility for $\Delta p = 0$, even when p and q are not zero. That is

$$\overline{W}_A = \overline{W}_a$$

in natural populations of this species. The fitnesses estimated for the three genotypes in the laboratory are

$$W_{ST/ST} = 0.89 \qquad W_{ST/CH} = 1.0 \qquad W_{CH/CH} = 0.41$$

Applying the formula for the equilibrium value $\hat{p}$ we obtain $\hat{p} = 0.85$, which agrees quite well with the observations in Figure 19-17.

Multiple Adaptive Peaks

We must avoid taking an overly simplified view of the consequences of selection. At the level of the gene, or even at the level of partial phenotype, the outcome of selection for a trait in a given environment is not unique. Selection to alter a trait (say, to increase size) may be successful in a number of ways. When F. Robertson and E. Reeve (1952) selected to change wing size in *Drosophila* in two different populations, they succeeded in both—but in one case the *number* of cells in the wing had changed, whereas in the other the *size* of the cells had changed. Two different genotypes have been selected, both causing a change in wing size. Which of these selections occurred depended on the initial state of the population when selection was started.

The way in which the same selection can lead to different outcomes can most easily be illustrated by a simple hypothetical case. Suppose there are two loci (there will usually be many more) whose variation influences a character and that (in a particular environment) intermediate phenotypes have the highest fitness. (For example, newborn babies have a higher chance of surviving birth if they are neither too big nor too small.) If the alleles act in a simple way in influencing the phenotype, then there are three genetic constitutions that would give high fitness: *Aa Bb*, *AA bb*, and *aa BB*, because all would be intermediate in phenotype. On the other hand, very low fitness will characterize the double homozygotes *AA BB* and *aa bb*. What will the result of selection be? We can predict the result by using the mean fitness $\overline{W}$ of a population. As previously discussed, selection acts in most simple cases to increase $\overline{W}$. Therefore, if we calculate $\overline{W}$ for every possible combination of gene frequencies at the two loci, we can find which combinations give high values of $\overline{W}$. Then we should be able to predict the course of selection, by following a curve of increasing $\overline{W}$. The surface of mean fitness for all possible combinations of allele frequency is called an **adaptive surface,** or an **adaptive landscape** (Figure 19-18). The figure is like a topographic map. The frequency of the allele A at one locus is plotted on one axis, and the frequency of allele B at the other locus is plotted on the other axis. The height above the plane (represented by topographic lines) is the value of $\overline{W}$ that the population would have for a particular combination of frequencies of A and B. According to the rule of increasing fitness, selection should carry the population from a low-fitness "valley" to a high-fitness "peak." However, Figure 19-18 shows that there are two adaptive peaks, corresponding to a fixed population of *AA bb* and a fixed population of *aa BB*, with an adaptive valley between them. Which peak the population will ascend, and therefore its final

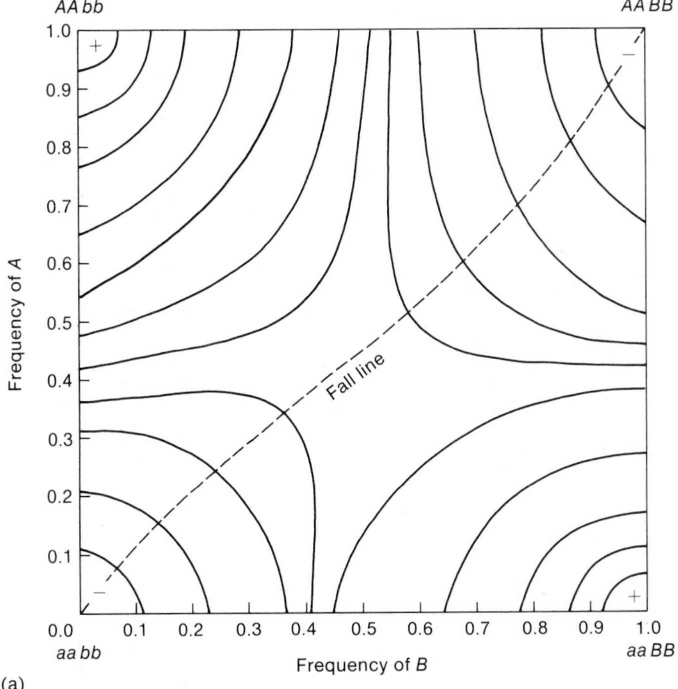

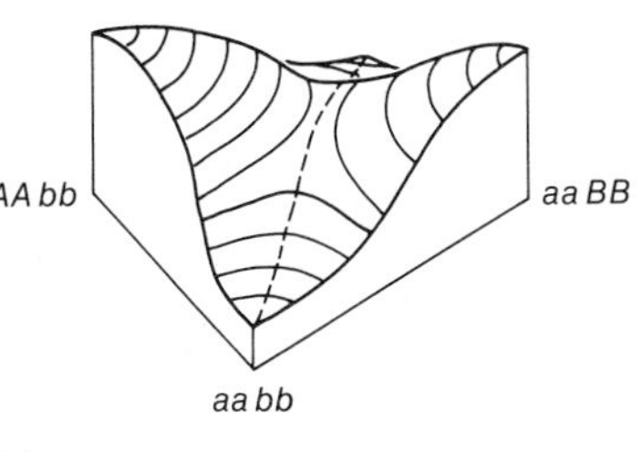

Figure 19-18. *An adaptive landscape with two adaptive peaks (+), two adaptive valleys (−), and a topographic saddle in the center of the landscape. The topographic lines are lines of equal mean fitness. If the genetic composition of a population always changes in such a way as to move the population "uphill" in the landscape, the final composition will depend on where the population began with respect to the fall line* (dashed line). *(a) Topographic map of the adaptive landscape.* *(b) A perspective sketch of the surface shown in the map.*

genetic composition, depends on whether the initial genetic composition of the population was on one side or the other of the "fall line" indicated by the dashed line.

Message
Under identical conditions of natural selection, two populations may arrive at two different genetic compositions as a direct result of natural selection

---

The existence of multiple adaptive peaks for a selective process means that some differences between species are the result of history and not of environmental differences. African rhinoceroses have two horns, and Indian rhinoceroses have one (Figure 19-19). We need not invent a special story to explain why it is better to have two horns on the African plains and one in India. It is

(a)

(b)

Figure 19-19. Differences in horn morphology in two geographically separated species of rhinoceroses. (a) The African rhinoceros. (b) The Indian rhinoceros. (Part a from Leonard Lee Rue, copyright © Tom Stack & Associates; part b copyright © Tom Stack & Associates.)

much more plausible that the trait of having horns was selected, but that having two long, slender horns or one short, stout horn are simply alternative adaptive peaks, and that historical accident differentiated the species. Explanations by natural selection do not require that every difference between species be differentially adaptive.

It is important to note that nothing in the theory of selection requires that the different adaptive peaks be of the same height. The kinetics of selection is such that $\overline{W}$ increases, not that it necessarily reaches the highest possible peak in the field of gene frequencies. Suppose, for example, that a population is near the peak *AA bb* in Figure 19-18 and that this peak is lower than the *aa BB* peak. Selection alone cannot carry the population to *aa BB* because that would require a temporary decrease in $\overline{W}$ as the population descended the *AA bb* slope, crossed the saddle, and ascended the other slope. Thus, the force of selection is myopic. It drives the population to a *local* maximum of $\overline{W}$ in the field of gene frequencies, not to a *global* one.

Artificial Selection

In contrast to the difficulties of finding simple, well-behaved cases in nature that will exemplify the simple formulas of natural selection, there is a vast record of the effectiveness of artificial selection in changing populations phenotypically. These changes have been produced by laboratory selection experiments and by selection of animals and plants in agriculture—for example, for increased milk production in cows or for rust resistance in wheat. No analysis of these experiments in terms of allelic frequencies is possible because individual loci have not been identified and followed. Nevertheless, it is clear that genetic changes have occurred in the populations and that some analysis of selected populations has been carried out by the methods of Chapter 18. Figure 19-20 shows, as an example, the large changes in average bristle number achieved in a selection

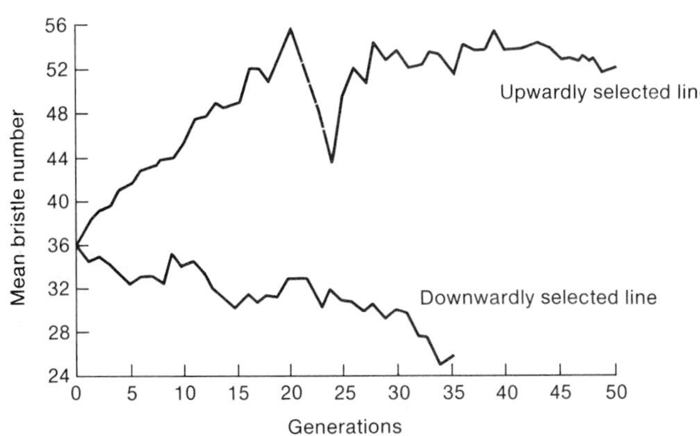

Figure 19-20. Changes in average bristle number obtained in two laboratory populations of Drosophila melanogaster *through artificial selection for high bristle number in one population and for low bristle number in the other. The dashed segment in the curve for the upwardly selected line indicates a period of five generations during which no selection was performed. (From K. Mather and B. J. Harrison, "The Manifold Effects of Selection,"* Heredity *3:1–52, 1949.)*

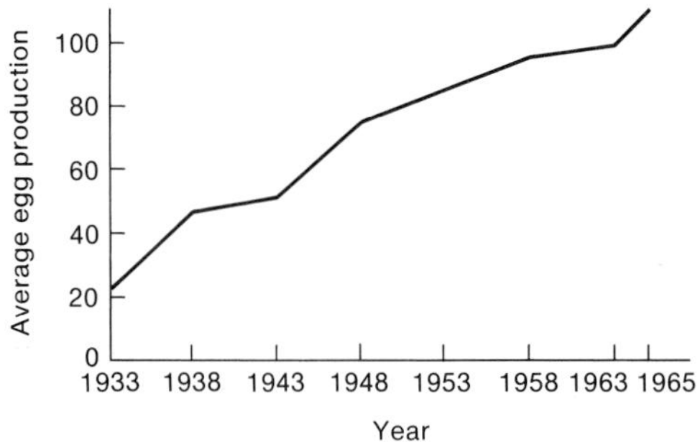

Figure 19-21. Changes in average egg production in a population selected for increase in egg-laying rate over a period of 30 years. (From Heredity, Evolution, and Society, *2nd. ed., by I. M. Lerner and W. J. Libby. W. H. Freeman and Company. Copyright © 1976. Data courtesy of D. C. Lowry.)*

experiment with *Drosophila melanogaster.* Figure 19-21 shows the change in the number of eggs laid per chicken as a consequence of 30 years of selection.

For characters of high heritability, the usual method of selection is **truncation selection.** The individuals in a given generation are pooled (irrespective of their families), a sample is measured, and only those individuals above (or below) a given phenotypic value (the truncation point) are chosen as parents for the next generation. This phenotypic value may be a fixed value over successive generations (**constant truncation**), or more commonly the highest (or lowest) K% of the population is chosen (**proportional truncation**). With constant truncation, the intensity of selection decreases with time, as more and more of the population exceeds the fixed truncation point. With proportional truncation, the intensity of selection is constant, but the truncation point moves upward as the population distribution moves. Figure 19-22 shows these two schemes.

For characters of low heritability, some sort of family selection is used. A number of mated pairs are separated, and their progeny are measured. Those families of progeny with the highest mean value are chosen for the next generation, even though some of the individuals within the family may fall below the selection criterion. If there is a low heritability, the mean of the family gives a much better estimate of the genetic differences between parental groups than do the individual phenotypes. Sometimes family and individual selection are combined, choosing the best families and then the best individuals within those families.

A common experience in artificial selection programs is that, as the population becomes more and more extreme, the viability and fertility decreases. As a result, no further progress under selection is possible, despite the presence of genetic variance for the character, because the selected individuals do not reproduce. The loss of fitness may be a direct phenotypic effect of the genes for the selected character, in which case nothing much can be done to improve the population further. Often, however, the loss of fitness comes from linked sterility genes that are carried along with the selected loci. In such cases, a number of

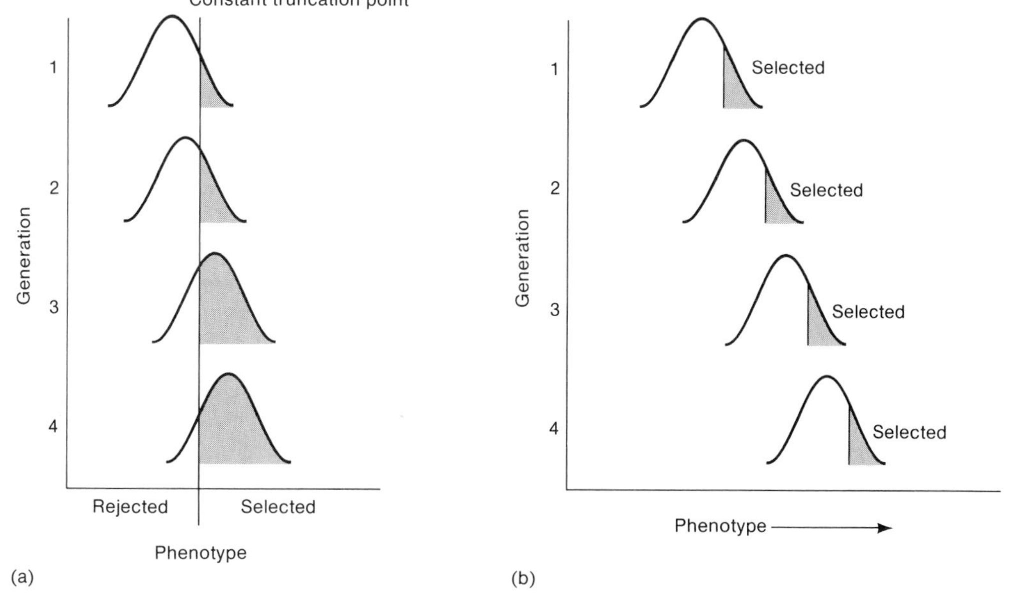

Figure 19-22. Two schemes of truncation selection for a continuously varying trait. (a) Constant truncation. (b) Proportional truncation.

generations without selection allow recombinants to be formed, and selection can then be continued.

One must be very careful in the interpretation of long-term agricultural selection programs. In the real world of agriculture, changes in cultivation methods, machinery, fertilizer, insecticides, herbicides, and so on are occurring along with the production of genetically improved varieties. Increases in average yields are consequences of all these changes. For example, the average yield of corn in the United States increased from 40 bushels to 80 bushels per acre between 1940 and 1970. But experiments with reconstruction of varieties and testing in common environments show that only about one-half of this increase is a direct result of new corn varieties, the other one-half being a result of improved farming techniques. Furthermore, the new varieties are most superior to the old ones at the high densities of modern planting for which they were selected.

Random Events

If a population is finite in size (as all populations are), and if a given pair of parents have only a small number of offspring, then, even in the absence of all selective forces, the frequency of a gene will not be exactly reproduced in the next generation. There is sampling error. If, in a population of 1000 individuals, the frequency of a is 0.5 in one generation, then in the next generation it may by chance be 0.493 or 0.505 because of the chance production of a few more or a few less progeny of each genotype. But in the second generation, there is again a sampling error based on the new gene frequency, so it may go from 0.505 to 0.511, or back to 0.498. This process of random fluctuation continues generation after generation, with no force pushing the frequency back to its initial state because the population has no "genetic memory" of its state many generations ago. Each generation is an independent event. The final result of this random change in allele frequency is that the population eventually drifts to $p = 1$ or $p = 0$, after which no further change is possible. It becomes homozygous. A different population, isolated from the first, also undergoes this **random genetic drift,** but it may become homozygous for allele A, whereas the first one drifted to homozygous a. As time goes on, isolated populations diverge from each other, each losing heterozyosity. The variation originally present *within* populations now appears as variation *between* populations.

The process just described should sound familiar. It is, in fact, another way of looking at the inbreeding effect in small populations discussed earlier. Whether regarded as inbreeding or as random sampling of genes, the effect is the same. Populations do not exactly reproduce their genetic constitutions; there is a random component of gene frequency change.

One result of random sampling is that most new mutations, even if they are not selected against, never succeed in entering the population. Suppose that a single individual is heterozygous for a new mutation. There is some chance that the individual in question will have no offspring at all. Even if it has one offspring, there is a chance of 1/2 that the new mutation will not be transmitted. If it has two offspring, the chance that neither carries the new mutation is 1/4, and so on. Suppose that the new mutation is successfully transmitted to an offspring. Then in the next generation the lottery is repeated, and again the allele may be lost. In fact, if a population is of size N, the chance that a new mutation is eventually lost by chance is $(2N - 1)/2N$. But if the new mutation is not lost, then the only thing that can happen to it in a finite population is that eventually it will sweep through the population and become fixed! This event has the probability of $1/2N$. In the absence of selection, then, the history of a population looks like Figure 19-23. For some period of time, it is homozygous. Then a new mutation appears. The new mutant allele will be lost immediately or very soon in most cases. Occasionally, however, a new mutant allele drifts through the population, and the population becomes homozygous for the new allele. The process then begins again.

Message

New mutations can become established in a population even though they are not favored by natural selection simply by a process of random genetic drift.

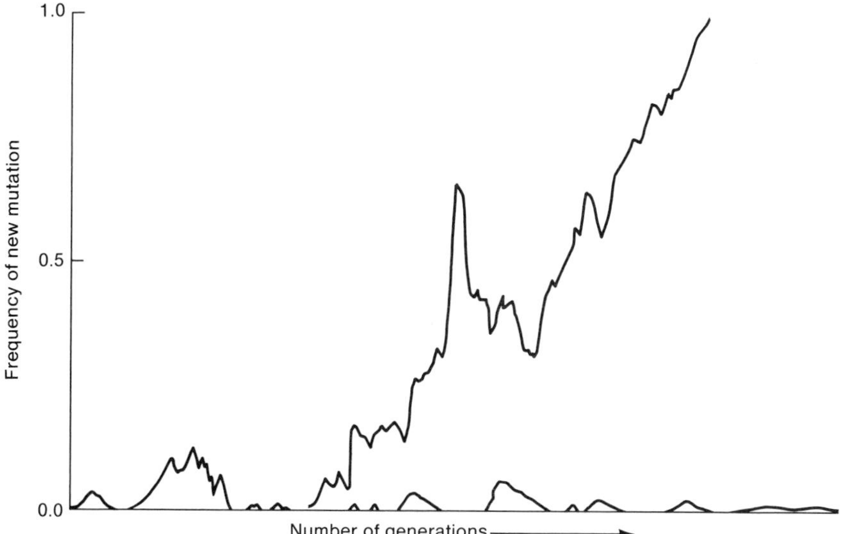

Figure 19-23. The appearance, loss, and eventual incorporation of new mutations during the life of a population. If random genetic drift does not cause the loss of a new mutation, then it must eventually cause the entire population to become homozygous for the mutation (in the absence of selection). (After J. Crow and M. Kimura, An Introduction to the Population Genetics Theory, *Harper & Row, 1970.)*

A Synthesis of Forces

The genetic variation within and between populations is a result of the interplay of the various evolutionary forces (Figure 19-24). Generally, as Table 19-13 shows, those forces that increase or maintain variation within populations prevent the differentiation of populations from each other, whereas the divergence of populations is a result of forces that make each population homozygous. Thus random drift (inbreeding) produces homozygosity while causing different populations to diverge. This divergence and homozygosity is counteracted by the constant flux of mutation and the migration between localities, which introduce variation into the population again and tend to make them more like each other.

The effects of selection are more variable. Directional selection pushes a population toward homozygosity, rejecting most new mutations as they are introduced but occasionally (if the mutation is advantageous) spreading a new allele through the population to create a new homozygous state. Whether or not such directional selection promotes differentiation of populations depends on the environment and on chance events. Two populations living in very similar environments may be kept genetically similar by directional selection, but, if there are environmental differences, selection may drive the populations to different compositions. Advantageous new mutations are rare, so that a given

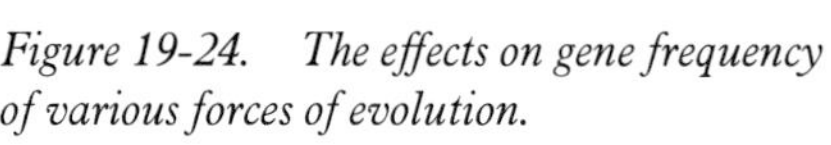

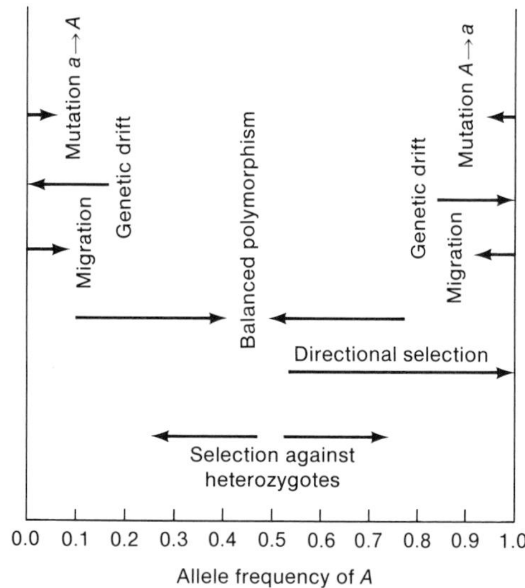

Figure 19-24. The effects on gene frequency of various forces of evolution.

Table 19-13. How the forces of evolution increase (+) or decrease (−) the variation within and between populations

Force	Variation within	Variation between
Inbreeding or genetic drift	−	+
Mutation	+	−
Migration	+	−
Selection:		
directional	−	+/−
balancing	+	−
incompatability	−	+

mutation may occur in one population but not (for a very long time) in others. Directional selection will then, temporarily, cause divergence of the population in which the mutation has appeared. Given enough time, of course, the mutation should be incorporated in all populations, especially if there is migration between them. But populations and species do not last forever, so directional selection operating on rare mutants may in fact be a cause of much divergence.

A particular case of interest, especially in human populations, is the interaction between mutation and directional selection in a very large population. New deleterious mutations are constantly arising spontaneously or as the result of the action of mutagens. These mutations may be completely recessive or partly dominant. Selection removes them from the population, but there will be an equilibrium between their appearance and removal. Let q be the frequency of the deleterious allele a, and let $p = 1 - q$ be the frequency of the normal allele.

The change in allele frequency due to the mutation rate μ is

$$\Delta q_{\text{mut}} = \mu p$$

For a recessive deleterious gene, a simple way to express the fitnesses is $W_{AA} = W_{Aa} = 1.0$ and $W_{aa} = 1 - s$, where s is the loss of fitness in homozygotes. We can now substitute these fitnesses in our general expression for allelic frequency change, $\Delta p = pq(W_A - W_a)/\overline{W}$ as follows:

$$W_A = pW_{AA} + qW_{Aa} = p(1) + q(1) = 1$$
$$W_a = pW_{Aa} + qW_{aa} = p(1) + q(1 - s) = p + q - sq = 1 - sq$$

so

$$\overline{W} = pW_A + qW_a = p(1) + q(1 - sq) = p + q - sq^2 = 1 - sq^2$$

Thus

$$\Delta p = \frac{pq(W_A - W_a)}{\overline{W}} = \frac{pq(sq)}{1 - sq^2}$$

but the change in q, the frequency of the a allele, must be equal in magnitude and opposite in sign to the change in p, the frequency of the A allele:

$$\Delta q = -\Delta p$$

Thus, we finally have

$$\Delta q_{\text{sel}} = \frac{-pq\,(sq)}{1 - sq^2}$$

The requirement for equilibrium is that

$$\Delta \hat{q}_{\text{mut}} + \Delta \hat{q}_{\text{sel}} = 0$$

Remembering that $\hat{q}$ at equilibrium will be quite small, so that $1 - s\hat{q}^2 \cong 1$, we have

$$\mu\hat{p} - \frac{s\hat{p}\hat{q}^2}{1 - s\hat{q}^2} \cong \mu\hat{p} - s\hat{p}\hat{q}^2 = 0$$

or

$$\hat{q} = \sqrt{\frac{\mu}{s}}$$

So, for example, a recessive lethal ($s = 1$) mutating at the rate $\mu = 10^{-6}$ will have an equilibrium frequency of 10^{-3}. Indeed, if we knew that a gene was a recessive lethal and had no heterozygous effects, we could estimate its mutation rate as the square of the allele frequency. One must be very careful, however. Sickle-cell anemia was once thought to be a recessive lethal with no heterozygous effects, which led to an estimated mutation rate in Africa of 0.1 for this locus!

For a partly dominant deleterious gene, let the fitnesses be $W_{AA} = 1.0$, $W_{Aa} = 1 - hs$, and $W_{aa} = 1 - s$. Then a similar calculation gives

$$\hat{q} = \frac{\mu}{hs}$$

where h is the degree of dominance of the deleterious allele. So, if $\mu = 10^{-6}$, and the lethal is not totally recessive but has a 5% deleterious effect in heterozygotes ($s = 1.0, h = 0.05$), then

$$\hat{q} = \frac{10^{-6}}{5 \times 10^{-2}} = 2 \times 10^{-5}$$

which is smaller by two orders of magnitude than the equilibrium frequency for the purely recessive case.

Selection favoring heterozygotes (balancing selection) will, for the most part, maintain more-or-less similar polymorphisms in different populations. However, again, if environment is different enough between them, the populations will show some divergence. The opposite of balancing selection is selection against heterozygotes, which produces unstable equilibria. Such selection will cause homozygosity and divergence between populations.

The Exploration of Adaptive Peaks

Random and selective forces should not be thought of as simple antagonists. The outcome of the evolutionary process is a result of their simultaneous operation. Random drift may counteract the force of selection, but it may enhance it as well. Figure 19-25 illustrates these possibilities. There are multiple adaptive peaks. Because of random drift, a population under selection does not ascend an adaptive peak smoothly. Instead, it takes an erratic course in the field of gene frequencies, like a drunken mountain climber. Pathway I shows a population history where adaptation has failed. The random fluctuations of gene frequency were sufficiently great that the population by chance went to fixation at an unfit genotype. In any population, some proportion of loci are fixed at a selectively unfavorable allele because the intensity of selection is insufficient to overcome the random drift to fixation. Very great skepticism should be maintained toward naive theories about evolution that assume that populations always or nearly always reach an optimal constitution under selection. The existence of multiple adaptive peaks and the random fixation of less fit alleles are integral features of the evolutionary process. Natural selection cannot be relied upon to produce the best of all possible worlds.

Pathway II in Figure 19-25, on the other hand, shows how random drift may improve adaptation. The population was originally in the sphere of influence of the lower adaptive peak, but, by random fluctuation in gene frequency, its composition passed over the adaptive saddle, and the population was captured by the higher, steeper adaptive peak. This passage from a lower to a higher adaptive

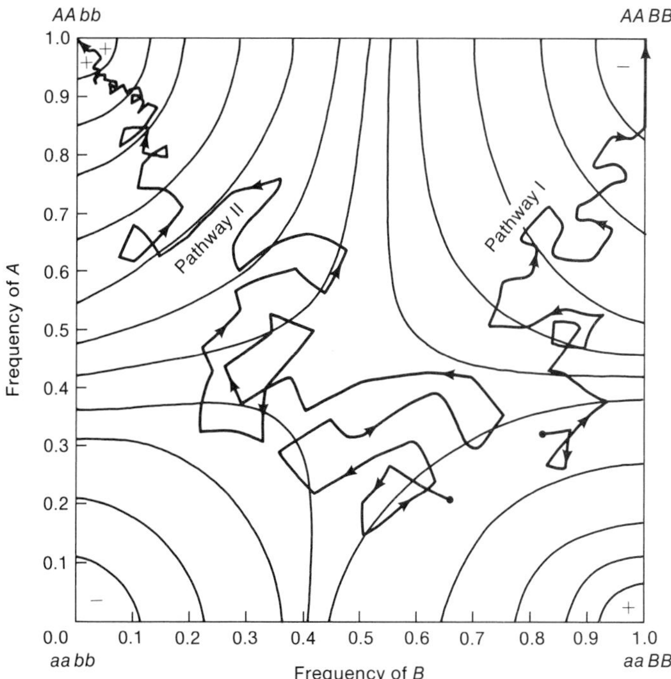

Figure 19-25. Selection and random drift can interact to produce different changes of gene frequency in an adaptive landscape. Without random drift, both populations would have moved toward aa BB *as a result of selection alone.*

stable state could never have occurred by selection in an infinite population, because, by selection alone, $\overline{W}$ could never decrease temporarily in order to cross from one slope to another.

Message
The interaction of selection and random drift makes possible the attainment of higher fitness states than would natural selection operating alone.

The Origin of Species

By a species (at least in sexually reproducing organisms), we mean a group of individuals biologically capable of interbreeding yet isolated genetically from other groups. The origin of a new species (**speciation**) is the origin of a group of individuals capable of making a living in a new way and at the same time acquiring some barrier to genetic exchange with the species from which it arose. The genetic differentiation of a population by inbreeding, genetic drift, and differential selection is always threatened by the reintroduction of genes from other groups by migration. The reduction of gene migration to a very low value is thus a prerequisite for speciation.

Generally, this reduction is the result of geographic isolation of the population as a consequence of chance historical events. A few long-distance migrants may reach a new island; a part of the mainland may be cut off by a rise in sea level; an insect vector that formerly passed pollen from one population to another may become locally extinct; the grassy plain that connected the feeding grounds of two grazers may, by a slight change in rainfall pattern, become a desert. Once the population is isolated physically, the processes of genetic differentiation will go on unimpeded until the genetic constitution of the isolated population is so different from its parental group that there is real difficulty in interbreeding.

If migration is reestablished before this critical period, speciation will not occur and the divergent population will once again converge. This has already happened in the human species, where genetic differentiation of geographical populations never proceeded beyond some superficial physical traits and a mixed differentiation of frequencies at polymorphic loci. On the other hand, if populations are very divergent before they come back in contact with each other, hybrid offspring will have genotypes with such low fitness that they do not survive or are sterile. At this stage of differentiation, there is a definite selective advantage for the newly forming species to avoid mating with each other and so avoid the wastage of gametes. New (secondary) barriers to interbreeding will then be selected, and the speciation process will be complete.

Beyond this generalized sketch of speciation, remarkably little can be said with certainty. Because species do not interbreed, it is difficult to analyze their differences genetically. A great deal must be made of the few cases where some hybrid offspring can be produced in the laboratory or garden. The methods of electrophoresis, immunology, and protein sequencing have made it possible to describe the differences in the proteins of species, but we have very few cases of species that have just recently separated. Thus, we do not know how much of the genome, and what part of it, is involved in the first divergence, nor do we know whether that divergence is often a consequence of diversifying selection or random drift. A detailed genetic analysis of the process of speciation remains one of the most important tasks for population genetics.

Summary

Charles Darwin revolutionized the study of biology when he constructed a theory of evolution based on the principles that variation existed within populations, that variation was heritable, and that the phenotype of the individuals in the population changed through generations because of natural selection. These basic tenets of evolution, put forward in *The Origin of Species* in 1859 prior to our knowledge of Mendelian genetics, have required only minor modification since that time. The study of changes within a population, or population genetics, relates the heritable changes in populations of organisms to the underlying individual processes of inheritance and development. Population genetics is the study of inherited variation and its modification in time and space.

Our identification of inherited variation within a population can be studied by observing morphological differences among individuals, or by examining the differences in specific amino-acid sequences of proteins, or even by examining, most recently, the differences in nucleotide sequences within the DNA. These kinds of observations have led to the conclusion that there is considerable polymorphism at many loci within a population. A measure of this variation is the amount of heterozygosity in a population. Population studies have shown that, in general, the genetic differences among individuals within human races is much greater than the average differences among races.

The ultimate source of all variation is mutation. However, within a population, the quantitative frequency of specific genotypes can be changed by recombination, immigration of genes, continued mutational events, and chance.

One property of Mendelian segregation is that, after one generation, random mating results in an equilibrium distribution of genotypes. However, inbreeding is one process that converts genetic variation within a population into differences between populations by making each separate population homozygous for a randomly chosen allele. On the other hand, for most populations, a balance is reached for any given environment between inbreeding, mutation from one allele to another, and immigration.

"Directed" changes of allelic frequencies within a population occur through natural selection of a favored genotype. In many cases such changes lead to homozygosity at a particular locus. On the other hand, the heterozygote may be more suited to a given environment than either of the homozygotes, leading to a balanced polymorphism.

Environmental selection of specific genotypes is rarely this simple, however. More often than not, phenotypes are determined by several interacting genes, and alleles at these different loci will be selected for at different rates. Furthermore, closely linked loci, unrelated to the phenotype in question, may have specific alleles carried along during the selection process. In general, genetic variation is the result of the interaction of evolutionary forces. For instance, a recessive deleterious mutant will never be totally eliminated from a population, because mutation constantly resupplies it to the population. Immigration can also reintroduce the undesirable allele into the population. And, indeed, a deleterious allele may, under environmental conditions of which we are unaware (including the remaining genetic makeup of the individual), be selected for.

Unless alternative alleles are in intermediate frequencies, selection, especially against recessives, is very slow, requiring many generations. In many populations, especially those of small size, new mutations can become established even though they are not favored by natural selection, simply by a process of random genetic drift. Such slow changes in different allele frequencies throughout the genome can lead eventually to the formation of new races and new species.

Problems

1. You are studying protein polymorphism in a natural population of a certain species of a sexually reproducing haploid organism. You isolate many strains from various parts of the test area and run extracts from each strain on electrophoretic gels. You stain the gels with a reagent specific for enzyme "X" and find that in the population there is a total of, say, five electrophoretic variants of enzyme X. You speculate that these variants represent various alleles of the structural gene for enzyme X.

   a. How would you demonstrate that this is so, both genetically and biochemically? (You can make crosses, make diploids, run gels, test enzyme activities, test amino-acid sequences, and so on.) Lay out the steps and conclusions precisely.

   b. Name at least one other possible way of generating the different electrophoretic variants, and say how you would distinguish this possibility from the one mentioned above.

2. A study made in 1958 in the mining town of Ashibetsu in Hokkaido province of Japan revealed the frequencies of MN blood-type genotypes shown in Table 19-14 (for individuals and for married couples).

Table 19-14.

Genotype	Number of individuals or couples
Individuals	
L^ML^M	406
L^ML^N	744
L^NL^N	332
Total	1482
Couples	
$L^ML^M \times L^ML^M$	58
$L^ML^M \times L^ML^N$	202
$L^ML^N \times L^ML^N$	190
$L^ML^M \times L^NL^N$	88
$L^ML^N \times L^NL^N$	162
$L^NL^N \times L^NL^N$	41
Total	741

   a. Show whether the population is in Hardy–Weinberg equilibrium with respect to the MN blood types.

   b. Show whether mating is random with respect to MN blood types.

   (Problem 2 is from J. Kuspira and G. W. Walker, *Genetics: Questions and Problems,* McGraw-Hill, 1973.)

3. For a sex-linked character in a species where the male is the heterogametic sex, suppose that the allelic frequencies at a locus are different for males and females.

a. Let the frequency of A be 0.8 in males and 0.2 in females. Show what happens to the allelic frequencies in successive generations in the two sexes.

b. Try to develop a general expression for the difference of frequency in males (p) and in females (P) in the nth generation, given that the initial values were p_0 and P_0.

4. Consider the populations whose genotypes are shown in Table 19-15.

Table 19-15.

Population	*AA*	*Aa*	*aa*
1	1.0	0.0	0.0
2	0.0	1.0	0.0
3	0.0	0.0	1.0
4	0.50	0.25	0.25
5	0.25	0.25	0.50
6	0.25	0.50	0.25
7	0.33	0.33	0.33
8	0.04	0.32	0.64
9	0.64	0.32	0.04
10	0.986049	0.013902	0.000049

a. Which of the populations are in Hardy–Weinberg equilibrium?

b. What are p and q in each population?

c. In population 10, it is discovered that the mutation rate from A to a is 5×10^{-6} and that reverse mutation is negligible. What must be the fitness of the *aa* phenotype?

d. In population 6, the a allele is detrimental, and furthermore the A allele is incompletely dominant so that *AA* is perfectly fit, *Aa* has a fitness of 0.8, and *aa* has a fitness of 0.6. If there is no mutation, what will p and q be in the next generation?

5. Color blindness is due to a sex-linked recessive allele. One male in ten is color blind.

a. What proportion of women are color blind?

b. By what factor is color blindness more common in men (or, how many color blind men are there for each color-blind woman)?

c. In what proportion of marriages would color blindness affect one-half of the children of each sex?

d. In what proportion of marriages would all children be normal?

e. In a population that is not in equilibrium, the frequency of the allele for color blindness is 0.2 in women and 0.6 in men. After one generation of random mating, what proportion of the female progeny will be color blind? What proportion of the male progeny?

f. What will the allele frequencies be in the male and in the female progeny in part e?

(Problem 5 courtesy of Clayton Person.)

6. In a wild population of beetles of species X, you notice that there is a 3:1 ratio of shiny to dull wing covers. Does this prove that shiny is dominant? (Assume that the two states are caused by the alleles of one gene.) If not, what does it prove? How would you elucidate the situation?

7. It seems clear that most new mutations are deleterious. Why?

8. Most mutations are recessive to wild type. Of those rare mutations that are dominant in *Drosophila,* for example, the majority turn out to be chromosomal aberrations or to be inseparable from chromosomal aberrations. Can you offer an explanation for why wild type is usually dominant?

9. Ten percent of the males of a large and randomly mating population are color blind. A representative group of 1000 from this population migrates to a South Pacific Island, where there are already 1000 inhabitants, and where 30% of the males are color blind. Assuming that Hardy–Weinberg conditions apply throughout (in the two original populations before emigration, and in the mixed population immediately following immigration), what fractions of males and females are expected to be color blind in the generation immediately following the arrival of the immigrants?

10. Using pedigree diagrams, find the probability of homozygosity by descent of the offspring of: (a) parent–offspring matings; (b) first-cousin matings; (c) aunt–nephew or uncle–niece matings.

11. In a survey of Indian tribes in Arizona and New Mexico, it was found that, in most groups, albinos were completely absent or very rare. (There is 1 albino per 20,000 North American caucasians.) However, in three populations, albino frequencies were exceptionally high: 1 per 277 Indians in Arizona, 1 per 140 Jemez Indians in New Mexico, and 1 per 247 Zuni Indians in New Mexico. All three of these populations were culturally, but not linguistically, related. What possible factors might explain the high incidence of albinos in these three tribes?

12. In an animal population, 20% of the individuals are *AA*, 60% are *Aa*, and 20% are *aa*. What are the allele frequencies? In this population, mating is always with *like phenotype* but is random within phenotype. What genotype and allele frequencies will prevail in the next generation? Such *assortative mating* is common in animal populations. Another type of assortative mating is that which occurs only between *unlike* phenotypes: answer the above question with this restriction imposed. What will the end result be after many generations of mating of both types?

13. In *Drosophila,* a stock isolated from nature has an average of 36 abdominal bristles. By selectively breeding only those flies with more bristles, the mean is raised to 56 in twenty generations! What would be the source of this genetic flexibility? The 56-bristle stock is very infertile, and so selection is relaxed for several generations and the bristle number drops to about 45. Why does it not drop to 36? When selection is reapplied, 56 bristles are soon attained, but this time the stock is *not* sterile. How could this situation arise?

14. The fitnesses of three genotypes are $W_{AA} = 0.9$; $W_{Aa} = 1.0$; and $W_{aa} = 0.7$.
    a. If the population starts at allele frequency $p = 0.5$, what is the value of p in the next generaton?
    b. What is the predicted equilibrium?

15. AA and Aa individuals are equally fertile. If 0.1% of the population is aa, what selection pressure exists against aa if the mutation rate $A \rightarrow a$ is 10^{-5}?

16. Gene B is a deleterious autosomal dominant. The frequency of affected individuals is 4.0×10^{-6}. Such individuals have a reproductive capacity about 30% that of normal individuals. Estimate μ, the rate at which b mutates to its deleterious allele B.

17. Of thirty-one children born of father–daughter matings, six died in infancy, twelve were very abnormal and died in childhood, and thirteen were normal. From this information, calculate roughly how many recessive lethal genes we have in our human genomes on the average. For example, if the answer is 1, then a daughter would stand a 50% chance of having it, and the probability of the union producing a lethal combination would be $1/2 \times 1/4 = 1/8$. (So, obviously, 1 is not the answer.) Consider also the possibility of undetected fatalities in utero in such matings. How would they affect your result?

18. Let us define the **total selection cost** to a population of a deleterious recessive gene as the loss of fitness per individual affected, s, times the frequency of affected individuals, q^2. That is,

$$\text{Genetic cost} = sq^2$$

    a. Suppose that a population is at equilibrium between mutation and selection for a deleterious recessive gene, where $s = 0.5$ and $\mu = 10^{-5}$. What is the equilibrium frequency of the gene? What is the genetic cost?

    b. Suppose that we now start irradiating people, so that the mutation rate doubles. What will be the new equilibrium frequency of the gene? What will be the genetic cost?

    c. Suppose that we do not change the mutation rate but instead lower the selection intensity to $s = 0.3$. What happens to the equilibrium frequency and genetic cost?

Further Reading

It is customary in introductory texts to include long lists of references to original research papers. We intend to break with this custom, because in our experience these bibliographies rarely attain the use that would merit their inclusion. Furthermore, the sophistication of modern library indexing and information retrieval systems (many computerized) enable the curious student to generate bibliographies in his own specific area of interest with remarkable ease. Nevertheless, some sources are so useful that they demand attention, and we have included a sampling of these.

Chapter 1

Clausen, J., D. D. Keck, and W. W. Hiesey. 1940. *Experimental Studies on the Nature of Species,* Vol. 1: *The Effect of Varied Environments on Western North American Plants.* Carnegie Institute of Washington, Publ. No. 520, 1–452. This publication and the following one by the same authors are the classic studies of norms of reaction of plants from natural populations.

Clausen, J., D. D. Keck, and W. W. Hiesey. 1958. *Experimental Studies on the Nature of Species,* Vol. 3: *Environmental Responses of Climatic Races of* Achillea. Carnegie Institute of Washington, Publ. No. 581, 1–129.

Schmalhausen, I. I. 1949. *Factors of Evolution: The Theory of Stabilizing Selection.* Philadelphia: Blakiston. The most general discussion of the relation of genotype and environment in the formation of phenotypic variation.

Chapter 2

Carlson, E. A. 1966. *The Gene: A Critical History.* Philadelphia: Saunders. A readable history of genetics.

Grant, V. 1975. *Genetics of Flowering Plants.* New York: Columbia University Press. One of the few texts on this subject.

Harpstead, D. 1971. "High-Lysine Corn." *Scientific American* (August). An account of the breeding of lines with increased amounts of normally limiting amino acids.

Hutt, F. B. 1964. *Animal Genetics.* New York: Ronald Press. A standard text on the subject with many interesting examples.

Jennings, P. R. 1976. "The Amplification of Agricultural Production." *Scientific American* (September). A discussion of genetics and the green revolution.

Olby, R. C. 1966. *Origins of Mendelism.* London: Constable. An enjoyable account of Mendel's work and the intellectual climate of his time.

Singer, S. 1978. *Human Genetics.* San Francisco: W. H. Freeman and Company. A short and readable treatment of the subject.

Stern, C., and E. R. Sherwood. 1966. *The Origin of Genetics. A Mendel Source Book.* San Francisco: W. H. Freeman and Company. A short collection of important early papers, including Mendel's papers and correspondence.

Sturtevant, A. H. 1965. *A History of Genetics.* New York: Harper & Row. Another useful historical text.

Todd, N. B. 1977. "Cats and Commerce." *Scientific American* (November). Includes some genetics of domestic cat coat colors and the use of this information to study cat migration throughout history.

Chapter 3

McLeish, J., and B. Snoad. 1958. *Looking at Chromosomes.* New York: Macmillan. A short classic book consisting of many superb photos of mitosis and meiosis.

Rick, C. M. 1978. "The Tomato." *Scientific American* (August). Includes an account of tomato genes and chromosomes and their role in breeding.

Stern, C. 1973. *Principles of Human Genetics,* 3rd ed. San Francisco: W. H. Freeman and Company. A standard text including many examples of the inheritance of human traits.

Chapter 4

Bodmer, W. F., and L. L. Cavalli-Sforza. 1976. *Genetics, Evolution and Man.* San Francisco: W. H. Freeman and Company. A very readable, well-illustrated book, including a clear account of HLA genetics.

Day, P. R. 1974. *Genetics of Host-Parasite Interaction.* San Francisco: W. H. Freeman and Company. A short technical account of the subject, which is of great importance to agriculture. It includes good examples of gene interaction.

Chapter 5

Peters, J. A., ed. 1959. *Classic Papers in Genetics.* Englewood Cliffs, N.J.: Prentice-Hall. A collection of important papers in the history of genetics.

Chapter 6

Finchman, J. R. S., P. R. Day, and A. Radford. 1979. *Fungal Genetics,* 3rd ed. London: Blackwell. A large, standard technical work. Good for tetrad analysis.

Kemp, R. 1970. *Cell Division and Heredity.* London: Edward Arnold. A short, clear introduction to genetics. Good for map functions and tetrad analysis.

Ruddle, F. H., and R. S. Kucherlapati. 1974. "Hybrid Cells and Human Genes." *Scientific American* (July). A popular account of the use of cell hybridization in mapping human genes.

Stahl, F. W. 1969. *The Mechanics of Inheritance,* 2nd ed. Englewood Cliffs, N.J.: Prentice-Hall. A short introduction to genetics, including some advanced material presented with a novel approach.

Chapter 7

Lawrence, C. W. 1971. *Cellular Radiobiology.* London: Edward Arnold. A short standard text, useful for target theory.

Chapter 8

Friedmann, T. 1971. "Prenatal Diagnosis of Genetic Disease." *Scientific American* (November). An early article on amniocentesis and its uses.

Fuchs, F. 1980. "Genetic Amniocentesis." *Scientific American* (August).

Hulse, J. H., and D. Spurgeon. 1974. "Triticale." *Scientific American* (August). An account of the development and possible benefits of this wheat–rye amphidiploid.

Lawrence, W. J. C. 1968. *Plant Breeding.* London: Edward Arnold (Studies in Biology No. 12). A short introduction to the subject.

Swanson, C. P., T. Mertz, and W. J. Young. 1967. *Cytogenetics.* Englewood Cliffs, N.J.: Prentice-Hall.

Weatherall, D. J., and J. B. Clegg. 1979. "Recent Developments in the Molecular Genetics of Human Haemoglobin." *Cell* 16: 467–478. A useful survey of recent research.

Chapter 9

Adelberg, E. A. 1966. *Papers on Bacterial Genetics.* Boston: Little, Brown.

Hayes, W. 1968. *The Genetics of Bacteria and Their Viruses,* 2nd ed. New York: Wiley. The standard and classic text, written by a pioneer in the subject.

Lewin, B. 1977. *Gene Expression,* Vol. 1: *Bacterial Genomes.* New York: Wiley. An excellent, up-to-date set of volumes, all of which are relevant to various sections of this text.

Lewin, B. 1977. *Gene Expression,* Vol. 3: *Plasmids and Phages.* New York: Wiley.

Stent, G. S., and R. Calendar. 1978. *Molecular Genetics,* 2nd ed. San Francisco: W. H. Freeman and Company. A lucidly written account of the development of our present understanding of the subject, based mainly on experiments in bacteria and phage.

Chapter 10

Benzer, S. 1962. "The Fine Structure of the Gene." *Scientific American* (January). A popular version of the author's pioneer experiments.

Watson, J. D. 1976. *The Molecular Biology of the Gene,* 3rd ed. Menlo Park, Calif: Benjamin/Cummings. A superb development of the subject, written in a highly readable style and well illustrated.

Chapter 11

Kornberg, A. 1980. *DNA Replication.* San Francisco: W. H. Freeman and Company. The definitive technical treatment of the subject, based on genetic and chemical analysis.

Watson, J. D. 1968. *The Double Helix.* New York: Atheneum. An enjoyable personal account of Watson and Crick's discovery, including the human dramas involved.

Chapter 12

Crick, F. H. C. 1962. "The Genetic Code." *Scientific American* (October).

Crick, F. H. C. 1966. "The Genetic Code: III." *Scientific American* (October). Popular accounts of code-cracking experiments.

Lane, C. 1976. "Rabbit Haemoglobin from Frog Eggs." *Scientific American* (August). This article describes experiments illustrating the universality of the genetic system.

Miller, O. L. 1973. "The Visualization of Genes in Action." *Scientific American* (March). A discussion of electron microscopy of transcription and translation.

Moore, P. B. 1976. "Neutron-Scattering Studies of the Ribosome." *Scientific American* (October). This article gives the details of ribosome substructure.

Nirenberg, M. W. 1963. "The Genetic Code: II." *Scientific American* (March). Another account of early code-cracking experiments.

Rich, A., and S. H. Kim. 1978. "The Three-Dimensional Structure of Transfer RNA." *Scientific American* (January). A presentation of the experimental evidence behind the structure described in this chapter.

Yanofsky, C. 1967. "Gene Structure and Protein Structure." *Scientific American* (May). This article gives the details of colinearity at the molecular level.

Chapter 13

Britten, R. J., and D. Kohne. 1968. "Repeated Sequences in DNA." *Science* 161: 529–540. One of the important summaries of the theoretical basis for distinguishing DNAs by renaturation.

Broda, P. 1979. *Plasmids.* San Francisco: W. H. Freeman and Company. One of the few technical books on the subject.

Brown, D. D. 1973. "The Isolation of Genes." *Scientific American* (August). Illustrates the power of focusing molecular techniques on one specific locus with special properties.

Cohen, S. 1975. "The Manipulation of Genes." *Scientific American* (July). A summary of recombinant DNA techniques by one of the main innovators.

Fiddes, J. C. 1977. "The Nucleotide Sequence of a Viral DNA." *Scientific American* (December). This is a review of a landmark, the DNA sequence of an entire virus genome with an unexpected discovery.

Gilbert, W., and L. Villa-Komaroff. 1980. "Useful Proteins from Recombinant Bacteria." *Scientific American* (April). A description of the method of DNA sequencing used most extensively. Also discusses the potential application of recombinant DNA techniques.

Itakura, K., et al. 1977. "Expression in *E. coli* of a Chemically Synthesized Gene for the Hormone Somatostatin." *Science* 198. A technical paper well worth reading for its historical significance. It represents the start of bioengineering—using DNA manipulation to modify cells to produce a medically useful human protein.

Khorana, H. G., et al. 1972. "Studies on Polynucleotides. CIII. Total Synthesis of the Structural Gene for an Alanine Transfer Ribonucleic Acid from Yeast." *J. Mol. Biol.* 72: 209–217. A classic technical paper.

Mertens, T. R. 1975. *Human Genetics: Readings on the Implications of Genetic Engineering.* New York: Wiley. A collection of popular articles.

Nathans, D., and H. O. Smith. 1975. "Restriction Endonucleases in the Analysis and Restructuring of DNA Molecules." *Ann. Rev. Biochem.* 44: 273–293. A technical review of restriction enzymes by two pioneers in the field.

Shapiro, J., et al. 1969. "Isolation of Pure *lac* Operon DNA." *Nature* 224: 768–774. An illustration of the elegance of genetic analysis.

Sinsheimer, R. L. 1977. "Recombinant DNA." *Ann. Rev. Biochem.* 46. A provocative article by a leading molecular biologist who has expressed concern about potential hazards of DNA manipulation.

Chapter 14

Britten, R. J., and D. E. Kohne. 1970. "Repeated Segments of DNA." *Scientific American* (April).

Brown, S. W. 1966. "Heterochromatin." *Science* 151: 417–425. A nice review of the classic cytological observations.

Davidson, E., and R. Britten. 1973. "Organization, Transcription and Regulation in the Animal Genome." *Quart. Rev. Biol.* 48: 565–613. The analysis of renaturation kinetics of DNA fragments provides insights into chromosome structure.

Dupraw, E. J. 1970. *DNA and Chromosomes.* New York: Holt, Rinehart & Winston. A useful book on chromosome substructure, containing excellent photographs by the author.

Edgar, R. S., and R. H. Epstein. 1965. "The Genetics of a Bacterial Virus." *Scientific* American (February). In analyzing a large number of mutations, the clustering of functionally related genes became apparent.

Glover, D. M., et al. 1975. "Characterization of 6 Cloned DNAs from *Drosophila melanogaster* Including One That Contains the Genes for rRNA." *Cell* 5. A technical paper showing how the isolation of random segments of *Drosophila* DNA in *E. coli* can be used to study chromosome structure.

Hayashi, S., et al. 1980. "Hybridization of tRNAs of *Drosophila melanogaster.*" *Chromosoma* 76: 65–84. A technical report showing how specific genes can be located cytologically by hybridization of labeled RNA to chromosomes in situ.

Herskowitz, I. 1973. "Control of Gene Expression in Bacteriophage Lambda." *Ann. Rev. Genetics* 7: 289–324. A technical review of the complex and well-analyzed regulation of lambda genes.

Jacob, F., and J. Monod. 1961. "Genetic Regulatory Mechanisms in the Synthesis of Proteins." *J. Mol. Biol.* 3: 318–356. A classic paper setting forth the elements of an operon and the experimental evidence.

Judd, B. H., M. W. Shen, and T. C. Kaufman. 1972. "The Anatomy and Function of a Segment of the X Chromosome of *Drosophila melanogaster.*" *Genetics* 71: 139–156. A beautiful example of genetic analysis leading to a significant insight into chromosome structure.

Kavenoff, R., and B. H. Zimm. 1973. "Chromosome-Sized DNA Molecules from *Drosophila.*" *Chromosoma* 41. An elegant experiment involving the study of chromosome aberrations with physiochemical techniques.

Kornberg, R. O. 1974. "Chromatin Structure: A Repeating Unit of Histones and DNA." *Science* 184: 868–871. A technical review of the evidence for nucleosomes.

Lewin, B. 1977. *Gene Expression,* Vol. 2: *Eucaryotic Chromosomes.* New York: Wiley.

Lucchesi, J. C. 1973. "Dosage Compensation in *Drosophila.*" *Ann. Rev. Genetics* 7: 225–237. A technical survey of the phenomenon and its possible mechanisms.

Lyon, M. L. 1962. "Sex Chromatin and Gene Action in the Mammalian X Chromosome." *Amer. J. Hum. Genet.* 14: 135–148. The first proposal that, in human females, one X is inactive, with supporting evidence of mosaic expression of sex-linked mutations.

Maniatis, T., and M. Ptashne. 1976. "A DNA Operator-Repressor System." *Scientific American* (January). This article discusses the molecular structures of the components of the *lac* operon.

Ptashne, M., and W. Gilbert. 1970. "Genetic Repressors." *Scientific American* (June). The exciting story of how repressors were identified and purified, thereby confirming the predictions of Jacob and Monod.

Stern, C. 1960. "Dosage Compensation—Development of a Concept and New Facts." *Canad. J. Genet. Cytol.* 2: 105–118. A delightful personal account of the history of dosage compensation by one of the main investigators.

Chapter 15

Borst, P., and L. A. Grivell. 1978. "The Mitochondrial Genome of Yeast." *Cell* 15: 705–723. A nontechnical review of the molecular biology of mtDNA.

Gillham, N. W. 1978. *Organelle Heredity.* New York: Raven Press. A rather technical but complete and up-to-date work on the subject.

Linnane, A. W., and P. Nagley. 1978. "Mitochondrial Genetics in Perspective: The Derivation of a Genetic and Physical Map of the Yeast Mitochondrial Genome." *Plasmid* 1: 324–345. An excellent review of the genetics and molecular biology of yeast mtDNA.

Sager, R. 1965. "Genes Outside Chromosomes." *Scientific American* (January). A popular description of early *Chlamydomonas* experiments.

Sager, R. 1972. *Cytoplasmic Genes and Organelles.* New York: Academic Press.

Chapter 16

Auerbach, C. 1976. *Mutation Research.* London: Chapman & Hall. Standard text by a pioneer researcher.

Bukhari, A. I., J. A. Shapiro, and S. L. Adhya, eds. 1977. *DNA Insertion Elements, Plasmids and Episomes.* Cold Spring Harbor Laboratory. An excellent large collection of short summary papers involving lower and higher life forms.

Cohen, S. N., and J. A. Shapiro. 1980. "Transposable Genetic Elements." *Scientific American* (February). A popular account stressing bacteria and phages.

Croce, C. M., and H. Koprowski. 1978. "The Genetics of Human Cancer." *Scientific American* (February). An excellent popular account.

Devoret, R. 1979. "Bacterial Tests for Potential Carcinogens." *Scientific American* (August). This article describes the use of mutation tests to screen for carcinogens and includes some details of DNA reactions.

Drake, J. W. 1970. *The Molecular Basis of Mutation.* San Francisco: Holden-Day. One of the few standard texts on the subject.

Stahl, F. W. 1979. *Genetic Recombination: Thinking About It in Phage and Fungi.* San Francisco: W. H. Freeman and Company. A rather technical short book on recombination models.

Whitehouse, H. L. K. 1973. *Towards an Understanding of the Mechanism of Heredity,* 3rd ed. London: Edward Arnold. An excellent general introduction to genetics stressing the historical approach and the pivotal experiments. It includes a good section on recombination models.

Chapter 17

Beermann, W., and U. Clever. 1964. "Chromosome Puffs." *Scientific American* (April). A description of the cytology of puffing.

Benzer, S. 1973. "The Genetic Dissection of Behavior." *Scientific American* (December). A beautiful illustration of the power of genetic analysis to probe development and behavior. Includes focus mapping.

DeRobertis, E. M., and J. B. Gurdon. 1979. "Gene Transplantation and the Analysis of Development." *Scientific American* (December).

Garcia-Bellido, A., P. A. Lawrence, and G. Morata. 1979. "Compartments in Animal Development." *Scientific American* (July). This article is not easy reading, but it points to an underlying principle in development that appears to be widespread.

Garcia-Bellido, A., and J. R. Merriam. 1969. "Cell Lineage of the Imaginal Discs of *Drosophila melanogaster.*" *J. Exp. Zool.* 170: 61–76. A clever genetic experiment to determine the end of gene activity in development.

Gurdon, J. B. 1968. "Transplanted Nuclei and Cell Differentiation." *Scientific American* (December). A review of nuclear transplantation and the successful cloning of a vertebrate.

Hadorn, E. 1968. "Transdetermination in Cells." *Scientific American* (November). A fascinating look at a puzzling phenomenon.

Hood, L. E., I. L. Weissman, and W. B. Wood. 1978. *Immunology.* Menlo Park, Calif.: Benjamin/Cummings.

Illmensee, K., and L. C. Stevens. 1979. "Teratomas and Chimeras." *Scientific American* (April). A review of the use of genetic mosaics to analyze development in mammals.

Jonathan, P., G. Butler, and A. Klug. 1978. "The Assembly of a Virus." *Scientific American* (November). A study of the assembly of RNA and coat protein to form a functional tobacco mosaic virion.

Markert, C. L., and H. Urpsprung. 1971. *Developmental Genetics.* Englewood Cliffs, N.J.: Prentice-Hall. A good short book on the subject.

Suzuki, D. T. 1970. "Temperature-Sensitive Mutations in *Drosophila melanogaster.*" *Science* 170: 695–706. A review of the usefulness of conditional mutations.

Wood, W. B., and R. S. Edgar. 1967. "Building a Bacterial Virus." *Scientific American* (July). A fascinating review of the discovery that virus particles assemble in a specific sequence.

Chapter 18

Bodmer, W. F., and L. L. Cavalli-Sforza. 1970. "Intelligence and Race." *Scientific American* (October). A popular treatment, including a discussion of heritability.

Briggs, D., and S. M. Walters. 1969. *Plant Variation and Evolution.* New York: McGraw-Hill. An excellent short text on the genetics of plant variation. Well illustrated.

Falconer, D. S. 1970. *Introduction to Quantitative Genetics.* New York: Ronald Press. A widely read text with a strong mathematical emphasis.

Feldman, M. W., and R. C. Lewontin. 1975. "The Heritability Hangup." *Science* 190: 1163–1168. A discussion of the meaning of heritability and its limitations, especially in relation to human intelligence.

Lewontin, R. C. 1974. "The Analysis of Variance and the Analysis of Causes." *Amer. J. Hum. Genet.* 26: 400–411. A discussion of the meaning of the analysis of variance in genetics as a method for determining the roles of heredity and environment in determining phenotype.

Chapter 19

Beadle, G. W. 1980. "The Ancestry of Corn." *Scientific American* (January). A popular account of the various clues that led to the modern version.

Cavalli-Sforza, L. L., and W. F. Bodmer. 1971. *The Genetics of Human Populations.* San Francisco: W. H. Freeman and Company. A comprehensive mathematical text useful for population genetics in general.

Clarke, B. 1975. "The Causes of Biological Diversity." *Scientific American* (August). A popular account emphasizing genetic polymorphism.

Crow, J. F. 1979. "Genes That Violate Mendel's Rules." *Scientific American* (February). An interesting article on a topic called segregation distortion (not treated in this text) and its effects in populations.

Crow, J. F., and M. Kimura. 1970. *An Introduction to Population Genetics Theory.* New York: Harper & Row. A standard mathematical text by two well-known population geneticists.

Dobzhansky, T. 1951. *Genetics and the Origin of Species.* New York: Columbia University Press. The classic synthesis of population genetics and the processes of evolution. The most influential book on evolution since Darwin's *Origin of Species.*

Ford, E. B. 1971. *Ecological Genetics,* 3rd ed. London: Chapman & Hall. A nonmathematical treatment of the role of genetic variation in nature, stressing morphological variation.

Futuyma, D. J. 1979. *Evolutionary Biology.* Sunderland, Mass.: Sinauer Associates. The best modern discussion of population genetics, ecology, and evolution from both a theoretical and an experimental point of view.

Lerner, I. M., and W. J. Libby. 1976. *Heredity, Evolution and Society,* 2nd ed. San Francisco: W. H. Freeman and Company. A text meant for non-science students.

Lewontin, R. C. 1974. *The Genetic Basis of Evolutionary Change.* New York: Columbia University Press. A discussion of the prevalence and role of genetic variation in natural populations. Both morphological and protein variations are considered.

Scientific American. September 1978. *Evolution.* This volume contains several articles relevant to population genetics.

Glossary

A Adenine, or adenosine.

abortive transduction The failure of a transducing DNA segment to be incorporated into the recipient chromosome.

acentric chromosome A chromosome having no centromere.

acrocentric chromosome A chromosome having the centromere located slightly nearer one end than the other.

active site The part of a protein that must be maintained in a specific shape if the protein is to be functional—for example, in an enzyme, the part to which the substrate binds.

adaptation In the evolutionary sense, some heritable feature of an individual's phenotype that improves its chances of survival and reproduction in the existing environment.

adaptive landscape The surface plotted in a three-dimensional graph, with all possible combinations of allele frequencies for different loci plotted in the plane, and mean fitness for each combination plotted in the third dimension.

adaptive peak A high point (perhaps one of several) on an adaptive landscape; selection tends to drive the genotype composition of the population toward a combination corresponding to an adaptive peak.

adaptive surface *See* **adaptive landscape.**

additive genetic variance Genetic variance associated with the average effects of substituting one allele for another.

adenine A purine base that pairs with thymine in the DNA double helix.

adenosine The nucleoside containing adenine as its base.

adenosine triphosphate *See* **ATP.**

adjacent segregation In a reciprocal translocation, the passage of a translocated and a normal chromosome to each of the poles.

ADP Adenosine diphosphate.

Ala Alanine (an amino acid).

alkylating agent A chemical agent that can add alkyl groups (for example, ethyl or methyl groups) to another molecule; many mutagens act through alkylation.

allele One of two or more forms that can exist at a single gene locus, distinguished by their differing effects on the phenotype.

allele exclusion The detection in a clone of immune cells of only one allele of an immunoglobulin gene for which the organism is heterozygotic.

allele frequency A measure of the commonness of an allele in a population; the proportion of all alleles of that gene in the population that are of this specific type.

allopolyploid *See* **amphidiploid.**

alternate segregation In a reciprocal translocation, the passage of both normal chromosomes to one pole and both translocated chromosomes to the other pole.

alternation of generations The alternation of gametophyte and sporophyte stages in the life cycle of a plant.

amber codon The codon UAG, a nonsense codon.

amber suppressor A mutant allele coding for tRNA whose anticodon is altered in such a way that the tRNA inserts an amino acid at an amber codon in translation.

Ames test A widely used test to detect possible chemical carcinogens; based on mutagenicity in the bacterium *Salmonella.*

amino acid A peptide; the basic building block of proteins (or polypeptides).

amniocentesis A technique for testing the genotype of an embryo or fetus in utero with minimal risk to the mother or the child.

AMP Adenosine monophosphate.

amphidiploid An allopolyploid; a polyploid formed from the union of two separate chromosome sets and their subsequent doubling.

amplification The production of many DNA copies from one master region of DNA.

aneuploid cell A cell having a chromosome number that differs from the normal chromosome number for the species by a small number of chromosomes.

angstrom unit (Å) A unit of length equal to 10^{-10} meter; many scientists now prefer to use the nanometer (1 nm $= 10^{-9}$ m $=$ 10 Å).

animal breeding The practical application of genetic analysis for development of pure-breeding lines of domestic animals suited to human purposes.

annealing Spontaneous alignment of two single DNA strands to form a double helix.

antibody A protein (immunoglobulin) molecule, produced by the immune system, that recognizes a particular foreign antigen and binds to it; if the antigen is on the surface of a cell, this binding leads to cell aggregation and subsequent destruction.

anticodon A nucleotide triplet in a tRNA molecule that aligns with a particular codon in mRNA under the influence of the ribosome, so that the peptide carried by the tRNA is inserted in a growing protein chain.

antigen A molecule (typically found in the surface of a cell) whose shape triggers the production of antibodies that will bind to the antigen.

antiparallel A term used to describe the opposite orientations of the two strands of a DNA double helix; the 5′ end of one strand aligns with the 3′ end of the other strand.

Arg Arginine (an amino acid).

ascospore A sexual spore from certain fungus species in which spores are found in a sac called an ascus.

ascus In fungi, a sac that encloses a tetrad or an octad of ascospores.

asexual spores *See* **spore.**

Asn Asparagine (an amino acid).

Asp Aspartate (an amino acid).

ATP Adenosine triphosphate, the "energy molecule" of cells, synthesized mainly in mitochondria and chloroplasts; energy from the breakdown of ATP drives many important reactions in the cell.

attachment point (*ap*) A hypothetical point on cpDNA in *Chlamydomonas*, proposed as an analogy to the centromere in nuclear chromosomes, in relation to which chloroplast genes may be mapped.

attenuator A region adjacent to the structural genes of the *trp* operon; this region acts in the presence of tryptophan to reduce the rate of transcription from the structural genes.

autonomous controlling element A controlling element that apparently has both regulator and receptor functions combined in the single unit, which enters a gene and makes it an unstable mutant.

autopolyploid A polyploid formed from the doubling of a single genome.

autoradiography A process in which radioactive materials are incorporated into cell structures, which are then placed next to a film of photographic emulsion, thus forming a pattern on the film corresponding to the location of the radioactive compounds within the cell.

autosome Any chromosome that is not a sex chromosome.

auxotroph A strain of microorganisms that will proliferate only when the medium is supplemented with some specific substance not required by wild-type organisms.

B cell An immature lymphocyte (in mammals, formed in the bone marrow) that is stimulated by the presence of a particular antigen to differentiate and divide, producing daughter cells (plasma cells) that release large amounts of the corresponding antibody.

B chromosomes Small plant chromosomes of variable number between individuals of a species, having no known phenotypic role.

bacteriophage (phage) A virus that infects bacteria.

balanced polymorphism Stable genetic polymorphism maintained by natural selection.

Balbiani ring A large chromosome puff.

Barr body A densely staining mass that represents an X chromosome inactivated by dosage compensation.

base analog A chemical whose molecular structure mimics that of a DNA base; because of this mimicry, the analog may act as a mutagen.

bead theory The disproved hypothesis that genes are arranged on the chromosome like beads on a necklace, with the chromosome structure serving merely as structural support.

Bence–Jones proteins Proteins excreted in the urine of multiple myeloma patients; now known to be light chains from immunoglobulins.

bimodal distribution A distribution having two modes.

binary fission The process in which a parent cell splits into two daughter cells of approximately equal size.

biparental zygote A *Chlamydomonas* zygote that contains cpDNA from both parents; such cells generally are rare.

blastoderm In an insect embryo, the layer of cells that completely surrounds an internal mass of yolk.

blending inheritance A discredited model of inheritance suggesting that the characteristics of an individual result from the smooth blending of fluidlike influences from its parents.

bridging cross A cross made to transfer alleles between two sexually isolated species by first transferring the alleles to an intermediate species that is sexually compatible with both.

broad heritability (H^2) The proportion of total phenotypic variance at the population level that is contributed by genetic variance.

bud A daughter cell formed by mitosis in yeast; one daughter cell retains the cell walls of the parent, and the other (the bud) forms new cell walls.

buoyant density A measure of the tendency of a substance to float in some other substance; large molecules are distinguished by their differing buoyant densities in some standard fluid. Measured by density-gradient ultracentrifugation.

C Cytosine, or cytidine.

cAMP Cyclic adenosine monophosphate; a molecule that plays a key role in the regulation of various processes within the cell.

canalized character A character whose phenotype is kept within narrow boundaries even in the presence of disturbing environments or mutations.

cancer A syndrome that involves the uncontrolled and abnormal division of eukaryotic cells.

CAP Catabolite activator protein; a protein whose presence is necessary for the activation of the *lac* operon.

carbon source A nutrient (such as sugar) that provides carbon "skeletons" needed in the organism's synthesis of organic molecules.

carcinogen A substance that causes cancer.

carrier An individual who possesses a mutant allele but does not express it in the phenotype because of a dominant allelic partner; thus, an individual of genotype *Aa* is a carrier of *a* if there is complete dominance of *A* over *a*.

cassette model A model to explain mating-type interconversion in yeast. Information for both *a* and α mating types is assumed to be present as silent "cassettes"; a copy of either type of cassette may be transposed to the mating-type locus, where it is "played" (transcribed).

catabolite activator protein *See* **CAP.**

catabolite repression The inactivation of an operon caused by the presence of large amounts of the metabolic end product of the operon.

cation A positively charged ion (such as K^+).

cDNA *See* **complementary DNA.**

cell division The process by which two cells are formed from one.

cell lineage A pedigree of cells related through asexual division.

cell-mediated system That part of the immune system in which the foreign particle is destroyed by contact with specialized host cells.

central dogma The hypothesis that information flows only from DNA to RNA to protein; although some exceptions are now known, the rule is generally valid.

centromere A kinetochore; the constricted region of a nuclear chromosome, to which the spindle fibers attach during division.

character Some attribute of individuals within a species for which various heritable differences can be defined.

character difference Alternative forms of the same attribute within a species.

chase *See* **pulse–chase experiment.**

chiasma (plural, **chiasmata**) A cross-shaped structure commonly observed between sister chromatids of a dyad during meiosis; the site of crossing over.

chi-square (χ^2) test A statistical procedure used to determine whether differences between two sets of values exceed those expected by chance if the two sets were randomly selected from a single large population.

chimera *See* **mosaic.**

chloroplast A chlorophyll-containing organelle in plants that is the site of photosynthesis.

chromatid One of the two side-by-side replicas produced by chromosome division.

chromatid conversion A type of gene conversion that is inferred from the existence of identical sister-spore pairs in a fungal octad that shows a non-Mendelian allele ratio.

chromatid inference A situation where the occurrence of a crossover between any two nonsister chromatids can be shown to affect the probability of those chromatids being involved in other crossovers in the same meiosis.

chromatin The substance of chromosomes; now known to include DNA, chromosomal proteins, and chromosomal RNA.

chromocenter The point at which the polytene chromosomes appear to be attached together.

chromomere A small beadlike structure visible on a chromosome during prophase of meiosis and mitosis.

chromosomal mutation *See* **chromosome aberration.**

chromosome A linear end-to-end arrangement of genes and other DNA, sometimes with associated protein and RNA.

chromosome aberration Any type of change in the chromosome structure or number.

chromosome loss Failure of a chromosome to become incorporated into a daughter nucleus at cell division.

chromosome map *See* **linkage map.**

chromosome puff A swelling at a site along the length of a polytene chromosome; the site of active transcription.

chromosome rearrangement A chromosome aberration involving new juxtapositions of chromosome parts.

chromosome set The group of different chromosomes that carries the basic set of genetic information for a particular species.

chromosome theory of inheritance The unifying theory stating that inheritance patterns may be generally explained by assuming that genes are located in specific sites on chromosomes.

***cis* conformation** In a heterozygote involving two mutant sites within a gene, the arrangement $a_1\,a_2/+\,+$.

***cis* dominance** The ability of a gene to affect genes next to it on the same chromosome.

***cis-trans* test** A test to determine whether two mutant sites of a gene are in the same cistron.

cistron A region of a gene within which two mutations cannot complement; a region coding for one polypeptide.

clonal selection A theory of antibody specificity, holding that an antigen selects the clone with an antibody specific to it from a vast number of such clones already present in the body.

clone (1) A group of genetically identical cells or individuals derived by asexual division from a common ancestor. (2) (*colloquial*) An individual formed by some asexual process so that it is genetically identical to its "parent." (3) *See* **DNA clone.**

code dictionary A listing of the 64 possible codons and their translational meanings (the corresponding amino acids).

codominance The situation in which a heterozygote shows the phenotypic effects of both alleles.

codon A section of DNA (three nucleotide pairs in length) that codes for a single amino acid.

coefficient of coincidence The ratio of the observed number of double recombinants to the expected number.

colinearity The correspondence between the location of a mutant site within a gene and the location of an amino-acid substitution within the polypeptide translated from that gene.

colony A visible clone of cells.

compartmentalization The existence of boundaries within the organism beyond which a specific clone of cells will never extend during development.

complementary DNA (cDNA) Synthetic DNA transcribed from a specific RNA through the action of the enzyme reverse transcriptase.

complementary gene action Intergenic complementation between mutant alleles at different loci to give wild-type phenotype.

complementary RNA (cRNA) Synthetic RNA produced by transcription from a specific DNA single-stranded template.

complementation The production of a wild-type phenotype when two different mutations are combined in a diploid or a heterokaryon.

complementation map A gene map in which each mutation is represented by a "bar" that overlaps the bars for other mutations with which it will not complement; each bar probably represents the region of the polypeptide that is distorted by an amino-acid substitution.

complementation test *See* ***cis–trans*** **test.**

conditional mutation A mutation that has the wild-type phenotype under certain (permissive) environmental conditions and a mutant phenotype under other (restrictive) conditions.

conjugation The union of two bacterial cells, during which chromosomal material is transferred from the donor to the recipient cell.

conjugation tube *See* **pilus.**

conservative replication A disproved model of DNA synthesis suggesting that one-half of the daughter DNA molecules should have both strands composed of newly polymerized nucleotides.

constant region A region of an antibody molecule that is nearly identical with the corresponding regions of antibodies of different specificities.

constitutive heterochromatin Specific regions of heterochromatin always present and in both homologs of a chromosome.

controlling element A mobile genetic element capable of producing an unstable mutant target gene; two types exist, the regulator and the receptor elements.

copy-choice model A model of the mechanism for crossing over, suggesting that crossing over occurs during chromosome division and can occur only between two supposedly "new" nonsister chromatids; the experimental evidence does not support this model.

correction The production (possibly by excision and repair) of a properly paired nucleotide pair from a sequence of hybrid DNA that contains an illegitimate pair.

correlation coefficient A statistical measure of the extent to which variations in one variable are related to variations in another.

cosegregation In *Chlamydomonas,* parallel behavior of different chloroplast markers in a cross, due to their close linkage on cpDNA.

Cot curve A graph of the fraction of DNA that has reannealed versus the normalized time, C_0t (where C_0 is the initial concentration of single-stranded DNA, and t is the elapsed time); this graph is used in studying the renaturation of denatured repetitive DNA.

cotransduction The simultaneous transduction of two bacterial marker genes.

cotransformation The simultaneous transformation of two bacterial marker genes.

coupling conformation Linked heterozygous gene pairs in the arrangement $A\ B/a\ b$.

covariance A statistical measure used in computing the correlation coefficient between two variables; the covariance is the sum of $(x - \bar{x})(y - \bar{y})$ over all pairs of values for the variables x and y, where $\bar{x}$ is the mean of the x values and $\bar{y}$ is the mean of the y values.

cpDNA Chloroplast DNA.

cri-du-chat syndrome An abnormal human condition caused by deletion of part of one homolog of chromosome 5.

criss-cross inheritance Transmission of a gene from male parent to female child to male grandchild—for example, X-linked inheritance.

cRNA *See* **complementary RNA.**

crossing over The exchange of corresponding chromosome parts between homologs by breakage and reunion.

crossover suppressor An inversion (usually complex) that makes pairing and crossing over impossible.

cruciform configuration A region of DNA with palindromic sequences in both strands, so that each strand pairs with itself to form a helix extending sideways from the main helix.

CSAR (cytoplasmic segregation and recombination) An acronym used in this book to describe the process whereby organelle-based genes assort and recombine in a cytohet.

culture Tissue or cells multiplying by asexual division, grown for experimentation.

cyclic adenosine monophosphate *See* **cAMP.**

Cys Cysteine (an amino acid).

cytidine The nucleoside containing cytosine as its base.

cytochromes A class of proteins, found in mitochondrial membranes, whose main function is oxidative phosphorylation of ADP to form ATP.

cytogenetics The cytological approach to genetics, mainly involving microscopic studies of chromosomes.

cytohet A cell containing two genetically distinct types of a specific organelle.

cytoplasm The material between the nuclear and cell membranes; includes fluid (cytosol), organelles, and various membranes.

cytoplasmic inheritance Inheritance via genes found in cytoplasmic organelles.

cytosine A pyrimidine base that pairs with guanine.

cytosol The fluid portion of the cytoplasm, outside the organelles.

Darwinian fitness The relative probability of survival and reproduction for a genotype.

degenerate code A genetic code in which some amino acids may be encoded by more than one codon each.

deletion Removal of a chromosomal segment from a chromosome set.

denaturation The separation of the two strands of a DNA double helix, or the severe disruption of the structure of any complex molecule without breaking the major bonds of its chains.

denaturation map A map of a stretch of DNA showing the locations of local denaturation loops, which correspond to regions of high AT content.

deoxyribonuclease *See* **DNase.**

deoxyribonucleic acid *See* **DNA.**

development The process whereby a single cell becomes a differentiated organism.

developmental noise The influence of random molecular proceses on development.

dicentric chromosome A chromosome with two centromeres.

dihybrid cross A cross between two individuals identically heterozygous at two loci—for example, $Aa\,Bb \times Aa\,Bb$.

dimorphism A "polymorphism" involving only two forms.

dioecious plant A plant species in which male and female organs appear on separate individuals.

diploid A cell having two chromosome sets, or an individual having two chromosome sets in each of its cells.

directional selection Selection that changes the frequency of an allele in a constant direction, either toward or away from fixation for that allele.

dispersive replication Disproved model of DNA synthesis suggesting more-or-less random interspersion of parental and new segments in daughter DNA molecules.

distribution *See* **statistical distribution.**

distribution function A graph of some precise quantitative measure of a character against its frequency of occurrence.

DNA Deoxyribonucleic acid; a double chain of linked nucleotides (having deoxyribose as their sugars); the fundamental substance of which genes are composed.

DNA clone A section of DNA that has been inserted into the chromosome of a bacterium, phage, or plasmid vector and has been replicated to form many copies.

DNA polymerase An enzyme that can synthesize new DNA strands using a DNA template; several such enzymes exist.

DNase Deoxyribonuclease; an enzyme that degrades DNA to nucleotides.

dominance variance Genetic variance at a single locus attributable to dominance of one allele over another.

dominant allele An allele that expresses its phenotypic effect even when heterozygous with a recessive allele; thus if A is dominant over a, then AA and Aa have the same phenotype.

dominant phenotype The phenotype of a genotype containing the dominant allele; the parental phenotype that is expressed in a heterozygote.

dosage compensation (1) Inactivation of X chromosomes in mammals so that no cell has more than one functioning X chromosome. (2) Regulation at some autosomal loci so that homozygous dominants do not produce twice as much product as the heterozygote.

dose *See* **gene dose.**

double crossover Two crossovers occurring in a chromosomal region under study.

double helix The structure of DNA first proposed by Watson and Crick, with two interlocking helices joined by hydrogen bonds between paired bases.

double infection Infection of a bacterium with two genetically different phages.

Down's syndrome A human phenotype involving several typical morphological characteristics, due to a trisomy of chromosome 21.

drift *See* **random genetic drift.**

duplicate genes Two identical allele pairs in one diploid individual.

duplication More than one copy of a particular chromosomal segment in a chromosome set.

dyad A pair of sister chromatids joined at the centromere, as in the first division of meiosis.

ecdysone A molting hormone in insects.

electrophoresis A technique for separating the components of a mixture of molecules (of protein or of DNAs) in an electric field within a gel.

endogenote *See* **merozygote.**

endopolyploidy An increase in the number of chromosome sets caused by replication without cell division.

enforced outbreeding Deliberate avoidance of mating between relatives.

enucleate cell A cell having no nucleus.

environment The combination of all the conditions external to the genome that potentially affect its expression and its structure.

environmental variance The variance due to environmental variation.

enzyme A protein that functions as a catalyst.

episome A genetic element in bacteria that can replicate free in the cytoplasm or can be inserted into the main bacterial chromosome and replicate with the chromosome.

epistasis A situation in which an allele of one gene obliterates the phenotypic expression of all allelic alternatives of another gene.

equational division A nuclear division that maintains the same ploidy level of the cell.

ethidium A molecule that can intercalate into DNA double helices when the helix is under torsional stress.

euchromatin A chromosomal region that stains normally; thought to contain the normally functioning genes.

eugenics Controlled human breeding based on notions of desirable and undesirable genotypes.

eukaryote An organism having eukaryotic cells.

eukaryotic cell A cell containing a nucleus.

euploid A cell having any number of complete chromosome sets, or an individual composed of such cells.

excision repair The repair of a DNA lesion by removal of the faulty DNA segment and its replacement with a wild-type segment.

exconjugant A female bacterial cell that has just been in conjugation with a male and that contains a fragment of male DNA.

exogenote *See* **merozygote.**

expressivity The degree to which a particular genotype is expressed in the phenotype.

extrapolation number (target number) In target theory, the intercept of the survival curve with a logarithmic vertical axis.

extron The portion of the gene other than the intron; the portion of the gene that is transcribed into mRNA and is translated into protein.

F^- cell In *E. coli,* a cell having no fertility factor; a female cell.

F^+ cell In *E. coli,* a cell having a free fertility factor; a male cell.

F factor *See* **fertility factor.**

F′ factor A fertility factor into which a portion of the bacterial chromosome has been incorporated.

F_1 generation The first filial generation, produced by crossing two parental lines.

F_2 generation The second filial generation, produced by selfing or intercrossing the F_1.

facultative heterochromatin Heterochromatin located in positions that are composed of euchromatin in other individuals of the same species, or even in the other homolog of a chromosome pair.

familial trait A trait shared by members of a family.

family selection A breeding technique of selecting a pair on the basis of the average performance of their progeny.

fate map A map of an embryo showing areas that are destined to develop into specific adult tissues and organs.

fertility (F) factor A bacterial episome whose presence confers donor ability (maleness).

filial generations Successive generations of progeny in a controlled series of crosses, starting with two specific parents (the P generation) and selfing or intercrossing the progeny of each new (F_1, F_2, . . .) generation.

filter enrichment A technique for recovering auxotrophic mutants in filamentous fungi.

fingerprint The characteristic spot pattern produced by electrophoresis of the polypeptide fragments obtained through denaturation of a particular protein with a proteolytic enzyme.

first-division segregation pattern A linear pattern of spore phenotypes within the ascus for a particular allele pair, produced when the alleles go into separate nuclei at the first meiotic division, showing that no crossover has occurred between that allele pair and the centromere.

fitness *See* **Darwinian fitness.**

fixed allele An allele for which all members of the population under study are homozygous, so that no other alleles for this locus exist in the population.

fixed breakage point According to the hybrid DNA recombination model, the point from which unwinding of the DNA double helices begins, as a prelude to formation of hybrid DNA.

fluctuation test A test used in microbes to establish the random nature of mutation, or to measure mutation rates.

fMet *See* **formylmethionine.**

focus map A fate map of areas of the *Drosophila* blastoderm destined to become specific adult structures, based on the frequencies of specific kinds of mosaics.

formylmethionine (fMet) A specialized amino acid that is the very first one incorporated into the polypeptide chain in the synthesis of proteins.

forward mutation A mutation that converts a wild-type allele to a mutant allele.

frameshift mutation The insertion or deletion of a nucleotide pair, causing a disruption of the translational reading frame.

frequency-dependent fitness Fitness differences whose intensity changes with changes in the relative frequency of genotypes in the population.

frequency-dependent selection Selection that involves frequency-dependent fitness.

frequency histogram A "step curve" in which the frequencies of various arbitrarily bounded classes are graphed.

frequency-independent fitness Fitness that is not determined by interactions with other individuals of the same species.

frequency-independent selection Selection in which the fitnesses of genotypes are independent of their relative frequency in the population.

fruiting body In fungi, the organ in which meiosis occurs and sexual spores are produced.

G Guanine, or guanosine.

gamete A specialized haploid cell that fuses with a gamete from the opposite sex or mating type to form a diploid zygote; in mammals, an egg or sperm.

gametophyte The haploid gamete-producing stage in the life cycle of plants; prominent and independent in some species but reduced and parasitic in others.

gene The fundamental physical unit of heredity, recognized through its variant alleles, which transmits a specification or specifications from one generation to the next; a segment of DNA coding for one function or several related functions.

gene conversion A meiotic process of directed change in which one allele directs the conversion of a partner allele to its own form.

gene dose The number of copies of a particular gene present in the genome.

gene frequency *See* **allele frequency.**

gene interaction The collaboration of several different genes in the production of one phenotypic character (or related group of characters).

gene locus The specific place on a chromosome where a gene is located.

gene map A linear designation of mutant sites within a gene, based upon the various frequencies of interallelic (intragenic) recombination.

gene mutation A point mutation that results from changes within the structure of a gene.

gene pair The two copies of a particular type of gene present in a diploid cell (one in each chromosome set).

generalized transduction The ability of certain phages to transduce any gene in the bacterial chromosome.

genetic code The set of correspondences between nucleotide pair triplets in DNA and amino acids in protein.

genetic dissection The use of recombination and mutation to piece together the various components of a given biological function.

genetic markers Alleles used as experimental probes to mark a nucleus, a chromosome, or a gene.

genetic variance The phenotypic variance due to the presence of different genotypes in the population.

genetics (1) The study of genes. (2) The study of inheritance.

genome The entire complement of genetic material in a cell.

genotype The specific allelic composition of a cell—either of the entire genome, or more commonly for a certain gene or a set of genes.

germinal mutations Mutations occurring in the cells that are destined to develop into gametes.

Gln Glutamine (an amino acid).

Glu Glutamate (an amino acid).

Gly Glycine (an amino acid).

gradient A gradual change in some quantitative property over a specific distance.

guanine A purine base that pairs with cytosine.

guanosine The nucleoside having guanine as its base.

gynandromorph A sexual mosaic.

Hardy–Weinberg equilibrium The presence in a population of the three genotypes *AA*, *Aa*, and *aa* in the frequencies p^2, $2pq$, and q^2, respectively (where p and q are the frequencies of the alleles *A* and *a*). At these genotype frequencies, there is no change from one generation to the next.

hemoglobin (Hb) The oxygen-transporting blood protein in animals.

half-chromatid conversion A type of gene conversion that is inferred from the existence of nonidentical sister spores in a fungal octad showing a non-Mendelian allele ratio.

haploid A cell having one chromosome set, or an organism composed of such cells.

haploidization Production of a haploid from a diploid by progressive chromosome loss; studied mainly in *Aspergillus.*

harlequin chromosomes Sister chromatids that stain differently, so that one appears dark and the other light.

hemizygous gene A gene present in only one copy in a diploid organism—for example, X-linked genes in a male mammal.

heredity The similarity among offspring and parents.

heritability in the broad sense *See* **broad heritability.**

heritability in the narrow sense (h^2) The proportion of phenotypic variance that can be attributed to additive genetic variance.

hermaphrodite (1) A plant species in which male and female organs occur in the same flower of a single individual (*compare* **monoecious plant**). (2) An animal with both male and female sex organs.

heterochromatin Densely staining chromosomal regions, believed to be for the most part genetically inert.

heteroduplex A DNA double helix formed by annealing single strands from different sources; if there is a structural difference between the strands, the heteroduplex may show such abnormalities as loops or buckles.

heterogametic sex The sex that has heteromorphic sex chromosomes (for example, XY) and hence produces two different kinds of gametes with respect to the sex chromosomes.

heterogeneous nuclear RNA (HnRNA) A diverse assortment of RNA types found in the nucleus, including mRNA precursors and other types of RNA.

heterokaryon A culture of cells composed of two different nuclear types in a common cytoplasm.

heterokaryon test A test for cytoplasmic mutations, based on new associations of phenotypes in cells derived from specially marked heterokaryons.

heteromorphic chromosomes A chromosome pair with some homology but differing in size or shape.

heterothallic fungus A fungus species in which two different mating types must unite to complete the sexual cycle.

heterozygosity A measure of the genetic variation in a population; with respect to one locus, stated as the frequency of heterozygotes for that locus.

heterozygote An individual having a heterozygous gene pair.

heterozygous gene pair A gene pair having different alleles in the two chromosome sets of the diploid individual, usually (but not necessarily) one dominant and one recessive–for example, Aa or A^1A^2.

hexaploid A cell having six chromosome sets, or an organism composed of such cells.

high-frequency recombination (Hfr) cell In *E. coli*, a cell having its fertility factor integrated into the bacterial chromosome; a donor (male) cell.

His Histidine (an amino acid).

histocompatibility antigens Antigens that determine the acceptance or rejection of a tissue graft.

histocompatibility genes The genes that code for the histocompatibility antigens.

HnRNA *See* **heterogeneous nuclear RNA.**

homeotic mutations Mutations that can change the fate of an imaginal disk.

homoeologous chromosomes Partially homologous chromosomes, usually indicating some original ancestral homology.

homogametic sex The sex with homologous sex chromosomes (for example, XX).

homolog A member of a pair of homologous chromosomes.

homologous chromosomes Chromosomes that are identical in shape, size, and function.

homothallic fungus A fungus species in which a single sexual spore can complete the entire sexual cycle (*compare* **heterothallic fungus**).

homozygote An individual having a homozygous gene pair.

homozygous gene pair A gene pair having identical alleles in both copies–for example, AA or A^1A^1.

host range The spectrum of strains of a given bacterial species that a given strain of phage can infect.

hot spot A part of a gene that shows a very high tendency to become a mutant site under the action of a particular mutagen.

humoral system That part of the immune system in which the foreign particle is inactivated by circulating antibody molecules.

hybrid (1) A heterozygote. (2) A progeny individual from any cross involving parents of differing genotypes.

hybrid DNA model A model that explains both crossing over and gene conversion by assuming the production of a short stretch of heteroduplex (hybrid) DNA (formed from both parental DNAs) in the vicinity of a chiasma.

hybridization in situ Finding the location of a gene by adding radioactive copies of the specific RNA transcripts of the gene and detecting the location of the radioactivity on the chromosome after the RNA hybridizes to the DNA.

hybridize (1) To form a hybrid by performing a cross. (2) To anneal nucleic acid strands from different sources.

hydrogen bond A weak bond involving the sharing of an electron with a hydrogen atom; hydrogen bonds are important in the specificity of base pairing in nucleic acids and in determining protein shape.

hydroxyapatite A form of calcium phosphate that binds double-stranded DNA.

hypervariable region The part of a variable region that actually determines the specificity of an antibody.

hypha (plural, **hyphae**) A threadlike structure (composed of cells attached end to end) that forms the main tissue in many fungus species.

Ig *See* **immunoglobulin.**

Ile Isoleucine (an amino acid).

imaginal disk A small group of cells in an insect larva that is destined to develop into an entire adult structure (for example, a leg).

immune system The animal cells and tissues involved in recognizing and attacking foreign substances within the body.

immune tolerance *See* **tolerance.**

immunoglobin (Ig) A general term for the kind of globular blood proteins that constitute antibodies.

immunological memory The ability of the immune system to respond more rapidly and more extensively to a second appearance of a given antigen.

in situ "In place"; *see* **hybridization in situ.**

in vitro In an experimental situation outside the organism (literally, "in glass").

in vivo In a living cell or organism.

inbreeding Mating between relatives more frequently than would be expected by chance.

inbreeding coefficient The probability of homozygosity that results because the zygote obtains copies of the *same* ancestral gene.

incomplete dominance The situation in which a heterozygote shows a phenotype quantitatively intermediate between the corresponding homozygote phenotypes.

independent assortment *See* **Mendel's second law.**

inducer An environmental agent that triggers transcription from an operon.

infectious transfer The rapid transmission of free episomes (plus any chromosomal genes they may carry) from donor to recipient cells in a bacterial population.

inosine A rare base that is important at the wobble position of some tRNA anticodons.

insertion sequence (IS) A mobile piece of bacterial DNA (several hundred nucleotide pairs in length) that is capable of inactivating a gene into which it inserts.

insertional translocation The insertion of a segment from one chromosome into another nonhomologous one.

intercalating agent A chemical that can insert itself between the stacked bases at the center of the DNA double helix, possibly causing a frameshift mutation.

interchromosomal recombination Recombination resulting from independent assortment.

interference A measure of the independence of crossovers from each other, calculated by subtracting the coefficient of coincidence from 1.

intergenic complementation Complementation involving mutations in two different genes.

interrupted mating A technique used to map bacterial genes by determining the sequence in which donor genes enter recipient cells.

interstitial region The chromosomal region between the centromere and the site of a rearrangement.

intervening sequence An intron; a segment of unknown function within a gene. This segment may be initially transcribed, but it is not found in the functional mRNA.

intrachromosomal recombination Recombination resulting from crossing over between two gene pairs.

intragenic complementation Complementation involving two mutations within the same gene.

intron *See* **intervening sequence.**

inversion A chromosomal mutation involving the removal of a chromosome segment, its rotation through 180°, and its reinsertion in the same location.

inverted repeat (IR) sequence A sequence found in identical (but inverted) form, for example, at the opposite ends of a transposon.

IR *See* **inverted repeat sequence.**

IS *See* **insertion sequence.**

isoaccepting tRNAs The various types of tRNA molecules that carry a specific amino acid.

isotope One of several forms of an atom having the same atomic number but differing atomic masses.

karyotype The entire chromosome complement of an individual or cell, as seen during mitotic metaphase.

killer cell Immune system cell type which destroys cells of incompatible tissue.

kineotochore *See* **centromere.**

Klinefelter's syndrome An abnormal human male phenotype involving an extra X chromosome (XXY).

λ (lambda) phage One kind ("species") of temperate bacteriophage.

λ*dgal* A λ phage carrying a *gal* bacterial gene and defective (*d*) for some phage function.

lampbrush chromosomes Large chromosomes found in amphibian eggs, with lateral DNA loops that produce a brushlike appearance under the microscope.

lawn A continuous layer of bacteria on the surface of an agar medium.

leaky mutant A mutant (typically, an auxotroph) that represents a partial rather than a complete inactivation of the wild-type function.

lesion A damaged area in a gene (a mutant site), a chromosome, or a protein.

lethal gene A gene whose expression results in the death of the individual expressing it.

Leu Leucine (an amino acid).

ligase An enzyme that can rejoin a broken phosphodiester bond in a nucleic acid.

line A group of identical pure-breeding diploid or polyploid organisms, distinguished from other individuals of the same species by some unique phenotype and genotype.

linear tetrad A tetrad that results from the occurrence of the meiotic and postmeiotic nuclear divisions in such a way that sister products remain adjacent to one another (with no passing of nuclei).

linkage The association of genes on the same chromosome.

linkage group A group of genes known to be linked; a chromosome.

linkage map A chromosome map; an abstract map of chromosomal loci, based on recombinant frequencies.

locus (plural, **loci**) *See* **gene locus.**

Lys Lysine (an amino acid).

lysis The rupture and death of a bacterial cell upon the release of phage progeny.

lysogenic bacterium A bacterial cell capable of spontaneous lysis due to the uncoupling of a prophage from the bacterial chromosome.

macromolecule A large polymer such as DNA, a protein, or a polysaccharide.

map unit (m.u.) The "distance" between two linked gene pairs where 1% of the products of meiosis are recombinant; a unit of distance in a linkage map.

mapping function A formula expressing the relationship between distance in a linkage map and recombinant frequency.

marker *See* **genetic marker.**

marker retention A technique used in yeast to test the degree of linkage between two mitochrondrial mutations.

mating types The equivalent in lower organisms of the sexes in higher organisms; the mating types typically differ only physiologically and not in physical form.

maternal effect The environmental influence of the mother's tissues on the phenotype of the offspring.

maternal inheritance A type of uniparental inheritance in which all progeny have the genotype and phenotype of the parent acting as the female parent.

matroclinous inheritance Inheritance in which all offspring have the phenotype of the mother.

mean The arithmetic average.

medium Any material on (or in) which experimental cultures are grown.

meiosis Two successive nuclear divisions (with corresponding cell divisions) that produce gametes (in animals) or sexual spores (in plants) having one-half of the genetic material of the original cell.

melting Denaturation of DNA.

Mendelian ratio A ratio of progeny phenotypes reflecting the operation of Mendel's laws.

Mendel's first law The two members of a gene pair segregate from each other during meiosis; each gamete has an equal probability of obtaining either member of the gene pair.

Mendel's second law The law of independent assortment; different segregating gene pairs behave independently. (This law is now known to apply only to unlinked or distantly linked pairs.)

merozygote A partially diploid *E. coli* cell formed from a complete female chromosome (the endogenote) plus a male fragment (the exogenote).

messenger RNA *See* **mRNA.**

Met Methionine (an amino acid).

metabolism The chemical reactions that occur in a living cell.

metacentric chromosome A chromosome having its centromere in the middle.

midparent value The mean of the values of a quantitative phenotype for two specific parents.

minimal medium A medium containing only inorganic salts, a carbon source, and water.

missense mutation A mutation that alters a codon so that it encodes a different amino acid.

mitochrondrial sex locus (ω) A locus on the mitochrondrial DNA whose heterozygosity or homozygosity can produce widely differing recombination values in mitochondrial "crosses."

mitochondrion A eukaryotic organelle that is the site of ATP synthesis and of the citric acid cycle.

mitosis A type of nuclear division (occurring at cell division) that produces two daughter nuclei identical to the parent nucleus.

mitotic crossover A crossover resulting from the pairing of homologs in a mitotic diploid.

mobile genetic element *See* **transposable genetic element.**

mode The single class in a statistical distribution having the greatest frequency.

modifier gene A gene that affects the phenotypic expression of another gene.

molecular genetics The study of the molecular processes underlying gene structure and function.

monoecious plant A plant species in which male and female organs are found on the same plant but in different flowers (for example, corn).

monohybrid cross A cross between two individuals identically heterozygous at one gene pair—for example, $Aa \times Aa$.

monoploid A cell having only one chromosome set (usually as an aberration), or an organism composed of such cells.

monosomic genome A genome that is basically diploid but that has only one copy of one particular chromosome type and thus has chromosome number $2n - 1$.

mosaic A chimera; a tissue containing two or more genetically distinct cell types, or an individual composed of such tissues.

mRNA (messenger RNA) An RNA molecule transcribed from the DNA of a gene, and from which a protein is translated by the action of ribosomes.

mtDNA Mitochrondrial DNA.

m.u. *See* **map unit.**

***mu* phage** A kind ("species") of phage with properties similar to those of insertion sequences, being able to insert, transpose, inactivate, and cause rearrangements.

multimeric structure A structure composed of several identical or different subunits held together by weak bonds.

multiple allelism The existence of several known alleles of a gene.

multiple-factor hypothesis A hypothesis to explain quantitative variation by assuming the interaction of a large number of genes (polygenes), each with a small additive effect on the character.

multiplicity of infection The average number of phage particles that infect a single bacterial cell in a specific experiment.

mutagen An agent that is capable of increasing the mutation rate.

mutant An organism or cell carrying a mutation.

mutant allele An allele differing from the allele found in the standard or wild type.

mutant hunt The process of accumulating different mutants showing abnormalities in a certain structure or function, as a preparation for mutational dissection of that function.

mutant site The damaged or altered area within a mutated gene.

mutation (1) The process that produces a gene or a chromosome set differing from the wild type. (2) The gene or chromosome set that results from such a process.

mutation breeding Use of mutagens to develop variants that can increase agricultural yield.

mutation event The actual occurrence of a mutation in time and space.

mutation frequency The frequency of mutants in a population.

mutation rate The number of mutation events per gene per unit of time (for example, per cell generation).

mutational dissection The study of the parts of a biological function through a study of mutations affecting that function.

muton The smallest part of a gene that can be involved in a mutation event; now known to be a nucleotide pair.

myeloma A cancer of the bone marrow.

narrow heritability *See* **heritability in the narrow sense.**

negative assortative mating Preferential mating between phenotypically unlike partners.

Neurospora A pink mold, commonly found growing on old food.

neutral equilibrium *See* **passive equilibrium.**

neutral mutation (1) A mutation that has no adaptive significance. (2) A mutation that has no phenotypic effect.

neutral petite A petite that produces all wild-type progeny when crossed with wild type.

neutrality *See* **selective neutrality.**

nicking Nuclease action to sever the sugar-phosphate backbone in one DNA strand at one specific site.

nitrocellulose filter A type of filter used to hold DNA for hybridization.

nitrogen bases Types of molecules that form important parts of nucleic acids, composed of nitrogen-containing ring structures; hydrogen bonds between bases link the two strands of a DNA double helix.

nondisjunction The failure of homologs (at meiosis) or sister chromatids (at mitosis) to separate properly to opposite poles.

nonlinear tetrad A tetrad in which the meiotic products are in no particular order.

non-Mendelian ratio An unusual ratio of progeny phenotypes that does not reflect the simple operation of Mendel's laws; for example, mutant:wild ratios of 3:5, 5:3, 6:2, or 2:6 in tetrads indicate that gene conversion has occurred.

nonparental ditype (NPD) A tetrad type containing two different genotypes, both of which are recombinant.

nonsense codon A codon for which no normal tRNA molecule exists; the presence of a nonsense codon causes termination of translation (the end of the polypeptide chain). The three nonsense codons are called amber, ocher, and opal codons.

nonsense mutation A mutation that alters a gene so as to insert a nonsense codon.

nonsense suppressor A mutation that produces a sequence coding for an altered tRNA that will insert an amino acid during translation in response to a nonsense codon.

norm of reaction The pattern of phenotypes produced by a given genotype under different environmental conditions.

NPD *See* **nonparental ditype.**

nu body *See* **nucleosome.**

nuclease An enzyme that can degrade DNA by breaking its phosphodiester bonds.

nucleoid A DNA mass within a chloroplast or mitochondrion.

nucleolar organizer A region (or regions) of the chromosome set physically associated with the nucleolus and containing rRNA genes.

nucleolus An organelle found in the nucleus, containing rRNA and amplified multiple copies of the genes coding for rRNA.

nucleoside A nitrogen base bound to a sugar molecule.

nucleosome A nu body; the basic unit of eukaryotic chromosome structure; a ball of eight histone molecules wrapped about by two coils of DNA.

nucleotide A molecule composed of a nitrogen base, a sugar, and a phosphate group; the basic building block of nucleic acids.

nucleotide pair A pair of nucleotides (one in each strand of DNA) that are joined by hydrogen bonds.

nucleotide-pair substitution The replacement of a specific nucleotide pair by a different pair; often mutagenic.

null allele An allele whose effect is either an absence of normal gene product at the molecular level or an absence of normal function at the phenotypic level.

nullisomic genome An abnormal genome with one chromosomal type missing, thus producing a chromosome number such as $n - 1$ or $2n - 2$.

ocher codon The codon UAA, a nonsense codon.

octad An ascus containing eight ascospores, produced in species where the tetrad normally undergoes a postmeiotic mitotic division.

opal codon The codon UGA, a nonsense codon.

operational definition A working definition; the recognition of an entity or a phenomenon on the basis of key features which appear during experimentation.

operator A DNA region at one end of an operon that acts as the binding site for repressor protein.

operon A set of adjacent structural genes whose mRNA is synthesized in one piece, plus the adjacent regulatory genes that affect transcription of the structural genes.

organelle A subcellular structure having a specialized function—for example, the mitochondrion, the chloroplast, or the spindle apparatus.

overdominance A phenotypic relation in which the phenotypic expression of the heterozygote is greater than that of either homozygote.

paracentric inversion An inversion not involving the centromere.

parental ditype (PD) A tetrad type containing two different genotypes, both of which are parental.

partial diploid *See* **merozygote.**

particulate inheritance The model proposing that genetic information is transmitted from one generation to the next in discrete units ("particles"), so that the character of the offspring is not a smooth blend of characters from the parents (*compare* **blending inheritance**).

pathogen An organism that causes disease in another organism.

patroclinous inheritance Inheritance in which all offspring have the phenotype of the father.

PD *See* **parental ditype.**

pedigree A "family tree," drawn with standard genetic symbols, showing inheritance patterns for specific phenotypic characters.

penetrance The proportion of individuals with a specific genotype who manifest that genotype at the phenotype level.

peptide *See* **amino acid.**

peptide bond A bond joining two amino acids.

pericentric inversion An inversion that involves the centromere.

permissive conditions Those environmental conditions under which a conditional mutant shows the wild-type phenotype.

petite A yeast mutation producing small colonies and altered mitochondrial functions. In cytoplasmic petites (neutral and suppressive petites), the mutation occurs in mitochrondrial DNA; in segregational petites, the mutation occurs in nuclear DNA.

phage *See* **bacteriophage.**

Phe Phenylalanine (an amino acid).

phenocopy An environmentally induced phenotype that resembles the phenotype produced by a mutation.

phenotype (1) The form taken by some character (or group of characters) in a specific individual. (2) The detectable outward manifestations of a specific genotype.

phenotypic sex determination Sex determination by nongenetic means.

phosphodiester bond A bond between a sugar group and a phosphate group; such bonds form the sugar–phosphate–sugar–phosphate backbone of DNA.

pilus (plural, **pili**) A conjugation tube; a hollow hairlike appendage of a donor *E. coli* cell that acts as a bridge for transmission of donor DNA to the recipient cell during conjugation.

plaque A clear area on a bacterial lawn, left by lysis of the bacteria through progressive infections by a phage and its descendants.

plant breeding The application of genetic analysis to development of plant lines better suited for human purposes.

plasma cell A cell that secretes a specific antibody.

plasmid Originally defined as a genetic element existing solely in the bacterial cytoplasm, but this term is now used as a synonym for *episome.*

plate (1) A flat dish used to culture microbes. (2) To spread cells over the surface of solid medium in a plate.

pleiotropic mutation A mutation that has effects on several different characters.

point mutation A mutation that can be mapped to one specific locus.

Poisson distribution A statistical equation that describes the process of sampling in situations where the number of events per sample is potentially very large but in practice is very small.

poky A slow-growing mitochondrial mutation in *Neurospora.*

polar gene conversion A gradient of conversion frequency along the length of a gene.

polar granules Granules at the anterior end of the cytoplasm of a *Drosophila* egg.

polar mutation A mutation that affects the transcription or translation of the part of the gene or operon only on one side of the mutant site—for example, nonsense mutations, frameshift mutations, and IS-induced mutations.

poly-A tail A string of adenine nucleotides added to mRNA after transcription.

polyacrilamide A material used to make electrophoretic gels for DNA separation.

polygenes *See* **multiple-factor hypothesis.**

polymorphism The occurrence in a population (or between populations) of several phenotypic forms associated with alleles of one gene or homologs of one chromosome.

polypeptide A chain of linked amino acids; a protein.

polyploid A cell having three or more chromosome sets, or an organism composed of such cells.

polysaccharide A biological polymer composed of sugar subunits—for example, starch or cellulose.

polytene chromosome A giant chromosome produced by an endomitotic process in which the multiple sets remain bound in a haploid number of chromosomes.

position effect Used to describe a situation where the phenotypic influence of a gene is altered by changes in the position of the gene within the genome.

position-effect variegation Variegation caused by the inactivation of a gene in some cells through its abnormal juxtaposition with heterochromatin.

positive assortative mating A situation in which like phenotypes mate more commonly than expected by chance.

primary structure of a protein The sequence of amino acids in the polypeptide chain.

Pro Proline (an amino acid).

probe Defined nucleic acid segment which can be used to identify specific DNA clones bearing the complementary sequence.

product rule The probability of two independent events occurring simultaneously is the product of the individual probabilities.

proflavin A mutagen that tends to produce frameshift mutations.

prokaryote An organism composed of a prokaryotic cell, such as bacteria and blue-green algae.

prokaryotic cell A cell having no nuclear membrane and hence no separate nucleus.

promoter A regulatory region at one end of an operon that acts as the binding site for RNA polymerase.

prophage A phage "chromosome" inserted as part of the linear structure of the DNA chromosome of a bacterium.

propositus In a human pedigree, the individual who first came to the attention of the geneticist.

protoplast A plant cell whose wall has been removed.

prototroph A strain of organisms that will proliferate on minimal medium (*contrast* **auxotroph**).

provirus A virus "chromosome" integrated into the DNA of the host cell.

pseudoalleles Alleles that do not complement, but between which a rare crossover may produce a wild-type chromosome. (This term is mainly of historical significance.)

pseudodominance The sudden appearance of a recessive phenotype in a pedigree, due to deletion of a masking dominant gene.

puff *See* **chromosome puff.**

pulse–chase experiment An experiment in which cells are grown in radioactive medium for a brief period (the pulse) and then transferred to nonradioactive medium for a longer period (the chase).

Punnett square A grid used as a graphic representation of the progeny zygotes resulting from different gamete fusions in a specific cross.

pure-breeding (true-breeding) line or strain A group of identical individuals that always produce offspring of the same phenotype when intercrossed.

purine A type of nitrogen base; the purine bases in DNA are adenine and guanine.

pyrimidine A type of nitrogen base; the pyrimidine bases in DNA are cytosine and thymine.

quantitative variation The existence of a range of phenotypes for a specific character, differing by degree rather than by distinct qualitative differences.

quaternary structure of a protein The multimeric constitution of the protein.

R plasmid A plasmid containing one or several transposons that bear resistance genes.

random genetic drift Changes in allele frequency that result because the genes appearing in offspring do not represent a perfectly representative sampling of the parental genes.

random mating Mating between individuals where the choice of a partner is not influenced by their genotypes (with respect to specific genes under study).

reading frame The codon sequence that is determined by reading nucleotides in groups of three from some specific fixed starting nucleotide.

reannealing Spontaneous realignment of two single DNA strands to re-form a DNA double helix that had been denatured.

receptor element A controlling element that can insert into a gene (making it a mutant) and can also exit (thus making the mutation unstable); both of these functions are nonautonomous, being under the influence of the regulator element.

recessive allele An allele whose phenotypic effect is not expressed in a heterozygote.

recessive phenotype The phenotype of a homozygote for the recessive allele; the parental phenotype that is not expressed in a heterozygote.

reciprocal crosses A pair of crosses of the type ♀ genotype A × ♂ genotype B and ♀ genotype B × ♂ genotype A.

reciprocal translocation A translocation in which part of one chromosome is exchanged with a part of a separate nonhomologous chromosome.

recombinant An individual or cell with a genotype produced by recombination.

recombinant DNA A novel DNA sequence formed by the combination of two nonhomologous DNA molecules.

recombinant frequency (RF) The proportion (or percentage) of recombinant cells or individuals.

recombination (1) In general, any process in a diploid or partially diploid cell that generates new gene or chromosomal combinations not found in that cell or in its progenitors. (2) At meiosis, the process that generates a haploid product of meiosis whose genotype is different from either of the two haploid genotypes that constituted the meiotic diploid.

recombination repair The repair of a DNA lesion through a process, similar to recombination, that uses recombination enzymes.

recon A region of a gene within which there can be no crossing over; now known to be a nucleotide pair.

reduction division A nuclear division that produces daughter nuclei each having one-half as many centromeres as the parental nucleus.

redundant DNA *See* **repetitive DNA.**

regulator element *See* **receptor element.**

regulatory genes Genes that are involved in turning on or off the transcription of structural genes.

repetitive DNA Redundant DNA; DNA sequences that are present in many copies per chromosome set.

replication DNA synthesis.

replicon A chromosomal region under the influence of one adjacent replication-initiation locus.

repressor protein A molecule that binds to the operator and prevents transcription of an operon.

repulsion conformation Two linked heterozygous gene pairs in the arrangement $A\,b/a\,B$.

resolving power The ability of an experimental technique to distinguish between two genetic conditions (typically discussed when one condition is rare and of particular interest).

restricted (specialized) transduction The situation in which a particular phage will transduce only specific regions of the bacterial chromosome.

restriction enzyme A nuclease that will recognize specific target nucleotide sequences in DNA and break the DNA chain at those points; a variety of these enzymes are known, and they are extensively used in genetic engineering.

restrictive conditions Environmental conditions under which a conditional mutant shows the mutant phenotype.

reverse transcriptase An enzyme that catalyzes the synthesis of a DNA strand from an RNA template.

reversion The production of a wild-type gene from a mutant gene.

RF *See* **recombinant frequency.**

ribonucleic acid *See* **RNA.**

ribosomal RNA *See* **rRNA.**

ribosome A complex organelle that catalyzes translation of messenger RNA into an amino-acid sequence. Composed of proteins plus rRNA.

RNA (ribonucleic acid) A single-stranded nucleic acid similar to DNA but having ribose sugar rather than deoxyribose sugar and uracil rather than thymine as one of the bases.

RNA polymerase An enzyme that catalyzes the synthesis of an RNA strand from a DNA template.

rRNA (ribosomal RNA) RNA molecules, coded in the nucleolar organizer, that have an integral (but poorly understood) role in ribosome structure and function.

S (Svedberg) A unit of sedimentation velocity, commonly used to describe molecular units of various sizes (because sedimentation velocity is related to size).

satellite A terminal section of a chromosome, separated from the main body of the chromosome by a narrow constriction.

satellite chromosomes Chromosomes that seem to be additions to the normal genome.

satellite DNA DNA that forms a separate band in a density gradient because of its different nucleotide composition.

scaffold The central core of a eukaryotic nuclear chromosome, from which DNA loops extend.

SCE *See* **sister-chromatid exchange.**

second-division segregation pattern A pattern of ascospore genotypes for a gene pair showing that the two alleles separate into different nuclei only at the second meiotic division, as a result of a crossover between that gene pair and its centromere; can only be detected in a linear ascus.

second-site mutation The second mutation of a double mutation within a gene; in many cases, the second-site mutation suppresses the first mutation, so that the double mutant has the wild-type phenotype.

secondary structure of a protein A spiral or zigzag arrangement of the polypeptide chain.

sector An area of tissue whose phenotype is detectably different from the surrounding tissue phenotype.

sedimentation The sinking of a molecule under the opposing forces of gravitation and buoyancy.

segregation (1) Cytologically, the separation of homologous structures. (2) Genetically, the production of two separate phenotypes, corresponding to two alleles of a gene, either in different individuals (meiotic segregation) or in different tissues (mitotic segregation).

segregational petite A petite that in a cross with wild type produces 1/2 petite and 1/2 wild-type progeny; caused by a nuclear mutation.

selection coefficient (*s*) The proportional excess or deficiency of fitness of one genotype in relation to another genotype.

selective neutrality A situation where different alleles of a certain gene confer equal fitness.

selective system An experimental technique that enhances the recovery of specific (usually rare) genotypes.

self To fertilize eggs with sperms from the same individual.

self-assembly The ability of certain multimeric biological structures to assemble from their component parts through random movements of the molecules and formation of weak chemical bonds between surfaces with complementary shapes.

semiconservative replication The established model of DNA replication in which each double-stranded molecule is composed of one parental strand and one newly polymerized strand.

semisterility The phenotype of individuals heterozygotic for certain types of chromosome aberration; expressed as a reduced number of viable gametes and hence reduced fertility.

Ser Serine (an amino acid).

sex chromosome A chromosome whose presence or absence is correlated with the sex of the bearer; a chromosome that plays a role in sex determination.

sex linkage The location of a gene on a sex chromosome.

sexduction Sexual transmission of donor *E. coli* chromosomal genes on the fertility factor.

sexual spores *See* **spore.**

shotgun technique Cloning a large number of different DNA fragments as a prelude to selecting one particular clone type for intensive study.

side-by-side helix A model of DNA structure (an alternative model to the Watson-Crick double helix) involving noninterlocking strands.

sister-chromatid exchange (SCE) An event similar to crossing over that can occur between sister chromatids at mitosis or at meiosis; detected in harlequin chromosomes.

S-9 mix A liver-derived supernatant used in the Ames test to activate or inactivate mutagens.

solenoid structure The supercoiled arrangement of DNA in eukaryotic nuclear chromosomes.

somatic cell A cell that is not destined to become a gamete; a "body cell," whose genes will not be passed on to future generations.

somatic mutation A mutation occurring in a somatic cell.

somatic-cell genetics Asexual genetics, involving study of cell fusion, somatic assortment, and somatic crossing over.

somatostatin A human growth hormone.

SOS repair The error-prone process whereby gross structural DNA damage is circumvented by allowing replication to proceed past the damage through imprecise polymerization.

spacer DNA Repetitive DNA found between genes; its function is unknown.

specialized transduction *See* **restricted transduction.**

specific-locus test A system for detecting recessive mutations in diploids. Normal individuals treated with mutagen are crossed to testers that are homozygous for the recessive alleles at a number of specific loci; the progeny are then screened for recessive phenotypes.

spindle The set of microtubular fibers that appear to move eukaryotic chromosomes during division.

spindle fiber One of the fibers making up the spindle.

spontaneous mutation A mutation occurring in the absence of mutagens, usually due to errors in the normal functioning of cellular enzymes.

spore (1) In plants and fungi, sexual spores are the haploid cells produced by meiosis. (2) In fungi, asexual spores are somatic cells that are cast off to act either as gametes or as the initial cells for new haploid individuals.

sporophyte The diploid sexual-spore-producing generation in the life cycle of plants—that is, the stage in which meiosis occurs.

stacking The packing of the flattish nitrogen bases at the center of the DNA double helix.

standard deviation The square root of the variance.

statistic A computed quantity characteristic of a population, such as the mean.

statistical distribution The array of frequencies of different quantitative or qualitative classes in a population.

strain A pure-breeding lineage, usually of haploid organisms, bacteria, or viruses.

structural gene A gene encoding the amino-acid sequence of a protein.

subvital gene A gene that causes the death of some proportion (but not all) of the individuals that express it.

sum rule The probability that one or the other of two mutually exclusive events will occur is the sum of their individual probabilities.

superinfection Phage infection of a cell that already harbors a prophage.

supersuppressor A mutation that can suppress a variety of other mutations; typically a nonsense suppressor.

suppressive petite A petite that in a cross with wild type produces progeny all of which are petite.

suppressor A mutation that can restore the wild-type phenotype to another mutation.

synapsis Close pairing of homologs at meiosis.

syncitium A single cell with many nuclei.

T (1) Thymine, or thymidine. (2) *See* **tetratype.**

T cell An immature lymphocyte (in mammals, produced in the thymus) that is stimulated by the presence of a particular antigen to differentiate and divide, producing daughter cells (killer cells) that attack and kill the cells bearing the antigen.

tandem duplication Adjacent identical chromosome segments.

target number *See* **extrapolation number.**

target theory A theory in which death from radiation is interpreted in terms of inactivation of specific radiation targets within the organism.

tautomeric shift The spontaneous isomerization of a nitrogen base to an alternative hydrogen-bonding condition, possibly resulting in a mutation.

telocentric chromosome A chromosome having the centromere at one end.

telomere The tip (or end) of a chromosome.

temperate phage A phage that can become a prophage.

temperature-sensitive mutation A conditional mutation that produces the mutant phenotype in one temperature range and the wild-type phenotype in another temperature range.

template A molecular "mold" that shapes the structure or sequence of another molecule; for example, the nucleotide sequence of DNA acts as a template to control the nucleotide sequence of RNA during transcription.

teratogen An agent that interferes with normal development.

teratoma A tumor composed of a chaotic array of different tissue types.

terminal redundancy In phage, a linear DNA molecule with single-stranded ends that are longer than is necessary to close the DNA circle.

tertiary structure of a protein The folding or coiling of the secondary structure to form a globular molecule.

testcross A cross of an individual of unknown genotype or a heterozygote (or a multiple heterozygote) to a tester individual.

tester An individual homozygous for one or more recessive alleles; used in a testcross.

testicular feminization The creation of an apparent female phenotype in an XY individual as a result of an X-linked mutation.

tetrad (1) Four homologous chromatids in a bundle in the first meiotic prophase and metaphase. (2) The four haploid product cells from a single meiosis.

tetrad analysis The use of tetrads (definition 2) to study the behavior of chromosomes and genes during meiosis.

tetraparental mouse A mouse that develops from an embryo created by the experimental fusion of two separate blastulas.

tetraploid A cell having four chromosome sets; an organism composed of such cells.

tetratype (T) A tetrad type containing four different genotypes, two parental and two recombinant.

Thr Threonine (an amino acid).

three-point testcross A testcross involving one parent with three heterozygous gene pairs.

thymidine The nucleoside having thymine as its base.

thymine A pyrimidine base that pairs with adenine.

thymine dimer A pair of chemically bonded adjacent thymine bases in DNA; the cellular processes that repair this lesion often make errors that create mutations.

tolerance (immune tolerance) The failure to produce antibodies to a foreign antigen due to previous exposure to the antigen very early in development.

totipotency The ability of a cell to proceed through all the stages of development and thus produce a normal adult.

***trans* conformation** In a heterozygote involving two mutant sites within one gene, the arrangement $a_1 + / + a_2$.

transcription The synthesis of RNA using a DNA template.

transdetermination A specific change in the fate of an imaginal disk that can occur when the disk is cultured.

transduction The movement of genes from a bacterial donor to a bacterial recipient using a phage as the vector.

transfer RNA *See* **tRNA.**

transformation The directed modification of a bacterial genome by the external application of DNA from a bacterial cell of different genotype.

transient diploid The stage of the life cycle of predominantly haploid fungi (and algae) during which meiosis occurs.

transition A type of nucleotide-pair substitution involving the replacement of a purine with another purine, or of a pyrimidine with another pyrimidine—for example, GC → AT.

translation The ribosome-mediated production of a polypeptide whose amino-acid sequence is derived from the codon sequence of an mRNA molecule.

translocation The relocation of a chromosomal segment in a different position in the genome.

transmission genetics The study of the mechanisms involved in the passage of a gene from one generation to the next.

transposable (mobile) genetic element A general term for any genetic unit that can insert into a chromosome, exit, and relocate; includes insertion sequences, transposons, some phages, and controlling elements.

transposon A mobile piece of phage or bacterial DNA that is flanked by inverted repeat sequences; typically bears genes of identifiable function such as drug-resistance genes.

transversion A type of nucleotide-pair substitution involving the replacement of a purine with a pyrimidine, or vice versa—for example, GC → TA.

triplet The three nucleotide pairs that comprise a codon.

triploid A cell having three chromosome sets, or an organism composed of such cells.

trisomic A basically diploid genome with an extra chromosome of one type, producing a chromosome number of the form $2n + 1$.

tritium A radioactive isotope of hydrogen.

tRNA (transfer RNA) Small RNA molecules that bear specific amino acids to the ribosome during translation; the amino acid is inserted into the growing polypeptide chain when the anticodon of the tRNA pairs with a codon on the mRNA being translated.

Trp Tryptophan (an amino acid).

true-breeding line or strain *See* **pure-breeding line or strain.**

truncation selection A breeding technique in which individuals above or below a certain phenotypic value (the truncation point) are selected as parents for the next generation.

Turner's syndrome An abnormal phenotype of the human female produced by the presence of only one X chromosome (XO).

twin spot A pair of mutant sectors within wild-type tissue, produced by a mitotic crossover in an individual of appropriate heterozygous genotype.

Tyr Tyrosine (an amino acid).

U Uracil, or uradine.

underdominance A situation where the phenotype of the heterozygote is less than that of either homozygote.

unequal crossover A crossover between homologs that are not perfectly aligned.

uniparental inheritance The transmission of certain phenotypes from one parental type to all the progeny; such inheritance is generally produced by organelle genes.

unstable mutation A mutation that has a high frequency of reversion; a mutation caused by the insertion of a controlling element, whose subsequent exit produces a reversion.

uracil A pyrimidine base that appears in RNA in place of thymine found in DNA.

uradine The nucleoside having uracil as its base.

Val Valine (an amino acid).

variable A property that may have different values in various cases.

variable region A region in an immunoglobin molecule that shows many sequence differences between antibodies of different specificities; the part of the antibody that binds to the antigen.

variance A measure of the variation around the central class of a distribution; the average squared deviation of the observations from their mean value.

variant An individual organism that is recognizably different from an arbitrary standard type in that species.

variate A specific numerical value of a variable.

variation The differences among parents and their offspring or among individuals in a population.

variegation The occurrence within a tissue of sectors with differing phenotypes.

vector In cloning, the plasmid or phage used to carry the cloned DNA segment.

viability The probability that a fertilized egg will survive and develop into an adult organism.

virulent phage A phage that cannot become a prophage; infection by such a phage always leads to lysis of the host cell.

wild type The genotype or phenotype that is found in nature or in the standard laboratory stock for a given organism.

wobble The ability of certain bases at the third position of an anticodon in tRNA to form hydrogen bonds in various ways, causing alignment with several possible codons.

X linkage The presence of a gene on the X chromosome but not on the Y.

X-and-Y linkage The presence of a gene on both the X and Y chromosomes (rare).

X-ray crystallography A technique for deducing molecular structure by aiming a beam of X rays at a crystal of the test compound and measuring the scatter of rays.

Y linkage The presence of a gene on the Y chromosome but not on the X.

zygote The cell formed by the fusion of an egg and a sperm; the unique diploid cell that will divide mitotically to create a differentiated diploid organism.

zygotic induction The sudden release of a lysogenic phage from an Hfr chromosome when the prophage enters the F^- cell, and the subsequent lysis of the recipient cell.

Answers to Selected Problems

Chapter 2

1. Charlie was not pure breeding but heterozygous, *Bb,* where *b* is a recessive allele causing red and white coat. His mates were *BB.* Half of his progeny were *Bb* and these, when inter-bred, gave $3/4\ B-$ and $1/4\ bb$.

2. Recessive.

6. Father probably *Hh* (disease is rare), so man has 1/2 chance of getting *H* from father.
   (a) 1/2
   (b) $1/2 \times 1/2 = 1/4$

7. (a) Dominant.
   (b) *Dd* and *Dd.*
   (c) 1/4 for normal, 3/4 for dwarf.

9. 91 spotted-leafed plants and 28 unspotted, approximately a 3:1 ratio.
   (a) Spots $= S$, no spots $= s$. Parental plant *Ss*, gave $3/4\ S-$ and $1/4\ ss$.
   (b) All unspotted plants shuld be pure breeding, and 1/3 of spotted plants should be pure breeding.

14. (Pedigree 1 answer only.) Must be recessive condition.
    Genotypes, line 1: $A-$, *aa*. Line 2: $Aa, Aa, Aa, A-, A-, Aa$. Line 3: Aa, Aa. Line 4: *aa.*

15. 1/4

16. 1/32

18. (a) $CcSs \times CcSs$
    (b) $CCSs \times CCss$
    (c) $CcSS \times ccSS$
    (d) $ccSs \times ccSs$
    (e) $Ccss \times Ccss$
    (f) $CCSs \times CCSs$
    (g) $CcSs \times Ccss$

20. (a) 2
(b) bow and knock are alleles of one; hairy and smooth are alleles of other.
(c) B = Bow, b = knock
H = Hairy, h = smooth
(d) $\frac{\text{Bb Hh,}}{\text{(p)}}$ $\frac{\text{Bb HH,}}{\text{(q)}}$ $\frac{\text{Bb hh,}}{\text{(r)}}$ $\frac{\text{bb hh,}}{\text{(s)}}$ $\frac{\text{BB Hh.}}{\text{(t)}}$
So that 1. is p × q *or* q × p
2. is r × s
3. is p × s
4. is p × t *or* t × p

22. *Phenotypically*
(a) $1/2 \times 3/4 \times 1 \times 3/4 \times 1/2 = 9/64$
(b) $1/2 \times 3/4 \times 1 \times 3/4 \times 1/2 = 9/64$
(c) $9/64 + 9/64 = 18/64 = 9/32$
(d) $1 - 9/32 = 23/32$

Genotypically
(a) $1/2 \times 1/2 \times 1/2 \times 1/2 \times 1/2 = 1/32$
(b) $1/2 \times 1/2 \times 1/2 \times 1/2 \times 1/2 = 1/32$
(c) $1/32 + 1/32 = 2/32 = 1/16$
(d) $1 - 1/16 = 15/16$

Chapter 3

2. (a) 92; (b) 92; (c) 46; (d) 46; (e) 23.
3. 5
5. For a child's karyotype to be the same as his, the man's sperm must contain a satellited 4, an abnormal 7, and a Y. The overall probability of this is $1/2 \times 1/2 \times 1/2 = 1/8$.
6. Probability $= 1 \times 1/2 \times 1/2 \times 1/2 \ldots\ldots\ldots 1/2 = (1/2)^{n-1}$
9. In schmoos, females are obviously the heterogametic sex. ♀ G × ♂ gg gives Gg male and g female offspring where G = graceful, g = gruesome.
10. 4
13. (a) Her mother must have been Dd, so there is a 1/2 chance that the woman got the d from her mother and is heterozygous herself. 1/2 of her children will be sons, and 1/2 of them will have muscular dystrophy.
Overall probability $= 1/2 \times 1/2 \times 1/2 = 1/8$.
(b) One of the grandparents had the gene so there is a 1/2 chance that your mother has, and a further 1/2 chance that you have it, if your are a ♂ or ♀.
Overall probability $= 1/2 \times 1/2 = 1/4$.
(c) Since your father does not have the disease, there is no way you can get it.
16. (a) Suppose the genes for black and yellow are alleles on the X chromosome, call them t^b and t^y, respectively. Then males are $X^{t^b}Y$ (black) and $X^{t^y}Y$ (yellow).

Females could be $X^{tb}X^{tb}$ (black), $X^{ty}X^{ty}$ (yellow), and $X^{tb}X^{ty}$ (tortoise shell). Females heterozygous for the two alleles have a phenotype different from either homozygote.

(b) $X^{ty}X^{ty}$ (yellow)♀ × X^{tb}Y (black)♂
↓
$X^{ty}X^{tb}$ (tortoise shell)♀ X^{ty}Y (yellow)♂

(c) $X^{tb}X^{tb}$ (black)♀ × X^{ty}Y (yellow)♂
↓
$X^{tb}X^{ty}$ (tortoise shell)♀ X^{tb}Y (black)♂

(d) tortoise shell ♀♀ and black ♂♂.

(e) tortoise shell ♀♀ and yellow ♂♂.

17. 1/4

20. (a) excluded; (b) consistent; (c) excluded; (d) excluded; (e) excluded.

21. (a) Bent is dominant (bent × bent gives normals in 6).
(b) X-linked
(c)

(1)	$bb \times B$	Bb	b
(2)	$Bb \times b$	$1/2Bb$ $1/2bb$	$1/2B$ $1/2b$
(3)	$BB \times b$	Bb	B
(4)	$bb \times b$	bb	b
(5)	$BB \times B$	BB	B
(6)	$Bb \times B$	$1/2BB$ $1/2Bb$	$1/2B$ $1/2b$

Chapter 4

2. Erminette is a case of codominance—the phenotype is a heterozygote (*Ee*), where *EE* = black and *ee* = white. Thus *Ee* selfed gives a 1:2:1 ratio. One possible test is to cross erminettes to white; a ratio of 1/2 erminette to 1/2 white is expected.

4. Progeny types: 12, 14, 23, 34. (a) 1/4; (b) 1/4; (c) 1/2.

6. e

8. Baby 1 is from cross 4, baby 2 is from cross 2, baby 3 is from cross 1, baby 4 is from cross 3.

10. Platinum phenotype due to an allele which is lethal when homozygous. Test by crossing platinum to normal; predict a 1:1 ratio. Or examine aborted fetuses in platinum × platinum; expect 1/4.

12. Short bristled flies are always females. Only half as many males are produced as females. Suppose the allele for short bristles is dominant (call it S). Hence S/S^+ females have short bristles, but S is also a recessive lethal so that S Y males never live. The first cross is

SS^+ (short bristle)♀ × S^+Y (long)♂
↓
1 SS^+(short)♀ : 1 S^+S^+ (long)♀ : 1 S^+Y (long)♂ : 1 SY ♂ dies

The second cross is

S^+S^+ (long)♀ × S^+Y (long)♂
↓
S^+S^+ (long)♀ S^+Y (long)♂

The third cross is the same as the first.

16. Recessive albino alleles at two independent genes. Cross is $A_1A_1a_2a_2 \times a_1a_1A_2A_2$, which gives all $A_1a_1A_2a_2$ (normal).

17. c^{sr} = sun red, c^{o} = orange, c^{p} = pink, allelic series with dominance in the order shown.

    1. $c^{sr}c^{sr} \times c^{p}c^{p}$ gives F_1 $c^{sr}c^{p}$, gives 3:1 in F_2.
    2. $c^{o}c^{o} \times c^{sr}c^{sr}$ gives F_1 $c^{o}c^{sr}$, gives 3:1 in F_2.
    3. $c^{o}c^{o} \times c^{p}c^{p}$ gives F_1 $c^{o}c^{p}$, gives 3:1 in F_2.

    Scarlet is an allele of a separate independent gene *s*.

    4. $c^{o}c^{o}SS \times CCss$

       F_1 $Cc^{o}Ss$, normal (yellow)
       F_2 $C-\ S-$, yellow 9
       $C-\ ss$, scarlet 3
       $c^{o}c^{o}S-$, orange 3
       $c^{o}c^{o}ss$, orange 1 (epistasis)

20. The 9:3:3:1 indicates that 2 pairs of genes are involved. The alleles for scarlet and brown eyes must be recessive since the F_1 are normal. Let the allele for scarlet eyes be *s* and for brown eyes be *b*.

$ss\,b^+b^+$ (scarlet) $\times$ $s^+s^+\,bb$ (brown)
↓
$s^+s\,b^+b$ (red) $\times$ $s^+s\,b^+b$ (red)
↓

$9\ s^+-\ b^+-$	red
$3\ ss\,b^+-$	scarlet
$3\ s^+-\ bb$	brown
$1\ ss\,bb$	white

23. (a) The genes causing scarlet eyes in strains A and B are different. They are recessive to their wild-type alleles. The gene for scarlet in strain B must be autosomal.

    (b) The gene for scarlet eyes in strain A is sex-linked.

    (c) Let the genes for scarlet in strain A be *a* and in strain B be *b* (the normal alleles are *A* and *B*, respectively).

$aYBB$ ♂ $\times$ $AA\,bb$ ♀
↓
$Aa\,Bb$ ♀ $\times$ $AYBb$ ♂
↓

1/4 AA	1/4 BB	—1/16 $AA\,BB$ normal ♀	
	1/2 Bb	—1/8 $AA\,Bb$ normal ♀	6/16
	1/4 bb	—1/16 $AA\,bb$ scarlet ♀	
1/4 Aa	1/4 BB	—1/16 $Aa\ BB$ normal ♀	
	1/2 Bb	—1/8 $Aa\ Bb$ normal ♀	2/16
	1/4 bb	—1/16 $Aa\ bb$ scarlet ♀	
1/4 AY	1/4 BB	—1/16 $AY\,BB$ normal ♂	
	1/2 Bb	—1/8 $AY\,Bb$ normal ♂	3/16
	1/4 bb	—1/16 $AY\,bb$ scarlet ♂	
1/4 aY	1/4 BB	—1/16 $aY\ BB$ scarlet ♂	
	1/2 Bb	—1/8 $aY\ Bb$ scarlet ♂	5/16
	1/4 bb	—1/16 $aY\ bb$ scarlet ♂	

26. Line 1 proves plant is either *Aa CC* or *AA Cc.*
Line 2 proves plant is *Aa Rr.*
Line 3 confirms *AA CC* alternative from line 1.
Plant is *Aa CC Rr.*

27. (a) *td* + (1/4)　　(b) 1/4:3/4 or 1:3
+ *su* (1/4)
+ + (1/4)
td su (1/4)

28.

	Cross 1	*Cross 2*
white	32/48	16/24
solid black	9/48	3/24
spotted black	3/48	3/24
solid chestnut	3/48	1/24
spotted chestnut	1/48	1/24

30. *For host:*

$F_2\ R_\alpha - R_\beta -$	9	no disease
$R_\alpha - r_\beta\, r_\beta$	3	disease
$r_\alpha r_\alpha\ R_\beta -$	3	disease
$r_\alpha r_\alpha\ r_\beta\, r_\beta$	1	disease

R_α and R_β are dominant host alleles at separate loci, conferring resistance to α and β respectively.

A is $r_\alpha r_\alpha R_\beta R_\beta$
B is $R_\alpha R_\alpha r_\beta r_\beta$
F_1 is $R_\alpha r_\alpha R_\beta r_\beta$

For parasite:
Hypothetis 1. Virulence alleles (*) at separate loci, that is, α is $V_A{}^*V_B$, β is $V_AV_B{}^*$.
Prediction: 4 genotypes from crossing $\alpha \times \beta$

	A	B
$V_A{}^*V_B$	+	−
$V_A\ V_B{}^*$	−	+
$V_A{}^*V_B{}^*$	+	+
$V_A\ V_B$	−	−

Hypothesis 2. Different virulence alleles at one locus, that is, α is V_A, β is V_B.
Prediction: 2 equally frequent genotypes from $\alpha \times \beta$

	A	B
1/2 V_A	+	−
1/2 V_B	−	+

31. A must have had the gene but with very low (or zero) penetrance, due either to suppressor genes or environmental suppression of gene action.

Chapter 5

2. The H/h locus is closely linked to the I/i locus, so that crossovers are very rare.

3. c

5. Two genes are linked 10 map units apart. The female parent was of the type AB/ab.

7. b

8. The results indicate linkage. The parents of the children must have been $\frac{P\,I^{A}}{p\,i}$

   (where P is the allele for nail-patella syndrome).
   If you make a grid of the combinations of parental and recombinant gametes, you will see that the $p\,i/p\,i$ (normal, 0 blood) is the only phenotypically unique class and is produced by two parentals. The frequency of $p\,i$ gametes must be the square root of the $p\,i/p\,i$ class $= \sqrt{0.16} = 0.4$.
   Assume the $P\,I$ gametes occur with the same frequency so the recombinants occur at a frequency of 0.2 and the distance between p and i is 20 map units.

	0.4 $P\,I^{A}$	0.4 $p\,i$	0.1 $P\,i$	0.1 $p\,I^{A}$
0.4 $P\,I^{A}$	0.16 $\frac{P\,I^{A}}{P\,I^{A}}$	0.16 $\frac{p\,i}{P\,I^{A}}$	0.04 $\frac{P\,i}{P\,I^{A}}$	0.04 $\frac{p\,I^{A}}{P\,I^{A}}$
0.4 $p\,i$	0.16 $\frac{P\,I^{A}}{p\,i}$	0.16 $\frac{p\,i}{p\,i}$	0.04 $\frac{P\,i}{p\,i}$	0.04 $\frac{p\,I^{A}}{p\,i}$
0.1 $P\,i$	0.04 $\frac{P\,I^{A}}{P\,i}$	0.04 $\frac{p\,i}{P\,i}$	0.01 $\frac{P\,i}{P\,i}$	0.01 $\frac{P\,I^{A}}{P\,i}$
0.1 $p\,I^{A}$	0.04 $\frac{P\,I^{A}}{p\,I^{A}}$	0.04 $\frac{p\,i}{p\,I^{A}}$	0.01 $\frac{P\,i}{p\,I^{A}}$	0.01 $\frac{p\,I^{A}}{p\,I^{A}}$

10. There were two linked genes, each with two alleles coding for an F and S allele. The conformation of the alleles was ADH^{F} PGM^{S}/ADH^{S} PGM^{F} (repulsion). The 21 + 19 individuals are recombinants, RF $= 40/237 = 0.17$.

12. (a) The cross is $a\,b\,c/+\,+\,+$♀ × $a\,b\,c/+\,+\,+$♂. Since there is no crossing over in males, two types of sperm are produced, $a\,b\,c$ and $+\,+\,+$. There should be as many $+\,+\,+/a\,b\,c$ as $a\,b\,c/a\,b\,c$. The data then are:

Parental ($a\,b\,c$ and $+\,+\,+$)	730
Single recombinants between a and b ($a\,+\,+$ and $+\,b\,c$)	91
Single recombinants between b and c ($a\,b\,+$ and $+\,+\,c$)	171
Double recombinants ($+\,b\,+$ and $a\,+\,c$)	9
Total	1001

$a - b = [(91 + 9)/1001] \times 100 = 10$ map units
$b - c = [(171 + 9)/1001] \times 100 = 18$ map units

(b) Expected number of double recombinants assuming no interference = $(0.1 \times 0.18)(1001) = 18$. Coefficient of coincidence = observed/expected = $9/18 = 0.5$.

14. Gene sequence is $lg - b - y$. Map distances are 28 m.u. (*lg* to *b*) and 18 m.u. (*b* to *y*). Coefficient of coincidence is about 0.8.

16. Cross $+ + + \times fat^- \; tail^- \; flag^-$

Progeny
$+ + +$
$+ + flag^-$
$fat^- + +$
$fat^- + flag^-$
$fat^- \; tail^- \; flag^-$
$fat^- \; tail^- +$
$+ \; tail^- \; flag^-$
$+ \; tail^- +$

Linkage map $+/fat^-$ — $+/tail^-$ — $+/flag^-$
9.3 mu 15.3 mu

19. (a) yes (b) probably dominant (c) yes, mother would be RE/re. The single crossover-derived child (*Rree*) provides a rough estimate of 10 m.u.

20. (a) $PAR = (0.85 \times 0.80) \div 2 = 0.34$
(b) $par = (0.85 \times 0.80) \div 2 = 0.34$
(c) $PaR = (0.15 \times 0.20) \div 2 = 0.015$
(d) $Par = (0.15 \times 0.80) \div 2 = 0.06$

22. 1. χ^2 (3 d.f.) = 2.12; p = > 50%; unlinked.
2. χ^2 (3 d.f.) = 6.6; p = > 10%; unlinked.
3. χ^2 (3 d.f.) = 66; p = < 0.5%; linked.
4. χ^2 (3 d.f.) = 11.6; p = ~ 1%; linked.

Chapter 6

1. (a) *ad* and *nic* linked, *leu* and *arg* linked. Parents were $ad^-nic^+ \; leu^+arg^- \times ad^+nic^- \; leu^-arg^+$.
(b) Crossover between *ad* and *nic* (the reciprocal type not found in this small sample).

3. (a) $0.2 = 1/2\,(1 - e^{-m})$. $m = 0.51 \equiv 0.51 \times 50 = 25.5$ m.u.
(b) Hypothesis: no linkage. Prediction 50:50:50:50.
$\chi^2 = 2.52$ (3 d.f.); p = 30 to 50%.
Therefore these data are a common deviation from a 1:1:1:1 ratio and there is no need to postulate linkage.

5. 1. A PD
2. Derived from a self of $y^- \; paba^+$
3. Derived from a self of $y^+ \; paba^-$

7. 

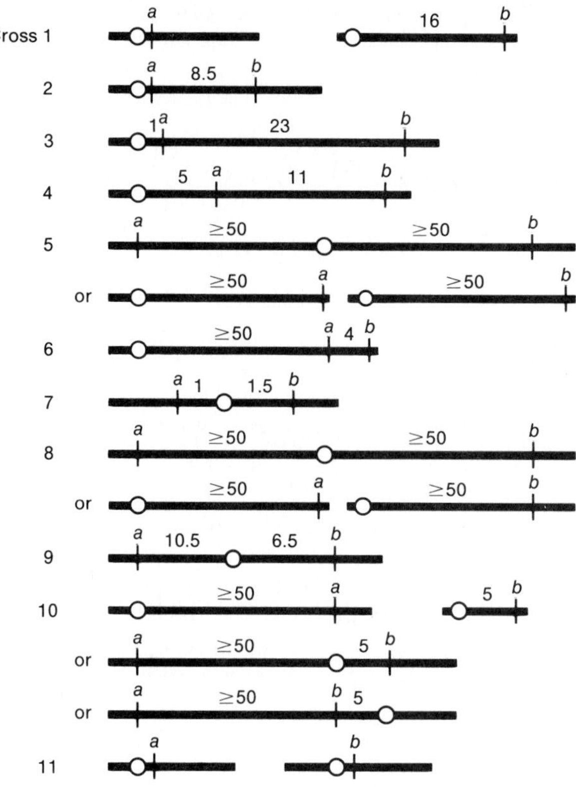

8.

$+/b$ \ $+/a$	M_I 90%	M_{II} 10%
M_I 80%	72%: 36% PD, 36% NPD	8% T
M_{II} 20%	18% T	2%: 0.5% PD, 0.5% NPD, 1.0% T

(a) 36.5%
(b) 36.5%
(c) 27.0%
(d) 50.0%
(e) 25.0%

10. 1. linked 2. unlinked 3. linked or unlinked 4. unlinked 5. unlinked 6. linked or unlinked.

12. *his-4*, because T requires exchange between centromere and either *ad-3* or *his-?*, and since *ad-3* is known to be close to its centromere, only *his-4* has the necessary distance to give 60% T.

13. First mutant w, second at separate locus t.
Crossed $w\,t^+ \times w^+\,t$

Asci			
	$w\,t^+$ (white)	$w\,t^+$ (white)	$w\,t$ (white)
	$w\,t^+$ (white)	$w\,t$ (white)	$w\,t$ (white)
	$w^+\,t$ (tan)	$w^+\,t^+$ (black)	$w^+\,t^+$ (black)
	$w^+\,t$ (tan)	$w^+\,t$ (tan)	$w^+\,t^+$ (black)
	(PD)	(T)	(NPD)

(Note w is epistatic to t.)

15. *Yy* (yellowish). Mitotic crossover produces a twin spot of genotype *YY* (normal) and *yy* (yellow).

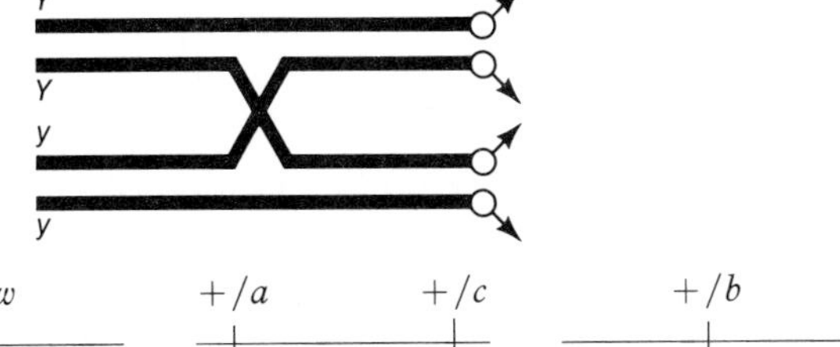

17.

$+/w$	$+/a$	$+/c$	$+/b$

19. (a) They indicate linkage of *pro* and *paba* to *fpa*, and their different frequencies reflect the sizes of linkage group intervals. The absence of a class requiring *pro* only means that *paba* is distal to *pro*.

●	*pro*		*paba*		*fpa*
6 (= 9/154)		71 (= 110/154)		23 (= 35/154)	

(c) *pro, paba, fpa.*

21. α on 7, β on 1, γ on 5, δ on 6, ϵ not on these eight.

Chapter 7

2. Make master plate on medium with arginine, replicate plate onto arginineless medium and look for failure to duplicate a colony.

4. Strain 2 carried a recessive *mei* mutant and passed this on to strains 4, 6, 7, and 8. Only crosses homozygous for *mei* show meiotic abnormalities. Perhaps *mei* inhibits pairing.

5. $e^{-\mu.10^6} = 0.37$, and $\mu = -\ln 0.37.10^{-6}. = 1$ per 10^6 cell division.
7. We have to make the assumption that chromosomes are genetically inert when they are heterochromatic. The data then suggest that in males the heterochromatic chromosomes are all of paternal origin. Hence, radiation of females produces lethal mutations which kill females (dominant) and males (recessive). Radiation of males kills daughters but has no effect on sons because their irradiated chromosomes are inert.

9. Study pollen grains (which are haploid) and look for red grains under the microscope.

13. It is possible (although not certain) that the achondroplasia mutation arose in the germ line of the father as a result of radiation exposure. However, the hemophilia allele *must* have been transmitted from the mother (father contributes Y), and plant cannot be held responsible.

14. There are $10 - 2 = 8$ new mutant alleles per $9{,}4075 \times 2$ alleles; this equals 4.25×10^{-5}, or about 1 in 24,000 alleles.

Chapter 8

2. Consider two hypotheses:
   (i) Deletion of left arm in nucleus 2.
   (ii) Mitotic crossover generates homozygosity of left arm of nucleus 1.

   Hypothesis (ii) is unlikely, since we should be able to recover nucleus 2 alleles in a cross. Although these alleles must be present in the HK to provide *ad-3A*$^+$ and *nic*$^+$ function, they cannot cross because their mating allele has been deleted.

4. Small deletions *within* the *ad-3B* gene.

5. (a) Deletions spanning the loci whose mutant alleles are expressed.
   (b) A deletion not spanning a mutant allele; or, alternatively, a dominant lethal point mutation.

7. e

9.

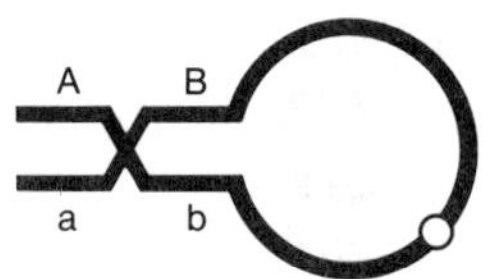

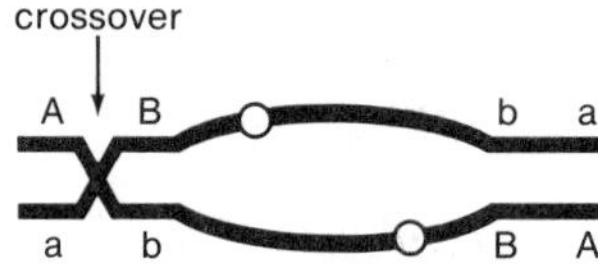

Another possible rare type

11. (a) Inversion from close to *f* to close to *y*.
    (b) Double crossovers; or single crossovers in the noninverted end of the *f*-*b* region.

12. (a) m = 0.22
    (b) (i) 0.8
    (ii) 0.18
    (iii) 0.02
    1.00 (rounded off)
    (c) (i) 8:0 (dark:light)
    (ii) 4:4
    (iii) 8:0, 0:8, and 4:4 in a 1:1:2 ratio.

(d)

# crossovers	8:0	0:8	4:4
0 (0.8)	0.8	–	–
1 (0.18)	–	–	0.18
2 (0.02)	0.005	0.005	0.01
Totals	0.805	0.005	0.19

13. (a) Classes 1 and 2 are parentals. Classes 3 to 10 are all double recombinants. We'd expect such results if one of the chromosomes carried a very large inversion spanning the entire marked region from *y* to *car*.
   (b) Class 11 is patroclinous in that the flies carry a paternal X chromosome. This could result from nondisjunction or a four-strand double-exchange tetrad which produces nullo X eggs.
   (c) If egg nuclei could be looked at cytologically, you'd expect to see double dicentric bridges if the four-strand doubles occur.

16. The wild type from nature was a reciprocal translocation for the chromosomes carrying the marked loci.

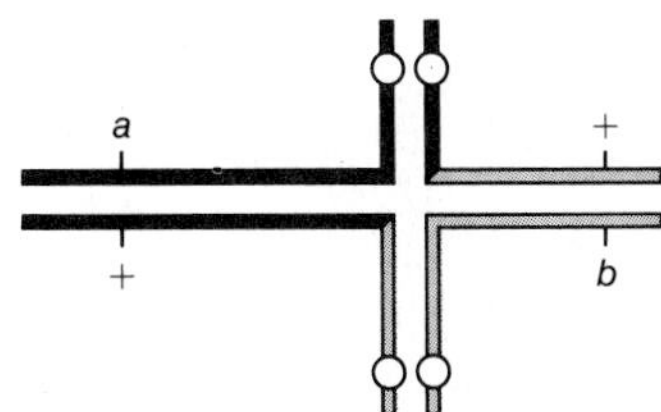

Alternate segregation guarantees *ab* and ++ genotypes (predominantly), with some *a*+ and +*b* resulting from exchange in the breakpoint to locus intervals.

20. Treat the translocation point as a locus and calculate RF in the usual way. This gives the value of 18.2%, which = 18.2 m.u.

21.

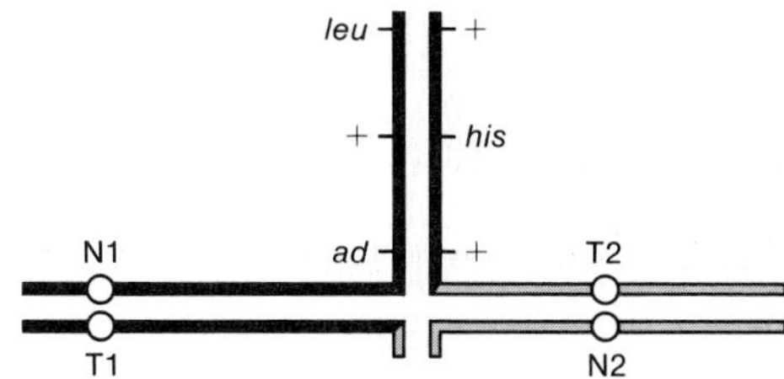

Since short arm nonessential, N1+T2 is now a viable duplication and is the only POM which will grow on medium lacking leucine, histidine, and adenine. (Double recombinants will be much rarer.)

24. species B
↓
species D—paracentric inversion of *x y* segment.
↓
species E—translocation of *z x y* to *k l m*.
↓
species A—translocation of *a b c* to *d e f*.
↓
species C—pericentric inversion of *b c d e*.

26. Solve this by putting *AA BB*, *AA bb*, *aa BB*, or *aa bb* as the female contributions, and fertilize each with one of four different male nuclei, *AB*, *Ab*, *aB*, or *ab*. (Use a Punnet square.)

28. 1/3 of the time, *B* pairs with *B*, and *b* with *b*. This produces asci of "type I." The other 2/3 of the time, *B* will pair with *b* at both homozygous pairs. Two equally frequent segregations will produce asci of "type I" and "type II." Hence, the answer is 2/3 asci type I and 1/3 type II.

	Bb	*BB*	
(type I)	*Bb*	*BB*	(type II)
	Bb	*bb*	
	Bb	*bb*	

30.

thurberi	═══ × 13	
herbaceum		═══ × 13
thurb × herb sterile hybrid	——— × 13	——— × 13
amphidiploid = hirsutum	═══ × 13	═══ × 13
hirs × thurb hybrid	═══ × 13	——— × 13
hirs × herb hybrid	——— × 13	═══ × 13

Test by reconstructing amphidiploid with colchicine.

31. (a) 1/36 *FFGG*, 4/36 *FFGg*, 1/36 *FFgg*, 4/36 *FfGG*, 16/36 *FfGg*, 4/36 *Ffgg*, 1/36 *ffGG*, 4/36 *ffGg*, 1/36 *ffgg*.
   (b) *FFFf* produced by *FF* + *Ff* or Ff + *FF* = 4/6 × 1/6 × 2 = 8/36
   GGgg produced by *GG* + *gg*, *gg* + GG, or *Gg* + *Gg*
   = 1/36 + 1/36 + 16/36 = 18/36
   Hence *FFFfGGgg* = 8/36 × 18/36 = 1/9.
   ffffgggg only produced by *ffgg* + *ffgg* = 1/36 × 1/36 = 1/1296.

35. One might expect half of the gametes to be 2 × 21 ($n + 1$) and half to be normal 1 × 21 (n). Thus, various pairings could give anywhere from 4 × 21 ($2n + 2$, probably lethal) to normal, 2 × 21 ($2n$).

40. (a) Female zygote X X (lost)
   (b) Klinefelter zygote X X Y
   (c) Female zygote X X
   (d) (i) two zygotes fused?
   (ii) fertilization of egg by one sperm (say X) and fertilization of polar body by another (Y), and fusion.
   (e) Same as (c) but at later mitosis so XX cells present.

42. Origin by meiotic nondisjunction, forming an $n + 1$ (disomic) P.O.M. of genotype

$$\frac{a \;\; + \;\; c \;\; + \;\; e}{+ \;\; b \;\; + \;\; d \;\; +}$$

The disomic was unstable and haploidized during cell growth to give haploid nuclei containing component chromosome types. One possibility:

$a \quad + \quad c \quad + \quad e \qquad + \quad b \quad + \quad d \quad +$

43. (a) mutation; (b) crossover; (c) nondisjunction.

Chapter 9

2. (a) F^+; (b) F^-; (c) F^-; (d) Hfr; (e) Hfr; (f) Hfr; (g) F^-; (h) F^+.
3. QWDMTPXACNBQ
6. (a) To select for met^+ to ensure a merozygote.
   (b) *met* – *pur* – *thi*
   (c) *met* – *pur* RF = 52/338 = 15.4%
   met – *thi* RF = 58/338 = 17.2%
   thi – *pur* RF = 17.2 – 15.4 = 1.8%
8. The order is ad-Z_2-Z_1.
   In the first cross, Z_2^+ is sometimes included in the double which inserts ad^+

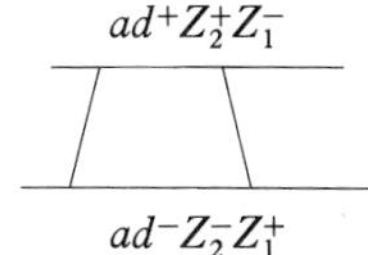

In the second cross, a separate insertion is needed, hence the combined frequency is very low.

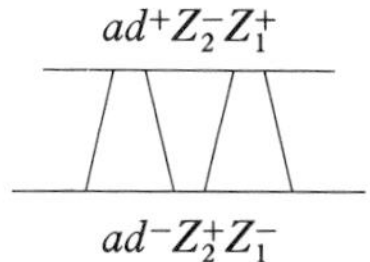

11. The pro^+ gene had been sexduced by an F′ episome which was transmitted infectiously for two generations.
15. (a) Multiple transformants are most rare whenever B is selected, so B is the distant gene.
    (b) ADC
17. (a) The distances are:

$$m - r = \frac{162 + 520 + 474 + 172}{10{,}342} \times 100 = 12.8 \text{ map units}$$

$$r - tu = \frac{853 + 162 + 172 + 965}{10{,}342} \times 100 = 20.8 \text{ map units}$$

$$m - tu = \frac{853 + 520 + 474 + 965}{10{,}342} \times 100 = 27.2 \text{ map units}$$

(looking only at *m* and *tu*)

(b) The linkage data suggest that m and tu are farthest apart so the order must be $m - r - tu$. Now we can include the double crossovers, so that distance $m - tu =$

$$\frac{853 + (2 \times 162) + 520 + 474 + (2 \times 172) + 965}{10{,}342} \times 100 = 33.7$$

(c) The coefficient of coincidence = observed doubles/expected doubles =

$$\frac{162 + 172}{275.1} = \frac{334}{275.1} = 1.2$$

There is an excess of doubles over the expected, so interference is negative.

20. (a) specialized transduction; (b) in the *cys–leu* region.

23. There is a group of genes in the order *cead.* Gene *b* is more distant as it is never cotransduced.

Chapter 10

2. Withhold lactose from the diet. Recessive.

4. (a) Main use is in diagnosing carriers. Parents of diseased individuals must be carriers, and these show intermediate enzyme levels. Also, using amniocentesis, the genotype of an unborn progeny may be determined (see next question).
   (b) One source of ambiguity is the range overlap shown in the galactosemia example, that is, an activity of 25-30 units could mean carrier or normal.
   (c) At the molecular level these genes show incomplete dominance, but at the phenotypic level they are dominant. Presumably a certain minimal level of activity (a threshold) is necessary for normal function and phenotype.

6. a

8. Development of wild-type eye color in *v* or *cu* disks transplanted to wild type hosts suggests a factor made by the wild cells can diffuse into the transplants which lack it. Failure of a *v* host to provide a factor to a *cu* implant, while a *cu* host does provide a factor to a *v* implant, can be explained by a sequential reaction controlled by the two genes.

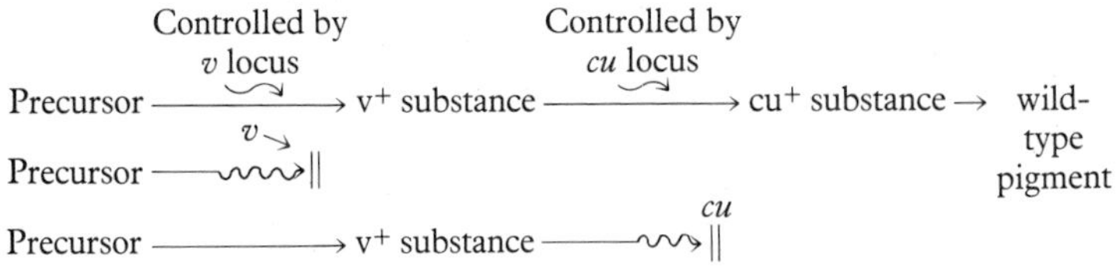

Controlled by *v* locus; Controlled by *cu* locus

Precursor ⟶ v^+ substance ⟶ cu^+ substance ⟶ wild-type pigment

Precursor ⟶ (*v*) ||

Precursor ⟶ v^+ substance ⟶ (*cu*) ||

Thus, *cu* mutants have v^+ substance, which can diffuse into *v* cells, which can convert it to pigment. A simple test would be to grind up wild-type cells and inject into *v* and *cu* hosts and look for wild-type eye color. The substance (s) could be identified using such an assay system.

10. (a) E ⟶ A ⟶ C ⟶ B ⟶ D ⟶ G
    (b) 5 4 2 1 3
    (c) yes, no, yes.

12. (a) m_2; (b) Purple; (c) 9 Purple:3 blue:3 red:1 white. c, because they lack functional enzyme.

14. A cistron V, W.
B cistron U, X, Y, Z.

16.

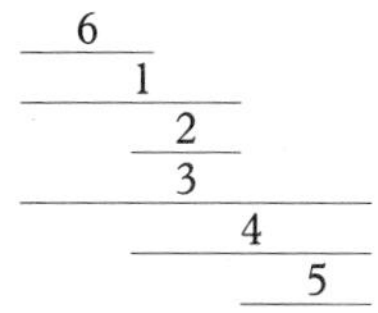

3 may be a deletion, or a mutation causing a large conformational distortion (or as we will see in later chapters, a frameshift or a nonsense mutation).

18. (a) 2 black:4 pale:2 colorless.
(b) 4 black:4 colorless. (Another segregation is possible, but this would give 8 pale spores, the usual ascus in such crosses.)

20.

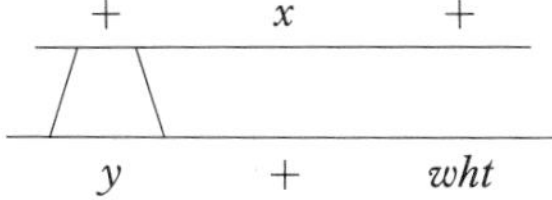

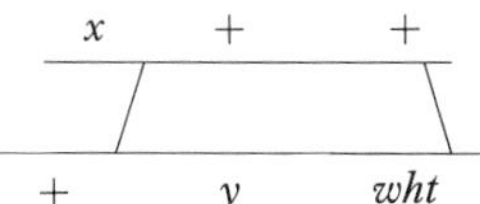

Therefore, high green/white means donor to left and low green/white means donor to right. Order then becomes —1—3—5—6—2—4—*wht*. Abortive transductions presumably occur in all crosses, but only where mutants are complementary do microcolonies form. There are three complementation groups (which may or may not be cistrons) as follows: (1)(3,5)(6,2,4).

22. (a) Line indicates extent of deletion

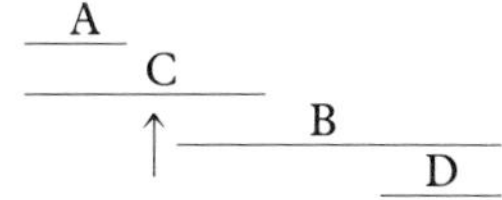

(b) Assuming deletions span entire gene, then arrow indicates position of point mutation

24. (a)

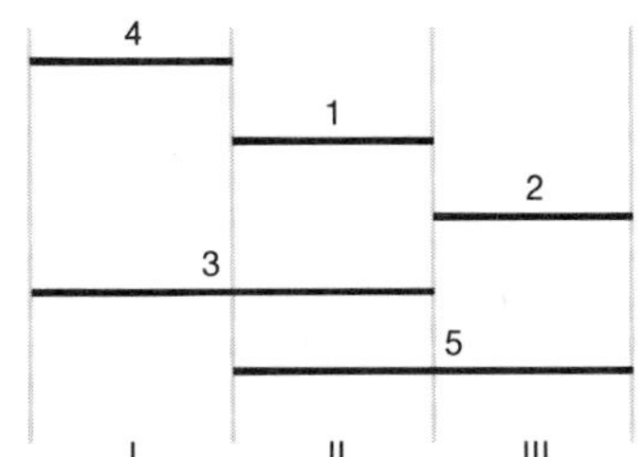

(b) I, II, III are discrete complementation regions (3)

(c)

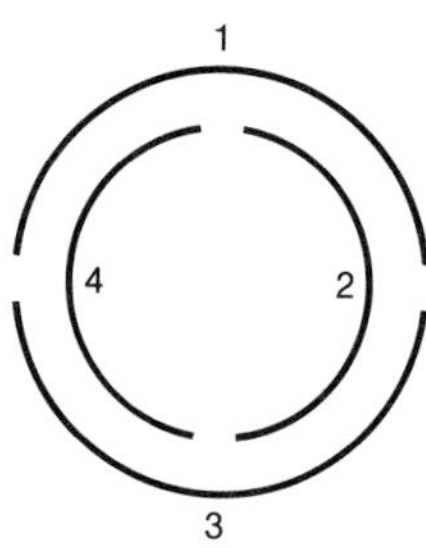

25. (a)

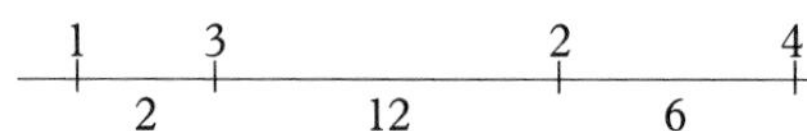

(b) No; homozygous crosses give no prototrophs.

28. S^n dominant. No. Mutations in other parts of genome could modify threshold to 40, perhaps by more efficient transport of the factor, or more efficient packaging into cell walls, etc., and hence S^f would become the dominant allele at generation *t*.

Chapter 11

1. 35%
2. Assume a diploid cell

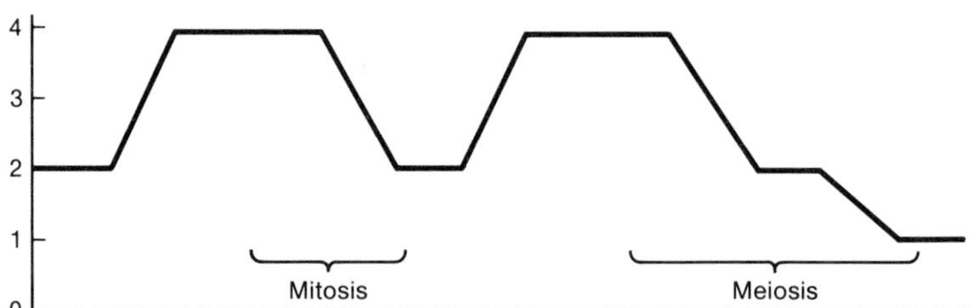

7. Because A does not equal T, nor does G equal C, the DNA must be single stranded. It replicates by synthesizing a complementary strand and uses that as a template.
12. The virus DNA has probably inserted itself into the mouse DNA like a prophage. This may be the manner in which the viruses cause cancer in mice (and possibly in humans).
14. From each harlequin pair of sister chromatids, another round of replication produces another harlequin pair of chromatids, plus an all-light pair.
15. (a) mutant; (b) wild; (c) mutant; (d) mutant; (e) mutant; (f) wild; (g) mutant; (h) wild; (i) mutant; (j) wild.

Chapter 12

2. (a) Both A+U/G+C ratios are compatible with the RNAs being transcribed off their primer DNA, but do *not* tell us if copied off one or both DNA strands since same A+U/G+C ratio would result in either case.

   (b) The purine/pyrimidine ratio of 0.8 proves the *E. coli* RNA is not double stranded. The 1.02 ratio could indicate double strandedness of *B. subtilis* RNA, or alternatively that one strand of primer DNA has equal numbers of purines and pyrimidines.

3. One nucleotide change should give rise to three adjacent amino-acid substitutions in protein.

6. (a) in 8 cases; (b) 3 cases, arg, ser, and leu.

8. (a) 1/8; (b) 1/4; (c) 1/8; (d) 1/8.

11. The codon changes in the RNA are as follows:

Mutant 1 – missense mutation
(Ser) AGU ⟶ (Arg) AG A/G
Mutant 2 – nonsense mutation
(Trp) UGG ⟶ (stop) UGA or UAG
Mutant 3 – two frameshift mutations

	⊖ ↓			⊕ ↓		
Ala — Pro —	Gly —	Val —	Lys —	Asn —	Cys —	His
GC(C/U/A/G) CC(C/U/A/G)	GGA	GUG	AAA	AA(U/C)	UG(U/C)	CA(U/C)

Mutant 4 – inversion

Ala — Pro — Trp — Phe — Phe — Thr — Cys — His
(inverted segment ↑ Phe — Phe — Thr ↑)
UUU UUC ACU

The wild-type DNA sequence is:

CGT(G/C/A)GG(G/C/A)T ACC TCA CTT TTT AC(A/G) GT(A/G)

13. e

15.

C	G	T	A	C	C	A	C	T	G	C	A
G	C	A	T	G	G	T	G	A	C	G	T
G	C	A	U	G	G	U	G	A	C	G	U
C	G	U	A	C	C	A	C	U	G	C	A

ala trp nothing (stop) nothing

17. (a) 120 nucleotides

(b)

		I	II	
1	Gln	C	A	A/G
2	Ser	A	G	U/C

Exchange in	I	II	I+II
	AA A/G = Lys	AG A/G = Arg	CG A/G = Arg
	CG U/C = Arg	CA U/C = His	AA U/C = Asn

Since Arg results from both singles, its frequency is twice His or Lys. Doubles are rare so Asn is least common. If markers are at adjacent nucleotides in DNA, the RF should be 4×10^{-7}.

Chapter 13

2.

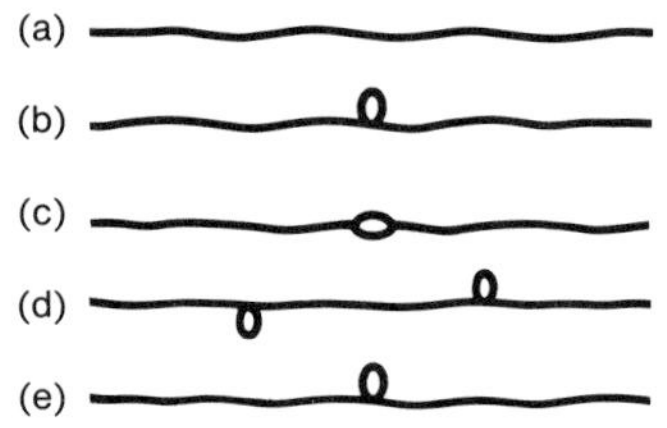

5. Palindrome DNA

7.

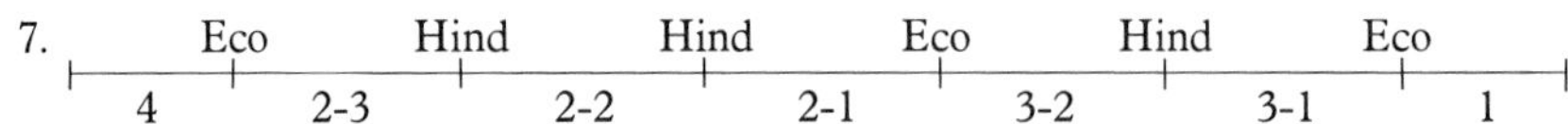

9. From the top (the ^{32}P end), the sequence of sites:

Hind-Hae-Hind-Hae-Hae-Hind-Hind-Hind-Hae-Hind-Hind-Hae-Eco.

Chapter 14

1. c

6. The recognition point for the migration mechanism is in the vicinity of the heterochromatic knob.

9. (a) The extent of denaturation of the DNA and the accessibility of the DNA to the RNA. The type of radioactive emission determines the size of the "spot" on the film.
The length of time the emulsion is exposed to the chromosomes.
The number of DNA copies in the chromosome.
The specific activity of the RNA (i.e., the amount of radioactivity per set amount of RNA).
(b) Add excess of cold RNA from a wide range of organisms and see if it will compete out the radioactive RNA.
(c) You infer that you've located the gene(s) responsible for the production of that RNA.

10. Perhaps they act as pairing regions to keep chromosomes together in the nucleus.
Perhaps they initiate rough chromosome alignments prior to meiosis.
Perhaps they play some unknown role in directing spindle fibers to the centromere.
Any reasonable speculation will do.

12. The loops are DNA. The loops are genetically different. Some contain sites for the specific restriction enzyme used.

14. (a) From line 1, $a^- = i^c$ or o^c.
From line 2, $c^- = i^c$ or o^c.
Therefore b must be the z gene.
From line 7, since i^c is recessive and we have a constitutive synthesis, the regulatory mutant producing the constitutive synthesis must be an o^c mutant. We know o^c mutants work only in cis, and a^- is the only candidate in cis conformation with a functional z^+ gene. Therefore a is the o gene and c is the i gene.

(b) Line 3, $c^- = i^c$ or i^s.
Line 4, $a^- = o^c$, $c^- = i^c$ or i^s.
Line 5, $a^- = o^c$ or o^o, $c^- = i^c$.
Line 6, $a^- = o^c$ or o^o, $c^- = i^c$.
Line 7, $a^- = o^c$, $c^- = i^c$ or i^s.

16. Type 1 Mutation probably in structural gene for E_1.
Type 2 Mutation probably in structural gene for E_2.
Type 3 Mutation in regulatory gene which controls function of both structural genes.

Points

(i) Normal product of regulatory gene is needed for E_1 and E_2 production.
(ii) Regulatory gene is on different chromosome.
(iii) Genes for sequential enzymatic steps are not adjacent.
(iv) Since regulatory gene must interact with structural loci there must be adjacent recognition sites, comparable to operators, for both. These might be found in an expanded mutant hunt, or even among types 1 and 2.

18. Assume the entire sequence of ADH is known. Isolate *adh*⁻ mutants by selecting survivors on pentenol. Look for immunologically cross-reacting protein with anti-ADH serum. Sequence all mutant proteins and determine the amino acid affected. This will allow you to select mutants at each end of the *adh* cistron. Select for mutants affecting the amount of ADH (for example, overproduction of ADH might give flies that can tolerate very high levels of ethanol). Select for mutants which alter the time or tissue of ADH production.

Chapter 15

1. Both crosses show predominantly maternal inheritance of the alternative phenotypes. The genes concerned are presumably carried on the cpDNA since a chlorophyll character is affected. (The minority classes are possibly due to a small male contribution of cytoplasm to the zygote. This is often encountered in cytoplasmic inheritance, for example in *poky Neurospora.*)

3. The results are explained by the maternal genotype influencing the phenotype of the zygote. Dwarfness is caused by a recessive nuclear gene *d*.

Original dwarf ♀	*dd*	×	*DD* ♂		
F_1		↓ *Dd*			(All dwarf because mother genetically dwarf)
F_2	1/4*DD*	1/2*Dd*	1/4*dd*	(All normal)	
F_3	↓ (Normal)	↓ (Normal)	↓ (dwarf)		

4. Continually backcross using pollen from plant B. In this way, the egg cytoplasm of plant A is maintained while the nucleus is "filled up" with genes from plant B. (See figure at right.)

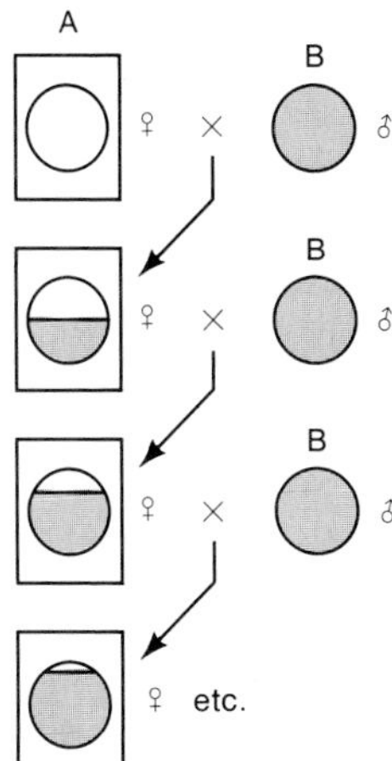

8. Provides evidence that cpDNA is the site of uniparentally inherited genes in *Chlamydomonas.*

10. (a) *ap* – *ant 1* – *ant 3* – *ant 5* – *ant 2* – *ant 4*
(b) The slopes could be used as map distances.

12. The positive heterokaryon test shows that the red phenotype must be caused by a cytoplasmic organelle mutation.

15. Define the *poky* gene as p and place it in brackets to show it is cytoplasmic, (p). A nuclear suppressor of *poky* will be called f. The ♀ parent is shown first.
    (a) $(+)+ \times (p)+ \rightarrow$ all $(+)+$
    (b) $(+)f \times (p)+ \rightarrow 1/2\,(+)f,\ 1/2\,(+)+$
    (c) $(p)+ \times (+)+ \rightarrow$ all $(p)+$
    (d) $(p)+ \times (+)f \rightarrow 1/2\,(p)+(=\mathrm{D}),\ 1/2\,(p)f(=\mathrm{E})$
    (e) $(p)f \times (+)f \rightarrow$ all $(p)f$
    (f) $(p)f \times (+)+ \rightarrow 1/2\,(p)+,\ 1/2\,(p)f.$

16. (a), (b). Each meiosis shows uniparental inheritance. This suggests cytoplasmic (mitochondrial) inheritance.
    (c) Since ant^r is probably on mtDNA, and petites represent loss of mtDNA, ant^r should be lost in some cases making ant^s petites.

18. Single gene losses were rarest for the *apt-cob* pair (45 total), most common for *apt-bar* (207), and intermediate for *bar-cob* (117).
    Rough map:

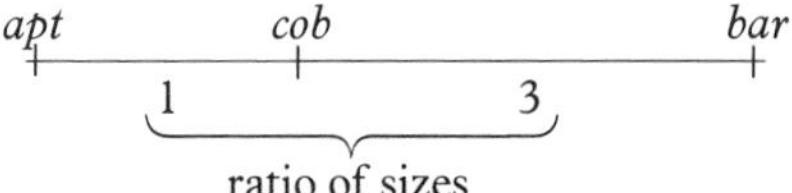

20. Some tetrads will show only strain 1-type DNA, other tetrads will show only strain 2-type DNA, and still others will show a recombinant DNA. (As an example of the latter, all the ascospores in one ascus might show recombinant DNA with 4 restriction sites, another ascus no restriction sites.)

Chapter 16

1. Possibly the agent cannot penetrate the cells. Test by adding together with an agent that loosens cell walls and enhances permeability.

3. The enzyme or protein damaged by the mutation still has some residual activity; for example, it might still be able to bind substrate in its newly shaped active site on rare occasions.

4. The wild type contained a gene which increases spontaneous mutation rate: call it a mutator gene m. m appears to be unlinked to *ad-3*.

A	×	Wild type
ad-3, +	↓	+, *m*

1. 1/4 *ad-3*, *m*
2. 1/4 *ad-3*, +
3. 1/4 +, *m*
4. 1/4 +, +

Prediction: Cross individual of type 2 to several *ad-3*$^+$ progeny; half the crosses should show a repeat of the above results. Also test m on other reverting mutants.

6. (a) UAA does not contain G or C so cannot be obtained this way.
    UAG and UGA could be derived by such transitions, but only from UAA, which is itself a nonsense codon.

(b) Yes, for example, UG*G* → UA*G*
(Trp)

(c) No, since G is only site acted on and UAA is always produced.
e.g., UA*G* → UA*A*

8. allele 1: a deletion or other chromosomal rearrangement

allele 2: a base-pair substitution

Prototroph A: a true revertant. *nic* → +
Prototroph B: mutation at an unlinked suppressor locus. + → *su* (tRNA?)
Prototroph C: mutation at a very closely linked site (0.8 m.u. away). Possibly intragenic suppressor altering mutant protein conformation.

10.

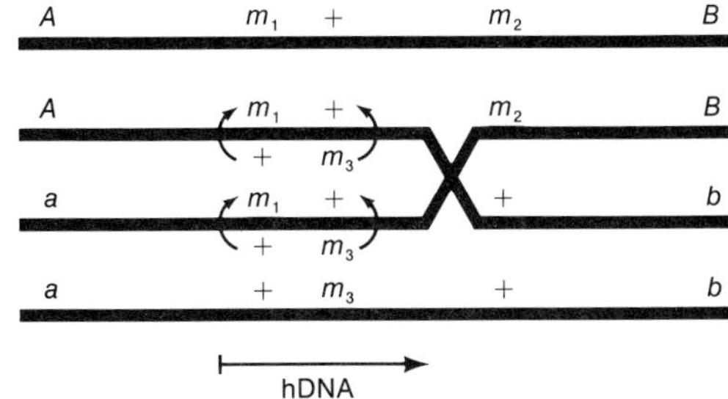

Order is $A/a - m_1 m_3 m_2 - B/b$ (or $m_3 m_1 m_2$)

Hybrid DNA entered m gene from left and ended between m_3 and m_2, and hence spanned m_1 and m_3.

One single excision/repair event corrected $+m_3$ to m_1+ on both hybrid DNA molecules.

12. (a), (b) Distortion in hybrid DNA is greater for addition/deletion and is corrected more efficiently by excision/repair system.

(c) Excision/repair system recognizes buckle but excises the *other* strand, e.g., for hybrid DNA between addition m and $+$.

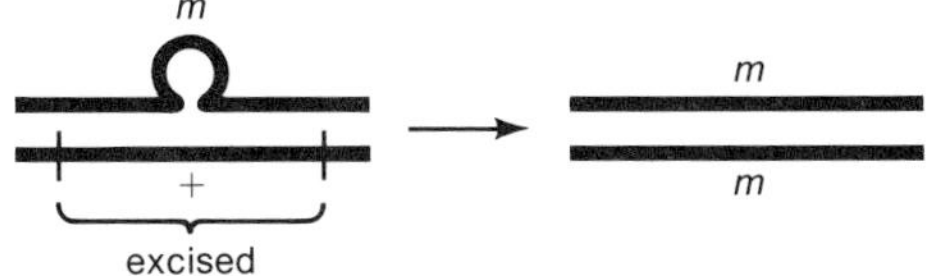

(d) When excision/repair system recognizes frameshift site, the excision and repair spans *both* the frameshift site and the substitution site, e.g., for addition

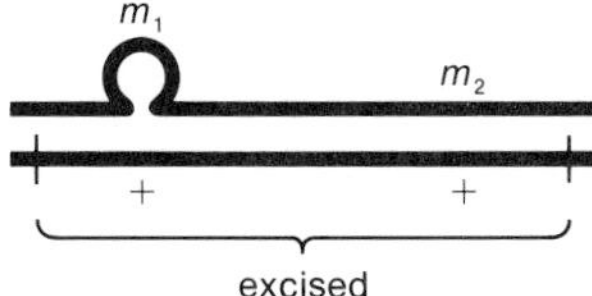

14. The second backcross is obviously a testcross since it shows the segregation of some genetic factor which affects recombinant frequency, call it *rec*. rec^+, borne by A,

his-1$_x$ parent, is a dominant inhibitor of recombination. *rec*$^-$, borne by *a*, *his-1*$_y$ parent, is a recessive allele which permits high recombinant frequency. The *rec* locus assorts independently of *his-1*, so half of the *A*, *his-1*$_x$ progeny are *rec*$^+$ and half are *rec*$^-$. The same applies to the *a*, *his-1*$_y$ progeny, but since the backcross is to *rec*$^+$ the segregation is not observed.
Test: 1/4 of the crosses between *A*, *his-1*$_x$ and *a*, *his-1*$_y$ progeny should be *rec*$^-$ $\times$ *rec*$^-$ and show high prototroph frequency. *rec* is a meiotic function occurring at meiosis in the diploid stage of the cycle.

17. Possibly caused by an insertion element in A.

20.

a	*a*	Hom	Hom
Hom	Hom	Hom	Hom

21. (a) Any recessive recombination-affecting mutation will immediately become homozygous in the ensuing meiosis.
    Step 1. Mutagenize many hundreds of disomic cells.
    Step 2. Isolate individual cells to generate separate disomic cultures.
    Step 3. Allow each culture to go through meiosis.
    Step 4. Plate products on leucineless medium and measure prototroph frequency. Select those cultures which produce abnormal prototroph frequencies (high or low); these probably have recombination-affecting mutations.

Chapter 17

1. Somatic mutation, mitotic chromosome loss, mitotic chromosome nondisjunction, mitotic crossing over, position-effect variegation, cytoplasmic mutation (and subsequent segregation), fusion of different zygotes.

6. (a) Regeneration must be controlled by the nucleus.
   (b) Hat forming substance is localized near the top. If enucleate, new substance cannot be synthesized.
   (c) No turnover of materials forming the hat.

10. For example, in the presence of excess S_1, P_1 is produced which would shut down operon 2 and permit operon 1 to stay active. When some of the S_1 is used up, and S_2 is in excess, reversal of the situation occurs.

12. The woman was heterozygous (Gd^+Gd^-), and dosage compensation has occurred randomly, inactivating either the malarial resistance or the sensitivity-determining X chromosome.

15. Femaleness

17. Petersen's studies suggest that muscular dystrophy results from defects in the nerves. The parabiosis studies show that the mutant effect is no longer operative in adults. Consequently, the nerve effects on muscles must be induced during development.

18. (a) B is before A. (B product builds up and is later used when A functions.)
    (b) A is before B.

(c) Both reactions have to be working simultaneously.

(d) The reactions of A and B are alternatives.

21. A tumor cells are recognized as bearing foreign antigens by the original mouse. These alien antigens were produced by a somatic mutation in the gene coding for the structure of the antigen (say, a cell surface protein). Antibodies are produced in response to this antigen, which make the cured mouse immune. Other mutations in this or in other antigen genes (e.g., those for tumor B) are not identical to those of A, and there is no cross-immunity.

25. The flies could have a structural defect in legs preventing normal movement. A simple example would be a muscle defect. The flies could have an alteration in their nerve wiring so that the wrong legs are signaled to move. If you could do physiological tests, action potentials of nerves and muscles could be tested and compared with wild type. Embryonic focus mapping using mosaics could indicate what cells and tissues are defective.

Chapter 18

1. mean = 4.783; variance = .323; standard deviation = .568.

2. (a) 1.0; (b) .83; (c) .66; (d) −.20.

3. By looking across the rows, we can find the effect of substituting a "low" for a "high" chromosome I. In the first row the differences are $25.1 - 22.2 = 2.9$ and $22.2 - 19.0 = 3.2$. In the second row these differences are 3.1 and 5.2, while in the third row they are 2.7 and 6.8. Averaging over all three rows, we see that the average difference between an h/h and h/l is $(2.9 + 3.2 + 2.7)/3 = 2.9$, while the average difference between an h/l and an l/l is $(3.2 + 5.2 + 6.8)/3 = 5.1$. Thus there is some dominance of the h genes, since the heterozygotes h/l are closer to h/h than to l/l.

   A similar calculation for the columns gives the average effect of a difference between h/h and h/l and between h/l and l/l for chromosome II. These are 2.9 and 11.5, respectively, so there is even more dominance on chromosome II. Finally, there is distinct epistasis. For example, the average effect of changing from h/l to l/l for chromosome I is 5.1, and the same change in chromosome II causes a difference of 11.5. We then expect the difference between $h/l\ h/l$ and $l/l\ l/l$ to be $5.1 + 11.5 = 16.6$, but we see that it is 17.6. Another way to see this is that the effect of a change from h/l to h/h on the first row is only 3.2 bristles, while in the third row it is 6.8 bristles.

4. (a) Homozygous at 1 locus $3(1/2)^3 = 3/8$
   Homozygous at 2 loci $3(1/2)^3 = 3/8$
   Homozygous at 3 loci $(1/2)^3 = 1/8$

   (b) Carrying 0 capital-letter alleles $= (1/2)^6 = 1/64$
   Carrying 1 capital-letter alleles = 6/64
   Carrying 2 capital-letter alleles = 15/64
   Carrying 3 capital-letter alleles = 20/64
   Carrying 4 capital-letter alleles = 15/64
   Carrying 5 capital-letter alleles = 6/64
   Carrying 6 capital-letter alleles = 1/64

5. Since the loci are on different chromosomes, it is simple to calculate the proportion of any genotype. For example, *AABbcc* will be in proportion (1/4)(1/2)(1/4)

= 2/64. Its phenotypic score will be 4 + 3 + 1 = 8. There are 27 possible genotypes. The frequency and score for each one can be calculated in this manner and the distribution of scores constructed. The distribution of scores is:

Score	*Proportion*
3	1/64
5	6/64
6	3/64
7	12/64
8	12/64
9	11/64
10	12/64
11	6/64
12	1/64

Note that the 3-bristle class, which comprises 19/64 of all flies, includes 7 genotypes ranging from *AaBbCC* to *AABBCC*, while the 1-bristle class includes only *aabbcc* individuals. All the rest of the genotypes, 19 of them, ranging from *aabbcc* to *AABBcc*, fall into the 2-bristle class.

11.

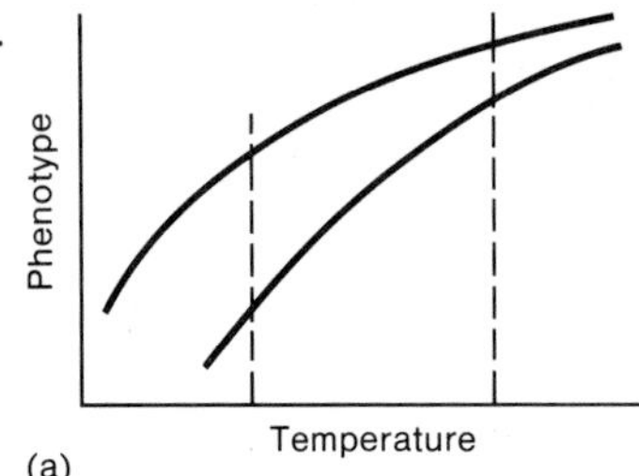

(a)

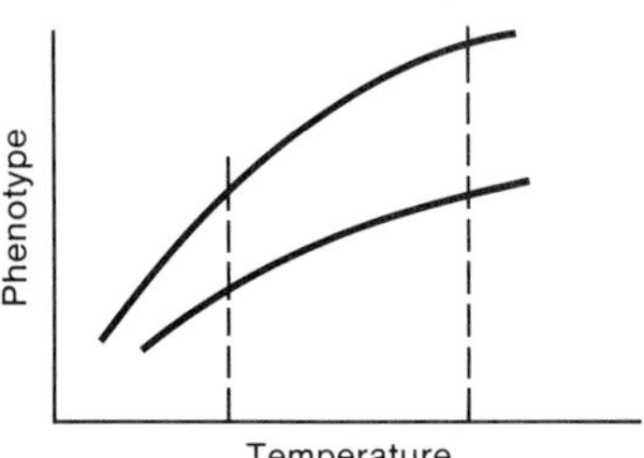

(b)

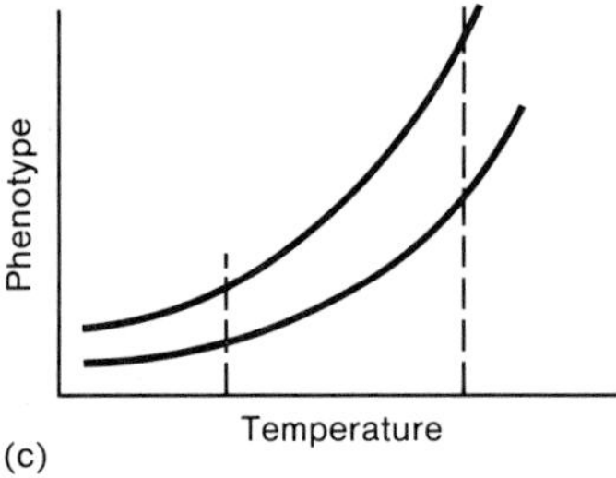

(c)

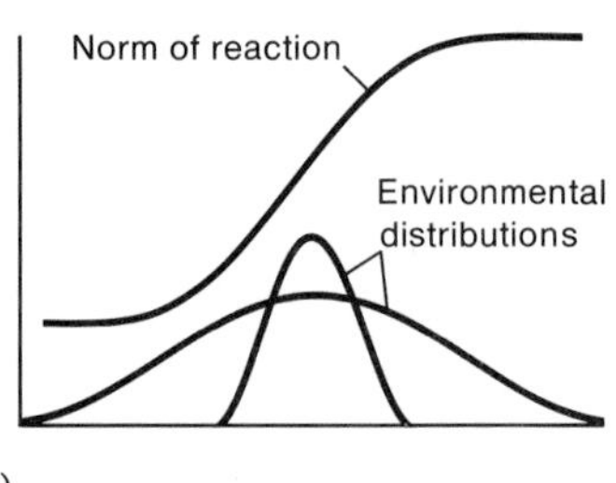

(d)

12. $H^2 = 4$ [correlation full sibs − half sibs]

$$= 4\left[\frac{2.7}{14.8} - \frac{.8}{14.8}\right] = .51$$

$H^2 = 4$ [correlation of full sibs] − 2 [parent-offspring correlation]

$$= 4\left(\frac{2.7}{14.8}\right) - 2\left(\frac{1.7}{14.8}\right) = .50$$

$$h^2 = 2\,(\text{parent-offspring correlation}) - 2\left(\frac{1.7}{14.8}\right) = .23$$

Chapter 19

2. (a) $p = \dfrac{406 + 372}{1482} = .525$. Therefore the expected members of the three genotypes are:

$$L^M L^M = (.525)^2\,(1482) = 408.5$$
$$L^M L^N = 2(.525)(.475)\,(1482) = 739.1$$
$$L^N L^N = (.475)^2\,(1482) = 334.4$$

so the agreement with HWE is excellent.

(b)

Mating	*Expected frequency*
$L^M L^M \times L^M L^M$	$(p^2)(p^2)(1482) = 56.3$
$L^M L^M \times L^M L^N$	$2(p^2)(2pq)(1482) = 203.7$
$L^M L^N \times L^M L^N$	$(2pq)(2pq)(1482) = 184.3$
$L^M L^M \times L^N L^N$	$2(p^2)(q^2)(1482) = 92.2$
$L^M L^N \times L^N L^N$	$2(2pq)(q^2)(1482) = 166.8$
$L^N L^N \times L^N L^N$	$(q^2)(q^2)(1482) = 37.7$

Again, the agreement is quite close.

3. (a)

Generation	p♂	P♀
0	.8	.2
1	.2	.5
2	.5	.35
3	.35	.425
⋮	⋮	⋮
n	$P_{(n-1)}$	$\dfrac{P_{(n-1)} + p_{(n-1)}}{2}$

Let $d = P - p$
then $d_{(n)} = -1/2\, d_{(n-1)}$
so $d_{(n)} = (-1/2)^n d_0$

4. (i); (iii); (vi); (viii); (ix); (x).

5. (a) .01
   (b) 10 times
   (c) A mating of a heterozygous female to a color-blind male. Proportion is $(2pq)(q) = (.18)(.1) = .018$.
   (d) All matings in which the female was homozygous normal. Proportion is $(.9)^2 = .81$.
   (e) Female progeny proportion color blind = .12
   Male progeny proportion color blind = .20
   (f) Female allele frequency = .40
   Male allele frequency = .20

9. Color-blind male offspring = .2
   Color-blind female offspring = .04

10\. (a)

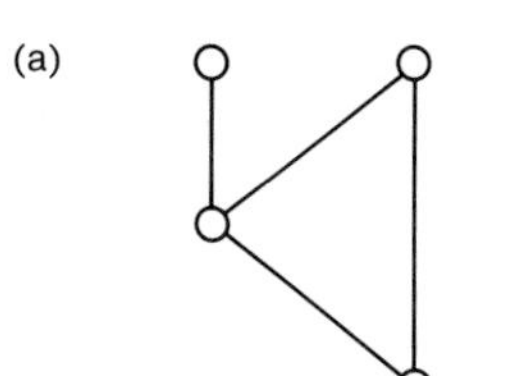

(b)

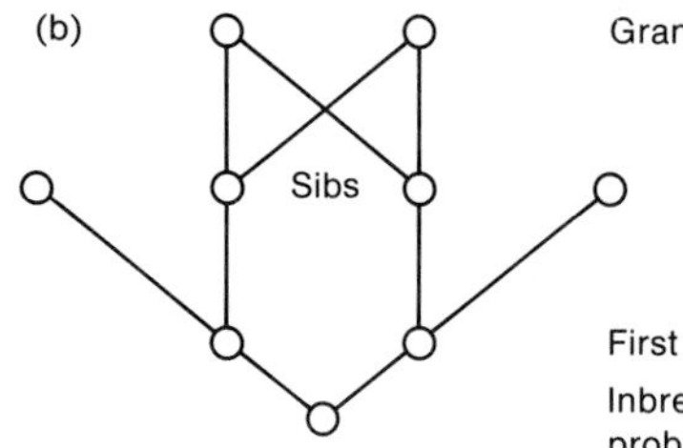

(c)

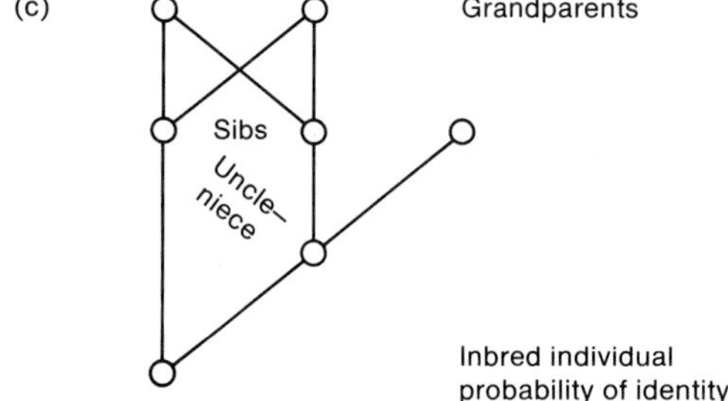

12\. Allele frequencies are $p(A) = q(a) = .50$. Assuming dominance of A over a, 80% of matings involve AA and Aa individuals, and 20% are $aa \times aa$. Among the 80% the frequencies are:

$$AA \times AA = \left(\frac{.2}{.8}\right)^2 = .0625$$

$$AA \times Aa = 2\left(\frac{.2}{.8}\right)\left(\frac{.6}{.8}\right) = .3750$$

$$Aa \times Aa = \left(\frac{.6}{.8}\right)^2 = .5625$$

Each of these frequencies is then multiplied by .80 to get the frequencies in the total population of matings. The genotype frequencies in the next generation are as follows:

AA: from $AA \times AA = (.0625)(.8) = .05$
from $AA \times Aa = 1/2(.3750)(.8) = .15$
from $Aa \times Aa = 1/4(.5625)(.8) = .1125$

Total .3125

Aa: from $AA \times Aa = 1/2(.3750)(.8) = .15$
from $Aa \times AA = 1/2(.5625)(.8) = .225$

Total .3750

$$aa\text{: from } Aa \times Aa = 1/4(.5625)(.8) = .1125$$
$$\text{from } aa \times aa = .20$$
$$\text{Total} \quad .3125$$

For negative assortative mating, using a similar form of analysis, after a generation the frequencies will be (in the next generation): $AA = 0$, $Aa = 2/3$, $aa = 1/3$. In the second generation the frequencies will be $Aa = .5$ and $aa = .5$ and will remain so in the future.

14. (a) $p^1 = .528$
    (b) $p = .75$

15. Equilibrium frequency of $a = \sqrt{\mu/s}$, where μ is the mutation rate and s the selection coefficient. Therefore aa frequency is $\mu/s = .001$, and

$$s = \frac{10^{-5}}{10^{-3}} = 10^{-2} = .01$$

16. Affected individuals are in frequency $2pq$, and since p is nearly unity,
    $q = 4.0 \times 10^{-6}/2 = 2 \times 10^{-6}$
    For a dominant deleterious gene, at equilibrium, $q = \mu/s$.
    Therefore $\mu = sq = (.3)\,(2 \times 10^{-6}) = 6 \times 10^{-7}$.

17. For each recessive lethal the chance of having a normal child is $7/8 = .875$. If there are n recessive lethals, the chance of escaping homozygosity for any of them is $(.875)^n$. Therefore $(.875)^n = 13/31$. Using logarithms we find that $n = 6.5$, so the number of recessive lethals is between 6 and 7.

18. (a)

$$q = \sqrt{\frac{\mu}{s}} = \sqrt{\frac{10^{-5}}{.5}} = 4.47 \times 10^{-3}$$

$$\text{genetic cost} = (.5)(2 \times 10^{-5}) = 10^{-5}$$

(b)

$$q = \sqrt{\frac{2 \times 10^{-5}}{.5}} = 6.32 \times 10^{-3}$$

$$\text{genetic cost} = (.5)(4 \times 10^{-5}) = 2 \times 10^{-5}$$

(c)

$$q = \sqrt{\frac{10^{-5}}{.3}} = 5.77 \times 10^{-3}$$

genetic cost $= (.3)(3.33 \times 10^{-5}) = 10^{-5}$, the same as in (a).

Index